CAMBRIDGE LIBRARY COLLECTION

Books of enduring scholarly value

Mathematics

From its pre-historic roots in simple counting to the algorithms powering modern desktop computers, from the genius of Archimedes to the genius of Einstein, advances in mathematical understanding and numerical techniques have been directly responsible for creating the modern world as we know it. This series will provide a library of the most influential publications and writers on mathematics in its broadest sense. As such, it will show not only the deep roots from which modern science and technology have grown, but also the astonishing breadth of application of mathematical techniques in the humanities and social sciences, and in everyday life.

Mathematical Tables

Prior to the advent of computers, no mathematician, physicist or engineer could do without a volume of tables of logarithmic and trigonometric functions. These tables made possible certain calculations which would otherwise be impossible. Unfortunately, carelessness and lazy plagiarism meant that the tables often contained serious errors. Those prepared by Charles Hutton (1737–1823) were notable for their reliability and remained the standard for a century. Hutton had risen, by mathematical ability, hard work and some luck, from humble beginnings to become a professor of mathematics at the Royal Military Academy. His mathematical work was distinguished by utility rather than originality, but his contributions to the teaching of the subject were substantial. This seventh edition was published in 1858 with additional material by Olinthus Gregory (1774–1841). The preliminary matter will be of interest to any modern-day reader who wishes to know how calculation was done before the electronic computer.

Mathematical Tables

Containing the Common,
Hyperbolic, and Logistic Logarithms

CHARLES HUTTON
OLINTHUS GREGORY

CAMBRIDGE
UNIVERSITY PRESS

CAMBRIDGE UNIVERSITY PRESS

Cambridge, New York, Melbourne, Madrid, Cape Town,
Singapore, São Paolo, Delhi, Mexico City

Published in the United States of America by Cambridge University Press, New York

www.cambridge.org
Information on this title: www.cambridge.org/9781108054027

© in this compilation Cambridge University Press 2012

This edition first published 1858
This digitally printed version 2012

ISBN 978-1-108-05402-7 Paperback

MATHEMATICAL TABLES:

CONTAINING THE

COMMON, HYPERBOLIC, AND LOGISTIC

LOGARITHMS,

ALSO

SINES, TANGENTS, SECANTS, AND VERSED SINES,

BOTH NATURAL AND LOGARITHMIC.

TOGETHER WITH

SEVERAL OTHER TABLES

USEFUL IN

MATHEMATICAL CALCULATIONS.

ALSO

THE COMPLETE DESCRIPTION AND USE OF THE TABLES.

BY

CHARLES HUTTON, LL.D. F.R.S. &c.,

FORMERLY PROFESSOR OF MATHEMATICS IN THE ROYAL MILITARY ACADEMY.

WITH SEVEN ADDITIONAL TABLES OF TRIGONOMETRICAL FORMULÆ,

BY

OLINTHUS GREGORY, LL.D.

Late Professor of Mathematics in the Royal Military Academy; Vice President of the Astronomical Society of London; Honorary Member of the Cambridge Philosophical Society, the Institution of Civil Engineers, &c.

NEW EDITION.

LONDON:

LONGMAN AND CO.; HAMILTON AND CO.; SIMPKIN AND CO.; WHITTAKER AND CO.; COWIE AND CO.; HOULSTON AND STONEMAN; F. AND J. RIVINGTON; SMITH, ELDER, AND CO.; T. BUMPUS; H. G. BOHN; J. CORNISH; ALLEN AND CO.; C. H. LAW; AND J. GREEN. CAMBRIDGE: DEIGHTON, BELL, AND CO.

1858.

CONTENTS.

I.—INTRODUCTION.

II.—TABLES.

PREFACE.

The very ample introduction, prefixed to the following collection of Mathematical Tables, supersedes the necessity of using many words here by way of preface, and leaves little more to be mentioned than the necessity and occasion of this work, with some account of the contents and mode of execution.

The undertaking was occasioned by the great incorrectness of all the editions of Sherwin's or Gardiner's Tables, and more especially by the bad arrangement in the fifth or last edition. Finding, as well from the report of others, as from my own experience, that those editions (to say nothing of the very improper alteration in the form of the table of sines, tangents, and secants in the last of them) were so very incorrectly printed, the errors being multiplied beyond all tolerable bounds, and no dependence to be placed on them for any thing of real practice, I was led to undertake the painful office of preparing a correct edition of another similar work. And I was lucky enough to meet with a bookseller of sufficient spirit to be at the great expense of printing the book, as well as to allow me what I demanded for my trouble in preparing it; which demand, however, was nothing adequate to the great labour attending it, as I was well aware that the profits of the book would not enable him fully to reward my pains.

I have, in the first place, therefore, used all the means in my power to render the work correct. I began by collating the third or best edition of Sherwin's tables, with some others of the most perfect works of the same kind, as Briggs's, Vlacq's, Gardiner's quarto book, &c.; by which means I detected many errors in each of them, which had not before been discovered; and of these, between twenty and thirty were in the two editions of Gardiner's quarto work, printed at London in 1742, and at Avignon in 1770. But, besides detecting many previously unknown errors in the said third edition of Sherwin, which was no more than was expected, I discovered, with no small surprise, that the last figures in the table of logarithms were not uniformly true to the nearest unit, except in a very few pages at the beginning and end of the table; though Mr. Gardiner, the editor of that edition, had made the table correct in that respect in his own quarto work before mentioned, which was also printed in the same year, 1742, with the said third edition of Sherwin! The errors from this cause, in that third edition, amounted to several thousands; and they have continued to run through all the editions of Sherwin ever since that time! But they are here corrected. Nor has less attention been employed in correcting the press, than in previously correcting the copy; every proof having been several times read over, and compared with the best of the books hitherto printed, by several persons attending to the reading of every proof-sheet.

But in giving this edition to the world, I was not satisfied with barely making it correct. I was aware that the materials themselves might be much improved; and I have accordingly enlarged, or otherwise greatly amended them, in various respects. Among the improvements of the old materials may be reckoned the following:—namely, in the large table of logarithms, the proportional parts, near the beginning, are more conveniently arranged, being now all placed in the same opening of the book where their corresponding differences occur; the logarithms to sixty-one figures are brought to their proper place in the book, and more conveniently disposed all in one page; the large table of sines, tangents, and secants, is more commodiously arranged, and rendered more distinct and convenient for use; the natural sines, tangents, secants, and versed sines, being all separated from the others, and placed all together on the left-hand pages, and the logarithmic ones facing them on the right-hand pages; the common differences, in both, set between the two columns to which each of them answers; and the versed sines here introduced into their proper place in the same pages with the sines, tangents, and secants. Besides these, there are some other alterations in the new tables here given, and the reader will find a number of very important improvements in the description and use of the whole; especially in the arithmetic of logarithms, and in the resolution of plane and spherical triangles, according to the present improved methods of calculation used by the Astronomer Royal, and other persons the most experienced in these matters.

The improvements in the tables, by the introduction of new matter, are both great and numerous. The tables numbered 2, 3, and 4, are here added, being an entire new set, with their differences, for finding numbers and logarithms to twenty places. The columns of common differences, in the pages of natural sines, &c., are now first introduced: as are also the tables of hyperbolic and logistic logarithms; the logarithmic sines and tangents for every second, in the first two degrees of the quadrant; together with a table of the length of arcs, a table to change common and hyperbolic logarithms from the one to the other, &c.,—the uses and exemplifications of the whole being very amply detailed.

Royal Mil. Acad. Woolwich,
   February, 1785.

⁎ In the large table of common logarithms, when the first of the last four figures in any logarithm changes from a 9 to a 0, in any line, in which case the first three or constant figures are prefixed to the next following line, instead of these three, it often happens that young beginners by mistake take out the three constant figures next above the said line. To guard against this error, the figures in this edition are so contrived, that where the said change happens, a bar is placed over the cipher, thus ō, through the remaining part of the line, in order to catch the eye, and remind the learner that the change there takes place. In this edition, too, the black rules formerly drawn across the pages, at the intervals of every five, or six, or ten lines, have been taken out, leaving thin white spaces across the pages instead of them. These improvements, besides that of new and better formed figures now introduced, and other attentions, contribute to render this edition of the tables more convenient and correct than either of the former ones.

December, 1800.   C. H.

In this *fifth edition* several of the tables have been much enlarged and improved, and some new ones introduced. Thus, the first large table of logarithms, which heretofore extended only to 100,000 numbers, is now enlarged by one whole sheet more, being continued to 108,000 numbers. Other tables are also extended to more numbers than formerly: and a new and extensive table of Hyperbolic Logarithms is introduced after the old one.

London, May, 1811.

The *seventh edition* is now presented to the public with still farther improvements. In order to reduce the price, the Historical Introduction, which occupied 124 pages, is omitted. The impartial, interesting, and valuable information, contained in that Introduction, is now to be found in Dr. Hutton's *Mathematical Tracts,* a work possessed by many purchasers of these Tables : so that the necessity for continuing its insertion in this volume no longer exists. The tables of logarithms to 61 places, with their differences, are also omitted, as of little, if any, use, to the ordinary purchasers of logarithm tables.

Seven tables of trigonometrical formulæ, relative to arcs and angles, their sums and differences, the differentials of trigonometrical lines, the sides and angles of plane and spherical triangles, and the solution of quadratic and cubic equations, are now added ; as well as the logarithms of many numbers often employed in mathematical computations. These are known and appreciated by the readers of *Cagnoli, Callet, Borda, Ursin,* &c.: but are, as yet, by no means so well known to English computers as they ought to be, notwithstanding the exhibition of tables nearly similar, by Mr. *F. Baily,* in his elegant and valuable collection for the use of astronomers.

The description and use of the tables have been, in some places, enlarged and improved. A few particulars have been transferred into this part from the Historical Introduction just mentioned ; and a page, exhibiting the surd values of the sines of arcs to every third degree in the quadrant. These, with the decimal values of the principal surds, as $\sqrt{3}$, $\sqrt{5}$, $\sqrt{(5 + \sqrt{5})}$, &c., arranged in immediate connection with them, will often serve to facilitate the labour of men of science.

Great care has been taken to render this edition as free from errata as possible ; and the editor cannot but hope that the improvements now introduced, will preserve to Hutton's Mathematical Tables the reputation, for accuracy and extensive utility, which they have maintained for nearly half a century.

OLINTHUS GREGORY.

Royal Mil. Acad. Woolwich,
    March, 1830.

LOGARITHMIC TABLES.

The Definition and Notation of Logarithms.

LOGARITHMS may be considered the indices or arithmetical series of numbers, adapted to the terms of a geometrical series, in such sort that 0 corresponds to 1, or is the index of it, in the geometricals.

Thus $\begin{cases} 0 & 1 & 2 & 3 & 4 & 5, \text{ \&c. indices or logarithms.} \\ 1 & 2 & 4 & 8 & 16 & 32, \text{ \&c. geometric progression.} \end{cases}$

or $\begin{cases} 0 & 1 & 2 & 3 & 4 & 5, \text{ \&c. indices or logarithms.} \\ 1 & 3 & 9 & 27 & 81 & 243, \text{ \&c. geometric series.} \end{cases}$

or $\begin{cases} 0 & 1 & 2 & 3 & 4 & 5, \text{ \&c. indices or logarithms.} \\ 1, 10, & 100, & 1000, & 10000, & 100000, \text{ \&c. geometric series.} \end{cases}$

Where the same indices serve equally for any geometric series; and from which it is evident, that there may be an endless variety of systems of logarithms to the same common numbers, by varying the 2d term, 2, or 3, or 10, &c. of the geometric series; as this will change the original series of terms, whose indices are the integer numbers, 1, 2, 3, &c.; then by interpolation the whole system of numbers may be made to enter the geometrical series, and receive their proportional logarithms, whether integers or decimals.

Or, the logarithm of any number is the index of that power of some other number, which is equal to the given number. So, if N be $= r^n$, then the logarithm of N is n, which may be either positive or negative, and r any number whatever, according to the different systems of logarithms. When N is 1, then $n = 0$, whatever the value of r is; and consequently the logarithm of 1 is always 0 in every system of logarithms. When n is $= 1$, then N is $= r$: consequently r is always the number whose logarithm is 1, in every system. When r is $= 2\cdot718281828459$, &c., the indices are the hyperbolic logarithms, such as in our 5th table: so that n is the hyperbolic logarithm of $(2\cdot718, \text{ \&c.})^n$. But in the common logarithms, r is $= 10$; so that the common logarithm of any number (10^n) is (n) the index of that power of 10 which is equal to the said number. So 1000, being the 3d power of 10, has 3 for its logarithm; and if 50 be $= 10^{1\cdot69897}$, then is 1.69897 the common logarithm of 50. And hence it follows, that this decuple series of terms

$$10^4 , \quad 10^3, \quad 10^2, 10^1, \quad 10^0, \quad 10^{-1}, \quad 10^{-2}, \quad 10^{-3}, \quad 10^{-4},$$
or 10000, 1000, 100, 10 , 1 , $\cdot$1 , $\cdot$01 , $\cdot$001 , $\cdot$0001 ,
have 4 , 3 , 2 , 1 , 0 ,—1 ,—2 , —3 , —4 ,
respectively for their logarithms.

The logarithm of a number comprehended between any two terms of the first series, is included between the two corresponding terms of the latter and therefore that logarithm will consist of the same index (whether positive or negative) as the less of those two terms, together with a decimal fraction, which will always be positive. So the number 50, falling between 10 and 100, its logarithm will fall between 1 and 2, and is $= 1\cdot69897$, the index of the less term, together with the same decimal $\cdot69897$ as before: also the number 05, falling between the terms 1 and $\cdot$01, its logarithm will fall between -1 and -2, and is indeed $= -2 + \cdot69897$. The index is also called the characteristic of the logarithms, and is always an integer either positive or negative, or else $= 0$; and it shows what place is occupied by the first significant figure of the given number, either above or below the place of units, being in the former case $+$ or positive, in the latter $-$ or negative.

When the characteristic of a logarithm is negative, the sign $-$ is commonly set over it, to distinguish it from the decimal part, which being the logarithm found in the tables is always positive: so $-2 + \cdot69897$, or the logarithm of $\cdot05$, is written thus $\bar{2}\cdot69897$. But on some occasions it is convenient to reduce the whole expression to a negative form; which is done by making the characteristic figure less by 1, and taking the arithmetical complement of the decimal, that is, beginning at the left hand, subtract each figure from 9, except the last significant figure, which subtract from 10; so shall the remainders form the logarithm entirely negative. Thus the logarithm of $\cdot05$, which is $\bar{2}\cdot69897$, or $-2 + \cdot69897$, is also expressed by $-1\cdot30103$, which is wholly negative. It is also sometimes thought more convenient to express such logarithms wholly as positive, namely, by only joining to the tabular decimal the complement of the index to 10: in which way the above logarithm is expressed by $8\cdot69897$; which is only increasing the indices in the scale by 10; and is now commonly done in the tables of logarithm sines, tangents, &c. It is also convenient, in many operations with logarithms, to take their arithmetical complements, which is done, by beginning at the left hand, and subtracting every figure from 9, but the last figure from 10. so the arithmetical complement

of $1\cdot69897$ $\{$ and of $\bar{2}\cdot69897$ $\}$ where the index -2, being negative, is added
is $8\cdot30103$ $\{$ is $11\cdot30103$ $\}$ to 9, and makes 11.

The Properties of Logarithms.

From the definition of logarithms, either as being the indices of a series of geometricals, or as the indices of the powers of the same root, it follows, that the multiplication of the numbers will answer to the addition of their logarithms; the division of numbers, to the subtraction of their logarithms; the raising of powers, to the multiplying the logarithm of the root by the index of the power; and the extracting of roots, to the dividing the logarithm of the given number by the index of the root required to be extracted. So, using L. to denote the logarithm of the quantity which follows it :—

1st.   L. ab or $a \times b$ is $=$ L. $a +$ L. b
     L. 18 or 3×6 is $=$ L. $3 +$ L. 6
     L. $5 \times 9 \times 73$ is $=$ L. $5 +$ L. $9 +$ L. 73

2d.   L. $a \div b$ is $=$ L. $a -$ L. b
     L. $18 \div 6$ is $=$ L. $18 -$ L. 6
     L. $79 \times 5 \div 9$ is $=$ L. $79 +$ L. $5 -$ L 9
     L. $\frac{1}{2}$ or $1 \div 2$ is $=$ L. $1 -$ L. $2 = 0 -$ L. $2 = -$ L. 2

     L. $\dfrac{1}{n}$ or $1 \div n$ is $= -$ L. n.

3d.   L. r^n is $= n$ L. r;   L. $r^{\frac{1}{n}}$ or L. $\sqrt[n]{r}$ is $= \dfrac{1}{n}$ L. r;   L. $r^{\frac{m}{n}}$ is $= \dfrac{m}{n}$ L. r.

     L 2^6 is $= 6$ L. 2;   L. $2^{\frac{1}{3}}$ or L. $\sqrt[3]{2}$ is $= \frac{1}{3}$ L. 2;   L. $2^{\frac{3}{5}}$ is $= \frac{3}{5}$ L. 2.

Thus, any number and its reciprocal have the same logarithm, but with contrary signs; and the sum of the logarithms of any number and its complement, is equal to 0.

To construct Logarithms.

It is shown in my *History of Logarithms*, that the logarithm of $\dfrac{b}{a}$ is $=$

$\dfrac{2}{m} \times : \dfrac{x}{z} + \dfrac{x^3}{3z^3} + \dfrac{x^5}{5z^5} + \dfrac{x^7}{7z^7}$ &c., where z is the sum and x the difference of a and b; also $m = 2\cdot302585092994$, &c., the hyp. logarithm of 10. Therefore if a and b be any two numbers differing only by unity, so that x or $b - a$ may be $= 1$; then shall the logarithm of b be $=$ L. $a + \dfrac{2}{m} \times :$

$\dfrac{1}{z} + \dfrac{1}{3z^3} + \dfrac{1}{5z^5}$ &c. Which gives this rule in words at length: call z the sum of any number (whose logarithm is sought) and the number next less by unity; divide $\cdot8685889638$, &c. (or $2 \div 2\cdot3025$, &c.) by z, and reserve the quotient: divide the reserved quotient by the square of z, and reserve this quotient: divide this last quotient also by the square of z, and again reserve this quotient: and thus proceed continually, dividing the last quotient by the square of z, as long as division can be made. Then write these quotients orderly under one another, the first uppermost, and divide them respectively by the uneven numbers 1, 3, 5, 7, 9, 11, &c., as long as division can be made: that is, divide the first reserved quotient by 1, the 2d by 3, the 3d by 5, the 4th by 7, &c. Add all these last quotients together, then the sum will be the logarithm of $b \div a$; and therefore to this logarithm adding also the logarithm of a the next less number, the sum will be the required logarithm of b the number proposed.

Example 1. To find the Log. of 2.

Here the next less number is 1, and $2 + 1 = 3 = z$, whose square is 9. Then

3)	·868588964	1)	·289529654	(·289529654
9)	·289529654	3)	32169962	(	10723321
9)	32169962	5)	3574440	(	714888
9)	3574440	7)	397160	(	56737
9)	397160	9)	44129	(	4903
9)	44129	11)	4903	(	446
9)	4903	13)	545	(	42
9)	545	15)	61	(	4
9)	61				

Log. $\frac{2}{1}$ .... ·301029995
Add Log. 1. .... ·000000000

Log. of 2. ..... ·301029995

Example 2. To find the Log. of 3.

Here the next less number is 2, and $2 + 3 = 5 = z$, whose square is 25, to divide by which always multiply by ·04. Then

5)	·868588964	1)	·173717793	(·173717793
25)	·173717793	3)	6948712	(	2316237
25)	6948712	5)	277948	(	55590
25)	277948	7)	11118	(	1588
25)	11118	9)	448	(	50
25)	445	11)	18	(	2
	18				

Log. $\frac{3}{2}$ .. .. ·176091260
Log. 2 add .... ·301029995

Log. 3 .... ·477121255

Then, because the sum of the logarithms of numbers gives the logarithm of their product, and the difference of the logarithms gives the logarithm of the quotient of the numbers, from the above two logarithms, and the logarithm of 10 which is 1, we may raise a great many other logarithms, thus:

Example 3.

Because 2 × 2 = 4, therefore to L. 2. . . ·301029995$\frac{2}{3}$
add L. 2. . . . . ·301029995$\frac{2}{3}$

sum is L. 4 . . . . . . ·602059991$\frac{1}{3}$

Example 4.

Because 2 × 3 = 6, therefore to L. 2. . . . ·301029995
add L. 3 . . . . ·477121255

sum is L. 6. . . . . ·778151250

Example 5.

Because 2³ = 8, therefore L. 2. . . . . . . ·301029995$\frac{2}{3}$
mult. by 3 3

gives L. 8 . . . . ·903089987

Example 6.

Because 3² = 9, therefore L. 3 . . . . . . ·477121254$\frac{7}{10}$
mult. by 2 2

gives L. 9 . . . . ·954242509

Example 7.

Because $\frac{10}{2}$ = 5, therefore from L. 10. . 1·000000000
take L. 2 . . . . ·301029995$\frac{2}{3}$

leaves L. 5 . . . . ·698970004$\frac{1}{3}$

Example 8.

Because 12 = 3 × 4, therefore to L. 3 ·477121255
add L. 4 . . . . . . ·602059991

gives L. 12 . . . . 1·079181246

And thus, computing by the general rule, the logarithms of the other prime numbers, 7, 11, 13, 17, 19, 23, &c. ; and then using composition and division, we may easily find as many logarithms as we please, or may speedily examine any logarithm in the table.

See, farther, the History of Logarithms, in my Tracts, vol. i. 8vo.

THE DESCRIPTION AND USE OF THE TABLES.

THE following collection consists of various tables, in the following order, viz. 1. A large table of logarithms to 7 places of figures; 2. A table for finding logarithms and numbers to 20 places; 3. Logarithms to 20 places, with their 1st, 2d, and 3d differences; 4. Another table of logarithms to 20 places, with their 1st, 2d, and 3d differences; 5. Hyperbolic logarithms; 6. Logistic logarithms; 7. Logarithmic sines and tangents to every second of the first 2 degrees; 8. Natural and logarithmic sines, tangents, secants, and versed sines, with their differences to every minute of the quadrant. After which follow several smaller tables: as, a table of the lengths of circular arcs; a traverse table, or table of difference of latitude and departure, to every degree and quarter point of the compass; a table for changing the common logarithms into hyperbolic logarithms; and a table of the names and number of degrees, &c., in every point of the compass; tables of useful trigonometrical formulæ, &c. Of each of which in their order.

Of the large Table of Logarithms.

The first is the large table of logarithms, to all numbers from 1 to 108000; by which may be found the logarithm to any number, and the number to any logarithm, to 7 places of figures. This table consists of two parts; the first contains, in 4 pages, the first 1000 numbers, with their corresponding logarithms in adjacent columns; the second contains all the 108000 numbers and their logarithms, with the differences and proportional parts, disposed as follows: in the first column of each page are the first 4 figures of the numbers, and along the top and bottom of the columns is the 5th figure, in which columns are placed all the logarithms, the first 3 figures of each logarithm being at the beginning of the lines in the first column of logarithms, signed 0 at the top and bottom, and the other 4 figures in the remaining columns. Sometimes the first three figures of the logarithms are found in the line next below the number, viz. when the fourth figures have changed from 9's to 0's, in which case, a bar is placed over the first ciphers in the whole of that line to catch the eye, thus $\bar{0}$. After the 10 columns of logarithms stands their column of differences, signed D; and lastly, after that, the column of proportional parts, signed Pro. Pts., showing what proportional part of each difference corresponds to 1, 2, 3, &c., the whole difference answering to 10; or showing the $\frac{1}{10}$, $\frac{2}{10}$, $\frac{3}{10}$, &c., of the differences.

Note, The logarithms in these columns are all supposed to be decimals, and their corresponding natural numbers may be either integers or decimals or mixt numbers; for the same figures, whatever be their denomination, have the same decimal logarithm, and these differ only in the index or characteristic, which is the integer number to be prefixed to the decimal part of the logarithm; and this is always the number which expresses the distance of the highest denomination, or left-hand figure, of the natural number, from the units place. So that, if the natural number consist of only one place of integers, the index of its log. will be 0: if of 2, 3, 4, 5, &c., the index of its logarithm will be respectively 1, 2, 3, 4, &c., being 1 less than the number of integer places: and the same figures made negative will give the index of the logarithm of a decimal, viz. if the natural number be a decimal, and its first significant figure be in the place of primes, 2ds, 3ds, 4ths, &c., the index of its logarithm will be respectively $\bar{1}$, $\bar{2}$, $\bar{3}$, $\bar{4}$, &c.; or the figure which expresses the distance of the first place of the natural number from the units place, but with a negative sign, as the number is below the place of units, the sign being written above the index instead of before it, as that part only of the logarithms is to be considered as negative, the decimal part of it being always affirmative. And in the arithmetical operations of addition and subtraction with logarithms, the negative indexes will have the contrary effect to that of the decimal part of the logarithm, viz. when the logarithm is to be added, the figure of the negative index must be subtracted, *et vice versa*

Hence, if 4234097 be the tabular or decimal part of the logarithm belonging to the figures 2651, without any regard to their particular denominations; then according as they are varied with respect to the number of decimals, as in the 1st annexed column, the index of their logarithm, and the complete logarithm, will vary as in the 2d column here annexed. And hence, like as when the natural number is given, we find the index of its logarithm by counting how far its first figure on the left hand is from the units

Number	Logar.
2651	3·4234097
265·1	2·4234097
26·51	1·4234097
2·651	0·4234097
·2651	$\bar{1}$·4234097
·02651	$\bar{2}$·4234097
·002651	$\bar{3}$·4234097

place; so when a logarithm is given the denominations of the figures in its natural number will be found by placing the decimal point so, that the number of integer places may be 1 more than that of the index when positive, or by setting the first significant figure in that decimal place, which is expressed by the number of the index when negative.

Of finding the Logarithm of a given Number, or the number to a given Logarithm.

1. To find the Logarithm of a Number consisting of 3 figures.

Find the number in the column of numbers in one of the first 4 pages of the table, and immediately on the right of it is its logarithm sought. So the logarithm of 72 is 1·8573325, and the logarithm of 3·33 is 0·5224442, when the proper index is supplied.

2. To find the Logarithm of a Number consisting of 4 Places.

In the first column (signed N) in some one of the pages of the table after the first four, find the given number, then against it in the 2d column (signed 0) is the logarithm sought. So the logarithm of 2254 is 3·3529539, and that of 31·32 is 1·4958218.

3. To find the Logarithm of a Number consisting of 5 Places.

Find the first 4 figures of the given number in the first column as before, and the 5th figure at the top or bottom; then the 7 figures of the logarithm are found in two columns on the line of the first 4 figures of the given number, viz. the first 3 figures of the logarithm are the first 3 common figures of the 2d column (signed 0), and the last 4 figures are on the same line, but in the column signed with the 5th figure of the given number. So the logarithm of 23204 is 4·3655629, and that of 746·40 is 2·8729716, and that of 083178 is $\bar{2}$·9200085.

Note, When the last four figures of the logarithm begin with a cipher, or any figure less than the last four in the 2d column begins with, then the first 3 common figures are those in the next lower line · so, in the last example, the first 3 common figures are 920, and not 919. This is indicated by the bar over the first cipher in $\bar{0}$085, as already explained.

4. To find the Logarithm of a Number of 6 Places.

Find the logarithm of the first five figures by the last article and take the difference between that logarithm and the next following logarithm, or (which is the same thing) find the difference nearest opposite in the last column but one, signed D; then under that difference in the last column (of proportional parts) and against the 6th figure of the given number, is the part to be added to the logarithm before found for the first 5 figures, the sum being the logarithm sought. So to find the logarithm of 340926 : the logarithm of 34092 the first 5 figures, being 5326525, and the common difference 127, under which and against 6 in the last column is 76, which being added to the former logarithm, and the proper index prefixed, i. e. 3 for 4 places of whole numbers, we have 3·5326601 for the whole logarithm required.

5. To find the Logarithm of a Number of 7 Places.

Find the logarithm of the first 5 figures by the 3d article, and of the sixth figure by the 4th article; then for the logarithm of the 7th figure, divide its proportional part by 10, that is, set it one place farther to the right hand than the last figure of the logarithm reaches; add all the three together, and their sum will be the logarithm required.

Thus, to find the logarithm of 3·409264. The several parts being taken out according to the rule, and placed as in the margin, the sum gives the whole logarithm sought.

Numb.	Logar.
34092 . .	5326525
6 .	76
4 .	5,1
3·409264 .	0·5326606

Note, In the same way we might take out the proportional part of an 8th figure, dividing its tabular part by 100, or setting it two places farther to the right hand than the first logarithm. Or the whole proportional part for any number of figures above five, may be found at once, by multiplying the common tabular difference of the logarithms, found as before, by all the figures after the 5th, cutting off from the product as many figures as we multiply by, and adding the rest to the logarithm of the first 5 figures before found. So in the last example above, having found the common difference 127, multiplying it by 64 the last two figures, cutting off two, add the rest to the logarithm of the first 5, as in the margin.

$$
\begin{array}{r}
127 \\
64 \\
\hline
508 \\
762 \\
\hline
81,28 \\
5326525 \\
\hline
0·5326606
\end{array}
$$

For another example, suppose we wanted the logarithm of the following 8 figures, 34092648. The operation by both methods will be as below.

34092	5326525		127
6	76		648
4	5,1		1016
8	1,02		508
			762
34092648	7·5326607		82,296
			5326525
			7·5326607 the same as the other.

6. To find the Logarithm of a Vulgar Fraction, or of a Mixt Number.

Either reduce the vulgar fraction to a decimal, and find its logarithm as above. Or else (having reduced the mixt number to an improper fraction),

subtract the logarithm of the denominator from the logarithm of the nume
rator, and the remainder will be the logarithm óf the fraction sought.

Example 1.

To find the log. of $\frac{3}{16}$ or $0\cdot1875$.

From log of 3	$0\cdot4771213$
Take log. of 16	$1\cdot2041200$
Rem. log. of $\frac{3}{16}$ or $\cdot1875$	$\overline{1}\cdot2730013$

Example 2.

To find the log. of $13\frac{3}{4}$ or $\frac{55}{4}$.

From log. of 55	$1\cdot7403627$
Take log. of 4	$0\cdot6020600$
Leaves log. of $\frac{55}{4}$ or $13\cdot75$	$1\cdot1383027$

7. *To find the Natural Number answering to any given Logarithm.*

Find the first 3 figures, next after the index of the given logarithm, in the
second column, signed 0, and the other 4 figures on the same line in one of
the nine following columns; if the figures of the logarithm be thus found
exactly, then on the same line in the first column are the first four figures of
the natural number, and the 5th is at the top or bottom of that column
in which the last four figures of the log. were found. So to find the number
answering to the logarithm $2\cdot5890108$. In p. 63 I find the first three
figures 589, and in column 6 of the line above are found the other four $\cdot0108$
(because the first three common figures are supposed to begin at that part of
the line above where they are placed): then on the same line in the column
of numbers stand the first four figures $388\cdot1$, and 6 at the top of the column,
making in all $388\cdot16$ for the number sought; having placed the decimal
point so as to make three integers, being 1 more than 2 the index of the
given logarithm.

But if the given logarithm be not found exactly in the table, subtract the
next less tabular logarithm from it, and look for the remainder in the proportional parts under the difference between the two tabular logarithms next
less and greater than the given logarithm, and against it, or the part next
less, is a 6th figure to be annexed to the five figures before found. And if
the remainder be not found exactly in the proportional parts, subtract the
next less part from it, and annex a cipher to this 2d remainder, then against
the nearest proportional part (either greater or less) is a 7th figure to be
annexed to the six before found. And that figure will be the nearest to the
truth in that place, either too much or too little.

Example.

To find the number answering to the logarithm $1\cdot2335678$
The next less tab. log. is the log. of 17122, viz. 2335545

		1st rem.	133
The difference is 254 $\Big\{$	$\cdot$ 5	for the part	127
and the table of pro. pts. gives $\Big\{$		2d rem.	60
	$\cdot$ 2	for the part	51

So that the number sought is $17\cdot12252$, making two integers for the index 1

Or the 6th and 7th figures may be found without the table of proportional parts, by dividing the first remainder by the tabular difference, annexing one cipher to the dividend for each figure to be found. So, in the last example, the remainder 133, with two ciphers annexed, being divided by the tabular difference 254, as in the margin, the quotient gives 52 for the 6th and 7th figures, the same as before. In like manner may be found the numbers to the following logarithms.

254) 133,00 (52
     127,0
     ─────
     600
     5о8

Logar.	1·2345678	3·7343003	1̄·0921406	2̄·3720468	4·6123004	3·2946809
Numb.	17·16200	5·423758	·1236348	·02355303	40954·39	1970·974

OF LOGARITHMICAL ARITHMETIC.

I. *Multiplication by Logarithms.*

Add together the logarithms of all the factors; then the sum is a logarithm, the natural number corresponding to which, being found in the table, will be the product required.

Observing to add, to the sum of the affirmative indices, what is carried from the sum of the decimal parts of the logarithms.

And that the difference between the affirmative and negative indices, is to be taken for the index to the logarithm of the product.

Example 1.

To multiply 23·14 by 5·062.

23·14 its log. is	1·3643634
5·062 its log. is	0·7043221
Product 117·1347	2·0686855

Example 2.

To multiply 2·581926 by 3·457291.

2·581926 its log. is	0·4119438
3·457291	0·5387359
Product 8·92647	0·9506797

Example 3.

To multiply 3·902, and 597·16, and ·0314728 all together.

3 902 its log. is	0·5912873
597·16	2·7760907
0314728	2̄·4979353
Product 73·33533	1·8653133

The 2̄ cancels the 2, and the 1 to carry from the decimals is set down.

Example 4.

To multiply 3·586 and 2·1046, and 0·8372, and 0·0294, all together.

3·586 its log. is	0·5546103
2·1046	0·3231696
0·8372	1̄·9228292
0·0294	2̄·4683473
Product ·1857618	1̄·2689564

Here the 2 to carry cancels the 2̄, and there remains the 1̄ to set down

b

II. *Division by Logarithms.*

From the logarithm of the dividend, subtract the logarithm of the divisor; the remainder is a logarithm, whose corresponding number will be the quotient required.

But first observe to change the sign of the index of the logarithm of the divisor, viz. from negative to affirmative, or from affirmative to negative; then take the sum of the indices if they be of the same kind, or their difference when of different kinds, with the sign of the greater, for the index to the logarithm of the quotient.

When 1 is borrowed in the left-hand place of the decimal part of the logarithm, add it to the index of the logarithm of the divisor when that index is affirmative, but subtract it when negative; then let the index thus found be changed, and worked with as before.

Example 1.

To divide 24163 by 4567.

Dividend 24163 its log.	4·3831509
Divisor 4567	3·6596310
Quotient 5·290782	0·7235199

Example 2

To divide 37·149 by 523·76.

Divid. 37·149 its log.	1·569947
Divis. 523·76	2·7191323
Quot. ·07092752	$\bar{2}$·8508148

Example 3.

To divide ·06314 by ·007241.

Divid. ·06314 its log.	$\bar{2}$·8003046
Divis. ·007241	$\bar{3}$·8597985
Quot. 8·719792	0·9405061

Here 1 carried from the decimals to the $\bar{3}$ makes it become $\bar{2}$, which taken from the other $\bar{2}$, leaves 0 remaining.

Example 4.

To divide ·7438 by 12·9476.

Divid. ·7438 its log.	$\bar{1}$·8714562
Divis. 12·9476	1·1121893
Quot. ·05744694	$\bar{2}$·7592669

Here the 1 taken from the $\bar{1}$ makes it become $\bar{2}$ to set down.

III. *The Rule of Three, or Proportion.*

Add the logarithms of the 2d and 3d terms together, and from their sum subtract the logarithm of the 1st, by the foregoing rules; the remainder will be the logarithm of the 4th term required.

Or in any compound proportion whatever, add together the logarithms of all the terms that are to be multiplied, and from that sum take the sum of the others; the remainder will be the logarithm of the term sought.

But instead of subtracting any logarithm, we may add its complement, and the result will be the same. By the complement is meant the logarithm of the reciprocal of the given number, or the remainder by taking the given logarithm from 0 or from 10, changing the radix from 0 to 10; the easiest method of doing which, is to begin at the left-hand, and subtract each figure from 9, except the last significant figure on the right hand, which must be subtracted from 10. But when the index is negative, add it to 9, and subtract the rest as before. And for every complement that is added, subtract 10 from the last sum of the indices.

Example 1.

To find a 4th proportional to 72·34, and 2.519, and 357·4862.

As 72·34 . comp. log. ....   8·1406215
To 2·519 ....   0·4012282
So 357·4862 .... ····   2·5532592

To 12·44827 ....   1·0951089

Example 2.

To find a 3d proportional to 12·796 and 3·24718.

As 12·796 . comp log. ....   8·8929258
To 3·24718 ....   0·5115064
So 3·24718 ....   0·5115064

To ·8240216 ....   1·9159386

Example 3.

To find a number in proportion to ·379145 as ·85132 is to ·0649.

As ·0649   comp. log. .....   11·1877553
To ·85132   .... $\overline{1}$·9300928
So ·379145   .... $\overline{1}$·5788054

To 4·973401 ....   0·6966535

Example 4.

If the interest of 100l. for a year or 365 days be 4·5l., what will be the interest of 279·25l. for 274 days?

As $\left\{\begin{array}{c}100\\365\end{array}\right\}$ comp. log. ...... $\left\{\begin{array}{c}8\cdot0000000\\7\cdot4377071\end{array}\right.$

To $\left\{\begin{array}{c}279\cdot25\\274\end{array}\right.$ ....   2·4459932
              ....   2·4377506

So 4·5 ....   0·6532125

To 9·433296 ....   0·9746634

IV. *Involution, or Raising of Powers.*

Multiply the logarithm of the number given by the proposed index of the power, and the product will be the logarithm of the power sought.

Note, In multiplying a logarithm with a negative index by any affirmative number, the product will be negative.—But what is to be carried from the decimal part of the logarithm will be affirmative.—Therefore the difference will be the index of the product; and it is to be accounted of the same kind with the greater

Example 1.

To find the 2d power of 2·5791.

Root 2·5791 its log. .... 0·4114682
        index..........   2

Power 6·651756 .... 0·8229364

Example 2.

To find the cube of 3·07146.

Root 3·07146 its log. .... 0·4873449
        index..........   3

Power 28·97575 .... 1·4620347

Example 3.

To find the 4th power of ·09163.

Root ·09163 its log. .... $\bar{2}$·9620377
        index   4

Power ·0000704938 .... $\bar{5}$·8481508

Here 4 times the negative index being 8, and 3 carried on, the difference 5 is the index of the product.

Example 4.

To find the 365th power of 1·0045.

Root 1·0045 its log. ... . 0·0019499
        index   365

                         97495
                      116994
                       58497

Power 5·148888 .... 0·7117135

V. *Evolution, or Extraction of Roots.*

Divide tne ogarithm of the power, or given number, by its index, and the quotient will be the logarithm of the root required.

Note, When the index of the logarithm is negative, and the divisor is not exactly contained in it without a remainder, increase it by such a number as will make it exactly divisible; and carry the units borrowed, as so many tens, to the left-hand place of the decimal part of the logarithm; then divide the results by the index of the root.

Example 1

To find the square root of 365.

Power 365 .... 2) 2·5622929
Root 19·10498 .... 1·2811465

Example 2.

To find the cube root of 12345.

Power 12345 .... 3) 4·0914911
Root 23·11162 .... 1·3638304

Example 3.

To find the 10th root of 2.

Power 2 .... 10) 0·3010300
Root 1·071773 .... 0·0301030

Example 4.

To find the 365th root of 1·045.

Power 1·045 ...... 365) 0·0191163
Root 1·000121 .... 0·0000524

Example 5.

To find the square root of ·093.

Power ·093 .... 2) $\bar{2}$·9684829
Root ·304959 .... $\bar{1}$·4842415

Here the divisor 2 is contained exactly once in $\bar{2}$ the negative index, there-fore the index of the quotient is $\bar{1}$.

Example 6.

To find the cube root of ·00048

Power ·00048 .... 3) $\bar{4}$·6812412
Root ·07829735 .... $\bar{2}$·8937471

Here the divisor 3 not being exactly contained in $\bar{4}$, augment it by 2, to make it become $\bar{6}$. in which the divisor is contained just $\bar{2}$ times ; and the 2 borrowed being carried to the other figures 6, &c., makes 2·6812412, which divided by 3 gives ·8937471.

OF THE TABLES FOR LOGARITHMS TO TWENTY PLACES.

THESE are tables 2d, 3d, and 4th. Of these, table 2 contains all numbers from 1 to 1000, and all uneven numbers from 1000 to 1161; with their logarithms to twenty places ; table 3 contains all numbers from 101000 to 101139, with their logarithms to twenty places, and the 1st, 2d, and 3d dif-ferences of those logarithms : and table 4 contains all logarithms regularly from 00001 to 00139, with their corresponding natural numbers to twenty places, as also the 1st, 2d, and 3d differences of those numbers. And by means of them may be found the logarithm to any other number, and the number to any other logarithm, to twenty places of figures.

(I.) *To find the Logarithms to given Numbers.*

CASE 1. If the given number *b* be found in any of these three tables; then its logarithm B is in the line even with it.

Case 2. If b is known to be the product or quotient of numbers found in these tables; then B is the sum or difference of the logarithms of those numbers.

Case 3. If a', the first six significant figures of a given number b', be found in table 3; let a' be an integer, A′ its logarithm; δ the remaining figures of b'; x the complement of δ to d' or 1; D′, D″, D‴, the 1st, 2d, 3d differences of the logarithms in the same line with A′; $f = \frac{1}{3}$ D‴ $\times$ $\overline{x+1}$ + D″ : Then B′ the logarithm of the number b' will be

$$(\text{D}' \times \delta) + \text{A}' \dots \dots \text{ to } 12$$
$$(\tfrac{1}{2} x \,\text{D}'' + \text{D}' \times \delta) + \text{A}' \dots \dots \text{ to } 17 \left.\right\} \text{places of figures nearly.}$$
$$(\tfrac{1}{2} x f + \text{D}' \times \delta) + \text{A}' \dots \dots \text{ to } 20$$

Example 1.

Given the number $b' = 0{\cdot}01010{,}26227{,}6351$, to find B′ its logarithm nearly to twelve places.

Here $a' = 101026$ A′ $= 00443{,}31579{,}747$

$\delta = 0{\cdot}2276351$ δ D′.... $+$ 9785,618 −

$\text{D}' = 429881746$ B′ $= \overline{2}{\cdot}00443{,}41365{,}365$ −

Example 2.

Given $b' = 0{\cdot}01010{,}26227{,}63509{,}626$, to find B′ its log. nearly to 17 places.   Here $a' = 101026$.

$\delta = 0{\cdot}22763{,}509626$; $x = 0{\cdot}772365$; D′ $= 42988{,}174579$; D″ $= 425510$.

Now $\tfrac{1}{2} x$ D″ $\dots \dots \dots \dots \dots$ 16432,45

         D′ $\dots \dots \dots \dots$ 42988,17457,86

$\tfrac{1}{2} x$ D″ $+$ D′ $\dots \dots \dots \dots$ 42988,33890,31

$\tfrac{1}{2} x$ D″ $+$ D′ $\times \delta$ $\dots \dots \dots$ 9785,65466,42

         A′ $\dots \dots$ 0$\cdot$0443,31579,74695,33

And $(\tfrac{1}{2} x$ D″ $+$ D′ $\times \delta) +$ A′, or B′ $\dots \dots$ $\overline{2}{\cdot}00443{,}41365{,}40161{,}75$

Example 3.

Given $b' = 0{\cdot}01010{,}26227{,}63509{,}62573{,}17345$, to find B′ its log. nearly to 20 places. $a' = 101026$.

$\delta = 0{\cdot}22763{,}50962{,}573173$; $x = 0{\cdot}77236{,}490374$; $x + 1 = 1{\cdot}772365$; D′ $= 42988{,}17457{,}86301$; D″ $= 42550{,}96343$; D‴ $= 84236$.

Now $\tfrac{1}{3}$ D‴ $\times \overline{x+1}$ $\dots \dots \dots$ 49766

         D″ $\dots \dots \dots$ 42550,96343

f $\dots \dots \dots \dots \dots$ 42551,46109

$\tfrac{1}{2} x f$ $\dots \dots \dots \dots \dots$ 16432,62757

         D′ $\dots \dots \dots$ 42988,17457,86301

$\tfrac{1}{2} x f +$ D′ $\dots \dots \dots$ 42988,33890,49058

$\tfrac{1}{2} x f +$ D′ $\times \delta$ $\dots \dots \dots$ 9785,65466,45604

         A′ $\dots \dots$ 00443,31579,74695,32791

And B′ $\dots \dots \dots \dots$ $\overline{2}{,}00443{,}41365{,}40161{,}78395$

Case 4. If the number b do not come under one of the preceding cases: put a for the first five figures of b; n for 101, the least, or some one, of the numbers in table 3; then $\dfrac{a}{n}$ or $\dfrac{n}{a} = a$ is to be had in table 2, with A its logarithm; let $b' = \dfrac{b}{a}$ or ba, and a' the first six significant figures of b' (found in table 3) be an integer, and a' its logarithm; put δ for the remaining figures of b'; x the complement of δ to d'; D', D'', E''', the 1st, 2d, 3d differences of the logarithms in the same line with A'; $f = \frac{1}{2} D''' \times \overline{x+1} + D''$. Then B the logarithm of the number b will be

$$\left. \begin{array}{l} \overline{D' \times \delta + A' \pm A = B' \pm A \text{ to } 12} \\ \overline{\frac{1}{2} x D'' + D' \times \delta + A' \pm A = B' \pm A \text{ to } 17} \\ \overline{\frac{1}{2} x f + D' \times \delta + A' \pm A = B' \pm A \text{ to } 20} \end{array} \right\} \text{places of figures nearly.}$$

Example.

Given $b' = 3{\cdot}14159,26535,89793,23846,26434$, to find B to twenty places.

Here $a = 31415$ \qquad\qquad Let $a = \dfrac{a}{n} = 311$.

Then $b' = \dfrac{b}{a} = 0{\cdot}01010,15840,95144,02970,57$; $a' = 101015$.

$\delta = 0\ 84095,14402,97057$; $\quad x = 0.15904,85597$; $\quad x + 1 = 1{\cdot}15905$;
$D' = 42992,85574,06337$; $\quad D'' = 42560,23099$; $\quad D''' = 84263$.

Now $\frac{1}{3} D''' \times \overline{x+1}$ 32555
$\qquad\qquad\qquad\qquad D''$ 42560,23099
f 42560,55654
$\frac{1}{2} x f$ 3384,59761
$\qquad\qquad D'$ 42992,85574,06337
$\frac{1}{2} x f + D'$ 42992,88958,66098
$\frac{1}{2} \overline{x f + D'} \times \delta$ 36154,93242,03919
$\qquad\qquad A'$ 00438,58681,74054,30961
$\qquad\qquad\qquad A$ 49276,03890,26837,50555
And B · .... 0·49714,98726,94133,85435

Or let $a = \dfrac{n}{a}\cdot = 3{\cdot}216 = 0{\cdot}536 \times 6$.

Then $b' = ba = 10{\cdot}10336,19739,44775,0549$; $a' = 101033$.
$\delta = 0{\cdot}61973,94477,50549$; $\quad x = 0{\cdot}38026,055225$; $\quad x + 1 = 1{\cdot}38026$;
$D' = 42985,19618,80760$; $\quad D'' = 42545,06747$; $\quad D''' = 84219$.

Now $\frac{1}{3} D''' \times \overline{x+1}$ 38748
$\qquad\qquad\qquad\qquad D''$ 42545,06747
f 42545,45495
$\frac{1}{2} x f$ 8089,17910
$\qquad\qquad D'$ 42985,19618,80760
$\frac{1}{2} x f + D'$ 42985,27707,98670
$\frac{1}{2} \overline{x f + D'} \times \delta$ 26639,67187,88811
$\qquad\qquad A'$ 00446,32488,03359,61854
$\qquad B'$ .... 1·00446,59127,70547,50665
$\qquad A$ .... 0·50731,60400,76413,65230
$B = B' - A$ .... 0·49714,98726,94133,85435

(II.) *To find the Numbers to given Logarithms.*

CASE 1. When the logarithm B is found in any of these three tables; then its number b is in the line even with it.

Case 2. If the first five figures (omitting the index) of a given logarithm B′, be between 00432 and 00492: take them as an integer, and put A′ and c′ for the logarithms, in table 3, next less and greater than B′, a′ and c′ their numbers; let D′ ($= c′ − A′$) and D″ be the 1st and 2d differences in the line with A′; $\Delta = B′ − A′$; $d′ = (c′ − a′ =) 1$; $x = \dfrac{D′ − \Delta}{D′}$; $\delta = \dfrac{\Delta}{D′ + \frac{1}{2} \times D″}$ then $b′ = a′ + \delta$, nearly true to 17 places of figures.

Example.

Given the logarithm B′ .... $= 5,00446,59127,70547,507$
to find b′ its number.   A′$= 5,00446,32488,03359,619$

   $a′ = 101033$ $\Delta = 0\cdot26639,67187,888$
   $\delta .... 0\cdot61973,944776$ D′ $= 0\cdot42985,19618,808$
   $b′ = 101033\cdot61973,944776$ D′ $- \Delta = 0\cdot16345,52430,920$
                              $x = 0\cdot38026$
                           D″ $= 0\cdot00000,42545$
                  $\frac{1}{2} \times$ D″ $= 0\cdot00000,08089,1$
        D′ $+ \frac{1}{2} \times$ D″ $= 0\cdot42985,27707,9$

But when any other logarithm B is given, subduct ·004321 from the first six figures of B: call the remainder R, and let A be the logarithm in table 2, next less than R, or next greater than the complement of R, and a its number: then B′ $=$ B $−$ A, or B′ $=$ B $+$ A, will be within the limits of table 3, and b will be found as in the preceding example; and if B′ $=$ B $−$ A, then $b = ab′$; or if B′ $=$ B $+$ A, then $b = \dfrac{b′}{a}$.

Case 3. If A′, the first five figures (omitting the index) of a given logarithm B′, be found in table 4: let a′ be its number; and put A′ as an integer, and Δ the remaining figures of B′, and x the complement of Δ to D′; d′, d″, d‴, the 1st, 2d, 3d differences of the numbers in the same line with a′; $f = d″ − \frac{1}{3} d‴ \times \overline{x + 1}$: then the number b′, whose logarithm is B′, will be

$$
\left.
\begin{array}{l}
(d′ \times \Delta) + a′ .... \text{to } 12 \\
\overline{(d′ − \frac{1}{2} \times d″} \times \Delta) + a′ .... \text{to } 17 \\
\overline{(d′ − \frac{1}{2} \times f} \times \Delta) + a′ .... \text{to } 20
\end{array}
\right\} \text{places of figures nearly.}
$$

Example.

Given the logarithm B′ $= 0\cdot00006,93311,37711,69929$, to find b′ its number to 20 places. Here A′ $= 00006$.

   $\Delta = 0\cdot93311,37711,69929$; $x = 0\cdot06688,622883$; $x + 1 = 1\cdot066886$;
   $d′ = 23029,29742,21293$; $d″ = 53027,52746$; $d‴ = 1\,22100$.

        Now $\frac{1}{3} d‴ \times \overline{x + 1}$ 43422
        $d″$ 53027,52746
      f 53027,09324
             $\frac{1}{2} \times f$ 1773,39115
      $d′$ 23029,29742,21293
   $d′ − \frac{1}{2} \times f$ 23029,27968,82178
   $d′ − \frac{1}{2} \times f \times \Delta$ 21488,93801,72000
              $a′$ 10001,38164,64943,57474
And b .... $1\cdot00015,96535,87452,9474$

CASE 4. If the logarithm B do not come under one of the preceding cases. Put A for the logarithm in table 2, next less than B, or next greater than the complement of B, and a its number; let $B' = B - A$, or $B' = B + A$; and A', the first five figures of B', may be had in table 4, with a' its number; put A' as an integer, and let Δ be the remaining figures of B'; x the complement of Δ to D'; d', d'', d''', the 1st, 2d, 3d differences of the numbers in the same line with a'; $f = d'' - \frac{1}{3}d''' \times \overline{x+1}$: then the number b', whose logarithm is B', will be

$$\left.\begin{array}{l} \overline{d' \times \Delta + a'} \times a = ab' \text{ to } 11 \\ \overline{d' - \frac{1}{2}x\,d' \times \Delta + a'} \times a = ab' \text{ to } 16 \\ \overline{d' - \frac{1}{2}x f \times \Delta + a'} \times a = ab' \text{ to } 19 \end{array}\right\} \text{places of figures nearly.}$$

Example.

Given $B = \overline{4}\cdot46372,61172,07184,15204$, to find b its number
Let $A = 1\cdot46239,79978,98956,08733.$ $a = 29.$

$B' = B - A = \overline{5}\cdot00132,81193,08228,06471.$ A' = 00132
$\Delta = 0\cdot81193,08228,06471$; x = $0\cdot18806,91772$; x + 1 = $1\cdot18807$;
$d' = 23096,20835,34589$; $d'' = 53181,59733$; $d''' = 1\cdot22457.$

Now $\frac{1}{3} d''' \times \overline{x+1}$	48496
d''	53181,59733
f	53181,11237
$\frac{1}{2}\times f$	5000,86402
d'	23096,20835,34589
$d' - \frac{1}{2}\times f$	23096,15834,48187
$d' - \frac{1}{2}\times f \times \Delta$	18752,48284,85771
a'	10030,44036,01963,96855
b'	10030,62788,50248,82626
$b = ab'$	0·00029,08882,08665,72159,6154

Or, given $B = \overline{4}\cdot46372,61172,07184,15204$, to find b.
Let $A = 2\cdot53655,84425,71530,11205.$ $a = 344.$

$B' = B + A = \overline{1}\cdot00028,45597,78714,26409.$ A' = 00028.
$\Delta = 0\cdot45597,78714,26409$; x = $0\cdot54402,21286$; x + 1 = $1\cdot54402$;
$d' = 23040,96629,91521$; $d'' = 53054,39634$; $d''' = 1\cdot22163$;

Now $\frac{1}{3} d''' \times \overline{x+1}$	62874
d''	53054,39634
f	53053,76760
$\frac{1}{2}\times f$	14431,21179
d'	23040,96629,91521
$d' - \frac{1}{2}\times f$	23040,82198,70342
$d' - \frac{1}{2}\times f \times \Delta$	10506,10496,55627
a'	10006,44931,70511,67281
b'	10006,55437,81008,22908
$b = \dfrac{b'}{a}$....	0 00029,08882,08665,72159,616

OF THE TABLE OF HYPERBOLIC LOGARITHMS.

This is table 5, in pages 219—223, which contain the series of numbers 1·01, 1·02, 1·03, &c., to 10·00, with their hyperbolic logarithms to seven places of figures. They are so called because they square the asymptotic spaces of the right-angled hyperbola; and they are very useful in finding fluents, and the sums of infinite series. The table, as well as the following rules, were first given at the end of Simpson's fluxions, but they were rendered much more correct in the French edition of Gardiner's tables, printed at Avignon in 1770, being very incorrect in the last figure in Simpson's book. But both those books are very erroneous in the example for finding logarithms by the table

1. *When the given Number is between 1 and 10.*

From the given number subtract the next less tabular number, divide the remainder by the said tabular number, increased by half the remainder; add the quotient to the logarithm of the said tabular number, and the sum will be the logarithm of the number proposed.

Example.

To find the hyperbolic logarithm of 3·45678

$$3\cdot45339 \,)\; \cdot00678 \;(\cdot0019633$$
$$1\cdot2383742$$
$$\text{log. } \overline{1\cdot2403375}$$

Here the next less number is 3·45, and its logarithm 1·2383742, the remainder or dividend ·00678, its half 339, which joined to the tabular number 3·45, gives the divisor; the quotient ·0019633 added to the tabular logarithm 1 2383742, gives 1·2403375, the required logarithm of 3·45678.

2. *When the given Number exceeds 10.*

Find the logarithm of the number as above, supposing all the figures after the first to be decimals, then to that logarithm add 2·3025851, or 4·6051702, or 6·9077553, &c., according as the given number contains 2, or 3, or 4, &c., places of integers. That is, add 2·302585092994 multiplied by the index of the power of 10, by which the given number was divided to bring it to one integer, or within the limits of the table.

Example.

To find the hyperbolic logarithm of 345·678.

$$1\cdot2403375$$
$$4\cdot6051702$$
$$\overline{5\cdot8455077}$$

This number divided by 100 or 10², to bring it within the limits of the table, or removing the decimal point two places, gives 3·45678, the logarithm of which as above found is 1·2403375, to which adding 4·6051702 the hyperbolic logarithm of 100, the sum is 5·8455077 the hyperbolic logarithm required of 345·678.

Note The hyperbolic logarithm of any number may be also found from Briggs's logarithms, viz. multiplying Briggs's logarithm of the same number by the hyperbolic logarithm of 10, viz.

Multiplying by . . . . 2·30258,50929,94045,68401.79914,
Or dividing by its reciprocal . ·43429,44819,03251,82765,11289.

Note 2. There is also now added a supplementary table of Hyperbolic Logarithms (Tab. 6.) of numbers from 1 to 1200.

OF THE LOGISTIC LOGARITHMS.

These are in table 7, pages 230—235, which contain the logistic logarithm of every second as far as the first 88′ or 5280″.

The logistic logarithm of any number of seconds is the difference between the logarithm of 3600″ and the logarithm of that number of seconds.

The chief use of the table of logistic logarithms is for the ready computing a proportional part in minutes and seconds, when two terms of the proportion are minutes and seconds, hours and minutes, or other numbers.

When two terms of the proportion are common numbers, their common logarithms may be used instead of their logistic logarithms, putting the logarithm where its complement should be, and the contrary.

1. *To find the Logistic Logarithm of any Number of Minutes and Seconds, within the Limits of the Table.*

At the top of the table find the minutes, and in the same column, even with the seconds on the left-hand side, is the logistic logarithm.

Note, When hours are made any terms of the proportion, they are to be taken as if they were minutes, and the minutes of an hour as if they were seconds.

2. *To find the Logistic Logarithm of any Number not exceeding 5280.*

In the 2d row, next the top of the table, find the number next less than that given; then in the same column, even with the difference on the left-hand side, is found the logistic logarithm.

When two given terms of the proportion are common numbers, one or both greater than 5280, take their halves, thirds, &c., instead of them. But when only one of the given terms is a common number, and that greater than 5280, take its half, third, &c., and multiply the 4th term by 2, 3, &c.

The logistic logarithms in this table are all affirmative, as well above as below 60′; but the index of those above 60′ is − 1; below 60′ down to 6′, the index is 0; and below 6′ the indices (being either 1, 2, or 3) are expressed in the table.

Examples.

		lo. log.			lo. log.			lo. log.
As 60′	.	0·0000	As 60′	..	0·0000	As 60′	.	0·0000
To 46′ 12″	..	0·1135	To 78′ 27″	..	I·8836	To 1531	..	0·3713
So 8 7	..	0·8688	So 13 53	..	0·6357	So 40′ 12″	..	0·1135
To 6 15	..	0 9823	To 18 9	..	0·5193	To 1179	..	0·4848
As 46′ 12″	co.	I·8865	As 78′ 27″	co.	0·1164	As 40′ 12″	co.	I·8865
To 60 0	..	0·0000	To 60 0	..	0·0000	To 1179	..	0·4848
So 6 15	..	0·9823	So 18 9	..	0·5193	So 60′ 0″	..	0 0000
To 8 7	..	0·8688	To 13 53	..	0·6357	To 1531	..	0·3713
As 60′	. co.	0·0000	As 24ʰ	.. co.	I·6021	As 24ʰ	.. co.	I·6021
To 4721	..	1·8823	To 46′ 11″	..	0·1137	To 76′ 34″	..	I·8941
So 37′ 28″	..	0·2045	So 8ʰ 7	..	0·8688	So 13ʰ 53′	..	0·6357
To 2948	..	0·0868	To 15′ 37″	.	0·5846	To 44′ 17″	..	0 1319
As 4721	.. co.	0·1177	As 46′ 11″	co.	I·8863	As 76′ 34″	co.	0·1059
To 60′ 0″	..	0·0000	To 24ʰ	..	0·3979	To 24ʰ	..	0·3979
So 2948	..	0·0868	So 15′ 37″	..	0·5846	So 44′ 17″	..	0·1319
To 37′ 28″	..	0·2045	To 8ʰ 7′	.	0·8688	To 13ʰ 53′	..	0·6357

The logistic logarithms may be used in trigonometrical operations, when two of the terms are small arcs, with the logarithmic sines or tangents of other arcs; observing, that instead of the logarithmic sine or tangent, to take the complement of their logistic logarithm; and the contrary.

But this may be as readily and more naturally done by the logarithmic sines and tangents themselves of such small arcs, as taken from the next following table of sines and tangents for every second of the first 2° or 120′.

OF THE LOGARITHMIC SINES AND TANGENTS TO EVERY SECOND.

Table 8, pages 238—267, contains the log. sines and tangents for every single second of the first 2 degrees of the quadrant; the sines being placed on the left-hand pages, and the tangents on the right. The degrees and minutes are placed at the top of the columns, and the seconds on the left-hand side of each page, the logarithmic sine or tangent being found in the common angle of meeting. So of 1° 52′ 54″ the log. sine is 8·5163420, and the log. tangent 8·5165762.

The same numbers are also the cosines and cotangents of the last 2 degrees of the quadrant, those degrees with their minutes being placed at the bottom of the columns, and their seconds ascending on the right-hand side of the pages. So the cosine of 88° 7′ 6″ is 8·5163420, and its cotangent 8·5165762.

When it is required to find the sine or tangent, &c. to 3ds, &c., or any other fractional part of a second, subtract the tabular sine or tangent of the complete seconds from the next to it in the table, and take the like proportional part of the difference; which part added to, or taken from, the said tabular sine or tangent, according as it is increasing or decreasing, will give the sine or tangent required.

Example.

To find the log. sine of 1° 52′ 54″ 25‴ or 1° 52′ 54″ $\frac{25}{60}$ or $\frac{5}{12}$.

$$
\begin{array}{ll}
\text{1° 52′ 54″ sine} & 8\text{·}5163420 \\
\text{1 52 55 ..} & 8\text{·}5164061 \\
\hline
\text{dif.} & 641 \\
& 5 \\
\hline
& 12)3205 \\
\hline
\text{pro. part.} & 267 \\
\text{1° 52′ 54″} & 8\text{·}5163420 \\
\hline
\text{1° 52′ 54″ 25‴} & 8\text{·}5163687
\end{array}
$$

Here the sine of 1° 52′ 54″ taken from the next leaves 641, which multiplied by 5 and divided by 12, or multiplied by 25 and divided by 60, gives 267 the pro. part; this added to the first sine gives that which was required.

On the contrary, if a sine or tangent be given, to find the corresponding arc; take the difference between it and the next less tabular number, and the difference between the next less and greater tabular numbers, so shall the less difference be the numerator, and the greater the denominator, of the fractional part to be added to the arc of the less tabular number; which fraction may also, if required, be either turned into a decimal, or into 3ds, &c., by multiplying the numerator by 60, and dividing by the denominator.

Example.

To find the arc whose sine is 8·5163900.

$$
\begin{array}{lll}
\text{1° 52′ 55″} & \text{....} & 8\text{·}5164061 \\
\text{1 52 54} & \text{....} & 8\text{·}5163420 \\
\text{1 52 54 45‴} & \text{....} & 8\text{·}5163900 \\
\hline
& \text{diff. ...} & 480 \\
& \text{diff. ...} & 641
\end{array}
$$

Finding the number is between the sines of 1° 52′ 55″ and 1° 52′ 54″, take the differences between the sines as in the margin, and the differences give $\frac{480}{641}$ for the fraction of a second, or $\frac{48}{64}$ nearly, which abbreviates to $\frac{3}{4}″ = 45‴$; and therefore the arc sought is 1° 52′ 54″ 45‴.

Where the 1st differences of the sines and tangents alter much, as near the beginning of the table, the 2d, 3d, &c. differences may be taken in, and then the logarithmic sine or tangent will be expressed by this series, viz. :—

$$
Q = A + x D' + x \cdot \frac{x-1}{2} D'' + x \cdot \frac{x-1}{2} \cdot \frac{x-2}{3} D''' \text{ &c., or nearly } A + (D' - \tfrac{1}{2} D'')\, x ;
$$

where A is the next less tabular logarithm, D′, D″, D‴, &c., the 1st, 2d, 3d, &c., differences of the tabular logarithms, and x the fractional part of the arc over the complete seconds.

Example.

To find the log. tangent of 5′ 1″ 12‴ 24⁗ or 5′ 1″ $\frac{6\,2}{3\,0\,0}$ or 5′ 1″ 206.

Tang.	D′	D″
5′ 0″ .. 7·1626964		
5 1 .. 7·1641417	14453	−49
5 2 .. 7·1655821	14404	−47
5 3 .. 7·1670178	14357	

Here A = 7·1641417; $x = \frac{6\,2}{3\,0\,0}$; D′ = 14404; and the mean 2d diff. D = − 48. Hence

$$A \dots \dots \dots \quad 7·1641417$$
$$x\,D′ \dots \dots \quad 2977$$
$$x\,.\frac{x-1}{2}\,D″ \dots \dots \quad 4$$

Therefore the tangent of 5′ 1″ 12‴ 24⁗ .... 7·1644398

And on the other hand, when the sine or tangent is given, and falls near the beginning of the table, from the same series we may find *x* the fractional part of a second. For suppose it be required to find the arc whose tangent is 7·1644398. This falling between the tangents of 5′ 1″ and 5′ 2″, take the differences, &c., as above, and the series gives 7·1644398 = 7·1641417 $+ x\,D′ + x\,.\frac{x-1}{2}\,D″$; or 2981 = 14404*x* − 24 (*x²* − *x*), or − 24*x²* + 14428*x* = 2981; which gives *x* = ·2067″ nearly = 12‴ 24⁗. Therefore the arc required is 5′ 1″ 12‴ 24⁗. Or rather the approximate value A + D′ − $\frac{1}{2}$D″ . *x* = Q, gives $x = \frac{Q-A}{D′-\frac{1}{2}D″} = \frac{2981}{14404+24} = \frac{2981}{14428} = ·2067$, the same as before.

OF THE LARGE TABLE OF NATURAL AND LOGARITHMIC SINES, TANGENTS, SECANTS, AND VERSED SINES.

Table 9, page 268—357, contains all the sines, tangents, secants, and versed sines, both natural and logarithmic, to every minute of the quadrant, the degrees at top, and minutes descending down the left-hand side as far as 45°, or the middle of the quadrant, and from thence returning with the degrees at the bottom, and the minutes ascending by the right-hand side to 90°, or the other half of the quadrant, in such sort, that any arc on the one side is on the same line with its complement on the other side; the respective sines, cosines, tangents, cotangents, &c., being on the same line with the minutes, and in the columns signed with their respective names, at top when the degrees are at top, but at the bottom when the degrees are at the bottom. The natural sines, tangents, &c., are placed all together on the left-hand pages, and the logarithmic ones all together, facing them, on the right-hand pages. Also in the naturals there are two columns of the common differences, and in the logarithmic 3 columns of common differences, each column of differences being placed between the two columns of numbers having the same differences; so that these differences serve for both their right-hand and left-hand adjacent columns: also each differential number is set opposite the space between the numbers whose difference it is. The numbers on the same line in those columns having such common differences, are mutually complements of each other; so that the sum of the decimal figures of any two such numbers is always 1 integer, with 0 in each place of decimals

All this will be evident by inspecting one page of each sort, as well as the method of taking out the sine, &c., to any degrees and complete minutes. It is, however, to be observed, that in all the log. sines, tangents, &c., and in such of the natural as have any significant figure for their index or characteristic, the indices are expressed in the table, and the separating point is placed between the index and the decimal part of the number; but in several columns of the natural sines, &c., having 0 for their integer or index, both the index and decimal separating point are omitted; and wherever this is the case, it is to be understood that all the figures in such columns are decimals, wanting before them only the separating point and index 0.

The sine, tangent, or secant of any arc, has the same value, or is expressed by the same number, as the sine, tangent, or secant of the supplement of that arc; for which reason the tables are carried only to a quadrant or 90 degrees. So that when an arc is greater than 90°, subtract it from 180°, and take the sine, tang. or secant of the remainder, for that of the arc given. But this property does not take place between the versed sines of arcs and their supplements; and to find the versed sine of an arc greater than 90°, proceed thus: in the natural versed sines, to radius add the natural cosine, the sum will be the natural versed sine; and in the log. versed sines, add 0·3010300 to twice the log. sine of half the arc, the sum, abating radius 10·0000000, will be the log. versed sine required.

1. *Given any Arc ; to find its Sine, Cosine, Tangent, &c.*

Seek the degrees at the top or bottom, and the minutes respectively on the left or right; then on the same line with these is the sine, &c. each in its proper column, the title being at the top or bottom, according as the degrees are.

But when the given arc contains any parts of a minute, intermediate to those found in the table, take the difference between the tabular sines, &c. of the given degrees and minutes, and of the minute next greater; then take the proportional part of that difference for the parts of the minute, and add to it the sine, tangent, secant, and versed sine, or subtract it from the cosine cotangent, cosecant, or coversed sine, of the given degrees and minutes; so shall the sum or remainder be the sine, &c. required.

Note, The proportional part is found thus, as 1′ is to the given intermediate part of a minute, so is the whole difference to the proportional part required; which therefore is found by multiplying the difference by the said intermediate part. Also that intermediate part may be expressed either by a vulgar fraction, or a decimal, or a sexagesimal in seconds, thirds, &c., and the fraction or sexagesimal may be first reduced to a decimal, if it be thought better so to do, by dividing the numerator of the fraction by the denominator, or by dividing the sexagesimal by 60.

Example 1.

To find the natural sine of 1° 48′ 28″ 12‴.

In the column of difference between the natural sines of 1° 48′ and 1° 49 is the difference 2907; and 28″ 12‴ being = 28·2″ = ·47′; therefore as

$$1 : 2907 :: ·47 : \text{the pro. part} + 1366$$
$$\text{to which add sin. } 1° 48′ \ldots 0314108$$
$$\overline{\text{makes sin. of } 1° 48′ 28″ 12‴ \quad 0315474}$$

Example 2.

To find the natural tangent of $8°\,9'\,10''\,24'''$.

  $8°\,10$ tang. 1435084

  8 9 1432115

            diff. 2969

$1:2969::(10''\,24'''=)\ \cdot 17'\tfrac{1}{3}:+515$

  $8°\,9'$ 1432115

  $8°\,9'\,10''\,24'''$ 1432630

Example 3.

To find the natural coversed sine of $4°\,6'\,5''\,40'''$

$\left.\begin{array}{l}1\ .\ 2902\ (\text{tab. dif.})::\tfrac{17'}{180}=\\[2pt]5''\,40''':\text{pro part.}......\end{array}\right\}-274$

  $4°\,6'$ covers............ 9285026

  $4°\,6'\,5''\,40''''$ 9284752

Example 4.

To find the logarithmic cosine of $6°\,8'\,42''$.

$1:136\ (\text{tab. dif.})::\cdot 7'=42'':\text{pr. pt.}-95$

  $6°\,8'$ cosine .... 9·9975069

  $6°\,8'\,42''$............. 9·9974974

Example 5.

To find the log. sec. of $7°\,12'\,50''$.

$1:160\ (\text{tab. dif.})::\tfrac{5'}{6}=50'':\text{pr. pt.}+133$

  $7°\,12'$ secant......... 10·0034381

  $7°\,12'\,50''$ .... 10·0034514

Example 6.

To find the logarithm cotangent of $39°\,4'\,12''\,20'''$

$\left.\begin{array}{l}1:2581\ (\text{tab. dif.})::\cdot 20\tfrac{5}{9}=\\[2pt]12''\,20''':\text{pro. part ...}\end{array}\right\}-531$

  $39°\,4'$ cotan......... 10·0905978

  $39°\,4'\,12''\,20'''$ ...... 10·0905447

The foregoing method of finding the proportional part of the tabular difference, to be added or subtracted, by one single proportion, is only true when those differences are nearly equal, and may do for all except for the tangents and secants of large arcs near the end of the quadrant in the natural sines, &c., and in the log. sines, &c., except the sines and versed sines of small arcs, the tangents of both large and small arcs, and the secants of large arcs. And when much accuracy is required, these excepted parts may be found by the series used in the last article, viz. $Q = A + xD' + x\ .\ \dfrac{x-1}{2}\ D''$

$+\ x\ .\ \dfrac{x-1}{2}\ .\ \dfrac{x-2}{3}\ D'''$ &c. or $= A + (D' - \tfrac{1}{2}D'')\ .\ x$ nearly; where A is the tabular number for the degrees and minutes, D', D'', D''', &c., the 1st, 2d, 3d, &c. tabular differences, and x the fractional part over the complete minutes,

&c.; at least it may be proper to find the tangents and secants of very large arcs from this series; but as to the log. sines, versed sines, and tangents of small arcs, they may also be found, perhaps easier, from their corresponding natural ones, viz. find the natural sine, versed sine, or tangent of the given small arc, and then find the log. of such natural number by the 1st or large table of logarithms, which will be the log. sine, &c. required. And the log. tangent and secant of large arcs will be also found by taking the difference between 20 and their log. cotangent and cosine respectively. And lastly, the natural tangents and secants of large arcs may also be found by first finding their log. tangent and secant, and then finding the corresponding number

Example 1.

To find the log. sine of 1° 48′ 28″ 12‴.

The natural sine, found in Ex. 1. above, is ·0315474; and the log. of this is 8·4989636, which is the log. sine required.

Example 2.

To find the log. versin. of 1° 48′ 28″ 12‴.

1° 48′ nat. vers. 0004934
1 : 92 tab. dif. :: ·47′ = 28″ 12‴ : + 43

1° 48′ 28″ 12‴ nat. vers. ·0004977
Its log. 1 48 28 12 log. vers. 6·6969676

Example 3.

To find the log. tang. of 2° 23′ 33″ 36‴.

2° 23′ its nat. tan. 0416210
1 : 2914 tab. dif. :: ·56′ = 33″ 36‴ : + 1632

2° 23′ 33″ 36‴ nat. tan. .. 0417842
Its log. 2 23 33 36 log. tang. 8·6210121

Example 4.

To find the log. tang. of 87° 36′ 26″ 24‴.

Its complement is .... 2° 23′ 33″ 36″
Whose log. tang. in Ex. 3 is .. 8·6210121
Taken from .... 20·0000000

Leaves log. tan. 87° 36′ 26″ 24‴ 11·3789879

Example 5.

To find the log. sec. of 88° 11′ 31″ 48‴.

Its complement is .... 1° 48′ 28″ 12‴
Its log. sine in Ex. 1 is .... 8·4989636
Which taken from .... 20·0000000

Leaves log. sec. 88° 11′ 31″ 48‴ 11·5010364

Example 6.

To find the nat. sec. of $88°.\ 11'\ 31''\ 48'''$.

	nat. sec.	D′	D″	D‴
88° 11′	31·544246	291979		
88 12	31·836225	297438	5459	
88 13	32·133663	303050	5612	153
88 14	32·436713			

Hence A $= 31·544246$; D′ $= 291979$; D″ $= 5459$; and D‴ $= 153$;

$x = ·53' = 31''\ 48'''$; $x \cdot \dfrac{x-1}{2} = -·12455$; $x \cdot \dfrac{x-1}{2} \cdot \dfrac{x-2}{3} = ·06125$.

$$\begin{array}{lr}
\text{Then A} \ldots\ldots\ldots\ldots\ldots & 31·544246 \\
x\ \text{D}' \ldots\ldots\ldots\ldots\ldots\ldots & 154748 \\
x \cdot \dfrac{x-1}{2}\ \text{D}'' \ldots\ldots\ldots\ldots & -680 \\
x \cdot \dfrac{x-1}{2} \cdot \dfrac{x-2}{3}\ \text{D}''' \ldots\ldots & 9 \\
\hline
& 31·698323
\end{array}$$

In the 6th example, the natural secant is found by the differential series to be 31·698324. But by taking the number to the logarithm of it, as found in the 5th example, it is 31·698329; the difference may be owing to the logs. not being far enough continued. But this method by the series seems to be, in many instances, more troublesome than finding the secant by dividing 1 by the cosine.

2. *Given any Sine, Tangent, &c. to find its Arc.*

Take the difference between the next less and greater tabular number of the same kind, and the difference between the given number and said next less or next greater tabular number, according as the given number is a sine, tangent, &c., or a cosine, cotangent, &c., noting its degrees and minutes; then the two differences will be the terms of a vulgar fraction of a minute, to be added to those minutes, to give the arc required.

And this vulgar fraction may also, if required, be reduced to a decimal by dividing the less or numerator by the denominator; or brought to sexagesimals, by multiplying by 60, &c. Also, where the tabular differences are printed, the subtraction of the less tabular number from the greater is saved.

Example 1.

To find the arc to the natural sine ·0315474.

$$\begin{array}{lr}
\text{Ans.} \ldots\ldots\ 1°\ 48'\ 28''\ 12'''\ \ldots & 0315474 \\
\text{Subtr.} \ldots\ldots\ 1\ \ 48'\ \text{next less} \ldots & 0314108 \\
\hline
& 1366 \\
& 60 \\
\hline
\text{Tab. difference} \ldots\ldots\ 2907\ )\ & 81960\ (\ 28'' \\
& 5814 \\
\hline
& 23820 \\
& 23256 \\
\hline
& 564 \\
& 60 \\
\hline
2907\ )\ & 33840\ (\ 12'''
\end{array}$$

Example 2.

To find the arc to natural tang. ·1432630.
Next greater 1435084
Answer 8° 9′ 10″ 24‴ 1432630
Next less, subt. fr. each. 1432115
                                          ———
                                          515
                                          60
                                          ———
Tab. difference . . . . . . 2969) 30900 (10″
                                    29690
                                    ———
                                    1210
                                    60
                                    ———
                                    72600 (24‴
                                     5938
                                    ———
                                    13220

Example 3.

To find the arc to logarithm cosine 9·9974974.
          6° 8′ . . . . 9·9975069
Answer 6° 8′ 42″ . . . . 9·9974974
                                        ———
                                        95
                                        60
                                        ———
Tab. difference . . . . 136) 5700
                                       544
                                       ———
                                       260

Example 4.

To find the arc to logarithm cot. 10·0905447.
          39° 4′ . . . . 10·0905978
Answer 39° 4′ 12″ 20‴ . . . . 10·0905447
                                       ———
                                       531
                                       60
                                       ———
Tab. difference . . . . . . 2581) 31860 (12″
                                    2581
                                    ———
                                    6050
                                    5162
                                    ———
                                    888
                                    60
                                    ———
                    2581) 53280 (20‴
                           5162
                           ———
                           1660

The above method of proportioning by the first difference alone, can only be true when the other differences are nothing, or very small; but other means must be used when they are large, viz. for the natural tangents and secants of very large arcs; and for the logarithmic sines, and versed sines of small arcs, also the log. secants of large arcs, with the log. tangents and cotangents both of small and large arcs. When the log. sine, versed sine, or tangent of a small arc is given, by means of the table of logarithms find the corresponding natural number, and then the arc answering to it in the table of natural sines, &c. But when the log. tangent or secant of a large arc is proposed, subtract it from 20, the remainder is the log. cotangent or cosine which will be the log. tangent or sine of a small arc which is the complement of that required, which complement will be found as in the last remark, by taking the corresponding natural number, and finding it in the natural tangents or sines; then subtracting that complemental arc from 90°, leaves the required large arc answering to the proposed log. tangent or secant. And when the natural tangent or secant of a large arc is proposed, change it into the log. tangent or secant of the same, by taking the log. of the proposed natural number; then proceed with it as above in the last remark.—Or, what relates to the log. sines and tangents of small arcs, or cosines and cotangents of large ones, will be best performed by the foregoing table for every second of the first 2 degrees.

Example 1.

To find the arc to natural tangent 50·0000000.

```
                                  20·0000000
    Given 50 0000000.... its log. 11·6989700
              ·02 ..............   8·3010300
          ·0197830 nat. tan. of 1° 8′
              2170
                60
    2910)  130200    (44″
           1164
           1380
           1164
            216
             60
          12960       (44‴
           1164
           1320
```

Hence from .... 90°   0′   0″   0‴
Take the comp. 1   8   44   44
Leaves arc required 88   51   15   16

Example 2.

To find the arc to natural secant 31·6983333.

```
                                  20·0000000
    Given 31·698½ its log...... 11·5010365
           ·0315474 .........   8·4989635
          ·0314108 nat. sine of 1° 48′
              1366
                60
    2907)  81960    (28″
           5814
          23820
          23256
            564
             60
          33840       (12‴
           2907
           4770
```

Hence from .... 90°   0′   0″   0‴
Take the comp. 1   48   28   12
Leaves arc required 88   11   31   48

The following rules in reference to the sines and tangents of small arcs, given by Dr. Maskelyne, in his introduction to *Taylor's Logarithms*, will often be useful :—

1. *To find the sine.* To the logarithm of the arc reduced into seconds, with the decimal annexed, add the constant quantity 4·6855749, and from the sum subtract one-third of the arithmetical complement of the log. cosine, the remainder will be the log. sine of the given arc.

2. *To find the tangent.* To the log. arc and the above constant quantity, add two-thirds of the arithmetical complement of the log. cosine, the sum is the log. tangent of the given arc.

3. *To find the arc from the sine.* To the given log. sine of a small arc 5·3144251, add ⅓ of arith. comp. of the log. cosine: subtract 10 from the index of the sum, the remainder will be the log. of the number of *seconds* and decimals in the arc sought.

4. *To find the arc from the tangent.* To the log. tangent add 5·3144251, and from the sum subtract ⅔ of the arith. comp. of log. cosine: take 10 from the index, and there will remain the logarithm of the number of *seconds* and decimals of a second in the given arc.

*** See, also, Dr. Maskelyne's remarks, before the directions for the use of the *Traverse Table.*

On the Construction of these Tables.

The computation of tables of sines, tangents, secants, &c. is evidently a work of great labour; but, considerable as it doubtless is, it has been much diminished by the employment of various trigonometrical theorems, discovered by different mathematicians. A few of these theorems, being of general utility in the doctrine of Trigonometry, are here presented.

Theorem 1. The square of the diameter of a circle is equal to the sum of the squares of the chord of an arc, and of the chord of its supplement to a semicircle.

2. The rectangle under the two diagonals of any quadrilateral inscribed in a circle, is equal to the sum of the two rectangles under the opposite sides.

3. The sum of the squares of the sine and cosine (often called the sine of the complement), is equal to the square of the radius.

4. The difference between the sines of two arcs that are equally distant from 60 degrees, or ⅙ of the whole circumference, the one as much greater as the other is less, is equal to the sine of half the difference of those arcs, or of the difference between either arc and the said arc of 60 degrees.

5. The sum of the cosine and versed sine is equal to the radius.

6. The sum of the squares of the sine and versed sine is equal to the square of the chord, or to the square of double the sine of half the arc.

7. The sine is a mean proportional between half the radius and the versed sine of double the arc.

8. A mean proportional, between the versed sine and half the radius, is equal to the sine of half the arc.

9. As radius is to the sine, so is twice the cosine to the sine of twice the arc.

10. As the chord of an arc is to the sum of the chords of the single and double arc, so is the difference of those chords to the chord of thrice the arc.

11. As the chord of an arc is to the sum of the chords of twice and thrice the arc, so is the difference of those chords to the chord of five times the arc.

12. And in general, as the chord of an arc is to the sum of the chords of n times and $n+1$ times the arc, so is the difference of those chords to the chord of $2n+1$ times the arc.

13. The sine of the sum of two arcs is equal to the sum of the products of the sine of each multiplied by the cosine of the other, and divided by the radius.

14. The sine of the difference of two arcs is equal to the difference of the said two products divided by radius.

15. The cosine of the sum of two arcs is equal to the difference between the products of their sines and of their cosines, divided by radius.

16. The cosine of the difference of two arcs is equal to the sum of the said products divided by radius.

17. A small arc is equal to its chord or sine, nearly.

18. As cosine is to sine, so is radius to tangent.

19. Radius is a mean proportional between the tangent and cotangent.

20. Half the difference between the tangent and cotangent of an arc is equal to the tangent of the difference between the arc and its complement. Or, the sum arising from the addition of double the tangent of an arc with the tangent of half its complement, is equal to the tangent of the sum of that arc and the said half complement.

21. The square of the secant of an arc is equal to the sum of the squares of the radius and tangent.

22. Radius is a mean proportional between the secant and cosine. Or, as cosine is to radius, so is radius to secant.

23. Radius is a mean proportional between the sine and cosecant.

24. The secant of an arc is equal to the sum of its tangent, and the tangent of half its complement. Or, the secant of the difference between an arc and its complement is equal to the tangent of the said difference added to the tangent of the less arc.

25. The secant of an arc is equal to the difference between the tangent of that arc and the tangent of the arc added to half its complement. Or the secant of the difference between an arc and its complement, is equal to the difference between the tangent of the said difference and the tangent of the greater arc.

26. Sir Isaac Newton also proposed a method of computing a table of sines by means of multiple angles; and in which radius, or 1, is the first term, and double the sine or cosine of the first angle is the 2d term of all the proportions by which the several successive multiple sines or cosines are found. The substance of the method is this: the best foundation for the construction of the tables of sines, is the continual addition of a given angle to itself or to another given angle. As if the angle A be to be added; inscribe

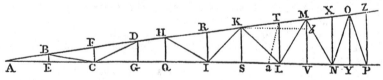

HI, IK, KL, LM, MN, NO, OP, &c., each equal to the radius AB; and to the opposite sides draw the perpendiculars BE, HQ. IR, KS, LT, MV, NX, OY, &c., so shall the angle A be the common difference of the angles HIQ, IKH, KLI, LMK, &c.; their sines HQ, IR, KS, &c.; and their cosines IQ, KR, LS, &c. Now let any one of them, LMK, be given, then the rest will be thus found:—
Draw Ta and Kb perpendicular to sv and MV; now because of the equiangular triangles ABE, TLa, KMb, ALT, AMV, &c., it will be AB : AE $\because$ KT : sa $(= \frac{1}{2}$LV $+ \frac{1}{2}$Ls$) \because$ LT : Ta $(= \frac{1}{2}$ MV $+ \frac{1}{2}$ KS,$)$ and AB : BE $\because$ LT : Ja $(= \frac{1}{2}$ LS $- \frac{1}{2}$ LV$) \because$ KT $(= \frac{1}{2}$KM$)$: $\frac{1}{2}$Mb $(= \frac{1}{2}$MV $- \frac{1}{2}$KS.$)$ Hence are given

the sines and cosines KS, MV, LS, LV. And the method of continuing the progressions is evident. Namely,

$$\text{as AB} : 2\text{AE} :: \begin{cases} \text{LV} : \text{MT} + \text{MX} :: \text{MX} : \text{NV} + \text{NY}, \&c. \\ \text{MV} : \text{NX} + \text{LT} :: \text{NX} : \text{OY} + \text{MV}, \&c. \end{cases}$$

$$\text{or AB} : 2\text{BE} :: \begin{cases} \text{LV} : \text{NX} - \text{LT} :: \text{MX} : \text{OY} - \text{MV}, \&c \\ \text{MV} : \text{MT} - \text{MX} :: \text{NX} : \text{NV} - \text{NY}, \&c. \end{cases}$$

And on the other hand, AB : 2AE :: LS : KT + KR, &c.
Therefore put AB = 1, and make BE × LT = La, AE × KT = sa, sa − La = LV,
2AE × LV − TM = MX, &c.

The sense of these general theorems is this, that if P be any one among a series of angles in arithmetical progression, the angle d being their common difference, then as radius or

$$1 : 2 \cos. d :: \begin{cases} \cos. P : \cos. (P+d) + \cos. (P-d) \\ \sin. P : \sin. (P+d) + \sin. (P-d) \end{cases}$$

$$1 : 2 \sin. d :: \begin{cases} \cos. P : \sin. (P+d) - \sin. (P-d) \\ \sin. P : \cos. (P+d) - \cos. (P-d) \end{cases}$$

where the 4th terms of these proportions are the sums or differences of the sines or cosines of the two angles next less and greater than any angle P in the series; and therefore subtracting the less extreme from the sum, or adding it to the difference, the result will be the greater extreme, or the next sine or cosine beyond that of the term P. And in the same manner are all the rest to be found. This method, it is evident, is equally applicable whether the common difference d, or angle A, be equal to one term of the series or not: when it *is* one of the terms, then the whole series of sines and cosines becomes thus, viz. as 1 : 2 cos. d ::

sin. d : sin. $2d$ :: sin. $2d$: sin. d + sin. $3d$ :: sin. $3d$: sin. $2d$ + sin. $4d$ :: sin. $4d$: sin. $3d$ + sin. $5d$, &c.

cos. d : 1 + cos. $2d$ :: cos. $2d$: cos. d + cos. $3d$ :: cos. $3d$: cos. $2d$ + cos. $4d$ :: cos. $4d$: cos. $3d$ + cos. $5d$, &c.

27. The following values of the natural sine for every third degree in the quadrant, have greatly contributed to facilitate the computations :—

sin. $0° = 0$

sin. $3° = \frac{1}{8}[\sqrt{(5+\sqrt{5})} + \sqrt{1\frac{1}{2}} + \sqrt{\frac{1}{2}} - \sqrt{(15+3\sqrt{5})} - \sqrt{\frac{1}{2}} - \sqrt{\frac{1}{2}}]$

sin. $6° = \frac{1}{8}[\sqrt{(30-6\sqrt{5})} - 1 - \sqrt{5}]$

sin. $9° = \frac{1}{4}[\sqrt{\frac{5}{2}} + \sqrt{\frac{1}{2}} - \sqrt{(5-\sqrt{5})}]$

sin. $12° = \frac{1}{8}[\sqrt{3} + \sqrt{(10+2\sqrt{5})} - \sqrt{15}]$

sin. $15° = \frac{1}{2}[\sqrt{\frac{3}{2}} - \sqrt{\frac{1}{2}}]$

sin. $18° = \frac{1}{4}[\sqrt{5} - 1]$

sin. $21° = \frac{1}{8}[\sqrt{(15-3\sqrt{5})} + \sqrt{(5-\sqrt{5})} + \sqrt{\frac{5}{2}} + \sqrt{\frac{1}{2}} - \sqrt{1\frac{1}{2}} - \sqrt{\frac{3}{2}}$

sin. $24° = \frac{1}{8}[\sqrt{15} + \sqrt{3} - \sqrt{(10-2\sqrt{5})}]$

sin. $27° = \frac{1}{4}[\sqrt{(5+\sqrt{5})} + \sqrt{\frac{1}{2}} - \sqrt{\frac{5}{2}}]$

sin. $30° = \frac{1}{2}$

sin. $33° = \frac{1}{8}[\sqrt{(15+3\sqrt{5})} + \sqrt{1\frac{1}{2}} + \sqrt{\frac{3}{2}} - \sqrt{(5+\sqrt{5})} - \sqrt{\frac{3}{2}} - \sqrt{\frac{1}{2}}]$

sin. $36° = \frac{1}{4}\sqrt{(10-2\sqrt{5})}$

sin. $39° = \frac{1}{8}[\sqrt{(5-\sqrt{5})} + \sqrt{1\frac{1}{2}} + \sqrt{\frac{5}{2}} + \sqrt{\frac{3}{2}} + \sqrt{\frac{1}{2}} - \sqrt{(15-3\sqrt{5})}]$

sin. $42° = \frac{1}{8}[1 + \sqrt{(30+6\sqrt{5})} - \sqrt{5}]$

sin. $45° = \sqrt{\frac{1}{2}}$

sin. $48° = \frac{1}{8}[\sqrt{15} + \sqrt{(10+2\sqrt{5})} - \sqrt{3}]$

sin. $51° = \frac{1}{8}[\sqrt{(15-3\sqrt{5})} + \sqrt{(5-\sqrt{5})} + \sqrt{1\frac{1}{2}} + \sqrt{\frac{3}{2}} - \sqrt{\frac{5}{2}} - \sqrt{\frac{1}{2}}]$

sin. $54° = \frac{1}{4}[\sqrt{5} + 1]$

$\sin. 57° = \frac{1}{8} \left[\surd (15 + 3\surd5) + \surd (5 + \surd5) + \surd\frac{5}{2} + \surd\frac{3}{2} - \surd\frac{15}{2} - \surd\frac{1}{2} \right]$

$\sin. 60° = \frac{1}{2}\surd3$

$\sin. 63° = \frac{1}{4} \left[\surd (5 + \surd5) + \surd\frac{5}{2} - \surd\frac{1}{2} \right]$

$\sin. 66° = \frac{1}{8} \left[1 + \surd (30 - 6\surd5) + \surd5 \right]$

$\sin. 69° = \frac{1}{8} \left[\surd (15 - 3\surd5) + \surd\frac{15}{2} + \surd\frac{5}{2} + \surd\frac{3}{2} + \surd\frac{1}{2} - \surd (5 - \surd5) \right]$

$\sin. 72° = \frac{1}{4}\surd (10 + 2\surd5)$

$\sin. 75° = \frac{1}{2} \left[\surd\frac{3}{2} + \surd\frac{1}{2} \right]$

$\sin. 78° = \frac{1}{8} \left[\surd5 + \surd (30 + 6\surd5) - 1 \right]$

$\sin. 81° = \frac{1}{4} \left[\surd\frac{5}{2} + \surd\frac{1}{2} + \surd (5 - \surd5) \right]$

$\sin. 84° = \frac{1}{8}.\left[\surd15 + \surd3 + \surd (10 - 2\surd5) \right]$

$\sin. 87° = \frac{1}{8} \left[(\surd5 + \surd5) + \surd (15 + 3\surd5) + \surd\frac{15}{2} + \surd\frac{1}{2} - \surd\frac{5}{2} - \surd\frac{3}{2} \right]$

$\sin. 90° = 1$

$\surd 3 = 1\cdot7320508076$ $\surd\frac{5}{2} = 1\cdot5811388301$

$\surd 5 = 2\cdot2360679775$ $\surd\frac{15}{2} = 2\cdot7386127875$

$\surd15 = 3\cdot8729833462$ $\surd (5 + \surd5) = 2\cdot6899940479$

$\surd\frac{1}{2} = 0\cdot7071067812$ $\surd (5 - \surd5) = 1\cdot6625077511$

$\surd\frac{3}{2} = 1\cdot2247448714$ $\surd (10 + 2\surd5) = 3\cdot8042260652$

$\surd (10 - 2\surd5) = 2\cdot3511410092$

$\surd (15 + 3\surd5) = 4\cdot6592063629$

$\surd (15 - 3\surd5) = 2\cdot8795478929$

$\surd (30 + 6\surd5) = 6\cdot5891128284$

$\surd (30 - 6\surd5) = 4\cdot0722956836$

28. The natural sines, tangents, secants, &c., being computed by means of these and other convenient theorems, are arranged in order on the left-hand pages of table 9; and the logarithmic sines, tangents, secants, &c., belonging to the same degrees and minutes respectively, stand on the opposite right-hand pages. They are, of course, easily ascertained by computing, one by one, the logarithms of the sines, tangents, &c., to which they belong. Thus, 9·3836752, the log. sine of 14°, is simply the log. of ·2419219, the natural sine of 14°; 9·3967711, the log. tangent of 14° is the logarithm of ·2493280, the natural tangent of 14°; and so on.

29. But, all this may indeed be accomplished with much greater facility. For, since sin. ÷ cos. = tan.; tan. : rad. :: rad. : cot.; sin. : rad. :: rad. : cosec.; cos. : rad. :: rad. : sec.; we may, after having found the logarithmic sines and cosines; or, in other words, all the sines, minute by minute, &c. to 90°, proceed thus:—

     From 10 + sin. take cos. there remains tan.

     From 20 take tan., there remains cot.

     From 20 take cos., there remains sec.

     From 20 take sin., there remains cosec.

Add ·3010300 to twice the log. sin. of half the arc, and take 10 from the index of the sum, the remainder is the log. versin.

Add ·3010300 to twice the log. sin. of half the comp. of the arc, and take 10 from the index of the sum, the remainder is the log. coversin.

TRIGONOMETRICAL RULES.

1. In a right-lined triangle, whose sides are A, B, C, and their opposite angles, a, b, c; having given any three of these, of which one is a side; to find the rest.

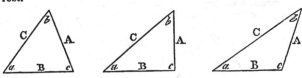

Put s for the sine, s′ the cosine, t the tangent, and t′ the cotangent, of an arch or angle, to the radius r; also L for a logarithm, and L′ its arithmetical complement. Then

Case 1. When three sides A, B, C, are given.

Put $P = \frac{1}{2} . \overline{A + B + C}$ or semiperimeter.

Then
$$s. \tfrac{1}{2}c = r \sqrt{\frac{(P-A) \times (P-B)}{A \times B}}.$$

And
$$s'. \tfrac{1}{2}c = r \sqrt{\frac{P \times (P-C)}{A \times B}}.$$

$$\text{L. s.} \tfrac{1}{2}c = \tfrac{1}{2}. (\text{L. } \overline{P-A} + \text{L. } \overline{P-B} + \text{L}' A + \text{L}' B).$$

$$\text{L}' \text{ s.} \tfrac{1}{2}c = \tfrac{1}{2} (\text{L. } P + \text{L. } \overline{P-C} + \text{L}' A + \text{L}' B).$$

Note, When A = B, then $s. \tfrac{1}{2}c = \frac{C}{A} \times \frac{r}{2}$. And $s' \tfrac{1}{2}c = r \sqrt{\frac{A^2 - \frac{1}{4}C^2}{A^2}}$.

Case 2. Given two sides A, B, and their included angle c.

Put $s = 90° - \tfrac{1}{2}c$, and t. $d = \frac{A-B}{A+B} \times t. s$; then $a = s + d$; and $b = s - d$.

And
$$c = \sqrt{\left(\frac{4 AB \times s^2 \frac{1}{2}c}{rr} + \overline{A-B} \right)^2}$$

Or in logarithms, putting L. Q = 2 L. (A − B), and L.R = L. 2 A + L. 2 B + 2 L. s. $\frac{1}{2}c - 20$,

then
$$\text{L. c} = \tfrac{1}{2} \text{L. } (Q + R).$$

If the angle c be right, or = 90°; then t. $a = \frac{A}{B} r$; t. $b = \frac{B}{A} r$;

$$c = \frac{r}{s. a} A, \text{ or } = \frac{r}{s. b} B, \text{ or } = \sqrt{A^2 + B^2}.$$

If A = B; then $a = b = 90° - \tfrac{1}{2}c$, and $c = \frac{s. \frac{1}{2}c}{r} \times 2 A$.

Case 3. When a side and its opposite angle are among the terms given.

Then $\frac{A}{s. a} = \frac{B}{s. b} = \frac{C}{s. c}$; from which equations any term wanted may be found.

When an angle, as a, is 90°, and A and c are given,

Then $$B = \sqrt{(A^2 - C^2)} = \sqrt{(A + C) \times (A - C)}.$$

And $$\text{L. } B = \tfrac{1}{2}(\text{L. } \overline{A + C} + \text{L. } \overline{A - C}).$$

Note, When two sides A, B, and an angle a opposite to one of them, are given; if A be less than B, then b, c, c, have each two values; otherwise only one value.

II.—In a spheric triangle, whose three sides are A, B, C, and their opposite angles, a, b, c; any three of these six terms being given, to find the rest.

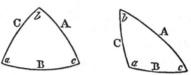

CASE 1.   Given the three sides A, B, C.

Calling 2 P the perim. or $P = \tfrac{1}{2}(A + B + C)$.

Then $$\text{s. } \tfrac{1}{2}c = r \sqrt{\frac{\text{s. } (P - A) \times \text{s. } (P - B)}{\text{s. } A \times \text{s. } B}}.$$

And $$\text{s}' \tfrac{1}{2}c = r \sqrt{\frac{\text{s. } P + \text{s. } (P - C)}{\text{s. } A \times \text{s. } B}}$$

$$\text{L. s. } \tfrac{1}{2}c = \tfrac{1}{2}(\text{L. s. } \overline{P - A} + \text{L s. } \overline{P - B} + \text{L's. } A + \text{L's. } B).$$

$$\text{L. s}' c = \tfrac{1}{2}(\text{L. s. } P + \text{L. s. } \overline{P - C} + \text{L's. } A + \text{L's. } B)$$

And the same for the other angles.

CASE 2.   Given the three angles.

Put $2 p = a + b + c$.

Then $$\text{s } \tfrac{1}{2}c = r \sqrt{\frac{\text{s}' p \times \text{s}' (p - c)}{\text{s. } a \times \text{s. } b}}$$

And $$\text{s}' \tfrac{1}{2}c = r \sqrt{\frac{\text{s}' (p - a) \times \text{s}' (p - b)}{\text{s. } a \times \text{s. } b}}.$$

$$\text{L. s. } \tfrac{1}{2}c = \tfrac{1}{2}(\text{L. s}' p + \text{L. s}' \overline{p - c} + \text{L's. } a + \text{L's. } b).$$

$$\text{L. s}' \tfrac{1}{2}c = \tfrac{1}{2}(\text{L. s}' \overline{p - a} + \text{L. s}' \overline{p - b} + \text{L's. } a + \text{L's. } b).$$

And the same for the other sides.

Note, The sign $>$ signifies greater than, and $<$ less than; also $\frown$ the difference.

CASE 3.   Given A, B, and included angle c.

To find an angle a opposite the side A, let $r : \text{s}' c :: \text{t. } A : \text{t. } M$, like or unlike A, as c is $>$ or $<$ 90°; also $N = B \frown M$: then s. N : s. M :: t. c : t. a, like or unlike c as M is $>$ or $<$ B. Or let $\text{s}' \tfrac{1}{2}. \overline{A + B} : \text{s}' \tfrac{1}{2}. \overline{A - B} :: \text{t}' \tfrac{1}{2}c$: t. M, which is $>$ or $<$ 90°, as A + B is $>$ or $<$ 180°, and s. $\tfrac{1}{2}. \overline{A + B}$: s. $A \frown B :: \text{t}' \tfrac{1}{2}c$: t. N, $>$ 90°, then $a = M + N$; and $b = M - N$.

Again, let $r : \text{s}' c :: \text{t. } A : \text{t. } M$, like or unlike A as c is $>$ or $<$ than 90°; and $N = B \sim M$. Then $\text{s}' M : \text{s}' N :: \text{s}' A : \text{s}' c$, like or unlike N as c is $>$ or $<$ 90°   Or,

$$\text{s. } \tfrac{1}{2}c = \sqrt{\frac{\text{s. } A \times \text{s. } B \times \text{s}^2 \tfrac{1}{2}c}{rr} + \text{s}^2 \tfrac{1}{2}. \overline{A \frown B}}.$$

In logarithms, put L. Q $=$ 2 L. s. $\frac{1}{2}$ $\overline{\text{A} \smile \text{B}}$; and L. R $=$ L. s. A $+$ L. s. B $+$ 2 L. s. $\overline{\frac{1}{2} \text{c} - 20}$; then L. s. $\frac{1}{2}$c $=$ $\frac{1}{2}$ L. (Q $+$ R).

CASE 4. Given a, b, and included side c.

First, let $r : s' \text{c} :: \text{t.} a : \text{t}' m$, like or unlike a as c is $>$ or $<$ 90°; also $n = b \smile m$. Then $s' n : s' m :: \text{t.}$ c $:$ t. A, like or unlike n as a is $>$ or $<$ 90°. Or, let $s' \frac{1}{2} . \overline{a + b} : s' \frac{1}{2} . \overline{a \smile b} :: \text{t.} \frac{1}{2} \text{c} : \text{t. M}$, $>$ or $<$ 90° as $a + b$ is $>$ or $<$ 180°; and s. $\frac{1}{2} \overline{a + b}$: s. $\frac{1}{2} \overline{a \smile b}' :: \text{t.} \frac{1}{2} \text{c} : \text{t. N}$, $>$ 90°; then A $=$ M $\pm$ N; and B $=$ M $\mp$ N.

Again, let $r : s'$ c $:: \text{t.} a : \text{t}' m$, like or unlike a as c is $>$ or $<$ 90°; and $n = b \smile m$: then s. $m :$ s. $n :: s' a : s'$ c, like or unlike a as m is $>$ or $<$ b

CASE 5. Given A, B, and an opposite angle a.

1st. s. A : s. $a :: $ s. B : s. b, $>$ or $<$ 90°.

2dly. Let $r : s'$ B $:: \text{t.} a : \text{t}' m$, like. or unlike B as a is $>$ or $<$ 90°; and t. A : t. B $:: s' m : s' n$, like or unlike A as a is $>$ or $<$ 90°; then c $=$ $m \pm n$, two values also.

3dly. Let $r : s' a :: \text{t.}$ B $:$ t. M, like or unlike B as a is $>$ or $<$ 90°; and s' B $: s'$ A $:: s'$ M $: s'$ N, like or unlike A as a is $>$ or $<$ 90°; then c $=$ M $\pm$ N, two values also.

But if A be equal to B, or to its supplement, or between B and its supplement; then is b like to B: also c is $= m \pm n$, and c $=$ M $\mp$ N, as B is like or unlike a.

CASE 6. Given a, b, and an opposite side A.

1st. s. $a :$ s. A. $::$ s. $b :$ s. B, $>$ or $<$ 90°.

2dly. Let $r : s' b :: \text{t.}$ A : t. M, like or unlike b as A is $>$ or $<$ 90°; and t. $a :$ t. $b :: $ s. M : s. N, $>$ or $<$ 90°; then c $=$ M $\pm$ N, as a is like or unlike b.

3dly. Let $r : s'$ A $:: \text{t.}$ $b : \text{t}' m$, like or unlike b as A $>$ or $<$ 90°; and $s' b : s' a :: $ s. $m :$ s. n, $>$ or $<$ 90°; then c $= m \pm n$, as a is like or unlike b.

But if A be equal to B, or to its supplement, or between B and its supplement; then B is unlike b, and only the less values of N, n, are possible.

Note, When two sides A, B, and their opposite angles a, b, are known; the third side c, and its opposite angle c, are readily found thus :—

$$\text{s.} \tfrac{1}{2} (a \smile b) : \text{s.} \tfrac{1}{2} . (a + b) :: \text{t.} \tfrac{1}{2} (\text{A} \smile \text{B}) : \text{t.} \tfrac{1}{2} \text{c.}$$
$$\text{s.} \tfrac{1}{2} . (\text{A} \smile \text{B}) : \text{s.} \tfrac{1}{2} . (\text{A} + \text{B}) :: \text{t.} \tfrac{1}{2} . (a \smile b) : \text{t.} \tfrac{1}{2} c.$$

III. In a right-angled spheric triangle, where H is the hypotenuse, or side opposite the right angle, B, P, the other two sides, and b, p, their opposite angles; any two of these five terms being given, to find the rest; the cases, with their solutions, are as in the following table.

The same table will also serve for the quadrantal triangle, or that which has one side $=$ 90°, H being the angle opposite to that side, B, P, the other two angles, and b, p, their opposite sides; observing, instead of H to take its supplement; or else mutually changing the terms *like* and *unlike* for each other where H is concerned, and its real value is taken.

Case	Given	Req^d	SOLUTIONS.
1	H B	b p P	s. H : r :: s.B : s.b, and is like B r : t'.H :: t.B : s'.p $\Big\}$, > or < 90° as H is like or un- s'.B : r :: s'.H : s.P $\Big\}$ like B
2	H b	B P p	r : s.H :: s.b : s.B, like b r : s'.b :: t.H : t.P $\Big\}$, > or < 90° as H is like or un- r : s'.H :: t.b : t'.p $\Big\}$ like b
3	B b	H P p	s.b : r :: s.B : s.H $\Big\}$ r : t.B :: t'.b : s.P $\Big\}$, each > or < 90°; both values s'.B': r :: s'.b : s.p $\Big\}$ true
4	B p	H b P	r : t'.B :: s'.p : t'.H, > or < 90° as B is like or unlike p r : s'.B :: s.p : s'.b, like B r : s.B :: t.p : t.P, like p
5	B P	H b p	r : s'.B :: s'.P : s'.H, > or < 90° as B is like or unlike P r : s.P :: t'.B : t'.b, like B r : s.B :: t' P : t'.p, like P
6	p b	H B P	r : t'.b :: t'.p : s'.H, > or < 90° as b is like or unlike p s.p : r :: s'.b : s'.B, like b s.b : r :: s'.p : s'.P, like p

The following Propositions and Remarks, concerning Spherical Triangles, (selected and communicated by the Rev. Nevil Maskelyne, D.D., Astronomer Royal, F.R.S.) will also render the calculation of them perspicuous, and free from ambiguity :—

" 1. A spherical triangle is equilateral, isosceler, or scalene, according as it has its three angles all equal, or two of them equal, or all three unequal; and *vice versâ*.

" 2. The greatest side is always opposite the greatest angle, and the smallest side opposite the smallest angle.

" 3. Any two sides, taken together, are greater than the third.

" 4. If the three angles are all acute, or all right, or all obtuse; the three sides will be, accordingly, all less than 90°, or equal to 90°, or greater than 90°; and *vice versâ*.

" 5. If from the three angles A, B, C, of a triangle ABC, as poles, there be described, upon the surface of the sphere, three arches of a great circle DE, DF, FE, forming by their intersections a new spherical triangle DEF; each side of the new triangle will be the supplement of the angle at its pole; and each angle of the same triangle, will be the supplement of the side opposite to it in the triangle A B C.

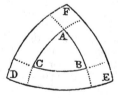

" 6. In any triangle A B C, or A b C, right angled
in A, 1st, The angles at the hypotenuse are always of
the same kind as their opposite sides ; 2dly, The
hypotenuse is less or greater than a quadrant according as the sides including the right angle are of
the same or different kinds ; that is to say, according
as these same sides are either both acute or both obtuse, or as one is acute and the other obtuse.   And,
vice versâ, 1st, The sides including the right angle, are always of the same
kind as their opposite angles ; 2dly, The sides including the right angle will
be of the same or different kinds, according as the hypotenuse is less or more
than 90° : but one at least of them will be of 90°, if the hypotenuse is so "

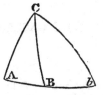

The values of the sine, cosine, &c. at the extremity of each quadrant, as
well as the changes of signs in passing through the different quadrants, may
be known from the following tables :—

ARC =	0°	90°	180°	270°	360°
Sin..	0	R	0	− R	0
Tan.. ...	0	∞	0	− ∞	0
Sec......	R	∞	− R	− ∞	R
Cos.	R	0	− R	0	R
Cot......	∞	0	− ∞	0	∞
Cosec..	∞	R	− ∞	− R	∞

The character ∞ denotes infinity.

The changes of signs are these :—

					sin.	cos.	tan.	cot.	sec.	cosec.
1st.	5th.	9th	13th.	} quadrants.	+	+	+	+	+	+
2d.	6th.	10th.	14th.		+	−	−	−	−	+
3d.	7th.	11th.	15th.		−	−	+	+	−	−
4th.	8th.	12th.	16th.		−	+	−	−	+	−

THE CASES OF PLANE TRIANGLES RESOLVED BY LOGARITHMS.

In this and the following solutions of spherical triangles, it is to be observed, that when we say the sine, tangent, &c. we mean the logarithmic sine, tangent, &c. as found by the table.

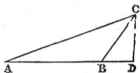

Prop. I.—*Having the angles, and one side; to find either of the other sides.*

Add the logarithm of the given side to the sine of the angle opposite to the side required, and from the sum subtract the sine of the angle opposed to the given side; the remainder will be the logarithm of the side required.

Example. In the triangle B C D, having the angle C D B 90°, C B D 51° 56′, B C D 38° 4′ and the side B D 197·3; to find the side C D.

$$
\begin{array}{l}
2\cdot2951271 \text{ log. of } 197\cdot3 \\
9\cdot8961369 \text{ sin. of } 51°\,56' \\
\hline
12\cdot1912640 \text{ sum} \\
9\cdot7899880 \text{ sin. of } 38°\,4' \\
\hline
2\cdot4012760 \text{ log. } 251\cdot9278 \text{ C D req.}
\end{array}
$$

Or you may add the complement of the sine of the angle opposed to the given side, to the two other logarithms, the sum (abating radius) is the logarithm of the side required; as shown in art. 3 of Log. Arith. And it is to be observed, that the complements of the sines in the table are to be found in the columns of the cosecants: for (passing over the first unit) the cosecants of the same arcs are the complements of the same sines. Also the complements of the tangents are the cotangents.

Example. The sine of 38° 4′ being 9·7899880, the cosecant of 38° 4′ is 10·2100120, which (omitting the first unit) is the complement of the said sine.

$$
\begin{array}{l}
0\cdot2100120 \text{ co. of sin. } 38°\,4' \\
2\cdot2951271 \text{ log. of } 197\cdot3 \\
9\cdot8961369 \text{ sin. of } 51°\,56' \\
\hline
2\cdot4012760 \text{ log. } 251\cdot9278, \text{ as before.}
\end{array}
$$

But if one side and the angles of a right-angled triangle be known, and you would have the other side, as in the former example, the operation will be easier thus:

Add the tangent of the angle opposite to the side required, to the logarithm of the given side, the sum (abating radius) is the logarithm of the side required.

$$
\begin{array}{l}
10\cdot1061489 \text{ tan. } 51°\,56' \\
2\cdot2951271 \text{ log. of } 197\cdot3 \\
\hline
2\cdot4012760 \text{ log. } 251\cdot9278 \text{ as before.}
\end{array}
$$

PROP. II.—*Having two sides, and an angle opposite to one of them; to find the other two angles, and the third side.*

Add the sine of the angle given to the logarithm of the side adjoining that angle, and from the sum subtract the logarithm of the side opposite to that angle, or add its arithmetical comp the remainder or sum will be the sine of the angle opposite to the adjoining side.

Example. In the triangle ABC, having the side AC 800, BC 320, and the angle ABC 128° 4′; to find the angles BAC, ACB, and the side AB.

$$
\begin{aligned}
&7{\cdot}0969100 \text{ ar. com. log. } 800\\
&2{\cdot}5051500 \text{ log. of } 320\\
&9{\cdot}8961369 \text{ sin. } 128°\ 4'\\
\hline
&9{\cdot}4981969 \text{ sin. } 18°\ 21'\ \text{BAC.}
\end{aligned}
$$

Having BAC and ABC the angle ACB is their supplement to 180°, viz. 33° 35′; and you may find the side AB by the first proposition.

PROP. III.—*Having two sides and the angle between them; to find the other two angles, and the third side.*

If the angle included be a right angle, add the radius to the logarithm of the less side, and from the sum subtract the logarithm of the greater side, or add its arith. comp.: the remainder or sum will be the tangent of the angle opposed to the less side.

Example. In the triangle BCD, having the side BE 197·3, and CD 251·9; to find the angles BCD, CBD, and the side CB.

$$
\begin{aligned}
&7{\cdot}5987728 \text{ ar. com. log. } 251{\cdot}9\\
&12{\cdot}2951271 \text{ rad. } + \text{ log. } 197{\cdot}3\\
\hline
&9{\cdot}8938989 \text{ tan. } 38°\ 4'\ \text{BCD.}
\end{aligned}
$$

But if the angle included be oblique, add the logarithm of the difference of the given sides to the tangent of half the sum of the unknown angles, and from the sum subtract the logarithm of the sum of the given sides, or add its complement; the remainder or sum will be the tangent of half their difference.

Example. In the triangle ABC, having the side AB 562, BC 320, and the angle ABC 128° 4′; to find the angles BAC, ACB, and the side AC.
The sum of the given sides is 882, and the difference 242, the half sum of the unknown angles is 25° 58′.

$$
\begin{aligned}
&7{\cdot}0545314 \text{ com. log. } 882\\
&2{\cdot}3838154 \text{ log. of } 242\\
&9{\cdot}6875402 \text{ tang. } 25°\ 58'\\
\hline
&9{\cdot}1258870 \text{ tang. } 7\ \ 37\\
&\phantom{9{\cdot}1258870 \text{ tang. } }25\ \ 58\\
\hline
\end{aligned}
$$

Angle ACB..... 33 35 sum.
Angle CAB..... 18 21 dif.

These 7° 37′ being added to 25° 58′ the half-sum of the angles unknown the sum is 33° 35′ for the greater angle ACB; and the same 7° 37′ being subtracted from 25° 58′, the remainder is 18° 21′ for the lesser angle CAB. Lastly, knowing the angles, and two sides, the third side may be found by the first proposition

Prop. IV.—*Having the three sides; to find any angle.*

Add the three sides together, and take half the sum, and the differences betwixt the half-sum and each side: then add the complements of the logarithms of the half-sum, and of the difference between the half-sum and the side opposite to the angle sought, to the logarithms of the differences of the half-sum, and the other side, half their sum will be the tangent of half the angle required.

Example. In the triangle ABC, having the side AB 562, AC 800, and BC 320; to find the angle ABC.

$$
\begin{array}{ll}
\text{AC} = 800\,| & \text{H} = 841 \dots\text{co.} \ 7\cdot0752040 \\
\text{AB} = 562\,| & \text{H} - \text{AC} = 41\ \text{co.}\ 8\cdot3872161 \\
\text{BC} = 320\,| & \text{H} - \text{AB} = 279\,..\ 2\cdot4456042 \\
\overline{\text{um}\ 1682}\,| & \text{H} - \text{BC} = 521\,..\ 2\cdot7168377 \\
\tfrac{1}{2}\ \text{sum}\ 841 = \text{H} & \quad\quad\text{sum}\ \overline{20\cdot6248620} \\
\end{array}
$$

Tang. of 64° 2′ = $\tfrac{1}{2}$ sum $\overline{10\cdot3124310}$
Whose double 128° 4′ is the angle ABC.

PLANE TRIANGLES RESOLVED BY THE NATURAL SINES, &c.

Although it is generally best to solve triangles by the table of log. sines, tangents, &c. yet, it will often happen that the results may be obtained with more rapidity by means of the natural sines, &c. than even by the aid of the logarithms.

Thus, with regard to Prop. II., when the sides are given in terms that are suited for easy multiplication and division, it will be best to work with the natural sines. As in the example already given. The sine of 128° 4′, or of its supplement 51° 56′, is ·7872939. Hence

$$800 : 320 :: 1 : \cdot4 :: \cdot7872939 : ?$$
$$\cdot4$$

nat. sin 18° 21′ $+$ $=$ $\overline{\cdot31491756}$

Here we have only to consult the table *twice;* while in the logarithmic method it must be done *four* times.

Again, for Prop. III. Suppose there are given AC = 450, BC = 540, and angle c = 80°; to find AB. By formula 7, Table 16, we have AB = $\sqrt{(\text{BC}^2 + \text{AC}^2 - 2\,\text{BC}.\text{AC}.\cos.\text{c})} = \sqrt{(540^2 + 450^2 - 2\,.\,540\,.\,450\,.\,1736482)} = 90\,\sqrt{(6^2 + 5^2 - 60 \times \cdot1736482)} = 90\,\sqrt{(61 - 10\cdot418892)} = 90\,\sqrt{50\cdot581108} = 640\cdot08$.

So, again, examples in Prop. IV. may be worked by a table of squares, and formula 43 for cos. A, in Table 16.

And the most useful cases in heights and distances may be readily solved by means of the natural tangents and secants; the base × tan. angle at base, giving the vertical altitude, and base × sec. angle at base, giving the sloping or hypotenusal distance.

THE CASES OF SPHERICAL TRIANGLES RESOLVED BY LOGARITHMS.

The resolution of spherical triangles is to be performed by the table of sines, tangents, and secants; which we shall show by the 28 propositions following; whereof 16 are of right-angled, and 12 are of oblique triangles; and first,

Of right-angled Triangles.

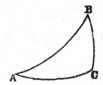

PROP. 1.—*Having the legs; to find the hypotenuse.*

Add the cosine of one leg, to the cosine of the other leg; the sum (abating radius) is the cosine of the hypotenuse required.

Example. In the right-angled triangle ABC, having AC 27° 54', and BC 11° 30'; to find AB the hypotenuse.

$$9\cdot9911927 \quad \text{cosin. } 11°\ 30'$$
$$9\cdot9463371 \quad \text{cosin. } 27\ \ 54$$
$$\overline{9\cdot9375298} \quad \text{cosin. } 30 \text{ AB req.}$$

PROP. II.—*Having the two legs; to find either of the angles.*

Add the sine of the leg next the angle sought, to the cotangent of the other leg: the sum (abating radius) is the cotangent of the angle required.

Example. In the right-angled triangle ABC, having AC 27° 54', and BC 11° 30'; to find the angle BAC.

$$9\cdot6701807 \quad \text{sin. next leg } 27°\ 54'$$
$$10\cdot6915374 \quad \text{cot. opp. leg } 11\ \ 30$$
$$\overline{10\cdot3617181} \quad \text{cotan. BAC } 23\ \ 30$$

PROP. III.—*Having the hypotenuse, and one of the angles; to find the other angle.*

Add the cosine of the hypotenuse to the tangent of the angle given; the sum (abating radius) is the cotangent of the angle required.

Example. In the right-angled triangle ABC, having the hypotenuse AB 30°, and the angle ABC 69° 22'; to find the angle BAC.

$$9\cdot9375306 \quad \text{cosin. hyp. AB } 30°\ 00'$$
$$10\cdot4241896 \quad \text{tang. ABC } \ldots 69\ \ 22$$
$$\overline{10\cdot3617202} \quad \text{cotan. BAC } .. \ 23\ \ 30$$

PROP. IV.—*Having the hypotenuse, and one of the angles; to find the leg next the given angle.*

Add the tangent of the hypotenuse to the cosine of the angle given; the sum (abating radius) is the tangent of the leg required.

d

Example. In the right-angled triangle ABC, having the hypotenuse AB 30°, and the angle ABC 69° 22′; to find the leg BC.

$$
\begin{array}{ll}
9\cdot7614393 & \text{tang. hyp. AB } 30°\ 00' \\
9\cdot5470188 & \text{cosin. ABC .. } 69\ 22 \\
\hline
9\cdot3084581 & \text{tang. BC .... } 11\ 30
\end{array}
$$

PROP. V.—*Having the hypotenuse, and one of the angles; to find the leg opposed to the given angle.*

Add the sine of the hypotenuse to the sine of the angle given; the sum (abating radius) is the sine of the leg required.

Example. In the right-angled triangle ABC, having the hypotenuse AB 30°, and the angle BAC 23° 30′; to find the leg BC.

$$
\begin{array}{ll}
9\cdot6989700 & \text{sin. hyp. AB .. } 30°\ 00' \\
9\cdot6006997 & \text{sin. BAC .... } 23\ 30 \\
\hline
9\cdot2996697 & \text{sin. BC ..... } 11\ 30
\end{array}
$$

PROP. VI.—*Having one of the legs, and the angle next it; to find the hypotenuse.*

Add the cotangent of the given leg to the cosine of the given angle; the sum (abating radius) is the cotangent of the hypotenuse required.

Example. In the right-angled triangle ABC, having the leg AC 27° 54′ and the angle BAC 23° 30′; to find the hypotenuse AB.

$$
\begin{array}{ll}
10\cdot2761563 & \text{cot. AC .... } 27°\ 54' \\
9\cdot9623977 & \text{cos. BAC ... } 23\ 30 \\
\hline
10\cdot2385540 & \text{cot. hyp. AB.. } 30\ 00
\end{array}
$$

PROP. VII.—*Having one of the legs, and the angle next it; to find the other leg.*

Add the sine of the leg given to the tangent of the angle given; the sum (abating radius) is the tangent of the leg required.

Example. In the right-angled triangle ABC, having the leg AC 27° 54′, and the angle BAC 23° 30′; to find the leg BC.

$$
\begin{array}{ll}
9\cdot6701807 & \text{sin. AC .. } 27°\ 54' \\
9\cdot6383019 & \text{tan. BAC } 23\ 30 \\
\hline
9\cdot3084826 & \text{tan. BC .. } 11\ 30
\end{array}
$$

PROP. VIII.—*Having one of the legs, and the angle next to it; to find the other angle.*

Add the cosine of the given leg to the sine of the given angle; the sum (abating radius) is the cosine of the angle required.

Example. In the right-angled triangle ABC, having the leg BC 11° 30′, and the angle ABC 69° 22′; to find the angle BAC.

$$
\begin{array}{ll}
9\cdot9911927 & \text{cos. BC } 11°\ 30 \\
9\cdot9712084 & \text{sin. ABC } 69\ 22 \\
\hline
9\cdot9624011 & \text{cos. BAC } 23\ 30
\end{array}
$$

PROP. IX.—*Having one of the legs, and the angle opposite to it; to find the hypotenuse.*

Add the radius to the sine of the given leg, and from the sum subtract the sine of the given angle, or add its cosecant; the remainder or sum is the sine of the hypotenuse required.

Example. In the right-angled triangle ABC, having the leg BC 11° 30′, and the angle BAC 23° 30′; to find the hypotenuse AB.

$$9{\cdot}2996553 \text{ sin. BC } 11° 30′$$
$$0{\cdot}3993003 \text{ cos. BAC } 23\ 30$$
$$\overline{9{\cdot}6989556} \text{ sin. AB } 30 \text{ required.}$$

PROP. X.—*Having one of the legs, and the angle opposite to it; to find the other leg.*

Add the tangent of the given leg to the cotangent of the given angle; the sum (abating radius) is the sine of the leg required.

Example. In the right-angled triangle ABC, having the leg BC 11° 30′, and the angle BAC 23° 30′; to find the leg AC.

$$9{\cdot}3084626 \text{ tang. BC } 11° 30′$$
$$10{\cdot}3616981 \text{ cot. BAC } 23\ 30$$
$$\overline{9{\cdot}6701607} \text{ sin. AC } 27\ 54$$

PROP. XI.—*Having one of the legs, and the angle opposite to it; to find the other angle.*

Add the radius to the cosine of the given angle, and from the sum subtract the cosine of the given leg, or add the secant; the remainder or sum is the sine of the angle required.

Example. In the right-angled triangle ABC, having the leg BC 11° 30′, and the angle BAC 23° 30′; to find the angle ABC.

$$9{\cdot}9623977 \text{ cos. BAC } 23° 30′$$
$$0{\cdot}0088073 \text{ sec. BC } 11\ 30$$
$$\overline{9{\cdot}9712050} \text{ sin. ABC } 69\ 22$$

PROP. XII.—*Having one of the legs, and the hypotenuse; to find the angle next the given leg.*

Add the tangent of the given leg to the cotangent of the hypotenuse, the sum (abating radius) is the cosine of the angle required.

Example. In the right-angled triangle ABC, having the leg AC 27° 54′, and the hypotenuse AB 30°; to find the angle BAC.

$$9{\cdot}7238436 \text{ tan. AC } 27° 54′$$
$$10{\cdot}2385606 \text{ cot. AB } 30\ 00$$
$$\overline{9{\cdot}9624042} \text{ cosi. BAC } 23\ 30$$

PROP. XIII.—*Having one of the legs, and the hypotenuse; to find the angle opposite to the given leg.*

Add the radius to the sine of the given leg, and from the sum subtract the sine of the hypotenuse, or add its cosecant; the remainder or sum will be the sine of the angle required.

d 2

Example. In the right-angled triangle ABC, having the leg BC 11° 30', ·and the hypotenuse AB 30°; to find the angle BAC.

$$
\begin{array}{lll}
9\text{·}2996553 & \text{sin. leg } \text{BC} & 11° 30' \\
0\text{·}3010300 & \text{cosec. hyp. } \text{AB } 30 & 00 \\
\hline
9\text{·}6006853 & \text{sine of } \text{BAC} & 23\ 30
\end{array}
$$

PROP. XIV.—*Having one of the legs, and the hypotenuse; to find the other leg.*

Add the radius to the cosine of the hypotenuse, and from the sum subtract the cosine of the given leg, or add its secant; the remainder or sum is the cosine of the leg required.

Example. In the right-angled triangle ABC, having the leg BC 11° 30', and the hypotenuse AB 30°; to find the leg AC.

$$
\begin{array}{lll}
9\text{·}9375306 & \text{cosin. } \text{AB} & 30°\ 00' \\
0\text{·}0088073 & \text{sec. } \text{BC} & 11\ 30 \\
\hline
9\text{·}9463379 & \text{cosin. } \text{AC} & 27\ 54
\end{array}
$$

PROP. XV.—*Having the angles; to find the hypotenuse.*

Add the cotangent of one oblique angle to the cotangent of the other oblique angle; the sum (abating radius) is the cosine of the hypotenuse required.

Example. In the right-angled triangle ABC, having the angle BAC 23° 30', and the angle ABC 59° 22'; to find the hypotenuse AB.

$$
\begin{array}{lll}
0\text{·}3616981 & \text{cot. } \text{BAC} & 23°\ 30' \\
9\text{·}5758104 & \text{cot. } \text{ABC} & 69\ 22 \\
\hline
9\text{·}9375085 & \text{cos. hyp. } \text{AB } 30 & 00
\end{array}
$$

PROP. XVI.—*Having the angles; to find either of the legs.*

Add the radius to the cosine of either oblique angle, and from the sum subtract the sine of the other oblique angle, or add its cosecant; the remainder or sum will be the cosine of the leg opposite to the angle whose cosine was taken.

Example. In the right-angled triangle ABC, having the angle BAC 23° 30', and the angle ABC 69° 22'; to find the leg BC.

$$
\begin{array}{lll}
9\text{·}9623977 & \text{cosin. } \text{BAC} & 23°\ 30' \\
0\text{·}0287916 & \text{cosec. } \text{ABC} & 69\ 22 \\
\hline
9\text{.}9911893 & \text{cosin. } \text{BC} & 11\ 30
\end{array}
$$

Of Oblique Triangles.

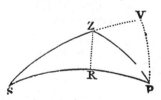

PROP. XVII.—*Having the three sides; to find any of the angles.*

Add the three sides together, and take half the sum; also the difference between the half-sum and the side opposite to the angle sought. Then add the cosecants, or the complements of the sines, of the other sides, to the sines of the half-sum and of the said difference; half the sum of these four logarithms is the cosine of half the angle required.

Example. In the triangle szp, having the side zs 40°, ps 70°, and pz 38° 30′; to find the angle zps.

ps =	70° 0′	cosec.	0·0270142
pz =	38 30	cosec.	0·2058505
zs =	40 0	sin. ½ sum	9·9833805
Sum	148 30	sin. dif.	9 7503579
½ sum	74 15	2)	19·9666031
zs =	40 0	cos.15° 47′	9·9833015
Diff.	34 15	zps 31 34 required.	

PROP. XVIII.—*Having the three angles; to find any of the sides.*

Let the angles be changed into sides, taking the supplement of one of them; then the operation will be the same as in the former proposition.

PROP. XIX.—*Having two angles, and a side opposite to one of them; to find the side opposed to the other angle.*

Add the sine of the side given to the sine of the angle opposite to the side required, and from the sum subtract the sine of the angle opposite to the side given, or add its cosecant; the remainder or sum will be the sine of the side required.

Example. In the triangle szp, having the angle szp 130° 3′ 12″, spz 31° 34′ 26″, and the side zs 40°; to find the side ps.

9·8080675	sin. zs	40° 0′ 0″		
9.8838294 ⎫	sin. szp	⎧ 49 56 .		
850 ⎭		⎩ . . 48		
0·2808858 ⎫	cos. spz	⎧ 31 35 .		
1165 ⎭		⎩ . −34		
9·9729842	sin. ps reqd. 70 0 0			

See page lxi following.

PROP. XX.—*Having two angles, and a side opposite to one of them; to find the side between the angles given.*

Let a perpendicular fall from the angle unknown, on its opposite side; then add the cosine of the given angle next the given side, to the tangent of the given side; the sum (abating radius) is the tangent of the first arc, comprehended between the given angle next the given side, and the segment of the side where the perpendicular falls.

And the second arc, comprehended between the same segment and the other angle, is to be found thus: add the sine of the arc found to the tangent of the given angle next the given side, and from the sum subtract the tangent of the other angle given, or add its cotangent; the remainder or sum will be the sine of the second arc.

The sum or difference of these two arcs will be the side required.

Example. In the triangle szp, having the angle zps 31° 34′ 26″, zsp 30° 28′ 12″, and the side pz 38° 30′; to find the side sp.

9·9303781 ⎫	cos. zps	⎧ 31° 35′ .″
440 ⎭		⎩ . −34
9·9006052	tan. pz	38 30 0
9·8310273	tan. pr 1st arc 34 7 30	

$$
\left.\begin{array}{r} 9\cdot7488698 \\ 932 \end{array}\right\} \text{ sin. } \textsc{pr} \quad \left\{\begin{array}{cc} 34^\circ & 7' \quad . \\ . & .\quad 30 \end{array}\right.
$$

$$
\left.\begin{array}{r} 9\cdot7884529 \\ 1227 \end{array}\right\} \text{ tan. } \textsc{zps} \quad \left\{\begin{array}{cc} 31 & 34 \quad . \\ . & .\quad 26 \end{array}\right.
$$

$$
\left.\begin{array}{r} 0\cdot2301404 \\ 2313 \end{array}\right\} \text{ cot. } \textsc{zsp} \quad \left\{\begin{array}{cc} 30 & 29 \quad . \\ . & -48 \end{array}\right.
$$

9·7679103 sin. sʀ 2d arc	35 52 30
add pʀ 1st arc	34 7 30
sum is sᴘ	70 0 0

See page lxi following.

But when the perpendicular falls out of the triangle, the difference of the two arcs will be the side required.

Pʀᴏᴘ. XXI.—*Having two angles, and a side opposite to one of them; to find the third angle.*

Let a perpendicular fall from the angle unknown, on its opposite side: then add the cosine of the given side to the tangent of the adjacent angle; the sum (abating radius) is the cotangent of the first angle to be found, comprehended by the given side and the perpendicular.

And the second angle, comprehended by the perpendicular and the side unknown, is to be found thus: add the sine of the angle found to the cosine of the given angle opposite to the given side, and from the sum subtract the cosine of the other angle given, or add its secant; the remainder or sum will be the sine of the second angle.

The sum or difference of these two angles will be the angle required.

Example. In the triangle sᴢᴘ, having the angle ᴢᴘs 31° 34′ 26″, ᴢsᴘ 30° 28′ 12″, and the side ᴘᴢ 38° 30′; to find the angle sᴢᴘ.

$$
\begin{array}{l} 9\cdot8935444 \quad \text{.cosin. } \textsc{pz} \quad 38^\circ 30'\ 0'' \\ \left.\begin{array}{r} 9\cdot7884529 \\ 1227 \end{array}\right\} \text{ tang. } \textsc{zps} \quad \left\{\begin{array}{cc} 31 & 34 \\ . & .\quad 26 \end{array}\right. \end{array}
$$

| 9·6821200 cot. 1st < pᴢʀ | 64 18 50 |

$$
\left.\begin{array}{r} 9\cdot9547619 \\ 507 \end{array}\right\} \text{ sin. } \textsc{pzr} \quad \left\{\begin{array}{cc} 64 & 18 \quad . \\ . & .\quad 50 \end{array}\right.
$$

$$
\left.\begin{array}{r} 9\cdot9353948 \\ 594 \end{array}\right\} \text{ cos. } \textsc{zsp} \quad \left\{\begin{array}{cc} 30 & 29 \quad . \\ . & -48 \end{array}\right.
$$

$$
\left.\begin{array}{r} 0\cdot0695443 \\ 336 \end{array}\right\} \text{ sec. } \textsc{zps} \quad \left\{\begin{array}{cc} 31 & 34 \quad . \\ . & .\quad 26 \end{array}\right.
$$

9·9598447 sin. 2d < sᴢʀ	65 44 21
then add 1st < pᴢʀ	64 18 50
the sum is sᴢᴘ	130 3 11

See page lxi following.

But when the perpendicular falls out of the triangle, the difference of the two angles will be the angle required.

Pʀᴏᴘ. XXII.—*Having two sides, and the angle between them; to find either of the other angles.*

Let a perpendicular fall from the unknown angle, which is not required, on its opposite side: then add the cosine of the given angle to the tangent of the given side opposite to the angle required; the sum (abating radius) is the tangent of the first arc, comprehended between the given angle and the segment of the given side where the perpendicular falls.

And the second arc is the difference of that side and the first arc, being comprehended between the same segment and the angle required.

Now add the sine of the first arc to the tangent of the given angle, and from the sum subtract the sine of the second arc, or add its cosecant; the remainder or sum will be the tangent of the angle required.

Example. In the triangle szp, having the side pz 38° 30′, ps 70°, and the angle zps 31° 34′ 26″; to find the angle zsp.

$$
\left.\begin{array}{r} 9\cdot9303781 \\ 440 \end{array}\right\} \text{ cosin. zps } \left\{\begin{array}{ccc} 31°34' & \cdot & '' \\ \cdot & \cdot & 26 \end{array}\right.
$$

9·9006052   tang. pz   38 30 0

9·8310273 tan. pr, 1st arc 34 7 30
        taken from ps 70 0 0

        leaves sr, 2d arc 35 52 30

$$
\left.\begin{array}{r} 9\cdot7488698 \\ 932 \end{array}\right\} \text{ sin. pr } \left\{\begin{array}{ccc} 34 & 7 & \cdot \\ \cdot & \cdot & 30 \end{array}\right.
$$

$$
\left.\begin{array}{r} 9\cdot7884529 \\ 1227 \end{array}\right\} \text{ tang. zps } \left\{\begin{array}{ccc} 31 & 34 & \cdot \\ \cdot & \cdot & 26 \end{array}\right.
$$

$$
\left.\begin{array}{r} 0\cdot2320011 \\ 873 \end{array}\right\} \text{ cosec. sr } \left\{\begin{array}{ccc} 35 & 53 & \cdot \\ \cdot & -30 \end{array}\right.
$$

9·7696270 tan. zps req. 30 28 12

See page lxi following.

To find both the unknown angles.

Add together the cosecant, or the complement of the sine, of half the sum of the given sides, the sine of half their difference, and the cotangent of half the angle given; the sum (abating radius) is the tangent of half the difference of the angles required.

Add also together the secant, or the complement of the cosine, of half the sum of the given sides, the cosine of half their difference, and the cotangent of half the angle given: the sum (abating radius) is the tangent of half the sum of the angles required.

Then add the half-difference of the angles required to their half-sum, and you will have the greater angle: and subtract the half-difference from the half-sum, and you will have the lesser angle required, the same as in the former operation.

ps =	70° 0′	Cosec. ½ sum	0·0906719	Sec. ½ sum	0·2334015
pz =	38 30	Sin. ½ diff.	9·4336746	Cosin. ½ diff.	9·9833805
Sum	108 30	Cot. ½ zps	10·5486352	Cot. ½ zps	10·5486352
Diff.	31 30	T. 49° 47′ 30″	10·0729817	T. 80° 15′ 42″	10·7654172
½ Sum	54 15	Half sum of angles required is..........			80° 15′ 42″
½ Diff.	15 45	Half the difference is			49 47 30
<zps =	31 34 26″	The greater angle szp is................			130 3 12
½<zps =	15 47 13	The lesser angle zsp is, as before			30 28 12

PROP. XXIII.—*Having two sides, and the angle between them; to find the third side.*

Let a perpendicular fall from either of the angles unknown, on its opposite side: then add the cosine of the given angle to the tangent of the side from whose end the perpendicular is let fall; the sum (abating radius) is the tangent of the first arc, comprehended between the given angle and the segment of the side where the perpendicular falls.

And the second arc is the difference of that side and the first arc, being comprehended between the same segment and the end of the side required.

Now add the cosine of the second arc to the cosine of the side from whose end the perpendicular falls, and from the sum subtract the cosine of the first arc found, or add its secant; the remainder or sum will be the cosine of the side required.

Example. In the triangle szp, having the side pz 38° 30′, ps 70°, and the angle zps 31° 34′ 26″; to find the side zs.

$$
\left.\begin{array}{l} 9 \cdot 9303781 \\ 440 \end{array}\right\} \text{cosin. zps} \quad \left\{\begin{array}{l} 31° 35′ \quad .'' \\ . \quad -34 \end{array}\right.
$$

9·9006052 tang. pz   38 30 0

9·8310273 tan. pr, 1st arc 34 7 30

taken from ps 70 0 0

leaves sr, 2d arc 35 52 30

$$
\left.\begin{array}{l} 9 \cdot 9085988 \\ 457 \end{array}\right\} \text{cosin. sr} \quad \left\{\begin{array}{l} 35 \ 53 \quad . \\ . \quad -30 \end{array}\right.
$$

9·8935444 cosin. pz   38 30 0

$$
\left.\begin{array}{l} 0 \cdot 0820236 \\ 428 \end{array}\right\} \text{sec. pr} \quad \left\{\begin{array}{l} 34 \ 7 \quad . \\ . \quad . \ 30 \end{array}\right.
$$

1·8842553 cosin. zs req. 40 0 0

See page lxi following.

PROP. XXIV.—*Having two sides, and the angle opposite to one of them; to find the angle opposed to the other side.*

Add the sine of the angle given to the sine of the side opposite to the angle required, and from the sum subtract the sine of the side opposite to the angle given, or add its cosecant; the remainder or sum will be the sine of the angle required.

Example. In the triangle szp, having the side ps 70°, zs 40°, and the angle szp 130° 3′ 12″; to find the angle zps.

$$
\left.\begin{array}{l} 9 \cdot 8838294 \\ 850 \end{array}\right\} \text{sin. sup. szp} \left\{\begin{array}{l} 49° 56′ \quad '' \\ . \quad . \ 48 \end{array}\right.
$$

9·8080675 sin. zs   40 0 0

0 0270142 cosec. ps   70 0 0

9·7189961 sin. zps req. 31 34 26

See page lxi following.

PROP. XXV.—*Having two sides, and the angle opposite to one of them; to find the third side.*

Let a perpendicular fall from the angle between the sides given, on its opposite side. then add the cosine of the angle given to the tangent of the given side next that angle; the sum (abating radius) is the tangent of the first arc, comprehended between the given angle and the segment of the side where the perpendicular falls.

Now the 2d arc, comprehended between the same segment, and the end of the side required, is to be found thus: add the cosine of the first arc to the cosine of the given side opposite to the angle given, and from the sum subtract the cosine of the other given side, or add its secant; the remainder or sum will be the cosine of the second arc.

The sum or difference of these two arcs will be the side required.

Example. In the triangle szP, having the side PZ 38°·30′, sz 40°, and the angle sPZ 31° 34′ 26″; to find the side PS.

9·9303781 }	cos. SPZ	{ 31° 35′	. ″
440 }		{ .	−34
9·9006052	tan. PZ	38 30	0
9·8310273	tang. PR, 1st arc	34 7	30

9·9178908 }	cosin. PR	{ 34 8	
428 }		{ .	−30
9·8842540	cosin. sz	40 0	0
0·1064556	sec PZ	38 30	0
9·9086432	cosin. SR, 2d arc	35 52	30
	add PR, 1st arc	34 7	30
	gives PS req.	70 0	0

See page lxi following.

But when the perpendicular falls out of the triangle, the difference of the two arcs will be the side required.

PROP. XXVI.—*Having two sides, and the angle opposite to one of them; to find the angle between them.*

Let a perpendicular fall from the angle between the sides given, on its opposite side: then add the cosine of the given side next the given angle, to the tangent of that angle; the sum (abating radius) is the cotangent of the first angle to be found, comprehended by the given side next the angle given, and by the perpendicular.

Now the second angle, comprehended by the perpendicular and the other given side, is to be found thus: add the cosine of the first angle found to the tangent of the given side next the angle given, and from the sum subtract the tangent of the other given side, or add its cotangent; the remainder or sum will be the cosine of the second angle to be found.

The sum or the difference of the first and second angles will be the angle required.

Example. In the triangle szP, having the side PZ 38° 30′, sz 40°, and the angle sPZ 31° 34′ 26″; to find the angle szP

9 8935444	cosin. PZ	38° 30′	0′
9·7884529 }	tang. szP	{ 31 34	.
1227 }		{ .	. 26
9·6821200	cotan. PZR, 1st <	64 18	50

9·6368859 }	cosin. PZR	{ 64 19	.
437 }		{ .	−10
9·9006052	tang. PZ	38 30	0
0·0761865	cotan. sz	40 0	0
9·6137213	cosin szR, 2d <	65 44	22
	add PZR, 1st <	64 18	50
	gives szP, req.	130 3	12

See page lxi following.

PROP. XXVII.—*Having two angles, and the side between them; to find either of the other sides.*

Let a perpendicular fall from the given angle, which is next the side required, upon its opposite side: then add the cosine of the given side to the tangent of the given angle opposite to the side required; the sum (abating radius) is the cotangent of the first angle to be found, comprehended by the given side and the perpendicular.

And the second angle is the difference between the first and the given angle next the required side, being comprehended by the perpendicular and that side.

Now add the cosine of the first angle found to the tangent of the side given, and from the sum subtract the cosine of the second angle, or add its secant; the remainder or sum will be the tangent of the side required.

Example. In the triangle SZP, having the angle SPZ 31° 34′ 26″, SZP 130° 3′ 12″, and the side PZ 38° 30′; to find the side SZ.

9·8935444	cosin. PZ	38° 30′ 0″		
9·7884529 ⎫	tang. SPZ	⎰ 31 34 .		
1227 ⎭		⎱ . . 26		
9·6821200	cot. PZR, 1st ∠	64 18 50		
	taken from SZP	130 3 12		
	leaves SZR, 2d ∠	65 44 22		

9·6368859 ⎫	cosin. PZR	⎰ 64 19 .	
437 ⎭		⎱ . . −10	
9·9006052	tang. PZ	38 30 0	
0·3861750 ⎫	sec. SZR	⎰ 65 44 .	
1028 ⎭		⎱ . . 22	
9·9238126	tan. SZ req.	40 0 0	

See page lxi following.

To find both the unknown sides.

Add together the cosecant, or the complement of the sine, of half the sum of the angles given, the sine of half their difference, and the tangent of half the given side; the sum (abating radius) is the tangent of half the difference of the sides required.

Add also together the secant, or the complement of the cosine, of half the sum of the given angles, the cosine of half their difference, and the tangent of half the given side: the sum (abating radius) is the tangent of half the sum of the sides required.

Then add half the difference of the sides required to their half-sum, and you will have the greater side: and subtract the half-difference from the half-sum, and you will have the lesser side required, the same as in the former operation.

SZP	130° 3′ 12″	Cosec. ½ sum	0·0056062	Sec. ½ sum	0·7968360
SPZ	31 34 26	Sin. ½ diff.	9·8793527	Cosin. ½ diff.	9·8148437
Sum	161 37 38	Tang. ½ PZ	9·5430936	Tang. ½ PZ	9·5430936
Diff.	98 28 46	Tang. of 15°	9·4280525	Tang. of 55°	10·1547733
½ Sum	80 48 49	Half sum of the sides required is			55°
½ Diff.	49 14 23	Half their difference is.			15
PZ	38 30 0	The greater side SP is.			70
½ PZ	19 15 0	Lesser side SZ is, as before			40

PROP. XXVIII.—*Having two angles, and the side between them ; to find the third angle.*

Let a perpendicular fall from either of the angles given, upon its opposite side: then add the cosine of the side given to the tangent of the given angle, from which the perpendicular does not fall ; the sum (abating radius) is the cotangent of the first angle, comprehended by the given side and the perpendicular.

And the second angle is the difference between the first and the given angle that the perpendicular fell from, being comprehended by the perpendicular and the side opposite to the other angle given.

Now add the sine of the second angle to the cosine of that given angle from which the perpendicular did not fall, and from the sum subtract the sine of the first angle found, or add its cosecant; the remainder or sum will be the cosine of the angle required.

Example. In the triangle szp, having the angle szp 130° 3′ 12″, spz 31° 34′ 26″, and the side pz 38° 30′ ; to find the angle psz

9·8935444	cosin. PZ	38° 30′ 0″
9·7884529 ⎱ 1227 ⎰	tang. SPZ	⎰ 31 34 . ⎱ . . 26
9·6821200	cot. PZR, 1st <	64 18 50
	taken from szp	130 3 12
	leaves SZR. 2d <	65 44 22

0·0451773 ⎱ 101 ⎰	cosec. PZR	⎰ 64 19 . ⎱ . −10
9·9303781 ⎱ 440 ⎰	cosin. SPZ	⎰ 31 35 . ⎱ . −34
9·9598246 ⎱ 209 ⎰	sin. SZR	⎰ 65 44 . ⎱ . . 22
9·9354550	cosin. PSZ req.	30 28 0

See page lxi following.

FOR THE USE OF THE VERSED SINES MAY BE ALSO ADDED THE FOLLOWING PROPOSITIONS.

Prop. I.—*Having two sides of a spheric triangle, with the angle between them ; to find the third side.*

Add together the log. versed sine of the contained angle, and the log. sines of the two sides; the sum (abating twice the radius) is the logarithm of a number to be found, which added to the natural versed sine of the difference of the two given sides, the sum will be the natural versed sine of the third side sought.

Or when the contained angle is above 90°, add the log. versed sine of its supplement, and the log. sines of the two sides together; the sum (abating twice the radius) is the logarithm of a number to be found, and subtracted from the natural versed sine of the sum of the two given sides, the remainder will be the natural versed sine of the third side sought.

Example 1. In the triangle szp, having the side pz 38° 30′, ps 70°, and the angle zps 31° 34′ 26″ ; to find the side zs.

```
9·1703625 log. ver. sine zsp  31° 34′ 26″
9·7941496 log. sine of pz     38  30   0
9·9729858 log. sine of ps     70   0   0
8·9374979 log. of the numb.   865960
Nat. vers. diff. sides 31° 30′  1473598
Nat. vers. zs 40° ........  2339558
```

Example 2. In the triangle szp, having the side pz 38° 30′, zs 40°, and the angle szp 130° 3′ 12″ ; to find the side ps
The angle vzp is the supplement of szp.

```
9·5520590 log. vers. vzp   49° 56′ 48″
9·7941496 log. sin. pz     38  30   0
9·8080675 log. sin. zs     40   0   0
9·1542761 log. of the num. 1426514
Nat. vers. sum sides 78° 30′ 8006321
Nat. vers. ps 70° ... ...  6579807
```

This proposition may be very useful in finding the distances of places on the earth, whose longitudes and latitudes are known; the distances of stars, whose declinations and right ascensions, or longitudes and latitudes, are known; and consequently the altitudes, or common altitude of two stars, or two altitudes of the sun, and time between the observations, or difference of azimuth, being taken, the latitude of the place may readily be found.

Prop. II —*Having two angles of a spheric triangle, and the side between them ; to find the third angle*

Let the angles be changed into sides, and the side into an angle; then proceed as in the former proposition, and the result will be the supplement of the third angle. But if one of the given angles exceed 90°, take its supplement, and the result will be the third angle.

The following remarks and directions, for rendering the proportional part of a logarithm always additive, and for using $c + t$, $c - t$, &c., for s or c, &c. in the foregoing propositions 20, 21, 22, 23, 25, 26, 27, 28, were communicated by the Rev. Nevil Maskelyne, D.D. astronomer royal, and F.R.S., the fourth case having been invented by him many years since, and delivered to the computers of the Nautical Ephemeris, as precepts necessary in computing the moon's distance from the stars in some cases; the rest he has now added on this occasion:—

" The result of trigonometrical calculations will be sometimes inaccurate, owing to the logarithms not being carried to a greater number of places in the table, as will sufficiently appear from the logarithmic differences being small. This will happen where the answer comes out in the cosine of a very small angle, or the sine of an angle near 90°. The greatness of the differences of the log. sines of small arcs, or cosines of large ones, will sometimes affect the accuracy of the result of the second part of the operation, unless the first arc be found to a small part of a minute or second: to prevent such error, and render the computation easier, putting t, t′, s, c for the tangent, cotangent, sine, and cosine of the 1st arc or angle, then in the 2d part of the work,

In Prop. 20, if the first arc   is very small,   for s use $c + t$
        21 .....   angle is very small,   for s use $c - t'$
        22 .....   arc   is very small,   for s use $c + t$
        23 .....   arc   is near 90°,   for -c use $t - s$
        25 .....   arc   is near 90°,   for c use $s - t$
        26 .....   angle is near 90°,   for c use $s + t'$
        27 .....   angle is near 90°,   for c use $s + t'$
        28 .....   angle is very small,   for -s use $t' - c$.

This obviates the necessity of finding the first arc to a very minute exactness, which otherwise would be necessary in taking out the sine or cosine of the same arc in the second part of the work.

Where the foregoing precepts direct to subtract a sine or cosine, it will be readier in practice to add a cosecant or secant; and where they direct to subtract a tangent (which is done in Prop. 26), it will be readier to add a cotangent. This method being used, if it be required to find the logarithmic sines, &c. to the exactness of a second, and the logarithm is increasing (as in the sines, tangents, and secants), write down the logarithm for the degree and minute without the seconds; and also write down the proportional part for the seconds; but if the logarithm is decreasing (as in the cosines, cotangents, and cosecants), write down the logarithm for the next greater minute, and also write down the proportional part for the complement of the seconds to 60; and proceed in like manner with every logarithmic sine, cosine, &c. used in the work; the sum of all the logarithms (abating one or two radii or tens in the index, according as 2 or 3 logarithmic sines, &c. are used in the part of the work in question) will be the logarithmic sine, cosine, tangent, or cotangent required.

Example 1. To find the log. sine of 34° 17′ 24″.

       Here log. sine of 34° 17′........ 9·7507287
       And as 60 : 24 or as 10 : 4 : : 1853 : .. 741
                              9 7508028

Example 2. To find the log. cos. of 55° 42′ 36″.

Log. cos. of 55° 43′ .... 9·7507287
60 : 24 (60 − 36), or 10 : 4 :: 1853 : .. 741

9·7508028

Example 3. In the triangle PLS, given

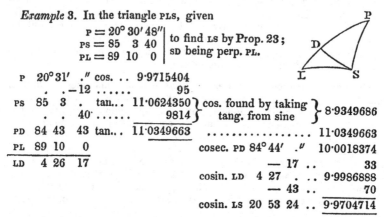

P = 20° 30′ 48″ ⎫
PS = 85   3 40   ⎬ to find LS by Prop. 23;
PL = 89 10   0   ⎭ SD being perp. PL.

P	20°31′	.″	cos. ..	9·9715404			
	.	.−12	......	95			
PS	85 3	.	tan...	11·0624350 ⎫ cos. found by taking ⎫			8·9349686
	. .	40·	......	9814 ⎭ tang. from sine ⎭			
PD	84 43	43	tan...	11·0349663	....	11·0349663	
PL	89 10	0			cosec. PD 84° 44′ .″	10·0018374	
LD	4 26	17			− 17 ..	33	
					cosin. LD 4 27 . ..	9·9986888	
					− 43 ..	70	
					cosin. LS 20 53 24 ..	9·9704714	

Here to avoid the trouble of finding the proportional part for the large logarithm difference of the cosine of PS, that cosine is found by subtracting the tangent of it (already found) from the sine, which is easily found, because the differences are small: and, for the same reason, the sum of the tangent and cosecant of PD, are used instead of its secant.

N.B.—The perpendicular should always be let fall from the end of the side, PS or PL, which differs most from 90°, over or under.

OF THE TRAVERSE TABLE.

This traverse table, or table of difference of latitude and departure, in pages 158, 159, is so contrived, as to have the whole in one view, and is so plainly titled as to want little or no explanation.

The distances 1, 2, 3, &c. at the top and bottom, may be accounted 10, 20, 30, &c., and the 10 as 100, if the minutes of latitude and departure, answering to the course, be increased in the same proportion; so that if the distance consists of two significant figures, the difference of latitude, and the departure, is each to be taken out at twice; and if of three figures, at thrice.

The chief design of this table is for the ready and exact working of traverses; but it may also be applied to the solution of the several cases of plain sailing, and to some other uses

PROP. I.—*Having the course and distance, to find the difference of latitude and departure.*

Seek the course on the left-hand of both pages downwards, if less than four points, or 45 degrees; or if greater, on the right-hand upwards; and even with it in the double column, signed at the top and bottom with the distance, is found both the difference of latitude and the departure.

Example 1. A ship sails ssw ¾ w 37 miles; the difference of latitude and the departure are required.

Find the course 2¾ points on the left-hand side of each page, and even with it in the double columns signed 3 and 7, the two figures of the distance the difference of latitude for 30 is 25·732, and for 7 is 6·004, the sum is 31·736 for the whole difference of latitude; and the departure for 30 is 15·423, and for 7 is 3·599, the sum is 19·022 for the whole departure. Thus,

Dist.	Diff. Lat.	Dep.
30	25·732 ..	15·423
7	6·004 ..	3·599
37 miles	31·736 .·.	19·022

Example 2. A ship sails SE 49° 148 miles; the difference of latitude and the departure are required.

Find the course 49 degrees on the right-hand side of each page, and even with it in the double columns signed 10, 4, and 8, the difference of latitude at 100 miles is 65·606, at 40 is 26·242, and at 8 is 5·248; the sum is 97·096 for the whole difference of latitude. And the departure at 100 miles is 75·471, at 40 is 30·188, and at 8 is 6·038; the sum is 111·697 for the whole departure. Thus,

Dist.	Diff. Lat.	Dep.
100	65·606 ..	75·471
40	26·242 .	30·188
8	5·248 ..	6·038
148 miles	97·096	111·697

PROP. II.—*Having several courses and distances; to find the difference of latitude and the departure.*

Make a table in the following manner, and put therein each course and distance; then find the difference of latitude and departure to each course by the preceding, and place them in the proper column; the difference of the sums of the northings and southings is the whole difference of latitude; and the difference of the sums of the eastings and westings is the whole departure.

Example. A ship from the latitude of 50° north sails according to the courses and distances set in the traverse table; the differences of latitude and the departure are found at the bottom.

THE TRAVERSE TABLE.

Courses.	Dist. Miles.	Diff. of Lat. North.	Diff. of Lat. South.	Departure. East.	Departure. West.
S S E ½ E	79		69·671	37·241	
S E ½ E	86		54·557	66·479	
S b W ¾ W	108		101·687		36·384
S 48° W	112		74·942		83·231
N 58° W	70	6·101			69·734
S 40° W	84		64·348		53·994
		6·101	365·205	103·720	243·343
			6·101		103·720
		Diff. la. 359·104		Depart.	139·623

This proposition may be applied in the surveying of large tracts of land, as a county, &c. and was made use of by Mr. Norwood in measuring the distance from York to London, as the road led him, observing the several bearings by his circumferentor, and finding by such a table his several differ-ences of latitude and departure, by which he obtained the distance between the parallels of London and York, pretty near the truth, so long ago as the year 1635; as may be seen in his *Seaman's Practice.*

Also in plotting the survey of a county thus taken, the circuit station-lines, though consisting of many hundreds, may be reduced to a few for the first closing, and the like for the intermediates of each line first plotted, by which every station may perhaps be more truly placed than by any other method: the distances in the table may be chains of 66, or 100 feet, as well as miles, or any other measure that the differences of latitude and departure would be had in.

PROP. III.—*Having the difference of latitude, and the departure; to find the course and distance.*

Seek the given difference of latitude and departure, taken together, in thei, columns, or the nearest numbers to them; and the course is even therewith at the side, and the distance at the top and bottom: but if the given difference of latitude and departure cannot be found nearly, take $\frac{1}{2}$, $\frac{1}{4}$, &c. part, or any equal multiple of them that can be found; then the course is even with them at the side, and such a part of the distance, as was taken of the difference of latitude and departure, at the top and bottom.

Example 1. Given the difference of latitude 59 miles s, and the departure 68 miles w; the course and distance are required.

In the double column over 9, even with 49° at the right-hand side, is found together the given difference of latitude and departure; therefore the course is 49° sw, and the distance 90 miles.

Example 2. Given the difference of latitude 30 miles N, and the departure 18 miles E; the course and distance are required.

Here the given difference of latitude and departure, or any numbers near them, are not to be found together in the table; therefore taking $\frac{1}{8}$ or the double of each, the course is found to be 31° NE, and the distance 35 miles.

Note. A table computed to every mile in the distance up to 100 miles would more readily solve this example.

PROP. IV.—*Having the departure and middle latitude; to find the difference of longitude, according to the method first given by* W. JONES, Esq. F.R.S.

Seek the given departure, or the next less number in the columns signed lat. even with the middle given latitude found among the courses, and at the top and bottom (signed dist.) is the difference of longitude sought; which, if not found directly at once, may be taken out at twice or thrice.

Example 1. Being yesterday noon in the latitude of 37° 17′ N, and this day noon in 38° 43′ N, and by the table the departure is found 70·921 E; the difference of longitude is required.

In the column signed lat. under 9, even with 38°, the middle latitude is found 7·0921; therefore 90 miles is the difference of longitude sought.

Example 2. Being yesterday noon in latitude 46° 25′ N, and this day at noon in 47° 35′ N, so that the middle latitude is 47° N, and the departure is found 112·53 miles w; required the difference of longitude.

In the column signed lat. over 10 at the bottom, even with 47 at the right-hand side, is 6·8200; therefore subducting 68·200 from 112·53, the remainder is 44·33; then over 6 is 4·0920, and 40·92 subducted from 44·33 leaves 3·41, which is found over 5; therefore the difference of longitude is 165 miles west.

If the middle latitude be not an even degree, but have odd minutes; find the difference of longitude, for the even degrees next less and greater, and add a proportional part of the difference between the two results to the lesser; the sum will be the difference of longitude sought.

Suppose the middle latitude in the last example had been 47° 20′ N; then, after finding the difference of longitude as before for 47°, find it also for 48°, which is 168 miles; then ⅓ of the difference being added to the former, gives the difference of longitude 166 miles west.

Note. Though this method is not in all cases near the truth, yet when the miles are geographical, it is sufficiently near for daily practice in any voyage, as well as easy, and very expeditious.

Prop. V.—*Having the latitudes and the longitudes of two places, to find the bearing and distance.*

Seek the complement of the middle latitude among the degrees, and the difference of longitude in minutes among the distances, the departure answering is found in its proper column; then with the difference of latitude and departure, find their bearing or course and distance by the third.

Example. Let the Lizard be given in the latitude of 49° 50′ N, and 5° 21′ w longitude, and Cape Ortegal in the latitude of 44° 10′ N, and 70° 43′ w longitude; to find the bearing and distance.

The difference of longitude is 142′; and in the columns signed dep. under 10, 4, and 2, even with 43° the co-middle latitude, are found 6·8200, 2·7280, and 1·3640; then increasing the two former as before shown, their sum is 96·844 miles w, for the departure; and the bearing, or course, answering to 340 miles difference of latitude, with 96·844 departure, is found about 16° sw; and the distance about 354 miles.

OF MERCATOR'S SAILING.

The uses of the table of meridional parts are fully supplied by the table of logarithmic tangents, as is demonstrated in N° 219 of the *Philosophical Transactions.* It is there proved, 1st. That the meridional line, or scale of Mercator's Chart, is a scale of the log. tangents of the half-complements of the latitude. 2dly. That such log. tangents of Mr. Briggs's form, are a scale of the differences of longitude, on the rumb which makes an angle of 51° 38′ 9″ with the meridian. And 3dly. That the differences of longitude on different rumbs, are to one another as the tangents of the angles of those rumbs with the meridian.

Hence it follows, that the difference of the log. tangents of the half complements of the latitudes, is to the difference of longitude a ship makes in sailing on any rumb from the one latitude to the other, as the tangent of 51° 38′ 9″ (whose logarithm is 10·1015104) to the tangent of the angle of the rumb or course with the meridian; so that:

I. If two latitudes, and the difference of longitude, be given, the course and distance are readily determined by this rule.

Take, by help of the tables, the difference of the log. tangents of the half-complements of the latitudes, esteeming the last three figures to be a decimal fraction; and add the complement of its logarithm to the logarithm of the difference of longitude reduced to minutes, and the constant log. 10·1015104 the sum (abating radius) shall be the log. tangent of the course. And to the log. secant of the course, add the logarithm of the difference of latitude

e

reduced to minutes, the sum (abating radius) shall be the logarithm of the distance in minutes.

Example. Given the Lizard to be in latitude 49° 55′ N, Barbadoes in 13° 10′ N, and their difference of longitude 53° 00′ or 3180′ w ; to find the course and distance.

$\frac{1}{2}$ Co. lat. $\left\{\begin{array}{l}\text{Barbadoes 38° 25′ } \quad \text{l. tan. 9·8993082} \qquad \text{l. 3180′} = \; 3·5024271 \\ \text{Lizard} \quad 20 \quad 2\frac{1}{2} \quad \text{l. tan. 9·5620477} \qquad \text{const. log. 10·1015104}\end{array}\right.$

diff. 3372·605   its co. log. 6 4720346

Log. tang. of the course 49° 59′ 10″ sw ....10·0759721

Log. sec. of the course 49 59 10......................10·1918067

Log. of 2205′ diff. of the latitudes ....,.............. 3·3434086

Log. of 3429·378 distance of Barbadoes from the Lizard ...... 3·5352153

II. If two latitudes and the course be given, the difference of longitude is obtained with the same ease : for as the tangent of 51° 38′ 9″ is to the tangent of the course, so is the difference of the log. tangents of the half-complements of the latitudes, to the difference of longitude sought. Therefore, to the complement of the constant log. 10·1015104, add the log. of the difference of the log. tangents of the half complements of the latitudes, and the log. tangent of the course, the sum (abating radius) will be the log. of the difference of longitude in minutes.

Example. Given the latitudes 49° 55′ and 13° 10′, and course 49° 59′ 10″ ; to find the difference of longitude.

Lat. 13° 10′ its $\frac{1}{2}$ co. lat. 38° 25′ l. tan. 9·8993082

Lat. 49 55.......... 20   2$\frac{1}{2}$ l. tan. 9·5620477 co. const. log. 9·8984896

diff. 3372·605...... its log. 3·5279654

Log. tang. of the course 49° 59′ 10″......................10·0759721

Log. of 3180′ = 53° for diff. of longitude .... 3·5024271

By this rule, having two good observations of the latitude, and the course duly steered, the reckoning of a ship's way is best ascertained, especially if you sail near the meridian.

III. If the latitude departed from, the course steered, and distance sailed. be given ; to find the ship's latitude, and difference of longitude.

First, the latitude is obtained from the consideration that the distance is to the difference of latitude, as radius to the cosine of the course, which is common to plain sailing. Therefore to the log. of the distance add the log. cosine of the course, the sum (abating radius) is the log. of the difference of latitudes ; which difference added to the lesser latitude, or subtracted from the greater, the sum or remainder is the present latitude · then having the two latitudes and the course, the difference of longitude is found by the second.

Example. Having sailed from the Lizard, in lat. 49° 55′ N, on a course 49° 59′ 10″ south-westerly 3429·378 miles : required what longitude and latitude the ship is found in.

Log. of 3429·378 the distance sailed....................... 3·5352153

Log. cosine of 49° 59′ 10″ the course .... 9·8081933

Log. of 2205′, or 36° 45′ diff. of the latitudes .... 3·3434086

Now subtracting 36° 45′ from 49° 55′, the remainder 13° 10′ N, is the latitude the ship is found in.

By which latitude, now known, the difference of log. tangents will be found 3372·605, and the further process in nothing differing from the second rule, by which the difference of longitude will be found 53° 00′.

Thus the dead reckoning by the log. line, and daily account of a ship's way, are duly kept, and the trouble very little more than by plain sailing.

These are all the cases that occur in practice; the rest, which are mostly speculative, are either easily reducible to these, or else not to be performed by logarithms, and therefore come not at present under our cognizance.

But it is to be noted, that both the complements of the latitudes are to be estimated from the same pole of the world; which may be from either; and therefore if one latitude be N, and the other s, to have their complements, you must add 90° to one of them, and subtract the other from 90°, and then the operation will be the same as in the preceding cases.

Example. Given St. Jago, one of the Cape-de-Verd Islands, in the latitude of 14° 56′ N; and the island St. Helena, in latitude 15° 45′ s, and their difference of longitude 30° 12′ E; to find the course and distance.

½ co. lat. $\begin{cases} \text{St. Jago} & 52° 28′. \text{ l. tan. } 10·1144965 \quad \text{l. } 1812′ \quad 3·2581582 \\ \text{St. Helena } 37 \ 7½. \text{ l. tan. } \ 9·8790845 \text{ const. log. } 10·1015104 \end{cases}$

$$2354·120 \text{ its co. log. } \quad 6·6281714$$

Log. tang. of the course 44° 11′ 53″ SE .... 9·9878400

Log. sec. of the course 44 11 53 .... 10·1445200

Log. of 1841′ diff. of the latitudes........................ 3·2650538

Log. of 2567·875 distance of St. Helena from St. Jago .... 3·4095738

Or if it be thought easier, when one latitude is N, and the other s, you may add 90° to each of them, the sum of the log. tangents of their halves (abating twice the radius) will be the same as the difference of the log. tangents of the former. For an example, take the same latitudes as in the preceding.

Then $90° + \begin{cases} 14° \ 56′ = 104° \ 56′ \\ 15 \ \ 45 = 105 \ \ 45 \end{cases}$ its half $\begin{cases} 52° \ 28′ \text{ l. tan. } 10·1144965 \\ 52 \ \ 52½ \text{ l. tan. } 10·1209155 \end{cases}$

The sum (abating twice the radius) equal to the former distance. . 2354·120

Also, when both latitudes are of the same name, that is both N or both s, you may add 90° to each of them, the difference of the log. tangents of half these sums will be the same as of the log tangents of half the complements of those latitudes.

TABLE FOR THE LENGTHS OF CIRCULAR ARCS.

This is Table 11, and constitutes page 360. It contains the lengths of every single degree up to 180, and of every minute, second, and third, each up to 60. The form of it is obvious; the length of each degree, minute, second, or third, immediately following it on the same line in the next column. And the two following examples will show the use of the table.

Example 1. To find the length of an arc of 57° 17′ 44″ 48‴.

Take out from their respective columns the lengths answering to each of these numbers singly, and add them all together, thus:

57° .... 0·9948377
17′ 49451
44″ 2133
48‴.... 39

the sum or 1·0000000 is the whole length, and is equal to the radius; that is, the length of an arc of 57° 17′ 44″ 48‴ is equal to the radius of the circle.

Example 2. To find the degrees, minutes, &c. in the arc 1, which is equal to the radius.

Subtract from it the next less tabular arc, and from the remainder the next less again, and so on till nothing remain; and opposite to the several numbers subtracted, will be the degree, minutes, &c.; thus:

Given length	1·0000000
57°	0·9948377
	51623
17′	49451
	2172
44″	2133
48‴	39

So that the arc, which is equal to the radius, contains 57° 17′ 44″ 48‴.

TABLE FOR COMPARING HYP. AND COMMON LOGS.

THIS is Table 12, and is the upper part of page 361. It contains the hyperbolic logs. answering to the first 100 common logs. and is very useful for speedily changing the one into the other.

Example 1. To find the hyp. log. answering to the common log. 0·9542425.

Beginning at the left-hand, and dividing the given number into periods of two figures each, including the index, take out the hyp. log. to each period, omitting two figures at the 2d period, four at the 3rd, and six at the 4th; then add them all together, thus:

com. log.		hyp. log.
09	......	2·0723266
54	......	1243396
24	. ....	5526
25	......	58
0·9542425		2·1972246 answer.

Example 2. To find the common log. answering to the hyp. log. 2·1972246.

Subtract continually each next less tabular hyp. log. from the given number, and from the remainders; and the several common logarithms answering to these tabular hyp. logs. joined together, will be the com. log. required, thus:

		hyp. log.
	given	2·1972246
09	......	2·0723266
		1248980
54	....	1243396
		5584
24	....	5526
		58
25	......	58
0·9542425 answer.		

The remaining tables are often useful in trigonometrical investigations, as well as in different researches in the higher mathematics; but they require no particular explanation.

N.	Log	N.	Log	N.	Log.	N.	Log.	N.	Log.
1	0·0000000	51	1·7075702	100	0000000	150	1760913	200	3010300
2	0·3010300	52	1·7160033	101	0043214	151	1789769	201	3031961
3	0·4771213	53	1·7242759	102	0086002	152	1818436	202	3053514
4	0·6020600	54	1·7323938	103	0128372	153	1846914	203	3074960
5	0·6989700	55	1·7403627	104	0170333	154	1875207	204	3096302
6	0·7781513	56	1·7481880	105	0211893	155	1903317	205	3117539
7	0·8450980	57	1·7558749	106	0253059	156	1931246	206	3138672
8	0·9030900	58	1·7634280	107	0293838	157	1958997	207	3159703
9	0·9542425	59	1·7708520	108	0334238	158	1986571	208	3180633
10	1·0000000	60	1·7781513	109	0374265	159	2013971	209	3201463
11	1·0413927	61	1·7853298	110	0413927	160	2041200	210	3222193
12	1·0791812	62	1·7923917	111	0453230	161	2068259	211	3242825
13	1·1139434	63	1·7993405	112	0492180	162	2095150	212	3263359
14	1·1461280	64	1·8061800	113	0530784	163	2121876	213	3283796
15	1·1760913	65	1·8129134	114	0569049	164	2148438	214	3304138
16	1·2041200	66	1·8195439	115	0606978	165	2174839	215	3324385
17	1·2304489	67	1·8260748	116	0644580	166	2201081	216	3344538
18	1·2552725	68	1·8325089	117	0681859	167	2227165	217	3364597
19	1·2787536	69	1·8388491	118	0718820	168	2253093	218	3384565
20	1·3010300	70	1·8450980	119	0755470	169	2278867	219	3404441
21	1·3222193	71	1·8512583	120	0791812	170	2304489	220	3424227
22	1·3424227	72	1·8573325	121	0827854	171	2329961	221	3443923
23	1·3617278	73	1·8633229	122	0863598	172	2355284	222	3463530
24	1·3802112	74	1·8692317	123	0899051	173	2380461	223	3483049
25	1·3979400	75	1·8750613	124	0934217	174	2405492	224	3502480
26	1·4149733	76	1·8808136	125	0969100	175	2430380	225	3521825
27	1·4313638	77	1·8864907	126	1003705	176	2455127	226	3541084
28	1·4471580	78	1·8920946	127	1038037	177	2479733	227	3560259
29	1·4623980	79	1·8976271	128	1072100	178	2504200	228	3579348
30	1·4771213	80	1·9030900	129	1105897	179	2528530	229	3598355
31	1·4913617	81	1·9084850	130	1139434	180	2552725	230	3617278
32	1·5051500	82	1·9138139	131	1172713	181	2576786	231	3636120
33	1·5185139	83	1·9190781	132	1205739	182	2600714	232	3654880
34	1·5314789	84	1·9242793	133	1238516	183	2624511	233	3673559
35	1·5440680	85	1·9294189	134	1271048	184	2648178	234	3692159
36	1·5563025	86	1·9344985	135	1303338	185	2671717	235	3710679
37	1·5682017	87	1·9395193	136	1335389	186	2695129	236	3729120
38	1·5797836	88	1·9444827	137	1367206	187	2718416	237	3747483
39	1·5910646	89	1·9493900	138	1398791	188	2741578	238	3765770
40	1·6020600	90	1·9542425	139	1430148	189	2764618	239	3783979
41	1·6127839	91	1·9590414	140	1461280	190	2787536	240	3802112
42	1·6232493	92	1·9637878	141	1492191	191	2810334	241	3820170
43	1·6334685	93	1·9684829	142	1522883	192	2833012	242	3838154
44	1·6434527	94	1·9731279	143	1553360	193	2855573	243	3856063
45	1·6532125	95	1·9777236	144	1583625	194	2878017	244	3873898
46	1·6627578	96	1·9822712	145	1613680	195	2900346	245	3891661
47	1·6720979	97	1·9867717	146	1643529	196	2922561	246	3909351
48	1·6812412	98	1·9912261	147	1673173	197	2944662	247	3926970
49	1·6901961	99	1·9956352	148	1702617	198	2966652	248	3944517
50	1·6989700	100	2·0000000	149	1731863	199	2988531	249	3961993
N.	Log	N.	Log.	N.	Log.	N.	Log.	N.	Log.

N.	Log.	N.	Log.	N.	Log.	N.	Log.	N.	Log.
250	3979400	300	4771213	350	5440680	400	6020600	450	6532125
251	3996737	301	4785665	351	5453071	401	6031444	451	6541765
252	4014005	302	4800069	352	5465427	402	6042261	452	6551384
253	4031205	303	4814426	353	5477747	403	6053050	453	6560982
254	4048337	304	4828736	354	5490033	404	6063814	454	6570559
255	4065402	305	4842998	355	5502284	405	6074550	455	6580114
256	4082400	306	4857214	356	5514500	406	6085260	456	6589648
257	4099331	307	4871384	357	5526682	407	6095944	457	6599162
258	4116197	308	4885507	358	5538830	408	6106602	458	6608655
259	4132998	309	4899585	359	5550944	409	6117233	459	6618127
260	4149733	310	4913617	360	5563025	410	6127839	460	6627578
261	4166405	311	4927604	361	5575072	411	6138418	461	6637009
262	4183013	312	4941546	362	5587086	412	6148972	462	6646420
263	4199557	313	4955443	363	5599066	413	6159501	463	6655810
264	4216039	314	4969296	364	5611014	414	6170003	464	6665180
265	4232459	315	4983106	365	5622929	415	6180481	465	6674530
266	4248816	316	4996871	366	5634811	416	6190933	466	6683859
267	4265113	317	5010593	367	5646661	417	6201361	467	6693169
268	4281348	318	5024271	368	5658478	418	6211763	468	6702459
269	4297523	319	5037907	369	5670264	419	6222140	469	6711728
270	4313638	320	5051500	370	5682017	420	6232493	470	6720979
271	4329693	321	5065050	371	5693739	421	6242821	471	6730209
272	4345689	322	5078559	372	5705429	422	6253125	472	6739420
273	4361626	323	5092025	373	5717088	423	6263404	473	6748611
274	4377506	324	5105450	374	5728716	424	6273659	474	6757783
275	4393327	325	5118834	375	5740313	425	6283889	475	6766936
276	4409091	326	5132176	376	5751878	426	6294096	476	6776070
277	4424798	327	5145478	377	5763414	427	6304279	477	6785184
278	4440448	328	5158738	378	5774918	428	6314438	478	6794279
279	4456042	329	5171959	379	5786392	429	6324573	479	6803355
280	4471580	330	5185139	380	5797836	430	6334685	480	6812412
281	4487063	331	5198280	381	5809250	431	6344773	481	6821451
282	4502491	332	5211381	382	5820634	432	6354837	482	6830470
283	4517864	333	5224442	383	5831988	433	6364879	483	6839471
284	4533183	334	5237465	384	5843312	434	6374897	484	6848454
285	4548449	335	5250448	385	5854607	435	6384893	485	6857417
286	4563660	336	5263393	386	5865873	436	6394865	486	6866363
287	4578819	337	5276299	387	5877110	437	6404814	487	6875290
288	4593925	338	5289167	388	5888317	438	6414741	488	6884198
289	4608978	339	5301997	389	5899496	439	6424645	489	6893089
290	4623980	340	5314789	390	5910646	440	6434527	490	6901961
291	4638930	341	5327544	391	5921768	441	6444386	491	6910815
292	4653829	342	5340261	392	5932861	442	6454223	492	6919651
293	4668676	343	5352941	393	5943926	443	6464037	493	6928469
294	4683473	344	5365584	394	5954962	444	6473830	494	6937269
295	4698220	345	5378191	395	5965971	445	6483600	495	6946052
296	4712917	346	5390761	396	5976952	446	6493349	496	6954817
297	4727564	347	5403295	397	5987905	447	6503075	497	6963564
298	4742163	348	5415792	398	5998831	448	6512780	498	6972293
299	4756712	349	5428254	399	6009729	449	6522463	499	6981005
N.	Log.	N.	Log.	N.	Log.	N.	Log.	N.	Log.

N.	Log.	N.	Log.	N.	Log.	N.	Log.	N.	Log.
500	6989700	550	7403627	600	7781513	650	8129134	700	8450980
501	6998877	551	7411516	601	7788745	651	8135810	701	8457180
502	7007037	552	7419391	602	7795965	652	8142476	702	8463371
503	7015680	553	7427251	603	7803173	653	8149132	703	8469553
504	7024305	554	7435098	604	7810369	654	8155777	704	8475727
505	7032914	555	7442930	605	7817554	655	8162413	705	8481891
506	7041505	556	7450748	606	7824726	656	8169038	706	8488047
507	7050080	557	7458552	607	7831887	657	8175654	707	8494194
508	7058637	558	7466342	608	7839036	658	8182259	708	8500333
509	7067178	559	7474118	609	7846173	659	8188854	709	8506462
510	7075702	560	7481880	610	7853298	660	8195439	710	8512583
511	7084209	561	7489629	611	7860412	661	8202015	711	8518696
512	7092700	562	7497363	612	7867514	662	8208580	712	8524800
513	7101174	563	7505084	613	7874605	663	8215135	713	8530895
514	7109631	564	7512791	614	7881684	664	8221681	714	8536982
515	7118072	565	7520484	615	7888751	665	8228216	715	8543060
516	7126497	566	7528164	616	7895807	666	8234742	716	8549130
517	7134905	567	7535831	617	7902852	667	8241258	717	8555192
518	7143298	568	7543483	618	7909885	668	8247765	718	8561244
519	7151674	569	7551123	619	7916906	669	8254261	719	8567289
520	7160033	570	7558749	620	7923917	670	8260748	720	8573325
521	7168377	571	7566361	621	7930916	671	8267225	721	8579353
522	7176705	572	7573960	622	7937904	672	8273693	722	8585372
523	7185017	573	7581546	623	7944880	673	8280151	723	8591383
524	7193313	574	7589119	624	7951846	674	8286599	724	8597386
525	7201593	575	7596678	625	7958800	675	8293038	725	8603380
526	7209857	576	7604225	626	7965743	676	8299467	726	8609366
527	7218106	577	7611758	627	7972675	677	8305887	727	8615344
528	7226339	578	7619278	628	7979596	678	8312297	728	8621314
529	7234557	579	7626786	629	7986506	679	8318698	729	8627275
530	7242759	580	7634280	630	7993405	680	8325089	730	8633229
531	7250945	581	7641761	631	8000294	681	8331471	731	8639174
532	7259116	582	7649230	632	8007171	682	8337844	732	8645111
533	7267272	583	7656686	633	8014037	683	8344207	733	8651040
534	7275413	584	7664128	634	8020893	684	8350561	734	8656961
535	7283538	585	7671559	635	8027737	685	8356906	735	8662873
536	7291648	586	7678976	636	8034571	686	8363241	736	8668778
537	7299743	587	7686381	637	8041394	687	8369567	737	8674675
538	7307823	588	7693773	638	8048207	688	8375884	738	8680564
539	7315888	589	7701153	639	8055009	689	8382192	739	8686444
540	7323938	590	7708520	640	8061800	690	8388491	740	8692317
541	7331973	591	7715875	641	8068580	691	8394780	741	8698182
542	7339993	592	7723217	642	8075350	692	8401061	742	8704039
543	7347998	593	7730547	643	8082110	693	8407332	743	8709888
544	7355989	594	7737864	644	8088859	694	8413595	744	8715729
545	7363965	595	7745170	645	8095597	695	8419848	745	8721563
546	7371926	596	7752463	646	8102325	696	8426092	746	8727388
547	7379873	597	7759743	647	8109043	697	8432328	747	8733206
548	7387806	598	7767012	648	8115750	698	8438554	748	8739016
549	7395723	599	7774268	649	8122447	699	8444772	749	8744818
N.	Log.	N.	Log.	N.	Log.	N.	Log.	N.	Log.

N.	Log.	N.	Log.	N.	Log.	N.	Log.	N.	Log.
750	8750613	800	9030900	850	9294189	900	9542425	950	9777236
751	8756399	801	9036325	851	9299296	901	9547248	951	9781805
752	8762178	802	9041744	852	9304396	902	9552065	952	9786369
753	8767950	803	9047155	853	9309490	903	9556878	953	9790929
754	8773713	804	9052560	854	9314579	904	9561684	954	9795484
755	8779470	805	9057959	855	9319661	905	9566486	955	9800034
756	8785218	806	9063350	856	9324738	906	9571282	956	9804579
757	8790959	807	9068735	857	9329808	907	9576073	957	9809119
758	8796692	808	9074114	858	9334873	908	9580858	958	9813655
759	8802418	809	9079485	859	9339932	909	9585639	959	9818186
760	8808136	810	9084850	860	9344985	910	9590414	960	9822712
761	8813847	811	9090209	861	9350032	911	9595184	961	9827234
762	8819550	812	9095560	862	9355073	912	9599948	962	9831751
763	8825245	813	9100905	863	9360108	913	9604708	963	9836263
764	8830934	814	9106244	864	9365137	914	9609462	964	9840770
765	8836614	815	9111576	865	9370161	915	9614211	965	9845273
766	8842288	816	9116902	866	9375179	916	9618955	966	9849771
767	8847954	817	9122221	867	9380191	917	9623693	967	9854265
768	8853612	818	9127533	868	9385197	918	9628427	968	9858754
769	8859263	819	9132839	869	9390198	919	9633155	969	9863238
770	8864907	820	9138139	870	9395193	920	9637878	970	9867717
771	8870544	821	9143432	871	9400182	921	9642596	971	9872192
772	8876173	822	9148718	872	9405165	922	9647309	972	9876663
773	8881795	823	9153998	873	9410142	923	9652017	973	9881128
774	8887410	824	9159272	874	9415114	924	9656720	974	9885590
775	8893017	825	9164539	875	9420081	925	9661417	975	9890046
776	8898617	826	9169800	876	9425041	926	9666110	976	9894498
777	8904210	827	9175055	877	9429996	927	9670797	977	9898946
778	8909796	828	9180303	878	9434945	928	9675480	978	9903389
779	8915375	829	9185545	879	9439889	929	9680157	979	9907827
780	8920946	830	9190781	880	9444827	930	9684829	980	9912261
781	8926510	831	9196010	881	9449759	931	9689497	981	9916690
782	8932068	832	9201233	882	9454686	932	9694159	982	9921115
783	8937618	833	9206450	883	9459607	933	9698816	983	9925535
784	8943161	834	9211661	884	9464523	934	9703469	984	9929951
785	8948697	835	9216865	885	9469433	935	9708116	985	9934362
786	8954225	836	9222063	886	9474337	936	9712758	986	9938769
787	8959747	837	9227255	887	9479236	937	9717396	987	9943172
788	8965262	838	9232440	888	9484130	938	9722028	988	9947569
789	8970770	839	9237620	889	9489018	939	9726656	989	9951963
790	8976271	840	9242793	890	9493900	940	9731279	990	9956352
791	8981765	841	9247960	891	9498777	941	9735896	991	9960737
792	8987252	842	9253121	892	9503649	942	9740509	992	9965117
793	8992732	843	9258276	893	9508515	943	9745117	993	9969492
794	8998205	844	9263424	894	9513375	944	9749720	994	9973864
795	9003671	845	9268567	895	9518230	945	9754318	995	9978231
796	9009131	846	9273704	896	9523080	946	9758911	996	9982593
797	9014583	847	9278834	897	9527924	947	9763500	997	9986952
798	9020029	848	9283959	898	9532763	948	9768083	998	9991305
799	9025468	849	9289077	899	9537597	949	9772662	999	9995655
N.	Log.	N.	Log	N.	Log.	N.	Log.	N.	Log.

N.	0	1	2	3	4	5	6	7	8	9
1000	0000000	0434	0869	1303	1737	2171	2605	3039	3473	3907
01	4341	4775	5208	5642	6076	6510	6943	7377	7810	8244
02	8677	9111	9544	9977	$\overline{0}$411	$\overline{0}$844	$\overline{1}$277	$\overline{1}$710	$\overline{2}$143	$\overline{2}$576
03	0013009	3442	3875	4308	4741	5174	5607	6039	6472	6905
04	7337	7770	8202	8635	9067	9499	9932	$\overline{0}$364	$\overline{0}$796	$\overline{1}$228
05	0021661	2093	2525	2957	3389	3821	4253	4685	5116	5548
06	5980	6411	6843	7275	7706	8138	8569	9001	9432	9863
07	0030295	0726	1157	1588	2019	2451	2882	3313	3744	4174
08	4605	5036	5467	5898	6328	6759	7190	7620	8051	8481
09	8912	9342	9772	$\overline{0}$203	$\overline{0}$633	$\overline{1}$063	$\overline{1}$493	$\overline{1}$924	$\overline{2}$354	$\overline{2}$784
1010	0043214	3644	4074	4504	4933	5363	5793	6223	6652	7082
11	7512	7941	8371	8800	9229	9659	$\overline{0}$089	$\overline{0}$517	$\overline{0}$947	$\overline{1}$376
12	0051805	2234	2663	3092	3521	3950	4379	4808	5237	5666
13	6094	6523	6952	7380	7809	8238	8666	9094	9523	9951
14	0060380	0808	1236	1664	2092	2521	2949	3377	3805	4233
15	4660	5088	5516	5944	6372	6799	7227	7655	8082	8510
16	8937	9365	9792	$\overline{0}$219	$\overline{0}$647	$\overline{1}$074	$\overline{1}$501	$\overline{1}$928	$\overline{2}$355	$\overline{2}$782
17	0073210	3637	4064	4490	4917	5344	5771	6198	6624	7051
18	7478	7904	8331	8757	9184	9610	$\overline{0}$037	$\overline{0}$463	$\overline{0}$889	$\overline{1}$316
19	0081742	2168	2594	3020	3446	3872	4298	4724	5150	5576
1020	6002	6427	6853	7279	7704	8130	8556	8981	9407	9832
21	0090257	0683	1108	1533	1959	2384	2809	3234	3659	4084
22	4509	4934	5359	5784	6208	6633	7058	7483	7907	8332
23	8756	9181	9605	$\overline{0}$030	$\overline{0}$454	$\overline{0}$878	$\overline{1}$303	$\overline{1}$727	$\overline{2}$151	$\overline{2}$575
24	0103000	3424	3848	4272	4696	5120	5544	5967	6391	6815
25	7239	7662	8086	8510	8933	9357	9780	$\overline{0}$204	$\overline{0}$627	$\overline{1}$050
26	0111474	1897	2320	2743	3166	3590	4013	4436	4859	5282
27	5704	6127	6550	6973	7396	7818	8241	8664	9086	9509
28	9931	$\overline{0}$354	$\overline{0}$776	$\overline{1}$198	$\overline{1}$621	$\overline{2}$043	$\overline{2}$465	$\overline{2}$887	$\overline{3}$310	$\overline{3}$732
29	0124154	4576	4998	5420	5842	6264	6685	7107	7529	7951
1030	8372	8794	9215	9637	$\overline{0}$059	$\overline{0}$480	$\overline{0}$901	$\overline{1}$323	$\overline{1}$744	$\overline{2}$165
31	0132587	3008	3429	3850	4271	4692	5113	5534	5955	6376
32	6797	7218	7639	8059	8480	8901	9321	9742	$\overline{0}$162	$\overline{0}$583
33	0141003	1424	1844	2264	2685	3105	3525	3945	4365	4785
34	5205	5625	6045	6465	6885	7305	7725	8144	8564	8984
35	9403	9823	$\overline{0}$243	$\overline{0}$662	$\overline{1}$082	$\overline{1}$501	$\overline{1}$920	$\overline{2}$340	$\overline{2}$759	$\overline{3}$178
36	0153598	4017	4436	4855	5274	5693	6112	6531	6950	7369
37	7788	8206	8625	9044	9462	9881	$\overline{0}$300	$\overline{0}$718	$\overline{1}$137	$\overline{1}$555
38	0161974	2392	2810	3229	3647	4065	4483	4901	5319	5737
39	6155	6573	6991	7409	7827	8245	8663	9080	9498	9916
1040	0170333	0751	1168	1586	2003	2421	2838	3256	3673	4090
41	4507	4924	5342	5759	6176	6593	7010	7427	7844	8260
42	8677	9094	9511	9927	$\overline{0}$344	$\overline{0}$761	$\overline{1}$177	$\overline{1}$594	$\overline{2}$010	$\overline{2}$427
43	0182843	3259	3676	4092	4508	4925	5341	5757	6173	6589
44	7005	7421	7837	8253	8669	9084	9500	9916	$\overline{0}$332	$\overline{0}$747
45	0191163	1578	1994	2410	2825	3240	3656	4071	4486	4902
46	5317	5732	6147	6562	6977	7392	7807	8222	8637	9052
47	9467	9882	$\overline{0}$296	$\overline{0}$711	$\overline{1}$126	$\overline{1}$540	$\overline{1}$955	$\overline{2}$369	$\overline{2}$784	$\overline{3}$198
48	0203613	4027	4442	4856	5270	5684	6099	6513	6927	7341
49	7755	8169	8583	8997	9411	9824	$\overline{0}$238	$\overline{0}$652	$\overline{1}$066	$\overline{1}$479
N.	0	1	2	3	4	5	6	7	8	9

Dif. & Pro. Pts.

	434	433	432
1	43	43	43
2	87	87	86
3	130	130	130
4	174	173	173
5	217	217	216
6	260	260	259
7	304	303	302
8	347	346	346
9	391	390	389

	431	430	429
1	43	43	43
2	86	86	86
3	129	129	129
4	172	172	172
5	216	215	215
6	259	258	257
7	302	301	300
8	345	344	343
9	388	387	386

	428	427	426
1	43	43	43
2	86	85	85
3	128	128	128
4	171	171	170
5	214	214	213
6	257	256	256
7	300	299	298
8	342	342	341
9	385	384	383

	425	424	423
1	43	42	42
2	85	85	85
3	128	127	127
4	170	170	169
5	213	212	212
6	255	254	254
7	298	297	296
8	340	339	338
9	383	382	381

	422	421	420
1	42	42	42
2	84	84	84
3	127	126	126
4	169	168	168
5	211	211	210
6	253	253	252
7	295	295	294
8	338	337	336
9	380	379	378

	419	418	417
1	42	42	42
2	84	84	83
3	126	125	125
4	168	167	167
5	210	209	209
6	251	251	250
7	293	293	292
8	335	334	334
9	377	376	375

Dif. & Pro. Pts.

N.	0	1	2	3	4	5	6	7	8	9
1050	0211893	2307	2720	3134	3547	3961	4374	4787	5201	5614
51	6027	6440	6854	7267	7680	8093	8506	8919	9332	9745
52	0221157	0570	0983	1396	1808	2221	2634	3046	3459	3871
53	4284	4696	5109	5521	5933	6345	6758	7170	7582	7994
54	8406	8818	9230	9642	0054	0466	0878	1289	1701	2113
55	0232525	2936	3348	3759	4171	4582	4994	5405	5817	6228
56	6639	7050	7462	7873	8284	8695	9106	9517	9928	0339
57	0240750	1161	1572	1982	2393	2804	3214	3625	4036	4446
58	4857	5267	5678	6088	6498	6909	7319	7729	8139	8549
59	8960	9370	9780	0190	0600	1010	1419	1829	2239	2649
1060	0253059	3468	3878	4288	4697	5107	5516	5926	6335	6744
61	7154	7563	7972	8382	8791	9200	9609	0018	0427	0836
62	0261245	1654	2063	2472	2881	3289	3698	4107	4515	4924
63	5333	5741	6150	6558	6967	7375	7783	8192	8600	9008
64	9416	9824	0233	0641	1049	1457	1865	2273	2680	3088
65	0273496	3904	4312	4719	5127	5535	5942	6350	6757	7165
66	7572	7979	8387	8794	9201	9609	0016	0423	0830	1237
67	0281644	2051	2458	2865	3272	3679	4086	4492	4899	5306
68	5713	6119	6526	6932	7339	7745	8152	8558	8964	9371
69	9777	0183	0590	0996	1402	1808	2214	2620	3026	3432
1070	0293838	4244	4649	5055	5461	5867	6272	6678	7084	7489
71	7895	8300	8706	9111	9516	9922	0327	0732	1138	1543
72	0301948	2353	2758	3163	3568	3973	4378	4783	5188	5592
73	5997	6402	6807	7211	7616	8020	8425	8830	9234	9638
74	0310043	0447	0851	1256	1660	2064	2468	2872	3277	3681
75	4085	4489	4893	5296	5700	6104	6508	6912	7315	7719
76	8123	8526	8930	9333	9737	0140	0544	0947	1350	1754
77	0322157	2560	2963	3367	3770	4173	4576	4979	5382	5785
78	6188	6590	6993	7396	7799	8201	8604	9007	9409	9812
79	0330214	0617	1019	1422	1824	2226	2629	3031	3433	3835
1080	4238	4640	5042	5444	5846	6248	6650	7052	7453	7855
81	8257	8659	9060	9462	9864	0265	0667	1068	1470	1871
82	0342273	2674	3075	3477	3878	4279	4680	5081	5482	5884
83	6285	6686	7087	7487	7888	8289	8690	9091	9491	9892
84	0350293	0693	1094	1495	1895	2296	2696	3096	3497	3897
85	4297	4698	5098	5498	5898	6298	6698	7098	7498	7898
86	8298	8698	9098	9498	9898	0297	0697	1097	1496	1896
87	0362295	2695	3094	3494	3893	4293	4692	5091	5491	5890
88	6289	6688	7087	7486	7885	8284	8683	9082	9481	9880
89	0370279	0678	1076	1475	1874	2272	2671	3070	3468	3867
1090	4265	4663	5062	5460	5858	6257	6655	7053	7451	7849
91	8248	8646	9044	9442	9839	0237	0635	1033	1431	1829
92	0382226	2624	3022	3419	3817	4214	4612	5009	5407	5804
93	6202	6599	6996	7393	7791	8188	8585	8982	9379	9776
94	0390173	0570	0967	1364	1761	2158	2554	2951	3348	3745
95	4141	4538	4934	5331	5727	6124	6520	6917	7313	7709
96	8106	8502	8898	9294	9690	0086	0482	0878	1274	1670
97	0402066	2462	2858	3254	3650	4045	4441	4837	5232	5628
98	6023	6419	6814	7210	7605	8001	8396	8791	9187	9582
99	9977	0372	0767	1162	1557	1952	2347	2742	3137	3532
N.	0	1	2	3	4	5	6	7	8	9

Dif. & Pro. Pts.

	416	415	414
1	42	42	41
2	83	83	83
3	125	125	124
4	166	166	166
5	208	208	207
6	250	249	248
7	291	291	290
8	333	332	331
9	374	374	373

	413	412	411
1	41	41	41
2	83	82	82
3	124	124	123
4	165	165	164
5	207	206	206
6	248	247	247
7	289	288	288
8	330	330	329
9	372	371	370

	410	409	408
1	41	41	41
2	82	82	82
3	123	123	122
4	164	164	163
5	205	205	204
6	246	245	245
7	287	286	286
8	328	327	326
9	369	368	367

	407	406	405
1	41	41	41
2	81	81	81
3	122	122	122
4	163	162	162
5	204	203	203
6	244	244	243
7	285	284	284
8	326	325	324
9	366	365	365

	404	403	402
1	40	40	40
2	81	81	80
3	121	121	121
4	162	161	161
5	202	202	201
6	242	242	241
7	283	282	281
8	323	322	322
9	364	363	362

	401	400	399
1	40	40	40
2	80	80	80
3	120	120	120
4	160	160	160
5	201	200	200
6	241	240	239
7	281	280	279
8	321	320	319
9	361	360	359

N.	0	1	2	3	4	5	6	7	8	9
1100	0413927	4322	4716	5111	5506	5900	6295	6690	7084	7479
01	7873	8268	8662	9056	9451	9845	0̄239	0̄633	1̄028	1̄422
02	0421816	2210	2604	2998	3392	3786	4180	4574	4968	5361
03	5755	6149	6543	6936	7330	7723	8117	8510	8904	9297
04	9691	0̄084	0̄477	0̄871	1̄264	1̄657	2̄050	2̄444	2̄837	3̄230
05	0433623	4016	4409	4802	5195	5587	5980	6373	6766	7159
06	7551	7944	8337	8729	9122	9514	9907	0̄299	0̄692	1̄084
07	0441476	1869	2261	2653	3045	3437	3829	4222	4614	5006
08	5398	5790	6181	6573	6965	7357	7749	8140	8532	8924
09	9315	9707	0̄099	0̄490	0̄882	1̄273	1̄664	2̄056	2̄447	2̄839
1110	0453230	3621	4012	4403	4795	5186	5577	5968	6359	6750
11	7141	7531	7922	8313	8704	9095	9485	9876	0̄267	0̄657
12	0461048	1438	1829	2219	2610	3000	3391	3781	4171	4561
13	4952	5342	5732	6122	6512	6902	7292	7682	8072	8462
14	8852	9242	9632	0̄021	0̄411	0̄801	1̄190	1̄580	1̄970	2̄359
15	0472749	3138	3528	3917	4306	4696	5085	5474	5864	6253
16	6642	7031	7420	7809	8198	8587	8976	9365	9754	0̄143
17	0480532	0921	1309	1698	2087	2475	2864	3253	3641	4030
18	4418	4806	5195	5583	5972	6360	6748	7136	7525	7913
19	8301	8689	9077	9465	9853	0̄241	0̄629	1̄017	1̄405	1̄792
1120	0492180	2568	2956	3343	3731	4119	4506	4894	5281	5669
21	6056	6444	6831	7218	7606	7993	8380	8767	9154	9541
22	9929	0̄316	0̄703	1̄090	1̄477	1̄863	2̄250	2̄637	3̄024	3̄411
23	0503798	4184	4571	4958	5344	5731	6117	6504	6890	7277
24	7663	8049	8436	8822	9208	9595	9981	0̄367	0̄753	1̄139
25	0511525	1911	2297	2683	3069	3455	3841	4227	4612	4998
26	5384	5770	6155	6541	6926	7312	7697	8083	8468	8854
27	9239	9624	0̄010	0̄395	0̄780	1̄166	1̄551	1̄936	2̄321	2̄706
28	0523091	3476	3861	4246	4631	5016	5400	5785	6170	6555
29	6939	7324	7709	8093	8478	8862	9247	9631	0̄016	0̄400
1130	0530784	1169	1553	1937	2321	2706	3090	3474	3858	4242
31	4626	5010	5394	5778	6162	6546	6929	7313	7697	8081
32	8464	8848	9232	9615	9999	0̄382	0̄766	1̄149	1̄532	1̄916
33	0542299	2682	3066	3449	3832	4215	4598	4981	5365	5748
34	6131	6514	6896	7279	7662	8045	8428	8811	9193	9576
35	9959	0̄341	0̄724	1̄106	1̄489	1̄871	2̄254	2̄636	3̄019	3̄401
36	0553783	4166	4548	4930	5312	5694	6077	6459	6841	7223
37	7605	7987	8369	8750	9132	9514	9896	0̄278	0̄659	1̄041
38	0561423	1804	2186	2567	2949	3330	3712	4093	4475	4856
39	5237	5619	6000	6381	6762	7143	7524	7905	8287	8668
1140	9049	9429	9810	0̄191	0̄572	0̄953	1̄334	1̄714	2̄095	2̄476
41	0572856	3237	3618	3998	4379	4759	5140	5520	5900	6281
42	6661	7041	7422	7802	8182	8562	8942	9322	9702	0̄082
43	0580462	0842	1222	1602	1982	2362	2741	3121	3501	3881
44	4260	4640	5019	5399	5778	6158	6537	6917	7296	7676
45	8055	8434	8813	9193	9572	9951	0̄330	0̄709	1̄088	1̄467
46	0591846	2225	2604	2983	3362	3741	4119	4498	4877	5256
47	5634	6013	6391	6770	7148	7527	7905	8284	8662	9041
48	9419	9797	0̄175	0̄554	0̄932	1̄310	1̄688	2̄066	2̄444	2̄822
49	0603200	3578	3956	4334	4712	5090	5468	5845	6223	6601
N.	0	1	2	3	4	5	6	7	8	9

Dif. & Pro. Pts.

	398	397	396
1	40	40	40
2	80	79	79
3	119	119	119
4	159	159	158
5	199	199	198
6	239	238	238
7	279	278	277
8	318	318	317
9	358	357	356

	395	394	393
1	40	39	39
2	79	79	79
3	119	118	118
4	158	158	157
5	198	197	197
6	237	236	236
7	277	276	275
8	316	315	314
9	356	355	354

	392	391	390
1	39	39	39
2	78	78	78
3	118	117	117
4	157	156	156
5	196	196	195
6	235	235	234
7	274	274	273
8	314	313	312
9	353	352	351

	389	388	387
1	39	39	39
2	78	78	77
3	117	116	116
4	156	155	155
5	195	194	194
6	233	233	232
7	272	272	271
8	311	310	310
9	350	349	348

	386	385	384
1	39	39	38
2	77	77	77
3	116	116	115
4	154	154	154
5	193	193	192
6	232	231	230
7	270	270	269
8	309	308	307
9	347	347	346

	383	382	381
1	38	38	38
2	77	76	76
3	115	115	114
4	153	153	152
5	192	191	191
6	230	229	229
7	268	267	267
8	306	306	305
9	345	344	343

N.	0	1	2	3	4	5	6	7	8	9
1150	0606978	7356	7734	8111	8489	8866	9244	9621	9999	0̄376
51	0610753	1131	1508	1885	2262	2639	3017	3394	3771	4148
52	4525	4902	5279	5656	6032	6409	6786	7163	7540	7916
53	8293	8670	9046	9423	9799	0̄176	0̄552	0̄929	1305	1̄682
54	0622058	2434	2811	3187	3563	3939	4316	4692	5068	5444
55	5820	6196	6572	6948	7324	7699	8075	8451	8827	9203
56	9578	9954	0̄330	0̄705	1̄081	1̄456	1̄832	2̄207	2̄583	2̄958
57	0633334	3709	4084	4460	4835	5210	5585	5960	6335	6711
58	7086	7461	7836	8211	8585	8960	9335	9710	0̄085	0̄460
59	0640834	1209	1584	1958	2333	2708	3082	3457	3831	4205
1160	4580	4954	5329	5703	6077	6451	6826	7200	7574	7948
61	8322	8696	9070	9444	9818	0̄192	0̄566	0̄940	1̄314	1̄688
62	0652061	2435	2809	3182	3556	3930	4303	4677	5050	5424
63	5797	6171	6544	6917	7291	7664	8037	8410	8784	9157
64	9530	9903	0̄276	0̄649	1̄022	1̄395	1̄768	2̄141	2̄514	2̄886
65	0663259	3632	4005	4377	4750	5123	5495	5868	6241	6613
66	6986	7358	7730	8103	8475	8847	9220	9592	9964	0̄336
67	0670709	1081	1453	1825	2197	2569	2941	3313	3685	4057
68	4428	4800	5172	5544	5915	6287	6659	7030	7402	7774
69	8145	8517	8888	9259	9631	0̄002	0̄374	0̄745	1̄116	1̄487
1170	0681859	2230	2601	2972	3343	3714	4085	4456	4827	5198
71	5569	5940	6311	6681	7052	7423	7794	8164	8535	8906
72	9276	9647	0̄017	0̄388	0̄758	1̄129	1̄499	1̄869	2̄240	2̄610
73	0692980	3350	3721	4091	4461	4831	5201	5571	5941	6311
74	6681	7051	7421	7791	8160	8530	8900	9270	9639	0̄009
75	0700379	0748	1118	1487	1857	2226	2596	2965	3335	3704
76	4073	4442	4812	5181	5550	5919	6288	6658	7027	7396
77	7765	8134	8503	8871	9240	9609	9978	0̄347	0̄715	1̄084
78	0711453	1822	2190	2559	2927	3296	3664	4033	4401	4770
79	5138	5506	5875	6243	6611	6979	7348	7716	8084	8452
1180	8820	9188	9556	9924	0̄292	0̄660	1̄028	1̄396	1̄763	2̄131
81	0722499	2867	3234	3602	3970	4337	4705	5072	5440	5807
82	6175	6542	6910	7277	7644	8011	8379	8746	9113	9480
83	9847	0̄215	0̄582	0̄949	1̄316	1̄683	2̄050	2̄416	2̄783	3̄150
84	0733517	3884	4251	4617	4984	5351	5717	6084	6450	6817
85	7184	7550	7916	8283	8649	9016	9382	9748	0̄114	0̄481
86	0740847	1213	1579	1945	2311	2677	3043	3409	3775	4141
87	4507	4873	5239	5605	5970	6336	6702	7068	7433	7799
88	8164	8530	8895	9261	9626	9992	0̄357	0̄723	1̄088	1̄453
89	0751819	2184	2549	2914	3279	3644	4010	4375	4740	5105
1190	5470	5835	6199	6564	6929	7294	7659	8024	8388	8753
91	9118	9482	9847	0̄211	0̄576	0̄940	1̄305	1̄669	2̄034	2̄398
92	0762763	3127	3491	3855	4220	4584	4948	5312	5676	6040
93	6404	6768	7132	7496	7860	8224	8588	8952	9316	9680
94	0770043	0407	0771	1134	1498	1862	2225	2589	2952	3316
95	3679	4042	4406	4769	5133	5496	5859	6222	6585	6949
96	7312	7675	8038	8401	8764	9127	9490	9853	0̄216	0̄579
97	0780942	1304	1667	2030	2393	2755	3118	3480	3843	4206
98	4568	4931	5293	5656	6018	6380	6743	7105	7467	7830
99	8192	8554	8916	9278	9640	0̄003	0̄365	0̄727	1̄089	1̄451
N.	0	1	2	3	4	5	6	7	8	9

Dif. & Pro. Pts.

	380	379	378
1	38	38	38
2	76	76	76
3	114	114	113
4	152	152	151
5	190	190	189
6	228	227	227
7	266	265	265
8	304	303	302
9	342	341	340

	377	376	375
1	38	38	38
2	75	75	75
3	113	113	113
4	151	150	150
5	189	188	188
6	226	226	225
7	264	263	263
8	302	301	300
9	339	338	338

	374	373	372
1	37	37	37
2	75	75	74
3	112	112	112
4	150	149	149
5	187	187	186
6	224	224	223
7	262	261	260
8	299	298	298
9	337	336	335

	371	370	369
1	37	37	37
2	74	74	74
3	111	111	111
4	148	148	148
5	186	185	185
6	223	222	221
7	260	259	258
8	297	296	295
9	334	333	332

	368	367	366
1	37	37	37
2	74	73	73
3	110	110	110
4	147	147	146
5	184	184	183
6	221	220	220
7	258	257	256
8	294	294	293
9	331	330	329

	365	364	363
1	37	36	36
2	73	73	73
3	110	109	109
4	146	146	145
5	183	182	182
6	219	218	218
7	256	255	254
8	292	291	290
9	329	328	327

Dif. & Pro. Pts.

N.	0	1	2	3	4	5	6	7	8	9
1200	0791812	2174	2536	2898	3260	3622	3983	4345	4707	5068
01	5430	5792	6153	6515	6876	7238	7599	7961	8322	8683
02	9045	9406	9767	0̄128	0̄490	0̄851	1̄212	1̄573	1̄934	2̄295
03	0802656	3017	3378	3739	4100	4461	4822	5183	5543	5904
04	6265	6626	6986	7347	7707	8068	8429	8789	9150	9510
05	9870	0̄231	0̄591	0̄952	1̄312	1̄672	2̄032	2̄393	2̄753	3̄113
06	0813473	3833	4193	4553	4913	5273	5633	5993	6353	6713
07	7073	7432	7792	8152	8512	8871	9231	9591	9950	0̄310
08	0820669	1029	1388	1748	2107	2467	2826	3185	3545	3904
09	4263	4622	4981	5341	5700	6059	6418	6777	7136	7495
1210	7854	8213	8571	8930	9289	9648	0̄007	0̄365	0̄724	1̄083
11	0831441	1800	2159	2517	2876	3234	3593	3951	4309	4668
12	5026	5385	5743	6101	6459	6817	7176	7534	7892	8250
13	8608	8966	9324	9682	0̄040	0̄398	0̄756	1̄114	1̄471	1̄829
14	0842187	2545	2902	3260	3618	3975	4333	4690	5048	5405
15	5763	6120	6478	6835	7192	7550	7907	8264	8621	8979
16	9336	9693	0̄050	0̄407	0̄764	1̄121	1̄478	1̄835	2̄192	2̄549
17	0852906	3263	3619	3976	4333	4690	5046	5403	5760	6116
18	6473	6829	7186	7542	7899	8255	8612	8968	9324	9681
19	0860037	0393	0750	1106	1462	1818	2174	2530	2886	3242
1220	3598	3954	4310	4666	5022	5378	5734	6089	6445	6801
21	7157	7512	7868	8224	8579	8935	9290	9646	0̄001	0̄357
22	0870712	1067	1423	1778	2133	2489	2844	3199	3554	3909
23	4265	4620	4975	5330	5685	6040	6395	6750	7104	7459
24	7814	8169	8524	8878	9233	9588	9943	0̄297	0̄652	1̄006
25	0881361	1715	2070	2424	2779	3133	3488	3842	4196	4550
26	4905	5259	5613	5967	6321	6676	7030	7384	7738	8092
27	8446	8800	9153	9507	9861	0̄215	0̄569	0̄923	1̄276	1̄630
28	0891984	2337	2691	3045	3398	3752	4105	4459	4812	5165
29	5519	5872	6226	6579	6932	7285	7639	7992	8345	8698
1230	9051	9404	9757	0̄110	0̄463	0̄816	1̄169	1̄522	1̄875	2̄228
31	0902581	2933	3286	3639	3991	4344	4697	5049	5402	5755
32	6107	6460	6812	7164	7517	7869	8222	8574	8926	9279
33	9631	9983	0̄335	0̄687	1̄039	1̄392	1̄744	2̄096	2̄448	2̄800
34	0913152	3504	3855	4207	4559	4911	5263	5614	5966	6318
35	6670	7021	7373	7724	8076	8427	8779	9130	9482	9833
36	0920185	0536	0887	1239	1590	1941	2292	2644	2995	3346
37	3697	4048	4399	4750	5101	5452	5803	6154	6505	6856
38	7206	7557	7908	8259	8609	8960	9311	9661	0̄012	0̄363
39	0930713	1064	1414	1764	2115	2465	2816	3166	3516	3867
1240	4217	4567	4917	5267	5618	5968	6318	6668	7018	7368
41	7718	8068	8418	8768	9117	9467	9817	0̄167	0̄517	0̄866
42	0941216	1566	1915	2265	2614	2964	3313	3663	4012	4362
43	4711	5061	5410	5759	6109	6458	6807	7156	7506	7855
44	8204	8553	8902	9251	9600	9949	0̄298	0̄647	0̄996	1̄345
45	0951694	2042	2391	2740	3089	3437	3786	4135	4483	4832
46	5180	5529	5877	6226	6574	6923	7271	7620	7968	8316
47	8665	9013	9361	9709	0̄057	0̄406	0̄754	1̄102	1̄450	1̄798
48	0962146	2494	2842	3190	3538	3885	4233	4581	4929	5277
49	5624	5972	6320	6667	7015	7363	7710	8058	8405	8753
N	0	1	2	3	4	5	6	7	8	9

Dif. & Pro. Pts.

	362	361	360
1	36	36	36
2	72	72	72
3	109	108	108
4	145	144	144
5	181	181	180
6	217	217	216
7	253	253	252
8	290	289	288
9	326	325	324

	359	358	357
1	36	36	36
2	72	72	71
3	108	107	107
4	144	143	143
5	180	179	179
6	215	215	214
7	251	251	250
8	287	286	286
9	323	322	321

	356	355	354
1	36	36	35
2	71	71	71
3	107	107	106
4	142	142	142
5	178	178	177
6	214	213	212
7	249	249	248
8	285	284	283
9	320	320	319

	353	352	351
1	35	35	35
2	71	70	70
3	106	106	105
4	141	141	140
5	177	176	176
6	212	211	211
7	247	246	246
8	282	282	281
9	318	317	316

	350	349	348
1	35	35	35
2	70	70	70
3	105	105	104
4	140	140	139
5	175	175	174
6	210	209	209
7	245	244	244
8	280	279	278
9	315	314	313

	347	346	345
1	35	35	35
2	69	69	69
3	104	104	104
4	139	138	138
5	174	173	173
6	208	208	207
7	243	242	242
8	278	277	276
9	312	311	311

Dif. & Pro. Pts.

N.	0	1	2	3	4	5	6	7	8	9
1250	0969100	9448	9795	ō142	ō490	ō837	Ī184	Ī531	Ī879	2226
51	0972573	2920	3267	3614	3962	4309	4656	5003	5349	5696
52	6043	6390	6737	7084	7431	7777	8124	8471	8817	9164
53	9511	9857	ō204	ō550	ō897	Ī243	Ī590	Ī936	2283	2629
54	0982975	3322	3668	4014	4360	4707	5053	5399	5745	6091
55	6437	6783	7129	7475	7821	8167	8513	8859	9205	9551
56	9896	ō242	ō588	ō934	Ī279	Ī625	Ī971	2316	2662	3007
57	0993353	3698	4044	4389	4735	5080	5425	5771	6116	6461
58	6806	7152	7497	7842	8187	8532	8877	9222	9567	9912
59	1000257	0602	0947	1292	1637	1982	2327	2671	3016	3361
1260	3705	4050	4395	4739	5084	5429	5773	6118	6462	6806
61	7151	7495	7840	8184	8528	8873	9217	9561	9905	ō249
62	1010594	0938	1282	1626	1970	2314	2658	3002	3346	3690
63	4034	4377	4721	5065	5409	5752	6096	6440	6784	7127
64	7471	7814	8158	8501	8845	9188	9532	9875	ō219	ō562
65	1020905	1249	1592	1935	2278	2621	2965	3308	3651	3994
66	4337	4680	5023	5366	5709	6052	6395	6738	7081	7423
67	7766	8109	8452	8794	9137	9480	9822	ō165	ō507	ō850
68	1031193	1535	1877	2220	2562	2905	3247	3589	3932	4274
69	4616	4958	5301	5643	5985	6327	6669	7011	7353	7695
1270	8037	8379	8721	9063	9405	9747	ō089	ō430	ō772	Ī114
71	1041456	1797	2139	2480	2822	3164	3505	3847	4188	4530
72	4871	5213	5554	5895	6237	6578	6919	7260	7602	7943
73	8284	8625	8966	9307	9648	9989	ō331	ō671	Ī012	Ī353
74	1051694	2035	2376	2717	3058	3398	3739	4080	4421	4761
75	5102	5442	5783	6124	6464	6805	7145	7486	7826	8166
76	8507	8847	9187	9528	9868	ō208	ō548	ō889	Ī229	Ī569
77	1061909	2249	2589	2929	3269	3609	3949	4289	4629	4969
78	5309	5648	5988	6328	6668	7007	7347	7687	8026	8366
79	8705	9045	9385	9724	ō063	ō403	ō742	Ī082	Ī421	Ī760
1280	1072100	2439	2778	3117	3457	3796	4135	4474	4813	5152
81	5491	5830	6169	6508	6847	7186	7525	7864	8203	8541
82	8880	9219	9558	9896	ō235	ō574	ō912	Ī251	Ī590	Ī928
83	1082267	2605	2944	3282	3620	3959	4297	4635	4974	5312
84	5650	5988	6327	6665	7003	7341	7679	8017	8355	8693
85	9031	9369	9707	ō045	ō383	ō721	Ī059	Ī396	Ī734	2072
86	1092410	2747	3085	3423	3760	4098	4435	4773	5111	5448
87	5785	6123	6460	6798	7135	7472	7810	8147	8484	8821
88	9159	9496	9833	ō170	ō507	ō844	Ī181	Ī518	Ī855	2192
89	1102529	2866	3203	3540	3877	4213	4550	4887	5224	5560
1290	5897	6234	6570	6907	7244	7580	7917	8253	8590	8926
91	9262	9599	9935	ō272	ō608	ō944	Ī280	Ī617	Ī953	2289
92	1112625	2961	3297	3633	3969	4306	4642	4977	5313	5649
93	5985	6321	6657	6993	7329	7664	8000	8336	8671	9007
94	9343	9678	ō014	ō350	ō685	Ī021	Ī356	Ī691	2027	2362
95	1122698	3033	3368	3704	4039	4374	4709	5045	5380	5715
96	6050	6385	6720	7055	7390	7725	8060	8395	8730	9065
97	9400	9735	ō069	ō404	ō739	Ī074	Ī408	Ī743	2078	2412
98	1132747	3081	3416	3751	4085	4420	4754	5088	5423	5757
99	6092	6426	6760	7094	7429	7763	8097	8431	8765	9099
N.	0	1	2	3	4	5	6	7	8	9

Differ.

	344	343
1	34	34
2	69	69
3	103	103
4	138	137
5	172	172
6	206	206
7	241	240
8	275	274
9	310	309

	342	341
1	34	34
2	68	68
3	103	102
4	137	136
5	171	171
6	205	205
7	239	239
8	274	273
9	308	307

	340	339
1	34	34
2	68	68
3	102	102
4	136	136
5	170	170
6	204	203
7	238	237
8	272	271
9	306	305

	338	337
1	34	34
2	68	67
3	101	101
4	135	135
5	169	169
6	203	202
7	237	236
8	270	270
9	304	303

	336	335
1	34	34
2	67	67
3	101	101
4	134	134
5	168	168
6	202	201
7	235	235
8	269	268
9	302	302

	334	333
1	33	33
2	67	67
3	100	100
4	134	133
5	167	167
6	200	200
7	234	233
8	267	266
9	301	300

Differ.

N.	0	1	2	3	4	5	6	7	8	9
1300	1139434	9768	0̄102	0̄436	0̄770	1̄104	1̄437	1̄771	2̄105	2̄439
01	1142773	3107	3441	3774	4108	4442	4775	.5109	5443	5776
02	6110	6443	6777	7110	7444	7777	8111	8444	8777	9111
03	9444	9777	0̄111	0̄444	0̄777	1̄110	1̄444	1̄777	2̄110	2̄443
04	1152776	3109	3442	3775	4108	4441	4774	5107	5439	5772
05	6105	6438	6771	7103	7436	7769	8101	8434	8767	9099
06	9432	9764	0̄097	0̄429	0̄762	1̄094	1̄427	1̄759	2̄091	2̄424
07	1162756	3088	3420	3753	4085	4417	4749	5081	5413	5745
08	6077	6409	6741	7073	7405	7737	8069	8401	8733	9065
09	9396	9728	0̄060	0̄392	0̄723	1̄055	1̄387	1̄718	2̄050	2̄381
1310	1172713	3044	3376	3707	4039	4370	4702	5033	5364	5696
11	6027	6358	6689	7021	7352	7683	8014	8345	8676	9007
12	9338	9669	0̄000	0̄331	0̄662	0̄993	1̄324	1̄655	1̄986	2̄316
13	1182647	2978	3309	3639	3970	4301	4631	4962	5293	5623
14	5954	6284	6615	6945	7276	7606	7936	8267	8597	8927
15	9258	9588	9918	0̄248	0̄578	0̄909	1̄239	1̄569	1̄899	2̄229
16	1192559	2889	3219	3549	3879	4209	4539	4868	5198	5528
17	5858	6187	6517	6847	7177	7506	7836	8165	8495	8825
18	9154	9484	9813	0̄143	0̄472	0̄801	1̄131	1̄460	1̄789	2̄119
19	1202448	2777	3106	3436	3765	4094	4423	4752	5081	5410
1320	5739	6068	6397	6726	7055	7384	7713	8042	8371	8699
21	9028	9357	9686	0̄014	0̄343	0̄672	1̄000	1̄329	1̄657	1̄986
22	1212315	2643	2972	3300	3628	3957	4285	4614	4942	5270
23	5598	5927	6255	6583	6911	7239	7568	7896	8224	8552
24	8880	9208	9536	9864	0̄192	0̄520	0̄848	1̄175	1̄503	1̄831
25	1222159	2487	2814	3142	3470	3797	4125	4453	4780	5108
26	5435	5763	6090	6418	6745	7073	7400	7727	8055	8382
27	8709	9036	9364	9691	0̄018	0̄345	0̄672	1̄000	1̄327	1̄654
28	1231981	2308	2635	2962	3289	3616	3942	4269	4596	4923
29	5250	5577	5903	6230	6557	6883	7210	7537	7863	8190
1330	8516	8843	9169	9496	9822	0̄149	0̄475	0̄802	1̄128	1̄454
31	1241781	2107	2433	2759	3086	3412	3738	4064	4390	4716
32	5042	5368	5694	6020	6346	6672	6998	7324	7650	7976
33	8301	8627	8953	9279	9605	9930	0̄256	0̄582	0̄907	1̄233
34	1251558	1884	2209	2535	2860	3186	3511	3837	4162	4487
35	4813	5138	5463	5788	6114	6439	6764	7089	7414	7739
36	8065	8390	8715	9040	9365	9690	0̄015	0̄339	0̄664	0̄989
37	1261314	1639	1964	2288	2613	2938	3263	3587	3912	4237
38	4561	4886	5210	5535	5859	6184	6508	6833	7157	7481
39	7806	8130	8454	8779	9103	9427	9751	0̄076	0̄400	0̄724
1340	1271048	1372	1696	2020	2344	2668	2992	3316	3640	3964
41	4288	4612	4935	5259	5583	5907	6230	6554	6878	7202
42	7525	7849	8172	8496	8819	9143	9466	9790	0̄113	0̄437
43	1280760	1083	1407	1730	2053	2377	2700	3023	3346	3670
44	3993	4316	4639	4962	5285	5608	5931	6254	6577	6900
45	7223	7546	7869	8191	8514	8837	9160	9483	9805	0̄128
46	1290451	0773	1096	1418	1741	2064	2386	2709	3031	3354
47	3676	3998	4321	4643	4965	5288	5610	5932	6255	6577
48	6899	7221	7543	7865	8187	8510	8832	9154	9476	9798
49	1300119	0441	0763	1085	1407	1729	2051	2372	2694	3016
N.	0	1	2	3	4	5	6	7	8	9

Differ.

	334	333
1	33	33
2	67	67
3	100	100
4	134	133
5	167	167
6	200	200
7	234	233
8	267	266
9	301	300

	332	331
1	33	33
2	66	66
3	100	99
4	133	132
5	166	166
6	199	199
7	232	232
8	266	265
9	299	298

	330	329
1	33	33
2	66	66
3	99	99
4	132	132
5	165	165
6	198	197
7	231	230
8	264	263
9	297	296

	328	327
1	33	33
2	66	65
3	98	98
4	131	131
5	164	164
6	197	196
7	230	229
8	262	262
9	295	294

	326	325
1	33	33
2	65	65
3	98	98
4	130	130
5	163	163
6	196	195
7	228	228
8	261	260
9	293	293

	324	323
1	32	32
2	65	65
3	97	97
4	130	129
5	162	162
6	194	194
7	227	226
8	259	258
9	292	291

Differ.

N.	0	1	2	3	4	5	6	7	8	9
1350	1303338	3659	3981	4303	4624	4946	5267	5589	5911	6232
51	6553	6875	7196	7518	7839	8161	8482	8803	9124	9446
52	9767	0088	0409	0730	1052	1373	1694	2015	2336	2657
53	1312978	3299	3620	3941	4262	4583	4903	5224	5545	5866
54	6187	6507	6828	7149	7469	7790	8111	8431	8752	9072
55	9393	9713	0034	0354	0675	0995	1316	1636	1956	2277
56	1322597	2917	3237	3558	3878	4198	4518	4838	5158	5478
57	5798	6119	6439	6758	7078	7398	7718	8038	8358	8678
58	8998	9317	9637	9957	0277	0596	0916	1236	1555	1875
59	1332195	2514	2834	3153	3473	3792	4112	4431	4750	5070
1360	5389	5708	6028	6347	6666	6985	7305	7624	7943	8262
61	8581	8900	9219	9538	9857	0176	0495	0814	1133	1452
62	1341771	2090	2409	2728	3046	3365	3684	4003	4321	4640
63	4959	5277	5596	5914	6233	6551	6870	7188	7507	7825
64	8144	8462	8780	9099	9417	9735	0054	0372	0690	1008
65	1351327	1645	1963	2281	2599	2917	3235	3553	3871	4189
66	4507	4825	5143	5461	5779	6096	6414	6732	7050	7367
67	7685	8003	8320	8638	8956	9273	9591	9908	0226	0543
68	1360861	1178	1496	1813	2131	2448	2765	3083	3400	3717
69	4034	4352	4669	4986	5303	5620	5937	6255	6572	6889
1370	7206	7523	7840	8157	8473	8790	9107	9424	9741	0058
71	1370375	0691	1008	1325	1641	1958	2275	2591	2908	3225
72	3541	3858	4174	4491	4807	5124	5440	5756	6073	6389
73	6705	7022	7338	7654	7970	8287	8603	8919	9235	9551
74	9867	0183	0499	0815	1131	1447	1763	2079	2395	2711
75	1383027	3343	3659	3974	4290	4606	4922	5237	5553	5869
76	6184	6500	6816	7131	7447	7762	8078	8393	8709	9024
77	9339	9655	9970	0285	0601	0916	1231	1547	1862	2177
78	1392492	2807	3122	3438	3753	4068	4383	4698	5013	5328
79	5643	5958	6272	6587	6902	7217	7532	7847	8161	8476
1380	8791	9106	9420	9735	0050	0364	0679	0993	1308	1622
81	1401937	2251	2566	2880	3195	3509	3823	4138	4452	4766
82	5080	5395	5709	6023	6337	6651	6966	7280	7594	7908
83	8222	8536	8850	9164	9478	9792	0106	0419	0733	1047
84	1411361	1675	1988	2302	2616	2930	3243	3557	3871	4184
85	4498	4811	5125	5438	5752	6065	6379	6692	7006	7319
86	7632	7946	8259	8572	8885	9199	9512	9825	0138	0451
87	1420765	1078	1391	1704	2017	2330	2643	2956	3269	3582
88	3895	4208	4520	4833	5146	5459	5772	6084	6397	6710
89	7022	7335	7648	7960	8273	8586	8898	9211	9523	9836
1390	1430148	0460	0773	1085	1398	1710	2022	2335	2647	2959
91	3271	3584	3896	4208	4520	4832	5144	5456	5768	6080
92	6392	6704	7016	7328	7640	7952	8264	8576	8888	9199
93	9511	9823	0135	0446	0758	1070	1381	1693	2005	2316
94	1442628	2939	3251	3562	3874	4185	4497	4808	5119	5431
95	5742	6053	6365	6676	6987	7298	7610	7921	8232	8543
96	8854	9165	9476	9787	0098	0409	0720	1031	1342	1653
97	1451964	2275	2586	2897	3207	3518	3829	4140	4450	4761
98	5072	5382	5693	6004	6314	6625	6935	7246	7556	7867
99	8177	8488	8798	9108	9419	9729	0039	0350	0660	0970
N.	0	1	2	3	4	5	6	7	8	9

Differ.

	322	321
1	32	32
2	64	64
3	97	96
4	129	128
5	161	161
6	193	193
7	225	225
8	258	257
9	290	289

	320	319
1	32	32
2	64	64
3	96	96
4	128	128
5	160	160
6	192	191
7	224	223
8	256	255
9	288	287

	318	317
1	32	32
2	64	63
3	95	95
4	127	127
5	159	159
6	191	190
7	223	222
8	254	254
9	286	285

	316	315
1	32	32
2	63	63
3	95	95
4	126	126
5	158	158
6	190	189
7	221	221
8	253	252
9	284	284

	314	313
1	31	31
2	63	63
3	94	94
4	126	125
5	157	157
6	188	188
7	220	219
8	251	250
9	283	282

	312	311
1	31	31
2	62	62
3	94	93
4	125	124
5	156	156
6	187	187
7	218	218
8	250	249
9	281	280

N.	0	1	2	3	4	5	6	7	8	9
1400	1461280	1591	1901	2211	2521	2831	3141	3451	3761	4071
01	4381	4691	5001	5311	5621	5931	6241	6551	6861	7170
02	7480	7790	8100	8409	8719	9029	9338	9648	9958	0̄267
03	1470577	0886	1196	1505	1815	2124	2434	2743	3052	3362
04	3671	3980	4290	4599	4908	5217	5527	5836	6145	6454
05	6763	7072	7381	7690	7999	8308	8617	8926	9235	9544
06	9853	0̄162	0̄471	0̄780	1̄089	1̄397	1̄706	2̄015	2̄324	2̄632
07	1482941	3250	3558	3867	4175	4484	4793	5101	5410	5718
08	6027	6335	6643	6952	7260	7569	7877	8185	8493	8802
09	9110	9418	9726	0̄035	0̄343	0̄651	0̄959	1̄267	1̄575	1̄883
1410	1492191	2499	2807	3115	3423	3731	4039	4347	4655	4962
11	5270	5578	5886	6193	6501	6809	7116	7424	7732	8039
12	8347	8655	8962	9270	9577	9885	0̄192	0̄499	0̄807	1̄114
13	1501422	1729	2036	2344	2651	2958	3265	3573	3880	4187
14	4494	4801	5108	5415	5722	6030	6337	6644	6951	7257
15	7564	7871	8178	8485	8792	9099	9406	9712	0̄019	0̄326
16	1510633	0939	1246	1553	1859	2166	2472	2779	3085	3392
17	3699	4005	4311	4618	4924	5231	5537	5843	6150	6456
18	6762	7069	7375	7681	7987	8293	8600	8906	9212	9518
19	9824	0̄130	0̄436	0̄742	1̄048	1̄354	1̄660	1̄966	2̄272	2̄578
1420	1522883	3189	3495	3801	4107	4412	4718	5024	5329	5635
21	5941	6246	6552	6858	7163	7469	7774	8080	8385	8691
22	8996	9301	9607	9912	0̄217	0̄523	0̄828	1̄133	1̄439	1̄744
23	1532049	2354	2659	2964	3270	3575	3880	4185	4490	4795
24	5100	5405	5710	6015	6320	6625	6929	7234	7539	7844
25	8149	8453	8758	9063	9368	9672	9977	0̄281	0̄586	0̄891
26	1541195	1500	1804	2109	2413	2718	3022	3327	3631	3935
27	4240	4544	4848	5153	5457	5761	6065	6370	6674	6978
28	7282	7586	7890	8194	8498	8802	9106	9410	9714	0̄018
29	1550322	0626	0930	1234	1538	1842	2145	2449	2753	3057
1430	3360	3664	3968	4271	4575	4879	5182	5486	5789	6093
31	6396	6700	7003	7307	7610	7914	8217	8520	8824	9127
32	9430	9733	0̄037	0̄340	0̄643	0̄946	1̄249	1̄553	1̄856	2̄159
33	1562462	2765	3068	3371	3674	3977	4280	4583	4886	5189
34	5492	5794	6097	6400	6703	7006	7308	7611	7914	8216
35	8519	8822	9124	9427	9729	0̄032	0̄334	0̄637	0̄939	1̄242
36	1571544	1847	2149	2452	2754	3056	3359	3661	3963	4265
37	4568	4870	5172	5474	5776	6079	6381	6683	6985	7287
38	7589	7891	8193	8495	8797	9099	9401	9702	0̄004	0̄306
39	1580608	0910	1212	1513	1815	2117	2418	2720	3022	3323
1440	3625	3927	4228	4530	4831	5133	5434	5736	6037	6338
41	6640	6941	7243	7544	7845	8146	8448	8749	9050	9351
42	9653	9954	0̄255	0̄556	0̄857	1̄158	1̄459	1̄760	2̄061	2̄362
43	1592663	2964	3265	3566	3867	4168	4469	4770	5070	5371
44	5672	5973	6273	6574	6875	7175	7476	7777	8077	8378
45	8678	8979	9280	9580	9881	0̄181	0̄481	0̄782	1̄082	1̄383
46	1601683	1983	2284	2584	2884	3184	3485	3785	4085	4385
47	4685	4985	5286	5586	5886	6186	6486	6786	7086	7386
48	7686	7986	8285	8595	8885	9185	9485	9785	0̄084	0̄384
49	1610684	0984	1283	1583	1883	2182	2482	2781	3081	3380
N.	0	1	2	3	4	5	6	7	8	9

Differ.

	310	309
1	31	31
2	62	62
3	93	93
4	124	124
5	155	155
6	186	185
7	217	216
8	248	247
9	279	278

	308	307
1	31	31
2	62	61
3	92	92
4	123	123
5	154	154
6	185	184
7	216	215
8	246	246
9	277	276

	306	305
1	31	31
2	61	61
3	92	92
4	122	122
5	153	153
6	184	183
7	214	214
8	245	244
9	275	275

	304	303
1	30	30
2	61	61
3	91	91
4	122	121
5	152	152
6	182	182
7	213	212
8	243	242
9	274	273

	302	301
1	30	30
2	60	60
3	91	90
4	121	120
5	151	151
6	181	181
7	211	211
8	242	241
9	272	271

	300	299
1	30	30
2	60	60
3	90	90
4	120	120
5	150	150
6	180	179
7	210	209
8	240	239
9	270	269

Differ.

N.	0	1	2	3	4	5	6	7	8	9
1450	1613680	3980	4279	4578	4878	5177	5477	5776	6075	6375
51	6674	6973	7273	7572	7871	8170	8470	8769	9068	9367
52	9666	9965	0264	0563	0862	1161	1460	1759	2058	2357
53	1622656	2955	3254	3553	3852	4150	4449	4748	5047	5345
54	5644	5943	6241	6540	6839	7137	7436	7734	8033	8331
55	8630	8928	9227	9525	9824	0122	0420	0719	1017	1315
56	1631614	1912	2210	2508	2807	3105	3403	3701	3999	4297
57	4596	4894	5192	5490	5788	6086	6384	6682	6979	7277
58	7575	7873	8171	8469	8767	9064	9362	9660	9958	0255
59	1640553	0851	1148	1446	1743	2041	2339	2636	2934	3231
1460	3529	3826	4123	4421	4718	5016	5313	5610	5908	6205
61	6502	6799	7097	7394	7691	7988	8285	8582	8880	9177
62	9474	9771	0068	0365	0662	0959	1256	1553	1850	2146
63	1652443	2740	3037	3334	3631	3927	4224	4521	4817	5114
64	5411	5707	6004	6301	6597	6894	7190	7487	7783	8080
65	8376	8673	8969	9265	9562	9858	0155	0451	0747	1043
66	1661340	1636	1932	2228	2525	2821	3117	3413	3709	4005
67	4301	4597	4893	5189	5485	5781	6077	6373	6669	6965
68	7261	7556	7852	8148	8444	8740	9035	9331	9627	9922
69	1670218	0514	0809	1105	1400	1696	1991	2287	2582	2878
1470	3173	3469	3764	4060	4355	4650	4946	5241	5536	5831
71	6127	6422	6717	7012	7308	7603	7898	8193	8488	8783
72	9078	9373	9668	9963	0258	0553	0848	1143	1438	1733
73	1682027	2322	2617	2912	3207	3501	3796	4091	4386	4680
74	4975	5269	5564	5859	6153	6448	6742	7037	7331	7626
75	7920	8215	8509	8803	9098	9392	9686	9981	0275	0569
76	1690864	1158	1452	1746	2040	2335	2629	2923	3217	3511
77	3805	4099	4393	4687	4981	5275	5569	5863	6157	6450
78	6744	7038	7332	7626	7920	8213	8507	8801	9094	9388
79	9682	9975	0269	0563	0856	1150	1443	1737	2030	2324
1480	1702617	2911	3204	3497	3791	4084	4377	4671	4964	5257
81	5551	5844	6137	6430	6723	7017	7310	7603	7896	8189
82	8482	8775	9068	9361	9654	9947	0240	0533	0826	1119
83	1711412	1704	1997	2290	2583	2876	3168	3461	3754	4046
84	4339	4632	4924	5217	5509	5802	6095	6387	6680	6972
85	7265	7557	7849	8142	8434	8727	9019	9311	9604	9896
86	1720188	0480	0773	1065	1357	1649	1941	2233	2526	2818
87	3110	3402	3694	3986	4278	4570	4862	5154	5446	5737
88	6029	6321	6613	6905	7197	7488	7780	8072	8364	8655
89	8947	9239	9530	9822	0113	0405	0697	0988	1280	1571
1490	1731863	2154	2446	2737	3028	3320	3611	3903	4194	4485
91	4776	5068	5359	5650	5941	6233	6524	6815	7106	7397
92	7688	7979	8270	8561	8852	9143	9434	9725	0016	0307
93	1740598	0889	1180	1471	1761	2052	2343	2634	2925	3215
94	3506	3797	4087	4378	4669	4959	5250	5540	5831	6121
95	6412	6702	6993	7283	7574	7864	8155	8445	8735	9026
96	9316	9606	9897	0187	0477	0767	1057	1348	1638	1928
97	1752218	2508	2798	3088	3378	3668	3958	4248	4538	4828
98	5118	5408	5698	5988	6278	6567	6857	7147	7437	7727
99	8016	8306	8596	8885	9175	9465	9754	0044	0333	0623
N	0	1	2	3	4	5	6	7	8	9

Differ.

	298	297
1	30	30
2	60	59
3	89	89
4	119	119
5	149	149
6	179	178
7	209	208
8	238	238
9	268	267

	296	295
1	30	30
2	59	59
3	89	89
4	118	118
5	148	148
6	178	177
7	207	207
8	237	236
9	266	266

	294	293
1	29	29
2	59	59
3	88	88
4	118	117
5	147	147
6	176	176
7	206	205
8	235	234
9	265	264

	292	291
1	29	29
2	58	58
3	88	87
4	117	116
5	146	146
6	175	175
7	204	204
8	234	233
9	263	262

	290	289
1	29	29
2	58	58
3	87	87
4	116	116
5	145	145
6	174	173
7	203	202
8	232	231
9	261	260

N.	0	1	2	3	4	5	6	7	8	9
1500	1760913	1202	1492	1781	2071	2360	2649	2939	3228	3518
01	3807	4096	4386	4675	4964	5253	5543	5832	6121	6410
02	6699	6988	7278	7567	7856	8145	8434	8723	9012	9301
03	9590	9879	0̄168	0̄457	0̄745	1̄034	1̄323	1̄612	1̄901	2̄190
04	1772478	2767	3056	3345	3633	3922	4211	4499	4788	5076
05	5365	5654	5942	6231	6519	6808	7096	7385	7673	7961
06	8250	8538	8826	9115	9403	9691	9980	0̄268	0̄556	0̄844
07	1781133	1421	1709	1997	2285	2573	2861	3149	3437	3725
08	4013	4301	4589	4877	5165	5453	5741	6029	6317	6605
09	6892	7180	7468	7756	8043	8331	8619	8907	9194	9482
1510	9769	0̄057	0̄345	0̄632	0̄920	1̄207	1̄495	1̄782	2̄070	2̄357
11	1792645	2932	3219	3507	3794	4082	4369	4656	4943	5231
12	5518	5805	6092	6380	6667	6954	7241	7528	7815	8102
13	8389	8676	8963	9250	9537	9824	0̄111	0̄398	0̄685	0̄972
14	1801259	1546	1832	2119	2406	2693	2980	3266	3553	3840
15	4126	4413	4700	4986	5273	5559	5846	6133	6419	6706
16	6992	7278	7565	7851	8138	8424	8711	8997	9283	9570
17	9856	0̄142	0̄428	0̄715	1̄001	1̄287	1̄573	1̄859	2̄145	2̄432
18	1812718	3004	3290	3576	3862	4148	4434	4720	5006	5292
19	5578	5864	6150	6435	6721	7007	7293	7579	7864	8150
1520	8436	8722	9007	9293	9579	9864	0̄150	0̄435	0̄721	1̄007
21	1821292	1578	1863	2149	2434	2720	3005	3290	3576	3861
22	4147	4432	4717	5002	5288	5573	5858	6143	6429	6714
23	6999	7284	7569	7854	8140	8425	8710	8995	9280	9565
24	9850	0̄135	0̄420	0̄704	0̄989	1̄274	1̄559	1̄844	2̄129	2̄414
25	1832698	2983	3268	3553	3837	4122	4407	4691	4976	5261
26	5545	5830	6114	6399	6684	6968	7253	7537	7822	8106
27	8390	8675	8959	9244	9528	9812	0̄096	0̄381	0̄665	0̄949
28	1841234	1518	1802	2086	2370	2654	2939	3223	3507	3791
29	4075	4359	4643	4927	5211	5495	5779	6063	6347	6630
1530	6914	7198	7482	7766	8050	8333	8617	8901	9185	9468
31	9752	0̄036	0̄319	0̄603	0̄886	1̄170	1̄454	1̄737	2̄021	2̄304
32	1852588	2871	3155	3438	3721	4005	4288	4572	4855	5138
33	5422	5705	5988	6271	6555	6838	7121	7404	7687	7970
34	8254	8537	8820	9103	9386	9669	9952	0̄235	0̄518	0̄801
35	1861084	1367	1650	1932	2215	2498	2781	3064	3347	3629
36	3912	4195	4478	4760	5043	5326	5608	5891	6174	6456
37	6739	7021	7304	7586	7869	8151	8434	8716	8999	9281
38	9563	9846	0̄128	0̄410	0̄693	0̄975	1̄257	1̄540	1̄822	2̄104
39	1872386	2668	2951	3233	3515	3797	4079	4361	4643	4925
1540	5207	5489	5771	6053	6335	6617	6899	7181	7463	7745
41	8026	8308	8590	8872	9154	9435	9717	9999	0̄280	0̄562
42	1880844	1125	1407	1689	1970	2252	2533	2815	3096	3378
43	3659	3941	4222	4504	4785	5066	5348	5629	5910	6192
44	6473	6754	7035	7317	7598	7879	8160	8441	8723	9004
45	9285	9566	9847	0̄128	0̄409	0̄690	0̄971	1̄252	1̄533	1̄814
46	1892095	2376	2657	2938	3218	3499	3780	4061	4342	4622
47	4903	5184	5465	5745	6026	6307	6597	6868	7148	7429
48	7710	7990	8271	8551	8832	9112	9393	9673	9953	0̄234
49	1900514	0795	1075	1355	1636	1916	2196	2476	2757	3037
N.	0	1	2	3'	4	5	6	7	8	9

Differ.

	290	289
1	29	29
2	58	58
3	87	87
4	116	116
5	145	145
6	174	173
7	203	202
8	232	231
9	261	260

	288	287
1	29	29
2	58	57
3	86	86
4	115	115
5	144	144
6	173	172
7	202	201
8	230	230
9	259	258

	286	285
1	29	29
2	57	57
3	86	86
4	114	114
5	143	143
6	172	171
7	200	200
8	229	228
9	257	257

	284	283
1	28	28
2	57	57
3	85	85
4	114	113
5	142	142
6	170	170
7	199	198
8	227	226
9	256	255

	282	281
1	28	28
2	56	56
3	85	84
4	113	112
5	141	141
6	169	169
7	197	197
8	226	225
9	254	253

Differ.

N.	0	1	2	3	4	5	6	7	8	9	Differ.
1550	1903317	3597	3877	4157	4438	4718	4998	5278	5558	5838	280 279
51	6118	6398	6678	6958	7238	7518	7798	8078	8357	8637	1 28 28
52	8917	9197	9477	9757	0036	0316	0596	0876	1155	1435	2 56 56
53	1911715	1994	2274	2553	2833	3113	3392	3672	3951	4231	3 84 84
54	4510	4790	5069	5348	5628	5907	6187	6466	6745	7025	4 112 112
55	7304	7583	7862	8142	8421	8700	8979	9259	9538	9817	5 140 140
56	1920096	0375	0654	0933	1212	1491	1770	2049	2328	2607	6 168 167
57	2886	3165	3444	3723	4002	4281	4559	4838	5117	5396	7 196 195
58	5675	5953	6232	6511	6789	7068	7347	7625	7904	8183	8 224 223
59	8461	8740	9018	9297	9575	9854	0132	0411	0689	0968	9 252 251
1560	1931246	1524	1803	2081	2359	2638	2916	3194	3473	3751	278 277
61	4029	4307	4585	4864	5142	5420	5698	5976	6254	6532	1 28 28
62	6810	7088	7366	7644	7922	8200	8478	8756	9034	9312	2 56 55
63	9590	9868	0145	0423	0701	0979	1257	1534	1812	2090	3 83 83
64	1942367	2645	2923	3200	3478	3756	4033	4311	4588	4866	4 111 111
65	5143	5421	5698	5976	6253	6531	6808	7086	7363	7640	5 139 139
66	7918	8195	8472	8749	9027	9304	9581	9858	0136	0413	6 167 166
67	1950690	0967	1244	1521	1798	2075	2353	2630	2907	3184	7 195 194
68	3461	3738	4014	4291	4568	4845	5122	5399	5676	5953	8 222 222
69	6229	6506	6783	7060	7336	7613	7890	8167	8443	8720	9 250 249
1570	8997	9273	9550	9826	0103	0379	0656	0932	1209	1485	276 275
71	1961762	2038	2315	2591	2867	3144	3420	3697	3973	4249	1 28 28
72	4525	4802	5078	5354	5630	5907	6183	6459	6735	7011	2 55 55
73	7287	7563	7839	8115	8391	8667	8943	9219	9495	9771	3 83 83
74	1970047	0323	0599	0875	1151	1427	1702	1978	2254	2530	4 110 110
75	2806	3081	3357	3633	3908	4184	4460	4735	5011	5287	5 138 138
76	5562	5838	6113	6389	6664	6940	7215	7491	7766	8042	6 166 165
77	8317	8592	8868	9143	9418	9694	9969	0244	0520	0795	7 193 193
78	1981070	1345	1620	1896	2171	2446	2721	2996	3271	3546	8 221 220
79	3821	4096	4371	4646	4921	5196	5471	5746	6021	6296	9 248 248
1580	6571	6846	7121	7395	7670	7945	8220	8495	8769	9044	274 273
81	9319	9593	9868	0143	0417	0692	0967	1241	1516	1790	1 27 27
82	1992065	2339	2614	2888	3163	3437	3712	3986	4260	4535	2 55 55
83	4809	5083	5358	5632	5906	6181	6455	6729	7003	7278	3 82 82
84	7552	7826	8100	8374	8648	8922	9197	9471	9745	0019	4 110 109
85	2000293	0567	0841	1115	1389	1662	1936	2210	2484	2758	5 137 137
86	3032	3306	3579	3853	4127	4401	4674	4948	5222	5496	6 164 164
87	5769	6043	6317	6590	6864	7137	7411	7664	7958	8231	7 192 191
88	8505	8778	9052	9325	9599	9872	0146	0419	0692	0966	8 219 218
89	2011239	1512	1786	2059	2332	2605	2879	3152	3425	3698	9 247 246
1590	3971	4244	4517	4791	5064	5337	5610	5883	6156	6429	
91	6702	6975	7248	7521	7794	8066	8339	8612	8885	9158	
92	9431	9703	9976	0249	0522	0794	1067	1340	1612	1885	272 271
93	2022158	2430	2703	2976	3248	3521	3793	4066	4338	4611	1 27 27
94	4883	5156	5428	5700	5973	6245	6518	6790	7062	7335	2 54 54
95	7607	7879	8151	8424	8696	8968	9240	9512	9785	0057	3 82 81
96	2030329	0601	0873	1145	1417	1689	1961	2233	2505	2777	4 109 108
97	3049	3321	3593	3865	4137	4409	4681	4952	5224	5496	5 136 136
98	5768	6040	6311	6583	6855	7126	7398	7670	7941	8213	6 163 163
99	8485	8756	9028	9299	9571	9842	0114	0385	0657	0928	7 190 190
											8 218 217
											9 245 244
N.	0	1	2	3	4	5	6	7	8	9	Differ

N.	0	1	2	3	4	5	6	7	8	9
1600	2041200	1471	1743	2014	2285	2557	2828	3099	3371	3642
01	3913	4185	4456	4727	4998	5269	5541	5812	6083	6354
02	6625	6896	7167	7438	7709	7980	8251	8522	8793	9064
03	9335	9606	9877	0̄148	0̄419	0̄690	0̄960	1̄231	1̄502	1̄773
04	2052044	2314	2585	2856	3127	3397	3668	3939	4209	4480
05	4750	5021	5292	5562	5833	6103	6374	6644	6915	7185
06	7455	7726	7996	8267	8537	8807	9078	9348	9618	9889
07	2060159	0429	0699	0969	1240	1510	1780	2050	2320	2590
08	2860	3131	3401	3671	3941	4211	4481	4751	5021	5291
09	5560	5830	6100	6370	6640	6910	7180	7449	7719	7989
1610	8259	8529	8798	9068	9338	9607	9877	0̄147	0̄416	0̄686
11	2070955	1225	1495	1764	2034	2303	2573	2842	3112	3381
12	3650	3920	4189	4459	4728	4997	5267	5536	5805	6074
13	6344	6613	6882	7151	7421	7690	7959	8228	8497	8766
14	9035	9304	9573	9842	0̄111	0̄380	0̄649	0̄918	1̄187	1̄456
15	2081725	1994	2263	2532	2801	3070	3338	3607	3876	4145
16	4414	4682	4951	5220	5488	5757	6026	6294	6563	6832
17	7100	7369	7637	7906	8174	8443	8711	8980	9248	9517
18	9785	0̄054	0̄322	0̄590	0̄859	1̄127	1̄395	1̄664	1̄932	2̄200
19	2092468	2737	3005	3273	3541	3810	4078	4346	4614	4882
1620	5150	5418	5686	5954	6222	6490	6758	7026	7294	7562
21	7830	8098	8366	8634	8902	9170	9437	9705	9973	0̄241
22	2100508	0776	1044	1312	1579	1847	2115	2382	2650	2918
23	3185	3453	3720	3988	4255	4523	4790	5058	5325	5593
24	5860	6128	6395	6662	6930	7197	7464	7732	7999	8266
25	8534	8801	9068	9335	9603	9870	0̄137	0̄404	0̄671	0̄938
26	2111205	1472	1740	2007	2274	2541	2808	3075	3342	3609
27	3876	4142	4409	4676	4943	5210	5477	5744	6010	6277
28	6544	6811	7078	7344	7611	7878	8144	8411	8678	8944
29	9211	9477	9744	0̄011	0̄277	0̄544	0̄810	1̄077	1̄343	1̄610
1630	2121876	2142	2409	2675	2942	3208	3474	3741	4007	4273
31	4540	4806	5072	5338	5605	5871	6137	6403	6669	6935
32	7202	7468	7734	8000	8266	8532	8798	9064	9330	9596
33	9862	0̄128	0̄394	0̄660	0̄926	1̄191	1̄457	1̄723	1̄989	2̄255
34	2132521	2786	3052	3318	3584	3849	4115	4381	4646	4912
35	5178	5443	5709	5974	6240	6505	6771	7037	7302	7568
36	7833	8098	8364	8629	8895	9160	9425	9691	9956	0̄221
37	2140487	0752	1017	1283	1548	1813	2078	2343	2609	2874
38	3139	3404	3669	3934	4199	4464	4730	4995	5260	5525
39	5790	6055	6319	6584	6849	7114	7379	7644	7909	8174
1640	8438	8703	8968	9233	9498	9762	0̄027	0̄292	0̄556	0̄821
41	2151086	1350	1615	1880	2144	2409	2673	2938	3203	3467
42	3732	3996	4260	4525	4789	5054	5318	5583	5847	6111
43	6376	6640	6904	7169	7433	7697	7961	8226	8490	8754
44	9018	9282	9546	9811	0̄075	0̄339	0̄603	0̄867	1̄131	1̄395
45	2161659	1923	2187	2451	2715	2979	3243	3507	3771	4034
46	4298	4562	4826	5090	5354	5617	5881	6145	6409	6672
47	6936	7200	7463	7727	7991	8254	8518	8781	9045	9309
48	9572	9836	0̄099	0̄363	0̄626	0̄890	1̄153	1̄416	1̄680	1̄943
49	2172207	2470	2733	2997	3260	3523	3786	4050	4313	4576
N.	0	1	2	3	4	5	6	7	8	9

Differ.

	272	271		270	269		268	267		266	265		264	263
1	27	27	1	27	27	1	27	27	1	·27	27	1	26	26
2	54	54	2	54	54	2	54	53	2	53	53	2	53	53
3	82	81	3	81	81	3	80	80	3	80	80	3	79	79
4	109	108	4	108	108	4	107	107	4	106	106	4	106	105
5	136	136	5	135	135	5	134	134	5	133	133	5	132	132
6	163	163	6	162	161	6	161	160	6	160	159	6	158	158
7	190	190	7	189	188	7	188	187	7	186	186	7	185	184
8	218	217	8	216	215	8	214	214	8	213	212	8	211	210
9	245	244	9	243	242	9	241	240	9	239	239	9	238	237

N.	0	1	2	3	4	5	6	7	8	9
1650	2174839	5103	5366	5629	5892	6155	6418	6682	6945	7208
51	7471	7734	7997	8260	8523	8786	9049	9312	9575	9838
52	2180100	0363	0626	0889	1152	1415	1677	1940	2203	2466
53	2729	2991	3254	3517	3779	4042	4305	4567	4830	5092
54	5355	5618	5880	6143	6405	6668	6930	7193	7455	7718
55	7980	8242	8505	8767	9030	9292	9554	9816	0̄079	0̄341
56	2190603	0866	1128	1390	1652	1914	2177	2439	2701	2963
57	3225	3487	3749	4011	4273	4535	4797	5059	5321	5583
58	5845	6107	6369	6631	6893	7155	7417	7678	7940	8202
59	8464	8726	8987	9249	9511	9773	0̄034	0̄296	0̄558	0̄819
1660	2201081	1342	1604	1866	2127	2389	2650	2912	3173	3435
61	3696	3958	4219	4481	4742	5003	5265	5526	5788	6049
62	6310	6571	6833	7094	7355	7617	7878	8139	8400	8661
63	8922	9184	9445	9706	9967	0̄228	0̄489	0̄750	1̄011	1̄272
64	2211533	1794	2055	2316	2577	2838	3099	3360	3621	3882
65	4142	4403	4664	4925	5186	5446	5707	5968	6229	6489
66	6750	7011	7271	7532	7793	8053	8314	8574	8835	9095
67	9356	9617	9877	0̄138	0̄398	0̄658	0̄919	1̄179	1̄440	1̄700
68	2221960	2221	2481	2741	3002	3262	3522	3783	4043	4303
69	4563	4824	5084	5344	5604	5864	6124	6384	6645	6905
1670	7165	7425	7685	7945	8205	8465	8725	8985	9245	9505
71	9764	0̄024	0̄284	0̄544	0̄804	1̄064	1̄324	1̄583	1̄843	2̄103
72	2232363	2622	2882	3142	3402	3661	3921	4181	4440	4700
73	4959	5219	5479	5738	5998	6257	6517	6776	7036	7295
74	7555	7814	8073	8333	8592	8852	9111	9370	9630	9889
75	2240148	0407	0667	0926	1185	1444	1704	1963	2222	2481
76	2740	2999	3258	3517	3777	4036	4295	4554	4813	5072
77	5331	5590	5849	6107	6366	6625	6884	7143	7402	7661
78	7920	8178	8437	8696	8955	9213	9472	9731	9990	0̄248
79	2250507	0766	1024	1283	1541	1800	2059	2317	2576	2834
1680	3093	3351	3610	3868	4127	4385	4644	4902	5160	5419
81	5677	5935	6194	6452	6710	6969	7227	7485	7743	8002
82	8260	8518	8776	9034	9293	9551	9809	0̄067	0̄325	0̄583
83	2260841	1099	1357	1615	1873	2131	2389	2647	2905	3163
84	3421	3679	3937	4194	4452	4710	4968	5226	5484	5741
85	5999	6257	6515	6772	7030	7288	7545	7803	8060	8318
86	8576	8833	9091	9348	9606	9863	0̄121	0̄378	0̄636	0̄893
87	2271151	1408	1666	1923	2180	2438	2695	2953	3210	3467
88	3724	3982	4239	4496	4753	5011	5268	5525	5782	6039
89	6296	6554	6811	7068	7325	7582	7839	8096	8353	8610
1690	8867	9124	9381	9638	9895	0̄152	0̄409	0̄666	0̄922	1̄179
91	2281436	1693	1950	2206	2463	2720	2977	3233	3490	3747
92	4004	4260	4517	4774	5030	5287	5543	5800	6057	6313
93	6570	6826	7083	7339	7596	7852	8108	8365	8621	8878
94	9134	9390	9647	9903	0̄159	0̄416	0̄672	0̄928	1̄185	1̄441
95	2291697	1953	2209	2466	2722	2978	3234	3490	3746	4002
96	4258	4515	4771	5027	5283	5539	5795	6051	6307	6562
97	6818	7074	7330	7586	7842	8098	8354	8609	8865	9121
98	9377	9633	9888	0̄144	0̄400	0̄656	0̄911	1̄167	1̄423	1̄678
99	2301934	2189	2445	2701	2956	3212	3467	3723	3978	4234
N.	0	1	2	3	4	5	6	7	8	9

Differ.

	264	263
1	26	26
2	53	53
3	79	79
4	106	105
5	132	132
6	158	158
7	185	184
8	211	210
9	238	237

	262	261
1	26	26
2	52	52
3	79	78
4	105	104
5	131	131
6	157	157
7	183	183
8	210	209
9	236	235

	260	259
1	26	26
2	52	52
3	78	78
4	104	104
5	130	130
6	156	155
7	182	181
8	208	207
9	234	233

	258	257
1	26	26
2	52	51
3	77	77
4	103	103
5	129	129
6	155	154
7	181	180
8	206	206
9	232	231

	256	255
1	26	26
2	51	51
3	77	77
4	102	102
5	128	128
6	154	153
7	179	179
8	205	204
9	230	230

N.	0	1	2	3	4	5	6	7	8	9	Differ.
1700	2304489	4745	5000	5256	5511	5766	6022	6277	6532	6788	
01	7043	7298	7554	7809	8064	8320	8575	8830	9085	9340	
02	9596	9851	0̄106	0̄361	0̄616	0̄871	1̄126	1̄381	1̄636	1̄891	
03	2312146	2401	2656	2911	3166	3421	3676	3931	4186	4441	
04	4696	4951	5206	5460	5715	5970	6225	6480	6734	6989	
05	7244	7499	7753	8008	8263	8517	8772	9026	9281	9536	
06	9790	0̄045	0̄299	0̄554	0̄808	1̄063	1̄317	1̄572	1̄826	2̄081	
07	2322335	2590	2844	3098	3353	3607	3861	4116	4370	4624	
08	4879	5133	5387	5641	5896	6150	6404	6658	6912	7166	
09	7421	7675	7929	8183	8437	8691	8945	9199	9453	9707	
1710	9961	0̄215	0̄469	0̄723	0̄977	1̄231	1̄485	1̄739	1̄992	2̄246	
11	2332500	2754	3008	3262	3515	3769	4023	4277	4530	4784	
12	5038	5291	5545	5799	6052	6306	6559	6813	7067	7320	
13	7574	7827	8081	8334	8588	8841	9095	9348	9601	9855	
14	2340108	0362	0615	0868	1122	1375	1628	1881	2135	2388	
15	2641	2894	3148	3401	3654	3907	4160	4414	4667	4920	
16	5173	5426	5679	5932	6185	6438	6691	6944	7197	7450	
17	7703	7956	8209	8462	8715	8967	9220	9473	9726	9979	
18	2350232	0484	0737	0990	1243	1495	1748	2001	2253	2506	
19	2759	3011	3264	3517	3769	4022	4274	4527	4779	5032	
1720	5284	5537	5789	6042	6294	6547	6799	7052	7304	7556	
21	7809	8061	8313	8566	8818	9070	9323	9575	9827	0̄079	
22	2360331	0584	0836	1088	1340	1592	1844	2097	2349	2601	
23	2853	3105	3357	3609	3861	4113	4365	4617	4869	5121	
24	5373	5625	5876	6128	6380	6632	6884	7136	7387	7639	
25	7891	8143	8394	8646	8898	9150	9401	9653	9905	0̄156	
26	2370408	0660	0911	1163	1414	1666	1917	2169	2420	2672	
27	2923	3175	3426	3678	3929	4181	4432	4683	4935	5186	
28	5437	5689	5940	6191	6443	6694	6945	7196	7448	7699	
29	7950	8201	8452	8703	8955	9206	9457	9708	9959	0̄210	
1730	2380461	0712	0963	1214	1465	1716	1967	2218	2469	2720	
31	2971	3222	3472	3723	3974	4225	4476	4727	4977	5228	
32	5479	5730	5980	6231	6482	6732	6983	7234	7484	7735	
33	7986	8236	8487	8737	8988	9238	9489	9739	9990	0̄240	
34	2390491	0741	0992	1242	1493	1743	1993	2244	2494	2744	
35	2995	3245	3495	3746	3996	4246	4496	4747	4997	5247	
36	5497	5747	5998	6248	6498	6748	6998	7248	7498	7748	
37	7998	8248	8498	8748	8998	9248	9498	9748	9998	0̄248	
38	2400498	0748	0997	1247	1497	1747	1997	2247	2496	2746	
39	2996	3246	3495	3745	3995	4244	4494	4744	4993	5243	
1740	5492	5742	5992	6241	6491	6740	6990	7239	7489	7738	
41	7988	8237	8487	8736	8985	9235	9484	9734	9983	0̄232	
42	2410482	0731	0980	1229	1479	1728	1977	2226	2476	2725	
43	2974	3223	3472	3721	3970	4220	4469	4718	4967	5216	
44	5465	5714	5963	6212	6461	6710	6959	7208	7457	7705	
45	7954	8203	8452	8701	8950	9199	9447	9696	9945	0̄194	
46	2420442	0691	0940	1189	1437	1686	1935	2183	2432	2680	
47	2929	3178	3426	3675	3923	4172	4420	4669	4917	5166	
48	5414	5663	5911	6160	6408	6656	6905	7153	7401	7650	
49	7898	8146	8395	8643	8891	9139	9388	9636	9884	0̄132	
N.	0	1	2	3	4	5	6	7	8	9	Differ.

Proportional parts (Differ.):

	256	255
1	26	26
2	51	51
3	77	77
4	102	102
5	128	128
6	154	153
7	179	179
8	205	204
9	230	230

	254	253
1	25	25
2	51	51
3	76	76
4	102	101
5	127	127
6	152	152
7	178	177
8	203	202
9	229	228

	252	251
1	25	25
2	50	50
3	76	75
4	101	100
5	126	126
6	151	151
7	176	176
8	202	201
9	227	226

	250	249
1	25	25
2	50	50
3	75	75
4	100	100
5	125	125
6	150	149
7	175	174
8	200	199
9	225	224

N.	0	1	2	3	4	5	6	7	8	9	Differ.
1750	2430380	0629	0877	1125	1373	1621	1869	2117	2365	2613	
51	2861	3109	3357	3605	3853	4101	4349	4597	4845	5093	
52	5341	5589	5837	6085	6332	6580	6828	7076	7324	7571	
53	7819	8067	8315	8562	8810	9058	9305	9553	9801	ō048	
54	2440296	0543	0791	1039	1286	1534	1781	2029	2276	2524	
55	2771	3019	3266	3514	3761	4008	4256	4503	4750	4998	
56	5245	5492	5740	5987	6234	6482	6729	6976	7223	7470	
57	7718	7965	8212	8459	8706	8953	9200	9448	9695	9942	
58	2450189	0436	0683	0930	1177	1424	1671	1918	2165	2411	
59	2658	2905	3152	3399	3646	3893	4140	4386	4633	4880	
1760	5127	5373	5620	5867	6114	6360	6607	6854	7100	7347	
61	7594	7840	8087	8333	8580	8826	9073	9320	9566	9813	
62	2460059	0306	0552	0798	1045	1291	1538	1784	2030	2277	
63	2523	2769	3016	3262	3508	3755	4001	4247	4493	4740	
64	4986	5232	5478	5724	5970	6217	6463	6709	6955	7201	
65	7447	7693	7939	8185	8431	8677	8923	9169	9415	9661	
66	9907	ō153	ō399	ō645	ō891	Ī136	Ī382	Ī628	Ī874	Ī2120	
67	2472365	2611	2857	3103	3349	3594	3840	4086	4331	4577	
68	4823	5068	5314	5559	5805	6051	6296	6542	6787	7033	
69	7278	7524	7769	8015	8260	8506	8751	8997	9242	9487	
1770	9733	9978	ō223	ō469	ō714	ō959	Ī205	Ī450	Ī695	Ī940	
71	2482186	2431	2676	2921	3166	3412	3657	3902	4147	4392	
72	4637	4882	5127	5372	5617	5862	6107	6352	6597	6842	
73	7087	7332	7577	7822	8067	8312	8557	8802	9047	9291	
74	9536	9781	ō026	ō271	ō515	ō760	Ī005	Ī249	Ī494	Ī739	
75	2491984	2228	2473	2718	2962	3207	3451	3696	3941	4185	
76	4430	4674	4919	5163	5408	5652	5897	6141	6385	6630	
77	6874	7119	7363	7607	7852	8096	8340	8585	8829	9073	
78	9318	9562	9806	ō050	ō294	ō539	ō783	Ī027	Ī271	Ī515	
79	2501759	2004	2248	2492	2736	2980	3224	3468	3712	3956	
1780	4200	4444	4688	4932	5176	5420	5664	5908	6151	6395	
81	6639	6883	7127	7371	7614	7858	8102	8346	8590	8833	
82	9077	9321	9564	9808	ō052	ō295	ō539	ō783	Ī026	Ī270	
83	2511513	1757	2001	2244	2488	2731	2975	3218	3462	3705	
84	3949	4192	4435	4679	4922	5166	5409	5652	5896	6139	
85	6382	6625	6869	7112	7355	7599	7842	8085	8328	8571	
86	8815	9058	9301	9544	9787	ō030	ō273	ō516	ō759	Ī002	
87	2521246	1489	1732	1975	2218	2461	2703	2946	3189	3432	
88	3675	3918	4161	4404	4647	4889	5132	5375	5618	5861	
89	6103	6346	6589	6832	7074	7317	7560	7802	8045	8288	
1790	8530	8773	9016	9258	9501	9743	9986	ō228	ō471	ō713	
91	2530956	1198	1441	1683	1926	2168	2411	2653	2895	3138	
92	3380	3622	3865	4107	4349	4592	4834	5076	5318	5561	
93	5803	6045	6287	6529	6772	7014	7256	7498	7740	7982	
94	8224	8466	8709	8951	9193	9435	9677	9919	ō161	ō403	
95	2540645	0886	1128	1370	1612	1854	2096	2338	2580	2822	
96	3063	3305	3547	3789	4030	4272	4514	4756	4997	5239	
97	5481	5722	5964	6206	6447	6689	6931	7172	7414	7655	
98	7897	8138	8380	8621	8863	9104	9346	9587	9829	ō070	
99	2550312	0553	0794	1036	1277	1519	1760	2001	2242	2484	
N.	0	1	2	3	4	5	6	7	8	9	Differ.

Proportional parts (Differ. column):

	248	247
1	25	25
2	50	49
3	74	74
4	99	99
5	124	124
6	149	148
7	174	173
8	198	198
9	223	222

	246	245
1	25	25
2	49	49
3	74	74
4	98	98
5	123	123
6	148	147
7	172	172
8	197	196
9	221	221

	244	243
1	24	24
2	49	49
3	73	73
4	98	97
5	122	122
6	146	146
7	171	170
8	195	194
9	220	219

	242	241
1	24	24
2	48	48
3	73	72
4	97	96
5	121	121
6	145	145
7	169	169
8	194	193
9	218	217

N.	0	1	2	3	4	5	6	7	8	9	D
1800	2552725	2966	3208	3449	3690	3931	4172	4414	4655	4896	
01	5137	5378	5619	5860	6102	6343	6584	6825	7066	7307	241
02	7548	7789	8030	8271	8512	8753	8994	9235	9475	9716	
03	9957	$\overline{0}$198	$\overline{0}$439	$\overline{0}$680	$\overline{0}$921	$\overline{1}$161	$\overline{1}$402	$\overline{1}$643	$\overline{1}$884	$\overline{2}$125	
04	2562365	2606	2847	3087	3328	3569	3810	4050	4291	4531	
05	4772	5013	5253	5494	5734	5975	6215	6456	6696	6937	
06	7177	7418	7658	7899	8139	8380	8620	8860	9101	9341	
07	9582	9822	$\overline{0}$062	$\overline{0}$302	$\overline{0}$543	$\overline{0}$783	$\overline{1}$023	$\overline{1}$264	$\overline{1}$504	$\overline{1}$744	
08	2571984	2224	2465	2705	2945	3185	3425	3665	3905	4146	
09	4386	4626	4866	5106	5346	5586	5826	6066	6306	6546	
1810	6786	7026	7266	7506	7745	7985	8225	8465	8705	8945	240
11	9185	9424	9664	9904	$\overline{0}$144	$\overline{0}$383	$\overline{0}$623	$\overline{0}$863	$\overline{1}$103	$\overline{1}$342	
12	2581582	1822	2061	2301	2541	2780	3020	3259	3499	3738	
13	3978	4218	4457	4697	4936	5176	5415	5655	5894	6133	
14	6373	6612	6852	7091	7330	7570	7809	8048	8288	8527	
15	8766	9006	9245	9484	9723	9963	$\overline{0}$202	$\overline{0}$441	$\overline{0}$680	$\overline{0}$919	
16	2591158	1398	1637	1876	2115	2354	2593	2832	3071	3310	
17	3549	3788	4027	4266	4505	4744	4983	5222	5461	5700	
18	5939	6178	6417	6655	6894	7133	7372	7611	7849	8088	239
19	8327	8566	8804	9043	9282	9521	9759	9998	$\overline{0}$237	$\overline{0}$475	
1820	2600714	0952	1191	1430	1668	1907	2145	2384	2622	2861	
21	3099	3338	3576	3815	4053	4292	4530	4769	5007	5245	
22	5484	5722	5960	6199	6437	6675	6914	7152	7390	7628	
23	7867	8105	8343	8581	8820	9058	9296	9534	9772	$\overline{0}$010	
24	2610248	0486	0725	0963	1201	1439	1677	1915	2153	2391	
25	2629	2867	3105	3343	3580	3818	4056	4294	4532	4770	238
26	5008	5246	5483	5721	5959	6197	6435	6672	6910	7148	
27	7385	7623	7861	8099	8336	8574	8811	9049	9287	9524	
28	9762	9999	$\overline{0}$237	$\overline{0}$475	$\overline{0}$712	$\overline{0}$950	$\overline{1}$187	$\overline{1}$425	$\overline{1}$662	$\overline{1}$900	
29	2622137	2374	2612	2849	3087	3324	3562	3799	4036	4274	
1830	4511	4748	4986	5223	5460	5697	5935	6172	6409	6646	
31	6883	7121	7358	7595	7832	8069	8306	8543	8781	9018	
32	9255	9492	9729	9966	$\overline{0}$203	$\overline{0}$440	$\overline{0}$677	$\overline{0}$914	$\overline{1}$151	$\overline{1}$388	
33	2631625	1862	2098	2335	2572	2809	3046	3283	3520	3757	237
34	3993	4230	4467	4704	4940	5177	5414	5651	5887	6124	
35	6361	6597	6834	7071	7307	7544	7780	8017	8254	8490	
36	8727	8963	9200	9436	9673	9909	$\overline{0}$146	$\overline{0}$382	$\overline{0}$619	$\overline{0}$855	
37	2641092	1328	1564	1801	2037	2273	2510	2746	2982	3219	
38	3455	3691	3928	4164	4400	4636	4873	5109	5345	5581	
39	5817	6053	6290	6526	6762	6998	7234	7470	7706	7942	
1840	8178	8414	8650	8886	9122	9358	9594	9830	$\overline{0}$066	$\overline{0}$302	236
41	2650538	0774	1010	1246	1481	1717	1953	2189	2425	2660	
42	2896	3132	3368	3604	3839	4075	4311	4546	4782	5018	
43	5253	5489	5725	5960	6196	6431	6667	6903	7138	7374	
44	7609	7845	8080	8316	8551	8787	9022	9257	9493	9728	
45	9964	$\overline{0}$199	$\overline{0}$434	$\overline{0}$670	$\overline{0}$905	$\overline{1}$140	$\overline{1}$376	$\overline{1}$611	$\overline{1}$846	$\overline{2}$082	
46	2662317	2552	2787	3023	3258	3493	3728	3963	4199	4434	
47	4669	4904	5139	5374	5609	5844	6080	6315	6550	6785	235
48	7020	7255	7490	7725	7960	8195	8429	8664	8899	9134	
49	9369	9604	9839	$\overline{0}$074	$\overline{0}$309	$\overline{0}$543	$\overline{0}$778	$\overline{1}$013	$\overline{1}$248	$\overline{1}$483	
N.	0	1	2	3	4	5	6	7	8	9	D

Pro. (Proportional parts)

	240	239	238	237	236	235
1	24	24	24	24	24	24
2	48	48	48	47	47	47
3	72	72	71	71	71	71
4	96	96	95	95	94	94
5	120	120	119	119	118	118
6	144	143	143	142	142	141
7	168	167	167	166	165	165
8	192	191	190	190	189	188
9	216	215	214	213	212	212

N.	0	1	2	3	4	5	6	7	8	9	D	Pro.
1850	2671717	1952	2187	2421	2656	2891	3126	3360	3595	3830		234
51	4064	4299	4533	4768	5003	5237	5472	5706	5941	6175		1 23
52	6410	6644	6879	7113	7348	7582	7817	8051	8285	8520		2 47
53	8754	8989	9223	9457	9692	9926	ō160	ō394	ō629	ō863		3 70
54	2681097	1332	1566	1800	2034	2268	2503	2737	2971	3205		4 94
55	3439	3673	3907	4141	4376	4610	4844	5078	5312	5546	234	5 117
56	5780	6014	6248	6482	6716	6950	7183	7417	7651	7885		6 140
57	8119	8353	8587	8821	9054	9288	9522	9756	9990	ō223		7 164
58	2690457	0691	0925	1158	1392	1626	1859	2093	2327	2560		8 187
59	2794	3028	3261	3495	3728	3962	4195	4429	4662	4896		9 211
1860	5129	5363	5596	5830	6063	6297	6530	6764	6997	7230		233
61	7464	7697	7930	8164	8397	8630	8864	9097	9330	9564		1 23
62	9797	ō030	ō263	ō496	ō730	ō963	Ī196	Ī429	Ī662	Ī895	233	2 47
63	2702129	2362	2595	2828	3061	3294	3527	3760	3993	4226		3 70
64	4459	4692	4925	5158	5391	5624	5857	6090	6323	6555		4 93
65	6788	7021	7254	7487	7720	7953	8185	8418	8651	8884		5 117
66	9116	9349	9582	9815	ō047	ō280	ō513	ō745	ō978	Ī211		6 140
67	2711443	1676	1908	2141	2374	2606	2839	3071	3304	3536		7 163
68	3769	4001	4234	4466	4699	4931	5163	5396	5628	5861		8 186
69	6093	6325	6558	6790	7022	7255	7487	7719	7952	8184		9 210
1870	8416	8648	8881	9113	9345	9577	9809	ō041	ō274	ō506	232	232
71	2720738	0970	1202	1434	1666	1898	2130	2362	2594	2826		1 23
72	3058	3290	3522	3754	3986	4218	4450	4682	4914	5146		2 46
73	5378	5610	5841	6073	6305	6537	6769	7001	7232	7464		3 70
74	7696	7928	8159	8391	8623	8854	9086	9318	9549	9781		4 93
75	2730013	0244	0476	0708	0939	1171	1402	1634	1865	2097		5 116
76	2328	2560	2791	3023	3254	3486	3717	3949	4180	4411		6 139
77	4643	4874	5105	5337	5568	5799	6031	6262	6493	6725		7 162
78	6956	7187	7418	7650	7881	8112	8343	8574	8806	9037		8 186
79	9268	9499	9730	9961	ō192	ō423	ō654	ō885	Ī116	Ī347		9 209
1880	2741578	1809	2040	2271	2502	2733	2964	3195	3426	3657		231
81	3888	4119	4350	4581	4811	5042	5273	5504	5735	5965		1 23
82	6196	6427	6658	6888	7119	7350	7581	7811	8042	8273		2 46
83	8503	8734	8964	9195	9426	9656	9887	ō117	ō348	ō578		3 69
84	2750809	1039	1270	1500	1731	1961	2192	2422	2653	2883	231	4 92
85	3114	3344	3574	3805	4035	4265	4496	4726	4956	5187		5 116
86	5417	5647	5877	6108	6338	6568	6798	7028	7259	7489		6 139
87	7719	7949	8179	8409	8640	8870	9100	9330	9560	9790		7 162
88	2760020	0250	0480	0710	0940	1170	1400	1630	1860	2090		8 185
89	2320	2549	2779	3009	3239	3469	3699	3929	4158	4388		9 208
1890	4618	4848	5078	5307	5537	5767	5997	6226	6456	6686		230
91	6915	7145	7375	7604	7834	8063	8293	8523	8752	8982		1 23
92	9211	9441	9670	9900	ō129	ō359	ō588	ō818	Ī047	Ī277		2 46
93	2771506	1736	1965	2194	2424	2653	2882	3112	3341	3570	230	3 69
94	3800	4029	4258	4488	4717	4946	5175	5405	5634	5863		4 92
95	6092	6321	6550	6780	7009	7238	7467	7696	7925	8154	229	5 115
96	8383	8612	8841	9070	9299	9528	9757	9986	ō215	ō444		6 138
97	2780673	0902	1131	1360	1589	1818	2047	2276	2504	2733		7 161
98	2962	3191	3420	3648	3877	4106	4335	4564	4792	5021		8 184
99	5250	5478	5707	5936	6164	6393	6622	6850	7079	7307		9 207
N.	0	1	2	3	4	5	6	7	8	9	D	Pts.

Pro. continued:

229
1 23
2 46
3 69
4 92
5 115
6 137
7 160
8 183
9 206

N.	0	1	2	3	4	5	6	7	8	9	D
1900	2787536	7765	7993	8222	8450	8679	8907	9136	9364	9593	
01	9821	$\overline{0050}$	$\overline{0278}$	$\overline{0506}$	$\overline{0735}$	$\overline{0963}$	$\overline{1}192$	$\overline{1}420$	$\overline{1}648$	$\overline{1}877$	
02	2792105	2333	2562	2790	3018	3247	3475	3703	3931	4160	
03	4388	4616	4844	5072	5301	5529	5757	5985	6213	6441	
04	6669	6898	7126	7354	7582	7810	8038	8266	8494	8722	228
05	8950	9178	9406	9634	9862	$\overline{0090}$	$\overline{0317}$	$\overline{0545}$	$\overline{0773}$	$\overline{1}001$	
06	2801229	1457	1685	1912	2140	2368	2596	2824	3051	3279	
07	3507	3735	3962	4190	4418	4645	4873	5101	5328	5556	
08	5784	6011	6239	6467	6694	6922	7149	7377	7604	7832	
09	8059	8287	8514	8742	8969	9197	9424	9651	9879	$\overline{0106}$	
1910	2810334	0561	0788	1016	1243	1470	1698	1925	2152	2380	
11	2607	2834	3061	3289	3516	3743	3970	4197	4425	4652	
12	4879	5106	5333	5560	5787	6014	6242	6469	6696	6923	227
13	7150	7377	7604	7831	8058	8285	8512	8739	8966	9192	
14	9419	9646	9873	$\overline{0100}$	$\overline{0327}$	$\overline{0554}$	$\overline{0781}$	$\overline{1}007$	$\overline{1}234$	$\overline{1}461$	
15	2821688	1915	2141	2368	2595	2822	3048	3275	3502	3728	
16	3955	4182	4408	4635	4862	5088	5315	5541	5768	5995	
17	6221	6448	6674	6901	7127	7354	7580	7807	8033	8260	
18	8486	8712	8939	9165	9392	9618	9844	$\overline{0071}$	$\overline{0297}$	$\overline{0523}$	
19	2830750	0976	1202	1429	1655	1881	2107	2334	2560	2786	226
1920	3012	3238	3465	3691	3917	4143	4369	4595	4821	5048	
21	5274	5500	5726	5952	6178	6404	6630	6856	7082	7308	
22	7534	7760	7986	8212	8438	8663	8889	9115	9341	9567	
23	9793	$\overline{0019}$	$\overline{0245}$	$\overline{0470}$	$\overline{0696}$	$\overline{0922}$	$\overline{1}148$	$\overline{1}373$	$\overline{1}599$	$\overline{1}825$	
24	2842051	2276	2502	2728	2953	3179	3405	3630	3856	4082	
25	4307	4533	4759	4984	5210	5435	5661	5886	6112	6337	
26	6563	6788	7014	7239	7465	7690	7916	8141	8366	8592	
27	8817	9043	9268	9493	9719	9944	$\overline{0169}$	$\overline{0394}$	$\overline{0620}$	$\overline{0845}$	
28	2851070	1296	1521	1746	1971	2196	2422	2647	2872	3097	
29	3322	3547	3773	3998	4223	4448	4673	4898	5123	5348	
1930	5573	5798	6023	6248	6473	6698	6923	7148	7373	7598	225
31	7823	8048	8273	8497	8722	8947	9172	9397	9622	9846	
32	2860071	0296	0521	0746	0970	1195	1420	1644	1869	2094	
33	2319	2543	2768	2993	3217	3442	3666	3891	4116	4340	
34	4565	4789	5014	5238	5463	5687	5912	6136	6361	6585	
35	6810	7034	7259	7483	7707	7932	8156	8381	8605	8829	
36	9054	9278	9502	9726	9951	$\overline{0175}$	$\overline{0399}$	$\overline{0624}$	$\overline{0848}$	$\overline{1}072$	
37	2871296	1520	1745	1969	2193	2417	2641	2865	3090	3314	
38	3538	3762	3986	4210	4434	4658	4882	5106	5330	5554	224
39	5778	6002	6226	6450	6674	6898	7122	7346	7570	7793	
1940	8017	8241	8465	8689	8913	9136	9360	9584	9808	$\overline{0032}$	
41	2880255	0479	0703	0927	1150	1374	1598	1821	2045	2269	
42	2492	2716	2939	3163	3387	3610	3834	4057	4281	4504	
43	4728	4952	5175	5399	5622	5845	6069	6292	6516	6739	223
44	6963	7186	7409	7633	7856	8079	8303	8526	8749	8973	
45	9196	9419	9643	9866	$\overline{0089}$	$\overline{0312}$	$\overline{0536}$	$\overline{0759}$	$\overline{0982}$	$\overline{1}205$	
46	2891428	1652	1875	2098	2321	2544	2767	2990	3213	3436	
47	3660	3883	4106	4329	4552	4775	4998	5221	5444	5667	
48	5890	6112	6335	6558	6781	7004	7227	7450	7673	7896	
49	8118	8341	8564	8787	9010	9232	9455	9678	9901	$\overline{0123}$	
N.	0	1	2	3	4	5	6	7	8	9	D

Pro. (Proportional parts)

228		227		226		225		224		223	
1	23	1	23	1	23	1	23	1	22	1	22
2	46	2	45	2	45	2	45	2	45	2	45
3	68	3	68	3	68	3	68	3	67	3	67
4	91	4	91	4	90	4	90	4	90	4	89
5	114	5	114	5	113	5	113	5	112	5	112
6	137	6	136	6	136	6	135	6	134	6	134
7	160	7	159	7	158	7	158	7	157	7	156
8	182	8	182	8	181	8	180	8	179	8	178
9	205	9	204	9	203	9	203	9	202	9	201

N.	0	1	2	3	4	5	6	7	8	9	D	Pro.
1950	2900346	0569	0792	1014	1237	1460	1682	1905	2127	2350		
51	2573	2795	3018	3240	3463	3686	3908	4131	4353	4576		222
52	4798	5021	5243	5466	5688	5910	6133	6355	6578	6800		1|22
53	7022	7245	7467	7690	7912	8134	8356	8579	8801	9023		2|44
54	9246	9468	9690	9912	ō135	ō357	ō579	ō801	Ī023	Ī245		3|67
												4|89
55	2911468	1690	1912	2134	2356	2578	2800	3022	3244	3466	222	5|111
56	3689	3911	4133	4355	4577	4799	5020	5242	5464	5686		6|133
57	5908	6130	6352	6574	6796	7018	7240	7461	7683	7905		7|155
58	8127	8349	8570	8792	9014	9236	9458	9679	9901	ō123		8|178
59	2920344	0566	0788	1009	1231	1453	1674	1896	2118	2339		9|200
1960	2561	2782	3004	3225	3447	3668	3890	4111	4333	4554		
61	4776	4997	5219	5440	5662	5883	6105	6326	6547	6769		221
62	6990	7211	7433	7654	7875	8097	8318	8539	8760	8982		1|22
63	9203	9424	9645	9867	ō088	ō309	ō530	ō751	ō973	Ī194		2|44
64	2931415	1636	1857	2078	2299	2520	2741	2962	3183	3405	221	3|66
												4|88
65	3626	3847	4068	4289	4510	4730	4951	5172	5393	5614		5|111
66	5835	6056	6277	6498	6719	6940	7160	7381	7602	7823		6|133
67	8044	8264	8485	8706	8927	9147	9368	9589	9810	ō030		7|155
68	2940251	0472	0692	0913	1134	1354	1575	1795	2016	2237		8|177
69	2457	2678	2898	3119	3339	3560	3780	4001	4221	4442		9|199
1970	4662	4883	5103	5324	5544	5764	5985	6205	6426	6646		
71	6866	7087	7307	7527	7748	7968	8188	8408	8629	8849		220
72	9069	9289	9510	9730	9950	ō170	ō390	ō610	ō831	Ī051		1|22
73	2951271	1491	1711	1931	2151	2371	2591	2811	3031	3251	220	2|44
74	3471	3691	3911	4131	4351	4571	4791	5011	5231	5451		3|66
												4|88
75	5671	5891	6111	6331	6550	6770	6990	7210	7430	7650		5|110
76	7869	8089	8309	8529	8748	8968	9188	9408	9627	9847		6|132
77	2960067	0286	0506	0726	0945	1165	1385	1604	1824	2043		7|154
78	2263	2482	2702	2922	3141	3361	3580	3800	4019	4238		8|176
79	4458	4677	4897	5116	5336	5555	5774	5994	6213	6433		9|198
1980	6652	6871	7091	7310	7529	7748	7968	8187	8406	8626		
81	8845	9064	9283	9502	9722	9941	ō160	ō379	ō598	ō817		219
82	2971037	1256	1475	1694	1913	2132	2351	2570	2789	3008	219	1|22
83	3227	3446	3665	3884	4103	4322	4541	4760	4979	5198		2|44
84	5417	5636	5854	6073	6292	6511	6730	6949	7168	7386		3|66
												4|88
85	7605	7824	8043	8261	8480	8699	8918	9136	9355	9574		5|110
86	9792	ō011	ō230	ō448	ō667	ō886	Ī104	Ī323	Ī542	Ī760		6|131
87	2981979	2197	2416	2634	2853	3071	3290	3508	3727	3945		7|153
88	4164	4382	4601	4819	5038	5256	5474	5693	5911	6129		8|175
89	6348	6566	6785	7003	7221	7439	7658	7876	8094	8313		9|197
1990	8531	8749	8967	9185	9404	9622	9840	ō058	ō276	ō494		
91	2990713	0931	1149	1367	1585	1803	2021	2239	2457	2675		218
92	2893	3111	3329	3547	3765	3983	4201	4419	4637	4855	218	1|22
93	5073	5291	5509	5727	5945	6162	6380	6598	6816	7034		2|44
94	7252	7469	7687	7905	8123	8340	8558	8776	8994	9211		3|65
												4|87
95	9429	9647	9864	ō082	ō300	ō517	ō735	ō953	Ī170	Ī388		5|109
96	3001605	1823	2041	2258	2476	2693	2911	3128	3346	3563		6|131
97	3781	3998	4216	4433	4650	4868	5085	5303	5520	5737		7|153
98	5955	6Ī72	6390	6607	6824	7042	7259	7476	7693	7911		8|174
99	8128	8345	8562	8780	8997	9214	9431	9648	9866	ō083		9|196
N.	0	1	2	3	4	5	6	7	8	9	D	Pts.

N.	0	1	2	3	4	5	6	7	8	9	D	Pro.
2000	3010300	0517	0734	0951	1168	1386	1603	1820	2037	2254		
01	2471	2688	2905	3122	3339	3556	3773	3990	4207	4424	217	217
02	4641	4858	5075	5291	5508	5725	5942	6159	6376	6593		1 22
03	6809	7026	7243	7460	7677	7893	8110	8327	8544	8760		2 43
04	8977	9194	9411	9627	9844	0̄061	0̄277	0̄494	0̄711	0̄927		3 65
												4 87
05	3021144	1360	1577	1794	2010	2227	2443	2660	2876	3093		5 109
06	3309	3526	3742	3959	4175	4392	4608	4825	5041	5257		6 130
07	5474	5690	5906	6123	6339	6556	6772	6988	7204	7421		7 152
08	7637	7853	8070	8286	8502	8718	8935	9151	9367	9583		8 174
09	9799	0̄016	0̄232	0̄448	0̄664	0̄880	1̄096	1̄312	1̄528	1̄745		9 195
2010	3031961	2177	2393	2609	2825	3041	3257	3473	3689	3905	216	
11	4121	4337	4553	4769	4984	5200	5416	5632	5848	6064		
12	6280	6496	6711	6927	7143	7359	7575	7790	8006	8222		216
13	8438	8653	8869	9085	9301	9516	9732	9948	0̄163	0̄379		1 22
14	3040595	0810	1026	1242	1457	1673	1888	2104	2319	2535		2 43
												3 65
15	2751	2966	3182	3397	3613	3828	4043	4259	4474	4690		4 86
16	4905	5121	5336	5552	5767	5982	6198	6413	6628	6844		5 108
17	7059	7274	7490	7705	7920	8135	8351	8566	8781	8996		6 130
18	9212	9427	9642	9857	0̄072	0̄288	0̄503	0̄718	0̄933	1̄148		7 151
19	3051363	1578	1793	2008	2224	2439	2654	2869	3084	3299		8 173
												9 194
2020	3514	3729	3944	4159	4374	4589	4803	5018	5233	5448	215	
21	5663	5878	6093	6308	6523	6737	6952	7167	7382	7597		
22	7812	8026	8241	8456	8671	8885	9100	9315	9529	9744		215
23	9959	0̄174	0̄388	0̄603	0̄817	1̄032	1̄247	1̄461	1̄676	1̄891		1 22
24	3062105	2320	2534	2749	2963	3178	3392	3607	3821	4036		2 43
												3 65
25	4250	4465	4679	4894	5108	5322	5537	5751	5966	6180		4 86
26	6394	6609	6823	7037	7252	7466	7680	7895	8109	8323		5 108
27	8537	8752	8966	9180	9394	9609	9823	0̄037	0̄251	0̄465		6 129
28	3070680	0894	1108	1322	1536	1750	1964	2178	2392	2606	214	7 151
29	2820	3035	3249	3463	3677	3891	4105	4319	4532	4746		8 172
												9 194
2030	4960	5174	5388	5602	5816	6030	6244	6458	6672	6885		
31	7099	7313	7527	7741	7954	8168	8382	8596	8810	9023		214
32	9237	9451	9664	9878	0̄092	0̄306	0̄519	0̄733	0̄947	1̄160		1 21
33	3081374	1587	1801	2015	2228	2442	2655	2869	3082	3296		2 43
34	3509	3723	3936	4150	4363	4577	4790	5004	5217	5431		3 64
												4 86
35	5644	5858	6071	6284	6498	6711	6924	7138	7351	7564		5 107
36	7778	7991	8204	8418	8631	8844	9057	9271	9484	9697		6 128
37	9910	0̄123	0̄337	0̄550	0̄763	0̄976	1̄189	1̄402	1̄616	1̄829		7 150
38	3092042	2255	2468	2681	2894	3107	3320	3533	3746	3959		8 171
39	4172	4385	4598	4811	5024	5237	5450	5663	5876	6089	213	9 193
2040	6302	6515	6727	6940	7153	7366	7579	7792	8004	8217		
41	8430	8643	8856	9068	9281	9494	9707	9919	0̄132	0̄345		213
42	3100557	0770	0983	1195	1408	1621	1833	2046	2258	2471		1 21
43	2684	2896	3109	3321	3534	3746	3959	4171	4384	4596		2 43
44	4809	5021	5234	5446	5659	5871	6084	6296	6508	6721		3 64
												4 85
45	6933	7145	7358	7570	7783	7995	8207	8419	8632	8844		5 107
46	9056	9269	9481	9693	9905	0̄117	0̄330	0̄542	0̄754	0̄966		6 128
47	3111178	1391	1603	1815	2027	2239	2451	2663	2875	3087		7 149
48	3300	3512	3724	3936	4148	4360	4572	4784	4996	5208	212	8 170
49	5420	5632	5843	6055	6267	6479	6691	6903	7115	7327		9 192
N.	0	1	2	3	4	5	6	7	8	9	D	Pts.

N	0	1	2	3	4	5	6	7	8	9
2050	3117539	7750	7962	8174	8386	8598	8810	9021	9233	9445
51	9657	9868	0̅080	0̅292	0̅504	0̅715	0̅927	1̅139	1̅350	1̅562
52	3121774	1985	2197	2408	2620	2832	3043	3255	3466	3678
53	3889	4101	4313	4524	4736	4947	5159	5370	5581	5793
54	6004	6216	6427	6639	6850	7061	7273	7484	7696	7907
55	8118	8330	8541	8752	8964	9175	9386	9597	9809	0̅020
56	3130231	0442	0654	0865	1076	1287	1498	1709	1921	2132
57	2343	2554	2765	2976	3187	3398	3610	3821	4032	4243
58	4454	4665	4876	5087	5298	5509	5720	5931	6142	6353
59	6563	6774	6985	7196	7407	7618	7829	8040	8251	8461
2060	8672	8883	9094	9305	9515	9726	9937	0̅148	0̅358	0̅569
61	3140780	0991	1201	1412	1623	1833	2044	2255	2465	2676
62	2887	3097	3308	3518	3729	3940	4150	4361	4571	4782
63	4992	5203	5413	5624	5834	6045	6255	6466	6676	6887
64	7097	7307	7518	7728	7939	8149	8359	8570	8780	8990
65	9201	9411	9621	9831	0̅042	0̅252	0̅462	0̅672	0̅883	1̅093
66	3151303	1513	1724	1934	2144	2354	2564	2774	2985	3195
67	3405	3615	3825	4035	4245	4455	4665	4875	5085	5295
68	5505	5715	5925	6135	6345	6555	6765	6975	7185	7395
69	7605	7815	8025	8235	8444	8654	8864	9074	9284	9494
2070	9703	9913	0̅123	0̅333	0̅543	0̅752	0̅962	1̅172	1̅382	1̅591
71	3161801	2011	2220	2430	2640	2849	3059	3269	3478	3688
72	3898	4107	4317	4526	4736	4945	5155	5364	5574	5784
73	5993	6203	6412	6621	6831	7040	7250	7459	7669	7878
74	8088	8297	8506	8716	8925	9134	9344	9553	9762	9972
75	3170181	0390	0600	0809	1018	1227	1437	1646	1855	2064
76	2273	2483	2692	2901	3110	3319	3528	3738	3947	4156
77	4365	4574	4783	4992	5201	5410	5619	5828	6037	6246
78	6455	6664	6873	7082	7291	7500	7709	7918	8127	8336
79	8545	8754	8963	9172	9380	9589	9798	0̅007	0̅216	0̅425
2080	3180633	0842	1051	1260	1468	1677	1886	2095	2303	2512
81	2721	2929	3138	3347	3556	3764	3973	4181	4390	4599
82	4807	5016	5224	5433	5642	5850	6059	6267	6476	6684
83	6893	7101	7310	7518	7727	7935	8143	8352	8560	8769
84	8977	9186	9394	9602	9811	0̅019	0̅227	0̅436	0̅644	0̅852
85	3191061	1269	1477	1685	1894	2102	2310	2518	2727	2935
86	3143	3351	3559	3768	3976	4184	4392	4600	4808	5016
87	5224	5433	5641	5849	6057	6265	6473	6681	6889	7097
88	7305	7513	7721	7929	8137	8345	8553	8761	8969	9176
89	9384	9592	9800	0̅008	0̅216	0̅424	0̅632	0̅839	1̅047	1̅255
2090	3201463	1671	1878	2086	2294	2502	2709	2917	3125	3333
91	3540	3748	3956	4163	4371	4579	4786	4994	5202	5409
92	5617	5824	6032	6240	6447	6655	6862	7070	7277	7485
93	7692	7900	8107	8315	8522	8730	8937	9145	9352	9559
94	9767	9974	0̅182	0̅389	0̅596	0̅804	1̅011	1̅218	1̅426	1̅633
95	3211840	2048	2255	2462	2669	2877	3084	3291	3498	3706
96	3913	4120	4327	4534	4742	4949	5156	5363	5570	5777
97	5984	6191	6398	6606	6813	7020	7227	7434	7641	7848
98	8055	8262	8469	8676	8883	9090	9297	9504	9711	9917
99	3220124	0331	0538	0745	0952	1159	1366	1572	1779	1986
N.	0	1	2	3	4	5	6	7	8	9

D (mean differences): 211, 210, 209, 208, 207

Pro. (proportional parts)

212		211		210		209		208	
1	21	1	21	1	21	1	21	1	21
2	42	2	42	2	42	2	42	2	42
3	64	3	63	3	63	3	63	3	62
4	85	4	84	4	84	4	84	4	83
5	106	5	106	5	105	5	105	5	104
6	127	6	127	6	126	6	125	6	125
7	148	7	148	7	147	7	146	7	146
8	170	8	169	8	168	8	167	8	166
9	191	9	190	9	189	9	188	9	187

N.	0	1	2	3	4	5	6	7	8	9	D
2100	3222193	2400	2607	2813	3020	3227	3434	3640	3847	4054	
01	4261	4467	4674	4881	5087	5294	5501	5707	5914	6121	
02	6327	6534	6740	6947	7153	7360	7567	7773	7980	8186	
03	8393	8599	8806	9012	9219	9425	9632	9838	0̄045	0̄251	
04	3230457	0664	0870	1077	1283	1489	1696	1902	2108	2315	
05	2521	2727	2934	3140	3346	3552	3759	3965	4171	4377	
06	4584	4790	4996	5202	5408	5615	5821	6027	6233	6439	
07	6645	6851	7058	7264	7470	7676	7882	8088	8294	8500	
08	8706	8912	9118	9324	9530	9736	9942	0̄148	0̄354	0̄560	206
09	3240766	0972	1178	1384	1589	1795	2001	2207	2413	2619	
2110	2825	3030	3236	3442	3648	3854	4059	4265	4471	4677	
11	4882	5088	5294	5499	5705	5911	6117	6322	6528	6734	
12	6939	7145	7350	7556	7762	7967	8173	8378	8584	8789	
13	8995	9201	9406	9612	9817	0̄023	0̄228	0̄433	0̄639	0̄844	
14	3251050	1255	1461	1666	1872	2077	2282	2488	2693	2898	
15	3104	3309	3514	3720	3925	4130	4336	4541	4746	4951	
16	5157	5362	5567	5772	5978	6183	6388	6593	6798	7003	
17	7209	7414	7619	7824	8029	8234	8439	8644	8849	9055	
18	9260	9465	9670	9875	0̄080	0̄285	0̄490	0̄695	0̄900	1̄105	
19	3261310	1515	1719	1924	2129	2334	2539	2744	2949	3154	205
2120	3359	3563	3768	3973	4178	4383	4588	4792	4997	5202	
21	5407	5611	5816	6021	6226	6430	6635	6840	7044	7249	
22	7454	7658	7863	8068	8272	8477	8682	8886	9091	9295	
23	9500	9705	9909	0̄114	0̄318	0̄523	0̄727	0̄932	1̄136	1̄341	
24	3271545	1750	1954	2158	2363	2567	2772	2976	3181	3385	
25	3589	3794	3998	4202	4407	4611	4815	5020	5224	5428	
26	5633	5837	6041	6245	6450	6654	6858	7062	7267	7471	
27	7675	7879	8083	8287	8492	8696	8900	9104	9308	9512	
28	9716	9920	0̄124	0̄328	0̄533	0̄737	0̄941	1̄145	1̄349	1̄553	204
29	3281757	1961	2165	2369	2572	2776	2980	3184	3388	3592	
2130	3796	4000	4204	4408	4612	4815	5019	5223	5427	5631	
31	5834	6038	6242	6446	6650	6853	7057	7261	7465	7668	
32	7872	8076	8279	8483	8687	8890	9094	9298	9501	9705	
33	9909	0̄112	0̄316	0̄519	0̄723	0̄926	1̄130	1̄334	1̄537	1̄741	
34	3291944	2148	2351	2555	2758	2962	3165	3369	3572	3775	
35	3979	4182	4386	4589	4792	4996	5199	5402	5606	5809	
36	6012	6216	6419	6622	6826	7029	7232	7436	7639	7842	
37	8045	8248	8452	8655	8858	9061	9264	9468	9671	9874	
38	3300077	0280	0483	0686	0889	1093	1296	1499	1702	1905	203
39	2108	2311	2514	2717	2920	3123	3326	3529	3732	3935	
2140	4138	4341	4544	4747	4949	5152	5355	5558	5761	5964	
41	6167	6370	6572	6775	6978	7181	7384	7586	7789	7992	
42	8195	8397	8600	8803	9006	9208	9411	9614	9816	0̄019	
43	3310222	0424	0627	0830	1032	1235	1437	1640	1843	2045	
44	2248	2450	2653	2855	3058	3261	3463	3666	3868	4070	
45	4273	4475	4678	4880	5083	5285	5488	5690	5892	6095	
46	6297	6500	6702	6904	7107	7309	7511	7714	7916	8118	
47	8320	8523	8725	8927	9129	9332	9534	9736	9938	0̄141	
48	3320343	0545	0747	0949	1151	1354	1556	1758	1960	2162	
49	2364	2566	2768	2970	3172	3374	3577	3779	3981	4183	202
N.	0	1	2	3	4	5	6	7	8	9	D

Pro.

207
1	21
2	41
3	62
4	83
5	104
6	124
7	145
8	166
9	186

206
1	21
2	41
3	62
4	82
5	103
6	124
7	144
8	165
9	185

205
1	21
2	41
3	62
4	82
5	103
6	123
7	144
8	164
9	185

204
1	20
2	41
3	61
4	82
5	102
6	122
7	143
8	163
9	184

203
1	20
2	41
3	61
4	81
5	102
6	122
7	143
8	163
9	184

203
1	20
2	41
3	61
4	81
5	102
6	122
7	142
8	162
9	183

Pts.

N.	0	1	2	3	4	5	6	7	8	9	D	Pro.
2150	3324385	4587	4789	4991	5193	5394	5596	5798	6000	6202	202	
51	6404	6606	6808	7010	7212	7414	7615	7817	8019	8221		202
52	8423	8624	8826	9028	9230	9432	9633	9835	ō037	ō239		1 20
53	3330440	0642	0844	1045	1247	1449	1650	1852	2054	2255		2 40
54	2457	2659	2860	3062	3263	3465	3667	3868	4070	4271		3 61
												4 81
55	4473	4674	4876	5077	5279	5480	5682	5883	6085	6286		5 101
56	6488	6689	6890	7092	7293	7495	7696	7897	8099	8300		6 121
57	8501	8703	8904	9105	9307	9508	9709	9911	ō112	ō313		7 141
58	3340514	0716	0917	1118	1319	1521	1722	1923	2124	2325		8 162
59	2526	2728	2929	3130	3331	3532	3733	3934	4135	4336		9 182
2160	4538	4739	4940	5141	5342	5543	5744	5945	6146	6347	201	
61	6548	6749	6950	7151	7351	7552	7753	7954	8155	8356		201
62	8557	8758	8959	9159	9360	9561	9762	9963	ō164	ō364		1 20
63	3350565	0766	0967	1168	1368	1569	1770	1970	2171	2372		2 40
64	2573	2773	2974	3175	3375	3576	3777	3977	4178	4378		3 60
												4 80
65	4579	4780	4980	5181	5381	5582	5782	5983	6183	6384		5 101
66	6585	6785	6986	7186	7386	7587	7787	7988	8188	8389		6 121
67	8589	8790	8990	9190	9391	9591	9791	9992	ō192	ō392		7 141
68	3360593	0793	0993	1194	1394	1594	1795	1995	2195	2395		8 161
69	2596	2796	2996	3196	3396	3597	3797	3997	4197	4397		9 181
2170	4597	4797	4998	5198	5398	5598	5798	5998	6198	6398	200	
71	6598	6798	6998	7198	7398	7598	7798	7998	8198	8398		200
72	8598	8798	8998	9198	9398	9598	9798	9998	ō198	ō397		1 20
73	3370597	0797	0997	1197	1397	1596	1796	1996	2196	2396		2 40
74	2595	2795	2995	3195	3394	3594	3794	3994	4193	4393		3 60
												4 80
75	4593	4792	4992	5192	5391	5591	5791	5990	6190	6389		5 100
76	6589	6788	6988	7188	7387	7587	7786	7986	8185	8385		6 120
77	8584	8784	8983	9183	9382	9582	9781	9981	ō180	ō379		7 140
78	3380579	0778	0978	1177	1376	1576	1775	1974	2174	2373		8 160
79	2572	2772	2971	3170	3369	3569	3768	3967	4166	4366		9 180
2180	4565	4764	4963	5163	5362	5561	5760	5959	6158	6358		
81	6557	6756	6955	7154	7353	7552	7751	7950	8149	8348	199	
82	8547	8746	8946	9145	9344	9543	9742	9940	ō139	ō338		199
83	3390537	0736	0935	1134	1333	1532	1731	1930	2129	2327		1 20
84	2526	2725	2924	3123	3322	3520	3719	3918	4117	4316		2 40
												3 60
85	4514	4713	4912	5111	5309	5508	5707	5906	6104	6303		4 80
86	6502	6700	6899	7098	7296	7495	7693	7892	8091	8289		5 100
87	8488	8686	8885	9084	9282	9481	9679	9878	ō076	ō275		6 119
88	3400473	0672	0870	1069	1267	1466	1664	1862	2061	2259		7 139
89	2458	2656	2854	3053	3251	3449	3648	3846	4045	4243		8 159
												9 179
2190	4441	4639	4838	5036	5234	5433	5631	5829	6027	6226		
91	6424	6622	6820	7018	7217	7415	7613	7811	8009	8207		
92	8405	8604	8802	9000	9198	9396	9594	9792	9990	ō188		198
93	3410386	0584	0782	0980	1178	1376	1574	1772	1970	2168	198	1 20
94	2366	2564	2762	2960	3158	3356	3554	3752	3950	4147		2 40
												3 59
95	4345	4543	4741	4939	5137	5334	5532	5730	5928	6126		4 79
96	6323	6521	6719	6917	7114	7312	7510	7708	7905	8103		5 99
97	8301	8498	8696	8894	9091	9289	9486	9684	9882	ō079		6 119
98	3420277	0474	0672	0870	1067	1265	1462	1660	1857	2055		7 139
99	2252	2450	2647	2845	3042	3240	3437	3635	3832	4029		8 158
												9 178
N.	0	1	2	3	4	5	6	7	8	9	D	Pts.

N.	0	1	2	3	4	5	6	7	8	9	D	Pro.
2200	3424227	4424	4622	4819	5016	5214	5411	5608	5806	6003		
01	,6200	6398	6595	6792	6990	7187	7384	7581	7779	7976		
02	8173	8370	8568	8765	8962	9159	9356	9554	9751	9948		**198**
03	3430145	0342	0539	0736	0933	1131	1328	1525	1722	1919		1|20 2|40
04	2116	2313	2510	2707	2904	3101	3298	3495	3692	3889	197	3|59 4|79 5|99
05	4086	4283	4480	4677	4874	5071	5268	5464	5661	5858		6|119
06	6055	6252	6449	6646	6842	7039	7236	7433	7630	7827		7|139
07	8023	8220	8417	8614	8810	9007	9204	9401	9597	9794		8|158
08	9991	0̄187	0̄384	0̄581	0̄777	0̄974	1̄171	1̄367	1̄564	1̄761		9|178
09	3441957	2154	2350	2547	2743	2940	3137	3333	3530	3726		
2210	3923	4119	4316	4512	4709	4905	5102	5298	5495	5691		
11	5887	6084	6280	6477	6673	6869	7066	7262	7459	7655		**197**
12	7851	8048	8244	8440	8636	8833	9029	9225	9422	9618		1|20 2|39
13	9814	0̄010	0̄207	0̄403	0̄599	0̄795	0̄991	1̄188	1̄384	1̄580		3|59 4|79
14	3451776	1972	2168	2365	2561	2757	2953	3149	3345	3541	196	5|99 6|118
15	3737	3933	4129	4325	4522	4718	4914	5110	5306	5502		7|138
16	5698	5894	6090	6285	6481	6677	6873	7069	7265	7461		8|158
17	7657	7853	8049	8245	8440	8636	8832	9028	9224	9420		9|177
18	9615	9811	0̄007	0̄203	0̄399	0̄594	0̄790	0̄986	1̄182	1̄377		
19	3461573	1769	1964	2160	2356	2551	2747	2943	3138	3334		
2220	3530	3725	3921	4117	4312	4508	4703	4899	5094	5290		
21	5486	5681	5877	6072	6268	6463	6659	6854	7050	7245		**196**
22	7441	7636	7831	8027	8222	8418	8613	8808	9004	9199		1|20 2|39
23	9395	9590	9785	9981	0̄176	0̄371	0̄567	0̄762	0̄957	1̄153		3|59 4|78
24	3471348	1543	1738	1934	2129	2324	2519	2715	2910	3105		5|98 6|118
25	3300	3495	3691	3886	4081	4276	4471	4666	4861	5056		7|137
26	5252	5447	5642	5837	6032	6227	6422	6617	6812	7007	195	8|157
27	7202	7397	7592	7787	7982	8177	8372	8567	8762	8957		9|176
28	9152	9347	9542	9737	9931	0̄126	0̄321	0̄516	0̄711	0̄906		
29	3481101	1296	1490	1685	1880	2075	2270	2464	2659	2854		
2230	3049	3243	3438	3633	3828	4022	4217	4412	4606	4801		
31	4996	5190	5385	5580	5774	5969	6164	6358	6553	6747		**195**
32	6942	7136	7331	7526	7720	7915	8109	8304	8498	8693		1|20 2|39
33	8887	9082	9276	9471	9665	9860	0̄054	0̄248	0̄443	0̄637		3|59 4|78
34	3490832	1026	1220	1415	1609	1804	1998	2192	2387	2581		5|98 6|117
35	2775	2970	3164	3358	3552	3747	3941	4135	4330	4524		7|137
36	4718	4912	5106	5301	5495	5689	5883	6077	6272	6466		8|156
37	6660	6854	7048	7242	7436	7630	7825	8019	8213	8407		9|176
38	8601	8795	8989	9183	9377	9571	9765	9959	0̄153	0̄347	194	
39	3500541	0735	0929	1123	1317	1511	1705	1898	2092	2286		
2240	2480	2674	2868	3062	3256	3449	3643	3837	4031	4225		
41	4419	4612	4806	5000	5194	5387	5581	5775	5969	6162		**194**
42	6356	6550	6743	6937	7131	7325	7518	7712	7905	8099		1|19 2|39
43	8293	8486	8680	8874	9067	9261	9454	9648	9841	0̄035		3|58 4|78
44	3510229	0422	0616	0809	1003	1196	1390	1583	1777	1970		5|97 6|116
45	2163	2357	2550	2744	2937	3131	3324	3517	3711	3904		7|136
46	4098	4291	4484	4678	4871	5064	5258	5451	5644	5837		8|155
47	6031	6224	6417	6611	6804	6997	7190	7383	7577	7770		9|175
48	7963	8156	8349	8543	8736	8929	9122	9315	9508	9701		
49	9895	0̄088	0̄281	0̄474	0̄667	0̄860	1̄053	1̄246	1̄439	1̄632	193	
N.	0	1	2	3	4	5	6	7	8	9	D	Pts.

N.	0	1	2	3	4	5	6	7	8	9	D	Pro.
2250	3521825	2018	2211	2404	2597	2790	2983	3176	3369	3562	193	
51	3755	3948	4141	4334	4527	4720	4912	5105	5298	5491		193
52	5684	5877	6070	6262	6455	6648	6841	7034	7226	7419		1|19
53	7612	7805	7997	8190	8383	8576	8768	8961	9154	9346		2|39
54	9539	9732	9924	0̄117	0̄310	0̄502	0̄695	0̄888	1̄080	1̄273		3|58
55	3531465	1658	1851	2043	2236	2428	2621	2813	3006	3198		4|77 5|97
56	3391	3583	3776	3968	4161	4353	4546	4738	4931	5123		6|116
57	5316	5508	5700	5893	6085	6278	6470	6662	6855	7047		7|135
58	7239	7432	7624	7816	8009	8201	8393	8586	8778	8970		8|154
59	9162	9355	9547	9739	9931	0̄123	0̄316	0̄508	0̄700	0̄892		9|174
2260	3541084	1277	1469	1661	1853	2045	2237	2429	2621	2814		
61	3006	3198	3390	3582	3774	3966	4158	4350	4542	4734	192	192
62	4926	5118	5310	5502	5694	5886	6078	6270	6462	6654		1|19
63	6846	7037	7229	7421	7613	7805	7997	8189	8381	8572		2|38
64	8764	8956	9148	9340	9531	9723	9915	0̄107	0̄299	0̄490		3|58
65	3550682	0874	1066	1257	1449	1641	1832	2024	2216	2407		4|77 5|96
66	2599	2791	2982	3174	3366	3557	3749	3940	4132	4324		6|115
67	4515	4707	4898	5090	5281	5473	5664	5856	6048	6239		7|134
68	6431	6622	6813	7005	7196	7388	7579	7771	7962	8154		8|154
69	8345	8536	8728	8919	9111	9302	9493	9685	9876	0̄067		9|173
2270	3560259	0450	0641	0832	1024	1215	1406	1598	1789	1980		
71	2171	2363	2554	2745	2936	3127	3319	3510	3701	3892		
72	4083	4274	4466	4657	4848	5039	5230	5421	5612	5803		191
73	5994	6185	6376	6568	6759	6950	7141	7332	7523	7714	191	1|19
74	7905	8096	8287	8478	8668	8859	9050	9241	9432	9623		2|38 3|57
75	9814	0̄005	0̄196	0̄387	0̄578	0̄768	0̄959	1̄150	1̄341	1̄532		4|76 5|96
76	3571723	1913	2104	2295	2486	2677	2867	3058	3249	3440		6|115
77	3630	3821	4012	4202	4393	4584	4775	4965	5156	5347		7|134
78	5537	5728	5918	6109	6300	6490	6681	6872	7062	7253		8|153
79	7443	7634	7824	8015	8205	8396	8586	8777	8967	9158		9|172
2280	9348	9539	9729	9920	0̄110	0̄301	0̄491	0̄682	0̄872	1̄062		
81	3581253	1443	1634	1824	2014	2205	2395	2585	2776	2966		
82	3156	3347	3537	3727	3918	4108	4298	4488	4679	4869		190
83	5059	5249	5440	5630	5820	6010	6200	6391	6581	6771		1|19
84	6961	7151	7341	7531	7722	7912	8102	8292	8482	8672	190	2|38 3|57
85	8862	9052	9242	9432	9622	9812	0̄002	0̄192	0̄382	0̄572		4|76 5|95
86	3590762	0952	1142	1332	1522	1712	1902	2092	2282	2472		6|114
87	2662	2852	3041	3231	3421	3611	3801	3991	4181	4370		7|133
88	4560	4750	4940	5130	5319	5509	5699	5889	6078	6268		8|152
89	6458	6648	6837	7027	7217	7406	7596	7786	7976	8165		9|171
2290	8355	8544	8734	8924	9113	9303	9493	9682	9872	0̄061		
91	3600251	0440	0630	0820	1009	1199	1388	1578	1767	1957		
92	2146	2336	2525	2715	2904	3093	3283	3472	3662	3851		189
93	4041	4230	4419	4609	4798	4987	5177	5366	5555	5745		1|19
94	5934	6123	6313	6502	6691	6881	7070	7259	7448	7638		2|38 3|57
95	7827	8016	8205	8395	8584	8773	8962	9151	9341	9530	189	4|76 5|95
96	9719	9908	0̄097	0̄286	0̄475	0̄664	0̄854	1̄043	1̄232	1̄421		6|113
97	3611610	1799	1988	2177	2366	2555	2744	2933	3122	3311		7|132
98	3500	3689	3878	4067	4256	4445	4634	4823	5012	5201		8|151
99	5390	5579	5768	5956	6145	6334	6523	6712	6901	7090		9|170
N.	0	1	2	3	4	5	6	7	8	9	D	Pts.

N.	0	1	2	3	4	5	6	7	8	9	D	Pro.
2300	3617278	7467	7656	7845	8034	8222	8411	8600	8789	8977		
01	9166	9355	9544	9732	9921	0̄110	0̄298	0̄487	0̄676	0̄865		189
02	3621053	1242	1430	1619	1808	1996	2185	2374	2562	2751		1|19
03	2939	3128	3317	3505	3694	3882	4071	4259	4448	4636		2|38
04	4825	5013	5202	5390	5579	5767	5956	6144	6332	6521		3|57
05	6709	6898	7086	7275	7463	7651	7840	8028	8216	8405		4|76
06	8593	8781	8970	9158	9346	9535	9723	9911	0̄099	0̄288		5|95 6|113
07	3630476	0664	0852	1041	1229	1417	1605	1794	1982	2170		7|132
08	2358	2546	2734	2923	3111	3299	3487	3675	3863	4051		8|151
09	4239	4427	4615	4804	4992	5180	5368	5556	5744	5932	188	9|170
2310	6120	6308	6496	6684	6872	7060	7248	7436	7624	7812		
11	7999	8187	8375	8563	8751	8939	9127	9315	9503	9690		188
12	9878	0̄066	0̄254	0̄442	0̄630	0̄817	1̄005	1̄193	1̄381	1̄569		1|19
13	3641756	1944	2132	2320	2507	2695	2883	3070	3258	3446		2|38
14	3634	3821	4009	4197	4384	4572	4759	4947	5135	5322		3|56
15	5510	5698	5885	6073	6260	6448	6635	6823	7010	7198		4|75
16	7386	7573	7761	7948	8136	8323	8511	8698	8885	9073		5|94 6|113
17	9260	9448	9635	9823	0̄010	0̄197	0̄385	0̄572	0̄760	0̄947		7|132
18	3651134	1322	1509	1696	1884	2071	2258	2446	2633	2820		8|150
19	3007	3195	3382	3569	3757	3944	4131	4318	4505	4693		9|169
2320	4880	5067	5254	5441	5629	5816	6003	6190	6377	6564		
21	6751	6939	7126	7313	7500	7687	7874	8061	8248	8435		
22	8622	8809	8996	9183	9370	9557	9744	9931	0̄118	0̄305	187	187
23	3660492	0679	0866	1053	1240	1427	1614	1801	1987	2174		1|19
24	2361	2548	2735	2922	3109	3296	3482	3669	3856	4043		2|37 3|56
25	4230	4416	4603	4790	4977	5163	5350	5537	5724	5910		4|75
26	6097	6284	6471	6657	6844	7031	7217	7404	7591	7777		5|94 6|112
27	7964	8150	8337	8524	8710	8897	9083	9270	9457	9643		7|131
28	9830	0̄016	0̄203	0̄389	0̄576	0̄762	0̄949	1̄135	1̄322	1̄508		8|150
29	3671695	1881	2068	2254	2441	2627	2814	3000	3186	3373		9|168
2330	3559	3746	3932	4118	4305	4491	4677	4864	5050	5236		
31	5423	5609	5795	5982	6168	6354	6540	6727	6913	7099		
32	7285	7472	7658	7844	8030	8217	8403	8589	8775	8961		186
33	9147	9334	9520	9706	9892	0̄078	0̄264	0̄450	0̄636	0̄822		1|19
34	3681009	1195	1381	1567	1753	1939	2125	2311	2497	2683	186	2|37 3|56
35	2869	3055	3241	3427	3613	3799	3985	4171	4357	4542		4|74
36	4728	4914	5100	5286	5472	5658	5844	6030	6215	6401		5|93 6|112
37	6587	6773	6959	7145	7330	7516	7702	7888	8074	8259		7|130
38	8445	8631	8817	9002	9188	9374	9559	9745	9931	0̄117		8|149
39	3690302	0488	0674	0859	1045	1230	1416	1602	1787	1973		9|167
2340	2159	2344	2530	2715	2901	3086	3272	3458	3643	3829		
41	4014	4200	4385	4571	4756	4942	5127	5313	5498	5683		
42	5869	6054	6240	6425	6611	6796	6981	7167	7352	7538		185
43	7723	7908	8094	8279	8464	8650	8835	9020	9205	9391		1|19
44	9576	9761	9947	0̄132	0̄317	0̄502	0̄688	0̄873	1̄058	1̄243		2|37 3|56
45	3701428	1614	1799	1984	2169	2354	2540	2725	2910	3095		4|74
46	3280	3465	3650	3835	4020	4206	4391	4576	4761	4946		5|93
47	5131	5316	5501	5686	5871	6056	6241	6426	6611	6796	185	6|111 7|130
48	6981	7166	7351	7536	7721	7906	8091	8275	8460	8645		8|148
49	8830	9015	9200	9385	9570	9754	9939	0̄124	0̄309	0̄494		9|167
N.	0	1	2	3	4	5	6	7	8	9	D	Pts.

N.	0	1	2	3	4	5	6	7	8	9	D
2350	3710679	0863	1048	1233	1418	1603	1787	1972	2157	2342	
51	2526	2711	2896	3080	3265	3450	3635	3819	4004	4189	
52	4373	4558	4742	4927	5112	5296	5481	5666	5850	6035	
53	6219	6404	6588	6773	6957	7142	7327	7511	7696	7880	
54	8065	8249	8434	8618	8802	8987	9171	9356	9540	9725	
55	9909	ō094	ō278	ō462	ō647	ō831	Ī015	Ī200	Ī384	Ī569	
56	3721753	1937	2122	2306	2490	2674	2859	3043	3227	3412	
57	3596	3780	3964	4149	4333	4517	4701	4885	5070	5254	
58	5438	5622	5806	5991	6175	6359	6543	6727	6911	7095	
59	7279	7464	7648	7832	8016	8200	8384	8568	8752	8936	
2360	9120	9304	9488	9672	9856	ō040	ō224	ō408	ō592	ō776	184
61	3730960	1144	1328	1512	1696	1879	2063	2247	2431	2615	
62	2799	2983	3167	3350	3534	3718	3902	4086	4270	4453	
63	4637	4821	5005	5189	5372	5556	5740	5924	6107	6291	
64	6475	6658	6842	7026	7210	7393	7577	7761	7944	8128	
65	8311	8495	8679	8862	9046	9230	9413	9597	9780	9964	
66	3740147	0331	0515	0698	0882	1065	1249	1432	1616	1799	
67	1983	2166	2350	2533	2716	2900	3083	3267	3450	3634	
68	3817	4000	4184	4367	4551	4734	4917	5101	5284	5467	
69	5651	5834	6017	6201	6384	6567	6750	6934	7117	7300	
2370	7483	7667	7850	8033	8216	8400	8583	8766	8949	9132	
71	9316	9499	9682	9865	ō048	ō231	ō414	ō598	ō781	ō964	183
72	3751147	1330	1513	1696	1879	2062	2245	2428	2611	2794	
73	2977	3160	3343	3526	3709	3892	4075	4258	4441	4624	
74	4807	4990	5173	5356	5539	5722	5905	6088	6270	6453	
75	6636	6819	7002	7185	7367	7550	7733	7916	8099	8282	
76	8464	8647	8830	9013	9195	9378	9561	9744	9926	ō109	
77	3760292	0475	0657	0840	1023	1205	1388	1571	1753	1936	
78	2119	2301	2484	2666	2849	3032	3214	3397	3579	3762	
79	3944	4127	4310	4492	4675	4857	5040	5222	5405	5587	
2380	5770	5952	6135	6317	6499	6682	6864	7047	7229	7412	
81	7594	7776	7959	8141	8323	8506	8688	8871	9053	9235	
82	9418	9600	9782	9965	ō147	ō329	ō511	ō694	ō876	Ī058	
83	3771240	1423	1605	1787	1969	2152	2334	2516	2698	2880	
84	3063	3245	3427	3609	3791	3973	4155	4338	4520	4702	
85	4884	5066	5248	5430	5612	5794	5976	6158	6340	6522	182
86	6704	6886	7068	7250	7432	7614	7796	7978	8160	8342	
87	8524	8706	8888	9070	9252	9434	9616	9798	9979	ō161	
88	3780343	0525	0707	0889	1071	1252	1434	1616	1798	1980	
89	2161	2343	2525	2707	2889	3070	3252	3434	3616	3797	
2390	3979	4161	4342	4524	4706	4887	5069	5251	5432	5614	
91	5796	5977	6159	6341	6522	6704	6885	7067	7249	7430	
92	7612	7793	7975	8156	8338	8519	8701	8882	9064	9245	
93	9427	9608	9790	9971	ō153	ō334	ō516	ō697	ō879	Ī060	
94	3791241	1423	1604	1786	1967	2148	2330	2511	2692	2874	
95	3055	3237	3418	3599	3780	3962	4143	4324	4506	4687	
96	4868	5049	5231	5412	5593	5774	5956	6137	6318	6499	
97	6680	6862	7043	7224	7405	7586	7767	7948	8130	8311	
98	8492	8673	8854	9035	9216	9397	9578	9759	9940	ō121	181
99	3800302	0484	0665	0846	1027	1208	1389	1570	1750	1931	
N.	0	1	2	3	4	5	6	7	8	9	D

Pro. (Proportional parts)

185	184	183	182	181
1 19	1 18	1 18	1 18	1 18
2 37	2 37	2 37	2 36	2 36
3 56	3 55	3 55	3 55	3 54
4 74	4 74	4 73	4 73	4 72
5 93	5 92	5 92	5 91	5 91
6 111	6 110	6 110	6 109	6 109
7 130	7 129	7 128	7 127	7 127
8 148	8 147	8 146	8 146	8 145
9 167	9 166	9 165	9 164	9 163

Pts.

N.	0	1	2	3	4	5	6	7	8	9	D
2400	3802112	2293	2474	2655	2836	3017	3198	3379	3560	3741	181
01	3922	4102	4283	4464	4645	4826	5007	5188	5368	5549	
02	5730	5911	6092	6272	6453	6634	6815	6995	7176	7357	
03	7538	7718	7899	8080	8261	8441	8622	8803	8983	9164	
04	9345	9525	9706	9887	0̄067	0̄248	0̄428	0̄609	0̄790	0̄970	
05	3811151	1331	1512	1693	1873	2054	2234	2415	2595	2776	
06	2956	3137	3317	3498	3678	3859	4039	4220	4400	4580	
07	4761	4941	5122	5302	5483	5663	5843	6024	6204	6384	
08	6565	6745	6926	7106	7286	7467	7647	7827	8007	8188	
09	8368	8548	8729	8909	9089	9269	9450	9630	9810	9990	
2410	3820170	0351	0531	0711	0891	1071	1252	1432	1612	1792	
11	1972	2152	2332	2512	2693	2873	3053	3233	3413	3593	180
12	3773	3953	4133	4313	4493	4673	4853	5033	5213	5393	
13	5573	5753	5933	6113	6293	6473	6653	6833	7013	7193	
14	7373	7553	7732	7912	8092	8272	8452	8632	8812	8992	
15	9171	9351	9531	9711	9891	0̄070	0̄250	0̄430	0̄610	0̄790	
16	3830969	1149	1329	1509	1688	1868	2048	2227	2407	2587	
17	2767	2946	3126	3306	3485	3665	3844	4024	4204	4383	
18	4563	4743	4922	5102	5281	5461	5640	5820	6000	6179	
19	6359	6538	6718	6897	7077	7256	7436	7615	7795	7974	
2420	8154	8333	8513	8692	8871	9051	9230	9410	9589	9769	
21	9948	0̄127	0̄307	0̄486	0̄665	0̄845	1̄024	1̄203	1̄383	1̄562	
22	3841741	1921	2100	2279	2459	2638	2817	2996	3176	3355	
23	3534	3713	3893	4072	4251	4430	4609	4789	4968	5147	
24	5326	5505	5684	5864	6043	6222	6401	6580	6759	6938	
25	7117	7297	7476	7655	7834	8013	8192	8371	8550	8729	179
26	8908	9087	9266	9445	9624	9803	9982	0̄161	0̄340	0̄519	
27	3850698	0877	1056	1235	1413	1592	1771	1950	2129	2308	
28	2487	2666	2845	3023	3202	3381	3560	3739	3918	4096	
29	4275	4454	4633	4812	4990	5169	5348	5527	5705	5884	
2430	6063	6241	6420	6599	6778	6956	7135	7314	7492	7671	
31	7850	8028	8207	8386	8564	8743	8921	9100	9279	9457	
32	9636	9814	9993	0̄171	0̄350	0̄528	0̄707	0̄886	1̄064	1̄243	
33	3861421	1600	1778	1957	2135	2314	2492	2670	2849	3027	
34	3206	3384	3563	3741	3919	4098	4276	4455	4633	4811	
35	4990	5168	5346	5525	5703	5881	6060	6238	6416	6595	
36	6773	6951	7129	7308	7486	7664	7842	8021	8199	8377	
37	8555	8733	8912	9090	9268	9446	9624	9803	9981	0̄159	
38	3870337	0515	0693	0871	1049	1228	1406	1584	1762	1940	178
39	2118	2296	2474	2652	2830	3008	3186	3364	3542	3720	
2440	3898	4076	4254	4432	4610	4788	4966	5144	5322	5500	
41	5678	5856	6034	6212	6389	6567	6745	6923	7101	7279	
42	7457	7634	7812	7990	8168	8346	8524	8701	8879	9057	
43	9235	9412	9590	9768	9946	0̄123	0̄301	0̄479	0̄657	0̄834	
44	3881012	1190	1367	1545	1723	1900	2078	2256	2433	2611	
45	2789	2966	3144	3321	3499	3677	3854	4032	4209	4387	
46	4565	4742	4920	5097	5275	5452	5630	5807	5985	6162	
47	6340	6517	6695	6872	7050	7227	7404	7582	7759	7937	
48	8114	8292	8469	8646	8824	9001	9178	9356	9533	9711	
49	9888	0̄065	0̄243	0̄420	0̄597	0̄774	0̄952	1̄129	1̄306	1̄484	
N.	0	1	2	3	4	5	6	7	8	9	D

Pro.

181
1 | 18
2 | 36
3 | 54
4 | 72
5 | 91
6 | 109
7 | 127
8 | 145
9 | 163

180
1 | 18
2 | 36
3 | 54
4 | 72
5 | 90
6 | 108
7 | 126
8 | 144
9 | 162

179
1 | 18
2 | 36
3 | 54
4 | 72
5 | 90
6 | 107
7 | 125
8 | 143
9 | 161

178
1 | 18
2 | 36
3 | 53
4 | 71
5 | 89
6 | 107
7 | 125
8 | 142
9 | 160

Pts.

N.	0	1	2	3	4	5	6	7	8	9	D
2450	3891661	1838	2015	2193	2370	2547	2724	2902	3079	3256	
51	3433	3610	3787	3965	4142	4319	4496	4673	4850	5028	
52	5205	5382	5559	5736	5913	6090	6267	6444	6621	6798	
53	6975	7153	7330	7507	7684	7861	8038	8215	8392	8569	177
54	8746	8923	9100	9276	9453	9630	9807	9984	0̄161	0̄338	
55	3900515	0692	0869	1046	1223	1399	1576	1753	1930	2107	
56	2284	2460	2637	2814	2991	3168	3344	3521	3698	3875	
57	4052	4228	4405	4582	4759	4935	5112	5289	5465	5642	
58	5819	5995	6172	6349	6525	6702	6879	7055	7232	7409	
59	7585	7762	7939	8115	8292	8468	8645	8821	8998	9175	
2460	9351	9528	9704	9881	0̄057	0̄234	0̄410	0̄587	0̄763	0̄940	
61	3911116	1293	1469	1646	1822	1998	2175	2351	2528	2704	
62	2880	3057	3233	3410	3586	3762	3940	4115	4291	4468	
63	4644	4820	4997	5173	5349	5526	5702	5878	6055	6231	
64	6407	6583	6760	6936	7112	7288	7464	7641	7817	7993	
65	8169	8345	8522	8698	8874	9050	9226	9402	9578	9755	
66	9931	0̄107	0̄283	0̄459	0̄635	0̄811	0̄987	1̄163	1̄339	1̄515	176
67	3921691	1868	2044	2220	2396	2572	2748	2924	3100	3276	
68	3452	3628	3803	3979	4155	4331	4507	4683	4859	5035	
69	5211	5387	5563	5739	5914	6090	6266	6442	6618	6794	
2470	6970	7145	7321	7497	7673	7849	8024	8200	8376	8552	
71	8727	8903	9079	9255	9430	9606	9782	9958	0̄133	0̄309	
72	3930485	0660	0836	1012	1187	1363	1539	1714	1890	2066	
73	2241	2417	2592	2768	2944	3119	3295	3470	3646	3821	
74	3997	4172	4348	4524	4699	4875	5050	5226	5401	5577	
75	5752	5928	6103	6278	6454	6629	6805	6980	7156	7331	
76	7506	7682	7857	8033	8208	8383	8559	8734	8909	9085	
77	9260	9435	9611	9786	9961	0̄137	0̄312	0̄487	0̄662	0̄838	
78	3941013	1188	1364	1539	1714	1889	2064	2240	2415	2590	
79	2765	2940	3116	3291	3466	3641	3816	3991	4167	4342	
2480	4517	4692	4867	5042	5217	5392	5567	5742	5918	6093	
81	6268	6443	6618	6793	6968	7143	7318	7493	7668	7843	175
82	8018	8193	8368	8543	8718	8893	9068	9242	9417	9592	
83	9767	9942	0̄117	0̄292	0̄467	0̄642	0̄817	0̄991	1̄166	1341	
84	3951516	1691	1866	2040	2215	2390	2565	2740	2914	3089	
85	3264	3439	3613	3788	3963	4138	4312	4487	4662	4837	
86	5011	5186	5361	5535	5710	5885	6059	6234	6409	6583	
87	6758	6932	7107	7282	7456	7631	7805	7980	8155	8329	
88	8504	8678	8853	9027	9202	9376	9551	9725	9900	0̄074	
89	3960249	0423	0598	0772	0947	1121	1296	1470	1645	1819	
2490	1993	2168	2342	2517	2691	2865	3040	3214	3389	3563	
91	3737	3912	4086	4260	4435	4609	4783	4958	5132	5306	
92	5480	5655	5829	6003	6177	6352	6526	6700	6874	7049	
93	7223	7397	7571	7745	7920	8094	8268	8442	8616	8790	
94	8964	9139	9313	9487	9661	9835	0̄009	0̄183	0̄357	0̄531	174
95	3970705	0880	1054	1228	1402	1576	1750	1924	2098	2272	
96	2446	2620	2794	2968	3142	3316	3490	3664	3838	4011	
97	4185	4359	4533	4707	4881	5055	5229	5403	5577	5750	
98	5924	6098	6272	6446	6620	6794	6967	7141	7315	7489	
99	7663	7836	8010	8184	8358	8531	8705	8879	9053	9226	
N.	0	1	2	3	4	5	6	7	8	9	D

Pro. Pts.

177
1 | 18
2 | 35
3 | 53
4 | 71
5 | 89
6 | 106
7 | 124
8 | 142
9 | 159

176
1 | 18
2 | 35
3 | 53
4 | 70
5 | 88
6 | 106
7 | 123
8 | 141
9 | 158

175
1 | 18
2 | 35
3 | 53
4 | 70
5 | 88
6 | 105
7 | 123
8 | 140
9 | 158

174
1 | 17
2 | 35
3 | 52
4 | 70
5 | 87
6 | 104
7 | 122
8 | 139
9 | 157

N.	0	1	2	3	4	5	6	7	8	9	D
2500	3979400	9574	9748	9921	$\overline{0}$095	$\overline{0}$269	$\overline{0}$442	$\overline{0}$616	$\overline{0}$790	$\overline{0}$963	
01	3981137	1311	1484	1658	1831	2005	2179	2352	2526	2699	
02	2873	3047	3220	3394	3567	3741	3914	4088	4261	4435	
03	4608	4782	4956	5129	5302	5476	5649	5823	5996	6170	
04	6343	6517	6690	6864	7037	7210	7384	7557	7731	7904	
05	8077	8251	8424	8597	8771	8944	9117	9291	9464	9637	
06	9811	9984	$\overline{0}$157	$\overline{0}$331	$\overline{0}$504	$\overline{0}$677	$\overline{0}$850	$\overline{1}$024	$\overline{1}$197	$\overline{1}$370	
07	3991543	1717	1890	2063	2236	2409	2583	2756	2929	3102	
08	3275	3448	3622	3795	3968	4141	4314	4487	4660	4834	
09	5007	5180	5353	5526	5699	5872	6045	6218	6391	6564	173
2510	6737	6910	7083	7256	7429	7602	7775	7948	8121	8294	
11	8467	8640	8813	8986	9159	9332	9505	9678	9851	$\overline{0}$023	
12	4000196	0369	0542	0715	0888	1061	1234	1406	1579	1752	
13	1925	2098	2271	2443	2616	2789	2962	3134	3307	3480	
14	3653	3825	3998	4171	4344	4516	4689	4862	5035	5207	
15	5380	5553	5725	5898	6071	6243	6416	6588	6761	6934	
16	7106	7279	7452	7624	7797	7969	8142	8314	8487	8660	
17	8832	9005	9177	9350	9522	9695	9867	$\overline{0}$040	$\overline{0}$212	$\overline{0}$385	
18	4010557	0730	0902	1075	1247	1420	1592	1764	1937	2109	
19	2282	2454	2626	2799	2971	3144	3316	3488	3661	3833	
2520	4005	4178	4350	4522	4695	4867	5039	5212	5384	5556	
21	5728	5901	6073	6245	6417	6590	6762	6934	7106	7279	
22	7451	7623	7795	7967	8140	8312	8484	8656	8828	9000	
23	9173	9345	9517	9689	9861	$\overline{0}$033	$\overline{0}$205	$\overline{0}$377	$\overline{0}$549	$\overline{0}$721	
24	4020894	1066	1238	1410	1582	1754	1926	2098	2270	2442	172
25	2614	2786	2958	3130	3302	3474	3646	3818	3990	4162	
26	4333	4505	4677	4849	5021	5193	5365	5537	5709	5881	
27	6052	6224	6396	6568	6740	6912	7083	7255	7427	7599	
28	7771	7942	8114	8286	8458	8630	8801	8973	9145	9317	
29	9488	9660	9832	$\overline{0}$003	$\overline{0}$175	$\overline{0}$347	$\overline{0}$519	$\overline{0}$690	$\overline{0}$862	$\overline{1}$034	
2530	4031205	1377	1549	1720	1892	2063	2235	2407	2578	2750	
31	2921	3093	3265	3436	3608	3779	3951	4122	4294	4465	
32	4637	4809	4980	5152	5323	5495	5666	5838	6009	6180	
33	6352	6523	6695	6866	7038	7209	7381	7552	7723	7895	
34	8066	8237	8409	8580	8752	8923	9094	9266	9437	9608	
35	9780	9951	$\overline{0}$122	$\overline{0}$294	$\overline{0}$465	$\overline{0}$636	$\overline{0}$807	$\overline{0}$979	$\overline{1}$150	$\overline{1}$321	
36	4041492	1664	1835	2006	2177	2349	2520	2691	2862	3033	
37	3205	3376	3547	3718	3889	4060	4232	4403	4574	4745	
38	4916	5087	5258	5429	5601	5772	5943	6114	6285	6456	
39	6627	6798	6969	7140	7311	7482	7653	7824	7995	8166	171
2540	8337	8508	8679	8850	9021	9192	9363	9534	9705	9876	
41	4050047	0218	0388	0559	0730	0901	1072	1243	1414	1585	
42	1755	1926	2097	2268	2439	2610	2780	2951	3122	3293	
43	3464	3634	3805	3976	4147	4317	4488	4659	4830	5000	
44	5171	5342	5512	5683	5854	6025	6195	6366	6537	6707	
45	6878	7049	7219	7390	7560	7731	7902	8072	8243	8413	
46	8584	8755	8925	9096	9266	9437	9607	9778	9948	$\overline{0}$119	
47	4060289	0460	0630	0801	0971	1142	1312	1483	1653	1824	
48	1994	2165	2335	2506	2676	2846	3017	3187	3358	3528	
49	3698	3869	4039	4209	4380	4550	4721	4891	5061	5231	
N.	0	1	2	3	4	5	6	7	8	9	D

Pro. (Proportional parts)

174	173	172	172	171
1 17	1 17	1 17	1 17	1 17
2 35	2 35	2 34	2 34	2 34
3 52	3 52	3 52	3 52	3 51
4 70	4 69	4 69	4 69	4 68
5 87	5 87	5 86	5 86	5 86
6 104	6 104	6 103	6 103	6 103
7 122	7 121	7 120	7 120	7 120
8 139	8 138	8 138	8 138	8 137
9 157	9 156	9 155	9 155	9 154

N.	0	1	2	3	4	5	6	7	8	9	D
2550	4065402	5572	5742	5913	6083	6253	6424	6594	6764	6934	
51	7105	7275	7445	7615	7786	7956	8126	8296	8466	8637	
52	8807	8977	9147	9317	9487	9658	9828	9998	ō168	ō338	
53	4070508	0678	0848	˙1018	1189	1359	1529	1699	1869	2039	
54	2209	2379	2549	2719	2889	3059	3229	3399	3569	3739	170
55	3909	4079	4249	4419	4589	4759	4929	5099	5269	5439	
56	5608	5778	5948	6118	6288	6458	6628	6798	6968	7137	
57	7307	7477	7647	7817	7987	8156	8326	8496	8666	8836	
58	9005	9175	9345	9515	9684	9854	ō024	ō194	ō363	ō533	
59	4080703	0873	1042	1212	1382	1551	1721	1891	2060	2230	
2560	2400	2569	2739	2909	3078	3248	3417	3587	3757	3926	
61	4096	4265	4435	4604	4774	4944	5113	5283	5452	5622	
62	5791	5961	6130	6300	6469	6639	6808	6978	7147	7317	
63	7486	7656	7825	7994	8164	8333	8503	8672	8841	9011	
64	9180	9350	9519	9688	9858	ō027	ō196	ō366	ō535	ō704	
65	4090874	1043	1212	1382	1551	1720	1889	2059	2228	2397	
66	2567	2736	2905	3074	3243	3413	3582	3751	3920	4089	
67	4259	4428	4597	4766	4935	5105	5274	5443	5612	5781	
68	5950	6119	6288	6458	6627	6796	6965	7134	7303	7472	169
69	7641	7810	7979	8148	8317	8486	8655	8824	8993	9162	
2570	9331	9500	9669	9838	ō007	ō176	ō345	ō514	ō683	ō852	
71	4101021	1190	1359	1527	1696	1865	2034	2203	2372	2541	
72	2710	2878	3047	3216	3385	3554	3723	3891	4060	4229	
73	4398	4567	4735	4904	5073	5242	5410	5579	5748	5917	
74	6085	6254	6423	6592	6760	6929	7098	7266	7435	7604	
75	7772	7941	8110	8278	8447	8616	8784	8953	9121	9290	
76	9459	9627	9796	9964	ō133	ō301	ō470	ō639	ō607	ō976	
77	4111144	1313	1481	1650	1818	1987	2155	2324	2492	2661	
78	2829	2998	3166	3334	3503	3671	3840	4008	4177	4345	
79	4513	4682	4850	5019	5187	5355	5524	5692	5860	6029	
2580	6197	6365	6534	6702	6870	7039	7207	7375	7544	7712	
81	7880	8048	8217	8385	8553	8721	8890	9058	9226	9394	
82	9562	9731	9899	ō067	ō235	ō403	ō571	ō740	ō908	ī076	
83	4121244	1412	1580	1748	1917	2085	2253	2421	2589	2757	
84	2925	3093	3261	3429	3597	3765	3933	4101	4269	4437	168
85	4605	4773	4941	5109	5277	5445	5613	5781	5949	6117	
86	6285	6453	6621	6789	6957	7125	7293	7461	7629	7796	
87	7964	8132	8300	8468	8636	8804	8971	9139	9307	9475	
88	9643	9811	9978	ō146	ō314	ō482	ō649	ō817	ō985	ī153	
89	4131321	1488	1656	1824	1991	2159	2327	2495	2662	2830	
2590	2998	3165	3333	3501	3668	3836	4004	4171	4339	4507	
91	4674	4842	5009	5177	5345	5512	5680	5847	6015	6182	
92	6350	6518	6685	6853	7020	7188	7355	7523	7690	7858	
93	8025	8193	8360	8528	8695	8863	9030	9197	9365	9532	
94	9700	9867	ō035	ō202	ō369	ō537	ō704	ō872	ī039	ī206	
95	4141374	1541	1708	1876	2043	2210	2378	2545	2712	2880	
96	3047	3214	3381	3549	3716	3883	4051	4218	4385	4552	
97	4719	4887	5054	5221	5388	5556	5723	5890	6057	6224	
98	6391	6559	6726	6893	7060	7227	7394	7561	7729	7896	
99	8063	8230	8397	8564	8731	8898	9065	9232	9399	9566	167
N.	0	1	2	3	4	5	6	7	8	9	D

Pro. (Proportional parts)

170	169	168	167
1 17	1 17	1 17	1 17
2 34	2 34	2 34	2 33
3 51	3 51	3 50	3 50
4 68	4 68	4 67	4 67
5 85	5 85	5 84	5 84
6 102	6 101	6 101	6 100
7 119	7 118	7 118	7 117
8 136	8 135	8 134	8 134
9 153	9 152	9 151	9 150

N.	0	1	2	3	4	5	6	7	8	9	D
2600	4149733	9901	ō068	ō235	ō402	ō569	ō736	ō903	ī070	ī237	167
01	4151404	1570	1737	1904	2071	2238	2405	2572	2739	2906	
02	3073	3240	3407	3574	3741	3907	4074	4241	4408	4575	
03	4742	4909	5075	5242	5409	5576	5743	5909	6076	6243	
04	6410	6577	6743	6910	7077	7244	7410	7577	7744	7911	
05	8077	8244	8411	8577	8744	8911	9077	9244	9411	9577	
06	9744	9911	ō077	ō244	ō411	ō577	ō744	ō911	ī077	ī244	
07	4161410	1577	1743	1910	2077	2243	2410	2576	2743	2909	
08	3076	3242	3409	3575	3742	3908	4075	4241	4408	4574	
09	4741	4907	5074	5240	5407	5573	5739	5906	6072	6239	
2610	6405	6571	6738	6904	7071	7237	7403	7570	7736	7902	
11	8069	8235	8401	8568	8734	8900	9067	9233	9399	9565	
12	9732	9898	ō064	ō231	ō397	ō563	ō729	ō895	ī062	ī228	
13	4171394	1560	1726	1893	2059	2225	2391	2557	2724	2890	
14	3056	3222	3388	3554	3720	3886	4053	4219	4385	4551	
15	4717	4883	5049	5215	5381	5547	5713	5879	6045	6211	166
16	6377	6543	6709	6875	7041	7207	7373	7539	7705	7871	
17	8037	8203	8369	8535	8701	8867	9033	9199	9365	9531	
18	9696	9862	ō028	ō194	ō360	ō526	ō692	ō857	ī023	ī189	
19	4181355	1521	1687	1852	2018	2184	2350	2516	2681	2847	
2620	3013	3179	3344	3510	3676	3842	4007	4173	4339	4505	
21	4670	4836	5002	5167	5333	5499	5664	5830	5996	6161	
22	6327	6493	6658	6824	6989	7155	7321	7486	7652	7817	
23	7983	8148	8314	8480	8645	8811	8976	9142	9307	9473	
24	9638	9804	9969	ō135	ō300	ō466	ō631	ō797	ō962	ī128	
25	4191293	1459	1624	1789	1955	2120	2286	2451	2616	2782	
26	2947	3113	3278	3443	3609	3774	3939	4105	4270	4435	
27	4601	4766	4931	5097	5262	5427	5593	5758	5923	6088	
28	6254	6419	6584	6749	6915	7080	7245	7410	7575	7741	
29	7906	8071	8236	8401	8567	8732	8897	9062	9227	9392	
2630	9557	9723	9888	ō053	ō218	ō383	ō548	ō713	ō878	ī043	
31	4201208	1374	1539	1704	1869	2034	2199	2364	2529	2694	165
32	2859	3024	3189	3354	3519	3684	3849	4014	4179	4344	
33	4509	4674	4838	5003	5168	5333	5498	5663	5828	5993	
34	6158	6323	6487	6652	6817	6982	7147	7312	7477	7641	
35	7806	7971	8136	8301	8465	8630	8795	8960	9125	9289	
36	9454	9619	9784	9948	ō113	ō278	ō442	ō607	ō772	ō937	
37	4211101	1266	1431	1595	1760	1925	2089	2254	2419	2583	
38	2748	2913	3077	3242	3406	3571	3736	3900	4065	4229	
39	4394	4558	4723	4888	5052	5217	5381	5546	5710	5875	
2640	6039	6204	6368	6533	6697	6862	7026	7191	7355	7520	
41	7684	7848	8013	8177	8342	8506	8671	8835	8999	9164	
42	9328	9493	9657	9821	9986	ō150	ō314	ō479	ō643	ō807	
43	4220972	1136	1300	1465	1629	1793	1957	2122	2286	2450	
44	2615	2779	2943	3107	3271	3436	3600	3764	3928	4093	
45	4257	4421	4585	4749	4913	5078	5242	5406	5570	5734	
46	5898	6063	6227	6391	6555	6719	6883	7047	7211	7375	
47	7539	7703	7868	8032	8196	8360	8524	8688	8852	9016	
48	9180	9344	9508	9672	9836	ō000	ō164	ō328	ō492	ō656	164
49	4230820	0984	1147	1311	1475	1639	1803	1967	2131	2295	
N.	0	1	2	3	4	5	6	7	8	9	D

Pro. (Proportional parts)

	167	166	165	164
1	17	17	17	16
2	33	33	33	33
3	50	50	50	49
4	67	66	66	66
5	84	83	83	82
6	100	100	99	98
7	117	116	116	115
8	134	133	132	131
9	150	149	149	148

N.	0	1	2	3	4	5	6	7	8	9	D	Pro.
2650	4232459	2623	2786	2950	3114	3278	3442	3606	3770	3933		
51	4097	4261	4425	4589	4753	4916	5080	5244	5408	5571		
52	5735	5899	6063	6226	6390	6554	6718	6881	7045	7209		**163**
53	7372	7536	7700	7864	8027	8191	8355	8518	8682	8846		1 16
54	9009	9173	9336	9500	9664	9827	9991	0̄154	0̄318	0̄482		2 33
55	4240645	0809	0972	1136	1300	1463	1627	1790	1954	2117		3 49
56	2281	2444	2608	2771	2935	3098	3262	3425	3589	3752		4 65
57	3916	4079	4242	4406	4569	4733	4896	5060	5223	5386		5 82
58	5550	5713	5877	6040	6203	6367	6530	6693	6857	7020		6 98
59	7183	7347	7510	7673	7837	8000	8163	8327	8490	8653		7 114
2660	8816	8980	9143	9306	9469	9633	9796	9959	0̄122	0̄286		8 130
61	4250449	0612	0775	0938	1102	1265	1428	1591	1754	1917		9 147
62	2081	2244	2407	2570	2733	2896	3059	3222	3385	3549	163	
63	3712	3875	4038	4201	4364	4527	4690	4853	5016	5179		
64	5342	5505	5668	5831	5994	6157	6320	6483	6646	6809		
65	6972	7135	7298	7461	7624	7787	7950	8113	8276	8439		
66	8601	8764	8927	9090	9253	9416	9579	9742	9904	0̄067		
67	4260230	0393	0556	0719	0881	1044	1207	1370	1533	1695		
68	1858	2021	2184	2347	2509	2672	2835	2998	3160	3323		
69	3486	3648	3811	3974	4137	4299	4462	4625	4787	4950		
2670	5113	5275	5438	5601	5763	5926	6088	6251	6414	6576		
71	6739	6901	7064	7227	7389	7552	7714	7877	8039	8202		162
72	8365	8527	8690	8852	9015	9177	9340	9502	9665	9827		1 16
73	9990	0̄152	0̄315	0̄477	0̄639	0̄802	0̄964	1̄127	1̄289	1̄452		2 32
74	4271614	1776	1939	2101	2264	2426	2588	2751	2913	3076		3 49
75	3238	3400	3563	3725	3887	4050	4212	4374	4536	4699		4 65
76	4861	5023	5186	5348	5510	5672	5835	5997	6159	6321		5 81
77	6484	6646	6808	6970	7133	7295	7457	7619	7781	7944		6 97
78	8106	8268	8430	8592	8754	8917	9079	9241	9403	9565		7 113
79	9727	9889	0̄051	0̄213	0̄376	0̄538	0̄700	0̄862	1̄024	1̄186		8 130
2680	4281348	1510	1672	1834	1996	2158	2320	2482	2644	2806	162	9 146
81	2968	3130	3292	3454	3616	3778	3940	4102	4264	4426		
82	4588	4750	4912	5073	5235	5397	5559	5721	5883	6045		
83	6207	6369	6530	6692	6854	7016	7178	7340	7501	7663		
84	7825	7987	8149	8311	8472	8634	8796	8958	9119	9281		
85	9443	9605	9766	9928	0̄090	0̄252	0̄413	0̄575	0̄737	0̄898		
86	4291060	1222	1383	1545	1707	1868	2030	2192	2353	2515		
87	2677	2838	3000	3162	3323	3485	3646	3808	3969	4131		
88	4293	4454	4616	4777	4939	5100	5262	5423	5585	5747		
89	5908	6070	6231	6393	6554	6715	6877	7038	7200	7361		
2690	7523	7684	7846	8007	8169	8330	8491	8653	8814	8976		
91	9137	9298	9460	9621	9782	9944	0̄105	0̄267	0̄428	0̄589		
92	4300751	0912	1073	1235	1396	1557	1718	1880	2041	2202		161
93	2364	2525	2686	2847	3009	3170	3331	3492	3653	3815		1 16
94	3976	4137	4298	4460	4621	4782	4943	5104	5265	5427		2 32
95	5588	5749	5910	6071	6232	6393	6554	6716	6877	7038	161	3 48
96	7199	7360	7521	7682	7843	8004	8165	8326	8487	8648		4 64
97	8809	8970	9132	9293	9454	9615	9776	9937	0̄098	0̄258		5 81
98	4310419	0580	0741	0902	1063	1224	1385	1546	1707	1868		6 97
99	2029	2190	2351	2512	2672	2833	2994	3155	3316	3477		7 113
												8 129
												9 145
N.	0	1	2	3	4	5	6	7	8	9	D	Pts.

N.	0	1	2	3	4	5	6	7	8	9	D
2700	4313638	3798	3959	4120	4281	4442	4603	4763	4924	5085	161
01	5246	5407	5567	5728	5889	6050	6210	6371	6532	6693	
02	6853	7014	7175	7336	7496	7657	7818	7978	8139	8300	
03	8460	8621	8782	8942	9103	9264	9424	9585	9746	9906	
04	4320067	0227	0388	0549	0709	0870	1030	1191	1352	1512	
05	1673	1833	1994	2154	2315	2475	2636	2796	2957	3117	
06	3278	3438	3599	3759	3920	4080	4241	4401	4562	4722	
07	4883	5043	5203	5364	5524	5685	5845	6005	6166	6326	
08	6487	6647	6807	6968	7128	7288	7449	7609	7769	7930	
09	8090	8250	8411	8571	8731	8892	9052	9212	9372	9533	
2710	9693	9853	0̄013	0̄174	0̄334	0̄494	0̄654	0̄815	0̄975	1̄135	
11	4331295	1455	1616	1776	1936	2096	2256	2416	2577	2737	
12	2897	3057	3217	3377	3537	3697	3858	4018	4178	4338	
13	4498	4658	4818	4978	5138	5298	5458	5618	5778	5938	160
14	6098	6258	6418	6578	6738	6898	7058	7218	7378	7538	
15	7698	7858	8018	8178	8338	8498	8658	8818	8978	9138	
16	9298	9458	9617	9777	9937	0̄097	0̄257	0̄417	0̄577	0̄737	
17	4340896	1056	1216	1376	1536	1696	1855	2015	2175	2335	
18	2495	2654	2814	2974	3134	3293	3453	3613	3773	3932	
19	4092	4252	4412	4571	4731	4891	5050	5210	5370	5529	
2720	5689	5849	6008	6168	6328	6487	6647	6807	6966	7126	
21	7285	7445	7605	7764	7924	8083	8243	8403	8562	8722	
22	8881	9041	9200	9360	9519	9679	9838	9998	0̄157	0̄317	
23	4350476	0636	0795	0955	1114	1274	1433	1593	1752	1912	
24	2071	2230	2390	2549	2709	2868	3028	3187	3346	3506	
25	3665	3824	3984	4143	4303	4462	4621	4781	4940	5099	
26	5259	5418	5577	5736	5896	6055	6214	6374	6533	6692	
27	6851	7011	7170	7329	7488	7648	7807	7966	8125	8284	
28	8444	8603	8762	8921	9080	9240	9399	9558	9717	9876	
29	4360035	0194	0354	0513	0672	0831	0990	1149	1308	1467	
2730	1626	1786	1945	2104	2263	2422	2581	2740	2899	3058	159
31	3217	3376	3535	3694	3853	4012	4171	4330	4489	4648	
32	4807	4966	5125	5284	5443	5602	5761	5920	6078	6237	
33	6396	6555	6714	6873	7032	7191	7350	7509	7667	7826	
34	7985	8144	8303	8462	8620	8779	8938	9097	9256	9415	
35	9573	9732	9891	0̄050	0̄208	0̄367	0̄526	0̄685	0̄843	1̄002	
36	4371161	1320	1478	1637	1796	1955	2113	2272	2431	2589	
37	2748	2907	3065	3224	3383	3541	3700	3859	4017	4176	
38	4334	4493	4652	4810	4969	5127	5286	5445	5603	5762	
39	5920	6079	6237	6396	6555	6713	6872	7030	7189	7347	
2740	7506	7664	7823	7981	8140	8298	8457	8615	8773	8932	
41	9090	9249	9407	9566	9724	9883	0̄041	0̄199	0̄358	0̄516	
42	4380675	0833	0991	1150	1308	1466	1625	1783	1941	2100	
43	2258	2416	2575	2733	2891	3050	3208	3366	3525	3683	
44	3841	3999	4158	4316	4474	4632	4791	4949	5107	5265	
45	5423	5582	5740	5898	6056	6214	6373	6531	6689	6847	
46	7005	7163	7322	7480	7638	7796	7954	8112	8270	8428	
47	8587	8745	8903	9061	9219	9377	9535	9693	9851	0̄009	
48	4390167	0325	0483	0641	0799	0957	1115	1273	1431	1589	158
49	1747	1905	2063	2221	2379	2537	2695	2853	3011	3169	
N.	0	1	2	3	4	5	6	7	8	9	D

Pro. (Proportional parts)

161
1	16
2	32
3	48
4	64
5	81
6	97
7	113
8	129
9	145

160
1	16
2	32
3	48
4	64
5	80
6	96
7	112
8	128
9	144

159
1	16
2	32
3	48
4	64
5	80
6	95
7	111
8	127
9	143

Pts.

N.	0	1	2	3	4	5	6	7	8	9	D	Pro.
2750	4393327	3485	3643	3801	3959	4116	4274	4432	4590	4748		
51	4906	5064	5222	5379	5537	5695	5853	6011	6169	6326		158
52	6484	6642	6800	6958	7115	7273	7431	7589	7747	7904		1|16
53	8062	8220	8378	8535	8693	8851	9009	9166	9324	9482		2|32
54	9639	9797	9955	ō112	ō270	ō428	ō585	ō743	ō901	ī058		3|47
												4|63
55	4401216	1374	1531	1689	1847	2004	2162	2319	2477	2635		5|79
56	2792	2950	3107	3265	3422	3580	3738	3895	4053	4210		6|95
57	4368	4525	4683	4840	4998	5155	5313	5470	5628	5785		7|111
58	5943	6100	6258	6415	6572	6730	6887	7045	7202	7360		8|126
59	7517	7674	7832	7989	8147	8304	8461	8619	8776	8933		9|142
2760	9091	9248	9406	9563	9720	9878	ō035	ō192	ō349	ō507		
61	4410664	0821	0979	1136	1293	1450	1608	1765	1922	2080		
62	2237	2394	2551	2708	2866	3023	3180	3337	3494	3652		
63	3809	3966	4123	4280	4438	4595	4752	4909	5066	5223		
64	5380	5538	5695	5852	6009	6166	6323	6480	6637	6794		
65	6951	7108	7265	7423	7580	7737	7894	8051	8208	8365	157	
66	8522	8679	8836	8993	9150	9307	9464	9621	9778	9935		
67	4420092	0249	0405	0562	0719	0876	1033	1190	1347	1504		
68	1661	1818	1975	2132	2288	2445	2602	2759	2916	3073		
69	3230	3386	3543	3700	3857	4014	4171	4327	4484	4641		
2770	4798	4954	5111	5268	5425	5582	5738	5895	6052	6209		
71	6365	6522	6679	6835	6992	7149	7306	7462	7619	7776		157
72	7932	8089	8246	8402	8559	8716	8872	9029	9185	9342		1|16
73	9499	9655	9812	9969	ō125	ō282	ō438	ō595	ō751	ō908		2|31
74	4431065	1221	1378	1534	1691	1847	2004	2160	2317	2473		3|47
												4|63
75	2630	2786	2943	3099	3256	3412	3569	3725	3882	4038		5|79
76	4195	4351	4507	4664	4820	4977	5133	5290	5446	5602		6|94
77	5759	5915	6072	6228	6384	6541	6697	6853	7010	7166		7|110
78	7322	7479	7635	7791	7948	8104	8260	8417	8573	8729		8|126
79	8885	9042	9198	9354	9511	9667	9823	9979	ō136	ō292		9|141
2780	4440448	0604	0760	0917	1073	1229	1385	1541	1698	1854		
81	2010	2166	2322	2478	2635	2791	2947	3103	3259	3415		
82	3571	3727	3883	4040	4196	4352	4508	4664	4820	4976		
83	5132	5288	5444	5600	5756	5912	6068	6224	6380	6536	156	
84	6692	6848	7004	7160	7316	7472	7628	7784	7940	8096		
85	8252	8408	8564	8720	8876	9032	9188	9343	9499	9655		
86	9811	9967	ō123	ō279	ō435	ō590	ō746	ō902	ī058	ī214		
87	4451370	1526	1681	1837	1993	2149	2305	2460	2616	2772		
88	2928	3083	3239	3395	3551	3706	3862	4018	4174	4329		
89	4485	4641	4797	4952	5108	5264	5419	5575	5731	5886		
2790	6042	6198	6353	6509	6665	6820	6976	7132	7287	7443		
91	7598	7754	7910	8065	8221	8376	8532	8687	8843	8999		
92	9154	9310	9465	9621	9776	9932	ō087	ō243	ō398	ō554		156
93	4460709	0865	1020	1176	1331	1487	1642	1798	1953	2109		1|16
94	2264	2419	2575	2730	2886	3041	3197	3352	3507	3663		2|31
												3|47
95	3818	3974	4129	4284	4440	4595	4750	4906	5061	5216		4|62
96	5372	5527	5682	5838	5993	6148	6304	6459	6614	6769		5|78
97	6925	7080	7235	7390	7546	7701	7856	8011	8167	8322		6|94
98	8477	8632	8788	8943	9098	9253	9408	9563	9719	9874		7|109
99	4470029	0184	0339	0494	0650	0805	0960	1115	1270	1425		8|125
												9|140
N.	0	1	2	3	4	5	6	7	8	9	D	Pts.

N.	0	1	2	3	4	5	6	7	8	9	D
2800	4471580	1735	1891	2046	2201	2356	2511	2666	2821	2976	
01	3131	3286	3441	3596	3751	3906	4061	4216	4371	4526	155
02	4681	4836	4991	5146	5301	5456	5611	5766	5921	6076	
03	6231	6386	6541	6696	6851	7006	7161	7315	7470	7625	
04	7780	7935	8090	8245	8400	8554	8709	8864	9019	9174	
05	9329	9483	9638	9793	9948	0̄103	0̄258	0̄412	0̄567	0̄722	
06	4480877	1031	1186	1341	1496	1650	1805	1960	2115	2269	
07	2424	2579	2734	2888	3043	3198	3352	3507	3662	3816	
08	3971	4126	4280	4435	4590	4744	4899	5054	5208	5363	
09	5517	5672	5827	5981	6136	6290	6445	6600	6754	6909	
2810	7063	7218	7372	7527	7681	7836	7990	8145	8299	8454	
11	8608	8763	8917	9072	9226	9381	9535	9690	9844	9999	
12	4490153	0308	0462	0616	0771	0925	1080	1234	1389	1543	
13	1697	1852	2006	2160	2315	2469	2624	2778	2932	3087	
14	3241	3395	3550	3704	3858	4013	4167	4321	4475	4630	
15	4784	4938	5093	5247	5401	5555	5710	5864	6018	6172	
16	6327	6481	6635	6789	6943	7098	7252	7406	7560	7714	
17	7868	8023	8177	8331	8485	8639	8793	8948	9102	9256	
18	9410	9564	9718	9872	0̄026	0̄180	0̄334	0̄489	0̄643	0̄797	
19	4500951	1105	1259	1413	1567	1721	1875	2029	2183	2337	154
2820	2491	2645	2799	2953	3107	3261	3415	3569	3723	3877	
21	4031	4185	4339	4493	4647	4801	4954	5108	5262	5416	
22	5570	5724	5878	6032	6186	6340	6493	6647	6801	6955	
23	7109	7263	7416	7570	7724	7878	8032	8186	8339	8493	
24	8647	8801	8954	9108	9262	9416	9570	9723	9877	0̄031	
25	4510185	0338	0492	0646	0799	0953	1107	1261	1414	1568	
26	1722	1875	2029	2183	2336	2490	2644	2797	2951	3104	
27	3258	3412	3565	3719	3873	4026	4180	4333	4487	4640	
28	4794	4948	5101	5255	5408	5562	5715	5869	6022	6176	
29	6329	6483	6636	6790	6943	7097	7250	7404	7557	7711	
2830	7864	8018	8171	8325	8478	8632	8785	8938	9092	9245	
31	9399	9552	9705	9859	0̄012	0̄166	0̄319	0̄472	0̄626	0̄779	
32	4520932	1086	1239	1393	1546	1699	1853	2006	2159	2312	
33	2466	2619	2772	2926	3079	3232	3385	3539	3692	3845	
34	3998	4152	4305	4458	4611	4765	4918	5071	5224	5377	
35	5531	5684	5837	5990	6143	6297	6450	6603	6756	6909	
36	7062	7215	7369	7522	7675	7828	7981	8134	8287	8440	
37	8593	8746	8900	9053	9206	9359	9512	9665	9818	9971	
38	4530124	0277	0430	0583	0736	0889	1042	1195	1348	1501	153
39	1654	1807	1960	2113	2266	2419	2572	2725	2878	3030	
2840	3183	3336	3489	3642	3795	3948	4101	4254	4407	4559	
41	4712	4865	5018	5171	5324	5477	5629	5782	5935	6088	
42	6241	6394	6546	6699	6852	7005	7158	7310	7463	7616	
43	7769	7921	8074	8227	8380	8532	8685	8838	8990	9143	
44	9296	9449	9601	9754	9907	0̄059	0̄212	0̄365	0̄517	0̄670	
45	4540823	0975	1128	1281	1433	1586	1739	1891	2044	2196	
46	2349	2502	2654	2807	2959	3112	3264	3417	3570	3722	
47	3875	4027	4180	4332	4485	4637	4790	4942	5095	5247	
48	5400	5552	5705	5857	6010	6162	6315	6467	6620	6772	
49	6924	7077	7229	7382	7534	7687	7839	7991	8144	8296	
N.	0	1	2	3	4	5	6	7	8	9	D

Pro. (Proportional Parts)

	155	154	153
1	16	15	15
2	31	31	31
3	47	46	46
4	62	62	61
5	78	77	77
6	93	92	92
7	109	108	107
8	124	123	122
9	140	139	138

N	0	1	2	3	4	5	6	7	8	9	D	Pro.
2850	4548449	8601	8753	8906	9058	9210	9363	9515	9668	9820		
51	9972	0125	0277	0429	0581	0734	0886	1038	1191	1343		
52	4551495	1647	1800	1952	2104	2257	2409	2561	2713	2865		152
53	3018	3170	3322	3474	3627	3779	3931	4083	4235	4388		1 15
54	4540	4692	4844	4996	5148	5300	5453	5605	5757	5909		2 30
55	6061	6213	6365	6517	6670	6822	6974	7126	7278	7430		3 46
56	7582	7734	7886	8038	8190	8342	8494	8646	8798	8950		4 61
57	9102	9254	9406	9558	9710	9862	0014	0166	0318	0470		5 76
58	4560622	0774	0926	1078	1230	1382	1534	1686	1838	1990	152	6 91
59	2142	2293	2445	2597	2749	2901	3053	3205	3357	3508		7 106
2860	3660	3812	3964	4116	4268	4420	4571	4723	4875	5027		8 122
61	5179	5330	5482	5634	5786	5938	6089	6241	6393	6545		9 137
62	6696	6848	7000	7152	7303	7455	7607	7758	7910	8062		
63	8213	8365	8517	8669	8820	8972	9124	9275	9427	9578		
64	9730	9882	0033	0185	0337	0488	0640	0791	0943	1095		
65	4571246	1398	1549	1701	1853	2004	2156	2307	2459	2610		
66	2762	2913	3065	3216	3368	3519	3671	3822	3974	4125		
67	4277	4428	4580	4731	4883	5034	5186	5337	5489	5640		
68	5791	5943	6094	6246	6397	6549	6700	6851	7003	7154		
69	7305	7457	7608	7760	7911	8062	8214	8365	8516	8668		
2870	8819	8970	9122	9273	9424	9576	9727	9878	0029	0181		151
71	4580332	0483	0634	0786	0937	1088	1239	1391	1542	1693		1 15
72	1844	1996	2147	2298	2449	2600	2752	2903	3054	3205		2 30
73	3356	3507	3659	3810	3961	4112	4263	4414	4565	4717		3 45
74	4868	5019	5170	5321	5472	5623	5774	5925	6076	6227		4 60
75	6378	6530	6681	6832	6983	7134	7285	7436	7587	7738		5 76
76	7889	8040	8191	8342	8493	8644	8795	8946	9097	9248	151	6 91
77	9399	9550	9701	9851	0002	0153	0304	0455	0606	0757		7 106
78	4590908	1059	1210	1361	1511	1662	1813	1964	2115	2266		8 121
79	2417	2567	2718	2869	3020	3171	3322	3472	3623	3774		9 136
2880	3925	4076	4226	4377	4528	4679	4830	4980	5131	5282		
81	5433	5583	5734	5895	6036	6186	6337	6488	6638	6789		
82	6940	7090	7241	7392	7542	7693	7844	7994	8145	8296		
83	8446	8597	8748	8898	9049	9200	9350	9501	9651	9802		
84	9953	0103	0254	0404	0555	0705	0856	1007	1157	1308		
85	4601458	1609	1759	1910	2060	2211	2361	2512	2662	2813		
86	2963	3114	3264	3415	3565	3716	3866	4017	4167	4317		
87	4468	4618	4769	4919	5070	5220	5370	5521	5671	5822		
88	5972	6122	6273	6423	6573	6724	6874	7024	7175	7325		
89	7475	7626	7776	7926	8077	8227	8377	8528	8678	8828		
2890	8978	9129	9279	9429	9579	9730	9880	0030	0180	0331		
91	4610481	0631	0781	0932	1082	1232	1382	1532	1683	1833		150
92	1983	2133	2283	2433	2584	2734	2884	3034	3184	3334		1 15
93	3484	3634	3785	3935	4085	4235	4385	4535	4685	4835		2 30
94	4985	5135	5285	5435	5585	5736	5886	6036	6186	6336		3 45
95	6486	6636	6786	6936	7086	7236	7386	7536	7686	7836	150	4 60
96	7986	8136	8285	8435	8585	8735	8885	9035	9185	9335		5 75
97	9485	9635	9785	9935	0085	0234	0384	0534	0684	0834		6 90
98	4620984	1134	1284	1433	1583	1733	1883	2033	2183	2332		7 105
99	2482	2632	2782	2932	3081	3231	3381	3531	3680	3830		8 120
												9 135
N.	0	1	2	3	4	5	6	7	8	9	D	Pts.

N.	0	1	2	3	4	5	6	7	8	9	D
2900	4623980	4130	4279	4429	4579	4729	4878	5028	5178	5328	
01	5477	5627	5777	5926	6076	6226	6375	6525	6675	6824	
02	6974	7124	7273	7423	7573	7722	7872	8022	8171	8321	
03	8470	8620	8770	8919	9069	9218	9368	9517	9667	9817	
04	9966	ō116	ō265	ō415	ō564	ō714	ō863	Ī013	Ī162	Ī312	
05	4631461	1611	1760	1910	2059	2209	2358	2508	2657	2807	
06	2956	3106	3255	3404	3554	3703	3853	4002	4152	4301	
07	4450	4600	4749	4898	5048	5197	5347	5496	5645	5795	
08	5944	6093	6243	6392	6541	6691	6840	6989	7139	7288	
09	7437	7587	7736	7885	8034	8184	8333	8482	8631	8781	
2910	8930	9079	9228	9378	9527	9676	9825	9974	ō124	ō273	
11	4640422	0571	0720	0870	1019	1168	1317	1466	1615	1765	
12	1914	2063	2212	2361	2510	2659	2808	2958	3107	3256	
13	3405	3554	3703	3852	4001	4150	4299	4448	4597	4746	
14	4895	5045	5194	5343	5492	5641	5790	5939	6088	6237	149
15	6386	6535	6684	6833	6981	7130	7279	7428	7577	7726	
16	7875	8024	8173	8322	8471	8620	8769	8918	9067	9215	
17	9364	9513	9662	9811	9960	ō109	ō258	ō406	ō555	ō704	
18	4650853	1002	1151	1299	1448	1597	1746	1895	2043	2192	
19	2341	2490	2639	2787	2936	3085	3234	3382	3531	3680	
2920	3829	3977	4126	4275	4423	4572	4721	4870	5018	5167	
21	5316	5464	5613	5762	5910	6059	6208	6356	6505	6653	
22	6802	6951	7099	7248	7397	7545	7694	7842	7991	8140	
23	8288	8437	8585	8734	8882	9031	9180	9328	9477	9625	
24	9774	9922	ō071	ō219	ō368	ō516	ō665	ō813	ō962	Ī110	
25	4661259	1407	1556	1704	1853	2001	2149	2298	2446	2595	
26	2743	2892	3040	3188	3337	3485	3634	3782	3930	4079	
27	4227	4376	4524	4672	4821	4969	5117	5266	5414	5562	
28	5711	5859	6007	6156	6304	6452	6601	6749	6897	7045	
29	7194	7342	7490	7639	7787	7935	8083	8232	8380	8528	
2930	8676	8824	8973	9121	9269	9417	9565	9714	9862	ō010	
31	4670158	0306	0455	0603	0751	0899	1047	1195	1343	1492	
32	1640	1788	1936	2084	2232	2380	2528	2676	2824	2973	
33	3121	3269	3417	3565	3713	3861	4009	4157	4305	4453	148
34	4601	4749	4897	5045	5193	5341	5489	5637	5785	5933	
35	6081	6229	6377	6525	6673	6821	6969	7117	7265	7413	
36	7561	7708	7856	8004	8152	8300	8448	8596	8744	8892	
37	9039	9187	9335	9483	9631	9779	9927	ō074	ō222	ō370	
38	4680518	0666	0814	0961	1109	1257	1405	1553	1700	1848	
39	1996	2144	2291	2439	2587	2735	2882	3030	3178	3326	
2940	3473	3621	3769	3916	4064	4212	4360	4507	4655	4803	
41	4950	5098	5246	5393	5541	5689	5836	5984	6131	6279	
42	6427	6574	6722	6870	7017	7165	7312	7460	7607	7755	
43	7903	8050	8198	8345	8493	8640	8788	8935	9083	9231	
44	9378	9526	9673	9821	9968	ō116	ō263	ō411	ō558	ō706	
45	4690853	1000	1148	1295	1443	1590	1738	1885	2033	2180	
46	2327	2475	2622	2770	2917	3064	3212	3359	3507	3654	
47	3801	3949	4096	4243	4391	4538	4685	4833	4980	5127	
48	5275	5422	5569	5717	5864	6011	6159	6306	6453	6600	
49	6748	6895	7042	7190	7337	7484	7631	7778	7926	8073	
N.	0	1	2	3	4	5	6	7	8	9	D

Pro.

150
1| 15
2| 30
3| 45
4| 60
5| 75
6| 90
7|105
8|120
9|135

149
1| 15
2| 30
3| 45
4| 60
5| 75
6| 89
7|104
8|119
9|134

148
1| 15
2| 30
3| 44
4| 59
5| 74
6| 89
7|104
8|118
9|133

Pts.

N.	0	1	2	3	4	5	6	7	8	9	D	Pro.
2950	4698220	8367	8515	8662	8809	8956	9103	9251	9398	9545		
51	9692	9839	9986	ō134	ō281	ō428	ō575	ō722	ō869	Ī016		147
52	4701164	1311	1458	1605	1752	1899	2046	2193	2340	2487		1 15
53	2634	2782	2929	3076	3223	3370	3517	3664	3811	3958	147	2 29
54	4105	4252	4399	4546	4693	4840	4987	5134	5281	5428		3 44
												4 59
55	5575	5722	5869	6016	6163	6310	6457	6604	6750	6897		5 74
56	7044	7191	7338	7485	7632	7779	7926	8073	8219	8366		6 88
57	8513	8660	8807	8954	9101	9248	9394	9541	9688	9835		7 103
58	9982	ō129	ō275	ō422	ō569	ō716	ō863	Ī009	Ī156	Ī303		8 118
59	4711450	1596	1743	1890	2037	2183	2330	2477	2624	2770		9 132
2960	2917	3064	3211	3357	3504	3651	3797	3944	4091	4237		
61	4384	4531	4677	4824	4971	5117	5264	5411	5557	5704		
62	5851	5997	6144	6290	6437	6584	6730	6877	7023	7170		
63	7317	7463	7610	7756	7903	8049	8196	8342	8489	8635		
64	8782	8929	9075	9222	9368	9515	9661	9808	9954	ō101		
65	4720247	0393	0540	0686	0833	0979	1126	1272	1419	1565		
66	1711	1858	2004	2151	2297	2444	2590	2736	2883	3029		
67	3175	3322	3468	3615	3761	3907	4054	4200	4346	4493		
68	4639	4785	4932	5078	5224	5371	5517	5663	5809	5956		
69	6102	6248	6395	6541	6687	6833	6980	7126	7272	7418		
2970	7564	7711	7857	8003	8149	8296	8442	8588	8734	8880		
71	9027	9173	9319	9465	9611	9757	9903	ō050	ō196	ō342		146
72	4730488	0634	0780	0926	1073	1219	1365	1511	1657	1803		1 15
73	1949	2095	2241	2387	2533	2679	2825	2972	3118	3264	146	2 29
74	3410	3556	3702	3848	3994	4140	4286	4432	4578	4724		3 41
												4 58
75	4870	5016	5162	5308	5454	5600	5746	5891	6037	6183		5 73
76	6329	6475	6621	6767	6913	7059	7205	7351	7497	7642		6 88
77	7788	7934	8080	8226	8372	8518	8664	8809	8955	9101		7 102
78	9247	9393	9539	9684	9830	9976	ō122	ō268	ō413	ō559		8 117
79	4740705	0851	0997	1142	1288	1434	1580	1725	1871	2017		9 131
2980	2163	2308	2454	2600	2746	2891	3037	3183	3328	3474		
81	3620	3765	3911	4057	4202	4348	4494	4639	4785	4931		
82	5076	5222	5368	5513	5659	5805	5950	6096	6241	6387		
83	6533	6678	6824	6969	7115	7260	7406	7552	7697	7843		
84	7988	8134	8279	8425	8570	8716	8861	9007	9152	9298		
85	9443	9589	9734	9880	ō025	ō171	ō316	ō462	ō607	ō753		
86	4750898	Ī043	1189	1334	1480	1625	1771	1916	2061	2207		
87	2352	2498	2643	2788	2934	3079	3225	3370	3515	3661		
88	3806	3951	4097	4242	4387	4533	4678	4823	4969	5114		
89	5259	5404	5550	5695	5840	5986	6131	6276	6421	6567		
2990	6712	6857	7002	7148	7293	7438	7583	7729	7874	8019		
91	8164	8309	8455	8600	8745	8890	9035	9180	9326	9471		
92	9616	9761	9906	ō051	ō196	ō342	ō487	ō632	ō777	ō922		145
93	4761067	1212	1357	1502	1648	1793	1938	2083	2228	2373	145	1 15
94	2518	2663	2808	2953	3098	3243	3388	3533	3678	3823		2 29
												3 44
95	3968	4113	4258	4403	4548	4693	4838	4983	5128	5273		4 58
96	5418	5563	5708	5853	5998	6143	6288	6433	6578	6723		5 73
97	6867	7012	7157	7302	7447	7592	7737	7882	8027	8171		6 87
98	8316	8461	8606	8751	8896	9041	9185	9330	9475	9620		7 102
99	9765	9909	ō054	ō199	ō344	ō489	ō633	ō778	ō923	Ī068		8 116
												9 131
N.	0	1	2	3	4	5	6	7	8	9	D	Pts.

N.	0	1	2	3	4	5	6	7	8	9	D	Pro.
3000	4771213	1357	1502	1647	1792	1936	2081	2226	2371	2515		
01	2660	2805	2949	3094	3239	3383	3528	3673	3818	3962		145
02	4107	4252	4396	4541	4686	4830	4975	5119	5264	5409		1|15
03	5553	5698	5843	5987	6132	6276	6421	6566	6710	6855		2|29
04	6999	7144	7288	7433	7578	7722	7867	8011	8156	8300		3|44
												4|58
05	8445	8589	8734	8878	9023	9167	9312	9456	9601	9745		5|73
06	9890	ō034	ō179	ō323	ō468	ō612	ō757	ō901	ī045	ī190		6|87
07	4781334	1479	1623	1768	1912	2056	2201	2345	2490	2634		7|102
08	2778	2923	3067	3211	3356	3500	3645	3789	3933	4078		8|116
09	4222	4366	4511	4655	4799	4943	5088	5232	5376	5521		9|131
3010	5665	5809	5954	6098	6242	6386	6531	6675	6819	6963		
11	7108	7252	7396	7540	7684	7829	7973	8117	8261	8405		
12	8550	8694	8838	8982	9126	9271	9415	9559	9703	9847		
13	9991	ō135	ō280	ō424	ō568	ō712	ō856	ī000	ī144	ī288		
14	4791432	1577	1721	1865	2009	2153	2297	2441	2585	2729		
15	2873	3017	3161	3305	3449	3593	3737	3881	4025	4169	144	
16	4313	4457	4601	4745	4889	5033	5177	5321	5465	5609		
17	5753	5897	6041	6185	6329	6473	6617	6761	6905	7048		
18	7192	7336	7480	7624	7768	7912	8056	8200	8343	8487		
19	8631	8775	8919	9063	9207	9350	9494	9638	9782	9926		
3020	4800069	0213	0357	0501	0645	0788	0932	1076	1220	1363		
21	1507	1651	1795	1939	2082	2226	2370	2513	2657	2801		144
22	2945	3088	3232	3376	3519	3663	3807	3950	4094	4238		1|14
23	4381	4525	4669	4812	4956	5100	5243	5387	5531	5674		2|29
24	5818	5961	6105	6249	6392	6536	6679	6823	6967	7110		3|43
25	7254	7397	7541	7684	7828	7972	8115	8259	8402	8546		4|58
26	8689	8833	8976	9120	9263	9407	9550	9694	9837	9981		5|72
27	4810124	0268	0411	0555	0698	0842	0985	1128	1272	1415		6|86
28	1559	1702	1846	1989	2132	2276	2419	2563	2706	2849		7|101
29	2993	3136	3279	3423	3566	3710	3853	3996	4140	4283		8|115
												9|130
3030	4426	4570	4713	4856	5000	5143	5286	5429	5573	5716		
31	5859	6003	6146	6289	6432	6576	6719	6862	7005	7149		
32	7292	7435	7578	7722	7865	8008	8151	8295	8438	8581		
33	8724	8867	9010	9154	9297	9440	9583	9726	9869	ō013		
34	4820156	0299	0442	0585	0728	0871	1015	1158	1301	1444		
35	1587	1730	1873	2016	2159	2302	2445	2589	2732	2875		
36	3018	3161	3304	3447	3590	3733	3876	4019	4162	4305		
37	4448	4591	4734	4877	5020	5163	5306	5449	5592	5735		
38	5878	6021	6164	6307	6449	6592	6735	6878	7021	7164	143	
39	7307	7450	7593	7736	7879	8021	8164	8307	8450	8593		
3040	8736	8879	9022	9164	9307	9450	9593	9736	9879	ō021		
41	4830164	0307	0450	0593	0735	0878	1021	1164	1307	1449		
42	1592	1735	1878	2020	2163	2306	2449	2591	2734	2877		143
43	3020	3162	3305	3448	3590	3733	3876	4018	4161	4304		1|14
44	4446	4589	4732	4874	5017	5160	5302	5445	5588	5730		2|29
												3|43
45	5873	6016	6158	6301	6443	6586	6729	6871	7014	7156		4|57
46	7299	7442	7584	7727	7869	8012	8154	8297	8439	8582		5|72
47	8725	8867	9010	9152	9295	9437	9580	9722	9865	ō007		6|86
48	4840150	0292	0435	0577	0720	0862	1004	1147	1289	1432		7|100
49	1574	1717	1859	2002	2144	2286	2429	2571	2714	2856		8|114
												9|129
N.	0	1	2	3	4	5	6	7	8	9	D	Pts.

N.	0	1	2	3	4	5	6	7	8	9	D
3050	4842998	3141	3283	3426	3568	3710	3853	3995	4137	4280	
51	4422	4564	4707	4849	4991	5134	5276	5418	5561	5703	
52	5845	5988	6130	6272	6414	6557	6699	6841	6984	7126	
53	7268	7410	7553	7695	7837	7979	8121	8264	8406	8548	
54	8690	8833	8975	9117	9259	9401	9543	9686	9828	9970	
55	4850112	0254	0396	0539	0681	0823	0965	1107	1249	1391	
56	1533	1676	1818	1960	2102	2244	2386	2528	2670	2812	142
57	2954	3096	3239	3381	3523	3665	3807	3949	4091	4233	
58	4375	4517	4659	4801	4943	5085	5227	5369	5511	5653	
59	5795	5937	6079	6221	6363	6505	6647	6788	6930	7072	
3060	7214	7356	7498	7640	7782	7924	8066	8208	8350	8491	
61	8633	8775	8917	9059	9201	9343	9484	9626	9768	9910	
62	4860052	0194	0336	0477	0619	0761	0903	1045	1186	1328	
63	1470	1612	1754	1895	2037	2179	2321	2462	2604	2746	
64	2888	3029	3171	3313	3455	3596	3738	3880	4021	4163	
65	4305	4446	4588	4730	4872	5013	5155	5297	5438	5580	
66	5722	5863	6005	6146	6288	6430	6571	6713	6855	6996	
67	7138	7279	7421	7563	7704	7846	7987	8129	8270	8412	
68	8554	8695	8837	8978	9120	9261	9403	9544	9686	9827	
69	9969	ō110	ō252	ō393	ō535	ō676	ō818	ō959	Ī101	Ī242	
3070	4871384	1525	1667	1808	1950	2091	2232	2374	2515	2657	
71	2798	2940	3081	3222	3364	3505	3647	3788	3929	4071	
72	4212	4353	4495	4636	4778	4919	5060	5202	5343	5484	
73	5626	5767	5908	6050	6191	6332	6473	6615	6756	6897	
74	7039	7180	7321	7462	7604	7745	7886	8027	8169	8310	
75	8451	8592	8734	8875	9016	9157	9299	9440	9581	9722	
76	9863	ō004	ō146	ō287	ō428	ō569	ō710	ō852	ō993	Ī134	
77	4881275	1416	1557	1698	1839	1981	2122	2263	2404	2545	
78	2686	2827	2968	3109	3251	3392	3533	3674	3815	3956	
79	4097	4238	4379	4520	4661	4802	4943	5084	5225	5366	141
3080	5507	5648	5789	5930	6071	6212	6353	6494	6635	6776	
81	6917	7058	7199	7340	7481	7622	7763	7904	8045	8185	
82	8326	8467	8608	8749	8890	9031	9172	9313	9454	9594	
83	9735	9876	ō017	ō158	ō299	ō440	ō580	ō721	ō862	Ī003	
84	4891144	1285	1425	1566	1707	1848	1989	2129	2270	2411	
85	2552	2692	2833	2974	3115	3256	3396	3537	3678	3818	
86	3959	4100	4241	4381	4522	4663	4804	4944	5085	5226	
87	5366	5507	5648	5788	5929	6070	6210	6351	6492	6632	
88	6773	6914	7054	7195	7335	7476	7617	7757	7898	8038	
89	8179	8320	8460	8601	8741	8882	9023	9163	9304	9444	
3090	9585	9725	9866	ō006	ō147	ō287	ō428	ō569	ō709	ō850	
91	4900990	1131	1271	1412	1552	1693	1833	1973	2114	2254	
92	2395	2535	2676	2816	2957	3097	3238	3378	3518	3659	
93	3799	3940	4080	4220	4361	4501	4642	4782	4922	5063	
94	5203	5343	5484	5624	5765	5905	6045	6186	6326	6466	
95	6607	6747	6887	7027	7168	7308	7448	7589	7729	7869	
96	8010	8150	8290	8430	8571	8711	8851	8991	9132	9272	
97	9412	9552	9693	9833	9973	ō113	ō253	ō394	ō534	ō674	
98	4910814	0954	1094	1235	1375	1515	1655	1795	1935	2076	
99	2216	2356	2496	2636	2776	2916	3057	3197	3337	3477	
N.	0	1	2	3	4	5	6	7	8	9	D

Pro.

142
1 14
2 28
3 43
4 57
5 71
6 85
7 99
8 114
9 128

141
1 14
2 28
3 42
4 56
5 71
6 85
7 99
8 113
9 127

140
1 14
2 28
3 42
4 56
5 70
6 84
7 98
8 112
9 126

Pts.

N.	0	1	2	3	4	5	6	7	8	9	D	Pro.
3100	4913617	3757	3897	4037	4177	4317	4457	4597	4738	4878		
01	5018	5158	5298	5438	5578	5718	5858	5998	6138	6278		140
02	6418	6558	6698	6838	6978	7118	7258	7398	7538	7678	140	1 14
03	7818	7958	8098	8238	8378	8517	8657	8797	8937	9077		2 28
04	9217	9357	9497	9637	9777	9917	0̄057	0̄196	0̄336	0̄476		3 42
												4 56
05	4920616	0756	0896	1036	1175	1315	1455	1595	1735	1875		5 70
06	2015	2154	2294	2434	2574	2714	2853	2993	3133	3273		6 84
07	3413	3552	3692	3832	3972	4111	4251	4391	4531	4670		7 98
08	4810	4950	5090	5229	5369	5509	5648	5788	5928	6068		8 112
09	6207	6347	6487	6626	6766	6906	7045	7185	7325	7464		9 126
3110	7604	7744	7883	8023	8162	8302	8442	8581	8721	8861		
11	9000	9140	9279	9419	9558	9698	9838	9977	0̄117	0̄256		
12	4930396	0535	0675	0815	0954	1094	1233	1373	1512	1652		
13	1791	1931	2070	2210	2349	2489	2628	2768	2907	3047		
14	3186	3326	3465	3604	3744	3883	4023	4162	4302	4441		
15	4581	4720	4859	4999	5138	5278	5417	5556	5696	5835		
16	5974	6114	6253	6393	6532	6671	6811	6950	7089	7229		
17	7368	7507	7647	7786	7925	8065	8204	8343	8483	8622		
18	8761	8900	9040	9179	9318	9457	9597	9736	9875	0̄015		
19	4940154	0293	0432	0571	0711	0850	0989	1128	1268	1407		
3120	1546	1685	1824	1964	2103	2242	2381	2520	2659	2799		
21	2938	3077	3216	3355	3494	3633	3773	3912	4051	4190		
22	4329	4468	4607	4746	4885	5024	5164	5303	5442	5581		139
23	5720	5859	5998	6137	6276	6415	6554	6693	6832	6971		1 14
24	7110	7249	7388	7527	7666	7805	7944	8083	8222	8361	139	2 28
												3 42
25	8500	8639	8778	8917	9056	9195	9334	9473	9612	9751		4 56
26	9890	0̄029	0̄168	0̄307	0̄445	0̄584	0̄723	0̄862	1̄001	1̄140		5 70
27	4951279	1418	1557	1695	1834	1973	2112	2251	2390	2529		6 83
28	2667	2806	2945	3084	3223	3362	3500	3639	3778	3917		7 97
29	4056	4194	4333	4472	4611	4750	4888	5027	5166	5305		8 111
												9 125
3130	5443	5582	5721	5860	5998	6137	6276	6415	6553	6692		
31	6831	6969	7108	7247	7385	7524	7663	7802	7940	8079		
32	8218	8356	8495	8634	8772	8911	9049	9188	9327	9465		
33	9604	9743	9881	0̄020	0̄158	0̄297	0̄436	0̄574	0̄713	0̄851		
34	4960990	1128	1267	1406	1544	1683	1821	1960	2098	2237		
35	2375	2514	2653	2791	2930	3068	3207	3345	3484	3622		
36	3761	3899	4038	4176	4314	4453	4591	4730	4868	5007		
37	5145	5284	5422	5560	5699	5837	5976	6114	6253	6391		
38	6529	6668	6806	6945	7083	7221	7360	7498	7636	7775		
39	7913	8052	8190	8328	8467	8605	8743	8882	9020	9158		
3140	9296	9435	9573	9711	9850	9988	0̄126	0̄265	0̄403	0̄541		
41	4970679	0818	0956	1094	1232	1371	1509	1647	1785	1924		
42	2062	2200	2338	2476	2615	2753	2891	3029	3167	3306		138
43	3444	3582	3720	3858	3996	4135	4273	4411	4549	4687		1 14
44	4825	4964	5102	5240	5378	5516	5654	5792	5930	6068		2 28
												3 41
45	6206	6345	6483	6621	6759	6897	7035	7173	7311	7449	138	4 55
46	7587	7725	7863	8001	8139	8277	8415	8553	8691	8829		5 69
47	8967	9105	9243	9381	9519	9657	9795	9933	0̄071	0̄209		6 83
48	4980347	0485	0623	0761	0899	1037	1175	1313	1451	1589		7 97
49	1727	1865	2002	2140	2278	2416	2554	2692	2830	2968		8 110
												9 124
N.	0	1	2	3	4	5	6	7	8	9	D	Pts.

N	0	1	2	3	4	5	6	7	8	9	D	Pro.
3150	4983106	3243	3381	3519	3657	3795	3933	4071	4208	4346		
51	4484	4622	4760	4897	5035	5173	5311	5449	5587	5724		138
52	5862	6000	6138	6275	6413	6551	6689	6826	6964	7102		1 14
53	7240	7377	7515	7653	7791	7928	8066	8204	8341	8479		2 28
54	8617	8755	8892	9030	9168	9305	9443	9581	9718	9856		3 41
55	9994	0131	0269	0407	0544	0682	0819	0957	1095	1232		4 55
56	4991370	1508	1645	1783	1920	2058	2196	2333	2471	2608		5 69
57	2746	2883	3021	3158	3296	3434	3571	3709	3846	3984		6 83
58	4121	4259	4396	4534	4671	4809	4946	5084	5221	5359		7 97
59	5496	5634	5771	5909	6046	6184	6321	6459	6596	6733		8 110
3160	6871	7008	7146	7283	7421	7558	7695	7833	7970	8108		9 124
61	8245	8382	8520	8657	8794	8932	9069	9207	9344	9481		
62	9619	9756	9893	0031	0168	0305	0443	0580	0717	0855		
63	5000992	1129	1267	1404	1541	1678	1816	1953	2090	2227		
64	2365	2502	2639	2777	2914	3051	3188	3325	3463	3600		
65	3737	3874	4012	4149	4286	4423	4560	4698	4835	4972		
66	5109	5246	5383	5521	5658	5795	5932	6069	6206	6344		
67	6481	6618	6755	6892	7029	7166	7303	7440	7578	7715		
68	7852	7989	8126	8263	8400	8537	8674	8811	8948	9085	137	
69	9222	9359	9496	9634	9771	9908	0045	0182	0319	0456		
3170	5010593	0730	0867	1004	1141	1278	1415	1552	1688	1825		
71	1962	2099	2236	2373	2510	2647	2784	2921	3058	3195		137
72	3332	3469	3606	3743	3879	4016	4153	4290	4427	4564		1 14
73	4701	4838	4974	5111	5248	5385	5522	5659	5796	5932		2 27
74	6069	6206	6343	6480	6617	6753	6890	7027	7164	7301		3 41
75	7437	7574	7711	7848	7984	8121	8258	8395	8531	8668		4 55
76	8805	8942	9078	9215	9352	9489	9625	9762	9899	0035		5 69
77	5020172	0309	0446	0582	0719	0856	0992	1129	1266	1402		6 82
78	1539	1676	1812	1949	2086	2222	2359	2495	2632	2769		7 96
79	2905	3042	3178	3315	3452	3588	3725	3861	3998	4135		8 110
3180	4271	4408	4544	4681	4817	4954	5091	5227	5364	5500		9 123
81	5637	5773	5910	6046	6183	6319	6456	6592	6729	6865		
82	7002	7138	7275	7411	7548	7684	7821	7957	8093	8230		
83	8366	8503	8639	8776	8912	9049	9185	9321	9458	9594		
84	9731	9867	0003	0140	0276	0413	0549	0685	0822	0958		
85	5031094	1231	1367	1503	1640	1776	1912	2049	2185	2321		
86	2458	2594	2730	2867	3003	3139	3276	3412	3548	3684		
87	3821	3957	4093	4229	4366	4502	4638	4774	4911	5047		
88	5183	5319	5456	5592	5728	5864	6000	6137	6273	6409		
89	6545	6681	6818	6954	7090	7226	7362	7498	7635	7771		
3190	7907	8043	8179	8315	8451	8587	8724	8860	8996	9132		
91	9268	9404	9540	9676	9812	9948	0085	0221	0357	0493		
92	5040629	0765	0901	1037	1173	1309	1445	1581	1717	1853		136
93	1989	2125	2261	2397	2533	2669	2805	2941	3077	3213	136	1 14
94	3349	3485	3621	3757	3893	4029	4165	4301	4437	4573		2 27
95	4709	4845	4980	5116	5252	5388	5524	5660	5796	5932		3 41
96	6068	6204	6339	6475	6611	6747	6883	7019	7155	7291		4 54
97	7426	7562	7698	7834	7970	8106	8241	8377	8513	8649		5 68
98	8785	8920	9056	9192	9328	9464	9599	9735	9871	0007		6 82
99	5050142	0278	0414	0550	0685	0821	0957	1093	1228	1364		7 95
												8 109
												9 122
N.	0	1	2	3	4	5	6	7	8	9	D	Pts.

E

N.	0	1	2	3	4	5	6	7	8	9	D	Pro.
3200	5051500	1635	1771	1907	2043	2178	2314	2450	2585	2721		
01	2857	2992	3128	3264	3399	3535	3671	3806	3942	4078		136
02	4213	4349	4485	4620	4756	4891	5027	5163	5298	5434		1 14
03	5569	5705	5841	5976	6112	6247	6383	6518	6654	6790		2 27
04	6925	7061	7196	7332	7467	7603	7738	7874	8009	8145		3 41 / 4 54
05	8280	8416	8551	8687	8822	8958	9093	9229	9364	9500		5 68
06	9635	9771	9906	0042	0177	0312	0448	0583	0719	0854		6 82
07	5060990	1125	1260	1396	1531	1667	1802	1937	2073	2208		7 95
08	2344	2479	2614	2750	2885	3020	3156	3291	3426	3562		8 109
09	3697	3833	3968	4103	4238	4374	4509	4644	4780	4915		9 122
3210	5050	5186	5321	5456	5591	5727	5862	5997	6133	6268		
11	6403	6538	6674	6809	6944	7079	7214	7350	7485	7620		
12	7755	7891	8026	8161	8296	8431	8567	8702	8837	8972		
13	9107	9242	9378	9513	9648	9783	9918	0053	0188	0324		
14	5070459	0594	0729	0864	0999	1134	1269	1405	1540	1675		
15	1810	1945	2080	2215	2350	2485	2620	2755	2890	3025		
16	3160	3295	3430	3566	3701	3836	3971	4106	4241	4376	135	
17	4511	4646	4781	4916	5051	5186	5321	5456	5590	5725		
18	5860	5995	6130	6265	6400	6535	6670	6805	6940	7075		
19	7210	7345	7480	7614	7749	7884	8019	8154	8289	8424		
3220	8559	8694	8828	8963	9098	9233	9366	9503	9638	9772		
21	9907	0042	0177	0312	0447	0581	0716	0851	0986	1121		135
22	5081255	1390	1525	1660	1794	1929	2064	2199	2334	2468		1 14
23	2603	2738	2873	3007	3142	3277	3411	3546	3681	3816		2 27
24	3950	4085	4220	4354	4489	4624	4758	4893	5028	5163		3 41 / 4 54
25	5297	5432	5567	5701	5836	5970	6105	6240	6374	6509		5 68
26	6644	6778	6913	7047	7182	7317	7451	7586	7720	7855		6 81
27	7990	8124	8259	8393	8528	8663	8797	8932	9066	9201		7 95
28	9335	9470	9604	9739	9873	0008	0142	0277	0411	0546		8 108
29	5090680	0815	0949	1084	1218	1353	1487	1622	1756	1891		9 122
3230	2025	2160	2294	2429	2563	2697	2832	2966	3101	3235		
31	3370	3504	3638	3773	3907	4042	4176	4310	4445	4579		
32	4714	4848	4982	5117	5251	5385	5520	5654	5788	5923		
33	6057	6191	6326	6460	6594	6729	6863	6997	7132	7266		
34	7400	7534	7669	7803	7937	8072	8206	8340	8474	8609		
35	8743	8877	9011	9146	9280	9414	9548	9682	9817	9951		
36	5100085	0219	0354	0488	0622	0756	0890	1024	1159	1293		
37	1427	1561	1695	1829	1964	2098	2232	2366	2500	2634		
38	2768	2903	3037	3171	3305	3439	3573	3707	3841	3975		
39	4109	4244	4378	4512	4646	4780	4914	5048	5182	5316		
3240	5450	5584	5718	5852	5986	6120	6254	6388	6522	6656	134	
41	6790	6924	7058	7192	7326	7460	7594	7728	7862	7996		
42	8130	8264	8398	8532	8666	8800	8934	9068	9202	9336		
43	9469	9603	9737	9871	0005	0139	0273	0407	0541	0675		134
44	5110808	0942	1076	1210	1344	1478	1612	1745	1879	2013		1 13 / 2 27
45	2147	2281	2415	2548	2682	2816	2950	3084	3218	3351		3 40
46	3485	3619	3753	3887	4020	4154	4288	4422	4555	4689		4 54
47	4823	4957	5090	5224	5358	5492	5625	5759	5893	6026		5 67 / 6 80
48	6160	6294	6428	6561	6695	6829	6962	7096	7230	7363		7 94
49	7497	7631	7764	7898	8032	8165	8299	8433	8566	8700		8 107 / 9 121
N.	0	1	2	3	4	5	6	7	8	9	D	Pts.

N.	0	1	2	3	4	5	6	7	8	9	D	Pro.
3250	5118834	8967	9101	9234	9368	9502	9635	9769	9903	ō036		
51	5120170	0303	0437	0570	0704	0838	0971	1105	1238	1372		134
52	1505	1639	1772	1906	2040	2173	2307	2440	2574	2707		1 13
53	2841	2974	3108	3241	3375	3508	3642	3775	3909	4042		2 27
54	4175	4309	4442	4576	4709	4843	4976	5110	5243	5377		3 40
55	5510	5643	5777	5910	6044	6177	6310	6444	6577	6711		4 54
56	6844	6977	7111	7244	7377	7511	7644	7778	7911	8044		5 67
57	8178	8311	8444	8578	8711	8844	8978	9111	9244	9377		6 80
58	9511	9644	9777	9911	ō044	ō177	ō311	ō444	ō577	ō710		7 94
59	5130844	0977	1110	1243	1377	1510	1643	1776	1910	2043		8 107
												9 121
3260	2176	2309	2442	2576	2709	2842	2975	3108	3242	3375	133	
61	3508	3641	3774	3908	4041	4174	4307	4440	4573	4706		
62	4840	4973	5106	5239	5372	5505	5638	5771	5905	6038		
63	6171	6304	6437	6570	6703	6836	6969	7102	7235	7368		
64	7502	7635	7768	7901	8034	8167	8300	8433	8566	8699		
65	8832	8965	9098	9231	9364	9497	9630	9763	9896	ō029		
66	5140162	0295	0428	0561	0694	0827	0960	1093	1225	1358		
67	1491	1624	1757	1890	2023	2156	2289	2422	2555	2688		
68	2820	2953	3086	3219	3352	3485	3618	3751	3883	4016		
69	4149	4282	4415	4548	4681	4813	4946	5079	5212	5345		
3270	5478	5610	5743	5876	6009	6142	6274	6407	6540	6673		
71	6805	6938	7071	7204	7336	7469	7602	7735	7867	8000		
72	8133	8266	8398	8531	8664	8797	8929	9062	9195	9327		133
73	9460	9593	9725	9858	9991	ō123	ō256	ō389	ō521	ō654		1 13
74	5150787	0919	1052	1185	1317	1450	1583	1715	1848	1980		2 27
75	2113	2246	2378	2511	2643	2776	2909	3041	3174	3306		3 40
76	3439	3571	3704	3837	3969	4102	4234	4367	4499	4632		4 53
77	4764	4897	5029	5162	5294	5427	5560	5692	5825	5957		5 67
78	6089	6222	6354	6487	6619	6752	6884	7017	7149	7282		6 80
79	7414	7547	7679	7811	7944	8076	8209	8341	8474	8606		7 93
3280	8738	8871	9003	9136	9268	9400	9533	9665	9798	9930		8 106
81	5160062	0195	0327	0459	0592	0724	0856	0989	1121	1253		9 120
82	1386	1518	1650	1783	1915	2047	2180	2312	2444	2577		
83	2709	2841	2973	3106	3238	3370	3502	3635	3767	3899		
84	4031	4164	4296	4428	4560	4693	4825	4957	5089	5222		
85	5354	5486	5618	5750	5883	6015	6147	6279	6411	6543		
86	6676	6808	6940	7072	7204	7336	7469	7601	7733	7865		
87	7997	8129	8261	8393	8526	8658	8790	8922	9054	9186		
88	9318	9450	9582	9714	9846	9978	ō111	ō243	ō375	ō507		
89	5170639	0771	0903	1035	1167	1299	1431	1563	1695	1827	132	
3290	1959	2091	2223	2355	2487	2619	2751	2883	3015	3147		
91	3279	3411	3543	3675	3807	3939	4071	4202	4334	4466		
92	4598	4730	4862	4994	5126	5258	5390	5522	5654	5785		132
93	5917	6049	6181	6313	6445	6577	6709	6840	6972	7104		1 13
94	7236	7368	7500	7631	7763	7895	8027	8159	8291	8422		2 26
95	8554	8686	8818	8950	9081	9213	9345	9477	9608	9740		3 40
96	9872	ō004	ō136	ō267	ō399	ō531	ō663	ō794	ō926	Ī058		4 53
97	5181189	1321	1453	1585	1716	1848	1980	2111	2243	2375		5 66
98	2507	2638	2770	2902	3033	3165	3297	3428	3560	3692		6 79
99	3823	3955	4086	4218	4350	4481	4613	4745	4876	5008		7 92
												8 106
												9 119
N.	0	1	2	3	4	5	6	7	8	9	D	Pts.

N.	0	1	2	3	4	5	6	7	8	9	D	Pro.
3300	5185139	5271	5403	5534	5666	5797	5929	6061	6192	6324		
01	6455	6587	6718	6850	6981	7113	7245	7376	7508	7639		132
02	7771	7902	8034	8165	8297	8428	8560	8691	8823	8954		1 13
03	9086	9217	9349	9480	9612	9743	9875	ō006	ō137	ō269		2 26
04	5190400	0532	0663	0795	0926	1058	1189	1320	1452	1583		3 40
05	1715	1846	1977	2109	2240	2372	2503	2634	2766	2897		4 53
06	3028	3160	3291	3423	3554	3685	3817	3948	4079	4211		5 66
07	4342	4474	4605	4736	4867	4999	5130	5261	5392	5524		6 79
08	5655	5786	5918	6049	6180	6311	6443	6574	6705	6836		7 92
09	6968	7099	7230	7361	7493	7624	7755	7886	8018	8149		8 106
3310	8280	8411	8542	8674	8805	8936	9067	9198	9329	9461		9 119
11	9592	9723	9854	9985	ō116	ō248	ō379	ō510	ō641	ō772		
12	5200903	1034	1166	1297	1428	1559	1690	1821	1952	2083		
13	2214	2345	2477	2608	2739	2870	3001	3132	3263	3394		
14	3525	3656	3787	3918	4049	4180	4311	4442	4573	4704	131	
15	4835	4966	5097	5228	5359	5490	5621	5752	5883	6014		
16	6145	6276	6407	6538	6669	6800	6931	7062	7193	7324		
17	7455	7586	7717	7847	7978	8109	8240	8371	8502	8633		
18	8764	8895	9026	9156	9287	9418	9549	9680	9811	9942		
19	5210073	0203	0334	0465	0596	0727	0858	0988	1119	1250		
3320	1381	1512	1642	1773	1904	2035	2166	2296	2427	2558		
21	2689	2820	2950	3081	3212	3343	3473	3604	3735	3866		131
22	3996	4127	4258	4388	4519	4650	4781	4911	5042	5173		1 13
23	5303	5434	5565	5695	5826	5957	6088	6218	6349	6479		2 26
24	6610	6741	6871	7002	7133	7263	7394	7525	7655	7786		3 39
25	7916	8047	8178	8308	8439	8570	8700	8831	8961	9092		4 52
26	9222	9353	9484	9614	9745	9875	ō006	ō136	ō267	ō397		5 66
27	5220528	0659	0789	0920	1050	1181	1311	1442	1572	1703		6 79
28	1833	1964	2094	2225	2355	2486	2616	2747	2877	3007		7 92
29	3138	3268	3399	3529	3660	3790	3921	4051	4181	4312		8 105
3330	4442	4573	4703	4834	4964	5094	5225	5355	5486	5616		9 118
31	5746	5877	6007	6137	6268	6398	6529	6659	6789	6920		
32	7050	7180	7311	7441	7571	7702	7832	7962	8093	8223		
33	8353	8483	8614	8744	8874	9005	9135	9265	9395	9526		
34	9656	9786	9916	ō047	ō177	ō307	ō437	ō568	ō698	ō828		
35	5230958	1089	1219	1349	1479	1609	1740	1870	2000	2130		
36	2260	2391	2521	2651	2781	2911	3041	3172	3302	3432		
37	3562	3692	3822	3952	4083	4213	4343	4473	4603	4733		
38	4863	4993	5124	5254	5384	5514	5644	5774	5904	6034		
39	6164	6294	6424	6554	6684	6814	6945	7075	7205	7335		
3340	7465	7595	7725	7855	7985	8115	8245	8375	8505	8635		
41	8765	8895	9025	9155	9285	9415	9545	9675	9805	9935		
42	5240064	0194	0324	0454	0584	0714	0844	0974	1104	1234		130
43	1364	1494	1624	1753	1883	2013	2143	2273	2403	2533		1 13
44	2663	2793	2922	3052	3182	3312	3442	3572	3702	3831		2 26
45	3961	4091	4221	4351	4481	4610	4740	4870	5000	5130		3 39
46	5259	5389	5519	5649	5779	5908	6038	6168	6298	6427		4 52
47	6557	6687	6817	6946	7076	7206	7336	7465	7595	7725	130	5 65
48	7854	7984	8114	8244	8373	8503	8633	8762	8892	9022		6 78
49	9151	9281	9411	9540	9670	9800	9929	ō059	ō189	ō318		7 91
												8 104
												9 117
N.	0	1	2	3	4	5	6	7	8	9	D	Pts.

N.	0	1	2	3	4	5	6	7	8	9	D
3350	5250448	0578	0707	0837	0967	1096	1226	1355	1485	1615	
51	1744	1874	2003	2133	2263	2392	2522	2651	2781	2911	
52	3040	3170	3299	3429	3558	3688	3817	3947	4076	4206	
53	4336	4465	4595	4724	4854	4983	5113	5242	5372	5501	
54	5631	5760	5890	6019	6148	6278	6407	6537	6666	6796	
55	6925	7055	7184	7314	7443	7572	7702	7831	7961	8090	
56	8220	8349	8478	8608	8737	8867	8996	9125	9255	9384	
57	9513	9643	9772	9902	ō031	ō160	ō290	ō419	ō548	ō678	
58	5260807	0936	1066	1195	1324	1454	1583	1712	1841	1971	
59	2100	2229	2359	2488	2617	2746	2876	3005	3134	3264	
3360	3393	3522	3651	3781	3910	4039	4168	4297	4427	4556	
61	4685	4814	4944	5073	5202	5331	5460	5590	5719	5848	
62	5977	6106	6235	6365	6494	6623	6752	6881	7010	7140	
63	7269	7398	7527	7656	7785	7914	8043	8173	8302	8431	
64	8560	8689	8818	8947	9076	9205	9334	9463	9593	9722	
65	9851	9980	ō109	ō238	ō367	ō496	ō625	ō754	ō883	Ī012	
66	5271141	1270	1399	1528	1657	1786	1915	2044	2173	2302	129
67	2431	2560	2689	2818	2947	3076	3205	3334	3463	3592	
68	3721	3850	3979	4108	4237	4366	4494	4623	4752	4881	
69	5010	5139	5268	5397	5526	5655	5783	5912	6041	6170	
3370	6299	6428	6557	6686	6814	6943	7072	7201	7330	7459	
71	7588	7716	7845	7974	8103	8232	8360	8489	8618	8747	
72	8876	9004	9133	9262	9391	9520	9648	9777	9906	ō035	
73	5280163	0292	0421	0550	0678	0807	0936	1065	1193	1322	
74	1451	1579	1708	1837	1966	2094	2223	2352	2480	2609	
75	2738	2866	2995	3124	3252	3381	3510	3638	3767	3896	
76	4024	4153	4282	4410	4539	4668	4796	4925	5053	5182	
77	5311	5439	5568	5696	5825	5954	6082	6211	6339	6468	
78	6596	6725	6854	6982	7111	7239	7368	7496	7625	7753	
79	7882	8010	8139	8267	8396	8525	8653	8782	8910	9039	
3380	9167	9295	9424	9552	9681	9809	9938	ō066	ō195	ō323	
81	5290452	0580	0709	0837	0965	1094	1222	1351	1479	1608	
82	1736	1864	1993	2121	2250	2378	2506	2635	2763	2892	
83	3020	3148	3277	3405	3533	3662	3790	3919	4047	4175	
84	4304	4432	4560	4689	4817	4945	5074	5202	5330	5458	
85	5587	5715	5843	5972	6100	6228	6356	6485	6613	6741	
86	6870	6998	7126	7254	7383	7511	7639	7767	7896	8024	
87	8152	8280	8408	8537	8665	8793	8921	9049	9178	9306	
88	9434	9562	9690	9819	9947	ō075	ō203	ō331	ō459	ō588	
89	5300716	0844	0972	1100	1228	1356	1485	1613	1741	1869	
3390	1997	2125	2253	2381	2509	2637	2766	2894	3022	3150	
91	3278	3406	3534	3662	3790	3918	4046	4174	4302	4430	128
92	4558	4686	4814	4943	5071	5199	5327	5455	5583	5711	
93	5839	5967	6095	6223	6351	6479	6607	6734	6862	6990	
94	7118	7246	7374	7502	7630	7758	7886	8014	8142	8270	
95	8398	8526	8654	8782	8909	9037	9165	9293	9421	9549	
96	9677	9805	9933	ō060	ō188	ō316	ō444	ō572	ō700	ō828	
97	5310955	1083	1211	1339	1467	1595	1722	1850	1978	2106	
98	2234	2362	2489	2617	2745	2873	3001	3128	3256	3384	
99	3512	3639	3767	3895	4023	4150	4278	4406	4534	4661	
N.	0	1	2	3	4	5	6	7	8	9	D

Pro.

	130		129		128
1	13	1	13	1	13
2	26	2	26	2	26
3	39	3	39	3	38
4	52	4	52	4	51
5	65	5	65	5	64
6	78	6	77	6	77
7	91	7	90	7	90
8	104	8	103	8	102
9	117	9	116	9	115

Pts.

N.	0	1	2	3	4	5	6	7	8	9	D	Pro.
3400	5314789	4917	5045	5172	5300	5428	5556	5683	5811	5939		
01	6066	6194	6322	6449	6577	6705	6832	6960	7088	7215		128
02	7343	7471	7598	7726	7854	7981	8109	8237	8364	8492		1 13
03	8619	8747	8875	9002	9130	9258	9385	9513	9640	9768		2 26
04	9896	0̄023	0̄151	0̄278	0̄406	0̄533	0̄661	0̄789	0̄916	1̄044		3 38
05	5321171	1299	1426	1554	1681	1809	1936	2064	2191	2319		4 51
06	2446	2574	2701	2829	2956	3084	3211	3339	3466	3594		5 64
07	3721	3849	3976	4104	4231	4359	4486	4614	4741	4868		6 77
08	4996	5123	5251	5378	5506	5633	5760	5888	6015	6143		7 90
09	6270	6397	6525	6652	6780	6907	7034	7162	7289	7416		8 102
3410	7544	7671	7799	7926	8053	8181	8308	8435	8563	8690		9 115
11	8817	8945	9072	9199	9326	9454	9581	9708	9836	9963		
12	5330090	0218	0345	0472	0599	0727	0854	0981	1108	1236		
13	1363	1490	1617	1745	1872	1999	2126	2254	2381	2508		
14	2635	2762	2890	3017	3144	3271	3398	3526	3653	3780		
15	3907	4034	4161	4289	4416	4543	4670	4797	4924	5051		
16	5179	5306	5433	5560	5687	5814	5941	6068	6196	6323		
17	6450	6577	6704	6831	6958	7085	7212	7339	7466	7594		
18	7721	7848	7975	8102	8229	8356	8483	8610	8737	8864		
19	8991	9118	9245	9372	9499	9626	9753	9880	0̄007	0̄134	127	
3420	5340261	0388	0515	0642	0769	0896	1023	1150	1277	1404		
21	1531	1658	1785	1912	2039	2165	2292	2419	2546	2673		127
22	2800	2927	3054	3181	3308	3435	3561	3688	3815	3942		1 13
23	4069	4196	4323	4450	4576	4703	4830	4957	5084	5211		2 25
24	5338	5464	5591	5718	5845	5972	6099	6225	6352	6479		3 38
25	6606	6733	6859	6986	7113	7240	7366	7493	7620	7747		4 51
26	7874	8000	8127	8254	8381	8507	8634	8761	8888	9014		5 64
27	9141	9268	9394	9521	9648	9775	9901	0̄028	0̄155	0̄281		6 76
28	5350408	0535	0662	0788	0915	1042	1168	1295	1422	1548		7 89
29	1675	1802	1928	2055	2181	2308	2435	2561	2688	2815		8 102
3430	2941	3068	3194	3321	3448	3574	3701	3827	3954	4081		9 114
31	4207	4334	4460	4587	4713	4840	4967	5093	5220	5346		
32	5473	5599	5726	5852	5979	6105	6232	6359	6485	6612		
33	6738	6865	6991	7118	7244	7371	7497	7623	7750	7876		
34	8003	8129	8256	8382	8509	8635	8762	8888	9015	9141		
35	9267	9394	9520	9647	9773	9900	0̄026	0̄152	0̄279	0̄405		
36	5360532	0658	0784	0911	1037	1163	1290	1416	1543	1669		
37	1795	1922	2048	2174	2301	2427	2553	2680	2806	2932		
38	3059	3185	3311	3438	3564	3690	3817	3943	4069	4195		
39	4322	4448	4574	4701	4827	4953	5079	5206	5332	5458		
3440	5584	5711	5837	5963	6089	6216	6342	6468	6594	6721		
41	6847	6973	7099	7225	7352	7478	7604	7730	7856	7982		
42	8109	8235	8361	8487	8613	8739	8866	8992	9118	9244		126
43	9370	9496	9622	9749	9875	0̄001	0̄127	0̄253	0̄379	0̄505		1 13
44	5370631	0758	0884	1010	1136	1262	1388	1514	1640	1766		2 25
45	1892	2018	2144	2270	2396	2523	2649	2775	2901	3027		3 38
46	3153	3279	3405	3531	3657	3783	3909	4035	4161	4287		4 50
47	4413	4539	4665	4791	4917	5043	5169	5295	5421	5547	126	5 63
48	5673	5799	5924	6050	6176	6302	6428	6554	6680	6806		6 76
49	6932	7058	7184	7310	7436	7561	7687	7813	7939	8065		7 88
												8 101
												9 113
N.	0	1	2	3	4	5	6	7	8	9	D	Pts.

N.	0	1	2	3	4	5	6	7	8	9	D
3450	5378191	8317	8443	8569	8694	8820	8946	9072	9198	9324	
51	9450	9575	9701	9827	9953	0̄079	0̄205	0̄330	0̄456	0̄582	
52	5380708	0834	0959	1085	1211	1337	1463	1588	1714	1840	
53	1966	2092	2217	2343	2469	2595	2720	2846	2972	3098	
54	3223	3349	3475	3601	3726	3852	3978	4103	4229	4355	
55	4481	4606	4732	4858	4983	5109	5235	5360	5486	5612	
56	5737	5863	5989	6114	6240	6366	6491	6617	6743	6868	
57	6994	7119	7245	7371	7496	7622	7747	7873	7999	8124	
58	8250	8375	8501	8627	8752	8878	9003	9129	9255	9380	
59	9506	9631	9757	9882	0̄008	0̄133	0̄259	0̄384	0̄510	0̄635	
3460	5390761	0887	1012	1138	1263	1389	1514	1640	1765	1891	
61	2016	2141	2267	2392	2518	2643	2769	2894	3020	3145	
62	3271	3396	3522	3647	3772	3898	4023	4149	4274	4400	
63	4525	4650	4776	4901	5027	5152	5277	5403	5528	5653	
64	5779	5904	6030	6155	6280	6406	6531	6656	6782	6907	
65	7032	7158	7283	7408	7534	7659	7784	7910	8035	8160	
66	8286	8411	8536	8661	8787	8912	9037	9163	9288	9413	
67	9538	9664	9789	9914	0̄039	0̄165	0̄290	0̄415	0̄540	0̄666	
68	5400791	0916	1041	1167	1292	1417	1542	1667	1793	1918	
69	2043	2168	2293	2419	2544	2669	2794	2919	3044	3170	
3470	3295	3420	3545	3670	3795	3920	4046	4171	4296	4421	
71	4546	4671	4796	4921	5047	5172	5297	5422	5547	5672	
72	5797	5922	6047	6172	6297	6423	6548	6673	6798	6923	
73	7048	7173	7298	7423	7548	7673	7798	7923	8048	8173	125
74	8298	8423	8548	8673	8798	8923	9048	9173	9298	9423	
75	9548	9673	9798	9923	0̄048	0̄173	0̄298	0̄423	0̄548	0̄673	
76	5410798	0923	1048	1172	1297	1422	1547	1672	1797	1922	
77	2047	2172	2297	2422	2546	2671	2796	2921	3046	3171	
78	3296	3421	3546	3670	3795	3920	4045	4170	4295	4419	
79	4544	4669	4794	4919	5044	5168	5293	5418	5543	5668	
3480	5792	5917	6042	6167	6292	6416	6541	6666	6791	6915	
81	7040	7165	7290	7415	7539	7664	7789	7913	8038	8163	
82	8288	8412	8537	8662	8787	8911	9036	9161	9285	9410	
83	9535	9659	9784	9909	0̄033	0̄158	0̄283	0̄407	0̄532	0̄657	
84	5420781	0906	1031	1155	1280	1405	1529	1654	1779	1903	
85	2028	2152	2277	2402	2526	2651	2775	2900	3025	3149	
86	3274	3398	3523	3648	3772	3897	4021	4146	4270	4395	
87	4519	4644	4769	4893	5018	5142	5267	5391	5516	5640	
88	5765	5889	6014	6138	6263	6387	6512	6636	6761	6885	
89	7010	7134	7259	7383	7508	7632	7756	7881	8005	8130	
3490	8254	8379	8503	8628	8752	8876	9001	9125	9250	9374	
91	9498	9623	9747	9872	9996	0̄120	0̄245	0̄369	0̄494	0̄618	
92	5430742	0867	0991	1115	1240	1364	1488	1613	1737	1862	
93	1986	2110	2235	2359	2483	2607	2732	2856	2980	3105	
94	3229	3353	3478	3602	3726	3850	3975	4099	4223	4348	
95	4472	4596	4720	4845	4969	5093	5217	5342	5466	5590	
96	5714	5838	5963	6087	6211	6335	6460	6584	6708	6832	
97	6956	7081	7205	7329	7453	7577	7701	7826	7950	8074	
98	8198	8322	8446	8571	8695	8819	8943	9067	9191	9315	
99	9439	9564	9688	9812	9936	0̄060	0̄184	0̄308	0̄432	0̄556	
N.	0	1	2	3	4	5	6	7	8	9	D

Pro. Pts.

126
1 | 13
2 | 25
3 | 38
4 | 50
5 | 63
6 | 76
7 | 88
8 | 101
9 | 113

125
1 | 13
2 | 25
3 | 38
4 | 50
5 | 63
6 | 75
7 | 88
8 | 100
9 | 113

124
1 | 12
2 | 25
3 | 37
4 | 50
5 | 62
6 | 74
7 | 87
8 | 99
9 | 112

N.	0	1	2	3	4	5	6	7	8	9	D	Pro.
3500	5440680	0805	0929	1053	1177	13(J)1	1425	1549	1673	1797		
01	1921	2045	2169	2293	2417	2541	2665	2789	2913	3037		124
02	3161	3285	3409	3533	3657	3781	3905	4029	4153	4277		1 12
03	4401	4525	4649	4773	4897	5021	5145	5269	5393	5517	124	2 25
04	5641	5765	5889	6013	6137	6261	6385	6508	6632	6756		3 37
												4 50
05	6880	7004	7128	7252	7376	7500	7624	7747	7871	7995		5 62
06	8119	8243	8367	8491	8615	8738	8862	8986	9110	9234		6 74
07	9358	9481	9605	9729	9853	9977	ō101	ō224	ō348	ō472		7 87
08	5450596	0720	0843	0967	1091	1215	1339	1462	1586	1710		8 99
09	1834	1957	2081	2205	2329	2452	2576	2700	2824	2947		9 112
3510	3071	3195	3319	3442	3566	3690	3813	3937	4061	4185		
11	4308	4432	4556	4679	4803	4927	5050	5174	5298	5421		
12	5545	5669	5792	5916	6040	6163	6287	6411	6534	6658		
13	6781	6905	7029	7152	7276	7400	7523	7647	7770	7894		
14	8018	8141	8265	8388	8512	8635	8759	8883	9006	9130		
15	9253	9377	9500	9624	9747	9871	9995	ō118	ō242	ō365		
16	5460489	0612	0736	0859	0983	1106	1230	1353	1477	1600		
17	1724	1847	1971	2094	2218	2341	2465	2588	2711	2835		
18	2958	3082	3205	3329	3452	3576	3699	3822	3946	4069		
19	4193	4316	4439	4563	4686	4810	4933	5056	5180	5303		
3520	5427	5550	5673	5797	5920	6043	6167	6290	6414	6537		
21	6660	6784	6907	7030	7154	7277	7400	7524	7647	7770		123
22	7894	8017	8140	8263	8387	8510	8633	8757	8880	9003		1 12
23	9126	9250	9373	9496	9620	9743	9866	9989	ō113	ō236		2 25
24	5470359	0482	0605	0729	0852	0975	1098	1222	1345	1468		3 37
												4 49
25	1591	1714	1838	1961	2084	2207	2330	2454	2577	2700		5 62
26	2823	2946	3069	3193	3316	3439	3562	3685	3808	3931		6 74
27	4055	4178	4301	4424	4547	4670	4793	4916	5040	5163		7 86
28	5286	5409	5532	5655	5778	5901	6024	6147	6270	6394		8 98
29	6517	6640	6763	6886	7009	7132	7255	7378	7501	7624		9 111
3530	7747	7870	7993	8116	8239	8362	8485	8608	8731	8854	123	
31	8977	9100	9223	9346	9469	9592	9715	9838	9961	ō084		
32	5480207	0330	0453	0576	0699	0822	0945	1068	1191	1313		
33	1436	1559	1682	1805	1928	2051	2174	2297	2420	2543		
34	2665	2788	2911	3034	3157	3280	3403	3526	3648	3771		
35	3894	4017	4140	4263	4386	4508	4631	4754	4877	5000		
36	5123	5245	5368	5491	5614	5737	5859	5982	6105	6228		
37	6351	6473	6596	6719	6842	6964	7087	7210	7333	7456		
38	7578	7701	7824	7947	8069	8192	8315	8437	8560	8683		
39	8806	8928	9051	9174	9296	9419	9542	9665	9787	9910		
3540	5490033	0155	0278	0401	0523	0646	0769	0891	1014	1137		
41	1259	1382	1505	1627	1750	1872	1995	2118	2240	2363		
42	2486	2608	2731	2853	2976	3099	3221	3344	3466	3589		122
43	3712	3834	3957	4079	4202	4324	4447	4569	4692	4815		1 12
44	4937	5060	5182	5305	5427	5550	5672	5795	5917	6040		2 24
												3 37
45	6162	6285	6407	6530	6652	6775	6897	7020	7142	7265		4 49
46	7387	7510	7632	7755	7877	8000	8122	8245	8367	8489		5 61
47	8612	8734	8857	8979	9102	9224	9346	9469	9591	9714		6 73
48	9836	9959	ō081	ō203	ō326	ō448	0570	ō693	ō815	ō938		7 85
49	5501060	1182	1305	1427	1549	1672	1794	1917	2039	2161		8 98
												9 110
N.	0	1	2	3	4	5	6	7	8	9	D	Pts.

N.	0	1	2	3	4	5	6	7	8	9	D	Pro.
3550	5502284	2406	2528	2651	2773	2895	3017	3140	3262	3384		
51	3507	3629	3751	3874	3996	4118	4240	4363	4485	4607		122
52	4730	4852	4974	5096	5219	5341	5463	5585	5708	5830		1 12
53	5952	6074	6197	6319	6441	6563	6685	6808	6930	7052		2 24
54	7174	7296	7419	7541	7663	7785	7907	8030	8152	8274		3 37
												4 49
55	8396	8518	8640	8763	8885	9007	9129	9251	9373	9495		5 61
56	9618	9740	9862	9984	ō106	ō228	ō350	ō472	ō594	ō717		6 73
57	5510839	0961	1083	1205	1327	1449	1571	1693	1815	1937		7 85
58	2059	2181	2304	2426	2548	2670	2792	2914	3036	3158		8 98
59	3280	3402	3524	3646	3768	3890	4012	4134	4256	4378	122	9 110
3560	4500	4622	4744	4866	4988	5110	5232	5354	5476	5598		
61	5720	5842	5964	6086	6208	6329	6451	6573	6695	6817		
62	6939	7061	7183	7305	7427	7549	7671	7793	7914	8036		
63	8158	8280	8402	8524	8646	8768	8890	9011	9133	9255		
64	9377	9499	9621	9743	9864	9986	ō108	ō230	ō352	ō474		
65	5520595	0717	0839	0961	1083	1204	1326	1448	1570	1692		
66	1813	1935	2057	2179	2301	2422	2544	2666	2788	2909		
67	3031	3153	3275	3396	3518	3640	3762	3883	4005	4127		
68	4248	4370	4492	4614	4735	4857	4979	5100	5222	5344		
69	5465	5587	5709	5831	5952	6074	6196	6317	6439	6561		
3570	6682	6804	6925	7047	7169	7290	7412	7534	7655	7777		
71	7899	8020	8142	8263	8385	8507	8628	8750	8871	8993		
72	9115	9236	9358	9479	9601	9722	9844	9965	ō087	ō209		
73	5530330	0452	0573	0695	0816	0938	1059	1181	1302	1424		
74	1545	1667	1789	1910	2032	2153	2275	2396	2517	2639		
75	2760	2882	3003	3125	3246	3368	3489	3611	3732	3854		
76	3975	4097	4218	4339	4461	4582	4704	4825	4947	5068		
77	5189	5311	5432	5554	5675	5796	5918	6039	6161	6282		
78	6403	6525	6646	6767	6889	7010	7132	7253	7374	7496		
79	7617	7738	7860	7981	8102	8224	8345	8466	8588	8709		
3580	8830	8952	9073	9194	9315	9437	9558	9679	9801	9922		
81	5540043	0164	0286	0407	0528	0650	0771	0892	1013	1135		
82	1256	1377	1498	1620	1741	1862	1983	2104	2226	2347		
83	2468	2589	2710	2832	2953	3074	3195	3316	3438	3559		
84	3680	3801	3922	4044	4165	4286	4407	4528	4649	4770		
85	4892	5013	5134	5255	5376	5497	5618	5740	5861	5982		
86	6103	6224	6345	6466	6587	6708	6829	6951	7072	7193		
87	7314	7435	7556	7677	7798	7919	8040	8161	8282	8403		
88	8524	8645	8766	8887	9008	9130	9251	9372	9493	9614		
89	9735	9856	9977	ō098	ō219	ō340	ō461	ō582	ō703	ō824	121	
3590	5550944	1065	1186	1307	1428	1549	1670	1791	1912	2033		
91	2154	2275	2396	2517	2638	2759	2880	3001	3121	3242		121
92	3363	3484	3605	3726	3847	3968	4089	4210	4330	4451		1 12
93	4572	4693	4814	4935	5056	5176	5297	5418	5539	5660		2 24
94	5781	5902	6022	6143	6264	6385	6506	6627	6747	6868		3 36
												4 48
95	6989	7110	7231	7351	7472	7593	7714	7835	7955	8076		5 61
96	8197	8318	8438	8559	8680	8801	8921	9042	9163	9284		6 73
97	9404	9525	9646	9767	9887	ō008	ō129	ō249	ō370	ō491		7 85
98	5560612	0732	0853	0974	1094	1215	1336	1456	1577	1698		8 97
99	1818	1939	2060	2180	2301	2422	2542	2663	2784	2904		9 109
N.	0	1	2	3	4	5	6	7	8	9	D	Pts.

N.	0	1	2	3	4	5	6	7	8	9	D	Pro.
3600	5563025	3146	3266	3387	3508	3628	3749	3869	3990	4111		121
01	4231	4352	4472	4593	4714	4834	4955	5075	5196	5317		
02	5437	5558	5678	5799	5919	6040	6160	6281	6402	6522		1 12
03	6643	6763	6884	7004	7125	7245	7366	7486	7607	7727		2. 24
04	7848	7968	8089	8209	8330	8450	8571	8691	8812	8932		3 36
												4 48
05	9053	9173	9294	9414	9535	9655	9775	9896	0016	0137		5 61
06	5570257	0378	0498	0619	0739	0859	0980	1100	1221	1341		6 73
07	1461	1582	1702	1823	1943	2063	2184	2304	2425	2545		7 85
08	2665	2786	2906	3026	3147	3267	3387	3508	3628	3748		8 97
09	3869	3989	4109	4230	4350	4470	4591	4711	4831	4952		9 109
3610	5072	5192	5313	5433	5553	5673	5794	5914	6034	6155		
11	6275	6395	6515	6636	6756	6876	6996	7117	7237	7357		
12	7477	7598	7718	7838	7958	8079	8199	8319	8439	8559		
13	8680	8800	8920	9040	9160	9281	9401	9521	9641	9761		
14	9881	0002	0122	0242	0362	0482	0602	0723	0843	0963		
15	5581083	1203	1323	1443	1564	1684	1804	1924	2044	2164		
16	2284	2404	2524	2645	2765	2885	3005	3125	3245	3365		
17	3485	3605	3725	3845	3965	4085	4205	4325	4446	4566	120	
18	4686	4806	4926	5046	5166	5286	5406	5526	5646	5766		
19	5886	6006	6126	6246	6366	6486	6606	6726	6846	6966		
3620	7086	7206	7326	7446	7566	7686	7805	7925	8045	8165		
21	8285	8405	8525	8645	8765	8885	9005	9125	9245	9365		
22	9484	9604	9724	9844	9964	0084	0204	0324	0444	0563		
23	5590683	0803	0923	1043	1163	1283	1403	1522	1642	1762		
24	1882	2002	2122	2241	2361	2481	2601	2721	2840	2960		
25	3080	3200	3320	3440	3559	3679	3799	3919	4038	4158		
26	4278	4398	4518	4637	4757	4877	4997	5116	5236	5356		
27	5476	5595	5715	5835	5954	6074	6194	6314	6433	6553		
28	6673	6792	6912	7032	7152	7271	7391	7511	7630	7750		
29	7870	7989	8109	8229	8348	8468	8588	8707	8827	8947		
3630	9066	9186	9306	9425	9545	9664	9784	9904	0023	0143		
31	5600262	0382	0502	0621	0741	0860	0980	1100	1219	1339		
32	1458	1578	1698	1817	1937	2056	2176	2295	2415	2534		
33	2654	2774	2893	3013	3132	3252	3371	3491	3610	3730		
34	3849	3969	4088	4208	4327	4447	4566	4686	4805	4925		
35	5044	5164	5283	5403	5522	5641	5761	5880	6000	6119		
36	6239	6358	6478	6597	6716	6836	6955	7075	7194	7314		
37	7433	7552	7672	7791	7911	8030	8149	8269	8388	8508		
38	8627	8746	8866	8985	9104	9224	9343	9463	9582	9701		
39	9821	9940	0059	0179	0298	0417	0537	0656	0775	0895		
3640	5611014	1133	1252	1372	1491	1610	1730	1849	1968	2088		
41	2207	2326	2445	2565	2684	2803	2922	3042	3161	3280		
42	3399	3519	3638	3757	3876	3996	4115	4234	4353	4472		120
43	4592	4711	4830	4949	5069	5188	5307	5426	5545	5665		1 12
44	5784	5903	6022	6141	6260	6380	6499	6618	6737	6856		2 24
												3 36
45	6975	7094	7214	7333	7452	7571	7690	7809	7928	8048		4 48
46	8167	8286	8405	8524	8643	8762	8881	9000	9119	9239		5 60
47	9358	9477	9596	9715	9834	9953	0072	0191	0310	0429		6 72
48	5620548	0667	0786	0905	1024	1144	1263	1382	1501	1620		7 84
49	1739	1858	1977	2096	2215	2334	2453	2572	2691	2810	119	8 96 / 9 108
N.	0	1	2	3	4	5	6	7	8	9	D	Pts.

N.	0	1	2	3	4	5	6	7	8	9	D	Pro.
3650	5622929	3048	3167	3286	3405	3524	3642	3761	3880	3999		119
51	4118	4237	4356	4475	4594	4713	4832	4951	5070	5189		
52	5308	5427	5546	5664	5783	5902	6021	6140	6259	6378		1 12
53	6497	6616	6734	6853	6972	7091	7210	7329	7448	7567		2 24
54	7685	7804	7923	8042	8161	8280	8398	8517	8636	8755		3 36
												4 48
55	8874	8993	9111	9230	9349	9468	9587	9705	9824	9943		5 60
56	5630062	0181	0299	0418	0537	0656	0775	0893	1012	1131		6 71
57	1250	1368	1487	1606	1725	1843	1962	2081	2200	2318		7 83
58	2437	2556	2674	2793	2912	3031	3149	3268	3387	3505		8 95
59	3624	3743	3861	3980	4099	4218	4336	4455	4574	4692		9 107
3660	4811	4930	5048	5167	5285	5404	5523	5641	5760	5879		
61	5997	6116	6235	6353	6472	6590	6709	6828	6946	7065		
62	7183	7302	7421	7539	7658	7776	7895	8013	8132	8251		
63	8369	8488	8606	8725	8843	8962	9081	9199	9318	9436		
64	9555	9673	9792	9910	0̄029	0̄147	0̄266	0̄384	0̄503	0̄621		
65	5640740	0858	0977	1095	1214	1332	1451	1569	1688	1806		
66	1925	2043	2162	2280	2398	2517	2635	2754	2872	2991		
67	3109	3228	3346	3464	3583	3701	3820	3938	4056	4175		
68	4293	4412	4530	4648	4767	4885	5004	5122	5240	5359		
69	5477	5595	5714	5832	5951	6069	6187	6306	6424	6542		
3670	6661	6779	6897	7016	7134	7252	7371	7489	7607	7726		
71	7844	7962	8080	8199	8317	8435	8554	8672	8790	8908		
72	9027	9145	9263	9382	9500	9618	9736	9855	9973	0̄091		
73	5650209	0̄328	0446	0564	0682	0800	0919	1037	1155	1273		
74	1392	1510	1628	1746	1864	1983	2101	2219	2337	2455		
75	2573	2692	2810	2928	3046	3164	3282	3401	3519	3637		
76	3755	3873	3991	4109	4228	4346	4464	4582	4700	4818		
77	4936	5054	5173	5291	5409	5527	5645	5763	5881	5999		
78	6117	6235	6353	6471	6590	6708	6826	6944	7062	7180	118	
79	7298	7416	7534	7652	7770	7888	8006	8124	8242	8360		
3680	8478	8596	8714	8832	8950	9068	9186	9304	9422	9540		
81	9658	9776	9894	0̄012	0̄130	0̄248	0̄366	0̄484	0̄602	0̄720		
82	5660838	0956	1074	1192	1310	1428	1545	1663	1781	1899		
83	2017	2135	2253	2371	2489	2607	2725	2843	2960	3078		
84	3196	3314	3432	3550	3668	3786	3903	4021	4139	4257		
85	4375	4493	4611	4728	4846	4964	5082	5200	5318	5435		
86	5553	5671	5789	5907	6025	6142	6260	6378	6496	6614		
87	6731	6849	6967	7085	7203	7320	7438	7556	7674	7791		
88	7909	8027	8145	8262	8380	8498	8616	8733	8851	8969		
89	9087	9204	9322	9440	9557	9675	9793	9911	0̄028	0̄146		
3690	5670264	0381	0499	0617	0734	0852	0970	1087	1205	1323		
91	1440	1558	1676	1793	1911	2029	2146	2264	2382	2499		
92	2617	2735	2852	2970	3087	3205	3323	3440	3558	3675		118
93	3793	3911	4028	4146	4263	4381	4499	4616	4734	4851		1 12
94	4969	5086	5204	5322	5439	5557	5674	5792	5909	6027		2 24
												3 35
95	6144	6262	6379	6497	6615	6732	6850	6967	7085	7202		4 47
96	7320	7437	7555	7672	7790	7907	8025	8142	8260	8377		5 59
97	8495	8612	8729	8847	8964	9082	9199	9317	9434	9552		6 71
98	9669	9787	9904	0̄021	0̄139	0̄256	0̄374	0̄491	0̄608	0̄726		7 83
99	5680843	0961	1078	1196	1313	1430	1548	1665	1782	1900		8 94
												9 106
N.	0	1	2	3	4	5	6	7	8	9	D	Pts.

N.	0	1	2	3	4	5	6	7	8	9	D	Pro.
3700	5682017	2135	2252	2369	2487	2604	2721	2839	2956	3074		
01	3191	3308	3426	3543	3660	3778	3895	4012	4130	4247		118
02	4364	4481	4599	4716	4833	4951	5068	5185	5303	5420		1 12
03	5537	5654	5772	5889	6006	6123	6241	6358	6475	6593		2 24
04	6710	6827	6944	7062	7179	7296	7413	7530	7648	7765		3 35
												4 47
05	7882	7999	8117	8234	8351	8468	8585	8703	8820	8937		5 59
06	9054	9171	9289	9406	9523	9640	9757	9874	9992	0̄109		6 71
07	5690226	0343	0460	0577	0694	0812	0929	1046	1163	1280		7 83
08	1397	1514	1631	1749	1866	1983	2100	2217	2334	2451		8 94
09	2568	2685	2803	2920	3037	3154	3271	3388	3505	3622		9 106
3710	3739	3856	3973	4090	4207	4324	4441	4558	4675	4793	117	
11	4910	5027	5144	5261	5378	5495	5612	5729	5846	5963		
12	6080	6197	6314	6431	6548	6665	6782	6899	7016	7133		
13	7249	7366	7483	7600	7717	7834	7951	8068	8185	8302		
14	8419	8536	8653	8770	8887	9004	9121	9237	9354	9471		
15	9588	9705	9822	9939	0̄056	0̄173	0̄290	0̄406	0̄523	0̄640		
16	5700757	0874	0991	1108	1225	1341	1458	1575	1692	1809		
17	1926	2042	2159	2276	2393	2510	2627	2743	2860	2977		
18	3094	3211	3327	3444	3561	3678	3795	3911	4028	4145		
19	4262	4379	4495	4612	4729	4846	4962	5079	5196	5313		
3720	5429	5546	5663	5780	5896	6013	6130	6247	6363	6480		
21	6597	6713	6830	6947	7064	7180	7297	7414	7530	7647		
22	7764	7880	7997	8114	8230	8347	8464	8580	8697	8814		
23	8930	9047	9164	9280	9397	9514	9630	9747	9863	9980		
24	5710097	0213	0330	0447	0563	0680	0796	0913	1030	1146		
25	1263	1379	1496	1613	1729	1846	1962	2079	2195	2312		
26	2429	2545	2662	2778	2895	3011	3128	3244	3361	3477		
27	3594	3710	3827	3943	4060	4177	4293	4410	4526	4643		
28	4759	4876	4992	5109	5225	5341	5458	5574	5691	5807		
29	5924	6040	6157	6273	6390	6506	6623	6739	6855	6972		
3730	7088	7205	7321	7438	7554	7670	7787	7903	8020	8136		
31	8252	8369	8485	8602	8718	8834	8951	9067	9184	9300		
32	9416	9533	9649	9765	9882	9998	0̄115	0̄231	0̄347	0̄464		
33	5720580	0696	0813	0929	1045	1162	1278	1394	1511	1627		
34	1743	1859	1976	2092	2208	2325	2441	2557	2674	2790		
35	2906	3022	3139	3255	3371	3487	3604	3720	3836	3952		
36	4069	4185	4301	4417	4534	4650	4766	4882	4999	5115		
37	5231	5347	5463	5580	5696	5812	5928	6044	6161	6277		
38	6393	6509	6625	6742	6858	6974	7090	7206	7322	7438		
39	7555	7671	7787	7903	8019	8135	8252	8368	8484	8600		
3740	8716	8832	8948	9064	9180	9297	9413	9529	9645	9761		
41	9877	9993	0̄109	0̄225	0̄341	0̄457	0̄574	0̄690	0̄806	0̄922		
42	5731038	1154	1270	1386	1502	1618	1734	1850	1966	2082		117
43	2198	2314	2430	2546	2662	2778	2894	3010	3126	3242		1 12
44	3358	3474	3590	3706	3822	3938	4054	4170	4286	4402	116	2 23
												3 35
45	4518	4634	4750	4866	4982	5098	5214	5330	5446	5562		4 47
46	5678	5794	5910	6026	6141	6257	6373	6489	6605	6721		5 59
47	6837	6953	7069	7185	7301	7416	7532	7648	7764	7880		6 70
48	7996	8112	8228	8343	8459	8575	8691	8807	8923	9039		7 82
49	9154	9270	9386	9502	9618	9734	9849	9965	0̄081	0̄197		8 94
												9 105
N	0	1	2	3	4	5	6	7	8	9	D	Pts.

N.	0	1	2	3	4	5	6	7	8	9	D	Pro.
3750	5740313	0428	0544	0660	0776	0892	1007	1123	1239	1355		
51	1471	1586	1702	1818	1934	2050	2165	2281	2397	2513		116
52	2628	2744	2860	2976	3091	3207	3323	3438	3554	3670		1 12
53	3786	3901	4017	4133	4248	4364	4480	4596	4711	4827		2 23
54	4943	5058	5174	5290	5405	5521	5637	5752	5868	5984		3 35
												4 46
55	6099	6215	6331	6446	6562	6678	6793	6909	7025	7140		5 58
56	7256	7371	7487	7603	7718	7834	7950	8065	8181	8296		6 70
57	8412	8528	8643	8759	8874	8990	9105	9221	9337	9452		7 81
58	9568	9683	9799	9914	0̄030	0̄146	0̄261	0̄377	0̄492	0̄608		8 93
59	5750723	0839	0954	1070	1185	1301	1416	1532	1647	1763		9 104
3760	1878	1994	2109	2225	2340	2456	2571	2687	2802	2918		
61	3033	3149	3264	3380	3495	3611	3726	3842	3957	4072		
62	4188	4303	4419	4534	4650	4765	4881	4996	5111	5227		
63	5342	5458	5573	5688	5804	5919	6035	6150	6265	6381		
64	6496	6612	6727	6842	6958	7073	7188	7304	7419	7534		
65	7650	7765	7881	7996	8111	8227	8342	8457	8573	8688		
66	8803	8918	9034	9149	9264	9380	9495	9610	9726	9841		
67	9956	0̄071	0̄187	0̄302	0̄417	0̄533	0̄648	0̄763	0̄878	0̄994		
68	5761109	1224	1339	1455	1570	1685	1800	1916	2031	2146		
69	2261	2377	2492	2607	2722	2837	2953	3068	3183	3298		
3770	3414	3529	3644	3759	3874	3989	4105	4220	4335	4450		
71	4565	4680	4796	4911	5026	5141	5256	5371	5487	5602		
72	5717	5832	5947	6062	6177	6292	6408	6523	6638	6753		
73	6868	6983	7098	7213	7328	7444	7559	7674	7789	7904		
74	8019	8134	8249	8364	8479	8594	8709	8824	8939	9055		
75	9170	9285	9400	9515	9630	9745	9860	9975	0̄090	0̄205		
76	5770320	0435	0550	0665	0780	0895	1010	1125	1240	1355		
77	1470	1585	1700	1815	1930	2045	2160	2275	2390	2505	115	
78	2620	2734	2849	2964	3079	3194	3309	3424	3539	3654		
79	3769	3884	3999	4114	4229	4343	4458	4573	4688	4803		
3780	4918	5033	5148	5263	5378	5492	5607	5722	5837	5952		
81	6067	6182	6296	6411	6526	6641	6756	6871	6986	7100		
82	7215	7330	7445	7560	7675	7789	7904	8019	8134	8249		
83	8363	8478	8593	8708	8823	8937	9052	9167	9282	9397		
84	9511	9626	9741	9856	9970	0̄085	0̄200	0̄315	0̄429	0̄544		
85	5780659	0774	0888	1003	1118	1233	1347	1462	1577	1691		
86	1806	1921	2036	2150	2265	2380	2494	2609	2724	2838		
87	2953	3068	3182	3297	3412	3526	3641	3756	3870	3985		
88	4100	4214	4329	4444	4558	4673	4788	4902	5017	5131		
89	5246	5361	5475	5590	5705	5819	5934	6048	6163	6278		
3790	6392	6507	6621	6736	6850	6965	7080	7194	7309	7423		
91	7538	7652	7767	7882	7996	8111	8225	8340	8454	8569		
92	8683	8798	8912	9027	9141	9256	9370	9485	9599	9714		115
93	9828	9943	0̄057	0̄172	0̄286	0̄401	0̄515	0̄630	0̄744	0̄859		1 12
94	5790973	1088	1202	1317	1431	1546	1660	1774	1889	2003		2 23
												3 35
95	2118	2232	2347	2461	2576	2690	2804	2919	3033	3148		4 46
96	3262	3376	3491	3605	3720	3834	3948	4063	4177	4292		5 58
97	4406	4520	4635	4749	4863	4978	5092	5207	5321	5435		6 69
98	5550	5664	5778	5893	6007	6121	6236	6350	6464	6579		7 81
99	6693	6807	6922	7036	7150	7264	7379	7493	7607	7722		8 92
												9 104
N.	0	1	2	3	4	5	6	7	8	9	D	Pts.

N.	0	1	2	3	4	5	6	7	8	9	D	Pro.
3800	5797836	7950	8065	8179	8293	8407	8522	8636	8750	8864		
01	8979	9093	9207	9321	9436	9550	9664	9778	9893	ō007		115
02	5800121	0235	0350	0464	0578	0692	0806	0921	1035	1149		1|12
03	1263	1377	1492	1606	1720	1834	1948	2063	2177	2291		2|23
04	2405	2519	2633	2748	2862	2976	3090	3204	3318	3432	114	3|35
												4|46
05	3547	3661	3775	3889	4003	4117	4231	4346	4460	4574		5|58
06	4688	4802	4916	5030	5144	5258	5372	5487	5601	5715		6|69
07	5829	5943	6057	6171	6285	6399	6513	6627	6741	6855		7|81
08	6969	7083	7197	7312	7426	7540	7654	7768	7882	7996		8|92
09	8110	8224	8338	8452	8566	8680	8794	8908	9022	9136		9|104
3810	9250	9364	9478	9592	9706	9820	9934	ō048	ō162	ō276		
11	5810389	0503	0617	0731	0845	0959	1073	1187	1301	1415		
12	1529	1643	1757	1871	1985	2099	2212	2326	2440	2554		
13	2668	2782	2896	3010	3124	3238	3351	3465	3579	3693		
14	3807	3921	4035	4148	4262	4376	4490	4604	4718	4832		
15	4945	5059	5173	5287	5401	5515	5628	5742	5856	5970		
16	6084	6197	6311	6425	6539	6653	6766	6880	6994	7108		
17	7222	7335	7449	7563	7677	7790	7904	8018	8132	8245		
18	8359	8473	8587	8700	8814	8928	9042	9155	9269	9383		
19	9497	9610	9724	9838	9951	ō065	ō179	ō293	ō406	ō520		
3820	5820634	0747	0861	0975	1088	1202	1316	1429	1543	1657		
21	1770	1884	1998	2111	2225	2339	2452	2566	2680	2793		
22	2907	3020	3134	3248	3361	3475	3589	3702	3816	3929		
23	4043	4157	4270	4384	4497	4611	4725	4838	4952	5065		
24	5179	5292	5406	5520	5633	5747	5860	5974	6087	6201		
25	6314	6428	6541	6655	6769	6882	6996	7109	7223	7336		
26	7450	7563	7677	7790	7904	8017	8131	8244	8358	8471		
27	8585	8698	8812	8925	9039	9152	9265	9379	9492	9606		
28	9719	9833	9946	ō060	ō173	ō287	ō400	ō513	ō627	ō740		
29	5830854	0967	1081	1194	1307	1421	1534	1648	1761	1874		
3830	1988	2101	2215	2328	2441	2555	2668	2781	2895	3008		
31	3122	3235	3348	3462	3575	3688	3802	3915	4028	4142		
32	4255	4368	4482	4595	4708	4822	4935	5048	5162	5275		
33	5388	5501	5615	5728	5841	5955	6068	6181	6295	6408		
34	6521	6634	6748	6861	6974	7087	7201	7314	7427	7540		
35	7654	7767	7880	7993	8107	8220	8333	8446	8560	8673		
36	8786	8899	9012	9126	9239	9352	9465	9578	9692	9805		
37	9918	ō031	ō144	ō258	ō371	ō484	ō597	ō710	ō823	ō937		
38	5841050	1163	1276	1389	1502	1615	1729	1842	1955	2068		
39	2181	2294	2407	2520	2634	2747	2860	2973	3086	3199		
3840	3312	3425	3538	3652	3765	3878	3991	4104	4217	4330		
41	4443	4556	4669	4782	4895	5008	5121	5234	5348	5461		114
42	5574	5687	5800	5913	6026	6139	6252	6365	6478	6591	113	1|11
43	6704	6817	6930	7043	7156	7269	7382	7495	7608	7721		2|23
44	7834	7947	8060	8173	8286	8399	8512	8625	8738	8850		3|34
												4|46
45	8963	9076	9189	9302	9415	9528	9641	9754	9867	9980		5|57
46	5850093	0206	0319	0432	0544	0657	0770	0883	0996	1109		6|68
47	1222	1335	1448	1561	1673	1786	1899	2012	2125	2238		7|80
48	2351	2463	2576	2689	2802	2915	3028	3141	3253	3366		8|91
49	3479	3592	3705	3818	3930	4043	4156	4269	4382	4494		9|103
N.	0	1	2	3	4	5	6	7	8	9	D	Pts.

N.	0	1	2	3	4	5	6	7	8	9	D	Pro.
3850	5854607	4720	4833	4946	5058	5171	5284	5397	5510	5622		113
51	5735	5848	5961	6073	6186	6299	6412	6525	6637	6750		1| 11
52	6863	6976	7088	7201	7314	7426	7539	7652	7765	7877		2| 23
53	7990	8103	8216	8328	8441	8554	8666	8779	8892	9004		3| 34
54	9117	9230	9342	9455	9568	9681	9793	9906	ō019	ō131		4| 45
55	5860244	0356	0469	0582	0694	0807	0920	1032	1145	1258		5| 57
56	1370	1483	1596	1708	1821	1933	2046	2159	2271	2384		6| 68
57	2496	2609	2722	2834	2947	3059	3172	3285	3397	3510		7| 79
58	3622	3735	3847	3960	4072	4185	4298	4410	4523	4635		8| 90
59	4748	4860	4973	5085	5198	5310	5423	5535	5648	5761		9|102
3860	5873	5986	6098	6211	6323	6436	6548	6661	6773	6886		
61	6998	7110	7223	7335	7448	7560	7673	7785	7898	8010		
62	8123	8235	8348	8460	8572	8685	8797	8910	9022	9135		
63	9247	9360	9472	9584	9697	9809	9922	ō034	ō146	ō259		
64	5870371	0484	0596	0708	0821	0933	1045	1158	1270	1383		
65	1495	1607	1720	1832	1944	2057	2169	2281	2394	2506		
66	2618	2731	2843	2955	3068	3180	3292	3405	3517	3629		
67	3742	3854	3966	4079	4191	4303	4416	4528	4640	4752		
68	4865	4977	5089	5201	5314	5426	5538	5651	5763	5875		
69	5987	6100	6212	6324	6436	6549	6661	6773	6885	6997		
3870	7110	7222	7334	7446	7559	7671	7783	7895	8007	8120		
71	8232	8344	8456	8568	8680	8793	8905	9017	9129	9241		
72	9353	9466	9578	9690	9802	9914	ō026	ō139	ō251	ō363		
73	5880475	0587	0699	0811	0923	1036	1148	1260	1372	1484		
74	1596	1708	1820	1932	2045	2157	2269	2381	2493	2605		
75	2717	2829	2941	3053	3165	3277	3389	3502	3614	3726	112	
76	3838	3950	4062	4174	4286	4398	4510	4622	4734	4846		
77	4958	5070	5182	5294	5406	5518	5630	5742	5854	5966		
78	6078	6190	6302	6414	6526	6638	6750	6862	6974	7086		
79	7198	7310	7422	7534	7646	7758	7870	7981	8093	8205		
3880	8317	8429	8541	8653	8765	8877	8989	9101	9213	9325		
81	9436	9548	9660	9772	9884	9996	ō108	ō220	ō332	ō443		
82	5890555	0667	0779	0891	1003	1115	1227	1338	1450	1562		
83	1674	1786	1898	2009	2121	2233	2345	2457	2569	2680		
84	2792	2904	3016	3128	3239	3351	3463	3575	3687	3798		
85	3910	4022	4134	4246	4357	4469	4581	4693	4804	4916		
86	5028	5140	5251	5363	5475	5587	5698	5810	5922	6034		
87	6145	6257	6369	6481	6592	6704	6816	6927	7039	7151		
88	7263	7374	7486	7598	7709	7821	7933	8044	8156	8268		
89	8379	8491	8603	8714	8826	8938	9049	9161	9273	9384		
3890	9496	9608	9719	9831	9943	ō054	ō166	ō277	ō389	ō501		
91	5900612	0724	0836	0947	1059	1170	1282	1394	1505	1617		112
92	1728	1840	1951	2063	2175	2286	2398	2509	2621	2732		1| 11
93	2844	2956	3067	3179	3290	3402	3513	3625	3736	3848		2| 22
94	3959	4071	4183	4294	4406	4517	4629	4740	4852	4963		3| 34
95	5075	5186	5298	5409	5521	5632	5744	5855	5967	6078		4| 45
96	6189	6301	6412	6524	6635	6747	6858	6970	7081	7193		5| 56
97	7304	7415	7527	7638	7750	7861	7973	8084	8196	8307		6| 67
98	8418	8530	8641	8753	8864	8975	9087	9198	9310	9421		7| 78
99	9532	9644	9755	9866	9978	ō089	ō201	ō312	ō423	ō535		8| 90
												9|101
N.	0	1	2	3	4	5	6	7	8	9	D	Pts.

N.	0	1	2	3	4	5	6	7	8	9	D	Pro.
3900	5910646	0757	0869	0980	1091	1203	1314	1426	1537	1648		
01	1760	1871	1982	2093	2205	2316	2427	2539	2650	2761		112
02	2873	2984	3095	3207	3318	3429	3540	3652	3763	3874		1 11
03	3986	4097	4208	4319	4431	4542	4653	4764	4876	4987		2 22
04	5098	5209	5321	5432	5543	5654	5765	5877	5988	6099		3 34
05	6210	6322	6433	6544	6655	6766	6878	6989	7100	7211		4 45
06	7322	7434	7545	7656	7767	7878	7989	8101	8212	8323		5 56
07	8434	8545	8656	8768	8879	8990	9101	9212	9323	9434		6 67
08	9546	9657	9768	9879	9990	0̄101	0̄212	0̄323	0̄434	0̄546		7 78
09	5920657	0768	0879	0990	1101	1212	1323	1434	1545	1656		8 90 / 9 101
3910	1769	1879	1990	2101	2212	2323	2434	2545	2656	2767		
11	2878	2989	3100	3211	3322	3433	3544	3655	3766	3877	111	
12	3988	4099	4210	4321	4433	4544	4655	4766	4876	4987		
13	5098	5209	5320	5431	5542	5653	5764	5875	5986	6097		
14	6208	6319	6430	6541	6652	6763	6874	6985	7096	7207		
15	7318	7429	7540	7650	7761	7872	7983	8094	8205	8316		
16	8427	8538	8649	8760	8870	8981	9092	9203	9314	9425		
17	9536	9647	9757	9868	9979	0̄090	0̄201	0̄312	0̄423	0̄533		
18	5930644	0755	0866	0977	1088	1199	1309	1420	1531	1642		
19	1753	1863	1974	2085	2196	2307	2417	2528	2639	2750		
3920	2861	2971	3082	3193	3304	3415	3525	3636	3747	3858		
21	3968	4079	4190	4301	4411	4522	4633	4744	4854	4965		
22	5076	5187	5297	5408	5519	5630	5740	5851	5962	6072		
23	6183	6294	6404	6515	6626	6737	6847	6958	7069	7179		
24	7290	7401	7511	7622	7733	7843	7954	8065	8175	8286		
25	8397	8507	8618	8729	8839	8950	9060	9171	9282	9392		
26	9503	9614	9724	9835	9945	0̄056	0̄167	0̄277	0̄388	0̄498		
27	5940609	0720	0830	0941	1051	1162	1273	1383	1494	1604		
28	1715	1825	1936	2046	2157	2268	2378	2489	2599	2710		
29	2820	2931	3041	3152	3262	3373	3483	3594	3704	3815		
3930	3926	4036	4147	4257	4368	4478	4588	4699	4809	4920		
31	5030	5141	5251	5362	5472	5583	5693	5804	5914	6025		
32	6135	6246	6356	6466	6577	6687	6798	6908	7019	7129		
33	7239	7350	7460	7571	7681	7792	7902	8012	8123	8233		
34	8344	8454	8564	8675	8785	8895	9006	9116	9227	9337		
35	9447	9558	9668	9778	9889	9999	0̄110	0̄220	0̄330	0̄441		
36	5950551	0661	0772	0882	0992	1103	1213	1323	1434	1544		
37	1654	1764	1875	1985	2095	2206	2316	2426	2537	2647		
38	2757	2867	2978	3088	3198	3308	3419	3529	3639	3750		
39	3860	3970	4080	4191	4301	4411	4521	4632	4742	4852		
3940	4962	5072	5183	5293	5403	5513	5624	5734	5844	5954		111
41	6064	6175	6285	6395	6505	6615	6725	6836	6946	7056		1 11
42	7166	7276	7387	7497	7607	7717	7827	7937	8047	8158		2 22
43	8268	8378	8488	8598	8708	8818	8929	9039	9149	9259		3 33
44	9369	9479	9589	9699	9810	9920	0̄030	0̄140	0̄250	0̄360		4 44
45	5960470	0580	0690	0800	0910	1020	1131	1241	1351	1461		5 56
46	1571	1681	1791	1901	2011	2121	2231	2341	2451	2561	110	6 67
47	2671	2781	2891	3001	3111	3221	3331	3441	3551	3661		7 78
48	3771	3881	3991	4101	4211	4321	4431	4541	4651	4761		8 89
49	4871	4981	5091	5201	5311	5421	5531	5641	5751	5861		9 100
N.	0	1	2	3	4	5	6	7	8	9	D	Pts.

N.	0	1	2	3	4	5	6	7	8	9	D	Pro.
3950	5965971	6081	6191	6301	6411	6521	6631	6741	6850	6960		
51	7070	7180	7290	7400	7510	7620	7730	7840	7950	8059		110
52	8169	8279	8389	8499	8609	8719	8829	8939	9048	9158		1| 11
53	9268	9378	9488	9598	9708	9817	9927	ō037	ō147	ō257		2| 22
54	5970367	0476	0586	0696	0806	0916	1026	1135	1245	1355		3| 33
												4| 44
55	1465	1575	1684	1794	1904	2014	2124	2233	2343	2453		5| 55
56	2563	2673	2782	2892	3002	3112	3221	3331	3441	3551		6| 66
57	3661	3770	3880	3990	4099	4209	4319	4429	4538	4648		7| 77
58	4758	4868	4977	5087	5197	5306	5416	5526	5636	5745		8| 88
59	5855	5965	6074	6184	6294	6403	6513	6623	6733	6842		9| 99
3960	6952	7062	7171	7281	7391	7500	7610	7719	7829	7939		
61	8048	8158	8268	8377	8487	8597	8706	8816	8925	9035		
62	9145	9254	9364	9474	9583	9693	9802	9912	ō022	ō131		
63	5980241	0350	0460	0569	0679	0789	0898	1008	1117	1227		
64	1336	1446	1556	1665	1775	1884	1994	2103	2213	2322		
65	2432	2541	2651	2761	2870	2980	3089	3199	3308	3418		
66	3527	3637	3746	3856	3965	4075	4184	4294	4403	4513		
67	4622	4731	4841	4950	5060	5169	5279	5388	5498	5607		
68	5717	5826	5936	6045	6154	6264	6373	6483	6592	6702		
69	6811	6920	7030	7139	7249	7358	7467	7577	7686	7796		
3970	7905	8014	8124	8233	8343	8452	8561	8671	8780	8890		
71	8999	9108	9218	9327	9436	9546	9655	9764	9874	9983		
72	5990092	0202	0311	0420	0530	0639	0748	0858	0967	1076		
73	1186	1295	1404	1514	1623	1732	1841	1951	2060	2169		
74	2279	2388	2497	2606	2716	2825	2934	3044	3153	3262		
75	3371	3481	3590	3699	3808	3918	4027	4136	4245	4355		
76	4464	4573	4682	4791	4901	5010	5119	5228	5338	5447		
77	5556	5665	5774	5884	5993	6102	6211	6320	6429	6539		
78	6648	6757	6866	6975	7084	7194	7303	7412	7521	7630		
79	7739	7849	7958	8067	8176	8285	8394	8503	8612	8722		
3980	8831	8940	9049	9158	9267	9376	9485	9594	9704	9813		
81	9922	ō031	ō140	ō249	ō358	ō467	ō576	ō685	ō794	ō903		
82	6001013	1122	1231	1340	1449	1558	1667	1776	1885	1994		
83	2103	2212	2321	2430	2539	2648	2757	2866	2975	3084	109	
84	3193	3302	3411	3520	3629	3738	3847	3956	4065	4174		
85	4283	4392	4501	4610	4719	4828	4937	5046	5155	5264		
86	5373	5482	5591	5700	5809	5918	6027	6136	6244	6353		
87	6462	6571	6680	6789	6898	7007	7116	7225	7334	7443		
88	7551	7660	7769	7878	7987	8096	8205	8314	8423	8531		
89	8640	8749	8858	8967	9076	9185	9294	9402	9511	9620		
3990	9729	9838	9947	ō055	ō164	ō273	ō382	ō491	ō600	ō708		
91	6010817	0926	1035	1144	1253	1361	1470	1579	1688	1797		
92	1905	2014	2123	2232	2340	2449	2558	2667	2776	2884		109
93	2993	3102	3211	3319	3428	3537	3646	3754	3863	3972		1| 11
94	4081	4189	4298	4407	4516	4624	4733	4842	4950	5059		2| 22
												3| 33
95	5168	5277	5385	5494	5603	5711	5820	5929	6037	6146		4| 44
96	6255	6363	6472	6581	6690	6798	6907	7016	7124	7233		5| 55
97	7341	7450	7559	7667	7776	7885	7993	8102	8211	8319		6| 65
98	8428	8537	8645	8754	8862	8971	9080	9188	9297	9405		7| 76
99	9514	9623	9731	9840	9948	ō057	ō166	ō274	ō383	ō491		8| 87
												9| 98
N.	0	1	2	3	4	5	6	7	8	9	D	Pts.

N.	0	1	2	3	4	5	6	7	8	9	D	Pro.
4000	6020600	0708	0817	0926	1034	1143	1251	1360	1468	1577		
01	1686	1794	1903	2011	2120	2228	2337	2445	2554	2662		109
02	2771	2879	2988	3096	3205	3313	3422	3530	3639	3747		1 11
03	3856	3964	4073	4181	4290	4398	4507	4615	4724	4832		2 22
04	4941	5049	5158	5266	5375	5483	5591	5700	5808	5917		3 33
												4 44
05	6025	6134	6242	6351	6459	6567	6676	6784	6893	7001		5 55
06	7109	7218	7326	7435	7543	7651	7760	7868	7977	8085		6 65
07	8193	8302	8410	8519	8627	8735	8844	8952	9060	9169		7 76
08	9277	9385	9494	9602	9711	9819	9927	ō036	ō144	ō252		8 87
09	6030361	0469	0577	0686	9794	0902	1010	1119	1227	1335		9 98
4010	1444	1552	1660	1769	1877	1985	2093	2202	2310	2418		
11	2527	2635	2743	2851	2960	3068	3176	3284	3393	3501		
12	3609	3717	3826	3934	4042	4150	4259	4367	4475	4583		
13	4692	4800	4908	5016	5124	5233	5341	5449	5557	5665		
14	5774	5882	5990	6098	6206	6315	6423	6531	6639	6747		
15	6855	6964	7072	7180	7288	7396	7504	7613	7721	7829		
16	7937	8045	8153	8261	837ʋ	8478	8586	8694	8802	8910		
17	9018	9126	9235	9343	9451	9559	9667	9775	9883	9991		
18	6040099	0207	0315	0424	0532	0640	0748	0856	0964	1072		
19	1180	1288	1396	1504	1612	1720	1828	1936	2044	2152		
4020	2261	2369	2477	2585	2693	2801	2909	3017	3125	3233	108	
21	3341	3449	3557	3665	3773	3881	3989	4097	4205	4313		
22	4421	4529	4637	4745	4853	4961	5068	5176	5284	5392		
23	5500	5608	5716	5824	5932	6040	6148	6256	6364	6472		
24	6580	6688	6796	6903	7011	7119	7227	7335	7443	7551		
25	7659	7767	7875	7983	8090	8198	8306	8414	8522	8630		
26	8738	8846	8953	9061	9169	9277	9385	9493	9601	9708		
27	9816	9924	ō032	ō140	ō248	ō355	ō463	ō571	ō679	ō787		
28	6050895	1002	1110	1218	1326	1434	1541	1649	1757	1865		
29	1973	2080	2188	2296	2404	2512	2619	2727	2835	2943		
4030	3050	3158	3266	3374	3482	3589	3697	3805	3912	4020		
31	4128	4236	4343	4451	4559	4667	4774	4882	4990	5098		
32	5205	5313	5421	5528	5636	5744	5851	5959	6067	6175		
33	6282	6390	6498	6605	6713	6821	6928	7036	7144	7251		
34	7359	7467	7574	7682	7790	7897	8005	8112	8220	8328		
35	8435	8543	8651	8758	8866	8974	9081	9189	9296	9404		
36	9512	9619	9727	9834	9942	ō050	ō157	ō265	ō372	ō480		
37	6060587	0695	0803	0910	1018	1125	1233	1340	1448	1556		
38	1663	1771	1878	1986	2093	2201	2308	2416	2523	2631		
39	2739	2846	2954	3061	3169	3276	3384	3491	3599	3706		
4040	3814	3921	4029	4136	4244	4351	4459	4566	4674	4781		
41	4889	4996	5103	5211	5318	5426	5533	5641	5748	5856		
42	5963	6071	6178	6285	6393	6500	6608	6715	6823	6930		108
43	7037	7145	7252	7360	7467	7574	7682	7789	7897	8004		1 11
44	8111	8219	8326	8434	8541	8648	8756	8863	8971	9078		2 22
												3 32
45	9185	9293	9400	9507	9615	9722	9829	9937	ō044	ō151		4 43
46	6070259	0366	0473	0581	0688	0795	0903	1010	1117	1225		5 54
47	1332	1439	1547	1654	1761	1869	1976	2083	2190	2298		6 65
48	2405	2512	2620	2727	2834	2941	3049	3156	3263	3371		7 76
49	3478	3585	3692	3800	3907	4014	4121	4229	4336	4443		8 86
												9 97
N.	0	1	2	3	4	5	6	7	8	9	D	Pts.

N.	0	1	2	3	4	5	6	7	8	9	D	Pro.
4050	6074550	4657	4765	4872	4979	5086	5194	5301	5408	5515		
51	5622	5730	5837	5944	6051	6158	6266	6373	6480	6587		107
52	6694	6802	6909	7016	7123	7230	7337	7445	7552	7659		1 11
53	7766	7873	7980	8087	8195	8302	8409	8516	8623	8730		2 21
54	8837	8945	9052	9159	9266	9373	9480	9587	9694	9801		3 32
												4 43
55	9909	0̄016	0̄123	0̄230	0̄337	0̄444	0̄551	0̄658	0̄765	0̄872		5 54
56	6080979	1087	1194	1301	1408	1515	1622	1729	1836	1943		6 64
57	2050	2157	2264	2371	2478	2585	2692	2799	2906	3013		7 75
58	3120	3227	3334	3441	3548	3656	3763	3870	3977	4084	107	8 86
59	4191	4298	4404	4511	4618	4725	4832	4939	5046	5153		9 96
4060	5260	5367	5474	5581	5688	5795	5902	6009	6116	6223		
61	6330	6437	6544	6651	6758	6865	6972	7078	7185	7292		
62	7399	7506	7613	7720	7827	7934	8041	8148	8254	8361		
63	8468	8575	8682	8789	8896	9003	9110	9216	9323	9430		
64	9537	9644	9751	9858	9964	0̄071	0̄178	0̄285	0̄392	0̄499		
65	6090605	0712	0819	0926	1033	1140	1246	1353	1460	1567		
66	1674	1781	1887	1994	2101	2208	2315	2421	2528	2635		
67	2742	2849	2955	3062	3169	3276	3382	3489	3596	3703		
68	3809	3916	4023	4130	4236	4343	4450	4557	4663	4770		
69	4877	4984	5090	5197	5304	5411	5517	5624	5731	5837		
4070	5944	6051	6157	6264	6371	6478	6584	6691	6798	6904		
71	7011	7118	7224	7331	7438	7544	7651	7758	7864	7971		
72	8078	8184	8291	8398	8504	8611	8718	8824	8931	9037		
73	9144	9251	9357	9464	9571	9677	9784	9890	9997	0̄104		
74	6100210	0317	0423	0530	0637	0743	0850	0956	1063	1170		
75	1276	1383	1489	1596	1702	1809	1916	2022	2129	2235		
76	2342	2448	2555	2661	2768	2874	2981	3088	3194	3301		
77	3407	3514	3620	3727	3833	3940	4046	4153	4259	4366		
78	4472	4579	4685	4792	4898	5005	5111	5218	5324	5431		
79	5537	5644	5750	5856	5963	6069	6176	6282	6389	6495		
4080	6602	6708	6815	6921	7027	7134	7240	7347	7453	7560		
81	7666	7772	7879	7985	8092	8198	8304	8411	8517	8624		
82	8730	8836	8943	9049	9156	9262	9368	9475	9581	9687		
83	9794	9900	0̄007	0̄113	0̄219	0̄326	0̄432	0̄538	0̄645	0̄751		
84	6110857	0964	1070	1176	1283	1389	1495	1602	1708	1814		
85	1921	2027	2133	2240	2346	2452	2558	2665	2771	2877		
86	2984	3090	3196	3302	3409	3515	3621	3728	3834	3940		
87	4046	4153	4259	4365	4471	4578	4684	4790	4896	5003		
88	5109	5215	5321	5428	5534	5640	5746	5852	5959	6065		
89	6171	6277	6384	6490	6596	6702	6808	6915	7021	7127		
4090	7233	7339	7445	7552	7658	7764	7870	7976	8082	8189		
91	8295	8401	8507	8613	8719	8826	8932	9038	9144	9250		
92	9356	9462	9569	9675	9781	9887	9993	0̄099	0̄205	0̄311		106
93	6120417	0524	0630	0736	0842	0948	1054	1160	1266	1372		1 11
94	1478	1584	1691	1797	1903	2009	2115	2221	2327	2433		2 21
												3 32
95	2539	2645	2751	2857	2963	3069	3175	3281	3387	3493		4 42
96	3599	3706	3812	3918	4024	4130	4236	4342	4448	4554	106	5 53
97	4660	4766	4872	4978	5084	5190	5296	5402	5508	5614		6 64
98	5720	5826	5931	6037	6143	6249	6355	6461	6567	6673		7 74
99	6779	6885	6991	7097	7203	7309	7415	7521	7627	7733		8 85
												9 95
N.	0	1	2	3	4	5	6	7	8	9	D	Pts.

N.	0	1	2	3	4	5	6	7	8	9	D	Pro.
4100	6127839	7944	8050	8156	8262	8368	8474	8580	8686	8792		
01	8898	9004	9109	9215	9321	9427	9533	9639	9745	9851		106
02	9957	$\bar{0}$062	$\bar{0}$168	$\bar{0}$274	$\bar{0}$380	$\bar{0}$486	$\bar{0}$592	$\bar{0}$698	$\bar{0}$803	$\bar{0}$909		1|11
03	6131015	1121	1227	1333	1439	1544	1650	1756	1862	1968		2|21
04	2074	2179	2285	2391	2497	2603	2708	2814	2920	3026		3|32 4|42
05	3132	3237	3343	3449	3555	3661	3766	3872	3978	4084		5|53
06	4189	4295	4401	4507	4613	4718	4824	4930	5036	5141		6|64
07	5247	5353	5459	5564	5670	5776	5881	5987	6093	6199		7|74
08	6304	6410	6516	6621	6727	6833	6939	7044	7150	7256		8|85
09	7361	7467	7573	7678	7784	7890	7996	8101	8207	8313		9|95
4110	8418	8524	8630	8735	8841	8947	9052	9158	9263	9369		
11	9475	9580	9686	9792	9897	$\bar{0}$003	$\bar{0}$109	$\bar{0}$214	$\bar{0}$320	$\bar{0}$425		
12	6140531	0637	0742	0848	0954	1059	1165	1270	1376	1482		
13	1587	1693	1798	1904	2009	2115	2221	2326	2432	2537		
14	2643	2748	2854	2960	3065	3171	3276	3382	3487	3593		
15	3698	3804	3909	4015	4121	4226	4332	4437	4543	4648		
16	4754	4859	4965	5070	5176	5281	5387	5492	5598	5703		
17	5809	5914	6020	6125	6231	6336	6442	6547	6652	6758		
18	6863	6969	7074	7180	7285	7391	7496	7602	7707	7812		
19	7918	8023	8129	8234	8340	8445	8550	8656	8761	8867		
4120	8972	9078	9183	9288	9394	9499	9605	9710	9815	9921		
21	6150026	0132	0237	0342	0448	0553	0658	0764	0869	0975		
22	1080	1185	1291	1396	1501	1607	1712	1817	1923	2028		
23	2133	2239	2344	2449	2555	2660	2765	2871	2976	3081		
24	3187	3292	3397	3502	3608	3713	3818	3924	4029	4134		
25	4240	4345	4450	4555	4661	4766	4871	4976	5082	5187		
26	5292	5397	5503	5608	5713	5818	5924	6029	6134	6239		
27	6345	6450	6555	6660	6766	6871	6976	7081	7186	7292		
28	7397	7502	7607	7712	7818	7923	8028	8133	8238	8344		
29	8449	8554	8659	8764	8870	8975	9080	9185	9290	9395		
4130	9501	9606	9711	9816	9921	$\bar{0}$026	$\bar{0}$131	$\bar{0}$237	$\bar{0}$342	$\bar{0}$447		
31	6160552	0657	0762	0867	0972	1078	1183	1288	1393	1498		
32	1603	1708	1813	1918	2024	2129	2234	2339	2444	2549		
33	2654	2759	2864	2969	3074	3179	3284	3390	3495	3600		
34	3705	3810	3915	4020	4125	4230	4335	4440	4545	4650		
35	4755	4860	4965	5070	5175	5280	5385	5490	5595	5700	105	
36	5805	5910	6015	6120	6225	6330	6435	6540	6645	6750		
37	6855	6960	7065	7170	7275	7380	7485	7590	7695	7800		
38	7905	8010	8115	8220	8325	8430	8535	8639	8744	8849		
39	8954	9059	9164	9269	9374	9479	9584	9689	9794	9899		
4140	6170003	0108	0213	0318	0423	0528	0633	0738	0843	0947		
41	1052	1157	1262	1367	1472	1577	1682	1786	1891	1996		
42	2101	2206	2311	2415	2520	2625	2730	2835	2940	3045		
43	3149	3254	3359	3464	3569	3673	3778	3883	3988	4093		105
44	4197	4302	4407	4512	4617	4721	4826	4931	5036	5141		1|11
45	5245	5350	5455	5560	5664	5769	5874	5979	6083	6188		2|21
46	6293	6398	6502	6607	6712	6817	6921	7026	7131	7236		3|32 4|42
47	7340	7445	7550	7655	7759	7864	7969	8073	8178	8283		5|53
48	8387	8492	8597	8702	8806	8911	9016	9120	9225	9330		6|63
49	9434	9539	9644	9748	9853	9958	$\bar{0}$062	$\bar{0}$167	$\bar{0}$272	$\bar{0}$376		7|74 8|84 9|95
N.	0	1	2	3	4	5	6	7	8	9	D	Pts.

N.	0	1	2	3	4	5	6	7	8	9	D	Pro.
4150	6180481	0586	0690	0795	0900	1004	1109	1213	1318	1423		
51	1527	1632	1737	1841	1946	2050	2155	2260	2364	2469		104
52	2573	2678	2783	2887	2992	3096	3201	3306	3410	3515		1| 10
53	3619	3724	3828	3933	4038	4142	4247	4351	4456	4560		2| 21
54	4665	4769	4874	4979	5083	5188	5292	5397	5501	5606		3| 31
												4| 42
55	5710	5815	5919	6024	6128	6233	6337	6442	6546	6651		5| 52
56	6755	6860	6964	7069	7173	7278	7382	7487	7591	7696		6| 62
57	7800	7905	8009	8114	8218	8323	8427	8531	8636	8740		7| 73
58	8845	8949	9054	9158	9263	9367	9471	9576	9680	9785		8| 83
59	9889	9994	0̄098	0̄202	0̄307	0̄411	0̄516	0̄620	0̄725	0̄829		9| 94
4160	6190933	1038	1142	1246	1351	1455	1560	1664	1768	1873		
61	1977	2082	2186	2290	2395	2499	2603	2708	2812	2916		
62	3021	3125	3229	3334	3438	3542	3647	3751	3855	3960		
63	4064	4168	4273	4377	4481	4586	4690	4794	4899	5003		
64	5107	5212	5316	5420	5524	5629	5733	5837	5942	6046		
65	6150	6254	6359	6463	6567	6671	6776	6880	6984	7088		
66	7193	7297	7401	7505	7610	7714	7818	7922	8027	8131		
67	8235	8339	8443	8548	8652	8756	8860	8964	9069	9173		
68	9277	9381	9485	9590	9694	9798	9902	0̄006	0̄111	0̄215		
69	6200319	0423	0527	0631	0736	0840	0944	1048	1152	1256		
4170	1361	1465	1569	1673	1777	1881	1985	2090	2194	2298		
71	2402	2506	2610	2714	2818	2922	3027	3131	3235	3339		
72	3443	3547	3651	3755	3859	3963	4068	4172	4276	4380		
73	4484	4588	4692	4796	4900	5004	5108	5212	5316	5420	104	
74	5524	5628	5733	5837	5941	6045	6149	6253	6357	6461		
75	6565	6669	6773	6877	6981	7085	7189	7293	7397	7501		
76	7605	7709	7813	7917	8021	8125	8229	8333	8437	8541		
77	8645	8749	8853	8957	9061	9165	9269	9373	9476	9580		
78	9684	9788	9892	9996	0̄100	0̄204	0̄308	0̄412	0̄516	0̄620		
79	6210724	0828	0932	1035	1139	1243	1347	1451	1555	1659		
4180	1763	1867	1971	2075	2178	2282	2386	2490	2594	2698		
81	2802	2906	3009	3113	3217	3321	3425	3529	3633	3736		
82	3840	3944	4048	4152	4256	4359	4463	4567	4671	4775		
83	4879	4982	5086	5190	5294	5398	5502	5605	5709	5813		
84	5917	6021	6124	6228	6332	6436	6540	6643	6747	6851		
85	6955	7058	7162	7266	7370	7473	7577	7681	7785	7888		
86	7992	8096	8200	8303	8407	8511	8615	8718	8822	8926		
87	9030	9133	9237	9341	9444	9548	9652	9756	9859	9963		
88	6220067	0170	0274	0378	0482	0585	0689	0793	0896	1000		
89	1104	1207	1311	1415	1518	1622	1726	1829	1933	2037		
4190	2140	2244	2348	2451	2555	2658	2762	2866	2969	3073		
91	3177	3280	3384	3487	3591	3695	3798	3902	4006	4109		103
92	4213	4316	4420	4524	4627	4731	4834	4938	5041	5145		1| 10
93	5249	5352	5456	5559	5663	5766	5870	5974	6077	6181		2| 21
94	6284	6388	6491	6595	6698	6802	6906	7009	7113	7216		3| 31
												4| 41
95	7320	7423	7527	7630	7734	7837	7941	8044	8148	8251		5| 52
96	8355	8458	8562	8665	8769	8872	8976	9079	9183	9286		6| 62
97	9390	9493	9597	9700	9804	9907	0̄011	0̄114	0̄217	0̄321		7| 72
98	6230424	0528	0631	0735	0838	0942	1045	1148	1252	1355		8| 82
99	1459	1562	1666	1769	1872	1976	2079	2183	2286	2389		9| 93
N.	0	1	2	3	4	5	6	7	8	9	D	Pts.

N.	0	1	2	3	4	5	6	7	8	9	D	Pro.
4200	6232493	2596	2700	2803	2906	3010	3113	3217	3320	3423		
01	3527	3630	3734	3837	3940	4044	4147	4250	4354	4457		103
02	4560	4664	4767	4871	4974	5077	5181	5284	5387	5491		1 10
03	5594	5697	5801	5904	6007	6111	6214	6317	6420	6524		2 21
04	6627	6730	6834	6937	7040	7144	7247	7350	7453	7557		3 31
												4 41
05	7660	7763	7867	7970	8073	8176	8280	8383	8486	8589		5 52
06	8693	8796	8899	9002	9106	9209	9312	9415	9519	9622		6 62
07	9725	9828	9932	ō035	ō138	ō241	ō344	ō448	ō551	ō654		7 72
08	6240757	0861	0964	1067	1170	1273	1377	1480	1583	1686		8 82
09	1789	1892	1996	2099	2202	2305	2408	2511	2615	2718		9 93
4210	2821	2924	3027	3130	3234	3337	3440	3543	3646	3749		
11	3852	3956	4059	4162	4265	4368	4471	4574	4677	4781		
12	4884	4987	5090	5193	5296	5399	5502	5605	5708	5812		
13	5915	6018	6121	6224	6327	6430	6533	6636	6739	6842		
14	6945	7048	7151	7254	7358	7461	7564	7667	7770	7873		
15	7976	8079	8182	8285	8388	8491	8594	8697	8800	8903		
16	9006	9109	9212	9315	9418	9521	9624	9727	9830	9933	103	
17	6250036	0139	0242	0345	0448	0551	0654	0757	0860	0963		
18	1066	1169	1272	1375	1478	1581	1683	1786	1889	1992		
19	2095	2198	2301	2404	2507	2610	2713	2816	2919	3022		
4220	3125	3227	3330	3433	3536	3639	3742	3845	3948	4051		
21	4154	4256	4359	4462	4565	4668	4771	4874	4977	5079		
22	5182	5285	5388	5491	5594	5697	5799	5902	6005	6108		
23	6211	6314	6416	6519	6622	6725	6828	6931	7033	7136		
24	7239	7342	7445	7548	7650	7753	7856	7959	8062	8164		
25	8267	8370	8473	8575	8678	8781	8884	8987	9089	9192		
26	9295	9398	9500	9603	9706	9809	9911	ō014	ō117	ō220		
27	6260322	0425	0528	0631	0733	0836	0939	1042	1144	1247		
28	1350	1453	1555	1658	1761	1863	1966	2069	2171	2274		
29	2377	2480	2582	2685	2788	2890	2993	3096	3198	3301		
4230	3404	3506	3609	3712	3814	3917	4020	4122	4225	4328		
31	4430	4533	4636	4738	4841	4943	5046	5149	5251	5354		
32	5457	5559	5662	5764	5867	5970	6072	6175	6277	6380		
33	6483	6585	6688	6790	6893	6996	7098	7201	7303	7406		
34	7509	7611	7714	7816	7919	8021	8124	8226	8329	8432		
35	8534	8637	8739	8842	8944	9047	9149	9252	9354	9457		
36	9560	9662	9765	9867	9970	ō072	ō175	ō277	ō380	ō482		
37	6270585	0687	0790	0892	0995	1097	1200	1302	1405	1507		
38	1610	1712	1814	1917	2019	2122	2224	2327	2429	2532		
39	2634	2737	2839	2942	3044	3146	3249	3351	3454	3556		
4240	3659	3761	3863	3966	4068	4171	4273	4376	4478	4580		
41	4683	4785	4888	4990	5092	5195	5297	5399	5502	5604		102
42	5707	5809	5911	6014	6116	6219	6321	6423	6526	6628		1 10
43	6730	6833	6935	7037	7140	7242	7344	7447	7549	7651		2 20
44	7754	7856	7958	8061	8163	8265	8368	8470	8572	8675		3 31
												4 41
45	8777	8879	8982	9084	9186	9288	9391	9493	9595	9698		5 51
46	9800	9902	ō004	ō107	ō209	ō311	ō414	ō516	ō618	ō720		6 61
47	6280823	0925	1027	1129	1232	1334	1436	1538	1641	1743		7 71
48	1845	1947	2050	2152	2254	2356	2458	2561	2663	2765		8 82
49	2867	2970	3072	3174	3276	3378	3481	3583	3685	3787		9 92
N.	0	1	2	3	4	5	6	7	8	9	D	Pts.

N.	0	1	2	3	4	5	6	7	8	9	D	Pro.
4250	6283889	3991	4094	4196	4298	4400	4502	4605	4707	4809		
51	4911	5013	5115	5218	5320	5422	5524	5626	5728	5830		102
52	5933	6035	6137	6239	6341	6443	6545	6647	6750	6852		1 | 10
53	6954	7056	7158	7260	7362	7464	7566	7669	7771	7873		2 | 20
54	7975	8077	8179	8281	8383	8485	8587	8689	8792	8894		3 | 31
												4 | 41
55	8996	9098	9200	9302	9404	9506	9608	9710	9812	9914		5 | 51
56	6290016	0118	0220	0322	0424	0526	0628	0730	0832	0934		6 | 61
57	1037	1139	1241	1343	1445	1547	1649	1751	1853	1955	102	7 | 71
58	2057	2159	2261	2363	2465	2567	2668	2770	2872	2974		8 | 82
59	3076	3178	3280	3382	3484	3586	3688	3790	3892	3994		9 | 92
4260	4096	4198	4300	4402	4504	4606	4708	4810	4911	5013		
61	5115	5217	5319	5421	5523	5625	5727	5829	5931	6033		
62	6134	6236	6338	6440	6542	6644	6746	6848	6950	7051		
63	7153	7255	7357	7459	7561	7663	7765	7866	7968	8070		
64	8172	8274	8376	8478	8579	8681	8783	8885	8987	9089		
65	9190	9292	9394	9496	9598	9699	9801	9903	0̄005	0̄107		
66	6300209	0310	0412	0514	0616	0717	0819	0921	1023	1125		
67	1226	1328	1430	1532	1634	1735	1837	1939	2041	2142		
68	2244	2346	2448	2549	2651	2753	2855	2956	3058	3160		
69	3262	3363	3465	3567	3668	3770	3872	3974	4075	4177		
4270	4279	4380	4482	4584	4686	4787	4889	4991	5092	5194		
71	5296	5397	5499	5601	5702	5804	5906	6007	6109	6211		
72	6312	6414	6516	6617	6719	6821	6922	7024	7126	7227		
73	7329	7431	7532	7634	7735	7837	7939	8040	8142	8244		
74	8345	8447	8548	8650	8752	8853	8955	9056	9158	9260		
75	9361	9463	9564	9666	9768	9869	9971	0̄072	0̄174	0̄275		
76	6310377	0479	0580	0682	0783	0885	0986	1088	1189	1291		
77	1393	1494	1596	1697	1799	1900	2002	2103	2205	2306		
78	2408	2509	2611	2712	2814	2915	3017	3118	3220	3321		
79	3423	3524	3626	3727	3829	3930	4032	4133	4235	4336		
4280	4438	4539	4641	4742	4844	4945	5046	5148	5249	5351		
81	5452	5554	5655	5757	5858	5959	6061	6162	6264	6365		
82	6467	6568	6669	6771	6872	6974	7075	7177	7278	7379		
83	7481	7582	7684	7785	7886	7988	8089	8190	8292	8393		
84	8495	8596	8697	8799	8900	9001	9103	9204	9306	9407		
85	9508	9610	9711	9812	9914	0̄015	0̄116	0̄218	0̄319	0̄420		
86	6320522	0623	0724	0826	0927	1028	1130	1231	1332	1434		
87	1535	1636	1737	1839	1940	2041	2143	2244	2345	2446		
88	2548	2649	2750	2852	2953	3054	3155	3257	3358	3459		
89	3560	3662	3763	3864	3965	4067	4168	4269	4370	4472		
4290	4573	4674	4775	4877	4978	5079	5180	5282	5383	5484		
91	5585	5686	5788	5889	5990	6091	6192	6294	6395	6496		101
92	6597	6698	6800	6901	7002	7103	7204	7305	7407	7508		1 | 10
93	7609	7710	7811	7912	8014	8115	8216	8317	8418	8519		2 | 20
94	8620	8722	8823	8924	9025	9126	9227	9328	9429	9531		3 | 30
												4 | 40
95	9632	9733	9834	9935	0̄036	0̄137	0̄238	0̄339	0̄441	0̄542		5 | 51
96	6330643	0744	0845	0946	1047	1148	1249	1350	1451	1552		6 | 61
97	1654	1755	1856	1957	2058	2159	2260	2361	2462	2563		7 | 71
98	2664	2765	2866	2967	3068	3169	3270	3371	3472	3573	101	8 | 81
99	3674	3775	3876	3978	4079	4180	4281	4382	4483	4584		9 | 91
N.	0	1	2	3	4	5	6	7	8	9	D	Pts.

N.	0	1	2	3	4	5	6	7	8	9	D	Pro.
4300	6334685	4786	4887	4988	5089	5190	5291	5391	5492	5593		
01	5694	5795	5896	5997	6098	6199	6300	6401	6502	6603		101
02	6704	6805	6906	7007	7108	7209	7310	7411	7512	7613		1 10
03	7713	7814	7915	8016	8117	8218	8319	8420	8521	8622		2 20
04	8723	8824	8924	9025	9126	9227	9328	9429	9530	9631		3 30
												4 40
05	9732	9832	9933	0̄034	0̄135	0̄236	0̄337	0̄438	0̄539	0̄639		5 51
06	6340740	0841	0942	1043	1144	1245	1345	1446	1547	1648		6 61
07	1749	1850	1950	2051	2152	2253	2354	2455	2555	2656		7 71
08	2757	2858	2959	3059	3160	3261	3362	3463	3563	3664		8 81
09	3765	3866	3967	4067	4168	4269	4370	4470	4571	4672		9 91
4310	4773	4873	4974	5075	5176	5276	5377	5478	5579	5679		
11	5780	5881	5982	6082	6183	6284	6385	6485	6536	6687		
12	6788	6888	6989	7090	7190	7291	7392	7492	7593	7694		
13	7795	7895	7996	8097	8197	8298	8399	8499	8600	8701		
14	8801	8902	9003	9103	9204	9305	9405	9506	9607	9707		
15	9808	9909	0̄009	0̄110	0̄211	0̄311	0̄412	0̄512	0̄613	0̄714		
16	6350814	0915	1016	1116	1217	1317	1418	1519	1619	1720		
17	1820	1921	2022	2122	2223	2323	2424	2525	2625	2726		
18	2826	2927	3028	3128	3229	3329	3430	3530	3631	3731		
19	3832	3933	4033	4134	4234	4335	4435	4536	4636	4737		
4320	4837	4938	5039	5139	5240	5340	5441	5541	5642	5742		
21	5843	5943	6044	6144	6245	6345	6446	6546	6647	6747		
22	6848	6948	7049	7149	7250	7350	7450	7551	7651	7752		
23	7852	7953	8053	8154	8254	8355	8455	8556	8656	8756		
24	8857	8957	9058	9158	9259	9359	9459	9560	9660	9761		
25	9861	9962	0̄062	0̄162	0̄263	0̄363	0̄464	0̄564	0̄664	0̄765		
26	6360865	0966	1066	1166	1267	1367	1467	1568	1668	1769		
27	1869	1969	2070	2170	2270	2371	2471	2571	2672	2772		
28	2873	2973	3073	3174	3274	3374	3475	3575	3675	3776		
29	3876	3976	4076	4177	4277	4377	4478	4578	4678	4779		
4330	4879	4979	5080	5180	5280	5380	5481	5581	5681	5782		
31	5882	5982	6082	6183	6283	6383	6483	6584	6684	6784		
32	6884	6985	7085	7185	7285	7386	7486	7586	7686	7787		
33	7887	7987	8087	8188	8288	8388	8488	8588	8689	8789		
34	8889	8989	9089	9190	9290	9390	9490	9590	9691	9791		
35	9891	9991	0̄091	0̄192	0̄292	0̄392	0̄492	0̄592	0̄692	0̄793		
36	6370893	0993	1093	1193	1293	1394	1494	1594	1694	1794		
37	1894	1994	2094	2195	2295	2395	2495	2595	2695	2795		
38	2895	2996	3096	3196	3296	3396	3496	3596	3696	3796		
39	3897	3997	4097	4197	4297	4397	4497	4597	4697	4797		
4340	4897	4997	5097	5197	5298	5398	5498	5598	5698	5798		
41	5898	5998	6098	6198	6298	6398	6498	6598	6698	6798		
42	6898	6998	7098	7198	7298	7398	7498	7598	7698	7798		
43	7898	7998	8098	8198	8298	8398	8498	8598	8698	8798	100	100
44	8898	8998	9098	9198	9298	9398	9498	9598	9698	9798		1 10
45	9898	9998	0̄098	0̄198	0̄298	0̄398	0̄497	0̄597	0̄697	0̄797		2 20
46	6380897	0997	1097	1197	1297	1397	1497	1597	1697	1796		3 30
47	1896	1996	2096	2196	2296	2396	2496	2596	2696	2795		4 40
48	2895	2995	3095	3195	3295	3395	3495	3594	3694	3794		5 50
49	3894	3994	4094	4194	4294	4393	4493	4593	4693	4793		6 60
												7 70
												8 80
												9 90
N.	0	1	2	3	4	5	6	7	8	9	D	Pts.

N	0	1	2	3	4	5	6	7	8	9	D	Pro.
4350	6384893	4992	5092	5192	5292	5392	5492	5591	5691	5791		
51	5891	5991	6090	6190	6290	6390	6490	6589	6689	6789		**99**
52	6889	6989	7088	7188	7288	7388	7488	7587	7687	7787		1 10
53	7887	7986	8086	8186	8286	8385	8485	8585	8685	8784		2 20
54	8884	8964	9084	9183	9283	9383	9483	9582	9682	9782		3 30
55	9682	9981	0̄081	0̄181	0̄280	0̄380	0̄480	0̄580	0̄679	0̄779		4 40 5 50
56	6390879	0978	1078	1178	1277	1377	1477	1577	1676	1776		6 59
57	1876	1975	2075	2175	2274	2374	2474	2573	2673	2773		7 69
58	2872	2972	3072	3171	3271	3371	3470	3570	3669	3769		8 79
59	3869	3968	4068	4168	4267	4367	4466	4566	4666	4765		9 89
4360	4865	4965	5064	5164	5263	5363	5463	5562	5662	5761		
61	5861	5960	6060	6160	6259	6359	6458	6558	6657	6757		
62	6857	6956	7056	7155	7255	7354	7454	7553	7653	7753		
63	7852	7952	8051	8151	8250	8350	8449	8549	8648	8748		
64	8847	8947	9046	9146	9245	9345	9444	9544	9643	9743		
65	9842	9942	0̄041	0̄141	0̄240	0̄340	0̄439	0̄539	0̄638	0̄738		
66	6400837	0937	1036	1136	1235	1335	1434	1534	1633	1732		
67	1832	1931	2031	2130	2230	2329	2429	2528	2627	2727		
68	2826	2926	3025	3125	3224	3323	3423	3522	3622	3721		
69	3820	3920	4019	4119	4218	4317	4417	4516	4616	4715		
4370	4814	4914	5013	5113	5212	5311	5411	5510	5609	5709		
71	5808	5907	6007	6106	6205	6305	6404	6504	6603	6702		
72	6802	6901	7000	7100	7199	7298	7398	7497	7596	7695		
73	7795	7894	7993	8093	8192	8291	8391	8490	8589	8688		
74	8788	8887	8986	9086	9185	9284	9383	9483	9582	9681		
75	9781	9880	9979	0̄078	0̄178	0̄277	0̄376	0̄475	0̄575	0̄674		
76	6410773	0872	0972	1071	1170	1269	1369	1468	1567	1666		
77	1765	1865	1964	2063	2162	2262	2361	2460	2559	2658		
78	2758	2857	2956	3055	3154	3254	3353	3452	3551	3650		
79	3749	3849	3948	4047	4146	4245	4344	4444	4543	4642		
4380	4741	4840	4939	5039	5138	5237	5336	5435	5534	5633		
81	5733	5832	5931	6030	6129	6228	6327	6426	6526	6625		
82	6724	6823	6922	7021	7120	7219	7318	7417	7517	7616		
83	7715	7814	7913	8012	8111	8210	8309	8408	8507	8606		
84	8705	8805	8904	9003	9102	9201	9300	9399	9498	9597		
85	9696	9795	9894	9993	0̄092	0̄191	0̄290	0̄389	0̄488	0̄587	99	
86	6420686	0785	0884	0983	1082	1181	1280	1379	1478	1577		
87	1676	1775	1874	1973	2072	2171	2270	2369	2468	2567		
88	2666	2765	2864	2963	3062	3161	3260	3359	3458	3557		
89	3656	3755	3854	3953	4052	4151	4249	4348	4447	4546		
4390	4645	4744	4843	4942	5041	5140	5239	5338	5437	5535		
91	5634	5733	5832	5931	6030	6129	6228	6327	6426	6524		
92	6623	6722	6821	6920	7019	7118	7217	7315	7414	7513		**98**
93	7612	7711	7810	7909	8007	8106	8205	8304	8403	8502		1 10
94	8601	8699	8798	8897	8996	9095	9194	9292	9391	9490		2 20 3 29
95	9589	9688	9786	9885	9984	0̄083	0̄182	0̄280	0̄379	0̄478		4 39
96	6430577	0676	0774	0873	0972	1071	1170	1268	1367	1466		5 49 6 59
97	1565	1663	1762	1861	1960	2058	2157	2256	2355	2454		7 69
98	2552	2651	2750	2848	2947	3046	3145	3243	3342	3441		8 78
99	3540	3638	3737	3836	3935	4033	4132	4231	4329	4428		9 88
N.	0	1	2	3	4	5	6	7	8	9	D	Pts.

N.	0	1	2	3	4	5	6	7	8	9	D	Pro.
4400	6434527	4625	4724	4823	4922	5020	5119	5218	5316	5415		
01	5514	5612	5711	5810	5908	6007	6106	6204	6303	6402		99
02	6500	6599	6698	6796	6895	6994	7092	7191	7290	7388		1 10
03	7487	7585	7684	7783	7881	7980	8079	8177	8276	8374		2 20
04	8473	8572	8670	8769	8868	8966	9065	9163	9262	9361		3 30
												4 40
05	9459	9558	9656	9755	9853	9952	0̄051	0̄149	0̄248	0̄346		5 50
06	6440445	0543	0642	0741	0839	0938	1036	1135	1233	1332		6 59
07	1431	1529	1628	1726	1825	1923	2022	2120	2219	2317		7 69
08	2416	2514	2613	2711	2810	2908	3007	3105	3204	3302		8 79
09	3401	3499	3598	3696	3795	3893	3992	4090	4189	4287		9 89
4410	4386	4484	4583	4681	4780	4878	4977	5075	5174	5272		
11	5371	5469	5567	5666	5764	5863	5961	6060	6158	6257		
12	6355	6453	6552	6650	6749	6847	6946	7044	7142	7241		
13	7339	7438	7536	7635	7733	7831	7930	8028	8127	8225		
14	8323	8422	8520	8618	8717	8815	8914	9012	9110	9209		
15	9307	9405	9504	9602	9701	9799	9897	9996	0̄094	0̄192		
16	6450291	0389	0487	0586	0684	0782	0881	0979	1077	1176		
17	1274	1372	1471	1569	1667	1766	1864	1962	2061	2159		
18	2257	2355	2454	2552	2650	2749	2847	2945	3043	3142		
19	3240	3338	3437	3535	3633	3731	3830	3928	4026	4124		
4420	4223	4321	4419	4517	4616	4714	4812	4910	5009	5107		
21	5205	5303	5402	5500	5598	5696	5795	5893	5991	6089		
22	6187	6286	6384	6482	6580	6678	6777	6875	6973	7071		
23	7169	7268	7366	7464	7562	7660	7758	7857	7955	8053		
24	8151	8249	8348	8446	8544	8642	8740	8838	8936	9035		
25	9133	9231	9329	9427	9525	9623	9722	9820	9918	0̄016		
26	6460114	0212	0310	0408	0507	0605	0703	0801	0899	0997		
27	1095	1193	1291	1390	1488	1586	1684	1782	1880	1978		
28	2076	2174	2272	2370	2468	2566	2665	2763	2861	2959		
29	3057	3155	3253	3351	3449	3547	3645	3743	3841	3939		
4430	4037	4135	4233	4331	4429	4527	4625	4723	4821	4919	98	
31	5018	5116	5214	5312	5410	5508	5606	5704	5802	5900		
32	5998	6096	6193	6291	6389	6487	6585	6683	6781	6879		
33	6977	7075	7173	7271	7369	7467	7565	7663	7761	7859		
34	7957	8055	8153	8251	8349	8447	8545	8642	8740	8838		
35	8936	9034	9132	9230	9328	9426	9524	9622	9720	9817		
36	9915	0̄013	0̄111	0̄209	0̄307	0̄405	0̄503	0̄601	0̄699	0̄796		
37	6470894	0992	1090	1188	1286	1384	1482	1579	1677	1775		
38	1873	1971	2069	2167	2264	2362	2460	2558	2656	2754		
39	2851	2949	3047	3145	3243	3341	3438	3536	3634	3732		
4440	3830	3928	4025	4123	4221	4319	4417	4514	4612	4710		
41	4808	4906	5003	5101	5199	5297	5394	5492	5590	5688		
42	5786	5883	5981	6079	6177	6274	6372	6470	6568	6665		98
43	6763	6861	6959	7056	7154	7252	7350	7447	7545	7643		1 10
44	7741	7838	7936	8034	8131	8229	8327	8425	8522	8620		2 20
												3 29
45	8718	8815	8913	9011	9108	9206	9304	9402	9499	9597		4 39
46	9695	9792	9890	9988	0̄085	0̄183	0̄281	0̄378	0̄476	0̄574		5 49
47	6480671	0769	0867	0964	1062	1160	1257	1355	1453	1550		6 59
48	1648	1745	1843	1941	2038	2136	2234	2331	2429	2526		7 69
49	2624	2722	2819	2917	3015	3112	3210	3307	3405	3503		8 78
												9 88
N.	0	1	2	3	4	5	6	7	8	9	D	Pts.

N.	0	1	2	3	4	5	6	7	8	9	D	Pro.
4450	6483600	3698	3795	3893	3990	4088	4186	4283	4381	4478		
51	4576	4674	4771	4869	4966	5064	5161	5259	5356	5454		**97**
52	5552	5649	5747	5844	5942	6039	6137	6234	6332	6429		1 10
53	6527	6624	6722	6820	6917	7015	7112	7210	7307	7405		2 19
54	7502	7600	7697	7795	7892	7990	8087	8185	8282	8380		3 29
												4 39
55	8477	8575	8672	8770	8867	8964	9062	9159	9257	9354		5 49
56	9452	9549	9647	9744	9842	9939	0̄037	0̄134	0̄231	0̄329		6 58
57	6490426	0524	0621	0719	0816	0914	1011	1108	1206	1303		7 68
58	1401	1498	1595	1693	1790	1888	1985	2083	2180	2277		8 78
59	2375	2472	2570	2667	2764	2862	2959	3056	3154	3251		9 87
4460	3349	3446	3543	3641	3738	3835	3933	4030	4128	4225		
61	4322	4420	4517	4614	4712	4809	4906	5004	5101	5198		
62	5296	5393	5490	5588	5685	5782	5880	5977	6074	6172		
63	6269	6366	6463	6561	6658	6755	6853	6950	7047	7145		
64	7242	7339	7436	7534	7631	7728	7826	7923	8020	8117		
65	8215	8312	8409	8506	8604	8701	8798	8895	8993	9090		
66	9187	9284	9382	9479	9576	9673	9771	9868	9965	0̄062		
67	6500160	0257	0354	0451	0548	0646	0743	0840	0937	1034		
68	1132	1229	1326	1423	1520	1618	1715	1812	1909	2006		
69	2104	2201	2298	2395	2492	2589	2687	2784	2881	2978		
4470	3075	3172	3270	3367	3464	3561	3658	3755	3852	3950		
71	4047	4144	4241	4338	4435	4532	4629	4727	4824	4921		
72	5018	5115	5212	5309	5406	5503	5601	5698	5795	5892		
73	5989	6086	6183	6280	6377	6474	6571	6669	6766	6863		
74	6960	7057	7154	7251	7348	7445	7542	7639	7736	7833		
75	7930	8027	8124	8222	8319	8416	8513	8610	8707	8804		
76	8901	8998	9095	9192	9289	9386	9483	9580	9677	9774	97	
77	9871	9968	0̄065	0̄162	0̄259	0̄356	0̄453	0̄550	0̄647	0̄744		
78	6510841	0938	1035	1132	1229	1326	1423	1520	1617	1714		
79	1811	1908	2005	2102	2198	2295	2392	2489	2586	2683		
4480	2780	2877	2974	3071	3168	3265	3362	3459	3556	3653		
81	3749	3846	3943	4040	4137	4234	4331	4428	4525	4622		
82	4719	4815	4912	5009	5106	5203	5300	5397	5494	5591		
83	5687	5784	5881	5978	6075	6172	6269	6365	6462	6559		
84	6656	6753	6850	6947	7043	7140	7237	7334	7431	7528		
85	7624	7721	7818	7915	8012	8109	8205	8302	8399	8496		
86	8593	8690	8786	8883	8980	9077	9174	9270	9367	9464		
87	9561	9657	9754	9851	9948	0̄045	0̄141	0̄238	0̄335	0̄432		
88	6520528	0625	0722	0819	0916	1012	1109	1206	1303	1399		
89	1496	1593	1690	1786	1883	1980	2076	2173	2270	2367		
4490	2463	2560	2657	2754	2850	2947	3044	3140	3237	3334		
91	3431	3527	3624	3721	3817	3914	4011	4107	4204	4301		
92	4397	4494	4591	4688	4784	4881	4978	5074	5171	5268		**96**
93	5364	5461	5558	5654	5751	5847	5944	6041	6137	6234		1 10
94	6331	6427	6524	6621	6717	6814	6910	7007	7104	7200		2 19
												3 29
95	7297	7394	7490	7587	7683	7780	7877	7973	8070	8166		4 38
96	8263	8360	8456	8553	8649	8746	8843	8939	9036	9132		5 48
97	9229	9325	9422	9519	9615	9712	9808	9905	0̄001	0̄098		6 58
98	6530195	0291	0388	0484	0581	0677	0774	0870	0967	1063		7 67
99	1160	1256	1353	1450	1546	1643	1739	1836	1932	2029		8 77
												9 86
N.	0	1	2	3	4	5	6	7	8	9	D	Pts.

N.	0	1	2	3	4	5	6	7	8	9	D	Pro.
4500	6532125	2222	2318	2415	2511	2608	2704	2801	2897	2994		
01	3090	3187	3283	3380	3476	3573	3669	3765	3862	3958		.97
02	4055	4151	4248	4344	4441	4537	4634	4730	4827	4923		1 10
03	5019	5116	5212	5309	5405	5502	5598	5695	5791	5887		2 19
04	5984	6080	6177	6273	6369	6466	6562	6659	6755	6852		3 29
												4 39
05	6948	7044	7141	7237	7334	7430	7526	7623	7719	7815		5 49
06	7912	8008	8105	8201	8297	8394	8490	8586	8683	8779		6 58
07	8876	8972	9068	9165	9261	9357	9454	9550	9646	9743		7 68
08	9839	9935	ō032	ō128	ō224	ō321	ō417	ō513	ō610	ō706		8 78
09	6540802	0899	0995	1091	1188	1284	1380	1477	1573	1669		9 87
4510	1765	1862	1958	2054	2151	2247	2343	2439	2536	2632		
11	2728	2825	2921	3017	3113	3210	3306	3402	3498	3595		
12	3691	3787	3883	3980	4076	4172	4268	4365	4461	4557		
13	4653	4750	4846	4942	5038	5134	5231	5327	5423	5519		
14	5616	5712	5808	5904	6000	6097	6193	6289	6385	6481		
15	6578	6674	6770	6866	6962	7058	7155	7251	7347	7443		
16	7539	7635	7732	7828	7924	8020	8116	8212	8309	8405		
17	8501	8597	8693	8789	8885	8982	9078	9174	9270	9366		
18	9462	9558	9655	9751	9847	9943	ō039	ō135	ō231	ō327		
19	6550423	0520	0616	0712	0808	0904	1000	1096	1192	1288		
4520	1384	1480	1577	1673	1769	1865	1961	2057	2153	2249		
21	2345	2441	2537	2633	2729	2825	2921	3017	3113	3210		
22	3306	3402	3498	3594	3690	3786	3882	3978	4074	4170		
23	4266	4362	4458	4554	4650	4746	4842	4938	5034	5130		
24	5226	5322	5418	5514	5610	5706	5802	5898	5994	6090	96	
25	6186	6282	6378	6474	6570	6666	6762	6858	6954	7050		
26	7145	7241	7337	7433	7529	7625	7721	7817	7913	8009		
27	8105	8201	8297	8393	8489	8585	8681	8776	8872	8968		
28	9064	9160	9256	9352	9448	9544	9640	9736	9831	9927		
29	6560023	0119	0215	0311	0407	0503	0599	0694	0790	0886		
4530	0982	1078	1174	1270	1365	1461	1557	1653	1749	1845		
31	1941	2036	2132	2228	2324	2420	2516	2612	2707	2803		
32	2899	2995	3091	3186	3282	3378	3474	3570	3666	3761		
33	3857	3953	4049	4145	4240	4336	4432	4528	4624	4719		
34	4815	4911	5007	5103	5198	5294	5390	5486	5581	5677		
35	5773	5869	5964	6060	6156	6252	6347	6443	6539	6635		
36	6730	6826	6922	7018	7113	7209	7305	7401	7496	7592		
37	7688	7784	7879	7975	8071	8166	8262	8358	8454	8549		
38	8645	8741	8836	8932	9028	9123	9219	9315	9410	9506		
39	9602	9698	9793	9889	9985	ō080	ō176	ō272	ō367	ō463		
4540	6570559	0654	0750	0845	0941	1037	1132	1228	1324	1419		
41	1515	1611	1706	1802	1898	1993	2089	2184	2280	2376		
42	2471	2567	2663	2758	2854	2949	3045	3141	3236	3332		96
43	3427	3523	3619	3714	3810	3905	4001	4096	4192	4288		1 10
44	4383	4479	4574	4670	4766	4861	4957	5052	5148	5243		2 19
45	5339	5434	5530	5626	5721	5817	5912	6008	6103	6199		3 29
46	6294	6390	6485	6581	6676	6772	6867	6963	7059	7154		4 38
47	7250	7345	7441	7536	7632	7727	7823	7918	8014	8109		5 48
48	8205	8300	8396	8491	8587	8682	8777	8873	8968	9064		6 58
49	9159	9255	9350	9446	9541	9637	9732	9828	9923	ō019		7 67
												8 77
												9 86
N.	0	1	2	3	4	5	6	7	8	9	D	Pts.

N.	0	1	2	3	4	5	6	7	8	9	D	Pro.
4550	6580114	0209	0305	0400	0496	0591	0687	0782	0877	0973		
51	1068	1164	1259	1355	1450	1545	1641	1736	1832	1927		95
52	2023	2118	2213	2309	2404	2500	2595	2690	2786	2881		1 10
53	2977	3072	3167	3263	3358	3453	3549	3644	3740	3835		2 19
54	3930	4026	4121	4216	4312	4407	4502	4598	4693	4788		3 29 / 4 38
55	4884	4979	5074	5170	5265	5361	5456	5551	5647	5742		5 48
56	5837	5932	6028	6123	6218	6314	6409	6504	6600	6695		6 57
57	6790	6886	6981	7076	7171	7267	7362	7457	7553	7648		7 67
58	7743	7838	7934	8029	8124	8220	8315	8410	8505	8601		8 76
59	8696	8791	8886	8982	9077	9172	9267	9363	9458	9553		9 86
4560	9648	9744	9839	9934	0̄029	0̄125	0̄220	0̄315	0̄410	0̄506		
61	6590601	0696	0791	0886	0982	1077	1172	1267	1362	1458		
62	1553	1648	1743	1838	1934	2029	2124	2219	2314	2410		
63	2505	2600	2695	2790	2885	2981	3076	3171	3266	3361		
64	3456	3552	3647	3742	3837	3932	4027	4122	4218	4313		
65	4408	4503	4598	4693	4788	4883	4979	5074	5169	5264		
66	5359	5454	5549	5644	5740	5835	5930	6025	6120	6215		
67	6310	6405	6500	6595	6690	6786	6881	6976	7071	7166		
68	7261	7356	7451	7546	7641	7736	7831	7926	8021	8117		
69	8212	8307	8402	8497	8592	8687	8782	8877	8972	9067		
4570	9162	9257	9352	9447	9542	9637	9732	9827	9922	0̄017		
71	6600112	0207	0302	0397	0492	0587	0682	0777	0872	0967	95	
72	1062	1157	1252	1347	1442	1537	1632	1727	1822	1917		
73	2012	2107	2202	2297	2392	2487	2582	2677	2772	2867		
74	2962	3057	3151	3246	3341	3436	3531	3626	3721	3816		
75	3911	4006	4101	4196	4291	4386	4481	4575	4670	4765		
76	4860	4955	5050	5145	5240	5335	5430	5524	5619	5714		
77	5809	5904	5999	6094	6189	6284	6379	6473	6568	6663		
78	6758	6853	6948	7042	7137	7232	7327	7422	7517	7612		
79	7706	7801	7896	7991	8086	8181	8275	8370	8465	8560		
4580	8655	8750	8844	8939	9034	9129	9224	9318	9413	9508		
81	9603	9698	9793	9887	9982	0̄077	0̄172	0̄266	0̄361	0̄456		
82	6610551	0646	0740	0835	0930	1025	1120	1214	1309	1404		
83	1499	1593	1688	1783	1878	1972	2067	2162	2257	2351		
84	2446	2541	2636	2730	2825	2920	3015	3109	3204	3299		
85	3393	3488	3583	3678	3772	3867	3962	4056	4151	4246		
86	4341	4435	4530	4625	4719	4814	4909	5003	5098	5193		
87	5287	5382	5477	5571	5666	5761	5855	5950	6045	6139		
88	6234	6329	6423	6518	6613	6707	6802	6897	6991	7086		
89	7181	7275	7370	7464	7559	7654	7748	7843	7938	8032		
4590	8127	8221	8316	8411	8505	8600	8695	8789	8884	8978		
91	9073	9168	9262	9357	9451	9546	9640	9735	9830	9924		
92	6620019	0113	0208	0303	0397	0492	0586	0681	0775	0870		94
93	0964	1059	1154	1248	1343	1437	1532	1626	1721	1815		1 9
94	1910	2004	2099	2194	2288	2383	2477	2572	2666	2761		2 19 / 3 28
95	2855	2950	3044	3139	3233	3328	3422	3517	3611	3706		4 38
96	3800	3895	3989	4084	4178	4273	4367	4462	4556	4651		5 47
97	4745	4840	4934	5028	5123	5217	5312	5406	5501	5595		6 56
98	5690	5784	5879	5973	6067	6162	6256	6351	6445	6540		7 66
99	6634	6729	6823	6917	7012	7106	7201	7295	7389	7484		8 75 / 9 85
N.	0	1	2	3	4	5	6	7	8	9	D	Pts.

N.	0	1	2	3	4	5	6	7	8	9	D	Pro.
4600	6627578	7673	7767	7862	7956	8050	8145	8239	8334	8428		
01	8522	8617	8711	8805	8900	8994	9089	9183	9277	9372		95
02	9466	9561	9655	9749	9844	9938	ō032	ō127	ō221	ō315		1 10
03	6630410	0504	0598	0693	0787	0881	0976	1070	1164	1259		2 19
04	1353	1447	1542	1636	1730	1825	1919	2013	2108	2202		3 29
												4 38
05	2296	2391	2485	2579	2674	2768	2862	2956	3051	3145		5 48
06	3239	3334	3428	3522	3616	3711	3805	3899	3994	4088		6 57
07	4182	4276	4371	4465	4559	4653	4748	4842	4936	5030		7 67
08	5125	5219	5313	5407	5502	5596	5690	5784	5879	5973		8 76
09	6067	6161	6256	6350	6444	6538	6632	6727	6821	6915		9 86
4610	7009	7103	7198	7292	7386	7480	7574	7669	7763	7857		
11	7951	8045	8140	8234	8328	8422	8516	8610	8705	8799		
12	8893	8987	9081	9175	9270	9364	9458	9552	9646	9740		
13	9835	9929	ō023	ō117	ō211	ō305	ō399	ō494	ō588	ō682		
14	6640776	0870	0964	1058	1152	1247	1341	1435	1529	1623		
15	1717	1811	1905	1999	2093	2188	2282	2376	2470	2564		
16	2658	2752	2846	2940	3034	3128	3222	3317	3411	3505		
17	3599	3693	3787	3881	3975	4069	4163	4257	4351	4445		
18	4539	4633	4727	4821	4915	5009	5104	5198	5292	5386		
19	5480	5574	5668	5762	5856	5950	6044	6138	6232	6326		
4620	6420	6514	6608	6702	6796	6890	6984	7078	7172	7266	94	
21	7360	7454	7548	7642	7736	7830	7924	8018	8111	8205		
22	8299	8393	8487	8581	8675	8769	8863	8957	9051	9145		
23	9239	9333	9427	9521	9615	9709	9803	9896	9990	ō084		
24	6650178	0272	0366	0460	0554	0648	0742	0836	0930	1023		
25	1117	1211	1305	1399	1493	1587	1681	1775	1869	1962		
26	2056	2150	2244	2338	2432	2526	2620	2713	2807	2901		
27	2995	3089	3183	3277	3370	3464	3558	3652	3746	3840		
28	3934	4027	4121	4215	4309	4403	4497	4590	4684	4778		
29	4872	4966	5059	5153	5247	5341	5435	5529	5622	5716		
4630	5810	5904	5998	6091	6185	6279	6373	6466	6560	6654		
31	6748	6842	6935	7029	7123	7217	7310	7404	7498	7592		
32	7686	7779	7873	7967	8061	8154	8248	8342	8436	8529		
33	8623	8717	8810	8904	8998	9092	9185	9279	9373	9467		
34	9560	9654	9748	9841	9935	ō029	ō123	ō216	ō310	ō404		
35	6660497	0591	0685	0778	0872	0966	1060	1153	1247	1341		
36	1434	1528	1622	1715	1809	1903	1996	2090	2184	2277		
37	2371	2465	2558	2652	2746	2839	2933	3027	3120	3214		
38	3307	3401	3495	3588	3682	3776	3869	3963	4056	4150		
39	4244	4337	4431	4525	4618	4712	4805	4899	4993	5086		
4640	5180	5273	5367	5461	5554	5648	5741	5835	5929	6022		
41	6116	6209	6303	6396	6490	6584	6677	6771	6864	6958		94
42	7051	7145	7238	7332	7426	7519	7613	7706	7S00	7893		1 9
43	7987	8080	8174	8267	8361	8454	8548	8642	8735	8829		2 19
44	8922	9016	9109	9203	9296	9390	9483	9577	9670	9764		3 28
45	9857	9951	ō044	ō138	ō231	ō325	ō418	ō512	ō605	ō699		4 38
46	6670792	0886	0979	1072	1166	1259	1353	1446	1540	1633		5 47
47	1727	1820	1914	2007	2101	2194	2287	2381	2474	2568		6 56
48	2661	2755	2848	2941	3035	3128	3222	3315	3409	3502		7 66
49	3595	3689	3782	3876	3969	4063	4156	4249	4343	4436		8 75
												9 85
N.	0	1	2	3	4	5	6	7	8	9	D	Pts.

N.	0	1	2	3	4	5	6	7	8	9	D	Pro.
4650	6674530	4623	4716	4810	4903	4996	5090	5183	5277	5370		
51	5463	5557	5650	5744	5837	5930	6024	6117	6210	6304		**93**
52	6397	6490	6584	6677	6770	6864	6957	7051	7144	7237		1 9
53	7331	7424	7517	7611	7704	7797	7891	7984	8077	8170		2 19
54	8264	8357	8450	8544	8637	8730	8824	8917	9010	9104		3 28
												4 37
55	9197	9290	9383	9477	9570	9663	9757	9850	9943	0̄036		5 47
56	6680130	0223	0316	0410	0503	0596	0689	0783	0876	0969		6 56
57	1062	1156	1249	1342	1435	1529	1622	1715	1808	1902		7 65
58	1995	2088	2181	2275	2368	2461	2554	2647	2741	2834		8 74
59	2927	3020	3114	3207	3300	3393	3486	3580	3673	3766		9 84
4660	3859	3952	4046	4139	4232	4325	4418	4511	4605	4698		
61	4791	4884	4977	5071	5164	5257	5350	5443	5536	5630		
62	5723	5816	5909	6002	6095	6188	6282	6375	6468	6561		
63	6654	6747	6840	6934	7027	7120	7213	7306	7399	7492		
64	7585	7679	7772	7865	7958	8051	8144	8237	8330	8423		
65	8516	8610	8703	8796	8889	8982	9075	9168	9261	9354		
66	9447	9540	9633	9727	9820	9913	0̄006	0̄099	0̄192	0̄285		
67	6690378	0471	0564	0657	0750	0843	0936	1029	1122	1215		
68	1308	1402	1495	1588	1661	1774	1867	1960	2053	2146		
69	2239	2332	2425	2518	2611	2704	2797	2890	2983	3076	93	
4670	3169	3262	3355	3448	3541	3634	3727	3820	3913	4006		
71	4099	4192	4285	4378	4471	4564	4656	4749	4842	4935		
72	5028	5121	5214	5307	5400	5493	5586	5679	5772	5865		
73	5958	6051	6144	6237	6330	6422	6515	6608	6701	6794		
74	6887	6980	7073	7166	7259	7352	7445	7537	7630	7723		
75	7816	7909	8002	8095	8188	8281	8373	8466	8559	8652		
76	8745	8838	8931	9024	9117	9209	9302	9395	9488	9581		
77	9674	9767	9859	9952	0̄045	0̄138	0̄231	0̄324	0̄416	0̄509		
78	6700602	0695	0788	0881	0974	1066	1159	1252	1345	1438		
79	1530	1623	1716	1809	1902	1995	2087	2180	2273	2366		
4680	2459	2551	2644	2737	2830	2922	3015	3108	3201	3294		
81	3386	3479	3572	3665	3758	3850	3943	4036	4129	4221		
82	4314	4407	4500	4592	4685	4778	4871	4963	5056	5149		
83	5242	5334	5427	5520	5613	5705	5798	5891	5983	6076		
84	6169	6262	6354	6447	6540	6632	6725	6818	6911	7003		
85	7096	7189	7281	7374	7467	7559	7652	7745	7837	7930		
86	8023	8116	8208	8301	8394	8486	8579	8672	8764	8857		
87	8950	9042	9135	9228	9320	9413	9505	9598	9691	9783		
88	9876	9969	0̄061	0̄154	0̄247	0̄339	0̄432	0̄524	0̄617	0̄710		
89	6710802	0895	0988	1080	1173	1265	1358	1451	1543	1636		
4690	1728	1821	1914	2006	2099	2191	2284	2377	2469	2562		
91	2654	2747	2839	2932	3025	3117	3210	3302	3395	3487		**92**
92	3580	3673	3765	3858	3950	4043	4135	4228	4320	4413		1 9
93	4506	4598	4691	4783	4876	4968	5061	5153	5246	5338		2 18
94	5431	5523	5616	5708	5801	5893	5986	6078	6171	6263		3 28
												4 37
95	6356	6448	6541	6633	6726	6818	6911	7003	7096	7188		5 46
96	7281	7373	7466	7558	7651	7743	7836	7928	8021	8113		6 55
97	8206	8298	8391	8483	8575	8668	8760	8853	8945	9038		7 64
98	9130	9223	9315	9407	9500	9592	9685	9777	9870	9962		8 74
99	6720054	0147	0239	0332	0424	0517	0609	0701	0794	0886		9 83
N.	0	1	2	3	4	5	6	7	8	9	D	Pts.

N.	0	1	2	3	4	5	6	7	8	9	D	Pro.
4700	6720979	1071	1163	1256	1348	1441	1533	1625	1718	1810		
01	1903	1995	2087	2180	2272	2364	2457	2549	2642	2734		93
02	2826	2919	3011	3103	3196	3288	3380	3473	3565	3657		1 9
03	3750	3842	3934	4027	4119	4211	4304	4396	4488	4581		2 19
04	4673	4765	4858	4950	5042	5135	5227	5319	5412	5504		3 28 4 37
05	5596	5689	5781	5873	5965	6058	6150	6242	6335	6427		5 47
06	6519	6612	6704	6796	6888	6981	7073	7165	7257	7350		6 56
07	7442	7534	7627	7719	7811	7903	7996	8088	8180	8272		7 65 8 74
08	8365	8457	8549	8641	8734	8826	8918	9010	9102	9195		9 84
09	9287	9379	9471	9564	9656	9748	9840	9932	ō025	ō117		
4710	6730209	0301	0393	0486	0578	0670	0762	0854	0947	1039		
11	1131	1223	1315	1408	1500	1592	1684	1776	1868	1961		
12	2053	2145	2237	2329	2421	2514	2606	2698	2790	2882		
13	2974	3067	3159	3251	3343	3435	3527	3619	3712	3804		
14	3896	3988	4080	4172	4264	4356	4449	4541	4633	4725		
15	4817	4909	5001	5093	5185	5277	5370	5462	5554	5646		
16	5738	5830	5922	6014	6106	6198	6290	6383	6475	6567		
17	6659	6751	6843	6935	7027	7119	7211	7303	7395	7487		
18	7579	7671	7763	7856	7948	8040	8132	8224	8316	8408		
19	8500	8592	8684	8776	8868	8960	9052	9144	9236	9328		
4720	9420	9512	9604	9696	9788	9880	9972	ō064	ō156	ō248	92	
21	6740340	0432	0524	0616	0708	0800	0892	0984	1076	1168		
22	1260	1352	1444	1536	1628	1720	1812	1904	1996	2088		
23	2179	2271	2363	2455	2547	2639	2731	2823	2915	3007		
24	3099	3191	3283	3375	3467	3559	3650	3742	3834	3926		
25	4018	4110	4202	4294	4386	4478	4570	4661	4753	4845		
26	4937	5029	5121	5213	5305	5397	5489	5580	5672	5764		
27	5856	5948	6040	6132	6224	6315	6407	6499	6591	6683		
28	6775	6867	6958	7050	7142	7234	7326	7418	7509	7601		
29	7693	7785	7877	7969	8060	8152	8244	8336	8428	8520		
4730	8611	8703	8795	8887	8979	9070	9162	9254	9346	9438		
31	9529	9621	9713	9805	9897	9988	ō080	ō172	ō264	ō356		
32	6750447	0539	0631	0723	0814	0906	0998	1090	1182	1273		
33	1365	1457	1549	1640	1732	1824	1916	2007	2099	2191		
34	2283	2374	2466	2558	2649	2741	2833	2925	3016	3108		
35	3200	3292	3383	3475	3567	3658	3750	3842	3934	4025		
36	4117	4209	4300	4392	4484	4575	4667	4759	4850	4942		
37	5034	5126	5217	5309	5401	5492	5584	5676	5767	5859		
38	5951	6042	6134	6226	6317	6409	6501	6592	6684	6775		
39	6867	6959	7050	7142	7234	7325	7417	7509	7600	7692		
4740	7783	7875	7967	8058	8150	8242	8333	8425	8516	8608		
41	8700	8791	8883	8974	9066	9158	9249	9341	9432	9524		
42	9615	9707	9799	9890	9982	ō073	ō165	ō257	ō348	ō440		
43	6760531	0623	0714	0806	0897	0989	1081	1172	1264	1355		92
44	1447	1538	1630	1721	1813	1905	1996	2088	2179	2271		1 9 2 18
45	2362	2454	2545	2637	2728	2820	2911	3003	3094	3186		3 28 4 37
46	3277	3369	3460	3552	3643	3735	3826	3918	4009	4101		5 46
47	4192	4284	4375	4467	4558	4650	4741	4833	4924	5016		6 55
48	5107	5199	5290	5382	5473	5564	5656	5747	5839	5930		7 64 8 74
49	6022	6113	6205	6296	6387	6479	6570	6662	6753	6845		9 83
N.	0	1	2	3	4	5	6	7	8	9	D	Pts.

N.	0	1	2	3	4	5	6	7	8	9	D	Pro.
4750	6766936	7028	7119	7210	7302	7393	7485	7576	7667	7759		
51	7850	7942	8033	8125	8216	8307	8399	8490	8582	8673		91
52	8764	8856	8947	9038	9130	9221	9313	9404	9495	9587		1　9
53	9678	9770	9861	9952	ō044	ō135	ō226	ō318	ō409	ō500		2　18
54	6770592	0683	0774	0866	0957	1049	1140	1231	1323	1414		3　27
												4　36
55	1505	1597	1688	1779	1871	1962	2053	2145	2236	2327		5　46
56	2418	2510	2601	2692	2784	2875	2966	3058	3149	3240		6　55
57	3332	3423	3514	3605	3697	3788	3879	3971	4062	4153		7　64
58	4244	4336	4427	4518	4609	4701	4792	4883	4975	5066		8　73
59	5157	5248	5340	5431	5522	5613	5705	5796	5887	5978		9　82
4760	6070	6161	6252	6343	6434	6526	6617	6708	6799	6891		
61	6982	7073	7164	7255	7347	7438	7529	7620	7712	7803		
62	7894	7985	8076	8168	8259	8350	8441	8532	8623	8715		
63	8806	8897	8988	9079	9171	9262	9353	9444	9535	9626		
64	9718	9809	9900	9991	ō082	ō173	ō264	ō356	ō447	ō538		
65	6780629	0720	0811	0902	0994	1085	1176	1267	1358	1449		
66	1540	1632	1723	1814	1905	1996	2087	2178	2269	2360		
67	2452	2543	2634	2725	2816	2907	2998	3089	3180	3271		
68	3362	3454	3545	3636	3727	3818	3909	4000	4091	4182		
69	4273	4364	4455	4546	4637	4729	4820	4911	5002	5093		
4770	5184	5275	5366	5457	5548	5639	5730	5821	5912	6003		
71	6094	6185	6276	6367	6458	6549	6640	6731	6822	6913		
72	7004	7095	7186	7277	7368	7459	7550	7641	7732	7823	91	
73	7914	8005	8096	8187	8278	8369	8460	8551	8642	8733		
74	8824	8915	9006	9097	9188	9279	9370	9461	9552	9643		
75	9734	9825	9916	ō007	ō098	ō188	ō279	ō370	ō461	ō552		
76	6790643	0734	0825	0916	1007	1098	1189	1280	1371	1461		
77	1552	1643	1734	1825	1916	2007	2098	2189	2280	2371		
78	2461	2552	2643	2734	2825	2916	3007	3098	3189	3279		
79	3370	3461	3552	3643	3734	3825	3916	4006	4097	4188		
4780	4279	4370	4461	4552	4642	4733	4824	4915	5006	5097		
81	5187	5278	5369	5460	5551	5642	5732	5823	5914	6005		
82	6096	6187	6277	6368	6459	6550	6641	6731	6822	6913		
83	7004	7095	7185	7276	7367	7458	7549	7639	7730	7821		
84	7912	8002	8093	8184	8275	8366	8456	8547	8638	8729		
85	8819	8910	9001	9092	9182	9273	9364	9455	9545	9636		
86	9727	9818	9908	9999	ō090	ō181	ō271	ō362	ō453	ō544		
87	6800634	0725	0816	0906	0997	1088	1179	1269	1360	1451		
88	1541	1632	1723	1814	1904	1995	2086	2176	2267	2358		
89	2448	2539	2630	2720	2811	2902	2992	3083	3174	3264		
4790	3355	3446	3536	3627	3718	3808	3899	3990	4080	4171		
91	4262	4352	4443	4534	4624	4715	4806	4896	4987	5077		
92	5168	5259	5349	5440	5531	5621	5712	5802	5893	5984		90
93	6074	6165	6256	6346	6437	6527	6618	6709	6799	6890		1　9
94	6980	7071	7161	7252	7343	7433	7524	7614	7705	7796		2　18
												3　27
95	7886	7977	8067	8158	8248	8339	8430	8520	8611	8701		4　36
96	8792	8882	8973	9063	9154	9244	9335	9426	9516	9607		5　45
97	9697	9788	9878	9969	ō059	ō150	ō240	ō331	ō421	ō512		6　54
98	6810602	0693	0783	0874	0964	1055	1145	1236	1327	1417		7　63
99	1507	1598	1688	1779	1869	1960	2050	2141	2231	2322		8　72
												9　81
N.	0	1	2	3	4	5.	6	7	8	9	D	Pts.

a

N.	0	1	2	3	4	5	6	7	8	9	D
4800	6812412	2503	2593	2684	2774	2865	2955	3046	3136	3227	
01	3317	3408	3498	3588	3679	3769	3860	3950	4041	4131	
02	4222	4312	4402	4493	4583	4674	4764	4855	4945	5035	
03	5126	5216	5307	5397	5488	5578	5668	5759	5849	5940	
04	6030	6120	6211	6301	6392	6482	6572	6663	6753	6844	
05	6934	7024	7115	7205	7295	7386	7476	7567	7657	7747	
06	7838	7928	8018	8109	8199	8289	8380	8470	8561	8651	
07	8741	8832	8922	9012	9103	9193	9283	9374	9464	9554	
08	9645	9735	9825	9916	0̄006	0̄096	0̄187	0̄277	0̄367	0̄457	
09	6820548	0638	0728	0819	0909	0999	1090	1180	1270	1360	
4810	1451	1541	1631	1722	1812	1902	1992	2083	2173	2263	
11	2354	2444	2534	2624	2715	2805	2895	2985	3076	3166	
12	3256	3346	3437	3527	3617	3707	3798	3888	3978	4068	
13	4159	4249	4339	4429	4520	4610	4700	4790	4880	4971	
14	5061	5151	5241	5331	5422	5512	5602	5692	5783	5873	
15	5963	6053	6143	6233	6324	6414	6504	6594	6684	6775	
16	6865	6955	7045	7135	7225	7316	7406	7496	7586	7676	
17	7766	7857	7947	8037	8127	8217	8307	8398	8488	8578	
18	8668	8758	8848	8938	9029	9119	9209	9299	9389	9479	
19	9569	9659	9750	9840	9930	0̄020	0̄110	0̄200	0̄290	0̄380	
4820	6830470	0560	0651	0741	0831	0921	1011	1101	1191	1281	
21	1371	1461	1551	1642	1732	1822	1912	2002	2092	2182	
22	2272	2362	2452	2542	2632	2722	2812	2902	2993	3083	
23	3173	3263	3353	3443	3533	3623	3713	3803	3893	3983	
24	4073	4163	4253	4343	4433	4523	4613	4703	4793	4883	90
25	4973	5063	5153	5243	5333	5423	5513	5603	5693	5783	
26	5873	5963	6053	6143	6233	6323	6413	6503	6593	6683	
27	6773	6863	6953	7043	7133	7223	7313	7403	7493	7583	
28	7673	7763	7853	7942	8032	8122	8212	8302	8392	8482	
29	8572	8662	8752	8842	8932	9022	9112	9202	9291	9381	
4830	9471	9561	9651	9741	9831	9921	0̄011	0̄101	0̄191	0̄280	
31	6840370	0460	0550	0640	0730	0820	0910	1000	1089	1179	
32	1269	1359	1449	1539	1629	1719	1808	1898	1988	2078	
33	2168	2258	2348	2438	2527	2617	2707	2797	2887	2977	
34	3066	3156	3246	3336	3426	3516	3605	3695	3785	3875	
35	3965	4055	4144	4234	4324	4414	4504	4594	4683	4773	
36	4863	4953	5043	5132	5222	5312	5402	5492	5581	5671	
37	5761	5851	5940	6030	6120	6210	6300	6389	6479	6569	
38	6659	6748	6838	6928	7018	7107	7197	7287	7377	7466	
39	7556	7646	7736	7825	7915	8005	8095	8184	8274	8364	
4840	8454	8543	8633	8723	8813	8902	8992	9082	9171	9261	
41	9351	9441	9530	9620	9710	9799	9889	9979	0̄068	0̄158	
42	6850248	0338	0427	0517	0607	0696	0786	0876	0965	1055	
43	1145	1234	1324	1414	1503	1593	1683	1772	1862	1952	
44	2041	2131	2221	2310	2400	2490	2579	2669	2759	2848	
45	2938	3027	3117	3207	3296	3386	3476	3565	3655	3744	
46	3834	3924	4013	4103	4193	4282	4372	4461	4551	4641	
47	4730	4820	4909	4999	5089	5178	5268	5357	5447	5537	
48	5626	5716	5805	5895	5984	6074	6164	6253	6343	6432	
49	6522	6611	6701	6791	6880	6970	7059	7149	7238	7328	

Pro.

91	
1	9
2	18
3	27
4	36
5	46
6	55
7	64
8	73
9	82

90	
1	9
2	18
3	27
4	36
5	45
6	54
7	63
8	72
9	81

N.	0	1	2	3	4	5	6	7	8	9	D	Pro.
4850	6857417	7507	7596	7686	7776	7865	7955	8044	8134	8223		
51	8313	8402	8492	8581	8671	8760	8850	8939	9029	9118		89
52	9208	9297	9387	9476	9566	9655	9745	9834	9924	ō013		1 9
53	6860103	0192	0282	0371	0461	0550	0640	0729	0819	0908		2 18
54	0998	1087	1177	1266	1356	1445	1535	1624	1713	1803		3 27
												4 36
55	1892	1982	2071	2161	2250	2340	2429	2518	2608	2697		5 45
56	2787	2876	2966	3055	3145	3234	3323	3413	3502	3592		6 53
57	3681	3770	3860	3949	4039	4128	4217	4307	4396	4486		7 62
58	4575	4665	4754	4843	4933	5022	5111	5201	5290	5380		8 71
59	5469	5558	5648	5737	5826	5916	6005	6095	6184	6273		9 80
4860	6363	6452	6541	6631	6720	6809	6899	6988	7078	7167		
61	7256	7346	7435	7524	7614	7703	7792	7882	7971	8060		
62	8150	8239	8328	8418	8507	8596	8685	8775	8864	8953		
63	9043	9132	9221	9311	9400	9489	9578	9668	9757	9846		
64	9936	ō025	ō114	ō204	ō293	ō382	ō471	ō561	ō650	ō739		
65	6870828	0918	1007	1096	1186	1275	1364	1453	1543	1632		
66	1721	1810	1900	1989	2078	2167	2257	2346	2435	2524		
67	2613	2703	2792	2881	2970	3060	3149	3238	3327	3416		
68	3506	3595	3684	3773	3863	3952	4041	4130	4219	4309		
69	4398	4487	4576	4665	4755	4844	4933	5022	5111	5200		
4870	5290	5379	5468	5557	5646	5735	5825	5914	6003	6092		
71	6181	6270	6360	6449	6538	6627	6716	6805	6895	6984		
72	7073	7162	7251	7340	7429	7518	7608	7697	7786	7875		
73	7964	8053	8142	8231	8321	8410	8499	8588	8677	8766		
74	8855	8944	9033	9123	9212	9301	9390	9479	9568	9657		
75	9746	9835	9924	ō013	ō103	ō192	ō281	ō370	ō459	ō548		
76	6880637	0726	0815	0904	0993	1082	1171	1260	1349	1439		
77	1528	1617	1706	1795	1884	1973	2062	2151	2240	2329		
78	2418	2507	2596	2685	2774	2863	2952	3041	3130	3219		
79	3308	3397	3486	3575	3664	3753	3842	3931	4020	4109	89	
4880	4198	4287	4376	4465	4554	4643	4732	4821	4910	4999		
81	5088	5177	5266	5355	5444	5533	5622	5711	5800	5889		
82	5978	6067	6156	6245	6334	6423	6511	6600	6689	6778		
83	6867	6956	7045	7134	7223	7312	7401	7490	7579	7668		
84	7757	7845	7934	8023	8112	8201	8290	8379	8468	8557		
85	8646	8735	8823	8912	9001	9090	9179	9268	9357	9446		
86	9535	9624	9712	9801	9890	9979	ō068	ō157	ō246	ō335		
87	6890423	0512	0601	0690	0779	0868	0957	1045	1134	1223		
88	1312	1401	1490	1579	1667	1756	1845	1934	2023	2112		
89	2200	2289	2378	2467	2556	2645	2733	2822	2911	3000		
4890	3089	3177	3266	3355	3444	3533	3621	3710	3799	3888		
91	3977	4065	4154	4243	4332	4421	4509	4598	4687	4776		
92	4664	4953	5042	5131	5220	5308	5397	5486	5575	5663		88
93	5752	5841	5930	6018	6107	6196	6285	6373	6462	6551		1 9
94	6640	6728	6817	6906	6995	7083	7172	7261	7350	7438		2 18
												3 26
95	7527	7616	7704	7793	7882	7971	8059	8148	8237	8325		4 35
96	8414	8503	8591	8680	8769	8858	8946	9035	9124	9212		5 44
97	9301	9390	9478	9567	9656	9744	9833	9922	ō010	ō099		6 53
98	6900188	0276	0365	0454	0542	0631	0720	0808	0897	0986		7 62
99	1074	1163	1252	1340	1429	1518	1606	1695	1784	1872		8 70
												9 79
N.	0	1	2	3	4	5	6	7	8	9	D	Pts.

N.	0	1	2	3	4	5	6	7	8	9	D	Pro.
4900	6901961	2049	2138	2227	2315	2404	2493	2581	2670	2758		
01	2847	2936	3024	3113	3201	3290	3379	3467	3556	3644		89
02	3733	3822	3910	3999	4087	4176	4265	4353	4442	4530		1 9
03	4619	4708	4796	4885	4973	5062	5150	5239	5327	5416		2 18
04	5505	5593	5682	5770	5859	5947	6036	6124	6213	6302		3 27 / 4 36
05	6390	6479	6567	6656	6744	6833	6921	7010	7098	7187		5 45
06	7275	7364	7452	7541	7630	7718	7807	7895	7984	8072		6 53
07	8161	8249	8338	8426	8515	8603	8692	8780	8869	8957		7 62
08	9046	9134	9223	9311	9399	9488	9576	9665	9753	9842		8 71
09	9930	0019	0107	0196	0284	0373	0461	0550	0638	0726		9 80
4910	6910815	0903	0992	1080	1169	1257	1346	1434	1522	1611		
11	1699	1788	1876	1965	2053	2141	2230	2318	2407	2495		
12	2584	2672	2760	2849	2937	3026	3114	3202	3291	3379		
13	3468	3556	3644	3733	3821	3910	3998	4086	4175	4263		
14	4352	4440	4528	4617	4705	4793	4882	4970	5058	5147		
15	5235	5324	5412	5500	5589	5677	5765	5854	5942	6030		
16	6119	6207	6295	6384	6472	6560	6649	6737	6825	6914		
17	7002	7090	7179	7267	7355	7444	7532	7620	7709	7797		
18	7885	7974	8062	8150	8238	8327	8415	8503	8592	8680		
19	8768	8857	8945	9033	9121	9210	9298	9386	9474	9563		
4920	9651	9739	9828	9916	0004	0092	0181	0269	0357	0445		
21	6920534	0622	0710	0798	0887	0975	1063	1151	1240	1328		
22	1416	1504	1593	1681	1769	1857	1945	2034	2122	2210		
23	2298	2387	2475	2563	2651	2739	2828	2916	3004	3092		
24	3180	3269	3357	3445	3533	3621	3710	3798	3886	3974		
25	4062	4151	4239	4327	4415	4503	4591	4680	4768	4856		
26	4944	5032	5120	5209	5297	5385	5473	5561	5649	5737		
27	5826	5914	6002	6090	6178	6266	6354	6443	6531	6619		
28	6707	6795	6883	6971	7059	7148	7236	7324	7412	7500		
29	7588	7676	7764	7853	7941	8029	8117	8205	8293	8381		
4930	8469	8557	8645	8733	8822	8910	8998	9086	9174	9262		
31	9350	9438	9526	9614	9702	9790	9878	9967	0055	0143		
32	6930231	0319	0407	0495	0583	0671	0759	0847	0935	1023		
33	1111	1199	1287	1375	1463	1551	1639	1727	1815	1903		
34	1991	2079	2167	2256	2344	2432	2520	2608	2696	2784	88	
35	2872	2960	3048	3136	3224	3312	3400	3488	3576	3664		
36	3752	3839	3927	4015	4103	4191	4279	4367	4455	4543		
37	4631	4719	4807	4895	4983	5071	5159	5247	5335	5423		
38	5511	5599	5687	5775	5863	5951	6039	6126	6214	6302		
39	6390	6478	6566	6654	6742	6830	6918	7006	7094	7182		
4940	7269	7357	7445	7533	7621	7709	7797	7885	7973	8061		
41	8149	8236	8324	8412	8500	8588	8676	8764	8852	8940		
42	9027	9115	9203	9291	9379	9467	9555	9643	9730	9818		
43	9906	9994	0082	0170	0258	0345	0433	0521	0609	0697		88
44	6940785	0872	0960	1048	1136	1224	1312	1399	1487	1575		1 9 / 2 18
45	1663	1751	1839	1926	2014	2102	2190	2278	2366	2453		3 26 / 4 35
46	2541	2629	2717	2805	2892	2980	3068	3156	3244	3331		5 44
47	3419	3507	3595	3682	3770	3858	3946	4034	4121	4209		6 53
48	4297	4385	4172	4560	4648	4736	4824	4911	4999	5087		7 62 / 8 70
49	5175	5262	5350	5438	5526	5613	5701	5789	5877	5964		9 79
N.	0	1	2	3	4	5	6	7	8	9	D	Pts.

N.	0	1	2	3	4	5	6	7	8	9	D	Pro.
4950	6946052	6140	6227	6315	6403	6491	6578	6666	6754	6842		
51	6929	7017	7105	7192	7280	7368	7456	7543	7631	7719		87
52	7806	7894	7982	8069	8157	8245	8333	8420	8508	8596		1 9
53	8683	8771	8859	8946	9034	9122	9209	9297	9385	9472		2 17
54	9560	9648	9735	9823	9911	9998	0086	0174	0261	0349	88	3 26 / 4 35
55	6950437	0524	0612	0700	0787	0875	0962	1050	1138	1225		5 44
56	1313	1401	1488	1576	1663	1751	1839	1926	2014	2102		6 52
57	2189	2277	2364	2452	2540	2627	2715	2802	2890	2978		7 61
58	3065	3153	3240	3328	3416	3503	3591	3678	3766	3854		8 70
59	3941	4029	4116	4204	4291	4379	4467	4554	4642	4729		9 78
4960	4817	4904	4992	5079	5167	5255	5342	5430	5517	5605		
61	5692	5780	5867	5955	6042	6130	6217	6305	6393	6480		
62	6568	6655	6743	6830	6918	7005	7093	7180	7268	7355		
63	7443	7530	7618	7705	7793	7880	7968	8055	8143	8230		
64	8318	8405	8493	8580	8668	8755	8843	8930	9018	9105		
65	9193	9280	9367	9455	9542	9630	9717	9805	9892	9980		
66	6960067	0155	0242	0330	0417	0504	0592	0679	0767	0854		
67	0942	1029	1116	1204	1291	1379	1466	1554	1641	1728		
68	1816	1903	1991	2078	2166	2253	2340	2428	2515	2603		
69	2690	2777	2865	2952	3040	3127	3214	3302	3389	3477		
4970	3564	3651	3739	3826	3913	4001	4088	4176	4263	4350		
71	4438	4525	4612	4700	4787	4874	4962	5049	5137	5224		
72	5311	5399	5486	5573	5661	5748	5835	5923	6010	6097		
73	6185	6272	6359	6447	6534	6621	6709	6796	6883	6970		
74	7058	7145	7232	7320	7407	7494	7582	7669	7756	7844		
75	7931	8018	8105	8193	8280	8367	8455	8542	8629	8716		
76	8804	8891	8978	9066	9153	9240	9327	9415	9502	9589		
77	9676	9764	9851	9938	0025	0113	0200	0287	0374	0462		
78	6970549	0636	0723	0811	0898	0985	1072	1160	1247	1334		
79	1421	1508	1596	1683	1770	1857	1945	2032	2119	2206		
4980	2293	2381	2468	2555	2642	2729	2817	2904	2991	3078		
81	3165	3253	3340	3427	3514	3601	3689	3776	3863	3950		
82	4037	4124	4212	4299	4386	4473	4560	4647	4735	4822		
83	4909	4996	5083	5170	5257	5345	5432	5519	5606	5693		
84	5780	5867	5955	6042	6129	6216	6303	6390	6477	6565		
85	6652	6739	6826	6913	7000	7087	7174	7261	7349	7436		
86	7523	7610	7697	7784	7871	7958	8045	8132	8220	8307		
87	8394	8481	8568	8655	8742	8829	8916	9003	9090	9177		
88	9264	9352	9439	9526	9613	9700	9787	9874	9961	0048		
89	6980135	0222	0309	0396	0483	0570	0657	0744	0831	0918	87	
4990	1005	1092	1180	1267	1354	1441	1528	1615	1702	1789		
91	1876	1963	2050	2137	2224	2311	2398	2485	2572	2659		
92	2746	2833	2920	3007	3094	3181	3268	3355	3442	3529		86
93	3616	3703	3790	3877	3964	4051	4138	4224	4311	4398		1 9
94	4485	4572	4659	4746	4833	4920	5007	5094	5181	5268		2 17 / 3 26
95	5355	5442	5529	5616	5703	5790	5877	5964	6050	6137		4 34
96	6224	6311	6398	6485	6572	6659	6746	6833	6920	7007		5 43
97	7093	7180	7267	7354	7441	7528	7615	7702	7789	7876		6 52
98	7963	8049	8136	8223	8310	8397	8484	8571	8658	8744		7 60
99	8831	8918	9005	9092	9179	9266	9353	9439	9526	9613		8 69 / 9 77
N.	0	1	2	3	4	5	6	7	8	9	D	Pts.

N.	0	1	2	3	4	5	6	7	8	9	D	Pro.
5000	6989700	9787	9874	9961	ō047	ō134	ō221	ō308	ō395	ō482		
01	6990569	0655	0742	0829	0916	1003	1090	1176	1263	1350		87
02	1437	1524	1611	1697	1784	1871	1958	2045	2131	2218		1 | 9
03	2305	2392	2479	2565	2652	2739	2826	2913	2999	3086	87	2 | 17
04	3173	3260	3347	3433	3520	3607	3694	3780	3867	3954		3 | 26
												4 | 35
05	4041	4128	4214	4301	4388	4475	4561	4648	4735	4822		5 | 44
06	4908	4995	5082	5169	5255	5342	5429	5516	5602	5689		6 | 52
07	5776	5863	5949	6036	6123	6210	6296	6383	6470	6556		7 | 61
08	6643	6730	6817	6903	6990	7077	7163	7250	7337	7424		8 | 70
09	7510	7597	7684	7770	7857	7944	8031	8117	8204	8291		9 | 78
5010	8377	8464	8551	8637	8724	8811	8897	8984	9071	9157		
11	9244	9331	9417	9504	9591	9677	9764	9851	9937	ō024		
12	7000111	0197	0284	0371	0457	0544	0630	0717	0804	0890		
13	0977	1064	1150	1237	1324	1410	1497	1583	1670	1757		
14	1843	1930	2017	2103	2190	2276	2363	2450	2536	2623		
15	2709	2796	2883	2969	3056	3142	3229	3316	3402	3489		
16	3575	3662	3748	3835	3922	4008	4095	4181	4268	4354		
17	4441	4528	4614	4701	4787	4874	4960	5047	5133	5220		
18	5307	5393	5480	5566	5653	5739	5826	5912	5999	6085		
19	6172	6258	6345	6432	6518	6605	6691	6778	6864	6951		
5020	7037	7124	7210	7297	7383	7470	7556	7643	7729	7816		
21	7902	7989	8075	8162	8248	8335	8421	8508	8594	8681		
22	8767	8854	8940	9027	9113	9199	9286	9372	9459	9545		
23	9632	9718	9805	9891	9978	ō064	ō151	ō237	ō323	ō410		
24	7010496	0583	0669	0756	0842	0929	1015	1101	1188	1274		
25	1361	1447	1534	1620	1706	1793	1879	1966	2052	2138		
26	2225	2311	2398	2484	2570	2657	2743	2830	2916	3002		
27	3089	3175	3262	3348	3434	3521	3607	3694	3780	3866		
28	3953	4039	4125	4212	4298	4385	4471	4557	4644	4730		
29	4816	4903	4989	5075	5162	5248	5334	5421	5507	5594		
5030	5680	5766	5853	5939	6025	6112	6198	6284	6371	6457		
31	6543	6629	6716	6802	6888	6975	7061	7147	7234	7320		
32	7406	7493	7579	7665	7752	7838	7924	8010	8097	8183		
33	8269	8356	8442	8528	8614	8701	8787	8873	8960	9046		
34	9132	9218	9305	9391	9477	9563	9650	9736	9822	9908		
35	9995	ō081	ō167	ō254	ō340	ō426	ō512	ō598	ō685	ō771		
36	7020857	0943	1030	1116	1202	1288	1375	1461	1547	1633		
37	1720	1806	1892	1978	2064	2151	2237	2323	2409	2495		
38	2582	2668	2754	2840	2926	3013	3099	3185	3271	3357		
39	3444	3530	3616	3702	3788	3874	3961	4047	4133	4219		
5040	4305	4392	4478	4564	4650	4736	4822	4909	4995	5081		
41	5167	5253	5339	5425	5512	5598	5684	5770	5856	5942		
42	6028	6115	6201	6287	6373	6459	6545	6631	6717	6804		
43	6890	6976	7062	7148	7234	7320	7406	7492	7579	7665		86
44	7751	7837	7923	8009	8095	8181	8267	8353	8440	8526		1 | 9
												2 | 17
45	8612	8698	8784	8870	8956	9042	9128	9214	9300	9386		3 | 26
46	9472	9559	9645	9731	9817	9903	9989	ō075	ō161	ō247		4 | 34
47	7030333	0419	0505	0591	0677	0763	0849	0935	1021	1107		5 | 43
48	1193	1279	1366	1452	1538	1624	1710	1796	1882	1968		6 | 52
49	2054	2140	2226	2312	2398	2484	2570	2656	2742	2828		7 | 60
											86	8 | 69
N.	0	1	2	3	4	5	6	7	8	9	D	9 | 77

Pts.

N.	0	1	2	3	4	5	6	7	8	9	D
5050	7032914	3000	3086	3172	3258	3344	3430	3516	3602	3688	
51	3774	3860	3946	4032	4118	4204	4290	4376	4461	4547	
52	4633	4719	4805	4891	4977	5063	5149	5235	5321	5407	
53	5493	5579	5665	5751	5837	5923	6009	6095	6181	6266	
54	6352	6438	6524	6610	6696	6782	6868	6954	7040	7126	
55	7212	7298	7383	7469	7555	7641	7727	7813	7899	7985	
56	8071	8157	8242	8328	8414	8500	8586	8672	8758	8844	
57	8930	9015	9101	9187	9273	9359	9445	9531	9617	9702	
58	9788	9874	9960	$\bar0$046	$\bar0$132	$\bar0$218	$\bar0$303	$\bar0$389	$\bar0$475	$\bar0$561	
59	7040647	0733	0818	0904	0990	1076	1162	1248	1334	1419	
5060	1505	1591	1677	1763	1848	1934	2020	2106	2192	2278	
61	2363	2449	2535	2621	2707	2792	2878	2964	3050	3136	
62	3221	3307	3393	3479	3565	3650	3736	3822	3908	3993	
63	4079	4165	4251	4337	4422	4508	4594	4680	4765	4851	
64	4937	5023	5108	5194	5280	5366	5452	5537	5623	5709	
65	5794	5880	5966	6052	6137	6223	6309	6395	6480	6566	
66	6652	6738	6823	6909	6995	7080	7166	7252	7338	7423	
67	7509	7595	7680	7766	7852	7938	8023	8109	8195	8280	
68	8366	8452	8537	8623	8709	8795	8880	8966	9052	9137	
69	9223	9309	9394	9480	9566	9651	9737	9823	9908	9994	
5070	7050080	0165	0251	0337	0422	0508	0594	0679	0765	0850	
71	0936	1022	1107	1193	1279	1364	1450	1536	1621	1707	
72	1792	1878	1964	2049	2135	2221	2306	2392	2477	2563	
73	2649	2734	2820	2905	2991	3077	3162	3248	3333	3419	
74	3505	3590	3676	3761	3847	3933	4018	4104	4189	4275	
75	4360	4446	4532	4617	4703	4788	4874	4959	5045	5131	
76	5216	5302	5387	5473	5558	5644	5729	5815	5901	5986	
77	6072	6157	6243	6328	6414	6499	6585	6670	6756	6841	
78	6927	7012	7098	7184	7269	7355	7440	7526	7611	7697	
79	7782	7868	7953	8039	8124	8210	8295	8381	8466	8552	
5080	8637	8723	8808	8894	8979	9065	9150	9236	9321	9406	
81	9492	9577	9663	9748	9834	9919	$\bar0$005	$\bar0$090	$\bar0$176	$\bar0$261	
82	7060347	0432	0518	0603	0688	0774	0859	0945	1030	1116	
83	1201	1287	1372	1457	1543	1628	1714	1799	1885	1970	
84	2055	2141	2226	2312	2397	2483	2568	2653	2739	2824	
85	2910	2995	3080	3166	3251	3337	3422	3507	3593	3678	
86	3764	3849	3934	4020	4105	4190	4276	4361	4447	4532	
87	4617	4703	4788	4873	4959	5044	5130	5215	5300	5386	
88	5471	5556	5642	5727	5812	5898	5983	6068	6154	6239	
89	6325	6410	6495	6581	6666	6751	6837	6922	7007	7092	
5090	7178	7263	7348	7434	7519	7604	7690	7775	7860	7946	
91	8031	8116	8202	8287	8372	8457	8543	8628	8713	8799	
92	8884	8969	9055	9140	9225	9310	9396	9481	9566	9651	
93	9737	9822	9907	9993	$\bar0$078	$\bar0$163	$\bar0$248	$\bar0$334	$\bar0$419	$\bar0$504	
94	7070589	0675	0760	0845	0930	1016	1101	1186	1271	1357	
95	1442	1527	1612	1698	1783	1868	1953	2039	2124	2209	
96	2294	2379	2465	2550	2635	2720	2805	2891	2976	3061	
97	3146	3232	3317	3402	3487	3572	3658	3743	3828	3913	
98	3998	4083	4169	4254	4339	4424	4509	4595	4680	4765	
99	4850	4935	5020	5106	5191	5276	5361	5446	5531	5617	86
N.	0	1	2	3	4	5	6	7	8	9	D

Pro. Pts.

86
1	9
2	17
3	26
4	34
5	43
6	52
7	60
8	69
9	77

85
1	9
2	17
3	26
4	34
5	43
6	51
7	60
8	68
9	77

N.	0	1	2	3	4	5	6	7	8	9	D	Pro.
5100	7075702	5797	5872	5957	6042	6128	6213	6298	6383	6468		
01	6553	6638	6724	6809	6894	6979	7064	7149	7234	7319		86
02	7405	7490	7575	7660	7745	7830	7915	8000	8085	8171		1 9
03	8256	8341	8426	8511	8596	8681	8766	8851	8936	9022		2 17
04	9107	9192	9277	9362	9447	9532	9617	9702	9787	9872		3 26
05	9957	0̄043	0̄128	0̄213	0̄298	0̄383	0̄468	0̄553	0̄638	0̄723		4 34
06	7080808	0893	0978	1063	1148	1233	1318	1403	1488	1574		5 43
07	1659	1744	1829	1914	1999	2084	2169	2254	2339	2424		6 52
08	2509	2594	2679	2764	2849	2934	3019	3104	3189	3274		7 60
09	3359	3444	3529	3614	3699	3784	3869	3954	4039	4124	85	8 69
5110	4209	4294	4379	4464	4549	4634	4719	4804	4889	4974		9 77
11	5059	5144	5229	5314	5399	5484	5569	5654	5739	5823		
12	5908	5993	6078	6163	6248	6333	6418	6503	6588	6673		
13	6758	6843	6928	7013	7098	7183	7268	7352	7437	7522		
14	7607	7692	7777	7862	7947	8032	8117	8202	8287	8371		
15	8456	8541	8626	8711	8796	8881	8966	9051	9136	9220		
16	9305	9390	9475	9560	9645	9730	9815	9900	9984	0̄069		
17	7090154	0239	0324	0409	0494	0579	0663	0748	0833	0918		
18	1003	1088	1173	1257	1342	1427	1512	1597	1682	1766		
19	1851	1936	2021	2106	2191	2275	2360	2445	2530	2615		
5120	2700	2784	2869	2954	3039	3124	3209	3293	3378	3463		
21	3548	3633	3717	3802	3887	3972	4057	4141	4226	4311		
22	4396	4481	4565	4650	4735	4820	4904	4989	5074	5159		
23	5244	5328	5413	5498	5583	5667	5752	5837	5922	6006		
24	6091	6176	6261	6345	6430	6515	6600	6684	6769	6854		
25	6939	7023	7108	7193	7278	7362	7447	7532	7617	7701		
26	7786	7871	7955	8040	8125	8210	8294	8379	8464	8548		
27	8633	8718	8803	8887	8972	9057	9141	9226	9311	9395		
28	9480	9565	9650	9734	9819	9904	9988	0̄073	0̄158	0̄242		
29	7100327	0412	0496	0581	0666	0750	0835	0920	1004	1089		
5130	1174	1258	1343	1428	1512	1597	1682	1766	1851	1936		
31	2020	2105	2189	2274	2359	2443	2528	2613	2697	2782		
32	2866	2951	3036	3120	3205	3290	3374	3459	3543	3628		
33	3713	3797	3882	3966	4051	4136	4220	4305	4389	4474		
34	4559	4643	4728	4812	4897	4982	5066	5151	5235	5320		
35	5404	5489	5574	5658	5743	5827	5912	5996	6081	6166		
36	6250	6335	6419	6504	6588	6673	6757	6842	6927	7011		
37	7096	7180	7265	7349	7434	7518	7603	7687	7772	7856		
38	7941	8026	8110	8195	8279	8364	8448	8533	8617	8702		
39	8786	8871	8955	9040	9124	9209	9293	9378	9462	9547		
5140	9631	9716	9800	9885	9969	0̄054	0̄138	0̄223	0̄307	0̄392		
41	7110476	0561	0645	0729	0814	0898	0983	1067	1152	1236		85
42	1321	1405	1490	1574	1659	1743	1827	1912	1996	2081		1 9
43	2165	2250	2334	2419	2503	2587	2672	2756	2841	2925		2 17
44	3010	3094	3178	3263	3347	3432	3516	3601	3685	3769		3 26
45	3854	3938	4023	4107	4191	4276	4360	4445	4529	4613		4 34
46	4698	4782	4867	4951	5035	5120	5204	5289	5373	5457		5 43
47	5542	5626	5710	5795	5879	5964	6048	6132	6217	6301		6 51
48	6385	6470	6554	6638	6723	6807	6892	6976	7060	7145		7 60
49	7229	7313	7398	7482	7566	7651	7735	7819	7904	7988		8 68
												9 77
N.	0	1	2	3	4	5	6	7	8	9	D	Pts.

N.	0	1	2	3	4	5	6	7	8	9	D	Pro.
5150	7118072	8157	8241	8325	8410	8494	8578	8663	8747	8831		
51	8915	9000	9084	9168	9253	9337	9421	9506	9590	9674		84
52	9759	9843	9927	ō011	ō096	ō180	ō264	ō349	ō433	ō517		1 8
53	7120601	0686	0770	0854	0939	1023	1107	1191	1276	1360		2 17
54	1444	1528	1613	1697	1781	1865	1950	2034	2118	2202		3 25 4 34
55	2287	2371	2455	2539	2624	2708	2792	2876	2961	3045		5 42
56	3129	3213	3298	3382	3466	3550	3634	3719	3803	3887		6 50
57	3971	4056	4140	4224	4308	4392	4477	4561	4645	4729		7 59
58	4813	4898	4982	5066	5150	5234	5319	5403	5487	5571		8 67
59	5655	5739	5824	5908	5992	6076	6160	6245	6329	6413		9 76
5160	6497	6581	6665	6750	6834	6918	7002	7086	7170	7254		
61	7339	7423	7507	7591	7675	7759	7843	7928	8012	8096		
62	8180	8264	8348	8432	8517	8601	8685	8769	8853	8937		
63	9021	9105	9189	9274	9358	9442	9526	9610	9694	9778		
64	9862	9946	ō031	ō115	ō199	ō283	ō367	ō451	ō535	ō619		
65	7130703	0787	0871	0956	1040	1124	1208	1292	1376	1460		
66	1544	1628	1712	1796	1880	1964	2048	2132	2217	2301		
67	2385	2469	2553	2637	2721	2805	2889	2973	3057	3141		
68	3225	3309	3393	3477	3561	3645	3729	3813	3897	3981		
69	4065	4149	4233	4317	4401	4485	4569	4653	4737	4821	84	
5170	4905	4989	5073	5157	5241	5325	5409	5493	5577	5661		
71	5745	5829	5913	5997	6081	6165	6249	6333	6417	6501		
72	6585	6669	6753	6837	6921	7005	7089	7173	7257	7341		
73	7425	7509	7593	7677	7761	7845	7928	8012	8096	8180		
74	8264	8348	8432	8516	8600	8684	8768	8852	8936	9020		
75	9104	9187	9271	9355	9439	9523	9607	9691	9775	9859		
76	9943	ō027	ō110	ō194	ō278	ō362	ō446	ō530	ō614	ō698		
77	7140782	0866	0949	1033	1117	1201	1285	1369	1453	1537		
78	1620	1704	1788	1872	1956	2040	2124	2208	2291	2375		
79	2459	2543	2627	2711	2795	2878	2962	3046	3130	3214		
5180	3298	3381	3465	3549	3633	3717	3801	3884	3968	4052		
81	4136	4220	4304	4387	4471	4555	4639	4723	4806	4890		
82	4974	5058	5142	5226	5309	5393	5477	5561	5645	5728		
83	5812	5896	5980	6063	6147	6231	6315	6399	6482	6566		
84	6650	6734	6817	6901	6985	7069	7153	7236	7320	7404		
85	7488	7571	7655	7739	7823	7906	7990	8074	8158	8241		
86	8325	8409	8493	8576	8660	8744	8828	8911	8995	9079		
87	9162	9246	9330	9414	9497	9581	9665	9749	9832	9916		
88	7150000	0083	0167	0251	0335	0418	0502	0586	0669	0753		
89	0837	0920	1004	1088	1171	1255	1339	1423	1506	1590		
5190	1674	1757	1841	1925	2008	2092	2176	2259	2343	2427		
91	2510	2594	2678	2761	2845	2929	3012	3096	3180	3263		
92	3347	3430	3514	3598	3681	3765	3849	3932	4016	4100		83
93	4183	4267	4350	4434	4518	4601	4685	4769	4852	4936		1 8
94	5019	5103	5187	5270	5354	5438	5521	5605	5688	5772		2 17 3 25
95	5856	5939	6023	6106	6190	6273	6357	6441	6524	6608		4 33
96	6691	6775	6859	6942	7026	7109	7193	7276	7360	7444		5 42
97	7527	7611	7694	7778	7861	7945	8029	8112	8196	8279		6 50
98	8363	8446	8530	8613	8697	8780	8864	8948	9031	9115		7 58 8 66
99	9198	9282	9365	9449	9532	9616	9699	9783	9866	9950		9 75
N.	0	1	2	3	4	5	6	7	8	9	D	Pts.

N.	0	1	2	3	4	5	6	7	8	9	D	Pro.
5200	7160033	0117	0200	0284	0367	0451	0535	0618	0702	0785		
01	0869	0952	1036	1119	1203	1286	1370	1453	1537	1620		84
02	1703	1787	1870	1954	2037	2121	2204	2288	2371	2455		1|8
03	2538	2622	2705	2789	2872	2956	3039	3123	3206	3289		2|17
04	3373	3456	3540	3623	3707	3790	3874	3957	4040	4124		3|25 4|34
05	4207	4291	4374	4458	4541	4625	4708	4791	4875	4958		5|42
06	5042	5125	5208	5292	5375	5459	5542	5626	5709	5792		6|50
07	5876	5959	6043	6126	6209	6293	6376	6460	6543	6626		7|59
08	6710	6793	6877	6960	7043	7127	7210	7293	7377	7460		8|67
09	7544	7627	7710	7794	7877	7960	8044	8127	8211	8294		9|76
5210	8377	8461	8544	8627	8711	8794	8877	8961	9044	9127		
11	9211	9294	9377	9461	9544	9627	9711	9794	9877	9961		
12	7170044	0127	0211	0294	0377	0461	0544	0627	0711	0794		
13	0877	0961	1044	1127	1210	1294	1377	1460	1544	1627		
14	1710	1794	1877	1960	2043	2127	2210	2293	2377	2460		
15	2543	2626	2710	2793	2876	2959	3043	3126	3209	3293		
16	3376	3459	3542	3626	3709	3792	3875	3959	4042	4125		
17	4208	4292	4375	4458	4541	4625	4708	4791	4874	4958		
18	5041	5124	5207	5290	5374	5457	5540	5623	5707	5790		
19	5873	5956	6039	6123	6206	6289	6372	6455	6539	6622		
5220	6705	6788	6871	6955	7038	7121	7204	7287	7371	7454		
21	7537	7620	7703	7786	7870	7953	8036	8119	8202	8286		
22	8369	8452	8535	8618	8701	8784	8868	8951	9034	9117		
23	9200	9283	9367	9450	9533	9616	9699	9782	9865	9949		
24	7180032	0115	0198	0281	0364	0447	0530	0614	0697	0780		
25	0863	0946	1029	1112	1195	1279	1362	1445	1528	1611		
26	1694	1777	1860	1943	2026	2110	2193	2276	2359	2442		
27	2525	2608	2691	2774	2857	2940	3023	3107	3190	3273		
28	3356	3439	3522	3605	3688	3771	3854	3937	4020	4103		
29	4186	4269	4353	4436	4519	4602	4685	4768	4851	4934		
5230	5017	5100	5183	5266	5349	5432	5515	5598	5681	5764		
31	5847	5930	6013	6096	6179	6262	6345	6428	6511	6594	83	
32	6677	6760	6843	6926	7009	7092	7175	7258	7341	7424		
33	7507	7590	7673	7756	7839	7922	8005	8088	8171	8254		
34	8337	8420	8503	8586	8669	8752	8835	8918	9001	9084		
35	9167	9250	9333	9416	9499	9582	9665	9748	9830	9913		
36	9996	$\overline{0}$079	$\overline{0}$162	$\overline{0}$245	$\overline{0}$328	$\overline{0}$411	$\overline{0}$494	$\overline{0}$577	$\overline{0}$660	$\overline{0}$743		
37	7190826	0909	0992	1075	1157	1240	1323	1406	1489	1572		
38	1655	1738	1821	1904	1987	2069	2152	2235	2318	2401		
39	2484	2567	2650	2733	2816	2898	2981	3064	3147	3230		
5240	3313	3396	3479	3562	3644	3727	3810	3893	3976	4059		
41	4142	4224	4307	4390	4473	4556	4639	4722	4804	4887		
42	4970	5053	5136	5219	5302	5384	5467	5550	5633	5716		83
43	5799	5881	5964	6047	6130	6213	6296	6378	6461	6544		1|8
44	6627	6710	6792	6875	6958	7041	7124	7207	7289	7372		2|17 3|25
45	7455	7538	7621	7703	7786	7869	7952	8034	8117	8200		4|33
46	8283	8366	8448	8531	8614	8697	8780	8862	8945	9028		5|42
47	9111	9193	9276	9359	9442	9524	9607	9690	9773	9856		6|50
48	9938	$\overline{0}$021	$\overline{0}$104	$\overline{0}$187	$\overline{0}$269	$\overline{0}$352	$\overline{0}$435	$\overline{0}$518	$\overline{0}$600	$\overline{0}$683		7|58
49	7200766	0848	0931	1014	1097	1179	1262	1345	1428	1510		8|66 9|75
N.	0	1	2	3	4	5	6	7	8	9	D	Pts.

N.	0	1	2	3	4	5	6	7	8	9	D	Pro.
5250	7201593	1676	1758	1841	1924	2007	2089	2172	2255	2337		
51	2420	2503	2586	2668	2751	2834	2916	2999	3082	3164		82
52	3247	3330	3413	3495	3578	3661	3743	3826	3909	3991		1 8
53	4074	4157	4239	4322	4405	4487	4570	4653	4735	4818		2 16
54	4901	4983	5066	5149	5231	5314	5397	5479	5562	5645		3 25
												4 33
55	5727	5810	5892	5975	6058	6140	6223	6306	6388	6471		5 41
56	6554	6636	6719	6801	6884	6967	7049	7132	7215	7297		6 49
57	7380	7462	7545	7628	7710	7793	7875	7958	8041	8123		7 57
58	8206	8288	8371	8454	8536	8619	8701	8784	8867	8949		8 66
59	9032	9114	9197	9279	9362	9445	9527	9610	9692	9775		9 74
5260	9857	9940	ō023	ō105	ō188	ō270	ō353	ō435	ō518	ō600		
61	7210683	0766	0848	0931	1013	1096	1178	1261	1343	1426		
62	1508	1591	1674	1756	1839	1921	2004	2086	2169	2251		
63	2334	2416	2499	2581	2664	2746	2829	2911	2994	3076		
64	3159	3241	3324	3406	3489	3571	3654	3736	3819	3901		
65	3984	4066	4149	4231	4314	4396	4479	4561	4644	4726		
66	4809	4891	4973	5056	5138	5221	5303	5386	5468	5551		
67	5633	5716	5798	5881	5963	6045	6128	6210	6293	6375		
68	6458	6540	6623	6705	6787	6870	6952	7035	7117	7200		
69	7282	7364	7447	7529	7612	7694	7777	7859	7941	8024		
5270	8106	8189	8271	8353	8436	8518	8601	8683	8765	8848		
71	8930	9013	9095	9177	9260	9342	9424	9507	9589	9672		
72	9754	9836	9919	ō001	ō084	ō166	ō248	ō331	ō413	ō495		
73	7220578	0660	0742	0825	0907	0990	1072	1154	1237	1319		
74	1401	1484	1566	1648	1731	1813	1895	1978	2060	2142		
75	2225	2307	2389	2472	2554	2636	2719	2801	2883	2966		
76	3048	3130	3212	3295	3377	3459	3542	3624	3706	3789		
77	3871	3953	4036	4118	4200	4282	4365	4447	4529	4612		
78	4694	4776	4858	4941	5023	5105	5188	5270	5352	5434		
79	5517	5599	5681	5763	5846	5928	6010	6092	6175	6257		
5280	6339	6421	6504	6586	6668	6750	6833	6915	6997	7079		
81	7162	7244	7326	7408	7491	7573	7655	7737	7820	7902		
82	7984	8066	8148	8231	8313	8395	8477	8559	8642	8724		
83	8806	8888	8971	9053	9135	9217	9299	9382	9464	9546		
84	9628	9710	9792	9875	9957	ō039	ō121	ō203	ō286	ō368		
85	7230450	0532	0614	0696	0779	0861	0943	1025	1107	1189		
86	1272	1354	1436	1518	1600	1682	1765	1847	1929	2011		
87	2093	2175	2257	2340	2422	2504	2586	2668	2750	2832		
88	2914	2997	3079	3161	3243	3325	3407	3489	3571	3654		
89	3736	3818	3900	3982	4064	4146	4228	4310	4393	4475		
5290	4557	4639	4721	4803	4885	4967	5049	5131	5213	5296		
91	5378	5460	5542	5624	5706	5788	5870	5952	6034	6116		81
92	6198	6280	6362	6445	6527	6609	6691	6773	6855	6937		1 8
93	7019	7101	7183	7265	7347	7429	7511	7593	7675	7757		2 16
94	7839	7921	8003	8086	8167	8250	8332	8414	8496	8578		3 24
												4 32
95	8660	8742	8824	8906	8988	9070	9152	9234	9316	9398	82	5 41
96	9480	9562	9644	9726	9808	9890	9972	ō054	ō136	ō218		6 49
97	7240300	0382	0464	0546	0628	0710	0792	0874	0956	1038		7 57
98	1120	1202	1283	1365	1447	1529	1611	1693	1775	1857		8 65
99	1939	2021	2103	2185	2267	2349	2431	2513	2595	2677		9 73
N.	0	1	2	3	4	5	6	7	8	9	D	Pts.

N.	0	1	2	3	4	5	6	7	8	9	D	Pro.
5300	7242759	2841	2923	3005	3086	3168	3250	3332	3414	3496		
01	3578	3660	3742	3824	3906	3988	4070	4151	4233	4315		82
02	4397	4479	4561	4643	4725	4807	4889	4971	5052	5134		1 8
03	5216	5298	5380	5462	5544	5626	5708	5790	5871	5953		2 16
04	6035	6117	6199	6281	6363	6445	6526	6608	6690	6772		3 25
												4 33
05	6854	6936	7018	7099	7181	7263	7345	7427	7509	7591		5 41
06	7672	7754	7836	7918	8000	8082	8164	8245	8327	8409		6 49
07	8491	8573	8655	8736	8818	8900	8982	9064	9146	9227		7 57
08	9309	9391	9473	9555	9636	9718	9800	9882	9964	0̄045		8 66
09	7250127	0209	0291	0373	0454	0536	0618	0700	0782	0863		9 74
5310	0945	1027	1109	1191	1272	1354	1436	1518	1599	1681		
11	1763	1845	1927	2008	2090	2172	2254	2335	2417	2499		
12	2581	2662	2744	2826	2908	2989	3071	3153	3235	3316		
13	3398	3480	3562	3643	3725	3807	3889	3970	4052	4134		
14	4216	4297	4379	4461	4542	4624	4706	4788	4869	4951		
15	5033	5114	5196	5278	5360	5441	5523	5605	5686	5768		
16	5850	5931	6013	6095	6176	6258	6340	6422	6503	6585		
17	6667	6748	6830	6912	6993	7075	7157	7238	7320	7402		
18	7483	7565	7647	7728	7810	7892	7973	8055	8137	8218		
19	8300	8382	8463	8545	8626	8708	8790	8871	8953	9035		
5320	9116	9198	9280	9361	9443	9524	9606	9688	9769	9851		
21	9933	0̄014	0̄096	0̄177	0̄259	0̄341	0̄422	0̄504	0̄585	0̄667		
22	7260749	0830	0912	0994	1075	1157	1238	1320	1401	1483		
23	1565	1646	1728	1809	1891	1973	2054	2136	2217	2299		
24	2380	2462	2544	2625	2707	2788	2870	2951	3033	3115		
25	3196	3278	3359	3441	3522	3604	3685	3767	3849	3930		
26	4012	4093	4175	4256	4338	4419	4501	4582	4664	4745		
27	4827	4908	4990	5072	5153	5235	5316	5398	5479	5561		
28	5642	5724	5805	5887	5968	6050	6131	6213	6294	6376		
29	6457	6539	6620	6702	6783	6865	6946	7028	7109	7191		
5330	7272	7354	7435	7517	7598	7679	7761	7842	7924	8005		
31	8087	8168	8250	8331	8413	8494	8576	8657	8739	8820		
32	8901	8983	9064	9146	9227	9309	9390	9472	9553	9634		
33	9716	9797	9879	9960	0̄042	0̄123	0̄204	0̄286	0̄367	0̄449		
34	7270530	0612	0693	0774	0856	0937	1019	1100	1181	1263		
35	1344	1426	1507	1588	1670	1751	1833	1914	1995	2077		
36	2158	2240	2321	2402	2484	2565	2647	2728	2809	2891		
37	2972	3053	3135	3216	3298	3379	3460	3542	3623	3704		
38	3786	3867	3948	4030	4111	4192	4274	4355	4437	4518		
39	4599	4681	4762	4843	4925	5006	5087	5169	5250	5331		
5340	5413	5494	5575	5657	5738	5819	5901	5982	6063	6144		
41	6226	6307	6388	6470	6551	6632	6714	6795	6876	6958		
42	7039	7120	7201	7283	7364	7445	7527	7608	7689	7770		81
43	7852	7933	8014	8096	8177	8258	8339	8421	8502	8583		1 8
44	8664	8746	8827	8908	8990	9071	9152	9233	9315	9396		2 16
45	9477	9558	9640	9721	9802	9883	9965	0̄046	0̄127	0̄208		3 24
												4 32
46	7280290	0371	0452	0533	0614	0696	0777	0858	0939	1021		5 41
47	1102	1183	1264	1346	1427	1508	1589	1670	1752	1833		6 49
48	1914	1995	2076	2158	2239	2320	2401	2482	2564	2645		7 57
49	2726	2807	2888	2970	3051	3132	3213	3294	3375	3457		8 65
												9 73
N.	0	1	2	3	4	5	6	7	8	9	D	Pts.

N	0	1	2	3	4	5	6	7	8	9	D	Pro.
5350	7283538	3619	3700	3781	3863	3944	4025	4106	4187	4268		
51	4350	4431	4512	4593	4674	4755	4836	4918	4999	5080		81
52	5161	5242	5323	5404	5486	5567	5648	5729	5810	5891		1 8
53	5972	6054	6135	6216	6297	6378	6459	6540	6621	6703		2 16
54	6784	6865	6946	7027	7108	7189	7270	7351	7433	7514		3 24
												4 32
55	7595	7676	7757	7838	7919	8000	8081	8162	8244	8325		5 41
56	8406	8487	8568	8649	8730	8811	8892	8973	9054	9135		6 49
57	9216	9298	9379	9460	9541	9622	9703	9784	9865	9946		7 57
58	7290027	0108	0189	0270	0351	0432	0513	0594	0675	0757		8 65
59	0838	0919	1000	1081	1162	1243	1324	1405	1486	1567		9 73
5360	1648	1729	1810	1891	1972	2053	2134	2215	2296	2377		
61	2458	2539	2620	2701	2782	2863	2944	3025	3106	3187		
62	3268	3349	3430	3511	3592	3673	3754	3835	3916	3997	81	
63	4078	4159	4240	4321	4402	4483	4564	4645	4726	4807		
64	4888	4969	5050	5131	5212	5292	5373	5454	5535	5616		
65	5697	5778	5859	5940	6021	6102	6183	6264	6345	6426		
66	6507	6588	6669	6749	6830	6911	6992	7073	7154	7235		
67	7316	7397	7478	7559	7640	7721	7801	7882	7963	8044		
68	8125	8206	8287	8368	8449	8530	8610	8691	8772	8853		
69	8934	9015	9096	9177	9258	9338	9419	9500	9581	9662		
5370	9743	9824	9905	9985	0̄066	0̄147	0̄228	0̄309	0̄390	0̄471		
71	7300552	0632	0713	0794	0875	0956	1037	1118	1198	1279		
72	1360	1441	1522	1603	1683	1764	1845	1926	2007	2088		
73	2168	2249	2330	2411	2492	2573	2653	2734	2815	2896		
74	2977	3057	3138	3219	3300	3381	3461	3542	3623	3704		
75	3785	3865	3946	4027	4108	4189	4269	4350	4431	4512		
76	4593	4673	4754	4835	4916	4997	5077	5158	5239	5320		
77	5400	5481	5562	5643	5723	5804	5885	5966	6046	6127		
78	6208	6289	6369	6450	6531	6612	6692	6773	6854	6935		
79	7015	7096	7177	7258	7338	7419	7500	7581	7661	7742		
5380	7823	7903	7984	8065	8146	8226	8307	8388	8468	8549		
81	8630	8711	8791	8872	8953	9033	9114	9195	9276	9356		
82	9437	9518	9598	9679	9760	9840	9921	0̄002	0̄082	0̄163		
83	7310244	0324	0405	0486	0567	0647	0728	0809	0889	0970		
84	1051	1131	1212	1292	1373	1454	1534	1615	1696	1776		
85	1857	1938	2018	2099	2180	2260	2341	2422	2502	2583		
86	2663	2744	2825	2905	2986	3067	3147	3228	3309	3389		
87	3470	3550	3631	3712	3792	3873	3953	4034	4115	4195		
88	4276	4356	4437	4518	4598	4679	4759	4840	4921	5001		
89	5082	5162	5243	5324	5404	5485	5565	5646	5727	5807		
5390	5888	5968	6049	6129	6210	6291	6371	6452	6532	6613		
91	6693	6774	6854	6935	7016	7096	7177	7257	7338	7419		
92	7499	7579	7660	7740	7821	7902	7982	8063	8143	8224		80
93	8304	8385	8465	8546	8626	8707	8787	8868	8948	9029		1 8
94	9109	9190	9270	9351	9431	9512	9592	9673	9753	9834		2 16
												3 24
95	9914	9995	0̄075	0̄156	0̄236	0̄317	0̄397	0̄478	0̄558	0̄639		4 32
96	7320719	0800	0880	0961	1041	1122	1202	1283	1363	1444		5 40
97	1524	1605	1685	1766	1846	1927	2007	2087	2168	2248		6 48
98	2329	2409	2490	2570	2651	2731	2812	2892	2972	3053		7 56
99	3133	3214	3294	3375	3455	3535	3616	3696	3777	3857		8 64
												9 72
N.	0	1	2	3	4	5	6	7	8	9	D	Pts.

N.	0	1	2	3	4	5	6	7	8	9	D	Pro.
5400	7323938	4018	4098	4179	4259	4340	4420	4501	4581	4661		
01	4742	4822	4903	4983	5063	5144	5224	5305	5385	5465		81
02	5546	5626	5707	5787	5867	5948	6028	6109	6189	6269		1 8
03	6350	6430	6510	6591	6671	6752	6832	6912	6993	7073		2 16
04	7153	7234	7314	7394	7475	7555	7636	7716	7796	7877		3 24
												4 32
05	7957	8037	8118	8198	8278	8359	8439	8519	8600	8680		5 41
06	8760	8841	8921	9001	9082	9162	9242	9323	9403	9483		6 49
07	9564	9644	9724	9805	9885	9965	0̄046	0̄126	0̄206	0̄287		7 57
08	7330367	0447	0527	0608	0688	0768	0849	0929	1009	1090		8 65
09	1170	1250	1330	1411	1491	1571	1652	1732	1812	1892		9 73
5410	1973	2053	2133	2213	2294	2374	2454	2535	2615	2695		
11	2775	2856	2936	3016	3096	3177	3257	3337	3417	3498		
12	3578	3658	3738	3819	3899	3979	4059	4140	4220	4300		
13	4380	4461	4541	4621	4701	4781	4862	4942	5022	5102		
14	5183	5263	5343	5423	5503	5584	5664	5744	5824	5904		
15	5985	6065	6145	6225	6305	6386	6466	6546	6626	6706		
16	6787	6867	6947	7027	7107	7187	7268	7348	7428	7508		
17	7588	7669	7749	7829	7909	7989	8069	8150	8230	8310		
18	8390	8470	8550	8630	8711	8791	8871	8951	9031	9111		
19	9192	9272	9352	9432	9512	9592	9672	9752	9833	9913		
5420	9993	0̄073	0̄153	0̄233	0̄313	0̄393	0̄474	0̄554	0̄634	0̄714		
21	7340794	0874	0954	1034	1115	1195	1275	1355	1435	1515		
22	1595	1675	1755	1835	1916	1996	2076	2156	2236	2316		
23	2396	2476	2556	2636	2716	2796	2877	2957	3037	3117		
24	3197	3277	3357	3437	3517	3597	3677	3757	3837	3917		
25	3997	4077	4158	4238	4318	4398	4478	4558	4638	4718		
26	4798	4878	4958	5038	5118	5198	5278	5358	5438	5518		
27	5598	5678	5758	5838	5918	5998	6078	6158	6238	6318		
28	6398	6478	6558	6638	6718	6798	6878	6958	7038	7118	80	
29	7198	7278	7358	7438	7518	7598	7678	7758	7838	7918		
5430	7998	8078	8158	8238	8318	8398	8478	8558	8638	8718		
31	8798	8878	8958	9038	9118	9198	9278	9358	9438	9518		
32	9598	9678	9758	9837	9917	9997	0̄077	0̄157	0̄237	0̄317		
33	7350397	0477	0557	0637	0717	0797	0877	0957	1036	1116		
34	1196	1276	1356	1436	1516	1596	1676	1756	1836	1916		
35	1995	2075	2155	2235	2315	2395	2475	2555	2635	2715		
36	2794	2874	2954	3034	3114	3194	3274	3354	3434	3513		
37	3593	3673	3753	3833	3913	3993	4073	4152	4232	4312		
38	4392	4472	4552	4632	4711	4791	4871	4951	5031	5111		
39	5191	5270	5350	5430	5510	5590	5670	5749	5829	5909		
5440	5989	6069	6149	6228	6308	6388	6468	6548	6628	6707		
41	6787	6867	6947	7027	7107	7186	7266	7346	7426	7506		
42	7585	7665	7745	7825	7905	7984	8064	8144	8224	8304		80
43	8383	8463	8543	8623	8702	8782	8862	8942	9022	9101		1 8
44	9181	9261	9341	9420	9500	9580	9660	9740	9819	9899		2 16
												3 24
45	9979	0̄059	0̄138	0̄218	0̄298	0̄378	0̄457	0̄537	0̄617	0̄697		4 32
46	7360776	0856	0936	1016	1095	1175	1255	1335	1414	1494		5 40
47	1574	1653	1733	1813	1893	1972	2052	2132	2212	2291		6 48
48	2371	2451	2530	2610	2690	2770	2849	2929	3009	3088		7 56
49	3168	3248	3327	3407	3487	3567	3646	3726	3806	3885		8 64
												9 72
N.	0	1	2	3	4	5	6	7	8	9	D	Pts.

N	0	1	2	3	4	5	6	7	8	9	D	Pro.
5450	7363965	4045	4124	4204	4284	4363	4443	4523	4602	4682		
51	4762	4841	4921	5001	5080	5160	5240	5319	5399	5479		80
52	5558	5638	5718	5797	5877	5957	6036	6116	6196	6275		1 8
53	6355	6435	6514	6594	6674	6753	6833	6912	6992	7072		2 16
54	7151	7231	7311	7390	7470	7549	7629	7709	7788	7868		3 24
												4 32
55	7948	8027	8107	8186	8266	8346	8425	8505	8584	8664		5 40
56	8744	8823	8903	8982	9062	9142	9221	9301	9380	9460		6 48
57	9540	9619	9699	9778	9858	9937	0̄017	0̄097	0̄176	0̄256		7 56
58	7370335	0415	0494	0̄574	0654	0733	0813	0892	0972	1051		8 64
59	1131	1210	1290	1370	1449	1529	1608	1688	1767	1847		9 72
5460	1926	2006	2086	2165	2245	2324	2404	2483	2563	2642		
61	2722	2801	2881	2960	3040	3119	3199	3278	3358	3437		
62	3517	3596	3676	3755	3835	3914	3994	4074	4153	4233		
63	4312	4392	4471	4550	4630	4709	4789	4868	4948	5027		
64	5107	5186	5266	5345	5425	5504	5584	5663	5743	5822		
65	5902	5981	6061	6140	6220	6299	6378	6458	6537	6617		
66	6696	6776	6855	6935	7014	7094	7173	7252	7332	7411		
67	7491	7570	7650	7729	7808	7888	7967	8047	8126	8206		
68	8285	8364	8444	8523	8603	8682	8762	8841	8920	9000		
69	9079	9159	9238	9317	9397	9476	9556	9635	9714	9794		
5470	9873	9953	0̄032	0̄111	0̄191	0̄270	0̄350	0̄429	0̄508	0̄588		
71	7380667	0747	0826	0905	0985	1064	1143	1223	1302	1382		
72	1461	1540	1620	1699	1778	1858	1937	2016	2096	2175		
73	2254	2334	2413	2493	2572	2651	2731	2810	2889	2969		
74	3048	3127	3207	3286	3365	3445	3524	3603	3683	3762		
75	3841	3921	4000	4079	4159	4238	4317	4396	4476	4555		
76	4634	4714	4793	4872	4952	5031	5110	5190	5269	5348		
77	5427	5507	5586	5665	5745	5824	5903	5982	6062	6141		
78	6220	6300	6379	6458	6537	6617	6696	6775	6854	6934		
79	7013	7092	7172	7251	7330	7409	7489	7568	7647	7726		
5480	7806	7885	7964	8043	8123	8202	8281	8360	8440	8519		
81	8598	8677	8756	8836	8915	8994	9073	9153	9232	9311		
82	9390	9470	9549	9628	9707	9786	9866	9945	0̄024	0̄103		
83	7390182	0262	0341	0420	0499	0578	0658	0737	0816	0895		
84	0974	1054	1133	1212	1291	1370	1450	1529	1608	1687		
85	1766	1845	1925	2004	2083	2162	2241	2321	2400	2479		
86	2558	2637	2716	2796	2875	2954	3033	3112	3191	3270		
87	3350	3429	3508	3587	3666	3745	3824	3904	3983	4062		
88	4141	4220	4299	4378	4458	4537	4616	4695	4774	4853		
89	4932	5011	5091	5170	5249	5328	5407	5486	5565	5644		
5490	5723	5803	5882	5961	6040	6119	6198	6277	6356	6435		
91	6514	6594	6673	6752	6831	6910	6989	7068	7147	7226		
92	7305	7384	7463	7543	7622	7701	7780	7859	7938	8017		79
93	8096	8175	8254	8333	8412	8491	8570	8649	8728	8808		1 8
94	8887	8966	9045	9124	9203	9282	9361	9440	9519	9598		2 16
												3 24
95	9677	9756	9835	9914	9993	0̄072	0̄151	0̄230	0̄309	0̄388		4 32
96	7400467	0546	0625	0704	0783	0862	0941	1020	1099	1178	79	5 40
97	1257	1336	1415	1494	1573	1652	1731	1810	1889	1968		6 47
98	2047	2126	2205	2284	2363	2442	2521	2600	2679	2758		7 55
												8 63
99	2837	2916	2995	3074	3153	3232	3311	3390	3469	3548		9 71
N	0	1	2	3	4	5	6	7	8	9	D	Pts.

N.	0	1	2	3	4	5	6	7	8	9	D	Pro.
5500	7403627	3706	3785	3864	3943	4022	4101	4180	4259	4338		
01	4416	4495	4574	4653	4732	4811	4890	4969	5048	5127		**79**
02	5206	5285	5364	5443	5522	5601	5679	5758	5837	5916		1 8
03	5995	6074	6153	6232	6311	6390	6469	6548	6626	6705		2 16
04	6784	6863	6942	7021	7100	7179	7258	7337	7415	7494		3 24
												4 32
05	7573	7652	7731	7810	7889	7968	8047	8125	8204	8283		5 40
06	8362	8441	8520	8599	8678	8756	8835	8914	8993	9072		6 47
07	9151	9230	9308	9387	9466	9545	9624	9703	9782	9860		7 55
08	9939	0̄018	0̄097	0̄176	0̄255	0̄334	0̄412	0̄491	0̄570	0̄649		8 63
09	7410728	0807	0885	0964	1043	1122	1201	1280	1358	1437		9 71
5510	1516	1595	1674	1752	1831	1910	1989	2068	2146	2225		
11	2304	2383	2462	2541	2619	2698	2777	2856	2935	3013		
12	3092	3171	3250	3328	3407	3486	3565	3644	3722	3801		
13	3880	3959	4037	4116	4195	4274	4353	4431	4510	4589		
14	4668	4746	4825	4904	4983	5061	5140	5219	5298	5376		
15	5455	5534	5613	5691	5770	5849	5928	6006	6085	6164		
16	6243	6321	6400	6479	6557	6636	6715	6794	6872	6951		
17	7030	7109	7187	7266	7345	7423	7502	7581	7660	7738		
18	7817	7896	7974	8053	8132	8210	8289	8368	8447	8525		
19	8604	8683	8761	8840	8919	8997	9076	9155	9233	9312		
5520	9391	9469	9548	9627	9705	9784	9863	9941	0̄020	0̄099		
21	7420177	0256	0335	0413	0492	0571	0649	0728	0807	0885		
22	0964	1043	1121	1200	1279	1357	1436	1515	1593	1672		
23	1750	1829	1908	1986	2065	2144	2222	2301	2379	2458		
24	2537	2615	2694	2773	2851	2930	3008	3087	3166	3244		
25	3323	3401	3480	3559	3637	3716	3794	3873	3952	4030		
26	4109	4187	4266	4345	4423	4502	4580	4659	4737	4816		
27	4895	4973	5052	5130	5209	5288	5366	5445	5523	5602		
28	5680	5759	5837	5916	5995	6073	6152	6230	6309	6387		
29	6466	6544	6623	6702	6780	6859	6937	7016	7094	7173		
5530	7251	7330	7408	7487	7565	7644	7722	7801	7880	7958		
31	8037	8115	8194	8272	8351	8429	8508	8586	8665	8743		
32	8822	8900	8979	9057	9136	9214	9293	9371	9450	9528		
33	9607	9685	9764	9842	9921	9999	0̄078	0̄156	0̄235	0̄313		
34	7430392	0470	0549	0627	0705	0784	0862	0941	1019	1098		
35	1176	1255	1333	1412	1490	1569	1647	1725	1804	1882		
36	1961	2039	2118	2196	2275	2353	2431	2510	2588	2667		
37	2745	2824	2902	2981	3059	3137	3216	3294	3373	3451		
38	3530	3608	3686	3765	3843	3922	4000	4078	4157	4235		
39	4314	4392	4470	4549	4627	4706	4784	4862	4941	5019		
5540	5098	5176	5254	5333	5411	5490	5568	5646	5725	5803		
41	5882	5960	6038	6117	6195	6273	6352	6430	6508	6587		
42	6665	6744	6822	6900	6979	7057	7135	7214	7292	7370		**78**
43	7449	7527	7605	7684	7762	7841	7919	7997	8076	8154		1 8
44	8232	8311	8389	8467	8546	8624	8702	8781	8859	8937		2 16
												3 23
45	9016	9094	9172	9250	9329	9407	9485	9564	9642	9720		4 31
46	9799	9877	9955	0̄034	0̄112	0̄190	0̄268	0̄347	0̄425	0̄503		5 39
47	7440582	0660	0738	0817	0895	0973	1051	1130	1208	1286		6 47
48	1365	1443	1521	1599	1678	1756	1834	1912	1991	2069		7 55
49	2147	2226	2304	2382	2460	2539	2617	2695	2773	2852		8 62
												9 70
N.	0	1	2	3	4	5	6	7	8	9	D	Pts.

N.	0	1	2	3	4	5	6	7	8	9	D	Pro.
5550	7442930	3008	3086	3165	3243	3321	3399	3478	3556	3634		
51	3712	3791	3869	3947	4025	4103	4182	4260	4338	4416		78
52	4495	4573	4651	4729	4807	4886	4964	5042	5120	5199		1 8
53	5277	5355	5433	5511	5590	5668	5746	5824	5902	5981		2 16
54	6059	6137	6215	6293	6372	6450	6528	6606	6684	6762		3 23
55	6841	6919	6997	7075	7153	7232	7310	7388	7466	7544		4 31
56	7622	7701	7779	7857	7935	8013	8091	8170	8248	8326		5 39
57	8404	8482	8560	8638	8717	8795	8873	8951	9029	9107		6 47
58	9185	9264	9342	9420	9498	9576	9654	9732	9810	9889		7 55
59	9967	0̄045	0̄123	0̄201	0̄279	0̄357	0̄435	0̄514	0̄592	0̄670		8 62
5560	7450748	0826	0904	0982	1060	1138	1217	1295	1373	1451		9 70
61	1529	1607	1685	1763	1841	1919	1998	2076	2154	2232		
62	2310	2388	2466	2544	2622	2700	2778	2856	2934	3013		
63	3091	3169	3247	3325	3403	3481	3559	3637	3715	3793		
64	3871	3949	4027	4105	4183	4261	4340	4418	4496	4574		
65	4652	4730	4808	4886	4964	5042	5120	5198	5276	5354		
66	5432	5510	5588	5666	5744	5822	5900	5978	6056	6134	78	
67	6212	6290	6368	6446	6524	6602	6680	6758	6836	6914		
68	6992	7070	7148	7226	7304	7382	7460	7538	7616	7694		
69	7772	7850	7928	8006	8084	8162	8240	8318	8396	8474		
5570	8552	8630	8708	8786	8864	8942	9020	9098	9176	9254		
71	9332	9410	9487	9565	9643	9721	9799	9877	9955	0̄033		
72	7460111	0189	0267	0345	0423	0501	0579	0657	0735	0813		
73	0890	0968	1046	1124	1202	1280	1358	1436	1514	1592		
74	1670	1748	1825	1903	1981	2059	2137	2215	2293	2371		
75	2449	2527	2605	2682	2760	2838	2916	2994	3072	3150		
76	3228	3306	3383	3461	3539	3617	3695	3773	3851	3929		
77	4006	4084	4162	4240	4318	4396	4474	4552	4629	4707		
78	4785	4863	4941	5019	5097	5174	5252	5330	5408	5486		
79	5564	5641	5719	5797	5875	5953	6031	6108	6186	6264		
5580	6342	6420	6498	6575	6653	6731	6809	6887	6965	7042		
81	7120	7198	7276	7354	7431	7509	7587	7665	7743	7821		
82	7898	7976	8054	8132	8210	8287	8365	8443	8521	8598		
83	8676	8754	8832	8910	8987	9065	9143	9221	9299	9376		
84	9454	9532	9610	9687	9765	9843	9921	9998	0̄076	0̄154		
85	7470232	0310	0387	0465	0543	0621	0698	0776	0854	0932		
86	1009	1087	1165	1243	1320	1398	1476	1554	1631	1709		
87	1787	1864	1942	2020	2098	2175	2253	2331	2409	2486		
88	2564	2642	2719	2797	2875	2953	3030	3108	3186	3263		
89	3341	3419	3497	3574	3652	3730	3807	3885	3963	4040		
5590	4118	4196	4273	4351	4429	4507	4584	4662	4740	4817		
91	4895	4973	5050	5128	5206	5283	5361	5439	5516	5594		
92	5672	5749	5827	5905	5982	6060	6138	6215	6293	6371		77
93	6448	6526	6603	6681	6759	6836	6914	6992	7069	7147		1 8
94	7225	7302	7380	7458	7535	7613	7690	7768	7846	7923		2 15
95	8001	8079	8156	8234	8311	8389	8467	8544	8622	8699		3 23
96	8777	8855	8932	9010	9087	9165	9243	9320	9398	9475		4 31
97	9553	9631	9708	9786	9863	9941	0̄019	0̄096	0̄174	0̄251		5 39
98	7480329	0407	0484	0562	0639	0717	0794	0872	0950	1027		6 46
99	1105	1182	1260	1337	1415	1492	1570	1648	1725	1803		7 54
												8 62
												9 69
N.	0	1	2	3	4	5	6	7	8	9	D	Pts.

H

N.	0	1	2	3	4	5	6	7	8	9	D	Pro.
5600	7481880	1958	2035	2113	2190	2268	2346	2423	2501	2578		
01	2656	2733	2811	2888	2966	3043	3121	3198	3276	3354		78
02	3431	3509	3586	3664	3741	3819	3896	3974	4051	4129		1 8
03	4206	4284	4361	4439	4516	4594	4671	4749	4826	4904		2 16
04	4981	5059	5136	5214	5291	5369	5446	5524	5601	5679		3 23
												4 31
05	5756	5834	5911	5989	6066	6144	6221	6299	6376	6453		5 39
06	6531	6608	6686	6763	6841	6918	6996	7073	7151	7228		6 47
07	7306	7383	7460	7538	7615	7693	7770	7848	7925	8003		7 55
08	8080	8157	8235	8312	8390	8467	8545	8622	8700	8777		8 62
09	8854	8932	9009	9087	9164	9242	9319	9396	9474	9551		9 70
5610	9629	9706	9783	9861	9938	0̄016	0̄093	0̄170	0̄248	0̄325		
11	7490403	0480	0557	0635	0712	0790	0867	0944	1022	1099		
12	1177	1254	1331	1409	1486	1564	1641	1718	1796	1873		
13	1950	2028	2105	2183	2260	2337	2415	2492	2569	2647		
14	2724	2801	2879	2956	3034	3111	3188	3266	3343	3420		
15	3498	3575	3652	3730	3807	3884	3962	4039	4116	4194		
16	4271	4348	4426	4503	4580	4658	4735	4812	4890	4967		
17	5044	5122	5199	5276	5353	5431	5508	5585	5663	5740		
18	5817	5895	5972	6049	6127	6204	6281	6358	6436	6513		
19	6590	6668	6745	6822	6899	6977	7054	7131	7209	7286		
5620	7363	7440	7518	7595	7672	7750	7827	7904	7981	8059		
21	8136	8213	8290	8368	8445	8522	8599	8677	8754	8831		
22	8908	8986	9063	9140	9217	9295	9372	9449	9526	9604		
23	9681	9758	9835	9913	9990	0̄067	0̄144	0̄221	0̄299	0̄376		
24	7500453	0530	0608	0685	0762	0839	0916	0994	1071	1148		
25	1225	1302	1380	1457	1534	1611	1688	1766	1843	1920		
26	1997	2074	2152	2229	2306	2383	2460	2538	2615	2692		
27	2769	2846	2924	3001	3078	3155	3232	3309	3387	3464		
28	3541	3618	3695	3772	3850	3927	4004	4081	4158	4235		
29	4312	4390	4467	4544	4621	4698	4775	4853	4930	5007		
5630	5084	5161	5238	5315	5392	5470	5547	5624	5701	5778		
31	5855	5932	6010	6087	6164	6241	6318	6395	6472	6549		
32	6626	6704	6781	6858	6935	7012	7089	7166	7243	7320		
33	7398	7475	7552	7629	7706	7783	7860	7937	8014	8091		
34	8168	8246	8323	8400	8477	8554	8631	8708	8785	8862		
35	8939	9016	9093	9170	9247	9325	9402	9479	9556	9633		
36	9710	9787	9864	9941	0̄018	0̄095	0̄172	0̄249	0̄326	0̄403		
37	7510480	0557	0634	0711	0789	0866	0943	1020	1097	1174		
38	1251	1328	1405	1482	1559	1636	1713	1790	1867	1944	77	
39	2021	2098	2175	2252	2329	2406	2483	2560	2637	2714		
5640	2791	2868	2945	3022	3099	3176	3253	3330	3407	3484		
41	3561	3638	3715	3792	3869	3946	4023	4100	4177	4254		
42	4331	4408	4485	4562	4639	4716	4793	4870	4947	5024		77
43	5101	5177	5254	5331	5408	5485	5562	5639	5716	5793		1 8
44	5870	5947	6024	6101	6178	6255	6332	6409	6486	6563		2 15
												3 23
45	6639	6716	6793	6870	6947	7024	7101	7178	7255	7332		4 31
46	7409	7486	7563	7639	7716	7793	7870	7947	8024	8101		5 39
47	8178	8255	8332	8409	8485	8562	8639	8716	8793	8870		6 46
48	8947	9024	9101	9178	9254	9331	9408	9485	9562	9639		7 54
49	9716	9793	9870	9946	0̄023	0̄100	0̄177	0̄254	0̄331	0̄408		8 62
												9 69
N.	0	1	2	3	4	5	6	7	8	9	D	Pts.

N.	0	1	2.	3	4	5	6	7	8	9	D	Pro.
5650	7520484	0561	0638	0715	0792	0869	0946	1023	1099	1176		
51	1253	1330	1407	1484	1560	1637	1714	1791	1868	1945		**77**
52	2022	2098	2175	2252	2329	2406	2483	2559	2636	2713		1 8
53	2790	2867	2944	3020	3097	3174	3251	3328	3404	3481		2 15
54	3558	3635	3712	3788	3865	3942	4019	4096	4172	4249		3 23
												4 31
55	4326	4403	4480	4556	4633	4710	4787	4864	4940	5017		5 39
56	5094	5171	5248	5324	5401	5478	5555	5631	5708	5785		6 46
57	5862	5939	6015	6092	6169	6246	6322	6399	6476	6553		7 54
58	6629	6706	6783	6860	6936	7013	7090	7167	7243	7320		8 62
59	7397	7474	7550	7627	7704	7781	7857	7934	8011	8088		9 69
5660	8164	8241	8318	8394	8471	8548	8625	8701	8778	8855		
61	8932	9008	9085	9162	9238	9315	9392	9469	9545	9622		
62	9699	9775	9852	9929	ō005	ō082	ō159	ō236	ō312	ō389		
63	7530466	0542	0619	0696	0772	0849	0926	1002	1079	1156		
64	1232	1309	1386	1462	1539	1616	1692	1769	1846	1922		
65	1999	2076	2152	2229	2306	2382	2459	2536	2612	2689		
66	2766	2842	2919	2996	3072	3149	3226	3302	3379	3455		
67	3532	3609	3685	3762	3839	3915	3992	4069	4145	4222		
68	4298	4375	4452	4528	4605	4682	4758	4835	4911	4988		
69	5065	5141	5218	5294	5371	5448	5524	5601	5677	5754		
5670	5831	5907	5984	6060	6137	6214	6290	6367	6443	6520		
71	6596	6673	6750	6826	6903	6979	7056	7133	7209	7286		
72	7362	7439	7515	7592	7668	7745	7822	7898	7975	8051		
73	8128	8204	8281	8357	8434	8511	8587	8664	8740	8817		
74	8893	8970	9046	9123	9199	9276	9353	9429	9506	9582		
75	9659	9735	9812	9888	9965	ō041	ō118	ō194	ō271	ō347		
76	7540424	0500	0577	0653	0730	0806	0883	0959	1036	1112		
77	1189	1265	1342	1418	1495	1571	1648	1724	1801	1877		
78	1954	2030	2107	2183	2260	2336	2413	2489	2566	2642		
79	2719	2795	2872	2948	3025	3101	3178	3254	3330	3407		
5680	3483	3560	3636	3713	3789	3866	3942	4019	4095	4171		
81	4248	4324	4401	4477	4554	4630	4707	4783	4859	4936		
82	5012	5089	5165	5242	5318	5394	5471	5547	5624	5700		
83	5777	5853	5929	6006	6082	6159	6235	6311	6388	6464		
84	6541	6617	6694	6770	6846	6923	6999	7076	7152	7228		
85	7305	7381	7457	7534	7610	7687	7763	7839	7916	7992		
86	8069	8145	8221	8298	8374	8450	8527	8603	8680	8756		
87	8832	8909	8985	9061	9138	9214	9290	9367	9443	9520		
88	9596	9672	9749	9825	9901	9978	ō054	ō130	ō207	ō283		
89	7550359	0436	0512	0588	0665	0741	0817	0894	0970	1046		
5690	1123	1199	1275	1352	1428	1504	1581	1657	1733	1810		
91	1886	1962	2038	2115	2191	2267	2344	2420	2496	2573		
92	2649	2725	2802	2878	2954	3030	3107	3183	3259	3336		**76**
93	3412	3488	3564	3641	3717	3793	3870	3946	4022	4098		1 8
94	4175	4251	4327	4403	4480	4556	4632	4709	4785	4861		2 15
												3 23
95	4937	5014	5090	5166	5242	5319	5395	5471	5547	5624		4 30
96	5700	5776	5852	5929	6005	6081	6157	6233	6310	6386		5 38
97	6462	6538	6615	6691	6767	6843	6920	6996	7072	7148		6 46
98	7224	7301	7377	7453	7529	7606	7682	7758	7834	7910		7 53
99	7987	8063	8139	8215	8291	8368	8444	8520	8596	8672		8 61
												9 68
N.	0	1	2	3	4	5	6	7	8	9	D	Pts.

N.	0	1	2	3	4	5	6	7	8	9	D	Pro.
5700	7558749	8825	8901	8977	9053	9130	9206	9282	9358	9434		
01	9510	9587	9663	9739	9815	9891	9967	ō044	ō120	ō196		
02	7560272	0348	0424	0501	0577	0653	0729	0805	0881	0958		**77**
03	1034	1110	1186	1262	1338	1414	1491	1567	1643	1719		1 | 8
04	1795	1871	1947	2024	2100	2176	2252	2328	2404	2480		2 | 15 3 | 23
05	2556	2633	2709	2785	2861	2937	3013	3089	3165	3242		4 | 31 5 | 39
06	3318	3394	3470	3546	3622	3698	3774	3850	3927	4003		6 | 46
07	4079	4155	4231	4307	4383	4459	4535	4611	4687	4764		7 | 54
08	4840	4916	4992	5068	5144	5220	5296	5372	5448	5524		8 | 62
09	5600	5677	5753	5829	5905	5981	6057	6133	6209	6285		9 | 69
5710	6361	6437	6513	6589	6665	6741	6817	6893	6970	7046		
11	7122	7198	7274	7350	7426	7502	7578	7654	7730	7806		
12	7882	7958	8034	8110	8186	8262	8338	8414	8490	8566		
13	8642	8718	8794	8870	8946	9022	9098	9174	9250	9326		
14	9402	9478	9554	9630	9706	9782	9858	9934	ō010	ō086	76	
15	7570162	0238	0314	0390	0466	0542	0618	0694	0770	0846		
16	0922	0998	1074	1150	1226	1302	1378	1454	1530	1606		
17	1682	1758	1834	1910	1986	2062	2138	2214	2290	2366		
18	2442	2517	2593	2669	2745	2821	2897	2973	3049	3125		
19	3201	3277	3353	3429	3505	3581	3657	3733	3808	3884		
5720	3960	4036	4112	4188	4264	4340	4416	4492	4568	4644		
21	4719	4795	4871	4947	5023	5099	5175	5251	5327	5403		
22	5479	5554	5630	5706	5782	5858	5934	6010	6086	6162		
23	6237	6313	6389	6465	6541	6617	6693	6769	6845	6920		
24	6996	7072	7148	7224	7300	7376	7451	7527	7603	7679		
25	7755	7831	7907	7982	8058	8134	8210	8286	8362	8438		
26	8513	8589	8665	8741	8817	8893	8968	9044	9120	9196		
27	9272	9348	9423	9499	9575	9651	9727	9803	9878	9954		
28	7580030	0106	0182	0258	0333	0409	0485	0561	0637	0712		
29	0788	0864	0940	1016	1091	1167	1243	1319	1395	1470		
5730	1546	1622	1698	1774	1849	1925	2001	2077	2153	2228		
31	2304	2380	2456	2531	2607	2683	2759	2835	2910	2986		
32	3062	3138	3213	3289	3365	3441	3516	3592	3668	3744		
33	3819	3895	3971	4047	4122	4198	4274	4350	4425	4501		
34	4577	4653	4728	4804	4880	4956	5031	5107	5183	5258		
35	5334	5410	5486	5561	5637	5713	5789	5864	5940	6016		
36	6091	6167	6243	6319	6394	6470	6546	6621	6697	6773		
37	6848	6924	7000	7076	7151	7227	7303	7378	7454	7530		
38	7605	7681	7757	7832	7908	7984	8060	8135	8211	8287		
39	8362	8438	8514	8589	8665	8741	8816	8892	8968	9043		
5740	9119	9195	9270	9346	9422	9497	9573	9649	9724	9800		
41	9875	9951	ō027	ō102	ō178	ō254	ō329	ō405	ō481	ō556		
42	7590632	0708	0783	0859	0934	1010	1086	1161	1237	1313		**76**
43	1388	1464	1539	1615	1691	1766	1842	1917	1993	2069		1 | 8
44	2144	2220	2296	2371	2447	2522	2598	2674	2749	2825		2 | 15 3 | 23
45	2900	2976	3052	3127	3203	3278	3354	3429	3505	3581		4 | 30
46	3656	3732	3807	3883	3959	4034	4110	4185	4261	4336		5 | 38
47	4412	4488	4563	4639	4714	4790	4865	4941	5016	5092		6 | 46
48	5168	5243	5319	5394	5470	5545	5621	5696	5772	5848		7 | 53 8 | 61
49	5923	5999	6074	6150	6225	6301	6376	6452	6527	6603		9 | 68
N.	0	1	2	3	4	5	6	7	8	9	D	Pts.

N.	0	1	2	3	4	5	6	7	8	9	D
5750	7596678	6754	6830	6905	6981	7056	7132	7207	7283	7358	
51	7434	7509	7585	7660	7736	7811	7887	7962	8038	8113	
52	8189	8264	8340	8415	8491	8566	8642	8717	8793	8868	
53	8944	9019	9095	9170	9246	9321	9397	9472	9548	9623	
54	9699	9774	9850	9925	ō000	ō076	ō151	ō227	ō302	ō378	
55	7600453	0529	0604	0680	0755	0831	0906	0981	1057	1132	
56	1208	1283	1359	1434	1510	1585	1661	1736	1811	1887	
57	1962	2038	2113	2189	2264	2339	2415	2490	2566	2641	
58	2717	2792	2867	2943	3018	3094	3169	3245	3320	3395	
59	3471	3546	3622	3697	3772	3848	3923	3999	4074	4149	
5760	4225	4300	4376	4451	4526	4602	4677	4753	4828	4903	
61	4979	5054	5130	5205	5280	5356	5431	5506	5582	5657	
62	5733	5808	5883	5959	6034	6109	6185	6260	6335	6411	
63	6486	6562	6637	6712	6788	6863	6938	7014	7089	7164	
64	7240	7315	7390	7466	7541	7616	7692	7767	7842	7918	
65	7993	8068	8144	8219	8294	8370	8445	8520	8596	8671	
66	8746	8822	8897	8972	9048	9123	9198	9274	9349	9424	
67	9500	9575	9650	9725	9801	9876	9951	ō027	ō102	ō177	
68	7610253	0328	0403	0478	0554	0629	0704	0780	0855	0930	
69	1005	1081	1156	1231	1307	1382	1457	1532	1608	1683	
5770	1758	1833	1909	1984	2059	2134	2210	2285	2360	2435	
71	2511	2586	2661	2737	2812	2887	2962	3037	3113	3188	
72	3263	3338	3414	3489	3564	3639	3715	3790	3865	3940	
73	4016	4091	4166	4241	4316	4392	4467	4542	4617	4693	
74	4768	4843	4918	4993	5069	5144	5219	5294	5369	5445	
75	5520	5595	5670	5745	5821	5896	5971	6046	6121	6197	
76	6272	6347	6422	6497	6573	6648	6723	6798	6873	6948	
77	7024	7099	7174	7249	7324	7400	7475	7550	7625	7700	
78	7775	7851	7926	8001	8076	8151	8226	8301	8377	8452	
79	8527	8602	8677	8752	8828	8903	8978	9053	9128	9203	
5780	9278	9354	9429	9504	9579	9654	9729	9804	9879	9955	
81	7620030	0105	0180	0255	0330	0405	0480	0556	0631	0706	
82	0781	0856	0931	1006	1081	1156	1232	1307	1382	1457	
83	1532	1607	1682	1757	1832	1907	1982	2058	2133	2208	
84	2283	2358	2433	2508	2583	2658	2733	2808	2883	2959	
85	3034	3109	3184	3259	3334	3409	3484	3559	3634	3709	
86	3784	3859	3934	4009	4085	4160	4235	4310	4385	4460	
87	4535	4610	4685	4760	4835	4910	4985	5060	5135	5210	
88	5285	5360	5435	5510	5585	5660	5735	5810	5885	5960	75
89	6035	6111	6186	6261	6336	6411	6486	6561	6636	6711	
5790	6786	6861	6936	7011	7086	7161	7236	7311	7386	7461	
91	7536	7611	7686	7761	7836	7911	7986	8061	8136	8211	
92	8286	8361	8435	8510	8585	8660	8735	8810	8885	8960	
93	9035	9110	9185	9260	9335	9410	9485	9560	9635	9710	
94	9785	9860	9935	ō010	ō085	ō160	ō235	ō310	ō385	ō459	
95	7630534	0609	0684	0759	0834	0909	0984	1059	1134	1209	
96	1284	1359	1434	1509	1583	1658	1733	1808	1883	1958	
97	2033	2108	2183	2258	2333	2408	2482	2557	2632	2707	
98	2782	2857	2932	3007	3082	3157	3232	3306	3381	3456	
99	3531	3606	3681	3756	3831	3906	3980	4055	4130	4205	
N.	0	1	2	3	4	5	6	7	8	9	D

Pro.

75	
1	8
2	15
3	23
4	30
5	38
6	45
7	53
8	60
9	68

74	
1	7
2	15
3	22
4	30
5	37
6	44
7	52
8	59
9	67

Pts.

N.	0	1	2	3	4	5	6	7	8	9	D	Pro.
5800	7634280	4355	4430	4505	4579	4654	4729	4804	4879	4954		
01	5029	5104	5178	5253	5328	5403	5478	5553	5628	5702		75
02	5777	5852	5927	6002	6077	6151	6226	6301	6376	6451		1 8
03	6526	6601	6675	6750	6825	6900	6975	7050	7124	7199		2 15
04	7274	7349	7424	7499	7573	7648	7723	7798	7873	7947	75	3 23 / 4 30
05	8022	8097	8172	8247	8321	8396	8471	8546	8621	8696		5 38
06	8770	8845	8920	8995	9070	9144	9219	9294	9369	9443		6 45
07	9518	9593	9668	9743	9817	9892	9967	ō042	ō117	ō191		7 53
08	7640266	0341	0416	0490	0565	0640	0715	0789	0864	0939		8 60
09	1014	1089	1163	1238	1313	1388	1462	1537	1612	1687		9 68
5810	1761	1836	1911	1986	2060	2135	2210	2285	2359	2434		
11	2509	2583	2658	2733	2808	2882	2957	3032	3107	3181		
12	3256	3331	3406	3480	3555	3630	3704	3779	3854	3929		
13	4003	4078	4153	4227	4302	4377	4451	4526	4601	4676		
14	4750	4825	4900	4974	5049	5124	5198	5273	5348	5423		
15	5497	5572	5647	5721	5796	5871	5945	6020	6095	6169		
16	6244	6319	6393	6468	6543	6617	6692	6767	6841	6916		
17	6991	7065	7140	7215	7289	7364	7439	7513	7588	7663		
18	7737	7812	7886	7961	8036	8110	8185	8260	8334	8409		
19	8484	8558	8633	8707	8782	8857	8931	9006	9081	9155		
5820	9230	9304	9379	9454	9528	9603	9678	9752	9827	9901		
21	9976	ō051	ō125	ō200	ō274	ō349	ō424	ō498	ō573	ō647		
22	7650722	0797	0871	0946	1020	1095	1170	1244	1319	1393		
23	1468	1542	1617	1692	1766	1841	1915	1990	2065	2139		
24	2214	2288	2363	2437	2512	2586	2661	2736	2810	2885		
25	2959	3034	3108	3183	3258	3332	3407	3481	3556	3630		
26	3705	3779	3854	3928	4003	4078	4152	4227	4301	4376		
27	4450	4525	4599	4674	4748	4823	4897	4972	5046	5121		
28	5195	5270	5344	5419	5493	5568	5643	5717	5792	5866		
29	5941	6015	6090	6164	6239	6313	6388	6462	6537	6611		
5830	6686	6760	6835	6909	6984	7058	7132	7207	7281	7356		
31	7430	7505	7579	7654	7728	7803	7877	7952	8026	8101		
32	8175	8250	8324	8399	8473	8547	8622	8696	8771	8845		
33	8920	8994	9069	9143	9218	9292	9366	9441	9515	9590		
34	9664	9739	9813	9888	9962	ō036	ō111	ō185	ō260	ō334		
35	7660409	0483	0557	0632	0706	0781	0855	0930	1004	1078		
36	1153	1227	1302	1376	1450	1525	1599	1674	1748	1823		
37	1897	1971	2046	2120	2195	2269	2343	2418	2492	2567		
38	2641	2715	2790	2864	2938	3013	3087	3162	3236	3310		
39	3385	3459	3534	3608	3682	3757	3831	3905	3980	4054		
5840	4128	4203	4277	4352	4426	4500	4575	4649	4723	4798		
41	4872	4946	5021	5095	5169	5244	5318	5393	5467	5541		
42	5616	5690	5764	5839	5913	5987	6062	6136	6210	6285		71
43	6359	6433	6508	6582	6656	6730	6805	6879	6953	7028		1 7
44	7102	7176	7251	7325	7399	7474	7548	7622	7697	7771		2 15 / 3 22
45	7845	7919	7994	8068	8142	8217	8291	8365	8440	8514	74	4 30
46	8588	8662	8737	8811	8885	8960	9034	9108	9182	9257		5 37
47	9331	9405	9479	9554	9628	9702	9777	9851	9925	9999		6 44 / 7 52
48	7670074	0148	0222	0296	0371	0445	0519	0593	0668	0742		8 59
49	0816	0890	0965	1039	1113	1187	1262	1336	1410	1484		9 67
N.	0	1	2	3	4	5	6	7	8	9	D	Pts.

N.	0	1	2	3	4	5	6	7	8	9	D	Pro.
5850	7671559	1633	1707	1781	1856	1930	2004	2078	2153	2227		
51	2301	2375	2449	2524	2598	2672	2746	2821	2895	2969		74
52	3043	3117	3192	3266	3340	3414	3488	3563	3637	3711		1 7
53	3785	3859	3934	4008	4082	4156	4230	4305	4379	4453		2 15
54	4527	4601	4676	4750	4824	4898	4972	5046	5121	5195		3 22
55	5269	5343	5417	5492	5566	5640	5714	5788	5862	5937		4 30
56	6011	6085	6159	6233	6307	6381	6456	6530	6604	6678		5 37
57	6752	6826	6901	6975	7049	7123	7197	7271	7345	7420		6 44
58	7494	7568	7642	7716	7790	7864	7938	8013	8087	8161		7 52
59	8235	8309	8383	8457	8531	8606	8680	8754	8828	8902		8 59
5860	8976	9050	9124	9198	9273	9347	9421	9495	9569	9643		9 67
61	9717	9791	9865	9940	0̄014	0̄088	0̄162	0̄236	0̄310	0̄384		
62	7680458	0532	0606	0680	0754	0829	0903	0977	1051	1125		
63	1199	1273	1347	1421	1495	1569	1643	1717	1791	1866		
64	1940	2014	2088	2162	2236	2310	2384	2458	2532	2606		
65	2680	2754	2828	2902	2976	3050	3124	3198	3273	3347		
66	3421	3495	3569	3643	3717	3791	3865	3939	4013	4087		
67	4161	4235	4309	4383	4457	4531	4605	4679	4753	4827	74	
68	4901	4975	5049	5123	5197	5271	5345	5419	5493	5567		
69	5641	5715	5789	5863	5937	6011	6085	6159	6233	6307		
5870	6381	6455	6529	6603	6677	6751	6825	6899	6973	7047		
71	7121	7195	7269	7343	7417	7491	7565	7639	7713	7787		
72	7860	7934	8008	8082	8156	8230	8304	8378	8452	8526		
73	8600	8674	8748	8822	8896	8970	9044	9118	9192	9265		
74	9339	9413	9487	9561	9635	9709	9783	9857	9931	0̄005		
75	7690079	0153	0227	0300	0374	0448	0522	0596	0670	0744		
76	0818	0892	0966	1040	1114	1187	1261	1335	1409	1483		
77	1557	1631	1705	1779	1852	1926	2000	2074	2148	2222		
78	2296	2370	2444	2517	2591	2665	2739	2813	2887	2961		
79	3035	3108	3182	3256	3330	3404	3478	3552	3626	3699		
5880	3773	3847	3921	3995	4069	4143	4216	4290	4364	4438		
81	4512	4586	4659	4733	4807	4881	4955	5029	5103	5176		
82	5250	5324	5398	5472	5546	5619	5693	5767	5841	5915		
83	5988	6062	6136	6210	6284	6358	6431	6505	6579	6653		
84	6727	6800	6874	6948	7022	7096	7169	7243	7317	7391		
85	7465	7538	7612	7686	7760	7834	7907	7981	8055	8129		
86	8203	8276	8350	8424	8498	8571	8645	8719	8793	8867		
87	8940	9014	9088	9162	9235	9309	9383	9457	9530	9604		
88	9678	9752	9826	9899	9973	0̄047	0̄121	0̄194	0̄268	0̄342		
89	7700416	0489	0563	0637	0711	0784	0858	0932	1005	1079		
5890	1153	1227	1300	1374	1448	1522	1595	1669	1743	1817		
91	1890	1964	2038	2111	2185	2259	2333	2406	2480	2554		
92	2627	2701	2775	2849	2922	2996	3070	3143	3217	3291		73
93	3364	3438	3512	3585	3659	3733	3807	3880	3954	4028		1 7
94	4101	4175	4249	4322	4396	4470	4543	4617	4691	4764		2 15
95	4838	4912	4985	5059	5133	5206	5280	5354	5427	5501		3 22
96	5575	5648	5722	5796	5869	5943	6017	6090	6164	6238		4 29
97	6311	6385	6459	6532	6606	6679	6753	6827	6900	6974		5 37
98	7048	7121	7195	7269	7342	7416	7489	7563	7637	7710		6 44
99	7784	7858	7931	8005	8078	8152	8226	8299	8373	8447		7 51
N.	0	1	2	3	4	5	6	7	8	9	D	Pts.

8 58
9 66

N.	0	1	2	3	4	5	6	7	8	9	D	Pro.
5900	7708520	8594	8667	8741	8815	8888	8962	9035	9109	9183		
01	9256	9330	9403	9477	9551	9624	9698	9771	9845	9918		**74**
02	9992	0̄066	0̄139	0̄213	0̄286	0̄360	0̄434	0̄507	0̄581	0̄654		1 7
03	7710728	0801	0875	0949	1022	1096	1169	1243	1316	1390		2 15
04	1463	1537	1611	1684	1758	1831	1905	1978	2052	2125		3 22 4 30
05	2199	2273	2346	2420	2493	2567	2640	2714	2787	2861		5 37
06	2934	3008	3081	3155	3229	3302	3376	3449	3523	3596		6 44
07	3670	3743	3817	3890	3964	4037	4111	4184	4258	4331		7 52
08	4405	4478	4552	4625	4699	4772	4846	4919	4993	5066		8 59
09	5140	5213	5287	5360	5434	5507	5581	5654	5728	5801		9 67
5910	5875	5948	6022	6095	6169	6242	6316	6389	6463	6536		
11	6610	6683	6757	6830	6903	6977	7050	7124	7197	7271		
12	7344	7418	7491	7565	7638	7712	7785	7858	7932	8005		
13	8079	8152	8226	8299	8373	8446	8519	8593	8666	8740		
14	8813	8887	8960	9034	9107	9180	9254	9327	9401	9474		
15	9547	9621	9694	9768	9841	9915	9988	0̄061	0̄135	0̄208		
16	7720282	0355	0428	0502	0575	0649	0722	0795	0869	0942		
17	1016	1089	1162	1236	1309	1383	1456	1529	1603	1676		
18	1750	1823	1896	1970	2043	2117	2190	2263	2337	2410		
19	2483	2557	2630	2704	2777	2850	2924	2997	3070	3144		
5920	3217	3290	3364	3437	3510	3584	3657	3731	3804	3877		
21	3951	4024	4097	4171	4244	4317	4391	4464	4537	4611		
22	4684	4757	4831	4904	4977	5051	5124	5197	5271	5344		
23	5417	5491	5564	5637	5711	5784	5857	5931	6004	6077		
24	6150	6224	6297	6370	6444	6517	6590	6664	6737	6810		
25	6884	6957	7030	7103	7177	7250	7323	7397	7470	7543		
26	7616	7690	7763	7836	7910	7983	8056	8129	8203	8276		
27	8349	8423	8496	8569	8642	8716	8789	8862	8935	9009		
28	9082	9155	9228	9302	9375	9448	9521	9595	9668	9741		
29	9815	9888	9961	0̄034	0̄107	0̄181	0̄254	0̄327	0̄400	0̄474		
5930	7730547	0620	0693	0767	0840	0913	0986	1060	1133	1206		
31	1279	1352	1426	1499	1572	1645	1719	1792	1865	1938		
32	2011	2085	2158	2231	2304	2377	2451	2524	2597	2670		
33	2743	2817	2890	2963	3036	3109	3183	3256	3329	3402		
34	3475	3549	3622	3695	3768	3841	3915	3988	4061	4134		
35	4207	4280	4354	4427	4500	4573	4646	4719	4793	4866		
36	4939	5012	5085	5158	5232	5305	5378	5451	5524	5597		
37	5670	5744	5817	5890	5963	6036	6109	6183	6256	6329		
38	6402	6475	6548	6621	6694	6768	6841	6914	6987	7060		
39	7133	7206	7280	7353	7426	7499	7572	7645	7718	7791		
5940	7864	7938	8011	8084	8157	8230	8303	8376	8449	8522		
41	8596	8669	8742	8815	8888	8961	9034	9107	9180	9253		
42	9326	9400	9473	9546	9619	9692	9765	9838	9911	9984		**73**
43	7740057	0130	0203	0277	0350	0423	0496	0569	0642	0715		1 7
44	0788	0861	0934	1007	1080	1153	1226	1299	1372	1446		2 15 3 22
45	1519	1592	1665	1738	1811	1884	1957	2030	2103	2176		4 29
46	2249	2322	2395	2468	2541	2614	2687	2760	2833	2906		5 37
47	2979	3052	3125	3198	3271	3345	3418	3491	3564	3637		6 44
48	3710	3783	3856	3929	4002	4075	4148	4221	4294	4367		7 51 8 58
49	4440	4513	4586	4659	4732	4805	4878	4951	5024	5097		9 66
N.	0	1	2	3	4	5	6	7	8	9	D	Pts.

N.	0	1	2	3	4	5	6	7	8	9	D	Pro.
5950	7745170	5243	5316	5389	5462	5535	5608	5681	5754	5827	73	
51	5900	5972	6045	6118	6191	6264	6337	6410	6483	6556		73
52	6629	6702	6775	6848	6921	6994	7067	7140	7213	7286		1 7
53	7359	7432	7505	7578	7651	7724	7797	7869	7942	8015		2 15
54	8088	8161	8234	8307	8380	8453	8526	8599	8672	8745		3 22
												4 29
55	8818	8891	8964	9036	9109	9182	9255	9328	9401	9474		5 37
56	9547	9620	9693	9766	9839	9911	9984	ō057	ō130	ō203		6 44
57	7750276	0349	0422	0495	0568	0641	0713	0786	0859	0932		7 51
58	1005	1078	1151	1224	1297	1369	1442	1515	1588	1661		8 58
59	1734	1807	1880	1952	2025	2098	2171	2244	2317	2390		9 66
5960	2463	2535	2608	2681	2754	2827	2900	2973	3046	3118		
61	3191	3264	3337	3410	3483	3555	3628	3701	3774	3847		
62	3920	3993	4065	4138	4211	4284	4357	4430	4502	4575		
63	4648	4721	4794	4867	4939	5012	5085	5158	5231	5304		
64	5376	5449	5522	5595	5668	5740	5813	5886	5959	6032		
65	6104	6177	6250	6323	6396	6469	6541	6614	6687	6760		
66	6832	6905	6978	7051	7124	7196	7269	7342	7415	7488		
67	7560	7633	7706	7779	7851	7924	7997	8070	8143	8215		
68	8288	8361	8434	8506	8579	8652	8725	8798	8870	8943		
69	9016	9089	9161	9234	9307	9380	9452	9525	9598	9671		
5970	9743	9816	9889	9962	ō034	ō107	ō180	ō253	ō325	ō398		
71	7760471	0543	0616	0689	0762	0834	0907	0980	1053	1125		
72	1198	1271	1343	1416	1489	1562	1634	1707	1780	1852		
73	1925	1998	2071	2143	2216	2289	2361	2434	2507	2579		
74	2652	2725	2798	2870	2943	3016	3088	3161	3234	3306		
75	3379	3452	3524	3597	3670	3743	3815	3888	3961	4033		
76	4106	4179	4251	4324	4397	4469	4542	4615	4687	4760		
77	4833	4905	4978	5051	5123	5196	5269	5341	5414	5486		
78	5559	5632	5704	5777	5850	5922	5995	6068	6140	6213		
79	6286	6358	6431	6503	6576	6649	6721	6794	6867	6939		
5980	7012	7084	7157	7230	7302	7375	7448	7520	7593	7665		
81	7738	7811	7883	7956	8028	8101	8174	8246	8319	8391		
82	8464	8537	8609	8682	8754	8827	8900	8972	9045	9117		
83	9190	9263	9335	9408	9480	9553	9626	9698	9771	9843		
84	9916	9988	ō061	ō134	ō206	ō279	ō351	ō424	ō496	ō569		
85	7770642	0714	0787	0859	0932	1004	1077	1149	1222	1295		
86	1367	1440	1512	1585	1657	1730	1802	1875	1947	2020		
87	2093	2165	2238	2310	2383	2455	2528	2600	2673	2745		
88	2818	2890	2963	3035	3108	3181	3253	3326	3398	3471		
89	3543	3616	3688	3761	3833	3906	3978	4051	4123	4196		
5990	4268	4341	4413	4486	4558	4631	4703	4776	4848	4921		
91	4993	5066	5138	5211	5283	5356	5428	5501	5573	5646		72
92	5718	5791	5863	5935	6008	6080	6153	6225	6298	6370		1 7
93	6443	6515	6588	6660	6733	6805	6878	6950	7022	7095		2 14
94	7167	7240	7312	7385	7457	7530	7602	7675	7747	7819		3 22
												4 29
95	7892	7964	8037	8109	8182	8254	8327	8399	8471	8544		5 36
96	8616	8689	8761	8834	8906	8978	9051	9123	9196	9268		6 43
97	9340	9413	9485	9558	9630	9703	9775	9847	9920	9992		7 50
98	7780065	0137	0209	0282	0354	0427	0499	0571	0644	0716		8 58
99	0789	0861	0933	1006	1078	1151	1223	1295	1368	1440		9 65
N.	0	1	2	3	4	5	6	7	8	9	D	Pts.

N.	0	1	2	3	4	5	6	7	8	9	D	Pro.
6000	7781513	1585	1657	1730	1802	1874	1947	2019	2092	2164		
01	2236	2309	2381	2453	2526	2598	2670	2743	2815	2888		73
02	2960	3032	3105	3177	3249	3322	3394	3466	3539	3611		1 7
03	3683	3756	3828	3900	3973	4045	4117	4190	4262	4335		2 15
04	4407	4479	4552	4624	4696	4768	4841	4913	4985	5058		3 22 / 4 29
05	5130	5202	5275	5347	5419	5492	5564	5636	5709	5781		5 37
06	5853	5926	5998	6070	6143	6215	6287	6359	6432	6504		6 44
07	6576	6649	6721	6793	6866	6938	7010	7082	7155	7227		7 51
08	7299	7372	7444	7516	7588	7661	7733	7805	7877	7950		8 58
09	8022	8094	8167	8239	8311	8383	8456	8528	8600	8672		9 66
6010	8745	8817	8889	8962	9034	9106	9178	9251	9323	9395		
11	9467	9540	9612	9684	9756	9829	9901	9973	ō045	ō117		
12	7790190	0262	0334	0406	0479	0551	0623	0695	0768	0840		
13	0912	0984	1056	1129	1201	1273	1345	1418	1490	1562		
14	1634	1706	1779	1851	1923	1995	2067	2140	2212	2284		
15	2356	2429	2501	2573	2645	2717	2790	2862	2934	3006		
16	3078	3150	3223	3295	3367	3439	3511	3584	3656	3728		
17	3800	3872	3944	4017	4089	4161	4233	4305	4377	4450		
18	4522	4594	4666	4738	4810	4883	4955	5027	5099	5171		
19	5243	5316	5388	5460	5532	5604	5676	5748	5821	5893		
6020	5965	6037	6109	6181	6253	6326	6398	6470	6542	6614		
21	6686	6758	6831	6903	6975	7047	7119	7191	7263	7335		
22	7408	7480	7552	7624	7696	7768	7840	7912	7984	8057		
23	8129	8201	8273	8345	8417	8489	8561	8633	8705	8778		
24	8850	8922	8994	9066	9138	9210	9282	9354	9426	9498		
25	9571	9643	9715	9787	9859	9931	ō003	ō075	ō147	ō219		
26	7800291	0363	0435	0507	0580	0652	0724	0796	0868	0940		
27	1012	1084	1156	1228	1300	1372	1444	1516	1588	1660		
28	1732	1804	1877	1949	2021	2093	2165	2237	2309	2381		
29	2453	2525	2597	2669	2741	2813	2885	2957	3029	3101		
6030	3173	3245	3317	3389	3461	3533	3605	3677	3749	3821		
31	3893	3965	4037	4109	4181	4253	4325	4397	4469	4541	72	
32	4613	4685	4757	4829	4901	4973	5045	5117	5189	5261		
33	5333	5405	5477	5549	5621	5693	5765	5837	5909	5981		
34	6053	6125	6197	6269	6341	6413	6485	6557	6629	6701		
35	6773	6845	6917	6989	7061	7133	7204	7276	7348	7420		
36	7492	7564	7636	7708	7780	7852	7924	7996	8068	8140		
37	8212	8284	8356	8428	8500	8571	8643	8715	8787	8859		
38	8931	9003	9075	9147	9219	9291	9363	9435	9506	9578		
39	9650	9722	9794	9866	9938	ō010	ō082	ō154	ō226	ō297		
6040	7810369	0441	0513	0585	0657	0729	0801	0873	0945	1016		
41	1088	1160	1232	1304	1376	1448	1520	1592	1663	1735		
42	1807	1879	1951	2023	2095	2167	2238	2310	2382	2454		72
43	2526	2598	2670	2742	2813	2885	2957	3029	3101	3173		1 7
44	3245	3316	3388	3460	3532	3604	3676	3748	3819	3891		2 14 / 3 22
45	3963	4035	4107	4179	4250	4322	4394	4466	4538	4610		4 29
46	4681	4753	4825	4897	4969	5041	5112	5184	5256	5328		5 36
47	5400	5472	5543	5615	5687	5759	5831	5902	5974	6046		6 43
48	6118	6190	6261	6333	6405	6477	6549	6620	6692	6764		7 50 / 8 58
49	6836	6908	6979	7051	7123	7195	7267	7338	7410	7482		9 65
N.	0	1	2	3	4	5	6	7	8	9	D	Pts

N.	0	1	2	3	4	5	6	7	8	9	D	Pro.
6050	7817554	7626	7697	7769	7841	7913	7984	8056	8128	8200		
51	8272	8343	8415	8487	8559	8630	8702	8774	8846	8917		72
52	8989	9061	9133	9204	9276	9348	9420	9491	9563	9635		1 7
53	9707	9778	9850	9922	9994	0065	0137	0209	0281	0352		2 14
54	7820424	0496	0568	0639	0711	0783	0855	0926	0998	1070		3 22
												4 29
55	1141	1213	1285	1357	1428	1500	1572	1644	1715	1787		5 36
56	1859	1930	2002	2074	2146	2217	2289	2361	2432	2504		6 43
57	2576	2647	2719	2791	2863	2934	3006	3078	3149	3221		7 50
58	3293	3364	3436	3508	3579	3651	3723	3794	3866	3938		8 58
59	4010	4081	4153	4225	4296	4368	4440	4511	4583	4655		9 65
6060	4726	4798	4870	4941	5013	5085	5156	5228	5300	5371		
61	5443	5514	5586	5658	5729	5801	5873	5944	6016	6088		
62	6159	6231	6303	6374	6446	6518	6589	6661	6732	6804		
63	6876	6947	7019	7091	7162	7234	7305	7377	7449	7520		
64	7592	7664	7735	7807	7878	7950	8022	8093	8165	8236		
65	8308	8380	8451	8523	8594	8666	8738	8809	8881	8952		
66	9024	9096	9167	9239	9310	9382	9454	9525	9597	9668		
67	9740	9812	9883	9955	0026	0098	0169	0241	0313	0384		
68	7830456	0527	0599	0670	0742	0814	0885	0957	1028	1100		
69	1171	1243	1314	1386	1458	1529	1601	1672	1744	1815		
6070	1887	1958	2030	2102	2173	2245	2316	2388	2459	2531		
71	2602	2674	2745	2817	2888	2960	3032	3103	3175	3246		
72	3318	3389	3461	3532	3604	3675	3747	3818	3890	3961		
73	4033	4104	4176	4247	4319	4390	4462	4533	4605	4676		
74	4748	4819	4891	4962	5034	5105	5177	5248	5320	5391		
75	5463	5534	5606	5677	5749	5820	5892	5963	6035	6106		
76	6178	6249	6321	6392	6464	6535	6606	6678	6749	6821		
77	6892	6964	7035	7107	7178	7250	7321	7393	7464	7536		
78	7607	7678	7750	7821	7893	7964	8036	8107	8179	8250		
79	8321	8393	8464	8536	8607	8679	8750	8821	8893	8964		
6080	9036	9107	9179	9250	9322	9393	9464	9536	9607	9679		
81	9750	9821	9893	9964	0036	0107	0179	0250	0321	0393		
82	7840464	0536	0607	0678	0750	0821	0893	0964	1035	1107		
83	1178	1250	1321	1392	1464	1535	1607	1678	1749	1821		
84	1892	1963	2035	2106	2178	2249	2320	2392	2463	2534		
85	2606	2677	2749	2820	2891	2963	3034	3105	3177	3248		
86	3319	3391	3462	3534	3605	3676	3748	3819	3890	3962		
87	4033	4104	4176	4247	4318	4390	4461	4532	4604	4675		
88	4746	4818	4889	4960	5032	5103	5174	5246	5317	5388		
89	5460	5531	5602	5674	5745	5816	5888	5959	6030	6102		
6090	6173	6244	6316	6387	6458	6529	6601	6672	6743	6815		
91	6886	6957	7029	7100	7171	7242	7314	7385	7456	7528		71
92	7599	7670	7742	7813	7884	7955	8027	8098	8169	8241		1 7
93	8312	8383	8454	8526	8597	8668	8739	8811	8882	8953		2 14
94	9024	9096	9167	9238	9310	9381	9452	9523	9595	9666		3 21
95	9737	9808	9880	9951	0022	0093	0165	0236	0307	0378		4 28
96	7850450	0521	0592	0663	0735	0806	0877	0948	1019	1091		5 36
97	1162	1233	1304	1376	1447	1518	1589	1661	1732	1803		6 43
98	1874	1945	2017	2088	2159	2230	2301	2373	2444	2515		7 50
99	2586	2658	2729	2800	2871	2942	3014	3085	3156	3227		8 57
												9 64
N.	0	1	2	3	4	5	6	7	8	9	D	Pts.

N.	0	1	2	3	4	5	6	7	8	9	D	Pro.
6100	7853298	3370	3441	3512	3583	3654	3726	3797	3868	3939		
01	4010	4081	4153	4224	4295	4366	4437	4509	4580	4651		72
02	4722	4793	4864	4936	5007	5078	5149	5220	5291	5363		1 7
03	5434	5505	5576	5647	5718	5789	5861	5932	6003	6074		2 14
04	6145	6216	6288	6359	6430	6501	6572	6643	6714	6786		3 22
												4 29
05	6857	6928	6999	7070	7141	7212	7283	7355	7426	7497		5 36
06	7568	7639	7710	7781	7852	7924	7995	8066	8137	8208		6 43
07	8279	8350	8421	8493	8564	8635	8706	8777	8848	8919		7 50
08	8990	9061	9132	9204	9275	9346	9417	9488	9559	9630		8 58
09	9701	9772	9843	9915	9986	0̄057	0̄128	0̄199	0̄270	0̄341		9 65
6110	7860412	0483	0554	0625	0696	0767	0839	0910	0981	1052		
11	1123	1194	1265	1336	1407	1478	1549	1620	1691	1762		
12	1833	1905	1976	2047	2118	2189	2260	2331	2402	2473		
13	2544	2615	2686	2757	2828	2899	2970	3041	3112	3183		
14	3254	3325	3396	3467	3538	3609	3681	3752	3823	3894		
15	3965	4036	4107	4178	4249	4320	4391	4462	4533	4604	71	
16	4675	4746	4817	4888	4959	5030	5101	5172	5243	5314		
17	5385	5456	5527	5598	5669	5740	5811	5882	5953	6024		
18	6095	6166	6237	6308	6379	6450	6521	6592	6663	6734		
19	6805	6876	6946	7017	7088	7159	7230	7301	7372	7443		
6120	7514	7585	7656	7727	7798	7869	7940	8011	8082	8153		
21	8224	8295	8366	8437	8508	8579	8649	8720	8791	8862		
22	8933	9004	9075	9146	9217	9288	9359	9430	9501	9572		
23	9643	9714	9784	9855	9926	9997	0̄068	0̄139	0̄210	0̄281		
24	7870352	0423	0494	0565	0635	0706	0777	0848	0919	0990		
25	1061	1132	1203	1274	1345	1415	1486	1557	1628	1699		
26	1770	1841	1912	1983	2053	2124	2195	2266	2337	2408		
27	2479	2550	2621	2691	2762	2833	2904	2975	3046	3117		
28	3188	3258	3329	3400	3471	3542	3613	3684	3754	3825		
29	3896	3967	4038	4109	4180	4250	4321	4392	4463	4534		
6130	4605	4676	4746	4817	4888	4959	5030	5101	5171	5242		
31	5313	5384	5455	5526	5596	5667	5738	5809	5880	5951		
32	6021	6092	6163	6234	6305	6376	6446	6517	6588	6659		
33	6730	6800	6871	6942	7013	7084	7155	7225	7296	7367		
34	7438	7509	7579	7650	7721	7792	7863	7933	8004	8075		
35	8146	8216	8287	8358	8429	8500	8570	8641	8712	8783		
36	8854	8924	8995	9066	9137	9207	9278	9349	9420	9490		
37	9561	9632	9703	9774	9844	9915	9986	0̄057	0̄127	0̄198		
38	7880269	0340	0410	0481	0552	0623	0693	0764	0835	0906		
39	0976	1047	1118	1189	1259	1330	1401	1472	1542	1613		
6140	1684	1754	1825	1896	1967	2037	2108	2179	2250	2320		
41	2391	2462	2532	2603	2674	2745	2815	2886	2957	3027		
42	3098	3169	3240	3310	3381	3452	3522	3593	3664	3734		71
43	3805	3876	3947	4017	4088	4159	4229	4300	4371	4441		1 7
44	4512	4583	4653	4724	4795	4865	4936	5007	5078	5148		2 14
												3 21
45	5219	5290	5360	5431	5502	5572	5643	5714	5784	5855		4 28
46	5926	5996	6067	6138	6208	6279	6350	6420	6491	6561		5 36
47	6632	6703	6773	6844	6915	6985	7056	7127	7197	7268		6 43
48	7339	7409	7480	7551	7621	7692	7762	7833	7904	7974		7 50
49	8045	8116	8186	8257	8327	8398	8469	8539	8610	8681		8 57
												9 64
N.	0	1	2	3	4	5	6	7	8	9	D	Pts.

N.	0	1	2	3	4	5	6	7	8	9	D	Pro.
6150	7888751	8822	8892	8963	9034	9104	9175	9245	9316	9387		71
51	9457	9528	9598	9669	9740	9810	9881	9951	ō022	ō093		
52	7890163	0234	0304	0375	0446	0516	0587	0657	0728	0799		1 7
53	0869	0940	1010	1081	1151	1222	1293	1363	1434	1504		2 14
54	1575	1645	1716	1787	1857	1928	1998	2069	2139	2210		3 21
												4 28
55	2281	2351	2422	2492	2563	2633	2704	2774	2845	2916		5 36
56	2986	3057	3127	3198	3268	3339	3409	3480	3550	3621		6 43
57	3692	3762	3833	3903	3974	4044	4115	4185	4256	4326		7 50
58	4397	4467	4538	4608	4679	4749	4820	4890	4961	5032		8 57
59	5102	5173	5243	5314	5384	5455	5525	5596	5666	5737		9 64
6160	5807	5878	5948	6019	6089	6160	6230	6301	6371	6442		
61	6512	6583	6653	6724	6794	6865	6935	7005	7076	7146		
62	7217	7287	7358	7428	7499	7569	7640	7710	7781	7851		
63	7922	7992	8063	8133	8204	8274	8344	8415	8485	8556		
64	8626	8697	8767	8838	8908	8979	9049	9119	9190	9260		
65	9331	9401	9472	9542	9613	9683	9753	9824	9894	9965		
66	7900035	0106	0176	0247	0317	0387	0458	0528	0599	0669		
67	0739	0810	0880	0951	1021	1092	1162	1232	1303	1373		
68	1444	1514	1584	1655	1725	1796	1866	1936	2007	2077		
69	2148	2218	2288	2359	2429	2500	2570	2640	2711	2781		
6170	2852	2922	2992	3063	3133	3204	3274	3344	3415	3485		
71	3555	3626	3696	3767	3837	3907	3978	4048	4118	4189		
72	4259	4330	4400	4470	4541	4611	4681	4752	4822	4892		
73	4963	5033	5103	5174	5244	5315	5385	5455	5526	5596		
74	5666	5737	5807	5877	5948	6018	6088	6159	6229	6299		
75	6370	6440	6510	6581	6651	6721	6792	6862	6932	7003		
76	7073	7143	7214	7284	7354	7424	7495	7565	7635	7706		
77	7776	7846	7917	7987	8057	8128	8198	8268	8338	8409		
78	8479	8549	8620	8690	8760	8831	8901	8971	9041	9112		
79	9182	9252	9323	9393	9463	9533	9604	9674	9744	9814		
6180	9885	9955	ō025	ō096	ō166	ō236	ō306	ō377	ō447	ō517		
81	7910587	0658	0728	0798	0868	0939	1009	1079	1150	1220		
82	1290	1360	1431	1501	1571	1641	1711	1782	1852	1922		
83	1992	2063	2133	2203	2273	2344	2414	2484	2554	2625		
84	2695	2765	2835	2905	2976	3046	3116	3186	3257	3327		
85	3397	3467	3537	3608	3678	3748	3818	3889	3959	4029		
86	4099	4169	4240	4310	4380	4450	4520	4591	4661	4731		
87	4801	4871	4942	5012	5082	5152	5222	5292	5363	5433		
88	5503	5573	5643	5714	5784	5854	5924	5994	6064	6135		
89	6205	6275	6345	6415	6486	6556	6626	6696	6766	6836		
6190	6906	6977	7047	7117	7187	7257	7327	7398	7468	7538		
91	7608	7678	7748	7818	7889	7959	8029	8099	8169	8239		70
92	8309	8380	8450	8520	8590	8660	8730	8800	8871	8941		
93	9011	9081	9151	9221	9291	9361	9432	9502	9572	9642		1 7
94	9712	9782	9852	9922	9992	ō063	ō133	ō203	ō273	ō343		2 14
												3 21
95	7920413	0483	0553	0623	0694	0764	0834	0904	0974	1044		4 28
96	1114	1184	1254	1324	1394	1465	1535	1605	1675	1745		5 35
97	1815	1885	1955	2025	2095	2165	2235	2306	2376	2446		6 42
98	2516	2586	2656	2726	2796	2866	2936	3006	3076	3146		7 49
99	3216	3286	3356	3427	3497	3567	3637	3707	3777	3847		8 56
												9 63
N.	0	1	2	3	4	5	6	7	8	9	D	Pts.

N.	0	1	2	3	4	5	6	7	8	9	D	Pro
6200	7923917	3987	4057	4127	4197	4267	4337	4407	4477	4547		
01	4617	4687	4757	4827	4897	4967	5038	5108	5178	5248		71
02	5318	5388	5458	5528	5598	5668	5738	5808	5878	5948		1 7
03	6018	6088	6158	6228	6298	6368	6438	6508	6578	6648	70	2 14
04	6718	6788	6858	6928	6998	7068	7138	7208	7278	7348		3 21
05	7418	7488	7558	7628	7698	7768	7838	7908	7978	8048		4 28
06	8118	8188	8258	8328	8398	8468	8538	8608	8678	8747		5 36
07	8817	8887	8957	9027	9097	9167	9237	9307	9377	9447		6 43
08	9517	9587	9657	9727	9797	9867	9937	0̄007	0̄077	0̄147		7 50
09	7930217	0287	0356	0426	0496	0566	0636	0706	0776	0846		8 57
												9 64
6210	0916	0986	1056	1126	1196	1266	1336	1406	1475	1545		
11	1615	1685	1755	1825	1895	1965	2035	2105	2175	2245		
12	2314	2384	2454	2524	2594	2664	2734	2804	2874	2944		
13	3014	3083	3153	3223	3293	3363	3433	3503	3573	3643		
14	3712	3782	3852	3922	3992	4062	4132	4202	4272	4341		
15	4411	4481	4551	4621	4691	4761	4831	4900	4970	5040		
16	5110	5180	5250	5320	5390	5459	5529	5599	5669	5739		
17	5809	5879	5948	6018	6088	6158	6228	6298	6367	6437		
18	6507	6577	6647	6717	6787	6856	6926	6996	7066	7136		
19	7206	7275	7345	7415	7485	7555	7625	7694	7764	7834		
6220	7904	7974	8043	8113	8183	8253	8323	8393	8462	8532		
21	8602	8672	8742	8811	8881	8951	9021	9091	9160	9230		
22	9300	9370	9440	9509	9579	9649	9719	9789	9858	9928		
23	9998	0̄068	0̄138	0̄207	0̄277	0̄347	0̄417	0̄487	0̄556	0̄626		
24	7940696	0766	0835	0905	0975	1045	1114	1184	1254	1324		
25	1394	1463	1533	1603	1673	1742	1812	1882	1952	2021		
26	2091	2161	2231	2300	2370	2440	2510	2579	2649	2719		
27	2789	2858	2928	2998	3068	3137	3207	3277	3347	3416		
28	3486	3556	3626	3695	3765	3835	3904	3974	4044	4114		
29	4183	4253	4323	4392	4462	4532	4602	4671	4741	4811		
6230	4880	4950	5020	5090	5159	5229	5299	5368	5438	5508		
31	5578	5647	5717	5787	5856	5926	5996	6065	6135	6205		
32	6274	6344	6414	6484	6553	6623	6693	6762	6832	6902		
33	6971	7041	7111	7180	7250	7320	7389	7459	7529	7598		
34	7668	7738	7807	7877	7947	8016	8086	8156	8225	8295		
35	8365	8434	8504	8574	8643	8713	8782	8852	8922	8991		
36	9061	9131	9200	9270	9340	9409	9479	9549	9618	9688		
37	9757	9827	9897	9966	0̄036	0̄106	0̄175	0̄245	0̄314	0̄384		
38	7950454	0523	0593	0663	0732	0802	0871	0941	1011	1080		
39	1150	1219	1289	1359	1428	1498	1567	1637	1707	1776		
6240	1846	1915	1985	2055	2124	2194	2263	2333	2403	2472		
41	2542	2611	2681	2751	2820	2890	2959	3029	3098	3168		
42	3238	3307	3377	3446	3516	3586	3655	3725	3794	3864		70
43	3933	4003	4072	4142	4212	4281	4351	4420	4490	4559		1 7
44	4629	4698	4768	4838	4907	4977	5046	5116	5185	5255		2 14
45	5324	5394	5464	5533	5603	5672	5742	5811	5881	5950		3 21
46	6020	6089	6159	6228	6298	6367	6437	6506	6576	6646		4 28
47	6715	6785	6854	6924	6993	7063	7132	7202	7271	7341		5 35
48	7410	7480	7549	7619	7688	7758	7827	7897	7966	8036		6 42
49	8105	8175	8244	8314	8383	8453	8522	8592	8661	8731		7 49
												8 56
												9 63
N.	0	1	2	3	4	5	6	7	8	9	D	Pts.

N	0	1	2	3	4	5	6	7	8	9	D	Pro.
6250	7958800	8870	8939	9009	9078	9148	9217	9287	9356	9426		
51	9495	9564	9634	9703	9773	9842	9912	9981	0̄051	0̄120		70
52	7960190	0259	0329	0398	0468	0537	0606	0676	0745	0815		1 7
53	0884	0954	1023	1093	1162	1232	1301	1370	1440	1509		2 14
54	1579	1648	1718	1787	1857	1926	1995	2065	2134	2204		3 21
55	2273	2343	2412	2481	2551	2620	2690	2759	2829	2898		4 28 / 5 35
56	2967	3037	3106	3176	3245	3314	3384	3453	3523	3592		6 42
57	3662	3731	3800	3870	3939	4009	4078	4147	4217	4286		7 49
58	4356	4425	4494	4564	4633	4703	4772	4841	4911	4980		8 56
59	5050	5119	5188	5258	5327	5396	5466	5535	5605	5674		9 63
6260	5743	5813	5882	5951	6021	6090	6160	6229	6298	6368		
61	6437	6506	6576	6645	6714	6784	6853	6923	6992	7061		
62	7131	7200	7269	7339	7408	7477	7547	7616	7685	7755		
63	7824	7893	7963	8032	8101	8171	8240	8309	8379	8448		
64	8517	8587	8656	8725	8795	8864	8933	9003	9072	9141		
65	9211	9280	9349	9419	9488	9557	9627	9696	9765	9835		
66	9904	9973	0̄043	0̄112	0̄181	0̄250	0̄320	0̄389	0̄458	0̄528		
67	7970597	0666	0736	0805	0874	0943	1013	1082	1151	1221		
68	1290	1359	1428	1498	1567	1636	1706	1775	1844	1913		
69	1983	2052	2121	2191	2260	2329	2398	2468	2537	2606		
6270	2675	2745	2814	2883	2952	3022	3091	3160	3229	3299		
71	3368	3437	3507	3576	3645	3714	3784	3853	3922	3991		
72	4060	4130	4199	4268	4337	4407	4476	4545	4614	4684		
73	4753	4822	4891	4961	5030	5099	5168	5237	5307	5376		
74	5445	5514	5584	5653	5722	5791	5860	5930	5999	6068		
75	6137	6207	6276	6345	6414	6483	6553	6622	6691	6760		
76	6829	6899	6968	7037	7106	7175	7245	7314	7383	7452		
77	7521	7590	7660	7729	7798	7867	7936	8006	8075	8144		
78	8213	8282	8351	8421	8490	8559	8628	8697	8766	8836		
79	8905	8974	9043	9112	9181	9251	9320	9389	9458	9527		
6280	9596	9666	9735	9804	9873	9942	0̄011	0̄080	0̄150	0̄219		
81	7980288	0357	0426	0495	0565	0634	0703	0772	0841	0910		
82	0979	1048	1118	1187	1256	1325	1394	1463	1532	1601		
83	1671	1740	1809	1878	1947	2016	2085	2154	2224	2293		
84	2362	2431	2500	2569	2638	2707	2776	2846	2915	2984		
85	3053	3122	3191	3260	3329	3398	3467	3536	3606	3675		
86	3744	3813	3882	3951	4020	4089	4158	4227	4296	4366		
87	4435	4504	4573	4642	4711	4780	4849	4918	4987	5056		
88	5125	5194	5263	5333	5402	5471	5540	5609	5678	5747		
89	5816	5885	5954	6023	6092	6161	6230	6299	6368	6437		
6290	6506	6575	6645	6714	6783	6852	6921	6990	7059	7128		
91	7197	7266	7335	7404	7473	7542	7611	7680	7749	7818		
92	7887	7956	8025	8094	8163	8232	8301	8370	8439	8508		69
93	8577	8646	8715	8784	8853	8922	8991	9060	9129	9198		1 7
94	9267	9336	9405	9474	9543	9612	9681	9750	9819	9888	69	2 14
95	9957	0̄026	0̄095	0̄164	0̄233	0̄302	0̄371	0̄440	0̄509	0̄578		3 21 / 4 28
96	7990647	0716	0785	0854	0923	0992	1061	1130	1199	1268		5 35
97	1337	1406	1475	1544	1613	1682	1751	1820	1889	1958		6 41
98	2027	2096	2164	2233	2302	2371	2440	2509	2578	2647		7 48
99	2716	2785	2854	2923	2992	3061	3130	3199	3268	3337		8 55 / 9 62
N.	0	1	2	3	4	5	6	7	8	9	D	Pts.

N.	0	1	2	3	4	5	6	7	8	9	D	Pro.
6300	7993405	3474	3543	3612	3681	3750	3819	3888	3957	4026		
01	4095	4164	4233	4302	4370	4439	4508	4577	4646	4715		69
02	4784	4853	4922	4991	5060	5129	5197	5266	5335	5404		1 7
03	5473	5542	5611	5680	5749	5818	5886	5955	6024	6093		2 14
04	6162	6231	6300	6369	6438	6506	6575	6644	6713	6782		3 21
												4 28
05	6851	6920	6989	7058	7126	7195	7264	7333	7402	7471		5 35
06	7540	7609	7677	7746	7815	7884	7953	8022	8091	8159		6 41
07	8228	8297	8366	8435	8504	8573	8641	8710	8779	8848		7 48
08	8917	8986	9055	9123	9192	9261	9330	9399	9468	9536		8 55
09	9605	9674	9743	9812	9881	9949	ō018	ō087	ō156	ō225		9 62
6310	8000294	0362	0431	0500	0569	0638	0707	0775	0844	0913		
11	0982	1051	1119	1188	1257	1326	1395	1463	1532	1601		
12	1670	1739	1808	1876	1945	2014	2083	2152	2220	2289		
13	2358	2427	2495	2564	2633	2702	2771	2839	2908	2977		
14	3046	3115	3183	3252	3321	3390	3458	3527	3596	3665		
15	3734	3802	3871	3940	4009	4077	4146	4215	4284	4352		
16	4421	4490	4559	4627	4696	4765	4834	4903	4971	5040		
17	5109	5178	5246	5315	5384	5453	5521	5590	5659	5727		
18	5796	5865	5934	6002	6071	6140	6209	6277	6346	6415		
19	6484	6552	6621	6690	6758	6827	6896	6965	7033	7102		
6320	7171	7239	7308	7377	7446	7514	7583	7652	7720	7789		
21	7858	7927	7995	8064	8133	8201	8270	8339	8408	8476		
22	8545	8614	8682	8751	8820	8888	8957	9026	9094	9163		
23	9232	9301	9369	9438	9507	9575	9644	9713	9781	9850		
24	9919	9987	ō056	ō125	ō193	ō262	ō331	ō399	ō468	ō537		
25	8010605	0674	0743	0811	0880	0949	1017	1086	1155	1223		
26	1292	1361	1429	1498	1566	1635	1704	1772	1841	1910		
27	1978	2047	2116	2184	2253	2322	2390	2459	2527	2596		
28	2665	2733	2802	2871	2939	3008	3076	3145	3214	3282		
29	3351	3420	3488	3557	3625	3694	3763	3831	3900	3968		
6330	4037	4106	4174	4243	4312	4380	4449	4517	4586	4655		
31	4723	4792	4860	4929	4998	5066	5135	5203	5272	5340		
32	5409	5478	5546	5615	5683	5752	5821	5889	5958	6026		
33	6095	6163	6232	6301	6369	6438	6506	6575	6643	6712		
34	6781	6849	6918	6986	7055	7123	7192	7261	7329	7398		
35	7466	7535	7603	7672	7740	7809	7878	7946	8015	8083		
36	8152	8220	8289	8357	8426	8494	8563	8631	8700	8769		
37	8837	8906	8974	9043	9111	9180	9248	9317	9385	9454		
38	9522	9591	9659	9728	9796	9865	9933	ō002	ō070	ō139		
39	8020208	0276	0345	0413	0482	0550	0619	0687	0756	0824		
6340	0893	0961	1030	1098	1167	1235	1304	1372	1441	1509		
41	1578	1646	1715	1783	1851	1920	1988	2057	2125	2194		
42	2262	2331	2399	2468	2536	2605	2673	2742	2810	2879		
43	2947	3016	3084	3153	3221	3289	3358	3426	3495	3563		68
44	3632	3700	3769	3837	3906	3974	4042	4111	4179	4248		1 7
												2 14
45	4316	4385	4453	4522	4590	4658	4727	4795	4864	4932		3 20
46	5001	5069	5138	5206	5274	5343	5411	5480	5548	5617		4 27
47	5685	5753	5822	5890	5959	6027	6096	6164	6232	6301		5 34
48	6369	6438	6506	6574	6643	6711	6780	6848	6916	6985		6 41
49	7053	7122	7190	7258	7327	7395	7464	7532	7600	7669		7 48
												8 54
												9 61
N.	0	1	2	3	4	5	6	7	8	9	D	Pts.

N.	0	1	2	3	4	5	6	7	8	9	D	Pro.
6350	8027737	7806	7874	7942	8011	8079	8148	8216	8284	8353		
51	8421	8490	8558	8626	8695	8763	8831	8900	8968	9037		69
52	9105	9173	9242	9310	9378	9447	9515	9583	9652	9720		1 7
53	9789	9857	9925	9994	ō062	ō130	ō199	ō267	ō335	ō404		2 14
54	8030472	0540	0609	0677	0745	0814	0882	0951	1019	1087		3 21
55	1156	1224	1292	1361	1429	1497	1566	1634	1702	1771		4 28
56	1839	1907	1976	2044	2112	2181	2249	2317	2385	2454		5 35
57	2522	2590	2659	2727	2795	2864	2932	3000	3069	3137		6 41
58	3205	3274	3342	3410	3478	3547	3615	3683	3752	3820		7 48
59	3888	3957	4025	4093	4161	4230	4298	4366	4435	4503		8 55
												9 62
6360	4571	4639	4708	4776	4844	4913	4981	5049	5117	5186		
61	5254	5322	5391	5459	5527	5595	5664	5732	5800	5868		
62	5937	6005	6073	6141	6210	6278	6346	6414	6483	6551		
63	6619	6687	6756	6824	6892	6960	7029	7097	7165	7233		
64	7302	7370	7438	7506	7575	7643	7711	7779	7848	7916		
65	7984	8052	8121	8189	8257	8325	8393	8462	8530	8598		
66	8666	8735	8803	8871	8939	9007	9076	9144	9212	9280		
67	9348	9417	9485	9553	9621	9690	9758	9826	9894	9962		
68	8040031	0099	0167	0235	0303	0372	0440	0508	0576	0644		
69	0712	0781	0849	0917	0985	1053	1122	1190	1258	1326		
6370	1394	1463	1531	1599	1667	1735	1803	1872	1940	2008		
71	2076	2144	2212	2281	2349	2417	2485	2553	2621	2690		
72	2758	2826	2894	2962	3030	3098	3167	3235	3303	3371		
73	3439	3507	3575	3644	3712	3780	3848	3916	3984	4052		
74	4121	4189	4257	4325	4393	4461	4529	4598	4666	4734		
75	4802	4870	4938	5006	5074	5143	5211	5279	5347	5415		
76	5483	5551	5619	5687	5756	5824	5892	5960	6028	6096		
77	6164	6232	6300	6368	6437	6505	6573	6641	6709	6777		
78	6845	6913	6981	7049	7118	7186	7254	7322	7390	7458		
79	7526	7594	7662	7730	7798	7866	7934	8003	8071	8139		
6380	8207	8275	8343	8411	8479	8547	8615	8683	8751	8819		
81	8887	8956	9024	9092	9160	9228	9296	9364	9432	9500		
82	9568	9636	9704	9772	9840	9908	9976	ō044	ō112	ō180		
83	8050248	0316	0385	0453	0521	0589	0657	0725	0793	0861		
84	0929	0997	1065	1133	1201	1269	1337	1405	1473	1541		
85	1609	1677	1745	1813	1881	1949	2017	2085	2153	2221		
86	2289	2357	2425	2493	2561	2629	2697	2765	2833	2901	68	
87	2969	3037	3105	3173	3241	3309	3377	3445	3513	3581		
88	3649	3717	3785	3853	3921	3989	4057	4125	4193	4261		
89	4329	4397	4465	4533	4601	4669	4737	4805	4873	4941		
6390	5009	5077	5145	5212	5280	5348	5416	5484	5552	5620		
91	5688	5756	5824	5892	5960	6028	6096	6164	6232	6300		
92	6368	6436	6504	6571	6639	6707	6775	6843	6911	6979		68
93	7047	7115	7183	7251	7319	7387	7455	7523	7590	7658		1 7
94	7726	7794	7862	7930	7998	8066	8134	8202	8270	8338		2 14
95	8405	8473	8541	8609	8677	8745	8813	8881	8949	9017		3 20
96	9085	9152	9220	9288	9356	9424	9492	9560	9628	9696		4 27
97	9764	9831	9899	9967	ō035	ō103	ō171	ō239	ō307	ō374		5 34
98	8060442	0510	0578	0646	0714	0782	0850	0917	0985	1053		6 41
99	1121	1189	1257	1325	1393	1460	1528	1596	1664	1732		7 48
												8 54
												9 61
N.	0	1	2	3	4	5	6	7	8	9	D	Pts.

1

N.	0	1	2	3	4	5	6	7	8	9	D	Pro.
6400	8061800	1868	1935	2003	2071	2139	2207	2275	2343	2410		
01	2478	2546	2614	2682	2750	2817	2885	2953	3021	3089		68
02	3157	3225	3292	3360	3428	3496	3564	3632	3699	3767		1 7
03	3835	3903	3971	4038	4106	4174	4242	4310	4378	4445		2 14
04	4513	4581	4649	4717	4784	4852	4920	4988	5056	5124		3 20
												4 27
05	5191	5259	5327	5395	5463	5530	5598	5666	5734	5802		5 34
06	5869	5937	6005	6073	6141	6208	6276	6344	6412	6479		6 41
07	6547	6615	6683	6751	6818	6886	6954	7022	7089	7157		7 48
08	7225	7293	7361	7428	7496	7564	7632	7699	7767	7835		8 54
09	7903	7970	8038	8106	8174	8242	8309	8377	8445	8513		9 61
6410	8580	8648	8716	8784	8851	8919	8987	9055	9122	9190		
11	9258	9326	9393	9461	9529	9596	9664	9732	9800	9867		
12	9935	0̄003	0̄071	0̄138	0̄206	0̄274	0̄342	0̄409	0̄477	0̄545		
13	8070612	0680	0748	0816	0883	0951	1019	1086	1154	1222		
14	1290	1357	1425	1493	1560	1628	1696	1764	1831	1899		
15	1967	2034	2102	2170	2237	2305	2373	2440	2508	2576		
16	2644	2711	2779	2847	2914	2982	3050	3117	3185	3253		
17	3320	3388	3456	3523	3591	3659	3726	3794	3862	3929		
18	3997	4065	4132	4200	4268	4335	4403	4471	4538	4606		
19	4674	4741	4809	4877	4944	5012	5080	5147	5215	5283		
6420	5350	5418	5486	5553	5621	5689	5756	5824	5891	5959		
21	6027	6094	6162	6230	6297	6365	6432	6500	6568	6635		
22	6703	6771	6838	6906	6974	7041	7109	7176	7244	7312		
23	7379	7447	7514	7582	7650	7717	7785	7853	7920	7988		
24	8055	8123	8191	8258	8326	8393	8461	8529	8596	8664		
25	8731	8799	8867	8934	9002	9069	9137	9204	9272	9340		
26	9407	9475	9542	9610	9678	9745	9813	9880	9948	0̄015		
27	8080083	0151	0218	0286	0353	0421	0488	0556	0624	0691		
28	0759	0826	0894	0961	1029	1096	1164	1232	1299	1367		
29	1434	1502	1569	1637	1704	1772	1840	1907	1975	2042		
6430	2110	2177	2245	2312	2380	2447	2515	2582	2650	2718		
31	2785	2853	2920	2988	3055	3123	3190	3258	3325	3393		
32	3460	3528	3595	3663	3730	3798	3865	3933	4000	4068		
33	4136	4203	4271	4338	4406	4473	4541	4608	4676	4743		
34	4811	4878	4946	5013	5081	5148	5216	5283	5351	5418		
35	5486	5553	5620	5688	5755	5823	5890	5958	6025	6093		
36	6160	6228	6295	6363	6430	6498	6565	6633	6700	6768		
37	6835	6903	6970	7037	7105	7172	7240	7307	7375	7442		
38	7510	7577	7645	7712	7780	7847	7914	7982	8049	8117		
39	8184	8252	8319	8387	8454	8521	8589	8656	8724	8791		
6440	8859	8926	8994	9061	9128	9196	9263	9331	9398	9466		
41	9533	9600	9668	9735	9803	9870	9938	0̄005	0̄072	0̄140		
42	8090207	0275	0342	0409	0477	0544	0612	0679	0747	0814		67
43	0881	0949	1016	1084	1151	1218	1286	1353	1421	1488		1 7
44	1555	1623	1690	1757	1825	1892	1960	2027	2094	2162		2 13
												3 20
45	2229	2297	2364	2431	2499	2566	2634	2701	2768	2836		4 27
46	2903	2970	3038	3105	3173	3240	3307	3375	3442	3509		5 34
47	3577	3644	3711	3779	3846	3914	3981	4048	4116	4183		6 40
48	4250	4318	4385	4452	4520	4587	4654	4722	4789	4856		7 47
49	4924	4991	5058	5126	5193	5260	5328	5395	5462	5530		8 54
												9 60
N.	0	1	2	3	4	5	6	7	8	9	D	Pts.

N.	0	1	2	3	4	5	6	7	8	9	D
6450	8095597	5664	5732	5799	5866	5934	6001	6068	6136	6203	
51	6270	6338	6405	6472	6540	6607	6674	6742	6809	6876	
52	6944	7011	7078	7146	7213	7280	7347	7415	7482	7549	
53	7617	7684	7751	7819	7886	7953	8020	8088	8155	8222	
54	8290	8357	8424	8491	8559	8626	8693	8761	8828	8895	
55	8962	9030	9097	9164	9232	9299	9366	9433	9501	9568	
56	9635	9702	9770	9837	9904	9972	ō039	ō106	ō173	ō241	
57	8100308	0375	0442	0510	0577	0644	0711	0779	0846	0913	
58	0980	1048	1115	1182	1249	1317	1384	1451	1518	1586	
59	1653	1720	1787	1855	1922	1989	2056	2123	2191	2258	
6460	2325	2392	2460	2527	2594	2661	2729	2796	2863	2930	
61	2997	3065	3132	3199	3266	3333	3401	3468	3535	3602	
62	3670	3737	3804	3871	3938	4006	4073	4140	4207	4274	
63	4342	4409	4476	4543	4610	4678	4745	4812	4879	4946	
64	5013	5081	5148	5215	5282	5349	5417	5484	5551	5618	
65	5685	5752	5820	5887	5954	6021	6088	6156	6223	6290	
66	6357	6424	6491	6558	6626	6693	6760	6827	6894	6961	
67	7029	7096	7163	7230	7297	7364	7432	7499	7566	7633	
68	7700	7767	7834	7902	7969	8036	8103	8170	8237	8304	
69	8372	8439	8506	8573	8640	8707	8774	8841	8909	8976	
6470	9043	9110	9177	9244	9311	9378	9446	9513	9580	9647	
71	9714	9781	9848	9915	9982	ō050	ō117	ō184	ō251	ō318	
72	8110385	0452	0519	0586	0653	0721	0788	0855	0922	0989	
73	1056	1123	1190	1257	1324	1392	1459	1526	1593	1660	
74	1727	1794	1861	1928	1995	2062	2129	2197	2264	2331	
75	2398	2465	2532	2599	2666	2733	2800	2867	2934	3001	
76	3068	3135	3203	3270	3337	3404	3471	3538	3605	3672	
77	3739	3806	3873	3940	4007	4074	4141	4208	4275	4342	
78	4409	4476	4544	4611	4678	4745	4812	4879	4946	5013	
79	5080	5147	5214	5281	5348	5415	5482	5549	5616	5683	
6480	5750	5817	5884	5951	6018	6085	6152	6219	6286	6353	67
81	6420	6487	6554	6621	6688	6755	6822	6889	6956	7023	
82	7090	7157	7224	7291	7358	7425	7492	7559	7626	7693	
83	7760	7827	7894	7961	8028	8095	8162	8229	8296	8363	
84	8430	8497	8564	8631	8698	8765	8832	8899	8966	9033	
85	9100	9167	9234	9301	9368	9435	9502	9569	9636	9702	
86	9769	9836	9903	9970	ō037	ō104	ō171	ō238	ō305	ō372	
87	8120439	0506	0573	0640	0707	0774	0841	0908	0975	1041	
88	1108	1175	1242	1309	1376	1443	1510	1577	1644	1711	
89	1778	1845	1912	1979	2045	2112	2179	2246	2313	2380	
6490	2447	2514	2581	2648	2715	2782	2848	2915	2982	3049	
91	3116	3183	3250	3317	3384	3451	3518	3584	3651	3718	
92	3785	3852	3919	3986	4053	4120	4186	4253	4320	4387	
93	4454	4521	4588	4655	4722	4788	4855	4922	4989	5056	
94	5123	5190	5257	5323	5390	5457	5524	5591	5658	5725	
95	5792	5858	5925	5992	6059	6126	6193	6260	6326	6393	
96	6460	6527	6594	6661	6728	6794	6861	6928	6995	7062	
97	7129	7196	7262	7329	7396	7463	7530	7597	7663	7730	
98	7797	7864	7931	7998	8064	8131	8198	8265	8332	8399	
99	8465	8532	8599	8666	8733	8799	8866	8933	9000	9067	
N.	0	1	2	3	4	5	6	7	8	9	D

Pro.

68		67	
1	7	1	7
2	14	2	13
3	20	3	20
4	27	4	27
5	34	5	34
6	41	6	40
7	48	7	47
8	54	8	54
9	61	9	60

Pts.

N.	0	1	2	3	4	5	6	7	8	9	D	Pro.
6500	8129134	9200	9267	9334	9401	9468	9534	9601	9668	9735		
01	9802	9868	9935	ō002	ō069	ō136	ō202	ō269	ō336	ō403		67
02	8130470	0536	0603	0670	0737	0804	0870	0937	1004	1071		1 7
03	1138	1204	1271	1338	1405	1471	1538	1605	1672	1739		2 13
04	1805	1872	1939	2006	2072	2139	2206	2273	2339	2406		3 20 / 4 27
05	2473	2540	2607	2673	2740	2807	2874	2940	3007	3074		5 34
06	3141	3207	3274	3341	3408	3474	3541	3608	3675	3741		6 40
07	3808	3875	3942	4008	4075	4142	4209	4275	4342	4409		7 47
08	4475	4542	4609	4676	4742	4809	4876	4943	5009	5076		8 54
09	5143	5209	5276	5343	5410	5476	5543	5610	5676	5743		9 60
6510	5810	5877	5943	6010	6077	6143	6210	6277	6344	6410		
11	6477	6544	6610	6677	6744	6810	6877	6944	7011	7077		
12	7144	7211	7277	7344	7411	7477	7544	7611	7677	7744		
13	7811	7877	7944	8011	8077	8144	8211	8278	8344	8411		
14	8478	8544	8611	8678	8744	8811	8878	8944	9011	9078		
15	9144	9211	9278	9344	9411	9477	9544	9611	9677	9744		
16	9811	9877	9944	ō011	ō077	ō144	ō211	ō277	ō344	ō411		
17	8140477	0544	0610	0677	0744	0810	0877	0944	1010	1077		
18	1144	1210	1277	1343	1410	1477	1543	1610	1677	1743		
19	1810	1876	1943	2010	2076	2143	2210	2276	2343	2409		
6520	2476	2543	2609	2676	2742	2809	2876	2942	3009	3075		
21	3142	3209	3275	3342	3408	3475	3542	3608	3675	3741		
22	3808	3875	3941	4008	4074	4141	4207	4274	4341	4407		
23	4474	4540	4607	4674	4740	4807	4873	4940	5006	5073		
24	5140	5206	5273	5339	5406	5472	5539	5605	5672	5739		
25	5805	5872	5938	6005	6071	6138	6204	6271	6338	6404		
26	6471	6537	6604	6670	6737	6803	6870	6937	7003	7070		
27	7136	7203	7269	7336	7402	7469	7535	7602	7668	7735		
28	7801	7868	7935	8001	8068	8134	8201	8267	8334	8400		
29	8467	8533	8600	8666	8733	8799	8866	8932	8999	9065		
6530	9132	9198	9265	9331	9398	9464	9531	9597	9664	9730		
31	9797	9863	9930	9996	ō063	ō129	ō196	ō262	ō329	ō395		
32	8150462	0528	0595	0661	0728	0794	0861	0927	0994	1060		
33	1127	1193	1260	1326	1392	1459	1525	1592	1658	1725		
34	1791	1858	1924	1991	2057	2124	2190	2257	2323	2389		
35	2456	2522	2589	2655	2722	2788	2855	2921	2988	3054		
36	3120	3187	3253	3320	3386	3453	3519	3586	3652	3718		
37	3785	3851	3918	3984	4051	4117	4183	4250	4316	4383		
38	4449	4516	4582	4648	4715	4781	4848	4914	4981	5047		
39	5113	5180	5246	5313	5379	5445	5512	5578	5645	5711		
6540	5777	5844	5910	5977	6043	6109	6176	6242	6309	6375		
41	6441	6508	6574	6641	6707	6773	6840	6906	6973	7039		
42	7105	7172	7238	7305	7371	7437	7504	7570	7636	7703		66
43	7769	7836	7902	7968	8035	8101	8167	8234	8300	8367		1 7
44	8433	8499	8566	8632	8698	8765	8831	8897	8964	9030		2 13 / 3 20
45	9097	9163	9229	9296	9362	9428	9495	9561	9627	9694		4 26
46	9760	9826	9893	9959	ō025	ō092	ō158	ō224	ō291	ō357		5 33
47	8160423	0490	0556	0622	0689	0755	0821	0888	0954	1020		6 40
48	1087	1153	1219	1286	1352	1418	1485	1551	1617	1684		7 46 / 8 53
49	1750	1816	1883	1949	2015	2081	2148	2214	2280	2347		9 59
N.	0	1	2	3	4	5	6	7	8	9	D	Pts.

N.	0	1	2	3	4	5	6	7	8	9	D	Pro.
6550	8162413	2479	2546	2612	2678	2745	2811	2877	2943	3010		
51	3076	3142	3209	3275	3341	3407	3474	3540	3606	3673		66
52	3739	3805	3871	3938	4004	4070	4137	4203	4269	4335		1 7
53	4402	4468	4534	4600	4667	4733	4799	4866	4932	4998		2 13
54	5064	5131	5197	5263	5329	5396	5462	5528	5594	5661		3 20
												4 26
55	5727	5793	5859	5926	5992	6058	6124	6191	6257	6323		5 33
56	6389	6456	6522	6588	6654	6721	6787	6853	6919	6986		6 40
57	7052	7118	7184	7251	7317	7383	7449	7515	7582	7648		7 46
58	7714	7780	7847	7913	7979	8045	8111	8178	8244	8310		8 53
59	8376	8443	8509	8575	8641	8707	8774	8840	8906	8972		9 59
6560	9038	9105	9171	9237	9303	9369	9436	9502	9568	9634		
61	9700	9767	9833	9899	9965	ō031	ō098	ō164	ō230	ō296		
62	8170362	0428	0495	0561	0627	0693	0759	0826	0892	0958		
63	1024	1090	1156	1223	1289	1355	1421	1487	1553	1620		
64	1686	1752	1818	1884	1950	2017	2083	2149	2215	2281		
65	2347	2413	2480	2546	2612	2678	2744	2810	2876	2943		
66	3009	3075	3141	3207	3273	3339	3406	3472	3538	3604		
67	3670	3736	3802	3869	3935	4001	4067	4133	4199	4265		
68	4331	4398	4464	4530	4596	4662	4728	4794	4860	4927		
69	4993	5059	5125	5191	5257	5323	5389	5455	5521	5588		
6570	5654	5720	5786	5852	5918	5984	6050	6116	6182	6249		
71	6315	6381	6447	6513	6579	6645	6711	6777	6843	6909		
72	6976	7042	7108	7174	7240	7306	7372	7438	7504	7570		
73	7636	7702	7768	7835	7901	7967	8033	8099	8165	8231		
74	8297	8363	8429	8495	8561	8627	8693	8759	8825	8892		
75	8958	9024	9090	9156	9222	9288	9354	9420	9486	9552		
76	9618	9684	9750	9816	9882	9948	ō014	ō080	ō146	ō212		
77	8180278	0344	0410	0477	0543	0609	0675	0741	0807	0873		
78	0939	1005	1071	1137	1203	1269	1335	1401	1467	1533	66	
79	1599	1665	1731	1797	1863	1929	1995	2061	2127	2193		
6580	2259	2325	2391	2457	2523	2589	2655	2721	2787	2853		
81	2919	2985	3051	3117	3183	3249	3315	3381	3447	3513		
82	3579	3645	3711	3777	3843	3909	3975	4041	4107	4173		
83	4239	4305	4370	4436	4502	4568	4634	4700	4766	4832		
84	4898	4964	5030	5096	5162	5228	5294	5360	5426	5492		
85	5558	5624	5690	5756	5822	5888	5953	6019	6085	6151		
86	6217	6283	6349	6415	6481	6547	6613	6679	6745	6811		
87	6877	6943	7008	7074	7140	7206	7272	7338	7404	7470		
88	7536	7602	7668	7734	7800	7866	7931	7997	8063	8129		
89	8195	8261	8327	8393	8459	8525	8591	8656	8722	8788		
6590	8854	8920	8986	9052	9118	9184	9250	9315	9381	9447		
91	9513	9579	9645	9711	9777	9843	9908	9974	ō040	ō106		
92	8190172	0238	0304	0370	0436	0501	0567	0633	0699	0765		65
93	0831	0897	0962	1028	1094	1160	1226	1292	1358	1424		1 7
94	1489	1555	1621	1687	1753	1819	1885	1950	2016	2082		2 13
												3 20
95	2148	2214	2280	2346	2411	2477	2543	2609	2675	2741		4 26
96	2806	2872	2938	3004	3070	3136	3202	3267	3333	3399		5 33
97	3465	3531	3597	3662	3728	3794	3860	3926	3991	4057		6 39
98	4123	4189	4255	4321	4386	4452	4518	4584	4650	4715		7 46
99	4781	4847	4913	4979	5045	5110	5176	5242	5308	5374		8 52
												9 59
N.	0	1	2	3	4	5	6	7	8	9	D	Pts.

N.	0	1	2	3	4	5	6	7	8	9	D	Pro.
6600	8195439	5505	5571	5637	5703	5768	5834	5900	5966	6032		
01	6097	6163	6229	6295	6360	6426	6492	6558	6624	6689		66
02	6755	6821	6887	6953	7018	7084	7150	7216	7281	7347		1 7
03	7413	7479	7545	7610	7676	7742	7808	7873	7939	8005		2 13
04	8071	8136	8202	8268	8334	8399	8465	8531	8597	8662		3 20
												4 26
05	8728	8794	8860	8925	8991	9057	9123	9188	9254	9320		5 33
06	9386	9451	9517	9583	9649	9714	9780	9846	9912	9977		6 40
07	8200043	0109	0175	0240	0306	0372	0437	0503	0569	0635		7 46
08	0̄700	0766	0832	0898	0963	1029	1095	1160	1226	1292		8 53
09	1358	1423	1489	1555	1620	1686	1752	1817	1883	1949		9 59
6610	2015	2080	2146	2212	2277	2343	2409	2474	2540	2606		
11	2672	2737	2803	2869	2934	3000	3066	3131	3197	3263		
12	3328	3394	3460	3525	3591	3657	3723	3788	3854	3920		
13	3985	4051	4117	4182	4248	4314	4379	4445	4511	4576		
14	4642	4708	4773	4839	4905	4970	5036	5102	5167	5233		
15	5298	5364	5430	5495	5561	5627	5692	5758	5824	5889		
16	5955	6021	6086	6152	6218	6283	6349	6414	6480	6546		
17	6611	6677	6743	6808	6874	6939	7005	7071	7136	7202		
18	7268	7333	7399	7464	7530	7596	7661	7727	7793	7858		
19	7924	7989	8055	8121	8186	8252	8317	8383	8449	8514		
6620	8580	8645	8711	8777	8842	8908	8973	9039	9105	9170		
21	9236	9301	9367	9433	9498	9564	9629	9695	9761	9826		
22	9892	9957	0̄023	0̄089	0̄154	0̄220	0̄285	0̄351	0̄416	0̄482		
23	8210548	0613	0679	0744	0810	0875	0941	1007	1072	1138		
24	1203	1269	1334	1400	1465	1531	1597	1662	1728	1793		
25	1859	1924	1990	2055	2121	2187	2252	2318	2383	2449		
26	2514	2580	2645	2711	2776	2842	2908	2973	3039	3104		
27	3170	3235	3301	3366	3432	3497	3563	3628	3694	3759		
28	3825	3891	3956	4022	4087	4153	4218	4284	4349	4415		
29	4480	4546	4611	4677	4742	4808	4873	4939	5004	5070		
6630	5135	5201	5266	5332	5397	5463	5528	5594	5659	5725		
31	5790	5856	5921	5987	6052	6118	6183	6249	6314	6380		
32	6445	6511	6576	6642	6707	6773	6838	6904	6969	7034		
33	7100	7165	7231	7296	7362	7427	7493	7558	7624	7689		
34	7755	7820	7886	7951	8017	8082	8147	8213	8278	8344		
35	8409	8475	8540	8606	8671	8737	8802	8867	8933	8998		
36	9064	9129	9195	9260	9326	9391	9456	9522	9587	9653		
37	9718	9784	9849	9914	9980	0̄045	0̄111	0̄176	0̄242	0̄307		
38	8220372	0438	0503	0569	0634	0700	0765	0830	0896	0961		
39	1027	1092	1158	1223	1288	1354	1419	1485	1550	1615		
6640	1681	1746	1812	1877	1942	2008	2073	2139	2204	2269		
41	2335	2400	2466	2531	2596	2662	2727	2793	2858	2923		
42	2989	3054	3119	3185	3250	3316	3381	3446	3512	3577		
43	3643	3708	3773	3839	3904	3969	4035	4100	4166	4231		65
44	4296	4362	4427	4492	4558	4623	4688	4754	4819	4884		1 7
												2 13
45	4950	5015	5081	5146	5211	5277	5342	5407	5473	5538		3 20
46	5603	5669	5734	5799	5865	5930	5995	6061	6126	6191		4 26
47	6257	6322	6387	6453	6518	6583	6649	6714	6779	6845		5 33
48	6910	6975	7041	7106	7171	7237	7302	7367	7433	7498		6 39
49	7563	7629	7694	7759	7825	7890	7955	8021	8086	8151		7 46
												8 52
												9 59
N.	0	1	2	3	4	5	6	7	8	9	D	Pts.

N.	0	1	2	3	4	5	6	7	8	9	D	Pro.
6650	8228216	8282	8347	8412	8478	8543	8608	8674	8739	8604		
51	8869	8935	9000	9065	9131	9196	9261	9327	9392	9457		65
52	9522	9588	9653	9718	9784	9849	9914	9979	0̄045	0̄110		1 7
53	8230175	0241	0306	0371	0436	0502	0567	0632	0697	0763		2 13
54	0628	0893	0958	1024	1089	1154	1220	1285	1350	1415		3 20 4 26
55	1481	1546	1611	1676	1742	1807	1872	1937	2003	2068		5 33
56	2133	2198	2264	2329	2394	2459	2525	2590	2655	2720		6 39
57	2786	2851	2916	2981	3047	3112	3177	3242	3307	3373		7 46
58	3438	3503	3568	3634	3699	3764	3829	3894	3960	4025		8 52
59	4090	4155	4221	4286	4351	4416	4481	4547	4612	4677		9 59
6660	4742	4808	4873	4938	5003	5068	5134	5199	5264	5329		
61	5394	5460	5525	5590	5655	5720	5786	5851	5916	5981		
62	6046	6111	6177	6242	6307	6372	6437	6503	6568	6633		
63	6698	6763	6828	6894	6959	7024	7089	7154	7220	7285		
64	7350	7415	7480	7545	7611	7676	7741	7806	7871	7936		
65	8002	8067	8132	8197	8262	8327	8392	8458	8523	8588		
66	8653	8718	8783	8849	8914	8979	9044	9109	9174	9239		
67	9305	9370	9435	9500	9565	9630	9695	9761	9826	9891		
68	9956	0̄021	0̄086	0̄151	0̄216	0̄282	0̄347	0̄412	0̄477	0̄542		
69	8240607	0672	0737	0803	0868	0933	0998	1063	1128	1193		
6670	1258	1323	1389	1454	1519	1584	1649	1714	1779	1844		
71	1909	1975	2040	2105	2170	2235	2300	2365	2430	2495		
72	2560	2625	2691	2756	2821	2886	2951	3016	3081	3146		
73	3211	3276	3341	3406	3472	3537	3602	3667	3732	3797		
74	3862	3927	3992	4057	4122	4187	4252	4318	4383	4448		
75	4513	4578	4643	4708	4773	4838	4903	4968	5033	5098		
76	5163	5228	5293	5358	5423	5489	5554	5619	5684	5749		
77	5814	5879	5944	6009	6074	6139	6204	6269	6334	6399		
78	6464	6529	6594	6659	6724	6789	6854	6919	6984	7049		
79	7114	7179	7244	7310	7375	7440	7505	7570	7635	7700		
6680	7765	7830	7895	7960	8025	8090	8155	8220	8285	8350		
81	8415	8480	8545	8610	8675	8740	8805	8870	8935	9000	65	
82	9065	9130	9195	9260	9325	9390	9455	9520	9585	9650		
83	9715	9780	9845	9910	9975	0̄040	0̄105	0̄169	0̄234	0̄299		
84	8250364	0429	0494	0559	0624	0689	0754	0819	0884	0949		
85	1014	1079	1144	1209	1274	1339	1404	1469	1534	1599		
86	1664	1729	1794	1859	1924	1988	2053	2118	2183	2248		
87	2313	2378	2443	2508	2573	2638	2703	2768	2833	2898		
88	2963	3028	3093	3157	3222	3287	3352	3417	3482	3547		
89	3612	3677	3742	3807	3872	3937	4002	4066	4131	4196		
6690	4261	4326	4391	4456	4521	4586	4651	4716	4780	4845		
91	4910	4975	5040	5105	5170	5235	5300	5365	5430	5494		
92	5559	5624	5689	5754	5819	5884	5949	6014	6078	6143		64
93	6208	6273	6338	6403	6468	6533	6598	6662	6727	6792		1 6
94	6857	6922	6987	7052	7117	7181	7246	7311	7376	7441		2 13 3 19
95	7506	7571	7636	7700	7765	7830	7895	7960	8025	8090		4 26
96	8154	8219	8284	8349	8414	8479	8544	8608	8673	8738		5 32
97	8803	8868	8933	8998	9062	9127	9192	9257	9322	9387		6 38
98	9451	9516	9581	9646	9711	9776	9840	9905	9970	0̄035		7 45
99	8260100	0165	0229	0294	0359	0424	0489	0554	0618	0683		8 51 9 58
N.	0	1	2	3	4	5	6	7	8	9	D	Pts.

N.	0	1	2	3	4	5	6	7	8	9	D	Pro.
6700	8260748	0813	0878	0942	1007	1072	1137	1202	1267	1331		
01	1396	1461	1526	1591	1655	1720	1785	1850	1915	1979		
02	2044	2109	2174	2239	2303	2368	2433	2498	2563	2627		65
03	2692	2757	2822	2887	2951	3016	3081	3146	3210	3275		1 7
04	3340	3405	3470	3534	3599	3664	3729	3794	3858	3923		2 13 3 20
05	3988	4053	4117	4182	4247	4312	4376	4441	4506	4571		4 26
06	4635	4700	4765	4830	4895	4959	5024	5089	5154	5218		5 33
07	5283	5348	5413	5477	5542	5607	5672	5736	5801	5866		6 39 7 46
08	5931	5995	6060	6125	6190	6254	6319	6384	6448	6513		8 52
09	6578	6643	6707	6772	6837	6902	6966	7031	7096	7160		9 59
6710	7225	7290	7355	7419	7484	7549	7614	7678	7743	7808		
11	7872	7937	8002	8067	8131	8196	8261	8325	8390	8455		
12	8519	8584	8649	8714	8778	8843	8908	8972	9037	9102		
13	9166	9231	9296	9361	9425	9490	9555	9619	9684	9749		
14	9813	9878	9943	ō007	ō072	ō137	ō201	ō266	ō331	ō395		
15	8270460	0525	0590	0654	0719	0784	0848	0913	0978	1042		
16	1107	1172	1236	1301	1366	1430	1495	1560	1624	1689		
17	1753	1818	1883	1947	2012	2077	2141	2206	2271	2335	64	
18	2400	2465	2529	2594	2659	2723	2788	2852	2917	2982		
19	3046	3111	3176	3240	3305	3370	3434	3499	3563	3628		
6720	3693	3757	3822	3887	3951	4016	4080	4145	4210	4274		
21	4339	4404	4468	4533	4597	4662	4727	4791	4856	4920		
22	4985	5050	5114	5179	5244	5308	5373	5437	5502	5567		
23	5631	5696	5760	5825	5889	5954	6019	6083	6148	6212		
24	6277	6342	6406	6471	6535	6600	6665	6729	6794	6858		
25	6923	6987	7052	7117	7181	7246	7310	7375	7439	7504		
26	7569	7633	7698	7762	7827	7891	7956	8021	8085	8150		
27	8214	8279	8343	8408	8473	8537	8602	8666	8731	8795		
28	8860	8924	8989	9053	9118	9183	9247	9312	9376	9441		
29	9505	9570	9634	9699	9763	9828	9893	9957	ō022	ō086		
6730	8280151	0215	0280	0344	0409	0473	0538	0602	0667	0731		
31	0796	0860	0925	0989	1054	1119	1183	1248	1312	1377		
32	1441	1506	1570	1635	1699	1764	1828	1893	1957	2022		
33	2086	2151	2215	2280	2344	2409	2473	2538	2602	2667		
34	2731	2796	2860	2925	2989	3054	3118	3183	3247	3312		
35	3376	3440	3505	3569	3634	3698	3763	3827	3892	3956		
36	4021	4085	4150	4214	4279	4343	4408	4472	4537	4601		
37	4665	4730	4794	4859	4923	4988	5052	5117	5181	5246		
38	5310	5375	5439	5503	5568	5632	5697	5761	5826	5890		
39	5955	6019	6083	6148	6212	6277	6341	6406	6470	6535		
6740	6599	6663	6728	6792	6857	6921	6986	7050	7114	7179		
41	7243	7308	7372	7437	7501	7565	7630	7694	7759	7823		
42	7887	7952	8016	8081	8145	8210	8274	8338	8403	8467		
43	8532	8596	8660	8725	8789	8854	8918	8982	9047	9111		64
44	9176	9240	9304	9369	9433	9498	9562	9626	9691	9755		1 6 2 13
45	9820	9884	9948	ō013	ō077	ō141	ō206	ō270	ō335	ō399		3 19
46	8290463	0528	0592	0656	0721	0785	0850	0914	0978	1043		4 26 5 32
47	1107	1171	1236	1300	1365	1429	1493	1558	1622	1686		6 38 7 45
48	1751	1815	1879	1944	2008	2073	2137	2201	2266	2330		8 51
49	2394	2459	2523	2587	2652	2716	2780	2845	2909	2973		9 58
N.	0	1	2	3	4	5	6	7	8	9	D	Pts.

N.	0	1	2	3	4	5	6	7	8	9	D	Pro.
6750	8293038	3102	3166	3231	3295	3359	3424	3488	3552	3617		
51	3681	3745	3810	3874	3938	4003	4067	4131	4196	4260		64
52	4324	4389	4453	4517	4582	4646	4710	4775	4839	4903		1 6
53	4967	5032	5096	5160	5225	5289	5353	5418	5482	5546		2 13
54	5611	5675	5739	5803	5868	5932	5996	6061	6125	6189		3 19
55	6254	6318	6382	6446	6511	6575	6639	6704	6768	6832		4 26
56	6896	6961	7025	7089	7154	7218	7282	7346	7411	7475		5 32
57	7539	7603	7668	7732	7796	7861	7925	7989	8053	8118		6 38
58	8182	8246	8310	8375	8439	8503	8567	8632	8696	8760		7 45
59	8824	8889	8953	9017	9081	9146	9210	9274	9338	9403		8 51
6760	9467	9531	9595	9660	9724	9788	9852	9917	9981	ō045		9 58
61	8300109	0174	0238	0302	0366	0431	0495	0559	0623	0687		
62	0752	0816	0880	0944	1009	1073	1137	1201	1265	1330		
63	1394	1458	1522	1587	1651	1715	1779	1843	1908	1972		
64	2036	2100	2164	2229	2293	2357	2421	2485	2550	2614		
65	2678	2742	2806	2871	2935	2999	3063	3127	3192	3256		
66	3320	3384	3448	3512	3577	3641	3705	3769	3833	3898		
67	3962	4026	4090	4154	4218	4283	4347	4411	4475	4539		
68	4604	4668	4732	4796	4860	4924	4988	5053	5117	5181		
69	5245	5309	5373	5438	5502	5566	5630	5694	5758	5823		
6770	5887	5951	6015	6079	6143	6207	6272	6336	6400	6464		
71	6528	6592	6656	6721	6785	6849	6913	6977	7041	7105		
72	7169	7234	7298	7362	7426	7490	7554	7618	7683	7747		
73	7811	7875	7939	8003	8067	8131	8195	8260	8324	8388		
74	8452	8516	8580	8644	8708	8772	8837	8901	8965	9029		
75	9093	9157	9221	9285	9349	9413	9478	9542	9606	9670		
76	9734	9798	9862	9926	9990	ō054	ō119	ō183	ō247	ō311		
77	8310375	0439	0503	0567	0631	0695	0759	0823	0887	0952		
78	1016	1080	1144	1208	1272	1336	1400	1464	1528	1592		
79	1656	1720	1784	1849	1913	1977	2041	2105	2169	2233		
6780	2297	2361	2425	2489	2553	2617	2681	2745	2809	2873		
81	2937	3001	3066	3130	3194	3258	3322	3386	3450	3514		
82	3578	3642	3706	3770	3834	3898	3962	4026	4090	4154		
83	4218	4282	4346	4410	4474	4538	4602	4666	4730	4794		
84	4858	4922	4986	5050	5114	5178	5242	5306	5371	5435		
85	5499	5563	5627	5691	5755	5819	5883	5947	6011	6075	64	
86	6139	6203	6267	6331	6395	6459	6523	6587	6651	6715		
87	6778	6842	6906	6970	7034	7098	7162	7226	7290	7354		
88	7418	7482	7546	7610	7674	7738	7802	7866	7930	7994		
89	8058	8122	8186	8250	8314	8378	8442	8506	8570	8634		
6790	8698	8762	8826	8890	8954	9018	9081	9145	9209	9273		
91	9337	9401	9465	9529	9593	9657	9721	9785	9849	9913		
92	9977	ō041	ō105	ō169	ō233	ō296	ō360	ō424	ō488	ō552		63
93	8320616	0680	0744	0808	0872	0936	1000	1064	1128	1192		1 6
94	1255	1319	1383	1447	1511	1575	1639	1703	1767	1831		2 13
												3 19
95	1895	1959	2022	2086	2150	2214	2278	2342	2406	2470		4 25
96	2534	2598	2662	2725	2789	2853	2917	2981	3045	3109		5 32
97	3173	3237	3300	3364	3428	3492	3556	3620	3684	3748		6 38
98	3812	3875	3939	4003	4067	4131	4195	4259	4323	4387		7 44
99	4450	4514	4578	4642	4706	4770	4834	4898	4961	5025		8 50
												9 57
N.	0	1	2	3	4	5	6	7	8	9	D	Pts.

N.	0	1	2	3	4	5	6	7	8	9	D	Pro.
6800	8325089	5153	5217	5281	5345	5408	5472	5536	5600	5664		
01	5728	5792	5855	5919	5983	6047	6111	6175	6239	6302		64
02	6366	6430	6494	6558	6622	6686	6749	6813	6877	6941		1 6
03	7005	7069	7132	7196	7260	7324	7388	7452	7515	7579		2 13
04	7643	7707	7771	7835	7898	7962	8026	8090	8154	8217		3 19
												4 26
05	8281	8345	8409	8473	8537	8600	8664	8728	8792	8856		5 32
06	8919	8983	9047	9111	9175	9238	9302	9366	9430	9494		6 38
07	9558	9621	9685	9749	9813	9877	9940	0̄004	0̄068	0̄132		7 45
08	8330195	0259	0323	0387	0451	0514	0578	0642	0706	0770		8 51
09	0833	0897	0961	1025	1088	1152	1216	1280	1344	1407		9 58
6810	1471	1535	1599	1662	1726	1790	1854	1918	1981	2045		
11	2109	2173	2236	2300	2364	2428	2491	2555	2619	2683		
12	2746	2810	2874	2938	3001	3065	3129	3193	3256	3320		
13	3384	3448	3511	3575	3639	3703	3766	3830	3894	3958		
14	4021	4085	4149	4212	4276	4340	4404	4467	4531	4595		
15	4659	4722	4786	4850	4913	4977	5041	5105	5168	5232		
16	5296	5360	5423	5487	5551	5614	5678	5742	5806	5869		
17	5933	5997	6060	6124	6188	6251	6315	6379	6443	6506		
18	6570	6634	6697	6761	6825	6888	6952	7016	7080	7143		
19	7207	7271	7334	7398	7462	7525	7589	7653	7716	7780		
6820	7844	7907	7971	8035	8098	8162	8226	8289	8353	8417		
21	8480	8544	8608	8672	8735	8799	8862	8926	8990	9053		
22	9117	9181	9244	9308	9372	9435	9499	9563	9626	9690		
23	9754	9817	9881	9945	0̄008	0̄072	0̄136	0̄199	0̄263	0̄327		
24	8340390	0454	0517	0581	0645	0708	0772	0836	0899	0963		
25	1027	1090	1154	1217	1281	1345	1408	1472	1536	1599		
26	1663	1726	1790	1854	1917	1981	2045	2108	2172	2235		
27	2299	2363	2426	2490	2553	2617	2681	2744	2808	2872		
28	2935	2999	3062	3126	3190	3253	3317	3380	3444	3508		
29	3571	3635	3698	3762	3826	3889	3953	4016	4080	4143		
6830	4207	4271	4334	4398	4461	4525	4589	4652	4716	4779		
31	4843	4906	4970	5034	5097	5161	5224	5288	5351	5415		
32	5479	5542	5606	5669	5733	5796	5860	5924	5987	6051		
33	6114	6178	6241	6305	6368	6432	6496	6559	6623	6686		
34	6750	6813	6877	6940	7004	7067	7131	7195	7258	7322		
35	7385	7449	7512	7576	7639	7703	7766	7830	7893	7957		
36	8021	8084	8148	8211	8275	8338	8402	8465	8529	8592		
37	8656	8719	8783	8846	8910	8973	9037	9100	9164	9227		
38	9291	9354	9418	9481	9545	9609	9672	9736	9799	9863		
39	9926	9990	0̄053	0̄117	0̄180	0̄244	0̄307	0̄371	0̄434	0̄498		
6840	8350561	0625	0688	0751	0815	0878	0942	1005	1069	1132		
41	1196	1259	1323	1386	1450	1513	1577	1640	1704	1767		
42	1831	1894	1958	2021	2085	2148	2212	2275	2338	2402		
43	2465	2529	2592	2656	2719	2783	2846	2910	2973	3037		63
44	3100	3163	3227	3290	3354	3417	3481	3544	3608	3671		1 6
												2 13
45	3735	3798	3861	3925	3988	4052	4115	4179	4242	4306		3 19
46	4369	4432	4496	4559	4623	4686	4750	4813	4876	4940		4 25
47	5003	5067	5130	5194	5257	5320	5384	5447	5511	5574		5 32
48	5638	5701	5764	5828	5891	5955	6018	6081	6145	6208		6 38
49	6272	6335	6398	6462	6525	6589	6652	6716	6779	6842		7 44
												8 50
												9 57
N.	0	1	2	3	4	5	6	7	8	9	D	Pts.

N.	0	1	2	3	4	5	6	7	8	9	D	Pro.
6850	8356906	6969	7033	7096	7159	7223	7286	7349	7413	7476		
51	7540	7603	7666	7730	7793	7857	7920	7983	8047	8110		63.
52	8174	8237	8300	8364	8427	8490	8554	8617	8681	8744		1 6
53	8807	8871	8934	8997	9061	9124	9188	9251	9314	9378		2 13
54	9441	9504	9568	9631	9694	9758	9821	9885	9948	0̄011		3 19
												4 25
55	8360075	0138	0201	0265	0328	0391	0455	0518	0581	0645		5 32
56	0708	0771	0835	0898	0961	1025	1088	1151	1215	1278		6 38
57	1341	1405	1468	1531	1595	1658	1721	1785	1848	1911		7 44
58	1975	2038	2101	2165	2228	2291	2355	2418	2481	2545		8 50
59	2608	2671	2735	2798	2861	2925	2988	3051	3115	3178		9 57
6860	3241	3304	3368	3431	3494	3558	3621	3684	3748	3811		
61	3874	3937	4001	4064	4127	4191	4254	4317	4381	4444		
62	4507	4570	4634	4697	4760	4824	4887	4950	5013	5077		
63	5140	5203	5267	5330	5393	5456	5520	5583	5646	5709		
64	5773	5836	5899	5963	6026	6089	6152	6216	6279	6342		
65	6405	6469	6532	6595	6658	6722	6785	6848	6911	6975		
66	7038	7101	7164	7228	7291	7354	7417	7481	7544	7607		
67	7670	7734	7797	7860	7923	7987	8050	8113	8176	8240		
68	8303	8366	8429	8493	8556	8619	8682	8745	8809	8872		
69	8935	8998	9062	9125	9188	9251	9314	9378	9441	9504		
6870	9567	9631	9694	9757	9820	9883	9947	0̄010	0̄073	0̄136		
71	8370199	0263	0326	0389	0452	0516	0579	0642	0705	0768		
72	0832	0895	0958	1021	1084	1147	1211	1274	1337	1400		
73	1463	1527	1590	1653	1716	1779	1843	1906	1969	2032		
74	2095	2158	2222	2285	2348	2411	2474	2538	2601	2664		
75	2727	2790	2853	2917	2980	3043	3106	3169	3232	3296		
76	3359	3422	3485	3548	3611	3674	3738	3801	3864	3927		
77	3990	4053	4117	4180	4243	4306	4369	4432	4495	4559		
78	4622	4685	4748	4811	4874	4937	5001	5064	5127	5190		
79	5253	5316	5379	5442	5506	5569	5632	5695	5758	5821		
6880	5884	5948	6011	6074	6137	6200	6263	6326	6389	6452		
81	6516	6579	6642	6705	6768	6831	6894	6957	7020	7084		
82	7147	7210	7273	7336	7399	7462	7525	7588	7652	7715		
83	7778	7841	7904	7967	8030	8093	8156	8219	8282	8346		
84	8409	8472	8535	8598	8661	8724	8787	8850	8913	8976		
85	9039	9103	9166	9229	9292	9355	9418	9481	9544	9607		
86	9670	9733	9796	9859	9922	9986	0̄049	0̄112	0̄175	0̄238		
87	8380301	0364	0427	0490	0553	0616	0679	0742	0805	0868		
88	0931	0994	1057	1121	1184	1247	1310	1373	1436	1499		
89	1562	1625	1688	1751	1814	1877	1940	2003	2066	2129		
6890	2192	2255	2318	2381	2444	2507	2570	2633	2696	2759		
91	2822	2886	2949	3012	3075	3138	3201	3264	3327	3390		62
92	3453	3516	3579	3642	3705	3768	3831	3894	3957	4020		1 6
93	4083	4146	4209	4272	4335	4398	4461	4524	4587	4650		2 12
94	4713	4776	4839	4902	4965	5028	5091	5154	5217	5280	63	3 19
95	5343	5406	5469	5532	5595	5658	5721	5784	5847	5910		4 25
96	5973	6036	6098	6161	6224	6287	6350	6413	6476	6539		5 31
97	6602	6665	6728	6791	6854	6917	6980	7043	7106	7169		6 37
98	7232	7295	7358	7421	7484	7547	7610	7673	7736	7798		7 43
99	7861	7924	7987	8050	8113	8176	8239	8302	8365	8428		8 50
												9 56
N.	0	1	2	3	4	5	6	7	8	9	D	Pts.

N.	0	1	2	3	4	5	6	7	8	9	D	Pro.
6900	8388491	8554	8617	8680	8743	8806	8869	8931	8994	9057		
01	9120	9183	9246	9309	9372	9435	9498	9561	9624	9687		63
02	9750	9812	9875	9938	0̄001	0̄064	0̄127	0̄190	0̄253	0̄316		
03	8390379	0442	0505	0567	0630	0693	0756	0819	0882	0945		1 6
04	1008	1071	1134	1197	1259	1322	1385	1448	1511	1574		2 13 3 19 4 25
05	1637	1700	1763	1826	1888	1951	2014	2077	2140	2203		5 32
06	2266	2329	2392	2454	2517	2580	2643	2706	2769	2832		6 38
07	2895	2957	3020	3083	3146	3209	3272	3335	3398	3460		7 44
08	3523	3586	3649	3712	3775	3838	3900	3963	4026	4089		8 50
09	4152	4215	4278	4341	4403	4466	4529	4592	4655	4718		9 57
6910	4780	4843	4906	4969	5032	5095	5158	5220	5283	5346		
11	5409	5472	5535	5597	5660	5723	5786	5849	5912	5974		
12	6037	6100	6163	6226	6289	6351	6414	6477	6540	6603		
13	6666	6728	6791	6854	6917	6980	7042	7105	7168	7231		
14	7294	7357	7419	7482	7545	7608	7671	7733	7796	7859		
15	7922	7985	8047	8110	8173	8236	8299	8361	8424	8487		
16	8550	8613	8675	8738	8801	8864	8927	8989	9052	9115		
17	9178	9241	9303	9366	9429	9492	9554	9617	9680	9743		
18	9806	9868	9931	9994	0̄057	0̄119	0̄182	0̄245	0̄308	0̄371		
19	8400433	0496	0559	0622	0684	0747	0810	0873	0935	0998		
6920	1061	1124	1186	1249	1312	1375	1437	1500	1563	1626		
21	1688	1751	1814	1877	1939	2002	2065	2128	2190	2253		
22	2316	2379	2441	2504	2567	2630	2692	2755	2818	2881		
23	2943	3006	3069	3132	3194	3257	3320	3382	3445	3508		
24	3571	3633	3696	3759	3821	3884	3947	4010	4072	4135		
25	4198	4260	4323	4386	4449	4511	4574	4637	4699	4762		
26	4825	4888	4950	5013	5076	5138	5201	5264	5326	5389		
27	5452	5515	5577	5640	5703	5765	5828	5891	5953	6016		
28	6079	6141	6204	6267	6330	6392	6455	6518	6580	6643		
29	6706	6768	6831	6894	6956	7019	7082	7144	7207	7270		
6930	7332	7395	7458	7520	7583	7646	7708	7771	7834	7896		
31	7959	8022	8084	8147	8210	8272	8335	8398	8460	8523		
32	8586	8648	8711	8773	8836	8899	8961	9024	9087	9149		
33	9212	9275	9337	9400	9463	9525	9588	9650	9713	9776		
34	9838	9901	9964	0̄026	0̄089	0̄152	0̄214	0̄277	0̄339	0̄402		
35	8410465	0527	0590	0653	0715	0778	0840	0903	0966	1028		
36	1091	1153	1216	1279	1341	1404	1467	1529	1592	1654		
37	1717	1780	1842	1905	1967	2030	2093	2155	2218	2280		
38	2343	2406	2468	2531	2593	2656	2719	2781	2844	2906		
39	2969	3031	3094	3157	3219	3282	3344	3407	3470	3532		
6940	3595	3657	3720	3782	3845	3908	3970	4033	4095	4158		
41	4220	4283	4346	4408	4471	4533	4596	4658	4721	4784		
42	4846	4909	4971	5034	5096	5159	5221	5284	5347	5409		
43	5472	5534	5597	5659	5722	5784	5847	5909	5972	6035		62
44	6097	6160	6222	6285	6347	6410	6472	6535	6597	6660		1 6 2 12 3 19
45	6723	6785	6848	6910	6973	7035	7098	7160	7223	7285		4 25
46	7348	7410	7473	7535	7598	7660	7723	7785	7848	7910		5 31
47	7973	8036	8098	8161	8225	8286	8348	8411	8473	8536		6 37
48	8598	8661	8723	8786	8848	8911	8973	9036	9098	9161		7 43 8 50
49	9223	9286	9348	9411	9473	9536	9598	9661	9723	9786		9 56
N.	0	1	2	3	4	5	6	7	8	9	D	Pts.

N.	0	1	2	3	4	5	6	7	8	9	D	Pro.
6950	8419848	9911	9973	ō036	ō098	ō160	ō223	ō285	ō348	ō410		
51	8420473	0535	0598	0660	0723	0785	0848	0910	0973	1035		63
52	1098	1160	1223	1285	1348	1410	1472	1535	1597	1660		1 6
53	1722	1785	1847	1910	1972	2035	2097	2160	2222	2284		2 13
54	2347	2409	2472	2534	2597	2659	2722	2784	2846	2909		3 19
												4 25
55	2971	3034	3096	3159	3221	3284	3346	3408	3471	3533		5 32
56	3596	3658	3721	3783	3845	3908	3970	4033	4095	4158		6 38
57	4220	4282	4345	4407	4470	4532	4595	4657	4719	4782		7 44
58	4844	4907	4969	5031	5094	5156	5219	5281	5344	5406		8 50
59	5468	5531	5593	5656	5718	5780	5843	5905	5968	6030		9 57
6960	6092	6155	6217	6280	6342	6404	6467	6529	6592	6654		
61	6716	6779	6841	6904	6966	7028	7091	7153	7215	7278		
62	7340	7403	7465	7527	7590	7652	7714	7777	7839	7902		
63	7964	8026	8089	8151	8213	8276	8338	8401	8463	8525		
64	8588	8650	8712	8775	8837	8899	8962	9024	9086	9149		
65	9211	9274	9336	9398	9461	9523	9585	9648	9710	9772		
66	9835	9897	9959	ō022	ō084	ō146	ō209	ō271	ō333	ō396		
67	8430458	0520	0583	0645	0707	0770	0832	0894	0957	1019		
68	1081	1144	1206	1268	1331	1393	1455	1518	1580	1642		
69	1705	1767	1829	1892	1954	2016	2079	2141	2203	2265		
6970	2328	2390	2452	2515	2577	2639	2702	2764	2826	2889		
71	2951	3013	3075	3138	3200	3262	3325	3387	3449	3511		
72	3574	3636	3698	3761	3823	3885	3948	4010	4072	4134		
73	4197	4259	4321	4383	4446	4508	4570	4633	4695	4757		
74	4819	4882	4944	5006	5069	5131	5193	5255	5318	5380		
75	5442	5504	5567	5629	5691	5753	5816	5878	5940	6002		
76	6065	6127	6189	6251	6314	6376	6438	6500	6563	6625		
77	6687	6749	6812	6874	6936	6998	7061	7123	7185	7247		
78	7310	7372	7434	7496	7559	7621	7683	7745	7808	7870		
79	7932	7994	8056	8119	8181	8243	8305	8368	8430	8492		
6980	8554	8616	8679	8741	8803	8865	8928	8990	9052	9114		
81	9176	9239	9301	9363	9425	9487	9550	9612	9674	9736		
82	9798	9861	9923	9985	ō047	ō109	ō172	ō234	ō296	ō358		
83	8440420	0483	0545	0607	0669	0731	0794	0856	0918	0980		
84	1042	1104	1167	1229	1291	1353	1415	1478	1540	1602		
85	1664	1726	1788	1851	1913	1975	2037	2099	2161	2224		
86	2286	2348	2410	2472	2534	2597	2659	2721	2783	2845		
87	2907	2970	3032	3094	3156	3218	3280	3343	3405	3467		
88	3529	3591	3653	3715	3778	3840	3902	3964	4026	4088		
89	4150	4213	4275	4337	4399	4461	4523	4585	4647	4710		
6990	4772	4834	4896	4958	5020	5082	5145	5207	5269	5331		
91	5393	5455	5517	5579	5642	5704	5766	5828	5890	5952		
92	6014	6076	6138	6201	6263	6325	6387	6449	6511	6573		62
93	6635	6697	6759	6822	6884	6946	7008	7070	7132	7194		1 6
94	7256	7318	7380	7443	7505	7567	7629	7691	7753	7815		2 12
												3 19
95	7877	7939	8001	8063	8126	8188	8250	8312	8374	8436		4 25
96	8498	8560	8622	8684	8746	8808	8870	8933	8995	9057		5 31
97	9119	9181	9243	9305	9367	9429	9491	9553	9615	9677		6 37
98	9739	9801	9863	9926	9988	ō050	ō112	ō174	ō236	ō298		7 43
99	8450360	0422	0484	0546	0608	0670	0732	0794	0856	0918		8 50
												9 56
N.	0	1	2	3	4	5	6	7	8	9	D	Pts.

N.	0	1	2	3	4	5	6	7	8	9	D	Pro.
7000	8450980	1042	1104	1167	1229	1291	1353	1415	1477	1539		
01	1601	1663	1725	1787	1849	1911	1973	2035	2097	2159		62
02	2221	2283	2345	2407	2469	2531	2593	2655	2717	2779		1 6
03	2841	2903	2965	3027	3089	3151	3213	3275	3337	3399		2 12
04	3461	3523	3585	3647	3709	3771	3833	3895	3957	4019		3 19 / 4 25
05	4081	4143	4205	4267	4329	4391	4453	4515	4577	4639	62	5 31
06	4701	4763	4825	4887	4949	5011	5073	5135	5197	5259		6 37
07	5321	5383	5445	5507	5569	5631	5693	5755	5817	5879		7 43
08	5941	6003	6065	6127	6189	6251	6313	6375	6437	6499		8 50
09	6561	6623	6685	6746	6808	6870	6932	6994	7056	7118		9 56
7010	7180	7242	7304	7366	7428	7490	7552	7614	7676	7738		
11	7800	7862	7924	7986	8047	8109	8171	8233	8295	8357		
12	8419	8481	8543	8605	8667	8729	8791	8853	8915	8976		
13	9038	9100	9162	9224	9286	9348	9410	9472	9534	9596		
14	9658	9720	9781	9843	9905	9967	0̄029	0̄091	0̄153	0̄215		
15	8460277	0339	0401	0462	0524	0586	0648	0710	0772	0834		
16	0896	0958	1020	1082	1143	1205	1267	1329	1391	1453		
17	1515	1577	1639	1700	1762	1824	1886	1948	2010	2072		
18	2134	2196	2257	2319	2381	2443	2505	2567	2629	2691		
19	2752	2814	2876	2938	3000	3062	3124	3186	3247	3309		
7020	3371	3433	3495	3557	3619	3680	3742	3804	3866	3928		
21	3990	4052	4113	4175	4237	4299	4361	4423	4485	4546		
22	4608	4670	4732	4794	4856	4917	4979	5041	5103	5165		
23	5227	5289	5350	5412	5474	5536	5598	5660	5721	5783		
24	5845	5907	5969	6031	6092	6154	6216	6278	6340	6401		
25	6463	6525	6587	6649	6711	6772	6834	6896	6958	7020		
26	7081	7143	7205	7267	7329	7391	7452	7514	7576	7638		
27	7700	7761	7823	7885	7947	8009	8070	8132	8194	8256		
28	8318	8379	8441	8503	8565	8626	8688	8750	8812	8874		
29	8935	8997	9059	9121	9183	9244	9306	9368	9430	9491		
7030	9553	9615	9677	9739	9800	9862	9924	9986	0̄047	0̄109		
31	8470171	0233	0295	0356	0418	0480	0542	0603	0665	0727		
32	0789	0850	0912	0974	1036	1097	1159	1221	1283	1344		
33	1406	1468	1530	1591	1653	1715	1777	1838	1900	1962		
34	2024	2085	2147	2209	2271	2332	2394	2456	2518	2579		
35	2641	2703	2764	2826	2888	2950	3011	3073	3135	3197		
36	3258	3320	3382	3443	3505	3567	3629	3690	3752	3814		
37	3876	3937	3999	4061	4122	4184	4246	4307	4369	4431		
38	4493	4554	4616	4678	4739	4801	4863	4925	4986	5048		
39	5110	5171	5233	5295	5356	5418	5480	5542	5603	5665		
7040	5727	5788	5850	5912	5973	6035	6097	6158	6220	6282		
41	6343	6405	6467	6528	6590	6652	6714	6775	6837	6899		
42	6960	7022	7084	7145	7207	7269	7330	7392	7454	7515		61
43	7577	7639	7700	7762	7824	7885	7947	8009	8070	8132		1 6
44	8193	8255	8317	8378	8440	8502	8563	8625	8687	8748		2 12 / 3 18
45	8810	8872	8933	8995	9057	9118	9180	9241	9303	9365		4 24
46	9426	9488	9550	9611	9673	9735	9796	9858	9919	9981		5 31
47	8480043	0104	0166	0228	0289	0351	0412	0474	0536	0597		6 37
48	0659	0721	0782	0844	0905	0967	1029	1090	1152	1213		7 43 / 8 49
49	1275	1337	1398	1460	1522	1583	1645	1706	1768	1830		9 55
N.	0	1	2	3	4	5	6	7	8	9	D	Pts.

N.	0	1	2	3	4	5	6	7	8	9	D	Pro.
7050	8481891	1953	2014	2076	2138	2199	2261	2322	2384	2446		
51	2507	2569	2630	2692	2754	2815	2877	2938	3000	3061		62
52	3123	3185	3246	3308	3369	3431	3493	3554	3616	3677		1 6
53	3739	3800	3862	3924	3985	4047	4108	4170	4231	4293		2 12
54	4355	4416	4478	4539	4601	4662	4724	4786	4847	4909		3 19
												4 25
55	4970	5032	5093	5155	5216	5278	5340	5401	5463	5524		5 31
56	5586	5647	5709	5770	5832	5893	5955	6017	6078	6140		6 37
57	6201	6263	6324	6386	6447	6509	6570	6632	6693	6755		7 43
58	6817	6878	6940	7001	7063	7124	7186	7247	7309	7370		8 50
59	7432	7493	7555	7616	7678	7739	7801	7862	7924	7985		9 56
7060	8047	8109	8170	8232	8293	8355	8416	8478	8539	8601		
61	8662	8724	8785	8847	8908	8970	9031	9093	9154	9216		
62	9277	9339	9400	9462	9523	9585	9646	9708	9769	9831		
63	9892	9954	ō015	ō077	ō138	ō199	ō261	ō322	ō384	ō445		
64	8490507	0568	0630	0691	0753	0814	0876	0937	0999	1060		
65	1122	1183	1245	1306	1368	1429	1490	1552	1613	1675		
66	1736	1798	1859	1921	1982	2044	2105	2167	2228	2289		
67	2351	2412	2474	2535	2597	2658	2720	2781	2843	2904		
68	2965	3027	3088	3150	3211	3273	3334	3396	3457	3518		
69	3580	3641	3703	3764	3826	3887	3948	4010	4071	4133		
7070	4194	4256	4317	4378	4440	4501	4563	4624	4686	4747		
71	4808	4870	4931	4993	5054	5115	5177	5238	5300	5361		
72	5423	5484	5545	5607	5668	5730	5791	5852	5914	5975		
73	6037	6098	6159	6221	6282	6344	6405	6466	6528	6589		
74	6651	6712	6773	6835	6896	6958	7019	7080	7142	7203		
75	7264	7326	7387	7449	7510	7571	7633	7694	7755	7817		
76	7878	7940	8001	8062	8124	8185	8246	8308	8369	8431		
77	8492	8553	8615	8676	8737	8799	8860	8922	8983	9044		
78	9106	9167	9228	9290	9351	9412	9474	9535	9596	9658		
79	9719	9780	9842	9903	9965	ō026	ō087	ō149	ō210	ō271		
7080	8500333	0394	0455	0517	0578	0639	0701	0762	0823	0885		
81	0946	1007	1069	1130	1191	1253	1314	1375	1437	1498		
82	1559	1621	1682	1743	1805	1866	1927	1988	2050	2111		
83	2172	2234	2295	2356	2418	2479	2540	2602	2663	2724		
84	2786	2847	2908	2969	3031	3092	3153	3215	3276	3337		
85	3399	3460	3521	3582	3644	3705	3766	3828	3889	3950		
86	4011	4073	4134	4195	4257	4318	4379	4440	4502	4563		
87	4624	4686	4747	4808	4869	4931	4992	5053	5115	5176		
88	5237	5298	5360	5421	5482	5543	5605	5666	5727	5788		
89	5850	5911	5972	6034	6095	6156	6217	6279	6340	6401		
7090	6462	6524	6585	6646	6707	6769	6830	6891	6952	7014		
91	7075	7136	7197	7259	7320	7381	7442	7504	7565	7626		
92	7687	7749	7810	7871	7932	7993	8055	8116	8177	8238		61
93	8300	8361	8422	8483	8545	8606	8667	8728	8789	8851		1 6
94	8912	8973	9034	9095	9157	9218	9279	9340	9402	9463		2 12
												3 18
95	9524	9585	9646	9708	9769	9830	9891	9952	ō014	ō075		4 24
96	8510136	0197	0258	0320	0381	0442	0503	0564	0626	0687		5 31
97	0748	0809	0870	0932	0993	1054	1115	1176	1238	1299		6 37
98	1360	1421	1482	1544	1605	1666	1727	1788	1849	1911		7 43
99	1972	2032	2094	2155	2216	2278	2339	2400	2461	2522		8 49
												9 55
N.	0	1	2	3	4	5	6	7	8	9	D	Pts.

N.	0	1	2	3	4	5	6	7	8	9	D	Pro.
7100	8512583	2645	2706	2767	2828	2889	2950	3012	3073	3134		
01	3195	3256	3317	3379	3440	3501	3562	3623	3684	3746		**62**
02	3807	3868	3929	3990	4051	4112	4174	4235	4296	4357		1 6
03	4418	4479	4540	4602	4663	4724	4785	4846	4907	4968		2 12
04	5030	5091	5152	5213	5274	5335	5396	5457	5519	5580		3 19
												4 25
05	5641	5702	5763	5824	5885	5946	6008	6069	6130	6191		5 31
06	6252	6313	6374	6435	6496	6558	6619	6680	6741	6802		6 37
07	6863	6924	6985	7046	7108	7169	7230	7291	7352	7413		7 43
08	7474	7535	7596	7657	7719	7780	7841	7902	7963	8024		8 50
09	8085	8146	8207	8268	8329	8391	8452	8513	8574	8635		9 56
7110	8696	8757	8818	8879	8940	9001	9062	9124	9185	9246		
11	9307	9368	9429	9490	9551	9612	9673	9734	9795	9856		
12	9917	9979	ō040	ō101	ō162	ō223	ō284	ō345	ō406	ō467		
13	8520528	0589	0650	0711	0772	0833	0894	0955	1017	1078		
14	1139	1200	1261	1322	1383	1444	1505	1566	1627	1688		
15	1749	1810	1871	1932	1993	2054	2115	2176	2237	2298		
16	2359	2420	2481	2542	2604	2665	2726	2787	2848	2909		
17	2970	3031	3092	3153	3214	3275	3336	3397	3458	3519		
18	3580	3641	3702	3763	3824	3885	3946	4007	4068	4129		
19	4190	4251	4312	4373	4434	4495	4556	4617	4678	4739		
7120	4800	4861	4922	4983	5044	5105	5166	5227	5288	5349	61	
21	5410	5471	5532	5593	5654	5715	5776	5837	5898	5959		
22	6020	6081	6142	6203	6264	6325	6386	6447	6508	6568		
23	6629	6690	6751	6812	6873	6934	6995	7056	7117	7178		
24	7239	7300	7361	7422	7483	7544	7605	7666	7727	7788		
25	7849	7910	7971	8032	8092	8153	8214	8275	8336	8397		
26	8458	8519	8580	8641	8702	8763	8824	8885	8946	9007		
27	9068	9129	9189	9250	9311	9372	9433	9494	9555	9616		
28	9677	9738	9799	9860	9921	9982	ō042	ō103	ō164	ō225		
29	8530286	0347	0408	0469	0530	0591	0652	0713	0773	0834		
7130	0895	0956	1017	1078	1139	1200	1261	1322	1383	1443		
31	1504	1565	1626	1687	1748	1809	1870	1931	1992	2052		
32	2113	2174	2235	2296	2357	2418	2479	2540	2600	2661		
33	2722	2783	2844	2905	2966	3027	3088	3148	3209	3270		
34	3331	3392	3453	3514	3575	3635	3696	3757	3818	3879		
35	3940	4001	4062	4122	4183	4244	4305	4366	4427	4488		
36	4548	4609	4670	4731	4792	4853	4914	4974	5035	5096		
37	5157	5218	5279	5340	5400	5461	5522	5583	5644	5705		
38	5765	5826	5887	5948	6009	6070	6130	6191	6252	6313		
39	6374	6435	6495	6556	6617	6678	6739	6800	6860	6921		
7140	6982	7043	7104	7165	7225	7286	7347	7408	7469	7530		
41	7590	7651	7712	7773	7834	7894	7955	8016	8077	8138		
42	8198	8259	8320	8381	8442	8502	8563	8624	8685	8746		
43	8807	8867	8928	8989	9050	9110	9171	9232	9293	9354		**61**
44	9414	9475	9536	9597	9658	9718	9779	9840	9901	9962		1 6
45	8540022	0083	0144	0205	0265	0326	0387	0448	0509	0569		2 12
46	0630	0691	0752	0812	0873	0934	0995	1056	1116	1177		3 18
47	1238	1299	1359	1420	1481	1542	1602	1663	1724	1785		4 24
48	1845	1906	1967	2028	2088	2149	2210	2271	2331	2392		5 31
49	2453	2514	2574	2635	2696	2757	2817	2878	2939	3000		6 37
												7 43
												8 49
												9 55
N.	0	1	2	3	4	5	6	7	8	9	D	Pts.

N.	0	1	2	3	4	5	6	7	8	9	D	Pro
7150	8543060	3121	3182	3243	3303	3364	3425	3486	3546	3607		
51	3668	3729	3789	3850	3911	3971	4032	4093	4154	4214		61
52	4275	4336	4397	4457	4518	4579	4639	4700	4761	4822		1 6
53	4882	4943	5004	5064	5125	5186	5247	5307	5368	5429		2 12
54	5489	5550	5611	5671	5732	5793	5854	5914	5975	6036		3 18
												4 24
55	6096	6157	6218	6278	6339	6400	6461	6521	6582	6643		5 31
56	6703	6764	6825	6885	6946	7007	7067	7128	7189	7249		6 37
57	7310	7371	7432	7492	7553	7614	7674	7735	7796	7856		7 43
58	7917	7978	8038	8099	8160	8220	8281	8342	8402	8463		8 49
59	8524	8584	8645	8706	8766	8827	8888	8948	9009	9070		9 55
7160	9130	9191	9252	9312	9373	9433	9494	9555	9615	9676		
61	9737	9797	9858	9919	9979	ō040	ō101	ō161	ō222	ō283		
62	8550343	0404	0464	0525	0586	0646	0707	0768	0828	0889		
63	0950	1010	1071	1131	1192	1253	1313	1374	1435	1495		
64	1556	1616	1677	1738	1798	1859	1919	1980	2041	2101		
65	2162	2223	2283	2344	2404	2465	2526	2586	2647	2707		
66	2768	2829	2889	2950	3010	3071	3132	3192	3253	3313		
67	3374	3435	3495	3556	3616	3677	3738	3798	3859	3919		
68	3980	4041	4101	4162	4222	4283	4343	4404	4465	4525		
69	4586	4646	4707	4768	4828	4889	4949	5010	5070	5131		
7170	5192	5252	5313	5373	5434	5494	5555	5616	5676	5737		
71	5797	5858	5918	5979	6039	6100	6161	6221	6282	6342		
72	6403	6463	6524	6584	6645	6706	6766	6827	6887	6948		
73	7008	7069	7129	7190	7250	7311	7372	7432	7493	7553		
74	7614	7674	7735	7795	7856	7916	7977	8037	8098	8159		
75	8219	8280	8340	8401	8461	8522	8582	8643	8703	8764		
76	8824	8885	8945	9006	9066	9127	9187	9248	9308	9369		
77	9429	9490	9550	9611	9672	9732	9793	9853	9914	9974		
78	8560035	0095	0156	0216	0277	0337	0398	0458	0519	0579		
79	0640	0700	0761	0821	0882	0942	1002	1063	1123	1184		
7180	1244	1305	1365	1426	1486	1547	1607	1668	1728	1789		
81	1849	1910	1970	2031	2091	2152	2212	2273	2333	2394		
82	2454	2514	2575	2635	2696	2756	2817	2877	2938	2998		
83	3059	3119	3180	3240	3301	3361	3421	3482	3542	3603		
84	3663	3724	3784	3845	3905	3965	4026	4086	4147	4207		
85	4268	4328	4389	4449	4509	4570	4630	4691	4751	4812		
86	4872	4933	4993	5053	5114	5174	5235	5295	5356	5416		
87	5476	5537	5597	5658	5718	5779	5839	5899	5960	6020		
88	6081	6141	6202	6262	6322	6383	6443	6504	6564	6624		
89	6685	6745	6806	6866	6926	6987	7047	7108	7168	7229		
7190	7289	7349	7410	7470	7531	7591	7651	7712	7772	7832		
91	7893	7953	8014	8074	8134	8195	8255	8316	8376	8436		
92	8497	8557	8618	8678	8738	8799	8859	8919	8980	9040		60
93	9101	9161	9221	9282	9342	9402	9463	9523	9584	9644		1 6
94	9704	9765	9825	9885	9946	ō006	ō067	ō127	ō187	ō248		2 12
												3 18
95	8570308	0368	0429	0489	0549	0610	0670	0730	0791	0851		4 24
96	0912	0972	1032	1093	1153	1213	1274	1334	1394	1455		5 30
97	1515	1575	1636	1696	1756	1817	1877	1937	1998	2058		6 36
98	2118	2179	2239	2299	2360	2420	2480	2541	2601	2661		7 42
99	2722	2782	2842	2903	2963	3023	3084	3144	3204	3265		8 48
												9 54
N.	0	1	2	3	4	5	6	7	8	9	D	Pts.

N.	0	1	2	3	4	5	6	7	8	9	D
7200	8573325	3385	3446	3506	3566	3627	3687	3747	3807	3868	
01	3928	3988	4049	4109	4169	4230	4290	4350	4411	4471	
02	4531	4591	4652	4712	4772	4833	4893	4953	5014	5074	
03	5134	5194	5255	5315	5375	5436	5496	5556	5616	5677	
04	5737	5797	5858	5918	5978	6038	6099	6159	6219	6280	
05	6340	6400	6460	6521	6581	6641	6701	6762	6822	6882	
06	6943	7003	7063	7123	7184	7244	7304	7364	7425	7485	
07	7545	7605	7666	7726	7786	7847	7907	7967	8027	8088	
08	8148	8208	8268	8329	8389	8449	8509	8570	8630	8690	
09	8750	8810	8871	8931	8991	9051	9112	9172	9232	9292	
7210	9353	9413	9473	9533	9594	9654	9714	9774	9835	9895	
11	9955	ō015	ō075	ō136	ō196	ō256	ō316	ō377	ō437	ō497	
12	8580557	0617	0678	0738	0798	0858	0918	0979	1039	1099	
13	1159	1220	1280	1340	1400	1460	1521	1581	1641	1701	
14	1761	1822	1882	1942	2002	2062	2123	2183	2243	2303	
15	2363	2424	2484	2544	2604	2664	2724	2785	2845	2905	
16	2965	3025	3086	3146	3206	3266	3326	3387	3447	3507	
17	3567	3627	3687	3748	3808	3868	3928	3988	4048	4109	
18	4169	4229	4289	4349	4409	4470	4530	4590	4650	4710	
19	4770	4831	4891	4951	5011	5071	5131	5192	5252	5312	
7220	5372	5432	5492	5552	5613	5673	5733	5793	5853	5913	
21	5973	6034	6094	6154	6214	6274	6334	6394	6455	6515	
22	6575	6635	6695	6755	6815	6876	6936	6996	7056	7116	
23	7176	7236	7296	7357	7417	7477	7537	7597	7657	7717	
24	7777	7837	7898	7958	8018	8078	8138	8198	8258	8318	
25	8379	8439	8499	8559	8619	8679	8739	8799	8859	8919	
26	8980	9040	9100	9160	9220	9280	9340	9400	9460	9520	
27	9581	9641	9701	9761	9821	9881	9941	ō001	ō061	ō121	
28	8590181	0242	0302	0362	0422	0482	0542	0602	0662	0722	
29	0782	0842	0902	0962	1023	1083	1143	1203	1263	1323	
7230	1383	1443	1503	1563	1623	1683	1743	1803	1863	1924	
31	1984	2044	2104	2164	2224	2284	2344	2404	2464	2524	
32	2584	2644	2704	2764	2824	2884	2944	3005	3065	3125	
33	3185	3245	3305	3365	3425	3485	3545	3605	3665	3725	
34	3785	3845	3905	3965	4025	4085	4145	4205	4265	4325	
35	4385	4445	4505	4565	4625	4685	4746	4806	4866	4926	
36	4986	5046	5106	5166	5226	5286	5346	5406	5466	5526	
37	5586	5646	5706	5766	5826	5886	5946	6006	6066	6126	60
38	6186	6246	6306	6366	6426	6486	6546	6606	6666	6726	
39	6786	6846	6906	6966	7026	7086	7146	7206	7266	7326	
7240	7386	7446	7506	7566	7626	7686	7746	7806	7866	7925	
41	7985	8045	8105	8165	8225	8285	8345	8405	8465	8525	
42	8585	8645	8705	8765	8825	8885	8945	9005	9065	9125	
43	9185	9245	9305	9365	9425	9485	9545	9605	9665	9724	
44	9784	9844	9904	9964	ō024	ō084	ō144	ō204	ō264	ō324	
45	8600384	0444	0504	0564	0624	0684	0744	0803	0863	0923	
46	0983	1043	1103	1163	1223	1283	1343	1403	1463	1523	
47	1583	1643	1702	1762	1822	1882	1942	2002	2062	2122	
48	2182	2242	2302	2362	2422	2481	2541	2601	2661	2721	
49	2781	2841	2901	2961	3021	3081	3140	3200	3260	3320	
N.	0	1	2	3	4	5	6	7	8	9	D

Pro.

60
1 | 6
2 | 12
3 | 18
4 | 24
5 | 30
6 | 36
7 | 42
8 | 48
9 | 54

59
1 | 6
2 | 12
3 | 18
4 | 24
5 | 30
6 | 35
7 | 41
8 | 47
9 | 53

Pts.

N.	0	1	2	3	4	5	6	7	8	9	D	Pro.
7250	8603380	3440	3500	3560	3620	3680	3739	3799	3859	3919		
51	3979	4039	4099	4159	4219	4279	4338	4398	4458	4518		60
52	4578	4638	4698	4758	4817	4877	4937	4997	5057	5117		1 6
53	5177	5237	5297	5356	5416	5476	5536	5596	5656	5716		2 12
54	5776	5835	5895	5955	6015	6075	6135	6195	6254	6314		3 18
												4 24
55	6374	6434	6494	6554	6614	6673	6733	6793	6853	6913		5 30
56	6973	7033	7092	7152	7212	7272	7332	7392	7452	7511		6 36
57	7571	7631	7691	7751	7811	7870	7930	7990	8050	8110		7 42
58	8170	8229	8289	8349	8409	8469	8529	8588	8648	8708		8 48
59	8768	8828	8888	8947	9007	9067	9127	9187	9247	9306		9 54
7260	9366	9426	9486	9546	9605	9665	9725	9785	9845	9905		
61	9964	0̄024	0̄084	0̄144	0̄204	0̄263	0̄323	0̄383	0̄443	0̄503		
62	8610562	0622	0682	0742	0802	0861	0921	0981	1041	1101		
63	1160	1220	1280	1340	1400	1459	1519	1579	1639	1699		
64	1758	1818	1878	1938	1997	2057	2117	2177	2237	2296		
65	2356	2416	2476	2536	2595	2655	2715	2775	2834	2894		
66	2954	3014	3073	3133	3193	3253	3313	3372	3432	3492		
67	3552	3611	3671	3731	3791	3850	3910	3970	4030	4089		
68	4149	4209	4269	4328	4388	4448	4508	4567	4627	4687		
69	4747	4806	4866	4926	4986	5045	5105	5165	5225	5284		
7270	5344	5404	5464	5523	5583	5643	5703	5762	5822	5882		
71	5941	6001	6061	6121	6180	6240	6300	6360	6419	6479		
72	6539	6598	6658	6718	6778	6837	6897	6957	7016	7076		
73	7136	7196	7255	7315	7375	7434	7494	7554	7614	7673		
74	7733	7793	7852	7912	7972	8031	8091	8151	8211	8270		
75	8330	8390	8449	8509	8569	8628	8688	8748	8808	8867		
76	8927	8987	9046	9106	9166	9225	9285	9345	9404	9464		
77	9524	9583	9643	9703	9762	9822	9882	9941	0̄001	0̄061		
78	8620121	0180	0240	0300	0359	0419	0479	0538	0598	0658		
79	0717	0777	0837	0896	0956	1016	1075	1135	1194	1254		
7280	1314	1373	1433	1493	1552	1612	1672	1731	1791	1851		
81	1910	1970	2030	2089	2149	2209	2268	2328	2387	2447		
82	2507	2566	2626	2686	2745	2805	2865	2924	2984	3043		
83	3103	3163	3222	3282	3342	3401	3461	3520	3580	3640		
84	3699	3759	3819	3878	3938	3997	4057	4117	4176	4236		
85	4296	4355	4415	4474	4534	4594	4653	4713	4772	4832		
86	4892	4951	5011	5070	5130	5190	5249	5309	5368	5428		
87	5488	5547	5607	5666	5726	5786	5845	5905	5964	6024		
88	6084	6143	6203	6262	6322	6382	6441	6501	6560	6620		
89	6680	6739	6799	6858	6918	6977	7037	7097	7156	7216		
7290	7275	7335	7394	7454	7514	7573	7633	7692	7752	7811		
91	7871	7931	7990	8050	8109	8169	8228	8288	8347	8407		
92	8467	8526	8586	8645	8705	8764	8824	8883	8943	9003		59
93	9062	9122	9181	9241	9300	9360	9419	9479	9539	9598		1 6
94	9658	9717	9777	9836	9896	9955	0̄015	0̄074	0̄134	0̄193		2 12
												3 18
95	8630253	0312	0372	0432	0491	0551	0610	0670	0729	0789		4 24
96	0848	0908	0967	1027	1086	1146	1205	1265	1324	1384		5 30
97	1443	1503	1562	1622	1682	1741	1801	1860	1920	1979		6 35
98	2039	2098	2158	2217	2277	2336	2396	2455	2515	2574		7 41
99	2634	2693	2753	2812	2872	2931	2991	3050	3110	3169		8 47
												9 53
N.	0	1	2	3	4	5	6	7	8	9	D	Pts.

к 2

N.	0	1	2	3	4	5	6	7	8	9	D	Pro.
7300	8633229	3288	3348	3407	3467	3526	3586	3645	3705	3764		
01	3823	3883	3942	4002	4061	4121	4180	4240	4299	4359		60
02	4418	4478	4537	4597	4656	4716	4775	4835	4894	4954		1 6
03	5013	5072	5132	5191	5251	5310	5370	5429	5489	5548		2 12
04	5608	5667	5727	5786	5845	5905	5964	6024	6083	6143		3 18
												4 24
05	6202	6262	6321	6381	6440	6499	6559	6618	6678	6737		5 30
06	6797	6856	6916	6975	7034	7094	7153	7213	7272	7332		6 36
07	7391	7451	7510	7569	7629	7688	7748	7807	7867	7926		7 42
08	7985	8045	8104	8164	8223	8283	8342	8401	8461	8520		8 48
09	8580	8639	8698	8758	8817	8877	8936	8996	9055	9114		9 54
7310	9174	9233	9293	9352	9411	9471	9530	9590	9649	9708		
11	9768	9827	9887	9946	0̄005	0̄065	0̄124	0̄184	0̄243	0̄302		
12	8640362	0421	0481	0540	0599	0659	0718	0778	0837	0896		
13	0956	1015	1075	1134	1193	1253	1312	1371	1431	1490		
14	1550	1609	1668	1728	1787	1846	1906	1965	2025	2084		
15	2143	2203	2262	2321	2381	2440	2500	2559	2618	2678		
16	2737	2796	2856	2915	2974	3034	3093	3152	3212	3271		
17	3331	3390	3449	3509	3568	3627	3687	3746	3805	3865		
18	3924	3983	4043	4102	4161	4221	4280	4339	4399	4458		
19	4517	4577	4636	4695	4755	4814	4873	4933	4992	5051		
7320	5111	5170	5229	5289	5348	5407	5467	5526	5585	5645		
21	5704	5763	5823	5882	5941	6001	6060	6119	6179	6238		
22	6297	6357	6416	6475	6534	6594	6653	6712	6772	6831		
23	6890	6950	7009	7068	7128	7187	7246	7305	7365	7424		
24	7483	7543	7602	7661	7721	7780	7839	7898	7958	8017		
25	8076	8136	8195	8254	8313	8373	8432	8491	8551	8610		
26	8669	8728	8788	8847	8906	8966	9025	9084	9143	9203		
27	9262	9321	9380	9440	9499	9558	9618	9677	9736	9795		
28	9855	9914	9973	0̄032	0̄092	0̄151	0̄210	0̄269	0̄329	0̄388		
29	8650447	0506	0566	0625	0684	0743	0803	0862	0921	0980		
7330	1040	1099	1158	1217	1277	1336	1395	1454	1514	1573		
31	1632	1691	1751	1810	1869	1928	1988	2047	2106	2165		
32	2225	2284	2343	2402	2461	2521	2580	2639	2698	2758		
33	2817	2876	2935	2995	3054	3113	3172	3231	3291	3350		
34	3409	3468	3527	3587	3646	3705	3764	3824	3883	3942		
35	4001	4060	4120	4179	4238	4297	4356	4416	4475	4534		
36	4593	4652	4712	4771	4830	4889	4948	5008	5067	5126		
37	5185	5244	5304	5363	5422	5481	5540	5600	5659	5718		
38	5777	5836	5895	5955	6014	6073	6132	6191	6251	6310		
39	6369	6428	6487	6546	6606	6665	6724	6783	6842	6901		
7340	6961	7020	7079	7138	7197	7256	7316	7375	7434	7493		
41	7552	7611	7671	7730	7789	7848	7907	7966	8025	8085		
42	8144	8203	8262	8321	8380	8440	8499	8558	8617	8676		
43	8735	8794	8854	8913	8972	9031	9090	9149	9208	9268		59
44	9327	9386	9445	9504	9563	9622	9681	9741	9800	9859		1 6
												2 12
45	9918	9977	0̄036	0̄095	0̄155	0̄214	0̄273	0̄332	0̄391	0̄450		3 18
46	8660509	0568	0627	0687	0746	0805	0864	0923	0982	1041		4 24
47	1100	1160	1219	1278	1337	1396	1455	1514	1573	1632		5 30
48	1691	1751	1810	1869	1928	1987	2046	2105	2164	2223		6 35
49	2282	2342	2401	2460	2519	2578	2637	2696	2755	2814		7 41
												8 47
												9 53
N.	0	1	2	3	4	5	6	7	8	9	D	Pts.

N	0	1	2	3	4	5	6	7	8	9	D	Pro
7350	8662873	2932	2992	3051	3110	3169	3228	3287	3346	3405		
51	3464	3523	3582	3641	3701	3760	3819	3878	3937	3996		59
52	4055	4114	4173	4232	4291	4350	4409	4468	4528	4587		1 6
53	4646	4705	4764	4823	4882	4941	5000	5059	5118	5177		2 12
54	5236	5295	5354	5413	5472	5532	5591	5650	5709	5768		3 18
												4 24
55	5827	5886	5945	6004	6063	6122	6181	6240	6299	6358		5 30
56	6417	6476	6535	6594	6653	6712	6771	6830	6889	6949		6 35
57	7008	7067	7126	7185	7244	7303	7362	7421	7480	7539		7 41
58	7598	7657	7716	7775	7834	7893	7952	8011	8070	8129		8 47
59	8188	8247	8306	8365	8424	8483	8542	8601	8660	8719		9 53
7360	8778	8837	8896	8955	9014	9073	9132	9191	9250	9309	59	
61	9368	9427	9486	9545	9604	9663	9722	9781	9840	9899		
62	9958	0̄017	0̄076	0̄135	0̄194	0̄253	0̄312	0̄371	0̄430	0̄489		
63	8670548	0607	0666	0725	0784	0843	0902	0961	1020	1079		
64	1138	1197	1256	1315	1374	1433	1492	1551	1610	1669		
65	1728	1786	1845	1904	1963	2022	2081	2140	2199	2258		
66	2317	2376	2435	2494	2553	2612	2671	2730	2789	2848		
67	2907	2966	3025	3084	3142	3201	3260	3319	3378	3437		
68	3496	3555	3614	3673	3732	3791	3850	3909	3968	4027		
69	4086	4145	4203	4262	4321	4380	4439	4498	4557	4616		
7370	4675	4734	4793	4852	4911	4970	5028	5087	5146	5205		
71	5264	5323	5382	5441	5500	5559	5618	5677	5735	5794		
72	5853	5912	5971	6030	6089	6148	6207	6266	6325	6383		
73	6442	6501	6560	6619	6678	6737	6796	6855	6914	6972		
74	7031	7090	7149	7208	7267	7326	7385	7444	7502	7561		
75	7620	7679	7738	7797	7856	7915	7974	8032	8091	8150		
76	8209	8268	8327	8386	8445	8503	8562	8621	8680	8739		
77	8798	8857	8916	8974	9033	9092	9151	9210	9269	9328		
78	9387	9445	9504	9563	9622	9681	9740	9799	9857	9916		
79	9975	0̄034	0̄093	0̄152	0̄211	0̄269	0̄328	0̄387	0̄446	0̄505		
7380	8680564	0622	0681	0740	0799	0858	0917	0976	1034	1093		
81	1152	1211	1270	1329	1387	1446	1505	1564	1623	1682		
82	1740	1799	1858	1917	1976	2035	2093	2152	2211	2270		
83	2329	2388	2446	2505	2564	2623	2682	2740	2799	2858		
84	2917	2976	3035	3093	3152	3211	3270	3329	3387	3446		
85	3505	3564	3623	3681	3740	3799	3858	3917	3975	4034		
86	4093	4152	4211	4269	4328	4387	4446	4505	4563	4622		
87	4681	4740	4799	4857	4916	4975	5034	5093	5151	5210		
88	5269	5328	5386	5445	5504	5563	5622	5680	5739	5798		
89	5857	5915	5974	6033	6092	6151	6209	6268	6327	6386		
7390	6444	6503	6562	6621	6679	6738	6797	6856	6915	6973		
91	7032	7091	7150	7208	7267	7326	7385	7443	7502	7561		
92	7620	7678	7737	7796	7855	7913	7972	8031	8090	8148		58
93	8207	8266	8325	8383	8442	8501	8560	8618	8677	8736		1 6
94	8794	8853	8912	8971	9029	9088	9147	9206	9264	9323		2 12
												3 17
95	9382	9441	9499	9558	9617	9675	9734	9793	9852	9910		4 23
96	9969	0̄028	0̄086	0̄145	0̄204	0̄263	0̄321	0̄380	0̄439	0̄497		5 29
97	8690556	0615	0674	0732	0791	0850	0908	0967	1026	1085		6 35
98	1143	1202	1261	1319	1378	1437	1495	1554	1613	1672		7 41
												8 46
99	1730	1789	1848	1906	1965	2024	2082	2141	2200	2259		9 52
N.	0	1	2	3	4	5	6	7	8	9	D	Pts.

N.	0	1	2	3	4	5	6	7	8	9	D	Pro.
7400	8692317	2376	2435	2493	2552	2611	2669	2728	2787	2845		
01	2904	2963	3021	3080	3139	3197	3256	3315	3373	3432		
02	3491	3549	3608	3667	3725	3784	3843	3901	3960	4019		59
03	4077	4136	4195	4253	4312	4371	4429	4488	4547	4605		1 6
04	4664	4723	4781	4840	4899	4957	5016	5075	5133	5192		2 12
05	5251	5309	5368	5427	5485	5544	5603	5661	5720	5778		3 18
06	5837	5896	5954	6013	6072	6130	6189	6248	6306	6365		4 24
07	6423	6482	6541	6599	6658	6717	6775	6834	6892	6951		5 30
08	7010	7068	7127	7186	7244	7303	7361	7420	7479	7537		6 35
09	7596	7655	7713	7772	7830	7889	7948	8006	8065	8123		7 41
7410	8182	8241	8299	8358	8417	8475	8534	8592	8651	8710		8 47
11	8768	8827	8885	8944	9003	9061	9120	9178	9237	9296		9 53
12	9354	9413	9471	9530	9588	9647	9706	9764	9823	9881		
13	9940	9999	ō057	ō116	ō174	ō233	ō292	ō350	ō409	ō467		!
14	8700526	0584	0643	0702	0760	0819	0877	0936	0994	1053		
15	1112	1170	1229	1287	1346	1404	1463	1522	1580	1639		
16	1697	1756	1814	1873	1931	1990	2049	2107	2166	2224		
17	2283	2341	2400	2458	2517	2576	2634	2693	2751	2810		
18	2868	2927	2985	3044	3102	3161	3220	3278	3337	3395		
19	3454	3512	3571	3629	3688	3746	3805	3863	3922	3981		
7420	4039	4098	4156	4215	4273	4332	4390	4449	4507	4566		
21	4624	4683	4741	4800	4858	4917	4975	5034	5092	5151		
22	5210	5268	5327	5385	5444	5502	5561	5619	5678	5736		
23	5795	5853	5912	5970	6029	6087	6146	6204	6263	6321		
24	6380	6438	6497	6555	6614	6672	6731	6789	6848	6906		
25	6965	7023	7082	7140	7199	7257	7316	7374	7432	7491		
26	7549	7608	7666	7725	7783	7842	7900	7959	8017	8076		
27	8134	8193	8251	8310	8368	8427	8485	8544	8602	8660		
28	8719	8777	8836	8894	8953	9011	9070	9128	9187	9245		
29	9304	9362	9421	9479	9537	9596	9654	9713	9771	9830		
7430	9888	9947	ō005	ō063	ō122	ō180	ō239	ō297	ō356	ō414		
31	8710473	0531	0589	0648	0706	0765	0823	0882	0940	0999		
32	1057	1115	1174	1232	1291	1349	1408	1466	1524	1583		
33	1641	1700	1758	1817	1875	1933	1992	2050	2109	2167		
34	2226	2284	2342	2401	2459	2518	2576	2634	2693	2751		
35	2810	2868	2927	2985	3043	3102	3160	3219	3277	3335		
36	3394	3452	3511	3569	3627	3686	3744	3803	3861	3919		
37	3978	4036	4095	4153	4211	4270	4328	4387	4445	4503		
38	4562	4620	4679	4737	4795	4854	4912	4970	5029	5087		
39	5146	5204	5262	5321	5379	5437	5496	5554	5613	5671		
7440	5729	5788	5846	5904	5963	6021	6080	6138	6196	6255		
41	6313	6371	6430	6488	6546	6605	6663	6722	6780	6838		
42	6897	6955	7013	7072	7130	7188	7247	7305	7363	7422		58
43	7480	7539	7597	7655	7714	7772	7830	7889	7947	8005		1 6
44	8064	8122	8180	8239	8297	8355	8414	8472	8530	8589		2 12
45	8647	8705	8764	8822	8880	8939	8997	9055	9114	9172		3 17
46	9230	9289	9347	9405	9464	9522	9580	9639	9697	9755		4 23
47	9814	9872	9930	9988	ō047	ō105	ō163	ō222	ō280	ō338		5 29
48	8720397	0455	0513	0572	0630	0688	0747	0805	0863	0921		6 35
49	0980	1038	1096	1155	1213	1271	1330	1388	1446	1504		7 41
												8 46
												9 52
N.	0	1	2	3	4	5	6	7	8	9	D	Pts.

N.	0	1	2	3	4	5	6	7	8	9	D	Pro.
7450	8721563	1621	1679	1738	1796	1854	1912	1971	2029	2087		
51	2146	2204	2262	2320	2379	2437	2495	2554	2612	2670		58
52	2728	2787	2845	2903	2962	3020	3078	3136	3195	3253		1 6
53	3311	3369	3428	3486	3544	3603	3661	3719	3777	3836		2 12
54	3894	3952	4010	4069	4127	4185	4243	4302	4360	4418		3 17
												4 23
55	4476	4535	4593	4651	4709	4768	4826	4884	4942	5001		5 29
56	5059	5117	5175	5234	5292	5350	5408	5467	5525	5583		6 35
57	5641	5700	5758	5816	5874	5933	5991	6049	6107	6166		7 41
58	6224	6282	6340	6398	6457	6515	6573	6631	6690	6748		8 46
59	6806	6864	6923	6981	7039	7097	7155	7214	7272	7330		9 52
7460	7388	7446	7505	7563	7621	7679	7738	7796	7854	7912		
61	7970	8029	8087	8145	8203	8261	8320	8378	8436	8494		
62	8552	8611	8669	8727	8785	8843	8902	8960	9018	9076		
63	9134	9193	9251	9309	9367	9425	9484	9542	9600	9658		
64	9716	9774	9833	9891	9949	0̄007	0̄065	0̄124	0̄182	0̄240		
65	8730298	0356	0414	0473	0531	0589	0647	0705	0764	0822		
66	0880	0938	0996	1054	1113	1171	1229	1287	1345	1403		
67	1462	1520	1578	1636	1694	1752	1810	1869	1927	1985		
68	2043	2101	2159	2218	2276	2334	2392	2450	2508	2566		
69	2625	2683	2741	2799	2857	2915	2973	3032	3090	3148		
7470	3206	3264	3322	3380	3439	3497	3555	3613	3671	3729		
71	3787	3845	3904	3962	4020	4078	4136	4194	4252	4311		
72	4369	4427	4485	4543	4601	4659	4717	4775	4834	4892		
73	4950	5008	5066	5124	5182	5240	5298	5357	5415	5473		
74	5531	5589	5647	5705	5763	5821	5880	5938	5996	6054		
75	6112	6170	6228	6286	6344	6402	6461	6519	6577	6635		
76	6693	6751	6809	6867	6925	6983	7041	7100	7158	7216		
77	7274	7332	7390	7448	7506	7564	7622	7680	7738	7797		
78	7855	7913	7971	8029	8087	8145	8203	8261	8319	8377		
79	8435	8493	8551	8610	8668	8726	8784	8842	8900	8958		
7480	9016	9074	9132	9190	9248	9306	9364	9422	9480	9538		
81	9597	9655	9713	9771	9829	9887	9945	0̄003	0̄061	0̄119		
82	8740177	0235	0293	0351	0409	0467	0525	0583	0641	0699		
83	0757	0815	0874	0932	0990	1048	1106	1164	1222	1280		
84	1338	1396	1454	1512	1570	1628	1686	1744	1802	1860		
85	1918	1976	2034	2092	2150	2208	2266	2324	2382	2440		
86	2498	2556	2614	2672	2730	2788	2846	2904	2962	3020	58	
87	3078	3136	3194	3252	3310	3368	3426	3484	3542	3600		
88	3658	3716	3774	3832	3890	3948	4006	4064	4122	4180		
89	4238	4296	4354	4412	4470	4528	4586	4644	4702	4760		
7490	4818	4876	4934	4992	5050	5108	5166	5224	5282	5340		
91	5398	5456	5514	5572	5630	5688	5746	5804	5862	5920		
92	5978	6036	6094	6152	6210	6268	6325	6383	6441	6499		57
93	6557	6615	6673	6731	6789	6847	6905	6963	7021	7079		1 6
94	7137	7195	7253	7311	7369	7427	7485	7543	7600	7658		2 11
												3 17
95	7716	7774	7832	7890	7948	8006	8064	8122	8180	8238		4 23
96	8296	8354	8412	8470	8528	8585	8643	8701	8759	8817		5 29
97	8875	8933	8991	9049	9107	9165	9223	9281	9339	9396		6 34
98	9454	9512	9570	9628	9686	9744	9802	9860	9918	9976		7 40
99	8750034	0091	0149	0207	0265	0323	0381	0439	0497	0555		8 46
												9 51
N.	0	1	2	3	4	5	6	7	8	9	D	Pts.

N.	0	1	2	3	4	5	6	7	8	9	D	Pro.
7500	8750613	0671	0728	0786	0844	0902	0960	1018	1076	1134		
01	1192	1250	1307	1365	1423	1481	1539	1597	1655	1713		58
02	1771	1828	1886	1944	2002	2060	2118	2176	2234	2292		1 6
03	2349	2407	2465	2523	2581	2639	2697	2755	2813	2870		2 12
04	2928	2986	3044	3102	3160	3218	3275	3333	3391	3449		3 17 / 4 23
05	3507	3565	3623	3681	3738	3796	3854	3912	3970	4028		5 29
06	4086	4143	4201	4259	4317	4375	4433	4491	4548	4606		6 35
07	4664	4722	4780	4838	4896	4953	5011	5069	5127	5185		7 41
08	5243	5300	5358	5416	5474	5532	5590	5648	5705	5763		8 46
09	5821	5879	5937	5995	6052	6110	6168	6226	6284	6342		9 52
7510	6399	6457	6515	6573	6631	6689	6746	6804	6862	6920		
11	6978	7035	7093	7151	7209	7267	7325	7382	7440	7498		
12	7556	7614	7671	7729	7787	7845	7903	7960	8018	8076		
13	8134	8192	8249	8307	8365	8423	8481	8539	8596	8654		
14	8712	8770	8828	8885	8943	9001	9059	9116	9174	9232		
15	9290	9348	9405	9463	9521	9579	9637	9694	9752	9810		
16	9868	9925	9983	$\bar{0}$041	$\bar{0}$099	$\bar{0}$157	$\bar{0}$214	$\bar{0}$272	$\bar{0}$330	$\bar{0}$388		
17	8760446	0503	0561	0619	0677	0734	0792	0850	0908	0965		
18	1023	1081	1139	1197	1254	1312	1370	1428	1485	1543		
19	1601	1659	1716	1774	1832	1890	1947	2005	2063	2121		
7520	2178	2236	2294	2352	2409	2467	2525	2583	2640	2698		
21	2756	2814	2871	2929	2987	3045	3102	3160	3218	3276		
22	3333	3391	3449	3506	3564	3622	3680	3737	3795	3853		
23	3911	3968	4026	4084	4142	4199	4257	4315	4372	4430		
24	4488	4546	4603	4661	4719	4776	4834	4892	4950	5007		
25	5065	5123	5180	5238	5296	5354	5411	5469	5527	5584		
26	5642	5700	5758	5815	5873	5931	5988	6046	6104	6161		
27	6219	6277	6335	6392	6450	6508	6565	6623	6681	6738		
28	6796	6854	6911	6969	7027	7085	7142	7200	7258	7315		
29	7373	7431	7488	7546	7604	7661	7719	7777	7834	7892		
7530	7950	8007	8065	8123	8180	8238	8296	8353	8411	8469		
31	8526	8584	8642	8699	8757	8815	8872	8930	8988	9045		
32	9103	9161	9218	9276	9334	9391	9449	9507	9564	9622		
33	9680	9737	9795	9853	9910	9968	$\bar{0}$026	$\bar{0}$083	$\bar{0}$141	$\bar{0}$199		
34	8770256	0314	0371	0429	0487	0544	0602	0660	0717	0775		
35	0833	0890	0948	1005	1063	1121	1178	1236	1294	1351		
36	1409	1467	1524	1582	1639	1697	1755	1812	1870	1928		
37	1985	2043	2100	2158	2216	2273	2331	2388	2446	2504		
38	2561	2619	2677	2734	2792	2849	2907	2965	3022	3080		
39	3137	3195	3253	3310	3368	3425	3483	3541	3598	3656		
7540	3713	3771	3829	3886	3944	4001	4059	4117	4174	4232		
41	4289	4347	4405	4462	4520	4577	4635	4693	4750	4808		
42	4865	4923	4980	5038	5096	5153	5211	5268	5326	5384		57
43	5441	5499	5556	5614	5671	5729	5787	5844	5902	5959		1 6
44	6017	6074	6132	6189	6247	6305	6362	6420	6477	6535		2 11 / 3 17
45	6592	6650	6708	6765	6823	6880	6938	6995	7053	7110		4 23
46	7168	7226	7283	7341	7398	7456	7513	7571	7628	7686		5 29
47	7743	7801	7859	7916	7974	8031	8089	8146	8204	8261		6 34
48	8319	8376	8434	8492	8549	8607	8664	8722	8779	8837		7 40 / 8 46
49	8894	8952	9009	9067	9124	9182	9239	9297	9354	9412		9 51
N.	0	1	2	3	4	5	6	7	8	9	D	Pts.

N.	0	1	2	3	4	5	6	7	8	9	D
7550	8779470	9527	9585	9642	9700	9757	9815	9872	9930	9987	
51	8780045	0102	0160	0217	0275	0332	0390	0447	0505	0562	
52	0620	0677	0735	0792	0850	0907	0965	1022	1080	1137	
53	1195	1252	1310	1367	1425	1482	1540	1597	1655	1712	
54	1770	1827	1885	1942	2000	2057	2115	2172	2230	2287	
55	2345	2402	2460	2517	2575	2632	2690	2747	2805	2862	
56	2919	2977	3034	3092	3149	3207	3264	3322	3379	3437	
57	3494	3552	3609	3667	3724	3782	3839	3896	3954	4011	
58	4069	4126	4184	4241	4299	4356	4414	4471	4529	4586	
59	4643	4701	4758	4816	4873	4931	4988	5046	5103	5161	
7560	5218	5275	5333	5390	5448	5505	5563	5620	5678	5735	
61	5792	5850	5907	5965	6022	6080	6137	6194	6252	6309	
62	6367	6424	6482	6539	6596	6654	6711	6769	6826	6884	
63	6941	6998	7056	7113	7171	7228	7286	7343	7400	7458	
64	7515	7573	7630	7687	7745	7802	7860	7917	7975	8032	
65	8089	8147	8204	8262	8319	8376	8434	8491	8549	8606	
66	8663	8721	8778	8836	8893	8950	9008	9065	9123	9180	
67	9237	9295	9352	9410	9467	9524	9582	9639	9696	9754	
68	9811	9869	9926	9983	ō041	ō098	ō156	ō213	ō270	ō328	
69	8790385	0442	0500	0557	0615	0672	0729	0787	0844	0901	
7570	0959	1016	1074	1131	1188	1246	1303	1360	1418	1475	
71	1532	1590	1647	1705	1762	1819	1877	1934	1991	2049	
72	2106	2163	2221	2278	2335	2393	2450	2508	2565	2622	
73	2680	2737	2794	2852	2909	2966	3024	3081	3138	3196	
74	3253	3310	3368	3425	3482	3540	3597	3654	3712	3769	
75	3826	3884	3941	3998	4056	4113	4170	4228	4285	4342	
76	4400	4457	4514	4572	4629	4686	4744	4801	4858	4916	
77	4973	5030	5088	5145	5202	5259	5317	5374	5431	5489	
78	5546	5603	5661	5718	5775	5833	5890	5947	6004	6062	
79	6119	6176	6234	6291	6348	6406	6463	6520	6577	6635	
7580	6692	6749	6807	6864	6921	6979	7036	7093	7150	7208	
81	7265	7322	7380	7437	7494	7551	7609	7666	7723	7781	
82	7838	7895	7952	8010	8067	8124	8181	8239	8296	8353	
83	8411	8468	8525	8582	8640	8697	8754	8811	8869	8926	
84	8983	9041	9098	9155	9212	9270	9327	9384	9441	9499	
85	9556	9613	9670	9728	9785	9842	9899	9957	ō014	ō071	
86	8800128	0186	0243	0300	0357	0415	0472	0529	0586	0644	
87	0701	0758	0815	0873	0930	0987	1044	1102	1159	1216	
88	1273	1330	1388	1445	1502	1559	1617	1674	1731	1788	
89	1846	1903	1960	2017	2074	2132	2189	2246	2303	2361	
7590	2418	2475	2532	2589	2647	2704	2761	2818	2875	2933	
91	2990	3047	3104	3162	3219	3276	3333	3390	3448	3505	
92	3562	3619	3676	3734	3791	3848	3905	3962	4020	4077	
93	4134	4191	4248	4306	4363	4420	4477	4534	4592	4649	
94	4706	4763	4820	4877	4935	4992	5049	5106	5163	5221	
95	5278	5335	5392	5449	5507	5564	5621	5678	5735	5792	
96	5850	5907	5964	6021	6078	6135	6193	6250	6307	6364	
97	6421	6478	6536	6593	6650	6707	6764	6821	6879	6936	
98	6993	7050	7107	7164	7222	7279	7336	7393	7450	7507	
99	7564	7622	7679	7736	7793	7850	7907	7964	8022	8079	
N.	0	1	2	3	4	5	6	7	8	9	D

Pro. / Pts.

	58		57
1	6	1	6
2	12	2	11
3	17	3	17
4	23	4	23
5	29	5	29
6	35	6	34
7	41	7	40
8	46	8	46
9	52	9	51

N.	0	1	2	3	4	5	6	7	8	9	D
7600	8808136	8193	8250	8307	8364	8422	8479	8536	8593	8650	
01	8707	8764	8822	8879	8936	8993	9050	9107	9164	9222	
02	9279	9336	9393	9450	9507	9564	9621	9679	9736	9793	
03	9850	9907	9964	ō021	ō078	ō136	ō193	ō250	ō307	ō364	
04	8810421	0478	0535	0592	0650	0707	0764	0821	0878	0935	
05	0992	1049	1106	1163	1221	1278	1335	1392	1449	1506	
06	1563	1620	1677	1735	1792	1849	1906	1963	2020	2077	
07	2134	2191	2248	2305	2363	2420	2477	2534	2591	2648	
08	2705	2762	2819	2876	2933	2990	3048	3105	3162	3219	
09	3276	3333	3390	3447	3504	3561	3618	3675	3732	3789	
7610	3847	3904	3961	4018	4075	4132	4189	4246	4303	4360	
11	4417	4474	4531	4588	4645	4703	4760	4817	4874	4931	
12	4988	5045	5102	5159	5216	5273	5330	5387	5444	5501	
13	5558	5615	5672	5729	5786	5844	5901	5958	6015	6072	
14	6129	6156	6243	6300	6357	6414	6471	6528	6585	6642	
15	6699	6756	6813	6870	6927	6984	7041	7098	7155	7212	
16	7269	7326	7383	7440	7497	7554	7611	7669	7726	7783	
17	7840	7897	7954	8011	8068	8125	8182	8239	8296	8353	
18	8410	8467	8524	8581	8638	8695	8752	8809	8866	8923	
19	8980	9037	9094	9151	9208	9265	9322	9379	9436	9493	57
7620	9550	9607	9664	9721	9778	9835	9892	9949	ō006	ō063	
21	8820120	0177	0234	0291	0348	0405	0462	0519	0575	0632	
22	0689	0746	0803	0860	0917	0974	1031	1088	1145	1202	
23	1259	1316	1373	1430	1487	1544	1601	1658	1715	1772	
24	1829	1886	1943	2000	2057	2114	2171	2228	2285	2342	
25	2398	2455	2512	2569	2626	2683	2740	2797	2854	2911	
26	2968	3025	3082	3139	3196	3253	3310	3367	3424	3481	
27	3537	3594	3651	3708	3765	3822	3879	3936	3993	4050	
28	4107	4164	4221	4278	4335	4392	4448	4505	4562	4619	
29	4676	4733	4790	4847	4904	4961	5018	5075	5132	5188	
7630	5245	5302	5359	5416	5473	5530	5587	5644	5701	5758	
31	5815	5871	5928	5985	6042	6099	6156	6213	6270	6327	
32	6384	6441	6497	6554	6611	6668	6725	6782	6839	6896	
33	6953	7010	7066	7123	7180	7237	7294	7351	7408	7465	
34	7522	7578	7635	7692	7749	7806	7863	7920	7977	8034	
35	8090	8147	8204	8261	8318	8375	8432	8489	8545	8602	
36	8659	8716	8773	8830	8887	8944	9000	9057	9114	9171	
37	9228	9285	9342	9399	9455	9512	9569	9626	9683	9740	
38	9797	9853	9910	9967	ō024	ō081	ō138	ō195	ō251	ō308	
39	8830365	0422	0479	0536	0593	0649	0706	0763	0820	0877	
7640	0934	0990	1047	1104	1161	1218	1275	1331	1388	1445	
41	1502	1559	1616	1673	1729	1786	1843	1900	1957	2014	
42	2070	2127	2184	2241	2298	2354	2411	2468	2525	2582	
43	2639	2695	2752	2809	2866	2923	2980	3036	3093	3150	
44	3207	3264	3320	3377	3434	3491	3548	3604	3661	3718	
45	3775	3832	3889	3945	4002	4059	4116	4173	4229	4286	
46	4343	4400	4457	4513	4570	4627	4684	4741	4797	4854	
47	4911	4968	5024	5081	5138	5195	5252	5308	5365	5422	
48	5479	5536	5592	5649	5706	5763	5819	5876	5933	5990	
49	6047	6103	6160	6217	6274	6330	6387	6444	6501	6558	
N.	0	1	2	3	4	5	6	7	8	9	D

Pro.

58
1 6
2 12
3 17
4 23
5 29
6 35
7 41
8 46
9 52

57
1 6
2 11
3 17
4 23
5 29
6 34
7 40
8 46
9 51

Pts.

N.	0	1	2	3	4	5	6	7	8	9	D	Pro.
7650	8836614	6671	6728	6785	6841	6898	6955	7012	7068	7125		
51	7182	7239	7296	7352	7409	7466	7523	7579	7636	7693		57
52	7750	7806	7863	7920	7977	8033	8090	8147	8204	8260		1| 6
53	8317	8374	8431	8487	8544	8601	8658	8714	8771	8828		2| 11
54	8885	8941	8998	9055	9112	9168	9225	9282	9338	9395		3| 17
												4| 23
55	9452	9509	9565	9622	9679	9736	9792	9849	9906	9963		5| 29
56	8840019	0076	0133	0189	0246	0303	0360	0416	0473	0530		6| 34
57	0586	0643	0700	0757	0813	0870	0927	0983	1040	1097		7| 40
58	1154	1210	1267	1324	1380	1437	1494	1551	1607	1664		8| 46
59	1721	1777	1834	1891	1948	2004	2061	2118	2174	2231		9| 51
7660	2288	2344	2401	2458	2514	2571	2628	2685	2741	2798		
61	2855	2911	2968	3025	3081	3138	3195	3251	3308	3365		
62	3421	3478	3535	3592	3648	3705	3762	3818	3875	3932		
63	3988	4045	4102	4158	4215	4272	4328	4385	4442	4498		
64	4555	4612	4668	4725	4782	4838	4895	4952	5008	5065		
65	5122	5178	5235	5292	5348	5405	5462	5518	5575	5631		
66	5688	5745	5801	5858	5915	5971	6028	6085	6141	6198		
67	6255	6311	6368	6425	6481	6538	6594	6651	6708	6764		
68	6821	6878	6934	6991	7048	7104	7161	7217	7274	7331		
69	7387	7444	7501	7557	7614	7671	7727	7784	7840	7897		
7670	7954	8010	8067	8124	8180	8237	8293	8350	8407	8463		
71	8520	8576	8633	8690	8746	8803	8860	8916	8973	9029		
72	9086	9143	9199	9256	9312	9369	9426	9482	9539	9595		
73	9652	9709	9765	9822	9878	9935	9992	0048	0105	0161		
74	8850218	0275	0331	0388	0444	0501	0557	0614	0671	0727		
75	0784	0840	0897	0954	1010	1067	1123	1180	1237	1293		
76	1350	1406	1463	1519	1576	1633	1689	1746	1802	1859		
77	1915	1972	2029	2085	2142	2198	2255	2311	2368	2425		
78	2481	2538	2594	2651	2707	2764	2820	2877	2934	2990		
79	3047	3103	3160	3216	3273	3329	3386	3443	3499	3556		
7680	3612	3669	3725	3782	3838	3895	3951	4008	4065	4121		
81	4178	4234	4291	4347	4404	4460	4517	4573	4630	4686		
82	4743	4800	4856	4913	4969	5026	5082	5139	5195	5252		
83	5308	5365	5421	5478	5534	5591	5647	5704	5761	5817		
84	5874	5930	5987	6043	6100	6156	6213	6269	6326	6382		
85	6439	6495	6552	6608	6665	6721	6778	6834	6891	6947		
86	7004	7060	7117	7173	7230	7286	7343	7399	7456	7512		
87	7569	7625	7682	7738	7795	7851	7908	7964	8021	8077		
88	8134	8190	8247	8303	8360	8416	8473	8529	8586	8642		
89	8699	8755	8812	8868	8925	8981	9037	9094	9150	9207		
7690	9263	9320	9376	9433	9489	9546	9602	9659	9715	9772		
91	9828	9885	9941	9998	0054	0110	0167	0223	0280	0336		
92	8860393	0449	0506	0562	0619	0675	0732	0788	0844	0901		56
93	0957	1014	1070	1127	1183	1240	1296	1352	1409	1465		1| 6
94	1522	1578	1635	1691	1748	1804	1860	1917	1973	2030		2| 11
												3| 17
95	2086	2143	2199	2256	2312	2368	2425	2481	2538	2594		4| 22
96	2651	2707	2763	2820	2876	2933	2989	3046	3102	3158		5| 28
97	3215	3271	3328	3384	3441	3497	3553	3610	3666	3723		6| 34
98	3779	3835	3892	3948	4005	4061	4118	4174	4230	4287		7| 39
99	4343	4400	4456	4512	4569	4625	4682	4738	4794	4851		8| 45
												9| 50
N.	0	1	2	3	4	5	6	7	8	9	D	Pts.

N.	0	1	2	3	4	5	6	7	8	9	D	Pro.
7700	8864907	4964	5020	5076	5133	5189	5246	5302	5358	5415		
01	5471	5528	5584	5640	5697	5753	5810	5866	5922	5979		
02	6035	6092	6148	6204	6261	6317	6373	6430	6486	6543		57
03	6599	6655	6712	6768	6824	6881	6937	6994	7050	7106		1|6
04	7163	7219	7275	7332	7388	7445	7501	7557	7614	7670		2|11
												3|17
05	7726	7783	7839	7896	7952	8008	8065	8121	8177	8234		4|23
06	8290	8346	8403	8459	8515	8572	8628	8685	8741	8797		5|29
07	8854	8910	8966	9023	9079	9135	9192	9248	9304	9361		6|34
08	9417	9473	9530	9586	9642	9699	9755	9811	9868	9924		7|40
09	9980	0̄037	0̄093	0̄149	0̄206	0̄262	0̄318	0̄375	0̄431	0̄487		8|46 9|51
7710	8870544	0600	0656	0713	0769	0825	0882	0938	0994	1051		
11	1107	1163	1220	1276	1332	1389	1445	1501	1558	1614		
12	1670	1727	1783	1839	1895	1952	2008	2064	2121	2177		
13	2233	2290	2346	2402	2459	2515	2571	2627	2684	2740		
14	2796	2853	2909	2965	3022	3078	3134	3190	3247	3303		
15	3359	3416	3472	3528	3584	3641	3697	3753	3810	3866		
16	3922	3978	4035	4091	4147	4204	4260	4316	4372	4429		
17	4485	4541	4598	4654	4710	4766	4823	4879	4935	4991		
18	5048	5104	5160	5217	5273	5329	5385	5442	5498	5554		
19	5610	5667	5723	5779	5835	5892	5948	6004	6060	6117		
7720	6173	6229	6286	6342	6398	6454	6511	6567	6623	6679		
21	6736	6792	6848	6904	6961	7017	7073	7129	7185	7242		
22	7298	7354	7410	7467	7523	7579	7635	7692	7748	7804		
23	7860	7917	7973	8029	8085	8142	8198	8254	8310	8366		
24	8423	8479	8535	8591	8648	8704	8760	8816	8872	8929		
25	8985	9041	9097	9154	9210	9266	9322	9378	9435	9491		
26	9547	9603	9659	9716	9772	9828	9884	9941	9997	0̄053		
27	8880109	0165	0222	0278	0334	0390	0446	0503	0559	0615		
28	0671	0727	0784	0840	0896	0952	1008	1064	1121	1177		
29	1233	1289	1345	1402	1458	1514	1570	1626	1683	1739		
7730	1795	1851	1907	1963	2020	2076	2132	2188	2244	2301		
31	2357	2413	2469	2525	2581	2638	2694	2750	2806	2862		
32	2918	2975	3031	3087	3143	3199	3255	3312	3368	3424		
33	3480	3536	3592	3649	3705	3761	3817	3873	3929	3986		
34	4042	4098	4154	4210	4266	4322	4379	4435	4491	4547		
35	4603	4659	4715	4772	4828	4884	4940	4996	5052	5108		
36	5165	5221	5277	5333	5389	5445	5501	5558	5614	5670		
37	5726	5782	5838	5894	5950	6007	6063	6119	6175	6231		
38	6287	6343	6400	6456	6512	6568	6624	6680	6736	6792		
39	6848	6905	6961	7017	7073	7129	7185	7241	7297	7353		
7740	7410	7466	7522	7578	7634	7690	7746	7802	7858	7915		
41	7971	8027	8083	8139	8195	8251	8307	8363	8419	8476		
42	8532	8588	8644	8700	8756	8812	8868	8924	8980	9037		56
43	9093	9149	9205	9261	9317	9373	9429	9485	9541	9597		1|6
44	9653	9710	9766	9822	9878	9934	9990	0̄046	0̄102	0̄158		2|11 3|17
45	8890214	0270	0326	0382	0439	0495	0551	0607	0663	0719		4|22
46	0775	0831	0887	0943	0999	1055	1111	1167	1223	1279		5|28
47	1336	1392	1448	1504	1560	1616	1672	1728	1784	1840		6|34
48	1896	1952	2008	2064	2120	2176	2232	2288	2345	2401		7|39 8|45
49	2457	2513	2569	2625	2681	2737	2793	2849	2905	2961		9|50
N.	0	1	2	3	4	5	6	7	8	9	D	Pts.

N.	0	1	2	3	4	5	6	7	8	9	D	Pro.
7750	8893017	3073	3129	3185	3241	3297	3353	3409	3465	3521		
51	3577	3633	3689	3745	3801	3858	3914	3970	4026	4082		56
52	4138	4194	4250	4306	4362	4418	4474	4530	4586	4642		1 6
53	4698	4754	4810	4866	4922	4978	5034	5090	5146	5202		2 11
54	5258	5314	5370	5426	5482	5538	5594	5650	5706	5762	56	3 17 / 4 22
55	5818	5874	5930	5986	6042	6098	6154	6210	6266	6322		5 28
56	6378	6434	6490	6546	6602	6658	6714	6770	6826	6882		6 34
57	6938	6994	7050	7106	7162	7218	7274	7330	7386	7442		7 39
58	7498	7554	7610	7666	7722	7778	7834	7890	7946	8002		8 45
59	8058	8113	8169	8225	8281	8337	8393	8449	8505	8561		9 50
7760	8617	8673	8729	8785	8841	8897	8953	9009	9065	9121		
61	9177	9233	9289	9345	9401	9457	9513	9569	9624	9680		
62	9736	9792	9848	9904	9960	0̄016	0̄072	0̄128	0̄184	0̄240		
63	8900296	0352	0408	0464	0520	0576	0632	0687	0743	0799		
64	0855	0911	0967	1023	1079	1135	1191	1247	1303	1359		
65	1415	1471	1526	1582	1638	1694	1750	1806	1862	1918		
66	1974	2030	2086	2142	2198	2253	2309	2365	2421	2477		
67	2533	2589	2645	2701	2757	2813	2869	2924	2980	3036		
68	3092	3148	3204	3260	3316	3372	3428	3484	3539	3595		
69	3651	3707	3763	3819	3875	3931	3987	4043	4098	4154		
7770	4210	4266	4322	4378	4434	4490	4546	4601	4657	4713		
71	4769	4825	4881	4937	4993	5049	5104	5160	5216	5272		
72	5328	5384	5440	5496	5551	5607	5663	5719	5775	5831		
73	5887	5943	5998	6054	6110	6166	6222	6278	6334	6389		
74	6445	6501	6557	6613	6669	6725	6781	6836	6892	6948		
75	7004	7060	7116	7172	7227	7283	7339	7395	7451	7507		
76	7563	7618	7674	7730	7786	7842	7898	7953	8009	8065		
77	8121	8177	8233	8289	8344	8400	8456	8512	8568	8624		
78	8679	8735	8791	8847	8903	8959	9014	9070	9126	9182		
79	9238	9294	9349	9405	9461	9517	9573	9629	9684	9740		
7780	9796	9852	9908	9963	0̄019	0̄075	0̄131	0̄187	0̄243	0̄298		
81	8910354	0410	0466	0522	0577	0633	0689	0745	0801	0856		
82	0912	0968	1024	1080	1135	1191	1247	1303	1359	1415		
83	1470	1526	1582	1638	1694	1749	1805	1861	1917	1972		
84	2028	2084	2140	2196	2251	2307	2363	2419	2475	2530		
85	2586	2642	2698	2754	2809	2865	2921	2977	3032	3088		
86	3144	3200	3256	3311	3367	3423	3479	3534	3590	3646		
87	3702	3758	3813	3869	3925	3981	4036	4092	4148	4204		
88	4259	4315	4371	4427	4482	4538	4594	4650	4706	4761		
89	4817	4873	4929	4984	5040	5096	5152	5207	5263	5319		
7790	5375	5430	5486	5542	5598	5653	5709	5765	5821	5876		
91	5932	5988	6044	6099	6155	6211	6266	6322	6378	6434		
92	6489	6545	6601	6657	6712	6768	6824	6880	6935	6991		55
93	7047	7102	7158	7214	7270	7325	7381	7437	7493	7548		1 6
94	7604	7660	7715	7771	7827	7883	7938	7994	8050	8105		2 11 / 3 17
95	8161	8217	8273	8328	8384	8440	8495	8551	8607	8663		4 22
96	8718	8774	8830	8885	8941	8997	9053	9108	9164	9220		5 28
97	9275	9331	9387	9442	9498	9554	9610	9665	9721	9777		6 33
98	9832	9888	9944	9999	0̄055	0̄111	0̄166	0̄222	0̄278	0̄334		7 39
99	8920389	0445	0501	0556	0612	0668	0723	0779	0835	0890		8 44 / 9 50
N.	0	1	2	3	4	5	6	7	8	9	D	Pts.

N.	0	1	2	3	4	5	6	7	8	9	D	Pro.
7800	8920946	1002	1057	1113	1169	1224	1280	1336	1391	1447		
01	1503	1558	1614	1670	1725	1781	1837	1892	1948	2004		56
02	2059	2115	2171	2226	2282	2338	2393	2449	2505	2560		1 6
03	2616	2672	2727	2783	2839	2894	2950	3006	3061	3117		2 11
04	3173	3228	3284	3340	3395	3451	3506	3562	3618	3673		3 17
05	3729	3785	3840	3896	3952	4007	4063	4119	4174	4230		4 22
06	4285	4341	4397	4452	4508	4564	4619	4675	4731	4786		5 28
07	4842	4897	4953	5009	5064	5120	5176	5231	5287	5342		6 34
08	5398	5454	5509	5565	5621	5676	5732	5787	5843	5899		7 39
09	5954	6010	6065	6121	6177	6232	6288	6344	6399	6455		8 45
7810	6510	6566	6622	6677	6733	6788	6844	6900	6955	7011		9 50
11	7066	7122	7178	7233	7289	7344	7400	7456	7511	7567		
12	7622	7678	7734	7789	7845	7900	7956	8011	8067	8123		
13	8178	8234	8289	8345	8401	8456	8512	8567	8623	8678		
14	8734	8790	8845	8901	8956	9012	9068	9123	9179	9234		
15	9290	9345	9401	9457	9512	9568	9623	9679	9734	9790		
16	9846	9901	9957	0̄012	0̄068	0̄123	0̄179	0̄234	0̄290	0̄346		
17	8930401	0457	0512	0568	0623	0679	0734	0790	0846	0901		
18	0957	1012	1068	1123	1179	1234	1290	1345	1401	1457		
19	1512	1568	1623	1679	1734	1790	1845	1901	1956	2012		
7820	2068	2123	2179	2234	2290	2345	2401	2456	2512	2567		
21	2623	2678	2734	2789	2845	2900	2956	3012	3067	3123		
22	3178	3234	3289	3345	3400	3456	3511	3567	3622	3678		
23	3733	3789	3844	3900	3955	4011	4066	4122	4177	4233		
24	4288	4344	4399	4455	4510	4566	4621	4677	4732	4788		
25	4843	4899	4954	5010	5065	5121	5176	5232	5287	5343		
26	5398	5454	5509	5565	5620	5676	5731	5787	5842	5898		
27	5953	6009	6064	6120	6175	6231	6286	6342	6397	6453		
28	6508	6564	6619	6675	6730	6786	6841	6897	6952	7007		
29	7063	7118	7174	7229	7285	7340	7396	7451	7507	7562		
7830	7618	7673	7729	7784	7839	7895	7950	8006	8061	8117		
31	8172	8228	8283	8339	8394	8450	8505	8560	8616	8671		
32	8727	8782	8838	8893	8949	9004	9059	9115	9170	9226		
33	9281	9337	9392	9448	9503	9558	9614	9669	9725	9780		
34	9836	9891	9947	0̄002	0̄057	0̄113	0̄168	0̄224	0̄279	0̄335		
35	8940390	0445	0501	0556	0612	0667	0723	0778	0833	0889		
36	0944	1000	1055	1111	1166	1221	1277	1332	1388	1443		
37	1498	1554	1609	1665	1720	1776	1831	1886	1942	1997		
38	2053	2108	2163	2219	2274	2330	2385	2440	2496	2551		
39	2607	2662	2717	2773	2828	2884	2939	2994	3050	3105		
7840	3161	3216	3271	3327	3382	3438	3493	3548	3604	3659		
41	3715	3770	3825	3881	3936	3991	4047	4102	4158	4213		
42	4268	4324	4379	4435	4490	4545	4601	4656	4711	4767		55
43	4822	4878	4933	4988	5044	5099	5154	5210	5265	5320		1 6
44	5376	5431	5487	5542	5597	5653	5708	5763	5819	5874		2 11
45	5929	5985	6040	6096	6151	6206	6262	6317	6372	6428		3 17
46	6483	6538	6594	6649	6704	6760	6815	6870	6926	6981		4 22
47	7037	7092	7147	7203	7258	7313	7369	7424	7479	7535		5 28
48	7590	7645	7701	7756	7811	7867	7922	7977	8033	8088		6 33
49	8143	8199	8254	8309	8365	8420	8475	8531	8586	8641		7 39
												8 44
												9 50
N.	0	1	2	3	4	5	6	7	8	9	D	Pts.

N.	0	1	2	3	4	5	6	7	8	9	D	Pro.
7850	8948697	8752	8807	8863	8918	8973	9028	9084	9139	9194		
51	9250	9305	9360	9416	9471	9526	9582	9637	9692	9748		56
52	9803	9858	9914	9969	ō024	ō079	ō135	ō190	ō245	ō301		1 6
53	8950356	0411	0467	0522	0577	0632	0688	0743	0798	0854		2 11
54	0909	0964	1020	1075	1130	1185	1241	1296	1351	1407		3 17
												4 22
55	1462	1517	1572	1628	1683	1738	1794	1849	1904	1959		5 28
56	2015	2070	2125	2181	2236	2291	2346	2402	2457	2512		6 34
57	2568	2623	2678	2733	2789	2844	2899	2954	3010	3065		7 39
58	3120	3176	3231	3286	3341	3397	3452	3507	3562	3618		8 45
59	3673	3728	3783	3839	3894	3949	4004	4060	4115	4170		9 50
7860	4225	4281	4336	4391	4446	4502	4557	4612	4667	4723		
61	4778	4833	4888	4944	4999	5054	5109	5165	5220	5275		
62	5330	5386	5441	5496	5551	5607	5662	5717	5772	5828		
63	5883	5938	5993	6048	6104	6159	6214	6269	6325	6380		
64	6435	6490	6545	6601	6656	6711	6766	6822	6877	6932		
65	6987	7042	7098	7153	7208	7263	7319	7374	7429	7484		
66	7539	7595	7650	7705	7760	7815	7871	7926	7981	8036		
67	8092	8147	8202	8257	8312	8368	8423	8478	8533	8588		
68	8644	8699	8754	8809	8864	8919	8975	9030	9085	9140		
69	9195	9251	9306	9361	9416	9471	9527	9582	9637	9692		
7870	9747	9803	9858	9913	9968	ō023	ō078	ō134	ō189	ō244		
71	8960299	0354	0409	0465	0520	0575	0630	0685	0741	0796		
72	0851	0906	0961	1016	1072	1127	1182	1237	1292	1347		
73	1403	1458	1513	1568	1623	1678	1733	1789	1844	1899		
74	1954	2009	2064	2120	2175	2230	2285	2340	2395	2450		
75	2506	2561	2616	2671	2726	2781	2837	2892	2947	3002		
76	3057	3112	3167	3222	3278	3333	3388	3443	3498	3553		
77	3608	3664	3719	3774	3829	3884	3939	3994	4050	4105		
78	4160	4215	4270	4325	4380	4435	4491	4546	4601	4656		
79	4711	4766	4821	4876	4931	4987	5042	5097	5152	5207		
7880	5262	5317	5372	5428	5483	5538	5593	5648	5703	5758		
81	5813	5868	5923	5979	6034	6089	6144	6199	6254	6309		
82	6364	6419	6475	6530	6585	6640	6695	6750	6805	6860		
83	6915	6970	7025	7081	7136	7191	7246	7301	7356	7411		
84	7466	7521	7576	7631	7686	7742	7797	7852	7907	7962		
85	8017	8072	8127	8182	8237	8292	8347	8403	8458	8513		
86	8568	8623	8678	8733	8788	8843	8898	8953	9008	9063		
87	9118	9173	9229	9284	9339	9394	9449	9504	9559	9614		
88	9669	9724	9779	9834	9889	9944	9999	ō054	ō109	ō165		
89	8970220	0275	0330	0385	0440	0495	0550	0605	0660	0715		
7890	0770	0825	0880	0935	0990	1045	1100	1155	1210	1265		
91	1320	1375	1431	1486	1541	1596	1651	1706	1761	1816		
92	1871	1926	1981	2036	2091	2146	2201	2256	2311	2366		55
93	2421	2476	2531	2586	2641	2696	2751	2806	2861	2916		1 6
94	2971	3026	3081	3136	3191	3246	3301	3356	3411	3466		2 11
												3 17
95	3521	3576	3631	3686	3741	3796	3851	3906	3961	4016	55	4 22
96	4071	4126	4181	4236	4291	4346	4401	4456	4511	4566		5 28
97	4621	4676	4731	4786	4841	4896	4951	5006	5061	5116		6 33
98	5171	5226	5281	5336	5391	5446	5501	5556	5611	5666		7 39
99	5721	5776	5831	5886	5941	5996	6051	6106	6161	6216		8 44
												9 50
N.	0	1	2	3	4	5	6	7	8	9	D	Pts.

N.	0	1	2	3	4	5	6	7	8	9	D	Pro.
7900	8976271	6326	6381	6436	6491	6546	6601	6656	6711	6766		
01	6821	6876	6931	6986	7040	7095	7150	7205	7260	7315		55
02	7370	7425	7480	7535	7590	7645	7700	7755	7810	7865		1 6
03	7920	7975	8030	8085	8140	8195	8250	8304	8359	8414		2 11
04	8469	8524	8579	8634	8689	8744	8799	8854	8909	8964		3 17
05	9019	9074	9129	9184	9238	9293	9348	9403	9458	9513		4 22
06	9568	9623	9678	9733	9788	9843	9898	9953	0̄008	0̄062		5 28
07	8980117	0172	0227	0282	0337	0392	0447	0502	0557	0612		6 33
08	0667	0722	0776	0831	0886	0941	0996	1051	1106	1161		7 39
09	1216	1271	1326	1380	1435	1490	1545	1600	1655	1710		8 44
												9 50
7910	1765	1820	1875	1930	1984	2039	2094	2149	2204	2259		
11	2314	2369	2424	2479	2533	2588	2643	2698	2753	2808		
12	2863	2918	2973	3027	3082	3137	3192	3247	3302	3357		
13	3412	3467	3521	3576	3631	3686	3741	3796	3851	3906		
14	3960	4015	4070	4125	4180	4235	4290	4345	4399	4454		
15	4509	4564	4619	4674	4729	4784	4838	4893	4948	5003		
16	5058	5113	5168	5222	5277	5332	5387	5442	5497	5552		
17	5606	5661	5716	5771	5826	5881	5936	5990	6045	6100		
18	6155	6210	6265	6320	6374	6429	6484	6539	6594	6649		
19	6703	6758	6813	6868	6923	6978	7032	7087	7142	7197		
7920	7252	7307	7361	7416	7471	7526	7581	7636	7690	7745		
21	7800	7855	7910	7965	8019	8074	8129	8184	8239	8294		
22	8348	8403	8458	8513	8568	8622	8677	8732	8787	8842		
23	8897	8951	9006	9061	9116	9171	9225	9280	9335	9390		
24	9445	9499	9554	9609	9664	9719	9774	9828	9883	9938		
25	9993	0̄048	0̄102	0̄157	0̄212	0̄267	0̄321	0̄376	0̄431	0̄486		
26	8990541	0595	0650	0705	0760	0815	0869	0924	0979	1034		
27	1089	1143	1198	1253	1308	1363	1417	1472	1527	1582		
28	1636	1691	1746	1801	1856	1910	1965	2020	2075	2129		
29	2184	2239	2294	2348	2403	2458	2513	2568	2622	2677		
7930	2732	2787	2841	2896	2951	3006	3060	3115	3170	3225	55	
31	3279	3334	3389	3444	3499	3553	3608	3663	3718	3772		
32	3827	3882	3937	3991	4046	4101	4156	4210	4265	4320		
33	4375	4429	4484	4539	4594	4648	4703	4758	4812	4867		
34	4922	4977	5031	5086	5141	5196	5250	5305	5360	5415		
35	5469	5524	5579	5634	5688	5743	5798	5852	5907	5962		
36	6017	6071	6126	6181	6235	6290	6345	6400	6454	6509		
37	6564	6619	6673	6728	6783	6837	6892	6947	7002	7056		
38	7111	7166	7220	7275	7330	7384	7439	7494	7549	7603		
39	7658	7713	7767	7822	7877	7932	7986	8041	8096	8150		
7940	8205	8260	8314	8369	8424	8479	8533	8588	8643	8697		
41	8752	8807	8861	8916	8971	9025	9080	9135	9189	9244		
42	9299	9354	9408	9463	9518	9572	9627	9682	9736	9791		54
43	9846	9900	9955	0̄010	0̄064	0̄119	0̄174	0̄228	0̄283	0̄338		1 5
44	9000392	0447	0502	0556	0611	0666	0720	0775	0830	0884		2 11
45	0939	0994	1048	1103	1158	1212	1267	1322	1376	1431		3 16
46	1486	1540	1595	1650	1704	1759	1814	1868	1923	1977		4 22
47	2032	2087	2141	2196	2251	2305	2360	2415	2469	2524		5 27
48	2579	2633	2688	2743	2797	2852	2906	2961	3016	3070		6 32
49	3125	3180	3234	3289	3344	3398	3453	3507	3562	3617		7 38
												8 43
												9 49
N.	0	1	2	3	4	5	6	7	8	9	D	Pts.

N.	0	1	2	3	4	5	6	7	8	9	D	Pro.
7950	9003671	3726	3781	3835	3890	3944	3999	4054	4108	4163		
51	4218	4272	4327	4381	4436	4491	4545	4600	4654	4709		55
52	4764	4818	4873	4928	4982	5037	5091	5146	5201	5255		1 6
53	5310	5364	5419	5474	5528	5583	5637	5692	5747	5801		2 11
54	5856	5910	5965	6020	6074	6129	6183	6238	6293	6347		3 17
												4 22
55	6402	6456	6511	6566	6620	6675	6729	6784	6839	6893		5 28
56	6948	7002	7057	7112	7166	7221	7275	7330	7384	7439		6 33
57	7494	7548	7603	7657	7712	7766	7821	7876	7930	7985		7 39
58	8039	8094	8148	8203	8258	8312	8367	8421	8476	8530		8 44
59	8585	8640	8694	8749	8803	8858	8912	8967	9022	9076		9 50
7960	9131	9185	9240	9294	9349	9403	9458	9513	9567	9622		
61	9676	9731	9785	9840	9894	9949	$\bar{0}$004	$\bar{0}$058	$\bar{0}$113	$\bar{0}$167		
62	9010222	0276	0331	0385	0440	0494	0549	0604	0658	0713		
63	0767	0822	0876	0931	0985	1040	1094	1149	1203	1258		
64	1313	1367	1422	1476	1531	1585	1640	1694	1749	1803		
65	1858	1912	1967	2021	2076	2130	2185	2239	2294	2349		
66	2403	2458	2512	2567	2621	2676	2730	2785	2839	2894		
67	2948	3003	3057	3112	3166	3221	3275	3330	3384	3439		
68	3493	3548	3602	3657	3711	3766	3820	3875	3929	3984		
69	4038	4093	4147	4202	4256	4311	4365	4420	4474	4529		
7970	4583	4638	4692	4747	4801	4856	4910	4965	5019	5074		
71	5128	5183	5237	5292	5346	5401	5455	5509	5564	5618		
72	5673	5727	5782	5836	5891	5945	6000	6054	6109	6163		
73	6218	6272	6327	6381	6436	6490	6544	6599	6653	6708		
74	6762	6817	6871	6926	6980	7035	7089	7144	7198	7252		
75	7307	7361	7416	7470	7525	7579	7634	7688	7743	7797		
76	7851	7906	7960	8015	8069	8124	8178	8233	8287	8341		
77	8396	8450	8505	8559	8614	8668	8723	8777	8831	8886		
78	8940	8995	9049	9104	9158	9212	9267	9321	9376	9430		
79	9485	9539	9594	9648	9702	9757	9811	9866	9920	9974		
7980	9020029	0083	0138	0192	0247	0301	0355	0410	0464	0519		
81	0573	0628	0682	0736	0791	0845	0900	0954	1008	1063		
82	1117	1172	1226	1280	1335	1389	1444	1498	1552	1607		
83	1661	1716	1770	1824	1879	1933	1988	2042	2096	2151		
84	2205	2260	2314	2368	2423	2477	2532	2586	2640	2695		
85	2749	2804	2858	2912	2967	3021	3076	3130	3184	3239		
86	3293	3347	3402	3456	3511	3565	3619	3674	3728	3782		
87	3837	3891	3946	4000	4054	4109	4163	4217	4272	4326		
88	4381	4435	4489	4544	4598	4652	4707	4761	4815	4870		
89	4924	4979	5033	5087	5142	5196	5250	5305	5359	5413		
7990	5468	5522	5577	5631	5685	5740	5794	5848	5903	5957		
91	6011	6066	6120	6174	6229	6283	6337	6392	6446	6500		
92	6555	6609	6663	6718	6772	6826	6881	6935	6989	7044		54
93	7098	7152	7207	7261	7315	7370	7424	7478	7533	7587		1 5
94	7641	7696	7750	7804	7859	7913	7967	8022	8076	8130		2 11
												3 16
95	8185	8239	8293	8348	8402	8456	8511	8565	8619	8674		4 22
96	8728	8782	8836	8891	8945	8999	9054	9108	9162	9217		5 27
97	9271	9325	9380	9434	9488	9542	9597	9651	9705	9760		6 32
98	9814	9868	9923	9977	$\bar{0}$031	$\bar{0}$085	$\bar{0}$140	$\bar{0}$194	$\bar{0}$248	$\bar{0}$303		7 38
99	9030357	0411	0466	0520	0574	0628	0683	0737	0791	0846		8 43 9 49
N.	0	1	2	3	4	5	6	7	8	9	D	Pts.

N.	0	1	2	3	4	5	6	7	8	9	D	Pro.
8000	9030900	0954	1008	1063	1117	1171	1226	1280	1334	1388		
01	1443	1497	1551	1606	1660	1714	1768	1823	1877	1931		55
02	1985	2040	2094	2148	2203	2257	2311	2365	2420	2474		1 6
03	2528	2582	2637	2691	2745	2799	2854	2908	2962	3017		2 11
04	3071	3125	3179	3234	3288	3342	3396	3451	3505	3559		3 17
												4 22
05	3613	3668	3722	3776	3830	3885	3939	3993	4047	4102		5 28
06	4156	4210	4264	4319	4373	4427	4481	4536	4590	4644		6 33
07	4698	4753	4807	4861	4915	4969	5024	5078	5132	5186		7 39
08	5241	5295	5349	5403	5458	5512	5566	5620	5674	5729		8 44
09	5783	5837	5891	5946	6000	6054	6108	6163	6217	6271		9 50
8010	6325	6379	6434	6488	6542	6596	6650	6705	6759	6813		
11	6867	6922	6976	7030	7084	7138	7193	7247	7301	7355		
12	7409	7464	7518	7572	7626	7680	7735	7789	7843	7897		
13	7951	8006	8060	8114	8168	8222	8277	8331	8385	8439		
14	8493	8548	8602	8656	8710	8764	8819	8873	8927	8981		
15	9035	9089	9144	9198	9252	9306	9360	9415	9469	9523		
16	9577	9631	9685	9740	9794	9848	9902	9956	0̄010	0̄065		
17	9040119	0173	0227	0281	0336	0390	0444	0498	0552	0606		
18	0661	0715	0769	0823	0877	0931	0985	1040	1094	1148		
19	1202	1256	1310	1365	1419	1473	1527	1581	1635	1690		
8020	1744	1798	1852	1906	1960	2014	2069	2123	2177	2231		
21	2285	2339	2393	2448	2502	2556	2610	2664	2718	2772		
22	2827	2881	2935	2989	3043	3097	3151	3206	3260	3314		
23	3368	3422	3476	3530	3584	3639	3693	3747	3801	3855		
24	3909	3963	4017	4072	4126	4180	4234	4288	4342	4396		
25	4450	4505	4559	4613	4667	4721	4775	4829	4883	4937		
26	4992	5046	5100	5154	5208	5262	5316	5370	5424	5479		
27	5533	5587	5641	5695	5749	5803	5857	5911	5965	6020		
28	6074	6128	6182	6236	6290	6344	6398	6452	6506	6560		
29	6615	6669	6723	6777	6831	6885	6939	6993	7047	7101		
8030	7155	7210	7264	7318	7372	7426	7480	7534	7588	7642		
31	7696	7750	7804	7858	7913	7967	8021	8075	8129	8183		
32	8237	8291	8345	8399	8453	8507	8561	8615	8670	8724		
33	8778	8832	8886	8940	8994	9048	9102	9156	9210	9264		
34	9318	9372	9426	9480	9534	9589	9643	9697	9751	9805		
35	9859	9913	9967	0̄021	0̄075	0̄129	0̄183	0̄237	0̄291	0̄345		
36	9050399	0453	0507	0561	0615	0669	0724	0778	0832	0886		
37	0940	0994	1048	1102	1156	1210	1264	1318	1372	1426		
38	1480	1534	1588	1642	1696	1750	1804	1858	1912	1966		
39	2020	2074	2128	2182	2236	2290	2344	2398	2452	2506		
8040	2560	2615	2669	2723	2777	2831	2885	2939	2993	3047		
41	3101	3155	3209	3263	3317	3371	3425	3479	3533	3587		54
42	3641	3695	3749	3803	3857	3911	3965	4019	4073	4127		1 5
43	4181	4235	4289	4343	4397	4451	4505	4559	4613	4667	54	2 11
44	4721	4775	4829	4883	4937	4991	5045	5099	5153	5207		3 16
45	5260	5314	5368	5422	5476	5530	5584	5638	5692	5746		4 22
46	5800	5854	5908	5962	6016	6070	6124	6178	6232	6286		5 27
47	6340	6394	6448	6502	6556	6610	6664	6718	6772	6826		6 32
48	6880	6934	6988	7042	7096	7149	7203	7257	7311	7365		7 38
49	7419	7473	7527	7581	7635	7689	7743	7797	7851	7905		8 43
												9 49
N	0	1	2	3	4	5	6	7	8·	9	D	Pts.

N.	0	1	2	3	4	5	6	7	8	9	D	Pro.
8050	9057959	8013	8067	8121	8175	8229	8282	8336	8390	8444		
51	8498	8552	8606	8660	8714	8768	8822	8876	8930	8984		54
52	9038	9092	9146	9199	9253	9307	9361	9415	9469	9523		1 5
53	9577	9631	9685	9739	9793	9847	9901	9954	0̄008	0̄062		2 11
54	9060116	0170	0224	0278	0332	0386	0440	0494	0548	0602		3 16
												4 22
55	0655	0709	0763	0817	0871	0925	0979	1033	1087	1141		5 27
56	1195	1248	1302	1356	1410	1464	1518	1572	1626	1680		6 32
57	1734	1788	1841	1895	1949	2003	2057	2111	2165	2219		7 38
58	2273	2327	2380	2434	2488	2542	2596	2650	2704	2758		8 43
59	2812	2865	2919	2973	3027	3081	3135	3189	3243	3297		9 49
8060	3350	3404	3458	3512	3566	3620	3674	3728	3781	3835		
61	3889	3943	3997	4051	4105	4159	4212	4266	4320	4374		
62	4428	4482	4536	4590	4643	4697	4751	4805	4859	4913		
63	4967	5020	5074	5128	5182	5236	5290	5344	5397	5451		
64	5505	5559	5613	5667	5721	5774	5828	5882	5936	5990		
65	6044	6098	6151	6205	6259	6313	6367	6421	6474	6528		
66	6582	6636	6690	6744	6798	6851	6905	6959	7013	7067		
67	7121	7174	7228	7282	7336	7390	7444	7497	7551	7605		
68	7659	7713	7767	7820	7874	7928	7982	8036	8090	8143		
69	8197	8251	8305	8359	8412	8466	8520	8574	8628	8682		
8070	8735	8789	8843	8897	8951	9004	9058	9112	9166	9220		
71	9273	9327	9381	9435	9489	9543	9596	9650	9704	9758		
72	9812	9865	9919	9973	0̄027	0̄081	0̄134	0̄188	0̄242	0̄296		
73	9070350	0403	0457	0511	0565	0618	0672	0726	0780	0834		
74	0887	0941	0995	1049	1103	1156	1210	1264	1318	1372		
75	1425	1479	1533	1587	1640	1694	1748	1802	1856	1909		
76	1963	2017	2071	2124	2178	2232	2286	2340	2393	2447		
77	2501	2555	2608	2662	2716	2770	2823	2877	2931	2985		
78	3038	3092	3146	3200	3254	3307	3361	3415	3469	3522		
79	3576	3630	3684	3737	3791	3845	3899	3952	4006	4060		
8080	4114	4167	4221	4275	4329	4382	4436	4490	4544	4597		
81	4651	4705	4759	4812	4866	4920	4974	5027	5081	5135		
82	5188	5242	5296	5350	5403	5457	5511	5565	5618	5672		
83	5726	5780	5833	5887	5941	5994	6048	6102	6156	6209		
84	6263	6317	6370	6424	6478	6532	6585	6639	6693	6747		
85	6800	6854	6908	6961	7015	7069	7123	7176	7230	7284		
86	7337	7391	7445	7498	7552	7606	7660	7713	7767	7821		
87	7874	7928	7982	8036	8089	8143	8197	8250	8304	8358		
88	8411	8465	8519	8573	8626	8680	8734	8787	8841	8895		
89	8948	9002	9056	9109	9163	9217	9270	9324	9378	9432		
8090	9485	9539	9593	9646	9700	9754	9807	9861	9915	9968		
91	9080022	0076	0129	0183	0237	0290	0344	0398	0451	0505		
92	0559	0612	0666	0720	0773	0827	0881	0934	0988	1042		53
93	1095	1149	1203	1256	1310	1364	1417	1471	1525	1578		1 5
94	1632	1686	1739	1793	1847	1900	1954	2008	2061	2115		2 11
95	2169	2222	2276	2329	2383	2437	2490	2544	2598	2651		3 16
96	2705	2759	2812	2866	2920	2973	3027	3080	3134	3188		4 21
97	3241	3295	3349	3402	3456	3510	3563	3617	3670	3724		5 27
98	3778	3831	3885	3939	3992	4046	4099	4153	4207	4260		6 32
99	4314	4368	4421	4475	4528	4582	4636	4689	4743	4797		7 37
												8 42
												9 48
N.	0	1	2	3	4	5	6	7	8	9	D	Pts.

N.	0	1	2	3	4	5	6	7	8	9	D	Pro.
8100	9084850	4904	4957	5011	5065	5118	5172	5225	5279	5333		
01	5386	5440	5494	5547	5601	5654	5708	5762	5815	5869		**54**
02	5922	5976	6030	6083	6137	6190	6244	6298	6351	6405		1 5
03	6458	6512	6566	6619	6673	6726	6780	6834	6887	6941		2 11
04	6994	7048	7102	7155	7209	7262	7316	7369	7423	7477		3 16
												4 22
05	7530	7584	7637	7691	7745	7798	7852	7905	7959	8012		5 27
06	8066	8120	8173	8227	8280	8334	8387	8441	8495	8548		6 32
07	8602	8655	8709	8762	8816	8870	8923	8977	9030	9084		7 38
08	9137	9191	9245	9298	9352	9405	9459	9512	9566	9619		8 43
09	9673	9727	9780	9834	9887	9941	9994	ō048	ō101	ō155		9 49
8110	9090209	0262	0316	0369	0423	0476	0530	0583	0637	0690		
11	0744	0798	0851	0905	0958	1012	1065	1119	1172	1226		
12	1279	1333	1386	1440	1494	1547	1601	1654	1708	1761		
13	1815	1868	1922	1975	2029	2082	2136	2189	2243	2297		
14	2350	2404	2457	2511	2564	2618	2671	2725	2778	2832		
15	2885	2939	2992	3046	3099	3153	3206	3260	3313	3367		
16	3420	3474	3527	3581	3634	3688	3741	3795	3848	3902		
17	3955	4009	4062	4116	4169	4223	4276	4330	4383	4437		
18	4490	4544	4597	4651	4704	4758	4811	4865	4918	4972		
19	5025	5079	5132	5186	5239	5293	5346	5400	5453	5507		
8120	5560	5614	5667	5721	5774	5828	5881	5935	5988	6042		
21	6095	6149	6202	6256	6309	6362	6416	6469	6523	6576		
22	6630	6683	6737	6790	6844	6897	6951	7004	7058	7111		
23	7165	7218	7271	7325	7378	7432	7485	7539	7592	7646		
24	7699	7753	7806	7860	7913	7966	8020	8073	8127	8180		
25	8234	8287	8341	8394	8447	8501	8554	8608	8661	8715		
26	8768	8822	8875	8929	8982	9035	9089	9142	9196	9249		
27	9303	9356	9409	9463	9516	9570	9623	9677	9730	9784		
28	9837	9890	9944	9997	ō051	ō104	ō158	ō211	ō264	ō318		
29	9100371	0425	0478	0532	0585	0638	0692	0745	0799	0852		
8130	0905	0959	1012	1066	1119	1173	1226	1279	1333	1386		
31	1440	1493	1546	1600	1653	1707	1760	1813	1867	1920		
32	1974	2027	2081	2134	2187	2241	2294	2348	2401	2454		
33	2508	2561	2615	2668	2721	2775	2828	2882	2935	2988		
34	3042	3095	3148	3202	3255	3309	3362	3415	3469	3522		
35	3576	3629	3682	3736	3789	3842	3896	3949	4003	4056		
36	4109	4163	4216	4270	4323	4376	4430	4483	4536	4590		
37	4643	4697	4750	4803	4857	4910	4963	5017	5070	5123		
38	5177	5230	5284	5337	5390	5444	5497	5550	5604	5657		
39	5710	5764	5817	5871	5924	5977	6031	6084	6137	6191		
8140	6244	6297	6351	6404	6457	6511	6564	6618	6671	6724		
41	6778	6831	6884	6938	6991	7044	7098	7151	7204	7258		
42	7311	7364	7418	7471	7524	7578	7631	7684	7738	7791		
43	7844	7898	7951	8004	8058	8111	8164	8218	8271	8324		**53**
44	8378	8431	8484	8538	8591	8644	8698	8751	8804	8858		1 5
												2 11
45	8911	8964	9018	9071	9124	9177	9231	9284	9337	9391		3 16
46	9444	9497	9551	9604	9657	9711	9764	9817	9871	9924		4 21
47	9977	ō030	ō084	ō137	ō190	ō244	ō297	ō350	ō404	ō457		5 27
48	9110510	0564	0617	0670	0723	0777	0830	0883	0937	0990		6 32
49	1043	1096	1150	1203	1256	1310	1363	1416	1470	1523		7 37
												8 42
												9 48
N.	0	1	2	3	4	5	6	7	8	9	D	Pts.

N.	0	1	2	3	4	5	6	7	8	9	.D	Pro.
8150	9111576	1629	1683	1736	1789	1843	1896	1949	2002	2056		
51	2109	2162	2215	2269	2322	2375	2429	2482	2535	2588		54
52	2642	2695	2748	2802	2855	2908	2961	3015	3068	3121		1 5
53	3174	3228	3281	3334	3387	3441	3494	3547	3601	3654		2 11
54	3707	3760	3814	3867	3920	3973	4027	4080	4133	4186		3 16
												4 22
55	4240	4293	4346	4399	4453	4506	4559	4612	4666	4719		5 27
56	4772	4825	4879	4932	4985	5038	5092	5145	5198	5251		6 32
57	5305	5358	5411	5464	5518	5571	5624	5677	5731	5784		7 38
58	5837	5890	5943	5997	6050	6103	6156	6210	6263	6316		8 43
59	6369	6423	6476	6529	6582	6635	6689	6742	6795	6848		9 49
8160	6902	6955	7008	7061	7114	7168	7221	7274	7327	7381		
61	7434	7487	7540	7593	7647	7700	7753	7806	7859	7913		
62	7966	8019	8072	8126	8179	8232	8285	8338	8392	8445		
63	8498	8551	8604	8658	8711	8764	8817	8870	8924	8977		
64	9030	9083	9136	9190	9243	9296	9349	9402	9456	9509		
65	9562	9615	9668	9721	9775	9828	9881	9934	9987	ō041		
66	9120094	0147	0200	0253	0306	0360	0413	0466	0519	0572		
67	0626	0679	0732	0785	0838	0891	0945	0998	1051	1104		
68	1157	1210	1264	1317	1370	1423	1476	1529	1583	1636		
69	1689	1742	1795	1848	1902	1955	2008	2061	2114	2167		
8170	2221	2274	2327	2380	2433	2486	2539	2593	2646	2699		
71	2752	2805	2858	2912	2965	3018	3071	3124	3177	3230		
72	3284	3337	3390	3443	3496	3549	3602	3656	3709	3762		
73	3815	3868	3921	3974	4028	4081	4134	4187	4240	4293		
74	4346	4399	4453	4506	4559	4612	4665	4718	4771	4824		
75	4878	4931	4984	5037	5090	5143	5196	5249	5303	5356		
76	5409	5462	5515	5568	5621	5674	5728	5781	5834	5887		
77	5940	5993	6046	6099	6152	6206	6259	6312	6365	6418		
78	6471	6524	6577	6630	6683	6737	6790	6843	6896	6949		
79	7002	7055	7108	7161	7214	7268	7321	7374	7427	7480		
8180	7533	7586	7639	7692	7745	7798	7852	7905	7958	8011		
81	8064	8117	8170	8223	8276	8329	8382	8436	8489	8542		
82	8595	8648	8701	8754	8807	8860	8913	8966	9019	9072		
83	9126	9179	9232	9285	9338	9391	9444	9497	9550	9603		
84	9656	9709	9762	9815	9868	9922	9975	ō028	ō081	ō134		
85	9130187	0240	0293	0346	0399	0452	0505	0558	0611	0664		
86	0717	0770	0824	0877	0930	0983	1036	1089	1142	1195		
87	1248	1301	1354	1407	1460	1513	1566	1619	1672	1725		
88	1778	1831	1884	1937	1990	2044	2097	2150	2203	2256		
89	2309	2362	2415	2468	2521	2574	2627	2680	2733	2786		
8190	2839	2892	2945	2998	3051	3104	3157	3210	3263	3316		
91	3369	3422	3475	3528	3581	3634	3687	3740	3793	3846		
92	3899	3952	4005	4058	4111	4165	4218	4271	4324	4377		53
93	4430	4483	4536	4589	4642	4695	4748	4801	4854	4907	53	1 5
94	4960	5013	5066	5119	5172	5225	5278	5331	5384	5437		2 11
												3 16
95	5490	5543	5596	5649	5702	5755	5808	5861	5914	5967		4 21
96	6019	6072	6125	6178	6231	6284	6337	6390	6443	6496		5 27
97	6549	6602	6655	6708	6761	6814	6867	6920	6973	7026		6 32
98	7079	7132	7185	7238	7291	7344	7397	7450	7503	7556		7 37
99	7609	7662	7715	7768	7821	7874	7927	7980	8033	8086		8 42
												9 48
N	0	1	2	3	4	5	6	7	8	9	D	Pts.

N.	0	1	2	3	4	5	6	7	8	9	D	Pro.
8200	9138139	8191	8244	8297	8350	8403	8456	8509	8562	8615		
01	8668	8721	8774	8827	8880	8933	8986	9039	9092	9145		53
02	9198	9251	9304	9356	9409	9462	9515	9568	9621	9674		1 5
03	9727	9780	9833	9886	9939	9992	ō045	ō098	ō151	ō204		2 11
04	9140257	0309	0362	0415	0468	0521	0574	0627	0680	0733		3 16
												4 21
05	0786	0839	0892	0945	0998	1050	1103	1156	1209	1262		5 27
06	1315	1368	1421	1474	1527	1580	1633	1686	1738	1791		6 32
07	1844	1897	1950	2003	2056	2109	2162	2215	2268	2321		7 37
08	2373	2426	2479	2532	2585	2638	2691	2744	2797	2850		8 42
09	2903	2955	3008	3061	3114	3167	3220	3273	3326	3379		9 48
8210	3432	3484	3537	3590	3643	3696	3749	3802	3855	3908		
11	3961	4013	4066	4119	4172	4225	4278	4331	4384	4437		
12	4489	4542	4595	4648	4701	4754	4807	4860	4912	4965		
13	5018	5071	5124	5177	5230	5283	5335	5388	5441	5494		
14	5547	5600	5653	5706	5758	5811	5864	5917	5970	6023		
15	6076	6129	6181	6234	6287	6340	6393	6446	6499	6551		
16	6604	6657	6710	6763	6816	6869	6921	6974	7027	7080		
17	7133	7186	7239	7291	7344	7397	7450	7503	7556	7609		
18	7661	7714	7767	7820	7873	7926	7978	8031	8084	8137		
19	8190	8243	8295	8348	8401	8454	8507	8560	8613	8665		
8220	8718	8771	8824	8877	8930	8982	9035	9088	9141	9194		
21	9246	9299	9352	9405	9458	9511	9563	9616	9669	9722		
22	9775	9828	9880	9933	9986	ō039	ō092	ō144	ō197	ō250		
23	9150303	0356	0409	0461	0514	0567	0620	0673	0725	0778		
24	0831	0884	0937	0989	1042	1095	1148	1201	1253	1306		
25	1359	1412	1465	1517	1570	1623	1676	1729	1781	1834		
26	1887	1940	1993	2045	2098	2151	2204	2257	2309	2362		
27	2415	2468	2521	2573	2626	2679	2732	2784	2837	2890		
28	2943	2996	3048	3101	3154	3207	3260	3312	3365	3418		
29	3471	3523	3576	3629	3682	3734	3787	3840	3893	3946		
8230	3998	4051	4104	4157	4209	4262	4315	4368	4420	4473		
31	4526	4579	4632	4684	4737	4790	4843	4895	4948	5001		
32	5054	5106	5159	5212	5265	5317	5370	5423	5476	5528		
33	5581	5634	5687	5739	5792	5845	5898	5950	6003	6056		
34	6109	6161	6214	6267	6320	6372	6425	6478	6531	6583	53	
35	6636	6689	6742	6794	6847	6900	6952	7005	7058	7111		
36	7163	7216	7269	7322	7374	7427	7480	7532	7585	7638		
37	7691	7743	7796	7849	7902	7954	8007	8060	8112	8165		
38	8218	8271	8323	8376	8429	8481	8534	8587	8640	8692		
39	8745	8798	8850	8903	8956	9009	9061	9114	9167	9219		
8240	9272	9325	9378	9430	9483	9536	9588	9641	9694	9746		
41	9799	9852	9905	9957	ō010	ō063	ō115	ō168	ō221	ō273		
42	9160326	0379	0431	0484	0537	0590	0642	0695	0748	0800		52
43	0853	0906	0958	1011	1064	1116	1169	1222	1274	1327		1 5
44	1380	1433	1485	1538	1591	1643	1696	1749	1801	1854		2 10
												3 16
45	1907	1959	2012	2065	2117	2170	2223	2275	2328	2381		4 21
46	2433	2486	2539	2591	2644	2697	2749	2802	2855	2907		5 26
47	2960	3013	3065	3118	3171	3223	3276	3329	3381	3434		6 31
48	3487	3539	3592	3644	3697	3750	3802	3855	3908	3960		7 36
49	4013	4066	4118	4171	4224	4276	4329	4382	4434	4487		8 42
												9 47
N.	0	1	2	3	4	5	6	7	8	9	D	Pts.

N.	0	1	2	3	4	5	6	7	8	9	D
8250	9164539	4592	4645	4697	4750	4803	4855	4908	4961	5013	
51	5066	5119	5171	5224	5276	5329	5382	5434	5487	5540	
52	5592	5645	5697	5750	5803	5855	5908	5961	6013	6066	
53	6118	6171	6224	6276	6329	6382	6434	6487	6539	6592	
54	6645	6697	6750	6802	6855	6908	6960	7013	7066	7118	
55	7171	7223	7276	7329	7381	7434	7486	7539	7592	7644	
56	7697	7749	7802	7855	7907	7960	8012	8065	8118	8170	
57	8223	8275	8328	8381	8433	8486	8538	8591	8644	8696	
58	8749	8801	8854	8907	8959	9012	9064	9117	9169	9222	
59	9275	9327	9380	9432	9485	9538	9590	9643	9695	9748	
8260	9800	9853	9906	9958	$\bar{0}$011	$\bar{0}$063	$\bar{0}$116	$\bar{0}$169	$\bar{0}$221	$\bar{0}$274	
61	9170326	0379	0431	0484	0537	0589	0642	0694	0747	0799	
62	0852	0904	0957	1010	1062	1115	1167	1220	1272	1325	
63	1378	1430	1483	1535	1588	1640	1693	1745	1798	1851	
64	1903	1956	2008	2061	2113	2166	2218	2271	2323	2376	
65	2429	2481	2534	2586	2639	2691	2744	2796	2849	2901	
66	2954	3007	3059	3112	3164	3217	3269	3322	3374	3427	
67	3479	3532	3584	3637	3690	3742	3795	3847	3900	3952	
68	4005	4057	4110	4162	4215	4267	4320	4372	4425	4477	
69	4530	4582	4635	4687	4740	4793	4845	4898	4950	5003	
8270	5055	5108	5160	5213	5265	5318	5370	5423	5475	5528	
71	5580	5633	5685	5738	5790	5843	5895	5948	6000	6053	
72	6105	6158	6210	6263	6315	6368	6420	6473	6525	6578	
73	6630	6683	6735	6788	6840	6893	6945	6998	7050	7103	
74	7155	7208	7260	7313	7365	7418	7470	7523	7575	7628	
75	7680	7733	7785	7837	7890	7942	7995	8047	8100	8152	
76	8205	8257	8310	8362	8415	8467	8520	8572	8625	8677	
77	8730	8782	8834	8887	8939	8992	9044	9097	9149	9202	
78	9254	9307	9359	9412	9464	9517	9569	9621	9674	9726	
79	9779	9831	9884	9936	9989	$\bar{0}$041	$\bar{0}$094	$\bar{0}$146	$\bar{0}$198	$\bar{0}$251	
8280	9180303	0356	0408	0461	0513	0566	0618	0671	0723	0775	
81	0828	0880	0933	0985	1038	1090	1143	1195	1247	1300	
82	1352	1405	1457	1510	1562	1614	1667	1719	1772	1824	
83	1877	1929	1981	2034	2086	2139	2191	2244	2296	2348	
84	2401	2453	2506	2558	2611	2663	2715	2768	2820	2873	
85	2925	2978	3030	3082	3135	3187	3240	3292	3344	3397	
86	3449	3502	3554	3607	3659	3711	3764	3816	3869	3921	
87	3973	4026	4078	4131	4183	4235	4288	4340	4393	4445	
88	4497	4550	4602	4655	4707	4759	4812	4864	4917	4969	
89	5021	5074	5126	5179	5231	5283	5336	5388	5441	5493	
8290	5545	5598	5650	5702	5755	5807	5860	5912	5964	6017	
91	6069	6122	6174	6226	6279	6331	6383	6436	6488	6541	
92	6593	6645	6698	6750	6802	6855	6907	6960	7012	7064	
93	7117	7169	7221	7274	7326	7378	7431	7483	7536	7588	
94	7640	7693	7745	7797	7850	7902	7954	8007	8059	8112	
95	8164	8216	8269	8321	8373	8426	8478	8530	8583	8635	
96	8687	8740	8792	8844	8897	8949	9002	9054	9106	9159	
97	9211	9263	9316	9368	9420	9473	9525	9577	9630	9682	
98	9734	9787	9839	9891	9944	9996	$\bar{0}$048	$\bar{0}$101	$\bar{0}$153	$\bar{0}$205	
99	9190258	0310	0362	0415	0467	0519	0572	0624	0676	0729	
N.	0	1	2	3	4	5	6	7	8	9	D

Pro.

53		52	
1	5	1	5
2	11	2	10
3	16	3	16
4	21	4	21
5	27	5	26
6	32	6	31
7	37	7	36
8	42	8	42
9	48	9	47

N.	0	1	2	3	4	5	6	7	8	9	D	Pro.
8300	9190781	0833	0886	0938	0990	1043	1095	1147	1200	1252		
01	1304	1356	1409	1461	1513	1566	1618	1670	1723	1775		53
02	1827	1880	1932	1984	2037	2089	2141	2193	2246	2298		1 5
03	2350	2403	2455	2507	2560	2612	2664	2717	2769	2821		2 11
04	2873	2926	2978	3030	3083	3135	3187	3239	3292	3344		3 16
												4 21
05	3396	3449	3501	3553	3606	3658	3710	3762	3815	3867		5 27
06	3919	3972	4024	4076	4128	4181	4233	4285	4338	4390		6 32
07	4442	4494	4547	4599	4651	4703	4756	4808	4860	4913		7 37
08	4965	5017	5069	5122	5174	5226	5279	5331	5383	5435		8 42
09	5488	5540	5592	5644	5697	5749	5801	5853	5906	5958		9 48
8310	6010	6062	6115	6167	6219	6272	6324	6376	6428	6481		
11	6533	6585	6637	6690	6742	6794	6846	6899	6951	7003		
12	7055	7108	7160	7212	7264	7317	7369	7421	7473	7526		
13	7578	7630	7682	7735	7787	7839	7891	7943	7996	8048		
14	8100	8152	8205	8257	8309	8361	8414	8466	8518	8570		
15	8623	8675	8727	8779	8831	8884	8936	8988	9040	9093		
16	9145	9197	9249	9301	9354	9406	9458	9510	9563	9615		
17	9667	9719	9771	9824	9876	9928	9980	0̄033	0̄085	0̄137		
18	9200189	0241	0294	0346	0398	0450	0502	0555	0607	0659		
19	0711	0763	0816	0868	0920	0972	1024	1077	1129	1181		
8320	1233	1285	1338	1390	1442	1494	1546	1599	1651	1703		
21	1755	1807	1860	1912	1964	2016	2068	2121	2173	2225		
22	2277	2329	2381	2434	2486	2538	2590	2642	2695	2747		
23	2799	2851	2903	2955	3008	3060	3112	3164	3216	3269		
24	3321	3373	3425	3477	3529	3582	3634	3686	3738	3790		
25	3842	3895	3947	3999	4051	4103	4155	4208	4260	4312		
26	4364	4416	4468	4521	4573	4625	4677	4729	4781	4833		
27	4886	4938	4990	5042	5094	5146	5199	5251	5303	5355		
28	5407	5459	5511	5564	5616	5668	5720	5772	5824	5876		
29	5929	5981	6033	6085	6137	6189	6241	6294	6346	6398		
8330	6450	6502	6554	6606	6659	6711	6763	6815	6867	6919		
31	6971	7023	7076	7128	7180	7232	7284	7336	7388	7440		
32	7493	7545	7597	7649	7701	7753	7805	7857	7910	7962		
33	8014	8066	8118	8170	8222	8274	8327	8379	8431	8483		
34	8535	8587	8639	8691	8743	8796	8848	8900	8952	9004		
35	9056	9108	9160	9212	9264	9317	9369	9421	9473	9525		
36	9577	9629	9681	9733	9785	9838	9890	9942	9994	0̄046		
37	9210098	0150	0202	0254	0306	0358	0411	0463	0515	0567		
38	0619	0671	0723	0775	0827	0879	0931	0983	1036	1088		
39	1140	1192	1244	1296	1348	1400	1452	1504	1556	1608		
8340	1661	1713	1765	1817	1869	1921	1973	2025	2077	2129		
41	2181	2233	2285	2337	2389	2442	2494	2546	2598	2650		
42	2702	2754	2806	2858	2910	2962	3014	3066	3118	3170		52
43	3222	3274	3327	3379	3431	3483	3535	3587	3639	3691		1 5
44	3743	3795	3847	3899	3951	4003	4055	4107	4159	4211		2 10
												3 16
45	4263	4315	4367	4420	4472	4524	4576	4628	4680	4732		4 21
46	4784	4836	4888	4940	4992	5044	5096	5148	5200	5252		5 26
47	5304	5356	5408	5460	5512	5564	5616	5668	5720	5772		6 31
48	5824	5876	5928	5980	6032	6085	6137	6189	6241	6293		7 36
49	6345	6397	6449	6501	6553	6605	6657	6709	6761	6813		8 42
												9 47
N.	0	1	2	3	4	5	6	7	8	9	D	Pts.

N	0	1	2	3	4	5	6	7	8	9	D	Pro.
8350	9216865	6917	6969	7021	7073	7125	7177	7229	7281	7333		
51	7385	7437	7489	7541	7593	7645	7697	7749	7801	7853		**52**
52	7905	7957	8009	8061	8113	8165	8217	8269	8321	8373		1 5
53	8425	8477	8529	8581	8633	8685	8737	8789	8841	8893	52	2 10
54	8945	8997	9049	9101	9153	9205	9257	9309	9361	9413		3 16
												4 21
55	9465	9517	9569	9620	9672	9724	9776	9828	9880	9932		5 26
56	9984	0036	0088	0140	0192	0244	0296	0348	0400	0452		6 31
57	9220504	0556	0608	0660	0712	0764	0816	0868	0920	0972		7 36
58	1024	1076	1128	1180	1232	1283	1335	1387	1439	1491		8 42
59	1543	1595	1647	1699	1751	1803	1855	1907	1959	2011		9 47
8360	2063	2115	2167	2219	2271	2323	2374	2426	2478	2530		
61	2582	2634	2686	2738	2790	2842	2894	2946	2998	3050		
62	3102	3154	3206	3257	3309	3361	3413	3465	3517	3569		
63	3621	3673	3725	3777	3829	3881	3933	3984	4036	4088		
64	4140	4192	4244	4296	4348	4400	4452	4504	4556	4608		
65	4659	4711	4763	4815	4867	4919	4971	5023	5075	5127		
66	5179	5231	5282	5334	5386	5438	5490	5542	5594	5646		
67	5698	5750	5801	5853	5905	5957	6009	6061	6113	6165		
68	6217	6269	6321	6372	6424	6476	6528	6580	6632	6684		
69	6736	6788	6839	6891	6943	6995	7047	7099	7151	7203		
8370	7255	7306	7358	7410	7462	7514	7566	7618	7670	7722		
71	7773	7825	7877	7929	7981	8033	8085	8137	8188	8240		
72	8292	8344	8396	8448	8500	8552	8603	8655	8707	8759		
73	8811	8863	8915	8967	9018	9070	9122	9174	9226	9278		
74	9330	9381	9433	9485	9537	9589	9641	9693	9744	9796		
75	9848	9900	9952	0004	0056	0107	0159	0211	0263	0315		
76	9230367	0419	0470	0522	0574	0626	0678	0730	0781	0833		
77	0885	0937	0989	1041	1093	1144	1196	1248	1300	1352		
78	1404	1455	1507	1559	1611	1663	1715	1766	1818	1870		
79	1922	1974	2026	2077	2129	2181	2233	2285	2337	2388		
8380	2440	2492	2544	2596	2647	2699	2751	2803	2855	2907		
81	2958	3010	3062	3114	3166	3217	3269	3321	3373	3425		
82	3477	3528	3580	3632	3684	3736	3787	3839	3891	3943		
83	3995	4046	4098	4150	4202	4254	4305	4357	4409	4461		
84	4513	4564	4616	4668	4720	4772	4823	4875	4927	4979		
85	5031	5082	5134	5186	5238	5290	5341	5393	5445	5497		
86	5549	5600	5652	5704	5756	5808	5859	5911	5963	6015		
87	6066	6118	6170	6222	6274	6325	6377	6429	6481	6532		
88	6584	6636	6688	6740	6791	6843	6895	6947	6998	7050		
89	7102	7154	7205	7257	7309	7361	7413	7464	7516	7568		
8390	7620	7671	7723	7775	7827	7878	7930	7982	8034	8085		
91	8137	8189	8241	8292	8344	8396	8448	8499	8551	8603		
92	8655	8707	8758	8810	8862	8913	8965	9017	9069	9120		**51**
93	9172	9224	9276	9327	9379	9431	9483	9534	9586	9638		1 5
94	9690	9741	9793	9845	9897	9948	0000	0052	0104	0155		2 10
												3 15
95	9240207	0259	0310	0362	0414	0466	0517	0569	0621	0673		4 20
96	0724	0776	0828	0879	0931	0983	1035	1086	1138	1190		5 26
97	1242	1293	1345	1397	1448	1500	1552	1604	1655	1707		6 31
98	1759	1810	1862	1914	1966	2017	2069	2121	2172	2224		7 36
99	2276	2328	2379	2431	2483	2534	2586	2638	2689	2741		8 41
												9 46
N.	0	1	2	3	4	5	6	7	8	9	D	Pts.

N.	0	1	2	3	4	5	6	7	8	9	D	Pro.
8400	9242793	2845	2896	2948	3000	3051	3103	3155	3206	3258		
01	3310	3362	3413	3465	3517	3568	3620	3672	3723	3775		52
02	3827	3878	3930	3982	4034	4085	4137	4189	4240	4292		1 5
03	4344	4395	4447	4499	4550	4602	4654	4705	4757	4809		2 10
04	4860	4912	4964	5015	5067	5119	5170	5222	5274	5326		3 16
05	5377	5429	5481	5532	5584	5636	5687	5739	5791	5842		4 21 / 5 26
06	5894	5946	5997	6049	6101	6152	6204	6255	6307	6359	51	6 31
07	6410	6462	6514	6565	6617	6669	6720	6772	6824	6875		7 36
08	6927	6979	7030	7082	7134	7185	7237	7289	7340	7392		8 42
09	7444	7495	7547	7598	7650	7702	7753	7805	7857	7908		9 47
8410	7960	8012	8063	8115	8167	8218	8270	8321	8373	8425		
11	8476	8528	8580	8631	8683	8734	8786	8838	8889	8941		
12	8993	9044	9096	9148	9199	9251	9302	9354	9406	9457		
13	9509	9561	9612	9664	9715	9767	9819	9870	9922	9973		
14	9250025	0077	0128	0180	0232	0283	0335	0386	0438	0490		
15	0541	0593	0644	0696	0748	0799	0851	0902	0954	1006		
16	1057	1109	1160	1212	1264	1315	1367	1418	1470	1522		
17	1573	1625	1676	1728	1780	1831	1883	1934	1986	2038		
18	2089	2141	2192	2244	2296	2347	2399	2450	2502	2554		
19	2605	2657	2708	2760	2811	2863	2915	2966	3018	3069		
8420	3121	3172	3224	3276	3327	3379	3430	3482	3534	3585		
21	3637	3688	3740	3791	3843	3895	3946	3998	4049	4101		
22	4152	4204	4256	4307	4359	4410	4462	4513	4565	4616		
23	4668	4720	4771	4823	4874	4926	4977	5029	5080	5132		
24	5184	5235	5287	5338	5390	5441	5493	5544	5596	5648		
25	5699	5751	5802	5854	5905	5957	6008	6060	6111	6163		
26	6215	6266	6318	6369	6421	6472	6524	6575	6627	6678		
27	6730	6781	6833	6885	6936	6988	7039	7091	7142	7194		
28	7245	7297	7348	7400	7451	7503	7554	7606	7657	7709		
29	7761	7812	7864	7915	7967	8018	8070	8121	8173	8224		
8430	8276	8327	8379	8430	8482	8533	8585	8636	8688	8739		
31	8791	8842	8894	8945	8997	9048	9100	9151	9203	9254		
32	9306	9357	9409	9460	9512	9563	9615	9667	9718	9770		
33	9821	9873	9924	9975	$\bar{0}$027	$\bar{0}$078	$\bar{0}$130	$\bar{0}$181	$\bar{0}$233	$\bar{0}$284		
34	9260336	0387	0439	0490	0542	0593	0645	0696	0748	0799		
35	0851	0902	0954	1005	1057	1108	1160	1211	1263	1314		
36	1366	1417	1469	1520	1572	1623	1675	1726	1778	1829		
37	1880	1932	1983	2035	2086	2138	2189	2241	2292	2344		
38	2395	2447	2498	2550	2601	2653	2704	2755	2807	2858		
39	2910	2961	3013	3064	3116	3167	3219	3270	3322	3373		
8440	3424	3476	3527	3579	3630	3682	3733	3785	3836	3888		
41	3939	3990	4042	4093	4145	4196	4248	4299	4351	4402		51
42	4453	4505	4556	4608	4659	4711	4762	4814	4865	4916		1 5
43	4968	5019	5071	5122	5174	5225	5277	5328	5379	5431		2 10
44	5482	5534	5585	5637	5688	5739	5791	5842	5894	5945		3 15 / 4 20
45	5997	6048	6099	6151	6202	6254	6305	6357	6408	6459		5 26
46	6511	6562	6614	6665	6716	6768	6819	6871	6922	6974		6 31
47	7025	7076	7128	7179	7231	7282	7333	7385	7436	7488		7 36
48	7539	7590	7642	7693	7745	7796	7847	7899	7950	8002		8 41
49	8053	8105	8156	8207	8259	8310	8362	8413	8464	8516		9 46
N.	0	1	2	3	4	5	6	7	8	9	D	Pts.

N.	0	1	2	3	4	5	6	7	8	9	D	Pro.
8450	9268567	8618	8670	8721	8773	8824	8875	8927	8978	9030		
51	9081	9132	9184	9235	9287	9338	9389	9441	9492	9543		52
52	9595	9646	9698	9749	9800	9852	9903	9955	ō006	ō057		1 5
53	9270109	0160	0211	0263	0314	0366	0417	0468	0520	0571		2 10
54	0622	0674	0725	0777	0828	0879	0931	0982	1033	1085		3 16
												4 21
55	1136	1187	1239	1290	1342	1393	1444	1496	1547	1598		5 26
56	1650	1701	1752	1804	1855	1907	1958	2009	2061	2112		6 31
57	2163	2215	2266	2317	2369	2420	2471	2523	2574	2625		7 36
58	2677	2728	2780	2831	2882	2934	2985	3036	3088	3139		8 42
59	3190	3242	3293	3344	3396	3447	3498	3550	3601	3652		9 47
8460	3704	3755	3806	3858	3909	3960	4012	4063	4114	4166		
61	4217	4268	4320	4371	4422	4474	4525	4576	4628	4679		
62	4730	4782	4833	4884	4935	4987	5038	5089	5141	5192		
63	5243	5295	5346	5397	5449	5500	5551	5603	5654	5705		
64	5757	5808	5859	5910	5962	6013	6064	6116	6167	6218		
65	6270	6321	6372	6424	6475	6526	6577	6629	6680	6731		
66	6783	6834	6885	6937	6988	7039	7090	7142	7193	7244		
67	7296	7347	7398	7449	7501	7552	7603	7655	7706	7757		
68	7808	7860	7911	7962	8014	8065	8116	8167	8219	8270		
69	8321	8373	8424	8475	8526	8578	8629	8680	8732	8783		
8470	8834	8885	8937	8988	9039	9090	9142	9193	9244	9296		
71	9347	9398	9449	9501	9552	9603	9654	9706	9757	9808		
72	9859	9911	9962	ō013	ō065	ō116	ō167	ō218	ō270	ō321		
73	9280372	0423	0475	0526	0577	0628	0680	0731	0782	0833		
74	0885	0936	0987	1038	1090	1141	1192	1243	1295	1346		
75	1397	1448	1500	1551	1602	1653	1705	1756	1807	1858		
76	1909	1961	2012	2063	2114	2166	2217	2268	2319	2371		
77	2422	2473	2524	2576	2627	2678	2729	2780	2832	2883		
78	2934	2985	3037	3088	3139	3190	3241	3293	3344	3395		
79	3446	3498	3549	3600	3651	3702	3754	3805	3856	3907		
8480	3959	4010	4061	4112	4163	4215	4266	4317	4368	4419		
81	4471	4522	4573	4624	4675	4727	4778	4829	4880	4931		
82	4983	5034	5085	5136	5187	5239	5290	5341	5392	5443		
83	5495	5546	5597	5648	5699	5751	5802	5853	5904	5955		
84	6007	6058	6109	6160	6211	6263	6314	6365	6416	6467		
85	6518	6570	6621	6672	6723	6774	6826	6877	6928	6979		
86	7030	7081	7133	7184	7235	7286	7337	7389	7440	7491		
87	7542	7593	7644	7696	7747	7798	7849	7900	7951	8003		
88	8054	8105	8156	8207	8258	8310	8361	8412	8463	8514		
89	8565	8616	8668	8719	8770	8821	8872	8923	8975	9026		
8490	9077	9128	9179	9230	9282	9333	9384	9435	9486	9537		
91	9588	9640	9691	9742	9793	9844	9895	9946	9998	ō049		
92	9290100	0151	0202	0253	0304	0356	0407	0458	0509	0560		51
93	0611	0662	0714	0765	0816	0867	0918	0969	1020	1071		1 5
94	1123	1174	1225	1276	1327	1378	1429	1480	1532	1583		2 10
												3 15
95	1634	1685	1736	1787	1838	1889	1941	1992	2043	2094		4 20
96	2145	2196	2247	2298	2350	2401	2452	2503	2554	2605		5 26
97	2656	2707	2758	2810	2861	2912	2963	3014	3065	3116		6 31
98	3167	3218	3269	3321	3372	3423	3474	3525	3576	3627		7 36
99	3678	3729	3780	3832	3883	3934	3985	4036	4087	4138		8 41
												9 46
N.	0	1	2	3	4	5	6	7	8	9	D	Pts.

N.	0	1	2	3	4	5	6	7	8	9	D	Pro.
8500	9294189	4240	4291	4343	4394	4445	4496	4547	4598	4649		
01	4700	4751	4802	4853	4905	4956	5007	5058	5109	5160		52
02	5211	5262	5313	5364	5415	5466	5517	5569	5620	5671		1 5
03	5722	5773	5824	5875	5926	5977	6028	6079	6130	6181		2 10
04	6233	6284	6335	6386	6437	6488	6539	6590	6641	6692		3 16 4 21
05	6743	6794	6845	6896	6947	6998	7050	7101	7152	7203		5 26
06	7254	7305	7356	7407	7458	7509	7560	7611	7662	7713		6 31
07	7764	7815	7866	7917	7969	8020	8071	8122	8173	8224		7 36
08	8275	8326	8377	8428	8479	8530	8581	8632	8683	8734		8 42
09	8785	8836	8887	8938	8989	9040	9091	9142	9194	9245		9 47
8510	9296	9347	9398	9449	9500	9551	9602	9653	9704	9755		
11	9806	9857	9908	9959	ō010	ō061	ō112	ō163	ō214	ō265		
12	9300316	0367	0418	0469	0520	0571	0622	0673	0724	0775		
13	0826	0877	0928	0979	1030	1081	1132	1183	1234	1285		
14	1336	1387	1438	1489	1540	1591	1643	1694	1745	1796		
15	1847	1898	1949	2000	2051	2102	2153	2204	2255	2306	51	
16	2357	2408	2459	2510	2561	2612	2663	2713	2764	2815		
17	2866	2917	2968	3019	3070	3121	3172	3223	3274	3325		
18	3376	3427	3478	3529	3580	3631	3682	3733	3784	3835		
19	3886	3937	3988	4039	4090	4141	4192	4243	4294	4345		
8520	4396	4447	4498	4549	4600	4651	4702	4753	4804	4855		
21	4906	4957	5008	5059	5110	5160	5211	5262	5313	5364		
22	5415	5466	5517	5568	5619	5670	5721	5772	5823	5874		
23	5925	5976	6027	6078	6129	6180	6231	6282	6333	6383		
24	6434	6485	6536	6587	6638	6689	6740	6791	6842	6893		
25	6944	6995	7046	7097	7148	7199	7250	7300	7351	7402		
26	7453	7504	7555	7606	7657	7708	7759	7810	7861	7912		
27	7963	8014	8064	8115	8166	8217	8268	8319	8370	8421		
28	8472	8523	8574	8625	8676	8727	8777	8828	8879	8930		
29	8981	9032	9083	9134	9185	9236	9287	9338	9388	9439		
8530	9490	9541	9592	9643	9694	9745	9796	9847	9898	9949		
31	9999	ō050	ō101	ō152	ō203	ō254	ō305	ō356	ō407	ō458		
32	9310508	0559	0610	0661	0712	0763	0814	0865	0916	0967		
33	1017	1068	1119	1170	1221	1272	1323	1374	1425	1475		
34	1526	1577	1628	1679	1730	1781	1832	1883	1933	1984		
35	2035	2086	2137	2188	2239	2290	2341	2391	2442	2493		
36	2544	2595	2646	2697	2748	2798	2849	2900	2951	3002		
37	3053	3104	3155	3205	3256	3307	3358	3409	3460	3511		
38	3562	3612	3663	3714	3765	3816	3867	3918	3968	4019		
39	4070	4121	4172	4223	4274	4324	4375	4426	4477	4528		
8540	4579	4630	4680	4731	4782	4833	4884	4935	4986	5036		
41	5087	5138	5189	5240	5291	5341	5392	5443	5494	5545		
42	5596	5647	5697	5748	5799	5850	5901	5952	6002	6053		51
43	6104	6155	6206	6257	6307	6358	6409	6460	6511	6562		1 5
44	6612	6663	6714	6765	6816	6867	6917	6968	7019	7070		2 10 3 15
45	7121	7171	7222	7273	7324	7375	7426	7476	7527	7578		4 20
46	7629	7680	7731	7781	7832	7883	7934	7985	8035	8086		5 26
47	8137	8188	8239	8289	8340	8391	8442	8493	8544	8594		6 31 7 36
48	8645	8696	8747	8798	8848	8899	8950	9001	9052	9102		8 41
49	9153	9204	9255	9306	9356	9407	9458	9509	9560	9610		9 46
N.	0	1	2	3	4	5	6	7	8	9	D	Pts.

N.	0	1	2	3	4	5	6	7	8	9	D	Pro.
8550	9319661	9712	9763	9814	9864	9915	9966	0̄017	0̄067	0̄118		
51	9320169	0220	0271	0321	0372	0423	0474	0525	0575	0626		51
52	0677	0728	0778	0829	0880	0931	0982	1032	1083	1134		1 5
53	1185	1235	1286	1337	1388	1439	1489	1540	1591	1642		2 10
54	1692	1743	1794	1845	1896	1946	1997	2048	2099	2149		3 15
												4 20
55	2200	2251	2302	2352	2403	2454	2505	2555	2606	2657		5 26
56	2708	2759	2809	2860	2911	2962	3012	3063	3114	3165		6 31
57	3215	3266	3317	3368	3418	3469	3520	3571	3621	3672		7 36
58	3723	3774	3824	3875	3926	3977	4027	4078	4129	4180		8 41
59	4230	4281	4332	4382	4433	4484	4535	4585	4636	4687		9 46
8560	4738	4788	4839	4890	4941	4991	5042	5093	5144	5194		
61	5245	5296	5346	5397	5448	5499	5549	5600	5651	5702		
62	5752	5803	5854	5904	5955	6006	6057	6107	6158	6209		
63	6259	6310	6361	6412	6462	6513	6564	6614	6665	6716		
64	6767	6817	6868	6919	6969	7020	7071	7122	7172	7223		
65	7274	7324	7375	7426	7476	7527	7578	7629	7679	7730		
66	7781	7831	7882	7933	7983	8034	8085	8136	8186	8237		
67	8288	8338	8389	8440	8490	8541	8592	8643	8693	8744		
68	8795	8845	8896	8947	8997	9048	9099	9149	9200	9251		
69	9301	9352	9403	9453	9504	9555	9606	9656	9707	9758		
8570	9808	9859	9910	9960	0̄011	0̄062	0̄112	0̄163	0̄214	0̄264		
71	9330315	0366	0416	0467	0518	0568	0619	0670	0720	0771		
72	0822	0872	0923	0974	1024	1075	1126	1176	1227	1278		
73	1328	1379	1430	1480	1531	1582	1632	1683	1733	1784		
74	1835	1885	1936	1987	2037	2088	2139	2189	2240	2291		
75	2341	2392	2443	2493	2544	2595	2645	2696	2746	2797		
76	2848	2898	2949	3000	3050	3101	3152	3202	3253	3303		
77	3354	3405	3455	3506	3557	3607	3658	3709	3759	3810		
78	3860	3911	3962	4012	4063	4114	4164	4215	4265	4316		
79	4367	4417	4468	4519	4569	4620	4670	4721	4772	4822		
8580	4873	4923	4974	5025	5075	5126	5177	5227	5278	5328		
81	5379	5430	5480	5531	5581	5632	5683	5733	5784	5834		
82	5885	5936	5986	6037	6088	6138	6189	6239	6290	6341		
83	6391	6442	6492	6543	6594	6644	6695	6745	6796	6846		
84	6897	6948	6998	7049	7099	7150	7201	7251	7302	7352		
85	7403	7454	7504	7555	7605	7656	7707	7757	7808	7858		
86	7909	7959	8010	8061	8111	8162	8212	8263	8313	8364		
87	8415	8465	8516	8566	8617	8668	8718	8769	8819	8870		
88	8920	8971	9021	9072	9123	9173	9224	9274	9325	9375		
89	9426	9477	9527	9578	9628	9679	9729	9780	9831	9881		
8590	9932	9982	0̄033	0̄083	0̄134	0̄184	0̄235	0̄286	0̄336	0̄387		
91	9340437	0488	0538	0589	0639	0690	0740	0791	0842	0892		
92	0943	0993	1044	1094	1145	1195	1246	1296	1347	1398		50
93	1448	1499	1549	1600	1650	1701	1751	1802	1852	1903		1 5
94	1953	2004	2055	2105	2156	2206	2257	2307	2358	2408		2 10
												3 15
95	2459	2509	2560	2610	2661	2711	2762	2812	2863	2914		4 20
96	2964	3015	3065	3116	3166	3217	3267	3318	3368	3419		5 25
97	3469	3520	3570	3621	3671	3722	3772	3823	3873	3924		6 30
98	3974	4025	4075	4126	4176	4227	4277	4328	4378	4429		7 35
99	4479	4530	4580	4631	4682	4732	4783	4833	4884	4934		8 40
												9 45
N.	0	1	2	3	4	5	6	7	8	9	D	Pts.

N.	0	1	2	3	4	5	6	7	8	9	D	Pro.
8600	9344985	5035	5086	5136	5187	5237	5287	5338	5388	5439		
01	5489	5540	5590	5641	5691	5742	5792	5843	5893	5944		51
02	5994	6045	6095	6146	6196	6247	6297	6348	6398	6449		1 5
03	6499	6550	6600	6651	6701	6752	6802	6853	6903	6954		2 10
04	7004	7054	7105	7155	7206	7256	7307	7357	7408	7458		3 15 / 4 20
05	7509	7559	7610	7660	7711	7761	7812	7862	7912	7963		5 26
06	8013	8064	8114	8165	8215	8266	8316	8367	8417	8468		6 31
07	8518	8568	8619	8669	8720	8770	8821	8871	8922	8972		7 36
08	9023	9073	9123	9174	9224	9275	9325	9376	9426	9477		8 41
09	9527	9578	9628	9678	9729	9779	9830	9880	9931	9981		9 46
8610	9350032	0082	0132	0183	0233	0284	0334	0385	0435	0485		
11	0536	0586	0637	0687	0738	0788	0838	0889	0939	0990		
12	1040	1091	1141	1191	1242	1292	1343	1393	1444	1494		
13	1544	1595	1645	1696	1746	1797	1847	1897	1948	1998		
14	2049	2099	2150	2200	2250	2301	2351	2402	2452	2502		
15	2553	2603	2654	2704	2754	2805	2855	2906	2956	3006		
16	3057	3107	3158	3208	3259	3309	3359	3410	3460	3511		
17	3561	3611	3662	3712	3763	3813	3863	3914	3964	4015		
18	4065	4115	4166	4216	4266	4317	4367	4418	4468	4518		
19	4569	4619	4670	4720	4770	4821	4871	4922	4972	5022		
8620	5073	5123	5173	5224	5274	5325	5375	5425	5476	5526		
21	5576	5627	5677	5728	5778	5828	5879	5929	5979	6030		
22	6080	6131	6181	6231	6282	6332	6382	6433	6483	6533		
23	6584	6634	6685	6735	6785	6836	6886	6936	6987	7037		
24	7087	7138	7188	7239	7289	7339	7390	7440	7490	7541		
25	7591	7641	7692	7742	7792	7843	7893	7943	7994	8044		
26	8095	8145	8195	8246	8296	8346	8397	8447	8497	8548		
27	8598	8648	8699	8749	8799	8850	8900	8950	9001	9051		
28	9101	9152	9202	9252	9303	9353	9403	9454	9504	9554		
29	9605	9655	9705	9756	9806	9856	9907	9957	ō007	ō058		
8630	9360108	0158	0209	0259	0309	0360	0410	0460	0511	0561		
31	0611	0661	0712	0762	0812	0863	0913	0963	1014	1064		
32	1114	1165	1215	1265	1316	1366	1416	1466	1517	1567		
33	1617	1668	1718	1768	1819	1869	1919	1970	2020	2070		
34	2120	2171	2221	2271	2322	2372	2422	2473	2523	2573		
35	2623	2674	2724	2774	2825	2875	2925	2975	3026	3076		
36	3126	3177	3227	3277	3327	3378	3428	3478	3529	3579		
37	3629	3679	3730	3780	3830	3881	3931	3981	4031	4082		
38	4132	4182	4233	4283	4333	4383	4434	4484	4534	4584		
39	4635	4685	4735	4786	4836	4886	4936	4987	5037	5087		
8640	5137	5188	5238	5288	5338	5369	5439	5489	5540	5590		
41	5640	5690	5741	5791	5841	5891	5942	5992	6042	6092		
42	6143	6193	6243	6293	6344	6394	6444	6494	6545	6595		50
43	6645	6695	6746	6796	6846	6896	6947	6997	7047	7097		1 5
44	7148	7198	7248	7298	7349	7399	7449	7499	7550	7600		2 10 / 3 15
45	7650	7700	7750	7801	7851	7901	7951	8002	8052	8102		4 20
46	8152	8203	8253	8303	8353	8403	8454	8504	8554	8604		5 25
47	8655	8705	8755	8805	8855	8906	8956	9006	9056	9107		6 30
48	9157	9207	9257	9307	9358	9408	9458	9508	9559	9609		7 35
49	9659	9709	9759	9810	9860	9910	9960	ō010	ō061	ō111		8 40 / 9 45
N.	0	1	2	3	4	5	6	7	8	9	D	Pts.

N.	0	1	2	3	4	5	6	7	8	9	D	Pro.
8650	9370161	0211	0261	0312	0362	0412	0462	0513	0563	0613		
51	0663	0713	0764	0814	0864	0914	0964	1015	1065	1115		50
52	1165	1215	1265	1316	1366	1416	1466	1516	1567	1617		1 5
53	1667	1717	1767	1818	1868	1918	1968	2018	2069	2119		2 10
54	2169	2219	2269	2319	2370	2420	2470	2520	2570	2621		3 15
												4 20
55	2671	2721	2771	2821	2871	2922	2972	3022	3072	3122		5 25
56	3172	3223	3273	3323	3373	3423	3474	3524	3574	3624		6 30
57	3674	3724	3775	3825	3875	3925	3975	4025	4075	4126		7 35
58	4176	4226	4276	4326	4376	4427	4477	4527	4577	4627		8 40
59	4677	4728	4778	4828	4878	4928	4978	5028	5079	5129		9 45
6660	5179	5229	5279	5329	5380	5430	5480	5530	5580	5630		
61	5680	5731	5781	5831	5881	5931	5981	6031	6082	6132		
62	6182	6232	6282	6332	6382	6432	6483	6533	6583	6633		
63	6683	6733	6783	6834	6884	6934	6984	7034	7084	7134		
64	7184	7235	7285	7335	7385	7435	7485	7535	7585	7636		
65	7686	7736	7786	7836	7886	7936	7986	8037	8087	8137		
66	8187	8237	8287	8337	8387	8437	8488	8538	8588	8638		
67	8688	8738	8788	8838	8888	8939	8989	9039	9089	9139		
68	9189	9239	9289	9339	9389	9440	9490	9540	9590	9640		
69	9690	9740	9790	9840	9890	9941	9991	ō041	ō091	ō141		
8670	9380191	0241	0291	0341	0391	0441	0492	0542	0592	0642		
71	0692	0742	0792	0842	0892	0942	0992	1042	1093	1143		
72	1193	1243	1293	1343	1393	1443	1493	1543	1593	1643		
73	1693	1744	1794	1844	1894	1944	1994	2044	2094	2144		
74	2194	2244	2294	2344	2394	2445	2495	2545	2595	2645		
75	2695	2745	2795	2845	2895	2945	2995	3045	3095	3145		
76	3195	3245	3296	3346	3396	3446	3496	3546	3596	3646		
77	3696	3746	3796	3846	3896	3946	3996	4046	4096	4146		
78	4196	4247	4297	4347	4397	4447	4497	4547	4597	4647		
79	4697	4747	4797	4847	4897	4947	4997	5047	5097	5147		
8680	5197	5247	5297	5347	5397	5447	5497	5547	5598	5648		
81	5698	5748	5798	5848	5898	5948	5998	6048	6098	6148		
82	6198	6248	6298	6348	6398	6448	6498	6548	6598	6648		
83	6698	6748	6798	6848	6898	6948	6998	7048	7098	7148		
84	7198	7248	7298	7348	7398	7448	7498	7548	7598	7648	50	
85	7698	7748	7798	7848	7898	7948	7998	8048	8098	8148		
86	8198	8248	8298	8348	8398	8448	8498	8548	8598	8648		
87	8698	8748	8798	8848	8898	8948	8998	9048	9098	9148		
88	9198	9248	9298	9348	9398	9448	9498	9548	9598	9648		
89	9698	9748	9798	9848	9898	9948	9998	ō048	ō098	ō148		
8690	9390198	0248	0298	0348	0398	0448	0498	0548	0598	0648		
91	0697	0747	0797	0847	0897	0947	0997	1047	1097	1147		
92	1197	1247	1297	1347	1397	1447	1497	1547	1597	1647		49
93	1697	1747	1797	1847	1897	1947	1997	2046	2096	2146		1 5
94	2196	2246	2296	2346	2396	2446	2496	2546	2596	2646		2 10
												3 15
95	2696	2746	2796	2846	2896	2946	2996	3045	3095	3145		4 20
96	3195	3245	3295	3345	3395	3445	3495	3545	3595	3645		5 25
97	3695	3745	3795	3845	3894	3944	3994	4044	4094	4144		6 29
98	4194	4244	4294	4344	4394	4444	4494	4544	4593	4643		7 34
99	4693	4743	4793	4843	4893	4943	4993	5043	5093	5143		8 39
												9 44
N.	0	1	2	3	4	5	6	7	8	9	D	Pts.

N.	0	1	2	3	4	5	6	7	8	9	D	Pro.
8700	9395193	5242	5292	5342	5392	5442	5492	5542	5592	5642		
91	5692	5742	5792	5841	5891	5941	5991	6041	6091	6141		
02	6191	6241	6291	6341	6390	6440	6490	6540	6590	6640		50
03	6690	6740	6790	6840	6889	6939	6989	7039	7089	7139		1 5
04	7189	7239	7289	7339	7388	7438	7488	7538	7588	7638		2 10
05	7688	7738	7788	7837	7887	7937	7987	8037	8087	8137		3 15
06	8187	8237	8286	8336	8386	8436	8486	8536	8586	8636		4 20
07	8685	8735	8785	8835	8885	8935	8985	9035	9084	9134		5 25
08	9184	9234	9284	9334	9384	9434	9483	9583	9583	9633		6 30
09	9683	9733	9783	9833	9882	9932	9982	ō032	ō082	ō132		7 35
8710	9400182	0231	0281	0331	0381	0431	0481	0531	0580	0630		8 40
11	0680	0730	0780	0830	0880	0929	0979	1029	1079	1129		9 45
12	1179	1229	1278	1328	1378	1428	1478	1528	1577	1627		
13	1677	1727	1777	1827	1877	1926	1976	2026	2076	2126		
14	2176	2225	2275	2325	2375	2425	2475	2524	2574	2624		
15	2674	2724	2774	2823	2873	2923	2973	3023	3073	3122		
16	3172	3222	3272	3322	3372	3421	3471	3521	3571	3621		
17	3670	3720	3770	3820	3870	3920	3969	4019	4069	4119		
18	4169	4218	4268	4318	4368	4418	4468	4517	4567	4617		
19	4667	4717	4766	4816	4866	4916	4966	5015	5065	5115		
8720	5165	5215	5264	5314	5364	5414	5464	5513	5563	5613		
21	5663	5713	5762	5812	5862	5912	5962	6011	6061	6111		
22	6161	6211	6260	6310	6360	6410	6460	6509	6559	6609		
23	6659	6709	6758	6808	6858	6908	6957	7007	7057	7107		
24	7157	7206	7256	7306	7356	7405	7455	7505	7555	7605		
25	7654	7704	7754	7804	7853	7903	7953	8003	8053	8102		
26	8152	8202	8252	8301	8351	8401	8451	8500	8550	8600		
27	8650	8700	8749	8799	8849	8899	8948	8998	9048	9098		
28	9147	9197	9247	9297	9346	9396	9446	9496	9545	9595		
29	9645	9695	9744	9794	9844	9894	9943	9993	ō043	ō093		
8730	9410142	0192	0242	0292	0341	0391	0441	0491	0540	0590		
31	0640	0690	0739	0789	0839	0889	0938	0988	1038	1088		
32	1137	1187	1237	1286	1336	1386	1436	1485	1535	1585		
33	1635	1684	1734	1784	1834	1883	1933	1983	2032	2082		
34	2132	2182	2231	2281	2331	2380	2430	2480	2530	2579		
35	2629	2679	2729	2778	2828	2878	2927	2977	3027	3077		
36	3126	3176	3226	3275	3325	3375	3425	3474	3524	3574		
37	3623	3673	3723	3772	3822	3872	3922	3971	4021	4071		
38	4120	4170	4220	4270	4319	4369	4419	4468	4518	4568		
39	4617	4667	4717	4766	4816	4866	4916	4965	5015	5065		
8740	5114	5164	5214	5263	5313	5363	5412	5462	5512	5562		
41	5611	5661	5711	5760	5810	5860	5909	5959	6009	6058		
42	6108	6158	6207	6257	6307	6356	6406	6456	6505	6555		49
43	6605	6654	6704	6754	6803	6853	6903	6952	7002	7052		1 5
44	7101	7151	7201	7250	7300	7350	7399	7449	7499	7548		2 10
45	7598	7648	7697	7747	7797	7846	7896	7946	7995	8045		3 15
46	8095	8144	8194	8244	8293	8343	8393	8442	8492	8542		4 20
47	8591	8641	8691	8740	8790	8840	8889	8939	8988	9038		5 25
48	9088	9137	9187	9237	9286	9336	9386	9435	9485	9535		6 29
49	9584	9634	9683	9733	9783	9832	9882	9932	9981	ō031		7 34
												8 39
												9 44
N	0	1	2	3	4	5	6	7	8	9	D	Pts.

N.	0	1	2	3	4	5	6	7	8	9	D	Pro.
8750	9420081	0130	0180	0229	0279	0329	0378	0428	0478	0527		
51	0577	0626	0676	0726	0775	0825	0875	0924	0974	1023		50
52	1073	1123	1172	1222	1272	1321	1371	1420	1470	1520		1 5
53	1569	1619	1669	1718	1768	1817	1867	1917	1966	2016		2 10
54	2065	2115	2165	2214	2264	2313	2363	2413	2462	2512		3 15
												4 20
55	2562	2611	2661	2710	2760	2810	2859	2909	2958	3008		5 25
56	3058	3107	3157	3206	3256	3306	3355	3405	3454	3504		6 30
57	3553	3603	3653	3702	3752	3801	3851	3901	3950	4000		7 35
58	4049	4099	4149	4198	4248	4297	4347	4397	4446	4496		8 40
59	4545	4595	4644	4694	4744	4793	4843	4892	4942	4991		9 45
8760	5041	5091	5140	5190	5239	5289	5339	5388	5438	5487		
61	5537	5586	5636	5686	5735	5785	5834	5884	5933	5983		
62	6032	6082	6132	6181	6231	6280	6330	6379	6429	6479		
63	6528	6578	6627	6677	6726	6776	6825	6875	6925	6974		
64	7024	7073	7123	7172	7222	7271	7321	7371	7420	7470		
65	7519	7569	7618	7668	7717	7767	7816	7866	7916	7965		
66	8015	8064	8114	8163	8213	8262	8312	8361	8411	8461		
67	8510	8560	8609	8659	8708	8758	8807	8857	8906	8956		
68	9005	9055	9104	9154	9204	9253	9303	9352	9402	9451		
69	9501	9550	9600	9649	9699	9748	9798	9847	9897	9946		
8770	9996	0̄045	0̄095	0̄144	0̄194	0̄244	0̄293	0̄343	0̄392	0̄442		
71	9430491	0541	0590	0640	0689	0739	0788	0838	0887	0937		
72	0986	1036	1085	1135	1184	1234	1283	1333	1382	1432		
73	1481	1531	1580	1630	1679	1729	1778	1828	1877	1927		
74	1976	2026	2075	2125	2174	2224	2273	2323	2372	2422		
75	2471	2521	2570	2620	2669	2719	2768	2818	2867	2917		
76	2966	3016	3065	3115	3164	3214	3263	3313	3362	3412		
77	3461	3510	3560	3609	3659	3708	3758	3807	3857	3906		
78	3956	4005	4055	4104	4154	4203	4253	4302	4352	4401		
79	4450	4500	4549	4599	4648	4698	4747	4797	4846	4896		
8780	4945	4995	5044	5094	5143	5192	5242	5291	5341	5390		
81	5440	5489	5539	5588	5638	5687	5737	5786	5835	5885		
82	5934	5984	6033	6083	6132	6182	6231	6280	6330	6379		
83	6429	6478	6528	6577	6627	6676	6726	6775	6824	6874		
84	6923	6973	7022	7072	7121	7170	7220	7269	7319	7368		
85	7418	7467	7517	7566	7615	7665	7714	7764	7813	7863		
86	7912	7961	8011	8060	8110	8159	8209	8258	8307	8357		
87	8406	8456	8505	8555	8604	8653	8703	8752	8802	8851		
88	8900	8950	8999	9049	9098	9148	9197	9246	9296	9345		
89	9395	9444	9493	9543	9592	9642	9691	9741	9790	9839		
8790	9889	9938	9988	0̄037	0̄086	0̄136	0̄185	0̄235	0̄284	0̄333		
91	9440383	0432	0482	0531	0580	0630	0679	0729	0778	0827		
92	0877	0926	0976	1025	1074	1124	1173	1223	1272	1321		49
93	1371	1420	1470	1519	1568	1618	1667	1716	1766	1815		1 5
94	1865	1914	1963	2013	2062	2112	2161	2210	2260	2309		2 10
												3 15
95	2358	2408	2457	2507	2556	2605	2655	2704	2753	2803		4 20
96	2852	2902	2951	3000	3050	3099	3148	3198	3247	3297		5 25
97	3346	3395	3445	3494	3543	3593	3642	3691	3741	3790		6 29
98	3840	3889	3938	3988	4037	4086	4136	4185	4234	4284		7 34
99	4333	4383	4432	4481	4531	4580	4629	4679	4728	4777		8 39
												9 44
N.	0	1	2	3	4	5	6	7	8	9	D	Pts.

M

N.	0	1	2	3	4	5	6	7	8	9	D	Pro.
8800	9444827	4876	4925	4975	5024	5073	5123	5172	5222	5271		
01	5320	5370	5419	5468	5518	5567	5616	5666	5715	5764		50
02	5814	5863	5912	5962	6011	6060	6110	6159	6208	6258		1 5
03	6307	6356	6406	6455	6504	6554	6603	6652	6702	6751		2 10
04	6800	6850	6899	6948	6998	7047	7096	7146	7195	7244		3 15
												4 20
05	7294	7343	7392	7442	7491	7540	7590	7639	7688	7737		5 25
06	7787	7836	7885	7935	7984	8033	8083	8132	8181	8231		6 30
07	8280	8329	8379	8428	8477	8527	8576	8625	8674	8724		7 35
08	8773	8822	8872	8921	8970	9020	9069	9118	9167	9217		8 40
09	9266	9315	9365	9414	9463	9513	9562	9611	9660	9710		9 45
8810	9759	9808	9858	9907	9956	0̄006	0̄055	0̄104	0̄153	0̄203		
11	9450252	0301	0351	0400	0449	0498	0548	0597	0646	0696		
12	0745	0794	0843	0893	0942	0991	1041	1090	1139	1188		
13	1238	1287	1336	1386	1435	1484	1533	1583	1632	1681		
14	1730	1780	1829	1878	1928	1977	2026	2075	2125	2174		
15	2223	2272	2322	2371	2420	2469	2519	2568	2617	2667		
16	2716	2765	2814	2864	2913	2962	3011	3061	3110	3159		
17	3208	3258	3307	3356	3405	3455	3504	3553	3602	3652		
18	3701	3750	3799	3849	3898	3947	3996	4046	4095	4144		
19	4193	4243	4292	4341	4390	4440	4489	4538	4587	4637		
20	4686	4735	4784	4834	4883	4932	4981	5031	5080	5129		
21	5178	5227	5277	5326	5375	5424	5474	5523	5572	5621		
22	5671	5720	5769	5818	5867	5917	5966	6015	6064	6114		
23	6163	6212	6261	6310	6360	6409	6458	6507	6557	6606		
24	6655	6704	6753	6803	6852	6901	6950	7000	7049	7098		
25	7147	7196	7246	7295	7344	7393	7442	7492	7541	7590		
26	7639	7688	7738	7787	7836	7885	7934	7984	8033	8082		
27	8131	8180	8230	8279	8328	8377	8426	8476	8525	8574		
28	8623	8672	8722	8771	8820	8869	8918	8968	9017	9066		
29	9115	9164	9214	9263	9312	9361	9410	9459	9509	9558		
8830	9607	9656	9705	9755	9804	9853	9902	9951	0̄000	0̄050		
31	9460099	0148	0197	0246	0296	0345	0394	0443	0492	0541		
32	0591	0640	0689	0738	0787	0836	0886	0935	0984	1033		
33	1082	1131	1181	1230	1279	1328	1377	1426	1476	1525		
34	1574	1623	1672	1721	1771	1820	1869	1918	1967	2016		
35	2066	2115	2164	2213	2262	2311	2360	2410	2459	2508		
36	2557	2606	2655	2705	2754	2803	2852	2901	2950	2999		
37	3049	3098	3147	3196	3245	3294	3343	3393	3442	3491		
38	3540	3589	3638	3687	3737	3786	3835	3884	3933	3982		
39	4031	4080	4130	4179	4228	4277	4326	4375	4424	4474		
8840	4523	4572	4621	4670	4719	4768	4817	4867	4916	4965		
41	5014	5063	5112	5161	5210	5260	5309	5358	5407	5456		
42	5505	5554	5603	5652	5702	5751	5800	5849	5898	5947		49
43	5996	6045	6094	6144	6193	6242	6291	6340	6389	6438		1 5
44	6487	6536	6586	6635	6684	6733	6782	6831	6880	6929		2 10
45	6978	7027	7077	7126	7.175	7224	7273	7322	7371	7420		3 15
46	7469	7518	7568	7617	7666	7715	7764	7813	7862	7911		4 20
47	7960	8009	8058	8108	8157	8206	8255	8304	8353	8402		5 25
48	8451	8500	8549	8598	8647	8697	8746	8795	8844	8893		6 29
49	8942	8991	9040	9089	9138	9187	9236	9285	9335	9384		7 34
												8 39
												9 44
N.	0	1	2	3	4	5	6	7	8	9	D	Pts.

N.	0	1	2	3	4	5	6	7	8	9	D	Pro.
8850	9469433	9482	9531	9580	9629	9678	9727	9776	9825	9874		
51	9923	9972	ō022	ō071	ō120	ō169	ō218	ō267	ō316	ō365		49
52	9470414	0463	0512	0561	0610	0659	0708	0757	0807	0856		1 5
53	0905	0954	1003	1052	1101	1150	1199	1248	1297	1346		2 10
54	1395	1444	1493	1542	1591	1640	1689	1739	1788	1837		3 15
												4 20
55	1886	1935	1984	2033	2082	2131	2180	2229	2278	2327		5 25
56	2376	2425	2474	2523	2572	2621	2670	2719	2768	2817		6 29
57	2866	2915	2965	3014	3063	3112	3161	3210	3259	3308		7 34
58	3357	3406	3455	3504	3553	3602	3651	3700	3749	3798		8 39
59	3847	3896	3945	3994	4043	4092	4141	4190	4239	4288		9 44
8860	4337	4386	4435	4484	4533	4582	4631	4680	4729	4778		
61	4827	4876	4925	4974	5023	5072	5121	5170	5219	5268		
62	5317	5366	5415	5464	5513	5562	5611	5660	5709	5758	49	
63	5807	5856	5905	5954	6003	6052	6101	6150	6199	6248		
64	6297	6346	6395	6444	6493	6542	6591	6640	6689	6738		
65	6787	6836	6885	6934	6983	7032	7081	7130	7179	7228		
66	7277	7326	7375	7424	7473	7522	7571	7620	7669	7718		
67	7767	7816	7865	7914	7963	8012	8061	8110	8159	8208		
68	8257	8306	8355	8404	8453	8502	8551	8600	8649	8698		
69	8747	8796	8844	8893	8942	8991	9040	9089	9138	9187		
8870	9236	9285	9334	9383	9432	9481	9530	9579	9628	9677		
71	9726	9775	9824	9873	9922	9971	ō020	ō068	ō117	ō166		
72	9480215	0264	0313	0362	0411	0460	0509	0558	0607	0656		
73	0705	0754	0803	0852	0901	0950	0998	1047	1096	1145		
74	1194	1243	1292	1341	1390	1439	1488	1537	1586	1635		
75	1684	1733	1781	1830	1879	1928	1977	2026	2075	2124		
76	2173	2222	2271	2320	2369	2418	2467	2515	2564	2613		
77	2662	2711	2760	2809	2858	2907	2956	3005	3054	3102		
78	3151	3200	3249	3298	3347	3396	3445	3494	3543	3592		
79	3641	3689	3738	3787	3836	3885	3934	3983	4032	4081		
8880	4130	4179	4227	4276	4325	4374	4423	4472	4521	4570		
81	4619	4668	4717	4765	4814	4863	4912	4961	5010	5059		
82	5108	5157	5205	5254	5303	5352	5401	5450	5499	5548		
83	5597	5646	5694	5743	5792	5841	5890	5939	5988	6037		
84	6085	6134	6183	6232	6281	6330	6379	6428	6477	6525		
85	6574	6623	6672	6721	6770	6819	6868	6916	6965	7014		
86	7063	7112	7161	7210	7259	7307	7356	7405	7454	7503		
87	7552	7601	7650	7698	7747	7796	7845	7894	7943	7992		
88	8040	8089	8138	8187	8236	8285	8334	8382	8431	8480		
89	8529	8578	8627	8676	8724	8773	8822	8871	8920	8969		
8890	9018	9066	9115	9164	9213	9262	9311	9360	9408	9457		
91	9506	9555	9604	9653	9701	9750	9799	9848	9897	9946		
92	9995	ō043	ō092	ō141	ō190	ō239	ō288	ō336	ō385	ō434		48
93	9490483	0532	0581	0629	0678	0727	0776	0825	0874	0922		1 5
94	0971	1020	1069	1118	1167	1215	1264	1313	1362	1411		2 10
												3 14
95	1460	1508	1557	1606	1655	1704	1752	1801	1850	1899		4 19
96	1948	1997	2045	2094	2143	2192	2241	2289	2338	2387		5 24
97	2436	2485	2534	2582	2631	2680	2729	2778	2826	2875		6 29
98	2924	2973	3022	3070	3119	3168	3217	3266	3314	3363		7 34
99	3412	3461	3510	3558	3607	3656	3705	3754	3802	3851		8 38
												9 43
N.	0	1	2	3	4	5	6	7	8	9	D	Pts.

N.	0	1	2	3	4	5	6	7	8	9	D	Pro.
8900	9493900	3949	3998	4046	4095	4144	4193	4242	4290	4339		
01	4388	4437	4486	4534	4583	4632	4681	4730	4778	4827		49
02	4876	4925	4973	5022	5071	5120	5169	5217	5266	5315		1 5
03	5364	5413	5461	5510	5559	5608	5656	5705	5754	5803		2 10
04	5852	5900	5949	5998	6047	6095	6144	6193	6242	6290		3 15
												4 20
05	6339	6388	6437	6486	6534	6583	6632	6681	6729	6778		5 25
06	6827	6876	6924	6973	7022	7071	7119	7168	7217	7266		6 29
07	7315	7363	7412	7461	7510	7558	7607	7656	7705	7753		7 34
08	7802	7851	7900	7948	7997	8046	8095	8143	8192	8241		8 39
09	8290	8338	8387	8436	8485	8533	8582	8631	8680	8728		9 44
8910	8777	8826	8875	8923	8972	9021	9069	9118	9167	9216		
11	9264	9313	9362	9411	9459	9508	9557	9606	9654	9703		
12	9752	9801	9849	9898	9947	9995	$\bar{0}$044	$\bar{0}$093	$\bar{0}$142	$\bar{0}$190		
13	9500239	0288	0337	0385	0434	0483	0531	0580	0629	0678		
14	0726	0775	0824	0872	0921	0970	1019	1067	1116	1165		
15	1213	1262	1311	1360	1408	1457	1506	1554	1603	1652		
16	1701	1749	1798	1847	1895	1944	1993	2042	2090	2139		
17	2188	2236	2285	2334	2382	2431	2480	2529	2577	2626		
18	2675	2723	2772	2821	2869	2918	2967	3016	3064	3113		
19	3162	3210	3259	3308	3356	3405	3454	3502	3551	3600		
8920	3649	3697	3746	3795	3843	3892	3941	3989	4038	4087		
21	4135	4184	4233	4281	4330	4379	4427	4476	4525	4574		
22	4622	4671	4720	4768	4817	4866	4914	4963	5012	5060		
23	5109	5158	5206	5255	5304	5352	5401	5450	5498	5547		
24	5596	5644	5693	5742	5790	5839	5888	5936	5985	6034		
25	6082	6131	6180	6228	6277	6326	6374	6423	6472	6520		
26	6569	6617	6666	6715	6763	6812	6861	6909	6958	7007		
27	7055	7104	7153	7201	7250	7299	7347	7396	7445	7493		
28	7542	7590	7639	7688	7736	7785	7834	7882	7931	7980		
29	8028	8077	8126	8174	8223	8271	8320	8369	8417	8466		
8930	8515	8563	8612	8660	8709	8758	8806	8855	8904	8952		
31	9001	9050	9098	9147	9195	9244	9293	9341	9390	9439		
32	9487	9536	9584	9633	9682	9730	9779	9827	9876	9925		
33	9973	$\bar{0}$022	$\bar{0}$071	$\bar{0}$119	$\bar{0}$168	$\bar{0}$216	$\bar{0}$265	$\bar{0}$314	$\bar{0}$362	$\bar{0}$411		
34	9510459	0508	0557	0605	0654	0703	0751	0800	0848	0897		
35	0946	0994	1043	1091	1140	1189	1237	1286	1334	1383		
36	1432	1480	1529	1577	1626	1675	1723	1772	1820	1869		
37	1918	1966	2015	2063	2112	2161	2209	2258	2306	2355		
38	2404	2452	2501	2549	2598	2646	2695	2744	2792	2841		
39	2889	2938	2987	3035	3084	3132	3181	3229	3278	3327		
8940	3375	3424	3472	3521	3569	3618	3667	3715	3764	3812		
41	3861	3910	3958	4007	4055	4104	4152	4201	4250	4298		
42	4347	4395	4444	4492	4541	4589	4638	4687	4735	4784		48
43	4832	4881	4929	4978	5027	5075	5124	5172	5221	5269		1 5
44	5318	5366	5415	5464	5512	5561	5609	5658	5706	5755		2 10
												3 14
45	5803	5852	5901	5949	5996	6046	6095	6143	6192	6240		4 19
46	6289	6337	6386	6435	6483	6532	6580	6629	6677	6726		5 24
47	6774	6823	6871	6920	6969	7017	7066	7114	7163	7211		6 29
48	7260	7308	7357	7405	7454	7502	7551	7599	7648	7697		7 34
49	7745	7794	7842	7891	7939	7988	8036	8085	8133	8182		8 38
												9 43
N.	0	1	2	3	4	5	6	7	8	9	D	Pts.

N.	0	1	2	3	4	5	6	7	8	9	D	Pro.
8950	9518230	8279	8327	8376	8424	8473	8521	8570	8619	8667		
51	8716	8764	8813	8861	8910	8958	9007	9055	9104	9152		49
52	9201	9249	9298	9346	9395	9443	9492	9540	9589	9637		1 5
53	9686	9734	9783	9831	9880	9928	9977	0̄025	0̄074	0̄122		2 10
54	9520171	0219	0268	0316	0365	0413	0462	0510	0559	0607		3 15
												4 20
55	0656	0704	0753	0801	0850	0898	0947	0995	1044	1092		5 25
56	1141	1189	1238	1286	1335	1383	1432	1480	1529	1577		6 29
57	1626	1674	1723	1771	1820	1868	1917	1965	2014	2062		7 34
58	2111	2159	2208	2256	2305	2353	2401	2450	2498	2547		8 39
59	2595	2644	2692	2741	2789	2838	2886	2935	2983	3032		9 44
8960	3080	3129	3177	3226	3274	3322	3371	3419	3468	3516		
61	3565	3613	3662	3710	3759	3807	3856	3904	3952	4001		
62	4049	4098	4146	4195	4243	4292	4340	4389	4437	4486		
63	4534	4582	4631	4679	4728	4776	4825	4873	4922	4970		
64	5018	5067	5115	5164	5212	5261	5309	5358	5406	5454		
65	5503	5551	5600	5648	5697	5745	5794	5842	5890	5939		
66	5987	6036	6084	6133	6181	6230	6278	6326	6375	6423		
67	6472	6520	6569	6617	6665	6714	6762	6811	6859	6908		
68	6956	7004	7053	7101	7150	7198	7247	7295	7343	7392		
69	7440	7489	7537	7586	7634	7682	7731	7779	7828	7876		
8970	7924	7973	8021	8070	8118	8167	8215	8263	8312	8360		
71	8409	8457	8505	8554	8602	8651	8699	8747	8796	8844		
72	8893	8941	8989	9038	9086	9135	9183	9231	9280	9328		
73	9377	9425	9473	9522	9570	9619	9667	9715	9764	9812		
74	9861	9909	9957	0̄006	0̄054	0̄103	0̄151	0̄199	0̄248	0̄296		
75	9530345	0393	0441	0490	0538	0587	0635	0683	0732	0780		
76	0828	0877	0925	0974	1022	1070	1119	1167	1215	1264		
77	1312	1361	1409	1457	1506	1554	1603	1651	1699	1748		
78	1796	1844	1893	1941	1989	2038	2086	2135	2183	2231		
79	2280	2328	2376	2425	2473	2522	2570	2618	2667	2715		
8980	2763	2812	2860	2908	2957	3005	3054	3102	3150	3199		
81	3247	3295	3344	3392	3440	3489	3537	3585	3634	3682		
82	3731	3779	3827	3876	3924	3972	4021	4069	4117	4166		
83	4214	4262	4311	4359	4407	4456	4504	4552	4601	4649		
84	4697	4746	4794	4842	4891	4939	4987	5036	5084	5132		
85	5181	5229	5277	5326	5374	5422	5471	5519	5567	5616		
86	5664	5712	5761	5809	5857	5906	5954	6002	6051	6099		
87	6147	6196	6244	6292	6341	6389	6437	6486	6534	6582		
88	6631	6679	6727	6776	6824	6872	6921	6969	7017	7065		
89	7114	7162	7210	7259	7307	7355	7404	7452	7500	7549		
8990	7597	7645	7694	7742	7790	7838	7887	7935	7983	8032		
91	8080	8128	8177	8225	8273	8321	8370	8418	8466	8515		
92	8563	8611	8660	8708	8756	8804	8853	8901	8949	8998		48
93	9046	9094	9143	9191	9239	9287	9336	9384	9432	9481		1 5
94	9529	9577	9625	9674	9722	9770	9819	9867	9915	9963		2 10
												3 14
95	9540012	0060	0108	0157	0205	0253	0301	0350	0398	0446		4 19
96	0494	0543	0591	0639	0688	0736	0784	0832	0881	0929		5 24
97	0977	1025	1074	1122	1170	1219	1267	1315	1363	1412		6 29
98	1460	1508	1556	1605	1653	1701	1749	1798	1846	1894		7 34
99	1943	1991	2039	2087	2136	2184	2232	2280	2329	2377		8 38
												9 43
N.	0	1	2	3	4	5	6	7	8	9	D	Pts.

N.	0	1	2	3	4	5	6	7	8	9	D	Pro.
9000	9542425	2473	2522	2570	2618	2666	2715	2763	2811	2859		
01	2908	2956	3004	3052	3101	3149	3197	3245	3294	3342		49
02	3390	3438	3487	3535	3583	3631	3680	3728	3776	3824		1 ·5
03	3873	3921	3969	4017	4065	4114	4162	4210	4258	4307		2 10
04	4355	4403	4451	4500	4548	4596	4644	4692	4741	4789		3 15
												4 20
05	4837	4885	4934	4982	5030	5078	5127	5175	5223	5271		5 25
06	5319	5368	5416	5464	5512	5561	5609	5657	5705	5753		6 29
07	5802	5850	5898	5946	5994	6043	6091	6139	6187	6236		7 34
08	6284	6332	6380	6428	6477	6525	6573	6621	6669	6718		8 39
09	6766	6814	6862	6910	6959	7007	7055	7103	7152	7200		9 44
9010	7248	7296	7344	7393	7441	7489	7537	7585	7634	7682		
11	7730	7778	7826	7874	7923	7971	8019	8067	8115	8164		
12	8212	8260	8308	8356	8405	8453	8501	8549	8597	8646		
13	8694	8742	8790	8838	8886	8935	8983	9031	9079	9127		
14	9176	9224	9272	9320	9368	9416	9465	9513	9561	9609		
15	9657	9705	9754	9802	9850	9898	9946	9995	ō043	ō091		
16	9550139	0187	0235	0284	0332	0380	0428	0476	0524	0573		
17	0621	0669	0717	0765	0813	0862	0910	0958	1006	1054		
18	1102	1150	1199	1247	1295	1343	1391	1439	1488	1536		
19	1584	1632	1680	1728	1776	1825	1873	1921	1969	2017		
9020	2065	2114	2162	2210	2258	2306	2354	2402	2451	2499		
21	2547	2595	2643	2691	2739	2798	2836	2884	2932	2980		
22	3028	3076	3125	3173	3221	3269	3317	3365	3413	3461		
23	3510	3558	3606	3654	3702	3750	3798	3846	3895	3943		
24	3991	4039	4087	4135	4183	4231	4280	4328	4376	4424		
25	4472	4520	4568	4616	4665	4713	4761	4809	4857	4905		
26	4953	5001	5050	5098	5146	5194	5242	5290	5338	5386		
27	5434	5483	5531	5579	5627	5675	5723	5771	5819	5867		
28	5916	5964	6012	6060	6108	6156	6204	6252	6300	6348		
29	6397	6445	6493	6541	6589	6637	6685	6733	6781	6829		
9030	6878	6926	6974	7022	7070	7118	7166	7214	7262	7310		
31	7358	7407	7455	7503	7551	7599	7647	7695	7743	7791		
32	7839	7887	7935	7984	8032	8080	8128	8176	8224	8272		
33	8320	8368	8416	8464	8512	8560	8609	8657	8705	8753		
34	8801	8849	8897	8945	8993	9041	9089	9137	9185	9234		
35	9282	9330	9378	9426	9474	9522	9570	9618	9666	9714		
36	9762	9810	9858	9906	9954	ō003	ō051	ō099	ō147	ō195		
37	9560243	0291	0339	0387	0435	0483	0531	0579	0627	0675		
38	0723	0771	0819	0868	0916	0964	1012	1060	1108	1156		
39	1204	1252	1300	1348	1396	1444	1492	1540	1588	1636		
9040	1684	1732	1780	1828	1876	1925	1973	2021	2069	2117		
41	2165	2213	2261	2309	2357	2405	2453	2501	2549	2597		48
42	2645	2693	2741	2789	2837	2885	2933	2981	3029	3077		1 5
43	3125	3173	3221	3269	3317	3365	3413	3461	3509	3558		2 10
44	3606	3654	3702	3750	3798	3846	3894	3942	3990	4038		3 14
45	4086	4134	4182	4230	4278	4326	4374	4422	4470	4518		4 19
46	4566	4614	4662	4710	4758	4806	4854	4902	4950	4998		5 24
47	5046	5094	5142	5190	5238	5286	5334	5382	5430	5478	48	6 29
48	5526	5574	5622	5670	5718	5766	5814	5862	5910	5958		7 34
49	6006	6054	6102	6150	6198	6246	6294	6342	6390	6438		8 38
												9 43
N.	0	1	2	3	4	5	6	7	8	9	D	Pts.

N.	0	1	2	3	4	5	6	7	8	9	D	Pro.
9050	9566486	6534	6582	6630	6678	6726	6774	6822	6870	6918		
51	6966	7014	7062	7110	7158	7206	7254	7302	7349	7397		48
52	7445	7493	7541	7589	7637	7685	7733	7781	7829	7877		1 5
53	7925	7973	8021	8069	8117	8165	8213	8261	8309	8357		2 10
54	8405	8453	8501	8549	8597	8645	8693	8741	8789	8837		3 14
												4 19
35	8885	8933	8980	9028	9076	9124	9172	9220	9268	9316		5 24
56	9364	9412	9460	9508	9556	9604	9652	9700	9748	9796		6 29
57	9844	9892	9940	9988	ō035	ō083	ō131	ō179	ō227	ō275		7 34
58	9570323	0371	0419	0467	0515	0563	0611	0659	0707	0755		8 38
59	0803	0851	0898	0946	0994	1042	1090	1138	1186	1234		9 43
9060	1282	1330	1378	1426	1474	1522	1570	1618	1665	1713		
61	1761	1809	1857	1905	1953	2001	2049	2097	2145	2193		
62	2241	2289	2336	2384	2432	2480	2528	2576	2624	2672		
63	2720	2768	2816	2864	2911	2959	3007	3055	3103	3151		
64	3199	3247	3295	3343	3391	3439	3486	3534	3582	3630		
65	3678	3726	3774	3822	3870	3918	3966	4013	4061	4109		
66	4157	4205	4253	4301	4349	4397	4445	4492	4540	4588		
67	4636	4684	4732	4780	4828	4876	4924	4971	5019	5067		
68	5115	5163	5211	5259	5307	5355	5402	5450	5498	5546		
69	5594	5642	5690	5738	5786	5833	5881	5929	5977	6025		
9070	6073	6121	6169	6217	6264	6312	6360	6408	6456	6504		
71	6552	6600	6647	6695	6743	6791	6839	6887	6935	6983		
72	7030	7078	7126	7174	7222	7270	7318	7366	7413	7461		
73	7509	7557	7605	7653	7701	7748	7796	7844	7892	7940		
74	7988	8036	8083	8131	8179	8227	8275	8323	8371	8418		
75	8466	8514	8562	8610	8658	8706	8753	8801	8849	8897		
76	8945	8993	9041	9088	9136	9184	9232	9280	9328	9376		
77	9423	9471	9519	9567	9615	9663	9710	9758	9806	9854		
78	9902	9950	9997	ō045	ō093	ō141	ō189	ō237	ō284	ō332		
79	9580380	0428	0476	0524	0571	0619	0667	0715	0763	0811		
9080	0858	0906	0954	1002	1050	1098	1145	1193	1241	1289		
81	1337	1385	1432	1480	1528	1576	1624	1672	1719	1767		
82	1815	1863	1911	1958	2006	2054	2102	2150	2198	2245		
83	2293	2341	2389	2437	2484	2532	2580	2628	2676	2723		
84	2771	2819	2867	2915	2962	3010	3058	3106	3154	3202		
85	3249	3297	3345	3393	3441	3488	3536	3584	3632	3680		
86	3727	3775	3823	3871	3919	3966	4014	4062	4110	4157		
87	4205	4253	4301	4349	4396	4444	4492	4540	4588	4635		
88	4683	4731	4779	4827	4874	4922	4970	5018	5065	5113		
89	5161	5209	5257	5304	5352	5400	5448	5495	5543	5591		
9090	5639	5687	5734	5782	5830	5878	5925	5973	6021	6069		
91	6117	6164	6212	6260	6308	6355	6403	6451	6499	6547		
92	6594	6642	6690	6738	6785	6833	6881	6929	6976	7024		47
93	7072	7120	7167	7215	7263	7311	7358	7406	7454	7502		1 5
94	7549	7597	7645	7693	7741	7788	7836	7884	7932	7979		2 9
												3 14
95	8027	8075	8123	8170	8218	8266	8314	8361	8409	8457		4 19
96	8505	8552	8600	8648	8695	8743	8791	8839	8886	8934		5 24
97	8982	9030	9077	9125	9173	9221	9268	9316	9364	9412		6 28
98	9459	9507	9555	9603	9650	9698	9746	9793	9841	9889		7 33
99	9937	9984	ō032	ō080	ō128	ō175	ō223	ō271	ō318	ō366		8 38
												9 42
N.	0	1	2	3	4	5	6	7	8	9	D	Pts.

N.	0	1	2	3	4	5	6	7	8	9	D	Pro.
9100	9590414	0462	0509	0557	0605	0653	0700	0748	0796	0843		
01	0891	0939	0987	1034	1082	1130	1177	1225	1273	1321		48
02	1368	1416	1464	1511	1559	1607	1655	1702	1750	1798		1 5
03	1845	1893	1941	1989	2036	2084	2132	2179	2227	2275		2 10
04	2322	2370	2418	2466	2513	2561	2609	2656	2704	2752		3 14
												4 19
05	2800	2847	2895	2943	2990	3038	3086	3133	3181	3229		5 24
06	3276	3324	3372	3420	3467	3515	3563	3610	365b	3706		6 29
07	3753	3801	3849	3896	3944	3992	4039	4087	4135	4183		7 34
08	4230	4278	4326	4373	4421	4469	4516	4564	4612	4659		8 38
09	4707	4755	4802	4850	4898	4945	4993	5041	5088	5136		9 43
9110	5184	5231	5279	5327	5374	5422	5470	5517	5565	5613		
11	5660	5708	5756	5803	5851	5899	5946	5994	6042	6089		
12	6137	6185	6232	6280	6328	6375	6423	6471	6518	6566		
13	6614	6661	6709	6757	6804	6852	6900	6947	6995	7043		
14	7090	7138	7186	7233	7281	7328	7376	7424	7471	7519		
15	7567	7614	7662	7710	7757	7805	7853	7900	7948	7996		
16	8043	8091	8138	8186	8234	8281	8329	8377	8424	8472		
17	8520	8567	8615	8662	8710	8758	8805	8853	8901	8948		
18	8996	9044	9091	9139	9186	9234	9282	9329	9377	9425		
19	9472	9520	9567	9615	9663	9710	9758	9806	9853	9901		
9120	9948	9996	0̄044	0̄091	0̄139	0̄186	0̄234	0̄282	0̄329	0̄377		
21	9600425	0472	0520	0567	0615	0663	0710	0758	0805	0853		
22	0901	0948	0996	1044	1091	1139	1186	1234	1282	1329		
23	1377	1424	1472	1520	1567	1615	1662	1710	1758	1805		
24	1853	1900	1948	1996	2043	2091	2138	2186	2234	2281		
25	2329	2376	2424	2472	2519	2567	2614	2662	2709	2757		
26	2805	2852	2900	2947	2995	3043	3090	3138	3185	3233		
27	3281	3328	3376	3423	3471	3518	3566	3614	3661	3709		
28	3756	3804	3851	3899	3947	3994	4042	4089	4137	4184		
29	4232	4280	4327	4375	4422	4470	4517	4565	4613	4660		
9130	4708	4755	4803	4850	4898	4946	4993	5041	5088	5136		
31	5183	5231	5279	5326	5374	5421	5469	5516	5564	5611		
32	5659	5707	5754	5802	5849	5897	5944	5992	6039	6087		
33	6135	6182	6230	6277	6325	6372	6420	6467	6515	6563		
34	6610	6658	6705	6753	6800	6848	6895	6943	6990	7038		
35	7086	7133	7181	7228	7276	7323	7371	7418	7466	7513		
36	7561	7608	7656	7704	7751	7799	7846	7894	7941	7989		
37	8036	8084	8131	8179	8226	8274	8321	8369	8416	8464		
38	8512	8559	8607	8654	8702	8749	8797	8844	8892	8939		
39	8987	9034	9082	9129	9177	9224	9272	9319	9367	9414		
9140	9462	9509	9557	9605	9652	9700	9747	9795	9842	9890		
41	9937	9985	0̄032	0̄080	0̄127	0̄175	0̄222	0̄270	0̄317	0̄365		
42	9610412	0460	0507	0555	0602	0650	0697	0745	0792	0840		47
43	0887	0935	0982	1030	1077	1125	1172	1220	1267	1315		1 5
44	1362	1410	1457	1505	1552	1600	1647	1695	1742	1790		2 9
												3 14
45	1837	1885	1932	1980	2027	2075	2122	2170	2217	2264		4 19
46	2312	2359	2407	2454	2502	2549	2597	2644	2692	2739		5 24
47	2787	2834	2882	2929	2977	3024	3072	3119	3167	3214		6 28
48	3262	3309	3357	3404	3451	3499	3546	3594	3641	3689		7 33
49	3736	3784	3831	3879	3926	3974	4021	4069	4116	4163		8 38
												9 42
N.	0	1	2	3	4	5	6	7	8	9	D	Pts.

N.	0	1	2	3	4	5	6	7	8	9	D
9150	9614211	4258	4306	4353	4401	4448	4496	4543	4591	4638	
51	4686	4733	4780	4828	4875	4923	4970	5018	5065	5113	
52	5160	5208	5255	5302	5350	5397	5445	5492	5540	5587	
53	5635	5682	5730	5777	5824	5872	5919	5967	6014	6062	
54	6109	6157	6204	6251	6299	6346	6394	6441	6489	6536	
55	6583	6631	6678	6726	6773	6821	6868	6916	6963	7010	
56	7058	7105	7153	7200	7248	7295	7342	7390	7437	7485	
57	7532	7580	7627	7674	7722	7769	7817	7864	7912	7959	
58	8006	8054	8101	8149	8196	8243	8291	8338	8386	8433	
59	8481	8528	8575	8623	8670	8718	8765	8812	8860	8907	
9160	8955	9002	9050	9097	9144	9192	9239	9287	9334	9381	
61	9429	9476	9524	9571	9618	9666	9713	9761	9608	9855	
62	9903	9950	9998	ō045	ō092	ō140	ō187	ō235	ō282	ō329	
63	9620377	0424	0472	0519	0566	0614	0661	0709	0756	0803	
64	0851	0898	0946	0993	1040	1088	1135	1183	1230	1277	
65	1325	1372	1419	1467	1514	1562	1609	1656	1704	1751	
66	1799	1846	1893	1941	1988	2035	2083	2130	2178	2225	
67	2272	2320	2367	2414	2462	2509	2557	2604	2651	2699	
68	2746	2793	2841	2888	2936	2983	3030	3078	3125	3172	
69	3220	3267	3314	3362	3409	3457	3504	3551	3599	3646	
9170	3693	3741	3788	3835	3883	3930	3978	4025	4072	4120	
71	4167	4214	4262	4309	4356	4404	4451	4498	4546	4593	
72	4640	4688	4735	4783	4830	4877	4925	4972	5019	5067	
73	5114	5161	5209	5256	5303	5351	5398	5445	5493	5540	
74	5587	5635	5682	5729	5777	5824	5871	5919	5966	6013	
75	6061	6108	6155	6203	6250	6297	6345	6392	6439	6487	
76	6534	6581	6629	6676	6723	6771	6818	6865	6913	6960	
77	7007	7055	7102	7149	7197	7244	7291	7339	7386	7433	
78	7481	7528	7575	7622	7670	7717	7764	7812	7859	7906	
79	7954	8001	8048	8096	8143	8190	8238	8285	8332	8380	
9180	8427	8474	8521	8569	8616	8663	8711	8758	8805	8853	
81	8900	8947	8994	9042	9089	9136	9184	9231	9278	9326	
82	9373	9420	9467	9515	9562	9609	9657	9704	9751	9799	
83	9846	9893	9940	9988	ō035	ō082	ō130	ō177	ō224	ō271	
84	9630319	0366	0413	0461	0508	0555	0602	0650	0697	0744	
85	0792	0839	0886	0933	0981	1028	1075	1123	1170	1217	
86	1264	1312	1359	1406	1454	1501	1548	1595	1643	1690	
87	1737	1784	1832	1879	1926	1974	2021	2068	2115	2163	
88	2210	2257	2304	2352	2399	2446	2493	2541	2588	2635	
89	2683	2730	2777	2824	2872	2919	2966	3013	3061	3108	
9190	3155	3202	3250	3297	3344	3391	3439	3486	3533	3580	
91	3628	3675	3722	3769	3817	3864	3911	3958	4006	4053	
92	4100	4147	4195	4242	4289	4336	4384	4431	4478	4525	
93	4573	4620	4667	4714	4762	4809	4856	4903	4951	4998	
94	5045	5092	5139	5187	5234	5281	5328	5376	5423	5470	
95	5517	5565	5612	5659	5706	5753	5801	5848	5895	5942	
96	5990	6037	6084	6131	6179	6226	6273	6320	6367	6415	
97	6462	6509	6556	6604	6651	6698	6745	6792	6840	6887	
98	6934	6981	7028	7076	7123	7170	7217	7265	7312	7359	
99	7406	7453	7501	7548	7595	7642	7689	7737	7784	7831	
N	0	1	2	3	4	5	6	7	8	9	D

Pro.

48
1	5
2	10
3	14
4	19
5	24
6	29
7	34
8	38
9	43

47
1	5
2	9
3	14
4	19
5	24
6	28
7	33
8	38
9	42

Pts.

N.	0	1	2	3	4	5	6	7	8	9	D
9200	9637878	7925	7973	8020	8067	8114	8161	8209	8256	8303	
01	8350	8398	8445	8492	8539	8586	8634	8681	8728	8775	
02	8822	8869	8917	8964	9011	9058	9105	9153	9200	9247	
03	9294	9341	9389	9436	9483	9530	9577	9625	9672	9719	
04	9766	9813	9860	9908	9955	ō002	ō049	ō096	ō144	ō191	
05	9640238	0285	0332	0379	0427	0474	0521	0568	0615	0663	
06	0710	0757	0804	0851	0898	0946	0993	1040	1087	1134	
07	1181	1229	1276	1323	1370	1417	1464	1512	1559	1606	
08	1653	1700	1747	1795	1842	1889	1936	1983	2030	2078	
09	2125	2172	2219	2266	2313	2361	2408	2455	2502	2549	
9210	2596	2643	2691	2738	2785	2832	2879	2926	2974	3021	
11	3068	3115	3162	3209	3256	3304	3351	3398	3445	3492	
12	3539	3586	3634	3681	3728	3775	3822	3869	3916	3964	
13	4011	4058	4105	4152	4199	4246	4294	4341	4388	4435	
14	4482	4529	4576	4623	4671	4718	4765	4812	4859	4906	
15	4953	5001	5048	5095	5142	5189	5236	5283	5330	5378	
16	5425	5472	5519	5566	5613	5660	5707	5755	5802	5849	
17	5896	5943	5990	6037	6084	6131	6179	6226	6273	6320	
18	6367	6414	6461	6508	6555	6603	6650	6697	6744	6791	
19	6838	6885	6932	6979	7027	7074	7121	7168	7215	7262	
9220	7309	7356	7403	7451	7498	7545	7592	7639	7686	7733	
21	7780	7827	7874	7922	7969	8016	8063	8110	8157	8204	
22	8251	8298	8345	8392	8440	8487	8534	8581	8628	8675	
23	8722	8769	8816	8863	8910	8958	9005	9052	9099	9146	
24	9193	9240	9287	9334	9381	9428	9475	9523	9570	9617	
25	9664	9711	9758	9805	9852	9899	9946	9993	ō040	ō087	
26	9650135	0182	0229	0276	0323	0370	0417	0464	0511	0558	
27	0605	0652	0699	0746	0793	0841	0888	0935	0982	1029	
28	1076	1123	1170	1217	1264	1311	1358	1405	1452	1499	
29	1546	1594	1641	1688	1735	1782	1829	1876	1923	1970	
9230	2017	2064	2111	2158	2205	2252	2299	2346	2393	2440	
31	2488	2535	2582	2629	2676	2723	2770	2817	2864	2911	
32	2958	3005	3052	3099	3146	3193	3240	3287	3334	3381	
33	3428	3475	3522	3569	3617	3664	3711	3758	3805	3852	
34	3899	3946	3993	4040	4087	4134	4181	4228	4275	4322	
35	4369	4416	4463	4510	4557	4604	4651	4698	4745	4792	
36	4839	4886	4933	4980	5027	5074	5121	5168	5215	5262	
37	5309	5356	5403	5450	5497	5545	5592	5639	5686	5733	
38	5780	5827	5874	5921	5968	6015	6062	6109	6156	6203	
39	6250	6297	6344	6391	6438	6485	6532	6579	6626	6673	
9240	6720	6767	6814	6861	6908	6955	7002	7049	7096	7143	47
41	7190	7237	7284	7331	7378	7425	7472	7519	7566	7613	
42	7660	7707	7754	7801	7848	7895	7942	7989	8036	5083	
43	8130	8177	8224	8270	8317	8364	8411	8458	8505	8552	
44	8599	8646	8693	8740	8787	8834	8881	8928	8975	9022	
45	9069	9116	9163	9210	9257	9304	9351	9398	9445	9492	
46	9539	9586	9633	9680	9727	9774	9821	9868	9915	9962	
47	9660009	0056	0103	0149	0196	0243	0290	0337	0384	0431	
48	0478	0525	0572	0619	0666	0713	0760	0807	0854	0901	
49	0948	0995	1042	1089	1136	1183	1230	1276	1323	1370	
N.	0	1	2	3	4	5	6	7	8	9	D

Pro.

48
1	5
2	10
3	14
4	19
5	24
6	29
7	34
8	38
9	43

47
1	5
2	9
3	14
4	19
5	24
6	28
7	33
8	38
9	42

Pts.

N.	0	1	2	3	4	5	6	7	8	9	D	Pro.
9250	9661417	1464	1511	1558	1605	1652	1699	1746	1793	1840		
51	1887	1934	1981	2028	2075	2122	2168	2215	2262	2309		47
52	2356	2403	2450	2497	2544	2591	2638	2685	2732	2779		1|5
53	2826	2873	2919	2966	3013	3060	3107	3154	3201	3248		2|9
54	3295	3342	3389	3436	3483	3530	3577	3623	3670	3717	47	3|14 4|19
55	3764	3811	3858	3905	3952	3999	4046	4093	4140	4187		5|24
56	4233	4280	4327	4374	4421	4468	4515	4562	4609	4656		6|28
57	4703	4750	4796	4843	4890	4937	4984	5031	5078	5125		7|33
58	5172	5219	5266	5312	5359	5406	5453	5500	5547	5594		8|38
59	5641	5688	5735	5782	5828	5875	5922	5969	6016	6063		9|42
9260	6110	6157	6204	6251	6297	6344	6391	6438	6485	6532		
61	6579	6626	6673	6720	6766	6813	6860	6907	6954	7001		
62	7048	7095	7142	7188	7235	7282	7329	7376	7423	7470		
63	7517	7564	7610	7657	7704	7751	7798	7845	7892	7939		
64	7985	8032	8079	8126	8173	8220	8267	8314	8360	8407		
65	8454	8501	8548	8595	8642	8689	8735	8782	8829	8876		
66	8923	8970	9017	9064	9110	9157	9204	9251	9298	9345		
67	9392	9438	9485	9532	9579	9626	9673	9720	9767	9813		
68	9860	9907	9954	0̄001	0̄048	0̄095	0̄141	0̄188	0̄235	0̄282		
69	9670329	0376	0423	0469	0516	0563	0610	0657	0704	0750		
9270	0797	0844	0891	0938	0985	1032	1078	1125	1172	1219		
71	1266	1313	1359	1406	1453	1500	1547	1594	1641	1687		
72	1734	1781	1828	1875	1922	1968	2015	2062	2109	2156		
73	2203	2249	2296	2343	2390	2437	2484	2530	2577	2624		
74	2671	2718	2765	2811	2858	2905	2952	2999	3046	3092		
75	3139	3186	3233	3280	3326	3373	3420	3467	3514	3561		
76	3607	3654	3701	3748	3795	3841	3888	3935	3982	4029		
77	4076	4122	4169	4216	4263	4310	4356	4403	4450	4497		
78	4544	4590	4637	4684	4731	4778	4825	4871	4918	4965		
79	5012	5059	5105	5152	5199	5246	5293	5339	5386	5433		
9280	5480	5527	5573	5620	5667	5714	5761	5807	5854	5901		
81	5948	5995	6041	6088	6135	6182	6228	6275	6322	6369		
82	6416	6462	6509	6556	6603	6650	6696	6743	6790	6837		
83	6884	6930	6977	7024	7071	7117	7164	7211	7258	7305		
84	7351	7398	7445	7492	7538	7585	7632	7679	7726	7772		
85	7819	7866	7913	7959	8006	8053	8100	8146	8193	8240		
86	8287	8334	8380	8427	8474	8521	8567	8614	8661	8708		
87	8754	8801	8848	8895	8942	8988	9035	9082	9129	9175		
88	9222	9269	9316	9362	9409	9456	9503	9549	9596	9643		
89	9690	9736	9783	9830	9877	9923	9970	0̄017	0̄064	0̄110		
9290	9650157	0204	0251	0297	0344	0391	0438	0484	0531	0578		
91	0625	0671	0718	0765	0812	0858	0905	0952	0999	1045		
92	1092	1139	1185	1232	1279	1326	1372	1419	1466	1513		46
93	1559	1606	1653	1700	1746	1793	1840	1886	1933	1980		1|5
94	2027	2073	2120	2167	2214	2260	2307	2354	2400	2447		2|9 3|14
95	2494	2541	2587	2634	2681	2728	2774	2821	2868	2914		4|18
96	2961	3008	3055	3101	3148	3195	3241	3288	3335	3382		5|23
97	3428	3475	3522	3568	3615	3662	3709	3755	3802	3849		6|28
98	3895	3942	3989	4036	4082	4129	4176	4222	4269	4316		7|32 8|37
99	4362	4409	4456	4503	4549	4596	4643	4689	4736	4783		9|41
N.	0	1	2	3	4	5	6	7	8	9	D	Pts.

N.	0	1	2	3	4	5	6	7	8	9	D	Pro.
9300	9684829	4876	4923	4970	5016	5063	5110	5156	5203	5250		
01	5296	5343	5390	5437	5483	5530	5577	5623	5670	5717		47
02	5763	5810	5857	5903	5950	5997	6043	6090	6137	6184		1 5
03	6230	6277	6324	6370	6417	6464	6510	6557	6604	6650		2 9
04	6697	6744	6790	6837	6884	6930	6977	7024	7070	7117		3 14
												4 19
05	7164	7210	7257	7304	7350	7397	7444	7490	7537	7584		5 24
06	7630	7677	7724	7770	7817	7864	7910	7957	8004	8050		6 28
07	8097	8144	8190	8237	8284	8330	8377	8424	8470	8517		7 33
08	8564	8610	8657	8704	8750	8797	8844	8890	8937	8984		8 38
09	9030	9077	9124	9170	9217	9264	9310	9357	9404	9450		9 42
9310	9497	9543	9590	9637	9683	9730	9777	9823	9870	9917		
11	9963	0̄010	0̄057	0̄103	0̄150	0̄196	0̄243	0̄290	0̄336	0̄383		
12	9690430	0476	0523	0570	0616	0663	0709	0756	0803	0849		
13	0896	0943	0989	1036	1083	1129	1176	1222	1269	1316		
14	1362	1409	1456	1502	1549	1595	1642	1689	1735	1782		
15	1829	1875	1922	1968	2015	2062	2108	2155	2202	2248		
16	2295	2341	2388	2435	2481	2528	2574	2621	2668	2714		
17	2761	2808	2854	2901	2947	2994	3041	3087	3134	3180		
18	3227	3274	3320	3367	3413	3460	3507	3553	3600	3647		
19	3693	3740	3786	3833	3880	3926	3973	4019	4066	4113		
9320	4159	4206	4252	4299	4346	4392	4439	4485	4532	4578		
21	4625	4672	4718	4765	4811	4858	4905	4951	4998	5044		
22	5091	5138	5184	5231	5277	5324	5371	5417	5464	5510		
23	5557	5603	5650	5697	5743	5790	5836	5883	5929	5976		
24	6023	6069	6116	6162	6209	6256	6302	6349	6395	6442		
25	6488	6535	6582	6628	6675	6721	6768	6814	6861	6908		
26	6954	7001	7047	7094	7140	7187	7234	7280	7327	7373		
27	7420	7466	7513	7559	7606	7653	7699	7746	7792	7839		
28	7885	7932	7978	8025	8072	8118	8165	8211	8258	8304		
29	8351	8397	8444	8491	8537	8584	8630	8677	8723	8770		
9330	8816	8863	8910	8956	9003	9049	9096	9142	9189	9235		
31	9282	9328	9375	9422	9468	9515	9561	9608	9654	9701		
32	9747	9794	9840	9887	9933	9980	0̄027	0̄073	0̄120	0̄166		
33	9700213	0259	0306	0352	0399	0445	0492	0538	0585	0631		
34	0678	0724	0771	0818	0864	0911	0957	1004	1050	1097		
35	1143	1190	1236	1283	1329	1376	1422	1469	1515	1562		
36	1608	1655	1701	1748	1794	1841	1888	1934	1981	2027		
37	2074	2120	2167	2213	2260	2306	2353	2399	2446	2492		
38	2539	2585	2632	2678	2725	2771	2818	2864	2911	2957		
39	3004	3050	3097	3143	3190	3236	3283	3329	3376	3422		
9340	3469	3515	3562	3608	3655	3701	3748	3794	3841	3887		
41	3934	3980	4027	4073	4120	4166	4213	4259	4306	4352		
42	4399	4445	4492	4538	4585	4631	4678	4724	4771	4817		46
43	4863	4910	4956	5003	5049	5096	5142	5189	5235	5282		1 5
44	5328	5375	5421	5468	5514	5561	5607	5654	5700	5747	46	2 9
45	5793	5840	5886	5932	5979	6025	6072	6118	6165	6211		3 14
46	6258	6304	6351	6397	6444	6490	6537	6583	6629	6676		4 18
47	6722	6769	6815	6862	6908	6955	7001	7048	7094	7141		5 23
48	7187	7233	7280	7326	7373	7419	7466	7512	7559	7605		6 28
49	7652	7698	7745	7791	7837	7884	7930	7977	8023	8070		7 32
												8 37
												9 41
N.	0	1	2	3	4	5	6	7	8	9	D	Pts.

N.	0	1	2	3	4	5	6	7	8	9	D	Pro.
9350	9708116	8163	8209	8255	8302	8348	8395	8441	8488	8534		
51	8581	8627	8673	8720	8766	8813	8859	8906	8952	8999		47
52	9045	9091	9138	9184	9231	9277	9324	9370	9416	9463		1 5
53	9509	9556	9602	9649	9695	9742	9788	9834	9881	9927		2 9
54	9974	ō020	ō067	ō113	ō159	ō206	ō252	ō299	ō345	ō391		3 14
												4 19
55	9710438	0484	0531	0577	0624	0670	0716	0763	0809	0856		5 24
56	0902	0949	0995	1041	1088	1134	1181	1227	1273	1320		6 28
57	1366	1413	1459	1506	1552	1598	1645	1691	1738	1784		7 33
58	1830	1877	1923	1970	2016	2062	2109	2155	2202	2248		8 38
59	2294	2341	2387	2434	2480	2526	2573	2619	2666	2712		9 42
9360	2758	2805	2851	2898	2944	2990	3037	3083	3130	3176		
61	3222	3269	3315	3362	3408	3454	3501	3547	3594	3640		
62	3686	3733	3779	3826	3872	3918	3965	4011	4057	4104		
63	4150	4197	4243	4289	4336	4382	4429	4475	4521	4568		
64	4614	4660	4707	4753	4800	4846	4892	4939	4985	5031		
65	5078	5124	5171	5217	5263	5310	5356	5402	5449	5495		
66	5542	5588	5634	5681	5727	5773	5820	5866	5912	5959		
67	6005	6052	6098	6144	6191	6237	6283	6330	6376	6422		
68	6469	6515	6562	6608	6654	6701	6747	6793	6840	6886		
69	6932	6979	7025	7071	7118	7164	7211	7257	7303	7350		
9370	7396	7442	7489	7535	7581	7628	7674	7720	7767	7813		
71	7859	7906	7952	7998	8045	8091	8137	8184	8230	8276		
72	8323	8369	8415	8462	8508	8554	8601	8647	8694	8740		
73	8786	8833	8879	8925	8972	9018	9064	9111	9157	9203		
74	9249	9296	9342	9388	9435	9481	9527	9574	9620	9666		
75	9713	9759	9805	9852	9898	9944	9991	ō037	ō083	ō130		
76	9720176	0222	0269	0315	0361	0408	0454	0500	0547	0593		
77	0639	0685	0732	0778	0824	0871	0917	0963	1010	1056		
78	1102	1149	1195	1241	1288	1334	1380	1426	1473	1519		
79	1565	1612	1658	1704	1751	1797	1843	1889	1936	1982		
9380	2028	2075	2121	2167	2214	2260	2306	2352	2399	2445		
81	2491	2538	2584	2630	2677	2723	2769	2815	2862	2908		
82	2954	3001	3047	3093	3139	3186	3232	3278	3325	3371		
83	3417	3463	3510	3556	3602	3649	3695	3741	3787	3834		
84	3880	3926	3973	4019	4065	4111	4158	4204	4250	4296		
85	4343	4389	4435	4482	4528	4574	4620	4667	4713	4759		
86	4805	4852	4898	4944	4991	5037	5083	5129	5176	5222		
87	5268	5314	5361	5407	5453	5500	5546	5592	5638	5685		
88	5731	5777	5823	5870	5916	5962	6008	6055	6101	6147		
89	6193	6240	6286	6332	6378	6425	6471	6517	6563	6610		
9390	6656	6702	6748	6795	6841	6887	6933	6980	7026	7072		
91	7118	7165	7211	7257	7303	7350	7396	7442	7488	7535		46
92	7581	7627	7673	7720	7766	7812	7858	7905	7951	7997		1 5
93	8043	8089	8136	8182	8228	8274	8321	8367	8413	8459		2 9
94	8506	8552	8598	8644	8690	8737	8783	8829	8875	8922		3 14
												4 18
95	8968	9014	9060	9107	9153	9199	9245	9291	9338	9384		5 23
96	9430	9476	9523	9569	9615	9661	9707	9754	9800	9846		6 28
97	9892	9938	9985	ō031	ō077	ō123	ō170	ō216	ō262	ō308		7 32
98	9730354	0401	0447	0493	0539	0585	0632	0678	0724	0770		8 37
99	0816	0863	0909	0955	1001	1048	1094	1140	1186	1232		9 41
N.	0	1	2	3	4	5	6	7	8	9	D	Pts.

N.	0	1	2	3	4	5	6	7	8	9	D
9400	9731279	1325	1371	1417	1463	1510	1556	1602	1648	1694	
01	1741	1787	1833	1879	1925	1972	2018	2064	2110	2156	
02	2202	2249	2295	2341	2387	2433	2480	2526	2572	2618	
03	2664	2711	2757	2803	2849	2895	2941	2988	3034	3080	
04	3126	3172	3219	3265	3311	3357	3403	3449	3496	3542	
05	3588	3634	3680	3727	3773	3819	3865	3911	3957	4004	
06	4050	4096	4142	4188	4234	4281	4327	4373	4419	4465	
07	4511	4558	4604	4650	4696	4742	4788	4835	4881	4927	
08	4973	5019	5065	5112	5158	5204	5250	5296	5342	5389	
09	5435	5481	5527	5573	5619	5665	5712	5758	5804	5850	
9410	5896	5942	5989	6035	6081	6127	6173	6219	6265	6312	
11	6358	6404	6450	6496	6542	6588	6635	6681	6727	6773	
12	6819	6865	6911	6958	7004	7050	7096	7142	7188	7234	
13	7281	7327	7373	7419	7465	7511	7557	7604	7650	7696	
14	7742	7788	7834	7880	7926	7973	8019	8065	8111	8157	
15	8203	8249	8295	8342	8388	8434	8480	8526	8572	8618	
16	8664	8711	8757	8803	8849	8895	8941	8987	9033	9080	
17	9126	9172	9218	9264	9310	9356	9402	9449	9495	9541	
18	9587	9633	9679	9725	9771	9817	9864	9910	9956	5002	
19	9740048	0094	0140	0186	0232	0279	0325	0371	0417	0463	
9420	0509	0555	0601	0647	0693	0740	0786	0832	0878	0924	
21	0970	1016	1062	1108	1154	1201	1247	1293	1339	1385	
22	1431	1477	1523	1569	1615	1661	1708	1754	1800	1846	
23	1892	1938	1984	2030	2076	2122	2168	2215	2261	2307	
24	2353	2399	2445	2491	2537	2583	2629	2675	2721	2768	
25	2814	2860	2906	2952	2998	3044	3090	3136	3182	3228	
26	3274	3320	3367	3413	3459	3505	3551	3597	3643	3689	
27	3735	3781	3827	3873	3919	3965	4011	4058	4104	4150	
28	4196	4242	4288	4334	4380	4426	4472	4518	4564	4610	
29	4656	4702	4748	4795	4841	4887	4933	4979	5025	5071	
9430	5117	5163	5209	5255	5301	5347	5393	5439	5485	5531	
31	5577	5623	5670	5716	5762	5808	5854	5900	5946	5992	
32	6038	6084	6130	6176	6222	6268	6314	6360	6406	6452	
33	6498	6544	6590	6636	6683	6729	6775	6821	6867	6913	
34	6959	7005	7051	7097	7143	7189	7235	7281	7327	7373	
35	7419	7465	7511	7557	7603	7649	7695	7741	7787	7833	
36	7879	7925	7971	8017	8063	8109	8155	8201	8246	8294	
37	8340	8386	8432	8478	8524	8570	8616	8662	8708	8754	
38	8800	8846	8892	8938	8984	9030	9076	9122	9168	9214	
39	9260	9306	9352	9398	9444	9490	9536	9582	9628	9674	
9440	9720	9766	9812	9858	9904	9950	9996	ō042	ō088	ō134	46
41	9750180	0226	0272	0318	0364	0410	0456	0502	0548	0594	
42	0640	0686	0732	0778	0824	0870	0916	0962	1008	1054	
43	1100	1146	1192	1238	1284	1330	1376	1422	1468	1514	
44	1560	1606	1652	1698	1744	1790	1836	1882	1928	1974	
45	2020	2066	2112	2158	2204	2250	2296	2341	2387	2433	
46	2479	2525	2571	2617	2663	2709	2755	2801	2847	2893	
47	2939	2985	3031	3077	3123	3169	3215	3261	3307	3353	
48	3399	3445	3491	3537	3583	3629	3675	3721	3767	3813	
49	3858	3904	3950	3996	4042	4088	4134	4180	4226	4272	
N.	0	1	2	3	4	5	6	7	8	9	D

Pro. Pts.

Proportional parts 47:

1	5
2	9
3	14
4	19
5	24
6	28
7	33
8	38
9	42

Proportional parts 46:

1	5
2	9
3	14
4	18
5	23
6	28
7	32
8	37
9	41

N.	0	1	2	3	4	5	6	7	8	9	D	Pro.
9450	9754318	4364	4410	4456	4502	4548	4594	4640	4686	4732		
51	4778	4824	4870	4915	4961	5007	5053	5099	5145	5191		46
52	5237	5283	5329	5375	5421	5467	5513	5559	5605	5651		1 5
53	5697	5743	5788	5834	5880	5926	5972	6018	6064	6110		2 9
54	6156	6202	6248	6294	6340	6386	6432	6478	6523	6569		3 14
												4 18
55	6615	6661	6707	6753	6799	6845	6891	6937	6983	7029		5 23
56	7075	7121	7166	7212	7258	7304	7350	7396	7442	7488		6 28
57	7534	7580	7626	7672	7718	7763	7809	7855	7901	7947		7 32
58	7993	8039	8085	8131	8177	8223	8269	8315	8360	8406		8 37
59	8452	8498	8544	8590	8636	8682	8728	8774	8820	8865		9 41
9460	8911	8957	9003	9049	9095	9141	9187	9233	9279	9325		
61	9370	9416	9462	9508	9554	9600	9646	9692	9738	9784		
62	9829	9875	9921	9967	0̄013	0̄059	0̄105	0̄151	0̄197	0̄243		
63	9760288	0334	0380	0426	0472	0518	0564	0610	0656	0701		
64	0747	0793	0839	0885	0931	0977	1023	1069	1114	1160		
65	1206	1252	1298	1344	1390	1436	1481	1527	1573	1619		
66	1665	1711	1757	1803	1849	1894	1940	1986	2032	2078		
67	2124	2170	2216	2261	2307	2353	2399	2445	2491	2537		
68	2582	2628	2674	2720	2766	2812	2858	2904	2949	2995		
69	3041	3087	3133	3179	3225	3270	3316	3362	3408	3454		
9470	3500	3546	3592	3637	3683	3729	3775	3821	3867	3913		
71	3958	4004	4050	4096	4142	4188	4233	4279	4325	4371		
72	4417	4463	4509	4554	4600	4646	4692	4738	4784	4830		
73	4875	4921	4967	5013	5059	5105	5150	5196	5242	5288		
74	5334	5380	5425	5471	5517	5563	5609	5655	5701	5746		
75	5792	5838	5884	5930	5976	6021	6067	6113	6159	6205		
76	6251	6296	6342	6388	6434	6480	6525	6571	6617	6663		
77	6709	6755	6800	6846	6892	6938	6984	7030	7075	7121		
78	7167	7213	7259	7305	7350	7396	7442	7488	7534	7579		
79	7625	7671	7717	7763	7808	7854	7900	7946	7992	8038		
9480	8083	8129	8175	8221	8267	8312	8358	8404	8450	8496		
81	8541	8587	8633	8679	8725	8770	8816	8862	8908	8954		
82	9000	9045	9091	9137	9183	9229	9274	9320	9366	9412		
83	9458	9503	9549	9595	9641	9686	9732	9778	9824	9870		
84	9915	9961	0̄007	0̄053	0̄099	0̄144	0̄190	0̄236	0̄282	0̄328		
85	9770373	0419	0465	0511	0556	0602	0648	0694	0740	0785		
86	0831	0877	0923	0969	1014	1060	1106	1152	1197	1243		
87	1289	1335	1381	1426	1472	1518	1564	1609	1655	1701		
88	1747	1793	1838	1884	1930	1976	2021	2067	2113	2159		
89	2204	2250	2296	2342	2388	2433	2479	2525	2571	2616		
9490	2662	2708	2754	2799	2845	2891	2937	2982	3028	3074		
91	3120	3165	3211	3257	3303	3349	3394	3440	3486	3532		
92	3577	3623	3669	3715	3760	3806	3852	3898	3943	3989		45
93	4035	4081	4126	4172	4218	4264	4309	4355	4401	4447		1 5
94	4492	4538	4584	4630	4675	4721	4767	4812	4858	4904		2 9
												3 14
95	4950	4995	5041	5087	5133	5178	5224	5270	5316	5361		4 18
96	5407	5453	5499	5544	5590	5636	5681	5727	5773	5819		5 23
97	5864	5910	5956	6002	6047	6093	6139	6184	6230	6276		6 27
98	6322	6367	6413	6459	6505	6550	6596	6642	6687	6733		7 32
99	6779	6825	6870	6916	6962	7007	7053	7099	7145	7190		8 36
												9 41
N.	0	1	2	3	4	5	6	7	8	9	D	Pts.

N.	0	1	2	3	4	5	6	7	8	9	D	Pro.
9500	9777236	7282	7327	7373	7419	7465	7510	7556	7602	7647		
01	7693	7739	7785	7830	7876	7922	7967	8013	8059	8105		46
02	8150	8196	8242	8287	8333	8379	8424	8470	8516	8562		1 5
03	8607	8653	8699	8744	8790	8836	8881	8927	8973	9019		2 9
04	9064	9110	9156	9201	9247	9293	9338	9384	9430	9476		3 14
												4 18
05	9521	9567	9613	9658	9704	9750	9795	9841	9887	9932		5 23
06	9978	$\overline{0}$024	$\overline{0}$069	$\overline{0}$115	$\overline{0}$161	$\overline{0}$207	$\overline{0}$252	$\overline{0}$298	$\overline{0}$344	$\overline{0}$389		6 28
07	9780435	0481	0526	0572	0618	0663	0709	0755	0800	0846		7 32
08	0892	0937	0983	1029	1074	1120	1166	1211	1257	1303		8 37
09	1348	1394	1440	1485	1531	1577	1622	1668	1714	1760		9 41
9510	1805	1851	1897	1942	1988	2033	2079	2125	2170	2216		
11	2262	2307	2353	2399	2444	2490	2536	2581	2627	2673		
12	2718	2764	2810	2855	2901	2947	2992	3038	3084	3129		
13	3175	3221	3266	3312	3358	3403	3449	3495	3540	3586		
14	3631	3677	3723	3768	3814	3860	3905	3951	3997	4042		
15	4088	4134	4179	4225	4270	4316	4362	4407	4453	4499		
16	4544	4590	4636	4681	4727	4773	4818	4864	4909	4955		
17	5001	5046	5092	5138	5183	5229	5274	5320	5366	5411		
18	5457	5503	5548	5594	5640	5685	5731	5776	5822	5868		
19	5913	5959	6005	6050	6096	6141	6187	6233	6278	6324		
9520	6369	6415	6461	6506	6552	6598	6643	6689	6734	6780		
21	6826	6871	6917	6962	7008	7054	7099	7145	7191	7236		
22	7282	7327	7373	7419	7464	7510	7555	7601	7647	7692		
23	7738	7783	7829	7875	7920	7966	8011	8057	8103	8148		
24	8194	8239	8285	8331	8376	8422	8467	8513	8559	8604		
25	8650	8695	8741	8787	8832	8878	8923	8969	9015	9060		
26	9106	9151	9197	9243	9288	9334	9379	9425	9470	9516		
27	9562	9607	9653	9698	9744	9790	9835	9881	9926	9972		
28	9790017	0063	0109	0154	0200	0245	0291	0337	0382	0428		
29	0473	0519	0564	0610	0656	0701	0747	0792	0838	0883		
9530	0929	0975	1020	1066	1111	1157	1202	1248	1294	1339		
31	1385	1430	1476	1521	1567	1613	1658	1704	1749	1795		
32	1840	1886	1931	1977	2023	2068	2114	2159	2205	2250		
33	2296	2341	2387	2433	2478	2524	2569	2615	2660	2706		
34	2751	2797	2843	2888	2934	2979	3025	3070	3116	3161		
35	3207	3253	3298	3344	3389	3435	3480	3526	3571	3617		
36	3662	3708	3754	3799	3845	3890	3936	3981	4027	4072		
37	4118	4163	4209	4254	4300	4346	4391	4437	4482	4528		
38	4573	4619	4664	4710	4755	4801	4846	4892	4937	4983		
39	5028	5074	5120	5165	5211	5256	5302	5347	5393	5438		
9540	5484	5529	5575	5620	5666	5711	5757	5802	5848	5893		
41	5939	5984	6030	6076	6121	6167	6212	6258	6303	6349		
42	6394	6440	6485	6531	6576	6622	6667	6713	6758	6804		45
43	6849	6895	6940	6986	7031	7077	7122	7168	7213	7259		1 5
44	7304	7350	7395	7441	7486	7532	7577	7623	7668	7714		2 9
												3 14
45	7759	7805	7850	7896	7941	7987	8032	8078	8123	8169		4 18
46	8214	8260	8305	8351	8396	8442	8487	8533	8578	8624		5 23
47	8669	8715	8760	8806	8851	8897	8942	8988	9033	9079		6 27
48	9124	9170	9215	9261	9306	9352	9397	9442	9488	9533		7 32
49	9579	9624	9670	9715	9761	9806	9852	9897	9943	9988		8 36
												9 41
N.	0	1	2	3	4	5	6	7	8	9	D	Pts.

N.	0	1	2	3	4	5	6	7	8	9	D	Pro.
9550	9800034	0079	0125	0170	0216	0261	0307	0352	0398	0443		
51	0488	0534	0579	0625	0670	0716	0761	0807	0852	0898		46
52	0943	0989	1034	1080	1125	1170	1216	1261	1307	1352		1|5
53	1398	1443	1489	1534	1580	1625	1671	1716	1761	1807		2|9
54	1852	1898	1943	1989	2034	2080	2125	2171	2216	2261		3|14
												4|18
55	2307	2352	2398	2443	2489	2534	2580	2625	2671	2716		5|23
56	2761	2807	2852	2898	2943	2989	3034	3080	3125	3170		6|28
57	3216	3261	3307	3352	3398	3443	3489	3534	3579	3625		7|32
58	3670	3716	3761	3807	3852	3897	3943	3988	4034	4079		8|37
59	4125	4170	4215	4261	4306	4352	4397	4443	4488	4533		9|41
9560	4579	4624	4670	4715	4761	4806	4851	4897	4942	4988		
61	5033	5079	5124	5169	5215	5260	5306	5351	5397	5442		
62	5487	5533	5578	5624	5669	5714	5760	5805	5851	5896		
63	5942	5987	6032	6078	6123	6169	6214	6259	6305	6350		
64	6396	6441	6486	6532	6577	6623	6668	6714	6759	6804		
65	6850	6895	6941	6986	7031	7077	7122	7168	7213	7258		
66	7304	7349	7395	7440	7485	7531	7576	7622	7667	7712		
67	7758	7803	7849	7894	7939	7985	8030	8075	8121	8166		
68	8212	8257	8302	8348	8393	8439	8484	8529	8575	8620		
69	8666	8711	8756	8802	8847	8892	8938	8983	9029	9074		
9570	9119	9165	9210	9256	9301	9346	9392	9437	9482	9528		
71	9573	9619	9664	9709	9755	9800	9845	9891	9936	9982		
72	9810027	0072	0118	0163	0208	0254	0299	0344	0390	0435		
73	0481	0526	0571	0617	0662	0707	0753	0798	0844	0889		
74	0934	0980	1025	1070	1116	1161	1206	1252	1297	1342		
75	1388	1433	1479	1524	1569	1615	1660	1705	1751	1796		
76	1841	1887	1932	1977	2023	2068	2113	2159	2204	2250		
77	2295	2340	2386	2431	2476	2522	2567	2612	2658	2703		
78	2748	2794	2839	2884	2930	2975	3020	3066	3111	3156		
79	3202	3247	3292	3338	3383	3428	3474	3519	3564	3610		
9580	3655	3700	3746	3791	3836	3882	3927	3972	4018	4063		
81	4108	4154	4199	4244	4290	4335	4380	4426	4471	4516		
82	4562	4607	4652	4698	4743	4788	4834	4879	4924	4970		
83	5015	5060	5106	5151	5196	5241	5287	5332	5377	5423		
84	5468	5513	5559	5604	5649	5695	5740	5785	5831	5876		
85	5921	5966	6012	6057	6102	6148	6193	6238	6284	6329		
86	6374	6420	6465	6510	6555	6601	6646	6691	6737	6782		
87	6827	6873	6918	6963	7008	7054	7099	7144	7190	7235		
88	7280	7326	7371	7416	7461	7507	7552	7597	7643	7688		
89	7733	7778	7824	7869	7914	7960	8005	8050	8095	8141		
9590	8186	8231	8277	8322	8367	8412	8458	8503	8548	8594		
91	8639	8684	8729	8775	8820	8865	8911	8956	9001	9046		
92	9092	9137	9182	9228	9273	9318	9363	9409	9454	9499		45
93	9544	9590	9635	9680	9726	9771	9816	9861	9907	9952		1|5
94	9997	0042	0088	0133	0178	0223	0269	0314	0359	0405		2|9
												3|14
95	9820450	0495	0540	0586	0631	0676	0721	0767	0812	0857		4|18
96	0902	0948	0993	1038	1083	1129	1174	1219	1264	1310		5|23
97	1355	1400	1445	1491	1536	1581	1626	1672	1717	1762		6|27
98	1807	1853	1898	1943	1988	2034	2079	2124	2169	2215		7|32
99	2260	2305	2350	2396	2441	2486	2531	2577	2622	2667		8|36
												9|41
N.	0	1	2	3	4	5	6	7	8	9	D	Pts.

N.	0	1	2	3	4	5	6	7	8	9	D	Pro.
9600	9822712	2758	2803	2848	2893	2939	2984	3029	3074	3119		
01	3165	3210	3255	3300	3346	3391	3436	3481	3527	3572		46
02	3617	3662	3707	3753	3798	3843	3888	3934	3979	4024		1 5
03	4069	4115	4160	4205	4250	4295	4341	4386	4431	4476		2 9
04	4522	4567	4612	4657	4702	4748	4793	4838	4883	4928		3 14
												4 18
05	4974	5019	5064	5109	5155	5200	5245	5290	5335	5381		5 23
06	5426	5471	5516	5561	5607	5652	5697	5742	5787	5833		6 28
07	5878	5923	5968	6014	6059	6104	6149	6194	6240	6285		7 32
08	6330	6375	6420	6466	6511	6556	6601	6646	6692	6737		8 37
09	6782	6827	6872	6918	6963	7008	7053	7098	7143	7189		9 41
9610	7234	7279	7324	7369	7415	7460	7505	7550	7595	7641		
11	7686	7731	7776	7821	7867	7912	7957	8002	8047	8092		
12	8138	8183	8228	8273	8318	8364	8409	8454	8499	8544		
13	8589	8635	8680	8725	8770	8815	8860	8906	8951	8996		
14	9041	9086	9132	9177	9222	9267	9312	9357	9403	9448		
15	9493	9538	9583	9628	9674	9719	9764	9809	9854	9899		
16	9945	9990	0̄035	0̄080	0̄125	0̄170	0̄216	0̄261	0̄306	0̄351		
17	9830396	0441	0486	0532	0577	0622	0667	0712	0757	0803		
18	0848	0893	0938	0983	1028	1073	1119	1164	1209	1254		
19	1299	1344	1390	1435	1480	1525	1570	1615	1660	1706		
9620	1751	1796	1841	1886	1931	1976	2022	2067	2112	2157		
21	2202	2247	2292	2338	2383	2428	2473	2518	2563	2608		
22	2654	2699	2744	2789	2834	2879	2924	2969	3015	3060		
23	3105	3150	3195	3240	3285	3331	3376	3421	3466	3511		
24	3556	3601	3646	3692	3737	3782	3827	3872	3917	3962		
25	4007	4053	4098	4143	4188	4233	4278	4323	4368	4413		
26	4459	4504	4549	4594	4639	4684	4729	4774	4819	4865		
27	4910	4955	5000	5045	5090	5135	5180	5225	5271	5316		
28	5361	5406	5451	5496	5541	5586	5631	5677	5722	5767		
29	5812	5857	5902	5947	5992	6037	6082	6128	6173	6218		
9630	6263	6308	6353	6398	6443	6488	6533	6579	6624	6669		
31	6714	6759	6804	6849	6894	6939	6984	7029	7075	7120		
32	7165	7210	7255	7300	7345	7390	7435	7480	7525	7571		
33	7616	7661	7706	7751	7796	7841	7886	7931	7976	8021		
34	8066	8111	8157	8202	8247	8292	8337	8382	8427	8472		
35	8517	8562	8607	8652	8697	8743	8788	8833	8878	8923		
36	8968	9013	9058	9103	9148	9193	9238	9283	9328	9374		
37	9419	9464	9509	9554	9599	9644	9689	9734	9779	9624		
38	9869	9914	9959	0̄004	0̄049	0̄095	0̄140	0̄185	0̄230	0̄275		
39	9840320	0365	0410	0455	0500	0545	0590	0635	0680	0725		
9640	0770	0815	0860	0905	0951	0996	1041	1086	1131	1176		
41	1221	1266	1311	1356	1401	1446	1491	1536	1581	1626		
42	1671	1716	1761	1806	1851	1896	1942	1987	2032	2077		45
43	2122	2167	2212	2257	2302	2347	2392	2437	2482	2527		1 5
44	2572	2617	2662	2707	2752	2797	2842	2887	2932	2977		2 9
												3 14
45	3022	3067	3112	3157	3202	3247	3292	3338	3383	3428		4 18
46	3473	3518	3563	3608	3653	3698	3743	3788	3833	3878		5 23
47	3923	3968	4013	4058	4103	4148	4193	4238	4283	4328		6 27
48	4373	4418	4463	4508	4553	4598	4643	4688	4733	4778		7 32
49	4823	4868	4913	4958	5003	5048	5093	5138	5183	5228		8 36
												9 41
N.	0	1	2	3	4	5	6	7	8	9	D	Pts

N.	0	1	2	3	4	5	6	7	8	9	D	Pro.
9650	9845273	5318	5363	5408	5453	5498	5543	5588	5633	5678		
51	5723	5768	5813	5858	5903	5948	5993	6038	6083	6128		45
52	6173	6218	6263	6308	6353	6398	6443	6488	6533	6578	45	1 5
53	6623	6668	6713	6758	6803	6848	6893	6938	6983	7028		2 9
54	7073	7118	7163	7208	7253	7298	7343	7388	7433	7478		3 14
												4 18
55	7523	7568	7613	7658	7703	7748	7793	7838	7883	7928		5 23
56	7973	8018	8063	8107	8152	8197	8242	8287	8332	8377		6 27
57	8422	8467	8512	8557	8602	8647	8692	8737	8782	8827		7 32
58	8872	8917	8962	9007	9052	9097	9142	9187	9232	9277		8 36
59	9322	9367	9412	9457	9502	9546	9591	9636	9681	9726		9 41
9660	9771	9816	9861	9906	9951	9996	ō041	ō086	ō131	ō176		
61	9850221	0266	0311	0356	0401	0446	0491	0535	0580	0625		
62	0670	0715	0760	0805	0850	0895	0940	0985	1030	1075		
63	1120	1165	1210	1255	1300	1345	1389	1434	1479	1524		
64	1569	1614	1659	1704	1749	1794	1839	1884	1929	1974		
65	2019	2064	2108	2153	2198	2243	2288	2333	2378	2423		
66	2468	2513	2558	2603	2648	2693	2737	2782	2827	2872		
67	2917	2962	3007	3052	3097	3142	3187	3232	3277	3321		
68	3366	3411	3456	3501	3546	3591	3636	3681	3726	3771		
69	3816	3861	3905	3950	3995	4040	4085	4130	4175	4220		
9670	4265	4310	4355	4399	4444	4489	4534	4579	4624	4669		
71	4714	4759	4804	4849	4893	4938	4983	5028	5073	5118		
72	5163	5208	5253	5298	5342	5387	5432	5477	5522	5567		
73	5612	5657	5702	5747	5791	5836	5881	5926	5971	6016		
74	6061	6106	6151	6196	6240	6285	6330	6375	6420	6465		
75	6510	6555	6600	6644	6689	6734	6779	6824	6869	6914		
76	6959	7003	7048	7093	7138	7183	7228	7273	7318	7363		
77	7407	7452	7497	7542	7587	7632	7677	7722	7766	7811		
78	7856	7901	7946	7991	8036	8081	8125	8170	8215	8260		
79	8305	8350	8395	8440	8484	8529	8574	8619	8664	8709		
9680	8754	8798	8843	8888	8933	8978	9023	9068	9112	9157		
81	9202	9247	9292	9337	9382	9426	9471	9516	9561	9606		
82	9651	9696	9740	9785	9830	9875	9920	9965	ō010	ō054		
83	9860099	0144	0189	0234	0279	0324	0368	0413	0458	0503		
84	0548	0593	0637	0682	0727	0772	0817	0862	0907	0951		
85	0996	1041	1086	1131	1176	1220	1265	1310	1355	1400		
86	1445	1489	1534	1579	1624	1669	1714	1758	1803	1848		
87	1893	1938	1983	2027	2072	2117	2162	2207	2252	2296		
88	2341	2366	2431	2476	2521	2565	2610	2655	2700	2745		
89	2790	2834	2879	2924	2969	3014	3058	3103	3148	3193		
9690	3238	3283	3327	3372	3417	3462	3507	3551	3596	3641		
91	3686	3731	3776	3820	3865	3910	3955	4000	4044	4089		
92	4134	4179	4224	4268	4313	4358	4403	4448	4493	4537		44
93	4582	4627	4672	4717	4761	4806	4851	4896	4941	4985		1 4
94	5030	5075	5120	5165	5209	5254	5299	5344	5389	5433		2 9
												3 13
95	5478	5523	5568	5613	5657	5702	5747	5792	5836	5881		4 18
96	5926	5971	6016	6060	6105	6150	6195	6240	6284	6329		5 22
97	6374	6419	6464	6508	6553	6598	6643	6687	6732	6777		6 26
98	6822	6867	6911	6956	7001	7046	7090	7135	7180	7225		7 31
99	7270	7314	7359	7404	7449	7493	7538	7583	7628	7673		8 35
												9 40
N.	0	1	2	3	4	5	6	7	8	9	D	Pts.

N.	0	1	2	3	4	5	6	7	8	9	D	Pro.
9700	9867717	7762	7807	7852	7896	7941	7986	8031	8076	8120		
01	8165	8210	8255	8299	8344	8389	8434	8478	8523	8568		
02	8613	8657	8702	8747	8792	8837	8881	8926	8971	9016		45
03	9060	9105	9150	9195	9239	9284	9329	9374	9418	9463		1 5
04	9508	9553	9597	9642	9687	9732	9776	9821	9866	9911		2 9 / 3 14
05	9955	ō000	ō045	ō090	ō134	ō179	ō224	ō269	ō313	ō358		4 18 / 5 23
06	9870403	0448	0492	0537	0582	0627	0671	0716	0761	0806		6 27
07	0850	0895	0940	0985	1029	1074	1119	1163	1208	1253		7 32
08	1298	1342	1387	1432	1477	1521	1566	1611	1656	1700		8 36
09	1745	1790	1834	1879	1924	1969	2013	2058	2103	2148		9 41
9710	2192	2237	2282	2326	2371	2416	2461	2505	2550	2595		
11	2640	2684	2729	2774	2818	2863	2908	2953	2997	3042		
12	3087	3131	3176	3221	3266	3310	3355	3400	3444	3489		
13	3534	3579	3623	3668	3713	3757	3802	3847	3892	3936		
14	3981	4026	4070	4115	4160	4205	4249	4294	4339	4383		
15	4428	4473	4517	4562	4607	4652	4696	4741	4786	4830		
16	4875	4920	4964	5009	5054	5099	5143	5188	5233	5277		
17	5322	5367	5411	5456	5501	5545	5590	5635	5680	5724		
18	5769	5814	5858	5903	5948	5992	6037	6082	6126	6171		
19	6216	6261	6305	6350	6395	6439	6484	6529	6573	6618		
9720	6663	6707	6752	6797	6841	6886	6931	6975	7020	7065		
21	7109	7154	7199	7243	7288	7333	7377	7422	7467	7511		
22	7556	7601	7646	7690	7735	7780	7824	7869	7914	7958		
23	8003	8048	8092	8137	8182	8226	8271	8316	8360	8405		
24	8450	8494	8539	8583	8628	8673	8717	8762	8807	8851		
25	8896	8941	8985	9030	9075	9119	9164	9209	9253	9298		
26	9343	9387	9432	9477	9521	9566	9611	9655	9700	9745		
27	9789	9834	9878	9923	9968	ō012	ō057	ō102	ō146	ō191		
28	9880236	0280	0325	0370	0414	0459	0503	0548	0593	0637		
29	0682	0727	0771	0816	0861	0905	0950	0994	1039	1084		
9730	1128	1173	1218	1262	1307	1352	1396	1441	1485	1530		
31	1575	1619	1664	1709	1753	1798	1842	1887	1932	1976		
32	2021	2066	2110	2155	2200	2244	2289	2333	2378	2423		
33	2467	2512	2556	2601	2646	2690	2735	2780	2824	2869		
34	2913	2958	3003	3047	3092	3136	3181	3226	3270	3315		
35	3360	3404	3449	3493	3538	3583	3627	3672	3716	3761		
36	3806	3850	3895	3939	3984	4029	4073	4118	4162	4207		
37	4252	4296	4341	4386	4430	4475	4519	4564	4609	4653		
38	4698	4742	4787	4831	4876	4921	4965	5010	5054	5099		
39	5144	5188	5233	5277	5322	5367	5411	5456	5500	5545		
9740	5590	5634	5679	5723	5768	5813	5857	5902	5946	5991		
41	6035	6080	6125	6169	6214	6258	6303	6348	6392	6437		
42	6481	6526	6570	6615	6660	6704	6749	6793	6838	6882		44
43	6927	6972	7016	7061	7105	7150	7194	7239	7284	7328		1 4
44	7373	7417	7462	7506	7551	7596	7640	7685	7729	7774		2 9 / 3 13
45	7818	7863	7908	7952	7997	8041	8086	8130	8175	8220		4 18 / 5 22
46	8264	8309	8353	8398	8442	8487	8531	8576	8621	8665		6 26
47	8710	8754	8799	8843	8888	8932	8977	9022	9066	9111		7 31
48	9155	9200	9244	9289	9333	9378	9423	9467	9512	9556		8 35
49	9601	9645	9690	9734	9779	9823	9868	9913	9957	ō002		9 40
N.	0	1	2	3	4	5	6	7	8	9	D	Pts.

N.	0	1	2	3	4	5	6	7	8	9	D	Pro.
9750	9890046	0091	0135	0180	0224	0269	0313	0358	0402	0447		
51	0492	0536	0581	0625	0670	0714	0759	0803	0848	0892		45
52	0937	0981	1026	1071	1115	1160	1204	1249	1293	1338		1 5
53	1382	1427	1471	1516	1560	1605	1649	1694	1738	1783		2 9
54	1828	1872	1917	1961	2006	2050	2095	2139	2184	2228		3 14
												4 18
55	2273	2317	2362	2406	2451	2495	2540	2584	2629	2673		5 23
56	2718	2762	2807	2851	2896	2940	2985	3030	3074	3119		6 27
57	3163	3208	3252	3297	3341	3386	3430	3475	3519	3564		7 32
58	3608	3653	3697	3742	3786	3831	3875	3920	3964	4009		8 36
59	4053	4098	4142	4187	4231	4276	4320	4365	4409	4454		9 41
9760	4498	4543	4587	4632	4676	4721	4765	4810	4854	4899		
61	4943	4988	5032	5077	5121	5166	5210	5255	5299	5344		
62	5388	5433	5477	5521	5566	5610	5655	5699	5744	5788		
63	5833	5877	5922	5966	6011	6055	6100	6144	6189	6233		
64	6278	6322	6367	6411	6456	6500	6545	6589	6634	6678		
65	6722	6767	6811	6856	6900	6945	6989	7034	7078	7123		
66	7167	7212	7256	7301	7345	7390	7434	7478	7523	7567		
67	7612	7656	7701	7745	7790	7834	7879	7923	7968	8012		
68	8057	8101	8145	8190	8234	8279	8323	8368	8412	8457		
69	8501	8546	8590	8634	8679	8723	8768	8812	8857	8901		
9770	8946	8990	9035	9079	9123	9168	9212	9257	9301	9346		
71	9390	9435	9479	9523	9568	9612	9657	9701	9746	9790		
72	9835	9879	9923	9968	0̄012	0̄057	0̄101	0̄146	0̄190	0̄235		
73	9900279	0323	0368	0412	0457	0501	0546	0590	0634	0679		
74	0723	0768	0812	0857	0901	0946	0990	1034	1079	1123		
75	1168	1212	1257	1301	1345	1390	1434	1479	1523	1568		
76	1612	1656	1701	1745	1790	1834	1878	1923	1967	2012		
77	2056	2101	2145	2189	2234	2278	2323	2367	2411	2456		
78	2500	2545	2589	2634	2678	2722	2767	2811	2856	2900		
79	2944	2989	3033	3078	3122	3167	3211	3255	3300	3344		
9780	3389	3433	3477	3522	3566	3611	3655	3699	3744	3788		
81	3833	3877	3921	3966	4010	4055	4099	4143	4188	4232		
82	4277	4321	4365	4410	4454	4499	4543	4587	4632	4676		
83	4721	4765	4809	4854	4898	4942	4987	5031	5076	5120		
84	5164	5209	5253	5298	5342	5386	5431	5475	5520	5564		
85	5608	5653	5697	5741	5786	5830	5875	5919	5963	6008		
86	6052	6096	6141	6185	6230	6274	6318	6363	6407	6452		
87	6496	6540	6585	6629	6673	6718	6762	6806	6851	6895		
88	6940	6984	7028	7073	7117	7161	7206	7250	7295	7339		
89	7383	7428	7472	7516	7561	7605	7649	7694	7738	7783		
9790	7827	7871	7916	7960	8004	8049	8093	8137	8182	8226		
91	8271	8315	8359	8404	8448	8492	8537	8581	8625	8670		
92	8714	8758	8803	8847	8891	8936	8980	9025	9069	9113		44
93	9158	9202	9246	9291	9335	9379	9424	9468	9512	9557		1 4
94	9601	9645	9690	9734	9778	9823	9867	9911	9956	0̄000		2 9
												3 13
95	9910044	0089	0133	0177	0222	0266	0310	0355	0399	0443		4 18
96	0488	0532	0576	0621	0665	0709	0754	0798	0842	0887		5 22
97	0931	0975	1020	1064	1108	1153	1197	1241	1286	1330		6 26
98	1374	1419	1463	1507	1552	1596	1640	1685	1729	1773		7 31
99	1818	1862	1906	1951	1995	2039	2083	2128	2172	2216		8 35
												9 40
N.	0	1	2	3	4	5	6	7	8	9	D	Pts.

N.	0	1	2	3	4	5	6	7	8	9	D	Pro.
9800	9912261	2305	2349	2394	2438	2482	2527	2571	2615	2660		
01	2704	2748	2793	2837	2881	2925	2970	3014	3058	3103		45
02	3147	3191	3236	3280	3324	3369	3413	3457	3501	3546		1 5
03	3590	3634	3679	3723	3767	3812	3856	3900	3944	3989		2 9
04	4033	4077	4122	4166	4210	4255	4299	4343	4387	4432		3 14
												4 18
05	4476	4520	4565	4609	4653	4697	4742	4786	4830	4875		5 23
06	4919	4963	5007	5052	5096	5140	5185	5229	5273	5317		6 27
07	5362	5406	5450	5495	5539	5583	5627	5672	5716	5760		7 32
08	5805	5849	5893	5937	5982	6026	6070	6115	6159	6203		8 36
09	6247	6292	6336	6380	6424	6469	6513	6557	6602	6646		9 41
9810	6690	6734	6779	6823	6867	6911	6956	7000	7044	7088		
11	7133	7177	7221	7266	7310	7354	7398	7443	7487	7531		
12	7575	7620	7664	7708	7752	7797	7841	7885	7929	7974		
13	8018	8062	8107	8151	8195	8239	8284	8328	8372	8416		
14	8461	8505	8549	8593	8638	8682	8726	8770	8815	8859		
15	8903	8947	8992	9036	9080	9124	9169	9213	9257	9301		
16	9345	9390	9434	9478	9522	9567	9611	9655	9699	9744		
17	9788	9832	9876	9921	9965	0̄009	0̄053	0̄098	0̄142	0̄186		
18	9920230	0275	0319	0363	0407	0451	0496	0540	0584	0628		
19	0673	0717	0761	0805	0850	0894	0938	0982	1026	1071		
9820	1115	1159	1203	1248	1292	1336	1380	1424	1469	1513		
21	1557	1601	1646	1690	1734	1778	1822	1867	1911	1955		
22	1999	2044	2088	2132	2176	2220	2265	2309	2353	2397		
23	2441	2486	2530	2574	2618	2662	2707	2751	2795	2839		
24	2884	2928	2972	3016	3060	3105	3149	3193	3237	3281		
25	3326	3370	3414	3458	3502	3547	3591	3635	3679	3723		
26	3768	3812	3856	3900	3944	3989	4033	4077	4121	4165		
27	4210	4254	4298	4342	4386	4431	4475	4519	4563	4607		
28	4651	4696	4740	4784	4828	4872	4917	4961	5005	5049		
29	5093	5138	5182	5226	5270	5314	5358	5403	5447	5491		
9830	5535	5579	5624	5668	5712	5756	5800	5844	5889	5933		
31	5977	6021	6065	6109	6154	6198	6242	6286	6330	6375		
32	6419	6463	6507	6551	6595	6640	6684	6728	6772	6816		
33	6860	6905	6949	6993	7037	7081	7125	7170	7214	7258		
34	7302	7346	7390	7435	7479	7523	7567	7611	7655	7699		
35	7744	7788	7832	7876	7920	7964	8009	8053	8097	8141		
36	8185	8229	8274	8318	8362	8406	8450	8494	8538	8583		
37	8627	8671	8715	8759	8803	8847	8892	8936	8980	9024		
38	9068	9112	9156	9201	9245	9289	9333	9377	9421	9465		
39	9510	9554	9598	9642	9686	9730	9774	9819	9863	9907		
9840	9951	9995	0̄039	0̄083	0̄128	0̄172	0̄216	0̄260	0̄304	0̄348		
41	9930392	0436	0481	0525	0569	0613	0657	0701	0745	0789		
42	0834	0878	0922	0966	1010	1054	1098	1142	1187	1231		44
43	1275	1319	1363	1407	1451	1495	1540	1584	1628	1672		1 4
44	1716	1760	1804	1848	1893	1937	1981	2025	2069	2113		2 9
												3 13
45	2157	2201	2245	2290	2334	2378	2422	2466	2510	2554		4 18
46	2598	2642	2687	2731	2775	2819	2863	2907	2951	2995		5 22
47	3039	3083	3128	3172	3216	3260	3304	3348	3392	3436		6 26
48	3480	3524	3569	3613	3657	3701	3745	3789	3833	3877		7 31
49	3921	3965	4010	4054	4098	4142	4186	4230	4274	4318		8 35
												9 40
N.	0	1	2	3	4	5	6	7	8	9	D	Pts.

N.	0	1	2	3	4	5	6	7	8	9	D
9850	9934362	4406	4450	4495	4539	4583	4627	4671	4715	4759	
51	4803	4847	4891	4935	4980	5024	5068	5112	5156	5200	
52	5244	5288	5332	5376	5420	5464	5509	5553	5597	5641	
53	5685	5729	5773	5817	5861	5905	5949	5993	6037	6082	
54	6126	6170	6214	6258	6302	6346	6390	6434	6478	6522	
55	6566	6610	6654	6698	6743	6787	6831	6875	6919	6963	
56	7007	7051	7095	7139	7183	7227	7271	7315	7359	7404	
57	7448	7492	7536	7580	7624	7668	7712	7756	7800	7844	
58	7888	7932	7976	8020	8064	8108	8152	8197	8241	8285	
59	8329	8373	8417	8461	8505	8549	8593	8637	8681	8725	
9860	8769	8813	8857	8901	8945	6989	9033	9077	9122	9166	
61	9210	9254	9298	9342	9386	9430	9474	9518	9562	9606	
62	9650	9694	9738	9782	9826	9870	9914	9958	ō002	ō046	
63	9940090	0134	0178	0222	0266	0310	0355	0399	0443	0487	
64	0531	0575	0619	0663	0707	0751	0795	0839	0883	0927	
65	0971	1015	1059	1103	1147	1191	1235	1279	1323	1367	
66	1411	1455	1499	1543	1587	1631	1675	1719	1763	1807	
67	1851	1895	1939	1983	2027	2071	2115	2159	2203	2247	
68	2291	2335	2379	2423	2467	2511	2555	2599	2643	2687	
69	2731	2775	2820	2864	2908	2952	2996	3040	3084	3128	44
9870	3172	3216	3260	3304	3348	3392	3436	3480	3524	3568	
71	3612	3656	3700	3744	3788	3831	3875	3919	3963	4007	
72	4051	4095	4139	4183	4227	4271	4315	4359	4403	4447	
73	4491	4535	4579	4623	4667	4711	4755	4799	4843	4887	
74	4931	4975	5019	5063	5107	5151	5195	5239	5283	5327	
75	5371	5415	5459	5503	5547	5591	5635	5679	5723	5767	
76	5811	5855	5899	5943	5987	6031	6075	6119	6163	6207	
77	6251	6295	6338	6382	6426	6470	6514	6558	6602	6646	
78	6690	6734	6778	6822	6866	6910	6954	6998	7042	7086	
79	7130	7174	7218	7262	7306	7350	7394	7438	7482	7525	
9880	7569	7613	7657	7701	7745	7789	7833	7877	7921	7965	
81	8009	8053	8097	8141	8185	8229	8273	8317	8361	8405	
82	8448	8492	8536	8580	8624	8668	8712	8756	8800	8844	
83	8888	8932	8976	9020	9064	9108	9152	9196	9239	9283	
84	9327	9371	9415	9459	9503	9547	9591	9635	9679	9723	
85	9767	9811	9855	9899	9942	9986	ō030	ō074	ō118	ō162	
86	9950206	0250	0294	0338	0382	0426	0470	0514	0557	0601	
87	0645	0689	0733	0777	0821	0865	0909	0953	0997	1041	
88	1085	1128	1172	1216	1260	1304	1348	1392	1436	1480	
89	1524	1568	1612	1656	1699	1743	1787	1831	1875	1919	
9890	1963	2007	2051	2095	2139	2182	2226	2270	2314	2358	
91	2402	2446	2490	2534	2578	2622	2665	2709	2753	2797	
92	2841	2885	2929	2973	3017	3061	3104	3148	3192	3236	
93	3280	3324	3368	3412	3456	3500	3543	3587	3631	3675	
94	3719	3763	3807	3851	3895	3939	3982	4026	4070	4114	
95	4158	4202	4246	4290	4334	4377	4421	4465	4509	4553	
96	4597	4641	4685	4729	4772	4816	4860	4904	4948	4992	
97	5036	5080	5123	5167	5211	5255	5299	5343	5387	5431	
98	5474	5518	5562	5606	5650	5694	5738	5782	5825	5869	
99	5913	5957	6001	6045	6089	6133	6176	6220	6264	6308	
N.	0	1	2	3	4	5	6	7	8	9	D

Pro.

44

1	4
2	9
3	13
4	18
5	22
6	26
7	31
8	35
9	40

43

1	4
2	9
3	13
4	17
5	22
6	26
7	30
8	34
9	39

Pts.

N.	0	1	2	3	4	5	6	7	8	9	D	Pro.
9900	9956352	6396	6440	6484	6527	6571	6615	6659	6703	6747		
01	6791	6834	6878	6922	6966	7010	7054	7098	7142	7185		44
02	7229	7273	7317	7361	7405	7449	7492	7536	7580	7624		1 4
03	7668	7712	7755	7799	7843	7887	7931	7975	8019	8062		2 9
04	8106	8150	8194	8238	8282	8326	8369	8413	8457	8501		3 13
												4 18
05	8545	8589	8632	8676	8720	8764	8808	8852	8896	8939		5 22
06	8983	9027	9071	9115	9159	9202	9246	9290	9334	9378		6 26
07	9422	9465	9509	9553	9597	9641	9685	9728	9772	9816		7 31
08	9860	9904	9948	9991	ō035	ō079	ō123	ō167	ō211	ō254		8 35
09	9960298	0342	0386	0430	0471	0517	0561	0605	0649	0693		9 40
9910	0737	0780	0824	0868	0912	0956	0999	1043	1087	1131		
11	1175	1219	1262	1306	1350	1394	1438	1481	1525	1569		
12	1613	1657	1701	1744	1788	1832	1876	1920	1963	2007		
13	2051	2095	2139	2182	2226	2270	2314	2358	2402	2445		
14	2489	2533	2577	2621	2664	2708	2752	2796	2840	2883		
15	2927	2971	3015	3059	3102	3146	3190	3234	3278	3321		
16	3365	3409	3453	3497	3540	3584	3628	3672	3716	3759		
17	3803	3847	3891	3935	3978	4022	4066	4110	4153	4197		
18	4241	4285	4329	4372	4416	4460	4504	4548	4591	4635		
19	4679	4723	4766	4810	4854	4898	4942	4985	5029	5073		
9920	5117	5161	5204	5248	5292	5336	5379	5423	5467	5511		
21	5554	5598	5642	5686	5730	5773	5817	5861	5905	5948		
22	5992	6036	6080	6124	6167	6211	6255	6299	6342	6386		
23	6430	6474	6517	6561	6605	6649	6693	6736	6780	6824		
24	6868	6911	6955	6999	7043	7086	7130	7174	7218	7261		
25	7305	7349	7393	7436	7480	7524	7568	7611	7655	7699		
26	7743	7786	7830	7874	7918	7961	8005	8049	8093	8136		
27	8180	8224	8268	8311	8355	8399	8443	8486	8530	8574		
28	8618	8661	8705	8749	8793	8836	8880	8924	8968	9011		
29	9055	9099	9143	9186	9230	9274	9318	9361	9405	9449		
9930	9492	9536	9580	9624	9667	9711	9755	9799	9842	9886		
31	9930	9974	ō017	ō061	ō105	ō148	ō192	ō236	ō280	ō323		
32	9970367	0411	0455	0498	0542	0586	0629	0673	0717	0761		
33	0804	0848	0892	0936	0979	1023	1067	1110	1154	1198		
34	1242	1285	1329	1373	1416	1460	1504	1548	1591	1635		
35	1679	1722	1766	1810	1854	1897	1941	1985	2028	2072		
36	2116	2160	2203	2247	2291	2334	2378	2422	2465	2509		
37	2553	2597	2640	2684	2728	2771	2815	2859	2903	2946		
38	2990	3034	3077	3121	3165	3208	3252	3296	3340	3383		
39	3427	3471	3514	3558	3602	3645	3689	3733	3776	3820		
9940	3864	3908	3951	3995	4039	4082	4126	4170	4213	4257		
41	4301	4344	4388	4432	4475	4519	4563	4607	4650	4694		
42	4738	4781	4825	4869	4912	4956	5000	5043	5087	5131		43
43	5174	5218	5262	5305	5349	5393	5436	5480	5524	5567		1 4
44	5611	5655	5699	5742	5786	5830	5873	5917	5961	6004		2 9
												3 13
45	6048	6092	6135	6179	6223	6266	6310	6354	6397	6441		4 17
46	6485	6528	6572	6616	6659	6703	6747	6790	6834	6878		5 22
47	6921	6965	7009	7052	7096	7139	7183	7227	7270	7314		6 26
48	7358	7401	7445	7489	7532	7576	7620	7663	7707	7751		7 30
												8 34
49	7794	7838	7882	7925	7969	8013	8056	8100	8144	8187		9 39
N.	0	1	2	3	4	5	6	7	8	9	D	Pts.

N.	0	1	2	3	4	5	6	7	8	9	D	Pro.
9950	9978231	8274	8318	8362	8405	8449	8493	8536	8580	8624		
51	8667	8711	8755	8798	8842	8885	8929	8973	9016	9060		44
52	9104	9147	9191	9235	9278	9322	9365	9409	9453	9496		1 4
53	9540	9584	9627	9671	9715	9758	9802	9845	9889	9933		2 9
54	9976	0̄020	0̄064	0̄107	0̄151	0̄195	0̄238	0̄282	0̄325	0̄369		3 13
												4 18
55	9980413	0456	0500	0544	0587	0631	0674	0718	0762	0805		5 22
56	0849	0893	0936	0980	1023	1067	1111	1154	1198	1241		6 26
57	1285	1329	1372	1416	1460	1503	1547	1590	1634	1678		7 31
58	1721	1765	1808	1852	1896	1939	1983	2026	2070	2114		8 35
59	2157	2201	2245	2288	2332	2375	2419	2463	2506	2550		9 40
9960	2593	2637	2681	2724	2768	2811	2855	2899	2942	2986		
61	3029	3073	3117	3160	3204	3247	3291	3335	3378	3422		
62	3465	3509	3553	3596	3640	3683	3727	3771	3814	3858		
63	3901	3945	3988	4032	4076	4119	4163	4206	4250	4294		
64	4337	4381	4424	4468	4512	4555	4599	4642	4686	4729		
65	4773	4817	4860	4904	4947	4991	5035	5078	5122	5165		
66	5209	5252	5296	5340	5383	5427	5470	5514	5557	5601		
67	5645	5688	5732	5775	5819	5862	5906	5950	5993	6037		
68	6080	6124	6167	6211	6255	6298	6342	6385	6429	6472		
69	6516	6560	6603	6647	6690	6734	6777	6821	6864	6908		
9970	6952	6995	7039	7082	7126	7169	7213	7256	7300	7344		
71	7387	7431	7474	7518	7561	7605	7648	7692	7736	7779		
72	7823	7866	7910	7953	7997	8040	8084	8128	8171	8215		
73	8258	8302	8345	8389	8432	8476	8519	8563	8607	8650		
74	8694	8737	8781	8824	8868	8911	8955	8998	9042	9086		
75	9129	9173	9216	9260	9303	9347	9390	9434	9477	9521		
76	9564	9608	9651	9695	9739	9782	9826	9869	9913	9956		
77	9990000	0043	0087	0130	0174	0217	0261	0304	0348	0391		
78	0435	0479	0522	0566	0609	0653	0696	0740	0783	0827		
79	0870	0914	0957	1001	1044	1088	1131	1175	1218	1262		
9980	1305	1349	1392	1436	1479	1523	1567	1610	1654	1697		
81	1741	1784	1828	1871	1915	1958	2002	2045	2089	2132		
82	2176	2219	2263	2306	2350	2393	2437	2480	2524	2567		
83	2611	2654	2698	2741	2785	2828	2872	2915	2959	3002		
84	3046	3089	3133	3176	3220	3263	3307	3350	3394	3437		
85	3481	3524	3568	3611	3655	3698	3742	3785	3829	3872		
86	3916	3959	4003	4046	4090	4133	4177	4220	4264	4307		
87	4350	4394	4437	4481	4524	4568	4611	4655	4698	4742		
88	4785	4829	4872	4916	4959	5003	5046	5090	5133	5177		
89	5220	5264	5307	5351	5394	5438	5481	5524	5568	5611		
9990	5655	5698	5742	5785	5829	5872	5916	5959	6003	6046		
91	6090	6133	6177	6220	6263	6307	6350	6394	6437	6481		
92	6524	6568	6611	6655	6698	6742	6785	6828	6872	6915		43
93	6959	7002	7046	7089	7133	7176	7220	7263	7307	7350		1 4
94	7393	7437	7480	7524	7567	7611	7654	7698	7741	7785		2 9
												3 13
95	7828	7871	7915	7958	8002	8045	8089	8132	8176	8219		4 17
96	8262	8306	8349	8393	8436	8480	8523	8567	8610	8653		5 22
97	8697	8740	8784	8827	8871	8914	8958	9001	9044	9088		6 26
98	9131	9175	9218	9262	9305	9349	9392	9435	9479	9522		7 30
99	9566	9609	9653	9696	9739	9783	9826	9870	9913	9957		8 34
												9 39
N.	0	1	2	3	4	5	6	7	8	9	D	Pts.

N.	0	1	2	3	4	5	6	7	8	9	D	Pro.
10000	00000000	0434	0869	1303	1737	2171	2606	3040	3474	3908		
01	4343	4777	5211	5645	6080	6514	6948	7382	7817	8251	435	435
02	8685	9119	9553	9988	ō422	ō856	1290	1724	2159	2593		1|44
03	00013027	3461	3895	4329	4764	5198	5632	6066	6500	6934		2|87
04	7368	7802	8237	8671	9105	9539	9973	ō407	ō841	1275		3|131 4|174
05	00021709	2143	2577	3012	3446	3880	4314	4748	5182	5616		5|218
06	6050	6484	6918	7352	7786	8220	8654	9088	9522	9956		6|261
07	00030390	0824	1258	1692	2126	2560	2994	3428	3862	4296		7|305
08	4730	5164	5598	6031	6465	6899	7333	7767	8201	8635		8|348
09	9069	9503	9937	ō371	ō805	1238	1672	2106	2540	2974		9|392
10010	00043408	3842	4275	4709	5143	5577	6011	6445	6878	7312		
11	7746	8180	8614	9048	9481	9915	ō349	ō783	1217	1650		
12	00052084	2518	2952	3385	3819	4253	4687	5120	5554	5988		
13	6422	6855	7289	7723	8157	8590	9024	9458	9891	ō325		
14	00060759	1192	1626	2060	2493	2927	3361	3794	4228	4662		
15	5095	5529	5963	6396	6830	7264	7697	8131	8564	8998		
16	9432	9865	ō299	ō732	1166	1600	2033	2467	2900	3334		
17	00073767	4201	4634	5068	5502	5935	6369	6802	7236	7669		
18	8103	8536	8970	9403	9837	ō270	ō704	1137	1571	2004		
19	00082438	2871	3305	3738	4172	4605	5038	5472	5905	6339		
10020	6772	7206	7639	8072	8506	8939	9373	9806	ō239	ō673		
21	00091106	1540	1973	2406	2840	3273	3706	4140	4573	5006		
22	5440	5873	6307	6740	7173	7606	8040	8473	8906	9346	434	434
23	9773	ō206	ō640	1073	1506	1939	2373	2806	3239	3673		1|43
24	00104106	4539	4972	5406	5839	6272	6705	7138	7572	8005		2|87 3|130
25	8438	8871	9305	9738	ō171	ō604	1037	1471	1904	2337		4|174
26	00112770	3203	3636	4070	4503	4936	5369	5802	6235	6668		5|217
27	7101	7535	7968	8401	8834	9267	9700	ō133	ō566	ō999		6|260 7|304
28	00121433	1866	2299	2732	3165	3598	4031	4464	4897	5330		8|347
29	5763	6196	6629	7062	7495	7928	8361	8794	9227	9660		9|391
10030	00130093	0526	0959	1392	1825	2258	2691	3124	3557	3990		
31	4423	4856	5289	5722	6155	6588	7021	7454	7887	8319		
32	8752	9185	9618	ō051	ō484	ō917	1350	1783	2215	2648		
33	00143081	3514	3947	4380	4813	5246	5678	6111	6544	6977		
34	7410	7842	8275	8708	9141	9574	ō007	ō439	ō872	1305		
35	00151738	2170	2603	3036	3469	3902	4334	4767	5200	5633		
36	6065	6498	6931	7363	7796	8229	8662	9094	9527	9960		
37	00160392	0825	1258	1690	2123	2556	2988	3421	3854	4286		
38	4719	5152	5584	6017	6450	6882	7315	7748	8180	8613		
39	9045	9478	9911	ō348	ō776	1208	1641	2074	2506	2939		
10040	00173371	3804	4236	4669	5102	5534	5967	6399	6832	7264		
41	7697	8129	8562	8994	9427	9859	ō292	ō724	1157	1589		
42	00182022	2454	2887	3319	3752	4184	4616	5049	5481	5914	433	433
43	6346	6779	7211	7644	8076	8508	8941	9373	9806	ō238		1|43
44	00190670	1103	1535	1968	2400	2832	3265	3697	4129	4562		2|87 3|130
45	4994	5426	5859	6291	6723	7156	7588	8020	8453	8885		4|173
46	9317	9750	ō182	ō614	1047	1479	1911	2343	2776	3208		5|217
47	00203640	4072	4505	4937	5369	5801	6234	6666	7098	7530		6|260 7|303
48	7963	8395	8827	9259	9691	ō124	ō556	ō988	1420	1852		8|346
49	00212285	2717	3149	3581	4013	4445	4878	5310	5742	6174		9|390
N.	0	1	2	3	4	5	6	7	8	9	D	Pts.

N.	0	1	2	3	4	5	6	7	8	9	D	Pro.
10050	00216606	7038	7470	7903	8335	8767	9199	9631	ō063	ō495		
51	00220927	1359	1791	2224	2656	3088	3520	3952	4384	4816	432	432
52	5248	5680	6112	6544	6976	7408	7840	8272	8704	9136		1 43
53	9568	ō000	ō432	ō864	ī296	ī728	2̄160	2̄592	3̄024	3̄456		2 86
54	00233888	4320	4752	5184	5616	6048	6480	6912	7344	7776		3 130
												4 173
55	8207	8639	9071	9503	9935	ō367	ō799	ī231	ī663	2̄095		5 216
56	00242526	2958	3390	3822	4254	4686	5118	5549	5981	6413		6 259
57	6845	7277	7709	8140	8572	9004	9436	9868	ō300	ō731		7 302
58	00251163	1595	2027	2458	2890	3322	3754	4186	4617	5049		8 346
59	5481	5913	6344	6776	7208	7639	8071	8503	8935	9366		9 389
10060	9798	ō230	ō661	ī093	ī525	ī957	2̄388	2̄820	3̄252	3̄683		
61	00264115	4547	4978	5410	5842	6273	6705	7136	7568	8000		
62	8431	8863	9295	9726	ō158	ō589	ī021	ī453	ī884	2̄316		
63	00272747	3179	3610	4042	4474	4905	5337	5768	6200	6631		
64	7063	7494	7926	8357	8789	9220	9652	ō083	ō515	ō946		
65	00281378	1809	2241	2672	3104	3535	3967	4398	4830	5261		
66	5693	6124	6555	6987	7418	7850	8281	8713	9144	9575		
67	00290007	0438	0870	1301	1732	2164	2595	3027	3458	3889		
68	4321	4752	5183	5615	6046	6477	6909	7340	7771	8203		
69	8634	9065	9497	9928	ō359	ō791	ī222	ī653	2̄084	2̄516		
10070	00302947	3378	3810	4241	4672	5103	5535	5966	6397	6828	431	431
71	7260	7691	8122	8553	8984	9416	9847	ō278	ō709	ī141		1 43
72	00311572	2003	2434	2865	3296	3728	4159	4590	5021	5452		2 86
73	5883	6315	6746	7177	7608	8039	8470	8901	9332	9764		3 129
74	00320195	0626	1057	1488	1919	2350	2781	3212	3643	4074		4 172
												5 216
75	4505	4937	5368	5799	6230	6661	7092	7523	7954	8385		6 259
76	8816	9247	9678	ō109	ō540	ō971	ī402	ī833	2̄264	2̄695		7 302
77	00333126	3557	3988	4419	4850	5281	5712	6143	6574	7004		8 345
78	7435	7866	8297	8728	9159	9590	ō021	ō452	ō883	ī314		9 388
79	00341745	2175	2606	3037	3468	3899	4330	4761	5192	5622		
10080	6053	6484	6915	7346	7777	8207	8638	9069	9500	9931		
81	00350361	0792	1223	1654	2085	2515	2946	3377	3808	4239		
82	4669	5100	5531	5962	6392	6823	7254	7685	8115	8546		
83	8977	9407	9838	ō269	ō700	ī130	ī561	ī992	2̄422	2̄853		
84	00363284	3714	4145	4576	5006	5437	5868	6298	6729	7160		
85	7590	8021	8452	8882	9313	9743	ō174	ō605	ī035	ī466		
86	00371896	2327	2758	3188	3619	4049	4480	4910	5341	5772		
87	6202	6633	7063	7494	7924	8355	8785	9216	9646	ō077		
88	00380507	0938	1368	1799	2229	2660	3090	3521	3951	4382		
89	4812	5243	5673	6104	6534	6964	7395	7825	8256	8686		
10090	9117	9547	9977	ō408	ō838	ī269	ī699	2̄129	2̄560	2̄990		
91	00393421	3851	4281	4712	5142	5572	6003	6433	6864	7294		
92	7724	8155	8585	9015	9445	9876	ō306	ō736	ī167	ī597	430	430
93	00402027	2458	2888	3318	3748	4179	4609	5039	5470	5900		1 43
94	6330	6760	7191	7621	8051	8481	8911	9342	9772	ō202		2 86
												3 129
95	00410632	1063	1493	1923	2353	2783	3213	3644	4074	4504		4 172
96	4934	5364	5795	6225	6655	7085	7515	7945	8375	8806		5 215
97	9236	9666	ō096	ō526	ō956	ī386	ī816	2̄246	2̄676	3̄107		6 258
98	00423537	3967	4397	4827	5257	5687	6117	6547	6977	7407		7 301
99	7837	8267	8697	9127	9557	9987	ō417	ō847	ī277	ī707		8 344
												9 387
N.	0	1	2	3	4	5	6	7	8	9	D	Pts.

N.	0	1	2	3	4	5	6	7	8	9	D	Pro.
10100	00432137	2567	2997	3427	3857	4287	4717	5147	5577	6007		
01	6437	6867	7297	7727	8157	8587	9017	9447	9877	ō307	430	430
02	00440736	1166	1596	2026	2456	2886	3316	3746	4176	4605		1 43
03	5035	5465	5895	6325	6755	7185	7614	8044	8474	8904		2 86
04	9334	9764	ō193	ō623	ī053	ī483	ī913	2̄342	2̄772	3̄202		3 129 / 4 172
05	00453632	4062	4491	4921	5351	5781	6210	6640	7070	7500		5 215
06	7929	8359	8789	9219	9648	ō078	ō508	ō937	ī367	ī797		6 258
07	00462227	2656	3086	3516	3945	4375	4805	5234	5664	6094		7 301
08	6523	6953	7383	7812	8242	8672	9101	9531	9960	ō390		8 344
09	00470820	1249	1679	2108	2538	2968	3397	3827	4256	4686		9 387
10110	5116	5545	5975	6404	6834	7263	7693	8122	8552	8982		
11	9411	9841	ō270	ō700	ī129	ī559	ī988	2̄418	2̄847	3̄277		
12	00483706	4136	4565	4995	5424	5853	6283	6712	7142	7571		
13	8001	8430	8860	9289	9718	ō148	ō577	ī007	ī436	ī866		
14	00492295	2724	3154	3583	4012	4442	4871	5301	5730	6159		
15	6589	7018	7447	7877	8306	8735	9165	9594	ō023	ō453		
16	00500882	1311	1741	2170	2599	3029	3458	3887	4316	4746		
17	5175	5604	6034	6463	6892	7321	7751	8180	8609	9038		
18	9468	9897	ō326	ō755	ī184	ī614	2̄043	2̄472	2̄901	3̄330		
19	00513760	4189	4618	5047	5476	5905	6335	6764	7193	7622		
10120	8051	8480	8910	9339	9768	ō197	ō626	ī055	ī484	ī913		
21	00522342	2772	3201	3630	4059	4488	4917	5346	5775	6204	429	429
22	6633	7062	7491	7920	8350	8779	9208	9637	ō066	ō495		1 43
23	00530924	1353	1782	2211	2640	3069	3498	3927	4356	4785		2 86
24	5214	5643	6072	6501	6930	7358	7787	8216	8645	9074		3 129 / 4 172
25	9503	9932	ō361	ō790	ī219	ī648	2̄077	2̄506	2̄935	3̄363		5 215
26	00543792	4221	4650	5079	5508	5937	6366	6794	7223	7652		6 257
27	8081	8510	8939	9368	9796	ō225	ō654	ī083	ī512	ī940		7 300
28	00552369	2798	3227	3656	4084	4513	4942	5371	5800	6228		8 343
29	6657	7086	7515	7943	8372	8801	9230	9658	ō087	ō516		9 386
10130	00560945	1373	1802	2231	2659	3088	3517	3945	4374	4803		
31	5232	5660	6089	6518	6946	7375	7804	8232	8661	9089		
32	9518	9947	ō375	ō804	ī233	ī661	2̄090	2̄518	2̄947	3̄376		
33	00573804	4233	4661	5090	5519	5947	6376	6804	7233	7661		
34	8090	8519	8947	9376	9804	ō233	ō661	ī090	ī518	ī947		
35	00582375	2804	3232	3661	4089	4518	4946	5375	5803	6232		
36	6660	7089	7517	7946	8374	8802	9231	9659	ō088	ō516		
37	00590945	1373	1801	2230	2658	3087	3515	3944	4372	4800		
38	5229	5657	6085	6514	6942	7371	7799	8227	8656	9084		
39	9512	9941	ō369	ō797	ī226	ī654	2̄082	2̄511	2̄939	3̄367		
10140	00603795	4224	4652	5080	5509	5937	6365	6793	7222	7650		
41	8078	8507	8935	9363	9791	ō219	ō648	ī076	ī504	ī932		
42	00612361	2789	3217	3645	4073	4502	4930	5358	5786	6214	428	428
43	6643	7071	7499	7927	8355	8783	9212	9640	ō068	ō496		1 43
44	00620924	1352	1780	2208	2637	3065	3493	3921	4349	4777		2 86 / 3 128
45	5205	5633	6061	6489	6917	7346	7774	8202	8630	9058		4 171
46	9486	9914	ō342	ō770	ī198	ī626	2̄054	2̄482	2̄910	3̄338		5 214
47	00633766	4194	4622	5050	5478	5906	6334	6762	7190	7618		6 257
48	8046	8474	8902	9330	9758	ō186	ō614	ī041	ī469	ī897		7 300
49	00642325	2753	3181	3609	4037	4465	4893	5321	5748	6176		8 342 / 9 385
N.	0	1	2	3	4	5	6	7	8	9	D	Pts.

N	0	1	2	3	4	5	6	7	8	9	D
10150	00646604	7032	7460	7888	8316	8744	9171	9599	0̄027	0̄455	
51	00650883	1311	1738	2166	2594	3022	3450	3878	4305	4733	427
52	5161	5589	6016	6444	6872	7300	7728	8155	8583	9011	
53	9439	9866	0̄294	0̄722	1̄150	1̄577	2̄005	2̄433	2̄860	3̄288	
54	00663716	4144	4571	4999	5427	5854	6282	6710	7137	7565	
55	7993	8420	8848	9276	9703	0̄131	0̄559	0̄986	1̄414	1̄842	
56	00672269	2697	3124	3552	3980	4407	4835	5262	5690	6118	
57	6545	6973	7400	7828	8256	8683	9111	9538	9966	0̄393	
58	00680821	1248	1676	2103	2531	2958	3386	3814	4241	4669	
59	5096	5524	5951	6379	6806	7233	7661	8088	8516	8943	
10160	9371	9798	0̄226	0̄653	1̄081	1̄508	1̄935	2̄363	2̄790	3̄218	
61	00693645	4073	4500	4927	5355	5782	6210	6637	7064	7492	
62	7919	8346	8774	9201	9629	0̄056	0̄483	0̄911	1̄338	1̄765	
63	00702193	2620	3047	3475	3902	4329	4756	5184	5611	6038	
64	6466	6893	7320	7747	8175	8602	9029	9457	9884	0̄311	
65	00710738	1166	1593	2020	2447	2874	3302	3729	4156	4583	
66	5011	5438	5865	6292	6719	7146	7574	8001	8428	8855	
67	9282	9710	0̄137	0̄564	0̄991	1̄418	1̄845	2̄272	2̄700	3̄127	
68	00723554	3981	4408	4835	5262	5689	6116	6543	6971	7398	
69	7825	8252	8679	9106	9533	9960	0̄387	0̄814	1̄241	1̄668	
10170	00732095	2522	2949	3376	3803	4230	4657	5084	5511	5938	
71	6365	6792	7219	7646	8073	8500	8927	9354	9781	0̄208	
72	00740635	1062	1489	1916	2343	2770	3197	3624	4051	4478	426
73	4904	5331	5758	6185	6612	7039	7466	7893	8320	8746	
74	9173	9600	0̄027	0̄454	0̄881	1̄308	1̄734	2̄161	2̄588	3̄015	
75	00753442	3869	4295	4722	5149	5576	6003	6429	6856	7283	
76	7710	8137	8563	8990	9417	9844	0̄270	0̄697	1̄124	1̄551	
77	00761977	2404	2831	3258	3684	4111	4538	4965	5391	5818	
78	6245	6671	7098	7525	7951	8378	8805	9231	9658	0̄085	
79	00770511	0938	1365	1791	2218	2645	3071	3498	3925	4351	
10180	4778	5204	5631	6058	6484	6911	7337	7764	8191	8617	
81	9044	9470	9897	0̄323	0̄750	1̄177	1̄603	2̄030	2̄456	2̄883	
82	00783309	3736	4162	4589	5015	5442	5868	6295	6721	7148	
83	7574	8001	8427	8854	9280	9707	0̄133	0̄560	0̄986	1̄413	
84	00791839	2266	2692	3118	3545	3971	4398	4824	5251	5677	
85	6103	6530	6956	7383	7809	8235	8662	9088	9514	9941	
86	00800367	0794	1220	1646	2073	2499	2925	3352	3778	4204	
87	4631	5057	5483	5910	6336	6762	7188	7615	8041	8467	
88	8894	9320	9746	0̄172	0̄599	1̄025	1̄451	1̄877	2̄304	2̄730	
89	00813156	3582	4009	4435	4861	5287	5714	6140	6566	6992	
10190	7418	7845	8271	8697	9123	9549	9976	0̄402	0̄828	1̄254	
91	00821680	2106	2532	2959	3385	3811	4237	4663	5089	5515	
92	5941	6368	6794	7220	7646	8072	8498	8924	9350	9776	425
93	00830202	0628	1055	1481	1907	2333	2759	3185	3611	4037	
94	4463	4889	5315	5741	6167	6593	7019	7445	7871	8297	
95	8723	9149	9575	0̄001	0̄427	0̄853	1̄279	1̄705	2̄131	2̄557	
96	00842983	3409	3835	4260	4686	5112	5538	5964	6390	6816	
97	7242	7668	8094	8520	8946	9371	9797	0̄223	0̄649	1̄075	
98	00851501	1927	2352	2778	3204	3630	4056	4482	4908	5333	
99	5759	6185	6611	7037	7462	7888	8314	8740	9166	9591	
N.	0	1	2	3	4	5	6	7	8	9	D

Pro. — Pts.

427
1	43
2	85
3	128
4	171
5	214
6	256
7	299
8	342
9	384

426
1	43
2	85
3	128
4	170
5	213
6	256
7	298
8	341
9	383

425
1	43
2	85
3	128
4	170
5	213
6	255
7	298
8	340
9	383

N.	0	1	2	3	4	5	6	7	8	9	D	Pro.
10200	00860017	0443	0869	1294	1720	2146	2572	2998	3423	3849		
01	4275	4700	5126	5552	5978	6403	6829	7255	7681	8106	426	426
02	8532	8958	9383	9809	0̄235	0̄660	Ī086	Ī512	Ī937	2̄363		1 43
03	00872789	3214	3640	4066	4491	4917	5343	5768	6194	6619		2 85
04	7045	7471	7896	8322	8747	9173	9599	0̄024	0̄450	0̄875		3 128 / 4 170
05	00881301	1726	2152	2578	3003	3429	3854	4280	4705	5131		5 213
06	5556	5982	6407	6833	7258	7684	8109	8535	8960	9386		6 256
07	9811	0̄237	0̄662	Ī088	Ī513	Ī939	2̄364	2̄790	3̄215	3̄641		7 298
08	00894066	4492	4917	5342	5768	6193	6619	7044	7470	7895		8 341
09	8320	8746	9171	9597	0̄022	0̄447	0̄873	Ī298	Ī723	2̄149		9 383
10210	00902574	2999	3425	3850	4276	4701	5126	5551	5977	6402		
11	6828	7253	7678	8103	8529	8954	9379	9804	0̄230	0̄655		
12	00911081	1506	1931	2356	2782	3207	3632	4057	4483	4908		
13	5333	5758	6184	6609	7034	7459	7885	8310	8735	9160		
14	9585	0̄010	0̄436	0̄861	Ī286	Ī711	2̄136	2̄561	2̄987	3̄412		
15	00923837	4262	4687	5112	5538	5963	6388	6813	7238	7663		
16	8088	8513	8939	9364	9789	0̄214	0̄639	Ī064	Ī489	Ī914		
17	00932339	2764	3189	3614	4040	4465	4890	5315	5740	6165		
18	6590	7015	7440	7865	8290	8715	9140	9565	9990	0̄415		
19	00940840	1265	1690	2115	2540	2965	3390	3815	4240	4665		
10220	5090	5515	5939	6364	6789	7214	7639	8064	8489	8914		
21	9339	9764	0̄189	0̄614	Ī038	Ī463	Ī888	2̄313	2̄738	3̄163	425	425
22	00953588	4013	4437	4862	5287	5712	6137	6562	6986	7411		1 43
23	7836	8261	8686	9111	9535	9960	0̄385	0̄810	Ī234	Ī659		2 85
24	00962084	2509	2934	3359	3783	4208	4633	5058	5482	5907		3 128 / 4 170
25	6332	6757	7181	7606	8031	8456	8880	9305	9729	0̄154		5 213
26	00970579	1004	1428	1853	2278	2703	3127	3552	3976	4401		6 255
27	4826	5251	5675	6100	6524	6949	7373	7798	8223	8648		7 298
28	9072	9497	9921	0̄346	0̄770	Ī195	Ī620	2̄045	2̄469	2̄894		8 340
29	00983318	3743	4167	4592	5016	5441	5865	6290	6714	7139		9 383
10230	7563	7988	8412	8837	9261	9686	0̄110	0̄535	0̄959	Ī384		
31	00991808	2233	2657	3082	3506	3931	4355	4780	5204	5629		
32	6053	6478	6902	7327	7751	8176	8600	9025	9449	9873		
33	01000297	0722	1146	1571	1995	2420	2844	3269	3693	4117		
34	4541	4966	5390	5815	6239	6663	7087	7512	7936	8361		
35	8785	9209	9633	0̄058	0̄482	0̄907	Ī331	Ī755	2̄179	2̄604		
36	01013028	3452	3876	4301	4725	5149	5573	5998	6422	6846		
37	7270	7695	8119	8543	8967	9392	9816	0̄240	0̄664	Ī088		
38	01021512	1937	2361	2785	3209	3634	4058	4482	4906	5330		
39	5754	6179	6603	7027	7451	7875	8299	8723	9147	9572		
10240	9996	0̄420	0̄844	Ī268	Ī692	2̄116	2̄540	2̄964	3̄388	3̄813		
41	01034237	4661	5085	5509	5933	6357	6781	7205	7629	8053		
42	8477	8901	9325	9749	0̄173	0̄597	Ī021	Ī445	Ī869	2̄293	424	424
43	01042717	3141	3565	3989	4413	4837	5261	5685	6109	6533		1 42
44	6957	7381	7805	8229	8653	9077	9501	9925	0̄348	0̄772		2 85 / 3 127
45	01051196	1620	2044	2468	2892	3316	3740	4164	4587	5011		4 170
46	5435	5859	6283	6707	7131	7555	7978	8402	8826	9250		5 212
47	9674	0̄098	0̄521	0̄945	Ī369	Ī793	2̄216	2̄640	3̄064	3̄488		6 254
48	01063912	4336	4759	5183	5607	6031	6454	6878	7302	7726		7 297
49	8149	8573	8997	9421	9844	0̄268	0̄692	Ī116	Ī539	Ī963		8 339 / 9 382
N.	0	1	2	3	4	5	6	7	8	9	D	Pts.

N.	0	1	•2	3	4	5	6	7	8	9	D	Pro.
10250	01072386	2810	3234	3658	4081	4505	4929	5353	5776	6200		
51	6623	7047	7471	7895	8318	8742	9165	9589	0̄012	0̄436	423	423
52	01080860	1284	1707	2131	2554	2978	3401	3825	4249	4673		1 42
53	5096	5520	5943	6367	6790	7214	7637	8061	8484	8908		2 85
54	9331	9755	0̄178	0̄602	1̄025	1̄449	1̄872	2̄296	2̄719	3̄143		3 127 / 4 169
55	01093566	3990	4413	4837	5260	5684	6107	6531	6954	7378		5 212
56	7801	8225	8648	9072	9495	9919	0̄342	0̄766	1̄189	1̄613		6 254
57	01102036	2459	2882	3306	3729	4153	4576	5000	5423	5846		7 296
58	6269	6693	7116	7540	7963	8387	8810	9233	9656	0̄080		8 338
59	01110503	0927	1350	1773	2196	2620	3043	3466	3889	4313		9 381
10260	4736	5160	5583	6006	6429	6853	7276	7699	8122	8546		
61	8969	9392	9815	0̄238	0̄662	1̄085	1̄508	1̄931	2̄355	2̄778		
62	01123201	3624	4047	4470	4894	5317	5740	6163	6587	7010		
63	7433	7856	8279	8702	9126	9549	9972	0̄395	0̄818	1̄241		
64	01131664	2087	2511	2934	3357	3780	4203	4626	5049	5472		
65	5895	6318	6742	7165	7588	8011	8434	8857	9280	9703		
66	01140126	0549	0972	1395	1818	2241	2664	3087	3510	3933		
67	4356	4779	5202	5625	6048	6471	6894	7317	7740	8163		
68	8586	9009	9432	9855	0̄278	0̄701	1̄124	1̄547	1̄970	2̄393		
69	01152815	3238	3661	4084	4507	4930	5353	5776	6199	6622		
10270	7044	7467	7890	8313	8736	9159	9582	0̄005	0̄427	0̄850	422	422
71	01161273	1696	2119	2542	2964	3387	3810	4233	4655	5078		1 42
72	5501	5924	6347	6770	7192	7615	8038	8461	8883	9306		2 84
73	9729	0̄152	0̄574	0̄997	1̄420	1̄843	2̄265	2̄688	3̄111	3̄534		3 127 / 4 169
74	01173956	4379	4802	5225	5647	6070	6492	6915	7338	7761		5 211
75	8183	8606	9028	9451	9874	0̄297	0̄719	1̄142	1̄564	1̄987		6 253
76	01182410	2833	3255	3678	4100	4523	4945	3568	5790	6213		7 295
77	6636	7059	7481	7904	8326	8749	9171	9594	0̄016	0̄439		8 338
78	01190861	1284	1706	2129	2552	2975	3397	3820	4242	4665		9 380
79	5087	5510	5932	6355	6777	7200	7622	8045	8467	8889		
10280	9311	9734	0̄156	0̄579	1̄001	1̄424	1̄846	2̄269	2̄691	3̄114		
81	01203536	3959	4381	4804	5226	5648	6070	6493	6915	7338		
82	7760	8183	8605	9027	9449	9872	0̄294	0̄717	1̄139	1̄562		
83	01211984	2406	2828	3251	3673	4096	4518	4940	5362	5785		
84	6207	6629	7051	7474	7896	8319	8741	9163	9585	0̄008		
85	01220430	0852	1274	1697	2119	2541	2963	3386	3808	4230		
86	4652	5074	5496	5919	6341	6763	7185	7608	8030	8452		
87	8874	9296	9718	0̄141	0̄563	0̄985	1̄407	1̄829	2̄251	2̄674		
88	01233096	3518	3940	4362	4784	5206	5628	6051	6473	6895		
89	7317	7739	8161	8583	9005	9427	9849	0̄271	0̄693	1̄115		
10290	01241537	1960	2382	2804	3226	3648	4070	4492	4914	5336		
91	5758	6180	6602	7024	7446	7868	8290	8712	9134	9556		
92	9978	0̄400	0̄822	1̄244	1̄666	2̄088	2̄510	2̄932	3̄353	3̄775	421	421
93	01254197	4619	5041	5463	5885	6307	6729	7151	7573	7995		1 42
94	8416	8838	9260	9682	0̄104	0̄526	0̄948	1̄370	1̄791	2̄213		2 84 / 3 126
95	01262635	3057	3479	3901	4322	4744	5166	5588	6010	6432		4 168
96	6853	7275	7697	8119	8541	8962	9384	9806	0̄228	0̄649		5 211
97	01271071	1493	1915	2336	2758	3180	3602	4023	4445	4867		6 253
98	5289	5710	6132	6554	6976	7397	7819	8241	8662	9084		7 295
99	9506	9928	0̄349	0̄771	1̄193	1̄614	2̄036	2̄458	2̄879	3̄301		8 337 / 9 379
N.	0	1	2	3	4	5	6	7	8	9	D	Pts.

N.	0	1	2	3	4	5	6	7	8	9	D
10300	01283723	4144	4566	4987	5409	5831	6252	6674	7095	7517	
01	7939	8360	8782	9204	9625	0̄047	0̄468	0̄890	1̄311	1̄733	422
02	01292155	2576	2998	3419	3841	4262	4684	5105	5527	5949	
03	6370	6792	7213	7635	8056	8478	8899	9321	9742	0̄164	
04	01300585	1006	1428	1849	2271	2692	3114	3535	3957	4378	
05	4800	5221	5642	6064	6485	6907	7328	7750	8171	8592	
06	9014	9435	9857	0̄278	0̄699	1̄121	1̄542	1̄964	2̄385	2̄806	
07	01313228	3649	4070	4492	4913	5334	5756	6177	6598	7020	
08	7441	7862	8284	8705	9126	9548	9969	0̄390	0̄811	1̄233	
09	01321654	2075	2497	2918	3339	3760	4182	4603	5024	5445	
10310	5867	6288	6709	7130	7551	7973	8394	8815	9236	9657	
11	01330079	0500	0921	1342	1763	2185	2606	3027	3448	3869	
12	4290	4712	5133	5554	5975	6396	6817	7238	7659	8081	
13	8502	8923	9344	9765	0̄186	0̄607	1̄028	1̄450	1̄871	2̄292	
14	01342713	3134	3555	3976	4397	4818	5239	5660	6081	6502	
15	6923	7344	7765	8186	8607	9028	9449	9870	0̄291	0̄712	
16	01351133	1554	1975	2396	2817	3238	3659	4080	4501	4922	
17	5343	5764	6185	6606	7027	7448	7869	8290	8711	9131	
18	9552	9973	0̄394	0̄615	1̄236	1̄657	2̄078	2̄499	2̄920	3̄340	
19	01363761	4182	4603	5024	5445	5866	6286	6707	7128	7549	
10320	7970	8391	8811	9232	9653	0̄074	0̄495	0̄915	1̄336	1̄757	
21	01372178	2599	3019	3440	3861	4282	4702	5123	5544	5965	421
22	6386	6806	7227	7648	8068	8489	8910	9331	9751	0̄172	
23	01380593	1013	1434	1855	2276	2696	3117	3538	3958	4379	
24	4800	5220	5641	6062	6482	6903	7324	7744	8165	8585	
25	9006	9427	9847	0̄268	0̄688	1̄109	1̄530	1̄950	2̄371	2̄791	
26	01393212	3633	4053	4474	4894	5315	5735	6156	6577	6997	
27	7418	7838	8259	8679	9100	9520	9941	0̄361	0̄782	1̄202	
28	01401623	2043	2464	2884	3305	3725	4146	4566	4987	5407	
29	5828	6248	6669	7089	7509	7930	8350	8771	9191	9612	
10330	01410032	0453	0873	1293	1714	2134	2555	2975	3395	3816	
31	4236	4656	5077	5497	5918	6338	6758	7179	7599	8019	
32	8440	8860	9280	9701	0̄121	0̄541	0̄962	1̄382	1̄802	2̄223	
33	01422643	3063	3484	3904	4324	4744	5165	5585	6005	6425	
34	6846	7266	7686	8106	8527	8947	9367	9787	0̄208	0̄628	
35	01431048	1468	1889	2309	2729	3149	3569	3990	4410	4830	
36	5250	5670	6090	6511	6931	7351	7771	8191	8611	9032	
37	9452	9872	0̄292	0̄712	1̄132	1̄552	1̄972	2̄393	2̄813	3̄233	
38	01443653	4073	4493	4913	5333	5753	6173	6593	7013	7433	
39	7854	8274	8694	9114	9534	9954	0̄374	0̄794	1̄214	1̄634	
10340	01452054	2474	2894	3314	3734	4154	4574	4994	5414	5834	
41	6254	6674	7094	7514	7934	8354	8774	9193	9613	0̄033	
42	01460453	0873	1293	1713	2133	2553	2973	3393	3813	4233	420
43	4653	5072	5492	5912	6332	6752	7172	7592	8012	8431	
44	8851	9271	9691	0̄111	0̄530	0̄950	1̄370	1̄790	2̄210	2̄630	
45	01473049	3469	3889	4309	4729	5149	5568	5988	6408	6828	
46	7247	7667	8087	8507	8926	9346	9766	0̄186	0̄605	1̄025	
47	01481445	1865	2284	2704	3124	3544	3963	4383	4803	5222	
48	5642	6062	6481	6901	7321	7740	8160	8580	8990	9419	
49	9839	0̄258	0̄678	1̄098	1̄517	1̄937	2̄357	2̄776	3̄196	3̄615	
N.	0	1	2	3	4	5	6	7	8	9	D

Pro.

422	421	420
1 42	1 42	1 42
2 84	2 84	2 84
3 127	3 126	3 126
4 169	4 168	4 168
5 211	5 211	5 210
6 253	6 253	6 252
7 295	7 295	7 294
8 338	8 337	8 336
9 380	9 379	9 378

Pts.

N.	0	1	2	3	4	5	6	7	8	9	D	Pro.
10350	01494035	4455	4874	5294	5713	6133	6553	6972	7392	7811		
51	8231	8651	9070	9490	9909	0̄329	0̄748	1̄168	1̄587	2̄007	419	419
52	01502426	2846	3265	3685	4104	4524	4943	5363	5782	6202		
53	6621	7041	7460	7880	8299	8719	9138	9558	9977	0̄397		
54	01510816	1236	1655	2074	2494	2913	3333	3752	4172	4591		
55	5010	5430	5849	6269	6688	7107	7527	7946	8366	8785		
56	9204	9624	0̄043	0̄462	0̄882	1̄301	1̄720	2̄140	2̄559	2̄978		
57	01523398	3817	4236	4656	5075	5494	5913	6333	6752	7171		
58	7591	8010	8429	8848	9268	9687	0̄106	0̄525	0̄945	1̄364		
59	01531783	2203	2622	3041	3460	3879	4299	4718	5137	5556		
10360	5976	6395	6814	7233	7652	8071	8491	8910	9329	9748		
61	01540167	0587	1006	1425	1844	2263	2682	3101	3520	3940		
62	4359	4778	5197	5616	6035	6454	6873	7293	7712	8131		
63	8550	8969	9388	9807	0̄226	0̄645	1̄064	1̄483	1̄902	2̄321		
64	01552740	3159	3578	3997	4416	4836	5255	5674	6093	6512		
65	6931	7350	7769	8188	8607	9026	9445	9864	0̄283	0̄702		
66	01561120	1539	1958	2377	2796	3215	3634	4053	4472	4891		
67	5310	5729	6148	6567	6985	7404	7823	8242	8661	9080		
68	9499	9918	0̄337	0̄755	1174	1̄593	2̄012	2̄431	2̄850	3̄269		
69	01573688	4106	4525	4944	5363	5782	6200	6619	7038	7457		
10370	7876	8294	8713	9132	9551	9970	0̄388	0̄807	1̄226	1̄645	418	418
71	01582063	2482	2901	3320	3738	4157	4576	4995	5413	5832		
72	6251	6670	7088	7507	7926	8344	8763	9182	9600	0̄019		
73	01590438	0857	1275	1694	2113	2531	2950	3369	3787	4206		
74	4625	5043	5462	5880	6299	6718	7136	7555	7973	8392		
75	8811	9229	9648	0̄066	0̄485	0̄903	1̄322	1̄741	2̄159	2̄578		
76	01602996	3415	3833	4252	4670	5089	5508	5926	6345	6763		
77	7182	7600	8019	8437	8856	9274	9693	0̄111	0̄530	0̄948		
78	01611367	1785	2204	2622	3041	3459	3877	4296	4714	5133		
79	5551	5970	6388	6806	7225	7643	8062	8480	8899	9317		
10380	9735	0̄154	0̄572	0̄990	1̄409	1̄827	2̄246	2̄664	3̄082	3̄501		
81	01623919	4337	4756	5174	5592	6011	6429	6847	7266	7684		
82	8102	8521	8939	9357	9776	0̄194	0̄612	1̄031	1̄449	1̄867		
83	01632285	2704	3122	3540	3959	4377	4795	5213	5632	6050		
84	6468	6886	7304	7723	8141	8559	8977	9395	9814	0̄232		
85	01640650	1068	1487	1905	2323	2741	3159	3577	3996	4414		
86	4832	5250	5668	6086	6504	6922	7341	7759	8177	8595		
87	9013	9431	9849	0̄268	0̄686	1̄104	1̄522	1̄940	2̄358	2̄776		
88	01653194	3612	4030	4448	4866	5284	5702	6120	6539	6957		
89	7375	7793	8211	8629	9047	9465	9883	0̄301	0̄719	1̄137		
10390	01661555	1973	2391	2809	3227	3645	4063	4481	4899	5317		
91	5735	6152	6570	6988	7406	7824	8242	8660	9078	9496		
92	9914	0̄332	0̄750	1̄168	1̄585	2̄003	2̄421	2̄839	3̄257	3̄675	417	417
93	01674093	4511	4928	5346	5764	6182	6600	7018	7436	7853		
94	8271	8689	9107	9525	9942	0̄360	0̄778	1̄196	1̄614	2̄031		
95	01682449	2867	3285	3703	4121	4538	4956	5374	5792	6209		
96	6627	7045	7463	7880	8298	8716	9134	9551	9969	0̄387		
97	01690804	1222	1640	2058	2475	2893	3311	3728	4146	4563		
98	4981	5399	5817	6234	6652	7070	7487	7905	8323	8740		
99	9158	9575	9993	0̄411	0̄828	1̄246	1̄663	2̄081	2̄499	2̄916		
N.	0	1	2	3	4	5	6	7	8	9	D	Pts.

Pro. 419

1	42
2	84
3	126
4	168
5	210
6	251
7	293
8	335
9	377

Pro. 418

1	42
2	84
3	125
4	167
5	209
6	251
7	293
8	334
9	376

Pro. 417

1	42
2	83
3	125
4	167
5	209
6	250
7	292
8	334
9	375

N.	0	1	2	3	4	5	6	7	8	9	D	Pro.
10400	01703334	3752	4169	4587	5004	5422	5839	6257	6675	7092		
01	7510	7927	8345	8762	9180	9597	0̄015	0̄432	0̄850	1̄267	418	418
02	01711685	2102	2520	2937	3355	3772	4190	4607	5025	5442		1 42
03	5860	6277	6695	7112	7530	7947	8365	8782	9199	9617		2 84
04	01720034	0452	0869	1287	1704	2121	2539	2956	3374	3791		3 125 / 4 167
05	4208	4626	5043	5461	5878	6295	6713	7130	7547	7965		5 209
06	8382	8800	9217	9634	0̄052	0̄469	0̄886	1̄304	1̄721	2̄138		6 251
07	01732556	2973	3390	3807	4225	4642	5059	5477	5894	6311		7 293
08	6728	7146	7563	7980	8397	8815	9232	9649	0̄066	0̄484		8 334
09	01740901	1318	1735	2152	2570	2987	3404	3821	4238	4656		9 376
10410	5073	5490	5907	6324	6742	7159	7576	7993	8410	8827		
11	9245	9662	0̄079	0̄496	0̄913	1̄330	1̄747	2̄165	2̄582	2̄999		
12	01753416	3833	4250	4667	5084	5501	5919	6336	6753	7170		
13	7587	8004	8421	8838	9255	9672	0̄089	0̄506	0̄923	1̄340		
14	01761757	2174	2591	3008	3425	3842	4259	4676	5093	5510		
15	5927	6344	6761	7178	7595	8012	8429	8846	9263	9680		
16	01770097	0514	0931	1348	1765	2182	2599	3016	3433	3850		
17	4266	4683	5100	5517	5934	6351	6768	7185	7602	8019		
18	8435	8852	9269	9686	0̄103	0̄520	0̄936	1̄353	1̄770	2̄187		
19	01782604	3021	3437	3854	4271	4688	5105	5521	5938	6355		
10420	6772	7189	7606	8022	8439	8856	9273	9689	0̄106	0̄523		
21	01790940	1356	1773	2190	2607	3023	3440	3857	4273	4690	417	417
22	5107	5524	5940	6357	6774	7190	7607	8024	8441	8857		1 42
23	9274	9690	0̄107	0̄524	0̄940	1̄357	1̄774	2̄190	2̄607	3̄024		2 83
24	01803440	3857	4274	4690	5107	5523	5940	6357	6773	7190		3 125 / 4 167
25	7606	8023	8440	8856	9273	9689	0̄106	0̄522	0̄939	1̄356		5 209
26	01811772	2189	2605	3022	3438	3855	4271	4688	5104	5521		6 250
27	5937	6354	6770	7187	7603	8020	8436	8853	9269	9686		7 292
28	01820102	0519	0935	1352	1768	2185	2601	3017	3434	3850		8 334
29	4267	4683	5100	5516	5932	6349	6765	7182	7598	8014		9 375
10430	8431	8847	9264	9680	0̄096	0̄513	0̄929	1̄345	1̄762	2̄178		
31	01832595	3011	3427	3844	4260	4676	5092	5509	5925	6342		
32	6758	7174	7590	8007	8423	8839	9256	9672	0̄088	0̄505		
33	01840921	1337	1753	2169	2586	3002	3418	3834	4251	4667		
34	5083	5499	5916	6332	6748	7164	7580	7997	8413	8829		
35	9245	9662	0̄078	0̄494	0̄910	1̄326	1̄742	2̄159	2̄575	2̄991		
36	01853407	3823	4239	4655	5072	5488	5904	6320	6736	7152		
37	7568	7984	8401	8817	9233	9649	0̄065	0̄481	0̄897	1̄313		
38	01861729	2145	2561	2977	3393	3809	4226	4642	5058	5474		
39	5890	6306	6722	7138	7554	7970	8386	8802	9218	9634		
10440	01870050	0466	0882	1298	1714	2130	2546	2962	3378	3794		
41	4210	4626	5041	5457	5873	6289	6705	7121	7537	7953		
42	8369	8785	9201	9617	0̄033	0̄448	0̄864	1̄280	1̄696	2̄112	416	416
43	01882528	2944	3360	3775	4191	4607	5023	5439	5855	6270		1 42
44	6686	7102	7518	7934	8350	8765	9181	9597	0̄013	0̄429		2 83 / 3 125
45	01890844	1260	1676	2092	2508	2923	3339	3755	4171	4586		4 166
46	5002	5418	5834	6249	6665	7081	7497	7912	8328	8744		5 208
47	9159	9575	9991	0̄407	0̄822	1̄238	1̄654	2̄069	2̄485	2̄901		6 250
48	01903316	3732	4148	4563	4979	5395	5810	6226	6642	7057		7 291
49	7473	7889	8304	8720	9135	9551	9967	0̄382	0̄798	1̄213		8 333 / 9 374
N.	0	1	2	3	4	5	6	7	8	9	D	Pts

N.	0	1	2	3	4	5	6	7	8	9	D	Pro.
10450	01911629	2045	2460	2876	3291	3707	4122	4538	4954	5369		
51	5785	6200	6616	7031	7447	7862	8278	8694	9109	9525	415	415
52	9940	0̄356	0̄771	1̄187	1̄602	2̄018	2̄433	2̄849	3̄264	3̄680		1 42
53	01924095	4510	4926	5341	5757	6172	6588	7003	7419	7834		2 83
54	8250	8665	9080	9496	9911	0̄327	0̄742	1̄157	1̄573	1̄988		3 125 / 4 166
55	01932404	2819	3235	3650	4065	4481	4896	5311	5727	6142		5 208
56	6557	6973	7388	7804	8219	8634	9050	9465	9880	0̄296		6 249
57	01940711	1126	1541	1957	2372	2787	3203	3618	4033	4449		7 291
58	4864	5279	5694	6109	6525	6940	7355	7770	8186	8601		8 332
59	9016	9432	9847	0̄262	0̄677	1̄092	1̄508	1̄923	2̄338	2̄753		9 374
10460	01953168	3584	3999	4414	4829	5244	5659	6075	6490	6905		
61	7320	7735	8150	8566	8981	9396	9811	0̄226	0̄641	1̄056		
62	01961472	1887	2302	2717	3132	3547	3962	4377	4792	5208		
63	5623	6038	6453	6868	7283	7698	8113	8528	8943	9358		
64	9773	0̄188	0̄603	1̄018	1̄433	1̄848	2̄263	2̄678	3̄093	3̄508		
65	01973923	4338	4753	5168	5583	5998	6413	6828	7243	7658		
66	8073	8488	8903	9318	9733	0̄148	0̄563	0̄978	1̄393	1̄807		
67	01982222	2637	3052	3467	3882	4297	4712	5127	5542	5957		
68	6371	6786	7201	7616	8031	8446	8861	9275	9690	0̄105		
69	01990520	0935	1350	1764	2179	2594	3009	3424	3838	4253		
10470	4668	5083	5498	5913	6327	6742	7157	7572	7987	8401	414	414
71	8816	9231	9645	0̄060	0̄475	0̄890	1̄304	1̄719	2̄134	2̄549		1 41
72	02002963	3378	3793	4207	4622	5037	5452	5866	6281	6696		2 83
73	7110	7525	7940	8354	8769	9184	9598	0̄013	0̄428	0̄842		3 124
74	02011257	1672	2086	2501	2916	3330	3745	4159	4574	4989		4 166 / 5 207
75	5403	5818	6232	6647	7062	7476	7891	8305	8720	9135		6 248
76	9549	9964	0̄378	0̄793	1̄207	1̄622	2̄036	2̄451	2̄865	3̄280		7 290
77	02023694	4109	4523	4938	5352	5767	6181	6596	7010	7425		8 331
78	7839	8254	8668	9083	9497	9912	0̄326	0̄741	1̄155	1̄570		9 373
79	02031984	2399	2813	3227	3642	4056	4471	4885	5299	5714		
10480	6128	6543	6957	7372	7786	8200	8615	9029	9444	9858		
81	02040272	0687	1101	1515	1930	2344	2758	3173	3587	4001		
82	4416	4830	5244	5658	6073	6487	6901	7316	7730	8144		
83	8559	8973	9387	9801	0̄216	0̄630	1̄044	1̄458	1̄873	2̄287		
84	02052701	3116	3530	3944	4358	4772	5187	5601	6015	6429		
85	6843	7258	7672	8086	8500	8915	9329	9743	0̄157	0̄571		
86	02060985	1400	1814	2228	2642	3056	3470	3884	4299	4713		
87	5127	5541	5955	6369	6783	7197	7612	8026	8440	8854		
88	9268	9682	0̄096	0̄510	0̄924	1̄338	1̄752	2̄166	2̄581	2̄995		
89	02073409	3823	4237	4651	5065	5479	5893	6307	6721	7135		
10490	7549	7963	8377	8791	9205	9619	0̄033	0̄447	0̄861	1̄275		
91	02081689	2103	2517	2931	3345	3759	4173	4587	5000	5414	413	413
92	5828	6242	6656	7070	7484	7898	8312	8726	9140	9553		1 41
93	9967	0̄381	0̄795	1̄209	1̄623	2̄037	2̄451	2̄864	3̄278	3̄692		2 83
94	02094106	4520	4934	5347	5761	6175	6589	7003	7417	7831		3 124 / 4 165
95	8244	8658	9072	9486	9900	0̄313	0̄727	1̄141	1̄555	1̄969		5 207
96	02102382	2796	3210	3624	4037	4451	4865	5279	5692	6106		6 248
97	6520	6934	7347	7761	8175	8588	9002	9416	9829	0̄243		7 289
98	02110657	1071	1484	1898	2312	2725	3139	3553	3966	4380		8 330
99	4794	5207	5621	6035	6448	6862	7275	7689	8103	8516		9 372
N.	0	1	2	3	4	5	6	7	8	9	D	Pts.

N.	0	1	2	3	4	5	6	7	8	9	D	Pro.
10500	02118930	9344	9757	0̄171	0̄584	0̄998	1̄412	1̄825	2̄239	2̄652		
01	02123066	3479	3893	4307	4720	5134	5547	5961	6374	6788	414	414
02	7201	7615	8028	8442	8855	9269	9682	0̄096	0̄509	0̄923		1 41
03	02131337	1750	2164	2577	2991	3404	3817	4231	4644	5058		2 83
04	5471	5885	6298	6712	7125	7539	7952	8365	8779	9192		3 124
05	9606	0̄019	0̄433	0̄846	1̄259	1̄673	2̄086	2̄500	2̄913	3̄326		4 166
06	02143740	4153	4566	4980	5393	5807	6220	6633	7047	7460		5 207
07	7873	8287	8700	9113	9526	9940	0̄353	0̄766	1̄180	1̄593		6 248
08	02152006	2420	2833	3246	3660	4073	4486	4899	5313	5726		7 290
09	6139	6553	6966	7379	7792	8205	8619	9032	9445	9858		8 331; 9 373
10510	02160272	0685	1098	1511	1924	2338	2751	3164	3577	3990		
11	4404	4817	5230	5643	6056	6469	6882	7296	7709	8122		
12	8535	8948	9361	9775	0̄188	0̄601	1̄014	1̄427	1̄840	2̄253		
13	02172666	3080	3493	3906	4319	4732	5145	5558	5971	6384		
14	6797	7210	7623	8036	8450	8863	9276	9689	0̄102	0̄515		
15	02180928	1341	1754	2167	2580	2993	3406	3819	4232	4645		
16	5058	5471	5884	6297	6710	7123	7535	7948	8361	8774		
17	9187	9600	0̄013	0̄426	0̄839	1̄252	1̄665	2̄078	2̄491	2̄904		
18	02193317	3730	4142	4555	4968	5381	5794	6207	6620	7033		
19	7446	7858	8271	8684	9097	9510	9923	0̄336	0̄748	1̄161		
10520	02201574	1987	2400	2812	3225	3638	4051	4464	4876	5289		
21	5702	6115	6528	6941	7353	7766	8179	8592	9004	9417	413	413
22	9830	0̄242	0̄655	1̄068	1̄481	1̄893	2̄306	2̄719	3̄132	3̄544		1 41
23	02213957	4370	4782	5195	5608	6021	6433	6846	7259	7671		2 83
24	8084	8497	8909	9322	9735	0̄147	0̄560	0̄973	1̄385	1̄798		3 124
25	02222210	2623	3036	3448	3861	4273	4686	5099	5511	5924		4 165
26	6337	6749	7162	7574	7987	8400	8812	9225	9637	0̄050		5 207
27	02230462	0875	1288	1700	2113	2525	2938	3350	3763	4175		6 248
28	4588	5000	5413	5825	6238	6650	7063	7475	7888	8300		7 289
29	8713	9125	9538	9950	0̄363	0̄775	1̄188	1̄600	2̄012	2̄425		8 330; 9 372
10530	02242837	3250	3662	4074	4487	4899	5312	5724	6137	6549		
31	6961	7374	7786	8199	8611	9023	9436	9848	0̄261	0̄673		
32	02251085	1497	1910	2322	2735	3147	3559	3972	4384	4796		
33	5208	5621	6033	6445	6858	7270	7682	8095	8507	8919		
34	9331	9744	0̄156	0̄568	0̄981	1̄393	1̄805	2̄217	2̄630	3042		
35	02263454	3866	4279	4691	5103	5515	5927	6340	6752	7164		
36	7576	7988	8401	8813	9225	9637	0̄049	0̄462	0̄874	1̄286		
37	02271698	2110	2522	2935	3347	3759	4171	4583	4995	5407		
38	5819	6231	6644	7056	7468	7880	8292	8704	9116	9528		
39	9940	0̄353	0̄765	1̄177	1̄589	2̄001	2̄413	2̄825	3̄237	3̄649		
10540	02284061	4473	4885	5297	5709	6121	6533	6945	7357	7769		
41	8181	8593	9005	9417	9829	0̄241	0̄653	1̄065	1̄477	1̄889		
42	02292301	2713	3125	3537	3949	4361	4773	5185	5597	6009	412	412
43	6421	6833	7245	7656	8068	8480	8892	9304	9716	0̄128		1 41
44	02300540	0952	1364	1775	2187	2599	3011	3423	3835	4247		2 82
45	4658	5070	5482	5894	6306	6718	7130	7541	7953	8365		3 124
46	8777	9189	9600	0̄012	0̄424	0̄836	1̄248	1̄659	2̄071	2̄483		4 165
47	02312895	3306	3718	4130	4542	4954	5365	5777	6189	6600		5 206
48	7012	7423	7835	8247	8659	9071	9482	9894	0̄306	0̄718		6 247; 7 288
49	02321129	1541	1953	2364	2776	3188	3599	4011	4423	4834		8 330; 9 371
N.	0	1	2	3	4	5	6	7	8	9	D	Pts.

N.	0	1	2	3	4	5	6	7	8	9	D	Pro.
10550	02325246	5657	6069	6481	6893	7304	7716	8127	8539	8951		
51	9362	9774	0̄186	0̄597	1̄009	1̄420	1̄832	2̄244	2̄655	3̄067	411	411
52	02333478	3890	4301	4713	5124	5536	5948	6359	6771	7182		1| 41
53	7594	8005	8417	8829	9240	9652	0̄063	0̄475	0̄886	1̄298		2| 82
54	02341709	2121	2532	2944	3355	3767	4178	4590	5001	5412		3|123
												4|164
55	5824	6235	6647	7058	7470	7881	8292	8704	9115	9527		5 205
56	9938	0̄350	0̄761	1̄172	1̄584	1̄995	2̄407	2̄818	3̄229	3̄641		6 247
57	02354052	4464	4875	5286	5698	6109	6520	6932	7343	7755		7 288
58	8166	8577	8989	9400	9811	0̄223	0̄634	1̄045	1̄456	1̄868		8 329
59	02362279	2690	3102	3513	3924	4336	4747	5158	5569	5981		9 370
10560	6392	6803	7214	7626	8037	8448	8859	9271	9682	0̄093		
61	02370504	0916	1327	1738	2149	2560	2972	3383	3794	4205		
62	4616	5028	5439	5850	6261	6672	7083	7495	7906	8317		
63	8728	9139	9550	9961	0̄373	0̄784	1̄195	1̄606	2̄017	2̄428		
64	02382839	3250	3661	4073	4484	4895	5306	5717	6128	6539		
65	6950	7361	7772	8183	8594	9005	9416	9828	0̄239	0̄650		
66	02391061	1472	1883	2294	2705	3116	3527	3938	4349	4760		
67	5171	5582	5993	6404	6815	7226	7637	8048	8459	8870		
68	9281	9692	0̄103	0̄514	0̄924	1̄335	1̄746	2̄157	2̄568	2̄979		
69	02403390	3801	4212	4623	5033	5444	5855	6266	6677	7088		
10570	7499	7910	8321	8731	9142	9553	9964	0̄375	0̄786	1̄196		
71	02411607	2018	2429	2840	3251	3662	4072	4483	4894	5305	410	410
72	5715	6126	6537	6948	7359	7769	8180	8591	9002	9413		1| 41
73	9823	0̄234	0̄645	1̄056	1̄466	1̄877	2̄288	2̄699	3̄109	3̄520		2| 82
74	02423931	4341	4752	5163	5573	5984	6395	6806	7216	7627		3|123
												4|164
75	8038	8448	8859	9270	9680	0̄091	0̄502	0̄912	1̄323	1̄734		5|205
76	02432144	2555	2966	3376	3787	4197	4608	5019	5429	5840		6|246
77	6250	6661	7072	7482	7893	8303	8714	9125	9535	9946		7|287
78	02440356	0767	1178	1588	1999	2409	2820	3230	3641	4051		8|328
79	4462	4872	5283	5693	6104	6514	6925	7335	7746	8156		9|369
10580	8567	8977	9388	9798	0̄209	0̄619	1̄030	1̄440	1̄851	2̄261		
81	02452671	3082	3492	3903	4313	4724	5134	5545	5955	6365		
82	6776	7186	7597	8007	8417	8828	9238	9649	0̄059	0̄469		
83	02460880	1290	1700	2111	2521	2932	3342	3752	4163	4573		
84	4983	5394	5804	6214	6624	7035	7445	7855	8266	8676		
85	9086	9497	9907	0̄317	0̄727	1̄138	1̄548	1̄958	2̄369	2̄779		
86	02473189	3599	4010	4420	4830	5240	5651	6061	6471	6881		
87	7291	7702	8112	8522	8932	9342	9753	0̄163	0̄573	0̄983		
88	02481393	1804	2214	2624	3034	3444	3854	4265	4675	5085		
89	5495	5905	6315	6725	7135	7546	7956	8366	8776	9186		
10590	9596	0̄006	0̄416	0̄826	1̄236	1̄647	2̄057	2̄467	2̄877	3̄287		
91	02493697	4107	4517	4927	5337	5747	6157	6567	6977	7387		
92	7797	8207	8617	9027	9437	9847	0̄257	0̄667	1̄077	1̄487	409	409
93	02501897	2307	2717	3127	3537	3947	4357	4767	5177	5587		1| 41
94	5997	6407	6817	7227	7637	8047	8457	8866	9276	9686		2| 82
												3|123
95	02510096	0506	0916	1326	1736	2146	2556	2965	3375	3785		4|164
96	4195	4605	5015	5425	5835	6245	6654	7064	7474	7884		5|205
97	8293	8703	9113	9523	9933	0̄343	0̄752	1̄162	1̄572	1̄982		6 245
98	02522392	2801	3211	3621	4031	4441	4850	5260	5670	6080		7 286
99	6489	6899	7309	7719	8128	8539	8948	9357	9767	0̄177		8 327
												9 368
N.	0	1	2	3	4	5	6	7	8	9	D	Pts.

N.	0	1	2	3	4	5	6	7	8	9	D	Pro.
10600	02530587	0996	1406	1816	2225	2635	3045	3454	3864	4274		
01	4683	5093	5503	5913	6322	6732	7142	7551	7961	8370	410	410
02	8780	9190	9599	0̄009	0̄419	0̄828	1̄238	1̄647	2̄057	2̄467		1|41
03	02542876	3286	3695	4105	4515	4924	5334	5743	6153	6562		2|82
04	6972	7382	7791	8201	8610	9020	9429	9839	0̄248	0̄658		3|123 4|164
05	02551067	1477	1886	2296	2705	3115	3524	3934	4343	4753		5|205
06	5162	5572	5981	6391	6800	7209	7619	8029	8438	8848		6|246
07	9257	9666	0̄076	0̄485	0̄895	1̄304	1̄714	2̄123	2̄532	2̄942		7|287
08	02563351	3761	4170	4579	4989	5398	5808	6217	6626	7036		8|328
09	7445	7854	8264	8673	9083	9492	9901	0̄310	0̄720	1̄129		9|369
10610	02571538	1948	2357	2766	3176	3585	3994	4404	4813	5222		
11	5631	6041	6450	6859	7269	7678	8087	8497	8906	9315		
12	9724	0̄133	0̄543	0̄952	1̄361	1̄770	2̄180	2̄589	2̄998	3̄407		
13	02583816	4226	4635	5044	5453	5862	6272	6681	7090	7499		
14	7908	8318	8727	9136	9545	9954	0̄363	0̄773	1̄182	1̄591		
15	02592000	2409	2818	3227	3636	4046	4455	4864	5273	5682		
16	6091	6500	6909	7318	7727	8137	8546	8955	9364	9773		
17	02600182	0591	1000	1409	1818	2227	2636	3045	3454	3863		
18	4272	4681	5090	5499	5908	6317	6726	7135	7544	7953		
19	8362	8771	9180	9589	9998	0̄407	0̄816	1̄225	1̄634	2̄043		
10620	02612452	2861	3270	3679	4088	4496	4905	5314	5723	6132	409	409
21	6541	6950	7359	7768	8177	8585	8994	9403	9812	0̄221		1|41
22	02620630	1039	1448	1856	2265	2674	3083	3492	3901	4309		2|82
23	4718	5127	5536	5945	6353	6762	7171	7580	7989	8397		3|123 4|104
24	8806	9215	9624	0̄033	0̄441	0̄850	1̄259	1̄668	2̄077	2̄485		5|205
25	02632894	3303	3712	4120	4529	4938	5346	5755	6164	6573		6|245
26	6981	7390	7799	8207	8616	9025	9433	9842	0̄251	0̄660		7|286
27	02641068	1477	1886	2294	2703	3112	3520	3929	4337	4746		8|327
28	5155	5563	5972	6381	6789	7198	7606	8015	8424	8832		9|368
29	9241	9649	0̄058	0̄467	0̄875	1̄284	1̄692	2̄101	2̄509	2̄918		
10630	02653326	3735	4144	4552	4961	5369	5778	6186	6595	7003		
31	7412	7820	8229	8637	9046	9454	9863	0̄271	0̄680	1̄088		
32	02661497	1905	2314	2722	3131	3539	3948	4356	4765	5173		
33	5581	5990	6398	6807	7215	7624	8032	8440	8849	9257		
34	9666	0̄074	0̄482	0̄891	1̄299	1̄708	2̄116	2̄524	2̄933	3̄341		
35	02673749	4158	4566	4975	5383	5791	6200	6608	7016	7425		
36	7833	8241	8650	9058	9466	9874	0̄283	0̄691	1̄099	1̄508		
37	02681916	2324	2733	3141	3549	3957	4366	4774	5182	5590		
38	5999	6407	6815	7224	7632	8040	8448	8856	9265	9673		
39	02690081	0489	0897	1306	1714	2122	2530	2938	3347	3755		
10640	4163	4571	4979	5387	5796	6204	6612	7020	7428	7836		
41	8244	8653	9061	9469	9877	0̄285	0̄693	1̄101	1̄509	1̄917		
42	02702326	2734	3142	3550	3958	4366	4774	5182	5590	5998	408	408
43	6406	6814	7222	7631	8039	8447	8855	9263	9671	0̄079		1|41
44	02710487	0895	1303	1711	2119	2527	2935	3343	3751	4159		2|82 3|122
45	4567	4975	5383	5791	6199	6607	7015	7423	7830	8238		4|163
46	8646	9054	9462	9870	0̄278	0̄686	1̄094	1̄502	1̄910	2̄318		5|204
47	02722725	3133	3541	3949	4357	4765	5173	5581	5989	6396		6|245
48	6804	7212	7620	8028	8436	8844	9252	9659	0̄067	0̄475		7|286 8|326
49	02730883	1291	1698	2106	2514	2922	3330	3737	4145	4553		9|367
N.	0	1	2	3	4	5	6	7	8	9	D	Pts.

N.	0	1	2	3	4	5	6	7	8	9	D	Pro.
10650	02734961	5369	5776	6184	6592	7000	7407	7815	8223	8631		
51	9039	9446	9854	0̄262	0̄669	1̄077	1̄485	1̄893	2̄300	2̄708	407	407
52	02743116	3524	3931	4339	4747	5154	5562	5970	6377	6785		1 41
53	7193	7600	8008	8416	8823	9231	9639	0̄046	0̄454	0̄862		2 81
54	02751269	1677	2085	2492	2900	3307	3715	4123	4530	4938		3 122 / 4 163
55	5345	5753	6161	6568	6976	7383	7791	8199	8606	9014		5 204
56	9421	9829	0̄236	0̄644	1̄051	1̄459	1̄866	2̄274	2̄682	3̄089		6 244 / 7 285
57	02763497	3904	4312	4719	5127	5534	5942	6349	6757	7164		8 326
58	7572	7979	8387	8794	9201	9609	0̄016	0̄424	0̄831	1̄239		9 366
59	02771646	2054	2461	2869	3276	3683	4091	4498	4906	5313		
10660	5720	6128	6535	6943	7350	7758	8165	8572	8980	9387		
61	9794	0̄202	0̄609	1̄016	1̄424	1̄831	2̄238	2̄646	3̄053	3̄460		
62	02783868	4275	4682	5090	5497	5904	6312	C719	7126	7534		
63	7941	8348	8756	9163	9570	9977	0̄385	0̄792	1̄199	1̄606		
64	02792014	2421	2828	3235	3643	4050	4457	4864	5271	5679		
65	6086	6493	6900	7308	7715	8122	8529	8936	9344	9751		
66	02800158	0565	0972	1379	1787	2194	2601	3008	3415	3822		
67	4230	4637	5044	5451	5858	6265	6672	7079	7487	7894		
68	8301	8708	9115	9522	9929	0̄336	0̄743	1̄150	1̄558	1̄965		
69	02812372	2779	3186	3593	4000	4407	4814	5221	5628	6035	406	406
10670	6442	6849	7256	7663	8070	8477	8884	9291	9698	0̄105		1 41
71	02820512	0919	1326	1733	2140	2547	2954	3361	3768	4175		2 81
72	4582	4989	5396	5803	6209	6616	7023	7430	7837	8244		3 122
73	8651	9058	9465	9872	0̄279	0̄685	1̄092	1̄499	1̄906	2̄313		4 162 / 5 203
74	02832720	3127	3534	3940	4347	4754	5161	5568	5975	6382		6 244 / 7 284
75	6788	7195	7602	8009	8416	8823	9229	9636	0̄043	0̄450		8 325
76	02840857	1263	1670	2077	2484	2891	3297	3704	4111	4518		9 365
77	4924	5331	5738	6145	6551	6958	7365	7772	8178	8585		
78	8992	9398	9805	0̄212	0̄618	1̄025	1̄432	1̄839	2̄245	2̄652		
79	02853059	3465	3872	4279	4685	5092	5499	5905	6312	6719		
10680	7125	7532	7939	8345	8752	9159	9565	9972	0̄378	0̄785		
81	02861192	1598	2005	2411	2818	3225	3631	4038	4444	4851		
82	5257	5664	6071	6477	6884	7290	7697	8103	8510	8916		
83	9323	9729	0̄136	0̄542	0̄949	1̄355	1̄762	2̄168	2̄575	2̄981		
84	02873388	3794	4201	4607	5014	5420	5827	6233	6640	7046		
85	7453	7859	8265	8672	9078	9485	9891	0̄298	0̄704	1̄111		
86	02881517	1923	2330	2736	3143	3549	3955	4362	4768	5175		
87	5581	5987	6394	6800	7206	7613	8019	8425	8832	9238		
88	9645	0̄051	0̄457	0̄864	1̄270	1̄676	2̄083	2̄489	2̄995	3̄301		
89	02893708	4114	4520	4927	5333	5739	6146	6552	6958	7364		
10690	7771	8177	8583	8989	9395	9802	0̄208	0̄614	1̄020	1̄427		
91	02901833	2239	2645	3052	3458	3864	4270	4676	5083	5489		
92	5895	6301	6707	7114	7520	7926	8332	8738	9144	9550	405	405
93	9957	0̄363	0̄769	1̄175	1̄581	1̄987	2̄394	2̄800	3̄206	3̄612		1 41
94	02914018	4424	4830	5236	5642	6049	6455	6861	7267	7673		2 81 / 3 122
95	8079	8485	8891	9297	9703	0̄109	0̄515	0̄921	1̄327	1̄733		4 162
96	02922139	2546	2952	3358	3764	4170	4576	4982	5388	5794		5 203
97	6200	6606	7012	7418	7824	8230	8635	9041	9447	9853		6 243 / 7 284
98	02930259	0665	1071	1477	1883	2289	2695	3101	3507	3913		8 324
99	4319	4725	5131	5536	5942	6348	6754	7160	7566	7972		9 365
N	0	1	2	3	4	5	6	7	8	9	D	Pts.

N.	0	1	2	3	4	5	6	7	8	9	D
10700	02938378	8784	9190	9595	0̄001	0̄407	0̄813	1̄219	1̄625	2̄031	
01	02942436	2842	3248	3654	4060	4465	4871	5277	5683	6089	406
02	6495	6901	7307	7712	8118	8524	8930	9335	9741	0̄147	
03	02950553	0958	1364	1770	2176	2581	2987	3393	3799	4205	
04	4610	5016	5422	5827	6233	6639	7044	7450	7856	8261	
05	8667	9073	9479	9884	0̄290	0̄696	1̄101	1̄507	1̄913	2̄318	
06	02962724	3130	3535	3941	4347	4752	5158	5563	5969	6375	
07	6780	7186	7592	7997	8403	8808	9214	9620	0̄025	0̄431	
08	02970836	1242	1647	2053	2458	2864	3270	3675	4081	4486	
09	4892	5298	5703	6109	6514	6920	7325	7731	8136	8542	
10710	8947	9353	9758	0̄164	0̄569	0̄975	1̄380	1̄786	2̄191	2̄597	
11	02983002	3407	3813	4218	4624	5029	5435	5840	6246	6651	
12	7056	7462	7867	8273	8678	9084	9489	9894	0̄300	0̄705	
13	02991111	1516	1921	2327	2732	3137	3543	3948	4354	4759	
14	5164	5569	5975	6380	6786	7191	7596	8002	8407	8812	
15	9218	9623	0̄028	0̄433	0̄839	1̄244	1̄649	2̄055	2̄460	2̄865	
16	03003271	3676	4081	4486	4892	5297	5702	6107	6513	6918	
17	7323	7728	8134	8539	8944	9349	9755	0̄160	0̄565	0̄970	
18	03011375	1781	2186	2591	2996	3401	3807	4212	4617	5022	
19	5427	5832	6238	6643	7048	7453	7858	8263	8668	9073	
10720	9479	9884	0̄289	0̄694	1̄099	1̄504	1̄909	2̄314	2̄719	3̄124	
21	03023529	3935	4340	4745	5150	5555	5960	6365	6770	7175	
22	7580	7985	8391	8796	9201	9606	0̄011	0̄416	0̄821	1̄226	405
23	03031631	2036	2441	2846	3251	3656	4061	4466	4871	5276	
24	5681	6086	6491	6896	7301	7706	8111	8515	8920	9325	
25	9730	0̄135	0̄540	0̄945	1̄350	1̄755	2̄160	2̄565	2̄970	3̄374	
26	03043779	4184	4589	4994	5399	5804	6209	6614	7019	7423	
27	7828	8233	8638	9043	9448	9853	0̄257	0̄662	1̄067	1̄472	
28	03051877	2281	2686	3091	3496	3901	4305	4710	5115	5520	
29	5925	6329	6734	7139	7544	7949	8353	8758	9163	9568	
10730	9972	0̄377	0̄782	1̄187	1̄591	1̄996	2̄401	2̄805	3̄210	3̄615	
31	03064020	4424	4829	5234	5638	6043	6448	6852	7257	7662	
32	8066	8471	8876	9281	9685	0̄090	0̄495	0̄899	1̄304	1̄708	
33	03072113	2518	2922	3327	3732	4136	4541	4945	5350	5755	
34	6159	6564	6968	7373	7777	8182	8587	8991	9396	9800	
35	03080205	0610	1014	1419	1823	2228	2632	3037	3441	3846	
36	4250	4655	5059	5464	5868	6273	6677	7082	7486	7891	
37	8295	8700	9104	9509	9913	0̄318	0̄722	1̄127	1̄531	1̄936	
38	03092340	2745	3149	3553	3958	4362	4767	5171	5575	5980	
39	6384	6789	7193	7597	8002	8406	8811	9215	9619	0̄024	
10740	03100428	0833	1237	1641	2046	2450	2854	3259	3663	4067	
41	4472	4876	5280	5685	6089	6493	6898	7302	7706	8111	
42	8515	8919	9323	9728	0̄132	0̄536	0̄941	1̄345	1̄749	2̄153	404
43	03112558	2962	3366	3771	4175	4579	4983	5387	5792	6196	
44	6600	7004	7408	7813	8217	8621	9025	9429	9834	0̄238	
45	03120642	1046	1450	1855	2259	2663	3067	3471	3875	4280	
46	4684	5088	5492	5896	6300	6704	7109	7513	7917	8321	
47	8725	9129	9533	9937	0̄341	0̄745	1̄150	1̄554	1̄958	2̄362	
48	03132766	3170	3574	3978	4382	4786	5190	5594	5998	6402	
49	6806	7210	7614	8018	8422	8826	9230	9634	0̄038	0̄442	
N.	0	1	2	3	4	5	6	7	8	9	D

Pro. (Proportional parts)

	406	405	404
1	41	41	40
2	81	81	81
3	122	122	121
4	162	162	162
5	203	203	202
6	244	243	242
7	284	284	283
8	325	324	323
9	365	365	364

N.	0	1	2	3	4	5	6	7	8	9	D	Pro.
10750	03140846	1250	1654	2058	2462	2866	3270	3674	4078	4482		404
51	4886	5290	5694	6098	6502	6906	7310	7714	8118	8522	404	404
52	8926	9330	9733	0̄137	0̄541	0̄945	1̄349	1̄753	2̄157	2̄561		1|40
53	03152965	3369	3772	4176	4580	4984	5388	5792	6196	6599		2|81
54	7003	7407	7811	8215	8619	9023	9426	9830	0̄234	0̄638		3|121
												4|162
55	03161042	1445	1849	2253	2657	3061	3464	3868	4272	4676		5|202
56	5080	5483	5887	6291	6695	7098	7502	7906	8310	8713		6|242
57	9117	9521	9924	0̄328	0̄732	1̄136	1̄539	1̄943	2̄347	2̄750		7|283
58	03173154	3558	3961	4365	4769	5173	5576	5980	6384	6787		8|323
59	7191	7594	7998	8402	8806	9209	9613	0̄016	0̄420	0̄824		9|364
10760	03181227	1631	2034	2438	2842	3245	3649	4052	4456	4860		
61	5263	5667	6070	6474	6878	7281	7685	8088	8492	8895		
62	9299	9702	0̄106	0̄510	0̄913	1̄317	1̄720	2̄124	2̄527	2̄931		
63	03193334	3738	4141	4545	4948	5352	5755	6159	6562	6966		
64	7369	7772	8176	8579	8983	9386	9790	0̄193	0̄597	1̄000		
65	03201403	1807	2210	2614	3017	3421	3824	4227	4631	5034		
66	5438	5841	6244	6648	7051	7455	7858	8261	8665	9068		
67	9471	9875	0̄278	0̄682	1̄085	1̄488	1̄892	2̄295	2̄698	3101		
68	03213505	3908	4311	4715	5118	5521	5925	6328	6731	7134		
69	7538	7941	8344	8748	9151	9554	9958	0̄361	0̄764	1167		
10770	03221570	1974	2377	2780	3183	3587	3990	4393	4796	5199		
71	5603	6006	6409	6812	7215	7619	8022	8425	8828	9231		
72	9635	0̄038	0̄441	0̄844	1̄247	1̄651	2̄054	2̄457	2̄860	3263	403	403
73	03233666	4069	4472	4875	5279	5682	6085	6488	6891	7294		1|40
74	7697	8100	8503	8906	9310	9713	0̄116	0̄519	0̄922	1̄325		2|81
												3|121
75	03241728	2131	2534	2937	3340	3743	4146	4549	4952	5355		4|161
76	5758	6161	6564	6967	7370	7773	8176	8579	8982	9385		5|202
77	9788	0̄191	0̄594	0̄997	1̄400	1̄803	2̄206	2̄609	3012	3415		6|242
78	03253818	4221	4624	5027	5430	5833	6236	6639	7041	7444		7|282
79	7847	8250	8653	9056	9459	9862	0̄265	0̄668	1̄070	1̄473		8|322
												9|363
10780	03261876	2279	2682	3085	3488	3891	4293	4696	5099	5502		
81	5905	6308	6710	7113	7516	7919	8322	8724	9127	9530		
82	9933	0̄336	0̄738	1̄141	1̄544	1̄947	2̄350	2̄752	3155	3558		
83	03273961	4363	4766	5169	5572	5974	6377	6780	7183	7585		
84	7988	8391	8793	9196	9599	0̄001	0̄404	0̄807	1̄210	1̄612		
85	03282015	2418	2820	3223	3626	4028	4431	4834	5236	5639		
86	6042	6444	6847	7250	7652	8055	8458	8860	9263	9665		
87	03290068	0471	0873	1276	1678	2081	2484	2886	3289	3691		
88	4094	4496	4899	5302	5704	6107	6509	6912	7314	7717		
89	8119	8522	8925	9327	9730	0̄132	0̄535	0̄937	1̄340	1̄742		
10790	03302145	2547	2950	3352	3755	4157	4560	4962	5365	5767		
91	6169	6572	6974	7377	7779	8182	8584	8987	9389	9791		
92	03310194	0596	0999	1401	1803	2206	2608	3011	3413	3815	402	402
93	4218	4620	5023	5425	5827	6230	6632	7035	7437	7839		1|40
94	8241	8644	9046	9449	9851	0̄253	0̄656	1̄058	1̄460	1̄862		2|80
												3|121
95	03322265	2667	3069	3472	3874	4276	4679	5061	5483	5885		4|161
96	6288	6690	7092	7495	7897	8299	8701	9104	9506	9908		5|201
97	03330310	0712	1115	1517	1919	2321	2724	3126	3528	3930		6|241
98	4332	4735	5137	5539	5941	6343	6746	7148	7550	7952		7|281
99	8354	8756	9159	9561	9963	0̄365	0̄767	1̄169	1̄571	1̄973		8|322
												9|362
N.	0	1	2	3	4	5	6	7	8	9	D	Pts.

TABLE II.

For finding Logarithms and Numbers to 20 Places of Figures.

N.	Logarithms.	N.	Logarithms.
1	00000 00000 00000 00000	51	70757 01760 97936 36584
2	30102 99956 63981 19521	52	71600 33436 34799 15963
3	47712 12547 19662 43730	53	72427 58696 00789 04563
4	60205 99913 27962 39043	54	73239 37598 22968 50710
5	69897 00043 36018 80479	55	74036 26894 94243 84554
6	77815 12503 83643 63251	56	74818 80270 06200 41635
7	84509 80400 14256 83071	57	75587 48556 72491 39883
8	90308 99869 91943 58564	58	76342 79935 62937 28255
9	95424 25094 39324 87459	59	77085 20116 42144 19026
10	00000 00000 00000 00000	60	77815 12503 83643 63251
11	04139 26851 58225 04075	61	78532 98350 10767 03389
12	07918 12460 47624 82772	62	79239 16894 98253 87488
13	11394 33523 06836 76921	63	79934 05494 53581 70530
14	14612 80356 78238 02593	64	80617 99739 83887 17128
15	17609 12590 55681 24208	65	81291 33566 42855 57399
16	20411 99826 55924 78085	66	81954 39355 41868 67326
17	23044 89213 78273 92854	67	82607 48027 00826 43415
18	25527 25051 03306 06980	68	83250 89127 06236 31897
19	27875 36009 52828 96154	69	83884 90907 37255 31616
20	30102 99956 63981 19521	70	84509 80400 14256 83071
21	32221 92947 33919 26801	71	85125 83487 19075 28609
22	34242 26808 22206 23596	72	85733 24964 31268 46023
23	36172 78360 17592 87887	73	86332 28601 20455 90107
24	38021 12417 11606 02294	74	86923 17197 30976 19202
25	39794 00086 72037 60957	75	87506 12633 91700 04687
26	41497 33479 70817 96442	76	88081 35922 80791 35196
27	43136 37641 58987 31189	77	88649 07251 72481 87146
28	44715 80313 42219 22114	78	89209 46026 90480 40172
29	46239 79978 98956 08733	79	89762 70912 90441 42799
30	47712 12547 19662 43730	80	90308 99869 91943 58564
31	49136 16938 34272 67967	81	90848 50188 78649 74918
32	50514 99783 19905 97607	82	91361 38523 83716 68972
33	51851 39398 77887 47805	83	91907 80923 76073 90383
34	53147 89170 42255 12375	84	92427 92860 61881 65843
35	54406 80443 50275 63550	85	92941 89257 14292 73333
36	55630 25007 67287 26502	86	93449 84512 43567 72162
37	56820 17240 66994 99681	87	93951 92526 18618 52463
38	57978 35966 16810 15675	88	94448 26721 50168 62639
39	59106 46070 26499 20650	89	94939 00066 44912 78472
40	60205 99913 27962 39043	90	95424 25094 39324 87459
41	61278 38567 19735 49451	91	95904 13923 21093 59992
42	62324 92903 97900 46322	92	96378 78273 45555 26930
43	63346 84555 79586 52641	93	96848 29485 53935 11696
44	64345 26764 86187 43118	94	97312 78535 99698 65963
45	65321 25137 75343 67938	95	97772 36052 88847 76632
46	66275 78316 81574 07408	96	98227 12330 39568 41336
47	67209 78579 35717 46441	97	98677 17342 66244 85178
48	68124 12373 75587 21815	98	99122 60756 92494 85664
49	69019 60800 28513 66142	99	99563 51945 97549 91534
50	69897 00043 36018 80479	100	00000 00000 00000 00000

Tab. 2.   LOGARITHMS TO 20 PLACES.   203

N.	Logarithms.				N.	Logarithms.			
101	00432	13737	82642	57428	151	17897	69472	93169	43687
102	00860	01717	61917	56105	152	18184	35879	44772	54718
103	01283	72247	05172	20517	153	18469	14308	17598	80313
104	01703	33392	98780	35485	154	18752	07208	36463	06668
105	02118	92990	69938	07279	155	19033	16981	70291	48445
106	02530	58652	64770	24085	156	19312	45983	54461	59693
107	02938	37776	85209	64083	157	19589	96524	09233	73676
108	03342	37554	86949	70231	158	19865	70869	54422	62321
109	03742	64979	40623	63520	159	20139	71243	20451	48293
110	04139	26851	58225	04075	160	20411	99826	55924	78085
111	04532	29787	86657	43410	161	20682	58760	31849	70958
112	04921	80226	70181	61157	162	20951	50145	42630	94439
113	05307	84434	83419	72280	163	21218	76044	03957	80764
114	05690	48513	36472	59405	164	21484	38480	47697	88494
115	06069	78403	53611	68365	165	21748	39442	13906	28283
116	06445	79892	26918	47776	166	22010	80880	40055	09905
117	06818	58617	46161	64380	167	22271	64711	47583	27998
118	07188	20073	06125	38547	168	22530	92817	25862	85365
119	07554	69613	92530	75925	169	22788	67046	13673	53841
120	07918	12460	47624	82772	170	23044	89213	78273	92654
121	08278	53703	16450	08150	171	23299	61103	92153	83613
122	08635	98306	74748	22910	172	23552	84469	07548	91683
123	08990	51114	39397	93180	173	23804	61031	28795	41456
124	09342	16851	62235	07009	174	24054	92482	82599	71984
125	09691	00130	08056	41436	175	24303	80486	86294	44028
126	10037	05451	17562	90052	176	24551	26678	14149	82161
127	10380	37209	55956	86425	177	24797	32663	61806	62756
128	10720	99696	47868	36650	178	25042	00023	08893	97994
129	11058	97102	99248	96370	179	25285	30309	79893	16957
130	11394	33523	06836	76921	180	25527	25051	03306	06980
131	11727	12956	55764	26081	181	25767	85748	69184	51029
132	12057	39312	05849	86847	182	26007	13879	85074	79513
133	12385	16409	67085	79225	183	26245	10897	30429	47118
134	12710	47983	64807	62936	184	26481	78230	09536	46451
135	13033	37684	95006	11667	185	26717	17284	03013	80159
136	13353	89083	70217	51418	186	26951	29442	17916	31218
137	13672	05671	56406	76856	187	27184	16065	36498	96929
138	13987	90864	01236	51138	188	27415	78492	63679	85484
139	14301	48002	54095	08046	189	27646	18041	73244	14260
140	14612	80356	78238	02593	190	27875	36009	52828	96154
141	14921	91126	55379	90171	191	28103	33672	47727	53764
142	15228	83443	83056	48131	192	28330	12287	03549	60858
143	15533	60374	65061	80996	193	28555	73090	07773	76060
144	15836	24920	95249	65545	194	28780	17299	30226	04700
145	16136	80022	34974	89212	195	29003	46113	62518	01129
146	16435	28557	84437	09629	196	29225	60713	56476	05185
147	16731	73347	48176	09872	197	29446	62261	61592	92737
148	17026	17153	94957	38724	198	29666	51902	61531	11035
149	17318	62684	12274	03826	199	29885	30764	09706	65010
150	17609	12590	55681	24208	200	30102	99956	63981	19521

N.	Logarithms.				N.	Logarithms.			
201	30319	60574	20488	87144	251	39967	37214	81038	13934
202	30535	13694	46623	76949	252	40140	05407	81544	09573
203	30749	60379	13212	91805	253	40312	05211	75817	91962
204	30963	01674	25898	75626	254	40483	37166	19938	05946
205	31175	38610	55754	29930	255	40654	01804	33955	17062
206	31386	72203	69153	40038	256	40823	99653	11849	56171
207	31597	03454	56917	75346	257	40993	31233	31294	53716
208	31806	33349	62761	55006	258	41161	97059	63230	15891
209	32014	62861	11054	00229	259	41329	97640	81251	82752
210	32221	92947	33919	26801	260	41497	33479	70817	96442
211	32428	24552	97692	66508	261	41664	05073	38280	96192
212	32633	58609	28751	43606	262	41830	12913	19745	45602
213	32837	96034	38737	72339	263	41995	57484	89757	86897
214	33041	37733	49190	83605	264	42160	39268	69831	06369
215	33243	84599	15605	33119	265	42324	58739	36807	85042
216	33445	37511	50930	89753	266	42488	16366	31066	98746
217	33645	97338	48529	51038	267	42651	12613	64575	22202
218	33845	64936	04604	83041	268	42813	47940	28788	82458
219	34044	41148	40118	33837	269	42975	22800	02407	98009
220	34242	26808	22206	23596	270	43136	37641	58987	31189
221	34439	22736	85110	69775	271	43296	92908	74405	72952
222	34635	29744	50638	62932	272	43456	89040	34198	70940
223	34830	48630	48160	67348	273	43616	26470	40756	03721
224	35024	80183	34162	80678	274	43775	05628	20387	96378
225	35218	25181	11362	48416	275	43933	26938	30262	65032
226	35410	84391	47400	.91801	276	44090	90820	65217	70659
227	35602	58571	93122	72010	277	44247	97690	64448	55378
228	35793	48470	00453	78926	278	44404	47959	18076	27567
229	35983	54823	39887	99413	279	44560	42032	73597	55426
230	36172	78360	17592	87887	280	44715	80313	42219	22114
231	36361	19798	92144	30876	281	44870	63199	05079	89286
232	36548	79848	90899	67297	282	45024	91083	19361	09692
233	36735	59210	26018	97219	283	45178	64355	24290	23556
234	36921	58574	10142	83901	284	45331	83400	47037	67652
235	37106	78622	71736	26920	285	45484	48600	08510	20362
236	37291	20029	70106	58069	286	45636	60331	29043	00517
237	37474	83460	10103	86529	287	45788	18967	33992	32522
238	37657	69570	56511	95447	288	45939	24877	59230	85066
239	37839	79009	48137	68500	289	46089	78427	56547	85708
240	38021	12417	11606	02294	290	46239	79978	98956	08733
241	38201	70425	74868	38408	291	46389	29889	85907	28908
242	38381	53659	80431	27671	292	46538	28514	48418	29150
243	38560	62735	98312	18648	293	46686	76203	54109	45624
244	38738	98263	38729	42431	294	46834	73304	12157	29393
245	38916	60843	64532	46621	295	46982	20159	78162	99505
246	39093	51071	03379	12702	296	47129	17110	58938	58245
247	39269	69532	59665	73074	297	47275	64493	17212	35264
248	39445	16808	26216	26531	298	47421	62640	76255	23347
249	39619	93470	95736	34113	299	47567	11883	24429	64807
250	39794	00086	72037	60957	300	47712	12547	19662	43730

Tab. 2. TO 20 PLACES. 205

N.	Logarithms.	N.	Logarithms.
301	47856 64955 93843 35712	351	54530 71164 65824 08109
302	48000 69429 57150 63208	352	54654 26634 78131 01682
303	48144 26285 02305 01157	353	54777 47053 87822 56550
304	48287 35836 08753 74239	354	54900 32620 25787 82277
305	48429 98393 46785 83867	355	55022 83530 55094 09088
306	48572 14264 81579 99834	356	55144 99979 72875 17515
307	48713 83754 77186 48475	357	55266 82161 12193 19655
308	48855 07165 00444 26189	358	55388 30266 43874 36478
309	48995 84794 24834 64247	359	55509 44485 78319 14782
310	49136 16938 34272 67967	360	55630 25007 67287 26502
311	49276 03890 26837 50555	361	55750 72019 05657 92307
312	49415 45940 18442 79214	362	55870 85705 33165 70550
313	49554 43375 46448 48481	363	55990 66250 36112 51880
314	49692 96480 73214 93198	364	56110 13836 49055 99035
315	49831 05537 89600 51009	365	56229 28644 56474 70586
316	49968 70826 18403 81842	366	56348 10853 94410 66639
317	50105 92622 17751 49455	367	56466 60642 52089 33799
318	50242 71199 84432 67814	368	56584 78186 73517 65972
319	50379 06830 57181 12808	369	56702 63661 59060 36910
320	50514 99783 19905 97607	370	56820 17240 66994 99681
321	50650 50324 04872 07813	371	56937 39096 15045 87635
322	50785 58716 95830 90479	372	57054 29398 81897 50739
323	50920 25223 31102 89008	373	57170 88318 08687 60551
324	51054 50102 06612 13961	374	57287 16022 00480 16450
325	51188 33609 78874 37878	375	57403 12677 27718 85165
326	51321 76000 67939 00285	376	57518 78449 27661 05006
327	51454 77526 60286 07250	377	57634 13502 05792 85654
328	51587 38437 11679 08015	378	57749 17998 37225 33781
329	51719 58979 49974 29513	379	57863 92099 68072 34193
330	51851 39398 77887 47805	380	57978 35966 16810 15675
331	51982 79937 75718 73861	381	58092 49756 75619 30154
332	52113 80837 04036 29426	382	58206 33629 11708 73285
333	52244 42335 06319 87140	383	58319 87739 68622 74038
334	52374 64668 11564 47520	384	58433 12243 67530 80379
335	52504 48070 36845 23894	385	58546 07295 08500 67625
336	52633 92773 89844 04886	386	58658 73046 71754 95581
337	52762 99008 71338 62619	387	58771 09650 18911 40100
338	52891 67002 77654 73363	388	58883 17255 94207 24221
339	53019 96982 03082 16009	389	58994 96013 25707 73624
340	53147 89170 42255 12375	390	59106 46070 26499 20650
341	53275 43789 92497 72042	391	59217 67573 95866 80741
342	53402 61060 56135 03134	392	59328 60670 20457 24707
343	53529 41200 42770 49214	393	59439 25503 75426 69811
344	53655 84425 71530 11205	394	59549 62218 25574 12259
345	53781 90950 73274 12095	395	59659 70956 26460 23278
346	53907 60987 92776 60977	396	59769 51859 25512 30577
347	54032 94747 90873 71854	397	59879 05067 63115 06588
348	54157 92439 46580 91506	398	59988 30720 73687 84531
349	54282 54269 59179 89654	399	60097 28956 86748 22954
350	54406 80443 50275 63550	400	60205 99913 27962 39043

N.	Logarithms.				N.	Logarithms.			
401	60314	43726	20182	30654	451	65417	65418	77960	53526
402	60422	60530	84470	06666	452	65513	84348	11382	11322
403	60530	50461	41109	44887	453	65609	82020	12831	87416
404	60638	13651	10604	96470	454	65705	58528	57103	91532
405	60745	50232	14668	55397	455	65801	13966	57112	40470
406	60852	60335	77194	11326	456	65896	48426	64434	98447
407	60959	44092	25220	03756	457	65991	62000	69850	22235
408	61066	01630	89879	95148	458	66086	54780	03869	18934
409	61172	33080	07341	80361	459	66181	26855	37261	24043
410	61278	38567	19735	49451	460	66275	78316	81574	07408
411	61384	18218	76069	20586	461	66370	09253	89648	14507
412	61489	72160	33134	59560	462	66464	19755	56125	50397
413	61595	00516	56401	02097	463	66558	09910	17953	13567
414	61700	03411	20898	94867	464	66651	79805	54880	86819
415	61804	80967	12092	70862	465	66745	29528	89953	92175
416	61909	33306	26742	74528	466	66838	59166	90000	16740
417	62013	60549	73757	51775	467	66931	68805	66112	16309
418	62117	62817	75035	19750	468	67024	58530	74124	03422
419	62221	40229	66295	30985	469	67117	28427	15083	26486
420	62324	92903	97900	46322	470	67209	78579	35717	46441
421	62428	20958	35668	30744	471	67302	09071	28896	17406
422	62531	24509	61673	86030	472	67394	19986	34087	77590
423	62634	03673	75042	33900	473	67486	11407	37811	56716
424	62736	58565	92732	63127	474	67577	83416	74085	06050
425	62838	89300	50311	53811	475	67669	36096	24866	57111
426	62940	95991	02718	91860	476	67760	69527	20493	14968
427	63042	78750	25023	86460	477	67851	83790	40113	92022
428	63144	37690	13172	03126	478	67942	78966	12118	88022
429	63245	72921	84724	24725	479	68033	55134	14563	22010
430	63346	84555	79586	52641	480	68124	12373	75587	21815
431	63447	72701	60731	60075	481	68214	50763	73831	76601
432	63548	37468	14912	09274	482	68304	70382	38849	57929
433	63648	78963	53365	44270	483	68394	71307	51512	14688
434	63748	97295	12510	70559	484	68484	53616	44412	47193
435	63848	92569	54637	32941	485	68574	17386	02263	65657
436	63948	64892	68586	02563	486	68663	62692	62293	38169
437	64048	14369	70421	84040	487	68752	89612	14634	33246
438	64147	41105	04099	53358	488	68841	98220	02710	61953
439	64246	45202	42121	37063	489	68930	88591	23620	24494
440	64345	26764	86187	43118	490	69019	60800	28513	66142
441	64443	85894	67838	53601	491	69108	14921	22968	47275
442	64542	22693	49091	89296	492	69196	51027	67360	32223
443	64640	37262	23069	56023	493	69284	69192	77230	01587
444	64738	29701	14619	82453	494	69372	69489	23646	92596
445	64836	00109	80931	58951	495	69460	51989	33568	72013
446	64933	48587	12141	86869	496	69548	16764	90197	46052
447	65030	75231	31936	47555	497	69635	63887	33332	11681
448	65127	80139	98144	00199	498	69722	93427	59717	53634
449	65224	63410	03323	17492	499	69810	05456	23389	91417
450	65321	25137	75343	67938	500	69897	00043	36018	80479

Tab. 2. TO 20 PLACES. 207

N.	Logarithms.				N.	Logarithms.			
501	69983	77258	67245	71728	551	74115	15988	51785	04887
502	70070	37171	45019	33455	552	74193	90777	29198	90180
503	70156	79850	55927	39710	553	74272	51313	04698	25871
504	70243	05364	45525	29094	554	74350	97647	28429	74899
505	70329	13781	18661	37906	555	74429	29831	22676	23889
506	70415	05168	39799	11483	556	74507	47915	82057	47088
507	70500	79593	33335	97571	557	74585	51951	73728	90044
508	70586	37122	83919	25467	558	74663	41989	37578	74947
509	70671	77823	36758	74657	559	74741	18078	86423	29561
510	70757	01760	97936	36584	560	74818	80270	06200	41635
511	70842	09001	34712	73179	561	74896	28612	56161	40659
512	70926	99609	75830	75692	562	74973	63155	69061	08808
513	71011	73651	11816	27342	563	75050	83948	51346	22909
514	71096	31189	95275	73238	564	75127	91039	83342	29214
515	71180	72290	41191	00996	565	75204	84478	19438	52758
516	71264	97016	27211	35413	566	75281	64311	88271	43077
517	71349	05430	93942	50516	567	75358	30588	92906	57989
518	71432	97597	45233	02273	568	75434	83357	11018	87173
519	71516	73578	48457	85186	569	75511	22663	95071	17229
520	71600	33436	34799	15963	570	75587	48556	72491	39883
521	71683	77232	99524	47424	571	75663	61082	45848	05004
522	71767	05030	02262	15714	572	75739	60287	93024	20038
523	71850	16888	67274	23926	573	75815	46219	67389	97493
524	71933	12869	83726	65124	574	75891	18923	97973	52044
525	72015	93034	05956	87758	575	75966	78446	89630	48844
526	72098	57441	53739	06419	576	76042	24834	23212	04587
527	72181	06152	12546	60821	577	76117	58131	55731	42849
528	72263	39225	33812	25890	578	76192	78384	20529	05229
529	72345	56720	35185	75774	579	76267	85637	27436	19789
530	72427	58696	00789	04563	580	76342	79935	62937	28255
531	72509	45210	81469	06485	581	76417	61323	90330	73454
532	72591	16322	95048	18268	582	76492	29846	49888	48429
533	72672	72090	26572	26372	583	76566	85547	59014	08638
534	72754	12570	28556	41723	584	76641	28471	12399	48672
535	72835	37820	21228	44562	585	76715	58660	82180	44858
536	72916	47896	92770	01979	586	76789	76160	18090	65146
537	72997	42856	99555	60687	587	76863	81012	47614	47606
538	73078	22756	66389	17530	588	76937	73260	76138	48915
539	73158	87651	86738	70217	589	77011	52947	87101	64120
540	73239	37598	22968	50710	590	77085	20116	42144	19026
541	73319	72651	06569	43688	591	77158	74808	81255	36467
542	73399	92865	38386	92473	592	77232	17067	22919	77766
543	73479	98295	88846	94758	593	77305	46933	64262	60640
544	73559	88996	98179	90461	594	77378	64449	81193	54785
545	73639	65022	76642	43999	595	77451	69657	28549	56404
546	73719	26427	04737	23243	596	77524	62597	40236	42868
547	73798	73263	33430	77381	597	77597	43311	29369	08740
548	73878	05584	84369	15899	598	77670	11839	88410	84329
549	73957	23444	50091	90848	599	77742	68223	89311	37983
550	74036	26894	94243	84554	600	77815	12503	83643	63251

N.	Logarithms.	N.	Logarithms.
601	77887 44720 02739 52089	651	81358 09885 68191 94767
602	77959 64912 57824 55233	652	81424 75957 31920 19807
603	78031 73121 40151 30874	653	81491 31812 75073 92143
604	78103 69386 21131 82730	654	81557 77483 24267 26771
605	78175 53746 52468 88629	655	81624 12999 91783 06560
606	78247 26241 66286 20678	656	81690 38393 75660 27536
607	78318 86910 75257 58096	657	81756 53695 59780 77566
608	78390 35792 72734 93761	658	81822 58936 13955 49034
609	78461 72926 32875 35534	659	81888 54145 94009 86128
610	78532 98350 10767 03389	660	81954 39355 41868 67326
611	78604 12102 42554 23362	661	82020 14594 85640 23665
612	78675 14221 45561 19356	662	82085 79894 39699 93392
613	78746 04745 18415 03774	663	82151 35284 04773 13504
614	78816 83711 41167 67997	664	82216 80793 68017 48947
615	78887 51157 75416 73659	665	82282 16453 03104 59703
616	78958 07121 64425 45710	666	82347 42291 70301 06661
617	79028 51640 33241 68205	667	82412 58339 16548 96620
618	79098 84750 88815 83768	668	82477 64624 75545 67041
619	79169 06490 20117 97680	669	82542 61177 67823 11077
620	79239 16894 98253 87488	670	82607 48027 90826 43415
621	79309 16001 76580 19075	671	82672 25201 68992 07464
622	79379 03846 90818 70077	672	82736 92730 53825 24408
623	79448 80466 59169 61544	673	82801 50642 23976 84648
624	79518 45896 82423 98736	674	82865 98965 35319 82140
625	79588 00173 44075 21915	675	82930 37728 31024 92146
626	79657 43332 10429 68002	676	82994 66959 41635 92884
627	79726 75408 30716 43958	677	83058 86686 85144 31601
628	79795 96437 37196 12719	678	83122 96938 67063 35530
629	79865 06454 45268 92535	679	83186 97742 80501 68250
630	79934 05494 53581 70530	680	83250 89127 06236 31897
631	80002 93592 44134 31302	681	83314 71119 12785 15740
632	80071 70782 82385 01364	682	83378 43746 56478 91563
633	80140 37100 17355 10238	683	83442 07036 81532 56340
634	80208 92578 81732 68977	684	83505 61017 20116 22655
635	80277 37252 91975 66903	685	83569 05714 92425 57335
636	80345 71156 48413 87336	686	83632 41157 06751 68735
637	80413 94323 35350 43063	687	83695 67370 59550 43142
638	80482 06787 21162 32330	688	83758 84382 35511 30726
639	80550 08581 58400 16068	689	83821 92219 07625 81434
640	80617 99739 83887 17128	690	83884 90907 37255 31616
641	80685 80295 18817 42225	691	83947 80473 74198 40758
642	80753 50280 68853 27334	692	84010 60944 56757 80499
643	80821 09729 24222 07249	693	84073 32346 11806 74605
644	80888 58673 59812 10001	694	84135 94704 54854 91375
645	80955 97146 35267 76849	695	84198 48045 90113 88524
646	81023 25179 95084 08529	696	84260 92396 10562 11027
647	81090 42806 68700 38446	697	84323 27780 98009 42305
648	81157 50058 70593 33482	698	84385 54226 23161 09175
649	81224 46968 00369 23101	699	84447 71757 45681 40948
650	81291 33566 42855 57399	700	84509 80400 14256 83071

Tab. 2.   TO 20 PLACES.   209

N.	Logarithms.	N.	Logarithms.
701	84571 80179 66658 65706	751	87563 99370 04168 38975
702	84633 71121 29805 27631	752	87621 78405 91642 24527
703	84695 53250 19823 95834	753	87679 49762 00700 57664
704	84757 26591 42112 21203	754	87737 13458 69774 05175
705	84818 91169 91398 70650	755	87794 69516 29188 24166
706	84880 47010 51803 76071	756	87852 17955 01206 53302
707	84941 94137 96899 40499	757	87909 58795 00072 75709
708	85003 32576 89769 01798	758	87966 92056 32053 53715
709	85064 62351 83066 54285	759	88024 17758 95480 35691
710	85125 83487 19075 28609	760	88081 35922 80791 35196
711	85186 96007 29766 30258	761	88138 46567 70572 82637
712	85247 99936 36856 37036	762	88195 49713 39600 49676
713	85308 95298 51865 55853	763	88252 45379 54880 46591
714	85369 82117 76174 39176	764	88309 33585 75689 92806
715	85430 60418 01080 61474	765	88366 14351 53617 60792
716	85491 30223 07855 56000	766	88422 87696 32603 93559
717	85551 91556 67800 12230	767	88479 53639 48980 95947
718	85612 44442 42300 34303	768	88536 12200 31511 99900
719	85672 88903 82882 60777	769	88592 63398 01431 03960
720	85733 24964 31268 46023	770	88649 07251 72481 87146
721	85793 52647 19429 03588	771	88705 43780 50956 97446
722	85853 71975 69639 11829	772	88761 73003 35736 15102
723	85913 82972 94530 82137	773	88817 94939 18324 90897
724	85973 85661 97146 90071	774	88874 09606 82892 59621
725	86033 80065 70993 69691	775	88930 17025 06310 28924
726	86093 66207 00093 71401	776	88986 17212 58188 43743
727	86153 44108 59037 83621	777	89042 10188 00914 26482
728	86213 13793 13037 18556	778	89097 95969 89688 93146
729	86272 75283 17974 62377	779	89153 74576 72564 45605
730	86332 28601 20455 90107	780	89209 46026 90480 40172
731	86391 73769 57860 45495	781	89265 10338 77300 32684
732	86451 10810 58391 86161	782	89320 67530 59848 00262
733	86510 39746 41127 94317	783	89376 17620 57943 39922
734	86569 60599 16070 53320	784	89431 60626 84438 44228
735	86628 73390 84194 90351	785	89486 96567 45252 54155
736	86687 78143 37498 85494	786	89542 25460 39407 89332
737	86746 74878 59051 47490	787	89597 47323 59064 55847
738	86805 63618 23041 56431	788	89652 62174 89555 31780
739	86864 44383 94825 73669	789	89707 70032 09420 30627
740	86923 17197 30976 19202	790	89762 70912 90441 42799
741	86981 82079 79328 16804	791	89817 64834 97676 55351
742	87040 39052 79027 07156	792	89872 51815 89493 50098
743	87098 88137 60575 29242	793	89927 31873 17603 80309
744	87157 29355 45878 70260	794	89982 05024 27096 26109
745	87215 62727 48292 84304	795	90036 71286 56470 28771
746	87273 88274 72668 80072	796	90091 30677 37669 04053
747	87332 06018 15398 77842	797	90145 83213 96112 34727
748	87390 15978 64461 35972	798	90200 28913 50729 42476
749	87448 18176 99466 47155	799	90254 67793 13991 39295
750	87506 12633 91700 04687	800	90308 99869 91943 58564

N.	Logarithms.	N.	Logarithms.
801	90363 25160 84237 65931	851	92992 95600 84587 87568
802	90417 43682 84163 50176	852	93043 95947 66700 11382
803	90471 55452 78680 94182	853	93094 90311 67523 03000
804	90525 60487 48451 26187	854	93145 78706 89005 05981
805	90579 58803 67868 51437	855	93196 61147 28172 64091
806	90633 50418 05090 64409	856	93247 37646 77153 22648
807	90687 35347 22070 41738	857	93298 08219 23198 16429
808	90741 13607 74586 15992	858	93348 72878 48705 44247
809	90794 85216 12272 30432	859	93399 31638 31242 30263
810	90848 50188 78649 74918	860	93449 84512 43567 72162
811	90902 08542 11156 03069	861	93500 31514 53654 76252
812	90955 60292 41175 30847	862	93550 72658 24712 79596
813	91009 05455 94068 16682	863	93601 07957 15209 59266
814	91062 44048 89201 23277	864	93651 37424 78893 28795
815	91115 76087 39976 61243	865	93701 61074 64814 21935
816	91169 01587 53861 14669	866	93751 78920 17346 63791
817	91222 20565 32415 48794	867	93801 90974 76210 29438
818	91275 33036 71322 99882	868	93851 97251 76491 90081
819	91328 39017 60418 47451	869	93901 97764 48666 46875
820	91381 38523 83716 68972	870	93951 92526 18618 52463
821	91434 31571 19440 77180	871	94001 81550 07663 20336
822	91487 18175 40050 40107	872	94051 64849 32567 22084
823	91539 98352 12269 83977	873	94101 42437 05569 72637
824	91592 72116 97115 79081	874	94151 14326 34403 03562
825	91645 39485 49925 08762	875	94200 80530 22313 24507
826	91698 00473 20382 21619	876	94250 41061 68080 72880
827	91750 55095 52546 67071	877	94299 95933 66040 51823
628	91803 03367 84880 14389	878	94349 45159 06102 56585
829	91855 45305 50273 55312	879	94398 88750 73771 89354
830	91907 80923 76073 90383	880	94448 26721 50168 62639
831	91960 10237 84110 99107	881	94497 59084 12047 91274
832	92012 33262 90723 94049	882	94546 85851 31819 73123
833	92064 50014 06787 58996	883	94596 07035 77568 58562
834	92116 60506 37738 71297	884	94645 22650 13073 08817
835	92168 64754 83602 08477	885	94694 32706 97825 43234
836	92220 62774 39016 39271	886	94743 37218 87050 75544
837	92272 54579 93259 99155	887	94792 36198 31726 39220
838	92324 40186 30276 50506	888	94841 29657 78601 01974
839	92376 19608 28700 27500	889	94890 17609 70213 69496
840	92427 92860 61881 65843	890	94939 00066 44912 78472
841	92479 59957 97912 17467	891	94987 77040 36874 78993
842	92531 20914 99649 50266	892	95036 48543 76123 06390
843	92582 75746 24742 33016	893	95085 14588 88546 42595
844	92634 24466 25655 05551	894	95133 75187 95917 67077
845	92685 67089 49692 34320	895	95182 30353 15911 97436
846	92737 03630 39023 53422	896	95230 80096 62125 19721
847	92788 34103 30706 91221	897	95279 24430 44092 08537
848	92839 58522 56713 82649	898	95327 63366 67304 37013
849	92890 76902 43952 67285	899	95375 96917 33228 76700
850	92941 89257 14292 73333	900	95424 25094 39324 87459

Tab. 2. TO 20 PLACES. 211

N.	Logarithms.	N.	Logarithms.
901	95472 47909 79062 97417	951	97818 05169 37413 93185
902	95520 65375 41941 73047	952	97863 69483 84474 34489
903	95568 77503 13505 79441	953	97909 29006 38326 40853
904	95616 84304 75363 30844	954	97954 83747 04095 11544
905	95664 85792 05203 31508	955	98000 33715 83746 34242
906	95712 81976 76813 06938	956	98045 78922 76100 07543
907	95760 72870 60095 25585	957	98091 19377 76843 56538
908	95808 58485 21085 11053	958	98136 55090 78544 41531
909	95856 38832 21967 44887	959	98181 86071 70663 59928
910	95904 13923 21093 59992	960	98227 12330 39568 41336
911	95951 83769 72998 24763	961	98272 33876 68545 35933
912	95999 48383 28416 17969	962	98317 50720 37812 96123
913	96047 07775 34298 94458	963	98362 62871 24534 51542
914	96094 61957 33831 41757	964	98407 70339 02830 77450
915	96142 10940 66448 27597	965	98452 73133 43792 56538
916	96189 54736 67850 38456	966	98497 71264 15493 34209
917	96236 93356 70021 09152	967	98542 64740 83001 67360
918	96284 26812 01242 43564	968	98587 53573 08393 66714
919	96331 55113 86111 26520	969	98632 37770 50765 32737
920	96378 78273 45555 26930	970	98677 17349 66244 85178
921	96425 96301 96848 92205	971	98721 92299 08004 86280
922	96473 09210 53629 34029	972	98766 62649 26274 57690
923	96520 17010 25912 05530	973	98811 28402 68351 91117
924	96567 19712 20106 69918	974	98855 89568 78615 52768
925	96614 17327 39032 60638	975	98900 46156 98536 81607
926	96661 09866 81934 33089	976	98944 98176 66691 81474
927	96707 97341 44497 07976	977	98989 45637 18773 07091
928	96754 79762 18862 06340	978	99033 88547 87601 44015
929	96801 57139 93641 76318	979	99078 26918 03137 82547
930	96848 29485 53935 11696	980	99122 60756 92494 85664
931	96894 96809 81342 62296	981	99166 90073 79948 50979
932	96941 59123 53981 36262	982	99211 14877 86949 66797
933	96988 16437 46499 94285	983	99255 35178 32135 62275
934	97034 68762 30093 35830	984	99299 50984 31341 51745
935	97081 16108 72517 77408	985	99343 62304 97611 73216
936	97127 58487 38105 22944	986	99387 69149 41211 21109
937	97173 95908 87778 26303	987	99431 71526 69636 73242
938	97220 28383 79064 46008	988	99475 69445 87628 12117
939	97266 55922 66110 92210	989	99519 62915 97179 40527
940	97312 78535 99698 65963	990	99563 51945 97549 91534
941	97358 96234 27256 90834	991	99607 36544 85275 32836
942	97405 09027 92877 36927	992	99651 16721 54178 65574
943	97451 16927 37328 37338	993	99694 92484 95381 17590
944	97497 19942 98068 97112	994	99738 63843 97313 31202
945	97543 18085 09262 94738	995	99782 30807 45725 45489
946	97589 11364 01792 76237	996	99825 93384 23698 73156
947	97634 99790 03273 41875	997	99869 51583 11655 71988
948	97680 83373 38066 25572	998	99913 05412 87371 1093°
949	97726 62124 27292 67028	999	99956 54882 25982 30869
950	97772 36052 88847 76632	1001	00043 40774 79318 64067

N.	Logarithms.	N.	Logarithms.
1003	00130 09330 20418 11880	1103	04257 55124 40190 59866
1005	00216 60617 56507 67623	1105	04336 22780 21129 50253
1007	00302 94705 53618 00717	1107	04414 76208 78722 80639
1009	00389 11662 36910 52172	1109	04493 15461 49160 06471
1011	00475 11555 91001 06349	1111	04571 40589 40867 61503
1013	00560 94453 60280 42845	1113	04649 51643 34708 31364
1015	00646 60422 49231 72283	1115	04727 48673 84179 47826
1017	00732 09529 22744 59739	1117	04805 31731 15609 05702
1019	00817 41840 06426 39490	1119	04883 00865 28350 04280
1021	00902 57420 86910 24725	1121	04960 56125 94973 15180
1023	00987 56337 12160 15771	1123	05037 97562 61457 78469
1025	01072 38653 91773 10408	1125	05115 25224 47381 28895
1027	01157 04435 97278 19720	1127	05192 39160 46106 54029
1029	01241 53747 62432 92943	1129	05269 39419 24967 86114
1031	01325 86652 83516 54691	1131	05346 26049 25455 29383
1033	01410 03215 19620 57904	1133	05422 99098 63397 24592
1035	01494 03497 92936 55824	1135	05499 58615 29141 52489
1037	01577 87563 89040 96243	1137	05576 04646 87734 77923
1039	01661 55475 57177 41240	1139	05652 37240 79100 36269
1041	01745 07295 10536 15583	1141	05728 56444 18214 63835
1043	01828 43084 26530 86897	1143	05804 62303 95281 73884
1045	01911 62904 47072 80707	1145	05880 54866 75906 79891
1047	01994 66816 78842 33384	1147	05956 34179 01267 67648
1049	02077 54881 93557 85991	1149	06032 00286 88285 17768
1051	02160 27160 28242 22008	1151	06107 53236 29791 80185
1053	02242 83711 85486 51839	1153	06182 93072 94699 02164
1055	02325 24596 33711 46987	1155	06258 19842 28163 11355
1057	02407 49873 07426 26753	1157	06333 33589 51749 55393
1059	02489 59601 07485 00279	1159	06408 34359 63595 99542
1061	02571 53839 01340 66612	1161	06483 22197 38573 83829
1063	02653 32645 23296 75697	1163	06557 97147 28448 41139
1065	02734 96077 74756 52817	1165	06632 59253 62037 77698
1067	02816 44194 24469 89253	1167	06707 08560 45370 17354
1069	02897 77052 08778 01749	1169	06781 45111 61840 11070
1071	02978 94708 31855 63385	1171	06855 68950 72363 12990
1073	03059 97219 65951 08414	1173	06929 80121 15529 24470
1075	03140 84642 51624 13598	1175	07003 78666 07755 07399
1077	03221 57032 97981 58511	1177	07077 64628 43434 68158
1079	03302 14446 82910 67304	1179	07151 38050 95089 13540
1081	03382 56939 53310 34328	1181	07224 98976 13514 79908
1083	03462 84566 25320 36037	1183	07298 47446 27930 36913
1085	03542 97381 84548 31517	1185	07371 83503 46122 67006
1087	03622 95440 86294 53993	1187	07445 07189 54591 22047
1089	03702 78797 55774 95610	1189	07518 18546 18691 58184
1091	03782 47505 88341 87761	1191	07591 17614 82777 50317
1093	03862 01619 49702 79227	1193	07664 04436 70341 87278
1095	03941 41191 76137 14316	1195	07736 79052 84156 48979
1097	04020 66275 74711 13222	1197	07809 41504 06410 66684
1099	04099 76924 23490 56747	1199	07881 91830 98848 67595
1101	04178 73189 71751 77528		

Num.	Logarithms	Differ. 1.	Diff. 2.	D. 3.
101000	00432 13737 82642 57428	42999 24078 66099	42572 87346	84301
101001	00432 56737 06721 23527	42998 81505 78753	42572 03045	84298
101002	00432 99735 88227 02280	42998 38933 75708	42571 18747	84295
101003	00433 42734 27160 77988	42997 96362 56961	42570 34452	84294
101004	00433 85732 23523 34949	42997 53792 22509	42569 50158	84290
101005	00434 28729 77315 57458	42997 11222 72351	42568 65868	84288
101006	00434 71726 88538 29809	42996 68654 06483	42567 81580	84286
101007	00435 14723 57192 36292	42996 26086 24903	42566 97294	84283
101008	00435 57719 83278 61195	42995 83519 27609	42566 13011	84281
101009	00436 00715 66797 88804	42995 40953 14598	42565 28730	84277
101010	00436 43711 07751 03402	42994 98387 85868	42564 44453	84277
101011	00436 86706 06138 89270	42994 55823 41415	42563 60176	84272
101012	00437 29700 61962 30685	42994 13259 81239	42562 75904	84271
101013	00437 72694 75222 11924	42993 70697 05335	42561 91633	84268
101014	00438 15688 45919 17259	42993 28135 13702	42561 07365	84266
101015	00438 58681 74054 30961	42992 85574 06337	42560 23099	84263
101016	00439 01674 59628 37298	42992 43013 83238	42559 38836	84260
101017	00439 44667 02642 20536	42992 00454 44402	42558 54576	84258
101018	00439 87659 03096 64938	42991 57895 89826	42557 70318	84257
101019	00440 30650 60992 54764	42991 15338 19508	42556 86061	84251
101020	00440 73641 76330 74272	42990 72781 33447	42556 01810	84252
101021	00441 16632 49112 07719	42990 30225 31637	42555 17558	84248
101022	00441 59622 79337 39356	42989 87670 14079	42554 33310	84245
101023	00442 02612 67007 53435	42989 45115 80769	42553 49065	84244
101024	00442 45602 12123 34204	42989 02562 31704	42552 64821	84239
101025	00442 88591 14685 65908	42988 60009 66883	42551 80582	84239
101026	00443 31579 74695 32791	42988 17457 86301	42550 96343	84236
101027	00443 74567 92153 19092	42987 74906 89958	42550 12107	84233
101028	00444 17555 67060 09050	42987 32356 77851	42549 27874	84230
101029	00444 60542 99416 86901	42986 89807 49977	42548 43644	84228
101030	00445 03529 89224 36878	42986 47259 06333	42547 59416	84225
101031	00445 46516 36483 43211	42986 04711 46917	42546 75191	84225
101032	00445 89502 41194 90128	42985 62164 71726	42545 90966	84219
101033	00446 32488 03359 61854	42985 19618 80760	42545 06747	84219
101034	00446 75473 22978 42614	42984 77073 74013	42544 22528	84214
101035	00447 18458 00052 16627	42984 34529 51485	42543 38314	84215
101036	00447 61442 34581 68112	42983 91986 13171	42542 54099	84209
101037	00448 04426 26567 81283	42983 49443 59072	42541 69890	84209
101038	00448 47409 76011 40355	42983 06901 89182	42540 85681	84205
101039	00448 90392 82913 29537	42982 64361 03501	42540 01476	84204
101040	00449 33375 47274 33038	42982 21821 02025	42539 17272	84199
101041	00449 76357 69095 35063	42981 79281 84753	42538 33073	84199
101042	00450 19339 48377 19816	42981 36743 51680	42537 48874	84196
101043	00450 62320 85120 71496	42980 94206 02806	42536 64678	84193
101044	00451 05301 79326 74302	42980 51669 38128	42535 80485	84189
101045	00451 48282 30996 12430	42980 09133 57643	42534 96296	84189
101046	00451 91262 40129 70073	42979 66598 61347	42534 12107	84187
101047	00452 34242 06728 31420	42979 24064 49240	42533 27920	84181
101048	00452 77221 30792 80660	42978 81531 21320	42532 43739	84181
101049	00453 20200 12324 01980	42978 38998 77581	42531 59558	84178

Num.	Logarithms.	Differ. 1.	Diff. 2.	D. 3.
101050	00453 63178 51322 79561	42977 96467 18023	42530 75380	84177
101051	00454 06156 47789 97584	42977 53936 42643	42529 91203	84172
101052	00454 49134 01726 40227	42977 11406 .51440	42529 07031	84170
101053	00454 92111 13132 91667	42976 68877 44409	42528 22861	84168
101054	00455 35087 82010 36076	42976 26349 21548	42527 38693	84167
101055	00455 78064 08359 57624	42975 83821 82855	42526 54526	84162
101056	00456 21039 92181 40479	42975 41295 28329	42525 70364	84161
101057	00456 64015 33476 68808	42974 98769 57965	42524 86203	84158
101058	00457 06990 32246 26773	42974 56244 71762	42524 02045	84155
101059	00457 49964 88490 98535	42974 13720 69717	42523 17890	84154
101060	00457 92939 02211 68252	42973 71197 51827	42522 33736	84150
101061	00458 35912 73409 20079	42973 28675 18091	42521 49586	84148
101062	00458 78886 02084 38170	42972 86153 68505	42520 65438	84146
101063	00459 21858 88238 06675	42972 43633 03067	42519 81292	84143
101064	00459 64831 31871 09742	42972 01113 21775	42518 97149	84140
101065	00460 07803 32984 31517	42971 58594 24626	42518 13009	84138
101066	00460 50774 91578 56143	42971 16076 11617	42517 28871	84137
101067	00460 93746 07654 67760	42970 73558 82746	42516 44734	84132
101068	00461 36716 81213 50506	42970 31042 38012	42515 60602	84130
101069	00461 79687 12255 88518	42969 88526 77410	42514 76472	84130
101070	00462 22657 00782 65928	42969 46012 00938	42513 92342	84124
101071	00462 65626 46794 66866	42969 03498 08596	42513 08218	84123
101072	00463 08595 50292 75462	42968 60985 00378	42512 24095	84122
101073	00463 51564 11277 75840	42968 18472 76283	42511 39973	84117
101074	00463 94532 29750 52123	42967 75961 36310	42510 55856	84116
101075	00464 37500 05711 88433	42967 33450 80454	42509 71740	84113
101076	00464 80467 39162 68887	42966 90941 08714	42508 87627	84111
101077	00465 23434 30103 77601	42966 48432 21087	42508 03516	84107
101078	00465 66400 78535 98688	42966 05924 17571	42507 19409	84107
101079	00466 09366 84460 16259	42965 63416 98162	42506 35302	84102
101080	00466 52332 47877 14421	42965 20910 62860	42505 51200	84102
101081	00466 95297 68787 77281	42964 78405 11660	42504 67098	84097
101082	00467 38262 47192 88941	42964 35900 44562	42503 83001	84095
101083	00467 81226 83093 33503	42963 93396 61561	42502 98906	84095
101084	00468 24190 76489 95064	42963 50893 62655	42502 14811	84090
101085	00468 67154 27383 57719	42963 08391 47844	42501 30721	84087
101086	00469 10117 35775 05563	42962 65890 17123	42500 46634	84087
101087	00469 53080 01665 22686	42962 23389 70489	42499 62547	84083
101088	00469 96042 25054 93175	42961 80890 07942	42498 78464	84080
101089	00470 39004 05945 01117	42961 38391 29478	42497 94384	84078
101090	00470 81965 44336 30595	42960 95893 35094	42497 10306	84077
101091	00471 24926 40229 65689	42960 53396 24788	42496 26229	84073
101092	00471 67886 93625 90477	42960 10899 98559	42495 42156	84069
101093	00472 10847 04525 89036	42959 68404 56403	42494 58087	84070
101094	00472 53806 72930 45439	42959 25909 98316	42493 74017	84065
101095	00472 96765 98840 43755	42958 83416 24299	42492 89952	84063
101096	00473 39724 82256 68054	42958 40923 34347	42492 05889	84060
101097	00473 82683 23180 02401	42957 98431 28458	42491 21829	84061
101098	00474 25641 21611 30859	42957 55940 06629	42490 37768	84052
101099	00474 68598 77551 37488	42957 13449 68861	42489 53716	84056

Tab. 3.   TO 20 PLACES.   215

Num.	Logarithms.				Differ. 1.			Diff. 2.		D.3.
101100	00475	11555	91001	06349	42956	70960	15145	42488	69660	84050
101101	00475	54512	61961	21494	42956	28471	45485	42487	85610	84048
101102	00475	97468	90432	66979	42955	85983	59875	42487	01562	84046
101103	00476	40424	76416	26854	42955	43496	58313	42486	17516	84043
101104	00476	83380	19912	85167	42955	01010	40797	42485	33473	84040
101105	00477	26335	20923	25964	42954	58525	07324	42484	49433	84040
101106	00477	69289	79448	33288	42954	16040	57891	42483	65393	84034
101107	00478	12243	95488	91179	42953	73556	92498	42482	81359	84035
101108	00478	55197	69045	83677	42953	31074	11139	42481	97324	84030
101109	00478	98151	00119	94816	42952	88592	13815	42481	13294	84028
101110	00479	41103	88712	08631	42952	46111	00521	42480	29266	84025
101111	00479	84056	34823	09152	42952	03630	71255	42479	45241	84026
101112	00480	27008	38453	80407	42951	61151	26014	42478	61215	84018
101113	00480	69959	99605	06421	42951	18672	64799	42477	77197	84020
101114	00481	12911	18277	71220	42950	76194	87602	42476	93177	84015
101115	00481	55861	94472	58822	42950	33717	94425	42476	09162	84013
101116	00481	98812	28190	53247	42949	91241	85263	42475	25149	84012
101117	00482	41762	19432	38510	42949	48766	60114	42474	41137	84008
101118	00482	84711	68198	98624	42949	06292	18977	42473	57129	84006
101119	00483	27660	74491	17601	42948	63818	61848	42472	73123	84002
101120	00483	70609	38309	79449	42948	21345	88725	42471	89121	84003
101121	00484	13557	59655	68174	42947	78873	99604	42471	05118	83996
101122	00484	56505	38529	67778	42947	36402	94486	42470	21122	83998
101123	00484	99452	74932	62264	42946	93932	73364	42469	37124	83993
101124	00485	42399	68865	35628	42946	51463	36240	42468	53131	83990
101125	00485	85346	20328	71868	42946	08994	83109	42467	69141	83988
101126	00486	28292	29323	54977	42945	66527	13968	42466	85153	83988
101127	00486	71237	95850	68945	42945	24060	28815	42466	01165	83982
101128	00487	14183	19910	97760	42944	81594	27650	42465	17183	83981
101129	00487	57128	01505	25410	42944	39129	10467	42464	33202	83978
101130	00488	00072	40634	35877	42943	96664	77265	42463	49224	83978
101131	00488	43016	37299	13142	42943	54201	28041	42462	65246	83971
101132	00488	85959	91500	41183	42943	11738	62795	42461	81275	83972
101133	00489	28903	03239	03978	42942	69276	81520	42460	97303	83969
101134	00489	71845	72515	85498	42942	26815	84217	42460	13334	83965
101135	00490	14787	99331	69715	42941	84355	70883	42459	29369	83964
101136	00490	57729	83637	40598	42941	41896	41514	42458	45405	83962
101137	00491	00671	25583	82112	42940	99437	96109	42457	61443	83957
101138	00491	43612	25021	78221	42940	56980	34666	42456	77486	83956
101139	00491	86552	82002	12887	42940	14523	57180	42455	93530	83954
101140	00492	29492	96525	70067	42939	72067	63650	42455	09576	83951
101141	00492	72432	68593	33717	42939	29612	54074	42454	25625	83949
101142	00493	15371	98205	87791	42938	87158	28449	42453	41676	83946
101143	00493	58310	85364	16240	42938	44704	86773	42452	57730	83944
101144	00494	01249	30069	03013	42938	02252	29043	42451	73786	83941
101145	00494	44187	32321	32056	42937	59800	55257	42450	89845	83938
101146	00494	87124	92121	87313	42937	17349	65412	42450	05907	83936
101147	00495	30062	09471	52725	42936	74899	59505	42449	21971	83934
101148	00495	72998	84371	12230	42936	32450	37534	42448	38037	83931
101149	00496	15935	16821	49764	42935	90001	99497	42447	54106	83929

TABLE IV.

Log.	Number.	Differ. 1.	Diff. 2.	D. 3.
00000	10000 00000 00000 00000	23026 11602 68807	53020 20192	1 22087
00001	10000 23026 11602 68807	23026 64622 88999	53021 42279	1 22085
00002	10000 46052 76225 57806	23027 17644 31278	53022 64364	1 22093
00003	10000 69079 93869 89084	23027 70666 95642	53023 86457	1 22093
00004	10000 92107 64536 84726	23028 23690 82099	53025 08550	1 22094
00005	10001 15135 88227 60825	23028 76715 90649	53026 30644	1 22102
00006	10001 38164 64943 57474	23029 29742 21293	53027 52746	1 22100
00007	10001 61193 94685 78767	23029 82769 74039	53028 74846	1 22106
00008	10001 84223 77455 52806	23030 35798 48885	53029 96952	1 22104
00009	10002 07254 13254 01691	23030 88828 45837	53031 19056	1 22114
00010	10002 30285 02082 47528	23031 41859 64893	53032 41170	1 22112
00011	10002 53316 43942 12421	23031 94892 06063	53033 63282	1 22115
00012	10002 76348 38834 18464	23032 47925 69345	53034 85397	1 22120
00013	10002 99380 86759 87829	23033 00960 54742	53036 07517	1 22120
00014	10003 22413 87720 42571	23033 53996 62259	53037 29637	1 22125
00015	10003 45447 41717 04830	23034 07033 91896	53038 51762	1 22128
00016	10003 68481 48750 96726	23034 60072 43658	53039 73890	1 22128
00017	10003 91516 08823 40384	23035 13112 17548	53040 96018	1 22134
00018	10004 14551 21935 57932	23035 66153 13566	53042 18152	1 22136
00019	10004 37586 88088 71498	23036 19195 31718	53043 40288	1 22137
00020	10004 60623 07284 03216	23036 72238 72006	53044 62425	1 22142
00021	10004 83659 79522 75222	23037 25283 34431	53045 84567	1 22144
00022	10005 06697 04806 09653	23037 78329 18998	53047 06711	1 22146
00023	10005 29734 83135 28651	23038 31376 25709	53048 28857	1 22151
00024	10005 52773 14511 54360	23038 84424 54566	53049 51008	1 22151
00025	10005 75811 98936 08926	23039 37474 05574	53050 73159	1 22156
00026	10005 98851 36410 14500	23039 90524 78733	53051 95315	1 22158
00027	10006 21891 26934 93233	23040 43576 74048	53053 17473	1 22161
00028	10006 44931 70511 67281	23040 96629 91521	53054 39634	1 22163
00029	10006 67972 67141 58802	23041 49684 31155	53055 61797	1 22167
00030	10006 91014 16825 89957	23042 02739 92952	53056 83964	1 22170
00031	10007 14056 19565 82909	23042 55796 76916	53058 06134	1 22170
00032	10007 37098 75362 59825	23043 08854 83050	53059 28304	1 22177
00033	10007 60141 84217 42875	23043 61914 11354	53060 50481	1 22177
00034	10007 83185 46131 54229	23044 14974 61835	53061 72658	1 22180
00035	10008 06229 61106 16064	23044 68036 34493	53062 94838	1 22185
00036	10008 29274 29142 50557	23045 21099 29331	53064 17023	1 22184
00037	10008 52319 50241 79888	23045 74163 46354	53065 39207	1 22190
00038	10008 75365 24405 26242	23046 27228 85561	53066 61397	1 22192
00039	10008 98411 51634 11803	23046 80295 46958	53067 83589	1 22195
00040	10009 21458 31929 58761	23047 33363 30547	53069 05784	1 22196
00041	10009 44505 65292 89308	23047 86432 36331	53070 27980	1 22202
00042	10009 67553 51725 25639	23048 39502 64311	53071 50182	1 22202
00043	10009 90601 91227 89950	23048 92574 14493	53072 72384	1 22206
00044	10010 13650 83802 04443	23049 45646 86877	53073 94590	1 22208
00045	10010 36700 29448 91320	23049 98720 81467	53075 16798	1 22213
00046	10010 59750 28169 72787	23050 51795 98265	53076 39011	1 22213
00047	10010 82800 79965 71052	23051 04872 37276	53077 61224	1 22218
00048	10011 05851 84838 08328	23051 57949 98500	53078 83442	1 22219
00049	10011 28903 42788 06828	23052 11028 81942	53080 05661	1 22224

Log.	Number.			Differ. 1.			Diff. 2.		D. 3.
00050	10011	51955	53816 88770	23052	64108	87603	53081	27885	1 22225
00051	10011	75008	17925 76373	23053	17190	15488	53082	50110	1 22228
00052	10011	98061	35115 91861	23053	70272	65598	53083	72338	1 22232
00053	10012	21115	05388 57459	23054	23356	37936	53084	94570	1 22233
00054	10012	44169	28744 95395	23054	76441	32506	53086	16803	1 22238
00055	10012	67224	05186 27901	23055	29527	49309	53087	39041	1 22238
00056	10012	90279	34713 77210	23055	82614	88350	53088	61279	1 22244
00057	10013	13335	17328 65560	23056	35703	49629	53089	83523	1 22245
00058	10013	36391	53032 15189	23056	88793	33152	53091	05768	1 22247
00059	10013	59448	41825 48341	23057	41884	38920	53092	28015	1 22252
00060	10013	82505	83709 87261	23057	94976	66935	53093	50267	1 22254
00061	10014	05563	78686 54196	23058	48070	17202	53094	72521	1 22255
00062	10014	28622	26756 71398	23059	01164	89723	53095	94776	1 22261
00063	10014	51681	27921 61121	23059	54260	84499	53097	17037	1 22262
00064	10014	74740	82182 45620	23060	07358	01536	53098	39299	1 22263
00065	10014	97800	89540 47156	23060	60456	40835	53099	61562	1 22270
00066	10015	20861	49996 87991	23061	13556	02397	53100	83832	1 22270
00067	10015	43922	63552 90388	23061	66656	86229	53102	06102	1 22273
00068	10015	66984	30209 76617	23062	19758	92331	53103	28375	1 22276
00069	10015	90046	49968 68948	23062	72862	20706	53104	50651	1 22280
00070	10016	13109	22830 89654	23063	25966	71357	53105	72931	1 22282
00071	10016	36172	48797 61011	23063	79072	44288	53106	95213	1 22284
00072	10016	59236	27870 05299	23064	32179	39501	53108	17497	1 22287
00073	10016	82300	60049 44800	23064	85287	56998	53109	39784	1 22290
00074	10017	05365	45337 01798	23065	38396	96782	53110	62074	1 22295
00075	10017	28430	83733 98580	23065	91507	58856	53111	84369	1 22295
00076	10017	51496	75241 57436	23066	44619	43225	53113	06664	1 22299
00077	10017	74563	19861 00661	23066	97732	49889	53114	28963	1 22301
00078	10017	97630	17593 50550	23067	50846	78852	53115	51264	1 22305
00079	10018	20697	68440 29402	23068	03962	30116	53116	73569	1 22306
00080	10018	43765	72402 59518	23068	57079	03685	53117	95875	1 22312
00081	10018	66834	29481 63203	23069	10196	99560	53119	18187	1 22312
00082	10018	89903	39678 62763	23069	63316	17747	53120	40499	1 22315
00083	10019	12973	02994 80510	23070	16436	58246	53121	62814	1 22318
00084	10019	36043	19431 38756	23070	69558	21060	53122	85132	1 22324
00085	10019	59113	88989 59816	23071	22681	06192	53124	07456	1 22321
00086	10019	82185	11670 66008	23071	75805	13648	53125	29777	1 22329
00087	10020	05256	87475 79656	23072	28930	43425	53126	52106	1 22329
00088	10020	28329	16406 23081	23072	82056	95531	53127	74435	1 22332
00089	10020	51401	98463 18612	23073	35184	69966	53128	96767	1 22336
00090	10020	74475	33647 88578	23073	88313	66733	53130	19103	1 22339
00091	10020	97549	21961 55311	23074	41443	85836	53131	41442	1 22339
00092	10021	20623	63405 41147	23074	94575	27278	53132	63781	1 22346
00093	10021	43698	57980 68425	23075	47707	91059	53133	86127	1 22345
00094	10021	66774	05688 59484	23076	00841	77186	53135	08472	1 22351
00095	10021	89850	06530 36670	23076	53976	85658	53136	30823	1 22350
00096	10022	12926	60507 22328	23077	07113	16481	53137	53173	1 22358
00097	10022	36003	67620 38809	23077	60250	69654	53138	75531	1 22355
00098	10022	59081	27871 08463	23078	13389	45185	53139	97886	1 22363
00099	10022	82159	41260 53648	23078	66529	43071	53141	20249	1 22362

Log.	Number.	Differ. 1.	Diff. 2.	D. 3.
00100	10023 05238 07789 96719	23079 19670 63320	53142 42611	1 22368
00101	10023 28317 27460 60039	23079 72813 05931	53143 64979	1 22367
00102	10023 51397 00273 65970	23080 25956 70910	53144 87346	1 22374
00103	10023 74477 26230 36880	23080 79101 58256	53146 09720	1 22374
00104	10023 97558 05331 95136	23081 32247 67976	53147 32094	1 22378
00105	10024 20639 37579 63112	23081 85395 00070	53148 54472	1 22380
00106	10024 43721 22974 63182	23082 38543 54542	53149 76852	1 22383
00107	10024 66803 61518 17724	23082 91693 31394	53150 99235	1 22386
00108	10024 89886 53211 49118	23083 44844 30629	53152 21621	1 22390
00109	10025 12969 98055 79747	23083 97996 52250	53153 44011	1 22392
00110	10025 36053 96052 31997	23084 51149 96261	53154 66403	1 22393
00111	10025 59138 47202 28258	23085 04304 62664	53155 88796	1 22398
00112	10025 82223 51506 90922	23085 57460 51460	53157 11194	1 22400
00113	10026 05309 08967 42382	23086 10617 62654	53158 33594	1 22403
00114	10026 28395 19585 05036	23086 63775 96248	53159 55997	1 22407
00115	10026 51481 83361 01284	23087 16935 52245	53160 78404	1 22407
00116	10026 74569 00296 53529	23087 70096 30649	53162 00811	1 22412
00117	10026 97656 70392 84178	23088 23258 31460	53163 23223	1 22415
00118	10027 20744 93651 15638	23088 76421 54683	53164 45638	1 22416
00119	10027 43833 70072 70321	23089 29586 00321	53165 68054	1 22421
00120	10027 66922 99658 70642	23089 82751 68375	53166 90475	1 22422
00121	10027 90012 82410 39017	23090 35918 58850	53168 12897	1 22425
00122	10028 13103 18328 97867	23090 89086 71747	53169 35322	1 22430
00123	10028 36194 07415 69614	23091 42256 07069	53170 57752	1 22430
00124	10028 59285 49671 76683	23091 95426 64821	53171 80182	1 22434
00125	10028 82377 45098 41504	23092 48598 45003	53173 02616	1 22438
00126	10029 05469 93696 86507	23093 01771 47619	53174 25054	1 22438
00127	10029 28562 95468 34126	23093 54945 72673	53175 47492	1 22443
00128	10029 51656 50414 06799	23094 08121 20165	53176 69935	1 22446
00129	10029 74750 58535 26964	23094 61297 90100	53177 92381	1 22448
00130	10029 97845 19833 17064	23095 14475 82481	53179 14829	1 22450
00131	10030 20940 34308 99545	23095 67654 97310	53180 37279	1 22454
00132	10030 44036 01963 96855	23096 20835 34589	53181 59733	1 22457
00133	10030 67132 22799 31444	23096 74016 94322	53182 82190	1 22459
00134	10030 90228 96816 25766	23097 27199 76512	53184 04649	1 22463
00135	10031 13326 24016 02278	23097 80383 81161	53185 27112	1 22464
00136	10031 36424 04399 83439	23098 33569 08273	53186 49576	1 22469
00137	10031 59522 37968 91712	23098 86755 57849	53187 72045	1 22469
00138	10031 82621 24724 49561	23099 39943 29894	53188 94514	1 22475
00139	10032 05720 64667 79455	23099 93132 24408	53190 16989	1 22476
00140	10032 28820 57800 03863	23100 46322 41397	53191 39465	1 22478
00141	10032 51921 04122 45260	23100 99513 80862	53192 61943	1 22483
00142	10032 75022 03636 26122	23101 52706 42805	53193 84426	1 22483
00143	10032 98123 56342 68927	23102 05900 27231	53195 06909	1 22484
00144	10033 21225 62242 96158	23102 59095 34140	53196 29393	1 22490
00145	10033 44328 21338 30298	23103 12291 63533	53197 51883	1 22495
00146	10033 67431 33629 93831	23103 65489 15416	53198 74378	1 22499
00147	10033 90534 99119 09247	23104 18687 89794	53199 96877	1 22501
00148	10034 13639 17806 99041	23104 71897 86671	53201 19378	1 22503
00149	10034 36743 89694 85712	23105 25089 06049	53202 41881	1 22506

Tab. 5. HYPERBOLIC LOGARITHMS. 219

N.	Logar.	N.	Logar.	N.	Logar.	N.	Logar.
1·01	0·0099503	1·51	0·4121097	2·01	0·6981347	2·51	0·9202828
1·02	0·0198026	1·52	0·4187103	2·02	0·7030975	2·52	0·9242589
1·03	0·0295588	1·53	0·4252677	2·03	0·7080358	2·53	0·9282193
1·04	0·0392207	1·54	0·4317824	2·04	0·7129498	2·54	0·9321641
1·05	0·0487902	1·55	0·4382549	2·05	0·7178398	2·55	0·9360934
1·06	0·0582689	1·56	0·4446858	2·06	0·7227060	2·56	0 9400073
1·07	0·0676586	1·57	0·4510756	2·07	0·7275486	2·57	0·9439059
1·08	0·0769610	1·58	0·4574248	2·08	0·7323679	2·58	0·9477894
1·09	0·0861777	1·59	0·4637340	2·09	0·7371641	2·59	0·9516579
1·10	0·0953102	1·60	0·4700036	2·10	0·7419373	2·60	0·9555114
1·11	0·1043600	1·61	0·4762342	2·11	0·7466879	2·61	0·9593502
1·12	0·1133287	1·62	0·4824261	2·12	0·7514161	2·62	0·9631743
1·13	0·1222176	1·63	0·4885800	2·13	0·7561220	2·63	0·9669838
1·14	0·1310283	1·64	0·4946962	2·14	0·7608058	2·64	0·9707789
1·15	0·1397619	1·65	0·5007753	2·15	0·7654678	2·65	0·9745596
1·16	0·1484200	1·66	0·5068176	2·16	0·7701082	2·66	0·9783261
1·17	0·1570037	1·67	0·5128236	2·17	0·7747272	2·67	0·9820785
1·18	0·1655144	1·68	0·5187938	2·18	0·7793249	2·68	0·9858168
1·19	0·1739533	1·69	0·5247285	2·19	0·7839015	2·69	0·9895412
1·20	0·1823216	1·70	0·5306283	2·20	0·7884574	2·70	0·9932518
1·21	0·1906204	1·71	0·5364934	2·21	0·7929925	2·71	0·9969486
1·22	0·1988509	1·72	0·5423243	2·22	0·7975072	2·72	1·0006319
1·23	0·2070142	1·73	0·5481214	2·23	0·8020016	2·73	1·0043016
1·24	0·2151114	1·74	0·5538851	2·24	0·8064759	2·74	1·0079579
1·25	0·2231436	1·75	0·5596158	2·25	0·8109302	2·75	1·0116009
1·26	0·2311117	1·76	0·5653138	2·26	0·8153648	2·76	1·0152307
1·27	0·2390169	1·77	0·5709795	2·27	0·8197798	2·77	1·0188473
1·28	0·2468601	1·78	0·5766134	2·28	0·8241754	2·78	1·0224509
1·29	0·2546422	1·79	0·5822156	2·29	0·8285518	2·79	1·0260416
1·30	0·2623643	1·80	0·5877867	2·30	0·8329091	2·80	1·0296194
1·31	0·2700271	1 81	0·5933268	2·31	0·8372475	2·81	1·0331845
1·32	0·2776317	1·82	0·5988365	2·32	0·8415672	2·82	1·0367369
1·33	0·2851789	1·83	0·6043160	2·33	0·8458683	2·83	1·0402767
1·34	0·2926696	1·84	0·6097656	2·34	0·8501509	2·84	1·0438041
1·35	0·3001046	1·85	0·6151856	2·35	0·8544153	2·85	1·0473190
1·36	0·3074847	1·86	0·6205765	2·36	0·8586616	2·86	1·0508217
1·37	0·3148107	1·87	0·6259384	2·37	0·8628899	2·87	1·0543121
1·38	0·3220835	1·88	0·6312718	2·38	0·8671005	2·88	1·0577903
1·39	0·3293037	1·89	0·6365768	2·39	0·8712933	2·89	1·0612565
1·40	0·3364722	1·90	0·6418539	2 40	0·8754687	2·90	1·0647107
1·41	0·3435897	1·91	0·6471032	2·41	0·8796267	2·91	1·0681531
1·42	0·3506569	1·92	0·6523252	2·42	0·8837675	2·92	1·0715836
1·43	0·3576745	1·93	0·6575200	2·43	0·8878913	2·93	1·0750024
1·44	0·3646431	1·94	0·6626880	2·44	0·8919980	2·94	1·0784096
1·45	0·3715636	1·95	0·6678294	2·45	0·8960880	2·95	1·0818052
1·46	0·3784364	1·96	0·6729445	2·46	0·9001613	2·96	1·0851893
1·47	0·3852624	1·97	0·6780335	2·47	0·9042182	2·97	1·0885619
1·48	0·3920421	1·98	0·6830968	2·48	0·9082586	2·98	1·0919233
1·49	0·3987761	1·99	0·6881346	2·49	0·9122827	2·99	1·0952734
1·50	0·4054651	2·00	0·6931472	2·50	0·9162907	3·00	1·0986123

N	Logar.	N.	Logar.	N.	Logar.	N.	Logar.
3·01	1·1019401	3·51	1·2556160	4·01	1·3887912	4·51	1·5062971
3·02	1·1052568	3·52	1·2584610	4·02	1·3912819	4·52	1·5085120
3·03	1·1085626	3·53	1·2612979	4·03	1·3937664	4·53	1·5107219
3·04	1·1118575	3·54	1·2641267	4·04	1·3962447	4·54	1·5129270
3·05	1·1151416	3·55	1·2669476	4·05	1·3987169	4 55	1·5151272
3·06	1·1184149	3·56	1·2697605	4·06	1·4011829	4·56	1·5173226
3·07	1·1216776	3·57	1·2725656	4·07	1·4036429	4·57	1·5195132
3·08	1·1249296	3·58	1·2753628	4·08	1·4060970	4·58	1·5216990
3·09	1·1281711	3·59	1·2781522	4·09	1·4085450	4·59	1·5238800
3·10	1·1314021	3·60	1·2809338	4·10	1·4109870	4·60	1·5260563
3·11	1·1346227	3·61	1·2837078	4·11	1·4134230	4·61	1·5282278
3·12	1·1378330	3·62	1·2864740	4·12	1·4158532	4·62	1·5303947
3·13	1·1410330	3·63	1·2892326	4·13	1·4182774	4·63	1·5325569
3·14	1·1442228	3·64	1·2919837	4·14	1·4206958	4·64	1·5347144
3·15	1·1474025	3·65	1·2947272	4·15	1·4231083	4·65	1·5368672
3·16	1·1505720	3·66	1·2974631	4·16	1·4255151	4·66	1·5390154
3·17	1·1537316	3·67	1·3001917	4·17	1·4279160	4·67	1·5411591
3·18	1·1568812	3·68	1·3029128	4·18	1·4303112	4·68	1·5432981
3·19	1·1600209	3·69	1·3056265	4·19	1·4327007	4·69	1·5454326
3·20	1·1631508	3·70	1·3083328	4·20	1·4350845	4·70	1·5475625
3·21	1·1662709	3·71	1·3110319	4·21	1·4374626	4·71	1·5496879
3·22	1·1693814	3·72	1·3137237	4·22	1 4398351	4·72	1·5518088
3·23	1·1724821	3·73	1·3164082	4·23	1·4422020	4·73	1·5539252
3·24	1·1755733	3·74	1·3190856	4·24	1·4445633	4·74	1·5560371
3·25	1·1786550	3·75	1·3217558	4·25	1·4469190	4 75	1·5581446
3·26	1·1817272	3 76	1·3244190	4·26	1·4492692	4·76	1·5602476
3·27	1·1847900	3·77	1·3270750	4·27	1·4516138	4·77	1·5623463
3·28	1·1878434	3·78	1 3297240	4·28	1·4539530	4·78	1·5644405
3·29	1·1908876	3·79	1·3323660	4·29	1·4562868	4·79	1·5665303
3·30	1·1939225	3·80	1·3350011	4·30	1·4586150	4·80	1·5686159
3 31	1·1969482	3·81	1·3376292	4·31	1·4609379	4·81	1·5706971
3 32	1·1999648	3·82	1·3402504	4·32	1·4632554	4·82	1·5727739
3·33	1·2029723	3·83	1·3428648	4·33	1·4655675	4·83	1·5748465
3·34	1·2059708	3·84	1·3454724	4·34	1·4678743	4·84	1·5769147
3·35	1·2089603	3·85	1·3480731	4·35	1·4701758	4·85	1·5789787
3·36	1·2119410	3·86	1·3506672	4·36	1·4724721	4·86	1·5810384
3·37	1·2149127	3·87	1·3532545	4·37	1·4747630	4·87	1·5830939
3·38	1·2178757	3·88	1·3558352	4·38	1·4770487	4·88	1·5851452
3·39	1·2208299	3·89	1·3584092	4·39	1·4793292	4·89	1·5871923
3·40	1·2237754	3·90	1·3609766	4·40	1·4816045	4·90	1·5892352
3·41	1·2267123	3·91	1·3635374	4·41	1·4838747	4 91	1·5912739
3·42	1·2296406	3·92	1·3660917	4·42	1·4861397	4·92	1·5933085
3·43	1·2325603	3·93	1·3686394	4·43	1·4883996	4·93	1·5953390
3·44	1 2354715	3·94	1·3711807	4·44	1·4906544	4·94	1·5973653
3·45	1·2383742	3·95	1·3737156	4·45	1·4929041	4·95	1·5993876
3·46	1·2412686	3·96	1·3762440	4·46	1·4953488	4·96	1·6014057
3·47	1·2441546	3·97	1·3787661	4·47	1·4973884	4·97	1·6034198
3·48	1·2470323	3·98	1·3812818	4·48	1·4996230	4·98	1·6054299
3·49	1·2499017	3·99	1·3837912	4·49	1·5018527	4·99	1·6074359
3·50	1·2527630	4·00	1·3862944	4·50	1·5040774	5·00	1·6094379

N.	Logar.	N.	Logar.	N.	Logar.	N.	Logar.
5·01	1·6114359	5·51	1·7065646	6·01	1·7934247	6·51	1·8733395
5·02	1·6134299	5·52	1·7083779	6·02	1·7950873	6·52	1·8748744
5·03	1·6154200	5·53	1·7101878	6·03	1·7967470	6·53	1·8764069
5·04	1·6174061	5·54	1·7119945	6·04	1·7984040	6·54	1·8779372
5·05	1·6193883	5·55	1·7137979	6·05	1·8000583	6·55	1·8794650
5·06	1·6213665	5·56	1·7155981	6·06	1·8017098	6·56	1·8809906
5·07	1·6233408	5·57	1·7173951	6·07	1·8033586	6·57	1·8825138
5·08	1·6253113	5·58	1·7191888	6·08	1·8050047	6·58	1·8840347
5·09	1·6272778	5·59	1·7209793	6 09	1·8066481	6·59	1·8855533
5·10	1·6292405	5·60	1·7227666	6·10	1·8082888	6·60	1·8870696
5·11	1·6311994	5·61	1·7245507	6·11	1·8099268	6·61	1·8885836
5·12	1·6331544	5·62	1·7263317	6·12	1·8115621	6·62	1·8900954
5·13	1·6351057	5·63	1·7281094	6·13	1·8131947	6·63	1·8916048
5·14	1·637053.	5·64	1·7298841	6·14	1·8148247	6·64	1·8931120
5·15	1·6389967	5·65	1·7316555	6·15	1·8164521	6·65	1·8946169
5·16	1·6409366	5·66	1·7334239	6·16	1·8180768	6·66	1·8961195
5·17	1·6428727	5·67	1·7351891	6·17	1·8196988	6·67	1·8976198
5·18	1·6448051	5 68	1·7369512	6·18	1·8213183	6·68	1·8991180
5·19	1·6467337	5 69	1·7387102	6·19	1·8229351	6·69	1·9006139
5·20	1·6486586	5·70	1 7404662	6·20	1·8245493	6·70	1·9021075
5·21	1·6505799	5·71	1·7422190	6·21	1·8261609	6·71	1·9035990
5 22	1·6524974	5·72	1·7439689	6·22	1·8277699	6·72	1·9050882
5·23	1·6544113	5·73	1·7457155	6·23	1·8293763	6·73	1 9065751
5·24	1·6563215	5·74	1·7474593	6·24	1·8309802	6·74	1·9080599
5·25	1·6582281	5·75	1·7491998	6·25	1·8325815	6·75	1·9095425
5·26	1·6601310	5·76	1·7509375	6·26	1·8341802	6·76	1·9110229
5·27	1·6620304	5·77	1·7526721	6·27	1·8357764	6·77	1·9125011
5·28	1·6639261	5·78	1·7544037	6·28	1·8373700	6·78	1·9139771
5·29	1·6658182	5·79	1·7561323	6·29	1·8389611	6·79	1·9154509
5·30	1·6677068	5·80	1·7578579	6·30	1·8405496	6·80	1·9169226
5·31	1·6695918	5·81	1·7595806	6·31	1·8421357	6·81	1 9183921
5·32	1·6714733	5·82	1·7613003	6·32	1 8437192	6·82	1·9198595
5·33	1·6733512	5·83	1·7630170	6·33	1·8453002	6·83	1·9213247
5·34	1·6752257	5·84	1·7647308	6·34	1·8468788	6·84	1·9227877
5·35	1·6770966	5·85	1·7664416	6·35	1·8484548	6·85	1·9242487
5·36	1·6789640	5·86	1·7681496	6 36	1·8500284	6·86	1·9257074
5·37	1·6808279	5·87	1·7698546	6·37	1·8515995	6·87	1·9271641
5·38	1·6826884	5·88	1·7715568	6·38	1·8531681	6·88	1·9286187
5·39	1·6845454	5·89	1·7732560	6·39	1·8547343	6·89	1·9300711
5·40	1·6863990	5·90	1·7749524	6·40	1·8562980	6·90	1·9315214
5·41	1·6882491	5·91	1·7766458	6·41	1·8578593	6·91	1·9329696
5·42	1·6900958	5·92	1·7783364	6·42	1·8594181	6·92	1·9344158
5·43	1·6919391	5·93	1·7800242	6·43	1·8609745	6·93	1·9358598
5·44	1·6937791	5·94	1·7817091	6·44	1·8625285	6·94	1·9373018
5·45	1·6956156	5·95	1·7833912	6·45	1·8640801	6·95	1·9387417
5·46	1·6974488	5·96	1·7850705	6·46	1·8656293	6·96	1·9401795
5·47	1·6992786	5·97	1·7867469	6·47	1·8671761	6·97	1·9416152
5·48	1·7011051	5·98	1·7884206	6·48	1·8687205	6·98	1·9430489
5·49	1·7029283	5·99	1·7900914	6·49	1·8702625	6·99	1·9444805
5·50	1·7047481	6·00	1·7917595	6·50	1·8718022	7·00	1·9459101

N.	Logar.	N.	Logar.	N.	Logar.	N.	Logar.
7·01	1·9473377	7·51	2·0162355	8·01	2·0806908	8·51	2·1412419
7·02	1·9487632	7·52	2·0175661	8·02	2·0819384	8·52	2·1424163
7·03	1·9501867	7·53	2·0188950	8·03	2·0831845	8·53	2·1435894
7·04	1·9516082	7·54	2·0202222	8·04	2·0844291	8·54	2·1447610
7·05	1·9530276	7·55	2·0215476	8·05	2·0856721	8·55	2·1459313
7·06	1·9544451	7·56	2·0228712	8·06	2·0869136	8·56	2·1471002
7·07	1·9558605	7·57	2·0241931	8·07	2·0881535	8·57	2·1482677
7·08	1·9572739	7·58	2·0255132	8·08	2·0893919	8·58	2·1494340
7·09	1·9586853	7·59	2·0268316	8·09	2·0906287	8·59	2·1505988
7·10	1·9600948	7·60	2·0281482	8·10	2·0918641	8·60	2·1517622
7·11	1·9615022	7·61	2·0294632	8·11	2·0930979	8·61	2·1529244
7·12	1·9629077	7·62	2·0307764	8·12	2·0943301	8·62	2·1540851
7·13	1·9643112	7·63	2·0320878	8·13	2·0955609	8·63	2·1552445
7·14	1·9657128	7·64	2·0333976	8·14	2·0967901	8·64	2·1564026
7·15	1·9671124	7·65	2·0347056	8·15	2·0980179	8·65	2·1575593
7·16	1·9685100	7·66	2·0360120	8·16	2·0992442	8·66	2·1587147
7·17	1·9699056	7·67	2·0373166	8·17	2·1004689	8·67	2·1598688
7·18	1·9712994	7·68	2·0386195	8·18	2·1016922	8·68	2·1610215
7·19	1·9726912	7·69	2·0399208	8·19	2·1029139	8·69	2·1621729
7·20	1·9740810	7·70	2·0412203	8·20	2·1041342	8·70	2·1633230
7·21	1·9754690	7·71	2·0425182	8·21	2·1053529	8·71	2·1644718
7·22	1·9768550	7·72	2·0438144	8·22	2·1065702	8·72	2·1656192
7·23	1·9782390	7·73	2·0451089	8·23	2·1077860	8·73	2·1667654
7·24	1·9796212	7·74	2·0464017	8·24	2·1090003	8·74	2·1679102
7·25	1·9810015	7·75	2·0476928	8·25	2·1102125	8·75	2·1690537
7·26	1·9823798	7·76	2·0489823	8·26	2·1114246	8·76	2·1701959
7·27	1·9837563	7·77	2·0502702	8·27	2·1126345	8·77	2·1713368
7·28	1·9851309	7·78	2·0515563	8·28	2·1138430	8·78	2·1724764
7·29	1·9865035	7·79	2·0528409	8·29	2·1150500	8·79	2·1736147
7·30	1·9878743	7·80	2·0541237	8·30	2·1162555	8·80	2·1747517
7·31	1·9892433	7·81	2·0554050	8·31	2·1174596	8·81	2·1758874
7·32	1·9906103	7·82	2·0566846	8·32	2·1186623	8·82	2·1770219
7·33	1·9919755	7·83	2·0579625	8·33	2·1198634	8·83	2·1781550
7·34	1·9933388	7·84	2·0592388	8·34	2·1210632	8·84	2·1792869
7·35	1·9947003	7·85	2·0605135	8·35	2·1222615	8·85	2·1804175
7·36	1·9960599	7·86	2·0617866	8·36	2·1234584	8·86	2·1815468
7·37	1·9974177	7·87	2·0630581	8·37	2·1246539	8·87	2·1826748
7·38	1·9987736	7·88	2·0643279	8·38	2·1258479	8·88	2·1838016
7·39	2·0001277	7·89	2·0655961	8·39	2·1270405	8·89	2·1849270
7·40	2·0014800	7·90	2·0668628	8·40	2·1282317	8·90	2·1860513
7·41	2·0028304	7·91	2·0681278	8·41	2·1294215	8·91	2·1871742
7·42	2·0041791	7·92	2·0693912	8·42	2·1306098	8·92	2·1882959
7·43	2·0055259	7·93	2·0706530	8·43	2·1317968	8·93	2·1894164
7·44	2·0068708	7·94	2·0719133	8·44	2·1329823	8·94	2·1905356
7·45	2·0082140	7·95	2·0731719	8·45	2·1341664	8·95	2·1916535
7·46	2·0095554	7·96	2·0744290	8·46	2·1353492	8·96	2·1927702
7·47	2·0108950	7·97	2·0756845	8·47	2·1365305	8·97	2·1938857
7·48	2·0122328	7·98	2·0769384	8·48	2·1377104	8·98	2·1949999
7·49	2·0135688	7·99	2·0781907	8·49	2·1388890	8·99	2·1961128
7·50	2·0149030	8·00	2·0794415	8·50	2·1400662	9·00	2·1972246

Tab. 5. HYPERBOLIC LOGARITHMS. 223

N.	Logar.	N.	Logar.	N.	Logar.
9·01	2·1983351	9·36	2·2364453	9·71	2·2731563
9·02	2·1994443	9·37	2·2375131	9·72	2·2741856
9·03	2·2005524	9·38	2·2385797	9·73	2·2752139
9·04	2·2016592	9·39	2·2396453	9·74	2·2762411
9·05	2·2027648	9·40	2·2407097	9·75	2·2772673
9·06	2·2038691	9·41	2·2417729	9·76	2·2782924
9·07	2·2049723	9·42	2·2428351	9 77	2·2793165
9·08	2·2060742	9·43	2·2438961	9·78	2·2803395
9·09	2·2071749	9·44	2·2449560	9·79	2·2813615
9·10	2·2082744	9·45	2·2460147	9·80	2·2823824
9·11	2·2093727	9·46	2·2470724	9·81	2·2834023
9·12	2·2104698	9·47	2·2481289	9·82	2·2844211
9·13	2·2115657	9·48	2·2491843	9·83	2·2854389
9·14	2·2126604	9·49	2·2502386	9·84	2·2864557
9·15	2·2137539	9·50	2·2512918	9·85	2·2874715
9·16	2·2148462	9·51	2·2523439	9·86	2·2884862
9·17	2·2159373	9 52	2·2533947	9·87	2·2894999
9·18	2·2170272	9·53	2·2544446	9·88	2 2905125
9·19	2·2181159	9·54	2·2554935	9·89	2·2915241
9·20	2·2192035	9·55	2·2565411	9·90	2·2925348
9·21	2·2202898	9·56	2·2575877	9·91	2 2935444
9·22	2·2213750	9·57	2·2586332	9·92	2·2945529
9·23	2·2224590	9·58	2·2596775	9·93	2·2955605
9·24	2·2235419	9·59	2·2607209	9·94	2·2965670
9·25	2·2246236	9·60	2·2617631	9·95	2·2975726
9·26	2·2257040	9·61	2·2628042	9·96	2·2985771
9·27	2·2267834	9·62	2·2638443	9·97	2·2995806
9·28	2·2278615	9·63	2·2648832	9·98	2·3005831
9·29	2·2289385	9·64	2·2659211	9·99	2·3015846
9·30	2·2300144	9·65	2·2669579	10·00	2·3025851
9·31	2·2310891	9·66	2·2679936	100·0	4·6051702
9·32	2·2321626	9·67	2·2690283	1000	6·9077553
9·33	2·2332350	9·68	2·2700619	10000	9·2103404
9·34	2·2343062	9·69	2·2710944	100000	11·51292546
9·35	2·2353763	9·70	2·2721259		

N.	Logar.	N.	Logar.	N.	Logar.	N.	Logar.
1	0·0000000	51	3·9318256	101	4·6151205	151	5·0172798
2	0·6931472	52	3·9512437	102	4·6249728	152	5·0238805
3	1·0986123	53	3·9702919	103	4·6347290	153	5·0304379
4	1·3862944	54	3·9889840	104	4·6443909	154	5·0369526
5	1·6094379	55	4·0073332	105	4·6539604	155	5·0434251
6	1·7917595	56	4·0253517	106	4·6634391	156	5·0498560
7	1·9459101	57	4·0430513	107	4·6728288	157	5·0562458
8	2·0794415	58	4·0604430	108	4·6821312	158	5·0625950
9	2·1972246	59	4·0775374	109	4·6913479	159	5·0689042
10	2·3025851	60	4·0943446	110	4·7004804	160	5·0751738
11	2·3978953	61	4·1108739	111	4·7095302	161	5·0814044
12	2·4849066	62	4·1271344	112	4·7184989	162	5 0875963
13	2·5649494	63	4·1431347	113	4·7273878	163	5·0937502
14	2·6390573	64	4·1588831	114	4·7351984	164	5·0998664
15	2·7080502	65	4·1743873	115	4·7449321	165	5·1059455
16	2·7725887	66	4·1896547	116	4·7535902	166	5·1119878
17	2·8332133	67	4·2046926	117	4·7621739	167	5·1179938
18	2·8903718	68	4·2195077	118	4·7706846	168	5·1239640
19	2·9444390	69	4·2341065	119	4·7791235	169	5·1298987
20	2·9957323	70	4·2484952	120	4·7874917	170	5·1357984
21	3·0445224	71	4·2626799	121	4·7957905	171	5·1416636
22	3·0910425	72	4·2766661	122	4·8040210	172	5·1474945
23	3·1354942	73	4·2904594	123	4·8121844	173	5·1532916
24	3·1780538	74	4·3040651	124	4·8202816	174	5·1590553
25	3·2188758	75	4·3174881	125	4·8283137	175	5·1647860
26	3·2580965	76	4·3307333	126	4·8362819	176	5·1704840
27	3·2958369	77	4·3438054	127	4·8441871	177	5·1761497
28	3·3322045	78	4·3567088	128	4·8520303	178	5·1817836
29	3·3672958	79	4·3694479	129	4·8598124	179	5·1873858
30	3·4011974	80	4·3820266	130	4·8675345	180	5·1929569
31	3·4339872	81	4·3944492	131	4·8751973	181	5·1984970
32	3·4657359	82	4·4067192	132	4·8828019	182	5·2040067
33	3·4965076	83	4·4188406	133	4·8903491	183	5·2094862
34	3·5263605	84	4·4308168	134	4·8978398	184	5·2149358
35	3·5553481	85	4·4426513	135	4·9052748	185	5·2203558
36	3·5835189	86	4·4543473	136	4 9126549	186	5·2257467
37	3·6109179	87	4·4659081	137	4·9199809	187	5·2311086
38	3·6375862	88	4·4773368	138	4·9272537	188	5·2364420
39	3·6635616	89	4·4886364	139	4·9344739	189	5·2417470
40	3·6888795	90	4·4998097	140	4·9416424	190	5·2470241
41	3·7135721	91	4·5108595	141	4·9487599	191	5·2522734
42	3·7376696	92	4·5217886	142	4·9558271	192	5·2574954
43	3·7612001	93	4·5325995	143	4·9628446	193	5·2626902
44	3·7841896	94	4·5432948	144	4·9698133	194	5·2678582
45	3·8066625	95	4·5538769	145	4·9767337	195	5·2729996
46	3·8286414	96	4·5643482	146	4·9836066	196	5·2781147
47	3·8501476	97	4·5747110	147	4·9904326	197	5·2832037
48	3·8712010	98	4·5849675	148	4·9972123	198	5·2882670
49	3·8918203	99	4·5951199	149	5·0039463	199	5·2933048
50	3·9120230	100	4·6051702	150	5·0106353	200	5·2983174

N.	Logar.	N.	Logar.	N.	Logar.	N.	Logar.
201	5·3033049	251	5·5254529	301	5·7071103	351	5·8607862
202	5·3082677	252	5·5294291	302	5·7104270	352	5·8636312
203	5·3132060	253	5·5333895	303	5·7137328	353	5·8664681
204	5·3181200	254	5·5373343	304	5·7170277	354	5·8692969
205	5·3230100	255	5·5412635	305	5·7203118	355	5·8721178
206	5·3278762	256	5 5451774	306	5·7235851	356	5·8749307
207	5·3327188	257	5·5490761	307	5·7268477	357	5·8777358
208	5·3375381	258	5·5529596	308	5·7300998	358	5·8805330
209	5·3423343	259	5·5568281	309	5·7333413	359	5·8833224
210	5·3471075	260	5·5606816	310	5·7365723	360	5·8961040
211	5·3518581	261	5·5645204	311	5·7397929	361	5·8888780
212	5·3565863	262	5·5683445	312	5·7430032	362	5·8916442
213	5·3612922	263	5·5721540	313	5·7462032	363	5·8944028
214	5·3659760	264	5·5759491	314	5·7493930	364	5·8971539
215	5·3706380	265	5·5797298	315	5·7525726	365	5·8998974
216	5·3752784	266	5·5834963	316	5·7557422	366	5·9026333
217	5·3798974	267	5·5872487	317	5·7589018	367	5·9053618
218	5·3844951	268	5·5909870	318	5·7620514	368	5·9080829
219	5·3890717	269	5·5947114	319	5·7651911	369	5 9107966
220	5·3936275	270	5·5984220	320	5·7683210	370	5·9135030
221	5·3981627	271	5·6021188	321	5·7714411	371	5·9162021
222	5·4026774	272	5·6058021	322	5·7745515	372	5·9188939
223	5·4071718	273	5·6094718	323	5·7776523	373	5·9215784
224	5·4116461	274	5·6131281	324	5·7807435	374	5·9242558
225	5·4161004	275	5·6167711	325	5·7838252	375	5·9269260
226	5·4205350	276	5·6204009	326	5·7868974	376	5·9295891
227	5·4249500	277	5·6240175	327	5·7899602	377	5·9322452
228	5·4293456	278	5·6276211	328	5·7930136	378	5·9348942
229	5·4337220	279	5·6312118	329	5·7960578	379	5 9375362
230	5·4380793	280	5·6347896	330	5·7990927	380	5·9401713
231	5·4424177	281	5·6383547	331	5·8021184	381	5·9427994
232	5·4467374	282	5·6419071	332	5·8051350	382	5·9454206
233	5·4510385	283	5·6454469	333	5·8081425	383	5·9480350
234	5·4553211	284	5·6489742	334	5·8111410	384	5·9506420
235	5·4595855	285	5·6524892	335	5·8141305	385	5·9532433
236	5·4638318	286	5·6559918	336	5·8171112	386	5·9558374
237	5·4680601	287	5·6594822	337	5·8200829	387	5·9584247
238	5·4722707	288	5·6629605	338	5·8230459	388	5·9610053
239	5·4764636	289	5·6664267	339	5·8260001	389	5 9635793
240	5·4806389	290	5·6698809	340	5·8289456	390	5·9661467
241	5·4847969	291	5·6733233	341	5·8318825	391	5·9687076
242	5·4889377	292	5·6767538	342	5·8348107	392	5·9712618
243	5·4930614	293	5·6801726	343	5·8377304	393	5·9738096
244	5·4971682	294	5·6835798	344	5·8406417	394	5 9763509
245	5·5012582	295	5·6869754	345	5·8435444	395	5·9788858
246	5 5053315	296	5·6903595	346	5·8464388	396	5·9814142
247	5·5093883	297	5·6937321	347	5·8493248	397	5·9839363
248	5·5134287	298	5·6970935	348	5·8522025	398	5·9864520
249	5·5174529	299	5·7004436	349	5·8550719	399	5·9889614
250	5·5214609	300	5·7037825	350	5·8579332	400	5 9914645

Q

N.	Logar.	N.	Logar.	N.	Logar.	N.	Logar.
401	5·9939614	451	6·1114673	501	6·2166061	551	6·3117348
402	5·9964521	452	6·1136822	502	6·2186001	552	6·3135480
403	5·9989366	453	6·1158921	503	6·2205902	553	6·3153580
404	6·0014149	454	6·1180972	504	6·2225763	554	6·3171647
405	6·0038871	455	6·1202974	505	6·2245584	555	6·3189681
406	6·0063532	456	6·1224928	506	6·2265367	556	6·3207683
407	6·0088132	457	6·1246834	507	6·2285110	557	6·3225652
408	6·0112672	458	6·1268692	508	6·2304814	558	6·3243590
409	6·0137152	459	6·1290502	509	6·2324480	559	6·3261495
410	6·0161572	460	6·1312265	510	6·2344107	560	6·3279368
411	6·0185932	461	6·1333980	511	6·2363696	561	6·3297209
412	6·0210233	462	6·1355649	512	6·2383246	562	6·3315018
413	6·0234476	463	6·1377271	513	6·2402758	563	6·3332796
414	6·0258660	464	6·1398846	514	6·2422233	564	6·3350543
415	6·0282785	465	6·1420374	515	6·2441669	565	6·3368257
416	6·0306853	466	6·1441856	516	6·2461068	566	6·3385941
417	6·0330862	467	6·1463293	517	6·2480429	567	6·3403593
418	6·0354814	468	6·1484683	518	6·2499752	568	6·3421214
419	6·0378709	469	6·1506028	519	6·2519039	569	6·3438804
420	6·0402547	470	6·1527327	520	6·2538288	570	6·3456364
421	6·0426328	471	6·1548581	521	6·2557500	571	6·3473892
422	6·0450053	472	6·1569790	522	6·2576676	572	6·3491390
423	6·0473722	473	6·1590954	523	6·2595815	573	6·3508857
424	6·0497335	474	6·1612073	524	6·2614917	574	6·3526294
425	6·0520892	475	6·1633148	525	6·2633983	575	6·3543700
426	6·0544393	476	6·1654179	526	6·2653012	576	6·3561077
427	6·0567840	477	6·1675165	527	6·2672005	577	6·3578423
428	6·0591232	478	6·1696107	528	6·2690963	578	6·3595739
429	6·0614569	479	6·1717006	529	6·2709884	579	6·3613025
430	6·0637852	480	6·1737861	530	6·2728770	580	6·3630281
431	6·0661081	481	6·1758673	531	6·2747620	581	6·3647508
432	6·0684256	482	6·1779441	532	6·2766435	582	6·3664704
433	6·0707377	483	6·1800167	533	6·2785214	583	6·3681872
434	6·0730445	484	6·1820849	534	6·2803958	584	6·3699010
435	6·0753460	485	6·1841489	535	6·2822667	585	6·3716118
436	6·0776422	486	6·1862086	536	6·2841342	586	6·3733198
437	6·0799332	487	6·1882641	537	6·2859981	587	6·3750248
438	6·0822189	488	6·1903154	538	6·2878586	588	6·3767269
439	6·0844994	489	6·1923625	539	6·2897156	589	6·3784262
440	6·0867747	490	6·1944054	540	6·2915691	590	6·3801225
441	6·0890449	491	6·1964441	541	6·2934193	591	6·3818160
442	6·0913099	492	6·1984787	542	6·2952660	592	6·3835066
443	6·0935698	493	6·2005092	543	6·2971093	593	6·3851944
444	6·0958246	494	6·2025355	544	6·2989492	594	6·3868793
445	6·0980743	495	6·2045578	545	6·3007858	595	6·3885614
446	6·1003190	496	6·2065759	546	6·3026190	596	6·3902407
447	6·1025586	497	6·2085900	547	6·3044488	597	6·3919171
448	6·1047932	498	6·2106001	548	6·3062753	598	6·3935908
449	6·1070229	499	6·2126061	549	6·3080984	599	6·3952616
450	6·1092476	500	6·2146081	550	6·3099183	600	6·3969297

Tab. 6.   HYPERBOLIC LOGARITHMS.   227

N.	Logar.	N.	Logar.	N.	Logar.	N.	Logar.
601	6·3985949	651	6·4785096	701	6·5525079	751	6·6214057
602	6·4002574	652	6·4800446	702	6·5539334	752	6 6227363
603	6·4019172	653	6·4815771	703	6·5553569	753	6·6240652
604	6·4035742	654	6·4831074	704	6·5567784	754	6·6253924
605	6·4052285	655	6·4846352	705	6·5581978	755	6 6267177
606	6·4068800	656	6·4861608	706	6·5596152	756	6·6280414
607	6·4085288	657	6·4876840	707	6·5610307	757	6·6293633
608	6·4101749	658	6·4892049	708	6·5624441	758	6·6306634
609	6·4118183	659	6 4907235	709	6·5638555	759	6·6320018
610	6·4134590	660	6·4922398	710	6 5652650	760	6·6333184
611	6·4150970	661	6·4937538	711	6·5666724	761	6·6346334
612	6·4167323	662	6·4952656	712	6·5680779	762	6·6359466
613	6·4183649	663	6·4967750	713	6·5694814	763	6·6372580
614	6·4199949	664	6·4982821	714	6·5708830	764	6·6385678
615	6·4216223	665	6 4997870	715	6·5722825	765	6·6398758
616	6·4232470	666	6·5012897	716	6·5736802	766	6 6411822
617	6·4248690	667	6·5027900	717	6·5750758	767	6·6424868
618	6·4264885	668	6·5042882	718	6·5764696	768	6·6437897
619	6·4281053	669	6·5057841	719	6·5778614	769	6·6450910
620	6·4297195	670	6·5072777	720	6·5792512	770	6·6463905
621	6·4313311	671	6·5087691	721	6·5806391	771	6 6476884
622	6·4329401	672	6·5102583	722	6·5820251	772	6·6489846
623	6·4345465	673	6·5117453	723	6·5834092	773	6·6502790
624	6·4361504	674	6.5132301	724	6·5847914	774	6·6515719
625	6·4377516	675	6·5147127	725	6·5861717	775	6·6528630
626	6·4393504	676	6·5161931	726	6·5875500	776	6·6541525
627	6·4409465	677	6·5176713	727	6·5889265	777	6·6554404
628	6·4425402	678	6·5191473	728	6·5903010	778	6·6567265
629	6·4441313	679	6·5206211	729	6·5916737	779	6·6580110
630	6·4457198	680	6·5220928	730	6·5930445	780	6·6592939
631	6·4473059	681	6·5235623	731	6·5944135	781	6·6605751
632	6·4488894	682	6·5250297	732	6·5957805	782	6·6618547
633	6·4504704	683	6·5264949	733	6·5971457	783	6·6631327
634	6·4520490	684	6·5279579	734	6·5985090	784	6·6644090
635	6 4536250	685	6·5294188	735	6·5998705	785	6·6656837
636	6·4551986	686	6·5308776	736	6 6012301	786	6·6669568
637	6·4567697	687	6·5323343	737	6·6025879	787	6·6682282
638	6·4583383	688	6 5337888	738	6·6039438	788	6·6694981
639	6·4599045	689	6·5352413	739	6 6052979	789	6·6707663
640	6·4614682	690	6·5366916	740	6·6066502	790	6·6720329
641	6·4630295	691	6·5381398	741	6·6080006	791	6·6732980
642	6·4645883	692	6·5395860	742	6·6093492	792	6·6745614
643	6·4661447	693	6·5410300	743	6·6106960	793	6 6758232
644	6·4676987	694	6·5124720	744	6·6120410	794	6·6770835
645	6·4692503	695	6·5439118	745	6·6133842	795	6·6783421
646	6 4707995	696	6·5453497	746	6·6147256	796	6·6795992
647	6·4723463	697	6·5467854	747	6·6160652	797	6·6808547
648	6·4738907	698	6·5482191	748	6·6174030	798	6·6821086
649	6 4754327	699	6·5496507	749	6 6187390	799	6·6833609
650	6·4769724	700	6·5510803	750	6·6200732	800	6·6846117

N.	Logar.	N.	Logar.	N.	Logar.	N.	Logar.
801	6·6858609	851	6·7464121	901	6 8035053	951	6·8575141
802	6·6871086	852	6·7475865	902	6·8046145	952	6·8585650
803	6·6883547	853	6·7487595	903	6·8057226	953	6·8596149
804	6·6895993	854	6·7499312	904	6·8068294	954	6·8606637
805	6·6908423	855	6·7511015	905	6·8079349	955	6 8617113
806	6·6920837	856	6·7522704	906	6·8090393	956	6·8627579
807	6·6933237	857	6·7534379	907	6·8101425	957	6·8638034
808	6·6945621	858	6·7546041	908	6·8112444	958	6·8648478
809	6·6957989	859	6·7557689	909	6·8123451	959	6·8658911
810	6·6970342	860	6·7569324	910	6·8134446	960	6·8669333
811	6·6982681	861	6·7580945	911	6·8145429	961	6·8679744
812	6·6995003	862	6:7592553	912	6·8156400	962	6·8690145
813	6·7007311	863	6·7604147	913	6·8167359	963	6·8700534
814	6·7019604	864	6·7615728	914	6·8178306	964	6·8710913
815	6·7031881	865	6·7627295	915	6·8189241	965	6·8721281
816	6·7044144	866	6·7638849	916	6·8200164	966	6·8731638
817	6·7056391	867	6·7650390	917	6·8211075	967	6·8741985
818	6·7068623	868	6·7661917	918	6·8221974	968	6·8752321
819	6·7080841	869	6·7673431	919	6·8232861	969	6·8762646
820	6·7093043	870	6·7684932	920	6·8243737	970	6·8772961
821	6 7105231	871	6·7696420	921	6·8254600	971	6·8783265
822	6·7117404	872	6·7707894	922	6·8265452	972	6·8793558
823	6·7129562	873	6·7719356	923	6·8276292	973	6·8803841
824	6·7141705	874	6·7730804	924	6·8287121	974	6·8814113
825	6·7153834	875	6·7742239	925	6·8297937	975	6·8824375
826	6·7165948	876	6·7753661	926	6·8308742	976	6·8834626
827	6·7178047	877	6·7765070	927	6·8319536	977	6·8844867
828	6·7190132	878	6·7776466	928	6·8330317	978	6·8855097
829	6·7202202	879	6·7787849	929	6·8341087	979	6·8865316
830	6·7214257	880	6·7799219	930	6·8351846	980	6·8875526
831	6·7226298	881	6·7810576	931	6·8362593	981	6·8885725
832	6·7238324	882	6·7821921	932	6·8373328	982	6·8895913
833	6·7250336	883	6·7833252	933	6·8384052	983	6·8906091
834	6 7262334	884	6·7844571	934	6·8394764	984	6·8916259
835	6·7274317	885	6·7855876	935	6·8405465	985	6·8926416
836	6·7286286	886	6·7867170	936	6·8416155	986	6·8936564
837	6·7298241	887	6·7878450	937	6·8426833	987	6·8946700
838	6·7310181	888	6·7889717	938	6·8437499	988	6·8956827
839	6·7322107	889	6·7900972	939	6·8448155	989	6·8966943
840	6·7334019	890	6·7912215	940	6·8458799	990	6·8977049
841	6·7345917	891	6·7923444	941	6·8469431	991	6·8987145
842	6·7357800	892	6·7934661	942	6·8480053	992	6·8997231
843	6·7369670	893	6·7945866	943	6·8490663	993	6·9007307
844	6·7381525	894	6·7957058	944	6·8501262	994	6·9017372
845	6·7393366	895	6·7968237	945	6·8511849	995	6·9027427
846	6·7405194	896	6·7979404	946	6·8522426	996	6·9037473
847	6·7417007	897	6 7990559	947	6·8532991	997	6·9047508
848	6·7428806	898	6·8001701	948	6·8543545	998	6·9057533
849	6·7440592	899	6·8012830	949	6·8554088	999	6·9067548
850	6·7452363	900	6 8023948	950	6·8564620	1000	6·9077553

N.	Logar.	N.	Logar.	N.	Logar.	N.	Logar
1001	6·9087548	1051	6·9574974	1101	7·0039741	1151	7·0483864
1002	6·9097533	1052	6·9584484	1102	7·0048820	1152	7·0492548
1003	6·9107508	1053	6·9593985	1103	7·0057890	1153	7·0501225
1004	6·9117473	1054	6·9603477	1104	7·0066952	1154	7·0509894
1005	6·9127428	1055	6·9612960	1105	7·0076006	1155	7·0518556
1006	6·9137374	1056	6·9622435	1106	7·0085052	1156	7·0527210
1007	6·9147309	1057	6·9631900	1107	7·0094089	1157	7·0535857
1008	6·9157234	1058	6·9641356	1108	7·0103119	1158	7·0544497
1009	6·9167150	1059	6·9650803	1109	7·0112140	1159	7·0553128
1010	6·9177056	1060	6·9660242	1110	7·0121153	1160	7·0561753
1011	6·9186952	1061	6·9669671	1111	7·0130158	1161	7·0570370
1012	6·9196838	1062	6·9679092	1112	7·0139155	1162	7·0578979
1013	6·9206715	1063	6·9688504	1113	7·0148144	1163	7·0587582
1014	6·9216582	1064	6·9697907	1114	7·0157124	1164	7·0596176
1015	6·9226439	1065	6·9707301	1115	7·0166097	1165	7·0604764
1016	6·9236286	1066	6·9716686	1116	7·0175061	1166	7·0613344
1017	6·9246124	1067	6·9726063	1117	7·0184018	1167	7·0621916
1018	6·9255952	1068	6·9735430	1118	7·0192967	1168	7·0630482
1019	6·9265770	1069	6·9744789	1119	7·0201907	1169	7·0639040
1020	6·9275579	1070	6·9754139	1120	7·0210840	1170	7·0647590
1021	6·9285378	1071	6·9763481	1121	7·0219764	1171	7·0656134
1022	6·9295168	1072	6·9772813	1122	7·0228681	1172	7·0664670
1023	6·9304948	1073	6·9782137	1123	7·0237590	1173	7·0673198
1024	6·9314718	1074	6·9791453	1124	7·0246490	1174	7·0681720
1025	6·9324479	1075	6·9800759	1125	7·0255383	1175	7·0690234
1026	6·9334230	1076	6·9810057	1126	7·0264268	1176	7·0698741
1027	6·9343972	1077	6·9819347	1127	7·0273145	1177	7·0707241
1028	6·9353704	1078	6·9828628	1128	7·0282014	1178	7·0715734
1029	6·9363427	1079	6·9837900	1129	7·0290876	1179	7·0724219
1030	6·9373141	1080	6·9847163	1130	7·0299729	1180	7·0732697
1031	6·9382845	1081	6·9856418	1131	7·0308575	1181	7·0741168
1032	6·9392539	1082	6·9865665	1132	7·0317413	1182	7·0749632
1033	6·9402225	1083	6·9874902	1133	7·0326243	1183	7·0758089
1034	6·9411901	1084	6·9884132	1134	7·0335065	1184	7·0766538
1035	6·9421567	1085	6·9893353	1135	7·0343879	1185	7·0774981
1036	6·9431224	1086	6·9902565	1136	7·0352686	1186	7·0783416
1037	6·9440872	1087	6·9911769	1137	7·0361485	1187	7·0791844
1038	6·9450511	1088	6·9920964	1138	7·0370276	1188	7·0800265
1039	6·9460140	1089	6·9930151	1139	7·0379060	1189	7·0808679
1040	6·9469760	1090	6·9939330	1140	7·0387835	1190	7·0817086
1041	6·9479371	1091	6·9948500	1141	7·0396603	1191	7·0825486
1042	6·9488972	1092	6·9957662	1142	7·0405364	1192	7·0833878
1043	6·9498565	1093	6·9966815	1143	7·0414117	1193	7·0842264
1044	6·9508148	1094	6·9975960	1144	7·0422862	1194	7·0850643
1045	6·9517722	1095	6·9985096	1145	7·0431599	1195	7·0859015
1046	6·9527286	1096	6·9994225	1146	7·0440329	1196	7·0867379
1047	6·9536842	1097	7·0003345	1147	7·0449051	1197	7·0875737
1048	6·9546389	1098	7·0012456	1148	7·0457766	1198	7·0884088
1049	6·9555926	1099	7·0021160	1149	7·0466473	1199	7·0892432
1050	6·9565454	1100	7·0030655	1150	7·0475172	1200	7·0900768

′ / ″	0	1	2	3	4	5	6	7	8	9	10	11	12
	0	60	120	180	240	300	360	420	480	540	600	660	720
0		1·7782	1·4771	1·3010	1·1761	1·0792	0000	9331	8751	8239	7782	7368	6990
1	3·5563	1·7710	1·4735	1·2986	1·1743	1·0777	9988	9320	8742	8231	7774	7361	6984
2	3·2553	1·7639	1·4699	1·2962	1·1725	1·0763	9976	9310	8733	8223	7767	7354	6978
3	3·0792	1·7570	1·4664	1·2939	1·1707	1·0749	9964	9300	8724	8215	7760	7348	6972
4	2·9542	1·7501	1·4629	1·2915	1·1689	1·0734	9952	9289	8715	8207	7753	7341	6966
5	2·8573	1·7434	1·4594	1·2891	1·1671	1·0720	9940	9279	8706	8199	7745	7335	6960
6	2·7782	1·7368	1·4559	1·2868	1·1654	1·0706	9928	9269	8697	8191	7738	7328	6954
7	2·7112	1·7302	1·4525	1·2845	1·1636	1·0692	9916	9259	8688	8183	7731	7322	6948
8	2·6532	1·7238	1·4491	1·2821	1·1619	1·0678	9905	9249	8679	8175	7724	7315	6942
9	2·6021	1·7175	1·4457	1·2798	1·1601	1·0663	9893	9238	8670	8167	7717	7309	6936
10	2·5563	1·7112	1·4424	1·2775	1·1584	1·0649	9881	9228	8661	8159	7710	7302	6930
11	2·5149	1·7050	1·4390	1·2753	1·1566	1·0635	9869	9218	8652	8152	7703	7296	6924
12	2·4771	1·6990	1·4357	1·2730	1·1549	1·0621	9858	9208	8643	8144	7696	7289	6918
13	2·4424	1·6930	1·4325	1·2707	1·1532	1·0608	9846	9198	8635	8136	7688	7283	6912
14	2·4102	1·6871	1·4292	1·2685	1·1515	1·0594	9834	9188	8626	8128	7681	7276	6906
15	2·3802	1·6812	1·4260	1·2663	1·1498	1·0580	9823	9178	8617	8120	7674	7270	6900
16	2·3522	1·6755	1·4228	1·2640	1·1481	1·0566	9811	9168	8608	8112	7667	7264	6894
17	2·3259	1·6698	1·4196	1·2618	1·1464	1·0552	9800	9158	8604	8104	7660	7257	6888
18	2·3010	1·6642	1·4165	1·2596	1·1447	1·0539	9788	9148	8591	8097	7653	7251	6882
19	2·2775	1·6587	1·4133	1·2574	1·1430	1·0525	9777	9138	8582	8089	7646	7244	6877
20	2·2553	1·6532	1·4102	1·2553	1·1413	1·0512	9765	9128	8573	8081	7639	7238	6871
21	2·2341	1·6478	1·4071	1·2531	1·1397	1·0498	9754	9119	8565	8073	7632	7232	6865
22	2·2139	1·6425	1·4040	1·2510	1·1380	1·0484	9742	9109	8556	8066	7625	7225	6859
23	2·1946	1·6372	1·4010	1·2488	1·1363	1·0471	9731	9099	8547	8058	7618	7219	6853
24	2·1761	1·6320	1·3979	1·2467	1·1347	1·0458	9720	9089	8539	8050	7611	7212	6847
25	2·1584	1·6269	1·3949	1·2445	1·1331	1·0444	9708	9079	8530	8043	7604	7206	6841
26	2·1413	1·6218	1·3919	1·2424	1·1314	1·0431	9697	9070	8522	8035	7597	7200	6836
27	2·1249	1·6168	1·3890	1·2403	1·1298	1·0418	9686	9060	8513	8027	7590	7193	6830
28	2·1091	1·6118	1·3860	1·2382	1·1282	1·0404	9675	9050	8504	8020	7583	7187	6824
29	2·0939	1·6069	1·3831	1·2362	1·1266	1·0391	9664	9041	8496	8012	7577	7181	6818
30	2·0792	1·6021	1·3802	1·2341	1·1249	1·0378	9652	9031	8487	8004	7570	7175	6812
31	2·0649	1·5973	1·3773	1·2320	1·1233	1·0365	9641	9021	8479	7997	7563	7168	6807
32	2·0512	1·5925	1·3745	1·2300	1·1217	1·0352	9630	9012	8470	7989	7556	7162	6801
33	2·0373	1·5878	1·3716	1·2279	1·1201	1·0339	9619	9002	8462	7981	7549	7156	6795
34	2·0248	1·5832	1·3688	1·2259	1·1186	1·0326	9608	8992	8453	7974	7542	7149	6789
35	2·0122	1·5786	1·3660	1·2239	1·1170	1·0313	9597	8983	8445	7966	7535	7143	6784
36	2·0000	1·5740	1·3632	1·2218	1·1154	1·0300	9586	8973	8437	7959	7528	7137	6778
37	1·9881	1·5695	1·3604	1·2198	1·1138	1·0287	9575	8964	8428	7951	7522	7131	6772
38	1·9765	1·5651	1·3576	1·2178	1·1123	1·0274	9564	8954	8420	7944	7515	7124	6766
39	1·9652	1·5607	1·3549	1·2159	1·1107	1·0261	9553	8945	8411	7936	7508	7118	6761
40	1·9542	1·5563	1·3522	1·2139	1·1091	1·0248	9542	8935	8403	7929	7501	7112	6755
41	1·9435	1·5520	1·3495	1·2119	1·1076	1·0235	9532	8926	8395	7921	7494	7106	6749
42	1·9331	1·5477	1·3468	1·2099	1·1061	1·0223	9521	8917	8386	7914	7488	7100	6743
43	1·9228	1·5435	1·3441	1·2080	1·1045	1·0210	9510	8907	8378	7906	7481	7093	6738
44	1·9128	1·5393	1·3415	1·2061	1·1030	1·0197	9499	8898	8370	7899	7474	7087	6732
45	1·9031	1·5351	1·3388	1·2041	1·1015	1·0185	9488	8888	8361	7891	7467	7081	6726
46	1·8935	1·5310	1·3362	1·2022	1·0999	1·0172	9478	8879	8353	7884	7461	7075	6721
47	1·8842	1·5269	1·3336	1·2003	1·0984	1·0160	9467	8870	8345	7877	7454	7069	6715
48	1·8751	1·5229	1·3310	1·1984	1·0969	1·0147	9456	8861	8337	7869	7447	7063	6709
49	1·8661	1·5189	1·3284	1·1965	1·0954	1·0135	9446	8851	8328	7862	7441	7057	6704
50	1·8573	1 5149	1·3259	1·1946	1·0939	1·0122	9435	8842	8320	7855	7434	7050	6698
51	1·8487	1·5110	1·3233	1·1927	1·0924	1·0110	9425	8833	8312	7847	7427	7044	6692
52	1·8403	1·5071	1·3208	1·1908	1·0909	1·0098	9414	8824	8304	7840	7421	7038	6687
53	1·8320	1·5032	1·3183	1·1889	1·0894	1·0085	9404	8814	8296	7832	7414	7032	6681
54	1·8239	1·4994	1·3158	1·1871	1·0880	1·0073	9393	8805	8288	7825	7407	7026	6676
55	1·8159	1·4956	1·3133	1·1852	1·0865	1·0061	9383	8796	8279	7818	7401	7020	6670
56	1·8081	1·4918	1·3108	1·1834	1·0850	1·0049	9372	8787	8271	7811	7394	7014	6664
57	1·8004	1·4881	1·3083	1·1816	1·0835	1·0036	9362	8778	8263	7803	7387	7008	6659
58	1·7929	1·4844	1·3059	1·1797	1·0821	1·0024	9351	8769	8255	7796	7381	7002	6653
59	1·7855	1·4808	1·3034	1·1779	1·0806	1·0012	9341	8760	8247	7789	7374	6996	6648
60	1·7782	1·4771	1·3010	1·1761	1·0792	1·0000	9331	8751	8239	7782	7368	6990	6642

′	13	14	15	16	17	18	19	20	21	22	23	24	25	26	27
″	780	840	900	960	1020	1080	1140	1200	1260	1320	1380	1440	1500	1560	1620
0	6642	6320	6021	5740	5477	5229	4994	4771	4559	4357	4164	3979	3802	3632	3468
1	6637	6315	6016	5736	5473	5225	4990	4768	4556	4354	4161	3976	3799	3629	3465
2	6631	6310	6011	5731	5469	5221	4986	4764	4552	4351	4158	3973	3796	3626	3463
3	6625	6305	6006	5727	5464	5217	4983	4760	4549	4347	4155	3970	3793	3623	3460
4	6620	6300	6001	5722	5460	5213	4979	4757	4546	4344	4152	3967	3791	3621	3457
5	6614	6294	5997	5718	5456	5209	4975	4753	4542	4341	4149	3964	3788	3618	3454
6	6609	6289	5992	5713	5452	5205	4971	4750	4539	4338	4145	3961	3785	3615	3452
7	6603	6284	5987	5709	5447	5201	4967	4746	4535	4334	4142	3958	3782	3612	3449
8	6598	6279	5982	5704	5443	5197	4964	4742	4532	4331	4139	3955	3779	3610	3446
9	6592	6274	5977	5700	5439	5193	4960	4739	4528	4328	4136	3952	3776	3607	3444
10	6587	6269	5973	5695	5435	5189	4956	4735	4525	4325	4133	3949	3773	3604	3441
11	6581	6264	5968	5691	5430	5185	4952	4732	4522	4321	4130	3946	3770	3601	3438
12	6576	6259	5963	5686	5426	5181	4949	4728	4518	4318	4127	3943	3768	3598	3436
13	6570	6254	5958	5682	5422	5177	4945	4724	4515	4315	4124	3940	3765	3596	3433
14	6565	6248	5954	5677	5418	5173	4941	4721	4511	4311	4120	3937	3762	3593	3431
15	6559	6243	5949	5673	5414	5169	4937	4717	4508	4308	4117	3934	3759	3590	3428
16	6554	6238	5944	5669	5409	5165	4933	4714	4505	4305	4114	3931	3756	3587	3425
17	6548	6233	5939	5664	5405	5161	4930	4710	4501	4302	4111	3928	3753	3585	3423
18	6543	6228	5935	5660	5401	5157	4926	4707	4498	4298	4108	3925	3750	3582	3420
19	6538	6223	5930	5655	5397	5153	4922	4703	4494	4295	4105	3922	3747	3579	3417
20	6532	6218	5925	5651	5393	5149	4918	4699	4491	4292	4102	3919	3745	3576	3415
21	6527	6213	5920	5646	5389	5145	4915	4696	4488	4289	4099	3917	3742	3574	3412
22	6521	6208	5916	5642	5384	5141	4911	4692	4484	4285	4096	3914	3739	3571	3409
23	6516	6203	5911	5637	5380	5137	4907	4689	4481	4282	4092	3911	3736	3568	3407
24	6510	6198	5906	5633	5376	5133	4903	4685	4477	4279	4089	3908	3733	3565	3404
25	6505	6193	5902	5629	5372	5129	4900	4682	4474	4276	4086	3905	3730	3563	3401
26	6500	6188	5897	5624	5368	5125	4896	4678	4471	4273	4083	3902	3727	3560	3399
27	6494	6183	5892	5620	5364	5122	4892	4675	4467	4269	4080	3899	3725	3557	3396
28	6489	6178	5888	5615	5359	5118	4889	4671	4464	4266	4077	3896	3722	3555	3393
29	6484	6173	5883	5611	5355	5114	4885	4668	4461	4263	4074	3893	3719	3552	3391
30	6478	6168	5878	5607	5351	5110	4881	4664	4457	4260	4071	3890	3716	3549	3388
31	6473	6163	5874	5602	5347	5106	4877	4660	4454	4256	4068	3887	3713	3546	3386
32	6467	6158	5869	5598	5343	5102	4874	4657	4450	4253	4065	3884	3710	3544	3383
33	6462	6153	5864	5594	5339	5098	4870	4653	4447	4250	4062	3881	3708	3541	3380
34	6457	6148	5860	5589	5335	5094	4866	4650	4444	4247	4059	3878	3705	3538	3378
35	6451	6143	5855	5585	5331	5090	4863	4646	4440	4244	4055	3875	3702	3535	3375
36	6446	6138	5850	5580	5326	5086	4859	4643	4437	4240	4052	3872	3699	3533	3372
37	6441	6133	5846	5576	5322	5082	4855	4639	4434	4237	4049	3869	3696	3530	3370
38	6435	6128	5841	5572	5318	5079	4852	4636	4430	4234	4046	3866	3693	3527	3367
39	6430	6123	5836	5567	5314	5075	4848	4632	4427	4231	4043	3863	3691	3525	3365
40	6425	6118	5832	5563	5310	5071	4844	4629	4424	4228	4040	3860	3688	3522	3362
41	6420	6113	5827	5559	5306	5067	4841	4625	4420	4224	4037	3857	3685	3519	3359
42	6414	6108	5823	5554	5302	5063	4837	4622	4417	4221	4034	3855	3682	3516	3357
43	6409	6103	5818	5550	5298	5059	4833	4618	4414	4218	4031	3852	3679	3514	3354
44	6404	6099	5813	5546	5294	5055	4830	4615	4410	4215	4028	3849	3677	3511	3351
45	6398	6094	5809	5541	5290	5051	4826	4611	4407	4212	4025	3846	3674	3508	3349
46	6393	6089	5804	5537	5285	5048	4822	4608	4404	4209	4022	3843	3671	3506	3346
47	6388	6084	5800	5533	5281	5044	4819	4604	4400	4205	4019	3840	3668	3503	3344
48	6383	6079	5795	5528	5277	5040	4815	4601	4397	4202	4016	3837	3665	3500	3341
49	6377	6074	5790	5524	5273	5036	4811	4597	4394	4199	4013	3834	3663	3497	3338
50	6372	6069	5786	5520	5269	5032	4808	4594	4390	4196	4010	3831	3660	3495	3336
51	6367	6064	5781	5516	5265	5028	4804	4590	4387	4193	4007	3828	3657	3492	3333
52	6362	6059	5777	5511	5261	5025	4800	4587	4384	4189	4004	3825	3654	3489	3331
53	6357	6055	5772	5507	5257	5021	4797	4584	4380	4186	4001	3822	3651	3487	3328
54	6351	6050	5768	5503	5253	5017	4793	4580	4377	4183	3998	3820	3649	3484	3325
55	6346	6045	5763	5498	5249	5013	4789	4577	4374	4180	3995	3817	3646	3481	3323
56	6341	6040	5758	5494	5245	5009	4786	4573	4370	4177	3991	3814	3643	3479	3320
57	6336	6035	5754	5490	5241	5005	4782	4570	4367	4174	3988	3811	3640	3476	3318
58	6331	6030	5749	5486	5237	5002	4778	4566	4364	4171	3985	3808	3637	3473	3315
59	6325	6025	5745	5481	5233	4998	4775	4563	4361	4167	3982	3805	3635	3471	3313
60	6320	6021	5740	5477	5229	4994	4771	4559	4357	4164	3979	3802	3632	3468	3310

′	28	29	30	31	32	33	34	35	36	37	38	39	40	41	42
″	1680	1740	1800	1860	1920	1980	2040	2100	2160	2220	2280	2340	2400	2460	2520
0	3310	3158	3010	2868	2730	2596	2467	2341	2218	2099	1984	1871	1761	1654	1549
1	3307	3155	3008	2866	2728	2594	2465	2339	2216	2098	1982	1869	1759	1652	1547
2	3305	3153	3005	2863	2725	2592	2462	2337	2214	2096	1980	1867	1757	1650	1546
3	3302	3150	3003	2861	2723	2590	2460	2335	2212	2094	1978	1865	1755	1648	1544
4	3300	3148	3001	2859	2721	2588	2458	2333	2210	2092	1976	1863	1754	1647	1542
5	3297	3145	2998	2856	2719	2585	2456	2331	2208	2090	1974	1862	1752	1645	1540
6	3294	3143	2996	2854	2716	2583	2454	2328	2206	2088	1972	1860	1750	1643	1539
7	3292	3140	2993	2852	2714	2581	2452	2326	2204	2086	1970	1858	1748	1641	1537
8	3289	3138	2991	2849	2712	2579	2450	2324	2202	2084	1968	1856	1746	1640	1535
9	3287	3135	2989	2847	2710	2577	2448	2322	2200	2082	1967	1854	1745	1638	1534
10	3284	3133	2986	2845	2707	2574	2445	2320	2198	2080	1965	1852	1743	1636	1532
11	3282	3130	2984	2842	2705	2572	2443	2318	2196	2078	1963	1850	1741	1634	1530
12	3279	3128	2981	2840	2703	2570	2441	2316	2194	2076	1961	1849	1739	1633	1528
13	3276	3125	2979	2838	2701	2568	2439	2314	2192	2074	1959	1847	1737	1631	1527
14	3274	3123	2977	2835	2698	2566	2437	2312	2190	2072	1957	1845	1736	1629	1525
15	3271	3120	2974	2833	2696	2564	2435	2310	2188	2070	1955	1843	1734	1627	1523
16	3269	3118	2972	2831	2694	2561	2433	2308	2186	2068	1953	1841	1732	1626	1522
17	3266	3115	2969	2828	2692	2559	2431	2306	2184	2066	1951	1839	1730	1624	1520
18	3264	3113	2967	2826	2689	2557	2429	2304	2182	2064	1950	1838	1728	1622	1518
19	3261	3110	2965	2824	2687	2555	2426	2302	2180	2062	1948	1836	1727	1620	1516
20	3259	3108	2962	2821	2685	2553	2424	2300	2178	2061	1946	1834	1725	1619	1515
21	3256	3105	2960	2819	2683	2551	2422	2298	2176	2059	1944	1832	1723	1617	1513
22	3253	3103	2958	2817	2681	2548	2420	2296	2174	2057	1942	1830	1721	1615	1511
23	3251	3101	2955	2815	2678	2546	2418	2294	2172	2055	1940	1828	1719	1613	1510
24	3248	3098	2953	2812	2676	2544	2416	2291	2170	2053	1938	1827	1718	1612	1508
25	3246	3096	2950	2810	2674	2542	2414	2289	2169	2051	1936	1825	1716	1610	1506
26	3243	3093	2948	2808	2672	2540	2412	2287	2167	2049	1934	1823	1714	1608	1504
27	3241	3091	2946	2805	2669	2538	2410	2285	2165	2047	1933	1821	1712	1606	1503
28	3238	3088	2943	2803	2667	2535	2408	2283	2163	2045	1931	1819	1711	1605	1501
29	3236	3086	2941	2801	2665	2533	2405	2281	2161	2043	1929	1817	1709	1603	1499
30	3233	3083	2939	2798	2663	2531	2403	2279	2159	2041	1927	1816	1707	1601	1498
31	3231	3081	2936	2796	2660	2529	2401	2277	2157	2039	1925	1814	1705	1599	1496
32	3228	3078	2934	2794	2658	2527	2399	2275	2155	2037	1923	1812	1703	1598	1494
33	3225	3076	2931	2792	2656	2525	2397	2273	2153	2035	1921	1810	1702	1596	1493
34	3223	3073	2929	2789	2654	2522	2395	2271	2151	2033	1919	1808	1700	1594	1491
35	3220	3071	2927	2787	2652	2520	2393	2269	2149	2032	1918	1806	1698	1592	1489
36	3218	3069	2924	2785	2649	2518	2391	2267	2147	2030	1916	1805	1696	1591	1487
37	3215	3066	2922	2782	2647	2516	2389	2265	2145	2028	1914	1803	1694	1589	1486
38	3213	3064	2920	2780	2645	2514	2387	2263	2143	2026	1912	1801	1693	1587	1484
39	3210	3061	2917	2778	2643	2512	2384	2261	2141	2024	1910	1799	1691	1585	1482
40	3208	3059	2915	2775	2640	2510	2382	2259	2139	2022	1908	1797	1689	1584	1481
41	3205	3056	2912	2773	2638	2507	2380	2257	2137	2020	1906	1795	1687	1582	1479
42	3203	3054	2910	2771	2636	2505	2378	2255	2135	2018	1904	1794	1686	1580	1477
43	3200	3052	2908	2769	2634	2503	2376	2253	2133	2016	1903	1792	1684	1578	1476
44	3198	3049	2905	2766	2632	2501	2374	2251	2131	2014	1901	1790	1682	1577	1474
45	3195	3047	2903	2764	2629	2499	2372	2249	2129	2012	1899	1788	1680	1575	1472
46	3193	3044	2901	2762	2627	2497	2370	2247	2127	2010	1897	1786	1678	1573	1470
47	3190	3042	2898	2760	2625	2494	2368	2245	2125	2009	1895	1785	1677	1571	1469
48	3188	3039	2896	2757	2623	2492	2366	2243	2123	2007	1893	1783	1675	1570	1467
49	3185	3037	2894	2755	2621	2490	2364	2241	2121	2005	1891	1781	1673	1568	1465
50	3183	3034	2891	2753	2618	2488	2362	2239	2119	2003	1889	1779	1671	1566	1464
51	3180	3032	2889	2750	2616	2486	2359	2237	2117	2001	1888	1777	1670	1565	1462
52	3178	3030	2887	2748	2614	2484	2357	2235	2115	1999	1886	1775	1668	1563	1460
53	3175	3027	2884	2746	2612	2482	2355	2233	2113	1997	1884	1774	1666	1561	1459
54	3173	3025	2882	2744	2610	2480	2353	2231	2111	1995	1882	1772	1664	1559	1457
55	3170	3022	2880	2741	2607	2477	2351	2229	2109	1993	1880	1770	1663	1558	1455
56	3168	3020	2877	2739	2605	2475	2349	2227	2107	1991	1878	1768	1661	1556	1454
57	3165	3018	2875	2737	2603	2473	2347	2225	2105	1989	1876	1766	1659	1554	1452
58	3163	3015	2873	2735	2601	2471	2345	2223	2103	1987	1875	1765	1657	1552	1450
59	3160	3013	2870	2732	2599	2469	2343	2220	2101	1986	1873	1763	1655	1551	1449
60	3158	3010	2868	2730	2596	2467	2341	2218	2099	1984	1871	1761	1654	1549	1447

Tab. 7. LOGISTIC LOGARITHMS. 233

′	43	44	45	46	47	48	49	50	51	52	53	54	55	56	57
″	2580	2640	2700	2760	2820	2880	2940	3000	3060	3120	3180	3240	3300	3360	3420
0	1447	1347	1249	1154	1061	0969	0880	0792	0706	0621	0539	0458	0378	0300	0223
1	1445	1345	1248	1152	1059	0968	0878	0790	0704	0620	0537	0456	0377	0298	0221
2	1443	1344	1246	1151	1057	0966	0877	0789	0703	0619	0536	0455	0375	0297	0220
3	1442	1342	1245	1149	1056	0965	0875	0787	0702	0617	0535	0454	0374	0296	0219
4	1440	1340	1243	1148	1054	0963	0874	0786	0700	0616	0533	0452	0373	0294	0218
5	1438	1339	1241	1146	1053	0962	0872	0785	0699	0615	0532	0451	0371	0293	0216
6	1437	1337	1240	1145	1051	0960	0871	0783	0697	0613	0531	0450	0370	0292	0215
7	1435	1335	1238	1143	1050	0959	0869	0782	0696	0612	0529	0448	0369	0291	0214
8	1433	1334	1237	1141	1048	0957	0868	0780	0694	0610	0528	0447	0367	0289	0213
9	1432	1332	1235	1140	1047	0956	0866	0779	0693	0609	0526	0446	0366	0288	0211
10	1430	1331	1233	1138	1045	0954	0865	0777	0692	0608	0525	0444	0365	0287	0210
11	1428	1329	1232	1137	1044	0953	0863	0776	0690	0606	0524	0443	0363	0285	0209
12	1427	1327	1230	1135	1042	0951	0862	0774	0689	0605	0522	0442	0362	0284	0208
13	1425	1326	1229	1134	1041	0950	0860	0773	0687	0603	0521	0440	0361	0283	0206
14	1423	1324	1227	1132	1039	0948	0859	0772	0686	0002	0520	0439	0359	0282	0205
15	1422	1322	1225	1130	1037	0947	0857	0770	0685	0601	0518	0438	0358	0280	0204
16	1420	1321	1224	1129	1036	0945	0856	0769	0683	0599	0517	0436	0357	0279	0202
17	1418	1319	1222	1127	1034	0944	0855	0767	0682	0598	0516	0435	0356	0278	0201
18	1417	1317	1221	1126	1033	0942	0853	0766	0680	0596	0514	0434	0354	0276	0200
19	1415	1316	1219	1124	1031	0941	0852	0764	0679	0595	0513	0432	0353	0275	0199
20	1413	1314	1217	1123	1030	0939	0850	0763	0678	0594	0512	0431	0352	0274	0197
21	1412	1313	1216	1121	1028	0938	0849	0762	0676	0592	0510	0430	0350	0273	0196
22	1410	1311	1214	1119	1027	0936	0847	0760	0675	0591	0509	0428	0349	0271	0195
23	1408	1309	1213	1118	1025	0935	0846	0759	0673	0590	0507	0427	0348	0270	0194
24	1407	1308	1211	1116	1024	0933	0844	0757	0672	0588	0506	0426	0346	0269	0192
25	1405	1306	1209	1115	1022	0932	0843	0756	0670	0587	0505	0424	0345	0267	0191
26	1403	1304	1208	1113	1021	0930	0841	0754	0669	0585	0503	0423	0344	0266	0190
27	1402	1303	1206	1112	1019	0929	0840	0753	0668	0584	0502	0422	0342	0265	0189
28	1400	1301	1205	1110	1018	0927	0838	0751	0666	0583	0501	0420	0341	0264	0187
29	1398	1300	1203	1109	1016	0926	0837	0750	0665	0581	0499	0419	0340	0262	0186
30	1397	1298	1201	1107	1015	0924	0835	0749	0663	0580	0498	0418	0339	0261	0185
31	1395	1296	1200	1105	1013	0923	0834	0747	0662	0579	0497	0416	0337	0260	0184
32	1393	1295	1198	1104	1012	0921	0833	0746	0661	0577	0495	0415	0336	0258	0182
33	1392	1293	1197	1102	1010	0920	0831	0744	0659	0576	0494	0414	0335	0257	0181
34	1390	1291	1195	1101	1008	0918	0830	0743	0658	0574	0493	0412	0333	0256	0180
35	1388	1290	1193	1099	1007	0917	0828	0741	0656	0573	0491	0411	0332	0255	0179
36	1387	1288	1192	1098	1005	0915	0827	0740	0655	0572	0490	0410	0331	0253	0177
37	1385	1287	1190	1096	1004	0914	0825	0739	0654	0570	0489	0408	0329	0252	0176
38	1383	1285	1189	1095	1002	0912	0824	0737	0652	0569	0487	0407	0328	0251	0175
39	1382	1283	1187	1093	1001	0911	0822	0736	0651	0568	0486	0406	0327	0250	0174
40	1380	1282	1186	1091	0999	0909	0821	0734	0649	0566	0484	0404	0326	0248	0172
41	1378	1280	1184	1090	0998	0908	0819	0733	0648	0565	0483	0403	0324	0247	0171
42	1377	1278	1182	1088	0996	0906	0818	0731	0647	0563	0482	0402	0323	0246	0170
43	1375	1277	1181	1087	0995	0905	0816	0730	0645	0562	0480	0400	0322	0244	0169
44	1373	1275	1179	1085	0993	0903	0815	0729	0644	0561	0479	0399	0320	0243	0167
45	1372	1274	1178	1084	0992	0902	0814	0727	0642	0559	0478	0398	0319	0242	0166
46	1370	1272	1176	1082	0990	0900	0812	0726	0641	0558	0476	0396	0318	0241	0165
47	1368	1270	1174	1081	0989	0899	0811	0724	0640	0557	0475	0395	0316	0239	0163
48	1367	1269	1173	1079	0987	0897	0809	0723	0638	0555	0474	0394	0315	0238	0162
49	1365	1267	1171	1078	0986	0896	0808	0721	0637	0554	0472	0392	0314	0237	0161
50	1363	1266	1170	1076	0984	0894	0806	0720	0635	0552	0471	0391	0313	0235	0160
51	1362	1264	1168	1074	0983	0893	0805	0719	0634	0551	0470	0390	0311	0234	0158
52	1360	1262	1167	1073	0981	0891	0803	0717	0633	0550	0468	0388	0310	0233	0157
53	1359	1261	1165	1071	0980	0890	0802	0716	0631	0548	0467	0387	0309	0232	0156
54	1357	1259	1163	1070	0978	0888	0801	0714	0630	0547	0466	0386	0307	0230	0155
55	1355	1257	1162	1068	0977	0887	0799	0713	0628	0545	0464	0384	0306	0229	0153
56	1354	1256	1160	1067	0975	0885	0798	0711	0627	0544	0463	0383	0305	0228	0152
57	1352	1254	1159	1065	0974	0884	0796	0710	0626	0543	0462	0382	0304	0227	0151
58	1350	1253	1157	1064	0972	0883	0795	0709	0624	0541	0460	0381	0302	0225	0150
59	1349	1251	1156	1062	0971	0881	0793	0707	0623	0540	0459	0379	0301	0224	0148
60	1347	1249	1154	1061	0969	0880	0792	0706	0621	0539	0458	0378	0300	0223	0147

′	58	59	60	61	62	63	64	65	66	67	68	69	70	71	72
″	3480	3540	3600	3660	3720	3780	3840	3900	3960	4020	4080	4140	4200	4260	4320
0	0147	0073	—	9928	9858	9788	9720	9652	9586	9521	9456	9393	9331	9269	9208
1	0146	0072	9999	9027	9856	9787	9719	9651	9585	9520	9455	9392	9329	9268	9207
2	0145	0071	9998	9926	9855	9786	9717	9650	9584	9519	9454	9391	9328	9267	9206
3	0143	0069	9996	9925	9854	9785	9716	9649	9583	9518	9453	9390	9327	9266	9205
4	0142	0068	9995	9923	9853	9784	9715	9648	9582	9516	9452	9389	9326	9265	9204
5	0141	0067	9994	9922	9852	9782	9714	9647	9581	9515	9451	9388	9325	9264	9203
6	0140	0066	9993	9921	9851	9781	9713	9646	9579	9514	9450	9387	9324	9263	9202
7	0139	0064	9992	9920	9849	9780	9712	9645	9578	9513	9449	9386	9323	9262	9201
8	0137	0063	9990	9919	9848	9779	9711	9643	9577	9512	9448	9385	9322	9261	9200
9	C136	0062	9989	9918	9847	9778	9710	9642	9576	9511	9447	9384	9321	9260	9199
10	0135	0061	9988	9916	9846	9777	9708	9641	9575	9510	9446	9383	9320	9259	9198
11	0134	0060	9987	9915	9844	9775	9707	9640	9574	9509	9445	9381	9319	9258	9197
12	0132	0058	9986	9914	9844	9774	9706	9639	9573	9508	9444	9380	9318	9257	9196
13	0131	0057	9984	9913	9842	9773	9705	9638	9572	9507	9443	9379	9317	9256	9195
14	0130	0056	9983	9912	9841	9772	9704	9637	9571	9506	9442	9378	9316	9255	9194
15	0129	0055	9982	9910	9840	9771	9703	9636	9570	9505	9440	9377	9315	9254	9193
16	0127	0053	9981	9909	9839	9770	9702	9635	9569	9504	9439	9376	9314	9253	9192
17	0126	0052	9980	9908	9838	9769	9701	9633	9567	9502	9438	9375	9313	9252	9191
18	0125	0051	9978	9907	9837	9767	9699	9632	9566	9501	9437	9374	9312	9251	9190
19	0124	0050	9977	9906	9835	9766	9698	9631	9565	9500	9436	9373	9311	9250	9189
20	0122	0049	9976	9905	9834	9765	9697	9630	9564	9499	9435	9372	9310	9249	9188
21	0121	0047	9975	9903	9833	9764	9696	9629	9563	9498	9434	9371	9309	9248	9187
22	0120	0046	9974	9902	9832	9763	9695	9628	9562	9497	9433	9370	9308	9247	9186
23	0119	0045	9972	9901	9831	9762	9694	9627	9561	9496	9432	9369	9307	9246	9185
24	0117	0044	9971	9900	9830	9761	9693	9626	9560	9495	9431	9368	9306	9245	9184
25	0116	0042	9970	9899	9829	9759	9692	9625	9559	9494	9430	9367	9305	9244	9183
26	0115	0041	9969	9897	9827	9758	9690	9624	9558	9493	9429	9366	9304	9243	9182
27	0114	0040	9968	9896	9826	9757	9689	9622	9557	9492	9428	9365	9303	9241	9181
28	0112	0039	9966	9895	9825	9756	9688	9621	9555	9491	9427	9364	9302	9240	9180
29	0111	0038	9965	9894	9824	9755	9687	9620	9554	9490	9426	9363	9301	9239	9179
30	0110	0036	9964	9893	9823	9754	9686	9619	9553	9488	9425	9362	9300	9238	9178
31	0109	0035	9963	9892	9822	9753	9685	9618	9552	9487	9424	9361	9299	9237	9177
32	0107	0034	9962	9890	9820	9751	9684	9617	9551	9486	9422	9360	9298	9236	9176
33	0106	0033	9960	9889	9819	9750	9683	9616	9550	9485	9421	9359	9297	9235	9175
34	0105	0031	9959	9888	9818	9749	9681	9615	9549	9484	9420	9358	9296	9234	9174
35	0104	0030	9958	9887	9817	9748	9680	9614	9548	9483	9419	9356	9294	9233	9173
36	0103	0029	9957	9836	9816	9747	9679	9612	9547	9482	9418	9355	9293	9232	9172
37	0101	0028	9956	9885	9815	9746	9678	9611	9546	9481	9417	9354	9292	9231	9171
38	0100	0027	9954	9883	9813	9745	9677	9610	9545	9480	9416	9353	9291	9230	9170
39	0099	0025	9953	9882	9812	9744	9676	9609	9544	9479	9415	9352	9290	9229	9169
40	0098	0024	9952	9881	9811	9742	9675	9608	9542	9478	9414	9351	9289	9228	9168
41	0096	0023	9951	9880	9810	9741	9674	9607	9541	9477	9413	9350	9288	9227	9167
42	0095	0022	9950	9879	9809	9740	9672	9606	9540	9476	9412	9349	9287	9226	9166
43	0094	0021	9948	9877	9808	9739	9671	9605	9539	9475	9411	9348	9286	9225	9165
44	0093	0019	9947	9876	9807	9738	9670	9604	9538	9473	9410	9347	9285	9224	9164
45	0091	0018	9946	9875	9805	9737	9669	9603	9537	9472	9409	9346	9284	9223	9163
46	0090	0017	9945	9874	9804	9736	9668	9601	9536	9471	9408	9345	9283	9222	9162
47	0089	0016	9944	9873	9803	9734	9667	9600	9535	9470	9407	9344	9282	9221	9161
48	0088	0015	9942	9872	9802	9733	9666	9599	9534	9469	9406	9343	9281	9220	9160
49	0087	0013	9941	9870	9801	9732	9665	9598	9533	9468	9405	9342	9280	9219	9159
50	0085	0012	9940	9869	9800	9731	9664	9597	9532	9467	9404	9341	9279	9218	9158
51	0084	0011	9939	9868	9798	9730	9662	9596	9530	9466	9402	9340	9278	9217	9157
52	0083	0010	9938	9867	9797	9729	9661	9595	9529	9465	9401	9339	9277	9216	9156
53	0082	0008	9937	9866	9796	9728	9660	9594	9528	9464	9400	9338	9276	9215	9155
54	0080	0007	9935	9865	9795	9727	9659	9593	9527	9463	9399	9337	9275	9214	9154
55	0079	0006	9934	9863	9794	9725	9658	9592	9526	9462	9398	9336	9274	9213	9153
56	0078	0005	9933	9862	9793	9724	9657	9590	9525	9461	9397	9335	9273	9212	9152
57	0077	0004	9932	9861	9792	9723	9656	9589	9524	9460	9396	9334	9272	9211	9151
58	0075	0002	9931	9860	9790	9722	9655	9588	9523	9459	9395	9333	9271	9210	9150
59	0074	0001	9929	9859	9789	9721	9653	9587	9522	9457	9394	9332	9270	9209	9149
60	0073	0000	9928	9858	9788	9720	9652	9586	9521	9456	9393	9331	9269	9208	9148

Tab. 7.　　　LOGISTIC LOGARITHMS.　　　235

′	73	74	75	76	77	78	79	80	81	82	83	84	85	86	87
″	4380	4440	4500	4560	4620	4680	4740	4800	4860	4920	4980	5040	5100	5160	5220
0	9148	9089	9031	8973	8917	8861	8805	8751	8697	8643	8591	8539	8487	8437	8386
1	9147	9088	9030	8972	8916	8860	8804	8750	8696	8642	8590	8538	8486	8436	8385
2	9146	9087	9029	8971	8915	8859	8803	8749	8695	8642	8589	8537	8486	8435	8385
3	9145	9086	9028	8971	8914	8858	8802	8748	8694	8641	8588	8536	8485	8434	8384
4	9144	9085	9027	8970	8913	8857	8802	8747	8693	8640	8587	8535	8484	8433	8383
5	9143	9084	9026	8969	8912	8856	8801	8746	8692	8639	8586	8534	8483	8432	8382
6	9142	9083	9025	8968	8911	8855	8800	8745	8691	8638	8585	8534	8482	8431	8381
7	9141	9082	9024	8967	8910	8854	8799	8744	8690	8637	8585	8533	8481	8431	8380
8	9140	9081	9023	8966	8909	8853	8798	8743	8689	8636	8584	8532	8481	8430	8380
9	9139	9080	9022	8965	8908	8852	8797	8743	8689	8635	8583	8531	8480	8429	8379
10	9138	9079	9021	8964	8907	8851	8796	8742	8688	8635	8582	8530	8479	8428	8378
11	9137	9078	9020	8963	8906	8850	8795	8741	8687	8634	8581	8529	8478	8427	8377
12	9136	9077	9019	8962	8905	8849	8794	8740	8686	8633	8580	8528	8477	8426	8376
13	9135	9076	9018	8961	8904	8849	8793	8739	8685	8632	8579	8528	8476	8426	8376
14	9134	9076	9017	8960	8903	8848	8792	8738	8684	8631	8579	8527	8475	8425	8375
15	9133	9075	9016	8959	8903	8847	8792	8737	8683	8630	8578	8526	8475	8424	8374
16	9132	9074	9015	8958	8902	8846	8791	8736	8682	8629	8577	8525	8474	8423	8373
17	9131	9073	9015	8957	8901	8845	8790	8735	8681	8628	8576	8524	8473	8422	8372
18	9130	9072	9014	8956	8900	8844	8789	8734	8681	8627	8575	8523	8472	8421	8371
19	9129	9071	9013	8955	8899	8843	8788	8733	8680	8627	8574	8522	8471	8421	8371
20	9128	9070	9012	8954	8898	8842	8787	8733	8679	8626	8573	8522	8470	8420	8370
21	9128	9069	9011	8953	8897	8841	8786	8732	8678	8625	8572	8521	8469	8419	8369
22	9127	9068	9010	8952	8896	8840	8785	8731	8677	8624	8572	8520	8469	8418	8368
23	9126	9067	9009	8952	8895	8839	8784	8729	8676	8623	8571	8519	8468	8417	8367
24	9125	9066	9008	8951	8894	8838	8783	8729	8675	8622	8570	8518	8467	8416	8366
25	9124	9065	9007	8950	8893	8837	8782	8728	8674	8621	8569	8517	8466	8416	8366
26	9123	9064	9006	8949	8892	8837	8781	8727	8673	8620	8568	8516	8465	8415	8365
27	9122	9063	9005	8948	8891	8836	8781	8726	8673	8620	8567	8515	8464	8414	8364
28	9121	9062	9004	8947	8890	8835	8780	8725	8672	8619	8566	8515	8464	8413	8363
29	9120	9061	9003	8946	8889	8834	8779	8724	8671	8618	8565	8514	8463	8412	8362
30	9119	9060	9002	8945	8888	8833	8778	8724	8670	8617	8565	8513	8462	8411	8361
31	9118	9059	9001	8944	8888	8832	8777	8723	8669	8616	8564	8512	8461	8410	8361
32	9117	9058	9000	8943	8887	8831	8776	8722	8668	8615	8563	8511	8460	8410	8360
33	9116	9057	8999	8942	8886	8830	8775	8721	8667	8614	8562	8510	8459	8409	8359
34	9115	9056	8998	8941	8885	8829	8774	8720	8666	8613	8561	8510	8458	8408	8358
35	9114	9055	8997	8940	8884	8828	8773	8719	8665	8613	8560	8509	8458	8407	8357
36	9113	9054	8996	8939	8883	8827	8772	8718	8665	8612	8559	8508	8457	8406	8356
37	9112	9053	8995	8938	8882	8826	8771	8717	8664	8611	8559	8507	8456	8405	8356
38	9111	9052	8994	8937	8881	8825	8771	8716	8663	8610	8558	8506	8455	8405	8355
39	9110	9051	8993	8936	8880	8825	8770	8715	8662	8609	8557	8505	8454	8404	8354
40	9109	9050	8992	8935	8879	8824	8769	8715	8661	8608	8556	8504	8453	8403	8353
41	9108	9049	8992	8935	8878	8823	8768	8714	8660	8607	8555	8504	8453	8402	8352
42	9107	9048	8991	8934	8877	8822	8767	8713	8659	8606	8554	8503	8452	8401	8351
43	9106	9047	8990	8933	8876	8821	8766	8712	8658	8606	8553	8501	8450	8400	8350
44	9105	9046	8989	8932	8875	8820	8765	8710	8657	8605	8553	8501	8450	8400	8350
45	9104	9045	8988	8931	8875	8819	8764	8710	8657	8604	8552	8500	8449	8399	8349
46	9103	9044	8987	8930	8874	8818	8763	8709	8656	8603	8551	8499	8448	8398	8348
47	9102	9043	8986	8929	8873	8817	8762	8708	8655	8602	8550	8498	8447	8397	8347
48	9101	9042	8985	8928	8872	8816	8761	8707	8654	8601	8549	8498	8447	8396	8347
49	9100	9042	8984	8927	8871	8815	8761	8706	8653	8600	8548	8497	8446	8395	8346
50	9099	9041	8983	8926	8870	8814	8760	8705	8652	8599	8547	8496	8445	8395	8345
51	9098	9040	8982	8925	8869	8813	8759	8704	8651	8599	8547	8495	8444	8394	8344
52	9097	9039	8981	8924	8868	8813	8758	8703	8650	8598	8546	8494	8443	8393	8343
53	9096	9038	8980	8923	8867	8812	8757	8703	8650	8597	8545	8493	8442	8392	8342
54	9095	9037	8979	8922	8866	8811	8756	8702	8649	8596	8544	8492	8441	8391	8341
55	9094	9036	8978	8921	8865	8810	8755	8701	8648	8595	8543	8492	8441	8390	8341
56	9093	9035	8977	8920	8864	8809	8754	8700	8647	8594	8542	8491	8440	8390	8340
57	9092	9034	8976	8919	8863	8808	8753	8699	8646	8593	8541	8490	8439	8389	8339
58	9091	9033	8975	8918	8862	8807	8752	8698	8645	8592	8540	8489	8438	8388	8338
59	9090	9032	8974	8918	8861	8806	8752	8697	8644	8592	8540	8488	8437	8387	8338
60	9089	9031	8973	8917	8861	8805	8751	8697	8643	8591	8539	8487	8437	8387	8237

"	0'	1'	2'	3'	4'	5'	6'	7'	"
0		6·4637261	6·7647561	6·9408473	7·0657860	7·1626960	7·2418771	7·3088239	60
1	4·6855749	6·4709047	6·7683602	6·9432534	7·0675918	7·1641412	7·2430818	7·3098567	59
2	4·9866049	6·4779665	6·7719347	6·9456462	7·0693001	7·1655817	7·2442832	7·3108870	58
3	5·1626961	6·4849154	6·7754300	6·9480259	7·0711810	7·1670173	7·2454813	7·3119149	57
4	5·2876349	6·4917548	6·7789965	6·9503926	7·0729646	7·1684483	7·2466760	7·3129404	56
5	5·3845449	6·4984882	6·7824849	6·9527465	7·0747408	7·1698745	7·2478675	7·3139635	55
6	5·4637261	6·5051188	6·7859454	6·9550878	7·0765099	7·1712961	7·2490557	7·3149842	54
7	5·5306729	6·5116497	6·7893786	6·9574164	7·0782717	7·1727131	7·2502407	7·3160024	53
8	5·5886649	6·5180838	6·7927848	6·9597327	7·0800264	7·1741254	7·2514225	7·3170183	52
9	5·6398174	6·5244239	6·7961645	6·9620366	7·0817741	7·1755332	7·2526010	7·3180318	51
10	5·6855749	6·5306729	6·7995182	6·9643284	7·0835148	7·1769364	7·2537764	7·3190430	50
11	5·7269676	6·5368332	6·8028461	6·9666082	7·0852485	7·1783351	7·2549485	7·3200518	49
12	5·7647561	6·5429074	6·8061488	6·9688760	7·0869753	7·1797293	7·2561176	7·3210583	48
13	5·7995182	6·5488977	6·8094265	6·9711321	7·0886953	7·1811190	7·2572835	7·3220624	47
14	5·8317029	6·5548066	6·8126796	6·9733765	7·0904085	7·1825043	7·2584462	7·3230643	46
15	5·8616661	6·5606361	6·8159086	6·9756094	7·0921149	7·1838853	7·2596059	7·3240638	45
16	5·8896948	6·5663884	6·8191137	6·9778309	7·0938147	7·1852618	7·2607625	7·3250610	44
17	5·9160238	6·5720656	6·8222954	6·9800410	7·0955079	7·1866340	7·2619160	7·3260560	43
18	5·9408474	6·5776695	6·8254539	6·9822400	7·0971945	7·1880018	7·2630664	7·3270487	42
19	5·9643285	6·5832019	6·8285896	6·9844279	7·0988745	7·1893654	7·2642138	7·3280391	41
20	5·9866049	6·5886648	6·8317029	6·9866048	7·1005481	7·1907247	7·2653582	7·3290272	40
21	6·0077942	6·5940599	6·8347939	6·9887709	7·1022153	7·1920797	7·2664996	7·3300131	39
22	6·0279975	6·5993887	6·8378632	6·9909262	7·1038760	7·1934306	7·2676380	7·3309968	38
23	6·0473027	6·6046529	6·8409109	6·9930708	7·1055305	7·1947772	7·2687734	7·3319783	37
24	6·0657861	6·6098541	6·8439373	6·9952050	7·1071787	7·1961197	7·2699058	7·3329575	36
25	6·0835149	6·6149938	6·8469428	6·9973287	7·1088206	7·1974580	7·2710353	7·3339345	35
26	6·1005482	6·6200733	6·8499277	6·9994420	7·1104564	7·1987923	7·2721619	7·3349094	34
27	6·1169386	6·6250941	6·8528922	7·0015451	7·1120860	7·2001224	7·2732856	7·3358821	33
28	6·1327329	6·6300575	6·8558365	7·0036381	7·1137095	7·2014485	7·2744063	7·3368525	32
29	6·1479729	6·6349649	6·8587611	7·0057211	7·1153270	7·2027706	7·2755242	7·3378209	31
30	6·1626961	6·6398174	6·8616661	7·0077841	7·1169385	7·2040886	7·2766392	7·3387870	30
31	6·1769366	6·6446162	6·8645518	7·0098572	7·1185440	7·2054027	7·2777514	7·3397511	29
32	6·1907248	6·6493627	6·8674184	7·0119107	7·1201436	7·2067128	7·2788607	7·3407130	28
33	6·2040888	6·6540578	6·8702663	7·0139544	7·1217374	7·2080189	7·2799672	7·3416727	27
34	6·2170538	6·6587027	6·8730955	7·0159886	7·1233253	7·2093211	7·2810708	7·3426304	26
35	6·2296429	6·6632985	6·8759065	7·0180132	7·1249074	7·2106195	7·2821717	7·3435859	25
36	6·2418774	6·6678461	6·8786994	7·0200285	7·1264838	7·2119140	7·2832698	7·3445394	24
37	6·2537766	6·6723466	6·8814745	7·0220345	7·1280545	7·2132046	7·2843651	7·3454907	23
38	6·2653585	6·6768009	6·8842319	7·0240313	7·1296195	7·2144914	7·2854577	7·3464400	22
39	6·2766395	6·6812100	6·8869719	7·0260189	7·1311789	7·2157744	7·2865475	7·3473872	21
40	6·2876349	6·6855748	6·8896948	7·0279975	7·1327328	7·2170536	7·2876346	7·3483323	20
41	6·2983587	6·6898962	6·8924007	7·0299671	7·1342811	7·2183290	7·2887190	7·3492754	19
42	6·3088242	6·6941750	6·8950898	7·0319278	7·1358238	7·2196008	7·2898006	7·3502165	18
43	6·3190433	6·6984121	6·8977624	7·0338796	7·1373612	7·2208688	7·2908796	7·3511555	17
44	6·3290275	6·7026082	6·9004187	7·0358228	7·1388931	7·2221331	7·2919560	7·3520925	16
45	6·3387874	6·7067641	6·9030588	7·0377573	7·1404196	7·2233938	7·2930296	7·3530275	15
46	6·3483327	6·7108807	6·9056829	7·0396832	7·1419408	7·2246508	7·2941006	7·3539604	14
47	6·3576727	6·7149586	6·9082913	7·0416006	7·1434566	7·2259041	7·2951690	7·3548914	13
48	6·3668161	6·7189986	6·9108841	7·0435096	7·1449672	7·2271539	7·2962347	7·3558203	12
49	6·3757709	6·7230013	6·9134615	7·0454103	7·1464726	7·2284001	7·2972979	7·3567473	11
50	6·3845449	6·7269675	6·9160237	7·0473026	7·1479727	7·2296427	7·2983584	7·3576723	10
51	6·3931450	6·7308978	6·9185709	7·0491868	7·1494677	7·2308818	7·2994164	7·3585954	9
52	6·4015782	6·7347929	6·9211033	7·0510628	7·1509576	7·2321173	7·3004718	7·3595165	8
53	6·4098507	6·7386533	6·9236209	7·0529307	7·1524423	7·2333494	7·3015246	7·3604356	7
54	6·4179686	6·7424797	6·9261241	7·0547906	7·1539221	7·2345779	7·3025749	7·3613528	6
55	6·4259376	6·7462727	6·9286129	7·0566426	7·1553967	7·2358030	7·3036227	7·3622681	5
56	6·4337629	6·7500328	6·9310875	7·0584868	7·1568664	7·2370246	7·3046679	7·3631814	4
57	6·4414497	6·7537607	6·9335481	7·0603231	7·1583312	7·2382429	7·3057106	7·3640929	3
58	6·4490029	6·7574569	6·9359948	7·0621517	7·1597910	7·2394577	7·3067509	7·3650024	2
59	6·4564269	6·7611218	6·9384278	7·0639727	7·1612459	7·2406691	7·3077886	7·3659100	1
60	6·4637261	6·7647561	6·9408473	7·0657860	7·1626960	7·2418771	7·3088239	7·3668157	0
"	59'	58'	57'	56'	55'	54'	53'	52'	"

"	0′	1′	2′	3′	4′	5′	6′	7′	"
0		6·4637261	6·7647562	6·9408475	7·0657863	7·1626964	7·2418778	7·3088248	60
1	4·6855749	6·4709047	6·7683603	6·9432536	7·0675921	7·1641417	7·2430825	7·3098576	59
2	4·9860049	6·4779066	6·7719347	6·9456464	7·0693904	7·1655821	7·2442839	7·3108879	58
3	5·1626961	6·4849154	6·7754800	6·9480261	7·0711813	7·1670178	7·2454819	7·3119158	57
4	5·2876349	6·4917549	6·7789966	6·9503928	7·0729649	7·1684488	7·2466767	7·3129413	56
5	5·3845449	6·4984882	6·7824849	6·9527467	7·0747412	7·1698750	7·2478682	7·3139644	55
6	5·4637261	6·5051188	6·7859455	6·9550879	7·0765102	7·1712966	7·2490564	7·3149851	54
7	5·5306729	6·5116497	6·7893786	6·9574166	7·0782720	7·1727136	7·2502414	7·3160034	53
8	5·5886049	6·5180838	6·7927849	6·9597328	7·0800268	7·1741259	7·2514231	7·3170193	52
9	5·6398174	6·5244240	6·7961646	6·9620368	7·0817744	7·1755337	7·2526017	7·3180328	51
10	5·6855749	6·5306729	6·7995183	6·9643286	7·0835151	7·1769369	7·2537771	7·3190440	50
11	5·7269676	6·5368332	6·8028462	6·9666084	7·0852488	7·1783356	7·2549492	7·3200528	49
12	5·7647561	6·5429074	6·8061489	6·9688762	7·0869756	7·1797298	7·2561183	7·3210592	48
13	5·7995182	6·5488977	6·8094266	6·9711323	7·0886956	7·1811195	7·2572842	7·3220634	47
14	5·8317029	6·5548066	6·8126797	6·9733767	7·0904088	7·1825049	7·2584469	7·3230652	46
15	5·8616661	6·5606361	6·8159087	6·9756096	7·0921153	7·1838858	7·2596066	7·3240648	45
16	5·8896948	6·5663885	6·8191138	6·9778311	7·0938151	7·1852623	7·2607632	7·3250620	44
17	5·9160238	6·5720056	6·8222955	6·9800412	7·0955082	7·1866345	7·2619167	7·3260570	43
18	5·9408474	6·5776695	6·8254540	6·9822402	7·0971948	7·1880023	7·2630672	7·3270496	42
19	5·9643285	6·5832020	6·8285897	6·9844281	7·0988749	7·1893659	7·2642146	7·3280400	41
20	5·9866049	6·5886649	6·8317030	6·9866050	7·1005484	7·1907252	7·2653590	7·3290282	40
21	6·0077942	6·5940599	6·8347940	6·9887711	7·1022156	7·1920802	7·2665003	7·3300141	39
22	6·0279975	6·5993887	6·8378633	6·9909264	7·1038764	7·1934311	7·2676387	7·3309978	38
23	6·0473027	6·6046530	6·8409110	6·9930710	7·1055309	7·1947777	7·2687741	7·3319793	37
24	6·0657861	6·6098542	6·8439374	6·9952052	7·1071790	7·1961202	7·2699066	7·3329585	36
25	6·0835149	6·6149938	6·8469429	6·9973289	7·1088210	7·1974586	7·2710361	7·3339356	35
26	6·1005482	6·6200733	6·8499278	6·9994422	7·1104567	7·1987928	7·2721627	7·3349104	34
27	6·1169386	6·6250941	6·8528923	7·0015454	7·1120864	7·2001230	7·2732863	7·3358831	33
28	6·1327329	6·6300576	6·8558367	7·0036383	7·1137099	7·2014491	7·2744071	7·3368536	32
29	6·1479729	6·6349649	6·8587612	7·0057213	7·1153274	7·2027711	7·2755250	7·3378219	31
30	6·1626961	6·6398174	6·8616662	7·0077943	7·1169389	7·2040892	7·2766400	7·3387881	30
31	6·1769366	6·6446163	6·8645519	7·0098575	7·1185444	7·2054032	7·2777521	7·3397521	29
32	6·1907248	6·6493627	6·8674185	7·0119109	7·1201440	7·2067133	7·2788615	7·3407140	28
33	6·2040888	6·6540578	6·8702664	7·0139546	7·1217378	7·2080195	7·2799679	7·3416738	27
34	6·2170538	6·6587027	6·8730957	7·0159888	7·1233257	7·2093217	7·2810716	7·3426314	26
35	6·2296429	6·6632985	6·8759066	7·0180135	7·1249078	7·2106201	7·2821725	7·3435870	25
36	6·2418774	6·6678461	6·8786995	7·0200288	7·1264842	7·2119145	7·2832706	7·3445404	24
37	6·2537766	6·6723466	6·8814746	7·0220348	7·1280549	7·2132052	7·2843659	7·3454918	23
38	6·2653585	6·6768010	6·8842320	7·0240315	7·1296199	7·2144920	7·2854585	7·3464411	22
39	6·2766395	6·6812101	6·8869721	7·0260191	7·1311793	7·2157750	7·2865483	7·3473883	21
40	6·2876349	6·6855749	6·8896949	7·0279977	7·1327332	7·2170542	7·2876354	7·3483334	20
41	6·2983587	6·6898963	6·8924008	7·0299673	7·1342815	7·2183296	7·2887198	7·3492765	19
42	6·3088242	6·6941751	6·8950900	7·0319280	7·1358242	7·2196014	7·2898015	7·3502176	18
43	6·3190433	6·6984121	6·8977626	7·0338799	7·1373616	7·2208694	7·2908805	7·3511566	17
44	6·3290275	6·7026082	6·9004188	7·0358231	7·1388935	7·2221337	7·2919568	7·3520936	16
45	6·3387874	6·7067642	6·9030589	7·0377576	7·1404200	7·2233944	7·2930304	7·3530286	15
46	6·3483327	6·7108808	6·9056830	7·0396835	7·1419412	7·2246514	7·2941015	7·3539615	14
47	6·3576727	6·7149587	6·9082914	7·0416009	7·1434570	7·2259048	7·2951698	7·3548925	13
48	6·3668161	6·7189987	6·9108842	7·0435099	7·1449676	7·2271545	7·2962356	7·3558215	12
49	6·3757709	6·7230014	6·9134617	7·0454105	7·1464730	7·2284007	7·2972987	7·3567485	11
50	6·3845449	6·7269676	6·9160239	7·0473029	7·1479732	7·2296433	7·2983593	7·3576735	10
51	6·3931450	6·7308979	6·9185711	7·0491870	7·1494681	7·2308824	7·2994173	7·3585965	9
52	6·4015782	6·7347929	6·9211034	7·0510630	7·1509580	7·2321180	7·3004727	7·3595176	8
53	6·4098507	6·7386534	6·9236211	7·0529310	7·1524428	7·2333500	7·3015255	7·3604368	7
54	6·4179686	6·7424798	6·9261242	7·0547909	7·1539225	7·2345786	7·3025758	7·3613540	6
55	6·4259376	6·7462728	6·9286130	7·0566429	7·1553972	7·2358936	7·3036235	7·3622692	5
56	6·4337629	6·7500329	6·9310876	7·0584871	7·1568669	7·2370253	7·3046688	7·3631826	4
57	6·4414497	6·7537608	6·9335482	7·0603234	7·1583316	7·2382435	7·3057115	7·3640940	3
58	6·4490029	6·7574570	6·9359950	7·0621520	7·1597914	7·2394583	7·3067517	7·3650035	2
59	6·4564260	6·7611219	6·9384280	7·0639730	7·1612464	7·2406698	7·3077895	7·3659112	1
60	6·4637261	6·7647562	6·9408475	7·0657863	7·1626964	7·2418778	7·3088248	7·3668169	0
"	59′	58′	57′	56′	55′	54′	53′	52′	"

"	8'	9'	10'	11'	12'	13'	14'	15'	"
0	7·3668157	7·4179681	7·4637255	7·5051181	7·5429065	7·5776684	7·6098530	7·6398160	60
1	7·3677195	7·4187716	7·4644487	7·5057756	7·5435092	7·5782249	7 6103697	7·6402983	59
2	7·3686216	7·4195737	7·4651707	7·5064321	7·5441112	7·5787806	7·6108858	7·6407800	58
3	7·3695216	7·4203742	7·4658916	7·5070876	7·5447123	7·5793356	7·6114012	7·6412612	57
4	7·3704198	7·4211733	7·4666112	7·5077422	7·5453125	7·5798899	7·6119161	7·6417419	56
5	7·3713162	7·4219709	7·4673296	7·5083958	7·5459120	7 5804435	7·6124304	7·6422221	55
6	7·3722107	7·4227670	7·4680469	7·5090483	7·5465106	7·5809964	7 6129440	7·6427017	54
7	7·3731034	7·4235617	7·4687629	7·5096999	7·5471084	7·5815485	7·6134571	7·6431808	53
8	7·3739943	7·4243549	7·4694778	7·5103506	7·5477053	7·5821000	7·6139695	7·6436593	52
9	7·3748832	7·4251467	7·4701915	7·5110002	7·5483015	7·5826508	7·6144813	7·6441373	51
10	7·3757705	7·4259370	7·4709041	7·5116489	7·5488968	7·5832009	7·6149926	7·6446149	50
11	7·3766559	7·4267259	7·4716154	7·5122966	7·5494913	7·5837503	7·6155032	7·6450918	49
12	7·3775396	7·4275134	7·4723257	7·5129434	7·5500850	7·5842990	7·6160132	7·6455683	48
13	7·3784214	7·4282995	7·4730347	7·5135892	7·5506779	7·5848470	7·6165227	7·6460442	47
14	7·3793014	7·4290841	7·4737426	7·5142340	7·5512700	7·5853943	7·6170315	7·6465136	46
15	7·3801796	7·4298673	7·4744493	7·5148779	7·5518613	7·5859409	7·6175397	7·6469945	45
16	7·3810561	7·4306491	7·4751549	7·5155208	7·5524518	7·5864869	7·6180474	7·6474689	44
17	7·3819308	7·4314295	7·4758594	7·5161628	7·5530414	7·5870321	7·6185544	7·6479428	43
18	7·3828038	7·4322085	7·4765627	7·5168038	7·5536303	7·5875767	7·6190609	7·6484161	42
19	7·3836750	7·4329861	7·4772649	7·5174439	7·5542184	7·5881206	7·6195668	7·6488889	41
20	7·3845444	7·4337624	7·4779659	7·5180830	7·5548057	7·5886638	7·6200721	7·6493613	40
21	7·3854122	7·4345372	7·4786658	7·5187212	7·5553921	7·5892063	7·6205768	7·6498331	39
22	7·3862782	7·4353106	7·4793646	7·5193585	7·5559778	7·5897481	7·6210809	7·6503043	38
23	7·3871424	7·4360827	7·4800623	7·5199948	7·5565627	7·5902893	7·6215844	7·6507751	37
24	7·3880050	7·4368534	7·4807588	7·5206302	7·5571469	7·5908298	7·6220873	7·6512454	36
25	7·3888658	7·4376228	7·4814542	7·5212646	7·5577302	7·5913696	7·6225897	7·6517151	35
26	7·3897249	7·4383908	7·4821485	7·5218982	7·5583127	7·5919088	7·6230915	7·6521844	34
27	7·3905824	7·4391574	7·4828417	7·5225308	7·5588945	7·5924473	7·6235927	7·6526531	33
28	7·3914381	7·4399227	7·4835338	7·5231625	7·5594755	7·5929851	7·6240933	7·6531214	32
29	7·3922922	7·4406866	7·4842248	7·5237933	7·5600557	7·5935223	7·6245934	7·6535891	31
30	7·3931446	7·4414492	7·4849147	7·5244231	7·5606352	7·5940588	7·6250928	7·6540563	30
31	7·3939953	7·4422104	7·4856035	7·5250521	7·5612138	7·5945946	7·6255917	7·6545231	29
32	7·3948444	7·4429703	7·4862913	7·5256801	7·5617917	7·5951298	7·6260901	7·6549893	28
33	7·3956918	7·4437289	7·4869779	7·5263073	7·5623689	7·5956643	7·6265878	7·6554550	27
34	7·3965375	7·4444862	7·4876634	7·5269335	7·5629452	7·5961981	7·6270850	7·6559203	26
35	7·3973816	7·4452422	7·4883479	7·5275588	7·5635208	7·5967313	7·6275816	7·6563850	25
36	7·3982241	7·4459968	7·4890313	7·5281833	7·5640957	7·5972639	7·6280777	7·6568492	24
37	7·3990650	7·4467501	7·4897136	7·5288068	7·5646698	7·5977958	7·6285732	7·6573130	23
38	7·3999042	7·4475021	7·4903949	7·5294295	7·5652431	7·5983270	7·6290681	7·6577762	22
39	7·4007418	7·4482529	7·4910750	7·5300512	7·5658157	7·5988576	7·6295624	7·6582390	21
40	7·4015778	7·4490023	7·4917541	7·5306721	7·5663875	7·5993876	7·6300562	7·6587012	20
41	7·4024121	7·4497504	7·4924322	7·5312920	7·5669585	7·5999169	7·6305495	7·6591630	19
42	7·4032449	7·4504973	7·4931092	7·5319111	7·5675289	7·6004455	7·6310421	7·6596243	18
43	7·4040761	7·4512428	7·4937851	7·5325294	7·5680984	7·6009735	7·6315342	7·6600850	17
44	7·4049057	7·4519871	7·4944600	7·5331467	7·5686672	7·6015009	7·6320258	7·6605453	16
45	7·4057337	7·4527302	7·4951339	7·5337631	7·5692353	7·6020277	7·6325168	7·6610052	15
46	7·4065601	7·4534719	7·4958067	7·5343787	7·5698026	7·6025538	7·6330073	7·6614645	14
47	7·4073850	7·4542124	7·4964784	7·5349934	7·5703692	7·6030792	7·6334971	7·6619233	13
48	7·4082083	7·4549516	7·4971492	7·5356073	7·5709351	7·6036040	7·6339865	7·6623817	12
49	7·4090301	7·4556896	7·4978188	7·5362202	7·5715002	7·6041282	7·6344753	7·6628395	11
50	7·4098503	7·4564263	7·4984875	7·5368324	7·5720646	7·6046518	7·6349635	7·6632969	10
51	7·4106689	7·4571618	7·4991551	7·5374436	7·5726282	7·6051747	7·6354512	7·6637538	9
52	7·4114860	7·4578960	7·4998217	7·5380540	7·5731912	7·6056970	7·6359384	7·6642103	8
53	7·4123016	7·4586290	7·5004873	7·5386635	7·5737533	7·6062187	7·6364250	7·6646662	7
54	7·4131156	7·4593607	7·5011519	7·5392722	7·5743148	7·6067397	7·6369110	7·6651217	6
55	7·4139282	7·4600912	7·5018154	7·5398800	7·5748755	7·6072602	7·6373965	7·6655767	5
56	7·4147392	7·4608205	7·5024780	7·5404870	7·5754356	7·6077800	7·6378815	7·6660312	4
57	7·4155487	7·4615486	7·5031395	7·5410931	7·5759949	7·6082991	7·6383659	7·6664852	3
58	7·4163567	7·4622754	7·5038000	7·5416984	7·5765534	7·6088177	7·6388498	7·6669388	2
59	7·4171631	7·4630011	7·5044595	7·5423029	7·5771113	7·6093356	7·6393332	7·6673919	1
60	7·4179681	7·4637255	7·5051181	7·5429065	7·5776684	7·6098530	7·6398160	7·6678445	0
"	51'	50'	49'	48'	47'	46'	45'	44'	"

"	8'	9'	10'	11'	12'	13'	14'	15'	"
0	7·3668169	7·4179696	7·4637273	7·5051203	7·5429091	7·5776715	7·6098566	7·6398201	60
1	7·3677207	7·4187731	7·4644506	7·5057778	7·5435119	7·5782280	7·6103733	7·6403024	59
2	7·3686227	7·4195752	7·4651726	7·5064343	7·5441138	7·5787837	7·6108894	7·6407842	58
3	7·3695228	7·4203757	7·4658934	7·5070899	7·5447149	7·5793387	7·6114049	7·6412654	57
4	7·3704210	7·4211748	7·4666130	7·5077444	7·5453152	7·5798930	7·6119197	7·6417461	56
5	7·3713174	7·4219724	7·4673315	7·5083980	7·5459147	7·5804466	7·6124340	7·6422262	55
6	7·3722119	7·4227685	7·4680487	7·5090506	7·5465133	7·5809995	7·6129477	7·6427059	54
7	7·3731046	7·4235632	7·4687648	7·5097022	7·5471111	7·5815517	7·6134607	7·6431850	53
8	7·3739955	7·4243564	7·4694797	7·5103528	7·5477080	7·5821032	7·6139732	7·6436635	52
9	7·3748845	7·4251482	7·4701934	7·5110025	7·5483042	7·5826540	7·6144850	7·6441416	51
10	7·3757718	7·4259386	7·4709060	7·5116512	7·5488995	7·5832041	7·6149963	7·6446191	50
11	7·3766572	7·4267275	7·4716173	7·5122989	7·5494941	7·5837535	7·6155069	7·6450961	49
12	7·3775408	7·4275150	7·4723276	7·5129457	7·5500878	7·5843022	7·6160169	7·6455725	48
13	7·3784226	7·4283010	7·4730366	7·5135915	7·5506807	7·5848502	7·6165264	7·6460485	47
14	7·3793026	7·4290857	7·4737445	7·5142363	7·5512728	7·5853975	7·6170352	7·6465239	46
15	7·3801809	7·4298689	7·4744513	7·5148802	7·5518640	7·5859441	7·6175435	7·6469988	45
16	7·3810574	7·4306507	7·4751569	7·5155231	7·5524545	7·5864901	7·6180511	7·6474732	44
17	7·3819321	7·4314311	7·4758613	7·5161651	7·5530442	7·5870353	7·6185582	7·6479471	43
18	7·3828051	7·4322101	7·4765646	7·5168061	7·5536331	7·5875799	7·6190647	7·6484204	42
19	7·3836763	7·4329877	7·4772668	7·5174462	7·5542212	7·5881238	7·6195705	7·6488933	41
20	7·3845457	7·4337640	7·4779679	7·5180854	7·5548084	7·5886670	7·6200758	7·6493656	40
21	7·3854134	7·4345388	7·4786678	7·5187236	7·5553949	7·5892096	7·6205805	7·6498374	39
22	7·3862794	7·4353123	7·4793666	7·5193608	7·5559806	7·5897514	7·6210847	7·6503087	38
23	7·3871437	7·4360843	7·4800642	7·5199972	7·5565656	7·5902926	7·6215882	7·6507795	37
24	7·3880063	7·4368551	7·4807608	7·5206326	7·5571497	7·5908331	7·6220911	7·6512497	36
25	7·3888671	7·4376244	7·4814562	7·5212670	7·5577330	7·5913730	7·6225935	7·6517195	35
26	7·3897263	7·4383924	7·4821505	7·5219006	7·5583156	7·5919121	7·6230953	7·6521888	34
27	7·3905837	7·4391590	7·4828437	7·5225332	7·5588974	7·5924506	7·6235965	7·6526575	33
28	7·3914395	7·4399243	7·4835359	7·5231649	7·5594784	7·5929884	7·6240972	7·6531258	32
29	7·3922935	7·4406882	7·4842269	7·5237957	7·5600586	7·5935256	7·6245972	7·6535935	31
30	7·3931459	7·4414508	7·4849168	7·5244256	7·5606380	7·5940621	7·6250967	7·6540608	30
31	7·3939967	7·4422121	7·4856056	7·5250545	7·5612167	7·5945980	7·6255956	7·6545275	29
32	7·3948457	7·4429720	7·4862933	7·5256826	7·5617946	7·5951331	7·6260939	7·6549937	28
33	7·3956931	7·4437306	7·4869799	7·5263097	7·5623718	7·5956677	7·6265917	7·6554595	27
34	7·3965389	7·4444879	7·4876655	7·5269360	7·5629481	7·5962015	7·6270889	7·6559247	26
35	7·3973830	7·4452438	7·4883500	7·5275613	7·5635238	7·5967347	7·6275855	7·6563895	25
36	7·3982255	7·4459985	7·4890334	7·5281858	7·5640986	7·5972673	7·6280816	7·6568537	24
37	7·3990663	7·4467518	7·4897157	7·5288093	7·5646727	7·5977992	7·6285771	7·6573174	23
38	7·3999055	7·4475038	7·4903969	7·5294319	7·5652460	7·5983304	7·6290720	7·6577807	22
39	7·4007431	7·4482546	7·4910771	7·5300537	7·5658186	7·5988611	7·6295664	7·6582435	21
40	7·4015791	7·4490040	7·4917562	7·5306746	7·5663904	7·5993910	7·6300602	7·6587057	20
41	7·4024135	7·4497521	7·4924343	7·5312946	7·5669615	7·5999203	7·6305534	7·6591675	19
42	7·4032463	7·4504990	7·4931113	7·5319137	7·5675318	7·6004490	7·6310461	7·6596288	18
43	7·4040775	7·4512446	7·4937872	7·5325319	7·5681014	7·6009770	7·6315382	7·6600896	17
44	7·4049071	7·4519889	7·4944621	7·5331492	7·5686702	7·6015044	7·6320298	7·6605499	16
45	7·4057351	7·4527319	7·4951360	7·5337657	7·5692383	7·6020311	7·6325208	7·6610097	15
46	7·4065616	7·4534737	7·4958088	7·5343813	7·5698056	7·6025572	7·6330113	7·6614690	14
47	7·4073864	7·4542141	7·4964806	7·5349960	7·5703722	7·6030827	7·6335012	7·6619279	13
48	7·4082097	7·4549534	7·4971513	7·5356008	7·5709381	7·6036075	7·6339905	7·6623863	12
49	7·4090315	7·4556913	7·4978210	7·5362228	7·5715032	7·6041317	7·6344793	7·6628441	11
50	7·4098517	7·4564281	7·4984897	7·5368349	7·5720676	7·6046553	7·6349676	7·6633015	10
51	7·4106703	7·4571635	7·4991573	7·5374462	7·5726313	7·6051782	7·6354553	7·6637585	9
52	7·4114875	7·4578978	7·4998239	7·5380566	7·5731942	7·6057005	7·6359424	7·6642149	8
53	7·4123030	7·4586308	7·5004895	7·5386661	7·5737564	7·6062222	7·6364290	7·6646709	7
54	7·4131171	7·4593625	7·5011541	7·5392748	7·5743179	7·6067433	7·6369151	7·6651263	6
55	7·4139296	7·4600930	7·5018176	7·5398826	7·5748786	7·6072637	7·6374006	7·6655813	5
56	7·4147406	7·4608223	7·5024802	7·5404896	7·5754386	7·6077835	7·6378856	7·6660359	4
57	7·4155501	7·4615504	7·5031417	7·5410958	7·5759979	7·6083027	7·6383700	7·6664899	3
58	7·4163581	7·4622773	7·5038022	7·5417011	7·5765565	7·6088213	7·6388539	7·6669435	2
59	7·4171646	7·4630030	7·5044618	7·5423055	7·5771144	7·6093392	7·6393373	7·6673966	1
60	7·4179696	7·4637273	7·5051203	7·5429091	7·5776715	7·6098566	7·6398201	7·6678492	0
"	51'	50'	49'	48'	47'	46'	45'	44'	"

"	16'	17'	18'	19'	20'	21'	22'	23'	"
0	7·6678445	7·6941733	7·7189966	7·7424775	7·7647537	7·7859427	7·8061458	7·8254507	60
1	7·6682967	7·6945988	7·7193986	7·7428583	7·7651154	7·7862872	7·8064747	7·8257653	59
2	7·6687484	7·6950240	7·7198001	7·7432388	7·7654769	7·7866315	7·8068033	7·8260797	58
3	7·6691996	7·6954487	7·7202013	7·7436189	7·7658380	7·7869755	7·8071317	7·8263938	57
4	7·6696503	7·6958730	7·7206021	7·7439987	7·7661989	7·7873192	7·8074599	7·8267077	56
5	7·6701006	7·6962969	7·7210026	7·7443781	7·7665594	7·7876627	7·8077878	7·8270214	55
6	7·6705504	7·6967204	7·7214027	7·7447573	7·7669197	7·7880058	7·8081154	7·8273348	54
7	7·6709998	7·6971435	7·7218024	7·7451360	7·7672797	7·7883488	7·8084428	7·8276481	53
8	7·6714486	7·6975662	7·7222017	7·7455145	7·7676393	7·7886914	7·8087699	7·8279611	52
9	7·6718970	7·6979884	7·7226007	7·7458926	7·7679987	7·7890337	7·8090968	7·8282738	51
10	7·6723450	7·6984103	7·7229993	7·7462705	7·7683577	7·7893758	7·8094235	7·8285864	50
11	7·6727925	7·6988317	7·7233976	7·7466479	7·7687165	7·7897177	7·8097499	7·8288987	49
12	7·6732395	7·6992528	7·7237955	7·7470251	7·7690750	7·7900592	7·8100761	7·8292108	48
13	7·6736861	7·6996734	7·7241930	7·7474019	7·7694332	7·7904005	7·8104020	7·8295227	47
14	7·6741322	7·7000936	7·7245902	7·7477784	7·7697910	7·7907415	7·8107277	7·8298343	46
15	7·6745779	7·7005134	7·7249869	7·7481546	7·7701486	7·7910823	7·8110531	7·8301458	45
16	7·6750231	7·7009328	7·7253834	7·7485304	7·7705059	7·7914228	7·8113783	7·8304570	44
17	7·6754678	7·7013518	7·7257794	7·7489059	7·7708629	7·7917630	7·8117032	7·8307680	43
18	7·6759121	7·7017704	7·7261752	7·7492811	7·7712196	7·7921029	7·8120279	7·8310787	42
19	7·6763559	7·7021886	7·7265705	7·7496560	7·7715760	7·7924426	7·8123524	7·8313893	41
20	7·6767993	7·7026064	7·7269655	7·7500306	7·7719322	7·7927820	7·8126766	7·8316996	40
21	7·6772422	7·7030238	7·7273601	7·7504048	7·7722880	7·7931212	7·8130006	7·8320097	39
22	7·6776847	7·7034407	7·7277544	7·7507787	7·7726435	7·7934601	7·8133243	7·8323195	38
23	7·6781267	7·7038573	7·7281483	7·7511523	7·7729988	7·7937987	7·8136478	7·8326292	37
24	7·6785683	7·7042735	7·7285419	7·7515255	7·7733537	7·7941371	7·8139711	7·8329386	36
25	7·6790094	7·7046893	7·7289351	7·7518985	7·7737084	7·7944752	7·8142941	7·8332478	35
26	7·6794501	7·7051047	7·7293279	7·7522711	7·7740628	7·7948130	7·8146168	7·8335568	34
27	7·6798904	7·7055197	7·7297204	7·7526434	7·7744169	7·7951506	7·8149394	7·8338656	33
28	7·6803302	7·7059343	7·7301125	7·7530154	7·7747707	7·7954879	7·8152617	7·8341741	32
29	7·6807695	7·7063485	7·7305043	7·7533871	7·7751242	7·7958250	7·8155837	7·8344825	31
30	7·6812084	7·7067623	7·7308957	7·7537584	7·7754774	7·7961617	7·8159055	7·8347906	30
31	7·6816469	7·7071757	7·7312868	7·7541294	7·7758303	7·7964983	7·8162271	7·8350985	29
32	7·6820849	7·7075887	7·7316776	7·7545001	7·7761830	7·7968345	7·8165484	7·8354062	28
33	7·6825224	7·7080014	7·7320679	7·7548705	7·7765354	7·7971705	7·8168695	7·8357136	27
34	7·6829596	7·7084136	7·7324579	7·7552406	7·7768874	7·7975063	7·8171904	7·8360209	26
35	7·6833963	7·7088254	7·7328476	7·7556104	7·7772392	7·7978418	7·8175110	7·8363279	25
36	7·6838325	7·7092369	7·7332369	7·7559798	7·7775907	7·7981770	7·8178314	7·8366347	24
37	7·6842683	7·7096480	7·7336259	7·7563490	7·7779420	7·7985120	7·8181516	7·8369413	23
38	7·6847037	7·7100586	7·7340145	7·7567178	7·7782929	7·7988467	7·8184715	7·8372477	22
39	7·6851387	7·7104689	7·7344028	7·7570863	7·7786436	7·7991811	7·8187912	7·8375538	21
40	7·6855732	7·7108788	7·7347908	7·7574545	7·7789939	7·7995153	7·8191106	7·8378598	20
41	7·6860072	7·7112883	7·7351783	7·7578224	7·7793440	7·7998493	7·8194298	7·8381655	19
42	7·6864409	7·7116975	7·7355656	7·7581900	7·7796938	7·8001830	7·8197488	7·8384710	18
43	7·6868741	7·7121062	7·7359525	7·7585572	7·7800434	7·8005164	7·8200676	7·8387763	17
44	7·6873069	7·7125146	7·7363390	7·7589242	7·7803926	7·8008496	7·8203861	7·8390814	16
45	7·6877392	7·7129225	7·7367252	7·7592908	7·7807416	7·8011825	7·8207043	7·8393863	15
46	7·6881711	7·7133301	7·7371111	7·7596572	7·7810903	7·8015151	7·8210224	7·8396909	14
47	7·6886026	7·7137373	7·7374966	7·7600232	7·7814387	7·8018475	7·8213402	7·8399954	13
48	7·6890337	7·7141442	7·7378818	7·7603889	7·7817868	7·8021797	7·8216578	7·8402996	12
49	7·6894643	7·7145506	7·7382666	7·7607543	7·7821347	7·8025116	7·8219751	7·8406036	11
50	7·6898945	7·7149567	7·7386511	7·7611194	7·7824822	7·8028432	7·8222922	7·8409074	10
51	7·6903243	7·7153624	7·7390353	7·7614842	7·7828295	7·8031746	7·8226091	7·8412110	9
52	7·6907536	7·7157677	7·7394191	7·7618487	7·7831765	7·8035058	7·8229258	7·8415144	8
53	7·6911826	7·7161726	7·7398026	7·7622129	7·7835233	7·8038367	7·8232422	7·8418176	7
54	7·6916111	7·7165772	7·7401857	7·7625768	7·7838697	7·8041673	7·8235584	7·8421205	6
55	7·6920392	7·7169814	7·7405685	7·7629403	7·7842159	7·8044977	7·8238743	7·8424233	5
56	7·6924668	7·7173852	7·7409510	7·7633036	7·7845618	7·8048278	7·8241901	7·8427258	4
57	7·6928941	7·7177886	7·7413331	7·7636666	7·7849075	7·8051577	7·8245056	7·8430281	3
58	7·6933209	7·7181917	7·7417149	7·7640292	7·7852528	7·8054873	7·8248209	7·8433302	2
59	7·6937473	7·7185943	7·7420964	7·7643916	7·7855979	7·8058167	7·8251359	7·8436321	1
60	7·6941733	7·7189966	7·7424775	7·7647537	7·7859427	7·8061458	7·8254507	7·8439338	0
"	43'	42'	41'	40'	39'	38'	37'	36'	"

"	16'	17'	18'	19'	20'	21'	22'	23'	"
0	7·6678492	7·6941786	7·7190026	7·7424841	7·7647610	7·7859508	7·8061547	7·8254604	60
1	7·6683014	7·6946042	7·7194045	7·7428649	7·7651228	7·7862954	7·8064836	7·8257750	59
2	7·6687531	7·6950293	7·7198061	7·7432454	7·7654843	7·7866396	7·8068123	7·8260894	58
3	7·6692043	7·6954541	7·7202073	7·7436255	7·7658454	7·7869836	7·8071407	7·8264036	57
4	7·6696551	7·6958784	7·7206081	7·7440053	7·7662063	7·7873274	7·8074688	7·8267175	56
5	7·6701053	7·6963023	7·7210086	7·7443848	7·7665669	7·7876708	7·8077967	7·8270312	55
6	7·6705552	7·6967258	7·7214087	7·7447640	7·7669271	7·7880140	7·8081244	7·8273446	54
7	7·6710045	7·6971489	7·7218084	7·7451428	7·7672871	7·7883569	7·8084518	7·8276579	53
8	7·6714534	7·6975716	7·7222078	7·7455212	7·7676468	7·7886996	7·8087789	7·8279709	52
9	7·6719018	7·6979938	7·7226068	7·7458994	7·7680061	7·7890420	7·8091059	7·8282837	51
10	7·6723498	7·6984157	7·7230054	7·7462772	7·7683652	7·7893841	7·8094325	7·8285962	50
11	7·6727973	7·6988371	7·7234037	7·7466547	7·7687240	7·7897259	7·8097590	7·8289086	49
12	7·6732443	7·6992582	7·7238016	7·7470319	7·7690825	7·7900675	7·8100851	7·8292207	48
13	7·6736909	7·6996788	7·7241991	7·7474087	7·7694407	7·7904088	7·8104111	7·8295326	47
14	7·6741371	7·7000990	7·7245963	7·7477852	7·7697986	7·7907498	7·8107368	7·8298443	46
15	7·6745827	7·7005189	7·7249931	7·7481614	7·7701562	7·7910906	7·8110622	7·8301557	45
16	7·6750279	7·7009383	7·7253895	7·7485372	7·7705135	7·7914311	7·8113874	7·8304669	44
17	7·6754727	7·7013573	7·7257856	7·7489128	7·7708705	7·7917713	7·8117124	7·8307779	43
18	7·6759170	7·7017759	7·7261813	7·7492880	7·7712272	7·7921113	7·8120371	7·8310887	42
19	7·6763608	7·7021941	7·7265767	7·7496629	7·7715836	7·7924510	7·8123615	7·8313992	41
20	7·6768042	7·7026119	7·7269717	7·7500374	7·7719398	7·7927904	7·8126858	7·8317096	40
21	7·6772471	7·7030293	7·7273663	7·7504117	7·7722956	7·7931296	7·8130098	7·8320197	39
22	7·6776896	7·7034463	7·7277606	7·7507856	7·7726512	7·7934685	7·8133335	7·8323296	38
23	7·6781317	7·7038629	7·7281545	7·7511592	7·7730064	7·7938071	7·8136570	7·8326392	37
24	7·6785733	7·7042791	7·7285481	7·7515325	7·7733614	7·7941455	7·8139803	7·8329467	36
25	7·6790144	7·7046949	7·7289413	7·7519054	7·7737161	7·7944836	7·8143033	7·8332579	35
26	7·6794551	7·7051103	7·7293342	7·7522780	7·7740705	7·7948215	7·8146261	7·8335669	34
27	7·6798953	7·7055253	7·7297267	7·7526504	7·7744246	7·7951590	7·8149486	7·8338757	33
28	7·6803351	7·7059399	7·7301188	7·7530224	7·7747784	7·7954964	7·8152709	7·8341843	32
29	7·6807745	7·7063541	7·7305106	7·7533940	7·7751319	7·7958334	7·8155930	7·8344926	31
30	7·6812134	7·7067679	7·7309020	7·7537654	7·7754851	7·7961702	7·8159148	7·8348007	30
31	7·6816519	7·7071813	7·7312931	7·7541364	7·7758381	7·7965068	7·8162364	7·8351087	29
32	7·6820899	7·7075944	7·7316839	7·7545072	7·7761907	7·7968431	7·8165578	7·8354163	28
33	7·6825275	7·7080070	7·7320742	7·7548776	7·7765431	7·7971791	7·8168789	7·8357238	27
34	7·6829646	7·7084193	7·7324643	7·7552477	7·7768952	7·7975148	7·8171998	7·8360311	26
35	7·6834013	7·7088311	7·7328540	7·7556174	7·7772470	7·7978503	7·8175204	7·8363381	25
36	7·6838376	7·7092426	7·7332433	7·7559869	7·7775985	7·7981856	7·8178408	7·8366449	24
37	7·6842734	7·7096537	7·7336323	7·7563560	7·7779498	7·7985206	7·8181610	7·8369515	23
38	7·6847088	7·7100643	7·7340209	7·7567249	7·7783007	7·7988553	7·8184809	7·8372579	22
39	7·6851438	7·7104746	7·7344092	7·7570934	7·7786514	7·7991898	7·8188006	7·8375641	21
40	7·6855783	7·7108846	7·7347972	7·7574616	7·7790018	7·7995240	7·8191201	7·8378701	20
41	7·6860124	7·7112941	7·7351848	7·7578295	7·7793519	7·7998579	7·8194393	7·8381758	19
42	7·6864460	7·7117032	7·7355720	7·7581971	7·7797017	7·8001916	7·8197583	7·8384813	18
43	7·6868792	7·7121120	7·7359589	7·7585644	7·7800513	7·8005251	7·8200770	7·8387867	17
44	7·6873120	7·7125203	7·7363455	7·7589313	7·7804005	7·8008582	7·8203956	7·8390918	16
45	7·6877444	7·7129283	7·7367317	7·7592980	7·7807495	7·8011912	7·8207139	7·8393966	15
46	7·6881763	7·7133359	7·7371176	7·7596643	7·7810982	7·8015238	7·8210319	7·8397013	14
47	7·6886078	7·7137432	7·7375031	7·7600304	7·7814466	7·8018563	7·8213497	7·8400058	13
48	7·6890389	7·7141500	7·7378883	7·7603961	7·7817948	7·8021884	7·8216673	7·8403100	12
49	7·6894695	7·7145565	7·7382731	7·7607615	7·7821426	7·8025203	7·8219847	7·8406140	11
50	7·6898997	7·7149625	7·7386577	7·7611266	7·7824902	7·8028520	7·8223018	7·8409179	10
51	7·6903295	7·7153682	7·7390418	7·7614915	7·7828375	7·8031834	7·8226187	7·8412215	9
52	7·6907589	7·7157736	7·7394257	7·7618560	7·7831845	7·8035146	7·8229354	7·8415249	8
53	7·6911878	7·7161785	7·7398091	7·7622202	7·7835313	7·8038455	7·8232518	7·8418280	7
54	7·6916163	7·7165831	7·7401923	7·7625840	7·7838778	7·8041761	7·8235680	7·8421310	6
55	7·6920444	7·7169873	7·7405751	7·7629476	7·7842240	7·8045065	7·8238840	7·8424338	5
56	7·6924721	7·7173911	7·7409576	7·7633109	7·7845699	7·8048366	7·8241997	7·8427363	4
57	7·6928993	7·7177945	7·7413397	7·7636739	7·7849155	7·8051665	7·8245153	7·8430387	3
58	7·6933262	7·7181976	7·7417215	7·7640366	7·7852609	7·8054962	7·8248305	7·8433408	2
59	7·6937526	7·7186003	7·7421030	7·7643989	7·7856060	7·8058256	7·8251456	7·8436427	1
60	7·6941786	7·7190026	7·7424841	7·7647610	7·7859508	7·8061547	7·8254604	7·8439444	0

| " | 43' | 42' | 41' | 40' | 39' | 38' | 37' | 36' | " |

"	24'	25'	26'	27'	28'	29'	30'	31'	"
0	7·8439338	7·8616623	7·8786953	7·8950854	7·9108793	7·9261190	7·9408419	7·9550819	60
1	7·8442353	7·8619517	7·8789736	7·8953534	7·9111378	7·9263685	7·9410831	7·9553153	59
2	7·8445366	7·8622410	7·8792517	7·8956212	7·9113960	7·9266179	7·9413241	7·9555486	58
3	7·8448377	7·8625300	7·8795297	7·8958889	7·9116542	7·9268671	7·9415651	7·9557818	57
4	7·8451385	7·8628189	7·8798075	7·8961564	7·9119121	7·9271162	7·9418059	7·9560149	56
5	7·8454392	7·8631075	7·8800850	7·8964237	7·9121699	7·9273651	7·9420465	7·9562478	55
6	7·8457396	7·8633960	7·8803625	7·8966909	7·9124276	7·9276139	7·9422871	7·9564806	54
7	7·8460398	7·8636843	7·8806397	7·8969579	7·9126851	7·9278626	7·9425275	7·9567133	53
8	7·8463399	7·8639723	7·8809167	7·8972248	7·9129425	7·9281111	7·9427677	7·9569458	52
9	7·8466397	7·8642602	7·8811936	7·8974914	7·9131997	7·9283595	7·9430070	7·9571782	51
10	7·8469393	7·8645479	7·8814703	7·8977580	7·9134567	7·9286077	7·9432479	7·9574105	50
11	7·8472387	7·8648354	7·8817469	7·8980243	7·9137136	7·9288558	7·9434877	7·9576427	49
12	7·8475379	7·8651228	7·8820232	7·8982905	7·9139704	7·9291037	7·9437275	7·9578747	48
13	7·8478369	7·8654099	7·8822994	7·8985565	7·9142269	7·9293516	7·9439671	7·9581067	47
14	7·8481357	7·8656968	7·8825754	7·8988224	7·9144834	7·9295992	7·9442066	7·9583385	46
15	7·8484343	7·8659836	7·8828512	7·8990881	7·9147397	7·9298467	7·9444459	7·9585702	45
16	7·8487326	7·8662702	7·8831269	7·8993536	7·9149958	7·9300941	7·9446851	7·9588017	44
17	7·8490308	7·8665565	7·8834023	7·8996190	7·9152518	7·9303414	7·9449242	7·9590331	43
18	7·8493288	7·8668427	7·8836776	7·8998842	7·9155076	7·9305885	7·9451631	7·9592645	42
19	7·8496265	7·8671287	7·8839528	7·9001493	7·9157633	7·9308354	7·9454019	7·9594956	41
20	7·8499241	7·8674145	7·8842277	7·9004141	7·9160189	7·9310823	7·9456406	7·9597267	40
21	7·8502215	7·8677001	7·8845025	7·9006789	7·9162743	7·9313289	7·9458792	7·9599576	39
22	7·8505186	7·8679856	7·8847771	7·9009434	7·9165295	7·9315755	7·9461176	7·9601885	38
23	7·8508156	7·8682708	7·8850515	7·9012078	7·9167846	7·9318219	7·9463559	7·9604192	37
24	7·8511123	7·8685559	7·8853258	7·9014721	7·9170395	7·9320682	7·9465940	7·9606497	36
25	7·8514088	7·8688408	7·8855999	7·9017362	7·9172943	7·9323143	7·9468321	7·9608802	35
26	7·8517052	7·8691254	7·8858738	7·9020001	7·9175489	7·9325603	7·9470700	7·9611105	34
27	7·8520013	7·8694099	7·8861475	7·9022639	7·9178034	7·9328061	7·9473077	7·9613407	33
28	7·8522973	7·8696942	7·8864211	7·9025275	7·9180578	7·9330518	7·9475454	7·9615708	32
29	7·8525930	7·8699784	7·8866945	7·9027909	7·9183120	7·9332974	7·9477829	7·9618008	31
30	7·8528885	7·8702623	7·8869677	7·9030542	7·9185660	7·9335428	7·9480203	7·9620306	30
31	7·8531839	7·8705461	7·8872407	7·9033173	7·9188199	7·9337881	7·9482575	7·9622603	29
32	7·8534790	7·8708296	7·8875136	7·9035803	7·9190736	7·9340332	7·9484946	7·9624899	28
33	7·8537739	7·8711130	7·8877863	7·9038431	7·9193272	7·9342783	7·9487316	7·9627194	27
34	7·8540687	7·8713962	7·8880589	7·9041057	7·9195807	7·9345231	7·9489685	7·9629487	26
35	7·8543632	7·8716792	7·8883312	7·9043682	7·9198340	7·9347679	7·9492052	7·9631780	25
36	7·8546575	7·8719621	7·8886034	7·9046305	7·9200871	7·9350125	7·9494418	7·9634071	24
37	7·8549517	7·8722447	7·8888754	7·9048927	7·9203401	7·9352569	7·9496783	7·9636361	23
38	7·8552456	7·8725272	7·8891473	7·9051547	7·9205930	7·9355012	7·9499146	7·9638649	22
39	7·8555393	7·8728095	7·8894190	7·9054166	7·9208457	7·9357454	7·9501508	7·9640937	21
40	7·8558329	7·8730916	7·8896905	7·9056783	7·9210983	7·9359895	7·9503869	7·9643223	20
41	7·8561262	7·8733735	7·8899618	7·9059398	7·9213507	7·9362334	7·9506229	7·9645508	19
42	7·8564193	7·8736552	7·8902330	7·9062012	7·9216030	7·9364772	7·9508587	7·9647792	18
43	7·8567123	7·8739367	7·8905040	7·9064624	7·9218551	7·9367208	7·9510944	7·9650075	17
44	7·8570050	7·8742181	7·8907749	7·9067235	7·9221071	7·9369643	7·9513300	7·9652356	16
45	7·8572976	7·8744993	7·8910455	7·9069844	7·9223589	7·9372077	7·9515654	7·9654637	15
46	7·8575899	7·8747803	7·8913160	7·9072451	7·9226106	7·9374509	7·9518008	7·9656916	14
47	7·8578821	7·8750611	7·8915864	7·9075057	7·9228621	7·9376940	7·9520360	7·9659194	13
48	7·8581740	7·8753417	7·8918565	7·9077662	7·9231135	7·9379369	7·9522710	7·9661470	12
49	7·8584658	7·8756222	7·8921265	7·9080265	7·9233648	7·9381798	7·9525060	7·9663746	11
50	7·8587574	7·8759025	7·8923963	7·9082866	7·9236159	7·9384224	7·9527408	7·9666020	10
51	7·8590487	7·8761826	7·8926660	7·9085466	7·9238668	7·9386650	7·9529755	7·9668293	9
52	7·8593399	7·8764625	7·8929355	7·9088064	7·9241177	7·9389074	7·9532100	7·9670565	8
53	7·8596309	7·8767422	7·8932048	7·9090660	7·9243683	7·9391497	7·9534444	7·9672836	7
54	7·8599217	7·8770218	7·8934740	7·9093256	7·9246188	7·9393918	7·9536787	7·9675106	6
55	7·8602123	7·8773011	7·8937430	7·9095849	7·9248692	7·9396338	7·9539129	7·9677374	5
56	7·8605027	7·8775803	7·8940118	7·9098441	7·9251195	7·9398757	7·9541470	7·9679641	4
57	7·8607929	7·8778594	7·8942804	7·9101031	7·9253696	7·9401175	7·9543809	7·9681907	3
58	7·8610829	7·8781382	7·8945489	7·9103620	7·9256195	7·9403591	7·9546147	7·9684172	2
59	7·8613727	7·8784168	7·8948173	7·9106208	7·9258693	7·9406005	7·9548484	7·9686436	1
60	7·8616623	7·8786953	7·8950854	7·9108793	7·9261190	7·9408419	7·9550819	7·9688698	0
"	35'	34'	33'	32'	31'	30'	29'	28'	"

LOG. COSINES.   89 Deg.

"	24'	25'	26'	27'	28'	29'	30'	31'	"
0	7·8439444	7·8616738	7·8787077	7·8950988	7·9108938	7·9261344	7·9408584	7·9550996	60
1	7·8442459	7·8619632	7·8789861	7·8953668	7·9111522	7·9263840	7·9410996	7·9553330	59
2	7·8445472	7·8622525	7·8792642	7·8956347	7·9114105	7·9266333	7·9413407	7·9555663	58
3	7·8448483	7·8625415	7·8795422	7·8959023	7·9116686	7·9268826	7·9415817	7·9557995	57
4	7·8451492	7·8628304	7·8798199	7·8961699	7·9119266	7·9271317	7·9418225	7·9560326	56
5	7·8454498	7·8631191	7·8800975	7·8964372	7·9121844	7·9273807	7·9420632	7·9562655	55
6	7·8457503	7·8634076	7·8803750	7·8967044	7·9124421	7·9276295	7·9423037	7·9564984	54
7	7·8460505	7·8636958	7·8806522	7·8969714	7·9126996	7·9278782	7·9425441	7·9567310	53
8	7·8463506	7·8639839	7·8809293	7·8972383	7·9129570	7·9281267	7·9427844	7·9569636	52
9	7·8466504	7·8642719	7·8812062	7·8975050	7·9132142	7·9283751	7·9430246	7·9571961	51
10	7·8469500	7·8645596	7·8814829	7·8977715	7·9134713	7·9286233	7·9432646	7 9574284	50
11	7·8472494	7·8648471	7·8817594	7·8980379	7·9137282	7·9288714	7·9435045	7·9576606	49
12	7·8475487	7·8651344	7·8820358	7·8983041	7·9139850	7·9291194	7·9437442	7·9578926	48
13	7·8478477	7·8654216	7·8823120	7·8985701	7·9142416	7·9293672	7·9439839	7·9581246	47
14	7·8481465	7·8657085	7·8825880	7·8988360	7·9144980	7·9296149	7·9442233	7·9583564	46
15	7·8484451	7·8659953	7·8828639	7·8991017	7·9147543	7·9298625	7·9444627	7·9585881	45
16	7·8487435	7·8662819	7·8831395	7·8993673	7·9150105	7·9301099	7·9447019	7·9588197	44
17	7·8490416	7·8665683	7·8834150	7·8996327	7·9152665	7·9303571	7·9449410	7·9590511	43
18	7·8493396	7·8668545	7·8836903	7·8998979	7·9155224	7·9306043	7·9451800	7·9592825	42
19	7·8496374	7·8671405	7·8839655	7·9001630	7·9157781	7·9308512	7·9454188	7·9595137	41
20	7·8499350	7·8674263	7·8842404	7·9004279	7·9160336	7·9310981	7·9456575	7·9597447	40
21	7·8502323	7·8677120	7·8845152	7·9006926	7·9162890	7·9313448	7·9458961	7·9599757	39
22	7·8505295	7·8679974	7·8847899	7·9009572	7·9165443	7·9315913	7·9461345	7·9602065	38
23	7·8508265	7·8682827	7·8850643	7·9012216	7·9167994	7·9318378	7·9463728	7·9604373	37
24	7·8511232	7·8685677	7·8853386	7·9014859	7·9170543	7·9320840	7·9466110	7·9606678	36
25	7·8514198	7·8688526	7·8856127	7·9017500	7·9173091	7·9323302	7·9468491	7·9608983	35
26	7·8517161	7·8691373	7·8858866	7·9020139	7·9175638	7·9325762	7·9470870	7·9611287	34
27	7·8520123	7·8694218	7·8861604	7·9022777	7·9178183	7·9328220	7·9473248	7·9613589	33
28	7·8523083	7·8697062	7·8864339	7·9025413	7·9180727	7·9330678	7·9475624	7·9615890	32
29	7·8526040	7·8699903	7·8867074	7·9028048	7·9183269	7·9333133	7·9478000	7·9618190	31
30	7·8528996	7·8702743	7·8869806	7·9030681	7·9185809	7·9335588	7·9480374	7·9620488	30
31	7·8531949	7·8705580	7·8872537	7·9033312	7·9188348	7·9338041	7·9482746	7·9622786	29
32	7·8534900	7·8708416	7·8875266	7·9035942	7·9190886	7·9340493	7·9485118	7·9625082	28
33	7·8537850	7·8711250	7·8877993	7·9038570	7·9193422	7·9342943	7·9487488	7·9627377	27
34	7·8540797	7·8714082	7·8880718	7·9041197	7·9195957	7·9345392	7·9489856	7·9629670	26
35	7·8543743	7·8716913	7·8883442	7·9043822	7·9198490	7·9347839	7·9492224	7·9631963	25
36	7·8546686	7·8719741	7·8886164	7·9046445	7·9201022	7·9350286	7·9494590	7·9634254	24
37	7·8549628	7·8722568	7·8888885	7·9049067	7·9203552	7·9352730	7·9496955	7·9636544	23
38	7·8552567	7·8725393	7·8891603	7·9051687	7·9206081	7·9355174	7·9499319	7·9638833	22
39	7·8555505	7·8728215	7·8894320	7·9054306	7·9208608	7·9357616	7·9501681	7·9641121	21
40	7·8558440	7·8731037	7·8897036	7·9056923	7·9211134	7·9360057	7·9504042	7·9643408	20
41	7·8561374	7·8733856	7·8899749	7·9059539	7·9213658	7·9362496	7·9506402	7·9645693	19
42	7·8564305	7·8736673	7·8902461	7·9062153	7·9216181	7·9364934	7·9508760	7·9647977	18
43	7·8567235	7·8739489	7·8905171	7·9064765	7·9218702	7·9367370	7·9511118	7·9650260	17
44	7·8570163	7·8742303	7·8907880	7·9067376	7·9221222	7·9369805	7·9513474	7·9652541	16
45	7·8573083	7·8745115	7·8910587	7·9069985	7·9223741	7·9372239	7·9515828	7·9654822	15
46	7·8576012	7·8747925	7·8913292	7·9072593	7·9226258	7·9374672	7·9518182	7·9657101	14
47	7·8578934	7·8750733	7·8915995	7·9075199	7·9228774	7·9377103	7·9520534	7·9659379	13
48	7·8581853	7·8753540	7·8918697	7·9077804	7·9231288	7·9379533	7·9522885	7·9661656	12
49	7·8584771	7·8756344	7·8921397	7·9080407	7·9233800	7·9381961	7·9525234	7·9663932	11
50	7·8587687	7·8759147	7·8924096	7·9083008	7·9236312	7·9384388	7·9527582	7·9666206	10
51	7·8590601	7·8761949	7·8926792	7·9085608	7·9238821	7·9386814	7·9529929	7·9668480	9
52	7·8593513	7·8764748	7·8929487	7·9088207	7·9241330	7·9389238	7·9532275	7·9670752	8
53	7·8596423	7·8767545	7·8932181	7·9090803	7·9243836	7·9391661	7·9534620	7·9673023	7
54	7·8599331	7·8770341	7·8934873	7·9093399	7·9246342	7·9394083	7·9536963	7·9675293	6
55	7·8602237	7·8773135	7·8937563	7·9095992	7·9248846	7·9396503	7·9539305	7·9677561	5
56	7·8605141	7·8775927	7·8940251	7·9098584	7·9251348	7·9398922	7·9541646	7·9679829	4
57	7·8608043	7·8778717	7·8942938	7·9101175	7·9253850	7·9401339	7·9543985	7·9682095	3
58	7·8610943	7·8781506	7·8945623	7·9103764	7·9256349	7·9403756	7·9546323	7·9684360	2
59	7·8613841	7·8784293	7·8948306	7·9106352	7·9258847	7·9406170	7·9548660	7·9686624	1
60	7·8616738	7·8787077	7·8950988	7·9108938	7·9261344	7·9408584	7·9550996	7·9688886	0
"	35'	34'	33'	32'	31'	30'	29'	28'	"

"	32'	33'	34'	35'	36'	37'	38'	39'	"
0	7·9688698	7·9822334	7·9951980	8·0077867	8·0200207	8·0319195	8·0435009	8·0547814	60
1	7·9690960	7·9824527	7·9954108	8·0079934	8·0202217	8·0321150	8·0436913	8·0549670	59
2	7·9693220	7·9826718	7·9956235	8·0082001	8·0204226	8·0323105	8·0438816	8·0551524	58
3	7·9695479	7·9828909	7·9958361	8·0084066	8·0206234	8·0325059	8·0440719	8·0553378	57
4	7·9697736	7·9831098	7·9960487	8·0086131	8·0208242	8·0327012	8·0442621	8·0555231	56
5	7·9699993	7·9833287	7·9962611	8·0088194	8·0210248	8·0328965	8·0444522	8·0557084	55
6	7·9702248	7·9835474	7·9964734	8·0090257	8·0212253	8·0330916	8·0446422	8·0558935	54
7	7·9704503	7·9837660	7·9966856	8·0092318	8·0214258	8·0332866	8·0448321	8·0560786	53
8	7·9706756	7·9839845	7·9968977	8·0094379	8·0216261	8·0334816	8·0450220	8·0562636	52
9	7·9709008	7·9842029	7·9971097	8·0096439	8·0218264	8·0336765	8·0452117	8·0564485	51
10	7·9711258	7·9844212	7·9973216	8·0098497	8·0220266	8·0338713	8·0454014	8·0566333	50
11	7·9713508	7·9846394	7·9975334	8·0100555	8·0222267	8·0340660	8·0455910	8·0568181	49
12	7·9715756	7·9848574	7·9977451	8·0102612	8·0224267	8·0342606	8·0457805	8·0570028	48
13	7·9718004	7·9850754	7·9979566	8·0104668	8·0226266	8·0344551	8·0459700	8·0571874	47
14	7·9720250	7·9852933	7·9981681	8·0106722	8·0228264	8·0346495	8·0461593	8·0573719	46
15	7·9722495	7·9855110	7·9983795	8·0108776	8·0230261	8·0348439	8·0463486	8·0575563	45
16	7·9724738	7·9857286	7·9985908	8·0110829	8·0232257	8·0350382	8·0465378	8·0577407	44
17	7·9726981	7·9859461	7·9988020	8·0112881	8·0234252	8·0352323	8·0467269	8·0579250	43
18	7·9729222	7·9861636	7·9990130	8·0114932	8·0236247	8·0354264	8·0469159	8·0581092	42
19	7·9731463	7·9863809	7·9992240	8·0116982	8·0238240	8·0356204	8·0471048	8·0582933	41
20	7·9733702	7·9865981	7·9994349	8·0119031	8·0240233	8·0358143	8·0472937	8·0584774	40
21	7·9735940	7·9868151	7·9996456	8·0121079	8·0242224	8·0360082	8·0474825	8·0586614	39
22	7·9738177	7·9870321	7·9998563	8·0123126	8·0244215	8·0362019	8·0476712	8·0588453	38
23	7·9740412	7·9872490	8·0000669	8·0125172	8·0246205	8·0363956	8·0478598	8·0590291	37
24	7·9742647	7·9874658	8·0002773	8·0127217	8·0248194	8·0365892	8·0480483	8·0592128	36
25	7·9744880	7·9876824	8·0004877	8·0129261	8·0250182	8·0367826	8·0482368	8·0593965	35
26	7·9747113	7·9878989	8·0006979	8·0131304	8·0252169	8·0369760	8·0484251	8·0595801	34
27	7·9749344	7·9881154	8·0009081	8·0133347	8·0254155	8·0371693	8·0486134	8·0597636	33
28	7·9751574	7·9883317	8·0011181	8·0135388	8·0256140	8·0373626	8·0488016	8·0599470	32
29	7·9753802	7·9885479	8·0013281	8·0137428	8·0258125	8·0375557	8·0489897	8·0601304	31
30	7·9756030	7·9887641	8·0015379	8·0139468	8·0260108	8·0377488	8·0491778	8·0603137	30
31	7·9758257	7·9889801	8·0017477	8·0141506	8·0262091	8·0379417	8·0493657	8·0604969	29
32	7·9760482	7·9891960	8·0019573	8·0143543	8·0264072	8·0381346	8·0495536	8·0606800	28
33	7·9762706	7·9894117	8·0021669	8·0145580	8·0266053	8·0383274	8·0497414	8·0608630	27
34	7·9764929	7·9896274	8·0023763	8·0147615	8·0268033	8·0385201	8·0499291	8·0610460	26
35	7·9767151	7·9898430	8·0025856	8·0149650	8·0270012	8·0387128	8·0501167	8·0612289	25
36	7·9769372	7·9900585	8·0027949	8·0151684	8·0271990	8·0389053	8·0503043	8·0614117	24
37	7·9771592	7·9902738	8·0030040	8·0153716	8·0273967	8·0390978	8·0504918	8·0615944	23
38	7·9773810	7·9904891	8·0032131	8·0155748	8·0275943	8·0392901	8·0506792	8·0617771	22
39	7·9776028	7·9907043	8·0034220	8·0157779	8·0277919	8·0394824	8·0508665	8·0619597	21
40	7·9778244	7·9909193	8·0036308	8·0159808	8·0279893	8·0396746	8·0510537	8·0621422	20
41	7·9780459	7·9911342	8·0038396	8·0161837	8·0281867	8·0398667	8·0512408	8·0623246	19
42	7·9782673	7·9913491	8·0040482	8·0163865	8·0283839	8·0400588	8·0514279	8·0625070	18
43	7·9784886	7·9915638	8·0042568	8·0165892	8·0285811	8·0402507	8·0516149	8·0626892	17
44	7·9787098	7·9917784	8·0044652	8·0167918	8·0287782	8·0404426	8·0518018	8·0628714	16
45	7·9789309	7·9919929	8·0046735	8·0169943	8·0289752	8·0406343	8·0519886	8·0630536	15
46	7·9791518	7·9922073	8·0048818	8·0171967	8·0291721	8·0408260	8·0521754	8·0632356	14
47	7·9793726	7·9924216	8·0050899	8·0173991	8·0293689	8·0410176	8·0523620	8·0634176	13
48	7·9795934	7·9926358	8·0052979	8·0176013	8·0295656	8·0412092	8·0525486	8·0635995	12
49	7·9798140	7·9928499	8·0055059	8·0178034	8·0297623	8·0414006	8·0527351	8·0637813	11
50	7·9800345	7·9930639	8·0057137	8·0180055	8·0299588	8·0415920	8·0529216	8·0639630	10
51	7·9802549	7·9932778	8·0059215	8·0182074	8·0301553	8·0417832	8·0531079	8·0641447	9
52	7·9804752	7·9934915	8·0061291	8·0184093	8·0303517	8·0419744	8·0532942	8·0643263	8
53	7·9806953	7·9937052	8·0063366	8·0186110	8·0305479	8·0421655	8·0534803	8·0645078	7
54	7·9809154	7·9939188	8·0065441	8·0188127	8·0307441	8·0423565	8·0536665	8·0646893	6
55	7·9811353	7·9941322	8·0067514	8·0190142	8·0309403	8·0425475	8·0538525	8·0648706	5
56	7·9813552	7·9943456	8·0069587	8·0192157	8·0311363	8·0427383	8·0540384	8·0650519	4
57	7·9815749	7·9945588	8·0071658	8·0194171	8·0313322	8·0429291	8·0542243	8·0652331	3
58	7·9817945	7·9947720	8·0073729	8·0196184	8·0315280	8·0431198	8·0544101	8·0654143	2
59	7·9820140	7·9949850	8·0075798	8·0198196	8·0317238	8·0433104	8·0545958	8·0655953	1
60	7·9822334	7·9951980	8·0077867	8·0200207	8·0319195	8·0435009	8·0547814	8·0657763	0
"	27'	26'	25'	24'	23'	22'	21'	20'	"

"	32′	33′	34′	35′	36′	37′	38′	39′	"
0	7·9688886	7·9822534	7·9952192	8·0078092	8·0200445	8·0319446	8·0435274	8·0548094	60
1	7·9691148	7·9824727	7·9954320	8·0080159	8·0202455	8·0321402	8·0437179	8·0549949	59
2	7·9693408	7·9826919	7·9956448	8·0082226	8·0204465	8·0323357	8·0439082	8·0551804	58
3	7·9695667	7·9829110	7·9958574	8·0084292	8·0206473	8·0325311	8·0440985	8·0553658	57
4	7·9697925	7·9831299	7·9960700	8·0086357	8·0208481	8·0327265	8·0442887	8·0555512	56
5	7·9700182	7·9833488	7·9962824	8·0088420	8·0210487	8·0329217	8·0444788	8·0557364	55
6	7·9702438	7·9835675	7·9964947	8·0090483	8·0212493	8·0331169	8·0446689	8·0559216	54
7	7·9704692	7·9837862	7·9967070	8·0092545	8·0214498	8·0333120	8·0448588	8·0561067	53
8	7·9706945	7·9840047	7·9969191	8·0094606	8·0216501	8·0335069	8·0450487	8·0562917	52
9	7·9709198	7·9842231	7·9971311	8·0096666	8·0218504	8·0337018	8·0452385	8·0564767	51
10	7·9711449	7·9844414	7·9973430	8·0098725	8·0220506	8·0338967	8·0454282	8·0566615	50
11	7·9713698	7·9846596	7·9975548	8·0100783	8·0222507	8·0340914	8·0456178	8·0568463	49
12	7·9715947	7·9848777	7·9977666	8·0102840	8·0224507	8·0342860	8·0458074	8·0570310	48
13	7·9718194	7·9850957	7·9979782	8·0104896	8·0226507	8·0344806	8·0459968	8·0572156	47
14	7·9720441	7·9853135	7·9981897	8·0106951	8·0228505	8·0346750	8·0461862	8·0574002	46
15	7·9722686	7·9855313	7·9984011	8·0109005	8·0230502	8·0348694	8·0463755	8·0575846	45
16	7·9724930	7·9857490	7·9986124	8·0111058	8·0232499	8·0350637	8·0465647	8·0577690	44
17	7·9727173	7·9859665	7·9988236	8·0113110	8·0234494	8·0352579	8·0467538	8·0579534	43
18	7·9729414	7·9861839	7·9990346	8·0115161	8·0236489	8·0354520	8·0469429	8·0581376	42
19	7·9731655	7·9864013	7·9992456	8·0117211	8·0238483	8·0356460	8·0471318	8·0583217	41
20	7·9733894	7·9866185	7·9994565	8·0119260	8·0240475	8·0358400	8·0473207	8·0585058	40
21	7·9736132	7·9868356	7·9996673	8·0121308	8·0242467	8·0360338	8·0475095	8·0586898	39
22	7·9738369	7·9870526	7·9998780	8·0123356	8·0244458	8·0362276	8·0476982	8·0588737	38
23	7·9740605	7·9872695	8·0000886	8·0125402	8·0246448	8·0364213	8·0478869	8·0590576	37
24	7·9742840	7·9874862	8·0002991	8·0127447	8·0248437	8·0366149	8·0480754	8·0592414	36
25	7·9745073	7·9877029	8·0005094	8·0129492	8·0250426	8·0368084	8·0482639	8·0594250	35
26	7·9747306	7·9879195	8·0007197	8·0131535	8·0252413	8·0370018	8·0484523	8·0596087	34
27	7·9749537	7·9881359	8·0009299	8·0133578	8·0254399	8·0371951	8·0486406	8·0597922	33
28	7·9751767	7·9883523	8·0011400	8·0135619	8·0256385	8·0373884	8·0488288	8·0599756	32
29	7·9753996	7·9885685	8·0013499	8·0137660	8·0258369	8·0375815	8·0490169	8·0601590	31
30	7·9756224	7·9887847	8·0015598	8·0139699	8·0260353	8·0377746	8·0492050	8·0603423	30
31	7·9758451	7·9890007	8·0017696	8·0141738	8·0262336	8·0379676	8·0493930	8·0605255	29
32	7·9760676	7·9892166	8·0019792	8·0143775	8·0264318	8·0381605	8·0495809	8·0607087	28
33	7·9762901	7·9894324	8·0021888	8·0145812	8·0266299	8·0383533	8·0497687	8·0608918	27
34	7·9765124	7·9896481	8·0023983	8·0147848	8·0268279	8·0385461	8·0499564	8·0610748	26
35	7·9767346	7·9898637	8·0026076	8·0149883	8·0270258	8·0387387	8·0501441	8·0612577	25
36	7·9769567	7·9900792	8·0028169	8·0151916	8·0272236	8·0389313	8·0503317	8·0614405	24
37	7·9771787	7·9902946	8·0030260	8·0153949	8·0274213	8·0391238	8·0505192	8·0616233	23
38	7·9774006	7·9905099	8·0032351	8·0155981	8·0276190	8·0393162	8·0507066	8·0618060	22
39	7·9776224	7·9907251	8·0034441	8·0158012	8·0278166	8·0395085	8·0508939	8·0619886	21
40	7·9778440	7·9909401	8·0036529	8·0160042	8·0280140	8·0397007	8·0510812	8·0621711	20
41	7·9780655	7·9911551	8·0038617	8·0162071	8·0282114	8·0398928	8·0512683	8·0623536	19
42	7·9782870	7·9913699	8·0040703	8·0164099	8·0284087	8·0400849	8·0514554	8·0625359	18
43	7·9785083	7·9915847	8·0042789	8·0166127	8·0286059	8·0402768	8·0516424	8·0627182	17
44	7·9787295	7·9917993	8·0044874	8·0168153	8·0288030	8·0404687	8·0518294	8·0629005	16
45	7·9789506	7·9920138	8·0046957	8·0170178	8·0290000	8·0406605	8·0520162	8·0630826	15
46	7·9791715	7·9922283	8·0049040	8·0172203	8·0291969	8·0408522	8·0522030	8·0632647	14
47	7·9793924	7·9924426	8·0051121	8·0174226	8·0293938	8·0410439	8·0523897	8·0634467	13
48	7·9796131	7·9926568	8·0053202	8·0176248	8·0295905	8·0412354	8·0525763	8·0636286	12
49	7·9798338	7·9928709	8·0055282	8·0178270	8·0297872	8·0414269	8·0527628	8·0638104	11
50	7·9800543	7·9930849	8·0057360	8·0180291	8·0299838	8·0416183	8·0529493	8·0639922	10
51	7·9802747	7·9932988	8·0059438	8·0182310	8·0301802	8·0418096	8·0531356	8·0641739	9
52	7·9804950	7·9935126	8·0061514	8·0184329	8·0303766	8·0420008	8·0533219	8·0643555	8
53	7·9807152	7·9937263	8·0063590	8·0186347	8·0305729	8·0421919	8·0535081	8·0645371	7
54	7·9809353	7·9939399	8·0065665	8·0188364	8·0307692	8·0423829	8·0536943	8·0647185	6
55	7·9811552	7·9941534	8·0067738	8·0190379	8·0309653	8·0425739	8·0538803	8·0648999	5
56	7·9813751	7·9943667	8·0069811	8·0192394	8·0311613	8·0427648	8·0540663	8·0650812	4
57	7·9815948	7·9945800	8·0071883	8·0194408	8·0313573	8·0429555	8·0542522	8·0652625	3
58	7·9818145	7·9947932	8·0073953	8·0196422	8·0315531	8·0431462	8·0544380	8·0654436	2
59	7·9820340	7·9950062	8·0076023	8·0198434	8·0317489	8·0433369	8·0546237	8·0656247	1
60	7·9822534	7·9952192	8·0078092	8·0200445	8·0319446	8·0435274	8·0548094	8·0658057	0
"	27′	26′	25′	24′	23′	22′	21′	20′	′

"	40'	41'	42'	43'	44'	45'	46'	47'	"
0	8·0657763	8·0764997	8·0869646	8·0971832	8·1071669	8·1169262	8·1264710	8·1358104	60
1	8·0659572	8·0766762	8·0871369	8·0973515	8·1073314	8·1170870	8·1266283	8·1359644	59
2	8·0661381	8·0768526	8·0873091	8·0975198	8·1074958	8·1172478	8·1267856	8·1361183	58
3	8·0663188	8·0770290	8·0874813	8·0976879	8·1076601	8·1174085	8·1269428	8·1362722	57
4	8·0664995	8·0772052	8·0876534	8·0978560	8·1078244	8·1175691	8·1270999	8·1364260	56
5	8·0666801	8·0773815	8·0878254	8·0980240	8·1079886	8·1177297	8·1272570	8·1365797	55
6	8·0668606	8·0775576	8·0879974	8·0981920	8·1081528	8·1178902	8·1274140	8·1367334	54
7	8·0670411	8·0777337	8·0881692	8·0983599	8·1083169	8·1180507	8·1275710	8·1368871	53
8	8·0672215	8·0779097	8·0883411	8·0985277	8·1084809	8·1182111	8·1277279	8·1370407	52
9	8·0674018	8·0780856	8·0885128	8·0986955	8·1086449	8·1183714	8·1278848	8·1371942	51
10	8·0675820	8·0782614	8·0886845	8·0988632	8·1088088	8·1185317	8·1280416	8·1373477	50
11	8·0677622	8·0784372	8·0888561	8·0990309	8·1089726	8·1186919	8·1281983	8·1375011	49
12	8·0679423	8·0786129	8·0890277	8·0991984	8·1091364	8·1188520	8·1283550	8·1376545	48
13	8·0681223	8·0787886	8·0891991	8·0993659	8·1093001	8·1190121	8·1285117	8·1378078	47
14	8·0683022	8·0789641	8·0893706	8·0995334	8·1094638	8·1191722	8·1286682	8·1379610	46
15	8·0684821	8·0791396	8·0895419	8·0997008	8·1096274	8·1193322	8·1288248	8·1381143	45
16	8·0686619	8·0793151	8·0897132	8·0998681	8·1097909	8·1194921	8·1289812	8·1382674	44
17	8·0688416	8·0794904	8·0898844	8·1000353	8·1099544	8·1196519	8·1291376	8·1384205	43
18	8·0690212	8·0796657	8·0900555	8·1002025	8·1101178	8·1198118	8·1292940	8·1385736	42
19	8·0692008	8·0798409	8·0902266	8·1003697	8·1102812	8·1199715	8·1294503	8·1387265	41
20	8·0693803	8·0800161	8·0903976	8·1005367	8·1104445	8·1201312	8·1296065	8·1388795	40
21	8·0695597	8·0801912	8·0905685	8·1007037	8·1106077	8·1202908	8·1297627	8·1390324	39
22	8·0697390	8·0803662	8·0907394	8·1008706	8·1107709	8·1204504	8·1299188	8·1391852	38
23	8·0699183	8·0805411	8·0909102	8·1010375	8·1109340	8·1206099	8·1300749	8·1393380	37
24	8·0700975	8·0807160	8·0910810	8·1012043	8·1110970	8·1207693	8·1302309	8·1394907	36
25	8·0702766	8·0808908	8·0912516	8·1013710	8·1112600	8·1209287	8·1303869	8·1396434	35
26	8·0704557	8·0810655	8·0914222	8·1015377	8·1114229	8·1210881	8·1305428	8·1397960	34
27	8·0706346	8·0812401	8·0915928	8·1017043	8·1115858	8·1212474	8·1306986	8·1399485	33
28	8·0708135	8·0814147	8·0917632	8·1018709	8·1117486	8·1214066	8·1308544	8·1401011	32
29	8·0709923	8·0815892	8·0919336	8·1020374	8·1119113	8·1215657	8·1310101	8·1402535	31
30	8·0711711	8·0817637	8·0921040	8·1022038	8·1120740	8·1217248	8·1311658	8·1404059	30
31	8·0713498	8·0819380	8·0922743	8·1023701	8·1122366	8·1218839	8·1313215	8·1405583	29
32	8·0715284	8·0821123	8·0924445	8·1025364	8·1123992	8·1220429	8·1314770	8·1407105	28
33	8·0717069	8·0822866	8·0926146	8·1027027	8·1125617	8·1222018	8·1316325	8·1408628	27
34	8·0718854	8·0824607	8·0927847	8·1028688	8·1127241	8·1223607	8·1317880	8·1410150	26
35	8·0720637	8·0826348	8·0929547	8·1030349	8·1128865	8·1225195	8·1319434	8·1411671	25
36	8·0722421	8·0828088	8·0931246	8·1032010	8·1130488	8·1226782	8·1320987	8·1413192	24
37	8·0724203	8·0829828	8·0932944	8·1033669	8·1132110	8·1228369	8·1322540	8·1414712	23
38	8·0725985	8·0831567	8·0934643	8·1035328	8·1133732	8·1229956	8·1324093	8·1416232	22
39	8·0727765	8·0833305	8·0936340	8·1036987	8·1135354	8·1231541	8·1325644	8·1417751	21
40	8·0729546	8·0835042	8·0938037	8·1038645	8·1136974	8·1233127	8·1327196	8·1419270	20
41	8·0731325	8·0836779	8·0939733	8·1040302	8·1138595	8·1234711	8·1328746	8·1420788	19
42	8·0733104	8·0838515	8·0941428	8·1041959	8·1140214	8·1236295	8·1330296	8·1422306	18
43	8·0734882	8·0840251	8·0943123	8·1043615	8·1141833	8·1237879	8·1331846	8·1423823	17
44	8·0736659	8·0841985	8·0944817	8·1045270	8·1143451	8·1239462	8·1333395	8·1425339	16
45	8·0738436	8·0843719	8·0946510	8·1046925	8·1145069	8·1241044	8·1334943	8·1426855	15
46	8·0740211	8·0845452	8·0948203	8·1048579	8·1146686	8·1242626	8·1336491	8·1428371	14
47	8·0741986	8·0847185	8·0949895	8·1050232	8·1148302	8·1244207	8·1338039	8·1429886	13
48	8·0743761	8·0848917	8·0951587	8·1051885	8·1149918	8·1245787	8·1339586	8·1431400	12
49	8·0745534	8·0850648	8·0953277	8·1053537	8·1151534	8·1247367	8·1341132	8·1432914	11
50	8·0747307	8·0852379	8·0954968	8·1055188	8·1153148	8·1248947	8·1342678	8·1434427	10
51	8·0749080	8·0854109	8·0956657	8·1056839	8·1154762	8·1250526	8·1344223	8·1435940	9
52	8·0750851	8·0855838	8·0958346	8·1058490	8·1156376	8·1252104	8·1345767	8·1437453	8
53	8·0752622	8·0857566	8·0960034	8·1060139	8·1157989	8·1253682	8·1347311	8·1438964	7
54	8·0754392	8·0859294	8·0961721	8·1061788	8·1159601	8·1255259	8·1348855	8·1440476	6
55	8·0756161	8·0861021	8·0963408	8·1063437	8·1161213	8·1256836	8·1350398	8·1441987	5
56	8·0757930	8·0862747	8·0965094	8·1065085	8·1162824	8·1258412	8·1351940	8·1443497	4
57	8·0759698	8·0864473	8·0966780	8·1066732	8·1164434	8·1259987	8·1353482	8·1445006	3
58	8·0761465	8·0866198	8·0968465	8·1068378	8·1166044	8·1261562	8·1355023	8·1446516	2
59	8·0763231	8·0867922	8·0970149	8·1070024	8·1167654	8·1263136	8·1356564	8·1448024	1
60	8·0764997	8·0869646	8·0971832	8·1071669	8·1169262	8·1264710	8·1358104	8·1449532	0
"	19'	18'	17'	16'	15'	14'	13'	12'	"

"	40'	41'	42'	43'	44'	45'	46'	47'	"
0	8·0658057	8·0765306	8·0869970	8·0972172	8·1072025	8·1169634	8·1265099	8·1358510	60
1	8·0659866	8·0767071	8·0871693	8·0973855	8·1073670	8·1171243	8·1266672	8·1360050	59
2	8·0661675	8·0768835	8·0873416	8·0975538	8·1075314	8·1172851	8·1268245	8·1361590	58
3	8·0663483	8·0770599	8·0875138	8·0977220	8·1076958	8·1174458	8·1269817	8·1363129	57
4	8·0665290	8·0772362	8·0876859	8·0978901	8·1078601	8·1176064	8·1271389	8·1364667	56
5	8·0667096	8·0774125	8·0878579	8·0980582	8·1080243	8·1177670	8·1272960	8·1366205	55
6	8·0668902	8·0775886	8·0880299	8·0982261	8·1081885	8·1179276	8·1274531	8·1367742	54
7	8·0670707	8·0777647	8·0882018	8·0983941	8·1083526	8·1180881	8·1276101	8·1369279	53
8	8·0672511	8·0779407	8·0883737	8·0985619	8·1085167	8·1182485	8·1277670	8·1370815	52
9	8·0674314	8·0781167	8·0885455	8·0987297	8·1086807	8·1184088	8·1279239	8·1372350	51
10	8·0676117	8·0782926	8·0887172	8·0988975	8·1088446	8·1185691	8·1280807	8·1373886	50
11	8·0677919	8·0784684	8·0888888	8·0990651	8·1090085	8·1187294	8·1282375	8·1375420	49
12	8·0679720	8·0786441	8·0890604	8·0992327	8·1091723	8·1188896	8·1283942	8·1376954	48
13	8·0681520	8·0788198	8·0892319	8·0994003	8·1093361	8·1190497	8·1285509	8·1378488	47
14	8·0683320	8·0789954	8·0894033	8·0995677	8·1094998	8·1192098	8·1287075	8·1380020	46
15	8·0685118	8·0791709	8·0895747	8·0997351	8·1096634	8·1193698	8·1288641	8·1381553	45
16	8·0686917	8·0793464	8·0897460	8·0999025	8·1098269	8·1195297	8·1290206	8·1383085	44
17	8·0688714	8·0795218	8·0899172	8·1000698	8·1099903	8·1196896	8·1291770	8·1384616	43
18	8·0690511	8·0796971	8·0900884	8·1002370	8·1101539	8·1198495	8·1293334	8·1386147	42
19	8·0692306	8·0798723	8·0902595	8·1004041	8·1103173	8·1200092	8·1294897	8·1387677	41
20	8·0694102	8·0800475	8·0904305	8·1005712	8·1104806	8·1201689	8·1296460	8·1389207	40
21	8·0695896	8·0802226	8·0906015	8·1007382	8·1106438	8·1203286	8·1298022	8·1390736	39
22	8·0697690	8·0803976	8·0907724	8·1009052	8·1108070	8·1204882	8·1299583	8·1392204	38
23	8·0699483	8·0805726	8·0909432	8·1010721	8·1109702	8·1206477	8·1301144	8·1393792	37
24	8·0701275	8·0807475	8·0911140	8·1012389	8·1111332	8·1208072	8·1302705	8·1395320	36
25	8·0703066	8·0809223	8·0912847	8·1014057	8·1112962	8·1209666	8·1304265	8·1396847	35
26	8·0704857	8·0810970	8·0914553	8·1015724	8·1114592	8·1211260	8·1305824	8·1398373	34
27	8·0706647	8·0812717	8·0916259	8·1017390	8·1116221	8·1212853	8·1307383	8·1399809	33
28	8·0708436	8·0814463	8·0917964	8·1019056	8·1117849	8·1214446	8·1308941	8·1401425	32
29	8·0710225	8·0816208	8·0919668	8·1020721	8·1119477	8·1216037	8·1310498	8·1402949	31
30	8·0712012	8·0817953	8·0921372	8·1022386	8·1121104	8·1217629	8·1312056	8·1404474	30
31	8·0713799	8·0819697	8·0923075	8·1024049	8·1122730	8·1219219	8·1313612	8·1405997	29
32	8·0715586	8·0821440	8·0924777	8·1025713	8·1124356	8·1220810	8·1315168	8·1407521	28
33	8·0717371	8·0823183	8·0926479	8·1027375	8·1125981	8·1222399	8·1316723	8·1409043	27
34	8·0719156	8·0824925	8·0928180	8·1029037	8·1127606	8·1223988	8·1318278	8·1410566	26
35	8·0720940	8·0826666	8·0929880	8·1030698	8·1129230	8·1225577	8·1319833	8·1412087	25
36	8·0722723	8·0828406	8·0931579	8·1032359	8·1130853	8·1227164	8·1321386	8·1413608	24
37	8·0724506	8·0830146	8·0933278	8·1034019	8·1132476	8·1228752	8·1322940	8·1415129	23
38	8·0726288	8·0831885	8·0934977	8·1035678	8·1134098	8·1230338	8·1324492	8·1416649	22
39	8·0728069	8·0833624	8·0936674	8·1037337	8·1135720	8·1231924	8·1326044	8·1418168	21
40	8·0729850	8·0835361	8·0938371	8·1038995	8·1137341	8·1233510	8·1327596	8·1419687	20
41	8·0731629	8·0837098	8·0940068	8·1040653	8·1138961	8·1235095	8·1329147	8·1421206	19
42	8·0733408	8·0838835	8·0941763	8·1042309	8·1140581	8·1236679	8·1330697	8·1422724	18
43	8·0735186	8·0840570	8·0943458	8·1043966	8·1142200	8·1238263	8·1332247	8·1424241	17
44	8·0736964	8·0842305	8·0945153	8·1045621	8·1143819	8·1239846	8·1333796	8·1425758	16
45	8·0738741	8·0844039	8·0946846	8·1047276	8·1145437	8·1241429	8·1335345	8·1427274	15
46	8·0740517	8·0845773	8·0948539	8·1048931	8·1147054	8·1243011	8·1336893	8·1428790	14
47	8·0742292	8·0847506	8·0950232	8·1050584	8·1148671	8·1244592	8·1338441	8·1430305	13
48	8·0744067	8·0849238	8·0951923	8·1052237	8·1150287	8·1246173	8·1339988	8·1431820	12
49	8·0745841	8·0850969	8·0953614	8·1053890	8·1151903	8·1247753	8·1341535	8·1433334	11
50	8·0747614	8·0852700	8·0955305	8·1055542	8·1153518	8·1249333	8·1343081	8·1434848	10
51	8·0749386	8·0854430	8·0956994	8·1057193	8·1155132	8·1250912	8·1344626	8·1436361	9
52	8·0751158	8·0856160	8·0958683	8·1058843	8·1156746	8·1252491	8·1346171	8·1437874	8
53	8·0752929	8·0857888	8·0960372	8·1060493	8·1158359	8·1254069	8·1347715	8·1439386	7
54	8·0754699	8·0859616	8·0962060	8·1062142	8·1159972	8·1255646	8·1349259	8·1440897	6
55	8·0756469	8·0861344	8·0963747	8·1063791	8·1161584	8·1257223	8·1350802	8·1442408	5
56	8·0758238	8·0863070	8·0965433	8·1065439	8·1163195	8·1258799	8·1352345	8·1443919	4
57	8·0760006	8·0864796	8·0967119	8·1067087	8·1164806	8·1260375	8·1353887	8·1445429	3
58	8·0761773	8·0866522	8·0968804	8·1068733	8·1166416	8·1261950	8·1355429	8·1446938	2
59	8·0763540	8·0868246	8·0970488	8·1070380	8·1168025	8·1263525	8·1356970	8·1448447	1
60	8·0765306	8·0869970	8·0972172	8·1072025	8·1169634	8·1265099	8·1358510	8·1449956	0
"	19'	18'	17'	16'	15'	14'	13'	12'	"

"	48'	49'	50'	51'	52'	53'	54'	55'	"
0	8·1449532	8·1539075	8·1626808	8·1712804	8·1797129	8·1879848	8·1961020	8·2040703	60
1	8·1451040	8·1540552	8·1628255	8·1714223	8·1798521	8·1881213	8·1962360	8·2042019	59
2	8·1452547	8·1542028	8·1629702	8·1715641	8·1799912	8·1882578	8·1963700	8·2043334	58
3	8·1454054	8·1543504	8·1631149	8·1717059	8·1801303	8·1883943	8·1965039	8·2044649	57
4	8·1455560	8·1544979	8·1632594	8·1718477	8·1802693	8·1885307	8·1966378	8·2045963	56
5	8·1457065	8·1546454	8·1634040	8·1719894	8·1804083	8·1886670	8·1967717	8·2047277	55
6	8·1458570	8·1547928	8·1635485	8·1721310	8·1805472	8·1888034	8·1969055	8·2048591	54
7	8·1460075	8·1549402	8·1636929	8·1722726	8·1806861	8·1889397	8·1970392	8·2049905	53
8	8·1461579	8·1550876	8·1638373	8·1724142	8·1808250	8·1890759	8·1971729	8·2051218	52
9	8·1463082	8·1552348	8·1639817	8·1725557	8·1809638	8·1892121	8·1973066	8·2052530	51
10	8·1464585	8·1553821	8·1641259	8·1726972	8·1811025	8·1893482	8·1974403	8·2053842	50
11	8·1466087	8·1555293	8·1642702	8·1728386	8·1812413	8·1894843	8·1975739	8·2055154	49
12	8·1467589	8·1556764	8·1644144	8·1729800	8·1813799	8·1896204	8·1977074	8·2056465	48
13	8·1469091	8·1558235	8·1645586	8·1731214	8·1815186	8·1897564	8·1978409	8·2057776	47
14	8·1470591	8·1559705	8·1647027	8·1732627	8·1816571	8·1898924	8·1979744	8·2059087	46
15	8·1472092	8·1561175	8·1648467	8·1734039	8·1817957	8·1900284	8·1981078	8·2060397	45
16	8·1473592	8·1562644	8·1649907	8·1735451	8·1819342	8·1901643	8·1982412	8·2061707	44
17	8·1475091	8·1564113	8·1651347	8·1736863	8·1820726	8·1903001	8·1983746	8·2063016	43
18	8·1476590	8·1565582	8·1652786	8·1738274	8·1822111	8·1904359	8·1985079	8·2064325	42
19	8·1478088	8·1567049	8·1654225	8·1739684	8·1823494	8·1905717	8·1986412	8·2065634	41
20	8·1479586	8·1568517	8·1655663	8·1741094	8·1824877	8·1907074	8·1987744	8·2066942	40
21	8·1481083	8·1569984	8·1657101	8·1742504	8·1826260	8·1908431	8·1989076	8·2068250	39
22	8·1482579	8·1571450	8·1658538	8·1743913	8·1827643	8·1909788	8·1990407	8·2069557	38
23	8·1484076	8·1572916	8·1659975	8·1745322	8·1829024	8·1911144	8·1991738	8·2070864	37
24	8·1485571	8·1574381	8·1661411	8·1746731	8·1830406	8·1912499	8·1993069	8·2072171	36
25	8·1487066	8·1575846	8·1662847	8·1748138	8·1831787	8·1913854	8·1994399	8·2073477	35
26	8·1488561	8·1577310	8·1664282	8·1749546	8·1833167	8·1915209	8·1995729	8·2074783	34
27	8·1490055	8·1578774	8·1665717	8·1750953	8·1834548	8·1916563	8·1997058	8·2076088	33
28	8·1491549	8·1580238	8·1667151	8·1752359	8·1835927	8·1917917	8·1998387	8·2077393	32
29	8·1493042	8·1581701	8·1668585	8·1753765	8·1837307	8·1919271	8·1999716	8·2078698	31
30	8·1494534	8·1583163	8·1670019	8·1755171	8·1838685	8·1920624	8·2001044	8·2080002	30
31	8·1496027	8·1584625	8·1671452	8·1756576	8·1840064	8·1921976	8·2002372	8·2081306	29
32	8·1497518	8·1586086	8·1672884	8·1757981	8·1841442	8·1923329	8·2003699	8·2082610	28
33	8·1499009	8·1587547	8·1674316	8·1759385	8·1842819	8·1924680	8·2005026	8·2083913	27
34	8·1500500	8·1589008	8·1675748	8·1760789	8·1844196	8·1926032	8·2006353	8·2085216	26
35	8·1501990	8·1590468	8·1677179	8·1762192	8·1845573	8·1927383	8·2007679	8·2086518	25
36	8·1503479	8·1591927	8·1678610	8·1763595	8·1846949	8·1928733	8·2009005	8·2087820	24
37	8·1504968	8·1593386	8·1680040	8·1764998	8·1848325	8·1930083	8·2010330	8·2089121	23
38	8·1506457	8·1594845	8·1681469	8·1766400	8·1849700	8·1931433	8·2011655	8·2090422	22
39	8·1507945	8·1596303	8·1682899	8·1767801	8·1851075	8·1932782	8·2012980	8·2091723	21
40	8·1509432	8·1597760	8·1684327	8·1769202	8·1852450	8·1934131	8·2014304	8·2093024	20
41	8·1510919	8·1599217	8·1685756	8·1770603	8·1853824	8·1935479	8·2015628	8·2094324	19
42	8·1512406	8·1600674	8·1687183	8·1772003	8·1855197	8·1936827	8·2016951	8·2095623	18
43	8·1513891	8·1602130	8·1688611	8·1773403	8·1856570	8·1938175	8·2018274	8·2096922	17
44	8·1515377	8·1603585	8·1690038	8·1774802	8·1857943	8·1939522	8·2019597	8·2098221	16
45	8·1516862	8·1605040	8·1691464	8·1776201	8·1859315	8·1940869	8·2020919	8·2099520	15
46	8·1518346	8·1606495	8·1692890	8·1777599	8·1860687	8·1942215	8·2022241	8·2100818	14
47	8·1519830	8·1607949	8·1694315	8·1778997	8·1862059	8·1943561	8·2023562	8·2102115	13
48	8·1521314	8·1609403	8·1695740	8·1780394	8·1863430	8·1944907	8·2024883	8·2103412	12
49	8·1522796	8·1610856	8·1697165	8·1781791	8·1864800	8·1946252	8·2026203	8·2104709	11
50	8·1524279	8·1612308	8·1698589	8·1783188	8·1866170	8·1947596	8·2027523	8·2106006	10
51	8·1525761	8·1613761	8·1700012	8·1784584	8·1867540	8·1948941	8·2028843	8·2107302	9
52	8·1527242	8·1615212	8·1701435	8·1785980	8·1868909	8·1950284	8·2030163	8·2108598	8
53	8·1528723	8·1616663	8·1702858	8·1787375	8·1870278	8·1951628	8·2031481	8·2109893	7
54	8·1530203	8·1618114	8·1704280	8·1788770	8·1871646	8·1952971	8·2032800	8·2111188	6
55	8·1531683	8·1619564	8·1705702	8·1790164	8·1873014	8·1954313	8·2034118	8·2112482	5
56	8·1533163	8·1621014	8·1707123	8·1791558	8·1874382	8·1955656	8·2035436	8·2113777	4
57	8·1534641	8·1622463	8·1708544	8·1792951	8·1875749	8·1956997	8·2036753	8·2115070	3
58	8·1536120	8·1623912	8·1709964	8·1794344	8·1877116	8·1958339	8·2038070	8·2116364	2
59	8·1537598	8·1625360	8·1711384	8·1795737	8·1878482	8·1959680	8·2039387	8·2117657	1
60	8·1539075	8·1626808	8·1712804	8·1797129	8·1879848	8·1961020	8·2040703	8·2118949	0
"	11'	10'	9'	8'	7'	6'	5'	4'	"

"	48'	49'	50'	51'	52'	53'	54'	55'	"
0	8·1449956	8·1539516	8·1627267	8·1713282	8·1797626	8·1880364	8·1961556	8·2041259	60
1	8·1451464	8·1540993	8·1628715	8·1714701	8·1799018	8·1881730	8·1962896	8·2042575	59
2	8·1452971	8·1542470	8·1630162	8·1716120	8·1800409	8·1883095	8·1964236	8·2043890	58
3	8·1454478	8·1543946	8·1631609	8·1717538	8·1801800	8·1884460	8·1965576	8·2045206	57
4	8·1455984	8·1545422	8·1633055	8·1718956	8·1803191	8·1885824	8·1966915	8·2046521	56
5	8·1457490	8·1546897	8·1634501	8·1720373	8·1804581	8·1887188	8·1968254	8·2047835	55
6	8·1458995	8·1548371	8·1635946	8·1721790	8·1805971	8·1888552	8·1969592	8·2049149	54
7	8·1460500	8·1549846	8·1637391	8·1723207	8·1807360	8·1889915	8·1970930	8·2050463	53
8	8·1462004	8·1551319	8·1638835	8·1724623	8·1808749	8·1891278	8·1972268	8·2051776	52
9	8·1463508	8·1552792	8·1640279	8·1726038	8·1810137	8·1892640	8·1973605	8·2053080	51
10	8·1465011	8·1554265	8·1641722	8·1727453	8·1811525	8·1894002	8·1974942	8·2054401	50
11	8·1466514	8·1555737	8·1643165	8·1728868	8·1812913	8·1895363	8·1976278	8·2055714	49
12	8·1468016	8·1557209	8·1644607	8·1730282	8·1814300	8·1896724	8·1977614	8·2057025	48
13	8·1469518	8·1558680	8·1646049	8·1731696	8·1815687	8·1898085	8·1978949	8·2058337	47
14	8·1471019	8·1560151	8·1647490	8·1733109	8·1817073	8·1899445	8·1980284	8·2059647	46
15	8·1472520	8·1561621	8·1648931	8·1734522	8·1818459	8·1900805	8·1981619	8·2060958	45
16	8·1474020	8·1563090	8·1650372	8·1735934	8·1819844	8·1902164	8·1982953	8·2062268	44
17	8·1475519	8·1564559	8·1651812	8·1737346	8·1821229	8·1903523	8·1984287	8·2063578	43
18	8·1477018	8·1566028	8·1653251	8·1738757	8·1822613	8·1904881	8·1985621	8·2064887	42
19	8·1478517	8·1567496	8·1654690	8·1740168	8·1823997	8·1906239	8·1986954	8·2066196	41
20	8·1480015	8·1568964	8·1656128	8·1741579	8·1825381	8·1907597	8·1988286	8·2067505	40
21	8·1481512	8·1570431	8·1657566	8·1742989	8·1826764	8·1908954	8·1989619	8·2068813	39
22	8·1483009	8·1571898	8·1659004	8·1744398	8·1828146	8·1910311	8·1990950	8·2070120	38
23	8·1484506	8·1573364	8·1660441	8·1745807	8·1829529	8·1911667	8·1992282	8·2071428	37
24	8·1486002	8·1574830	8·1661878	8·1747216	8·1830910	8·1913023	8·1993613	8·2072735	36
25	8·1487497	8·1576295	8·1663314	8·1748624	8·1832292	8·1914379	8 1994943	8·2074041	35
26	8·1488992	8·1577759	8·1664749	8·1750032	8·1833673	8·1915734	8·1996273	8·2075348	34
27	8·1490487	8·1579224	8·1666185	8·1751439	8·1835053	8·1917088	8·1997603	8·2076653	33
28	8·1491980	8·1580687	8·1667619	8·1752846	8·1836433	8·1918442	8·1998933	8·2077959	32
29	8·1493474	8·1582151	8·1669054	8·1754252	8·1837813	8·1919796	8·2000262	8·2079264	31
30	8·1494967	8·1583613	8·1670487	8·1755658	8·1839192	8·1921150	8·2001590	8·2080568	30
31	8·1496459	8·1585076	8·1671921	8·1757064	8·1840571	8·1922503	8·2002918	8·2081873	29
32	8·1497951	8·1586537	8·1673353	8·1758469	8·1841949	8·1923855	8·2004246	8·2083176	28
33	8·1499442	8·1587999	8·1674786	8·1759873	8·1843327	8·1925207	8·2005573	8·2084480	27
34	8·1500933	8·1589459	8·1676218	8·1761278	8·1844704	8·1926559	8·2006900	8·2085783	26
35	8·1502423	8·1590920	8·1677649	8·1762681	8·1846081	8·1927910	8·2008227	8·2087086	25
36	8.1503913	8·1592379	8·1679080	8·1764084	8·1847458	8·1929261	8·2009553	8·2088388	24
37	8·1505402	8·1593839	8·1680510	8·1765487	8·1848834	8·1930611	8·2010879	8·2089690	23
38	8·1506891	8·1595297	8·1681940	8·1766889	8·1850209	8·1931961	8·2012204	8·2090901	22
39	8·1508380	8·1596756	8·1683370	8·1768291	8·1851585	8·1933311	8·2013529	8·2092092	21
40	8·1509867	8·1598213	8·1684799	8·1769693	8·1852959	8·1934660	8·2014853	8·2093593	20
41	8·1511355	8·1599671	8·1686228	8·1771094	8·1854334	8·1936009	8·2016177	8·2094893	19
42	8·1512841	8 1601128	8·1687656	8·1772494	8·1855708	8·1937357	8·2017501	8·2096193	18
43	8·1514328	8·1602584	8·1689083	8·1773894	8·1857081	8·1938705	8·2018824	8·2097493	17
44	8·1515813	8·1604040	8·1690510	8·1775294	8·1858454	8·1940053	8·2020147	8·2098792	16
45	8·1517299	8·1605495	8·1691937	8·1776693	8·1859827	8·1941400	8·2021470	8·2100091	15
46	8·1518783	8·1606950	8·1693363	8·1778091	8·1861199	8·1942746	8·2022792	8·2101389	14
47	8·1520267	8·1608404	8·1694789	8·1779490	8·1862571	8·1944093	8·2024113	8·2102687	13
48	8·1521751	8·1609858	8·1696214	8·1780887	8·1863942	8·1945439	8·2025435	8·2103985	12
49	8·1523234	8·1611312	8·1697639	8·1782285	8·1865313	8·1946784	8·2026756	8·2105282	11
50	8·1524717	8·1612765	8·1699064	8·1783682	8·1866683	8·1948129	8·2028076	8·2106579	10
51	8·1526199	8·1614217	8·1700487	8·1785078	8·1868053	8·1949473	8·2029396	8·2107875	9
52	8·1527681	8·1615669	8·1701911	8·1786474	8·1869423	8·1950818	8·2030716	8·2109171	8
53	8·1529162	8·1617121	8·1703334	8·1787870	8·1870792	8·1952161	8·2032035	8·2110467	7
54	8·1530643	8·1618572	8·1704756	8·1789265	8·1872161	8·1953505	8·2033354	8·2111762	6
55	8·1532123	8·1620022	8·1706178	8·1790659	8·1873529	8·1954848	8·2034672	8·2113057	5
56	8·1533603	8·1621472	8·1707600	8·1792054	8·1874897	8·1956190	8·2035990	8·2114351	4
57	8·1535082	8·1622922	8·1709021	8·1793447	8·1876264	8·1957532	8·2037308	8·2115646	3
58	8·1536560	8·1624371	8·1710442	8·1794841	8·1877631	8·1958874	8·2038625	8·2116939	2
59	8·1538038	8·1625819	8·1711862	8·1796233	8·1878998	8·1960215	8·2039942	8·2118233	1
60	8·1539516	8·1627267	8·1713282	8·1797626	8·1880364	8·1961556	8·2041259	8·2119526	0
"	11'	10'	9'	8'	7'	6'	5'	4'	"

"	56'	57'	58'	59'	0'	1'	2'	3'	"
0	8·2118949	8·2195811	8·2271335	8·2345568	8·2418553	8·2490332	8·2560943	8·2630424	60
1	8·2120242	8·2197080	8·2272583	8·2346795	8·2419759	8·2491518	8·2562110	8·2631572	59
2	8·2121533	8·2198349	8·2273830	8·2348021	8·2420965	8·2492704	8·2563277	8·2632721	58
3	8·2122825	8·2199618	8·2275077	8·2349247	8·2422170	8·2493890	8·2564443	8·2633869	57
4	8·2124116	8·2200887	8·2276324	8·2350472	8·2423376	8·2495075	8·2565609	8·2635016	56
5	8·2125407	8·2202155	8·2277570	8·2351697	8·2424580	8·2496260	8·2566775	8·2636164	55
6	8·2126697	8·2203423	8·2278816	8·2352922	8·2425785	8·2497445	8·2567941	8·2637311	54
7	8·2127987	8·2204690	8·2280061	8·2354147	8·2426989	8·2498629	8·2569106	8·2638458	53
8	8·2129277	8·2205957	8·2281306	8·2355371	8·2428192	8·2499813	8·2570271	8·2639604	52
9	8·2130566	8·2207223	8·2282551	8·2356594	8·2429396	8·2500997	8·2571436	8·2640750	51
10	8·2131854	8·2208490	8·2283796	8·2357818	8·2430599	8·2502180	8·2572600	8·2641896	50
11	8·2133143	8·2209756	8·2285040	8·2359041	8·2431802	8·2503363	8·2573764	8·2643042	49
12	8·2134431	8·2211021	8·2286284	8·2360264	8·2433004	8·2504546	8·2574928	8·2644187	48
13	8·2135719	8·2212286	8·2287527	8·2361486	8·2434206	8·2505728	8·2576091	8·2645332	47
14	8·2137006	8·2213551	8·2288770	8·2362708	8·2435408	8·2506911	8·2577255	8·2646477	46
15	8·2138293	8·2214815	8·2290013	8·2363930	8·2436609	8·2508092	8·2578417	8·2647621	45
16	8·2139579	8·2216079	8·2291255	8·2365151	8·2437810	8·2509274	8·2579580	8·2648766	44
17	8·2140865	8·2217343	8·2292497	8·2366372	8·2439011	8·2510455	8·2580742	8·2649909	43
18	8·2142151	8·2218606	8·2293739	8·2367593	8·2440212	8·2511636	8·2581904	8·2651053	42
19	8·2143436	8·2219869	8·2294980	8·2368813	8·2441412	8·2512816	8·2583065	8·2652196	41
20	8·2144721	8·2221132	8·2296221	8·2370033	8·2442611	8·2513996	8·2584227	8·2653339	40
21	8·2146006	8·2222394	8·2297461	8·2371253	8·2443811	8·2515176	8·2585388	8·2654482	39
22	8·2147290	8·2223656	8·2298701	8·2372472	8·2445010	8·2516356	8·2586648	8·2655624	38
23	8·2148574	8·2224917	8·2299941	8·2373691	8·2446209	8·2517535	8·2587709	8·2656766	37
24	8·2149857	8·2226178	8·2301181	8·2374910	8·2447407	8·2518714	8·2588869	8·2657908	36
25	8·2151140	8·2227439	8·2302420	8·2376128	8·2448605	8·2519893	8·2590028	8·2659049	35
26	8·2152423	8·2228699	8·2303659	8·2377346	8·2449803	8·2521071	8·2591188	8·2660190	34
27	8·2153705	8·2229959	8·2304897	8·2378563	8·2451000	8·2522249	8·2592347	8·2661331	33
28	8·2154987	8·2231219	8·2306135	8·2379781	8·2452198	8·2523426	8·2593505	8·2662471	32
29	8·2156269	8·2232478	8·2307373	8·2380997	8·2453394	8·2524604	8·2594664	8·2663612	31
30	8·2157550	8·2233737	8·2308610	8·2382214	8·2454591	8·2525781	8·2595822	8·2664751	30
31	8·2158831	8·2234996	8·2309847	8·2383430	8·2455787	8·2526957	8·2596980	8·2665891	29
32	8·2160111	8·2236254	8·2311084	8·2384646	8·2456983	8·2528134	8·2598137	8·2667030	28
33	8·2161391	8·2237512	8·2312320	8·2385862	8·2458178	8·2529310	8·2599295	8·2668169	27
34	8·2162671	8·2238769	8·2313556	8·2387077	8·2459373	8·2530485	8·2600452	8·2669308	26
35	8·2163950	8·2240026	8·2314792	8·2388292	8·2460568	8·2531661	8·2601608	8·2670446	25
36	8·2165229	8·2241283	8·2316027	8·2389506	8·2461762	8·2532836	8·2602764	8·2671585	24
37	8·2166508	8·2242539	8·2317262	8·2390720	8·2462957	8·2534011	8·2603920	8·2672722	23
38	8·2167786	8·2243795	8·2318496	8·2391934	8·2464150	8·2535185	8·2605076	8·2673860	22
39	8·2169064	8·2245051	8·2319731	8·2393148	8·2465344	8·2536359	8·2606232	8·2674997	21
40	8·2170341	8·2246306	8·2320965	8·2394361	8·2466537	8·2537533	8·2607387	8·2676134	20
41	8·2171618	8·2247561	8·2322198	8·2395574	8·2467730	8·2538706	8·2608541	8·2677271	19
42	8·2172895	8·2248815	8·2323431	8·2396786	8·2468922	8·2539880	8·2609696	8·2678407	18
43	8·2174171	8·2250070	8·2324664	8·2397998	8·2470115	8·2541052	8·2610850	8·2679543	17
44	8·2175447	8·2251323	8·2325896	8·2399210	8·2471306	8·2542225	8·2612004	8·2680679	16
45	8·2176723	8·2252577	8·2327128	8·2400422	8·2472498	8·2543397	8·2613157	8·2681814	15
46	8·2177998	8·2253830	8·2328360	8·2401633	8·2473689	8·2544569	8·2614311	8·2682949	14
47	8·2179273	8·2255083	8·2329592	8·2402844	8·2474880	8·2545741	8·2615463	8·2684084	13
48	8·2180547	8·2256335	8·2330823	8·2404054	8·2476071	8·2546912	8·2616616	8·2685219	12
49	8·2181821	8·2257587	8·2332053	8·2405264	8·2477261	8·2548083	8·2617768	8·2686353	11
50	8·2183095	8·2258839	8·2333284	8·2406474	8·2478451	8·2549254	8·2618920	8·2687487	10
51	8·2184368	8·2260090	8·2334514	8·2407683	8·2479640	8·2550424	8·2620072	8·2688620	9
52	8·2185641	8·2261341	8·2335743	8·2408892	8·2480829	8·2551594	8·2621223	8·2689754	8
53	8·2186913	8·2262591	8·2336973	8·2410101	8·2482018	8·2552764	8·2622375	8·2690887	7
54	8·2188186	8·2263841	8·2338202	8·2411310	8·2483207	8·2553933	8·2623525	8·2692020	6
55	8·2189457	8·2265091	8·2339430	8·2412518	8·2484395	8·2555102	8·2624676	8·2693152	5
56	8·2190729	8·2266341	8·2340659	8·2413725	8·2485583	8·2556271	8·2625826	8·2694284	4
57	8·2192000	8·2267590	8·2341886	8·2414933	8·2486771	8·2557439	8·2626976	8·2695416	3
58	8·2193270	8·2268839	8·2343114	8·2416140	8·2487958	8·2558607	8·2628125	8·2696548	2
59	8·2194541	8·2270087	8·2344341	8·2417347	8·2489145	8·2559775	8·2629275	8·2697679	1
60	8·2195811	8·2271335	8·2345568	8·2418553	8·2490332	8·2560943	8·2630424	8·2698810	0
"	3'	2'	1'	0'	59'	58'	57'	56'	"

"	56′	57′	58′	59′	0′	1′	2′	3′	"
0	8·2119526	8·2196408	8·2271953	8·2346208	8·2419215	8·2491015	8·2561649	8·2631153	60
1	8·2120818	8·2197678	8·2273201	8·2347435	8·2420421	8·2492202	8·2562817	8·2632302	59
2	8·2122110	8·2198947	8·2274449	8·2348661	8·2421627	8·2493388	8·2563984	8·2633451	58
3	8·2123402	8·2200216	8·2275696	8·2349887	8·2422833	8·2494574	8·2565151	8·2634599	57
4	8·2124694	8·2201485	8·2276943	8·2351113	8·2424038	8·2495760	8·2566317	8·2635747	56
5	8·2125985	8·2202754	8·2278190	8·2352339	8·2425244	8·2496946	8·2567484	8·2636895	55
6	8·2127275	8·2204022	8·2279436	8·2353564	8·2426448	8·2498131	8·2568650	8·2638043	54
7	8·2128566	8·2205289	8·2280682	8·2354789	8·2427653	8·2499315	8·2569815	8·2639190	53
8	8·2129855	8·2206557	8·2281927	8·2356013	8·2428857	8·2500500	8·2570981	8·2640337	52
9	8·2131145	8·2207824	8·2283173	8·2357237	8·2430061	8·2501684	8·2572146	8·2641483	51
10	8·2132434	8·2209090	8·2284417	8·2358461	8·2431264	8·2502868	8·2573310	8·2642630	50
11	8·2133723	8·2210356	8·2285662	8·2359684	8·2432467	8·2504051	8·2574475	8·2643776	49
12	8·2135011	8·2211622	8·2286906	8·2360908	8·2433670	8·2505234	8·2575639	8·2644921	48
13	8·2136299	8·2212888	8·2288150	8·2362130	8·2434872	8·2506417	8·2576803	8·2646067	47
14	8·2137587	8·2214153	8·2289393	8·2363353	8·2436075	8·2507600	8·2577966	8·2647212	46
15	8·2138874	8·2215418	8·2290636	8·2364575	8·2437276	8·2508782	8·2579129	8·2648357	45
16	8·2140161	8·2216682	8·2291879	8·2365796	8·2438478	8·2509964	8·2580292	8·2649501	44
17	8·2141447	8·2217946	8·2293121	8·2367018	8·2439679	8·2511145	8·2581455	8·2650645	43
18	8·2142733	8·2219210	8·2294363	8·2368239	8·2440880	8·2512326	8·2582617	8·2651789	42
19	8·2144019	8·2220473	8·2295605	8·2369460	8·2442080	8·2513507	8·2583779	8·2652933	41
20	8·2145304	8·2221736	8·2296846	8·2370680	8·2443280	8·2514688	8·2584941	8·2654076	40
21	8·2146589	8·2222998	8·2298087	8·2371900	8·2444480	8·2515868	8·2586102	8·2655219	39
22	8·2147874	8·2224260	8·2299327	8·2373120	8·2445680	8·2517048	8·2587263	8·2656362	38
23	8·2149158	8·2225522	8·2300568	8·2374339	8·2446879	8·2518227	8·2588424	8·2657504	37
24	8·2150442	8·2226784	8·2301807	8·2375558	8·2448077	8·2519407	8·2589584	8·2658646	36
25	8·2151725	8·2228045	8·2303047	8·2376776	8·2449276	8·2520586	8·2590744	8·2659788	35
26	8·2153008	8·2229305	8·2304286	8·2377995	8·2450474	8·2521764	8·2591904	8·2660929	34
27	8·2154291	8·2230566	8·2305525	8·2379213	8·2451672	8·2522943	8·2593063	8·2662071	33
28	8·2155573	8·2231826	8·2306763	8·2380430	8·2452869	8·2524121	8·2594223	8·2663212	32
29	8·2156855	8·2233085	8·2308001	8·2381648	8·2454066	8·2525298	8·2595381	8·2664352	31
30	8·2158137	8·2234345	8·2309239	8·2382865	8·2455263	8·2526476	8·2596540	8·2665492	30
31	8·2159418	8·2235604	8·2310476	8·2384081	8·2456460	8·2527653	8·2597698	8·2666632	29
32	8·2160699	8·2236862	8·2311713	8·2385297	8·2457656	8·2528829	8·2598856	8·2667772	28
33	8·2161979	8·2238120	8·2312950	8·2386513	8·2458852	8·2530006	8·2600014	8·2668911	27
34	8·2163259	8·2239378	8·2314186	8·2387729	8·2460047	8·2531182	8·2601171	8·2670051	26
35	8·2164539	8·2240635	8·2315422	8·2388944	8·2461242	8·2532358	8·2602328	8·2671189	25
36	8·2165818	8·2241892	8·2316658	8·2390159	8·2462437	8·2533533	8·2603485	8·2672328	24
37	8·2167097	8·2243149	8·2317893	8·2391373	8·2463632	8·2534708	8·2604641	8·2673466	23
38	8·2168375	8·2244405	8·2319128	8·2392588	8·2464826	8·2535883	8·2605797	8·2674604	22
39	8·2169653	8·2245661	8·2320363	8·2393802	8·2466020	8·2537058	8·2606953	8·2675742	21
40	8·2170931	8·2246917	8·2321597	8·2395015	8·2467213	8·2538232	8·2608108	8·2676879	20
41	8·2172209	8·2248172	8·2322831	8·2396228	8·2468407	8·2539406	8·2609263	8·2678016	19
42	8·2173486	8·2249427	8·2324064	8·2397441	8·2469599	8·2540579	8·2610418	8·2679153	18
43	8·2174762	8·2250682	8·2325297	8·2398654	8·2470792	8·2541752	8·2611573	8·2680290	17
44	8·2176038	8·2251936	8·2326530	8·2399866	8·2471984	8·2542925	8·2612727	8·2681425	16
45	8·2177314	8·2253190	8·2327763	8·2401078	8·2473176	8·2544098	8·2613881	8·2682561	15
46	8·2178590	8·2254443	8·2328995	8·2402289	8·2474368	8·2545270	8·2615034	8·2683696	14
47	8·2179865	8·2255696	8·2330227	8·2403500	8·2475559	8·2546442	8·2616188	8·2684832	13
48	8·2181140	8·2256949	8·2331458	8·2404711	8·2476750	8·2547614	8·2617341	8·2685967	12
49	8·2182414	8·2258201	8·2332689	8·2405922	8·2477940	8·2548785	8·2618493	8·2687101	11
50	8·2183688	8·2259453	8·2333920	8·2407132	8·2479131	8·2549956	8·2619646	8·2688236	10
51	8·2184962	8·2260705	8·2335150	8·2408342	8·2480321	8·2551127	8·2620798	8·2689370	9
52	8·2186235	8·2261956	8·2336380	8·2409551	8·2481510	8·2552297	8·2621950	8·2690503	8
53	8·2187508	8·2263207	8·2337610	8·2410760	8·2482699	8·2553467	8·2623101	8·2691637	7
54	8·2188780	8·2264457	8·2338839	8·2411969	8·2483888	8·2554637	8·2624252	8·2692770	6
55	8·2190053	8·2265708	8·2340068	8·2413177	8·2485077	8·2555806	8·2625403	8·2693903	5
56	8·2191324	8·2266957	8·2341297	8·2414386	8·2486265	8·2556976	8·2626554	8·2695035	4
57	8·2192596	8·2268207	8·2342525	8·2415593	8·2487453	8·2558144	8·2627704	8·2696168	3
58	8·2193867	8·2269456	8·2343753	8·2416801	8·2488641	8·2559313	8·2628854	8·2697300	2
59	8·2195137	8·2270705	8·2344980	8·2418008	8·2489828	8·2560481	8·2630004	8·2698431	1
60	8·2196408	8·2271953	8·2346208	8·2419215	8·2491015	8·2561649	8·2631153	8·2699563	0
"	3′	2′	1′	0′	59′	58′	57′	56′	"

"	4'	5'	6'	7'	8'	9'	10'	11'	"
0	8·2698810	8·2766136	8·2832434	8·2897734	8·2962067	8·3025460	8·3087941	8·3149536	60
1	8·2699941	8·2767249	8·2833530	8·2898814	8·2963131	8·3026509	8·3088975	8·3150555	59
2	8·2701071	8·2768362	8·2834626	8·2899894	8·2964195	8·3027558	8·3090009	8·3151574	58
3	8·2702201	8·2769475	8·2835722	8·2900974	8·2965259	8·3028606	8·3091042	8·3152593	57
4	8·2703331	8·2770587	8·2836818	8·2902053	8·2966322	8·3029654	8·3092075	8·3153611	56
5	8·2704461	8·2771700	8·2837913	8·2903132	8·2967385	8·3030702	8·3093108	8·3154630	55
6	8·2705590	8·2772811	8·2839008	8·2904211	8·2968448	8·3031749	8·3094140	8·3155648	54
7	8·2706719	8·2773923	8·2840103	8·2905289	8·2969511	8·3032796	8·3095173	8·3156665	53
8	8·2707847	8·2775034	8·2841197	8·2906367	8·2970573	8·3033843	8·3096205	8·3157683	52
9	8·2708976	8·2776145	8·2842292	8·2907445	8·2971635	8·3034890	8·3097237	8·3158700	51
10	8·2710104	8·2777256	8·2843386	8·2908523	8·2972697	8·3035937	8·3098268	8·3159717	50
11	8·2711232	8·2778367	8·2844479	8·2909600	8·2973759	8·3036983	8·3099299	8·3160734	49
12	8·2712359	8·2779477	8·2845573	8·2910677	8·2974820	8·3038029	8·3100330	8·3161751	48
13	8·2713486	8·2780587	8·2846666	8·2911754	8·2975881	8·3039075	8·3101361	8·3162767	47
14	8·2714613	8·2781696	8·2847759	8·2912831	8·2976942	8·3040120	8·3102392	8·3163783	46
15	8·2715740	8·2782806	8·2848851	8·2913907	8·2978002	8·3041165	8·3103422	8·3164799	45
16	8·2716866	8·2783915	8·2849943	8·2914983	8·2979063	8·3042210	8·3104452	8·3165815	44
17	8·2717992	8·2785023	8·2851035	8·2916059	8·2980123	8·3043255	8·3105482	8·3166830	43
18	8·2719118	8·2786132	8·2852127	8·2917134	8·2981183	8·3044299	8·3106512	8·3167845	42
19	8·2720243	8·2787240	8·2853219	8·2918210	8·2982242	8·3045344	8·3107541	8·3168860	41
20	8·2721368	8·2788348	8·2854310	8·2919285	8·2983301	8·3046388	8·3108570	8·3169875	40
21	8·2722493	8·2789456	8·2855401	8·2920359	8·2984360	8·3047431	8·3109599	8·3170889	39
22	8·2723618	8·2790563	8·2856491	8·2921434	8·2985419	8·3048475	8·3110628	8·3171903	38
23	8·2724742	8·2791670	8·2857582	8·2922508	8·2986477	8·3049518	8·3111656	8·3172917	37
24	8·2725866	8·2792777	8·2858672	8·2923582	8·2987536	8·3050561	8·3112684	8·3173931	36
25	8·2726990	8·2793883	8·2859762	8·2924656	8·2988594	8·3051604	8·3113712	8·3174945	35
26	8·2728113	8·2794989	8·2860851	8·2925729	8·2989651	8·3052646	8·3114740	8·3175958	34
27	8·2729236	8·2796095	8·2861941	8·2926802	8·2990709	8·3053688	8·3115767	8·3176971	33
28	8·2730359	8·2797201	8·2863030	8·2927875	8·2991766	8·3054730	8·3116794	8·3177984	32
29	8·2731481	8·2798306	8·2864118	8·2928948	8·2992823	8·3055772	8·3117821	8·3178996	31
30	8·2732604	8·2799411	8·2865207	8·2930020	8·2993879	8·3056813	8·3118848	8·3180008	30
31	8·2733725	8·2800516	8·2866295	8·2931092	8·2994936	8·3057855	8·3119874	8·3181021	29
32	8·2734847	8·2801621	8·2867383	8·2932164	8·2995992	8·3058896	8·3120901	8·3182032	28
33	8·2735968	8·2802725	8·2868471	8·2933235	8·2997048	8·3059936	8·3121927	8·3183044	27
34	8·2737089	8·2803829	8·2869558	8·2934306	8·2998104	8·3060977	8·3122952	8·3184055	26
35	8·2738210	8·2804933	8·2870645	8·2935378	8·2999159	8·3062017	8·3123978	8·3185067	25
36	8·2739331	8·2806036	8·2871732	8·2936448	8·3000214	8·3063057	8·3125003	8·3186077	24
37	8·2740451	8·2807139	8·2872818	8·2937519	8·3001269	8·3064097	8·3126028	8·3187088	23
38	8·2741571	8·2808242	8·2873905	8·2938589	8·3002324	8·3065136	8·3127053	8·3188098	22
39	8·2742690	8·2809345	8·2874991	8·2939659	8·3003378	8·3066175	8·3128077	8·3189109	21
40	8·2743810	8·2810447	8·2876076	8·2940729	8·3004432	8·3067214	8·3129101	8·3190119	20
41	8·2744929	8·2811549	8·2877162	8·2941798	8·3005486	8·3068253	8·3130125	8·3191128	19
42	8·2746048	8·2812650	8·2878247	8·2942867	8·3006539	8·3069291	8·3131149	8·3192138	18
43	8·2747166	8·2813752	8·2879332	8·2943936	8·3007593	8·3070330	8·3132173	8·3193147	17
44	8·2748284	8·2814853	8·2880417	8·2945005	8·3008646	8·3071368	8·3133196	8·3194156	16
45	8·2749402	8·2815954	8·2881501	8·2946073	8·3009699	8·3072405	8·3134219	8·3195165	15
46	8·2750520	8·2817055	8·2882585	8·2947141	8·3010751	8·3073443	8·3135242	8·3196173	14
47	8·2751637	8·2818155	8·2883669	8·2948209	8·3011804	8·3074480	8·3136264	8·3197182	13
48	8·2752754	8·2819255	8·2884752	8·2949277	8·3012856	8·3075517	8·3137287	8·3198190	12
49	8·2753871	8·2820355	8·2885836	8·2950344	8·3013907	8·3076554	8·3138309	8·3199198	11
50	8·2754987	8·2821454	8·2886919	8·2951411	8·3014959	8·3077590	8·3139331	8·3200205	10
51	8·2756103	8·2822553	8·2888002	8·2952478	8·3016010	8·3078626	8·3140352	8·3201213	9
52	8·2757219	8·2823652	8·2889084	8·2953544	8·3017061	8·3079662	8·3141374	8·3202220	8
53	8·2758335	8·2824751	8·2890166	8·2954611	8·3018112	8·3080698	8·3142395	8·3203227	7
54	8·2759450	8·2825849	8·2891248	8·2955677	8·3019163	8·3081734	8·3143416	8·3204233	6
55	8·2760565	8·2826947	8·2892330	8·2956742	8·3020213	8·3082769	8·3144436	8·3205240	5
56	8·2761680	8·2828045	8·2893411	8·2957808	8·3021263	8·3083804	8·3145457	8·3206246	4
57	8·2762794	8·2829143	8·2894492	8·2958873	8·3022313	8·3084839	8·3146477	8·3207252	3
58	8·2763909	8·2830240	8·2895573	8·2959938	8·3023362	8·3085873	8·3147497	8·3208258	2
59	8·2765022	8·2831337	8·2896654	8·2961003	8·3024411	8·3086907	8·3148516	8·3209263	1
60	8·2766136	8·2832434	8·2897734	8·2962067	8·3025460	8·3087941	8·3149536	8·3210269	0
"	55'	54'	53'	52'	51'	50'	49'	48'	"

"	4'	5'	6'	7'	8'	9'	10'	11'	"
0	8·2699563	8·2766012	8·2833234	8·2898559	8·2962017	8·3026335	8·3088842	8·3150462	60
1	8·2700694	8·2768026	8·2834331	8·2899640	8·2963981	8·3027385	8·3089876	8·3151482	59
2	8·2701825	8·2769139	8·2835428	8·2900720	8·2965046	8·3028433	8·3090910	8·3152501	58
3	8·2702955	8·2770253	8·2836524	8·2901800	8·2966110	8·3029482	8·3091944	8·3153520	57
4	8·2704085	8·2771365	8·2837620	8·2902879	8·2967174	8·3030531	8·3092977	8·3154539	56
5	8·2705215	8·2772478	8·2838716	8·2903959	8·2968237	8·3031579	8·3094010	8·3155558	55
6	8·2706345	8·2773590	8·2839811	8·2905038	8·2969300	8·3032627	8·3095043	8·3156576	54
7	8·2707474	8·2774702	8·2840906	8·2906117	8·2970363	8·3033674	8·3096076	8·3157595	53
8	8·2708603	8·2775814	8·2842001	8·2907195	8·2971426	8·3034722	8·3097109	8·3158613	52
9	8·2709732	8·2776925	8·2843096	8·2908274	8·2972489	8·3035769	8·3098141	8·3159630	51
10	8·2710860	8·2778036	8·2844190	8·2909352	8·2973551	8·3036816	8·3099173	8·3160648	50
11	8·2711989	8·2779147	8·2845284	8·2910430	8·2974613	8·3037862	8·3100205	8·3161665	49
12	8·2713116	8·2780258	8·2846378	8·2911507	8·2975675	8·3038909	8·3101236	8·3162682	48
13	8·2714244	8·2781368	8·2847471	8·2912584	8·2976736	8·3039955	8·3102267	8·3163699	47
14	8·2715371	8·2782478	8·2848565	8·2913661	8·2977797	8·3041001	8·3103298	8·3164715	46
15	8·2716498	8·2783588	8·2849658	8·2914738	8·2978858	8·3042046	8·3104329	8·3165732	45
16	8·2717625	8·2784697	8·2850750	8·2915815	8·2979919	8·3043092	8·3105360	8·3166748	44
17	8·2718751	8·2785806	8·2851843	8·2916891	8·2980980	8·3044137	8·3106390	8·3167764	43
18	8·2719877	8·2786915	8·2852935	8·2917967	8·2982040	8·3045182	8·3107420	8·3168779	42
19	8·2721003	8·2788024	8·2854027	8·2919042	8·2983100	8·3046226	8·3108450	8·3169795	41
20	8·2722129	8·2789132	8·2855118	8·2920118	8·2984159	8·3047271	8·3109479	8·3170810	40
21	8·2723254	8·2790240	8·2856210	8·2921193	8·2985219	8·3048315	8·3110508	8·3171825	39
22	8·2724379	8·2791348	8·2857301	8·2922268	8·2986278	8·3049359	8·3111538	8·3172830	38
23	8·2725504	8·2792455	8·2858392	8·2923342	8·2987337	8·3050403	8·3112566	8·3173854	37
24	8·2726628	8·2793563	8·2859482	8·2924417	8·2988395	8·3051446	8·3113595	8·3174868	36
25	8·2727752	8·2794670	8·2860572	8·2925491	8·2989454	8·3052489	8·3114623	8·3175882	35
26	8·2728876	8·2795776	8·2861662	8·2926565	8·2990512	8·3053532	8·3115651	8·3176895	34
27	8·2729999	8·2796882	8·2862752	8·2927638	8·2991570	8·3054575	8·3116679	8·3177909	33
28	8·2731122	8·2797988	8·2863841	8·2928711	8·2992627	8·3055617	8·3117707	8·3178922	32
29	8·2732245	8·2799094	8·2864931	8·2929784	8·2993685	8·3056659	8·3118734	8·3179935	31
30	8·2733368	8·2800200	8·2866019	8·2930857	8·2994742	8·3057701	8·3119761	8·3180948	30
31	8·2734490	8·2801305	8·2867108	8·2931930	8·2995799	8·3058743	8·3120788	8·3181960	29
32	8·2735612	8·2802410	8·2868196	8·2933002	8·2996855	8·3059784	8·3121815	8·3182973	28
33	8·2736734	8·2803515	8·2869284	8·2934074	8·2997911	8·3060825	8·3122841	8·3183985	27
34	8·2737856	8·2804619	8·2870372	8·2935145	8·2998967	8·3061866	8·3123867	8·3184997	26
35	8·2738977	8·2805723	8·2871460	8·2936217	8·3000023	8·3062907	8·3124893	8·3186008	25
36	8·2740098	8·2806827	8·2872547	8·2937288	8·3001079	8·3063947	8·3125919	8·3187019	24
37	8·2741218	8·2807930	8·2873634	8·2938359	8·3002134	8·3064987	8·3126944	8·3188031	23
38	8·2742338	8·2809034	8·2874720	8·2939429	8·3003189	8·3066027	8·3127969	8·3189041	22
39	8·2743458	8·2810136	8·2875807	8·2940500	8·3004244	8·3067067	8·3128994	8·3190052	21
40	8·2744578	8·2811239	8·2876893	8·2941570	8·3005298	8·3068106	8·3130019	8·3191062	20
41	8·2745698	8·2812342	8·2877979	8·2942640	8·3006353	8·3069145	8·3131043	8·3192073	19
42	8·2746817	8·2813444	8·2879065	8·2943709	8·3007407	8·3070184	8·3132068	8·3193083	18
43	8·2747936	8·2814545	8·2880150	8·2944779	8·3008460	8·3071223	8·3133092	8·3194092	17
44	8·2749054	8·2815647	8·2881235	8·2945848	8·3009514	8·3072261	8·3134115	8·3195102	16
45	8·2750173	8·2816748	8·2882320	8·2946916	8·3010567	8·3073299	8·3135139	8·3196111	15
46	8·2751291	8·2817849	8·2883404	8·2947985	8·3011620	8·3074337	8·3136162	8·3197120	14
47	8·2752408	8·2818950	8·2884488	8·2949053	8·3012673	8·3075375	8·3137185	8·3198129	13
48	8·2753526	8·2820051	8·2885572	8·2950121	8·3013725	8·3076412	8·3138208	8·3199137	12
49	8·2754643	8·2821151	8·2886656	8·2951189	8·3014778	8·3077449	8·3139230	8·3200145	11
50	8·2755760	8·2822251	8·2887740	8·2952256	8·3015830	8·3078486	8·3140253	8·3201154	10
51	8·2756876	8·2823350	8·2888823	8·2953324	8·3016881	8·3079523	8·3141275	8·3202161	9
52	8·2757992	8·2824450	8·2889906	8·2954391	8·3017933	8·3080559	8·3142296	8·3203169	8
53	8·2759108	8·2825549	8·2890988	8·2955457	8·3018984	8·3081596	8·3143318	8·3204176	7
54	8·2760224	8·2826647	8·2892071	8·2956524	8·3020035	8·3082631	8·3144339	8·3205183	6
55	8·2761340	8·2827746	8·2893153	8·2957590	8·3021086	8·3083667	8·3145360	8·3206100	5
56	8·2762455	8·2828844	8·2894235	8·2958656	8·3022136	8·3084703	8·3146381	8·3207197	4
57	8·2763570	8·2829942	8·2895316	8·2959721	8·3023186	8·3085738	8·3147402	8·3208203	3
58	8·2764684	8·2831040	8·2896397	8·2960787	8·3024236	8·3086773	8·3148422	8·3209210	2
59	8·2765798	8·2832137	8·2897478	8·2961852	8·3025286	8·3087807	8·3149442	8·3210215	1
60	8·2766912	8·2833234	8·2898559	8·2962917	8·3026335	8·3088842	8·3150462	8·3211221	0
"	55'	54'	53'	52'	51'	50'	49'	48'	"

"	12'	13'	14'	15'	16'	17'	18'	19'	"
0	8·3210269	8·3270163	8·3329243	8·3387529	8·3445043	8·3501805	8·3557835	8·3613150	60
1	8·3211274	8·3271155	8·3330221	8·3388494	8·3445995	8·3502745	8·3558762	8·3614066	59
2	8·3212278	8·3272146	8·3331199	8·3389459	8·3446947	8·3503685	8·3559690	8·3614982	58
3	8·3213283	8·3273137	8·3332176	8·3390423	8·3447899	8·3504624	8·3560617	8·3615897	57
4	8·3214287	8·3274127	8·3333153	8·3391387	8·3448851	8·3505563	8·3561544	8·3616813	56
5	8·3215292	8·3275118	8·3334130	8·3392351	8·3449802	8·3506502	8·3562471	8·3617728	55
6	8·3216295	8·3276108	8·3335107	8·3393315	8·3450753	8·3507441	8·3563398	8·3618643	54
7	8·3217299	8·3277098	8·3336084	8·3394279	8·3451704	8·3508379	8·3564324	8·3619558	53
8	8·3218303	8·3278087	8·3337060	8·3395242	8·3452655	8·3509318	8·3565251	8·3620472	52
9	8·3219306	8·3279077	8·3338036	8·3396205	8·3453605	8·3510256	8·3566177	8·3621387	51
10	8·3220309	8·3280066	8·3339012	8·3397168	8·3454555	8·3511194	8·3567103	8·3622301	50
11	8·3221311	8·3281055	8·3339988	8·3398131	8·3455505	8·3512132	8·3568029	8·3623215	49
12	8·3222314	8·3282044	8·3340963	8·3399093	8·3456455	8·3513069	8·3568954	8·3624129	48
13	8·3223316	8·3283032	8·3341938	8·3400055	8·3457405	8·3514006	8·3569880	8·3625042	47
14	8·3224318	8·3284021	8·3342913	8·3401018	8·3458354	8·3514944	8·3570805	8·3625956	46
15	8·3225320	8·3285009	8·3343888	8·3401979	8·3459304	8·3515881	8·3571730	8·3626869	45
16	8·3226322	8·3285997	8·3344863	8·3402941	8·3460253	8·3516817	8·3572654	8·3627782	44
17	8·3227323	8·3286984	8·3345837	8·3403902	8·3461201	8·3517754	8·3573579	8·3628695	43
18	8·3228324	8·3287972	8·3346811	8·3404864	8·3462150	8·3518690	8·3574503	8·3629608	42
19	8·3229325	8·3288959	8·3347785	8·3405825	8·3463098	8·3519626	8·3575427	8·3630520	41
20	8·3230326	8·3289946	8·3348759	8·3406785	8·3464047	8·3520562	8·3576351	8·3631433	40
21	8·3231326	8·3290933	8·3349732	8·3407746	8·3464995	8·3521498	8·3577275	8·3632345	39
22	8·3232326	8·3291919	8·3350706	8·3408706	8·3465942	8·3522433	8·3578199	8·3633257	38
23	8·3233326	8·3292906	8·3351679	8·3409666	8·3466890	8·3523369	8·3579122	8·3634169	37
24	8·3234326	8·3293892	8·3352651	8·3410626	8·3467837	8·3524304	8·3580045	8·3635080	36
25	8·3235325	8·3294878	8·3353624	8·3411586	8·3468784	8·3525239	8·3580968	8·3635991	35
26	8·3236325	8·3295863	8·3354597	8·3412546	8·3469731	8·3526173	8·3581891	8·3636903	34
27	8·3237324	8·3296849	8·3355569	8·3413505	8·3470678	8·3527108	8·3582814	8·3637814	33
28	8·3238322	8·3297834	8·3356541	8·3414464	8·3471625	8·3528042	8·3583736	8·3638724	32
29	8·3239321	8·3298819	8·3357512	8·3415423	8·3472571	8·3528976	8·3584658	8·3639635	31
30	8·3240319	8·3299804	8·3358484	8·3416382	8·3473517	8·3529910	8·3585580	8·3640545	30
31	8·3241317	8·3300788	8·3359455	8·3417340	8·3474463	8·3530844	8·3586502	8·3641456	29
32	8·3242315	8·3301773	8·3360426	8·3418298	8·3475409	8·3531778	8·3587424	8·3642366	28
33	8·3243313	8·3302757	8·3361397	8·3419256	8·3476354	8·3532711	8·3588345	8·3643275	27
34	8·3244310	8·3303740	8·3362368	8·3420214	8·3477300	8·3533644	8·3589266	8·3644185	26
35	8·3245308	8·3304724	8·3363338	8·3421172	8·3478245	8·3534577	8·3590187	8·3645095	25
36	8·3246305	8·3305708	8·3364309	8·3422129	8·3479189	8·3535510	8·3591108	8·3646004	24
37	8·3247301	8·3306691	8·3365279	8·3423086	8·3480134	8·3536442	8·3592029	8·3646913	23
38	8·3248298	8·3307674	8·3366248	8·3424043	8·3481079	8·3537374	8·3592949	8·3647822	22
39	8·3249294	8·3308656	8·3367218	8·3425000	8·3482023	8·3538306	8·3593870	8·3648730	21
40	8·3250290	8·3309639	8·3368187	8·3425957	8·3482967	8·3539238	8·3594790	8·3649639	20
41	8·3251286	8·3310621	8·3369156	8·3426913	8·3483911	8·3540170	8·3595709	8·3650547	19
42	8·3252282	8·3311603	8·3370125	8·3427869	8·3484854	8·3541102	8·3596629	8·3651455	18
43	8·3253277	8·3312585	8·3371094	8·3428825	8·3485798	8·3542033	8·3597549	8·3652363	17
44	8·3254272	8·3313567	8·3372063	8·3429781	8·3486741	8·3542964	8·3598468	8·3653271	16
45	8·3255267	8·3314548	8·3373031	8·3430736	8·3487684	8·3543895	8·3599387	8·3654179	15
46	8·3256262	8·3315529	8·3373999	8·3431691	8·3488627	8·3544826	8·3600306	8·3655086	14
47	8·3257256	8·3316510	8·3374967	8·3432646	8·3489570	8·3545756	8·3601225	8·3655993	13
48	8·3258250	8·3317491	8·3375934	8·3433601	8·3490512	8·3546686	8·3602143	8·3656900	12
49	8·3259244	8·3318472	8·3376902	8·3434556	8·3491454	8·3547617	8·3603061	8·3657807	11
50	8·3260238	8·3319452	8·3377869	8·3435510	8·3492396	8·3548546	8·3603979	8·3658713	10
51	8·3261232	8·3320432	8·3378836	8·3436465	8·3493338	8·3549476	8·3604897	8·3659620	9
52	8·3262225	8·3321412	8·3379803	8·3437419	8·3494280	8·3550406	8·3605815	8·3660526	8
53	8·3263218	8·3322392	8·3380769	8·3438372	8·3495221	8·3551335	8·3606733	8·3661432	7
54	8·3264211	8·3323371	8·3381736	8·3439326	8·3496162	8·3552264	8·3607650	8·3662338	6
55	8·3265204	8·3324350	8·3382702	8·3440279	8·3497103	8·3553193	8·3608567	8·3663244	5
56	8·3266196	8·3325329	8·3383668	8·3441233	8·3498044	8·3554122	8·3609484	8·3664149	4
57	8·3267188	8·3326308	8·3384633	8·3442186	8·3498985	8·3555050	8·3610401	8·3665054	3
58	8·3268180	8·3327287	8·3385599	8·3443138	8·3499925	8·3555979	8·3611317	8·3665959	2
59	8·3269172	8·3328265	8·3386564	8·3444091	8·3500865	8·3556907	8·3612234	8·3666864	1
60	8·3270163	8·3329243	8·3387529	8·3445043	8·3501805	8·3557835	8·3613150	8·3667769	0
"	47'	46'	45'	44'	43'	42'	41'	40'	"

"	12′	13′	14′	15′	16′	17′	18′	19′	"
0	8·3211221	8·3271143	8·3330249	8·3388563	8·3446105	8·3502895	8·3558953	8·3614297	60
1	8·3212227	8·3272134	8·3331228	8·3389528	8·3447057	8·3503835	8·3559881	8·3615213	59
2	8·3213232	8·3273126	8·3332206	8·3390493	8·3448010	8·3504775	8·3560809	8·3616129	58
3	8·3214237	8·3274117	8·3333184	8·3391458	8·3448962	8·3505715	8·3561737	8·3617045	57
4	8·3215242	8·3275108	8·3334161	8·3392423	8·3449914	8·3506655	8·3562664	8·3617961	56
5	8·3216246	8·3276099	8·3335139	8·3393387	8·3450866	8·3507594	8·3563592	8·3618877	55
6	8·3217251	8·3277090	8·3336116	8·3394351	8·3451817	8·3508533	8·3564519	8·3619793	54
7	8·3218255	8·3278080	8·3337093	8·3395316	8·3452769	8·3509472	8·3565446	8·3620708	53
8	8·3219259	8·3279070	8·3338070	8·3396279	8·3453720	8·3510411	8·3566373	8·3621623	52
9	8·3220262	8·3280060	8·3339046	8·3397243	8·3454671	8·3511350	8·3567299	8·3622538	51
10	8·3221266	8·3281050	8·3340023	8·3398206	8·3455621	8·3512288	8·3568226	8·3623453	50
11	8·3222269	8·3282039	8·3340999	8·3399169	8·3456572	8·3513226	8·3569152	8·3624367	49
12	8·3223272	8·3283028	8·3341975	8·3400132	8·3457522	8·3514164	8·3570078	8·3625281	48
13	8·3224274	8·3284017	8·3342950	8·3401095	8·3458472	8·3515102	8·3571004	8·3626196	47
14	8·3225277	8·3285006	8·3343926	8·3402058	8·3459422	8·3516040	8·3571929	8·3627110	46
15	8·3226279	8·3285995	8·3344901	8·3403020	8·3460372	8·3516977	8·3572855	8·3628023	45
16	8·3227281	8·3286983	8·3345876	8·3403982	8·3461321	8·3517914	8·3573780	8·3628937	44
17	8·3228283	8·3287971	8·3346963	8·3404944	8·3462271	8·3518851	8·3574705	8·3629850	43
18	8·3229285	8·3288959	8·3347826	8·3405906	8·3463220	8·3519788	8·3575630	8·3630763	42
19	8·3230286	8·3289947	8·3348800	8·3406867	8·3464169	8·3520725	8·3576555	8·3631676	41
20	8·3231287	8·3290934	8·3349774	8·3407828	8·3465117	8·3521661	8·3577479	8·3632589	40
21	8·3232288	8·3291921	8·3350748	8·3408789	8·3466066	8·3522597	8·3578403	8·3633502	39
22	8·3233288	8·3292908	8·3351722	8·3409750	8·3467014	8·3523533	8·3579327	8·3634414	38
23	8·3234289	8·3293895	8·3352695	8·3410711	8·3467962	8·3524469	8·3580251	8·3635327	37
24	8·3235289	8·3294882	8·3353669	8·3411671	8·3468910	8·3525405	8·3581175	8·3636239	36
25	8·3236289	8·3295868	8·3354642	8·3412631	8·3469857	8·3526340	8·3582098	8·3637150	35
26	8·3237289	8·3296854	8·3355615	8·3413591	8·3470805	8·3527275	8·3583022	8·3638062	34
27	8·3238288	8·3297840	8·3356587	8·3414551	8·3471752	8·3528210	8·3583945	8·3638974	33
28	8·3239287	8·3298826	8·3357560	8·3415511	8·3472699	8·3529145	8·3584868	8·3639885	32
29	8·3240286	8·3299811	8·3358532	8·3416470	8·3473646	8·3530080	8·3585790	8·3640796	31
30	8·3241285	8·3300796	8·3359504	8·3417429	8·3474592	8·3531014	8·3586713	8·3641707	30
31	8·3242284	8·3301781	8·3360476	8·3418388	8·3475539	8·3531948	8·3587635	8·3642617	29
32	8·3243282	8·3302766	8·3361447	8·3419347	8·3476485	8·3532882	8·3588557	8·3643528	28
33	8·3244280	8·3303751	8·3362419	8·3420305	8·3477431	8·3533816	8·3589479	8·3644438	27
34	8·3245278	8·3304735	8·3363390	8·3421263	8·3478377	8·3534750	8·3590401	8·3645348	26
35	8·3246276	8·3305719	8·3364361	8·3422221	8·3479322	8·3535683	8·3591322	8·3646258	25
36	8·3247273	8·3306703	8·3365331	8·3423179	8·3480268	8·3536616	8·3592243	8·3647168	24
37	8·3248270	8·3307687	8·3366302	8·3424137	8·3481213	8·3537549	8·3593165	8·3648078	23
38	8·3249267	8·3308670	8·3367272	8·3425094	8·3482158	8·3538482	8·3594086	8·3648987	22
39	8·3250264	8·3309653	8·3368242	8·3426052	8·3483103	8·3539414	8·3595006	8·3649896	21
40	8·3251260	8·3310636	8·3369212	8·3427009	8·3484047	8·3540347	8·3595927	8·3650805	20
41	8·3252257	8·3311619	8·3370181	8·3427965	8·3484991	8·3541279	8·3596847	8·3651714	19
42	8·3253253	8·3312601	8·3371151	8·3428922	8·3485936	8·3542211	8·3597767	8·3652623	18
43	8·3254249	8·3313584	8·3372120	8·3429878	8·3486879	8·3543143	8·3598687	8·3653531	17
44	8·3255244	8·3314566	8·3373089	8·3430835	8·3487823	8·3544074	8·3599607	8·3654439	16
45	8·3256240	8·3315548	8·3374058	8·3431791	8·3488767	8·3545006	8·3600527	8·3655347	15
46	8·3257235	8·3316529	8·3375026	8·3432746	8·3489710	8·3545937	8·3601446	8·3656255	14
47	8·3258230	8·3317511	8·3375994	8·3433702	8·3490653	8·3546868	8·3602365	8·3657163	13
48	8·3259224	8·3318492	8·3376963	8·3434657	8·3491596	8·3547799	8·3603284	8·3658070	12
49	8·3260219	8·3319473	8·3377930	8·3435612	8·3492539	8·3548729	8·3604203	8·3658978	11
50	8·3261213	8·3320454	8·3378898	8·3436567	8·3493481	8·3549660	8·3605121	8·3659885	10
51	8·3262207	8·3321434	8·3379866	8·3437522	8·3494423	8·3550590	8·3606040	8·3660792	9
52	8·3263201	8·3322415	8·3380833	8·3438476	8·3495365	8·3551520	8·3606958	8·3661698	8
53	8·3264194	8·3323395	8·3381800	8·3439431	8·3496307	8·3552450	8·3607876	8·3662605	7
54	8·3265188	8·3324375	8·3382767	8·3440385	8·3497249	8·3553379	8·3608794	8·3663511	6
55	8·3266181	8·3325354	8·3383733	8·3441339	8·3498191	8·3554309	8·3609711	8·3664417	5
56	8·3267173	8·3326334	8·3384700	8·3442292	8·3499132	8·3555238	8·3610629	8·3665323	4
57	8·3268166	8·3327313	8·3385666	8·3443246	8·3500073	8·3556167	8·3611546	8·3666229	3
58	8·3269158	8·3328292	8·3386632	8·3444199	8·3501014	8·3557096	8·3612463	8·3667135	2
59	8·3270151	8·3329271	8·3387597	8·3445152	8·3501954	8·3558024	8·3613380	8·3668040	1
60	8·3271143	8·3330249	8·3388563	8·3446105	8·3502895	8·3558953	8·3614297	8·3668945	0
"	47′	46′	45′	44′	43′	42′	41′	40′	"

″	20′	21′	22′	23′	24′	25′	26′	27′	″
0	8·3667769	8·3721710	8·3774988	8·3827620	8·3879622	8·3931008	8·3981793	8·4031990	60
1	8·3668674	8·3722603	8·3775870	8·3828492	8·3880483	8·3931859	8·3982634	8·4032822	59
2	8·3669578	8·3723496	8·3776753	8·3829364	8·3881345	8·3932710	8·3983475	8·4033653	58
3	8·3670482	8·3724389	8·3777635	8·3830235	8·3882206	8·3933561	8·3984316	8·4034485	57
4	8·3671386	8·3725282	8·3778517	8·3831106	8·3883067	8·3934412	8·3985157	8·4035316	56
5	8·3672290	8·3726174	8·3779398	8·3831978	8·3883927	8·3935263	8·3985998	8·4036147	55
6	8·3673193	8·3727067	8·3780280	8·3832848	8·3884788	8·3936113	8·3986839	8·4036978	54
7	8·3674097	8·3727959	8·3781161	8·3833719	8·3885648	8·3936964	8·3987679	8·4037809	53
8	8·3675000	8·3728851	8·3782042	8·3834590	8·3886509	8·3937814	8·3988519	8·4038639	52
9	8·3675903	8·3729743	8·3782924	8·3835460	8·3887369	8·3938664	8·3989359	8·4039470	51
10	8·3676806	8·3730635	8·3783804	8·3836330	8·3888229	8·3939513	8·3990199	8·4040300	50
11	8·3677708	8·3731526	8·3784685	8·3837201	8·3889088	8·3940363	8·3991039	8·4041130	49
12	8·3678611	8·3732418	8·3785566	8·3838070	8·3889948	8·3941213	8·3991879	8·4041960	48
13	8·3679513	8·3733309	8·3786446	8·3838940	8·3890807	8·3942062	8·3992718	8·4042790	47
14	8·3680415	8·3734200	8·3787326	8·3839810	8·3891666	8·3942911	8·3993557	8·4043620	46
15	8·3681317	8·3735091	8·3788206	8·3840679	8·3892526	8·3943760	8·3994397	8·4044449	45
16	8·3682219	8·3735981	8·3789086	8·3841548	8·3893384	8·3944609	8·3995236	8·4045279	44
17	8·3683120	8·3736872	8·3789965	8·3842417	8·3894243	8·3945457	8·3996074	8·4046108	43
18	8·3684022	8·3737762	8·3790845	8·3843286	8·3895102	8·3946306	8·3996913	8·4046937	42
19	8·3684923	8·3738652	8·3791724	8·3844155	8·3895960	8·3947154	8·3997751	8·4047766	41
20	8·3685824	8·3739542	8·3792603	8·3845023	8·3896818	8·3948002	8·3998590	8·4048594	40
21	8·3686725	8·3740431	8·3793482	8·3845892	8·3897676	8·3948850	8·3999428	8·4049423	39
22	8·3687625	8·3741321	8·3794361	8·3846760	8·3898534	8·3949698	8·4000266	8·4050251	38
23	8·3688526	8·3742210	8·3795239	8·3847628	8·3899392	8·3950546	8·4001104	8·4051080	37
24	8·3689426	8·3743099	8·3796117	8·3848496	8·3900249	8·3951393	8·4001941	8·4051908	36
25	8·3690326	8·3743988	8·3796996	8·3849363	8·3901107	8·3952240	8·4002779	8·4052736	35
26	8·3691226	8·3744877	8·3797874	8·3850231	8·3901964	8·3953088	8·4003616	8·4053563	34
27	8·3692125	8·3745766	8·3798751	8·3851098	8·3902821	8·3953935	8·4004453	8·4054391	33
28	8·3693025	8·3746654	8·3799629	8·3851965	8·3903678	8·3954781	8·4005290	8·4055218	32
29	8·3693924	8·3747542	8·3800507	8·3852832	8·3904534	8·3955628	8·4006127	8·4056046	31
30	8·3694823	8·3748430	8·3801384	8·3853699	8·3905391	8·3956475	8·4006964	8·4056873	30
31	8·3695722	8·3749318	8·3802261	8·3854565	8·3906247	8·3957321	8·4007801	8·4057700	29
32	8·3696621	8·3750206	8·3803138	8·3855432	8·3907103	8·3958167	8·4008637	8·4058527	28
33	8·3697519	8·3751094	8·3804015	8·3856298	8·3907959	8·3959013	8·4009473	8·4059353	27
34	8·3698418	8·3751981	8·3804891	8·3857164	8·3908815	8·3959859	8·4010309	8·4060180	26
35	8·3699316	8·3752868	8·3805768	8·3858030	8·3909671	8·3960705	8·4011145	8·4061006	25
36	8·3700214	8·3753755	8·3806644	8·3858896	8·3910526	8·3961550	8·4011981	8·4061832	24
37	8·3701111	8·3754642	8·3807520	8·3859761	8·3911382	8·3962395	8·4012816	8·4062658	23
38	8·3702009	8·3755528	8·3808396	8·3860627	8·3912237	8·3963241	8·4013652	8·4063484	22
39	8·3702907	8·3756415	8·3809271	8·3861492	8·3913092	8·3964086	8·4014487	8·4064310	21
40	8·3703804	8·3757301	8·3810147	8·3862357	8·3913947	8·3964930	8·4015322	8·4065135	20
41	8·3704701	8·3758187	8·3811022	8·3863222	8·3914801	8·3965775	8·4016157	8·4065961	19
42	8·3705598	8·3759073	8·3811897	8·3864087	8·3915656	8·3966620	8·4016992	8·4066786	18
43	8·3706494	8·3759959	8·3812772	8·3864951	8·3916510	8·3967464	8·4017826	8·4067611	17
44	8·3707391	8·3760844	8·3813647	8·3865816	8·3917364	8·3968308	8·4018661	8·4068436	16
45	8·3708287	8·3761729	8·3814522	8·3866680	8·3918218	8·3969152	8·4019495	8·4069261	15
46	8·3709183	8·3762615	8·3815396	8·3867544	8·3919072	8·3969996	8·4020329	8·4070085	14
47	8·3710079	8·3763500	8·3816271	8·3868408	8·3919926	8·3970840	8·4021163	8·4070910	13
48	8·3710975	8·3764384	8·3817145	8·3869271	8·3920779	8·3971683	8·4021997	8·4071734	12
49	8·3711870	8·3765269	8·3818019	8·3870135	8·3921633	8·3972527	8·4022831	8·4072558	11
50	8·3712766	8·3766153	8·3818892	8·3870998	8·3922486	8·3973370	8·4023664	8·4073382	10
51	8·3713661	8·3767038	8·3819766	8·3871861	8·3923339	8·3974213	8·4024497	8·4074206	9
52	8·3714556	8·3767922	8·3820639	8·3872724	8·3924191	8·3975056	8·4025331	8·4075030	8
53	8·3715451	8·3768806	8·3821513	8·3873587	8·3925044	8·3975898	8·4026164	8·4075853	7
54	8·3716346	8·3769689	8·3822386	8·3874450	8·3925897	8·3976741	8·4026996	8·4076677	6
55	8·3717240	8·3770573	8·3823258	8·3875312	8·3926749	8·3977583	8·4027829	8·4077500	5
56	8·3718134	8·3771456	8·3824131	8·3876174	8·3927601	8·3978425	8·4028662	8·4078323	4
57	8·3719028	8·3772339	8·3825004	8·3877037	8·3928453	8·3979268	8·4029494	8·4079146	3
58	8·3719922	8·3773222	8·3825876	8·3877898	8·3929305	8·3980109	8·4030326	8·4079969	2
59	8·3720816	8·3774105	8·3826748	8·3878760	8·3930156	8·3980951	8·4031158	8·4080791	1
60	8·3721710	8·3774988	8·3827620	8·3879622	8·3931008	8·3981793	8·4031990	8·4081614	0
″	39′	38′	37′	36′	35′	34′	33′	32′	″

"	20'	21'	22'	23'	24'	25'	26'	27'	"
0	8·3668945	8·3722915	8·3776223	8·3828886	8·3880918	8·3932336	8·3983152	8·4033381	60
1	8·3669850	8·3723809	8·3777106	8·3829758	8·3881780	8·3933187	8·3983994	8·4034213	59
2	8·3670755	8·3724703	8·3777989	8·3830631	8·3882642	8·3934039	8·3984835	8·4035045	58
3	8·3671660	8·3725596	8·3778872	8·3831503	8·3883504	8·3934891	8·3985677	8·4035877	57
4	8·3672564	8·3726489	8·3779754	8·3832374	8·3884365	8·3935742	8·3986519	8·4036709	56
5	8·3673468	8·3727383	8·3780636	8·3833246	8·3885227	8·3936593	8·3987360	8·4037541	55
6	8·3674372	8·3728275	8·3781519	8·3834117	8·3886088	8·3937444	8·3988201	8·4038372	54
7	8·3675276	8·3729168	8·3782400	8·3834989	8·3886949	8·3938295	8·3989042	8·4039203	53
8	8·3676180	8·3730061	8·3783282	8·3835860	8·3887809	8·3939145	8·3989883	8·4040035	52
9	8·3677083	8·3730953	8·3784164	8·3836731	8·3888670	8·3939996	8·3990723	8·4040866	51
10	8·3677987	8·3731845	8·3785045	8·3837601	8·3889530	8·3940846	8·3991564	8·4041696	50
11	8·3678890	8·3732737	8·3785926	8·3838472	8·3890391	8·3941696	8·3992404	8·4042527	49
12	8·3679793	8·3733629	8·3786807	8·3839342	8·3891251	8·3942546	8·3993244	8·4043358	48
13	8·3680696	8·3734521	8·3787688	8·3840213	8·3892111	8·3943396	8·3994084	8·4044188	47
14	8·3681598	8·3735412	8·3788569	8·3841083	8·3892970	8·3944246	8·3994924	8·4045018	46
15	8·3682501	8·3736304	8·3789449	8·3841953	8·3893830	8·3945095	8·3995764	8·4045848	45
16	8·3683403	8·3737195	8·3790329	8·3842822	8·3894689	8·3945945	8·3996603	8·4046678	44
17	8·3684305	8·3738086	8·3791209	8·3843692	8·3895548	8·3946794	8·3997442	8·4047508	43
18	8·3685207	8·3738976	8·3792089	8·3844561	8·3896408	8·3947643	8·3998282	8·4048337	42
19	8·3686108	8·3739867	8·3792969	8·3845430	8·3897266	8·3948492	8·3999121	8·4049167	41
20	8·3687010	8·3740757	8·3793849	8·3846299	8·3898125	8·3949340	8·3999959	8·4049996	40
21	8·3687911	8·3741647	8·3794728	8·3847168	8·3898984	8·3950189	8·4000798	8·4050825	39
22	8·3688812	8·3742538	8·3795607	8·3848037	8·3899842	8·3951037	8·4001637	8·4051654	38
23	8·3689713	8·3743427	8·3796486	8·3848905	8·3900700	8·3951885	8·4002475	8·4052483	37
24	8·3690614	8·3744317	8·3797365	8·3849774	8·3901558	8·3952733	8·4003313	8·4053311	36
25	8·3691514	8·3745206	8·3798244	8·3850642	8·3902416	8·3953581	8·4004151	8·4054140	35
26	8·3692414	8·3746096	8·3799122	8·3851510	8·3903274	8·3954429	8·4004989	8·4054968	34
27	8·3693315	8·3746985	8·3800001	8·3852378	8·3904131	8·3955276	8·4005827	8·4055796	33
28	8·3694215	8·3747874	8·3800879	8·3853245	8·3904989	8·3956124	8·4006664	8·4056624	32
29	8·3695114	8·3748762	8·3801757	8·3854113	8·3905846	8·3956971	8·4007502	8·4057452	31
30	8·3696014	8·3749651	8·3802634	8·3854980	8·3906703	8·3957818	8·4008339	8·4058280	30
31	8·3696913	8·3750539	8·3803512	8·3855847	8·3907560	8·3958665	8·4009176	8·4059107	29
32	8·3697812	8·3751428	8·3804390	8·3856714	8·3908417	8·3959511	8·4010013	8·4059935	28
33	8·3698711	8·3752316	8·3805267	8·3857581	8·3909273	8·3960358	8·4010850	8·4060762	27
34	8·3699610	8·3753203	8·3806144	8·3858448	8·3910129	8·3961204	8·4011686	8·4061589	26
35	8·3700509	8·3754091	8·3807021	8·3859314	8·3910986	8·3962050	8·4012523	8·4062416	25
36	8·3701407	8·3754979	8·3807898	8·3860180	8·3911842	8·3962897	8·4013359	8·4063242	24
37	8·3702306	8·3755866	8·3808774	8·3861046	8·3912697	8·3963742	8·4014195	8·4064069	23
38	8·3703204	8·3756753	8·3809650	8·3861912	8·3913553	8·3964588	8·4015031	8·4064895	22
39	8·3704102	8·3757640	8·3810527	8·3862778	8·3914409	8·3965434	8·4015867	8·4065722	21
40	8·3704999	8·3758527	8·3811403	8·3863643	8·3915264	8·3966279	8·4016702	8·4066548	20
41	8·3705897	8·3759413	8·3812278	8·3864509	8·3916119	8·3967124	8·4017538	8·4067374	19
42	8·3706794	8·3760299	8·3813154	8·3865374	8·3916974	8·3967969	8·4018373	8·4068199	18
43	8·3707692	8·3761186	8·3814030	8·3866239	8·3917829	8·3968814	8·4019208	8·4069025	17
44	8·3708589	8·3762072	8·3814905	8·3867104	8·3918684	8·3969659	8·4020043	8·4069850	16
45	8·3709485	8·3762958	8·3815780	8·3867969	8·3919538	8·3970503	8·4020878	8·4070676	15
46	8·3710382	8·3763843	8·3816655	8·3868833	8·3920393	8·3971348	8·4021713	8·4071501	14
47	8·3711278	8·3764729	8·3817530	8·3869698	8·3921247	8·3972192	8·4022547	8·4072326	13
48	8·3712175	8·3765614	8·3818404	8·3870562	8·3922101	8·3973036	8·4023381	8·4073151	12
49	8·3713071	8·3766499	8·3819279	8·3871426	8·3922955	8·3973880	8·4024216	8·4073975	11
50	8·3713967	8·3767384	8·3820153	8·3872290	8·3923808	8·3974724	8·4025050	8·4074800	10
51	8·3714862	8·3768269	8·3821027	8·3873153	8·3924662	8·3975567	8·4025884	8·4075624	9
52	8·3715758	8·3769153	8·3821901	8·3874017	8·3925515	8·3976411	8·4026717	8·4076449	8
53	8·3716653	8·3770038	8·3822775	8·3874880	8·3926368	8·3977254	8·4027551	8·4077273	7
54	8·3717548	8·3770922	8·3823648	8·3875743	8·3927221	8·3978097	8·4028384	8·4078097	6
55	8·3718443	8·3771806	8·3824522	8·3876606	8·3928074	8·3978940	8·4029217	8·4078920	5
56	8·3719338	8·3772690	8·3825395	8·3877469	8·3928927	8·3979782	8·4030050	8·4079744	4
57	8·3720232	8·3773574	8·3826268	8·3878332	8·3929779	8·3980625	8·4030883	8·4080567	3
58	8·3721127	8·3774457	8·3827141	8·3879194	8·3930631	8·3981467	8·4031716	8·4081391	2
59	8·3722021	8·3775340	8·3828014	8·3880056	8·3931484	8·3982310	8·4032549	8·4082214	1
60	8·3722915	8·3776223	8·3828886	8·3880918	8·3932336	8·3983152	8·4033381	8·4083037	0
"	39'	38'	37'	36'	35'	34'	33'	32'	"

"	28'	29'	30'	31'	32'	33'	34'	35'	"
0	8·4081614	8·4130676	8·4179190	8·4227168	8·4274621	8·4321561	8·4367999	8·4413944	60
1	8·4082436	8·4131489	8·4179994	8·4227963	8·4275408	8·4322339	8·4368768	8·4414706	59
2	8·4083258	8·4132302	8·4180798	8·4228758	8·4276194	8·4323117	8·4369538	8·4415468	58
3	8·4084080	8·4133115	8·4181602	8·4229553	8·4276980	8·4323895	8·4370307	8·4416229	57
4	8·4084902	8·4133927	8·4182405	8·4230348	8·4277766	8·4324672	8·4371077	8·4416990	56
5	8·4085723	8·4134740	8·4183209	8·4231142	8·4278552	8·4325450	8·4371846	8·4417751	55
6	8·4086545	8·4135552	8·4184012	8·4231937	8·4279338	8·4326227	8·4372615	8·4418512	54
7	8·4087366	8·4136364	8·4184815	8·4232731	8·4280124	8·4327004	8·4373384	8·4419273	53
8	8·4088187	8·4137176	8·4185618	8·4233525	8·4280909	8·4327781	8·4374153	8·4420034	52
9	8·4089008	8·4137988	8·4186421	8·4234319	8·4281694	8·4328558	8·4374921	8·4420795	51
10	8·4089829	8·4138800	8·4187223	8·4235113	8·4282480	8·4329335	8·4375690	8·4421555	50
11	8·4090650	8·4139611	8·4188026	8·4235907	8·4283265	8·4330112	8·4376458	8·4422315	49
12	8·4091471	8·4140422	8·4188828	8·4236700	8·4284050	8·4330888	8·4377227	8·4423076	48
13	8·4092291	8·4141234	8·4189630	8·4237494	8·4284835	8·4331665	8·4377995	8·4423836	47
14	8·4093111	8·4142045	8·4190432	8·4238287	8·4285619	8·4332441	8·4378763	8·4424596	46
15	8·4093931	8·4142856	8·4191234	8·4239080	8·4286404	8·4333217	8·4379531	8·4425355	45
16	8·4094751	8·4143666	8·4192036	8·4239873	8·4287188	8·4333993	8·4380298	8·4426115	44
17	8·4095571	8·4144477	8·4192838	8·4240666	8·4287972	8·4334769	8·4381066	8·4426875	43
18	8·4096391	8·4145287	8·4193639	8·4241458	8·4288756	8·4335544	8·4381833	8·4427634	42
19	8·4097210	8·4146098	8·4194441	8·4242251	8·4289540	8·4336320	8·4382601	8·4428393	41
20	8·4098029	8·4146908	8·4195242	8·4243043	8·4290324	8·4337095	8·4383368	8·4429152	40
21	8·4098849	8·4147718	8·4196043	8·4243836	8·4291108	8·4337871	8·4384135	8·4429911	39
22	8·4099668	8·4148528	8·4196844	8·4244628	8·4291891	8·4338646	8·4384902	8·4430670	38
23	8·4100486	8·4149337	8·4197644	8·4245420	8·4292675	8·4339421	8·4385669	8·4431429	37
24	8·4101305	8·4150147	8·4198445	8·4246211	8·4293458	8·4340196	8·4386435	8·4432187	36
25	8·4102124	8·4150956	8·4199245	8·4247003	8·4294241	8·4340970	8·4387202	8·4432946	35
26	8·4102942	8·4151765	8·4200046	8·4247795	8·4295024	8·4341745	8·4387968	8·4433704	34
27	8·4103760	8·4152575	8·4200846	8·4248586	8·4295807	8·4342519	8·4388734	8·4434462	33
28	8·4104578	8·4153383	8·4201646	8·4249377	8·4296590	8·4343294	8·4389501	8·4435221	32
29	8·4105396	8·4154192	8·4202446	8·4250168	8·4297372	8·4344068	8·4390266	8·4435978	31
30	8·4106214	8·4155001	8·4203245	8·4250959	8·4298154	8·4344842	8·4391032	8·4436736	30
31	8·4107032	8·4155809	8·4204045	8·4251750	8·4298937	8·4345616	8·4391798	8·4437494	29
32	8·4107849	8·4156618	8·4204844	8·4252541	8·4299719	8·4346389	8·4392564	8·4438251	28
33	8·4108667	8·4157426	8·4205644	8·4253331	8·4300501	8·4347163	8·4393329	8·4439009	27
34	8·4109484	8·4158234	8·4206443	8·4254122	8·4301283	8·4347937	8·4394094	8·4439766	26
35	8·4110301	8·4159042	8·4207242	8·4254912	8·4302064	8·4348710	8·4394859	8·4440523	25
36	8·4111118	8·4159850	8·4208040	8·4255702	8·4302846	8·4349483	8·4395624	8·4441280	24
37	8·4111934	8·4160657	8·4208839	8·4256492	8·4303627	8·4350256	8·4396389	8·4442037	23
38	8·4112751	8·4161465	8·4209638	8·4257282	8·4304409	8·4351029	8·4397154	8·4442794	22
39	8·4113567	8·4162272	8·4210436	8·4258071	8·4305190	8·4351802	8·4397919	8·4443551	21
40	8·4114383	8·4163079	8·4211234	8·4258861	8·4305971	8·4352574	8·4398683	8·4444307	20
41	8·4115200	8·4163886	8·4212032	8·4259650	8·4306751	8·4353347	8·4399447	8·4445063	19
42	8·4116015	8·4164693	8·4212830	8·4260439	8·4307532	8·4354119	8·4400212	8·4445820	18
43	8·4116831	8·4165499	8·4213628	8·4261229	8·4308313	8·4354892	8·4400976	8·4446576	17
44	8·4117647	8·4166306	8·4214426	8·4262018	8·4309093	8·4355664	8·4401740	8·4447332	16
45	8·4118462	8·4167112	8·4215223	8·4262806	8·4309873	8·4356436	8·4402503	8·4448087	15
46	8·4119278	8·4167919	8·4216020	8·4263595	8·4310654	8·4357207	8·4403267	8·4448843	14
47	8·4120093	8·4168725	8·4216818	8·4264383	8·4311434	8·4357979	8·4404031	8·4449599	13
48	8·4120908	8·4169531	8·4217615	8·4265172	8·4312213	8·4358751	8·4404794	8·4450354	12
49	8·4121723	8·4170336	8·4218412	8·4265960	8·4312993	8·4359522	8·4405557	8·4451109	11
50	8·4122537	8·4171142	8·4219208	8·4266748	8·4313773	8·4360293	8·4406321	8·4451865	10
51	8·4123352	8·4171948	8·4220005	8·4267536	8·4314552	8·4361064	8·4407083	8·4452620	9
52	8·4124166	8·4172753	8·4220801	8·4268324	8·4315332	8·4361835	8·4407846	8·4453375	8
53	8·4124981	8·4173558	8·4221598	8·4269111	8·4316111	8·4362606	8·4408609	8·4454129	7
54	8·4125795	8·4174363	8·4222394	8·4269899	8·4316890	8·4363377	8·4409372	8·4454884	6
55	8·4126609	8·4175168	8·4223190	8·4270686	8·4317669	8·4364148	8·4410134	8·4455638	5
56	8·4127422	8·4175973	8·4223986	8·4271474	8·4318447	8·4364918	8·4410896	8·4456393	4
57	8·4128236	8·4176777	8·4224782	8·4272261	8·4319226	8·4365688	8·4411659	8·4457147	3
58	8·4129050	8·4177582	8·4225577	8·4273048	8·4320004	8·4366459	8·4412421	8·4457901	2
59	8·4129863	8·4178386	8·4226373	8·4273834	8·4320783	8·4367229	8·4413183	8·4458655	1
60	8·4130676	8·4179190	8·4227168	8·4274621	8·4321561	8·4367999	8·4413944	8·4459409	0
"	31'	30'	29'	28'	27'	26'	25'	24'	"

"	28′	29′	30′	31′	32′	33′	34′	35′	"
0	8·4083037	8·4132132	8·4180679	8·4228690	8·4276176	8·4323150	8·4369622	8·4415603	60
1	8·4083859	8·4132945	8·4181483	8·4229485	8·4276963	8·4323929	8·4370393	8·4416365	59
2	8·4084682	8·4133759	8·4182288	8·4230281	8·4277750	8·4324707	8·4371163	8·4417127	58
3	8·4085505	8·4134572	8·4183092	8·4231076	8·4278537	8·4325486	8·4371933	8·4417889	57
4	8·4086327	8·4135385	8·4183896	8·4231872	8·4279324	8·4326264	8·4372703	8·4418651	56
5	8·4087149	8·4136198	8·4184700	8·4232667	8·4280110	8·4327042	8·4373473	8·4419413	55
6	8·4087971	8·4137011	8·4185504	8·4233462	8·4280897	8·4327820	8·4374242	8·4420174	54
7	8·4088793	8·4137823	8·4186307	8·4234257	8·4281683	8·4328598	8·4375012	8·4420936	53
8	8·4089615	8·4138636	8·4187111	8·4235051	8·4282469	8·4329375	8·4375781	8·4421697	52
9	8·4090436	8·4139448	8·4187914	8·4235846	8·4283255	8·4330153	8·4376550	8·4422458	51
10	8·4091258	8·4140261	8·4188717	8·4236640	8·4284041	8·4330930	8·4377320	8·4423219	50
11	8·4092079	8·4141073	8·4189520	8·4237434	8·4284826	8·4331707	8·4378089	8·4423980	49
12	8·4092900	8·4141885	8·4190323	8·4238229	8·4285612	8·4332484	8·4378857	8·4424741	48
13	8·4093721	8·4142696	8·4191126	8·4239023	8·4286397	8·4333261	8·4379626	8·4425502	47
14	8·4094542	8·4143508	8·4191929	8·4239816	8·4287182	8·4334038	8·4380395	8·4426262	46
15	8·4095362	8·4144319	8·4192731	8·4240610	8·4287968	8·4334815	8·4381163	8·4427023	45
16	8·4096183	8·4145131	8·4193533	8·4241404	8·4288752	8·4335591	8·4381931	8·4427783	44
17	8·4097003	8·4145942	8·4194336	8·4242197	8·4289537	8·4336368	8·4382700	8·4428543	43
18	8·4097823	8·4146753	8·4195138	8·4242990	8·4290322	8·4337144	8·4383468	8·4429303	42
19	8·4098643	8·4147564	8·4195940	8·4243783	8·4291106	8·4337920	8·4384235	8·4430063	41
20	8·4099463	8·4148374	8·4196741	8·4244576	8·4291891	8·4338696	8·4385003	8·4430822	40
21	8·4100283	8·4149185	8·4197543	8·4245369	8·4292675	8·4339472	8·4385771	8·4431582	39
22	8·4101103	8·4149995	8·4198344	8·4246162	8·4293459	8·4340248	8·4386538	8·4432341	38
23	8·4101922	8·4150805	8·4199146	8·4246954	8·4294243	8·4341023	8·4387306	8·4433101	37
24	8·4102741	8·4151616	8·4199947	8·4247747	8·4295027	8·4341799	8·4388073	8·4433860	36
25	8·4103560	8·4152425	8·4200748	8·4248539	8·4295811	8·4342574	8·4388840	8·4434619	35
26	8·4104379	8·4153235	8·4201549	8·4249331	8·4296594	8·4343349	8·4389607	8·4435378	34
27	8·4105198	8·4154045	8·4202349	8·4250123	8·4297377	8·4344124	8·4390374	8·4436137	33
28	8·4106017	8·4154854	8·4203150	8·4250916	8·4298161	8·4344899	8·4391140	8·4436895	32
29	8·4106835	8·4155664	8·4203950	8·4251706	8·4298944	8·4345674	8·4391907	8·4437654	31
30	8·4107653	8·4156473	8·4204750	8·4252498	8·4299727	8·4346448	8·4392673	8·4438412	30
31	8·4108472	8·4157282	8·4205550	8·4253289	8·4300510	8·4347223	8·4393440	8·4439171	29
32	8·4109290	8·4158091	8·4206350	8·4254080	8·4301292	8·4347997	8·4394206	8·4439929	28
33	8·4110107	8·4158900	8·4207150	8·4254872	8·4302075	8·4348771	8·4394972	8·4440687	27
34	8·4110925	8·4159708	8·4207950	8·4255662	8·4302857	8·4349545	8·4395738	8·4441444	26
35	8·4111743	8·4160517	8·4208749	8·4256453	8·4303639	8·4350319	8·4396503	8·4442202	25
36	8·4112560	8·4161325	8·4209549	8·4257244	8·4304422	8·4351093	8·4397269	8·4442960	24
37	8·4113377	8·4162133	8·4210348	8·4258034	8·4305204	8·4351867	8·4398034	8·4443717	23
38	8·4114194	8·4162941	8·4211147	8·4258825	8·4305985	8·4352640	8·4398800	8·4444475	22
39	8·4115011	8·4163749	8·4211946	8·4259615	8·4306767	8·4353413	8·4399565	8·4445232	21
40	8·4115828	8·4164556	8·4212745	8·4260405	8·4307549	8·4354187	8·4400330	8·4445989	20
41	8·4116645	8·4165364	8·4213543	8·4261195	8·4308330	8·4354960	8·4401095	8·4446746	19
42	8·4117461	8·4166171	8·4214342	8·4261985	8·4309111	8·4355733	8·4401860	8·4447503	18
43	8·4118278	8·4166979	8·4215140	8·4262774	8·4309892	8·4356506	8·4402624	8·4448259	17
44	8·4119094	8·4167786	8·4215938	8·4263564	8·4310673	8·4357278	8·4403389	8·4449016	16
45	8·4119910	8·4168593	8·4216736	8·4264353	8·4311454	8·4358051	8·4404153	8·4449772	15
46	8·4120726	8·4169399	8·4217534	8·4265142	8·4312235	8·4358823	8·4404918	8·4450529	14
47	8·4121541	8·4170206	8·4218332	8·4265932	8·4313016	8·4359595	8·4405682	8·4451285	13
48	8·4122357	8·4171012	8·4219130	8·4266720	8·4313796	8·4360367	8·4406446	8·4452041	12
49	8·4123172	8·4171819	8·4219927	8·4267509	8·4314576	8·4361139	8·4407209	8·4452797	11
50	8·4123988	8·4172625	8·4220725	8·4268298	8·4315356	8·4361911	8·4407973	8·4453552	10
51	8·4124803	8·4173431	8·4221522	8·4269086	8·4316136	8·4362683	8·4408737	8·4454308	9
52	8·4125618	8·4174237	8·4222319	8·4269875	8·4316916	8·4363455	8·4409500	8·4455063	8
53	8·4126432	8·4175043	8·4223116	8·4270663	8·4317696	8·4364226	8·4410263	8·4455819	7
54	8·4127247	8·4175848	8·4223912	8·4271451	8·4318476	8·4364997	8·4411027	8·4456574	6
55	8·4128062	8·4176654	8·4224709	8·4272239	8·4319255	8·4365768	8·4411790	8·4457329	5
56	8·4128876	8·4177459	8·4225505	8·4273027	8·4320034	8·4366540	8·4412553	8·4458084	4
57	8·4129690	8·4178264	8·4226302	8·4273814	8·4320814	8·4367310	8·4413315	8·4458839	3
58	8·4130504	8·4179069	8·4227098	8·4274602	8·4321593	8·4368081	8·4414078	8·4459594	2
59	8·4131318	8·4179874	8·4227894	8·4275389	8·4322372	8·4368852	8·4414841	8·4460348	1
60	8·4132132	8·4180679	8·4228690	8·4276176	8·4323150	8·4369622	8·4415603	8·4461103	0
"	31′	30′	29′	28′	27′	26′	25′	24′	"

"	36'	37'	38'	39'	40'	41'	42'	43'	"
0	8·4459409	8·4504402	8·4548934	8·4593013	8·4636649	8·4679850	8·4722026	8·4764984	60
1	8·4460163	8·4505148	8·4549672	8·4593744	8·4637372	8·4680567	8·4723335	8·4765686	59
2	8·4460916	8·4505894	8·4550410	8·4594474	8·4638096	8·4681283	8·4724044	8·4766388	58
3	8·4461670	8·4506640	8·4551148	8·4595205	8·4638819	8·4681999	8·4724753	8·4767091	57
4	8·4462423	8·4507385	8·4551886	8·4595936	8·4639542	8·4682715	8·4725462	8·4767793	56
5	8·4463176	8·4508131	8·4552624	8·4596666	8·4640265	8·4683431	8·4726171	8·4768495	55
6	8·4463929	8·4508876	8·4553362	8·4597396	8·4640988	8·4684147	8·4726880	8·4769197	54
7	8·4464682	8·4509621	8·4554099	8·4598126	8·4641711	8·4684862	8·4727589	8·4769899	53
8	8·4465435	8·4510366	8·4554837	8·4598856	8·4642434	8·4685578	8·4728297	8·4770600	52
9	8·4466188	8·4511111	8·4555574	8·4599586	8·4643156	8·4686293	8·4729006	8·4771302	51
10	8·4466940	8·4511856	8·4556311	8·4600316	8·4643879	8·4687009	8·4729714	8·4772003	50
11	8·4467693	8·4512601	8·4557048	8·4601046	8·4644601	8·4687724	8·4730422	8·4772705	49
12	8·4468445	8·4513345	8·4557785	8·4601775	8·4645323	8·4688439	8·4731130	8·4773406	48
13	8·4469197	8·4514090	8·4558522	8·4602505	8·4646046	8·4689154	8·4731838	8·4774107	47
14	8·4469949	8·4514834	8·4559259	8·4603234	8·4646768	8·4689869	8·4732546	8·4774808	46
15	8·4470701	8·4515578	8·4559996	8·4603963	8·4647489	8·4690584	8·4733254	8·4775509	45
16	8·4471453	8·4516322	8·4560732	8·4604692	8·4648211	8·4691298	8·4733962	8·4776210	44
17	8·4472205	8·4517066	8·4561468	8·4605421	8·4648933	8·4692013	8·4734669	8·4776910	43
18	8·4472956	8·4517810	8·4562205	8·4606150	8·4649654	8·4692727	8·4735377	8·4777611	42
19	8·4473707	8·4518553	8·4562941	8·4606878	8·4650376	8·4693441	8·4736084	8·4778311	41
20	8·4474459	8·4519297	8·4563677	8·4607607	8·4651097	8·4694156	8·4736791	8·4779012	40
21	8·4475210	8·4520040	8·4564412	8·4608335	8·4651818	8·4694870	8·4737498	8·4779712	39
22	8·4475961	8·4520784	8·4565148	8·4609064	8·4652539	8·4695583	8·4738205	8·4780412	38
23	8·4476712	8·4521527	8·4565884	8·4609792	8·4653260	8·4696297	8·4738912	8·4781112	37
24	8·4477462	8·4522270	8·4566619	8·4610520	8·4653981	8·4697011	8·4739618	8·4781812	36
25	8·4478213	8·4523013	8·4567354	8·4611248	8·4654702	8·4697725	8·4740325	8·4782511	35
26	8·4478963	8·4523755	8·4568090	8·4611976	8·4655422	8·4698438	8·4741032	8·4783211	34
27	8·4479714	8·4524498	8·4568825	8·4612703	8·4656143	8·4699151	8·4741738	8·4783911	33
28	8·4480464	8·4525240	8·4569560	8·4613431	8·4656863	8·4699865	8·4742444	8·4784610	32
29	8·4481214	8·4525983	8·4570295	8·4614158	8·4657583	8·4700578	8·4743150	8·4785309	31
30	8·4481964	8·4526725	8·4571029	8·4614886	8·4658303	8·4701291	8·4743856	8·4786009	30
31	8·4482714	8·4527467	8·4571764	8·4615613	8·4659023	8·4702003	8·4744562	8·4786708	29
32	8·4483463	8·4528209	8·4572498	8·4616340	8·4659743	8·4702716	8·4745268	8·4787407	28
33	8·4484213	8·4528951	8·4573233	8·4617067	8·4660463	8·4703429	8·4745974	8·4788105	27
34	8·4484962	8·4529693	8·4573967	8·4617794	8·4661182	8·4704141	8·4746679	8·4788804	26
35	8·4485712	8·4530434	8·4574701	8·4618520	8·4661902	8·4704854	8·4747385	8·4789503	25
36	8·4486461	8·4531176	8·4575435	8·4619247	8·4662621	8·4705566	8·4748090	8·4790201	24
37	8·4487210	8·4531917	8·4576169	8·4619973	8·4663340	8·4706278	8·4748795	8·4790900	23
38	8·4487959	8·4532659	8·4576902	8·4620700	8·4664059	8·4706990	8·4749500	8·4791598	22
39	8·4488708	8·4533400	8·4577636	8·4621426	8·4664778	8·4707702	8·4750205	8·4792296	21
40	8·4489456	8·4534141	8·4578369	8·4622152	8·4665497	8·4708414	8·4750910	8·4792994	20
41	8·4490205	8·4534881	8·4579103	8·4622878	8·4666216	8·4709126	8·4751615	8·4793692	19
42	8·4490953	8·4535622	8·4579836	8·4623604	8·4666935	8·4709837	8·4752320	8·4794390	18
43	8·4491701	8·4536363	8·4580569	8·4624330	8·4667653	8·4710549	8·4753024	8·4795088	17
44	8·4492450	8·4537103	8·4581302	8·4625055	8·4668372	8·4711260	8·4753729	8·4795785	16
45	8·4493198	8·4537844	8·4582035	8·4625781	8·4669090	8·4711971	8·4754433	8·4796483	15
46	8·4493945	8·4538584	8·4582768	8·4626506	8·4669808	8·4712682	8·4755137	8·4797180	14
47	8·4494693	8·4539324	8·4583500	8·4627231	8·4670526	8·4713393	8·4755841	8·4797878	13
48	8·4495441	8·4540064	8·4584233	8·4627957	8·4671244	8·4714104	8·4756545	8·4798575	12
49	8·4496188	8·4540804	8·4584965	8·4628682	8·4671962	8·4714815	8·4757249	8·4799272	11
50	8·4496936	8·4541543	8·4585697	8·4629406	8·4672680	8·4715526	8·4757953	8·4799969	10
51	8·4497683	8·4542283	8·4586429	8·4630131	8·4673397	8·4716236	8·4758656	8·4800666	9
52	8·4498430	8·4543023	8·4587161	8·4630856	8·4674115	8·4716947	8·4759360	8·4801362	8
53	8·4499177	8·4543762	8·4587893	8·4631580	8·4674832	8·4717657	8·4760063	8·4802059	7
54	8·4499924	8·4544501	8·4588625	8·4632305	8·4675549	8·4718367	8·4760766	8·4802755	6
55	8·4500671	8·4545240	8·4589357	8·4633029	8·4676266	8·4719077	8·4761470	8·4803452	5
56	8·4501417	8·4545979	8·4590088	8·4633753	8·4676983	8·4719787	8·4762173	8·4804148	4
57	8·4502164	8·4546718	8·4590819	8·4634477	8·4677700	8·4720497	8·4762876	8·4804844	3
58	8·4502910	8·4547457	8·4591551	8·4635201	8·4678417	8·4721207	8·4763578	8·4805540	2
59	8·4503656	8·4548195	8·4592282	8·4635925	8·4679134	8·4721916	8·4764281	8·4806236	1
60	8·4504402	8·4548934	8·4593013	8·4636649	8·4679850	8·4722626	8·4764984	8·4806932	0
"	23'	22'	21'	20'	19'	18'	17'	16'	"

″	36′	37′	38′	39′	40′	41′	42′	43′	″
0	8·4461103	8·4506131	8·4550699	8·4594814	8·4638486	8·4681725	8·4724538	8·4766933	60
1	8·4461857	8·4506878	8·4551438	8·4595545	8·4639211	8·4682442	8·4725248	8·4767636	59
2	8·4462611	8·4507624	8·4552176	8·4596277	8·4639935	8·4683159	8·4725957	8·4768339	58
3	8·4463365	8·4508371	8·4552915	8·4597008	8·4640659	8·4683875	8·4726667	8·4769042	57
4	8·4464119	8·4509117	8·4553654	8·4597739	8·4641382	8·4684592	8·4727377	8·4769745	56
5	8·4464873	8·4509863	8·4554392	8·4598470	8·4642106	8·4685309	8·4728086	8·4770448	55
6	8·4465627	8·4510609	8·4555130	8·4599201	8·4642830	8·4686025	8·4728796	8·4771150	54
7	8·4466380	8·4511354	8·4555868	8·4599932	8·4643553	8·4686741	8·4729505	8·4771853	53
8	8·4467133	8·4512100	8·4556607	8·4600662	8·4644276	8·4687458	8·4730214	8·4772555	52
9	8·4467887	8·4512846	8·4557344	8·4601393	8·4645000	8·4688174	8·4730923	8·4773257	51
10	8·4468640	8·4513591	8·4558082	8·4602123	8·4645723	8·4688890	8·4731632	8·4773959	50
11	8·4469393	8·4514336	8·4558820	8·4602853	8·4646446	8·4689605	8·4732341	8·4774661	49
12	8·4470146	8·4515081	8·4559558	8·4603584	8·4647168	8·4690321	8·4733050	8·4775363	48
13	8·4470898	8·4515826	8·4560295	8·4604314	8·4647891	8·4691037	8·4733758	8·4776065	47
14	8·4471651	8·4516571	8·4561032	8·4605043	8·4648614	8·4691752	8·4734467	8·4776766	46
15	8·4472404	8·4517316	8·4561769	8·4605773	8·4649336	8·4692468	8·4735175	8·4777468	45
16	8·4473156	8·4518061	8·4562506	8·4606503	8·4650059	8·4693183	8·4735884	8·4778169	44
17	8·4473908	8·4518805	8·4563243	8·4607232	8·4650781	8·4693898	8·4736592	8·4778871	43
18	8·4474660	8·4519549	8·4563980	8·4607962	8·4651503	8·4694613	8·4737300	8·4779572	42
19	8·4475412	8·4520294	8·4564717	8·4608691	8·4652225	8·4695328	8·4738008	8·4780273	41
20	8·4476164	8·4521038	8·4565453	8·4609420	8·4652947	8·4696043	8·4738715	8·4780974	40
21	8·4476916	8·4521782	8·4566190	8·4610149	8·4653669	8·4696757	8·4739423	8·4781675	39
22	8·4477667	8·4522526	8·4566926	8·4610878	8·4654390	8·4697472	8·4740131	8·4782375	38
23	8·4478419	8·4523269	8·4567662	8·4611607	8·4655112	8·4698186	8·4740838	8·4783076	37
24	8·4479170	8·4524013	8·4568398	8·4612336	8·4655833	8·4698900	8·4741545	8·4783776	36
25	8·4479921	8·4524757	8·4569134	8·4613064	8·4656555	8·4699615	8·4742253	8·4784477	35
26	8·4480672	8·4525500	8·4569870	8·4613792	8·4657276	8·4700329	8·4742960	8·4785177	34
27	8·4481423	8·4526243	8·4570606	8·4614521	8·4657997	8·4701043	8·4743667	8·4785577	33
28	8·4482174	8·4526986	8·4571341	8·4615249	8·4658718	8·4701756	8·4744374	8·4785577	32
29	8·4482925	8·4527729	8·4572077	8·4615977	8·4659439	8·4702470	8·4745080	8·4787277	31
30	8·4483675	8·4528472	8·4572812	8·4616705	8·4660159	8·4703184	8·4745787	8·4787977	30
31	8·4484426	8·4529215	8·4573547	8·4617433	8·4660880	8·4703897	8·4746494	8·4788677	29
32	8·4485176	8·4529957	8·4574282	8·4618160	8·4661600	8·4704611	8·4747200	8·4789376	28
33	8·4485926	8·4530700	8·4575017	8·4618888	8·4662321	8·4705324	8·4747906	8·4790076	27
34	8·4486676	8·4531442	8·4575752	8·4619615	8·4663041	8·4706037	8·4748612	8·4790775	26
35	8·4487426	8·4532184	8·4576487	8·4620343	8·4663761	8·4706750	8·4749319	8·4791475	25
36	8·4488176	8·4532926	8·4577221	8·4621070	8·4664481	8·4707463	8·4750025	8·4792174	24
37	8·4488925	8·4533668	8·4577956	8·4621797	8·4665201	8·4708176	8·4750730	8·4792873	23
38	8·4489675	8·4534410	8·4578690	8·4622524	8·4665921	8·4708888	8·4751436	8·4793572	22
39	8·4490424	8·4535152	8·4579424	8·4623251	8·4666640	8·4709601	8·4752142	8·4794271	21
40	8·4491173	8·4535893	8·4580158	8·4623978	8·4667360	8·4710313	8·4752847	8·4794969	20
41	8·4491923	8·4536635	8·4580892	8·4624704	8·4668079	8·4711026	8·4753553	8·4795668	19
42	8·4492672	8·4537376	8·4581626	8·4625431	8·4668798	8·4711738	8·4754258	8·4796366	18
43	8·4493420	8·4538117	8·4582360	8·4626157	8·4669517	8·4712450	8·4754963	8·4797065	17
44	8·4494169	8·4538859	8·4583094	8·4626883	8·4670236	8·4713162	8·4755668	8·4797763	16
45	8·4494918	8·4539599	8·4583827	8·4627609	8·4670955	8·4713874	8·4756373	8·4798461	15
46	8·4495666	8·4540340	8·4584560	8·4628335	8·4671674	8·4714586	8·4757078	8·4799159	14
47	8·4496415	8·4541081	8·4585293	8·4629061	8·4672393	8·4715297	8·4757783	8·4799857	13
48	8·4497163	8·4541822	8·4586027	8·4629787	8·4673111	8·4716009	8·4758487	8·4800555	12
49	8·4497911	8·4542562	8·4586760	8·4630512	8·4673830	8·4716720	8·4759192	8·4801252	11
50	8·4498659	8·4543302	8·4587492	8·4631238	8·4674548	8·4717431	8·4759896	8·4801950	10
51	8·4499407	8·4544043	8·4588225	8·4631963	8·4675266	8·4718142	8·4760600	8·4802648	9
52	8·4500154	8·4544783	8·4588958	8·4632689	8·4675984	8·4718853	8·4761304	8·4803345	8
53	8·4500902	8·4545523	8·4589690	8·4633414	8·4676702	8·4719564	8·4762008	8·4804042	7
54	8·4501649	8·4546262	8·4590422	8·4634139	8·4677420	8·4720275	8·4762712	8·4804739	6
55	8·4502397	8·4547002	8·4591155	8·4634864	8·4678138	8·4720986	8·4763416	8·4805436	5
56	8·4503144	8·4547742	8·4591887	8·4635588	8·4678855	8·4721696	8·4764120	8·4806133	4
57	8·4503891	8·4548481	8·4592619	8·4636313	8·4679573	8·4722407	8·4764823	8·4806830	3
58	8·4504638	8·4549220	8·4593351	8·4637038	8·4680290	8·4723117	8·4765527	8·4807527	2
59	8·4505385	8·4549960	8·4594082	8·4637762	8·4681008	8·4723827	8·4766230	8·4808223	1
60	8·4506131	8·4550699	8·4594814	8·4638486	8·4681725	8·4724538	8·4766933	8·4808920	0
″	23′	22′	21′	20′	19′	18′	17′	16′	″

| | LOG. COTANGENTS. | | | | | | | 88 Deg. |

"	44′	45′	46′	47′	48′	49′	50′	51′	"
0	8·4806932	8·4848479	8·4889632	8·4930398	8·4970784	8·5010798	8·5050447	8·5089736	60
1	8·4807628	8·4849168	8·4890314	8·4931074	8·4971454	8·5011462	8·5051105	8·5090388	59
2	8·4808323	8·4849857	8·4890997	8·4931750	8·4972124	8·5012126	8·5051762	8·5091040	58
3	8·4809019	8·4850546	8·4891679	8·4932426	8·4972794	8·5012790	8·5052420	8·5091691	57
4	8·4809714	8·4851235	8·4892361	8·4933102	8·4973463	8·5013453	8·5053077	8·5092343	56
5	8·4810410	8·4851923	8·4893043	8·4933778	8·4974133	8·5014116	8·5053735	8·5092994	55
6	8·4811105	8·4852612	8·4893726	8·4934453	8·4974802	8·5014780	8·5054392	8·5093646	54
7	8·4811800	8·4853300	8·4894407	8·4935129	8·4975472	8·5015443	8·5055049	8·5094297	53
8	8·4812495	8·4853989	8·4895089	8·4935804	8·4976141	8·5016106	8·5055706	8·5094948	52
9	8·4813190	8·4854677	8·4895771	8·4936480	8·4976810	8·5016769	8·5056363	8·5095599	51
10	8·4813884	8·4855365	8·4896453	8·4937155	8·4977479	8·5017432	8·5057020	8·5096250	50
11	8·4814579	8·4856053	8·4897134	8·4937830	8·4978148	8·5018095	8·5057677	8·5096901	49
12	8·4815273	8·4856741	8·4897816	8·4938505	8·4978817	8·5018757	8·5058333	8·5097552	48
13	8·4815968	8·4857429	8·4898497	8·4939180	8·4979485	8·5019420	8·5058990	8·5098202	47
14	8·4816662	8·4858116	8·4899178	8·4939855	8·4980154	8·5020082	8·5059646	8·5098853	46
15	8·4817356	8·4858804	8·4899859	8·4940530	8·4980823	8·5020745	8·5060303	8·5099503	45
16	8·4818050	8·4859491	8·4900540	8·4941204	8·4981491	8·5021407	8·5060959	8·5100154	44
17	8·4818744	8·4860179	8·4901221	8·4941879	8·4982159	8·5022069	8·5061615	8·5100804	43
18	8·4819438	8·4860866	8·4901902	8·4942553	8·4982827	8·5022731	8·5062271	8·5101454	42
19	8·4820132	8·4861553	8·4902582	8·4943228	8·4983495	8·5023393	8·5062927	8·5102104	41
20	8·4820825	8·4862240	8·4903263	8·4943902	8·4984163	8·5024055	8·5063583	8·5102754	40
21	8·4821519	8·4862927	8·4903943	8·4944576	8·4984831	8·5024717	8·5064239	8·5103404	39
22	8·4822212	8·4863614	8·4904624	8·4945250	8·4985499	8·5025378	8·5064894	8·5104054	38
23	8·4822905	8·4864300	8·4905304	8·4945924	8·4986167	8·5026040	8·5065550	8·5104703	37
24	8·4823599	8·4864987	8·4905984	8·4946597	8·4986834	8·5026701	8·5066205	8·5105353	36
25	8·4824292	8·4865673	8·4906664	8·4947271	8·4987502	8·5027363	8·5066861	8·5106002	35
26	8·4824985	8·4866360	8·4907344	8·4947945	8·4988169	8·5028024	8·5067516	8·5106652	34
27	8·4825677	8·4867046	8·4908024	8·4948618	8·4988836	8·5028685	8·5068171	8·5107301	33
28	8·4826370	8·4867732	8·4908703	8·4949292	8·4989504	8·5029346	8·5068826	8·5107950	32
29	8·4827063	8·4868418	8·4909383	8·4949965	8·4990171	8·5030007	8·5069481	8·5108599	31
30	8·4827755	8·4869104	8·4910063	8·4950638	8·4990838	8·5030668	8·5070136	8·5109248	30
31	8·4828448	8·4869790	8·4910742	8·4951311	8·4991504	8·5031329	8·5070791	8·5109897	29
32	8·4829140	8·4870476	8·4911421	8·4951984	8·4992171	8·5031989	8·5071446	8·5110546	28
33	8·4829832	8·4871161	8·4912100	8·4952657	8·4992838	8·5032650	8·5072100	8·5111195	27
34	8·4830524	8·4871847	8·4912779	8·4953330	8·4993504	8·5033310	8·5072755	8·5111843	26
35	8·4831216	8·4872532	8·4913458	8·4954002	8·4994171	8·5033971	8·5073409	8·5112492	25
36	8·4831908	8·4873217	8·4914137	8·4954675	8·4994837	8·5034631	8·5074063	8·5113140	24
37	8·4832600	8·4873903	8·4914816	8·4955347	8·4995503	8·5035291	8·5074717	8·5113789	23
38	8·4833291	8·4874588	8·4915495	8·4956020	8·4996169	8·5035951	8·5075371	8·5114437	22
39	8·4833983	8·4875273	8·4916173	8·4956692	8·4996835	8·5036611	8·5076025	8·5115085	21
40	8·4834674	8·4875957	8·4916852	8·4957364	8·4997501	8·5037271	8·5076679	8·5115733	20
41	8·4835365	8·4876642	8·4917530	8·4958036	8·4998167	8·5037931	8·5077333	8·5116381	19
42	8·4836057	8·4877327	8·4918208	8·4958708	8·4998833	8·5038590	8·5077987	8·5117029	18
43	8·4836748	8·4878011	8·4918886	8·4959380	8·4999499	8·5039250	8·5078640	8·5117676	17
44	8·4837439	8·4878696	8·4919564	8·4960051	8·5000164	8·5039909	8·5079294	8·5118324	16
45	8·4838129	8·4879380	8·4920242	8·4960723	8·5000829	8·5040569	8·5079947	8·5118972	15
46	8·4838820	8·4880064	8·4920920	8·4961394	8·5001495	8·5041228	8·5080601	8·5119619	14
47	8·4839511	8·4880748	8·4921598	8·4962066	8·5002160	8·5041887	8·5081254	8·5120266	13
48	8·4840201	8·4881432	8·4922275	8·4962737	8·5002825	8·5042546	8·5081907	8·5120914	12
49	8·4840892	8·4882116	8·4922953	8·4963408	8·5003490	8·5043205	8·5082560	8·5121561	11
50	8·4841582	8·4882800	8·4923630	8·4964079	8·5004155	8·5043864	8·5083213	8·5122208	10
51	8·4842272	8·4883484	8·4924307	8·4964750	8·5004820	8·5044523	8·5083866	8·5122855	9
52	8·4842962	8·4884167	8·4924984	8·4965421	8·5005485	8·5045181	8·5084518	8·5123502	8
53	8·4843652	8·4884851	8·4925661	8·4966092	8·5006149	8·5045840	8·5085171	8·5124148	7
54	8·4844342	8·4885534	8·4926338	8·4966763	8·5006814	8·5046498	8·5085823	8·5124795	6
55	8·4845032	8·4886217	8·4927015	8·4967433	8·5007478	8·5047157	8·5086476	8·5125442	5
56	8·4845721	8·4886900	8·4927692	8·4968104	8·5008142	8·5047815	8·5087128	8·5126088	4
57	8·4846411	8·4887583	8·4928368	8·4968774	8·5008806	8·5048473	8·5087780	8·5126735	3
58	8·4847100	8·4888266	8·4929045	8·4969444	8·5009471	8·5049131	8·5088432	8·5127381	2
59	8·4847790	8·4888949	8·4929721	8·4970114	8·5010135	8·5049789	8·5089084	8·5128027	1
60	8·4848479	8·4889632	8·4930398	8·4970784	8·5010798	8·5050447	8·5089736	8·5128673	0
"	15′	14′	13′	12′	11′	10′	9′	8′	"

"	44'	45'	46'	47'	48'	49'	50'	51'	"
0	8·4808920	8·4850505	8·4891696	8·4932502	8·4972928	8·5012982	8·5052671	8·5092001	60
1	8·4809616	8·4851195	8·4892380	8·4933179	8·4973598	8·5013646	8·5053329	8·5092653	59
2	8·4810312	8·4851884	8·4893063	8·4933855	8·4974269	8·5014311	8·5053987	8·5093305	58
3	8·4811008	8·4852574	8·4893746	8·4934532	8·4974939	8·5014975	8·5054646	8·5093958	57
4	8·4811704	8·4853263	8·4894429	8·4935208	8·4975610	8·5015639	8·5055304	8·5094610	56
5	8·4812400	8·4853953	8·4895112	8·4935885	8·4976280	8·5016303	8·5055962	8·5095262	55
6	8·4813096	8·4854642	8·4895794	8·4936561	8·4976950	8·5016967	8·5056620	8·5095914	54
7	8·4813792	8·4855331	8·4896477	8·4937237	8·4977620	8·5017631	8·5057277	8·5096566	53
8	8·4814487	8·4856020	8·4897159	8·4937914	8·4978290	8·5018295	8·5057935	8·5097218	52
9	8·4815183	8·4856709	8·4897842	8·4938590	8·4978959	8·5018958	8·5058593	8·5097870	51
10	8·4815878	8·4857397	8·4898524	8·4939266	8·4979629	8·5019622	8·5059250	8·5098521	50
11	8·4816574	8·4858086	8·4899206	8·4939941	8·4980299	8·5020285	8·5059908	8·5099173	49
12	3·4817269	8·4858775	8·4899888	8·4940617	8·4980968	8·5020949	8·5060565	8·5099824	48
13	8·4817964	8·4859463	8·4900570	8·4941293	8·4981638	8·5021612	8·5061222	8·5100475	47
14	8·4818659	8·4860151	8·4901252	8·4941968	8·4982307	8·5022275	8·5061879	8·5101127	46
15	8·4819353	8·4860839	8·4901934	8·4942643	8·4982976	8·5022938	8·5062536	8·5101778	45
16	8·4820048	8·4861528	8·4902615	8·4943319	8·4983645	8·5023601	8·5063193	8·5102429	44
17	8·4820743	8·4862216	8·4903297	8·4943994	8·4984314	8·5024264	8·5063850	8·5103080	43
18	8·4821437	8·4862903	8·4903978	8·4944669	8·4984983	8·5024927	8·5064507	8·5103731	42
19	8·4822131	8·4863591	8·4904660	8·4945344	8·4985652	8·5025589	8·5065164	8·5104381	41
20	8·4822826	8·4864279	8·4905341	8·4946019	8·4986320	8·5026252	8·5065820	8·5105032	40
21	8·4823520	8·4864966	8·4906022	8·4946694	8·4986989	8·5026914	8·5066477	8·5105683	39
22	8·4824214	8·4865654	8·4906703	8·4947368	8·4987657	8·5027576	8·5067133	8·5106333	38
23	8·4824908	8·4866341	8·4907384	8·4948043	8·4988325	8·5028239	8·5067789	8·5106983	37
24	8·4825602	8·4867028	8·4908065	8·4948717	8·4988994	8·5028901	8·5068445	8·5107634	36
25	8·4826295	8·4867716	8·4908745	8·4949392	8·4989662	8·5029563	8·5069101	8·5108284	35
26	8·4826989	8·4868403	8·4909426	3·4950066	8·4990330	8·5030225	8·5069757	8·5108934	34
27	8·4827682	8·4869089	8·4910106	8·4950740	8·4990998	8·5030887	8·5070413	8·5109584	33
28	8·4828376	8·4869776	8·4910787	8·4951414	8·4991666	8·5031548	8·5071069	8·5110234	32
29	8·4829069	8·4870463	8·4911467	8·4952088	8·4992333	8·5032210	8·5071724	8·5110883	31
30	8·4829762	8·4871149	8·4912147	8·4952762	8·4993001	8·5032871	8·5072380	8·5111533	30
31	8·4830455	8·4871836	8·4912827	8·4953435	8·4993668	8·5033533	8·5073035	8·5112183	29
32	8·4831148	8·4872522	8·4913507	8·4954109	8·4994336	8·5034194	8·5073691	8·5112832	28
33	8·4831841	8·4873209	8·4914187	8·4954782	8·4995003	8·5034855	8·5074346	8·5113482	27
34	8·4832533	8·4873895	8·4914866	8·4955456	8·4995670	8·5035517	8·5075001	8·5114131	26
35	8·4833226	8·4874581	8·4915546	8·4956129	8·4996337	8·5036178	8·5075656	8·5114780	25
36	8·4833919	8·4875267	8·4916226	8·4956802	8·4997004	8·5036838	8·5076311	8·5115429	24
37	8·4834611	8·4875952	8·4916905	8·4957476	8·4997671	8·5037499	8·5076966	8·5116078	23
38	8·4835303	8·4876638	8·4917584	8·4958148	8·4998338	8·5038160	8·5077621	8·5116727	22
39	8·4835995	8·4877324	8·4918263	8·4958821	8·4999005	8·5038821	8·5078275	8·5117376	21
40	8·4836687	8·4878009	8·4918942	8·4959494	8·4999671	8·5039481	8·5078930	8·5118025	20
41	8·4837379	8·4878695	8·4919621	8·4960167	8·5000338	8·5040142	8·5079584	8·5118673	19
42	8·4838071	8·4879380	8·4920300	8·4960839	8·5001004	8·5040802	8·5080239	8·5119322	18
43	8·4838763	8·4880065	8·4920979	8·4961512	8·5001671	8·5041462	8·5080893	8·5119970	17
44	8·4839454	8·4880750	8·4921658	8·4962184	8·5002337	8·5042122	8·5081547	8·5120618	16
45	8·4840146	8·4881435	8·4922336	8·4962856	8·5003003	8·5042782	8·5082201	8·5121267	15
46	8·4840837	8·4882120	8·4923015	8·4963529	8·5003669	8·5043442	8·5082855	8·5121915	14
47	8·4841528	8·4882805	8·4923693	8·4964201	8·5004335	8·5044102	8·5083509	8·5122563	13
48	8·4842220	8·4883489	8·4924371	8·4964873	8·5005000	8·5044762	8·5084163	8·5123211	12
49	8·4842911	8·4884174	8·4925049	8·4965544	8·5005666	8·5045421	8·5084817	8·5123859	11
50	8·4843602	8·4884858	8·4925727	8·4966216	8·5006332	8·5046081	8·5085470	8·5124506	10
51	8·4844292	8·4885543	8·4926405	8·4966888	8·5006997	8·5046740	8·5086124	8·5125154	9
52	8·4844983	8·4886227	8·4927083	8·4967559	8·5007663	8·5047400	8·5086777	8·5125801	8
53	8·4845674	8·4886911	8·4927761	8·4968231	8·5008328	8·5048059	8·5087430	8·5126449	7
54	8·4846364	8·4887595	8·4928438	8·4968902	8·5008993	8·5048718	8·5088084	8·5127096	6
55	8·4847055	8·4888279	8·4929116	8·4969573	8·5009658	8·5049377	8·5088737	8·5127743	5
56	8·4847745	8·4888962	8·4929793	8·4970244	8·5010323	8·5050036	8·5089390	8·5128391	4
57	8·4848435	8·4889646	8·4930471	8·4970915	8·5010988	8·5050695	8·5090042	8·5129038	3
58	8·4849125	8·4890330	8·4931148	8·4971586	8·5011653	8·5051353	8·5090695	8·5129685	2
59	8·4849815	8·4891013	8·4931825	8·4972257	8·5012317	8·5052012	8·5091348	8·5130332	1
60	8·4850505	8·4891696	8·4932502	8·4972928	8·5012982	8·5052671	8·5092001	8·5130978	0
"	15'	14'	13'	12'	11'	10'	9'	8'	"

"	52'	53'	54'	55'	56'	57'	58'	59'	"
0	8·5128673	8·5167264	8·5205514	8·5243430	8·5281017	8·5318281	8·5355228	8·5391863	60
1	8·5129319	8·5167904	8·5206148	8·5244059	8·5281641	8·5318900	8·5355842	8·5392471	59
2	8·5129965	8·5168544	8·5206783	8·5244688	8·5282264	8·5319518	8·5356455	8·5393079	58
3	8·5130611	8·5169184	8·5207417	8·5245317	8·5282888	8·5320136	8·5357068	8·5393687	57
4	8·5131256	8·5169824	8·5208052	8·5245946	8·5283511	8·5320754	8·5357680	8·5394295	56
5	8·5131902	8·5170464	8·5208686	8·5246574	8·5284135	8·5321372	8·5358293	8·5394902	55
6	8·5132548	8·5171104	8·5209320	8·5247203	8·5284758	8·5321990	8·5358906	8·5395510	54
7	8·5133193	8·5171743	8·5209954	8·5247832	8·5285381	8·5322608	8·5359518	8·5396117	53
8	8·5133838	8·5172383	8·5210588	8·5248460	8·5286004	8·5323226	8·5360131	8·5396725	52
9	8·5134484	8·5173023	8·5211222	8·5249088	8·5286627	8·5323844	8·5360743	8·5397332	51
10	8·5135129	8·5173662	8·5211856	8·5249717	8·5287250	8·5324461	8·5361356	8·5397939	50
11	8·5135774	8·5174301	8·5212490	8·5250345	8·5287873	8·5325079	8·5361968	8·5398546	49
12	8·5136419	8·5174941	8·5213123	8·5250973	8·5288495	8·5325696	8·5362580	8·5399153	48
13	8·5137064	8·5175580	8·5213757	8·5251601	8·5289118	8·5326313	8·5363192	8·5399760	47
14	8·5137708	8·5176219	8·5214390	8·5252229	8·5289741	8·5326931	8·5363804	8·5400367	46
15	8·5138353	8·5176858	8·5215024	8·5252857	8·5290363	8·5327548	8·5364416	8·5400974	45
16	8·5138997	8·5177497	8·5215657	8·5253485	8·5290985	8·5328165	8·5365028	8·5401581	44
17	8·5139642	8·5178135	8·5216290	8·5254112	8·5291608	8·5328782	8·5365640	8·5402187	43
18	8·5140286	8·5178774	8·5216923	8·5254740	8·5292230	8·5329399	8·5366251	8·5402794	42
19	8·5140931	8·5179413	8·5217556	8·5255367	8·5292852	8·5330015	8·5366863	8·5403400	41
20	8·5141575	8·5180051	8·5218189	8·5255995	8·5293474	8·5330632	8·5367474	8·5404007	40
21	8·5142219	8·5180689	8·5218822	8·5256622	8·5294096	8·5331249	8·5368086	8·5404613	39
22	8·5142863	8·5181328	8·5219455	8·5257249	8·5294718	8·5331865	8·5368697	8·5405219	38
23	8·5143507	8·5181966	8·5220087	8·5257877	8·5295339	8·5332482	8·5369308	8·5405825	37
24	8·5144150	8·5182604	8·5220720	8·5258504	8·5295961	8·5333098	8·5369920	8·5406431	36
25	8·5144794	8·5183242	8·5221352	8·5259131	8·5296583	8·5333714	8·5370531	8·5407037	35
26	8·5145438	8·5183880	8·5221985	8·5259757	8·5297204	8·5334330	8·5371142	8·5407643	34
27	8·5146081	8·5184518	8·5222617	8·5260384	8·5297826	8·5334946	8·5371752	8·5408249	33
28	8·5146725	8·5185156	8·5223249	8·5261011	8·5298447	8·5335562	8·5372363	8·5408854	32
29	8·5147368	8·5185793	8·5223881	8·5261637	8·5299068	8·5336178	8·5372974	8·5409460	31
30	8·5148011	8·5186431	8·5224513	8·5262264	8·5299689	8·5336794	8·5373585	8·5410066	30
31	8·5148654	8·5187068	8·5225145	8·5262890	8·5300310	8·5337410	8·5374195	8·5410671	29
32	8·5149297	8·5187706	8·5225777	8·5263517	8·5300931	8·5338026	8·5374806	8·5411276	28
33	8·5149940	8·5188343	8·5226408	8·5264143	8·5301552	8·5338641	8·5375416	8·5411882	27
34	8·5150583	8·5188980	8·5227040	8·5264769	8·5302173	8·5339257	8·5376026	8·5412487	26
35	8·5151226	8·5189617	8·5227672	8·5265395	8·5302793	8·5339872	8·5376636	8·5413092	25
36	8·5151869	8·5190254	8·5228303	8·5266021	8·5303414	8·5340487	8·5377247	8·5413697	24
37	8·5152511	8·5190891	8·5228934	8·5266647	8·5304034	8·5341103	8·5377857	8·5414302	23
38	8·5153154	8·5191528	8·5229566	8·5267273	8·5304655	8·5341718	8·5378466	8·5414907	22
39	8·5153796	8·5192164	8·5230197	8·5267898	8·5305275	8·5342333	8·5379076	8·5415511	21
40	8·5154438	8·5192801	8·5230828	8·5268524	8·5305895	8·5342948	8·5379686	8·5416116	20
41	8·5155080	8·5193438	8·5231459	8·5269149	8·5306516	8·5343563	8·5380296	8·5416721	19
42	8·5155722	8·5194074	8·5232090	8·5269775	8·5307136	8·5344177	8·5380905	8·5417325	18
43	8·5156364	8·5194710	8·5232720	8·5270400	8·5307756	8·5344792	8·5381515	8·5417929	17
44	8·5157006	8·5195347	8·5233351	8·5271025	8·5308375	8·5345407	8·5382124	8·5418534	16
45	8·5157648	8·5195983	8·5233982	8·5271651	8·5308995	8·5346021	8·5382734	8·5419138	15
46	8·5158290	8·5196619	8·5234612	8·5272276	8·5309615	8·5346636	8·5383343	8·5419742	14
47	8·5158931	8·5197255	8·5235243	8·5272901	8·5310235	8·5347250	8·5383952	8·5420346	13
48	8·5159573	8·5197891	8·5235873	8·5273525	8·5310854	8·5347864	8·5384561	8·5420950	12
49	8·5160214	8·5198526	8·5236503	8·5274150	8·5311473	8·5348478	8·5385170	8·5421554	11
50	8·5160856	8·5199162	8·5237133	8·5274775	8·5312093	8·5349092	8·5385779	8·5422158	10
51	8·5161497	8·5199798	8·5237763	8·5275400	8·5312712	8·5349706	8·5386388	8·5422762	9
52	8·5162138	8·5200433	8·5238393	8·5276024	8·5313331	8·5350320	8·5386997	8·5423365	8
53	8·5162779	8·5201069	8·5239023	8·5276648	8·5313950	8·5350934	8·5387605	8·5423969	7
54	8·5163420	8·5201704	8·5239653	8·5277273	8·5314569	8·5351548	8·5388214	8·5424572	6
55	8·5164061	8·5202339	8·5240283	8·5277897	8·5315188	8·5352161	8·5388822	8·5425176	5
56	8·5164701	8·5202974	8·5240912	8·5278521	8·5315807	8·5352775	8·5389431	8·5425779	4
57	8·5165342	8·5203609	8·5241542	8·5279145	8·5316426	8·5353389	8·5390039	8·5426382	3
58	8·5165983	8·5204244	8·5242171	8·5279769	8·5317044	8·5354002	8·5390647	8·5426986	2
59	8·5166623	8·5204879	8·5242800	8·5280393	8·5317663	8·5354615	8·5391255	8·5427589	1
60	8·5167264	8·5205514	8·5243430	8·5281017	8·5318281	8·5355228	8·5391863	8·5428192	0
"	7'	6'	5'	4'	3'	2'	1'	0'	"

`"`	52′	53′	54′	55′	56′	57′	58′	59′	`"`
0	8·5130978	8·5169610	8·5207902	8·5245860	8·5283490	8·5320797	8·5357787	8·5394466	60
1	8·5131625	8·5170251	8·5208537	8·5246490	8·5284114	8·5321416	8·5358401	8·5395075	59
2	8·5132272	8·5170892	8·5209173	8·5247120	8·5284739	8·5322035	8·5359015	8·5395683	58
3	8·5132918	8·5171533	8·5209808	8·5247749	8·5285363	8·5322654	8·5359629	8·5396292	57
4	8·5133564	8·5172173	8·5210443	8·5248379	8·5285987	8·5323273	8·5360242	8·5396900	56
5	8·5134211	8·5172814	8·5211078	8·5249008	8·5286611	8·5323892	8·5360856	8·5397509	55
6	8·5134857	8·5173455	8·5211713	8·5249638	8·5287235	8·5324510	8·5361469	8·5398117	54
7	8·5135503	8·5174095	8·5212348	8·5250267	8·5287859	8·5325129	8·5362082	8·5398725	53
8	8·5136149	8·5174735	8·5212982	8·5250896	8·5288483	8·5325747	8·5362696	8·5399333	52
9	8·5136795	8·5175375	8·5213617	8·5251525	8·5289106	8·5326366	8·5363309	8·5399941	51
10	8·5137441	8·5176016	8·5214251	8·5252154	8·5289730	8·5326984	8·5363922	8·5400549	50
11	8·5138087	8·5176656	8·5214886	8·5252783	8·5290353	8·5327602	8·5364535	8·5401157	49
12	8·5138732	8·5177296	8·5215520	8·5253412	8·5290977	8·5328220	8·5365148	8·5401765	48
13	8·5139378	8·5177935	8·5216154	8·5254041	8·5291600	8·5328838	8·5365761	8·5402372	47
14	8·5140023	8·5178575	8·5216789	8·5254669	8·5292223	8·5329456	8·5366373	8·5402980	46
15	8·5140668	8·5179215	8·5217423	8·5255298	8·5292847	8·5330074	8·5366986	8·5403587	45
16	8·5141314	8·5179854	8·5218057	8·5255926	8·5293470	8·5330692	8·5367599	8·5404195	44
17	8·5141959	8·5180494	8·5218690	8·5256555	8·5294093	8·5331310	8·5368211	8·5404802	43
18	8·5142604	8·5181133	8·5219324	8·5257183	8·5294716	8·5331927	8·5368823	8·5405409	42
19	8·5143249	8·5181772	8·5219958	8·5257811	8·5295338	8·5332545	8·5369436	8·5406017	41
20	8·5143894	8·5182412	8·5220591	8·5258439	8·5295961	8·5333162	8·5370048	8·5406624	40
21	8·5144539	8·5183051	8·5221225	8·5259067	8·5296584	8·5333779	8·5370660	8·5407231	39
22	8·5145183	8·5183690	8·5221858	8·5259695	8·5297206	8·5334397	8·5371272	8·5407838	38
23	8·5145828	8·5184329	8·5222492	8·5260323	8·5297829	8·5335014	8·5371884	8·5408445	37
24	8·5146472	8·5184967	8·5223125	8·5260951	8·5298451	8·5335631	8·5372496	8·5409051	36
25	8·5147117	8·5185606	8·5223758	8·5261579	8·5299073	8·5336248	8·5373108	8·5409658	35
26	8·5147761	8·5186245	8·5224391	8·5262206	8·5299696	8·5336865	8·5373719	8·5410264	34
27	8·5148405	8·5186883	8·5225024	8·5262834	8·5300318	8·5337482	8·5374331	8·5410871	33
28	8·5149049	8·5187522	8·5225657	8·5263461	8·5300940	8·5338098	8·5374942	8·5411477	32
29	8·5149693	8·5188160	8·5226290	8·5264088	8·5301562	8·5338715	8·5375554	8·5412084	31
30	8·5150337	8·5188798	8·5226922	8·5264716	8·5302183	8·5339331	8·5376165	8·5412690	30
31	8·5150981	8·5189436	8·5227555	8·5265343	8·5302805	8·5339948	8·5376777	8·5413296	29
32	8·5151625	8·5190074	8·5228187	8·5265970	8·5303427	8·5340564	8·5377388	8·5413902	28
33	8·5152268	8·5190712	8·5228820	8·5266597	8·5304048	8·5341181	8·5377999	8·5414508	27
34	8·5152912	8·5191350	8·5229452	8·5267223	8·5304670	8·5341797	8·5378610	8·5415114	26
35	8·5153555	8·5191988	8·5230084	8·5267850	8·5305291	8·5342413	8·5379221	8·5415720	25
36	8·5154199	8·5192626	8·5230717	8·5268477	8·5305912	8·5343029	8·5379832	8·5416326	24
37	8·5154842	8·5193263	8·5231349	8·5269103	8·5306534	8·5343645	8·5380442	8·5416931	23
38	8·5155485	8·5193901	8·5231980	8·5269730	8·5307155	8·5344261	8·5381053	8·5417537	22
39	8·5156128	8·5194538	8·5232612	8·5270356	8·5307776	8·5344876	8·5381664	8·5418142	21
40	8·5156771	8·5195175	8·5233244	8·5270983	8·5308397	8·5345492	8·5382274	8·5418748	20
41	8·5157414	8·5195813	8·5233876	8·5271609	8·5309018	8·5346108	8·5382884	8·5419353	19
42	8·5158057	8·5196450	8·5234507	8·5272235	8·5309638	8·5346723	8·5383495	8·5419958	18
43	8·5158699	8·5197087	8·5235139	8·5272861	8·5310259	8·5347339	8·5384105	8·5420563	17
44	8·5159342	8·5197724	8·5235770	8·5273487	8·5310880	8·5347954	8·5384715	8·5421168	16
45	8·5159984	8·5198361	8·5236401	8·5274113	8·5311500	8·5348569	8·5385325	8·5421773	15
46	8·5160627	8·5198997	8·5237033	8·5274739	8·5312121	8·5349184	8·5385935	8·5422378	14
47	8·5161269	8·5199634	8·5237664	8·5275364	8·5312741	8·5349799	8·5386545	8·5422983	13
48	8·5161911	8·5200271	8·5238295	8·5275990	8·5313361	8·5350414	8·5387155	8·5423588	12
49	8·5162553	8·5200907	8·5238926	8·5276615	8·5313981	8·5351029	8·5387765	8·5424193	11
50	8·5163195	8·5201543	8·5239557	8·5277241	8·5314601	8·5351644	8·5388374	8·5424797	10
51	8·5163837	8·5202180	8·5240187	8·5277866	8·5315221	8·5352259	8·5388984	8·5425402	9
52	8·5164479	8·5202816	8·5240818	8·5278491	8·5315841	8·5352873	8·5389593	8·5426006	8
53	8·5165121	8·5203452	8·5241449	8·5279116	8·5316461	8·5353488	8·5390203	8·5426610	7
54	8·5165762	8·5204088	8·5242079	8·5279741	8·5317081	8·5354102	8·5390812	8·5427214	6
55	8·5166404	8·5204724	8·5242709	8·5280366	8·5317700	8·5354717	8·5391421	8·5427819	5
56	8·5167045	8·5205360	8·5243340	8·5280991	8·5318320	8·5355331	8·5392030	8·5428423	4
57	8·5167687	8·5205995	8·5243970	8·5281616	8·5318939	8·5355945	8·5392639	8·5429027	3
58	8·5168328	8·5206631	8·5244600	8·5282241	8·5319559	8·5356559	8·5393248	8·5429631	2
59	8·5168969	8·5207267	8·5245230	8·5282865	8·5320178	8·5357173	8·5393857	8·5430234	1
60	8·5169610	8·5207902	8·5245860	8·5283490	8·5320797	8·5357787	8·5394466	8·5430838	0
`"`	7′	6′	5′	4′	3′	2′	1′	0′	`"`

′	Sine.	Dif.	Covers	Cosec.	Tang.	Cotang.	Secant	Vers.	D.	Cosine	′
0	0000000	2909	1·000000	Infinite.	0000000	Infinite.	1·0000000	0000000	0	1·000000	60
1	0002909	2909	9997091	3437·7468	0002909	3437·7467	1·0000000	0000000	2	1·000000	59
2	0005818	2909	9994182	1718·8735	0005818	1718·8732	1·0000002	0000002	2	9999998	58
3	0008727	2909	9991273	1145·9157	0008727	1145·9153	1·0000004	0000004	3	9999996	57
4	0011636	2908	9988364	859·43689	0011636	859·43630	1·0000007	0000007	3	9999993	56
5	0014544	2909	9985456	687·54960	0014544	687·54887	1·0000011	0000011	4	9999989	55
6	0017453	2909	9982547	572·95809	0017453	572·95721	1·0000015	0000015	4	9999985	54
7	0020362	2909	9979638	491·10702	0020362	491·10600	1·0000021	0000021	6	9999979	53
8	0023271	2909	9976729	429·71873	0023271	429·71757	1·0000027	0000027	6	9999973	52
9	0026180	2909	9973820	381·97230	0026180	381·97099	1·0000034	0000034	7	9999966	51
10	0029089	2909	9970911	343·77516	0029089	343·77371	1·0000042	0000042	8	9999958	50
11	0031998	2909	9968002	312·52297	0031998	312·52137	1·0000051	0000051	9	9999949	49
12	0034907	2908	9965093	286·47948	0034907	286·47773	1·0000061	0000061	10	9999939	48
13	0037815	2909	9962185	264·44269	0037816	264·44080	1·0000072	0000072	11	9999928	47
14	0040724	2909	9959276	245·55402	0040725	245·55198	1·0000083	0000083	11	9999917	46
15	0043633	2909	9956367	229·18385	0043634	229·18166	1·0000095	0000095	12	9999905	45
16	0046542	2909	9953458	214·85995	0046542	214·85762	1·0000108	0000108	13	9999892	44
17	0049451	2909	9950549	202·22122	0049451	202·21875	1·0000122	0000122	14	9999878	43
18	0052360	2908	9947640	190·98680	0052360	190·98419	1·0000137	0000137	16	9999863	42
19	0055268	2909	9944732	180·93496	0055269	180·93220	1·0000153	0000153	16	9999847	41
20	0058177	2909	9941823	171·88831	0058178	171·88540	1·0000169	0000169	18	9999831	40
21	0061086	2909	9938914	163·70325	0061087	163·70019	1·0000187	0000187	18	9999813	39
22	0063995	2909	9936005	156·26228	0063996	156·25908	1·0000205	0000205	19	9999795	38
23	0066904	2909	9933096	149·46837	0066905	149·46502	1·0000224	0000224	20	9999776	37
24	0069813	2908	9930187	143·24061	0069814	143·23712	1·0000244	0000244	20	9999756	36
25	0072721	2909	9927279	137·51108	0072723	137·50745	1·0000264	0000264	22	9999736	35
26	0075630	2909	9924370	132·22229	0075632	132·21851	1·0000286	0000286	22	9999714	34
27	0078539	2909	9921461	127·32526	0078541	127·32134	1·0000308	0000308	24	9999692	33
28	0081448	2909	9918552	122·77803	0081450	122·77396	1·0000332	0000332	24	9999668	32
29	0084357	2908	9915643	118·54440	0084360	118·54018	1·0000356	0000356	25	9999644	31
30	0087265	2909	9912735	114·59301	0087269	114·58865	1·0000381	0000381	26	9999619	30
31	0090174	2909	9909826	110·89656	0090178	110·89205	1·0000407	0000407	26	9999593	29
32	0093083	2909	9906917	107·43114	0093087	107·42648	1·0000433	0000433	28	9999567	28
33	0095992	2909	9904008	104·17574	0095996	104·17094	1·0000461	0000461	28	9999539	27
34	0098900	2909	9901100	101·11185	0098905	101·10690	1·0000489	0000489	29	9999511	26
35	0101809	2909	9898191	98·223033	0101814	98·217943	1·0000518	0000518	30	9999482	25
36	0104718	2909	9895282	95·494711	0104724	95·489475	1·0000548	0000548	31	9999452	24
37	0107627	2908	9892373	92·913869	0107633	92·908487	1·0000579	0000579	32	9999421	23
38	0110535	2909	9889465	90·468863	0110542	90·463336	1·0000611	0000611	32	9999389	22
39	0113444	2909	9886556	88·149244	0113451	88·143572	1·0000644	0000643	34	9999357	21
40	0116353	2908	9883647	85·945609	0116361	85·939791	1·0000677	0000677	34	9999323	20
41	0119261	2909	9880739	83·849470	0119270	83·843507	1·0000711	0000711	35	9999289	19
42	0122170	2909	9877830	81·853150	0122179	81·847041	1·0000746	0000746	36	9999254	18
43	0125079	2908	9874921	79·949684	0125088	79·943430	1·0000782	0000782	37	9999218	17
44	0127987	2909	9872013	78·132742	0127998	78·126342	1·0000819	0000819	38	9999181	16
45	0130896	2909	9869104	76·396554	0130907	76·390009	1·0000857	0000857	38	9999143	15
46	0133805	2908	9866195	74·735856	0133817	74·729165	1·0000895	0000895	40	9999105	14
47	0136713	2909	9863287	73·145827	0136726	73·138991	1·0000935	0000935	40	9999065	13
48	0139622	2908	9860378	71·622052	0139635	71·615070	1·0000975	0000975	41	9999025	12
49	0142530	2909	9857470	70·160474	0142545	70·153346	1·0001016	0001016	42	9998984	11
50	0145439	2909	9854561	68·757360	0145454	68·750086	1·0001058	0001058	42	9998942	10
51	0148348	2908	9851652	67·409272	0148364	67·401854	1·0001101	0001100	44	9998900	9
52	0151256	2909	9848744	66·113036	0151273	66·105473	1·0001144	0001144	44	9998856	8
53	0154165	2908	9845835	64·865716	0154183	64·858008	1·0001189	0001188	46	9998812	7
54	0157073	2909	9842927	63·664595	0157093	63·656741	1·0001234	0001234	46	9998766	6
55	0159982	2908	9840018	62·507153	0160000	62·499154	1·0001280	0001280	47	9998720	5
56	0162890	2909	9837110	61·391050	0162912	61·382905	1·0001327	0001327	48	9998673	4
57	0165799	2908	9834201	60·314110	0165821	60·305820	1·0001375	0001375	48	9998625	3
58	0168707	2909	9831293	59·274308	0168731	59·265872	1 0001423	0001423	50	9998577	2
59	0171616	2908	9828384	58·269755	0171641	58·261174	1·0001473	0001473	50	9998527	1
60	0174524		9825476	57·298688	0174551	57·289962	1·0001523	0001523		9998477	0

| ′ | Cosine | Dif. | Vers. | Secant. | Cotan. | Tang. | Cosec. | Covers | D. | Sine. | ′ |

'	Sine.	Diff.	Cosec.	Verseds.	Tang.	Diff.	Cotang.	Covers.	Secant.	D	Cosine.	'
0	Inf. Neg.	Infin.	Infinite.	Inf. Neg.	Inf. Neg	Infin.	Infinite.	10·000000	10·0000000		10·000000	60
1	6·4637261	3010300	13·5362739	2·6264222	6·4637261	3010301	13·5362739	9·9998737	10·0000000	0	10·000000	59
2	6·7647561	1760012	13·2352439	3·2284822	6·7647562	1760013	13·2352438	9·9997473	10·0000001	1	9·9999999	58
3	6·9408473	1249387	13·0591527	3·5806647	6·9408475	1249388	13·0591525	9·9996208	10·0000003	1	9·9999997	57
4	7·0657860	969100	12·9342140	3·8305422	7·0657863	960101	12·9342137	9·9994944	10·0000003	1	9·9999997	56
5	7·1620960	791811	12·8373040	4·0243622	7·1620964	791814	12·8373036	9·9993679	10·0000005	2	9·9999995	55
6	7·2418771	669468	12·7581229	4·1827246	7·2418778	669470	12·7581222	9·9992414	10·0000007	2	9·9999993	54
7	7·3088239	579918	12·6911761	4·3166182	7·3088248	579921	12·6911752	9·9991148	10·0000009	3	9·9999991	53
8	7·3668157	511524	12·6331843	4·4326020	7·3668169	511527	12·6331831	9·9989882	10·0000012	3	9·9999988	52
9	7·4179681	457574	12·5820319	4·5349070	7·4179696	457577	12·5820304	9·9988615	10·0000015	3	9·9999985	51
10	7·4637255	413926	12·5362745	4·6264219	7·4637273	413930	12·5362727	9·9987348	10·0000018	3	9·9999982	50
11	7·5051181	377884	12·4948819	4·7092072	7·5051203	377888	12·4948797	9·9986081	10·0000022	4	9·9999978	49
12	7·5429065	347619	12·4570935	4·7847843	7·5429091	347624	12·4570909	9·9984814	10·0000026	4	9·9999974	48
13	7·5776684	321846	12·4223316	4·8543084	7·5776715	321851	12·4223285	9·9983546	10·0000031	5	9·9999969	47
14	7·6098530	299630	12·3901470	4·9186777	7·6098566	299635	12·3901434	9·9982278	10·0000036	5	9·9999964	46
15	7·6398160	280285	12·3601840	4·9786041	7·6398201	280291	12·3601799	9·9981009	10·0000041	5	9·9999959	45
16	7·6678445	263288	12·3321555	5·0346614	7·6678492	263294	12·3321508	9·9979740	10·0000047	6	9·9999953	44
17	7·6941733	248233	12·3058267	5·0873192	7·6941786	248240	12·3058214	9·9978471	10·0000053	7	9·9999947	43
18	7·7189966	234809	12·2810034	5·1369663	7·7190026	234815	12·2809974	9·9977201	10·0000060	6	9·9999940	42
19	7·7424775	222762	12·2575225	5·1839283	7·7424841	222769	12·2575159	9·9975931	10·0000066	7	9·9999934	41
20	7·7647537	211890	12·2352463	5·2284810	7·7647610	211898	12·2352390	9·9974660	10·0000073	7	9·9999927	40
21	7·7859427	202031	12·2140573	5·2708595	7·7859508	202039	12·2140492	9·9973389	10·0000081	8	9·9999919	39
22	7·8061458	193049	12·1938542	5·3112661	7·8061547	193057	12·1938453	9·9972118	10·0000089	8	9·9999911	38
23	7·8254507	184831	12·1745493	5·3498763	7·8254604	184840	12·1745396	9·9970846	10·0000097	9	9·9999903	37
24	7·8439338	177285	12·1560662	5·3868430	7·8439444	177294	12·1560556	9·9969574	10·0000106		9·9999894	36
25	7·8616623	170330	12·1383377	5·4223003	7·8616738	170339	12·1383262	9·9968302	10·0000115	9	9·9999885	35
26	7·8786953	163901	12·1213047	5·4563669	7·8787077	163911	12·1212923	9·9967029	10·0000124	10	9·9999876	34
27	7·8950854	157939	12·1049146	5·4891475	7·8950988	157950	12·1049012	9·9965756	10·0000134	10	9·9999866	33
28	7·9108793	152397	12·0891207	5·5207359	7·9108938	152406	12·0891062	9·9964483	10·0000144	11	9·9999856	32
29	7·9261190	147229	12·0738810	5·5512157	7·9261344	147240	12·0738656	9·9963209	10·0000155	11	9·9999845	31
30	7·9408419	142400	12·0591581	5·5806620	7·9408584	142412	12·0591416	9·9961935	10·0000165	12	9·9999835	30
31	7·9550819	137879	12·0449181	5·6091427	7·9550996	137890	12·0449004	9·9960660	10·0000177	11	9·9999823	29
32	7·9688698	133636	12·0311302	5·6367191	7·9688886	133648	12·0311114	9·9959385	10·0000188	12	9·9999812	28
33	7·9822334	129646	12·0177666	5·6634468	7·9822534	129658	12·0177466	9·9958110	10·0000200	12	9·9999800	27
34	7·9951980	125887	12·0048020	5·6893765	7·9952192	125900	12·0047808	9·9956834	10·0000212	13	9·9999788	26
35	8·0077867	122340	11·9922133	5·7145546	8·0078092	122353	11·9921908	9·9955558	10·0000225	13	9·9999775	25
36	8·0200207	118988	11·9799793	5·7390233	8·0200445	119001	11·9799555	9·9954282	10·0000238	14	9·9999762	24
37	8·0319195	115814	11·9680805	5·7628215	8·0319446	115828	11·9680554	9·9953005	10·0000252	14	9·9999748	23
38	8·0435009	112805	11·9564991	5·7859850	8·0435274	112820	11·9564726	9·9951728	10·0000265	14	9·9999735	22
39	8·0547814	109949	11·9452186	5·8085468	8·0548094	109963	11·9451906	9·9950450	10·0000279	15	9·9999721	21
40	8·0657763	107234	11·9342237	5·8305373	8·0658057	107249	11·9341943	9·9949172	10·0000294	15	9·9999706	20
41	8·0764997	104649	11·9235003	5·8519848	8·0765306	104664	11·9234694	9·9947894	10·0000309	15	9·9999691	19
42	8·0869646	102186	11·9130354	5·8729154	8·0869970	102202	11·9130030	9·9946615	10·0000324	16	9·9999676	18
43	8·0971832	99837	11·9028168	5·8933535	8·0972172	99853	11·9027828	9·9945336	10·0000340	16	9·9999660	17
44	8·1071669	97593	11·8928331	5·9133927	8·1072025	97609	11·8927975	9·9944057	10·0000356	16	9·9999644	16
45	8·1169262	95448	11·8830738	5·9328411	8·1169634	95465	11·8830366	9·9942777	10·0000372	17	9·9999628	15
46	8·1264710	93394	11·8735290	5·9519314	8·1265099	93411	11·8734901	9·9941497	10·0000389	17	9·9999611	14
47	8·1358104	91428	11·8641896	5·9706112	8·1358510	91446	11·8641490	9·9940217	10·0000406	17	9·9999594	13
48	8·1449532	89543	11·8550468	5·9888977	8·1449956	89560	11·8550044	9·9938936	10·0000423	18	9·9999577	12
49	8·1539075	87733	11·8460925	6·0068070	8·1539516	87751	11·8460484	9·9937654	10·0000441	18	9·9999559	11
50	8·1626808	85996	11·8373192	6·0243546	8·1627267	86015	11·8372733	9·9936373	10·0000459	19	9·9999541	10
51	8·1712804	84325	11·8287196	6·0415546	8·1713282	84344	11·8286718	9·9935091	10·0000478	19	9·9999522	9
52	8·1797129	82719	11·8202871	6·0584206	8·1797626	82738	11·8202374	9·9933808	10·0000497	19	9·9999503	8
53	8·1879848	81172	11·8120152	6·0749654	8·1880364	81192	11·8119636	9·9932526	10·0000516	20	9·9999484	7
54	8·1961020	79683	11·8038980	6·0912008	8·1961556	79703	11·8038444	9·9931243	10·0000536	20	9·9999464	6
55	8·2040703	78246	11·7959297	6·1071384	8·2041259	78267	11·7958741	9·9929959	10·0000556	20	9·9999444	5
56	8·2118949	76862	11·7881051	6·1227887	8·2119526	76882	11·7880474	9·9928675	10·0000576	21	9·9999424	4
57	8·2195811	75524	11·7804189	6·1381620	8·2196408	75545	11·7803592	9·9927391	10·0000597	21	9·9999403	3
58	8·2271335	74233	11·7728665	6·1532679	8·2271953	74255	11·7728047	9·9926106	10·0000618	22	9·9999382	2
59	8·2345568	72985	11·7654432	6·1681156	8·2346208	73007	11·7653792	9·9924821	10·0000640	22	9·9999360	1
60	8·2418553		11·7581447	6·1827137	8·2419215		11·7580785	9·9923536	10·0000662		9·9999338	0

| ' | Cosine. | Diff. | Secant. | Covers. | Cotang. | Diff. | Tang. | Verseds. | Cosec. | D | Sine. | ' |

′	Sine.	Dif.	Covers	Cosec.	Tang.	Cotang.	Secant.	Vers.	D.	Cosine	′
0	0174524	2908	9825476	57·298688	0174551	57·289962	1·0001523	0001523	51	9998477	60
1	0177432	2909	9822568	56·359462	0177460	56·350590	1·0001574	0001574	52	9998426	59
2	0180341	2908	9819659	55·450534	0180370	55·441517	1·0001627	0001626	53	9998374	58
3	0183249	2909	9816751	54·570464	0183280	54·561300	1·0001679	0001679	54	9998321	57
4	0186158	2908	9813842	53·717896	0186190	53·708587	1·0001733	0001733	54	9998267	56
5	0189066	2908	9810934	52·891564	0189100	52·882109	1·0001788	0001787	56	9998213	55
6	0191974	2909	9808026	52·090272	0192010	52·080673	1·0001843	0001843	56	9998157	54
7	0194883	2908	9805117	51·312902	0194920	51·303157	1·0001900	0001899	57	9998161	53
8	0197791	2908	9802209	50·558396	0197830	50·548506	1·0001957	0001956	58	9998044	52
9	0200699	2909	9799301	49·825762	0200740	49·815726	1·0002015	0002014	59	9997986	51
10	0203608	2908	9796392	49·114062	0203650	49·103881	1·0002073	0002073	60	9997927	50
11	0206516	2908	9793484	48·422411	0206560	48·412084	1·0002133	0002133	60	9997867	49
12	0209424	2908	9790576	47·749974	0209470	47·739501	1·0002194	0002193	62	9997807	48
13	0212332	2909	9787668	47·095961	0212380	47·085343	1·0002255	0002255	62	9997745	47
14	0215241	2908	9784759	46·459625	0215291	46·448862	1·0002317	0002317	63	9997683	46
15	0218149	2908	9781851	45·840260	0218201	45·829351	1·0002380	0002380	64	9997620	45
16	0221057	2908	9778943	45·237195	0221111	45·226141	1·0002444	0002444	64	9997556	44
17	0223965	2908	9776035	44·649795	0224021	44·638596	1·0002509	0002508	66	9997492	43
18	0226873	2908	9773127	44·077458	0226932	44·066113	1·0002575	0002574	66	9997426	42
19	0229781	2909	9770219	43·519612	0229842	43·508122	1·0002641	0002640	68	9997360	41
20	0232690	2908	9767310	42·975713	0232753	42·964077	1·0002708	0002708	68	9997292	40
21	0235598	2908	9764402	42·445245	0235663	42·433464	1·0002776	0002776	68	9997224	39
22	0238506	2908	9761494	41·927717	0238574	41·915790	1·0002845	0002844	70	9997156	38
23	0241414	2908	9758586	41·422660	0241484	41·410588	1·0002915	0002914	71	9997086	37
24	0244322	2908	9755678	40·929630	0244395	40·917412	1·0002986	0002985	72	9997015	36
25	0247230	2908	9752770	40·448201	0247305	40·435837	1·0003058	0003057	72	9996943	35
26	0250138	2908	9749862	39·977969	0250216	39·965460	1·0003130	0003129	73	9996871	34
27	0253046	2908	9746954	39·518549	0253127	39·505895	1·0003203	0003202	74	9996798	33
28	0255954	2908	9744046	39·069571	0256038	39·056771	1·0003277	0003276	75	9996724	32
29	0258862	2907	9741138	38·630683	0258948	38·617738	1·0003352	0003351	76	9996649	31
30	0261769	2908	9738231	38·201550	0261859	38·188459	1·0003428	0003427	76	9996573	30
31	0264677	2908	9735323	37·781849	0264770	37·768613	1·0003505	0003503	78	9996497	29
32	0267585	2908	9732415	37·371273	0267681	37·357892	1·0003582	0003581	78	9996419	28
33	0270493	2908	9729507	36·969528	0270592	36·956001	1·0003660	0003659	79	9996341	27
34	0273401	2908	9726599	36·576332	0273503	36·562659	1·0003739	0003738	80	9996262	26
35	0276309	2907	9723691	36·191414	0276414	36·177596	1·0003820	0003818	81	9996182	25
36	0279216	2908	9720784	35·814517	0279325	35·800553	1·0003900	0003899	81	9996101	24
37	0282124	2908	9717876	35·445391	0282236	35·431282	1·0003982	0003980	83	9996020	23
38	0285032	2908	9714968	35·083800	0285148	35·069546	1·0004065	0004063	83	9995937	22
39	0287940	2907	9712060	34·729515	0288059	34·715115	1·0004148	0004146	84	9995854	21
40	0290847	2908	9709153	34·382316	0290970	34·367771	1·0004232	0004230	86	9995770	20
41	0293755	2907	9706245	34·041994	0293882	34·027303	1·0004317	0004316	85	9995684	19
42	0296662	2908	9703338	33·708345	0296793	33·693509	1·0004403	0004401	87	9995599	18
43	0299570	2908	9700430	33·381176	0299705	33·366194	1·0004490	0004488	88	9995512	17
44	0302478	2907	9697522	33·060300	0302616	33·045173	1·0004578	0004576	88	9995424	16
45	0305385	2908	9694615	32·745537	0305528	32·730264	1·0004666	0004664	89	9995336	15
46	0308293	2907	9691707	32·436713	0308439	32·421295	1·0004756	0004753	90	9995247	14
47	0311200	2908	9688800	32·133663	0311351	32·118099	1·0004846	0004843	91	9995157	13
48	0314108	2907	9685892	31·836225	0314263	31·820516	1·0004937	0004934	92	9995066	12
49	0317015	2907	9682985	31·544246	0317174	31·528392	1·0005029	0005026	93	9994974	11
50	0319922	2908	9680078	31·257577	0320086	31·241577	1·0005121	0005119	93	9994881	10
51	0322830	2907	9677170	30·976074	0322998	30·959928	1·0005215	0005212	95	9994788	9
52	0325737	2907	9674263	30·699598	0325910	30·683307	1·0005309	0005307	95	9994693	8
53	0328644	2908	9671356	30·428017	0328822	30·411580	1·0005405	0005402	96	9994598	7
54	0331552	2907	9668448	30·161201	0331734	30·144619	1·0005501	0005498	97	9994502	6
55	0334459	2907	9665541	29·899026	0334646	29·882299	1·0005598	0005595	97	9994405	5
56	0337366	2908	9662634	29·641373	0337558	29·624499	1·0005696	0005692	99	9994308	4
57	0340274	2907	9659726	29·388124	0340471	29·371106	1·0005795	0005791	99	9994209	3
58	0343181	2907	9656819	29·139169	0343383	29·122005	1·0005894	0005890	101	9994110	2
59	0346088	2907	9653912	28·894398	0346295	28·877089	1·0005994	0005991	101	9994009	1
60	0348995		9651005	28·653708	0349208	28·636253	1·0006095	0006092		9993908	0

′	Cosine	Dif.	Vers.	Secant.	Cotan.	Tang.	Cosec.	Covers	D.	Sine.	′

'	Sine.	Diff.	Cosec.	Verseds.	Tang.	Diff.	Cotang.	Covers.	Secant.	D.	Cosine.	'
0	8·2418553	71779	11·7581447	6·1827137	8·2419215	71800	11·7580785	9·9923536	10·0000662	22	9·9999338	60
1	8·2490332	70611	11·7509668	6·1970705	8·2491015	70634	11·7508985	9·9922250	10·0000684	22	9·9999316	59
2	8·2560943	69481	11·7439057	6·2111938	8·2561649	69504	11·7438351	9·9920964	10·0000706	23	9·9999294	58
3	8·2630424	68386	11·7369576	6·2250012	8·2631153	68410	11·7368847	9·9919678	10·0000729	24	9·9999271	57
4	8·2698810	67326	11·7301190	6·2387696	8·2699563	67349	11·7300437	9·9918391	10·0000753	24	9·9999247	56
5	8·2766136	66298	11·7233864	6·2522360	8·2766912	66322	11·7233088	9·9917104	10·0000776	25	9·9999224	55
6	8·2832434	65300	11·7167566	6·2654968	8·2833234	65325	11·7166766	9·9915816	10·0000800	25	9·9999200	54
7	8·2897734	64333	11·7102266	6·2785581	8·2898559	64358	11·7101441	9·9914528	10·0000825	25	9·9999175	53
8	8·2962067	63393	11·7037933	6·2914259	8·2962917	63418	11·7037083	9·9913240	10·0000850	25	9·9999150	52
9	8·3025400	62481	11·6974540	6·3041058	8·3026335	62507	11·6973665	9·9911951	10·0000875	26	9·9999125	51
10	8·3087941	61595	11·6912059	6·3166033	8·3088842	61620	11·6911158	9·9910662	10·0000900	27	9·9999100	50
11	8·3149536	60733	11·6850464	6·3289235	8·3150462	60759	11·6849538	9·9909372	10·0000926	26	9·9999074	49
12	8·3210269	59894	11·6789731	6·3410714	8·3211221	59922	11·6788779	9·9908082	10·0000953	27	9·9999047	48
13	8·3270163	59080	11·6729837	6·3530516	8·3271143	59106	11·6728857	9·9906792	10·0000979	28	9·9999021	47
14	8·3329243	58286	11·6670757	6·3648689	8·3330249	58314	11·6669751	9·9905501	10·0001006	27	9·9998994	46
15	8·3387529	57514	11·6612471	6·3765275	8·3388563	57542	11·6611437	9·9904210	10·0001034	28	9·9998966	45
16	8·3445043	56762	11·6554957	6·3880317	8·3446105	56790	11·6553895	9·9902919	10·0001061	29	9·9998939	44
17	8·3501805	56030	11·6498195	6·3993855	8·3502895	56058	11·6497105	9·9901627	10·0001089	29	9·9998911	43
18	8·3557835	55315	11·6442165	6·4105928	8·3558953	55344	11·6441047	9·9900335	10·0001118	29	9·9998882	42
19	8·3613150	54619	11·6386850	6·4216573	8·3614297	54648	11·6385703	9·9899043	10·0001147	30	9·9998853	41
20	8·3667769	53941	11·6332231	6·4325826	8·3668945	53970	11·6331055	9·9897750	10·0001176	30	9·9998824	40
21	8·3721710	53278	11·6278290	6·4433722	8·3722915	53308	11·6277085	9·9896457	10·0001206	30	9·9998794	39
22	8·3774988	52632	11·6225012	6·4540294	8·3776223	52663	11·6223777	9·9895163	10·0001236	31	9·9998764	38
23	8·3827620	52002	11·6172380	6·4645573	8·3828886	52032	11·6171114	9·9893869	10·0001266	31	9·9998734	37
24	8·3879622	51386	11·6120378	6·4749592	8·3880918	51418	11·6119082	9·9892575	10·0001297	31	9·9998703	36
25	8·3931008	50785	11·6068992	6·4852380	8·3932336	50816	11·6067664	9·9891280	10·0001328	32	9·9998672	35
26	8·3981793	50197	11·6018207	6·4953965	8·3983152	50229	11·6016848	9·9889985	10·0001359	32	9·9998641	34
27	8·4031990	49624	11·5968010	6·5054376	8·4033381	49656	11·5966619	9·9888689	10·0001391	32	9·9998609	33
28	8·4081614	49062	11·5918386	6·5153639	8·4083037	49095	11·5916963	9·9887393	10·0001423	33	9·9998577	32
29	8·4130676	48514	11·5869324	6·5251780	8·4132132	48547	11·5867868	9·9886097	10·0001456	34	9·9998544	31
30	8·4179190	47978	11·5820810	6·5348825	8·4180679	48011	11·5819321	9·9884801	10·0001488	33	9·9998512	30
31	8·4227168	47453	11·5772832	6·5444707	8·4228690	47486	11·5771310	9·9883503	10·0001522	34	9·9998478	29
32	8·4274621	46940	11·5725379	6·5539720	8·4276176	46974	11·5723824	9·9882206	10·0001555	35	9·9998445	28
33	8·4321561	46438	11·5678439	6·5633616	8·4323150	46472	11·5676850	9·9880908	10·0001589	34	9·9998411	27
34	8·4367999	45945	11·5632001	6·5726509	8·4369622	45981	11·5630378	9·9879610	10·0001624	35	9·9998376	26
35	8·4413944	45465	11·5586056	6·5818418	8·4415603	45500	11·5584397	9·9878312	10·0001658	36	9·9998342	25
36	8·4459409	44993	11·5540591	6·5909365	8·4461103	45028	11·5538897	9·9877013	10·0001694	36	9·9998306	24
37	8·4504402	44532	11·5495598	6·5999369	8·4506131	44568	11·5493869	9·9875713	10·0001729	37	9·9998271	23
38	8·4548934	44079	11·5451066	6·6088450	8·4550699	44115	11·5449301	9·9874414	10·0001765	37	9·9998235	22
39	8·4593013	43636	11·5406987	6·6176626	8·4594814	43672	11·5405186	9·9873114	10·0001801	37	9·9998199	21
40	8·4636649	43201	11·5363351	6·6263916	8·4638486	43239	11·5361514	9·9871813	10·0001838	38	9·9998162	20
41	8·4679850	42776	11·5320150	6·6350337	8·4681725	42813	11·5318275	9·9870513	10·0001875	38	9·9998125	19
42	8·4722626	42358	11·5277374	6·6435907	8·4724538	42395	11·5275462	9·9869211	10·0001912	38	9·9998088	18
43	8·4764984	41948	11·5235016	6·6520642	8·4766933	41987	11·5233067	9·9867910	10·0001950	39	9·9998050	17
44	8·4806932	41547	11·5193068	6·6604558	8·4808920	41585	11·5191080	9·9866608	10·0001988	39	9·9998012	16
45	8·4848479	41153	11·5151521	6·6687671	8·4850505	41191	11·5149495	9·9865306	10·0002026	40	9·9997974	15
46	8·4889632	40766	11·5110368	6·6769996	8·4891696	40806	11·5108304	9·9864003	10·0002065	39	9·9997935	14
47	8·4930398	40386	11·5069602	6·6851547	8·4932502	40426	11·5067498	9·9862700	10·0002104	41	9·9997896	13
48	8·4970784	40014	11·5029216	6·6932340	8·4972928	40054	11·5027072	9·9861396	10·0002144	41	9·9997856	12
49	8·5010798	39649	11·4989202	6·7012389	8·5012982	39689	11·4987018	9·9860093	10·0002183	41	9·9997817	11
50	8·5050447	39289	11·4949553	6·7091706	8·5052671	39330	11·4947329	9·9858788	10·0002224	42	9·9997776	10
51	8·5089736	38937	11·4910264	6·7170305	8·5092001	38977	11·4907999	9·9857484	10·0002264	41	9·9997736	9
52	8·5128673	38591	11·4871327	6·7248199	8·5130978	38632	11·4869022	9·9856179	10·0002305	42	9·9997695	8
53	8·5167264	38250	11·4832736	6·7325400	8·5169610	38292	11·4830390	9·9854873	10·0002347	43	9·9997653	7
54	8·5205514	37916	11·4794486	6·7401921	8·5207902	37958	11·4792098	9·9853568	10·0002388	43	9·9997612	6
55	8·5243430	37587	11·4756570	6·7477774	8·5245860	37630	11·4754140	9·9852262	10·0002430	43	9·9997570	5
56	8·5281017	37264	11·4718983	6·7552970	8·5283490	37307	11·4716510	9·9850955	10·0002473	43	9·9997527	4
57	8·5318281	36947	11·4681719	6·7627520	8·5320797	36990	11·4679203	9·9849648	10·0002516	43	9·9997484	3
58	8·5355228	36635	11·4644772	6·7701436	8·5357787	36679	11·4642213	9·9848341	10·0002559	44	9·9997441	2
59	8·5391863	36329	11·4608137	6·7774728	8·5394466	36372	11·4605534	9·9847033	10·0002602		9·9997397	1
60	8·5428192		11·4571808	6·7847406	8·5430838		11·4569162	9·9845725	10·0002646		9·9997354	0
'	Cosine.	Diff.	Secant.	Covers.	Cotang.	Diff.	Tang.	Verseds.	Cosec.	D.	Sine.	'

	Sine.	Dif.	Covers	Cosec.	Tang.	Cotang.	Secant.	Vers.	D.	Cosine	
0	0348995	2907	9651005	28·653708	0349208	28·636253	1·0006095	0006092	102	9993908	60
1	0351902	2907	9648098	28·416997	0352120	28·399397	1·0006198	0006194	102	9993806	59
2	0354809	2907	9645191	28·184168	0355033	28·166422	1·0006300	0006296	102	9993704	58
3	0357716	2907	9642284	27·955125	0357945	27·937233	1·0006404	0006400	104	9993600	57
4	0360623	2907	9639377	27·729777	0360858	27·711740	1·0006509	0006505	105	9993495	56
5	0363530	2907	9636470	27·508035	0363771	27·489853	1·0006614	0006610	105	9993390	55
6	0366437	2907	9633563	27·289814	0366683	27·271486	1·0006721	0006716	106	9993284	54
7	0369344	2907	9630656	27·075030	0369596	27·056557	1·0006828	0006823	107	9993177	53
8	0372251	2907	9627749	26·863603	0372509	26·844984	1·0006936	0006931	108	9993069	52
9	0375158	2907	9624842	26·655455	0375422	26·636690	1·0007045	0007040	109	9992960	51
10	0378065	2907	9621935	26·450510	0378335	26·431600	1·0007154	0007149	109	9992851	50
11	0380971	2906	9619029	26·248694	0381248	26·229638	1·0007265	0007260	111	9992740	49
12	0383878	2907	9616122	26·049937	0384161	26·030736	1·0007376	0007371	111	9992629	48
13	0386785	2907	9613215	25·854169	0387074	25·834823	1·0007489	0007483	112	9992517	47
14	0389692	2906	9610308	25·661324	0389988	25·641832	1·0007602	0007596	113	9992404	46
15	0392598	2907	9607402	25·471337	0392901	25·451700	1·0007716	0007710	114	9992290	45
16	0395505	2906	9604495	25·284144	0395814	25·264361	1·0007830	0007824	114	9992176	44
17	0398411	2907	9601589	25·099685	0398728	25·079757	1·0007946	0007940	116	9992060	43
18	0401318	2906	9598682	24·917900	0401641	24·897826	1·0008063	0008056	116	9991944	42
19	0404224	2907	9595776	24·738731	0404555	24·718512	1·0008180	0008173	117	9991827	41
20	0407131	2906	9592869	24·562123	0407469	24·541758	1·0008298	0008291	118	9991709	40
21	0410037	2907	9589963	24·388020	0410383	24·367509	1·0008417	0008410	119	9991590	39
22	0412944	2906	9587056	24·216370	0413296	24·195714	1·0008537	0008530	120	9991470	38
23	0415850	2907	9584150	24·047121	0416210	24·026320	1·0008658	0008650	120	9991350	37
24	0418757	2906	9581243	23·880224	0419124	23·859277	1·0008779	0008772	122	9991228	36
25	0421663	2906	9578337	23·715630	0422038	23·694537	1·0008902	0008894	123	9991106	35
26	0424569	2906	9575431	23·553291	0424952	23·532052	1·0009025	0009017	124	9990983	34
27	0427475	2907	9572525	23·393161	0427866	23·371777	1·0009149	0009141	125	9990859	33
28	0430382	2906	9569618	23·235196	0430781	23·213666	1·0009274	0009266	125	9990734	32
29	0433288	2906	9566712	23·079351	0433695	23·057677	1·0009400	0009391	127	9990609	31
30	0436194	2906	9563806	22·925586	0436609	22·903766	1·0009527	0009518	127	9990482	30
31	0439100	2906	9560900	22·773857	0439524	22·751892	1·0009654	0009645	128	9990355	29
32	0442006	2906	9557994	22·624126	0442438	22·602015	1·0009783	0009773	129	9990227	28
33	0444912	2906	9555088	22·476353	0445353	22·454096	1·0009912	0009902	130	9990098	27
34	0447818	2906	9552182	22·330499	0448268	22·308097	1·0010042	0010032	131	9989968	26
35	0450724	2906	9549276	22·186528	0451183	22·163980	1·0010173	0010163	131	9989837	25
36	0453630	2906	9546370	22·044403	0454097	22·021710	1·0010305	0010294	133	9989706	24
37	0456536	2906	9543464	21·904090	0457012	21·881251	1·0010438	0010427	133	9989573	23
38	0459442	2905	9540558	21·765553	0459927	21·742569	1·0010571	0010560	134	9989440	22
39	0462347	2906	9537653	21·628759	0462842	21·605630	1·0010705	0010694	135	9989306	21
40	0465253	2906	9534747	21·493676	0465757	21·470401	1·0010841	0010829	136	9989171	20
41	0468159	2906	9531841	21·360272	0468673	21·336851	1·0010977	0010965	136	9989035	19
42	0471065	2905	9528935	21·228515	0471588	21·204949	1·0011114	0011101	138	9988899	18
43	0473970	2906	9526030	21·098376	0474503	21·074664	1·0011251	0011239	138	9988761	17
44	0476876	2905	9523124	20·969824	0477419	20·945966	1·0011390	0011377	139	9988623	16
45	0479781	2906	9520219	20·842830	0480334	20·818828	1·0011529	0011516	140	9988484	15
46	0482687	2905	9517313	20·717368	0483250	20·693220	1·0011670	0011656	141	9988344	14
47	0485592	2906	9514408	20·593409	0486166	20·569115	1·0011811	0011797	142	9988203	13
48	0488498	2905	9511502	20·470926	0489082	20·446486	1·0011953	0011939	142	9988061	12
49	0491403	2905	9508597	20·349893	0491997	20·325308	1·0012096	0012081	144	9987919	11
50	0494308	2906	9505692	20·230284	0494913	20·205553	1·0012239	0012225	144	9987775	10
51	0497214	2905	9502786	20·112075	0497829	20·087199	1·0012384	0012369	145	9987631	9
52	0500119	2905	9499881	19·995241	0500746	19·970219	1·0012529	0012514	146	9987486	8
53	0503024	2905	9496976	19·879758	0503662	19·854591	1·0012676	0012660	146	9987340	7
54	0505929	2906	9494071	19·765604	0506578	19·740291	1·0012823	0012806	148	9987194	6
55	0508835	2905	9491165	19·652754	0509495	19·627296	1·0012971	0012954	148	9987046	5
56	0511740	2905	9488260	19·541187	0512411	19·515584	1·0013120	0013102	150	9986898	4
57	0514645	2905	9485355	19·430882	0515328	19·405133	1·0013269	0013252	150	9986748	3
58	0517550	2905	9482450	19·321816	0518244	19·295922	1·0013420	0013402	151	9986598	2
59	0520455	2905	9479545	19·213970	0521161	19·187930	1·0013571	0013553	152	9986447	1
60	0523360		9476640	19·107323	0524078	19·081137	1·0013723	0013705		9986295	0
'	Cosine	Dif.	Vers.	Secant.	Cotan.	Tang.	Cosec.	Covers	D.	Sine.	'

87 Deg.

'	Sine.	Diff.	Cosec.	Versed s.	Tang.	Diff.	Cotang.	Covers.	Secant.	D.	Cosine.	'
0	8·5428192	36026	11·4571808	6·7847406	8·5430838	36071	11·4569162	9·9845725	10·0002646	45	9·9997354	60
1	8·5464218	35730	11·4535782	6·7919481	8·5466909	35774	11·4533091	9·9844417	10·0002691	44	9·9997309	59
2	8·5499948	35438	11·4500052	6·7990963	8·5502683	35483	11·4497317	9·9843108	10·0002735	45	9·9997265	58
3	8·5535386	35150	11·4464614	6·8061861	8·5538166	35196	11·4461834	9·9841799	10·0002780	46	9·9997220	57
4	8·5570536	34868	11·4429464	6·8132185	8·5573362	34914	11·4426638	9·9840490	10·0002826	46	9·9997174	56
5	8·5605404	34590	11·4394596	6·8201944	8·5608276	34636	11·4391724	9·9839180	10·0002872	46	9·9997128	55
6	8·5639994	34316	11·4360006	6·8271147	8·5642912	34363	11·4357088	9·9837869	10·0002918	46	9·9997082	54
7	8·5674310	34047	11·4325690	6·8339803	8·5677275	34093	11·4322725	9·9836559	10·0002964	47	9·9997036	53
8	8·5708357	33782	11·4291643	6·8407920	8·5711368	33829	11·4288632	9·9835248	10·0003011	47	9·9996989	52
9	8·5742139	33521	11·4257861	6·8475507	8·5745197	33569	11·4254803	9·9833936	10·0003058	48	9·9996942	51
10	8·5775660	33263	11·4224340	6·8542572	8·5778766	33311	11·4221234	9·9832624	10·0003106	48	9·9996894	50
11	8·5808923	33010	11·4191077	6·8609123	8·5812077	33059	11·4187923	9·9831312	10·0003154	48	9·9996846	49
12	8·5841933	32761	11·4158067	6·8675167	8·5845136	32809	11·4154864	9·9830000	10·0003202	49	9·9996798	48
13	8·5874694	32515	11·4125306	6·8740714	8·5877945	32564	11·4122055	9·9828687	10·0003251	49	9·9996749	47
14	8·5907209	32274	11·4092791	6·8805768	8·5910509	32323	11·4089491	9·9827373	10·0003300	50	9·9996700	46
15	8·5939483	32034	11·4060517	6·8870340	8·5942832	32085	11·4057168	9·9826060	10·0003350	49	9·9996650	45
16	8·5971517	31800	11·4028483	6·8934434	8·5974917	31850	11·4025083	9·9824745	10·0003399	51	9·9996601	44
17	8·6003317	31569	11·3996683	6·8998059	8·6006767	31619	11·3993233	9·9823431	10·0003450	50	9·9996550	43
18	8·6034886	31340	11·3965114	6·9061221	8·6038386	31391	11·3961614	9·9822116	10·0003500	51	9·9996500	42
19	8·6066226	31115	11·3933774	6·9123927	8·6069777	31166	11·3930223	9·9820801	10·0003551	51	9·9996449	41
20	8·6097341	30894	11·3902659	6·9186183	8·6100943	30946	11·3899057	9·9819485	10·0003602	52	9·9996398	40
21	8·6128235	30675	11·3871765	6·9247996	8·6131889	30727	11·3868111	9·9818169	10·0003654	52	9·9996346	39
22	8·6158910	30459	11·3841090	6·9309372	8·6162616	30511	11·3837384	9·9816853	10·0003706	52	9·9996294	38
23	8·6189369	30247	11·3810631	6·9370317	8·6193127	30300	11·3806873	9·9815536	10·0003758	53	9·9996242	37
24	8·6219616	30037	11·3780384	6·9430837	8·6223427	30091	11·3776573	9·9814219	10·0003811	53	9·9996189	36
25	8·6249653	29831	11·3750347	6·9490939	8·6253518	29884	11·3746482	9·9812901	10·0003864	54	9·9996136	35
26	8·6279484	29627	11·3720516	6·9550627	8·6283402	29681	11·3716598	9·9811583	10·0003918	54	9·9996082	34
27	8·6309111	29426	11·3690889	6·9609907	8·6313083	29480	11·3686917	9·9810265	10·0003972	54	9·9996028	33
28	8·6338537	29227	11·3661463	6·9668786	8·6342563	29282	11·3657437	9·9808946	10·0004026	55	9·9995974	32
29	8·6367764	29032	11·3632236	6·9727268	8·6371845	29086	11·3628155	9·9807627	10·0004081	54	9·9995919	31
30	8·6396796	28838	11·3603204	6·9785359	8·6400031	28894	11·3599069	9·9806308	10·0004135	56	9·9995865	30
31	8·6425634	28648	11·3574366	6·9843063	8·6429825	28703	11·3570175	9·9804988	10·0004191	56	9·9995809	29
32	8·6454282	28460	11·3545718	6·9900387	8·6458528	28516	11·3541472	9·9803668	10·0004247	56	9·9995753	28
33	8·6482742	28274	11·3517258	6·9957334	8·6487044	28331	11·3512956	9·9802347	10·0004303	56	9·9995697	27
34	8·6511016	28091	11·3488984	7·0013911	8·6515375	28147	11·3484625	9·9801026	10·0004359	57	9·9995641	26
35	8·6539107	27910	11·3460893	7·0070121	8·6543522	27968	11·3456478	9·9799704	10·0004416	57	9·9995584	25
36	8·6567017	27731	11·3432983	7·0125969	8·6571490	27789	11·3428510	9·9798383	10·0004473	58	9·9995527	24
37	8·6594748	27555	11·3405252	7·0181461	8·6599279	27612	11·3400721	9·9797061	10·0004531	58	9·9995469	23
38	8·6622303	27381	11·3377697	7·0236600	8·6626891	27440	11·3373109	9·9795738	10·0004589	58	9·9995411	22
39	8·6649684	27209	11·3350316	7·0291391	8·6654331	27267	11·3345669	9·9794415	10·0004647	58	9·9995353	21
40	8·6676893	27039	11·3323107	7·0345838	8·6681598	27099	11·3318402	9·9793092	10·0004705	59	9·9995295	20
41	8·6703932	26872	11·3296068	7·0399946	8·6708697	26931	11·3291303	9·9791768	10·0004764	60	9·9995236	19
42	8·6730804	26706	11·3269196	7·0453719	8·6735628	26765	11·3264372	9·9790444	10·0004824	60	9·9995176	18
43	8·6757510	26542	11·3242490	7·0507161	8·6762393	26603	11·3237607	9·9789119	10·0004881	60	9·9995116	17
44	8·6784052	26381	11·3215948	7·0560276	8·6788996	26441	11·3211004	9·9787795	10·0004944	60	9·9995056	16
45	8·6810433	26221	11·3189567	7·0613068	8·6815437	26282	11·3184563	9·9786469	10·0005004	61	9·9994996	15
46	8·6836654	26064	11·3163346	7·0665540	8·6841719	26125	11·3158201	9·9785144	10·0005065	61	9·9994935	14
47	8·6862718	25907	11·3137282	7·0717698	8·6867844	25969	11·3132156	9·9783818	10·0005126	62	9·9994874	13
48	8·6888625	25754	11·3111375	7·0769544	8·6893813	25816	11·3106187	9·9782491	10·0005188	62	9·9994812	12
49	8·6914379	25601	11·3085621	7·0821082	8·6919629	25663	11·3080371	9·9781164	10·0005250	62	9·9994750	11
50	8·6939980	25451	11·3060020	7·0872316	8·6945292	25514	11·3054708	9·9779837	10·0005312	63	9·9994688	10
51	8·6965431	25303	11·3034569	7·0923249	8·6970806	25366	11·3029194	9·9778510	10·0005375	63	9·9994625	9
52	8·6990734	25155	11·3009266	7·0973885	8·6996172	25220	11·3003828	9·9777182	10·0005438	64	9·9994562	8
53	8·7015889	25010	11·2984111	7·1024228	8·7021390	25075	11·2978610	9·9775853	10·0005502	63	9·9994498	7
54	8·7040899	24867	11·2959101	7·1074280	8·7046465	24930	11·2953535	9·9774525	10·0005565	65	9·9994435	6
55	8·7065766	24724	11·2934234	7·1124045	8·7071395	24790	11·2928605	9·9773195	10·0005630	64	9·9994370	5
56	8·7090490	24585	11·2909510	7·1173527	8·7096185	24649	11·2903815	9·9771866	10·0005694	65	9·9994306	4
57	8·7115075	24445	11·2884925	7·1222728	8·7120834	24511	11·2879166	9·9770536	10·0005759	65	9·9994241	3
58	8·7139520	24309	11·2860480	7·1271652	8·7145345	24374	11·2854655	9·9769206	10·0005824	66	9·9994176	2
59	8·7163829	24173	11·2836171	7·1320302	8·7169719	24239	11·2830281	9·9767875	10·0005890	66	9·9994110	1
60	8·7188002		11·2811998	7·1368680	8·7193958		11·2806042	9·9766544	10·0005956		9·9994044	0
'	Cosine.	Diff.	Secant.	Covers.	Cotang.	Diff.	Tang.	Versed s.	Cosec.	D.	Sine.	'

T

′	Sine.	Dif.	Covers	Cosec.	Tang.	Cotang.	Secant.	Vers.	D.	Cosine	′
0	0523360	2904	9476640	19·107323	0524078	19·081137	1·0018723	0013705	152	9986295	60
1	0526264	2905	9473736	19·001854	0526995	18·975523	1·0013877	0013857	154	9986143	59
2	0529169	2905	9470831	18·897545	0529912	18·871068	1·0014030	0014011	154	9985989	58
3	0532074	2905	9467926	18·794377	0532829	18·767754	1·0014185	0014165	155	9985835	57
4	0534979	2904	9465021	18·692330	0535746	18·665562	1·0014341	0014320	156	9985680	56
5	0537883	2905	9462117	18·591387	0538663	18·564473	1·0014497	0014476	157	9985524	55
6	0540788	2905	9459212	18·491530	0541581	18·464471	1·0014655	0014633	158	9985367	54
7	0543693	2904	9456307	18·392742	0544498	18·365537	1·0014813	0014791	159	9985209	53
8	0546597	2905	9453403	18·295005	0547416	18·267654	1·0014972	0014950	159	9985050	52
9	0549502	2904	9450498	18·198303	0550333	18·170807	1·0015132	0015109	160	9984891	51
10	0552406	2905	9447594	18·102619	0553251	18·074977	1·0015293	0015269	161	9984731	50
11	0555311	2904	9444689	18·007937	0556169	17·980150	1·0015454	0015430	162	9984570	49
12	0558215	2904	9441785	17·914243	0559087	17·886310	1·0015617	0015592	163	9984408	48
13	0561119	2905	9438881	17·821520	0562005	17·793442	1·0015780	0015755	164	9984245	47
14	0564024	2904	9435976	17·729753	0564923	17·701529	1·0015944	0015919	164	9984081	46
15	0566928	2904	9433072	17·638928	0567841	17·610559	1·0016109	0016083	166	9983917	45
16	0569832	2904	9430168	17·549030	0570759	17·520516	1·0016275	0016249	166	9983751	44
17	0572736	2904	9427264	17·460046	0573678	17·431385	1·0016442	0016415	167	9983585	43
18	0575640	2904	9424360	17·371960	0576596	17·343155	1·0016609	0016582	168	9983418	42
19	0578544	2904	9421456	17·284761	0579515	17·255809	1·0016778	0016750	168	9983250	41
20	0581448	2904	9418552	17·198434	0582434	17·169337	1·0016947	0016918	170	9983082	40
21	0584352	2904	9415648	17·112966	0585352	17·083724	1·0017117	0017088	170	9982912	39
22	0587256	2904	9412744	17·028346	0588271	16·998957	1·0017288	0017258	172	9982742	38
23	0590160	2904	9409840	16·944559	0591190	16·915025	1·0017460	0017430	172	9982570	37
24	0593064	2903	9406936	16·861594	0594109	16·831915	1·0017633	0017602	173	9982398	36
25	0595967	2904	9404033	16·779439	0597029	16·749614	1·0017806	0017775	173	9982225	35
26	0598871	2904	9401129	16·698082	0599948	16·668112	1·0017981	0017948	175	9982052	34
27	0601775	2903	9398225	16·617512	0602867	16·587396	1·0018156	0018123	176	9981877	33
28	0604678	2904	9395322	16·537717	0605787	16·507456	1·0018332	0018299	176	9981701	32
29	0607582	2903	9392418	16·458686	0608706	16·428279	1·0018509	0018475	177	9981525	31
30	0610485	2904	9389515	16·380408	0611626	16·349855	1·0018687	0018652	178	9981348	30
31	0613389	2903	9386611	16·302873	0614546	16·272174	1·0018866	0018830	179	9981170	29
32	0616292	2904	9383708	16·226069	0617466	16·195225	1·0019045	0019009	180	9980991	28
33	0619196	2903	9380804	16·149987	0620386	16·118998	1·0019225	0019189	180	9980811	27
34	0622099	2903	9377901	16·074617	0623306	16·043482	1·0019407	0019369	181	9980631	26
35	0625002	2903	9374998	15·999948	0626226	15·968667	1·0019589	0019550	183	9980450	25
36	0627905	2903	9372095	15·925971	0629147	15·894545	1·0019772	0019733	183	9980267	24
37	0630808	2903	9369192	15·852676	0632067	15·821105	1·0019956	0019916	184	9980084	23
38	0633711	2903	9366289	15·780054	0634988	15·748337	1·0020140	0020100	184	9979900	22
39	0636614	2903	9363386	15·708096	0637908	15·676233	1·0020326	0020284	186	9979716	21
40	0639517	2903	9360483	15·636793	0640829	15·604784	1·0020512	0020470	187	9979530	20
41	0642420	2903	9357580	15·566135	0643750	15·533981	1·0020699	0020657	187	9979343	19
42	0645323	2903	9354677	15·496114	0646671	15·463814	1·0020887	0020844	188	9979156	18
43	0648226	2903	9351774	15·426721	0649592	15·394276	1·0021076	0021032	189	9978968	17
44	0651129	2902	9348871	15·357949	0652513	15·325358	1·0021266	0021221	190	9978779	16
45	0654031	2903	9345969	15·289788	0655435	15·257052	1·0021457	0021411	190	9978589	15
46	0656934	2902	9343066	15·222231	0658356	15·189349	1·0021648	0021601	192	9978399	14
47	0659836	2903	9340164	15·155270	0661278	15·122242	1·0021841	0021793	192	9978207	13
48	0662739	2902	9337261	15·088896	0664199	15·055723	1·0022034	0021985	194	9978015	12
49	0665641	2903	9334359	15·023103	0667121	14·989784	1·0022228	0022179	194	9977821	11
50	0668544	2902	9331456	14·957882	0670043	14·924417	1·0022423	0022373	194	9977627	10
51	0671446	2903	9328554	14·893226	0672965	14·859616	1·0022619	0022567	196	9977433	9
52	0674349	2902	9325651	14·829128	0675887	14·795372	1·0022815	0022763	197	9977237	8
53	0677251	2902	9322749	14·765580	0678809	14·731679	1·0023013	0022960	197	9977040	7
54	0680153	2902	9319847	14·702576	0681732	14·668529	1·0023211	0023157	198	9976843	6
55	0683055	2902	9316945	14·640109	0684654	14·605916	1·0023410	0023355	200	9976645	5
56	0685957	2902	9314043	14·578172	0687577	14·543833	1·0023610	0023555	200	9976445	4
57	0688859	2902	9311141	14·516757	0690499	14·482273	1·0023811	0023755	200	9976245	3
58	0691761	2902	9308239	14·455859	0693422	14·421230	1·0024013	0023955	202	9976045	2
59	0694663	2902	9305337	14·395471	0696345	14·360696	1·0024216	0024157	202	9975843	1
60	0697565		9302435	14·335587	0699268	14·300666	1·0024419	0024359		9975641	0

′	Cosine	Dif.	Vers.	Secant.	Cotan.	Tang.	Cosec.	Covers	D.	Sine.	′

′	Sine.	Diff.	Cosec.	Verseds.	Tang.	Diff.	Cotang.	Covers.	Secant.	D.	Cosine.	′
0	8·7188002	24038	11·2811998	7·1368680	8·7193958	24105	11·2806042	9·9766544	10·0005956	66	9·9994044	60
1	8·7212040	23906	11·2787960	7·1416791	8·7218063	23972	11·2781937	9·9765213	10·0006022	67	9·9993978	59
2	8·7235946	23775	11·2764054	7·1464636	8·7242035	23842	11·2757965	9·9763881	10·0006089	67	9·9993911	58
3	8·7259721	23645	11·2740279	7·1512219	8·7265877	23712	11·2734123	9·9762549	10·0006156	68	9·9993844	57
4	8·7283366	23516	11·2716634	7·1559542	8·7289589	23585	11·2710411	9·9761216	10·0006224	68	9·9993776	56
5	8·7306882	23390	11·2693118	7·1606609	8·7313174	23457	11·2686826	9·9759883	10·0006292	68	9·9993708	55
6	8·7330272	23263	11·2669728	7·1653422	8·7336631	23333	11·2663369	9·9758550	10·0006360	68	9·9993640	54
7	8·7353535	23140	11·2646465	7·1699984	8·7359964	23208	11·2640036	9·9757216	10·0006428	69	9·9993572	53
8	8·7376675	23016	11·2623325	7·1746297	8·7383172	23086	11·2616828	9·9755882	10·0006497	70	9·9993503	52
9	8·7399691	22895	11·2600309	7·1792365	8·7406258	22964	11·2593742	9·9754547	10·0006567	69	9·9993433	51
10	8·7422586	22774	11·2577414	7·1838189	8·7429222	22845	11·2570778	9·9753212	10·0006636	71	9·9993364	50
11	8·7445360	22655	11·2554640	7·1883773	8·7452067	22725	11·2547933	9·9751877	10·0006707	70	9·9993293	49
12	8·7468015	22538	11·2531985	7·1929118	8·7474792	22608	11·2525208	9·9750541	10·0006777	71	9·9993223	48
13	8·7490553	22420	11·2509447	7·1974228	8·7497400	22492	11·2502600	9·9749205	10·0006848	71	9·9993152	47
14	8·7512973	22305	11·2487027	7·2019104	8·7519892	22377	11·2480108	9·9747868	10·0006919	72	9·9993081	46
15	8·7535278	22191	11·2464722	7·2063750	8·7542269	22262	11·2457731	9·9746532	10·0006991	71	9·9993009	45
16	8·7557469	22077	11·2442531	7·2108167	8·7564531	22150	11·2435469	9·9745194	10·0007062	73	9·3092938	44
17	8·7579546	21966	11·2420454	7·2152358	8·7586681	22038	11·2413319	9·9743857	10·0007135	72	9·9992865	43
18	8·7601512	21854	11·2398488	7·2196326	8·7608719	21928	11·2391281	9·9742519	10·0007207	73	9·9992793	42
19	8·7623366	21745	11·2376634	7·2240071	8·7630647	21818	11·2369353	9·9741180	10·0007280	74	9·9992720	41
20	8·7645111	21636	11·2354889	7·2283597	8·7652465	21710	11·2347535	9·9739841	10·0007354	74	9·9992646	40
21	8·7666747	21528	11·2333253	7·2326906	8·7674175	21602	11·2325825	9·9738502	10·0007428	74	9·9992572	39
22	8·7688275	21422	11·2311725	7·2370000	8·7695777	21497	11·2304223	9·9737162	10·0007502	74	9·9992498	38
23	8·7709697	21317	11·2290303	7·2412881	8·7717274	21391	11·2282726	9·9735822	10·0007576	75	9·9992424	37
24	8·7731014	21212	11·2268986	7·2455551	8·7738665	21287	11·2261335	9·9734482	10·0007651	75	9·9992349	36
25	8·7752226	21108	11·2247774	7·2498013	8·7759952	21184	11·2240048	9·9733141	10·0007726	76	9·9992274	35
26	8·7773334	21006	11·2226666	7·2540267	8·7781136	21082	11·2218864	9·9731800	10·0007802	76	9·9992198	34
27	8·7794340	20904	11·2205660	7·2582317	8·7802218	20981	11·2197782	9·9730458	10·0007878	76	9·9992122	33
28	8·7815244	20804	11·2184756	7·2624164	8·7823199	20880	11·2176801	9·9729117	10·0007954	77	9·9992046	32
29	8·7836048	20705	11·2163952	7·2665810	8·7844079	20782	11·2155921	9·9727774	10·0008031	77	9·9991969	31
30	8·7856753	20606	11·2143247	7·2707258	8·7864861	20683	11·2135139	9·9726431	10·0008108	77	9·9991892	30
31	8·7877359	20508	11·2122641	7·2748508	8·7885544	20586	11·2114456	9·9725088	10·0008185	78	9·9991815	29
32	8·7897867	20411	11·2102133	7·2789563	8·7906130	20490	11·2093870	9·9723745	10·0008263	78	9·9991737	28
33	8·7918278	20316	11·2081722	7·2830425	8·7926620	20394	11·2073380	9·9722401	10·0008341	79	9·9991659	27
34	8·7938594	20220	11·2061406	7·2871095	8·7947014	20299	11·2052986	9·9721057	10·0008420	79	9·9991580	26
35	8·7958814	20127	11·2041186	7·2911576	8·7967313	20206	11·2032687	9·9719712	10·0008499	79	9·9991501	25
36	8·7978941	20033	11·2021059	7·2951869	8·7987519	20113	11·2012481	9·9718367	10·0008578	80	9·9991422	24
37	8·7998974	19941	11·2001026	7·2991975	8·8007632	20021	11·1992368	9·9717021	10·0008658	80	9·9991342	23
38	8·8018915	19849	11·1981085	7·3031897	8·8027653	19930	11·1972347	9·9715675	10·0008738	80	9·9991262	22
39	8·8038764	19759	11·1961236	7·3071636	8·8047583	19839	11·1952417	9·9714329	10·0008818	81	9·9991182	21
40	8·8058523	19669	11·1941477	7·3111194	8·8067422	19750	11·1932578	9·9712982	10·0008899	81	9·9991101	20
41	8·8078192	19580	11·1921808	7·3150572	8·8087172	19662	11·1912828	9·9711635	10·0008980	82	9·9991020	19
42	8·8097772	19492	11·1902228	7·3189773	8·8106834	19573	11·1893166	9·9710288	10·0009062	82	9·9990938	18
43	8·8117264	19404	11·1882736	7·3228797	8·8126407	19487	11·1873593	9·9708940	10·0009144	82	9·9990856	17
44	8·8136668	19317	11·1863332	7·3267646	8·8145894	19400	11·1854106	9·9707592	10·0009226	83	9·9990774	16
45	8·8155985	19232	11·1844015	7·3306322	8·8165294	19314	11·1834706	9·9706243	10·0009309	83	9·9990691	15
46	8·8175217	19146	11·1824783	7·3344827	8·8184608	19230	11·1815392	9·9704894	10·0009392	83	9·9990608	14
47	8·8194363	19062	11·1805637	7·3383161	8·8203838	19146	11·1796162	9·9703545	10·0009475	84	9·9990525	13
48	8·8213425	18979	11·1786575	7·3421327	8·8222984	19062	11·1777016	9·9702195	10·0009559	84	9·9990441	12
49	8·8232404	18895	11·1767596	7·3459326	8·8242046	18980	11·1757954	9·9700845	10·0009643	84	9·9990357	11
50	8·8251299	18813	11·1748701	7·3497159	8·8261026	18898	11·1738974	9·9699494	10·0009727	85	9·9990273	10
51	8·8270112	18732	11·1729888	7·3534828	8·8279924	18817	11·1720076	9·9698143	10·0009812	85	9·9990188	9
52	8·8288844	18651	11·1711156	7·3572334	8·8298741	18737	11·1701259	9·9696792	10·0009897	86	9·9990103	8
53	8·8307495	18571	11·1692505	7·3609678	8·8317478	18656	11·1682522	9·9695440	10·0009983	86	9·9990017	7
54	8·8326066	18491	11·1673934	7·3646863	8·8336134	18578	11·1663866	9·9694088	10·0010069	86	9·9989931	6
55	8·8344557	18412	11·1655443	7·3683888	8·8354712	18499	11·1645288	9·9692735	10·0010155	87	9·9989845	5
56	8·8362969	18335	11·1637031	7·3720757	8·8373211	18422	11·1626789	9·9691382	10·0010242	87	9·9989758	4
57	8·8381304	18257	11·1618696	7·3757469	8·8391633	18344	11·1608367	9·9690029	10·0010329	87	9·9989671	3
58	8·8399561	18180	11·1600439	7·3794027	8·8409977	18268	11·1590023	9·9688675	10·0010416	88	9·9989584	2
59	8·8417741	18104	11·1582259	7·3830431	8·8428245	18192	11·1571755	9·9687321	10·0010504	88	9·9989496	1
60	8·8435845		11·1564155	7·3866683	8·8446437		11·1553563	9·9685967	10·0010592		9·9989408	0

| ′ | Cosine. | Diff. | Secant. | Covers. | Cotang. | Diff. | Tang. | Verseds. | Cosec. | D. | Sine. | ′ |

′	Sine.	Dif.	Covers	Cosec.	Tang.	Cotang.	Secant.	Vers.	D.	Cosine	′
0	0697565	2902	9302435	14·335587	0699268	14·300666	1·0024419	0024359	204	9975641	60
1	0700467	2901	9299533	14·276200	0702191	14·241134	1·0024623	0024563	204	9975437	59
2	0703368	2902	9296632	14·217304	0705115	14·182092	1·0024829	0024767	204	9975233	58
3	0706270	2901	9293730	14·158894	0708038	14·123536	1·0025035	0024972	205	9975028	57
4	0709171	2902	9290829	14·100963	0710961	14·065459	1·0025241	0025178	206	9974822	56
5	0712073	2901	9287927	14·043504	0713885	14·007856	1·0025449	0025385	207	9974615	55
6	0714974	2902	9285026	13·986514	0716809	13·950719	1·0025658	0025592	207	9974408	54
7	0717876	2901	9282124	13·929985	0719733	13·894045	1·0025867	0025801	209	9974199	53
8	0720777	2901	9279223	13·873913	0722657	13·837827	1·0026078	0026010	209	9973990	52
9	0723678	2902	9276322	13·818291	0725581	13·782060	1·0026289	0026220	210	9973780	51
10	0726580	2901	9273420	13·763115	0728505	13·726738	1·0026501	0026431	211	9973569	50
11	0729481	2901	9270519	13·708379	0731430	13·671856	1·0026714	0026643	212	9973357	49
12	0732382	2901	9267618	13·654077	0734354	13·617409	1·0026928	0026855	212	9973145	48
13	0735283	2901	9264717	13·600205	0737279	13·563391	1·0027142	0027069	214	9972931	47
14	0738184	2901	9261816	13·546758	0740203	13·509799	1·0027358	0027283	214	9972717	46
15	0741085	2901	9258915	13·493731	0743128	13·456625	1·0027574	0027498	215	9972502	45
16	0743986	2901	9256014	13·441118	0746053	13·403867	1·0027791	0027714	216	9972286	44
17	0746887	2900	9253113	13·388914	0748979	13·351518	1·0028009	0027931	217	9972069	43
18	0749787	2901	9250213	13·337116	0751904	13·299574	1·0028228	0028149	218	9971851	42
19	0752688	2901	9247312	13·285719	0754829	13·248031	1·0028448	0028367	218	9971633	41
20	0755589	2900	9244411	13·234717	0757755	13·196883	1·0028669	0028587	220	9971413	40
21	0758489	2901	9241511	13·184106	0760680	13·146127	1·0028890	0028807	220	9971193	39
22	0761390	2900	9238610	13·133882	0763606	13·095757	1·0029112	0029028	221	9970972	38
23	0764290	2900	9235710	13·084040	0766532	13·045769	1·0029336	0029250	222	9970750	37
24	0767190	2901	9232810	13·034576	0769458	12·996160	1·0029560	0029472	222	9970528	36
25	0770091	2900	9229909	12·985486	0772384	12·946924	1·0029785	0029696	224	9970304	35
26	0772991	2900	9227009	12·936765	0775311	12·898058	1·0030010	0029920	224	9970080	34
27	0775891	2900	9224109	12·888410	0778237	12·849557	1·0030146	0030146	226	9969854	33
28	0778791	2900	9221209	12·840416	0781164	12·801417	1·0030464	0030372	226	9969628	32
29	0781691	2900	9218309	12·792779	0784090	12·753634	1·0030693	0030599	227	9969401	31
30	0784591	2900	9215409	12·745495	0787017	12·706205	1·0030922	0030827	228	9969173	30
31	0787491	2900	9212509	12·698560	0789944	12·659125	1·0031152	0031055	228	9968945	29
32	0790391	2899	9209609	12·651971	0792871	12·612390	1·0031383	0031285	230	9968715	28
33	0793290	2900	9206710	12·605724	0795798	12·565997	1·0031615	0031515	230	9968485	27
34	0796190	2900	9203810	12·559815	0798726	12·519942	1·0031847	0031746	231	9968254	26
35	0799090	2899	9200910	12·514240	0801653	12·474221	1·0032081	0031978	232	9968022	25
36	0801989	2900	9198011	12·468995	0804581	12·428331	1·0032315	0032211	233	9967789	24
37	0804889	2899	9195111	12·424078	0807509	12·383768	1·0032550	0032445	234	9967555	23
38	0807788	2899	9192212	12·379484	0810437	12·339028	1·0032787	0032679	234	9967321	22
39	0810687	2900	9189313	12·335210	0813365	12·294609	1·0033024	0032915	236	9967085	21
40	0813587	2899	9186413	12·291252	0816293	12·250505	1·0033261	0033151	236	9966849	20
41	0816486	2899	9183514	12·247608	0819221	12·206716	1·0033500	0033389	237	9966612	19
42	0819385	2899	9180615	12·204274	0822150	12·163236	1·0033740	0033626	238	9966374	18
43	0822284	2899	9177716	12·161246	0825078	12·120062	1·0033980	0033865	239	9966135	17
44	0825183	2899	9174817	12·118522	0828007	12·077192	1·0034221	0034105	240	9965895	16
45	0828082	2899	9171918	12·076098	0830936	12·034622	1·0034463	0034345	240	9965655	15
46	0830981	2899	9169019	12·033970	0833865	11·992349	1·0034706	0034586	241	9965414	14
47	0833880	2898	9166120	11·992137	0836794	11·950370	1·0034950	0034828	242	9965172	13
48	0836778	2899	9163222	11·950595	0839723	11·908682	1·0035195	0035071	243	9964929	12
49	0839677	2899	9160323	11·909340	0842653	11·867282	1·0035440	0035315	244	9964685	11
50	0842576	2898	9157424	11·868370	0845583	11·826167	1·0035687	0035560	245	9964440	10
51	0845474	2899	9154526	11·827683	0848512	11·785333	1·0035934	0035805	245	9964195	9
52	0848373	2898	9151627	11·787274	0851442	11·744779	1·0036182	0036052	247	9963948	8
53	0851271	2898	9148729	11·747141	0854372	11·704500	1·0036431	0036299	247	9963701	7
54	0854169	2838	9145831	11·707282	0857302	11·664495	1·0036681	0036547	248	9963453	6
55	0857067	2899	9142933	11·667693	0860233	11·624761	1·0036932	0036796	249	9963204	5
56	0859966	2898	9140034	11·628372	0863163	11·585294	1·0037183	0037046	250	9962954	4
57	0862864	2898	9137136	11·589316	0866094	11·546093	1·0037436	0037296	250	9962704	3
58	0865762	2898	9134238	11·550523	0869025	11·507154	1·0037689	0037548	252	9962452	2
59	0868660	2897	9131340	11·511990	0871956	11·468474	1·0037943	0037800	252	9962200	1
60	0871557		9128443	11·473713	0874887	11·430052	1·0038198	0038053	253	9961947	0

′	Cosine	Dif.	Vers.	Secant.	Cotan.	Tang.	Cosec.	Covers	D.	Sine.	′

′	Sine.	Diff.	Cosec.	Verseds.	Tang.	Diff.	Cotang.	Covers.	Secant.	D.	Cosine.	′
0	8·8435845	18029	11·1564155	7·3866683	8·8446437	18117	11·1553563	9·9635967	10·0010592	89	9·9989408	60
1	8·8453874	17953	11·1546126	7·3902785	8·8464554	18043	11·1535446	9·9684612	10·0010681	89	9·9989319	59
2	8·8471827	17880	11·1528173	7·3938736	8·8482597	17969	11·1517403	9·9683256	10·0010770	89	9·9989230	58
3	8·8489707	17805	11·1510293	7·3974539	8·8500566	17895	11·1499434	9·9681901	10·0010859	89	9·9989141	57
4	8·8507512	17733	11·1492488	7·4010196	8·8518461	17822	11·1481539	9·9680544	10·0010948	90	9·9989052	56
5	8·8525245	17660	11·1474755	7·4045706	8·8536283	17751	11·1463717	9·9679188	10·0011038	91	9·9988962	55
6	8·8542905	17588	11·1457095	7·4081071	8·8554034	17679	11·1445966	9·9677831	10·0011129	91	9·9988871	54
7	8·8560493	17517	11·1439507	7·4116293	8·8571713	17608	11·1428287	9·9676474	10·0011220	91	9·9988780	53
8	8·8578010	17447	11·1421900	7·4151372	8·8589321	17538	11·1410679	9·9675116	10·0011311	91	9·9988689	52
9	8·8595457	17376	11·1404543	7·4186311	8·8606859	17468	11·1393141	9·9673758	10·0011402	92	9·9988598	51
10	8·8612833	17306	11·1387167	7·4221109	8·8624327	17398	11·1375673	9·9672399	10·0011494	92	9·9988506	50
11	8·8630139	17237	11·1369861	7·4255767	8·8641725	17330	11·1358275	9·9671041	10·0011586	93	9·9988414	49
12	8·8647376	17169	11·1352624	7·4290288	8·8659055	17262	11·1340945	9·9669681	10·0011679	93	9·9988321	48
13	8·8664545	17101	11·1335455	7·4324673	8·8676317	17194	11·1323683	9·9668322	10·0011772	93	9·9988228	47
14	8·8681646	17034	11·1318354	7·4358921	8·8693511	17127	11·1306489	9·9666961	10·0011865	94	9·9988135	46
15	8·8698680	16966	11·1301320	7·4393035	8·8710638	17061	11·1289362	9·9665601	10·0011959	94	9·9988041	45
16	8·8715646	16900	11·1284354	7·4427015	8·8727699	16995	11·1272301	9·9664240	10·0012053	94	9·9987947	44
17	8·8732546	16835	11·1267454	7·4460862	8·8744694	16929	11·1255306	9·9662879	10·0012147	95	9·9987853	43
18	8·8749381	16769	11·1250619	7·4494578	8·8761623	16864	11·1238377	9·9661517	10·0012242	95	9·9987758	42
19	8·8766150	16704	11·1233850	7·4528163	8·8778487	16799	11·1221513	9·9660155	10·0012337	96	9·9987663	41
20	8·8782854	16639	11·1217146	7·4561619	8·8795286	16736	11·1204714	9·9658793	10·0012433	96	9·9987567	40
21	8·8799493	16576	11·1200507	7·4594946	8·8812022	16672	11·1187978	9·9657430	10·0012529	96	9·9987471	39
22	8·8816069	16512	11·1183931	7·4628146	8·8828694	16609	11·1171306	9·9656067	10·0012625	97	9·9987375	38
23	8·8832581	16450	11·1167419	7·4661219	8·8845303	16547	11·1154697	9·9654703	10·0012722	97	9·9987278	37
24	8·8849031	16387	11·1150969	7·4694166	8·8861850	16484	11·1138150	9·9653339	10·0012819	97	9·9987181	36
25	8·8865418	16325	11·1134582	7·4726989	8·8878334	16423	11·1121666	9·9651974	10·0012916	98	9·9987084	35
26	8·8881743	16264	11·1118257	7·4759688	8·8894757	16362	11·1105243	9·9650610	10·0013014	98	9·9986986	34
27	8·8898007	16202	11·1101993	7·4792264	8·8911119	16301	11·1088881	9·9649244	10·0013112	98	9·9986888	33
28	8·8914209	16142	11·1085791	7·4824719	8·8927420	16240	11·1072580	9·9647879	10·0013210	99	9·9986790	32
29	8·8930351	16082	11·1069649	7·4857052	8·8943660	16182	11·1056340	9·9646513	10·0013309	99	9·9986691	31
30	8·8946433	16022	11·1053567	7·4889265	8·8959842	16121	11·1040158	9·9645146	10·0013409	99	9·9986591	30
31	8·8962455	15963	11·1037545	7·4921359	8·8975963	16063	11·1024037	9·9643779	10·0013508	100	9·9986492	29
32	8·8978418	15904	11·1021582	7·4953335	8·8992026	16004	11·1007974	9·9642412	10·0013608	100	9·9986392	28
33	8·8994322	15846	11·1005678	7·4985193	8·9008030	15947	11·0991970	9·9641044	10·0013708	101	9·9986292	27
34	8·9010168	15787	11·0989832	7·5016934	8·9023977	15889	11·0976023	9·9639676	10·0013809	101	9·9986191	26
35	8·9025955	15730	11·0974045	7·5048560	8·9039866	15831	11·0960134	9·9638308	10·0013910	102	9·9986090	25
36	8·9041685	15673	11·0958315	7·5080071	8·9055697	15775	11·0944303	9·9636939	10·0014012	102	9·9985988	24
37	8·9057358	15617	11·0942642	7·5111468	8·9071472	15718	11·0928528	9·9635570	10·0014114	102	9·9985886	23
38	8·9072975	15560	11·0927025	7·5142751	8·9087190	15663	11·0912810	9·9634200	10·0014216	102	9·9985784	22
39	8·9088535	15504	11·0911465	7·5173923	8·9102853	15607	11·0897147	9·9632830	10·0014318	103	9·9985682	21
40	8·9104039	15448	11·0895961	7·5204982	8·9118460	15552	11·0881540	9·9631460	10·0014421	104	9·9985579	20
41	8·9119487	15394	11·0880513	7·5235931	8·9134012	15497	11·0865988	9·9630089	10·0014525	103	9·9985475	19
42	8·9134881	15338	11·0865119	7·5266769	8·9149509	15443	11·0850491	9·9628718	10·0014628	104	9·9985372	18
43	8·9150219	15285	11·0849781	7·5297498	8·9164952	15388	11·0835048	9·9627346	10·0014734	105	9·9985268	17
44	8·9165504	15230	11·0834496	7·5328119	8·9180340	15335	11·0819660	9·9625974	10·0014837	105	9·9985163	16
45	8·9180734	15177	11·0819266	7·5358632	8·9195675	15282	11·0804325	9·9624602	10·0014942	105	9·9985058	15
46	8·9195911	15123	11·0804089	7·5389038	8·9210957	15229	11·0789043	9·9623229	10·0015047	105	9·9984953	14
47	8·9211034	15071	11·0788966	7·5419338	8·9226186	15177	11·0773814	9·9621856	10·0015152	106	9·9984848	13
48	8·9226105	15018	11·0773895	7·5449532	8·9241363	15124	11·0758637	9·9620482	10·0015258	106	9·9984742	12
49	8·9241123	14966	11·0758877	7·5479621	8·9256487	15073	11·0743513	9·9619108	10·0015364	107	9·9984636	11
50	8·9256089	14914	11·0743911	7·5509607	8·9271560	15021	11·0728440	9·9617733	10·0015471	107	9·9984529	10
51	8·9271003	14863	11·0728997	7·5539489	8·9286581	14971	11·0713419	9·9616359	10·0015585	107	9·9984422	9
52	8·9285866	14812	11·0714134	7·5569268	8·9301552	14919	11·0698448	9·9614983	10·0015685	108	9·9984315	8
53	8·9300678	14761	11·0699322	7·5598946	8·9316471	14869	11·0683529	9·9613608	10·0015793	108	9·9984207	7
54	8·9315439	14711	11·0684561	7·5628522	8·9331340	14820	11·0668660	9·9612232	10·0015901	109	9·9984099	6
55	8·9330150	14661	11·0669850	7·5657998	8·9346160	14769	11·0653840	9·9610855	10·0016010	109	9·9983990	5
56	8·9344811	14611	11·0655189	7·5687373	8·9360929	14721	11·0639071	9·9609478	10·0016119	109	9·9983881	4
57	8·9359422	14561	11·0640578	7·5716650	8·9375650	14671	11·0624350	9·9608101	10·0016228	109	9·9983772	3
58	8·9373983	14513	11·0626017	7·5745828	8·9390321	14623	11·0609679	9·9606723	10·0016337	110	9·9983663	2
59	8·9388496	14464	11·0611504	7·5774908	8·9404944	14574	11·0595056	9·9605345	10·0016447	111	9·9983553	1
60	8·9402960		11·0597040	7·5803891	8·9419518		11·0580482	9·9603967	10·0016558		9·9983442	0
′	Cosine.	Diff.	Secant.	Covers.	Cotang.	Diff.	Tang.	Verseds.	Cosec.	D.	Sine.	′

′	Sine.	Dif	Covers	Cosec.	Tang.	Cotang.	Secant.	Vers.	D.	Cosine	′
0	0871557	2898	9128443	11·473713	0874687	11·430052	1·0038198	0038053	254	9961947	60
1	0874455	2898	9125545	11·435692	0877818	11·391885	1·0038454	0038307	255	9961693	59
2	0877353	2898	9122647	11·397922	0880749	11·353970	1·0038711	0038562	255	9961438	58
3	0880251	2897	9119749	11·360402	0883681	11·316304	1·0038969	0038817	257	9961183	57
4	0883148	2898	9116852	11·323129	0886612	11·278885	1·0039227	0039074	257	9960926	56
5	0886046	2897	9113954	11·286101	0889544	11·241712	1·0039486	0039331	258	9960669	55
6	0888943	2897	9111057	11·249316	0892476	11·204780	1·0039747	0039589	259	9960411	54
7	0891840	2898	9108160	11·212770	0895408	11·168089	1·0040008	0039848	260	9960152	53
8	0894738	2897	9105262	11·176462	0898341	11·131635	1·0040270	0040108	261	9959892	52
9	0897635	2897	9102365	11·140389	0901273	11·095416	1·0040533	0040369	261	9959631	51
10	0900532	2897	9099468	11·104549	0904206	11·059431	1·0040796	0040630	263	9959370	50
11	0903429	2897	9096571	11·068940	0907138	11·023676	1·0041061	0040893	263	9959107	49
12	0906326	2897	9093674	11·033560	0910071	10·988150	1·0041326	0041156	264	9958844	48
13	0909223	2896	9090777	10·998406	0913004	10·952850	1·0041592	0041420	265	9958580	47
14	0912119	2897	9087881	10·963476	0915938	10·917775	1·0041859	0041685	266	9958315	46
15	0915016	2897	9084984	10·928768	0918871	10·882921	1·0042127	0041951	266	9958049	45
16	0917913	2896	9082087	10·894281	0921804	10·848288	1·0042396	0042217	268	9957783	44
17	0920809	2897	9079191	10·860011	0924738	10·813872	1·0042666	0042485	268	9957515	43
18	0923706	2896	9076294	10·825957	0927672	10·779673	1·0042937	0042753	269	9957247	42
19	0926602	2897	9073398	10·792117	0930606	10·745687	1·0043208	0043022	270	9956978	41
20	0929499	2896	9070501	10·758488	0933540	10·711913	1·0043480	0043292	271	9956708	40
21	0932395	2896	9067605	10·725070	0936474	10·678348	1·0043753	0043563	272	9956437	39
22	0935291	2896	9064709	10·691859	0939409	10·644992	1·0044028	0043835	272	9956165	38
23	0938187	2896	9061813	10·658854	0942344	10·611841	1·0044302	0044107	273	9955893	37
24	0941083	2896	9058917	10·626054	0945278	10·578895	1·0044578	0044380	275	9955620	36
25	0943979	2896	9056021	10·593455	0948213	10·546151	1·0044855	0044655	275	9955345	35
26	0946875	2896	9053125	10·561057	0951148	10·513607	1·0045132	0044930	275	9955070	34
27	0949771	2895	9050229	10·528857	0954084	10·481261	1·0045411	0045205	277	9954795	33
28	0952666	2896	9047334	10·496854	0957019	10·449112	1·0045690	0045482	278	9954518	32
29	0955562	2896	9044438	10·465046	0959955	10·417158	1·0045970	0045760	278	9954240	31
30	0958458	2895	9041542	10·433431	0962890	10·385397	1·0046251	0046038	279	9953962	30
31	0961353	2895	9038647	10·402007	0965826	10·353827	1·0046533	0046317	280	9953683	29
32	0964248	2896	9035752	10·370772	0968763	10·322447	1·0046815	0046597	281	9953403	28
33	0967144	2895	9032856	10·339726	0971699	10·291255	1·0047099	0046878	282	9953122	27
34	0970039	2895	9029961	10·308866	0974635	10·260249	1·0047383	0047160	283	9952840	26
35	0972934	2895	9027066	10·278190	0977572	10·229428	1·0047669	0047443	283	9952557	25
36	0975829	2895	9024171	10·247697	0980509	10·198789	1·0047955	0047726	284	9952274	24
37	0978724	2895	9021276	10·217386	0983446	10·168332	1·0048242	0048010	285	9951990	23
38	0981619	2895	9018381	10·187254	0986383	10·138054	1·0048530	0048295	286	9951705	22
39	0984514	2894	9015486	10·157300	0989320	10·107954	1·0048819	0048581	287	9951419	21
40	0987408	2895	9012592	10·127522	0992257	10·078031	1·0049108	0048868	288	9951132	20
41	0990303	2894	9009697	10·097920	0995194	10·048283	1·0049399	0049156	288	9950844	19
42	0993197	2895	9006803	10·068491	0998133	10·018708	1·0049690	0049444	290	9950556	18
43	0996092	2894	9003908	10·039234	1001071	9·9893050	1·0049982	0049734	290	9950266	17
44	0998986	2895	9001014	10·010147	1004009	9·9600724	1·0050275	0050024	291	9949976	16
45	1001881	2894	8998119	9·9812291	1006947	9·9310088	1·0050569	0050315	292	9949685	15
46	1004775	2894	8995225	9·9524787	1009886	9·9021125	1·0050864	0050607	292	9949393	14
47	1007669	2894	8992331	9·9238943	1012824	9·8733823	1·0051160	0050899	294	9949101	13
48	1010563	2894	8989437	9·8954744	1015763	9·8448166	1·0051456	0051193	294	9948807	12
49	1013457	2894	8986543	9·8672176	1018702	9·8164140	1·0051754	0051487	296	9948513	11
50	1016351	2894	8983649	9·8391227	1021641	9·7881732	1·0052052	0051783	296	9948217	10
51	1019245	2893	8980755	9·8111880	1024580	9·7600927	1·0052351	0052079	296	9947921	9
52	1022138	2894	8977862	9·7834124	1027520	9·7321713	1·0052651	0052375	298	9947625	8
53	1025032	2893	8974968	9·7557944	1030460	9·7044075	1·0052952	0052673	299	9947327	7
54	1027925	2894	8972075	9·7283327	1033399	9·6768000	1·0053254	0052972	299	9947028	6
55	1030819	2893	8969181	9·7010260	1036340	9·6493475	1·0053557	0053271	301	9946729	5
56	1033712	2893	8966288	9·6738730	1039280	9·6220486	1·0053860	0053572	301	9946428	4
57	1036605	2894	8963395	9·6468724	1042220	9·5949022	1·0054164	0053873	302	9946127	3
58	1039499	2893	8960501	9·6200229	1045161	9·5679068	1·0054470	0054175	302	9945825	2
59	1042392	2893	8957608	9·5933233	1048101	9·5410613	1·0054776	0054477	304	9945523	1
60	1045285		8954715	9·5667722	1051042	9·5143645	1·0055083	0054781		9945219	0

| ′ | Cosine | Dif. | Vers. | Secant. | Cotan. | Tang. | Cosec. | Covers | D. | Sine. | ′ |

'	Sine.	Diff.	Cosec.	Verseds.	Tang.	Diff.	Cotang.	Covers.	Secant.	D.	Cosine.	'
0	8·9402960	14416	11·0597040	7·5803891	8·9419518	14520	11·0580482	9·9603967	10·0016558	110	9·9983442	60
1	8·9417376	14367	11·0582624	7·5832778	8·9434044	14479	11·0565956	9·9602588	10·0016668	112	9·9983332	59
2	8·9431743	14320	11·0568257	7·5861568	8·9448523	14431	11·0551477	9·9601209	10·0016780	111	9·9983220	58
3	8·9446063	14272	11·0553937	7·5890263	8·9462954	14384	11·0537046	9·9599829	10·0016891	112	9·9983109	57
4	8·9460335	14226	11·0539665	7·5918864	8·9477338	14338	11·0522662	9·9598449	10·0017003	112	9·9982997	56
5	8·9474561	14178	11·0525439	7·5947370	8·9491676	14291	11·0508324	9·9597069	10·0017115	113	9·9982885	55
6	8·9488739	14132	11·0511261	7·5975783	8·9505907	14244	11·0494033	9·9595688	10·0017228	112	9·9982772	54
7	8·9502871	14086	11·0497129	7·6004103	8·9520211	14199	11·0479789	9·9594306	10·0017340	114	9·9982660	53
8	8·9516957	14039	11·0483043	7·6032331	8·9534410	14154	11·0465500	9·9592925	10·0017454	113	9·9982546	52
9	8·9530996	13995	11·0469004	7·6060468	8·9548564	14108	11·0451436	9·9591543	10·0017567	115	9·9982433	51
10	8·9544991	13949	11·0455009	7·6088513	8·9562672	14063	11·0437328	9·9590160	10·0017682	114	9·9982318	50
11	8·9558940	13903	11·0441060	7·6116468	8·9576735	14019	11·0423265	9·9588777	10·0017796	115	9·9982204	49
12	8·9572843	13860	11·0427157	7·6144333	8·9590754	13974	11·0409246	9·9587304	10·0017911	115	9·9982089	48
13	8·9586703	13814	11·0413297	7·6172109	8·9604728	13931	11·0395272	9·9586010	10·0018026	115	9·9981974	47
14	8·9600517	13771	11·0399483	7·6199796	8·9618659	13886	11·0381341	9·9584626	10·0018141	116	9·9981859	46
15	8·9614288	13726	11·0385712	7·6227395	8·9632545	13843	11·0367455	9·9583242	10·0018257	117	9·9981743	45
16	8·9628014	13683	11·0371986	7·6254906	8·9646388	13800	11·0353612	9·9581857	10·0018374	116	9·9981626	44
17	8·9641697	13640	11·0358303	7·6282330	8·9660188	13756	11·0339812	9·9580471	10·0018490	117	9·9981510	43
18	8·9655323	13597	11·0344663	7·6309668	8·9673944	13714	11·0326056	9·9579086	10·0018607	118	9·9981393	42
19	8·9668934	13553	11·0331066	7·6336920	8·9687658	13672	11·0312342	9·9577699	10·0018725	117	9·9981275	41
20	8·9682487	13512	11·0317513	7·6364086	8·9701330	13629	11·0298670	9·9576313	10·0018842	118	9·9981158	40
21	8·9695999	13469	11·0304001	7·6391167	8·9714959	13588	11·0285041	9·9574926	10·0018960	119	9·9981040	39
22	8·9709468	13427	11·0290532	7·6418164	8·9728523	13545	11·0271453	9·9573539	10·0019079	119	9·9980921	38
23	8·9722895	13385	11·0277105	7·6445078	8·9742092	13505	11·0257908	9·9572151	10·0019198	119	9·9980802	37
24	8·9736280	13344	11·0263720	7·6471908	8·9755597	13463	11·0244403	9·9570763	10·0019317	120	9·9980683	36
25	8·9749624	13302	11·0250376	7·6498655	8·9769060	13423	11·0230940	9·9569374	10·0019437	120	9·9980563	35
26	8·9762926	13262	11·0237074	7·6525320	8·9782483	13382	11·0217517	9·9567985	10·0019557	120	9·9980443	34
27	8·9776188	13220	11·0223812	7·6551903	8·9795865	13341	11·0204135	9·9566596	10·0019677	121	9·9980323	33
28	8·9789408	13181	11·0210592	7·6578404	8·9809206	13301	11·0190794	9·9565206	10·0019798	121	9·9980202	32
29	8·9802589	13140	11·0197411	7·6604825	8·9822507	13262	11·0177493	9·9563816	10·0019919	121	9·9980081	31
30	8·9815729	13100	11·0184271	7·6631166	8·9835769	13222	11·0164231	9·9562425	10·0020040	122	9·9979960	30
31	8·9828829	13060	11·0171171	7·6657427	8·9848991	13182	11·0151009	9·9561034	10·0020162	122	9·9979838	29
32	8·9841889	13021	11·0158111	7·6683608	8·9862173	13144	11·0137827	9·9559643	10·0020284	123	9·9979716	28
33	8·9854910	12981	11·0145090	7·6709711	8·9875317	13104	11·0124683	9·9558251	10·0020407	123	9·9979593	27
34	8·9867891	12943	11·0132109	7·6735735	8·9888421	13066	11·0111579	9·9556859	10·0020530	123	9·9979470	26
35	8·9880834	12903	11·0119166	7·6761682	8·9901487	13027	11·0098513	9·9555466	10·0020653	124	9·9979347	25
36	8·9893737	12865	11·0106263	7·6787550	8·9914514	12980	11·0085486	9·9554073	10·0020777	124	9·9979223	24
37	8·9906602	12827	11·0093398	7·6813342	8·9927503	12951	11·0072497	9·9552680	10·0020901	124	9·9979099	23
38	8·9919429	12788	11·0080571	7·6839058	8·9940454	12913	11·0059546	9·9551286	10·0021025	125	9·9978975	22
39	8·9932217	12751	11·0067783	7·6864697	8·9953367	12876	11·0046633	9·9549892	10·0021150	125	9·9978850	21
40	8·9944968	12713	11·0055032	7·6890260	8·9966243	12838	11·0033757	9·9548497	10·0021275	126	9·9978725	20
41	8·9957681	12675	11·0042319	7·6915749	8·9979081	12802	11·0020919	9·9547102	10·0021401	126	9·9978599	19
42	8·9970356	12638	11·0029644	7·6941162	8·9991883	12764	11·0008117	9·9545706	10·0021527	126	9·9978473	18
43	8·9982994	12601	11·0017006	7·6966502	9·0004647	12728	10·9995353	9·9544311	10·0021653	127	9·9978347	17
44	8·9995595	12565	11·0004405	7·6991767	9·0017375	12691	10·9982625	9·9542914	10·0021780	127	9·9978220	16
45	9·0008160	12527	10·9991840	7·7016959	9·0030066	12655	10·9969934	9·9541518	10·0021907	127	9·9978093	15
46	9·0020687	12492	10·9979313	7·7042078	9·0042721	12619	10·9957279	9·9540120	10·0022034	128	9·9977966	14
47	9·0033179	12455	10·9966821	7·7067124	9·0055340	12584	10·9944660	9·9538723	10·0022162	128	9·9977838	13
48	9·0045634	12419	10·9954366	7·7092098	9·0067924	12547	10·9932076	9·9537325	10·0022290	128	9·9977710	12
49	9·0058053	12383	10·9941947	7·7117001	9·0080471	12513	10·9919529	9·9535927	10·0022418	129	9·9977582	11
50	9·0070436	12348	10·9929564	7·7141832	9·0092984	12477	10·9907016	9·9534528	10·0022547	130	9·9977453	10
51	9·0082784	12312	10·9917216	7·7166592	9·0105461	12442	10·9894539	9·9533129	10·0022677	129	9·9977323	9
52	9·0095096	12278	10·9904904	7·7191281	9·0117903	12407	10·9882097	9·9531729	10·0022806	130	9·9977194	8
53	9·0107374	12242	10·9892626	7·7215900	9·0130310	12372	10·9869690	9·9530320	10·0022936	131	9·9977064	7
54	9·0119616	12207	10·9880384	7·7240450	9·0142682	12339	10·9857318	9·9528929	10·0023067	130	9·9976933	6
55	9·0131823	12173	10·9868177	7·7264930	9·0155021	12304	10·9844979	9·9527528	10·0023197	131	9·9976803	5
56	9·0143996	12139	10·9856004	7·7289341	9·0167325	12269	10·9832675	9·9526127	10·0023328	132	9·9976672	4
57	9·0156135	12104	10·9843865	7·7313683	9·0179594	12237	10·9820406	9·9524725	10·0023460	132	9·9976540	3
58	9·0168239	12070	10·9831761	7·7337958	9·0191831	12202	10·9808169	9·9523323	10·0023592	132	9·9976408	2
59	9·0180309	12037	10·9819691	7·7362164	9·0204033	12169	10·9795967	9·9521921	10·0023724	133	9·9976276	1
60	9·0192346		10·9807654	7·7386303	9·0216202		10·9783798	9·9520518	10·0023857		9·9976143	0
'	Cosine.	Diff.	Secant.	Covers.	Cotang.	Diff.	Tang.	Verseds.	Cosec.	D.	Sine.	'

′	Sine.	Dif.	Covers	Cosec.	Tang.	Cotang.	Secant.	Vers.	D.	Cosine	′
0	1045285	2893	8954715	9·5667722	1051042	9·5143645	1·0055083	0054781	305	9945219	60
1	1048178	2892	8951822	9·5403686	1053983	9·4878149	1·0055391	0055086	305	9944914	59
2	1051070	2893	8948930	9·5141110	1056925	9·4614116	1·0055699	0055391	306	9944609	58
3	1053963	2893	8946037	9·4879984	1059866	9·4351531	1·0056009	0055697	307	9944303	57
4	1056856	2892	8943144	9·4620296	1062808	9·4090384	1·0056319	0056004	308	9943996	56
5	1059748	2893	8940252	9·4362033	1065750	9·3830663	1·0056631	0056312	309	9943688	55
6	1062641	2892	8937359	9·4105184	1068692	9·3572355	1·0056943	0056621	309	9943379	54
7	1065533	2892	8934467	9·3849738	1071634	9·3315450	1·0057256	0056930	310	9943070	53
8	1068425	2893	8931575	9·3595682	1074576	9·3059936	1·0057570	0057240	312	9942760	52
9	1071318	2892	8928682	9·3343006	1077519	9·2805802	1·0057885	0057552	312	9942448	51
10	1074210	2892	8925790	9·3091699	1080462	9·2553035	1·0058200	0057864	313	9942136	50
11	1077102	2892	8922898	9·2841749	1083405	9·2301627	1·0058517	0058177	313	9941823	49
12	1079994	2891	8920006	9·2593145	1086348	9·2051564	1·0058834	0058490	315	9941510	48
13	1082885	2892	8917115	9·2345877	1089291	9·1802833	1·0059153	0058805	315	9941195	47
14	1085777	2892	8914223	9·2099934	1092234	9·1555436	1·0059472	0059120	317	9940880	46
15	1088669	2891	8911331	9·1855305	1095178	9·1309348	1·0059792	0059437	317	9940563	45
16	1091560	2892	8908440	9·1611980	1098122	9·1064564	1·0060113	0059754	318	9940246	44
17	1094452	2891	8905548	9·1369949	1101066	9·0821074	1·0060435	0060072	318	9939928	43
18	1097343	2891	8902657	9·1129200	1104010	9·0578867	1·0060757	0060390	320	9939610	42
19	1100234	2892	8899766	9·0889725	1106955	9·0337933	1·0061081	0060710	321	9939290	41
20	1103126	2891	8896874	9·0651512	1109899	9·0098261	1·0061405	0061031	321	9938969	40
21	1106017	2891	8893983	9·0414553	1112844	8·9859843	1·0061731	0061352	322	9938648	39
22	1108908	2891	8891092	9·0178837	1115789	8·9622668	1·0062057	0061674	323	9938326	38
23	1111799	2890	8888201	8·9944354	1118734	8·9386726	1·0062384	0061997	324	9938003	37
24	1114689	2891	8885311	8·9711095	1121680	8·9152009	1·0062712	0062321	324	9937679	36
25	1117580	2891	8882420	8·9479051	1124625	8·8918505	1·0063040	0062645	326	9937355	35
26	1120471	2890	8879529	8·9248211	1127571	8·8686206	1·0063370	0062971	326	9937029	34
27	1123361	2891	8876639	8·9018567	1130517	8·8455103	1·0063701	0063297	328	9936703	33
28	1126252	2890	8873748	8·8790109	1133463	8·8225186	1·0064032	0063625	328	9936375	32
29	1129142	2890	8870858	8·8562828	1136410	8·7996446	1·0064364	0063953	328	9936047	31
30	1132032	2890	8867968	8·8336715	1139356	8·7768874	1·0064697	0064281	330	9935719	30
31	1134922	2890	8865078	8·8111761	1142303	8·7542461	1·0065031	0064611	331	9935389	29
32	1137812	2890	8862188	8·7887957	1145250	8·7317198	1·0065366	0064942	331	9935058	28
33	1140702	2890	8859298	8·7665295	1148197	8·7093077	1·0065702	0065273	332	9934727	27
34	1143592	2890	8856408	8·7443766	1151144	8·6870088	1·0066039	0065605	333	9934395	26
35	1146482	2890	8853518	8·7223361	1154092	8·6648223	1·0066376	0065938	334	9934062	25
36	1149372	2889	8850628	8·7004071	1157039	8·6427475	1·0066714	0066272	335	9933728	24
37	1152261	2890	8847739	8·6785889	1159987	8·6207833	1·0067054	0066607	336	9933393	23
38	1155151	2889	8844849	8·6568805	1162936	8·5989290	1·0067394	0066943	336	9933057	22
39	1158040	2889	8841960	8·6352812	1165884	8·5771838	1·0067735	0067279	337	9932721	21
40	1160929	2889	8839071	8·6137901	1168832	8·5555468	1·0068077	0067616	339	9932384	20
41	1163818	2889	8836182	8·5924065	1171781	8·5340172	1·0068419	0067955	339	9932045	19
42	1166707	2889	8833293	8·5711295	1174730	8·5125943	1·0068763	0068294	339	9931706	18
43	1169596	2889	8830404	8·5499584	1177679	8·4912772	1·0069108	0068633	341	9931367	17
44	1172485	2889	8827515	8·5288923	1180628	8·4700651	1·0069453	0068974	341	9931026	16
45	1175374	2889	8824626	8·5079304	1183578	8·4489573	1·0069799	0069315	343	9930685	15
46	1178263	2888	8821737	8·4870721	1186528	8·4279531	1·0070146	0069658	343	9930342	14
47	1181151	2889	8818849	8·4663165	1189478	8·4070515	1·0070494	0070001	344	9929999	13
48	1184040	2888	8815960	8·4456629	1192428	8·3862519	1·0070843	0070345	345	9929655	12
49	1186928	2888	8813072	8·4251105	1195378	8·3655536	1·0071193	0070690	345	9929310	11
50	1189816	2888	8810184	8·4046586	1198329	8·3449558	1·0071544	0071035	347	9928965	10
51	1192704	2889	8807296	8·3843065	1201279	8·3244577	1·0071895	0071382	347	9928618	9
52	1195593	2888	8804407	8·3640534	1204230	8·3040586	1·0072248	0071729	349	9928271	8
53	1198481	2887	8801519	8·3438986	1207182	8·2837579	1·0072601	0072078	349	9927922	7
54	1201368	2888	8798632	8·3238415	1210133	8·2635547	1·0072955	0072427	349	9927573	6
55	1204256	2888	8795744	8·3038812	1213085	8·2434485	1·0073310	0072776	351	9927224	5
56	1207144	2887	8792856	8·2840171	1216036	8·2234384	1·0073666	0073127	352	9926873	4
57	1210031	2888	8789969	8·2642485	1218988	8·2035239	1·0074023	0073479	352	9926521	3
58	1212919	2887	8787081	8·2445748	1221941	8·1837041	1·0074380	0073831	353	9926169	2
59	1215806	2887	8784194	8·2249952	1224893	8·1639786	1·0074739	0074184	354	9925816	1
60	1218693		8781307	8·2055000	1227846	8·1443464	1·0075098	0074538		9925462	0

′	Cosine	Dif.	Vers.	Secant.	Cotan.	Tang.	Cosec.	Covers	D.	Sine.	′

′	Sine.	Diff.	Cosec.	Verseds.	Tang.	Diff.	Cotang.	Covers.	Secant.	D.	Cosine.	′
0	9·0192346	12002	10·9807654	7·7386303	9·0216202	12136	10·9783798	9·9520518	10·0023857	132	9·9976143	60
1	9·0204348	11970	10·9795652	7·7410375	9·0228338	12103	10·9771662	9·9519115	10·0023989	134	9·9976011	59
2	9·0216318	11936	10·9783682	7·7434380	9·0240441	12069	10·9759559	9·9517711	10·0024123	134	9·9975877	58
3	9·0228254	11903	10·9771746	7·7458319	9·0252510	12038	10·9747490	9·9516307	10·0024257	134	9·9975743	57
4	9·0240157	11870	10·9759843	7·7482192	9·0264548	12004	10·9735452	9·9514902	10·0024391	134	9·9975609	56
5	9·0252027	11838	10·9747973	7·7505999	9·0276552	11972	10·9723448	9·9513497	10·0024525	135	9·9975475	55
6	9·0263865	11804	10·9736135	7·7529742	9·0288524	11940	10·9711476	9·9512092	10·0024660	135	9·9975340	54
7	9·0275669	11773	10·9724331	7·7553419	9·0300464	11909	10·9699536	9·9510686	10·0024795	136	9·9975205	53
8	9·0287442	11740	10·9712558	7·7577031	9·0312373	11876	10·9687627	9·9509280	10·0024931	136	9·9975069	52
9	9·0299182	11708	10·9700818	7·7600580	9·0324249	11844	10·9675751	9·9507874	10·0025067	136	9·9974933	51
10	9·0310890	11677	10·9689110	7·7624064	9·0336093	11813	10·9663907	9·9506467	10·0025203	137	9·9974797	50
11	9·0322567	11645	10·9677433	7·7647485	9·0347906	11782	10·9652094	9·9505059	10·0025340	137	3·9974660	49
12	9·0334212	11613	10·9665788	7·7670843	9·0359688	11751	10·9640312	9·9503652	10·0025477	137	9·9974523	48
13	9·0345825	11582	10·9654175	7·7694138	9·0371439	11720	10·9628561	9·9502243	10·0025614	138	9·9974386	47
14	9·0357407	11551	10·9642593	7·7717371	9·0383159	11689	10·9616841	9·9500835	10·0025752	138	9·9974248	46
15	9·0368958	11519	10·9631042	7·7740541	9·0394848	11658	10·9605152	9·9499426	10·0025890	139	9·9974110	45
16	9·0380477	11439	10·9619523	7·7763649	9·0406506	11628	10·9593494	9·9498016	10·0026029	138	9·9973971	44
17	9·0391966	11458	10·9608034	7·7786696	9·0418134	11597	10·9581866	9·9496607	10·0026167	140	9·9973833	43
18	9·0403424	11428	10·9596576	7·7809682	9·0429731	11568	10·9570269	9·9495196	10·0026307	139	9·9973693	42
19	9·0414852	11397	10·9585148	7·7832607	9·0441290	11537	10·9558701	9·9493786	10·0026446	140	9·9973554	41
20	9·0426249	11368	10·9573751	7·7855472	9·0452836	11507	10·9547164	9·9492375	10·0026586	141	9·9973414	40
21	9·0437617	11337	10·9562383	7·7878276	9·0464343	11478	10·9535657	9·9490963	10·0026727	141	9·9973273	39
22	9·0448954	11307	10·9551046	7·7901020	9·0475821	11449	10·9524179	9·9489551	10·0026868	141	9·9973132	38
23	9·0460261	11277	10·9539739	7·7923705	9·0487270	11419	10·9512730	9·9488139	10·0027009	141	9·9972991	37
24	9·0471538	11248	10·9528462	7·7946331	9·0498689	11380	10·9501311	9·9486726	10·0027150	142	9·9972850	36
25	9·0482786	11219	10·9517214	7·7968897	9·0510078	11361	10·9489922	9·9485313	10·0027292	142	9·9972708	35
26	9·0494005	11189	10·9505995	7·7991405	9·0521439	11332	10·9478561	9·9483899	10·0027434	143	9·9972566	34
27	9·0505194	11160	10·9494806	7·8013855	9·0532771	11303	10·9467229	9·9482486	10·0027577	143	9·9972423	33
28	9·0516354	11131	10·9483646	7·8036246	9·0544074	11275	10·9455926	9·9481071	10·0027720	143	9·9972280	32
29	9·0527485	11103	10·9472515	7·8058580	9·0555349	11246	10·9444651	9·9479656	10·0027863	144	9·9972137	31
30	9·0538588	11073	10·9461412	7·8080856	9·0566595	11218	10·9433405	9·9478241	10·0028007	144	9·9971993	30
31	9·0549661	11045	10·9450339	7·8103075	9·0577813	11189	10·9422187	9·9476825	10·0028151	145	9·9971849	29
32	9·0560706	11017	10·9439294	7·8125237	9·0589002	11162	10·9410998	9·9475409	10·0028296	145	9·9971704	28
33	9·0571723	10988	10·9428277	7·8147343	9·0600164	11133	10·9399836	9·9473993	10·0028441	145	9·9971559	27
34	9·0582711	10961	10·9417289	7·8169392	9·0611297	11106	10·9388703	9·9472576	10·0028586	146	9·9971414	26
35	9·0593672	10932	10·9406328	7·8191386	9·0622403	11079	10·9377597	9·9471159	10·0028732	146	9·9971268	25
36	9·0604604	10905	10·9395396	7·8213324	9·0633482	11051	10·9366518	9·9469741	10·0028878	146	9·9971122	24
37	9·0615509	10877	10·9384491	7·8235205	9·0644533	11023	10·9355467	9·9468323	10·0029024	147	9·9970976	23
38	9·0626386	10849	10·9373614	7·8257032	9·0655556	10997	10·9344444	9·9466904	10·0029171	147	9·9970829	22
39	9·0637235	10822	10·9362765	7·8278804	9·0666553	10969	10·9333447	9·9465486	10·0029318	147	9·9970682	21
40	9·0648057	10795	10·9351943	7·8300522	9·0677522	10943	10·9322478	9·9464066	10·0029465	148	9·9970535	20
41	9·0658852	10767	10·9341148	7·8322185	9·0688465	10916	10·9311535	9·9462646	10·0029613	148	9·9970387	19
42	9·0669619	10741	10·9330381	7·8343794	9·0699381	10889	10·9300619	9·9461226	10·0029761	149	9·9970239	18
43	9·0680360	10714	10·9319640	7·8365349	9·0710270	10863	10·9289730	9·9459806	10·0029910	149	9·9970090	17
44	9·0691074	10687	10·9308926	7·8386851	9·0721133	10836	10·9278867	9·9458385	10·0030059	149	9·9969941	16
45	9·0701761	10660	10·9298239	7·8408299	9·0731969	10810	10·9268031	9·9456963	10·0030208	150	9·9969792	15
46	9·0712421	10634	10·9287579	7·8429695	9·0742779	10784	10·9257221	9·9455541	10·0030358	150	9·9969642	14
47	9·0723055	10608	10·9276945	7·8451037	9·0753563	10758	10·9246437	9·9454119	10·0030508	150	9·9969492	13
48	9·0733663	10581	10·9266337	7·8472327	9·0764321	10732	10·9235679	9·9452696	10·0030658	151	9·9969342	12
49	9·0744244	10555	10·9255756	7·8493565	9·0775053	10707	10·9224947	9·9451273	10·0030809	151	9·9969191	11
50	9·0754799	10530	10·9245201	7·8514751	9·0785760	10681	10·9214240	9·9449850	10·0030960	152	9·9969040	10
51	9·0765329	10503	10·9234671	7·8535885	9·0796441	10655	10·9203559	9·9448426	10·0031112	152	9·9968888	9
52	9·0775832	10478	10·9224168	7·8556968	9·0807096	10630	10·9192904	9·9447001	10·0031264	152	9·9968736	8
53	9·0786310	10452	10·9213690	7·8577999	9·0817726	10605	10·9182274	9·9445577	10·0031416	153	9·9968584	7
54	9·0796762	10427	10·9203238	7·8598980	9·0828331	10580	10·9171669	9·9444151	10·0031569	153	9·9968431	6
55	9·0807189	10401	10·9192811	7·8619910	9·0838911	10555	10·9161089	9·9442726	10·0031722	153	9·9968278	5
56	9·0817590	10376	10·9182410	7·8640789	9·0849466	10530	10·9150534	9·9441300	10·0031875	154	9·9968125	4
57	9·0827966	10351	10·9172034	7·8661618	9·0859996	10505	10·9140004	9·9439873	10·0032029	154	9·9967971	3
58	9·0838317	10326	10·9161683	7·8682397	9·0870501	10480	10·9129499	9·9438446	10·0032183	155	9·9967817	2
59	9·0848643	10302	10·9151357	7·8703126	9·0880981	10457	10·9119019	9·9437019	10·0032338	155	9·9967662	1
60	9·0858945		10·9141055	7·8723806	9·0891438		10·9108562	9·9435591	10·0032493	155	9·9967507	0
′	Cosine.	Diff.	Secant.	Covers.	Cotang.	Diff.	Tang.	Verseds.	Cosec.	D.	Sine.	′

′	Sine.	Dif.	Covers	Cosec.	Tang.	Cotang.	Secant.	Vers.	D.	Cosine	′
0	1218693	2688	8781307	8·2055090	1227846	8·1443464	1·0075098	0074538	355	9925462	60
1	1221581	2887	8778419	8·1861157	1230798	8·1248071	1·0075459	0074893	356	9925107	59
2	1224468	2887	8775532	8·1668145	1233752	8·1053599	1·0075820	0075249	357	9924751	58
3	1227355	2886	8772645	8·1476048	1236705	8·0860042	1·0076182	0075606	357	9924394	57
4	1230241	2887	8769759	8·1284860	1239668	8·0667394	1·0076545	0075963	358	9924037	56
5	1233128	2887	8766872	8·1094573	1242612	8·0475647	1·0076908	0076321	360	9923679	55
6	1236015	2886	8763985	8·0905182	1245560	8·0284796	1·0077273	0076681	360	9923319	54
7	1238901	2887	8761099	8·0716681	1248520	8·0094835	1·0077639	0077041	360	9922959	53
8	1241788	2886	8758212	8·0529062	1251474	7·9905756	1·0078005	0077401	362	9922599	52
9	1244674	2886	8755326	8·0342321	1254429	7·9717555	1·0078372	0077763	363	9922237	51
10	1247560	2886	8752440	8·0156450	1257384	7·9530224	1·0078741	0078126	363	9921874	50
11	1250446	2886	8749554	7·9971445	1260339	7·9343758	1·0079110	0078489	364	9921511	49
12	1253332	2886	8746668	7·9787298	1263294	7·9158151	1·0079480	0078853	365	9921147	48
13	1256218	2886	8743782	7·9604003	1266249	7·8973396	1·0079851	0079218	366	9920782	47
14	1259104	2886	8740896	7·9421556	1269205	7·8789489	1·0080222	0079584	367	9920416	46
15	1261990	2885	8738010	7·9239950	1272161	7·8606423	1·0080595	0079951	367	9920049	45
16	1264875	2886	8735125	7·9059179	1275117	7·8424191	1·0080968	0080318	368	9919682	44
17	1267761	2885	8732239	7·8879238	1278073	7·8242790	1·0081343	0080686	370	9919314	43
18	1270646	2885	8729354	7·8700120	1281030	7·8062212	1·0081718	0081056	370	9918944	42
19	1273531	2885	8726469	7·8521821	1283986	7·7882453	1·0082094	0081426	370	9918574	41
20	1276416	2886	8723584	7·8344335	1286943	7·7703506	1·0082471	0081796	372	9918204	40
21	1279302	2884	8720698	7·8167656	1289900	7·7525366	1·0082849	0082168	373	9917832	39
22	1282186	2885	8717814	7·7991778	1292858	7·7348028	1·0083228	0082541	373	9917459	38
23	1285071	2885	8714929	7·7816697	1295815	7·7171486	1·0083607	0082914	374	9917086	37
24	1287956	2885	8712044	7·7642406	1298773	7·6995735	1·0083988	0083288	375	9916712	36
25	1290841	2884	8709159	7·7468901	1301731	7·6820769	1·0084369	0083663	376	9916337	35
26	1293725	2884	8706275	7·7296176	1304690	7·6646584	1·0084752	0084039	377	9915961	34
27	1296609	2885	8703391	7·7124227	1307648	7·6473174	1·0085135	0084416	378	9915584	33
28	1299494	2884	8700506	7·6953047	1310607	7·6300533	1·0085519	0084794	378	9915206	32
29	1302378	2884	8697622	7·6782631	1313566	7·6128657	1·0085904	0085172	379	9914828	31
30	1305262	2884	8694738	7·6612976	1316525	7·5957541	1·0086290	0085551	380	9914449	30
31	1308146	2884	8691854	7·6444075	1319484	7·5787179	1·0086676	0085931	381	9914069	29
32	1311030	2883	8688970	7·6275923	1322444	7·5617567	1·0087064	0086312	382	9913688	28
33	1313913	2884	8686087	7·6108516	1325404	7·5448699	1·0087452	0086694	383	9913306	27
34	1316797	2884	8683203	7·5941849	1328364	7·5280571	1·0087842	0087077	383	9912923	26
35	1319681	2883	8680319	7·5775916	1331324	7·5113178	1·0088232	0087460	385	9912540	25
36	1322564	2883	8677436	7·5610713	1334285	7·4946514	1·0088623	0087845	385	9912155	24
37	1325447	2883	8674553	7·5446236	1337246	7·4780576	1·0089015	0088230	386	9911770	23
38	1328330	2883	8671670	7·5282478	1340207	7·4615357	1·0089408	0088616	387	9911384	22
39	1331213	2883	8668787	7·5119437	1343168	7·4450855	1·0089802	0089003	387	9910997	21
40	1334096	2883	8665904	7·4957106	1346129	7·4287064	1·0090196	0089390	389	9910610	20
41	1336979	2883	8663021	7·4795482	1349091	7·4123978	1·0090592	0089779	389	9910221	19
42	1339862	2882	8660138	7·4634560	1352053	7·3961595	1·0090988	0090168	390	9909832	18
43	1342744	2883	8657256	7·4474335	1355015	7·3799909	1·0091386	0090558	391	9909442	17
44	1345627	2882	8654373	7·4314803	1357978	7·3638916	1·0091784	0090949	392	9909051	16
45	1348509	2883	8651491	7·4155959	1360940	7·3478610	1·0092183	0091341	393	9908659	15
46	1351392	2882	8648608	7·3997798	1363903	7·3318989	1·0092583	0091734	393	9908266	14
47	1354274	2882	8645726	7·3840318	1366866	7·3160047	1·0092984	0092127	395	9907873	13
48	1357156	2882	8642844	7·3683512	1369830	7·3001780	1·0093386	0092522	395	9907478	12
49	1360038	2881	8639962	7·3527377	1372793	7·2844184	1·0093788	0092917	396	9907083	11
50	1362919	2882	8637081	7·3371909	1375757	7·2687255	1·0094192	0093313	397	9906687	10
51	1365801	2882	8634199	7·3217102	1378721	7·2530987	1·0094596	0093710	397	9906290	9
52	1368683	2881	8631317	7·3062954	1381685	7·2375378	1·0095001	0094107	399	9905893	8
53	1371564	2881	8628436	7·2909460	1384650	7·2220422	1·0095408	0094506	399	9905494	7
54	1374445	2882	8625555	7·2756616	1387615	7·2066116	1·0095815	0094905	401	9905095	6
55	1377327	2881	8622673	7·2604417	1390580	7·1912456	1·0096223	0095306	401	9904694	5
56	1380208	2881	8619792	7·2452859	1393545	7·1759437	1·0096631	0095707	402	9904293	4
57	1383089	2881	8616911	7·2301940	1396510	7·1607056	1·0097041	0096109	402	9903891	3
58	1385970	2880	8614030	7·2151653	1399476	7·1455308	1·0097452	0096511	404	9903489	2
59	1388850	2881	8611150	7·2001996	1402442	7·1304190	1·0097863	0096915	404	9903085	1
60	1391731		8608269	7·1852965	1405408	7·1153697	1·0098276	0097319		9902681	0

| ′ | Cosine | Dif. | Vers. | Secant. | Cotan. | Tang. | Cosec. | Covers | D. | Sine. | |

′	Sine.	Diff.	Cosec.	Verseds.	Tang.	Diff.	Cotang.	Covers.	Secant.	D.	Cosine.	′
0	9·0858945	10276	10·9141055	7·8723806	9·0891438	10431	10·9108562	9·9435591	10·0032493	155	9·9967507	60
1	9·0869221	10252	10·9130779	7·8744436	9·0901869	10408	10·9098131	9·9434163	10·0032648	156	9·9967352	59
2	9·0879473	10227	10·9120527	7·8765017	9·0912277	10383	10·9087723	9·9432735	10·0032804	156	9·9967196	58
3	9·0889700	10203	10·9110300	7·8785550	9·0922660	10360	10·9077340	9·9431306	10·0032960	156	9·9967040	57
4	9·0899903	10179	10·9100097	7·8806033	9·0933020	10335	10·9066980	9·9429876	10·0033116	157	9·9966884	56
5	9·0910082	10155	10·9089918	7·8826469	9·0943355	10312	10·9056645	9·9428447	10·0033273	157	9·9966727	55
6	9·0920237	10130	10·9079763	7·8846856	9·0953667	10288	10·9046333	9·9427016	10·0033430	158	9·9966570	54
7	9·0930367	10107	10·9069633	7·8867196	9·0963955	10264	10·9036045	9·9425586	10·0033588	158	9·9966412	53
8	9·0940474	10082	10·9059526	7·8887487	9·0974219	10241	10·9025781	9·9424155	10·0033746	158	9·9966254	52
9	9·0950556	10059	10·9049444	7·8907732	9·0984460	10218	10·9015540	9·9422723	10·0033904	159	9·9966096	51
10	9·0960615	10036	10·9039385	7·8927928	9·0994678	10194	10·9005322	9·9421291	10·0034063	159	9·9965937	50
11	9·0970651	10011	10·9029349	7·8948078	9·1004872	10172	10·8995128	9·9419859	10·0034222	159	9·9965778	49
12	9·0980662	9989	10·9019338	7·8968181	9·1015044	10148	10·8984956	9·9418426	10·0034381	160	9·9965619	48
13	9·0990651	9965	10·9009349	7·8988238	9·1025192	10125	10·8974808	9·9416993	10·0034541	160	9·9965459	47
14	9·1000616	9942	10·8999384	7·9008248	9·1035317	10103	10·8964683	9·9415560	10·0034701	161	9·9965299	46
15	9·1010558	9919	10·8989442	7·9028212	9·1045420	10080	10·8954580	9·9414126	10·0034862	161	9·9965138	45
16	9·1020477	9896	10·8979523	7·9048130	9·1055500	10057	10·8944500	9·9412691	10·0035023	161	9·9964977	44
17	9·1030373	9873	10·8969627	7·9068002	9·1065557	10034	10·8934443	9·9411256	10·0035184	161	9·9964816	43
18	9·1040246	9850	10·8959754	7·9087829	9·1075591	10013	10·8924409	9·9409821	10·0035345	162	9·9964655	42
19	9·1050096	9828	10·8949904	7·9107610	9·1085604	9990	10·8914396	9·9408385	10·0035507	163	9·9964493	41
20	9·1059924	9805	10·8940076	7·9127346	9·1095594	9968	10·8904406	9·9406949	10·0035670	163	9·9964330	40
21	9·1069729	9783	10·8930271	7·9147038	9·1105562	9946	10·8894438	9·9405513	10·0035833	163	9·9964167	39
22	9·1079512	9760	10·8920488	7·9166684	9·1115508	9923	10·8884492	9·9404076	10·0035996	163	9·9964004	38
23	9·1089272	9738	10·8910728	7·9186286	9·1125431	9902	10·8874569	9·9402638	10·0036161	164	9·9963841	37
24	9·1099010	9716	10·8900990	7·9205844	9·1135333	9880	10·8864667	9·9401201	10·0036323	164	9·9963677	36
25	9·1108726	9694	10·8891274	7·9225358	9·1145213	9859	10·8854787	9·9399762	10·0036487	165	9·9963513	35
26	9·1118420	9672	10·8881580	7·9244827	9·1155072	9837	10·8844928	9·9398324	10·0036652	165	9·9963348	34
27	9·1128092	9650	10·8871908	7·9264253	9·1164909	9815	10·8835091	9·9396885	10·0036817	165	9·9963183	33
28	9·1137742	9628	10·8862258	7·9283636	9·1174724	9794	10·8825276	9·9395445	10·0036982	166	9·9963018	32
29	9·1147370	9607	10·8852630	7·9302975	9·1184518	9773	10·8815482	9·9394005	10·0037148	166	9·9962852	31
30	9·1156977	9585	10·8843023	7·9322271	9·1194291	9752	10·8805709	9·9392565	10·0037314	167	9·9962686	30
31	9·1166562	9563	10·8833438	7·9341523	9·1204043	9730	10·8795957	9·9391124	10·0037481	167	9·9962519	29
32	9·1176125	9542	10·8823875	7·9360734	9·1213773	9709	10·8786227	9·9389683	10·0037648	167	9·9962352	28
33	9·1185667	9521	10·8814333	7·9379901	9·1223482	9689	10·8776518	9·9388242	10·0037815	168	9·9962185	27
34	9·1195188	9500	10·8804812	7·9399027	9·1233171	9668	10·8766829	9·9386800	10·0037983	168	9·9962017	26
35	9·1204688	9479	10·8795312	7·9418110	9·1242839	9647	10·8757161	9·9385357	10·0038151	168	9·9961849	25
36	9·1214167	9457	10·8785833	7·9437151	9·1252486	9626	10·8747514	9·9383914	10·0038319	169	9·9961681	24
37	9·1223624	9437	10·8776376	7·9456150	9·1262112	9606	10·8737888	9·9382471	10·0038488	169	9·9961512	23
38	9·1233061	9416	10·8766939	7·9475107	9·1271718	9585	10·8728282	9·9381027	10·0038657	169	9·9961343	22
39	9·1242477	9395	10·8757523	7·9494023	9·1281303	9565	10·8718697	9·9379583	10·0038826	170	9·9961174	21
40	9·1251872	9374	10·8748128	7·9512898	9·1290868	9545	10·8709132	9·9378139	10·0038996	170	9·9961004	20
41	9·1261246	9354	10·8738754	7·9531732	9·1300413	9524	10·8699587	9·9376694	10·0039166	171	9·9960834	19
42	9·1270600	9334	10·8729400	7·9550525	9·1309937	9505	10·8690063	9·9375248	10·0039337	171	9·9960663	18
43	9·1279934	9313	10·8720066	7·9569276	9·1319442	9484	10·8680558	9·9373802	10·0039508	171	9·9960492	17
44	9·1289247	9292	10·8710753	7·9587988	9·1328926	9465	10·8671074	9·9372356	10·0039679	172	9·9960321	16
45	9·1298539	9273	10·8701461	7·9606659	9·1338391	9444	10·8661609	9·9370909	10·0039851	172	9·9960149	15
46	9·1307812	9252	10·8692188	7·9625292	9·1347835	9425	10·8652165	9·9369462	10·0040023	173	9·9959977	14
47	9·1317064	9233	10·8682936	7·9643880	9·1357260	9405	10·8642740	9·9368015	10·0040196	173	9·9959804	13
48	9·1326297	9212	10·8673703	7·9662431	9·1366665	9386	10·8633335	9·9366567	10·0040369	173	9·9959631	12
49	9·1335509	9193	10·8664491	7·9680942	9·1376051	9366	10·8623949	9·9365119	10·0040542	174	9·9959458	11
50	9·1344702	9173	10·8655299	7·9699414	9·1385417	9347	10·8614583	9·9363670	10·0040716	174	9·9959284	10
51	9·1353875	9153	10·8646125	7·9717846	9·1394764	9328	10·8605236	9·9362220	10·0040889	175	9·9959111	9
52	9·1363028	9133	10·8636972	7·9736239	9·1404092	9308	10·8595908	9·9360771	10·0041064	175	9·9958936	8
53	9·1372161	9114	10·8627839	7·9754593	9·1413400	9289	10·8586600	9·9359321	10·0041239	175	9·9958761	7
54	9·1381275	9095	10·8618725	7·9772908	9·1422689	9270	10·8577311	9·9357870	10·0041414	175	9·9958586	6
55	9·1390370	9075	10·8609630	7·9791184	9·1431959	9251	10·8568041	9·9356419	10·0041589	176	9·9958411	5
56	9·1399445	9056	10·8600555	7·9809422	9·1441210	9232	10·8558790	9·9354968	10·0041765	176	9·9958235	4
57	9·1408501	9036	10·8591499	7·9827621	9·1450442	9213	10·8549558	9·9353516	10·0041941	177	9·9958059	3
58	9·1417537	9018	10·8582463	7·9845782	9·1459655	9194	10·8540345	9·9352064	10·0042118	177	9·9957882	2
59	9·1426555	8998	10·8573445	7·9863905	9·1468849	9176	10·8531151	9·9350611	10·0042295	177	9·9957705	1
60	9·1435553		10·8564447	7·9881990	9·1478025		10·8521975	9·9349158	10·0042472		9·9957528	0
′	Cosine.	Diff.	Secant.	Covers.	Cotang.	Diff.	Tang.	Verseds.	Cosec.	D.	Sine.	′

	Sine.	Dif.	Covers	Cosec.	Tang.	Cotang.	Secant.	Vers.	D.	Cosine	'
0	1391731	2881	8608263	7·1852965	1405408	7·1153697	1·0098276	0097319	406	9902681	60
1	1394612	2880	8605388	7·1704556	1408375	7·1003826	1·0098689	0097725	406	9902275	59
2	1397492	2880	8602508	7·1556764	1411342	7·0854573	1·0099103	0098131	407	9901869	58
3	1400372	2880	8599628	7·1409587	1414308	7·0705934	1·0099518	0098538	407	9901462	57
4	1403252	2880	8596748	7·1263019	1417276	7·0557905	1·0099934	0098945	409	9901055	56
5	1406132	2880	8593868	7·1117059	1420243	7·0410482	1·0100351	0099354	409	9900646	55
6	1409012	2880	8590988	7·0971700	1423211	7·0263662	1·0100769	0099763	411	9900237	54
7	1411892	2880	8588108	7·0826941	1426179	7·0117441	1·0101187	0100174	411	9899826	53
8	1414772	2880	8585228	7·0682777	1429147	6·9971806	1·0101607	0100585	412	9899415	52
9	1417651	2879	8582349	7·0539205	1432115	6·9826781	1·0102027	0100997	413	9899003	51
10	1420531	2880	8579469	7·0396220	1435084	6·9682335	1·0102449	0101410	413	9898590	50
11	1423410	2879	8576590	7·0253820	1438053	6·9538473	1·0102871	0101823	415	9898177	49
12	1426289	2879	8573711	7·0112001	1441022	6·9395192	1·0103294	0102238	415	9897762	48
13	1429168	2879	8570832	6·9970760	1443991	6·9252489	1·0103718	0102653	416	9897347	47
14	1432047	2879	8567953	6·9830092	1446961	6·9110359	1·0104143	0103069	417	9896931	46
15	1434926	2879	8565074	6·9689994	1449931	6·8968799	1·0104568	0103486	418	9896514	45
16	1437805	2879	8562195	6·9550464	1452901	6·8827807	1·0104995	0103904	419	9896096	44
17	1440684	2878	8559316	6·9411496	1455872	6·8687378	1·0105422	0104323	419	9895677	43
18	1443562	2878	8556438	6·9273089	1458842	6·8547508	1·0105851	0104742	420	9895258	42
19	1446440	2879	8553560	6·9135239	1461813	6·8408196	1·0106280	0105162	422	9894838	41
20	1449319	2878	8550681	6·8997942	1464784	6·8269437	1·0106710	0105584	422	9894416	40
21	1452197	2878	8547803	6·8861195	1467756	6·8131227	1·0107141	0106006	422	9893994	39
22	1455075	2878	8544925	6·8724995	1470727	6·7993565	1·0107573	0106428	424	9893572	38
23	1457953	2877	8542047	6·8589338	1473699	6·7856446	1·0108006	0106852	425	9893148	37
24	1460830	2878	8539170	6·8454222	1476672	6·7919867	1·0108440	0107277	425	9892723	36
25	1463708	2877	8536292	6·8319642	1479644	6·7583826	1·0108875	0107702	426	9892298	35
26	1466585	2877	8533415	6·8185597	1482617	6·7448318	1·0109310	0108128	427	9891872	34
27	1469463	2877	8530537	6·8052082	1485590	6·7313341	1·0109747	0108555	428	9891445	33
28	1472340	2877	8527660	6·7919095	1488563	6·7178891	1·0110184	0108983	429	9891017	32
29	1475217	2877	8524783	6·7786632	1491536	6·7044966	1·0110622	0109412	429	9890588	31
30	1478094	2877	8521906	6·7654691	1494510	6·6911562	1·0111061	0109841	431	9890159	30
31	1480971	2877	8519029	6·7523268	1497484	6·6778677	1·0111501	0110272	431	9889728	29
32	1483848	2876	8516152	6·7392360	1500458	6·6646307	1·0111942	0110703	432	9889297	28
33	1486724	2877	8513276	6·7261965	1503433	6·6514449	1·0112384	0111135	433	9888865	27
34	1489601	2876	8510399	6·7132079	1506408	6·6383100	1·0112827	0111568	434	9888432	26
35	1492477	2876	8507523	6·7002699	1509383	6·6252258	1·0113270	0112002	434	9887998	25
36	1495353	2877	8504647	6·6873822	1512358	6·6121919	1·0113715	0112436	436	9887564	24
37	1498230	2876	8501770	6·6745446	1515333	6·5992080	1·0114160	0112872	436	9887128	23
38	1501106	2875	8498894	6·6617568	1518309	6·5862739	1·0114606	0113308	437	9886692	22
39	1503981	2876	8496019	6·6490184	1521285	6·5733892	1·0115054	0113745	438	9886255	21
40	1506857	2876	8493143	6·6363293	1524262	6·5605538	1·0115502	0114183	439	9885817	20
41	1509733	2875	8490267	6·6236890	1527238	6·5477672	1·0115951	0114622	439	9885378	19
42	1512608	2876	8487392	6·6110973	1530215	6·5350293	1·0116400	0115061	441	9884939	18
43	1515484	2875	8484516	6·5985540	1533192	6·5223396	1·0116851	0115502	441	9884498	17
44	1518359	2875	8481641	6·5860587	1536170	6·5096981	1·0117303	0115943	442	9884057	16
45	1521234	2875	8478766	6·5736112	1539147	6·4971043	1·0117755	0116385	443	9883615	15
46	1524109	2875	8475891	6·5612113	1542125	6·4845581	1·0118209	0116828	444	9883172	14
47	1526984	2874	8473016	6·5488586	1545103	6·4720591	1·0118663	0117272	444	9882728	13
48	1529858	2875	8470142	6·5365528	1548082	6·4596070	1·0119118	0117716	446	9882284	12
49	1532733	2874	8467267	6·5242938	1551061	6·4472017	1·0119575	0118162	446	9881838	11
50	1535607	2875	8464393	6·5120812	1554040	6·4348428	1·0120032	0118608	447	9881392	10
51	1538482	2874	8461518	6·4999148	1557019	6·4225301	1·0120489	0119055	448	9880945	9
52	1541356	2874	8458644	6·4877944	1559998	6·4102633	1·0120948	0119503	449	9880497	8
53	1544230	2874	8455770	6·4757195	1562978	6·3980422	1·0121408	0119952	449	9880048	7
54	1547104	2874	8452896	6·4636901	1565958	6·3858665	1·0121869	0120401	451	9879599	6
55	1549978	2873	8450022	6·4517059	1568939	6·3737359	1·0122330	0120852	451	9879148	5
56	1552851	2874	8447149	6·4397666	1571919	6·3616502	1·0122793	0121303	452	9878697	4
57	1555725	2873	8444275	6·4278719	1574900	6·3496092	1·0123256	0121755	453	9878245	3
58	1558598	2874	8441402	6·4160216	1577881	6·3376126	1·0123720	0122208	454	9877792	2
59	1561472	2873	8438528	6·4042154	1580863	6·3256601	1·0124185	0122662	455	9877338	1
60	1564345		8435655	6·3924532	1583844	6·3137515	1·0124651	0123117		9876883	0
'	Cosine	Dif.	Vers.	Secant.	Cotan.	Tang.	Cosec.	Covers	D.	Sine.	'

′	Sine.	Dif.	Cosec.	Verseds.	Tang.	Dif.	Cotang.	Covers.	Secant.	D.	Cosine.	′
0	9·1435553	8979	10·8564447	7·9881390	9·1478025	9157	10·8521975	9·9349158	10·0042472	178	9·9957528	60
1	9·1444532	8961	10·8555468	7·9900038	9·1487182	9139	10·8512818	9·9347705	10·0042650	178	9·9957350	59
2	9·1453493	8942	10·8546507	7·9918047	9·1496321	9120	10·8503679	9·9346251	10·0042828	179	9·9957172	58
3	9·1462435	8923	10·8537565	7·9936020	9·1505441	9102	10·8494559	9·9344797	10·0043007	178	9·9956993	57
4	9·1471358	8904	10·8528642	7·9953955	9·1514543	9084	10·8485457	9·9343342	10·0043185	180	9·9956815	56
5	9·1480262	8886	10·8519738	7·9971853	9·1523627	9065	10·8476373	9·9341887	10·0043365	179	9·9956635	55
6	9·1489148	8867	10·8510852	7·9989713	9·1532692	9047	10·8467308	9·9340431	10·0043544	180	9·9956456	54
7	9·1498015	8849	10·8501985	8·0007537	9·1541739	9030	10·8458261	9·9338975	10·0043724	181	9·9956276	53
8	9·1506864	8830	10·8493136	8·0025325	9·1550769	9011	10·8449231	9·9337518	10·0043905	180	9·9956095	52
9	9·1515694	8813	10·8484306	8·0043076	9·1559780	8993	10·8440220	9·9336062	10·0044085	181	9·9955915	51
10	9·1524507	8794	10·8475493	8·0060790	9·1568773	8975	10·8431227	9·9334604	10·0044266	182	9·9955734	50
11	9·1533301	8775	10·8466699	8·0078468	9·1577748	8958	10·8422252	9·9333146	10·0044448	182	9·9955552	49
12	9·1542076	8758	10·8457924	8·0096110	9·1586706	8940	10·8413294	9·9331688	10·0044630	182	9·9955370	48
13	9·1550834	8740	10·8449166	8·0113716	9·1595646	8923	10·8404354	9·9330230	10·0044812	183	9·9955188	47
14	9·1559574	8722	10·8440426	8·0131287	9·1604569	8904	10·8395431	9·9328771	10·0044995	183	9·9955005	46
15	9·1568296	8704	10·8431704	8·0148822	9·1613473	8888	10·8386527	9·9327311	10·0045178	183	9·9954822	45
16	9·1577000	8686	10·8423000	8·0166321	9·1622361	8870	10·8377639	9·9325851	10·0045361	184	9·9954639	44
17	9·1585686	8668	10·8414314	8·0183785	9·1631231	8852	10·8368769	9·9324391	10·0045545	184	9·9954455	43
18	9·1594354	8651	10·8405646	8·0201213	9·1640083	8836	10·8359917	9·9322930	10·0045729	184	9·9954271	42
19	9·1603005	8634	10·8396995	8·0218607	9·1648919	8818	10·8351081	9·9321469	10·0045913	185	9·9954087	41
20	9·1611639	8615	10·8388361	8·0235965	9·1657737	8801	10·8342263	9·9320007	10·0046098	185	9·9953902	40
21	9·1620254	8599	10·8379746	8·0253289	9·1666538	8784	10·8333462	9·9318545	10·0046283	186	9·9953717	39
22	9·1628853	8581	10·8371147	8·0270578	9·1675322	8767	10·8324678	9·9317083	10·0046469	186	9·9953531	38
23	9·1637434	8564	10·8362566	8·0287833	9·1684089	8750	10·8315911	9·9315620	10·0046655	186	9·9953345	37
24	9·1645998	8546	10·8354002	8·0305053	9·1692839	8733	10·8307161	9·9314156	10·0046841	187	9·9953159	36
25	9·1654544	8530	10·8345456	8·0322239	9·1701572	8717	10·8298428	9·9312693	10·0047028	187	9·9952972	35
26	9·1663074	8512	10·8336926	8·0339391	9·1710289	8700	10·8289711	9·9311228	10·0047215	188	9·9952785	34
27	9·1671586	8495	10·8328414	8·0356508	9·1718989	8683	10·8281011	9·9309764	10·0047403	188	9·9952597	33
28	9·1680081	8478	10·8319919	8·0373592	9·1727672	8666	10·8272328	9·9308299	10·0047591	188	9·9952409	32
29	9·1688559	8462	10·8311441	8·0390643	9·1736338	8650	10·8263662	9·9306833	10·0047779	188	9·9952221	31
30	9·1697021	8444	10·8302979	8·0407659	9·1744988	8634	10·8255012	9·9305367	10·0047967	189	9·9952033	30
31	9·1705465	8428	10·8294535	8·0424642	9·1753622	8617	10·8246378	9·9303901	10·0048156	190	9·9951844	29
32	9·1713893	8412	10·8286107	8·0441592	9·1762239	8601	10·8237761	9·9302434	10·0048346	190	9·9951654	28
33	9·1722305	8394	10·8277695	8·0458509	9·1770840	8585	10·8229160	9·9300967	10·0048536	190	9·9951464	27
34	9·1730699	8378	10·8269301	8·0475393	9·1779425	8568	10·8220575	9·9299499	10·0048726	190	9·9951274	26
35	9·1739077	8362	10·8260923	8·0492243	9·1787993	8553	10·8212007	9·9298031	10·0048916	191	9·9951084	25
36	9·1747439	8345	10·8252561	8·0509061	9·1796546	8536	10·8203454	9·9296563	10·0049107	191	9·9950893	24
37	9·1755784	8328	10·8244216	8·0525846	9·1805082	8520	10·8194918	9·9295094	10·0049298	192	9·9950702	23
38	9·1764112	8313	10·8235888	8·0542599	9·1813602	8504	10·8186398	9·9293624	10·0049490	192	9·9950510	22
39	9·1772305	8296	10·8227575	8·0559319	9·1822106	8489	10·8177894	9·9292155	10·0049682	192	9·9950318	21
40	9·1780721	8280	10·8219279	8·0576007	9·1830595	8473	10·8169405	9·9290684	10·0049874	193	9·9950126	20
41	9·1789001	8264	10·8210999	8·0592663	9·1839068	8457	10·8160932	9·9289214	10·0050067	193	9·9949933	19
42	9·1797265	8247	10·8202735	8·0609286	9·1847525	8441	10·8152475	9·9287743	10·0050260	194	9·9949740	18
43	9·1805512	8232	10·8194488	8·0625878	9·1855966	8426	10·8144034	9·9286271	10·0050454	194	9·9949546	17
44	9·1813744	8216	10·8186256	8·0642438	9·1864392	8410	10·8135608	9·9284799	10·0050648	194	9·9949352	16
45	9·1821960	8200	10·8178040	8·0658966	9·1872802	8394	10·8127198	9·9283327	10·0050842	194	9·9949158	15
46	9·1830160	8184	10·8169840	8·0675463	9·1881196	8379	10·8118804	9·9281854	10·0051036	195	9·9948964	14
47	9·1838344	8168	10·8161656	8·0691928	9·1889575	8364	10·8110425	9·9280380	10·0051231	196	9·9948769	13
48	9·1846512	8153	10·8153488	8·0708362	9·1897939	8348	10·8102061	9·9278907	10·0051427	196	9·9948573	12
49	9·1854665	8137	10·8145335	8·0724764	9·1906287	8334	10·8093713	9·9277433	10·0051623	196	9·9948377	11
50	9·1862802	8121	10·8137198	8·0741136	9·1914621	8318	10·8085379	9·9275958	10·0051819	196	9·9948181	10
51	9·1870923	8106	10·8129077	8·0757476	9·1922939	8302	10·8077061	9·9274483	10·0052015	197	9·9947985	9
52	9·1879029	8091	10·8120971	8·0773786	9·1931241	8288	10·8068759	9·9273008	10·0052212	197	9·9947788	8
53	9·1887120	8075	10·8112880	8·0790065	9·1939529	8273	10·8060471	9·9271532	10·0052409	198	9·9947591	7
54	9·1895195	8059	10·8104805	8·0806313	9·1947802	8257	10·8052198	9·9270055	10·0052607	198	9·9947393	6
55	9·1903254	8045	10·8096746	8·0822531	9·1956059	8243	10·8043941	9·9268579	10·0052805	198	9·9947195	5
56	9·1911299	8029	10·8088701	8·0838718	9·1964302	8228	10·8035698	9·9267101	10·0053003	199	9·9946997	4
57	9·1919328	8014	10·8080672	8·0854875	9·1972530	8213	10·8027470	9·9265624	10·0053202	199	9·9946798	3
58	9·1927342	7999	10·8072658	8·0871002	9·1980743	8198	10·8019257	9·9264146	10·0053401	200	9·9946599	2
59	9·1935341	7983	10·8064659	8·0887099	9·1988941	8184	10·8011059	9·9262667	10·0053601	200	9·9946399	1
60	9·1943324		10·8056676	8·0903166	9·1997125		10·8002875	9·9261188	10·0053801		9·9946199	0
′	Cosine.	Dif.	Secant.	Covers.	Cotang.	Dif.	Tang.	Verseds.	Cosec.	D.	Sine.	′

'	Sine.	Dif.	Covers	Cosec.	Tang.	Cotang.	Secant.	Vers.	D.	Cosine	'
0	1564345	2873	8435655	6·3924532	1583844	6·3137515	1·0124651	0123117	455	9876883	60
1	1567218	2873	8432782	6·3807347	1586826	6·3018866	1·0125118	0123572	456	9876428	59
2	1570091	2872	8429909	6·3690595	1589809	6·2900651	1·0125586	0124028	458	9875972	58
3	1572963	2873	8427037	6·3574276	1592791	6·2782868	1·0126055	0124486	457	9875514	57
4	1575836	2872	8424164	6·3458386	1595774	6·2665515	1·0126524	0124943	459	9875057	56
5	1578708	2873	8421292	6·3342923	1598757	6·2548588	1·0126995	0125402	460	9874598	55
6	1581581	2872	8418419	6·3227884	1601740	6·2432086	1·0127466	0125862	460	9874138	54
7	1584453	2872	8415547	6·3113269	1604724	6·2316007	1·0127939	0126322	462	9873678	53
8	1587325	2872	8412675	6·2999073	1607708	6·2200347	1·0128412	0126784	462	9873216	52
9	1590197	2872	8409803	6·2885295	1610692	6·2085106	1·0128886	0127246	463	9872754	51
10	1593069	2871	8406931	6·2771933	1613677	6·1970279	1·0129361	0127709	464	9872291	50
11	1595940	2872	8404060	6·2658984	1616662	6·1855867	1·0129837	0128173	464	9871827	49
12	1598812	2871	8401188	6·2546446	1619647	6·1741865	1·0130314	0128637	466	9871363	48
13	1601683	2872	8398317	6·2434316	1622632	6·1628272	1·0130791	0129103	466	9870897	47
14	1604555	2871	8395445	6·2322594	1625618	6·1515085	1·0131270	0129569	467	9870431	46
15	1607426	2871	8392574	6·2211275	1628603	6·1402303	1·0131750	0130036	468	9869964	45
16	1610297	2870	8389703	6·2100359	1631590	6·1289923	1·0132230	0130504	469	9869496	44
17	1613167	2871	8386833	6·1989843	1634576	6·1177943	1·0132711	0130973	470	9869027	43
18	1616038	2871	8383962	6·1879725	1637563	6·1066360	1·0133194	0131443	470	9868557	42
19	1618909	2870	8381091	6·1770003	1640550	6·0955174	1·0133677	0131913	472	9868087	41
20	1621779	2871	8378221	6·1660674	1643537	6·0844381	1·0134161	0132385	472	9867615	40
21	1624650	2870	8375350	6·1551736	1646525	6·0733979	1·0134646	0132857	473	9867143	39
22	1627520	2870	8372480	6·1443189	1649513	6·0623967	1·0135132	0133330	474	9866670	38
23	1630390	2870	8369610	6·1335028	1652501	6·0514343	1·0135618	0133804	474	9866196	37
24	1633260	2869	8366740	6·1227253	1655489	6·0405103	1·0136106	0134278	476	9865722	36
25	1636129	2870	8363871	6·1119861	1658478	6·0296247	1·0136595	0134754	476	9865246	35
26	1638999	2869	8361001	6·1012850	1661467	6·0187772	1·0137084	0135230	477	9864770	34
27	1641868	2870	8358132	6·0906219	1664456	6·0079676	1·0137574	0135707	478	9864293	33
28	1644738	2869	8355262	6·0799964	1667446	5·9971957	1·0138066	0136185	479	9863815	32
29	1647607	2869	8352393	6·0694085	1670436	5·9864614	1·0138558	0136664	480	9863336	31
30	1650476	2869	8349524	6·0588580	1673426	5·9757644	1·0139051	0137144	481	9862856	30
31	1653345	2869	8346655	6·0483445	1676417	5·9651045	1·0139545	0137625	481	9862375	29
32	1656214	2868	8343786	6·0378680	1679407	5·9544815	1·0140040	0138106	482	9861894	28
33	1659082	2869	8340918	6·0274282	1682398	5·9438952	1·0140536	0138588	483	9861412	27
34	1661951	2868	8338049	6·0170250	1685390	5·9333455	1·0141032	0139071	484	9860929	26
35	1664819	2868	8335181	6·0066581	1688381	5·9228322	1·0141530	0139555	485	9860445	25
36	1667687	2869	8332313	5·9963274	1691373	5·9123550	1·0142029	0140040	485	9859960	24
37	1670556	2867	8329444	5·9860326	1694366	5·9019138	1·0142528	0140525	487	9859475	23
38	1673423	2868	8326577	5·9757737	1697358	5·8915084	1·0143028	0141012	487	9858988	22
39	1676291	2868	8323709	5·9655504	1700351	5·8811386	1·0143530	0141499	488	9858501	21
40	1679159	2867	8320841	5·9553625	1703344	5·8708042	1·0144035	0141987	489	9858013	20
41	1682026	2868	8317974	5·9452098	1706338	5·8605051	1·0144535	0142476	489	9857524	19
42	1684894	2867	8315106	5·9350922	1709331	5·8502410	1·0145039	0142965	491	9857035	18
43	1687761	2867	8312239	5·9250095	1712325	5·8400117	1·0145544	0143456	491	9856544	17
44	1690628	2867	8309372	5·9149614	1715320	5·8298172	1·0146050	0143947	492	9856053	16
45	1693495	2867	8306505	5·9049479	1718314	5·8196572	1·0146556	0144439	493	9855561	15
46	1696362	2866	8303638	5·8949688	1721309	5·8095315	1·0147064	0144932	494	9855068	14
47	1699228	2867	8300772	5·8850238	1724304	5·7994400	1·0147572	0145426	495	9854574	13
48	1702095	2866	8297905	5·8751128	1727300	5·7893825	1·0148082	0145921	496	9854079	12
49	1704961	2867	8295039	5·8652356	1730296	5·7793588	1·0148592	0146417	496	9853583	11
50	1707828	2866	8292172	5·8553921	1733292	5·7693688	1·0149103	0146913	497	9853087	10
51	1710694	2866	8289306	5·8455820	1736288	5·7594122	1·0149616	0147410	498	9852590	9
52	1713560	2865	8286440	5·8358053	1739285	5·7494889	1·0150129	0147908	499	9852092	8
53	1716425	2866	8283575	5·8260617	1742282	5·7395988	1·0150643	0148407	500	9851593	7
54	1719291	2865	8280709	5·8163510	1745279	5·7297416	1·0151158	0148907	500	9851093	6
55	1722156	2866	8277844	5·8066732	1748277	5·7199173	1·0151673	0149407	502	9850593	5
56	1725022	2865	8274978	5·7970280	1751275	5·7101256	1·0152190	0149909	502	9850091	4
57	1727887	2865	8272113	5·7874153	1754273	5·7003663	1·0152708	0150411	503	9849589	3
58	1730752	2865	8269248	5·7778350	1757272	5·6906394	1·0153226	0150914	503	9849086	2
59	1733617	2865	8266383	5·7682867	1760271	5·6809446	1·0153746	0151418	504	9848582	1
60	1736482		8263518	5·7587705	1763270	5·6712818	1·0154266	0151922	504	9848078	0
'	Cosine	Dif.	Vers.	Secant.	Cotan.	Tang.	Cosec.	Covers	D.	Sine.	'

′	Sine.	Dif.	Cosec.	Verseds.	Tang.	Dif.	Cotang.	Covers.	Secant.	D.	Cosine.	′
0	9·1943324	7969	10·8056676	8·0903166	9·1997125	8169	10·8002875	9·9261188	10·0053801	200	9·9946199	60
1	9·1951293	7954	10·8048707	8·0919203	9·2005294	8155	10·7994706	9·9259709	10·0054001	201	9·9945999	59
2	9·1959247	7939	10·8040753	8·0935210	9·2013449	8139	10·7986551	9·9258229	10·0054202	201	9·9945798	58
3	9·1967186	7924	10·8032814	8·0951188	9·2021588	8126	10·7978412	9·9256749	10·0054403	201	9·9945597	57
4	9·1975110	7909	10·8024890	8·0967136	9·2029714	8111	10·7970286	9·9255268	10·0054604	202	9·9945396	56
5	9·1983019	7894	10·8016981	8·0983055	9·2037825	8097	10·7962175	9·9253787	10·0054806	202	9·9945194	55
6	9·1990913	7880	10·8009087	8·0998944	9·2045922	8082	10·7954078	9·9252306	10·0055008	203	9·9944992	54
7	9·1998793	7865	10·8001207	8·1014804	9·2054004	8068	10·7945996	9·9250824	10·0055211	202	9·9944789	53
8	9·2006658	7851	10·7993342	8·1030635	9·2062072	8054	10·7937928	9·9249341	10·0055413	204	9·9944587	52
9	9·2014509	7836	10·7985491	8·1046437	9·2070126	8039	10·7929874	9·9247858	10·0055617	203	9·9944383	51
10	9·2022345	7822	10·7977655	8·1062211	9·2078165	8026	10·7921835	9·9246375	10·0055820	205	9·9944180	50
11	9·2030167	7807	10·7969833	8·1077955	9·2086191	8012	10·7913809	9·9244891	10·0056025	204	9·9943975	49
12	9·2037974	7792	10·7962026	8·1093671	9·2094203	7997	10·7905797	9·9243407	10·0056229	205	9·9943771	48
13	9·2045766	7779	10·7954234	8·1109358	9·2102200	7984	10·7897800	9·9241922	10·0056434	205	9·9943566	47
14	9·2053545	7764	10·7946455	8·1125017	9·2110184	7969	10·7889816	9·9240437	10·0056639	205	9·9943361	46
15	9·2061309	7750	10·7938691	8·1140647	9·2118153	7956	10·7881847	9·9238952	10·0056844	206	9·9943156	45
16	9·2069059	7736	10·7930941	8·1156249	9·2126109	7942	10·7873891	9·9237466	10·0057050	207	9·9942950	44
17	9·2076795	7721	10·7923205	8·1171823	9·2134051	7929	10·7865949	9·9235980	10·0057257	206	9·9942743	43
18	9·2084516	7708	10·7915484	8·1187369	9·2141980	7914	10·7858020	9·9234493	10·0057463	207	9·9942537	42
19	9·2092224	7693	10·7907776	8·1202887	9·2149894	7901	10·7850106	9·9233006	10·0057670	208	9·9942330	41
20	9·2099917	7680	10·7900083	8·1218377	9·2157795	7888	10·7842205	9·9231518	10·0057878	208	9·9942122	40
21	9·2107597	7666	10·7892403	8·1233840	9·2165683	7873	10·7834317	9·9230030	10·0058086	208	9·9941914	39
22	9·2115263	7651	10·7884737	8·1249274	9·2173556	7861	10·7826444	9·9228541	10·0058294	208	9·9941706	38
23	9·2122914	7638	10·7877086	8·1264681	9·2181417	7847	10·7818583	9·9227052	10·0058503	209	9·9941498	37
24	9·2130552	7624	10·7869448	8·1280061	9·2189264	7833	10·7810736	9·9225563	10·0058711	210	9·9941289	36
25	9·2138176	7611	10·7861824	8·1295413	9·2197097	7820	10·7802903	9·9224073	10·0058921	209	9·9941079	35
26	9·2145787	7597	10·7854213	8·1310738	9·2204917	7807	10·7795083	9·9222583	10·0059130	211	9·9940870	34
27	9·2153384	7583	10·7846616	8·1326036	9·2212724	7794	10·7787276	9·9221092	10·0059341	210	9·9940659	33
28	9·2160967	7569	10·7839033	8·1341307	9·2220518	7780	10·7779482	9·9219601	10·0059553	211	9·9940447	32
29	9·2168536	7556	10·7831464	8·1356551	9·2228298	7767	10·7771702	9·9218109	10·0059762	211	9·9940238	31
30	9·2176092	7543	10·7823908	8·1371768	9·2236065	7754	10·7763935	9·9216617	10·0059973	212	9·9940027	30
31	9·2183635	7529	10·7816365	8·1386958	9·2243819	7742	10·7756181	9·9215125	10·0060185	212	9·9939815	29
32	9·2191164	7516	10·7808836	8·1402121	9·2251561	7728	10·7748439	9·9213632	10·0060397	212	9·9939603	28
33	9·2198680	7502	10·7801320	8·1417258	9·2259289	7715	10·7740711	9·9212138	10·0060609	213	9·9939391	27
34	9·2206182	7489	10·7793818	8·1432368	9·2267004	7702	10·7732996	9·9210644	10·0060822	213	9·9939178	26
35	9·2213671	7476	10·7786329	8·1447452	9·2274706	7689	10·7725294	9·9209150	10·0061035	213	9·9938965	25
36	9·2221147	7462	10·7778853	8·1462510	9·2282395	7676	10·7717605	9·9207656	10·0061248	214	9·9938752	24
37	9·2228609	7450	10·7771391	8·1477541	9·2290071	7664	10·7709929	9·9206160	10·0061462	214	9·9938538	23
38	9·2236059	7436	10·7763941	8·1492546	9·2297735	7651	10·7702265	9·9204665	10·0061676	215	9·9938324	22
39	9·2243495	7423	10·7756505	8·1507525	9·2305386	7638	10·7694614	9·9203169	10·0061891	215	9·9938109	21
40	9·2250918	7410	10·7749082	8·1522478	9·2313024	7626	10·7686976	9·9201672	10·0062106	215	9·9937894	20
41	9·2258328	7397	10·7741672	8·1537405	9·2320650	7612	10·7679350	9·9200175	10·0062321	216	9·9937679	19
42	9·2265725	7385	10·7734275	8·1552307	9·2328262	7601	10·7671738	9·9198678	10·0062537	216	9·9937463	18
43	9·2273110	7371	10·7726890	8·1567182	9·2335863	7588	10·7664137	9·9197180	10·0062753	217	9·9937247	17
44	9·2280481	7358	10·7719519	8·1582032	9·2343451	7575	10·7656549	9·9195682	10·0062970	217	9·9937030	16
45	9·2287839	7346	10·7712161	8·1596857	9·2351026	7563	10·7648974	9·9194183	10·0063187	217	9·9936813	15
46	9·2295185	7333	10·7704815	8·1611656	9·2358589	7550	10·7641411	9·9192684	10·0063404	218	9·9936596	14
47	9·2302518	7320	10·7697482	8·1626430	9·2366139	7539	10·7633861	9·9191185	10·0063622	218	9·9936378	13
48	9·2309838	7307	10·7690162	8·1641178	9·2373678	7525	10·7626322	9·9189685	10·0063840	218	9·9936160	12
49	9·2317145	7295	10·7682855	8·1655902	9·2381203	7514	10·7618797	9·9188184	10·0064058	219	9·9935942	11
50	9·2324440	7282	10·7675560	8·1670600	9·2388717	7501	10·7611283	9·9186683	10·0064277	219	9·9935723	10
51	9·2331722	7270	10·7668278	8·1685273	9·2396218	7490	10·7603782	9·9185182	10·0064496	219	9·9935504	9
52	9·2338992	7257	10·7661008	8·1699921	9·2403708	7477	10·7596292	9·9183680	10·0064715	220	9·9935285	8
53	9·2346249	7245	10·7653751	8·1714545	9·2411185	7465	10·7588815	9·9182178	10·0064935	221	9·9935065	7
54	9·2353494	7232	10·7646506	8·1729144	9·2418650	7453	10·7581350	9·9180675	10·0065156	220	9·9934844	6
55	9·2360726	7220	10·7639274	8·1743718	9·2426103	7440	10·7573897	9·9179172	10·0065376	221	9·9934624	5
56	9·2367946	7207	10·7632054	8·1758267	9·2433543	7429	10·7566457	9·9177669	10·0065597	222	9·9934403	4
57	9·2375153	7196	10·7624847	8·1772792	9·2440972	7417	10·7559028	9·9176165	10·0065819	222	9·9934181	3
58	9·2382349	7183	10·7617651	8·1787292	9·2448389	7405	10·7551611	9·9174660	10·0066041	222	9·9933959	2
59	9·2389532	7170	10·7610468	8·1801768	9·2455794	7394	10·7544206	9·9173155	10·0066263	222	9·9933737	1
60	9·2396702		10·7603298	8·1816220	9·2463188		10·7536812	9·9171650	10·0066485		9·9933515	0

′	Cosine.	Dif.	Secant.	Covers.	Cotang.	Dif.	Tang.	Verseds.	Cosec.	D.	Sine.	′

'	Sine.	Dif.	Covers	Cosec.	Tang.	Cotang.	Secant.	Vers.	D.	Cosine	'
0	1736482	2864	8263518	5·7587705	1763270	5·6712818	1·0154266	0151922	506	9848078	60
1	1739346	2865	8260654	5·7492861	1766269	5·6616509	1·0154787	0152428	506	9847572	59
2	1742211	2864	8257789	5·7398333	1769269	5·6520516	1·0155310	0152934	508	9847066	58
3	1745075	2864	8254925	5·7304121	1772269	5·6424838	1·0155833	0153442	508	9846558	57
4	1747939	2864	8252061	5·7210223	1775270	5·6329474	1·0156357	0153950	508	9846050	56
5	1750803	2864	8249197	5·7116636	1778270	5·6234421	1·0156882	0154458	510	9845542	55
6	1753667	2864	8246333	5·7023360	1781271	5·6139680	1·0157408	0154968	511	9845032	54
7	1756531	2864	8243469	5·6930393	1784273	5·6045247	1·0157934	0155479	511	9844521	53
8	1759395	2863	8240605	5·6837734	1787274	5·5951121	1·0158462	0155990	512	9844010	52
9	1762258	2863	8237742	5·6745380	1790276	5·5857302	1·0158991	0156502	513	9843498	51
10	1765121	2863	8234879	5·6653331	1793279	5·5763786	1·0159520	0157015	514	9842985	50
11	1767984	2863	8232016	5·6561584	1796281	5·5670574	1·0160050	0157529	515	9842471	49
12	1770847	2863	8229153	5·6470140	1799284	5·5577663	1·0160582	0158044	515	9841956	48
13	1773710	2863	8226290	5·6378995	1802287	5·5485052	1·0161114	0158559	517	9841441	47
14	1776573	2862	8223427	5·6288148	1805291	5·5392740	1·0161647	0159076	517	9840924	46
15	1779435	2863	8220565	5·6197539	1808295	5·5300724	1·0162181	0159593	518	9840407	45
16	1782298	2862	8217702	5·6107345	1811299	5·5209005	1·0162716	0160111	519	9839889	44
17	1785160	2862	8214840	5·6017386	1814303	5·5117579	1·0163252	0160630	520	9839370	43
18	1788022	2862	8211978	5·5927719	1817308	5·5026446	1·0163789	0161150	520	9838850	42
19	1790884	2862	8209116	5·5838343	1820313	5·4935604	1·0164327	0161670	522	9838330	41
20	1793746	2861	8206254	5·5749258	1823319	5·4845052	1·0164865	0162192	522	9837808	40
21	1796607	2862	8203393	5·5660460	1826324	5·4754788	1·0165405	0162714	523	9837286	39
22	1799469	2861	8200531	5·5571951	1829330	5·4664812	1·0165946	0163237	524	9836763	38
23	1802330	2861	8197670	5·5483726	1832337	5·4575121	1·0166487	0163761	524	9836239	37
24	1805191	2861	8194809	5·5395786	1835343	5·4485715	1·0167029	0164285	526	9835715	36
25	1808052	2861	8191948	5·5308129	1838350	5·4396592	1·0167573	0164811	526	9835189	35
26	1810913	2861	8189087	5·5220754	1841358	5·4307750	1·0168117	0165337	527	9834663	34
27	1813774	2861	8186226	5·5133659	1844365	5·4219188	1·0168662	0165864	528	9834136	33
28	1816635	2860	8183365	5·5046843	1847373	5·4130906	1·0169208	0166392	529	9833608	32
29	1819495	2860	8180505	5·4960305	1850382	5·4042901	1·0169755	0166921	530	9833079	31
30	1822355	2860	8177645	5·4874043	1853390	5·3955172	1·0170303	0167451	530	9832549	30
31	1825215	2860	8174785	5·4788056	1856399	5·3867718	1·0170851	0167981	532	9832019	29
32	1828075	2860	8171925	5·4702342	1859409	5·3780538	1·0171401	0168513	532	9831487	28
33	1830935	2860	8169065	5·4616901	1862418	5·3693630	1·0171952	0169045	533	9830955	27
34	1833795	2859	8166205	5·4531731	1865428	5·3606993	1·0172503	0169578	534	9830422	26
35	1836654	2860	8163346	5·4446831	1868439	5·3520626	1·0173056	0170112	535	9829888	25
36	1839514	2859	8160486	5·4362199	1871449	5·3434527	1·0173609	0170647	535	9829353	24
37	1842373	2859	8157627	5·4277835	1874460	5·3348696	1·0174163	0171182	536	9828818	23
38	1845232	2859	8154768	5·4193737	1877471	5·3263131	1·0174719	0171718	538	9828282	22
39	1848091	2858	8151909	5·4109903	1880483	5·3177830	1·0175275	0172256	538	9827744	21
40	1850949	2859	8149051	5·4026333	1883495	5·3092793	1·0175832	0172794	538	9827206	20
41	1853808	2858	8146192	5·3943026	1886507	5·3008018	1·0176390	0173332	540	9826668	19
42	1856666	2858	8143334	5·3859979	1889520	5·2923505	1·0176949	0173872	541	9826128	18
43	1859524	2858	8140476	5·3777192	1892533	5·2839251	1·0177509	0174413	541	9825587	17
44	1862382	2858	8137618	5·3694664	1895546	5·2755255	1·0178060	0174954	542	9825046	16
45	1865240	2858	8134760	5·3612393	1898559	5·2671517	1·0178631	0175496	543	9824504	15
46	1868098	2858	8131902	5·3530379	1901573	5·2588035	1·0179194	0176039	544	9823961	14
47	1870956	2857	8129044	5·3448620	1904587	5·2504809	1·0179757	0176583	544	9823417	13
48	1873813	2857	8126187	5·3367114	1907602	5·2421836	1·0180321	0177127	546	9822873	12
49	1876670	2858	8123330	5·3285861	1910617	5·2339116	1·0180887	0177673	546	9822327	11
50	1879528	2857	8120472	5·3204860	1913632	5·2256647	1·0181453	0178219	547	9821781	10
51	1882385	2856	8117615	5·3124109	1916648	5·2174428	1·0182020	0178766	548	9821234	9
52	1885241	2857	8114759	5·3043608	1919664	5·2092459	1·0182588	0179314	549	9820686	8
53	1888098	2856	8111902	5·2963354	1922680	5·2010738	1·0183158	0179863	550	9820137	7
54	1890954	2857	8109046	5·2883347	1925696	5·1929264	1·0183728	0180413	550	9819587	6
55	1893811	2856	8106189	5·2803587	1928713	5·1848035	1·0184298	0180963	552	9819037	5
56	1896667	2856	8103333	5·2724070	1931731	5·1767051	1·0184870	0181515	552	9818485	4
57	1899523	2856	8100477	5·2644798	1934748	5·1686311	1·0185443	0182067	553	9817933	3
58	1902379	2855	8097621	5·2565768	1937766	5·1605813	1·0186017	0182620	554	9817380	2
59	1905234	2856	8094766	5·2486979	1940784	5·1525557	1·0186591	0183174	554	9816826	1
60	1908090		8091910	5·2408431	1943803	5·1445540	1·0187167	0183728		9816272	0
'	Cosine	Dif.	Vers.	Secant.	Cotan.	Tang.	Cosec.	Covers	D	Sine.	'

′	Sine.	Dif.	Cosec.	Versed s.	Tang.	Dif.	Cotang.	Covers.	Secant.	D.	Cosine.	′
0	9·2396702	7159	10·7603298	8·1816220	9·2463188	7381	10·7536812	9·9171650	10·0066485	223	9·9933515	60
1	9·2403861	7146	10·7596139	8·1830648	9·2470569	7370	10·7529431	9·9170144	10·0066708	224	9·9933292	59
2	9·2411007	7134	10·7588993	8·1845051	9·2477939	7358	10·7522061	9·9168638	10·0066932	223	9·9933068	58
3	9·2418141	7123	10·7581859	8·1859431	9·2485297	7346	10·7514703	9·9167131	10·0067155	224	9·9932845	57
4	9·2425264	7110	10·7574736	8·1873786	9·2492643	7335	10·7507357	9·9165624	10·0067379	225	9·9932621	56
5	9·2432374	7098	10·7567626	8·1888118	9·2499978	7323	10·7500022	9·9164117	10·0067604	225	9·9932396	55
6	9·2439472	7086	10·7560528	8·1902426	9·2507301	7311	10·7492699	9·9162609	10·0067829	225	9·9932171	54
7	9·2446558	7074	10·7553442	8·1916710	9·2514612	7300	10·7485388	9·9161100	10·0068054	226	9·9931946	53
8	9·2453632	7063	10·7546368	8·1930971	9·2521912	7288	10·7478088	9·9159591	10·0068280	226	9·9931720	52
9	9·2460695	7051	10·7539305	8·1945208	9·2529200	7277	10·7470800	9·9158082	10·0068506	226	9·9931494	51
10	9·2467746	7038	10·7532254	8·1959421	9·2536477	7266	10·7463523	9·9156572	10·0068732	227	9·9931268	50
11	9·2474784	7027	10·7525216	8·1973611	9·2543743	7254	10·7456257	9·9155062	10·0068959	227	9·9931041	49
12	9·2481811	7016	10·7518189	8·1987778	9·2550997	7243	10·7449003	9·9153551	10·0069186	227	9·9930814	48
13	9·2488827	7003	10·7511173	8·2001921	9·2558240	7232	10·7441760	9·9152040	10·0069413	228	9·9930587	47
14	9·2495830	6992	10·7504170	8·2016042	9·2565472	7220	10·7434528	9·9150528	10·0069641	228	9·9930359	46
15	9·2502822	6981	10·7497178	8·2030139	9·2572692	7209	10·7427308	9·9149016	10·0069869	229	9·9930131	45
16	9·2509803	6969	10·7490197	8·2044213	9·2579901	7198	10·7420099	9·9147504	10·0070098	229	9·9929902	44
17	9·2516772	6957	10·7483228	8·2058264	9·2587099	7186	10·7412901	9·9145991	10·0070327	229	9·9929673	43
18	9·2523729	6946	10·7476271	8·2072293	9·2594285	7176	10·7405715	9·9144478	10·0070556	230	9·9929444	42
19	9·2530675	6934	10·7469325	8·2086298	9·2601461	7164	10·7398539	9·9142964	10·0070786	230	9·9929214	41
20	9·2537609	6923	10·7462391	8·2100281	9·2608625	7154	10·7391375	9·9141450	10·0071016	231	9·9928984	40
21	9·2544532	6912	10·7455468	8·2114241	9·2615779	7142	10·7384221	9·9139935	10·0071247	231	9·9928753	39
22	9·2551444	6900	10·7448556	8·2128179	9·2622921	7132	10·7377079	9·9138420	10·0071478	231	9·9928522	38
23	9·2558344	6889	10·7441656	8·2142094	9·2630053	7120	10·7369947	9·9136904	10·0071709	232	9·9928291	37
24	9·2565233	6877	10·7434767	8·2155987	9·2637173	7110	10·7362827	9·9135388	10·0071941	232	9·9928059	36
25	9·2572110	6867	10·7427890	8·2169857	9·2644283	7099	10·7355717	9·9133872	10·0072173	232	9·9927827	35
26	9·2578977	6855	10·7421023	8·2183705	9·2651382	7088	10·7348618	9·9132355	10·0072405	233	9·9927595	34
27	9·2585832	6844	10·7414168	8·2197531	9·2658470	7077	10·7341530	9·9130837	10·0072638	233	9·9927362	33
28	9·2592676	6833	10·7407324	8·2211334	9·2665547	7066	10·7334453	9·9129319	10·0072871	234	9·9927129	32
29	9·2599509	6821	10·7400491	8·2225116	9·2672613	7056	10·7327387	9·9127801	10·0073105	234	9·9926895	31
30	9·2606330	6811	10·7393670	8·2238875	9·2679669	7045	10·7320331	9·9126282	10·0073339	234	9·9926661	30
31	9·2613141	6800	10·7386859	8·2252613	9·2686714	7035	10·7313286	9·9124763	10·0073573	235	9·9926427	29
32	9·2619941	6788	10·7380059	8·2266329	9·2693749	7023	10·7306251	9·9123244	10·0073808	235	9·9926192	28
33	9·2626729	6778	10·7373271	8·2280023	9·2700772	7014	10·7299228	9·9121723	10·0074043	235	9·9925957	27
34	9·2633507	6767	10·7366493	8·2293695	9·2707786	7002	10·7292214	9·9120203	10·0074278	236	9·9925722	26
35	9·2640274	6756	10·7359726	8·2307345	9·2714788	6992	10·7285212	9·9118682	10·0074514	236	9·9925486	25
36	9·2647030	6745	10·7352970	8·2320974	9·2721780	6982	10·7278220	9·9117161	10·0074750	237	9·9925250	24
37	9·2653775	6734	10·7346225	8·2334581	9·2728762	6971	10·7271238	9·9115639	10·0074987	237	9·9925013	23
38	9·2660509	6723	10·7339491	8·2348167	9·2735733	6961	10·7264267	9·9114116	10·0075224	237	9·9924776	22
39	9·2667232	6713	10·7332768	8·2361732	9·2742694	6950	10·7257306	9·9112593	10·0075461	238	9·9924539	21
40	9·2673945	6702	10·7326055	8·2375275	9·2749644	6940	10·7250356	9·9111070	10·0075699	238	9·9924301	20
41	9·2680647	6691	10·7319353	8·2388797	9·2756584	6930	10·7243416	9·9109547	10·0075937	239	9·9924063	19
42	9·2687338	6681	10·7312662	8·2402297	9·2763514	6920	10·7236486	9·9108022	10·0076176	239	9·9923824	18
43	9·2694019	6670	10·7305981	8·2415777	9·2770434	6909	10·7229566	9·9106498	10·0076415	239	9·9923585	17
44	9·2700689	6659	10·7299311	8·2429235	9·2777343	6899	10·7222657	9·9104973	10·0076654	240	9·9923346	16
45	9·2707348	6649	10·7292652	8·2442673	9·2784242	6889	10·7215758	9·9103447	10·0076894	240	9·9923106	15
46	9·2713997	6638	10·7286003	8·2456089	9·2791131	6878	10·7208869	9·9101921	10·0077131	240	9·9922866	14
47	9·2720635	6628	10·7279365	8·2469485	9·2798009	6869	10·7201991	9·9100395	10·0077374	241	9·9922626	13
48	9·2727263	6617	10·7272737	8·2482860	9·2804878	6858	10·7195122	9·9098868	10·0077615	241	9·9922385	12
49	9·2733880	6607	10·7266120	8·2496214	9·2811736	6849	10·7188264	9·9097341	10·0077856	242	9·9922144	11
50	9·2740487	6596	10·7259513	8·2509547	9·2818585	6838	10·7181415	9·9095813	10·0078098	242	9·9921902	10
51	9·2747083	6586	10·7252917	8·2522860	9·2825423	6828	10·7174577	9·9094285	10·0078340	242	9·9921660	9
52	9·2753669	6576	10·7246331	8·2536152	9·2832251	6819	10·7167749	9·9092756	10·0078582	243	9·9921418	8
53	9·2760245	6566	10·7239755	8·2549424	9·2839070	6808	10·7160930	9·9091227	10·0078825	243	9·9921175	7
54	9·2766811	6555	10·7233189	8·2562675	9·2845878	6799	10·7154122	9·9089697	10·0079068	243	9·9920932	6
55	9·2773366	6545	10·7226634	8·2575906	9·2852677	6789	10·7147323	9·9088167	10·0079311	244	9·9920689	5
56	9·2779911	6534	10·7220089	8·2589117	9·2859466	6779	10·7140534	9·9086637	10·0079555	244	9·9920445	4
57	9·2786445	6525	10·7213555	8·2602307	9·2866245	6769	10·7133755	9·9085106	10·0079799	245	9·9920201	3
58	9·2792970	6514	10·7207030	8·2615477	9·2873014	6759	10·7126986	9·9083575	10·0080044	245	9·9919956	2
59	9·2799484	6504	10·7200516	8·2628627	9·2879773	6750	10·7120227	9·9082043	10·0080289	245	9·9919711	1
60	9·2805988		10·7194012	8·2641757	9·2886523		10·7113477	9·9080510	10·0080534		9·9919466	0
	Cosine.	Dif.	Secant.	Covers.	Cotang.	Dif.	Tang.	Versed s.	Cosec.	D.	Sine.	′

′	Sine.	Dif.	Covers	Cosec.	Tang.	Cotang.	Secant.	Vers.	D.	Cosine	′
0	1908090	2855	8091910	5·2408431	1943803	5·1445540	1·0187167	0183728	556	9816272	60
1	1910945	2856	8089055	5·2330121	1946822	5·1365763	1·0187743	0184284	556	9815716	59
2	1913801	2855	8086199	5·2252050	1949841	5·1286224	1·0188321	0184840	557	9815160	58
3	1916656	2855	8083344	5·2174216	1952861	5·1206921	1·0188899	0185397	558	9814603	57
4	1919510	2854	8080490	5·2096618	1955881	5·1127855	1·0189478	0185955	559	9814045	56
5	1922365	2855	8077635	5·2019254	1958901	5·1049024	1·0190059	0186514	559	9813486	55
6	1925220	2854	8074780	5·1942125	1961922	5·0970426	1·0190640	0187073	561	9812927	54
7	1928074	2854	8071926	5·1865228	1964943	5·0892061	1·0191222	0187634	561	9812366	53
8	1930928	2854	8069072	5·1788563	1967964	5·0813928	1·0191805	0188195	562	9811805	52
9	1933782	2854	8066218	5·1712128	1970986	5·0736025	1·0192389	0188757	563	9811243	51
10	1936636	2854	8063364	5·1635924	1974008	5·0658352	1·0192973	0189320	564	9810680	50
11	1939490	2854	8060510	5·1559948	1977031	5·0580907	1·0193559	0189884	564	9810116	49
12	1942344	2853	8057656	5·1484199	1980053	5·0503690	1·0194146	0190448	566	9809552	48
13	1945197	2853	8054803	5·1408677	1983076	5·0426700	1·0194734	0191014	566	9808986	47
14	1948050	2853	8051950	5·1333381	1986100	5·0349935	1·0195322	0191580	567	9808420	46
15	1950903	2853	8049097	5·1258309	1989124	5·0273395	1·0195912	0192147	568	9807853	45
16	1953756	2853	8046244	5·1183461	1992148	5·0197078	1·0196502	0192715	569	9807285	44
17	1956609	2852	8043391	5·1108835	1995172	5·0120984	1·0197093	0193284	569	9806716	43
18	1959461	2853	8040539	5·1034431	1998197	5·0045111	1·0197686	0193853	571	9806147	42
19	1962314	2852	8037686	5·0960248	2001222	4·9969459	1·0198279	0194424	571	9805576	41
20	1965166	2852	8034834	5·0886284	2004248	4·9894027	1·0198873	0194995	572	9805005	40
21	1968018	2852	8031982	5·0812539	2007274	4·9818813	1·0199468	0195567	573	9804433	39
22	1970870	2852	8029130	5·0739012	2010300	4·9743817	1·0200064	0196140	574	9803860	38
23	1973722	2851	8026278	5·0665701	2013327	4·9669037	1·0200661	0196714	574	9803286	37
24	1976573	2852	8023427	5·0592606	2016354	4·9594474	1·0201259	0197288	576	9802712	36
25	1979425	2851	8020575	5·0519726	2019381	4·9520125	1·0201858	0197864	576	9802136	35
26	1982276	2851	8017724	5·0447060	2022409	4·9445990	1·0202457	0198440	577	9801560	34
27	1985127	2851	8014873	5·0374607	2025437	4·9372068	1·0203058	0199017	578	9800983	33
28	1987978	2851	8012022	5·0302367	2028465	4·9298358	1·0203660	0199595	578	9800405	32
29	1990829	2850	8009171	5·0230337	2031494	4·9224859	1·0204262	0200173	580	9799827	31
30	1993679	2851	8006321	5·0158517	2034523	4·9151570	1·0204866	0200753	580	9799247	30
31	1996530	2850	8003470	5·0086907	2037552	4·9078491	1·0205470	0201333	581	9798667	29
32	1999380	2850	8000620	5·0015505	2040582	4·9005620	1·0206075	0201914	582	9798086	28
33	2002230	2850	7997770	4·9944311	2043612	4·8932956	1·0206682	0202496	583	9797504	27
34	2005080	2850	7994920	4·9873323	2046643	4·8860499	1·0207289	0203079	584	9796921	26
35	2007930	2850	7992070	4·9802541	2049674	4·8788248	1·0207897	0203663	585	9796337	25
36	2010779	2849	7989221	4·9731964	2052705	4·8716201	1·0208506	0204248	585	9795752	24
37	2013629	2849	7986371	4·9661591	2055737	4·8644359	1·0209116	0204833	586	9795167	23
38	2016478	2849	7983522	4·9591421	2058769	4·8572719	1·0209727	0205419	587	9794581	22
39	2019327	2849	7980673	4·9521453	2061801	4·8501282	1·0210339	0206006	588	9793994	21
40	2022176	2848	7977824	4·9451687	2064834	4·8430045	1·0210952	0206594	588	9793406	20
41	2025024	2849	7974976	4·9382120	2067867	4·8359010	1·0211566	0207182	590	9792818	19
42	2027873	2848	7972127	4·9312754	2070900	4·8288174	1·0212180	0207772	590	9792228	18
43	2030721	2848	7969279	4·9243586	2073934	4·8217536	1·0212796	0208362	591	9791638	17
44	2033569	2849	7966431	4·9174616	2076968	4·8147096	1·0213413	0208953	592	9791047	16
45	2036418	2847	7963582	4·9105844	2080003	4·8076854	1·0214030	0209545	593	9790455	15
46	2039265	2848	7960735	4·9037267	2083038	4·8006808	1·0214649	0210138	594	9789862	14
47	2042113	2848	7957887	4·8968886	2086073	4·7936957	1·0215268	0210732	594	9789268	13
48	2044961	2847	7955039	4·8900700	2089109	4·7867300	1·0215888	0211326	595	9788674	12
49	2047808	2847	7952192	4·8832707	2092145	4·7797837	1·0216510	0211921	596	9788079	11
50	2050655	2847	7949345	4·8764907	2095181	4·7728568	1·0217132	0212517	597	9787483	10
51	2053502	2847	7946498	4·8697299	2098218	4·7659490	1·0217755	0213114	598	9786886	9
52	2056349	2846	7943651	4·8629883	2101255	4·7590603	1·0218379	0213712	599	9786288	8
53	2059195	2847	7940805	4·8562657	2104293	4·7521907	1·0219004	0214311	599	9785689	7
54	2062042	2846	7937958	4·8495621	2107331	4·7453401	1·0219630	0214910	600	9785090	6
55	2064888	2846	7935112	4·8428774	2110369	4·7385083	1·0220257	0215510	601	9784490	5
56	2067734	2846	7932266	4·8362114	2113407	4·7316954	1·0220885	0216111	602	9783889	4
57	2070580	2846	7929420	4·8295643	2116446	4·7249012	1·0221514	0216713	603	9783287	3
58	2073426	2846	7926574	4·8229357	2119486	4·7181256	1·0222144	0217316	604	9782684	2
59	2076272	2845	7923728	4·8163258	2122525	4·7113686	1·0222774	0217920	604	9782080	1
60	2079117		7920883	4·8097343	2125566	4·7046301	1·0223406	0218524		9781476	0
	Cosine	Dif.	Vers.	Secant.	Cotan.	Tang.	Cosec.	Covers	D.	Sine.	′

′	Sine.	Dif.	Cosec.	Verseds.	Tang.	Dif.	Cotang.	Covers.	Secant.	D.	Cosine.	′
0	9·2805988	6495	10·7194012	8·2641757	9·2886523	6740	10·7113477	9·9080510	10·0080534	246	9·9919466	60
1	9·2812483	6484	10·7187517	8·2654867	9·2893263	6730	10·7106737	9·9078978	10·0080780	246	9·9919220	59
2	9·2818967	6474	10·7181033	8·2667957	9·2899993	6720	10·7100007	9·9077445	10·0081026	247	9·9918974	58
3	9·2825441	6464	10·7174559	8·2681028	9·2906713	6711	10·7093287	9·9075911	10·0081273	247	9·9918727	57
4	9·2831905	6454	10·7168095	8·2694078	9·2913424	6702	10·7086576	9·9074377	10·0081520	247	9·9918480	56
5	9·2838359	6444	10·7161641	8·2707109	9·2920126	6691	10·7079874	9·9072842	10·0081767	247	9·9918233	55
6	9·2844803	6434	10·7155197	8·2720119	9·2926817	6683	10·7073183	9·9071307	10·0082014	249	9·9917986	54
7	9·2851237	6424	10·7148763	8·2733111	9·2933500	6672	10·7066500	9·9069772	10·0082263	248	9·9917737	53
8	9·2857661	6415	10·7142339	8·2746082	9·2940172	6664	10·7059828	9·9068236	10·0082511	249	9·9917489	52
9	9·2864076	6404	10·7135924	8·2759035	9·2946836	6653	10·7053164	9·9066699	10·0082760	249	9·9917240	51
10	9·2870480	6395	10·7129520	8·2771967	9·2953489	6645	10·7046511	9·9065163	10·0083009	250	9·9916991	50
11	9·2876875	6385	10·7123125	8·2784880	9·2960134	6635	10·7039866	9·9063625	10·0083259	249	9·9916741	49
12	9·2883260	6376	10·7116740	8·2797774	9·2966769	6626	10·7033231	9·9062087	10·0083508	251	9·9916492	48
13	9·2889636	6365	10·7110364	8·2810649	9·2973395	6616	10·7026605	9·9060549	10·0083759	251	9·9916241	47
14	9·2896001	6356	10·7103999	8·2823504	9·2980011	6607	10·7019989	9·9059011	10·0084010	251	9·9915990	46
15	9·2902357	6347	10·7097643	8·2836341	9·2986618	6598	10·7013382	9·9057471	10·0084261	251	9·9915739	45
16	9·2908704	6336	10·7091296	8·2849158	9·2993216	6588	10·7006784	9·9055932	10·0084512	252	9·9915488	44
17	9·2915040	6327	10·7084960	8·2861956	9·2999804	6579	10·7000196	9·9054392	10·0084764	252	9·9915236	43
18	9·2921367	6318	10·7078633	8·2874735	9·3006383	6571	10·6993617	9·9052851	10·0085016	253	9·9914984	42
19	9·2927685	6308	10·7072315	8·2887495	9·3012954	6560	10·6987046	9·9051310	10·0085269	253	9·9914731	41
20	9·2933993	6298	10·7066007	8·2900236	9·3019514	6552	10·6980486	9·9049769	10·0085522	253	9·9914478	40
21	9·2940291	6289	10·7059709	8·2912958	9·3026066	6543	10·6973934	9·9048227	10·0085775	254	9·9914225	39
22	9·2946580	6279	10·7053420	8·2925661	9·3032609	6534	10·6967391	9·9046685	10·0086029	254	9·9913971	38
23	9·2952859	6270	10·7047141	8·2938346	9·3039143	6524	10·6960857	9·9045142	10·0086283	255	9·9913717	37
24	9·2959129	6261	10·7040871	8·2951012	9·3045667	6516	10·6954333	9·9043599	10·0086538	255	9·9913462	36
25	9·2965390	6251	10·7034610	8·2963660	9·3052183	6506	10·6947817	9·9042055	10·0086793	255	9·9913207	35
26	9·2971641	6242	10·7028359	8·2976289	9·3058689	6498	10·6941311	9·9040511	10·0087048	256	9·9912952	34
27	9·2977883	6233	10·7022117	8·2988899	9·3065187	6488	10·6934813	9·9038966	10·0087304	256	9·9912696	33
28	9·2984116	6223	10·7015884	8·3001491	9·3071675	6480	10·6928325	9·9037421	10·0087560	256	9·9912440	32
29	9·2990339	6214	10·7009661	8·3014064	9·3078155	6471	10·6921845	9·9035876	10·0087816	257	9·9912184	31
30	9·2996553	6205	10·7003447	8·3026619	9·3084626	6462	10·6915374	9·9034330	10·0088073	257	9·9911927	30
31	9·3002758	6195	10·6997242	8·3039156	9·3091088	6453	10·6908912	9·9032783	10·0088330	258	9·9911670	29
32	9·3008953	6187	10·6991047	8·3051675	9·3097541	6444	10·6902459	9·9031236	10·0088588	258	9·9911412	28
33	9·3015140	6177	10·6984860	8·3064175	9·3103985	6436	10·6896015	9·9029689	10·0088846	258	9·9911154	27
34	9·3021317	6168	10·6978663	8·3076657	9·3110421	6427	10·6889579	9·9028141	10·0089104	259	9·9910896	26
35	9·3027485	6159	10·6972515	8·3089122	9·3116848	6418	10·6883152	9·9026593	10·0089363	259	9·9910637	25
36	9·3033644	6150	10·6966356	8·3101568	9·3123266	6409	10·6876734	9·9025044	10·0089622	259	9·9910378	24
37	9·3039794	6140	10·6960206	8·3113996	9·3129675	6401	10·6870325	9·9023495	10·0089881	260	9·9910119	23
38	9·3045934	6132	10·6954066	8·3126406	9·3136076	6392	10·6863924	9·9021945	10·0090141	261	9·9909859	22
39	9·3052066	6123	10·6947934	8·3138798	9·3142468	6383	10·6857532	9·9020395	10·0090402	260	9·9909598	21
40	9·3058189	6114	10·6941811	8·3151172	9·3148851	6375	10·6851149	9·9018845	10·0090662	261	9·9909338	20
41	9·3064303	6104	10·6935697	8·3163529	9·3155226	6366	10·6844774	9·9017294	10·0090923	262	9·9909077	19
42	9·3070407	6096	10·6929593	8·3175868	9·3161592	6358	10·6838408	9·9015742	10·0091185	262	9·9908815	18
43	9·3076503	6087	10·6923497	8·3188189	9·3167950	6349	10·6832050	9·9014190	10·0091447	262	9·9908553	17
44	9·3082590	6078	10·6917410	8·3200493	9·3174299	6341	10·6825701	9·9012638	10·0091709	262	9·9908291	16
45	9·3088668	6069	10·6911332	8·3212779	9·3180640	6332	10·6819360	9·9011085	10·0091971	263	9·9908029	15
46	9·3094737	6061	10·6905263	8·3225047	9·3186972	6323	10·6813028	9·9009531	10·0092234	264	9·9907766	14
47	9·3100798	6051	10·6899202	8·3237298	9·3193295	6316	10·6806705	9·9007978	10·0092498	263	9·9907502	13
48	9·3106849	6043	10·6893151	8·3249532	9·3199611	6307	10·6800389	9·9006423	10·0092761	265	9·9907239	12
49	9·3112892	6034	10·6887108	8·3261748	9·3205918	6298	10·6794082	9·9004869	10·0093026	264	9·9906974	11
50	9·3118926	6025	10·6881074	8·3273947	9·3212216	6290	10·6787784	9·9003313	10·0093290	265	9·9906710	10
51	9·3124951	6017	10·6875049	8·3286128	9·3218506	6282	10·6781494	9·9001758	10·0093555	265	9·9906445	9
52	9·3130968	6008	10·6869032	8·3298292	9·3224788	6273	10·6775212	9·9000202	10·0093820	266	9·9906180	8
53	9·3136976	5999	10·6863024	8·3310439	9·3231061	6266	10·6768939	9·8998645	10·0094086	266	9·9905914	7
54	9·3142975	5990	10·6857025	8·3322569	9·3237327	6257	10·6762673	9·8997088	10·0094352	266	9·9905648	6
55	9·3148965	5982	10·6851035	8·3334682	9·3243584	6248	10·6756416	9·8995531	10·0094618	267	9·9905382	5
56	9·3154947	5974	10·6845053	8·3346778	9·3249832	6241	10·6750168	9·8993973	10·0094885	267	9·9905115	4
57	9·3160921	5964	10·6839079	8·3358857	9·3256073	6232	10·6743927	9·8992414	10·0095152	268	9·9904848	3
58	9·3166885	5956	10·6833115	8·3370918	9·3262305	6224	10·6737695	9·8990855	10·0095420	268	9·9904580	2
59	9·3172841	5948	10·6827159	8·3382963	9·3268529	6216	10·6731471	9·8989296	10·0095688	268	9·9904312	1
60	9·3178789		10·6821211	8·3394991	9·3274745		10·6725255	9·8987736	10·0095956		9·9904044	0
′	Cosine.	Dif.	Secant.	Covers.	Cotang.	Dif.	Tang.	Verseds.	Cosec.	D.	Sine.	′

′	Sine.	Dif.	Covers	Cosec.	Tang.	Cotang.	Secant.	Vers.	D.	Cosine	′
0	2079117	2845	7920883	4·8097343	2125566	4·7046301	1·0223406	0218524	605	9781476	60
1	2081962	2845	7918038	4·8031613	2128606	4·6979100	1·0224039	0219129	606	9780871	59
2	2084807	2845	7915193	4·7966066	2131647	4·6912083	1·0224672	0219735	607	9780265	58
3	2087652	2845	7912348	4·7900702	2134688	4·6845248	1·0225307	0220342	608	9779658	57
4	2090497	2844	7909503	4·7835520	2137730	4·6778595	1·0225942	0220950	608	9779050	56
5	2093341	2845	7906659	4·7770519	2140772	4·6712124	1·0226578	0221558	610	9778442	55
6	2096186	2844	7903814	4·7705699	2143814	4·6645832	1·0227216	0222168	610	9777832	54
7	2099030	2844	7900970	4·7641058	2146857	4·6579721	1·0227854	0222778	611	9777222	53
8	2101874	2844	7898126	4·7576596	2149900	4·6513788	1·0228493	0223389	612	9776611	52
9	2104718	2843	7895282	4·7512312	2152944	4·6448034	1·0229133	0224001	612	9775999	51
10	2107561	2844	7892439	4·7448206	2155988	4·6382457	1·0229774	0224613	614	9775387	50
11	2110405	2843	7889595	4·7384277	2159032	4·6317056	1·0230416	0225227	614	9774773	49
12	2113248	2843	7886752	4·7320524	2162077	4·6251832	1·0231059	0225841	615	9774159	48
13	2116091	2843	7883909	4·7256945	2165122	4·6186783	1·0231703	0226456	616	9773544	47
14	2118934	2843	7881066	4·7193542	2168167	4·6121908	1·0232348	0227072	617	9772928	46
15	2121777	2842	7878223	4·7130313	2171213	4·6057207	1·0232994	0227689	618	9772311	45
16	2124619	2843	7875381	4·7067256	2174259	4·5992680	1·0233641	0228307	618	9771694	44
17	2127462	2842	7872538	4·7004372	2177306	4·5928325	1·0234288	0228925	619	9771075	43
18	2130304	2842	7869696	4·6941660	2180353	4·5864141	1·0234937	0229544	620	9770456	42
19	2133146	2842	7866854	4·6879119	2183400	4·5800129	1·0235587	0230164	621	9769836	41
20	2135988	2841	7864012	4·6816748	2186448	4·5736287	1·0236237	0230785	622	9769215	40
21	2138829	2842	7861171	4·6754548	2189496	4·5672615	1·0236889	0231407	623	9768593	39
22	2141671	2841	7858329	4·6692516	2192544	4·5609111	1·0237541	0232030	623	9767970	38
23	2144512	2841	7855488	4·6630652	2195593	4·5545776	1·0238195	0232653	624	9767347	37
24	2147353	2841	7852647	4·6568956	2198643	4·5482608	1·0238849	0233277	625	9766723	36
25	2150194	2841	7849806	4·6507427	2201692	4·5419608	1·0239504	0233902	626	9766098	35
26	2153035	2841	7846965	4·6446064	2204742	4·5356773	1·0240161	0234528	627	9765472	34
27	2155876	2840	7844124	4·6384867	2207793	4·5294105	1·0240818	0235155	627	9764845	33
28	2158716	2840	7841284	4·6323835	2210844	4·5231601	1·0241476	0235782	629	9764218	32
29	2161556	2840	7838444	4·6262967	2213895	4·5169261	1·0242135	0236411	629	9763589	31
30	2164396	2840	7835604	4·6202263	2216947	4·5107085	1·0242795	0237040	630	9762960	30
31	2167236	2840	7832764	4·6141722	2219999	4·5045072	1·0243456	0237670	631	9762330	29
32	2170076	2839	7829924	4·6081343	2223051	4·4983221	1·0244118	0238301	631	9761699	28
33	2172915	2839	7827085	4·6021126	2226104	4·4921532	1·0244781	0238932	633	9761068	27
34	2175754	2839	7824246	4·5961070	2229157	4·4860004	1·0245445	0239565	633	9760435	26
35	2178593	2839	7821407	4·5901174	2232211	4·4798636	1·0246110	0240198	634	9759802	25
36	2181432	2839	7818568	4·5841439	2235265	4·4737428	1·0246776	0240832	635	9759168	24
37	2184271	2839	7815729	4·5781862	2238319	4·4676379	1·0247442	0241467	636	9758533	23
38	2187110	2838	7812890	4·5722444	2241374	4·4615489	1·0248110	0242103	637	9757897	22
39	2189948	2838	7810052	4·5663183	2244429	4·4554756	1·0248779	0242740	637	9757260	21
40	2192786	2838	7807214	4·5604080	2247485	4·4494181	1·0249448	0243377	638	9756623	20
41	2195624	2838	7804376	4·5545134	2250541	4·4433762	1·0250119	0244015	640	9755985	19
42	2198462	2838	7801538	4·5486344	2253597	4·4373500	1·0250790	0244655	639	9755345	18
43	2201300	2837	7798700	4·5427709	2256654	4·4313392	1·0251463	0245294	641	9754706	17
44	2204137	2837	7795863	4·5369229	2259711	4·4253439	1·0252136	0245935	642	9754065	16
45	2206974	2837	7793026	4·5310903	2262769	4·4193641	1·0252811	0246577	642	9753423	15
46	2209811	2837	7790189	4·5252730	2265827	4·4133996	1·0253486	0247219	643	9752781	14
47	2212648	2837	7787352	4·5194711	2268885	4·4074504	1·0254162	0247862	644	9752138	13
48	2215485	2836	7784515	4·5136844	2271944	4·4015164	1·0254839	0248506	645	9751494	12
49	2218321	2837	7781679	4·5079129	2275003	4·3955977	1·0255518	0249151	646	9750849	11
50	2221158	2836	7778842	4·5021565	2278063	4·3896940	1·0256197	0249797	646	9750203	10
51	2223994	2836	7776006	4·4964152	2281123	4·3838054	1·0256877	0250444	647	9749556	9
52	2226830	2836	7773170	4·4906889	2284184	4·3779317	1·0257558	0251091	647	9748909	8
53	2229666	2835	7770334	4·4849775	2287244	4·3720731	1·0258240	0251739	648	9748261	7
54	2232501	2836	7767499	4·4792810	2290306	4·3662293	1·0258923	0252388	649	9747612	6
55	2235337	2835	7764663	4·4735993	2293367	4·3604003	1·0259607	0253038	650	9746962	5
56	2238172	2835	7761828	4·4679324	2296429	4·3545861	1·0260292	0253689	651	9746311	4
57	2241007	2835	7758993	4·4622803	2299492	4·3487866	1·0260978	0254340	651	9745660	3
58	2243842	2834	7756158	4·4566428	2302555	4·3430018	1·0261665	0254992	652	9745008	2
59	2246676	2835	7753324	4·4510198	2305618	4·3372316	1·0262352	0255645	653	9744355	1
60	2249511		7750489	4·4454115	2308682	4·3314759	1·0263041	0256299	654	9743701	0

′	Cosine	Dif.	Vers.	Secant.	Cotan.	Tang.	Cosec.	Covers	D.	Sine.	′

′	Sine.	Dif.	Cosec.	Verseds.	Tang.	Dif.	Cotang.	Covers.	Secant.	D.	Cosine.	′
0	9·3178789	5939	10·6821211	8·3394991	9·3274745	6208	10·6725255	9·8987736	10·0095956	269	9·9904044	60
1	9·3184728	5931	10·6815272	8·3407002	9·3280953	6200	10·6719047	9·8986176	10·0096225	269	9·9903775	59
2	9·3190659	5922	10·6809341	8·3418997	9·3287153	6192	10·6712847	9·8984615	10·0096494	269	9·9903506	58
3	9·3196581	5914	10·6803419	8·3430975	9·3293345	6183	10·6706655	9·8983054	10·0096763	270	9·9903237	57
4	9·3202495	5905	10·6797505	8·3442936	9·3299528	6176	10·6700472	9·8981492	10·0097033	270	9·9902967	56
5	9·3208400	5897	10·6791600	8·3454880	9·3305704	6168	10·6694296	9·8979930	10·0097303	271	9·9902697	55
6	9·3214297	5889	10·6785703	8·3466808	9·3311872	6159	10·6688128	9·8978367	10·0097574	271	9·9902426	54
7	9·3220186	5880	10·6779814	8·3478719	9·3318031	6152	10·6681969	9·8976804	10·0097845	272	9·9902155	53
8	9·3226066	5872	10·6773934	8·3490614	9·3324183	6144	10·6675817	9·8975241	10·0098117	271	9·9901883	52
9	9·3231938	5864	10·6768062	8·3502492	9·3330327	6136	10·6669673	9·8973677	10·0098388	273	9·9901612	51
10	9·3237802	5855	10·6762198	8·3514354	9·3336463	6128	10·6663537	9·8972112	10·0098661	272	9·9901339	50
11	9·3243657	5848	10·6756343	8·3526200	9·3342591	6120	10·6657409	9·8970547	10·0098933	273	9·9901067	49
12	9·3249505	5839	10·6750495	8·3538029	9·3348711	6112	10·6651289	9·8968982	10·0099206	273	9·9900794	48
13	9·3255344	5830	10·6744656	8·3549842	9·3354823	6104	10·6645177	9·8967416	10·0099479	274	9·9900521	47
14	9·3261174	5823	10·6738826	8·3561639	9·3360927	6097	10·6639073	9·8965850	10·0099753	274	9·9900247	46
15	9·3266997	5814	10·6733003	8·3573419	9·3367024	6089	10·6632976	9·8964283	10·0100027	275	9·9899973	45
16	9·3272811	5806	10·6727189	8·3585184	9·3373113	6081	10·6626887	9·8962716	10·0100302	275	9·9899698	44
17	9·3278617	5799	10·6721383	8·3596932	9·3379194	6073	10·6620806	9·8961148	10·0100577	275	9·9899423	43
18	9·3284416	5790	10·6715584	8·3608664	9·3385267	6066	10·6614733	9·8959580	10·0100852	275	9·9899148	42
19	9·3290206	5782	10·6709794	8·3620381	9·3391333	6058	10·6608667	9·8958011	10·0101127	276	9·9898873	41
20	9·3295988	5773	10·6704012	8·3632081	9·3397391	6050	10·6602609	9·8956442	10·0101403	277	9·9898597	40
21	9·3301761	5766	10·6698239	8·3643765	9·3403441	6043	10·6596559	9·8954872	10·0101680	277	9·9898320	39
22	9·3307527	5758	10·6692473	8·3655434	9·3409484	6035	10·6590516	9·8953302	10·0101957	277	9·9898043	38
23	9·3313285	5750	10·6686715	8·3667086	9·3415519	6027	10·6584481	9·8951732	10·0102234	277	9·9897766	37
24	9·3319035	5742	10·6680965	8·3678723	9·3421546	6020	10·6578454	9·8950161	10·0102511	278	9·9897489	36
25	9·3324777	5734	10·6675223	8·3690344	9·3427566	6012	10·6572434	9·8948589	10·0102789	279	9·9897211	35
26	9·3330511	5726	10·6669489	8·3701950	9·3433578	6005	10·6566422	9·8947017	10·0103068	278	9·9896932	34
27	9·3336237	5718	10·6663763	8·3713539	9·3439583	5997	10·6560417	9·8945445	10·0103346	280	9·9896654	33
28	9·3341955	5710	10·6658045	8·3725114	9·3445580	5990	10·6554420	9·8943872	10·0103626	279	9·9896374	32
29	9·3347665	5703	10·6652335	8·3736672	9·3451570	5982	10·6548430	9·8942299	10·0103905	280	9·9896095	31
30	9·3353368	5694	10·6646632	8·3748215	9·3457552	5975	10·6542448	9·8940725	10·0104185	280	9·9895815	30
31	9·3359062	5687	10·6640938	8·3759743	9·3463527	5967	10·6536473	9·8939150	10·0104465	281	9·9895535	29
32	9·3364749	5679	10·6635251	8·3771255	9·3469494	5960	10·6530506	9·8937576	10·0104746	281	9·9895254	28
33	9·3370428	5671	10·6629572	8·3782751	9·3475454	5953	10·6524546	9·8936000	10·0105027	281	9·9894973	27
34	9·3376099	5663	10·6623901	8·3794232	9·3481407	5945	10·6518593	9·8934425	10·0105308	282	9·9894692	26
35	9·3381762	5656	10·6618238	8·3805698	9·3487352	5938	10·6512648	9·8932849	10·0105590	282	9·9894410	25
36	9·3387418	5647	10·6612582	8·3817149	9·3493290	5930	10·6506710	9·8931272	10·0105872	283	9·9894128	24
37	9·3393065	5641	10·6606935	8·3828584	9·3499220	5923	10·6500780	9·8929695	10·0106155	283	9·9893845	23
38	9·3398706	5632	10·6601294	8·3840004	9·3505143	5916	10·6494857	9·8928117	10·0106438	283	9·9893562	22
39	9·3404338	5625	10·6595662	8·3851409	9·3511059	5909	10·6488941	9·8926539	10·0106721	284	9·9893279	21
40	9·3409963	5617	10·6590037	8·3862799	9·3516968	5901	10·6483032	9·8924961	10·0107005	284	9·9892995	20
41	9·3415580	5610	10·6584420	8·3874174	9·3522869	5894	10·6477131	9·8923382	10·0107289	284	9·9892711	19
42	9·3421190	5602	10·6578810	8·3885533	9·3528763	5887	10·6471237	9·8921802	10·0107573	285	9·9892427	18
43	9·3426792	5594	10·6573208	8·3896878	9·3534650	5880	10·6465350	9·8920222	10·0107858	286	9·9892142	17
44	9·3432386	5587	10·6567614	8·3908207	9·3540530	5872	10·6459470	9·8918642	10·0108144	285	9·9891856	16
45	9·3437973	5579	10·6562027	8·3919522	9·3546402	5865	10·6453598	9·8917061	10·0108429	286	9·9891571	15
46	9·3443552	5572	10·6556448	8·3930822	9·3552267	5859	10·6447733	9·8915480	10·0108715	287	9·9891285	14
47	9·3449124	5564	10·6550876	8·3942107	9·3558126	5851	10·6441874	9·8913898	10·0109002	287	9·9890998	13
48	9·3454688	5557	10·6545312	8·3953377	9·3563977	5844	10·6436023	9·8912316	10·0109289	287	9·9890711	12
49	9·3460245	5549	10·6539755	8·3964632	9·3569821	5837	10·6430179	9·8910733	10·0109576	287	9·9890424	11
50	9·3465794	5542	10·6534206	8·3975873	9·3575658	5829	10·6424342	9·8909150	10·0109863	288	9·9890137	10
51	9·3471336	5534	10·6528664	8·3987098	9·3581487	5823	10·6418513	9·8907566	10·0110151	289	9·9889849	9
52	9·3476870	5527	10·6523130	8·3998310	9·3587310	5816	10·6412690	9·8905982	10·0110440	289	9·9889560	8
53	9·3482397	5520	10·6517603	8·4009506	9·3593126	5809	10·6406874	9·8904397	10·0110729	289	9·9889271	7
54	9·3487917	5512	10·6512083	8·4020688	9·3598935	5801	10·6401065	9·8902812	10·0111018	289	9·9888982	6
55	9·3493429	5505	10·6506571	8·4031855	9·3604736	5795	10·6395264	9·8901226	10·0111307	290	9·9888693	5
56	9·3498934	5498	10·6501066	8·4043008	9·3610531	5788	10·6389469	9·8899640	10·0111597	290	9·9888403	4
57	9·3504432	5490	10·6495568	8·4054147	9·3616319	5781	10·6383681	9·8898054	10·0111887	291	9·9888113	3
58	9·3509922	5483	10·6490078	8·4065270	9·3622100	5774	10·6377900	9·8896467	10·0112178	291	9·9887822	2
59	9·3515405	5475	10·6484595	8·4076380	9·3627874	5767	10·6372126	9·8894879	10·0112469	292	9·9887531	1
60	9·3520880		10·6479120	8·4087475	9·3633641		10·6366359	9·8893291	10·0112761		9·9887239	0
′	Cosine.	Dif.	Secant.	Covers.	Cotang.	Dif.	Tang.	Verseds.	Cosec.	D.	Sine.	′

′	Sine.	Dif.	Covers	Cosec.	Tang.	Cotang.	Secant.	Vers.	D.	Cosine	′
0	2249511	2834	7750489	4·4454115	2308682	4·3314759	1·0263041	0256299	655	9743701	60
1	2252345	2834	7747655	4·4398176	2311746	4·3257347	1·0263731	0256954	656	9743046	59
2	2255179	2834	7744821	4·4342382	2314811	4·3200079	1·0264421	0257610	656	9742390	58
3	2258013	2833	7741987	4·4286731	2317876	4·3142955	1·0265113	0258266	657	9741734	57
4	2260846	2834	7739154	4·4231224	2320941	4·3085974	1·0265806	0258923	658	9741077	56
5	2263680	2833	7736320	4·4175859	2324007	4·3029136	1·0266499	0259581	659	9740419	55
6	2266513	2833	7733487	4·4120637	2327073	4·2972440	1·0267194	0260240	660	9739760	54
7	2269346	2833	7730654	4·4065556	2330140	4·2915885	1·0267889	0260900	661	9739100	53
8	2272179	2833	7727821	4·4010616	2333207	4·2859472	1·0268586	0261561	661	9738439	52
9	2275012	2833	7724988	4·3955817	2336274	4·2803199	1·0269283	0262222	662	9737778	51
10	2277844	2832	7722156	4·3901158	2339342	4·2747066	1·0269982	0262884	663	9737116	50
11	2280677	2832	7719323	4·3846638	2342410	4·2691072	1·0270681	0263547	664	9736463	49
12	2283509	2832	7716491	4·3792257	2345479	4·2635218	1·0271381	0264211	665	9735789	48
13	2286341	2831	7713659	4·3738015	2348548	4·2579501	1·0272082	0264876	665	9735124	47
14	2289172	2832	7710828	4·3683910	2351617	4·2523923	1·0272785	0265541	666	9734459	46
15	2292004	2831	7707996	4·3629943	2354687	4·2468482	1·0273488	0266207	668	9733793	45
16	2294835	2831	7705165	4·3576113	2357758	4·2413177	1·0274192	0266875	667	9733125	44
17	2297666	2831	7702334	4·3522419	2360829	4·2358009	1·0274897	0267542	669	9732458	43
18	2300497	2831	7699503	4·3468861	2363900	4·2302977	1·0275603	0268211	670	9731789	42
19	2303328	2831	7696672	4·3415438	2366971	4·2248080	1·0276310	0268881	670	9731119	41
20	2306159	2830	7693841	4·3362150	2370044	4·2193318	1·0277018	0269551	672	9730449	40
21	2308989	2830	7691011	4·3308996	2373116	4·2138690	1·0277727	0270223	672	9729777	39
22	2311819	2830	7688181	4·3255977	2376189	4·2084196	1·0278437	0270895	673	9729105	38
23	2314649	2830	7685351	4·3203090	2379262	4·2029835	1·0279148	0271568	673	9728432	37
24	2317479	2830	7682521	4·3150336	2382336	4·1975606	1·0279860	0272241	675	9727759	36
25	2320309	2829	7679691	4·3097715	2385410	4·1921510	1·0280573	0272916	675	9727084	35
26	2323138	2829	7676862	4·3045225	2388485	4·1867546	1·0281287	0273591	676	9726409	34
27	2325967	2829	7674033	4·2992867	2391560	4·1813713	1·0282002	0274267	677	9725733	33
28	2328796	2829	7671204	4·2940640	2394635	4·1760011	1·0282717	0274944	678	9725056	32
29	2331625	2829	7668375	4·2888543	2397711	4·1706440	1·0283434	0275622	679	9724378	31
30	2334454	2828	7665546	4·2836576	2400788	4·1652998	1·0284152	0276301	679	9723699	30
31	2337282	2828	7662718	4·2784738	2403864	4·1599685	1·0284871	0276980	681	9723020	29
32	2340110	2828	7659890	4·2733029	2406942	4·1546501	1·0285590	0277661	681	9722339	28
33	2342938	2828	7657062	4·2681449	2410019	4·1493446	1·0286311	0278342	682	9721658	27
34	2345766	2828	7654234	4·2629996	2413097	4·1440519	1·0287033	0279024	682	9720976	26
35	2348594	2827	7651406	4·2578671	2416176	4·1387719	1·0287755	0279706	684	9720294	25
36	2351421	2827	7648579	4·2527474	2419255	4·1335046	1·0288479	0280390	684	9719610	24
37	2354248	2827	7645752	4·2476402	2422334	4·1282499	1·0289203	0281074	686	9718926	23
38	2357075	2827	7642925	4·2425457	2425414	4·1230079	1·0289929	0281760	686	9718240	22
39	2359902	2827	7640098	4·2374637	2428494	4·1177784	1·0290655	0282446	687	9717554	21
40	2362729	2826	7637271	4·2323943	2431575	4·1125614	1·0291383	0283133	687	9716867	20
41	2365555	2826	7634445	4·2273373	2434656	4·1073569	1·0292111	0283820	689	9716180	19
42	2368381	2826	7631619	4·2222928	2437737	4·1021649	1·0292840	0284509	689	9715491	18
43	2371207	2826	7628793	4·2172606	2440819	4·0969852	1·0293571	0285198	690	9714802	17
44	2374033	2826	7625967	4·2122408	2443902	4·0918178	1·0294302	0285888	690	9714112	16
45	2376859	2825	7623141	4·2072333	2446984	4·0866627	1·0295034	0286579	691	9713421	15
46	2379684	2826	7620316	4·2022380	2450068	4·0815199	1·0295768	0287271	692	9712729	14
47	2382510	2825	7617490	4·1972549	2453151	4·0763892	1·0296502	0287964	693	9712036	13
48	2385335	2824	7614665	4·1922840	2456236	4·0712707	1·0297237	0288657	693	9711343	12
49	2388159	2825	7611841	4·1873252	2459320	4·0661643	1·0297973	0289351	694	9710649	11
50	2390984	2824	7609016	4·1823785	2462405	4·0610700	1·0298711	0290047	696	9709953	10
51	2393808	2825	7606192	4·1774438	2465491	4·0559877	1·0299449	0290742	695	9709258	9
52	2396633	2824	7603367	4·1725210	2468577	4·0509174	1·0300188	0291439	697	9708561	8
53	2399457	2823	7600543	4·1676102	2471663	4·0458590	1·0300928	0292137	698	9707863	7
54	2402280	2824	7597720	4·1627114	2474750	4·0408125	1·0301669	0292835	698	9707165	6
55	2405104	2823	7594896	4·1578243	2477837	4·0357779	1·0302411	0293534	699	9706466	5
56	2407927	2824	7592073	4·1529491	2480925	4·0307550	1·0303154	0294234	700	9705766	4
57	2410751	2823	7589249	4·1480856	2484013	4·0257440	1·0303898	0294935	701	9705065	3
58	2413574	2822	7586426	4·1432339	2487102	4·0207446	1·0304643	0295637	702	9704363	2
59	2416396	2823	7583604	4·1383939	2490191	4·0157570	1·0305389	0296339	702	9703661	1
60	2419219		7580781	4·1335655	2493280	4·0107809	1·0306136	0297043	704	9702957	0

′	Cosine	Dif.	Vers.	Secant.	Cotan.	Tang.	Cosec.	Covers	D.	Sine.	′

′	Sine.	Dif.	Cosec.	Verseds.	Tang.	Dif.	Cotang.	Covers.	Secant.	D.	Cosine.	′
0	9·3520880	5409	10·6479120	8·4087475	9·3633641	5760	10·6366359	9·8893291	10·0112761	292	9·9887239	60
1	9·3526349	5461	10·6473651	8·4098556	9·3639401	5754	10·6360599	9·8891703	10·0113053	292	9·9886947	59
2	9·3531810	5454	10·6468190	8·4109622	9·3645155	5746	10·6354845	9·8890114	10·0113345	292	9·9886655	58
3	9·3537264	5446	10·6462736	8·4120675	9·3650901	5740	10·6349099	9·8888525	10·0113637	293	9·9886363	57
4	9·3542710	5440	10·6457290	8·4131713	9·3656641	5733	10·6343359	9·8886935	10·0113930	294	9·9886070	56
5	9·3548150	5432	10·6451850	8·4142736	9·3662374	5726	10·6337626	9·8885344	10·0114224	294	9·9885776	55
6	9·3553582	5425	10·6446418	8·4153746	9·3668100	5719	10·6331900	9·8883754	10·0114518	294	9·9885482	54
7	9·3559007	5419	10·6440993	8·4164741	9·3673819	5713	10·6326181	9·8882162	10·0114812	294	9·9885188	53
8	9·3564426	5410	10·6435574	8·4175723	9·3679532	5706	10·6320468	9·8880571	10·0115106	295	9·9884894	52
9	9·3569836	5404	10·6430164	8·4186690	9·3685238	5699	10·6314762	9·8878978	10·0115401	296	9·9884599	51
10	9·3575240	5397	10·6424760	8·4197644	9·3690937	5692	10·6309063	9·8877386	10·0115697	295	9·9884303	50
11	9·3580637	5390	10·6419363	8·4208583	9·3696629	5686	10·6303371	9·8875792	10·0115992	296	9·9884008	49
12	9·3586027	5382	10·6413973	8·4219508	9·3702315	5679	10·6297685	9·8874199	10·0116288	297	9·9883712	48
13	9·3591409	5376	10·6408591	8·4230420	9·3707994	5673	10·6292006	9·8872605	10·0116585	297	9·9883415	47
14	9·3596785	5369	10·6403215	8·4241318	9·3713667	5666	10·6286333	9·8871010	10·0116882	297	9·9883118	46
15	9·3602154	5361	10·6397846	8·4252201	9·3719333	5659	10·6280667	9·8869415	10·0117179	298	9·9882821	45
16	9·3607515	5355	10·6392485	8·4263072	9·3724992	5653	10·6275008	9·8867819	10·0117477	298	9·9882523	44
17	9·3612870	5347	10·6387130	8·4273928	9·3730645	5646	10·6269355	9·8866223	10·0117775	298	9·9882225	43
18	9·3618217	5341	10·6381783	8·4284770	9·3736291	5639	10·6263709	9·8864627	10·0118073	299	9·9881927	42
19	9·3623558	5334	10·6376442	8·4295599	9·3741930	5633	10·6258070	9·8863030	10·0118372	299	9·9881628	41
20	9·3628892	5327	10·6371108	8·4306414	9·3747563	5627	10·6252437	9·8861432	10·0118671	300	9·9881329	40
21	9·3634219	5320	10·6365781	8·4317216	9·3753190	5620	10·6246810	9·8859834	10·0118971	300	9·9881029	39
22	9·3639539	5313	10·6360461	8·4328004	9·3758810	5613	10·6241190	9·8858236	10·0119271	300	9·9880729	38
23	9·3644852	5306	10·6355148	8·4338778	9·3764423	5607	10·6235577	9·8856637	10·0119571	301	9·9880429	37
24	9·3650158	5300	10·6349842	8·4349539	9·3770030	5601	10·6229970	9·8855038	10·0119872	301	9·9880128	36
25	9·3655458	5292	10·6344542	8·4360286	9·3775631	5594	10·6224369	9·8853438	10·0120173	302	9·9879827	35
26	9·3660750	5286	10·6339250	8·4371020	9·3781225	5588	10·6218775	9·8851837	10·0120475	302	9·9879525	34
27	9·3666036	5279	10·6333964	8·4381740	9·3786813	5581	10·6213187	9·8850236	10·0120777	302	9·9879223	33
28	9·3671315	5272	10·6328685	8·4392447	9·3792394	5575	10·6207606	9·8848635	10·0121079	303	9·9878921	32
29	9·3676587	5266	10·6323413	8·4403141	9·3797969	5568	10·6202031	9·8847033	10·0121382	303	9·9878618	31
30	9·3681853	5258	10·6318147	8·4413821	9·3803537	5563	10·6196463	9·8845431	10·0121685	303	9·9878315	30
31	9·3687111	5252	10·6312889	8·4424488	9·3809100	5555	10·6190900	9·8843828	10·0121988	304	9·9878012	29
32	9·3692363	5245	10·6307637	8·4435142	9·3814655	5549	10·6185345	9·8842225	10·0122292	304	9·9877708	28
33	9·3697608	5239	10·6302392	8·4445783	9·3820205	5543	10·6179795	9·8840621	10·0122596	305	9·9877404	27
34	9·3702847	5232	10·6297153	8·4456410	9·3825748	5537	10·6174252	9·8839017	10·0122901	305	9·9877099	26
35	9·3708079	5225	10·6291921	8·4467024	9·3831285	5531	10·6168715	9·8837413	10·0123206	306	9·9876794	25
36	9·3713304	5219	10·6286696	8·4477625	9·3836816	5524	10·6163184	9·8835807	10·0123512	305	9·9876488	24
37	9·3718523	5212	10·6281477	8·4488213	9·3842340	5518	10·6157660	9·8834202	10·0123817	307	9·9876183	23
38	9·3723735	5205	10·6276265	8·4498788	9·3847858	5512	10·6152142	9·8832596	10·0124124	306	9·9875876	22
39	9·3728940	5199	10·6271060	8·4509350	9·3853370	5506	10·6146630	9·8830989	10·0124430	307	9·9875570	21
40	9·3734139	5192	10·6265861	8·4519898	9·3858876	5500	10·6141124	9·8829382	10·0124737	308	9·9875264	20
41	9·3739331	5186	10·6260669	8·4530434	9·3864376	5493	10·6135624	9·8827775	10·0125045	307	9·9874955	19
42	9·3744517	5179	10·6255483	8·4540957	9·3869869	5487	10·6130131	9·8826167	10·0125352	309	9·9874648	18
43	9·3749696	5172	10·6250304	8·4551467	9·3875356	5481	10·6124644	9·8824558	10·0125661	308	9·9874339	17
44	9·3754868	5166	10·6245132	8·4561964	9·3880837	5475	10·6119163	9·8822949	10·0125969	309	9·9874031	16
45	9·3760034	5160	10·6239966	8·4572448	9·3886312	5469	10·6113688	9·8821340	10·0126278	309	9·9873722	15
46	9·3765194	5153	10·6234806	8·4582920	9·3891781	5463	10·6108219	9·8819730	10·0126587	310	9·9873413	14
47	9·3770347	5146	10·6229653	8·4593378	9·3897244	5456	10·6102756	9·8818119	10·0126897	310	9·9873103	13
48	9·3775493	5140	10·6224507	8·4603824	9·3902700	5451	10·6097300	9·8816508	10·0127207	311	9·9872793	12
49	9·3780633	5134	10·6219367	8·4614257	9·3908151	5444	10·6091849	9·8814897	10·0127518	311	9·9872482	11
50	9·3785767	5127	10·6214233	8·4624677	9·3913595	5439	10·6086405	9·8813285	10·0127829	311	9·9872171	10
51	9·3790894	5121	10·6209106	8·4635085	9·3919034	5432	10·6080966	9·8811673	10·0128140	311	9·9871860	9
52	9·3796015	5114	10·6203985	8·4645480	9·3924466	5427	10·6075534	9·8810060	10·0128451	313	9·9871549	8
53	9·3801129	5108	10·6198871	8·4655863	9·3929893	5420	10·6070107	9·8808446	10·0128764	312	9·9871236	7
54	9·3806237	5102	10·6193763	8·4666233	9·3935313	5414	10·6064687	9·8806833	10·0129076	313	9·9870924	6
55	9·3811339	5095	10·6188661	8·4676590	9·3940727	5409	10·6059273	9·8805218	10·0129389	313	9·9870611	5
56	9·3816434	5089	10·6183566	8·4686935	9·3946136	5402	10·6053864	9·8803604	10·0129702	314	9·9870298	4
57	9·3821523	5082	10·6178477	8·4697267	9·3951538	5397	10·6048462	9·8801988	10·0130016	314	9·9869984	3
58	9·3826605	5077	10·6173395	8·4707587	9·3956935	5391	10·6043065	9·8800372	10·0130330	314	9·9869670	2
59	9·3831682	5070	10·6168318	8·4717894	9·3962326	5385	10·6037674	9·8798756	10·0130644	315	9·9869356	1
60	9·3836752		10·6163248	8·4728189	9·3967711		10·6032289	9·8797140	10·0130959		9·9869041	0

| ′ | Cosine. | Dif. | Secant. | Covers. | Cotang. | Dif. | Tang. | Verseds. | Cosec. | D. | Sine. | ′ |

76 Deg.

'	Sine.	Dif	Covers	Cosec.	Tang.	Cotang.	Secant.	Vers.	D.	Cosine	'
0	2419219	2822	7580781	4·1335655	2493280	4·0107809	1·0306136	0297043	704	9702957	60
1	2422041	2822	7577959	4·1287487	2496370	4·0058165	1·0306884	0297747	705	9702253	59
2	2424863	2822	7575137	4·1239435	2499460	4·0008636	1·0307633	0298452	706	9701548	58
3	2427685	2822	7572315	4·1191498	2502551	3·9959223	1·0308383	0299158	706	9700842	57
4	2430507	2822	7569493	4·1143675	2505642	3·9909924	1·0309134	0299864	708	9700136	56
5	2433329	2821	7566671	4·1095967	2508734	3·9860739	1·0309886	0300572	708	9699428	55
6	2436150	2821	7563850	4·1048374	2511826	3·9811669	1·0310639	0301280	709	9698720	54
7	2438971	2821	7561029	4·1000893	2514919	3·9762712	1·0311393	0301989	710	9698071	53
8	2441792	2821	7558208	4·0953526	2518012	3·9713868	1·0312147	0302699	710	9697301	52
9	2444613	2820	7555387	4·0906272	2521106	3·9665137	1·0312903	0303409	712	9696591	51
10	2447433	2821	7552567	4·0859130	2524200	3·9616518	1·0313660	0304121	712	9695879	50
11	2450254	2820	7549746	4·0812100	2527294	3·9568011	1·0314418	0304833	714	9695167	49
12	2453074	2820	7546926	4·0765181	2530389	3·9519615	1·0315177	0305547	713	9694453	48
13	2455894	2819	7544106	4·0718374	2533484	3·9471331	1·0315936	0306260	715	9693740	47
14	2458713	2820	7541287	4·0671677	2536580	3·9423157	1·0316697	0306975	716	9693025	46
15	2461533	2819	7538467	4·0625091	2539676	3·9375094	1·0317459	0307691	716	9692309	45
16	2464352	2819	7535648	4·0578615	2542773	3·9327141	1·0318222	0308407	718	9691593	44
17	2467171	2819	7532829	4·0532249	2545870	3·9279297	1·0318985	0309125	718	9690875	43
18	2469990	2819	7530010	4·0485992	2548968	3·9231563	1·0319750	0309843	719	9690157	42
19	2472809	2818	7527191	4·0439844	2552066	3·9183937	1·0320516	0310562	719	9689438	41
20	2475627	2818	7524373	4·0393804	2555165	3·9136420	1·0321282	0311281	721	9688719	40
21	2478445	2818	7521555	4·0347872	2558264	3·9089011	1·0322050	0312002	721	9687998	39
22	2481263	2818	7518737	4·0302048	2561363	3·9041710	1·0322818	0312723	722	9687277	38
23	2484081	2818	7515919	4·0256332	2564463	3·8994516	1·0323588	0313445	723	9686555	37
24	2486899	2817	7513101	4·0210722	2567564	3·8947429	1·0324359	0314168	724	9685832	36
25	2489716	2817	7510284	4·0165219	2570664	3·8900448	1·0325130	0314892	725	9685108	35
26	2492533	2817	7507467	4·0119823	2573766	3·8853574	1·0325903	0315617	725	9684383	34
27	2495350	2817	7504650	4·0074532	2576868	3·8806805	1·0326676	0316342	727	9683658	33
28	2498167	2817	7501833	4·0029347	2579970	3·8760142	1·0327451	0317069	727	9682931	32
29	2500984	2816	7499016	3·9984267	2583073	3·8713584	1·0328227	0317796	728	9682204	31
30	2503800	2816	7496200	3·9939292	2586176	3·8667131	1·0329003	0318524	728	9681476	30
31	2506616	2816	7493384	3·9894421	2589280	3·8620782	1·0329781	0319252	730	9680748	29
32	2509432	2816	7490568	3·9849654	2592384	3·8574537	1·0330559	0319982	730	9680018	28
33	2512248	2815	7487752	3·9804991	2595488	3·8528396	1·0331339	0320712	731	9679288	27
34	2515063	2816	7484937	3·9760431	2598593	3·8482358	1·0332119	0321443	732	9678557	26
35	2517879	2815	7482121	3·9715975	2601699	3·8436424	1·0332901	0322175	733	9677825	25
36	2520694	2814	7479306	3·9671621	2604805	3·8390591	1·0333683	0322908	734	9677092	24
37	2523508	2815	7476492	3·9627369	2607911	3·8344861	1·0334467	0323642	734	9676358	23
38	2526323	2814	7473677	3·9583219	2611018	3·8299233	1·0335251	0324376	736	9675624	22
39	2529137	2815	7470863	3·9539171	2614126	3·8253707	1·0336037	0325112	736	9674888	21
40	2531952	2814	7468048	3·9495224	2617234	3·8208281	1·0336823	0325848	737	9674152	20
41	2534766	2813	7465234	3·9451379	2620342	3·8162957	1·0337611	0326585	737	9673415	19
42	2537579	2814	7462421	3·9407633	2623451	3·8117733	1·0338399	0327322	739	9672678	18
43	2540393	2813	7459607	3·9363988	2626560	3·8072609	1·0339188	0328061	739	9671939	17
44	2543206	2813	7456794	3·9320443	2629670	3·8027585	1·0339979	0328800	741	9671200	16
45	2546019	2813	7453981	3·9276997	2632780	3·7982661	1·0340770	0329541	741	9670459	15
46	2548832	2813	7451168	3·9233651	2635891	3·7937835	1·0341563	0330282	741	9669718	14
47	2551645	2813	7448355	3·9190403	2639002	3·7893109	1·0342356	0331023	743	9668977	13
48	2554458	2812	7445542	3·9147254	2642114	3·7848481	1·0343151	0331766	744	9668234	12
49	2557270	2812	7442730	3·9104203	2645226	3·7803951	1·0343946	0332510	744	9667490	11
50	2560082	2812	7439918	3·9061250	2648339	3·7759519	1·0344743	0333254	745	9666746	10
51	2562894	2811	7437106	3·9018395	2651452	3·7715185	1·0345540	0333999	746	9666001	9
52	2565705	2812	7434295	3·8975637	2654566	3·7670947	1·0346338	0334745	747	9665255	8
53	2568517	2811	7431483	3·8932976	2657680	3·7626807	1·0347138	0335492	747	9664508	7
54	2571328	2811	7428672	3·8890411	2660794	3·7582763	1·0347938	0336239	749	9663761	6
55	2574139	2811	7425861	3·8847943	2663909	3·7538815	1·0348740	0336988	749	9663012	5
56	2576950	2810	7423050	3·8805570	2667025	3·7494963	1·0349542	0337737	750	9662263	4
57	2579760	2810	7420240	3·8763293	2670141	3·7451207	1·0350346	0338487	751	9661513	3
58	2582570	2811	7417430	3·8721112	2673257	3·7407546	1·0351150	0339238	751	9660762	2
59	2585381	2809	7414619	3·8679025	2676374	3·7363980	1·0351955	0339989	753	9660011	1
60	2588190		7411810	3·8637033	2679492	3·7320508	1·0352762	0340742		9659258	0
'	Cosine	Dif.	Vers.	Secant.	Cotan.	Tang.	Cosec.	Covers	D.	Sine.	'

′	Sine.	Dif.	Cosec.	Verseds.	Tang.	Dif.	Cotang.	Covers.	Secant.	D.	Cosine.	′
0	9·3836752	5063	10·6163248	8·4728189	9·3967711	5378	10·6032289	9·8797140	10·0130959	315	9·9869041	60
1	9·3841815	5058	10·6158185	8·4738472	9·3973089	5374	10·6026911	9·8795522	10·0131274	316	9·9868726	59
2	9·3846873	5051	10·6153127	8·4748742	9·3978463	5367	10·6021537	9·8793905	10·0131590	316	9·9868410	58
3	9·3851924	5045	10·6148076	8·4759000	9·3583830	5361	10·6016170	9·8792286	10·0131906	316	9·9868094	57
4	9·3856969	5039	10·6143031	8·4769246	9·3989191	5356	10·6010809	9·8790668	10·0132222	317	9·9867778	56
5	9·3862008	5032	10·6137992	8·4779480	9·3994547	5349	10·6005453	9·8789049	10·0132539	317	9·9867461	55
6	9·3867040	5027	10·6132960	8·4789701	9·3999896	5344	10·6000104	9·8787429	10·0132856	317	9·9867144	54
7	9·3872067	5020	10·6127933	8·4799910	9·4005240	5338	10·5994760	9·8785809	10·0133173	318	9·9866827	53
8	9·3877087	5014	10·6122913	8·4810107	9·4010578	5332	10·5989422	9·8784188	10·0133491	318	9·9866509	52
9	9·3882101	5008	10·6117899	8·4820291	9·4015910	5327	10·5984090	9·8782567	10·0133809	319	9·9866191	51
10	9·3887109	5002	10·6112891	8·4830464	9·4021237	5321	10·5978763	9·8780946	10·0134128	319	9·9865872	50
11	9·3892111	4995	10·6107889	8·4840625	9·4026558	5315	10·5973442	9·8779324	10·0134447	320	9·9865553	49
12	9·3897106	4990	10·6102894	8·4850773	9·4031873	5309	10·5968127	9·8777701	10·0134767	320	9·9865233	48
13	9·3902096	4983	10·6097904	8·4860910	9·4037182	5304	10·5962818	9·8776078	10·0135087	320	9·9864913	47
14	9·3907079	4978	10·6092921	8·4871034	9·4042486	5298	10·5957514	9·8774454	10·0135407	320	9·9864593	46
15	9·3912057	4971	10·6087943	8·4881147	9·4047784	5292	10·5952216	9·8772830	10·0135727	321	9·9864273	45
16	9·3917028	4965	10·6082972	8·4891247	9·4053076	5287	10·5946924	9·8771206	10·0136048	322	9·9863952	44
17	9·3921993	4959	10·6078007	8·4901336	9·4058363	5281	10·5941637	9·8769581	10·0136370	322	9·9863630	43
18	9·3926952	4953	10·6073048	8·4911412	9·4063644	5275	10·5936356	9·8767955	10·0136692	322	9·9863308	42
19	9·3931905	4947	10·6068095	8·4921477	9·4068919	5270	10·5931081	9·8766329	10·0137014	323	9·9862986	41
20	9·3936852	4942	10·6063148	8·4931530	9·4074189	5264	10·5925811	9·8764703	10·0137337	323	9·9862663	40
21	9·3941794	4935	10·6058206	8·4941572	9·4079453	5259	10·5920547	9·8763076	10·0137660	323	9·9862340	39
22	9·3946729	4929	10·6053271	8·4951601	9·4084712	5253	10·5915288	9·8761449	10·0137983	324	9·9862017	38
23	9·3951658	4923	10·6048342	8·4961619	9·4089965	5247	10·5910035	9·8759821	10·0138307	324	9·9861693	37
24	9·3956581	4918	10·6043419	8·4971625	9·4095212	5242	10·5904788	9·8758192	10·0138631	324	9·9861369	36
25	9·3961499	4911	10·6038501	8·4981619	9·4100454	5236	10·5899546	9·8756563	10·0138955	325	9·9861045	35
26	9·3966410	4905	10·6033590	8·4991602	9·4105690	5231	10·5894310	9·8754934	10·0139280	326	9·9860720	34
27	9·3971315	4900	10·6028685	8·5001573	9·4110921	5225	10·5889079	9·8753304	10·0139606	325	9·9860394	33
28	9·3976215	4894	10·6023785	8·5011532	9·4116146	5220	10·5883854	9·8751674	10·0139931	327	9·9860069	32
29	9·3981109	4887	10·6018891	8·5021480	9·4121366	5215	10·5878634	9·8750043	10·0140258	326	9·9859742	31
30	9·3985996	4882	10·6014004	8·5031416	9·4126581	5208	10·5873419	9·8748412	10·0140584	327	9·9859416	30
31	9·3990878	4876	10·6009122	8·5041341	9·4131789	5204	10·5868211	9·8746780	10·0140011	327	9·9859089	29
32	9·3995754	4871	10·6004246	8·5051254	9·4136993	5198	10·5863007	9·8745147	10·0141238	328	9·9858762	28
33	9·4000625	4864	10·5999375	8·5061156	9·4142191	5192	10·5857809	9·8743515	10·0141566	328	9·9858434	27
34	9·4005489	4859	10·5994511	8·5071046	9·4147383	5187	10·5852617	9·8741881	10·0141894	329	9·9858106	26
35	9·4010348	4853	10·5989652	8·5080925	9·4152570	5182	10·5847430	9·8740248	10·0142223	328	9·9857777	25
36	9·4015201	4847	10·5984799	8·5090792	9·4157752	5176	10·5842248	9·8738613	10·0142551	330	9·9857449	24
37	9·4020048	4841	10·5979952	8·5100648	9·4162928	5171	10·5837072	9·8736978	10·0142881	329	9·9857119	23
38	9·4024889	4835	10·5975111	8·5110493	9·4168099	5166	10·5831901	9·8735343	10·0143210	330	9·9856790	22
39	9·4029724	4830	10·5970276	8·5120326	9·4173265	5160	10·5826735	9·8733707	10·0143540	331	9·9856460	21
40	9·4034554	4824	10·5965446	8·5130148	9·4178425	5155	10·5821575	9·8732071	10·0143871	331	9·9856129	20
41	9·4039378	4818	10·5960622	8·5139959	9·4183580	5149	10·5816420	9·8730434	10·0144202	331	9·9855798	19
42	9·4044196	4813	10·5955804	8·5149758	9·4188729	5145	10·5811271	9·8728797	10·0144533	332	9·9855467	18
43	9·4049009	4807	10·5950991	8·5159546	9·4193874	5139	10·5806126	9·8727159	10·0144865	332	9·9855135	17
44	9·4053816	4801	10·5946184	8·5169324	9·4199013	5133	10·5800987	9·8725521	10·0145197	332	9·9854803	16
45	9·4058617	4796	10·5941383	8·5179089	9·4204146	5129	10·5795854	9·8723883	10·0145529	333	9·9854471	15
46	9·4063413	4790	10·5936587	8·5188844	9·4209275	5123	10·5790725	9·8722243	10·0145862	333	9·9854138	14
47	9·4068203	4784	10·5931797	8·5198588	9·4214398	5117	10·5785602	9·8720604	10·0146195	334	9·9853805	13
48	9·4072987	4779	10·5927013	8·5208320	9·4219515	5113	10·5780485	9·8718963	10·0146529	333	9·9853471	12
49	9·4077766	4773	10·5922234	8·5218042	9·4224628	5107	10·5775372	9·8717323	10·0146862	335	9·9853138	11
50	9·4082539	4767	10·5917461	8·5227752	9·4229735	5103	10·5770265	9·8715682	10·0147197	335	9·9852803	10
51	9·4087306	4762	10·5912694	8·5237451	9·4234838	5097	10·5765162	9·8714040	10·0147532	335	9·9852468	9
52	9·4092068	4756	10·5907932	8·5247140	9·4239935	5091	10·5760065	9·8712398	10·0147867	335	9·9852133	8
53	9·4096824	4751	10·5903176	8·5256817	9·4245026	5087	10·5754974	9·8710755	10·0148202	336	9·9851798	7
54	9·4101575	4745	10·5898425	8·5266484	9·4250113	5081	10·5749887	9·8709112	10·0148538	337	9·9851462	6
55	9·4106320	4739	10·5893680	8·5276139	9·4255194	5077	10·5744806	9·8707468	10·0148875	336	9·9851125	5
56	9·4111059	4734	10·5888941	8·5285784	9·4260271	5071	10·5739729	9·8705824	10·0149211	337	9·9850789	4
57	9·4115793	4729	10·5884207	8·5295417	9·4265342	5066	10·5734658	9·8704179	10·0149348	338	9·9850452	3
58	9·4120522	4723	10·5879478	8·5305040	9·4270408	5061	10·5729592	9·8702534	10·0149886	338	9·9850114	2
59	9·4125245	4717	10·5874755	8·5314652	9·4275469	5056	10·5724531	9·8700889	10·0150224	338	9·9849776	1
60	9·4129962		10·5870038	8·5324253	9·4280525		10·5719475	9·8699243	10·0150562		9·9849438	0
′	Cosine.	Dif.	Secant.	Covers.	Cotang.	Dif.	Tang.	Verseds.	Cosec.	D.	Sine.	′

′	Sine.	Dif.	Covers	Cosec.	Tang.	Cotang.	Secant.	Vers.	D.	Cosine	′
0	2588190	2810	7411810	3·8637033	2679492	3·7320508	1·0352762	0340742	753	9659258	60
1	2591000	2810	7409000	3·8595135	2682610	3·7277131	1·0353569	0341495	754	9658505	59
2	2593810	2809	7406190	3·8553332	2685728	3·7233847	1·0354378	0342249	755	9657751	58
3	2596619	2809	7403381	3·8511622	2688847	3·7190658	1·0355187	0343004	756	9656996	57
4	2599428	2809	7400572	3·8470006	2691967	3·7147561	1·0355998	0343760	756	9656240	56
5	2602237	2808	7397763	3·8428482	2695087	3·7104558	1·0356809	0344516	758	9655484	55
6	2605045	2808	7394955	3·8387052	2698207	3·7061648	1·0357621	0345274	758	9654726	54
7	2607853	2809	7392147	3·8345713	2701328	3·7018830	1·0358435	0346032	759	9653968	53
8	2610662	2807	7389338	3·8304467	2704449	3·6976104	1·0359249	0346791	760	9653209	52
9	2613469	2808	7386531	3·8263313	2707571	3·6933469	1·0360065	0347551	760	9652449	51
10	2616277	2808	7383723	3·8222251	2710694	3·6890927	1·0360881	0348311	762	9651689	50
11	2619085	2807	7380915	3·8181280	2713817	3·6848475	1·0361699	0349073	762	9650927	49
12	2621892	2807	7378108	3·8140399	2716940	3·6806115	1·0362517	0349835	763	9650165	48
13	2624699	2807	7375301	3·8099610	2720064	3·6763845	1·0363337	0350598	764	9649402	47
14	2627506	2806	7372494	3·8058911	2723188	3·6721665	1·0364157	0351362	765	9648638	46
15	2630312	2806	7369688	3·8018301	2726313	3·6679575	1·0364979	0352127	765	9647873	45
16	2633118	2807	7366882	3·7977782	2729438	3·6637575	1·0365801	0352892	767	9647108	44
17	2635925	2805	7364075	3·7937352	2732564	3·6595665	1·0366625	0353659	767	9646341	43
18	2638730	2806	7361270	3·7897011	2735690	3·6553844	1·0367449	0354426	768	9645574	42
19	2641536	2806	7358464	3·7856760	2738817	3·6512111	1·0368275	0355194	769	9644806	41
20	2644342	2805	7355658	3·7816596	2741945	3·6470467	1·0369101	0355963	769	9644037	40
21	2647147	2805	7352853	3·7776522	2745072	3·6428911	1·0369929	0356732	771	9643268	39
22	2649952	2805	7350048	3·7736535	2748201	3·6387444	1·0370757	0357503	771	9642497	38
23	2652757	2804	7347243	3·7696636	2751330	3·6346064	1·0371587	0358274	772	9641726	37
24	2655561	2805	7344430	3·7656824	2754459	3·6304771	1·0372417	0359046	773	9640954	36
25	2658366	2804	7341634	3·7617100	2757589	3·6263566	1·0373249	0359819	774	9640181	35
26	2661170	2803	7338830	3·7577462	2760719	3·6222447	1·0374082	0360593	774	9639407	34
27	2663973	2804	7336027	3·7537911	2763850	3·6181415	1·0374915	0361367	775	9638633	33
28	2666777	2804	7333223	3·7498447	2766981	3·6140469	1·0375750	0362142	775	9637858	32
29	2669581	2803	7330419	3·7459068	2770113	3·6099609	1·0376585	0362919	776	9637081	31
30	2672384	2803	7327616	3·7419775	2773245	3·6058835	1·0377422	0363695	778	9636305	30
31	2675187	2802	7324813	3·7380568	2776378	3·6018146	1·0378260	0364473	779	9635527	29
32	2677989	2803	7322011	3·7341446	2779512	3·5977543	1·0379098	0365252	779	9634748	28
33	2680792	2802	7319208	3·7302409	2782646	3·5937024	1·0379938	0366031	780	9633969	27
34	2683594	2802	7316406	3·7263457	2785780	3·5896590	1·0380779	0366811	781	9633189	26
35	2686396	2802	7313604	3·7224589	2788915	3·5856241	1·0381621	0367592	782	9632408	25
36	2689198	2802	7310802	3·7185805	2792050	3·5815975	1·0382463	0368374	783	9631626	24
37	2692000	2801	7308000	3·7147105	2795186	3·5775794	1·0383307	0369157	783	9630843	23
38	2694801	2801	7305199	3·7108489	2798322	3·5735696	1·0384152	0369940	785	9630060	22
39	2697602	2801	7302398	3·7069956	2801459	3·5695681	1·0384998	0370725	785	9629275	21
40	2700403	2801	7299597	3·7031506	2804597	3·5655749	1·0385844	0371510	786	9628490	20
41	2703204	2800	7296796	3·6993139	2807735	3·5615900	1·0386692	0372296	787	9627704	19
42	2706004	2801	7293996	3·6954854	2810873	3·5576133	1·0387541	0373083	787	9626917	18
43	2708805	2800	7291195	3·6916652	2814012	3·5536449	1·0388391	0373870	788	9626130	17
44	2711605	2799	7288395	3·6878532	2817152	3·5496846	1·0389242	0374658	790	9625342	16
45	2714404	2800	7285596	3·6840493	2820292	3·5457325	1·0390094	0375448	790	9624552	15
46	2717204	2799	7282796	3·6802536	2823432	3·5417886	1·0390947	0376238	790	9623762	14
47	2720003	2799	7279997	3·6764660	2826573	3·5378528	1·0391800	0377028	792	9622972	13
48	2722802	2799	7277198	3·6726865	2829715	3·5339251	1·0392655	0377820	793	9622180	12
49	2725601	2799	7274399	3·6689151	2832857	3·5300054	1·0393511	0378613	793	9621387	11
50	2728400	2798	7271600	3·6651518	2835999	3·5260938	1·0394368	0379406	794	9620594	10
51	2731198	2799	7268802	3·6613964	2839143	3·5221902	1·0395226	0380200	795	9619800	9
52	2733997	2797	7266003	3·6576491	2842286	3·5182946	1·0396085	0380995	795	9619005	8
53	2736794	2798	7263206	3·6539097	2845430	3·5144070	1·0396945	0381790	797	9618210	7
54	2739592	2798	7260408	3·6501783	2848575	3·5105273	1·0397806	0382587	797	9617413	6
55	2742390	2797	7257610	3·6464548	2851720	3·5066555	1·0398669	0383384	798	9616616	5
56	2745187	2797	7254813	3·6427392	2854866	3·5027916	1·0399532	0384182	799	9615818	4
57	2747984	2797	7252016	3·6390315	2858012	3·4989356	1·0400396	0384981	800	9615019	3
58	2750781	2796	7249219	3·6353316	2861159	3·4950874	1·0401261	0385781	801	9614219	2
59	2753577	2797	7246423	3·6316395	2864306	3·4912470	1·0402127	0386582	801	9613418	1
60	2756374	2797	7243626	3·6279553	2867454	3·4874144	1·0402994	0387383	801	9612617	0

| ′ | Cosine | Dif. | Vers. | Secant. | Cotan. | Tang. | Cosec. | Covers | D. | Sine. | ′ |

′	Sine	Dif	Cosec	Verseds	Tang	Dif	Cotang	Covers	Secant	D	Cosine	′
0	9·4129962	4712	10·5870038	8·5324253	9·4280525	5050	10·5719475	9·8699243	10·0150562	339	9·9849438	60
1	9·4134674	4707	10·5865326	8·5333844	9·4285575	5046	10·5714425	9·8697596	10·0150901	339	9·9849099	59
2	9·4139381	4701	10·5860619	8·5343423	9·4290621	5040	10·5709379	9·8695949	10·0151240	340	9·9848760	58
3	9·4144082	4696	10·5855918	8·5352992	9·4295661	5036	10·5704339	9·8694301	10·0151580	339	9·9848420	57
4	9·4148778	4690	10·5851222	8·5362551	9·4300697	5030	10·5699303	9·8692653	10·0151919	341	9·9848081	56
5	9·4153468	4684	10·5846532	8·5372098	9·4305727	5026	10·5694273	9·8691004	10·0152260	340	9·9847740	55
6	9·4158152	4680	10·5841848	8·5381635	9·4310753	5020	10·5689247	9·8689355	10·0152600	341	9·9847400	54
7	9·4162832	4674	10·5837168	8·5391161	9·4315773	5016	10·5684227	9·8687706	10·0152941	342	9·9847059	53
8	9·4167506	4668	10·5832494	8·5400677	9·4320789	5010	10·5679211	9·8686056	10·0153283	342	9·9846717	52
9	9·4172174	4663	10·5827826	8·5410182	9·4325799	5005	10·5674201	9·8684405	10·0153625	342	9·9846375	51
10	9·4176837	4658	10·5823163	8·5419676	9·4330804	5001	10·5669196	9·8682754	10·0153967	343	9·9846033	50
11	9·4181495	4653	10·5818505	8·5429160	9·4335805	4995	10·5664195	9·8681102	10·0154310	343	9·9845690	49
12	9·4186148	4647	10·5813852	8·5438633	9·4340800	4991	10·5659200	9·8679450	10·0154653	343	9·9845347	48
13	9·4190795	4641	10·5809205	8·5448096	9·4345791	4985	10·5654209	9·8677798	10·0154996	344	9·9845004	47
14	9·4195436	4637	10·5804564	8·5457548	9·4350776	4981	10·5649224	9·8676145	10·0155340	344	9·9844660	46
15	9·4200073	4631	10·5799927	8·5466990	9·4355757	4976	10·5644243	9·8674491	10·0155684	345	9·9844316	45
16	9·4204704	4626	10·5795296	8·5476422	9·4360733	4971	10·5639267	9·8672837	10·0156029	345	9·9843971	44
17	9·4209330	4620	10·5790670	8·5485843	9·4365704	4966	10·5634296	9·8671182	10·0156374	345	9·9843626	43
18	9·4213950	4616	10·5786050	8·5495253	9·4370670	4961	10·5629330	9·8669527	10·0156719	346	9·9843281	42
19	9·4218566	4610	10·5781434	8·5504654	9·4375631	4956	10·5624369	9·8667872	10·0157065	346	9·9842935	41
20	9·4223176	4604	10·5776824	8·5514044	9·4380587	4951	10·5619413	9·8666216	10·0157411	347	9·9842589	40
21	9·4227780	4600	10·5772220	8·5523423	9·4385538	4947	10·5614462	9·8664559	10·0157758	347	9·9842242	39
22	9·4232380	4594	10·5767620	8·5532793	9·4390485	4941	10·5609515	9·8662902	10·0158105	347	9·9841895	38
23	9·4236974	4589	10·5763026	8·5542152	9·4395426	4937	10·5604574	9·8661244	10·0158452	348	9·9841548	37
24	9·4241563	4584	10·5758437	8·5551500	9·4400363	4932	10·5599637	9·8659586	10·0158800	348	9·9841200	36
25	9·4246147	4579	10·5753853	8·5560839	9·4405295	4927	10·5594705	9·8657928	10·0159148	349	9·9840852	35
26	9·4250726	4573	10·5749274	8·5570167	9·4410222	4923	10·5589778	9·8656269	10·0159497	349	9·9840503	34
27	9·4255299	4568	10·5744701	8·5579485	9·4415145	4917	10·5584855	9·8654609	10·0159846	349	9·9840154	33
28	9·4259867	4563	10·5740133	8·5588793	9·4420062	4913	10·5579938	9·8652949	10·0160195	350	9·9839805	32
29	9·4264430	4558	10·5735570	8·5598091	9·4424975	4908	10·5575025	9·8651288	10·0160545	350	9·9839455	31
30	9·4268988	4553	10·5731012	8·5607379	9·4429883	4903	10·5570117	9·8649627	10·0160895	350	9·9839105	30
31	9·4273541	4548	10·5726459	8·5616656	9·4434786	4899	10·5565214	9·8647966	10·0161245	351	9·9838755	29
32	9·4278089	4542	10·5721911	8·5625924	9·4439685	4894	10·5560315	9·8646303	10·0161596	352	9·9838404	28
33	9·4282631	4538	10·5717369	8·5635181	9·4444579	4889	10·5555421	9·8644641	10·0161948	351	9·9838052	27
34	9·4287169	4532	10·5712831	8·5644429	9·4449468	4884	10·5550532	9·8642978	10·0162299	353	9·9837701	26
35	9·4291701	4527	10·5708299	8·5653666	9·4454352	4880	10·5545648	9·8641314	10·0162652	352	9·9837348	25
36	9·4296228	4522	10·5703772	8·5662894	9·4459232	4875	10·5540768	9·8639650	10·0163004	353	9·9836996	24
37	9·4300750	4517	10·5699250	8·5672111	9·4464107	4871	10·5535893	9·8637985	10·0163357	353	9·9836643	23
38	9·4305267	4512	10·5694733	8·5681318	9·4468978	4865	10·5531022	9·8636320	10·0163710	354	9·9836290	22
39	9·4309779	4507	10·5690221	8·5690516	9·4473843	4861	10·5526157	9·8634655	10·0164064	354	9·9835936	21
40	9·4314286	4502	10·5685714	8·5699704	9·4478704	4857	10·5521296	9·8632989	10·0164418	355	9·9835582	20
41	9·4318788	4497	10·5681212	8·5708881	9·4483561	4852	10·5516439	9·8631322	10·0164773	355	9·9835227	19
42	9·4323285	4492	10·5676715	8·5718049	9·4488413	4847	10·5511587	9·8629655	10·0165128	355	9·9834872	18
43	9·4327777	4487	10·5672223	8·5727207	9·4493260	4842	10·5506740	9·8627987	10·0165483	356	9·9834517	17
44	9·4332264	4482	10·5667736	8·5736355	9·4498102	4838	10·5501898	9·8626319	10·0165839	356	9·9834161	16
45	9·4336746	4477	10·5663254	8·5745494	9·4502940	4834	10·5497060	9·8624651	10·0166195	356	9·9833805	15
46	9·4341223	4471	10·5658777	8·5754622	9·4507774	4828	10·5492226	9·8622981	10·0166551	357	9·9833449	14
47	9·4345694	4467	10·5654306	8·5763741	9·4512602	4825	10·5487398	9·8621312	10·0166908	357	9·9833092	13
48	9·4350161	4462	10·5649839	8·5772850	9·4517427	4819	10·5482573	9·8619642	10·0167265	358	9·9832735	12
49	9·4354623	4457	10·5645377	8·5781950	9·4522246	4815	10·5477754	9·8617971	10·0167623	358	9·9832377	11
50	9·4359080	4452	10·5640920	8·5791039	9·4527061	4811	10·5472939	9·8616300	10·0167981	358	9·9832019	10
51	9·4363532	4448	10·5636468	8·5800119	9·4531872	4806	10·5468128	9·8614628	10·0168339	359	9·9831661	9
52	9·4367980	4442	10·5632020	8·5809189	9·4536678	4801	10·5463322	9·8612956	10·0168698	360	9·9831302	8
53	9·4372422	4437	10·5627578	8·5818250	9·4541479	4797	10·5458521	9·8611283	10·0169058	359	9·9830942	7
54	9·4376859	4433	10·5623141	8·5827301	9·4546276	4793	10·5453724	9·8609610	10·0169417	360	9·9830583	6
55	9·4381292	4427	10·5618708	8·5836342	9·4551069	4788	10·5448931	9·8607936	10·0169777	361	9·9830223	5
56	9·4385719	4423	10·5614281	8·5845374	9·4555857	4784	10·5444143	9·8606262	10·0170138	361	9·9829862	4
57	9·4390142	4418	10·5609858	8·5854396	9·4560641	4779	10·5439359	9·8604588	10·0170499	361	9·9829501	3
58	9·4394560	4413	10·5605440	8·5863409	9·4565420	4774	10·5434580	9·8602912	10·0170860	362	9·9829140	2
59	9·4398973	4408	10·5601027	8·5872412	9·4570194	4770	10·5429806	9·8601237	10·0171222	362	9·9828778	1
60	9·4403381		10·5596619	8·5881406	9·4574964		10·5425036	9·8599560	10·0171584		9·9828416	0
′	Cosine	Dif	Secant	Covers	Cotang	Dif	Tang	Verseds	Cosec	D	Sine	′

′	Sine.	Dif.	Covers	Cosec.	Tang.	Cotang.	Secant.	Vers.	D.	Cosine	′
0	2756374	2796	7243626	3·6279553	2867454	3·4874144	1·0402994	0387383	802	9612617	60
1	2759170	2795	7240830	3·6242788	2870602	3·4835896	1·0403863	0388185	803	9611815	59
2	2761965	2796	7238035	3·6206101	2873751	3·4797726	1·0404732	0388988	804	9611012	58
3	2764761	2795	7235239	3·6169490	2876900	3·4759632	1·0405602	0389792	805	9610208	57
4	2767556	2796	7232444	3·6132957	2880050	3·4721616	1·0406473	0390597	805	9609403	56
5	2770352	2795	7229648	3·6096501	2883201	3·4683676	1·0407346	0391402	806	9608598	55
6	2773147	2794	7226853	3·6060121	2886352	3·4645813	1·0408219	0392208	808	9607792	54
7	2775941	2795	7224059	3·6023818	2889503	3·4608026	1·0409094	0393016	807	9606984	53
8	2778736	2794	7221264	3·5987590	2892655	3·4570315	1·0409969	0393823	809	9606177	52
9	2781530	2794	7218470	3·5951439	2895808	3·4532679	1·0410845	0394632	810	9605368	51
10	2784324	2794	7215676	3·5915363	2898961	3·4495120	1·0411723	0395442	810	9604558	50
11	2787118	2793	7212882	3·5879362	2902114	3·4457635	1·0412601	0396252	811	9603748	49
12	2789911	2793	7210089	3·5843437	2905269	3·4420226	1·0413481	0397063	812	9602937	48
13	2792704	2793	7207296	3·5807586	2908423	3·4382891	1·0414362	0397875	813	9602125	47
14	2795497	2793	7204503	3·5771810	2911578	3·4345631	1·0415243	0398688	813	9601312	46
15	2798290	2793	7201710	3·5736108	2914734	3·4308446	1·0416126	0399501	815	9600499	45
16	2801083	2792	7198917	3·5700481	2917890	3·4271334	1·0417009	0400316	815	9599684	44
17	2803875	2792	7196125	3·5664928	2921047	3·4234297	1·0417894	0401131	816	9598869	43
18	2806667	2792	7193333	3·5629448	2924205	3·4197333	1·0418780	0401947	817	9598053	42
19	2809459	2792	7190541	3·5594042	2927363	3·4160443	1·0419667	0402764	818	9597236	41
20	2812251	2791	7187749	3·5558710	2930521	3·4123626	1·0420554	0403582	818	9596418	40
21	2815042	2791	7184958	3·5523450	2933680	3·4086882	1·0421443	0404400	819	9595600	39
22	2817833	2791	7182167	3·5488263	2936839	3·4050210	1·0422333	0405219	820	9594781	38
23	2820624	2791	7179376	3·5453149	2939999	3·4013612	1·0423224	0406039	821	9593961	37
24	2823415	2790	7176585	3·5418107	2943160	3·3977085	1·0424116	0406860	822	9593140	36
25	2826205	2790	7173795	3·5383138	2946321	3·3940631	1·0425009	0407682	822	9592318	35
26	2828995	2790	7171005	3·5348240	2949483	3·3904249	1·0425903	0408504	824	9591496	34
27	2831785	2790	7168215	3·5313414	2952645	3·3867938	1·0426798	0409328	824	9590672	33
28	2834575	2789	7165425	3·5278660	2955808	3·3831699	1·0427694	0410152	825	9589848	32
29	2837364	2789	7162636	3·5243977	2958071	3·3795531	1·0428591	0410977	826	9589023	31
30	2840153	2789	7159847	3·5209365	2962135	3·3759434	1·0429489	0411803	826	9588197	30
31	2842942	2789	7157058	3·5174824	2965299	3·3723408	1·0430388	0412629	828	9587371	29
32	2845731	2789	7154269	3·5140354	2968464	3·3687453	1·0431289	0413457	828	9586543	28
33	2848520	2788	7151480	3·5105954	2971630	3·3651568	1·0432190	0414285	829	9585715	27
34	2851308	2788	7148692	3·5071625	2974796	3·3615753	1·0433092	0415114	830	9584886	26
35	2854096	2788	7145904	3·5037365	2977962	3·3580008	1·0433995	0415944	830	9584056	25
36	2856884	2787	7143116	3·5003175	2981129	3·3544333	1·0434900	0416774	832	9583226	24
37	2859671	2787	7140329	3·4969055	2984297	3·3508728	1·0435805	0417606	832	9582394	23
38	2862458	2788	7137542	3·4935004	2987465	3·3473191	1·0436712	0418438	833	9581562	22
39	2865246	2786	7134754	3·4901023	2990634	3·3437724	1·0437619	0419271	834	9580729	21
40	2868032	2787	7131968	3·4867110	2993803	3·3402326	1·0438528	0420105	835	9579895	20
41	2870819	2786	7129181	3·4833267	2996973	3·3366997	1·0439437	0420940	835	9579060	19
42	2873605	2786	7126395	3·4799492	3000144	3·3331736	1·0440348	0421775	836	9578225	18
43	2876301	2786	7123609	3·4765785	3003315	3·3296543	1·0441259	0422611	837	9577389	17
44	2879177	2786	7120823	3·4732146	3006486	3·3261419	1·0442172	0423448	838	9576552	16
45	2881963	2786	7118037	3·4698576	3009658	3·3226362	1·0443086	0424286	839	9575714	15
46	2884748	2785	7115252	3·4665073	3012831	3·3191373	1·0444001	0425125	840	9574875	14
47	2887533	2785	7112467	3·4631637	3016004	3·3156452	1·0444917	0425965	840	9574035	13
48	2890318	2785	7109682	3·4598269	3019178	3·3121598	1·0445833	0426805	841	9573195	12
49	2893103	2784	7106897	3·4564969	3022352	3·3086811	1·0446751	0427646	842	9572354	11
50	2895887	2784	7104113	3·4531735	3025527	3·3052091	1·0447670	0428488	843	9571512	10
51	2898671	2784	7101329	3·4498568	3028703	3·3017438	1·0448590	0429331	844	9570669	9
52	2901455	2784	7098545	3·4465467	3031879	3·2982851	1·0449511	0430175	844	9569825	8
53	2904239	2783	7095761	3·4432433	3035055	3·2948330	1·0450433	0431019	845	9568981	7
54	2907022	2783	7092978	3·4399465	3038232	3·2913876	1·0451357	0431864	846	9568136	6
55	2909805	2783	7090195	3·4366563	3041410	3·2879487	1·0452281	0432710	847	9567290	5
56	2912588	2783	7087412	3·4333727	3044588	3·2845164	1·0453206	0433557	847	9566443	4
57	2915371	2782	7084629	3·4300956	3047767	3·2810907	1·0454132	0434405	848	9565595	3
58	2918153	2782	7081847	3·4268251	3050946	3·2776715	1·0455060	0435253	848	9564747	2
59	2920935	2782	7079065	3·4235611	3054126	3·2742588	1·0455988	0436102	849	9563898	1
60	2923717		7076283	3·4203036	3057307	3·2708526	1·0456918	0436952	850	9563048	0

′	Cosine	Dif.	Vers.	Secant.	Cotan.	Tang.	Cosec.	Covers	D.	Sine.	′

'	Sine.	Dif.	Cosec.	Verseds.	Tang.	Dif.	Cotang.	Covers.	Secant.	D.	Cosine.	'
0	9·4403381	4403	10·5596619	8·5881406	9·4574964	4766	10·5425036	9·8599560	10·0171584	362	9·9828416	60
1	9·4407784	4398	10·5592216	8·5890390	9·4579730	4761	10·5420270	9·8597884	10·0171946	363	9·9828054	59
2	9·4412182	4394	10·5587818	8·5899365	9·4584491	4757	10·5415509	9·8596206	10·0172309	363	9·9827691	58
3	9·4416576	4389	10·5583424	8·5908330	9·4589248	4753	10·5410752	9·8594529	10·0172672	364	9·9827328	57
4	9·4420965	4384	10·5579035	8·5917286	9·4594001	4748	10·5405999	9·8592851	10·0173036	364	9·9826964	56
5	9·4425349	4379	10·5574651	8·5926233	9·4598749	4743	10·5401251	9·8591172	10·0173400	364	9·9826600	55
6	9·4429728	4375	10·5570272	8·5935170	9·4603492	4740	10·5396508	9·8589492	10·0173764	365	9·9826236	54
7	9·4434103	4369	10·5565897	8·5944097	9·4608232	4735	10·5391768	9·8587813	10·0174129	365	9·9825871	53
8	9·4438472	4365	10·5561528	8·5953016	9·4612967	4730	10·5387033	9·8586132	10·0174494	365	9·9825506	52
9	9·4442837	4360	10·5557163	8·5961925	9·4617697	4726	10·5382303	9·8584452	10·0174860	366	9·9825140	51
10	9·4447197	4356	10·5552803	8·5970824	9·4622423	4722	10·5377577	9·8582770	10·0175226	366	9·9824774	50
11	9·4451553	4351	10·5548447	8·5979715	9·4627145	4718	10·5372855	9·8581089	10·0175592	367	9·9824408	49
12	9·4455904	4346	10·5544096	8·5988596	9·4631863	4713	10·5368137	9·8579406	10·0175959	367	9·9824041	48
13	9·4460250	4341	10·5539750	8·5997468	9·4636576	4709	10·5363424	9·8577723	10·0176326	368	9·9823674	47
14	9·4464591	4336	10·5535409	8·6006330	9·4641285	4705	10·5358715	9·8576040	10·0176694	368	9·9823306	46
15	9·4468927	4332	10·5531073	8·6015184	9·4645990	4700	10·5354010	9·8574356	10·0177062	369	9·9822938	45
16	9·4473259	4327	10·5526741	8·6024028	9·4650690	4696	10·5349310	9·8572672	10·0177431	368	9·9822569	44
17	9·4477586	4323	10·5522414	8·6032863	9·4655386	4692	10·5344614	9·8570987	10·0177799	370	9·9822201	43
18	9·4481909	4318	10·5518091	8·6041689	9·4660078	4687	10·5339922	9·8569302	10·0178169	369	9·9821831	42
19	9·4486227	4313	10·5513773	8·6050506	9·4664765	4683	10·5335235	9·8567616	10·0178538	370	9·9821462	41
20	9·4490540	4309	10·5509460	8·6059313	9·4669448	4679	10·5330552	9·8565929	10·0178908	370	9·9821092	40
21	9·4494849	4304	10·5505151	8·6068112	9·4674127	4675	10·5325873	9·8564242	10·0179279	371	9·9820721	39
22	9·4499153	4299	10·5500847	8·6076901	9·4678802	4671	10·5321198	9·8562555	10·0179649	372	9·9820351	38
23	9·4503452	4295	10·5496548	8·6085681	9·4683473	4666	10·5316527	9·8560867	10·0180021	371	9·9819979	37
24	9·4507747	4290	10·5492253	8·6094453	9·4688139	4662	10·5311861	9·8559179	10·0180392	372	9·9819608	36
25	9·4512037	4285	10·5487963	8·6103215	9·4692801	4658	10·5307199	9·8557490	10·0180764	373	9·9819236	35
26	9·4516322	4281	10·5483678	8·6111968	9·4697459	4653	10·5302541	9·8555800	10·0181137	373	9·9818863	34
27	9·4520603	4276	10·5479397	8·6120712	9·4702112	4650	10·5297888	9·8554110	10·0181510	373	9·9818490	33
28	9·4524879	4272	10·5475121	8·6129448	9·4706762	4645	10·5293238	9·8552420	10·0181883	373	9·9818117	32
29	9·4529151	4267	10·5470849	8·6138174	9·4711407	4641	10·5288593	9·8550729	10·0182256	374	9·9817744	31
30	9·4533418	4263	10·5466582	8·6146891	9·4716048	4637	10·5283952	9·8549037	10·0182630	375	9·9817370	30
31	9·4537681	4258	10·5462319	8·6155600	9·4720685	4633	10·5279315	9·8547345	10·0183005	375	9·9816995	29
32	9·4541939	4253	10·5458061	8·6164299	9·4725318	4629	10·5274682	9·8545653	10·0183380	375	9·9816620	28
33	9·4546192	4249	10·5453808	8·6172990	9·4729947	4625	10·5270053	9·8543959	10·0183755	375	9·9816245	27
34	9·4550441	4245	10·5449559	8·6181672	9·4734572	4620	10·5265428	9·8542266	10·0184130	376	9·9815870	26
35	9·4554686	4240	10·5445314	8·6190345	9·4739192	4616	10·5260808	9·8540572	10·0184506	377	9·9815494	25
36	9·4558926	4235	10·5441074	8·6199009	9·4743808	4613	10·5256192	9·8538877	10·0184883	377	9·9815117	24
37	9·4563161	4231	10·5436839	8·6207664	9·4748421	4608	10·5251579	9·8537182	10·0185260	377	9·9814740	23
38	9·4567392	4226	10·5432608	8·6216311	9·4753029	4604	10·5246971	9·8535486	10·0185637	377	9·9814363	22
39	9·4571618	4222	10·5428382	8·6224948	9·4757633	4600	10·5242367	9·8533790	10·0186014	378	9·9813986	21
40	9·4575840	4218	10·5424160	8·6233577	9·4762233	4596	10·5237767	9·8532094	10·0186392	379	9·9813608	20
41	9·4580058	4213	10·5419942	8·6242197	9·4766829	4592	10·5233171	9·8530396	10·0186771	379	9·9813229	19
42	9·4584271	4209	10·5415729	8·6250809	9·4771421	4588	10·5228579	9·8528699	10·0187150	379	9·9812850	18
43	9·4588480	4204	10·5411520	8·6259412	9·4776009	4583	10·5223991	9·8527001	10·0187529	380	9·9812471	17
44	9·4592684	4200	10·5407316	8·6268006	9·4780592	4580	10·5219408	9·8525302	10·0187909	380	9·9812091	16
45	9·4596884	4195	10·5403116	8·6276591	9·4785172	4576	10·5214828	9·8523603	10·0188289	380	9·9811711	15
46	9·4601079	4191	10·5398921	8·6285168	9·4789748	4571	10·5210252	9·8521903	10·0188669	381	9·9811331	14
47	9·4605270	4186	10·5394730	8·6293736	9·4794319	4568	10·5205681	9·8520203	10·0189050	381	9·9810950	13
48	9·4609456	4182	10·5390544	8·6302295	9·4798887	4564	10·5201113	9·8518502	10·0189431	382	9·9810569	12
49	9·4613638	4178	10·5386362	8·6310846	9·4803451	4560	10·5196549	9·8516800	10·0189813	382	9·9810187	11
50	9·4617816	4173	10·5382184	8·6319388	9·4808011	4555	10·5191989	9·8515099	10·0190195	382	9·9809805	10
51	9·4621989	4169	10·5378011	8·6327922	9·4812566	4552	10·5187434	9·8513396	10·0190577	383	9·9809423	9
52	9·4626158	4165	10·5373842	8·6336447	9·4817118	4548	10·5182882	9·8511693	10·0190960	383	9·9809040	8
53	9·4630323	4160	10·5369677	8·6344964	9·4821666	4544	10·5178334	9·8509990	10·0191343	384	9·9808657	7
54	9·4634483	4156	10·5365517	8·6353472	9·4826210	4540	10·5173790	9·8508286	10·0191727	384	9·9808273	6
55	9·4638639	4151	10·5361361	8·6361971	9·4830750	4536	10·5169250	9·8506582	10·0192111	384	9·9807889	5
56	9·4642790	4148	10·5357210	8·6370462	9·4835286	4532	10·5164714	9·8504877	10·0192495	385	9·9807505	4
57	9·4646938	4143	10·5353062	8·6378945	9·4839818	4528	10·5160182	9·8503171	10·0192880	385	9·9807120	3
58	9·4651081	4138	10·5348919	8·6387419	9·4844346	4524	10·5155654	9·8501465	10·0193265	386	9·9806735	2
59	9·4655219	4134	10·5344781	8·6395884	9·4848870	4520	10·5151130	9·8499759	10·0193651	386	9·9806349	1
60	9·4659353		10·5340647	8·6404342	9·4853390		10·5146610	9·8498052	10·0194037		9·9805963	0
'	Cosine.	Dif.	Secant.	Covers.	Cotang.	Dif.	Tang.	Verseds.	Cosec.	D.	Sine.	'

′	Sine.	Dif.	Covers	Cosec.	Tang.	Cotang.	Secant.	Vers.	D.	Cosine	′
0	2923717	2782	7076283	3·4203036	3057307	3·2708526	1·0456918	0436952	851	9563048	60
1	2926499	2781	7073501	3·4170526	3060488	3·2674529	1·0457848	0437803	852	9562197	59
2	2929280	2781	7070720	3·4138080	3063670	3·2640596	1·0458780	0438655	853	9561345	58
3	2932061	2781	7067939	3·4105699	3066852	3·2606728	1·0459712	0439508	853	9560492	57
4	2934842	2781	7065158	3·4073382	3070034	3·2572924	1·0460646	0440361	854	9559639	56
5	2937623	2780	7062377	3·4041130	3073218	3·2539184	1·0461581	0441215	855	9558785	55
6	2940403	2780	7059597	3·4008941	3076402	3·2505508	1·0462516	0442070	856	9557930	54
7	2943183	2780	7056817	3·3976816	3079586	3·2471895	1·0463453	0442926	856	9557074	53
8	2945963	2780	7054037	3·3944754	3082771	3·2438346	1·0464391	0443782	857	9556218	52
9	2948743	2779	7051257	3·3912755	3085957	3·2404860	1·0465330	0444639	859	9555361	51
10	2951522	2780	7048478	3·3880820	3089143	3·2371438	1·0466270	0445498	859	9554502	50
11	2954302	2779	7045698	3·3848948	3092330	3·2338078	1·0467211	0446357	859	9553643	49
12	2957081	2778	7042919	3·3817138	3095517	3·2304780	1·0468153	0447216	861	9552784	48
13	2959859	2779	7040141	3·3785391	3098705	3·2271546	1·0469096	0448077	861	9551923	47
14	2962638	2778	7037362	3·3753707	3101893	3·2238373	1·0470040	0448938	863	9551062	46
15	2965416	2778	7034584	3·3722084	3105083	3·2205263	1·0470986	0449801	863	9550199	45
16	2968194	2777	7031806	3·3690524	3108272	3·2172215	1·0471932	0450664	863	9549336	44
17	2970971	2778	7029029	3·3659026	3111462	3·2139228	1·0472879	0451527	865	9548473	43
18	2973749	2777	7026251	3·3627589	3114653	3·2106304	1·0473828	0452392	865	9547608	42
19	2976526	2777	7023474	3·3596214	3117845	3·2073440	1·0474777	0453257	867	9546743	41
20	2979303	2776	7020697	3·3564900	3121036	3·2040638	1·0475728	0454124	867	9545876	40
21	2982079	2777	7017921	3·3533647	3124229	3·2007897	1·0476679	0454991	868	9545009	39
22	2984856	2776	7015144	3·3502455	3127422	3·1975217	1·0477632	0455859	868	9544141	38
23	2987632	2776	7012368	3·3471324	3130616	3·1942598	1·0478586	0456727	870	9543273	37
24	2990408	2776	7009592	3·3440254	3133810	3·1910039	1·0479540	0457597	870	9542403	36
25	2993184	2775	7006816	3·3409244	3137005	3·1877540	1·0480496	0458467	871	9541533	35
26	2995959	2775	7004041	3·3378294	3140200	3·1845102	1·0481453	0459338	872	9540662	34
27	2998734	2775	7001266	3·3347405	3143396	3·1812724	1·0482411	0460210	873	9539790	33
28	3001509	2775	6998491	3·3316575	3146593	3·1780406	1·0483370	0461083	873	9538917	32
29	3004284	2774	6995716	3·3285805	3149790	3·1748147	1·0484330	0461956	874	9538044	31
30	3007058	2774	6992942	3·3255095	3152988	3·1715948	1·0485291	0462830	876	9537170	30
31	3009832	2774	6990168	3·3224444	3156186	3·1683808	1·0486253	0463706	876	9536294	29
32	3012606	2774	6987394	3·3193853	3159385	3·1651728	1·0487217	0464582	876	9535418	28
33	3015380	2773	6984620	3·3163320	3162585	3·1619706	1·0488181	0465458	878	9534542	27
34	3018153	2773	6981847	3·3132847	3165785	3·1587744	1·0489146	0466336	878	9533664	26
35	3020926	2773	6979074	3·3102432	3168986	3·1555840	1·0490113	0467214	879	9532786	25
36	3023699	2772	6976301	3·3072076	3172187	3·1523994	1·0491080	0468093	880	9531907	24
37	3026471	2773	6973529	3·3041778	3175389	3·1492207	1·0492049	0468973	881	9531027	23
38	3029244	2772	6970756	3·3011539	3178591	3·1460478	1·0493019	0469854	882	9530146	22
39	3032016	2772	6967984	3·2981357	3181794	3·1428807	1·0493989	0470736	882	9529264	21
40	3034788	2771	6965212	3·2951234	3184998	3·1397194	1·0494961	0471618	883	9528382	20
41	3037559	2772	6962441	3·2921168	3188202	3·1365639	1·0495934	0472501	884	9527499	19
42	3040331	2771	6959669	3·2891160	3191407	3·1334141	1·0496908	0473385	885	9526615	18
43	3043102	2770	6956898	3·2861209	3194613	3·1302701	1·0497883	0474270	886	9525730	17
44	3045872	2771	6954128	3·2831316	3197819	3·1271317	1·0498859	0475156	886	9524844	16
45	3048643	2770	6951357	3·2801479	3201025	3·1239991	1·0499836	0476042	887	9523958	15
46	3051413	2770	6948587	3·2771700	3204232	3·1208722	1·0500815	0476929	888	9523071	14
47	3054183	2770	6945817	3·2741977	3207440	3·1177509	1·0501794	0477817	889	9522183	13
48	3056953	2770	6943047	3·2712311	3210649	3·1146353	1·0502774	0478706	890	9521294	12
49	3059723	2769	6940277	3·2682702	3213858	3·1115254	1·0503756	0479596	890	9520404	11
50	3062492	2769	6937508	3·2653149	3217067	3·1084210	1·0504738	0480486	891	9519514	10
51	3065261	2769	6934739	3·2623652	3220278	3·1053223	1·0505722	0481377	892	9518623	9
52	3068030	2768	6931970	3·2594211	3223489	3·1022291	1·0506706	0482269	893	9517731	8
53	3070798	2768	6929202	3·2564825	3226700	3·0991416	1·0507692	0483162	894	9516838	7
54	3073566	2768	6926434	3·2535496	3229912	3·0960596	1·0508679	0484056	894	9515944	6
55	3076334	2768	6923666	3·2506222	3233125	3·0929831	1·0509667	0484950	896	9515050	5
56	3079102	2767	6920898	3·2477003	3236338	3·0899122	1·0510656	0485846	896	9514154	4
57	3081869	2767	6918131	3·2447840	3239552	3·0868468	1·0511646	0486742	897	9513258	3
58	3084636	2767	6915364	3·2418732	3242766	3·0837869	1·0512637	0487639	897	9512361	2
59	3087403	2767	6912597	3·2389678	3245981	3·0807325	1·0513629	0488536	899	9511464	1
60	3090170		6909830	3·2360680	3249197	3·0776835	1·0514622	0489435		9510565	0

′	Cosine	Dif.	Vers.	Secant.	Cotan.	Tang.	Cosec.	Covers	D.	Sine.	′

′	Sine.	Dif.	Cosec.	Verseds.	Tang.	Dif.	Cotang.	Covers.	Secant.	D.	Cosine.	′
0	9·4659353	4130	10·5340647	8·6404342	9·4853390	4517	10·5146610	9·8498052	10·0194037	386	9·9805963	60
1	9·4663483	4126	10·5336517	8·6412791	9·4857907	4512	10·5142093	9·8496344	10·0194423	387	9·9805577	59
2	9·4667609	4121	10·5332391	8·6421231	9·4862419	4509	10·5137581	9·8494636	10·0194810	387	9·9805190	58
3	9·4671730	4118	10·5328270	8·6429663	9·4866928	4505	10·5133072	9·8492928	10·0195197	388	9·9804803	57
4	9·4675848	4112	10·5324152	8·6438087	9·4871433	4500	10·5128567	9·8491219	10·0195585	388	9·9804415	56
5	9·4679960	4109	10·5320040	8·6446502	9·4875933	4497	10·5124067	9·8489509	10·0195973	388	9·9804027	55
6	9·4684069	4104	10·5315931	8·6454909	9·4880430	4494	10·5119570	9·8487799	10·0196361	389	9·9803639	54
7	9·4688173	4100	10·5311827	8·6463308	9·4884924	4489	10·5115076	9·8486088	10·0196750	390	9·9803250	53
8	9·4692273	4096	10·5307727	8·6471698	9·4889413	4485	10·5110587	9·8484377	10·0197140	389	9·9802860	52
9	9·4696369	4092	10·5303631	8·6480080	9·4893898	4482	10·5106102	9·8482665	10·0197529	390	9·9802471	51
10	9·4700461	4087	10·5299539	8·6488454	9·4898380	4478	10·5101620	9·8480953	10·0197919	391	9·9802081	50
11	9·4704548	4083	10·5295452	8·6496820	9·4902858	4474	10·5097142	9·8479240	10·0198310	391	9·9801690	49
12	9·4708631	4079	10·5291369	8·6505177	9·4907332	4470	10·5092668	9·8477527	10·0198701	391	9·9801299	48
13	9·4712710	4075	10·5287290	8·6513526	9·4911802	4467	10·5088198	9·8475813	10·0199092	392	9·9800908	47
14	9·4716785	4071	10·5283215	8·6521867	9·4916269	4462	10·5083731	9·8474099	10·0199484	392	9·9800516	46
15	9·4720856	4066	10·5279144	8·6530200	9·4920731	4459	10·5079269	9·8472384	10·0199876	392	9·9800124	45
16	9·4724922	4063	10·5275078	8·6538524	9·4925190	4456	10·5074810	9·8470669	10·0200268	393	9·9799732	44
17	9·4728985	4058	10·5271015	8·6546841	9·4929646	4451	10·5070354	9·8468953	10·0200661	393	9·9799339	43
18	9·4733043	4054	10·5266957	8·6555149	9·4934097	4448	10·5065903	9·8467237	10·0201054	394	9·9798946	42
19	9·4737097	4049	10·5262903	8·6563449	9·4938545	4443	10·5061455	9·8465520	10·0201448	394	9·9798552	41
20	9·4741146	4046	10·5258854	8·6571741	9·4942988	4441	10·5057012	9·8463802	10·0201842	394	9·9798158	40
21	9·4745192	4042	10·5254808	8·6580025	9·4947429	4436	10·5052571	9·8462084	10·0202236	395	9·9797764	39
22	9·4749234	4037	10·5250766	8·6588301	9·4951865	4433	10·5048135	9·8460366	10·0202631	396	9·9797369	38
23	9·4753271	4033	10·5246729	8·6596569	9·4956298	4429	10·5043702	9·8458647	10·0203027	395	9·9796973	37
24	9·4757304	4030	10·5242696	8·6604829	9·4960727	4425	10·5039273	9·8456927	10·0203422	396	9·9796578	36
25	9·4761334	4025	10·5238666	8·6613081	9·4965152	4422	10·5034848	9·8455207	10·0203818	397	9·9796182	35
26	9·4765359	4021	10·5234641	8·6621324	9·4969574	4417	10·5030426	9·8453487	10·0204215	397	9·9795785	34
27	9·4769380	4016	10·5230620	8·6629560	9·4973991	4415	10·5026009	9·8451766	10·0204612	397	9·9795388	33
28	9·4773396	4013	10·5226604	8·6637788	9·4978406	4410	10·5021594	9·8450044	10·0205009	398	9·9794991	32
29	9·4777409	4009	10·5222591	8·6646008	9·4982816	4407	10·5017184	9·8448322	10·0205407	398	9·9794593	31
30	9·4781418	4005	10·5218582	8·6654220	9·4987223	4403	10·5012777	9·8446599	10·0205805	399	9·9794195	30
31	9·4785423	4000	10·5214577	8·6662424	9·4991626	4400	10·5008374	9·8444876	10·0206204	399	9·9793796	29
32	9·4789423	3997	10·5210577	8·6670620	9·4996026	4396	10·5003974	9·8443152	10·0206602	400	9·9793398	28
33	9·4793420	3992	10·5206580	8·6678808	9·5000422	4392	10·4999578	9·8441428	10·0207002	399	9·9792998	27
34	9·4797412	3989	10·5202588	8·6686988	9·5004814	4389	10·4995186	9·8439703	10·0207401	401	9·9792599	26
35	9·4801401	3984	10·5198599	8·6695160	9·5009203	4385	10·4990797	9·8437978	10·0207802	401	9·9792198	25
36	9·4805385	3981	10·5194615	8·6703324	9·5013588	4381	10·4986412	9·8436252	10·0208202	401	9·9791798	24
37	9·4809366	3976	10·5190634	8·6711481	9·5017969	4378	10·4982031	9·8434526	10·0208603	401	9·9791397	23
38	9·4813342	3973	10·5186658	8·6719630	9·5022347	4374	10·4977653	9·8432799	10·0209004	402	9·9790996	22
39	9·4817315	3968	10·5182685	8·6727771	9·5026721	4371	10·4973279	9·8431072	10·0209406	402	9·9790594	21
40	9·4821283	3965	10·5178717	8·6735904	9·5031092	4367	10·4968008	9·8429344	10·0209808	403	9·9790192	20
41	9·4825248	3960	10·5174752	8·6744029	9·5035459	4363	10·4964541	9·8427615	10·0210211	403	9·9789789	19
42	9·4829208	3957	10·5170792	8·6752147	9·5039822	4360	10·4960178	9·8425886	10·0210614	403	9·9789386	18
43	9·4833165	3952	10·5166835	8·6760256	9·5044182	4356	10·4955818	9·8424157	10·0211017	404	9·9788983	17
44	9·4837117	3949	10·5162883	8·6768358	9·5048538	4353	10·4951462	9·8422427	10·0211421	404	9·9788579	16
45	9·4841066	3944	10·5158934	8·6776453	9·5052891	4349	10·4947109	9·8420696	10·0211825	405	9·9788175	15
46	9·4845010	3941	10·5154990	8·6784539	9·5057240	4346	10·4942760	9·8418965	10·0212230	405	9·9787770	14
47	9·4848951	3937	10·5151049	8·6792618	9·5061586	4342	10·4938414	9·8417233	10·0212635	405	9·9787365	13
48	9·4852888	3932	10·5147112	8·6800689	9·5065928	4339	10·4934072	9·8415501	10·0213040	406	9·9786960	12
49	9·4856820	3929	10·5143180	8·6808753	9·5070267	4335	10·4929733	9·8413768	10·0213446	406	9·9786554	11
50	9·4860749	3925	10·5139251	8·6816809	9·5074602	4331	10·4925398	9·8412035	10·0213852	407	9·9786148	10
51	9·4864674	3921	10·5135326	8·6824857	9·5078933	4328	10·4921067	9·8410301	10·0214259	407	9·9785741	9
52	9·4868595	3917	10·5131405	8·6832897	9·5083261	4325	10·4916739	9·8408567	10·0214666	407	9·9785334	8
53	9·4872512	3914	10·5127488	8·6840930	9·5087586	4321	10·4912414	9·8406832	10·0215073	408	9·9784927	7
54	9·4876426	3909	10·5123574	8·6848956	9·5091907	4317	10·4908093	9·8405097	10·0215481	408	9·9784519	6
55	9·4880335	3905	10·5119665	8·6856973	9·5096224	4315	10·4903776	9·8403361	10·0215889	409	9·9784111	5
56	9·4884240	3902	10·5115760	8·6864934	9·5100539	4310	10·4899461	9·8401625	10·0216298	409	9·9783702	4
57	9·4888142	3898	10·5111858	8·6872986	9·5104849	4307	10·4895151	9·8399888	10·0216707	410	9·9783293	3
58	9·4892040	3894	10·5107960	8·6880981	9·5109156	4304	10·4890844	9·8398150	10·0217117	409	9·9782883	2
59	9·4895934	3890	10·5104066	8·6888969	9·5113460	4300	10·4886540	9·8396412	10·0217526	411	9·9782474	1
60	9·4899824		10·5100176	8·6896949	9·5117760		10·4882240	9·8394674	10·0217937		9·9782063	0

| ′ | Cosine. | Dif. | Secant. | Covers. | Cotang. | Dif. | Tang. | Verseds. | Cosec. | D. | Sine. | |

′	Sine.	Dif.	Covers	Cosec.	Tang.	Cotang.	Secant.	Vers.	D.	Cosine	′
0	3090170	2766	6909830	3·2360680	3249197	3·0776835	1·0514622	0489435	899	9510565	60
1	3092936	2766	6907064	3·2331736	3252413	3·0746400	1·0515617	0490334	900	9509666	59
2	3095702	2766	6904298	3·2302846	3255630	3·0716020	1·0516612	0491234	901	9508766	58
3	3098468	2766	6901532	3·2274011	3258848	3·0685694	1·0517608	0492135	902	9507865	57
4	3101234	2765	6898766	3·2245230	3262066	3·0655421	1·0518606	0493037	902	9506963	56
5	3103999	2765	6896001	3·2216503	3265284	3·0625203	1·0519605	0493939	904	9506061	55
6	3106764	2765	6893236	3·2187830	3268504	3·0595038	1·0520604	0494843	904	9505157	54
7	3109529	2765	6890471	3·2159210	3271724	3·0564928	1·0521605	0495747	905	9504253	53
8	3112294	2764	6887706	3·2130644	3274944	3·0534870	1·0522607	0496652	905	9503348	52
9	3115058	2764	6884942	3·2102132	3278165	3·0504866	1·0523610	0497557	907	9502443	51
10	3117822	2764	6882178	3·2073673	3281387	3·0474915	1·0524614	0498464	907	9501536	50
11	3120586	2763	6879414	3·2045266	3284610	3·0445018	1·0525619	0499371	908	9500629	49
12	3123349	2763	6876651	3·2016913	3287833	3·0415173	1·0526625	0500279	909	9499721	48
13	3126112	2763	6873888	3·1988613	3291056	3·0385381	1·0527633	0501188	910	9498812	47
14	3128875	2763	6871125	3·1960365	3294281	3·0355641	1·0528641	0502098	911	9497902	46
15	3131638	2762	6868362	3·1932170	3297505	3·0325954	1·0529651	0503009	911	9496991	45
16	3134400	2763	6865600	3·1904028	3300731	3·0296320	1·0530661	0503920	912	9496080	44
17	3137163	2762	6862837	3·1875937	3303957	3·0266737	1·0531673	0504832	913	9495168	43
18	3139925	2761	6860075	3·1847899	3307184	3·0237207	1·0532686	0505745	914	9494255	42
19	3142686	2762	6857314	3·1819913	3310411	3·0207728	1·0533699	0506650	915	9493341	41
20	3145448	2761	6854552	3·1791978	3313639	3·0178301	1·0534714	0507574	915	9492426	40
21	3148209	2760	6851791	3·1764095	3316868	3·0148926	1·0535730	0508489	916	9491511	39
22	3150969	2761	6849031	3·1736264	3320097	3·0119603	1·0536747	0509405	917	9490595	38
23	3153730	2760	6846270	3·1708484	3323327	3·0090330	1·0537765	0510322	918	9489678	37
24	3156490	2760	6843510	3·1680756	3326557	3·0061109	1·0538785	0511240	918	9488760	36
25	3159250	2760	6840750	3·1653078	3329788	3·0031939	1·0539805	0512158	920	9487842	35
26	3162010	2760	6837990	3·1625452	3333020	3·0002820	1·0540826	0513078	920	9486922	34
27	3164770	2759	6835230	3·1597876	3336252	2·9973751	1·0541849	0513998	921	9486002	33
28	3167529	2759	6832471	3·1570351	3339485	2·9944734	1·0542873	0514919	922	9485081	32
29	3170288	2759	6829712	3·1542877	3342719	2·9915766	1·0543897	0515841	922	9484159	31
30	3173047	2758	6826953	3·1515453	3345953	2·9886850	1·0544923	0516763	924	9483237	30
31	3175805	2758	6824195	3·1488079	3349188	2·9857983	1·0545950	0517687	924	9482313	29
32	3178563	2758	6821437	3·1460756	3352424	2·9829167	1·0546978	0518611	925	9481389	28
33	3181321	2758	6818679	3·1433483	3355660	2·9800400	1·0548007	0519536	926	9480464	27
34	3184079	2757	6815921	3·1406259	3358896	2·9771683	1·0549037	0520462	926	9479538	26
35	3186836	2757	6813164	3·1379086	3362134	2·9743016	1·0550068	0521388	928	9478612	25
36	3189593	2757	6810407	3·1351962	3365372	2·9714399	1·0551101	0522316	928	9477684	24
37	3192350	2756	6807650	3·1324887	3368610	2·9685831	1·0552134	0523244	929	9476756	23
38	3195106	2757	6804894	3·1297862	3371850	2·9657312	1·0553169	0524173	930	9475327	22
39	3197863	2756	6802137	3·1270886	3375090	2·9628842	1·0554204	0525103	931	9474897	21
40	3200619	2756	6799381	3·1243959	3378330	2·9600422	1·0555241	0526034	931	9473966	20
41	3203374	2756	6796626	3·1217081	3381571	2·9572050	1·0556279	0526965	932	9473035	19
42	3206130	2755	6793870	3·1190252	3384813	2·9543727	1·0557318	0527897	933	9472103	18
43	3208885	2755	6791115	3·1163472	3388056	2·9515453	1·0558358	0528830	934	9471170	17
44	3211640	2755	6788360	3·1136740	3391299	2·9487227	1·0559399	0529764	935	9470236	16
45	3214395	2754	6785605	3·1110057	3394543	2·9459050	1·0560441	0530699	935	9469301	15
46	3217149	2754	6782851	3·1083422	3397787	2·9430921	1·0561485	0531634	936	9468366	14
47	3219903	2754	6780097	3·1056835	3401032	2·9402840	1·0562529	0532570	937	9467430	13
48	3222657	2754	6777343	3·1030296	3404278	2·9374807	1·0563575	0533507	938	9466493	12
49	3225411	2753	6774589	3·1003805	3407524	2·9346822	1·0564621	0534445	930	9465555	11
50	3228164	2753	6771836	3·0977363	3410771	2·9318885	1·0565669	0535384	930	9464616	10
51	3230917	2753	6769083	3·0950967	3414019	2·9290995	1·0566718	0536323	941	9463677	9
52	3233670	2752	6766330	3·0924620	3417267	2·9263152	1·0567768	0537264	941	9462736	8
53	3236422	2752	6763578	3·0898319	3420516	2·9235358	1·0568819	0538205	941	9461795	7
54	3239174	2752	6760826	3·0872066	3423765	2·9207610	1·0569871	0539146	943	9460854	6
55	3241926	2752	6758074	3·0845860	3427015	2·9179909	1·0570924	0540089	943	9459911	5
56	3244678	2751	6755322	3·0819702	3430266	2·9152256	1·0571978	0541032	945	9458968	4
57	3247429	2751	6752571	3·0793590	3433518	2·9124649	1·0573034	0541977	945	9458023	3
58	3250180	2751	6749820	3·0767525	3436770	2·9097089	1·0574090	0542922	946	9457078	2
59	3252931	2751	6747069	3·0741507	3440023	2·9069576	1·0575148	0543868	946	9456132	1
60	3255682		6744318	3·0715535	3443276	2·9042109	1·0576207	0544814		9455186	0

| ′ | Cosine | Dif | Vers. | Secant. | Cotan. | Tang. | Cosec. | Covers | D. | Sine. | ′ |

71 Deg.

′	Sine.	Dif.	Cosec.	Verseds.	Tang.	Dif.	Cotang.	Covers.	Secant.	D.	Cosine.	′
0	9·4899824	3886	10·5100176	8·6896949	9·5117760	4297	10·4882240	9·8394674	10·0217937	410	9·9782063	60
1	9·4903710	3882	10·5096290	8·6904921	9·5122057	4294	10·4877943	9·8392935	10·0218347	412	9·9781653	59
2	9·4907592	3879	10·5092408	8·6912886	9·5126351	4290	10·4873649	9·8391195	10·0218759	411	9·9781241	58
3	9·4911471	3874	10·5088529	8·6920844	9·5130641	4286	10·4869359	9·8389455	10·0219170	412	9·9780830	57
4	9·4915345	3871	10·5084655	8·6928794	9·5134927	4283	10·4865073	9·8387714	10·0219582	412	9·9780418	56
5	9·4919216	3867	10·5080784	8·6936736	9·5139210	4280	10·4860790	9·8385973	10·0219994	413	9·9780006	55
6	9·4923083	3863	10·5076917	8·6944672	9·5143490	4276	10·4856510	9·8384231	10·0220407	413	9·9779593	54
7	9·4926946	3860	10·5073054	8·6952599	9·5147766	4273	10·4852234	9·8382489	10·0220820	414	9·9779180	53
8	9·4930806	3855	10·5069194	8·6960520	9·5152030	4270	10·4847961	9·8380746	10·0221234	413	9·9778766	52
9	9·4934661	3852	10·5065339	8·6968432	9·5156309	4266	10·4843691	9·8379003	10·0221647	415	9·9778353	51
10	9·4938513	3848	10·5061487	8·6976338	9·5160575	4263	10·4839425	9·8377259	10·0222062	415	9·9777938	50
11	9·4942361	3844	10·5057639	8·6984236	9·5164838	4259	10·4835162	9·8375515	10·0222477	415	9·9777523	49
12	9·4946205	3841	10·5053795	8·6992127	9·5169097	4256	10·4830903	9·8373770	10·0222892	415	9·9777108	48
13	9·4950046	3837	10·5049954	8·7000010	9·5173353	4253	10·4826647	9·8372024	10·0223307	416	9·9776693	47
14	9·4953883	3833	10·5046117	8·7007886	9·5177606	4249	10·4822394	9·8370278	10·0223723	417	9·9776277	46
15	9·4957716	3829	10·5042284	8·7015755	9·5181855	4246	10·4818145	9·8368532	10·0224140	416	9·9775860	45
16	9·4961545	3825	10·5038455	8·7023617	9·5186101	4243	10·4813899	9·8366785	10·0224556	418	9·9775444	44
17	9·4965370	3822	10·5034630	8·7031471	9·5190344	4239	10·4809656	9·8365037	10·0224974	417	9·9775026	43
18	9·4969192	3818	10·5030808	8·7039318	9·5194583	4236	10·4805417	9·8363289	10·0225391	418	9·9774609	42
19	9·4973010	3814	10·5026990	8·7047158	9·5198819	4233	10·4801181	9·8361540	10·0225809	419	9·9774191	41
20	9·4976824	3811	10·5023176	8·7054990	9·5203052	4230	10·4796948	9·8359791	10·0226228	418	9·9773772	40
21	9·4980635	3807	10·5019365	8·7062815	9·5207282	4226	10·4792718	9·8358041	10·0226646	420	9·9773354	39
22	9·4984442	3803	10·5015558	8·7070633	9·5211508	4222	10·4788492	9·8356291	10·0227066	419	9·9772934	38
23	9·4988245	3800	10·5011755	8·7078444	9·5215730	4220	10·4784270	9·8354540	10·0227485	420	9·9772515	37
24	9·4992045	3795	10·5007955	8·7086247	9·5219950	4216	10·4780050	9·8352789	10·0227905	421	9·9772095	36
25	9·4995840	3793	10·5004160	8·7094044	9·5224166	4213	10·4775834	9·8351037	10·0228326	421	9·9771674	35
26	9·4999633	3788	10·5000367	8·7101833	9·5228379	4210	10·4771621	9·8349285	10·0228747	421	9·9771253	34
27	9·5003421	3785	10·4996579	8·7109615	9·5232589	4206	10·4767411	9·8347532	10·0229168	422	9·9770832	33
28	9·5007206	3781	10·4992794	8·7117390	9·5236795	4204	10·4763205	9·8345778	10·0229590	422	9·9770410	32
29	9·5010987	3777	10·4989013	8·7125157	9·5240999	4200	10·4759001	9·8344024	10·0230012	423	9·9769988	31
30	9·5014764	3774	10·4985236	8·7132918	9·5245199	4196	10·4754801	9·8342269	10·0230434	423	9·9769566	30
31	9·5018538	3770	10·4981462	8·7140671	9·5249395	4194	10·4750605	9·8340514	10·0230857	423	9·9769143	29
32	9·5022308	3767	10·4977692	8·7148418	9·5253589	4190	10·4746411	9·8338759	10·0231280	424	9·9768720	28
33	9·5026075	3763	10·4973925	8·7156157	9·5257779	4187	10·4742221	9·8337002	10·0231704	424	9·9768296	27
34	9·5029838	3759	10·4970162	8·7163889	9·5261966	4184	10·4738034	9·8335246	10·0232128	425	9·9767872	26
35	9·5033597	3756	10·4966403	8·7171614	9·5266150	4181	10·4733850	9·8333488	10·0232553	425	9·9767447	25
36	9·5037353	3752	10·4962647	8·7179332	9·5270331	4177	10·4729669	9·8331731	10·0232978	425	9·9767022	24
37	9·5041105	3748	10·4958895	8·7187044	9·5274508	4174	10·4725492	9·8329972	10·0233403	426	9·9766597	23
38	9·5044853	3745	10·4955147	8·7194748	9·5278682	4171	10·4721318	9·8328213	10·0233829	426	9·9766171	22
39	9·5048598	3741	10·4951402	8·7202445	9·5282853	4168	10·4717147	9·8326454	10·0234255	427	9·9765745	21
40	9·5052339	3738	10·4947661	8·7210135	9·5287021	4165	10·4712979	9·8324694	10·0234682	427	9·9765318	20
41	9·5056077	3734	10·4943923	8·7217818	9·5291186	4161	10·4708814	9·8322933	10·0235109	427	9·9764891	19
42	9·5059811	3731	10·4940189	8·7225494	9·5295347	4158	10·4704653	9·8321172	10·0235536	428	9·9764464	18
43	9·5063542	3727	10·4936458	8·7233163	9·5299505	4156	10·4700495	9·8319411	10·0235964	428	9·9764036	17
44	9·5067269	3723	10·4932731	8·7240825	9·5303661	4152	10·4696339	9·8317649	10·0236392	429	9·9763608	16
45	9·5070992	3720	10·4929008	8·7248480	9·5307813	4148	10·4692187	9·8315886	10·0236821	429	9·9763179	15
46	9·5074712	3716	10·4925288	8·7256129	9·5311961	4146	10·4688039	9·8314123	10·0237250	429	9·9762750	14
47	9·5078428	3713	10·4921572	8·7263770	9·5316107	4143	10·4683893	9·8312359	10·0237679	430	9·9762320	13
48	9·5082141	3709	10·4917859	8·7271404	9·5320250	4139	10·4679750	9·8310595	10·0238109	430	9·9761891	12
49	9·5085850	3706	10·4914150	8·7279032	9·5324389	4137	10·4675611	9·8308830	10·0238539	431	9·9761461	11
50	9·5089556	3702	10·4910444	8·7286653	9·5328526	4133	10·4671474	9·8307064	10·0238970	431	9·9761030	10
51	9·5093258	3698	10·4906742	8·7294267	9·5332659	4130	10·4667341	9·8305299	10·0239401	432	9·9760599	9
52	9·5096956	3695	10·4903044	8·7301874	9·5336789	4127	10·4663211	9·8303532	10·0239833	431	9·9760170	8
53	9·5100651	3692	10·4899349	8·7309474	9·5340916	4124	10·4659084	9·8301765	10·0240264	433	9·9759736	7
54	9·5104343	3688	10·4895657	8·7317067	9·5345040	4121	10·4654960	9·8299997	10·0240697	433	9·9759303	6
55	9·5108031	3685	10·4891969	8·7324654	9·5349161	4117	10·4650839	9·8298229	10·0241130	433	9·9758870	5
56	9·5111716	3681	10·4888284	8·7332233	9·5353278	4115	10·4646722	9·8296461	10·0241563	433	9·9758437	4
57	9·5115397	3677	10·4884603	8·7339806	9·5357393	4112	10·4642607	9·8294692	10·0241996	434	9·9758004	3
58	9·5119074	3675	10·4880926	8·7347373	9·5361505	4108	10·4638495	9·8292922	10·0242430	434	9·9757570	2
59	9·5122749	3670	10·4877251	8·7354932	9·5365613	4106	10·4634387	9·8291152	10·0242865	435	9·9757135	1
60	9·5126419		10·4873581	8·7362485	9·5369719		10·4630281	9·8289381	10·0243299	434	9·9756701	0

′	Cosine.	Dif.	Secant.	Covers.	Cotang.	Dif.	Tang.	Verseds.	Cosec.	D.	Sine.	′

′	Sine	Dif.	Covers	Cosec.	Tang.	Cotang.	Secant.	Vers.	D.	Cosine	′
0	3255682	2750	6744318	3·0715535	3443276	2·9042109	1·0576207	0544814	948	9455186	60
1	3258432	2750	6741568	3·0689610	3446530	2·9014688	1·0577267	0545762	948	9454238	59
2	3261182	2750	6738818	3·0663731	3449785	2·8987314	1·0578328	0546710	949	9453290	58
3	3263932	2749	6736068	3·0637898	3453040	2·8959986	1·0579390	0547659	950	9452341	57
4	3266681	2749	6733319	3·0612111	3456296	2·8932704	1·0580453	0548609	950	9451301	56
5	3269430	2749	6730570	3·0586370	3459553	2·8905467	1·0581517	0549559	952	9450441	55
6	3272179	2749	6727821	3·0560675	3462810	2·8878277	1·0582583	0550511	952	9449489	54
7	3274928	2748	6725072	3·0535026	3466068	2·8851132	1·0583649	0551463	953	9448537	53
8	3277676	2748	6722324	3·0509423	3469327	2·8824033	1·0584717	0552416	954	9447584	52
9	3280424	2748	6719576	3·0483864	3472586	2·8796979	1·0585786	0553370	955	9446630	51
10	3283172	2747	6716828	3·0458352	3475846	2·8769970	1·0586855	0554325	955	9445675	50
11	3285919	2747	6714081	3·0432884	3479107	2·8743007	1·0587926	0555280	956	9444720	49
12	3288666	2747	6711334	3·0407462	3482368	2·8716088	1·0588999	0556236	957	9443764	48
13	3291413	2747	6708587	3·0382084	3485630	2·8689215	1·0590072	0557193	958	9442807	47
14	3294160	2746	6705840	3·0356752	3488893	2·8662386	1·0591146	0558151	959	9441849	46
15	3296906	2747	6703094	3·0331464	3492156	2·8635602	1·0592221	0559110	959	9440890	45
16	3299653	2745	6700347	3·0306221	3495420	2·8608863	1·0593298	0560069	960	9439931	44
17	3302398	2746	6697602	3·0281023	3498685	2·8582168	1·0594376	0561029	961	9438971	43
18	3305144	2745	6694856	3·0255868	3501950	2·8555517	1·0595454	0561990	962	9438010	42
19	3307889	2745	6692111	3·0230759	3505216	2·8528911	1·0596534	0562952	963	9437048	41
20	3310634	2745	6689366	3·0205693	3508483	2·8502349	1·0597615	0563915	963	9436085	40
21	3313379	2744	6686621	3·0180672	3511750	2·8475831	1·0598697	0564878	965	9435122	39
22	3316123	2744	6683877	3·0155694	3515018	2·8449356	1·0599781	0565843	965	9434157	38
23	3318867	2744	6681133	3·0130760	3518287	2·8422926	1·0600865	0566808	965	9433192	37
24	3321611	2744	6678389	3·0105870	3521556	2·8396539	1·0601951	0567773	967	9432227	36
25	3324355	2743	6675645	3·0081024	3524826	2·8370196	1·0603037	0568740	967	9431260	35
26	3327098	2743	6672902	3·0056221	3528096	2·8343896	1·0604125	0569707	969	9430293	34
27	3329841	2743	6670159	3·0031462	3531368	2·8317639	1·0605214	0570676	969	9429324	33
28	3332584	2742	6667416	3·0006746	3534640	2·8291426	1·0606304	0571645	969	9428355	32
29	3335326	2743	6664674	2·9982073	3537912	2·8265256	1·0607395	0572614	971	9427386	31
30	3338069	2741	6661931	2·9957443	3541186	2·8239129	1·0608487	0573585	971	9426415	30
31	3340810	2742	6659190	2·9932856	3544460	2·8213045	1·0609580	0574556	973	9425444	29
32	3343552	2741	6656448	2·9908312	3547734	2·8187003	1·0610675	0575529	973	9424471	28
33	3346293	2741	6653707	2·9883811	3551010	2·8161004	1·0611770	0576502	973	9423498	27
34	3349034	2741	6650966	2·9859352	3554286	2·8135048	1·0612867	0577475	975	9422525	26
35	3351775	2741	6648225	2·9834936	3557562	2·8109134	1·0613965	0578450	975	9421550	25
36	3354516	2740	6645484	2·9810563	3560840	2·8083263	1·0615064	0579425	977	9420575	24
37	3357256	2740	6642744	2·9786231	3564118	2·8057433	1·0616164	0580402	977	9419598	23
38	3359996	2739	6640004	2·9761942	3567397	2·8031646	1·0617265	0581379	977	9418621	22
39	3362735	2740	6637265	2·9737695	3570676	2·8005901	1·0618367	0582356	979	9417644	21
40	3365475	2739	6634525	2·9713490	3573956	2·7980198	1·0619471	0583335	979	9416665	20
41	3368214	2739	6631786	2·9689327	3577237	2·7954537	1·0620575	0584314	981	9415686	19
42	3370953	2738	6629047	2·9665205	3580518	2·7928917	1·0621681	0585295	981	9414705	18
43	3373691	2738	6626309	2·9641125	3583801	2·7903339	1·0622788	0586276	981	9413724	17
44	3376429	2738	6623571	2·9617087	3587083	2·7877802	1·0623896	0587257	983	9412743	16
45	3379167	2738	6620833	2·9593090	3590367	2·7852307	1·0625005	0588240	983	9411760	15
46	3381905	2737	6618095	2·9569135	3593651	2·7826853	1·0626115	0589223	984	9410777	14
47	3384642	2737	6615358	2·9545221	3596936	2·7801440	1·0627227	0590207	985	9409793	13
48	3387379	2737	6612621	2·9521348	3600222	2·7776069	1·0628339	0591192	986	9408808	12
49	3390116	2736	6609884	2·9497516	3603508	2·7750738	1·0629453	0592178	987	9407822	11
50	3392852	2737	6607148	2·9473725	3606795	2·7725448	1·0630568	0593165	987	9406835	10
51	3395589	2736	6604411	2·9449975	3610082	2·7700199	1·0631684	0594152	988	9405848	9
52	3398325	2735	6601675	2·9426265	3613371	2·7674990	1·0632801	0595140	989	9404860	8
53	3401060	2736	6598940	2·9402597	3616660	2·7649822	1·0633919	0596129	990	9403871	7
54	3403796	2735	6596204	2·9378968	3619949	2·7624695	1·0635038	0597119	990	9402881	6
55	3406531	2734	6593469	2·9355380	3623240	2·7599608	1·0636158	0598109	992	9401891	5
56	3409265	2735	6590735	2·9331833	3626531	2·7574561	1·0637280	0599101	992	9400899	4
57	3412000	2734	6588000	2·9308326	3629823	2·7549554	1·0638403	0600093	992	9399907	3
58	3414734	2734	6585266	2·9284858	3633115	2·7524588	1·0639527	0601086	993	9398914	2
59	3417468	2733	6582532	2·9261431	3636408	2·7499661	1·0640652	0602079	993	9397921	1
60	3420201		6579799	2·9238044	3639702	2·7474774	1·0641778	0603074	995	9396926	0

′	Cosine	Dif.	Vers.	Secunt.	Cotan.	Tang.	Cosec.	Covers	D.	Sine.	′

′	Sine.	Dif.	Cosec.	Verseds.	Tang.	Dif.	Cotang.	Covers.	Secant.	D.	Cosine.	′
0	9·5126419	3667	10·4873581	8·7362485	9·5369719	4102	10·4630281	9·8289381	10·0243299	436	9·9756701	60
1	9·5130086	3664	10·4869914	8·7370030	9·5373821	4099	10·4626179	9·8287609	10·0243735	435	9·9756265	59
2	9·5133750	3660	10·4866250	8·7377570	9·5377920	4097	10·4622080	9·8285837	10·0244170	436	9·9755830	58
3	9·5137410	3657	10·4862590	8·7385102	9·5382017	4093	10·4617983	9·8284065	10·0244606	437	9·9755394	57
4	9·5141067	3654	10·4858933	8·7392628	9·5386110	4090	10·4613890	9·8282292	10·0245043	436	9·9754957	56
5	9·5144721	3650	10·4855279	8·7400147	9·5390200	4087	10·4609800	9·8280518	10·0245479	438	9·9754521	55
6	9·5148371	3646	10·4851629	8·7407659	9·5394287	4084	10·4605713	9·8278744	10·0245917	437	9·9754083	54
7	9·5152017	3643	10·4847983	8·7415165	9·5398371	4082	10·4601629	9·8276970	10·0246354	438	9·9753646	53
8	9·5155660	3640	10·4844340	8·7422664	9·5402453	4078	10·4597547	9·8275194	10·0246792	439	9·9753208	52
9	9·5159300	3636	10·4840700	8·7430156	9·5406531	4075	10·4593469	9·8273419	10·0247231	439	9·9752769	51
10	9·5162936	3633	10·4837064	8·7437642	9·5410606	4072	10·4589394	9·8271642	10·0247670	439	9·9752330	50
11	9·5166569	3629	10·4833431	8·7445121	9·5414678	4069	10·4585322	9·8269866	10·0248109	440	9·9751891	49
12	9·5170198	3626	10·4829802	8·7452593	9·5418747	4066	10·4581253	9·8268088	10·0248549	440	9·9751451	48
13	9·5173824	3623	10·4826176	8·7460059	9·5422813	4064	10·4577187	9·8266310	10·0248989	441	9·9751011	47
14	9·5177447	3619	10·4822553	8·7467518	9·5426877	4060	10·4573123	9·8264532	10·0249430	441	9·9750570	46
15	9·5181066	3616	10·4818934	8·7474971	9·5430937	4057	10·4569063	9·8262753	10·0249871	441	9·9750129	45
16	9·5184682	3613	10·4815318	8·7482417	9·5434994	4054	10·4565006	9·8260973	10·0250312	442	9·9749688	44
17	9·5188295	3609	10·4811705	8·7489857	9·5439048	4052	10·4560952	9·8259193	10·0250754	442	9·9749246	43
18	9·5191904	3606	10·4808096	8·7497290	9·5443100	4048	10·4556900	9·8257412	10·0251196	443	9·9748804	42
19	9·5195510	3602	10·4804490	8·7504716	9·5447148	4045	10·4552852	9·8255631	10·0251639	443	9·9748361	41
20	9·5199112	3599	10·4800888	8·7512136	9·5451193	4043	10·4548807	9·8253849	10·0252082	443	9·9747918	40
21	9·5202711	3596	10·4797289	8·7519549	9·5455236	4040	10·4544764	9·8252067	10·0252525	444	9·9747475	39
22	9·5206307	3592	10·4793693	8·7526956	9·5459276	4036	10·4540724	9·8250284	10·0252969	444	9·9747031	38
23	9·5209899	3589	10·4790101	8·7534357	9·5463312	4034	10·4536688	9·8248501	10·0253413	445	9·9746587	37
24	9·5213488	3586	10·4786512	8·7541751	9·5467346	4031	10·4532654	9·8246717	10·0253858	445	9·9746142	36
25	9·5217074	3582	10·4782926	8·7549138	9·5471377	4028	10·4528623	9·8244932	10·0254303	445	9·9745697	35
26	9·5220656	3579	10·4779344	8·7556519	9·5475405	4025	10·4524595	9·8243147	10·0254748	446	9·9745252	34
27	9·5224235	3576	10·4775765	8·7563894	9·5479430	4022	10·4520570	9·8241362	10·0255134	447	9·9744806	33
28	9·5227811	3572	10·4772189	8·7571262	9·5483452	4019	10·4516548	9·8239576	10·0255593	446	9·9744359	32
29	9·5231383	3570	10·4768617	8·7578623	9·5487471	4016	10·4512529	9·8237789	10·0256087	447	9·9743913	31
30	9·5234953	3565	10·4765047	8·7585979	9·5491487	4013	10·4508513	9·8236002	10·0256534	448	9·9743466	30
31	9·5238518	3563	10·4761482	8·7593327	9·5495500	4011	10·4504500	9·8234214	10·0256982	448	9·9743018	29
32	9·5242081	3559	10·4757919	8·7600670	9·5499511	4008	10·4500489	9·8232425	10·0257430	448	9·9742570	28
33	9·5245640	3556	10·4754360	8·7608006	9·5503519	4004	10·4496481	9·8230636	10·0257878	449	9·9742122	27
34	9·5249196	3553	10·4750804	8·7615336	9·5507523	4002	10·4492477	9·8228847	10·0258327	449	9·9741673	26
35	9·5252749	3549	10·4747251	8·7622659	9·5511525	3999	10·4488475	9·8227057	10·0258776	450	9·9741224	25
36	9·5256298	3546	10·4743702	8·7629976	9·5515524	3997	10·4484476	9·8225266	10·0259226	450	9·9740774	24
37	9·5259844	3543	10·4740156	8·7637286	9·5519521	3993	10·4480479	9·8223475	10·0259676	451	9·9740324	23
38	9·5263387	3540	10·4736613	8·7644591	9·5523514	3990	10·4476486	9·8221684	10·0260127	451	9·9739873	22
39	9·5266927	3536	10·4733073	8·7651889	9·5527504	3988	10·4472496	9·8219891	10·0260578	451	9·9739422	21
40	9·5270463	3534	10·4729537	8·7659180	9·5531492	3985	10·4468508	9·8218099	10·0261029	452	9·9738971	20
41	9·5273997	3529	10·4726003	8·7666466	9·5535477	3982	10·4464523	9·8216305	10·0261481	452	9·9738519	19
42	9·5277526	3527	10·4722474	8·7673745	9·5539459	3979	10·4460541	9·8214511	10·0261933	452	9·9738067	18
43	9·5281053	3524	10·4718947	8·7681018	9·5543434	3977	10·4456562	9·8212717	10·0262385	453	9·9737615	17
44	9·5284577	3520	10·4715423	8·7688284	9·5547415	3973	10·4452585	9·8210922	10·0262838	453	9·9737162	16
45	9·5288097	3517	10·4711903	8·7695544	9·5551388	3971	10·4448612	9·8209126	10·0263291	454	9·9736709	15
46	9·5291614	3514	10·4708386	8·7702798	9·5555359	3968	10·4444641	9·8207330	10·0263745	454	9·9736255	14
47	9·5295128	3510	10·4704872	8·7710046	9·5559327	3965	10·4440673	9·8205533	10·0264199	455	9·9735801	13
48	9·5298638	3508	10·4701362	8·7717288	9·5563292	3963	10·4436708	9·8203736	10·0264654	455	9·9735346	12
49	9·5302146	3504	10·4697854	8·7724523	9·5567255	3959	10·4432745	9·8201938	10·0265109	456	9·9734891	11
50	9·5305650	3501	10·4694350	8·7731752	9·5571214	3957	10·4428786	9·8200140	10·0265565	455	9·9734435	10
51	9·5309151	3498	10·4690849	8·7738975	9·5575171	3954	10·4424829	9·8198341	10·0266020	457	9·9733980	9
52	9·5312649	3494	10·4687351	8·7746192	9·5579125	3952	10·4420875	9·8196542	10·0266477	456	9·9733523	8
53	9·5316143	3492	10·4683857	8·7753403	9·5583077	3948	10·4416923	9·8194742	10·0266933	457	9·9733067	7
54	9·5319635	3488	10·4680365	8·7760607	9·5587025	3946	10·4412975	9·8192941	10·0267390	458	9·9732610	6
55	9·5323123	3485	10·4676877	8·7767805	9·5590971	3943	10·4409029	9·8191140	10·0267848	458	9·9732152	5
56	9·5326608	3482	10·4673392	8·7774997	9·5594914	3940	10·4405086	9·8189338	10·0268306	458	9·9731694	4
57	9·5330090	3479	10·4669910	8·7782183	9·5598854	3938	10·4401146	9·8187536	10·0268764	459	9·9731236	3
58	9·5333569	3475	10·4666431	8·7789363	9·5602792	3935	10·4397208	9·8185733	10·0269223	459	9·9730777	2
59	9·5337044	3473	10·4662956	8·7796537	9·5606727	3932	10·4393273	9·8183930	10·0269682	4C0	9·9730318	1
60	9·5340517		10·4659483	8·7803705	9·5610659		10·4389341	9·8182126	10·0270142		9·9729858	0
′	Cosine.	Dif.	Secant.	Covers.	Cotang.	Dif.	Tang.	Verseds.	Cosec.	D.	Sine.	′

′	Sine.	Dif.	Covers	Cosec.	Tang.	Cotang.	Secant.	Vers.	Dif.	Cosine	′
0	3420201	2734	6579799	2·9238044	3639702	2·7474774	1·0641778	0603074	995	9396926	60
1	3422935		6577065	2·9214697	3642997	2·7449927	1·0642905	0604069	996	9395931	59
2	3425668	2733	6574332	2·9191389	3646292	2·7425120	1·0644033	0605065	997	9394935	58
3	3428400	2732	6571600	2·9168121	3649588	2·7400352	1·0645163	0606062	997	9393938	57
4	3431133	2733	6568867	2·9144892	3652885	2·7375623	1·0646294	0607060	998	9392940	56
5	3433865	2732	6566135	2·9121703	3656182	2·7350934	1·0647425	0608058	998	9391942	55
6	3436597	2732	6563403	2·9098553	3659480	2·7326284	1·0648558	0609057	999	9390943	54
7	3439329	2731	6560671	2·9075443	3662779	2·7301674	1·0649693	0610057	1000	9389943	53
8	3442060	2731	6557940	2·9052372	3666079	2·7277102	1·0650828	0611058	1001	9388942	52
9	3444791	2731	6555209	2·9029339	3669379	2·7252569	1·0651964	0612060	1002	9387940	51
10	3447521	2730	6552479	2·9006346	3672680	2·7228076	1·0653102	0613062	1002	9386938	50
11	3450252	2730	6549748	2·8983391	3675981	2·7203620	1·0654240	0614066	1004	9385934	49
12	3452982	2730	6547018	2·8960475	3679284	2·7179204	1·0655380	0615070	1004	9384930	48
13	3455712	2729	6544288	2·8937598	3682587	2·7154826	1·0656521	0616075	1005	9383925	47
14	3458441	2730	6541559	2·8914760	3685890	2·7130487	1·0657663	0617080	1005	9382920	46
15	3461171	2729	6538829	2·8891960	3689195	2·7106186	1·0658807	0618087	1007	9381913	45
16	3463900	2728	6536100	2·8869198	3692500	2·7081923	1·0659951	0619094	1007	9380906	44
17	3466628	2729	6533372	2·8846474	3695806	2·7057699	1·0661097	0620102	1008	9379898	43
18	3469357	2728	6530643	2·8823789	3699112	2·7033513	1·0662243	0621111	1009	9378889	42
19	3472085	2727	6527915	2·8801142	3702420	2·7009364	1·0663391	0622120	1009	9377880	41
20	3474812	2728	6525188	2·8778532	3705728	2·6985254	1·0664540	0623131	1011	9376869	40
21	3477540	2727	6522460	2·8755961	3709036	2·6961181	1·0665690	0624142	1011	9375858	39
22	3480267	2727	6519733	2·8733428	3712346	2·6937147	1·0666842	0625154	1012	9374846	38
23	3482994	2726	6517006	2·8710932	3715656	2·6913149	1·0667994	0626167	1013	9373833	37
24	3485720	2727	6514280	2·8688474	3718967	2·6889190	1·0669148	0627180	1013	9372820	36
25	3488447	2726	6511553	2·8666053	3722278	2·6865267	1·0670302	0628194	1014	9371806	35
26	3491173	2725	6508827	2·8643670	3725590	2·6841383	1·0671458	0629210	1016	9370790	34
27	3493898	2726	6506102	2·8621324	3728903	2·6817535	1·0672615	0630226	1016	9369774	33
28	3496624	2725	6503376	2·8599015	3732217	2·6793725	1·0673774	0631242	1016	9368758	32
29	3499349	2725	6500651	2·8576744	3735532	2·6769951	1·0674933	0632260	1018	9367740	31
30	3502074	2724	6497926	2·8554510	3738847	2·6746215	1·0676094	0633278	1018	9366722	30
31	3504798	2725	6495202	2·8532312	3742163	2·6722516	1·0677255	0634297	1019	9365703	29
32	3507523	2723	6492477	2·8510152	3745479	2·6698853	1·0678418	0635317	1020	9364683	28
33	3510246	2724	6489754	2·8488028	3748797	2·6675227	1·0679582	0636338	1021	9363662	27
34	3512970	2723	6487030	2·8465941	3752115	2·6651638	1·0680747	0637359	1021	9362641	26
35	3515693	2723	6484307	2·8443891	3755433	2·6628085	1·0681914	0638382	1023	9361618	25
36	3518416	2723	6481584	2·8421877	3758753	2·6604569	1·0683081	0639405	1023	9360595	24
37	3521139	2723	6478861	2·8399899	3762073	2·6581089	1·0684250	0640429	1024	9359571	23
38	3523862	2722	6476138	2·8377958	3765394	2·6557645	1·0685420	0641453	1024	9358547	22
39	3526584	2722	6473416	2·8356054	3768716	2·6534238	1·0686591	0642479	1026	9357521	21
40	3529306	2721	6470694	2·8334185	3772038	2·6510867	1·0687763	0643505	1026	9356495	20
41	3532027	2721	6467973	2·8312353	3775361	2·6487531	1·0688936	0644532	1027	9355468	19
42	3534748	2721	6465252	2·8290556	3778685	2·6464232	1·0690110	0645560	1028	9354440	18
43	3537469	2721	6462531	2·8268796	3782010	2·6440969	1·0691286	0646588	1028	9353412	17
44	3540190	2720	6459810	2·8247071	3785335	2·6417741	1·0692463	0647618	1030	9352382	16
45	3542910	2720	6457090	2·8225382	3788661	2·6394549	1·0693641	0648648	1030	9351352	15
46	3545630	2720	6454370	2·8203729	3791988	2·6371392	1·0694820	0649679	1031	9350321	14
47	3548350	2720	6451650	2·8182111	3795315	2·6348271	1·0696000	0650711	1032	9349289	13
48	3551070	2719	6448930	2·8160529	3798644	2·6325186	1·0697182	0651743	1032	9348257	12
49	3553789	2719	6446211	2·8138982	3801973	2·6302136	1·0698364	0652777	1034	9347223	11
50	3556508	2718	6443492	2·8117471	3805302	2·6279121	1·0699548	0653811	1034	9346189	10
51	3559226	2718	6440774	2·8095995	3808633	2·6256141	1·0700733	0654846	1035	9345154	9
52	3561944	2718	6438056	2·8074554	3811964	2·6233196	1·0701919	0655881	1035	9344119	8
53	3564662	2718	6435338	2·8053148	3815296	2·6210286	1·0703106	0656918	1037	9343082	7
54	3567380	2717	6432620	2·8031777	3818629	2·6187411	1·0704295	0657955	1037	9342045	6
55	3570097	2717	6429903	2·8010441	3821962	2·6164571	1·0705484	0658993	1038	9341007	5
56	3572814	2717	6427186	2·7989140	3825296	2·6141766	1·0706675	0660032	1039	9339968	4
57	3575531	2717	6424469	2·7967873	3828631	2·6118995	1·0707867	0661072	1040	9338928	3
58	3578248	2716	6421752	2·7946641	3831967	2·6096259	1·0709060	0662112	1040	9337888	2
59	3580964	2715	6419036	2·7925444	3835303	2·6073558	1·0710254	0663154	·042	9336846	1
60	3583679		6416321	2·7904281	3838640	2·6050891	1·0711450	0664196	1042	9335804	0

′	Cosine	Dif.	Vers.	Secant.	Cotan.	Tang.	Cosec.	Covers	Dif.	Sine.	′

′	Sine.	Dif.	Cosec.	Verseds.	Tang.	Dif.	Cotang.	Covers.	Secant.	D.	Cosine.	′
0	9·5340517	3469	10·4659483	8·7803705	9·5610059	3929	10·4389341	9·8182126	10·0270142	460	9·9729858	60
1	9·5343986	3466	10·4656014	8·7810866	9·5614588	3927	10·4385412	9·8180322	10·0270602	460	9·9729398	59
2	9·5347452	3463	10·4652548	8·7818022	9·5618515	3924	10·4381485	9·8178516	10·0271062	461	9·9728938	58
3	9·5350915	3460	10·4649085	8·7825171	9·5622439	3921	10·4377561	9·8176711	10·0271523	461	9·9728477	57
4	9·5354375	3457	10·4645625	8·7832314	9·5626360	3918	10·4373640	9·8174905	10·0271984	462	9·9728016	56
5	9·5357832	3454	10·4642168	8·7839452	9·5630278	3916	10·4369722	9·8173098	10·0272446	462	9·9727554	55
6	9·5361286	3451	10·4638714	8·7846583	9·5634194	3913	10·4365806	9·8171291	10·0272908	463	9·9727092	54
7	9·5364737	3447	10·4635263	8·7853708	9·5638107	3911	10·4361893	9·8169483	10·0273371	463	9·9726629	53
8	9·5368184	3445	10·4631816	8·7860827	9·5642018	3907	10·4357982	9·8167675	10·0273834	463	9·9726166	52
9	9·5371629	3441	10·4628371	8·7867940	9·5645925	3907	10·4354075	9·8165866	10·0274297	464	9·9725703	51
10	9·5375070	3438	10·4624930	8·7875047	9·5649831	3906	10·4350169	9·8164056	10·0274761	464	9·9725239	50
11	9·5378508	3435	10·4621492	8·7882149	9·5653733	3902	10·4346267	9·8162246	10·0275225	465	9·9724775	49
12	9·5381943	3432	10·4618057	8·7889244	9·5657633	3900	10·4342367	9·8160435	10·0275690	465	9·9724310	48
13	9·5385375	3429	10·4614625	8·7896333	9·5661530	3897	10·4338470	9·8158624	10·0276155	465	9·9723845	47
14	9·5388804	3426	10·4611196	8·7903416	9·5665424	3894	10·4334576	9·8156812	10·0276620	466	9·9723380	46
15	9·5392230	3423	10·4607770	8·7910494	9·5669316	3892	10·4330684	9·8155000	10·0277086	466	9·9722914	45
16	9·5395653	3420	10·4604347	8·7917565	9·5673205	3889	10·4326795	9·8153187	10·0277552	467	9·9722448	44
17	9·5399073	3416	10·4600927	8·7924630	9·5677091	3886	10·4322909	9·8151374	10·0278019	467	9·9721981	43
18	9·5402489	3414	10·4597511	8·7931690	9·5680975	3881	10·4319025	9·8149560	10·0278486	467	9·9721514	42
19	9·5405903	3411	10·4594097	8·7938743	9·5684856	3879	10·4315144	9·8147745	10·0278953	468	9·9721047	41
20	9·5409314	3407	10·4590686	8·7945791	9·5688735	3876	10·4311265	9·8145930	10·0279421	469	9·9720579	40
21	9·5412721	3405	10·4587279	8·7952833	9·5692611	3873	10·4307389	9·8144114	10·0279890	468	9·9720110	39
22	9·5416126	3401	10·4583874	8·7959869	9·5696484	3871	10·4303516	9·8142298	10·0280358	470	9·9719642	38
23	9·5419527	3399	10·4580473	8·7966899	9·5700355	3868	10·4299645	9·8140481	10·0280828	469	9·9719172	37
24	9·5422926	3395	10·4577074	8·7973923	9·5704223	3865	10·4295777	9·8138664	10·0281297	470	9·9718703	36
25	9·5426321	3392	10·4573679	8·7980941	9·5708088	3863	10·4291912	9·8136846	10·0281767	471	9·9718233	35
26	9·5429713	3390	10·4570287	8·7987953	9·5711951	3860	10·4288049	9·8135027	10·0282238	471	9·9717762	34
27	9·5433103	3386	10·4566897	8·7994960	9·5715811	3858	10·4284189	9·8133208	10·0282709	471	9·9717291	33
28	9·5436489	3384	10·4563511	8·8001961	9·5719669	3855	10·4280331	9·8131389	10·0283181	472	9·9716820	32
29	9·5439873	3380	10·4560127	8·8008956	9·5723524	3853	10·4276476	9·8129569	10·0283652	472	9·9716348	31
30	9·5443253	3377	10·4556747	8·8015945	9·5727377	3850	10·4272623	9·8127748	10·0284124	472	9·9715876	30
31	9·5446630	3375	10·4553370	8·8022928	9·5731227	3847	10·4268773	9·8125926	10·0284596	473	9·9715404	29
32	9·5450005	3371	10·4549995	8·8029906	9·5735074	3845	10·4264926	9·8124104	10·0285069	474	9·9714931	28
33	9·5453376	3369	10·4546624	8·8036878	9·5738919	3842	10·4261081	9·8122282	10·0285543	473	9·9714457	27
34	9·5456745	3365	10·4543255	8·8043843	9·5742761	3840	10·4257239	9·8120459	10·0286016	474	9·9713984	26
35	9·5460110	3362	10·4539890	8·8050803	9·5746601	3837	10·4253399	9·8118635	10·0286491	474	9·9713509	25
36	9·5463472	3360	10·4536528	8·8057758	9·5750438	3834	10·4249562	9·8116811	10·0286965	475	9·9713035	24
37	9·5466832	3357	10·4533168	8·8064707	9·5754272	3832	10·4245728	9·8114986	10·0287440	476	9·9712560	23
38	9·5470189	3353	10·4529811	8·8071649	9·5758104	3830	10·4241896	9·8113161	10·0287916	476	9·9712084	22
39	9·5473542	3351	10·4526458	8·8078587	9·5761934	3827	10·4238066	9·8111335	10·0288392	476	9·9711608	21
40	9·5476893	3347	10·4523107	8·8085518	9·5765761	3824	10·4234239	9·8109509	10·0288868	477	9·9711132	20
41	9·5480240	3345	10·4519760	8·8092444	9·5769585	3822	10·4230415	9·8107682	10·0289345	477	9·9710655	19
42	9·5483585	3342	10·4516415	8·8099364	9·5773407	3819	10·4226593	9·8105854	10·0289822	477	9·9710178	18
43	9·5486927	3339	10·4513073	8·8106278	9·5777226	3817	10·4222774	9·8104026	10·0290299	478	9·9709701	17
44	9·5490266	3336	10·4509734	8·8113187	9·5781043	3815	10·4218957	9·8102197	10·0290777	478	9·9709223	16
45	9·5493602	3333	10·4506398	8·8120090	9·5784858	3811	10·4215142	9·8100368	10·0291256	479	9·9708744	15
46	9·5496935	3330	10·4503065	8·8126988	9·5788669	3810	10·4211331	9·8098538	10·0291735	479	9·9708265	14
47	9·5500265	3327	10·4499735	8·8133879	9·5792479	3807	10·4207521	9·8096708	10·0292214	480	9·9707786	13
48	9·5503592	3324	10·4496408	8·8140765	9·5796286	3804	10·4203714	9·8094877	10·0292694	480	9·9707306	12
49	9·5506916	3321	10·4493084	8·8147646	9·5800090	3802	10·4199910	9·8093045	10·0293174	480	9·9706826	11
50	9·5510237	3319	10·4489763	8·8154521	9·5803892	3799	10·4196108	9·8091213	10·0293654	481	9·9706346	10
51	9·5513556	3315	10·4486444	8·8161390	9·5807691	3797	10·4192309	9·8089380	10·0294135	482	9·9705865	9
52	9·5516871	3313	10·4483129	8·8168253	9·5811488	3794	10·4188512	9·8087547	10·0294617	482	9·9705383	8
53	9·5520184	3310	10·4479816	8·8175111	9·5815282	3792	10·4184718	9·8085713	10·0295098	483	9·9704902	7
54	9·5523494	3307	10·4476506	8·8181964	9·5819074	3790	10·4180926	9·8083879	10·0295581	482	9·9704419	6
55	9·5526801	3304	10·4473199	8·8188810	9·5822864	3787	10·4177136	9·8082044	10·0296063	483	9·9703937	5
56	9·5530105	3301	10·4469895	8·8195652	9·5826651	3784	10·4173349	9·8080208	10·0296546	484	9·9703454	4
57	9·5533406	3298	10·4466594	8·8202487	9·5830435	3782	10·4169565	9·8078372	10·0297030	484	9·9702970	3
58	9·5536704	3295	10·4463296	8·8209317	9·5834217	3780	10·4165783	9·8076536	10·0297514	484	9·9702486	2
59	9·5539999	3293	10·4460001	8·8216142	9·5837997	3777	10·4162003	9·8074698	10·0297998	485	9·9702002	1
60	9·5543292		10·4456708	8·8222961	9·5841774		10·4158226	9·8072860	10·0298483		9·9701517	0

′	Cosine.	Dif.	Secant.	Covers.	Cotang.	Dif.	Tang.	Verseds.	Cosec.	D.	Sine.	′

′	Sine.	Dif.	Covers	Cosec.	Tang.	Cotang.	Secant.	Vers.	Dif.	Cosine	′
0	3583679	2716	6416321	2·7904281	3838640	2·6050891	1·0711450	0664196	1043	9335804	60
1	3586395	2715	6413605	2·7883153	3841978	2·6028258	1·0712647	0665239	1043	9334761	59
2	3589110	2715	6410890	2·7862059	3845317	2·6005659	1·0713844	0666282	1045	9333718	58
3	3591825	2715	6408175	2·7840999	3848656	2·5983095	1·0715043	0667327	1045	9332673	57
4	3594540	2714	6405460	2·7819973	3851996	2·5960564	1·0716244	0668372	1046	9331628	56
5	3597254	2714	6402746	2·7798982	3855337	2·5938068	1·0717445	0669418	1047	9330582	55
6	3599968	2714	6400032	2·7778024	3858679	2·5915606	1·0718647	0670465	1047	9329535	54
7	3602682	2713	6397318	2·7757100	3862021	2·5893177	1·0719851	0671512	1049	9328488	53
8	3605395	2713	6394605	2·7736211	3865364	2·5870782	1·0721056	0672561	1049	9327439	52
9	3608108	2713	6391892	2·7715355	3868708	2·5848421	1·0722262	0673610	1050	9326390	51
10	3610821	2713	6389179	2·7694532	3872053	2·5826094	1·0723469	0674660	1050	9325340	50
11	3613534	2712	6386466	2·7673744	3875398	2·5803800	1·0724678	0675710	1052	9324290	49
12	3616246	2712	6383754	2·7652988	3878744	2·5781539	1·0725887	0676762	1052	9323238	48
13	3618958	2711	6381042	2·7632267	3882091	2·5759312	1·0727098	0677814	1053	9322186	47
14	3621669	2711	6378331	2·7611578	3885439	2·5737118	1·0728310	0678867	1054	9321133	46
15	3624380	2711	6375620	2·7590923	3888787	2·5714957	1·0729523	0679921	1055	9320079	45
16	3627091	2711	6372909	2·7570301	3892136	2·5692830	1·0730737	0680976	1055	9319024	44
17	3629802	2710	6370198	2·7549712	3895486	2·5670735	1·0731953	0682031	1057	9317969	43
18	3632512	2710	6367488	2·7529157	3898837	2·5648674	1·0733170	0683088	1057	9316912	42
19	3635222	2710	6364778	2·7508634	3902189	2·5626645	1·0734388	0684145	1058	9315855	41
20	3637932	2709	6362068	2·7488144	3905541	2·5604649	1·0735607	0685203	1058	9314797	40
21	3640641	2710	6359359	2·7467687	3908894	2·5582686	1·0736827	0686261	1060	9313739	39
22	3643351	2708	6356649	2·7447263	3912247	2·5560756	1·0738048	0687321	1060	9312679	38
23	3646059	2709	6353941	2·7426871	3915602	2·5538858	1·0739271	0688381	1061	9311619	37
24	3648768	2708	6351232	2·7406512	3918957	2·5516992	1·0740495	0689442	1062	9310558	36
25	3651476	2708	6348524	2·7386186	3922313	2·5495160	1·0741720	0690504	1062	9309496	35
26	3654184	2707	6345816	2·7365892	3925670	2·5473359	1·0742946	0691566	1064	9308434	34
27	3656891	2708	6343109	2·7345630	3929027	2·5451591	1·0744173	0692630	1064	9307370	33
28	3659599	2707	6340401	2·7325400	3932386	2·5429855	1·0745402	0693694	1065	9306306	32
29	3662306	2706	6337694	2·7305203	3935745	2·5408151	1·0746631	0694759	1065	9305241	31
30	3665012	2707	6334988	2·7285038	3939105	2·5386479	1·0747862	0695824	1067	9304176	30
31	3667719	2706	6332281	2·7264905	3942465	2·5364839	1·0749095	0696891	1067	9303109	29
32	3670425	2705	6329575	2·7244804	3945827	2·5343231	1·0750328	0697958	1068	9302042	28
33	3673130	2706	6326870	2·7224735	3949189	2·5321655	1·0751562	0699026	1069	9300974	27
34	3675836	2705	6324164	2·7204698	3952552	2·5300111	1·0752798	0700095	1070	9299905	26
35	3678541	2705	6321459	2·7184693	3955916	2·5278598	1·0754035	0701165	1070	9298835	25
36	3681246	2704	6318754	2·7164719	3959280	2·5257117	1·0755273	0702235	1071	9297765	24
37	3683950	2704	6316050	2·7144777	3962645	2·5235667	1·0756512	0703306	1072	9296694	23
38	3686654	2704	6313346	2·7124866	3966011	2·5214249	1·0757753	0704378	1073	9295622	22
39	3689358	2703	6310642	2·7104987	3969378	2·5192863	1·0758995	0705451	1074	9294549	21
40	3692061	2704	6307939	2·7085139	3972746	2·5171507	1·0760237	0706525	1074	9293475	20
41	3694765	2703	6305235	2·7065323	3976114	2·5150183	1·0761481	0707599	1075	9292401	19
42	3697468	2702	6302532	2·7045538	3979483	2·5128890	1·0762727	0708674	1076	9291326	18
43	3700170	2702	6299830	2·7025784	3982853	2·5107629	1·0763973	0709750	1077	9290250	17
44	3702872	2702	6297128	2·7006061	3986224	2·5086398	1·0765221	0710827	1077	9289173	16
45	3705574	2702	6294426	2·6986370	3989595	2·5065198	1·0766470	0711904	1079	9288096	15
46	3708276	2701	6291724	2·6966709	3992968	2·5044029	1·0767720	0712983	1079	9287017	14
47	3710977	2701	6289023	2·6947079	3996341	2·5022891	1·0768971	0714062	1080	9285938	13
48	3713678	2701	6286322	2·6927480	3999715	2·5001784	1·0770224	0715142	1080	9284858	12
49	3716379	2700	6283621	2·6907912	4003089	2·4980707	1·0771477	0716222	1082	9283778	11
50	3719079	2701	6280921	2·6888374	4006465	2·4959661	1·0772732	0717304	1082	9282696	10
51	3721780	2699	6278220	2·6868867	4009841	2·4938645	1·0773988	0718386	1083	9281614	9
52	3724479	2700	6275521	2·6849391	4013218	2·4917660	1·0775246	0719469	1084	9280531	8
53	3727179	2699	6272821	2·6829945	4016596	2·4896706	1·0776504	0720553	1084	9279447	7
54	3729878	2699	6270122	2·6810530	4019974	2·4875781	1·0777764	0721637	1086	9278363	6
55	3732577	2698	6267423	2·6791145	4023354	2·4854887	1·0779025	0722723	1086	9277277	5
56	3735275	2698	6264725	2·6771790	4026734	2·4834023	1·0780287	0723809	1087	9276191	4
57	3737973	2698	6262027	2·6752465	4030115	2·4813190	1·0781550	0724896	1088	9275104	3
58	3740671	2698	6259329	2·6733171	4033496	2·4792386	1·0782815	0725984	1088	9274016	2
59	3743369	2697	6256631	2·6713906	4036879	2·4771612	1·0784080	0727072	1089	9272928	1
60	3746066		6253934	2·6694672	4040262	2·4750869	1·0785347	0728161		9271839	0
′	Cosine	Dif.	Vers.	Secant.	Cotan.	Tang.	Cosec.	Covers	Dif.	Sine.	′

'	Sine.	Dif.	Cosec.	Verseds.	Tang.	Dif.	Cotang.	Covers.	Secant.	D.	Cosine.	
0	9·5543292	3289	10·4456708	8·8222061	9·5841774	3775	10·4158226	9·8072860	10·0298483	485	9·9701517	60
1	9·5546581	3287	10·4453419	8·8229774	9·5845549	3772	10·4154451	9·8071022	10·0298968	485	9·9701032	59
2	9·5549868	3284	10·4450132	8·8236582	9·5849321	3770	10·4150679	9·8069183	10·0299453	486	9·9700547	58
3	9·5553152	3281	10·4446848	8·8243385	9·5853091	3768	10·4146909	9·8067344	10·0299939	486	9·9700061	57
4	9·5556433	3278	10·4443567	8·8250182	9·5856859	3765	10·4143141	9·8065503	10·0300426	487	9·9699574	56
5	9·5559711	3276	10·4440289	8·8256973	9·5860624	3762	10·4139376	9·8063663	10·0300913	487	9·9699087	55
6	9·5562987	3272	10·4437013	8·8263759	9·5864386	3761	10·4135614	9·8061821	10·0301400	488	9·9698600	54
7	9·5566259	3270	10·4433741	8·8270539	9·5868147	3757	10·4131853	9·8059980	10·0301888	488	9·9698112	53
8	9·5569529	3267	10·4430471	8·8277314	9·5871904	3756	10·4128096	9·8058137	10·0302376	488	9·9697624	52
9	9·5572796	3264	10·4427204	8·8284084	9·5875660	3753	10·4124340	9·8056294	10·0302864	488	9·9697136	51
10	9·5576060	3261	10·4423940	8·8290848	9·5879413	3750	10·4120587	9·8054451	10·0303353	489	9·9696647	50
11	9·5579321	3258	10·4420679	8·8297606	9·5883163	3749	10·4116837	9·8052606	10·0303842	490	9·9696158	49
12	9·5582579	3256	10·4417421	8·8304360	9·5886912	3745	10·4113088	9·8050762	10·0304332	491	9·9695668	48
13	9·5585835	3253	10·4414165	8·8311107	9·5890657	3744	10·4109343	9·8048916	10·0304823	490	9·9695177	47
14	9·5589088	3250	10·4410912	8·8317850	9·5894401	3741	10·4105599	9·8047070	10·0305313	490	9·9694687	46
15	9·5592338	3247	10·4407662	8·8324587	9·5898142	3739	10·4101858	9·8045224	10·0305804	492	9·9694196	45
16	9·5595585	3244	10·4404415	8·8331318	9·5901881	3736	10·4098119	9·8043377	10·0306296	492	9·9693704	44
17	9·5598829	3242	10·4401171	8·8338044	9·5905617	3734	10·4094383	9·8041529	10·0306788	492	9·9693212	43
18	9·5602071	3239	10·4397929	8·8344765	9·5909351	3731	10·4090649	9·8039681	10·0307280	493	9·9692720	42
19	9·5605310	3236	10·4394690	8·8351480	9·5913082	3730	10·4086918	9·8037832	10·0307773	493	9·9692227	41
20	9·5608546	3233	10·4391454	8·8358190	9·5916812	3727	10·4083188	9·8035983	10·0308266	493	9·9691734	40
21	9·5611779	3231	10·4388221	8·8364895	9·5920539	3724	10·4079461	9·8034133	10·0308759	495	9·9691241	39
22	9·5615010	3227	10·4384990	8·8371594	9·5924263	3722	10·4075737	9·8032283	10·0309254	494	9·9690746	38
23	9·5618237	3225	10·4381763	8·8378288	9·5927985	3720	10·4072015	9·8030432	10·0309748	495	9·9690252	37
24	9·5621462	3223	10·4378538	8·8384976	9·5931705	3718	10·4068295	9·8028580	10·0310243	495	9·9689757	36
25	9·5624685	3219	10·4375315	8·8391660	9·5935423	3715	10·4064577	9·8026728	10·0310738	496	9·9689262	35
26	9·5627904	3217	10·4372096	8·8398337	9·5939138	3713	10·4060862	9·8024875	10·0311234	496	9·9688766	34
27	9·5631121	3214	10·4368879	8·8405010	9·5942851	3713	10·4057149	9·8023021	10·0311730	497	9·9688270	33
28	9·5634335	3211	10·4365665	8·8411677	9·5946561	3710	10·4053439	9·8021167	10·0312227	497	9·9687773	32
29	9·5637546	3208	10·4362454	8·8418339	9·5950269	3708	10·4049731	9·8019313	10·0312724	497	9·9687276	31
30	9·5640754	3206	10·4359246	8·8424996	9·5953975	3706	10·4046025	9·8017458	10·0313221	498	9·9686779	30
31	9·5643960	3203	10·4356040	8·8431647	9·5957679	3704	10·4042321	9·8015602	10·0313719	498	9·9686281	29
32	9·5647163	3200	10·4352837	8·8438294	9·5961380	3701	10·4038620	9·8013746	10·0314217	499	9·9685783	28
33	9·5650363	3198	10·4349637	8·8444934	9·5965079	3699	10·4034921	9·8011889	10·0314716	499	9·9685284	27
34	9·5653561	3195	10·4346439	8·8451570	9·5968776	3697	10·4031224	9·8010031	10·0315215	499	9·9684785	26
35	9·5656756	3192	10·4343244	8·8458200	9·5972470	3694	10·4027530	9·8008173	10·0315714	500	9·9684286	25
36	9·5659948	3189	10·4340052	8·8464826	9·5976162	3692	10·4023838	9·8006315	10·0316214	500	9·9683786	24
37	9·5663137	3187	10·4336863	8·8471445	9·5979852	3690	10·4020148	9·8004456	10·0316715	501	9·9683285	23
38	9·5666324	3184	10·4333676	8·8478060	9·5983540	3685	10·4016460	9·8002596	10·0317216	501	9·9682784	22
39	9·5669508	3181	10·4330492	8·8484670	9·5987225	3683	10·4012775	9·8000735	10·0317717	502	9·9682283	21
40	9·5672689	3179	10·4327311	8·8491274	9·5990908	3680	10·4009092	9·7998875	10·0318219	502	9·9681781	20
41	9·5675868	3176	10·4324132	8·8497873	9·5994588	3679	10·4005412	9·7997013	10·0318721	502	9·9681279	19
42	9·5679044	3173	10·4320956	8·8504467	9·5998267	3676	10·4001733	9·7995151	10·0319223	503	9·9680777	18
43	9·5682217	3170	10·4317783	8·8511055	9·6001943	3674	10·3998057	9·7993288	10·0319726	503	9·9680274	17
44	9·5685387	3168	10·4314613	8·8517639	9·6005617	3672	10·3994383	9·7991425	10·0320229	504	9·9679771	16
45	9·5688555	3166	10·4311445	8·8524217	9·6009289	3669	10·3990711	9·7989561	10·0320733	504	9·9679267	15
46	9·5691721	3162	10·4308279	8·8530790	9·6012958	3667	10·3987042	9·7987697	10·0321237	505	9·9678763	14
47	9·5694883	3160	10·4305117	8·8537358	9·6016625	3665	10·3983375	9·7985832	10·0321742	505	9·9678258	13
48	9·5698043	3157	10·4301957	8·8543921	9·6020290	3663	10·3979710	9·7983966	10·0322247	506	9·9677753	12
49	9·5701200	3155	10·4298800	8·8550479	9·6023953	3660	10·3976047	9·7982100	10·0322753	506	9·9677247	11
50	9·5704355	3151	10·4295645	8·8557032	9·6027613	3658	10·3972387	9·7980233	10·0323259	506	9·9676741	10
51	9·5707506	3151	10·4292494	8·8563579	9·6031271	3656	10·3968729	9·7978366	10·0323765	507	9·9676235	9
52	9·5710656	3150	10·4289344	8·8570121	9·6034927	3654	10·3965073	9·7976498	10·0324272	507	9·9675728	8
53	9·5713802	3146	10·4286198	8·8576659	9·6038581	3652	10·3961419	9·7974629	10·0324779	508	9·9675221	7
54	9·5716946	3144	10·4283054	8·8583191	9·6042233	3649	10·3957767	9·7972760	10·0325287	508	9·9674713	6
55	9·5720087	3141	10·4279913	8·8589718	9·6045882	3647	10·3954118	9·7970890	10·0325795	508	9·9674205	5
56	9·5723226	3139	10·4276774	8·8596240	9·6049529	3645	10·3950471	9·7969020	10·0326303	508	9·9673697	4
57	9·5726362	3136	10·4273638	8·8602757	9·6053174	3643	10·3946826	9·7967149	10·0326812	509	9·9673188	3
58	9·5729495	3133	10·4270505	8·8609268	9·6056817	3640	10·3943183	9·7965278	10·0327321	509	9·9672679	2
59	9·5732626	3131	10·4267374	8·8615775	9·6060457	3639	10·3939543	9·7963406	10·0327831	510	9·9672169	1
60	9·5735754	3128	10·4264246	8·8622277	9·6064096		10·3935904	9·7961533	10·0328341	510	9·9671659	0
	Cosine.	Dif.	Secant.	Covers.	Cotang.	Dif.	Tang.	Verseds.	Cosec.	D.	Sine.	'

′	Sine.	Dif.	Covers	Cosec.	Tang.	Cotang.	Secant.	Vers.	Dif.	Cosine	′
0	3746066	2697	6253934	2·6694672	4040262	2·4750869	1·0785347	0728161	1091	9271839	60
1	3748763	2696	6251237	2·6675467	4043646	2·4730155	1·0786616	0729252	1090	9270748	59
2	3751459	2697	6248541	2·6656292	4047031	2·4709470	1·0787885	0730342	1092	9269658	58
3	3754156	2696	6245844	2·6637148	4050417	2·4688816	1·0789156	0731434	1092	9268566	57
4	3756852	2695	6243148	2·6618033	4053804	2·4668191	1·0790427	0732526	1094	9267474	56
5	3759547	2696	6240453	2·6598947	4057191	2·4647596	1·0791700	0733620	1094	9266380	55
6	3762243	2695	6237757	2·6579891	4060579	2·4627030	1·0792975	0734714	1004	9265286	54
7	3764938	2694	6235062	2·6560865	4063968	2·4606494	1·0794250	0735808	1096	9264192	53
8	3767632	2695	6232368	2·6541868	4067358	2·4585987	1·0795527	0736904	1096	9263096	52
9	3770327	2694	6229673	2·6522901	4070748	2·4565510	1·0796805	0738000	1098	9262000	51
10	3773021	2693	6226979	2·6503962	4074139	2·4545061	1·0798084	0739098	1097	9260902	50
11	3775714	2694	6224286	2·6485054	4077531	2·4524642	1·0799364	0740195	1099	9259805	49
12	3778408	2693	6221592	2·6466174	4080924	2·4504252	1·0800646	0741294	1100	9258706	48
13	3781101	2693	6218899	2·6447323	4084318	2·4483891	1·0801928	0742394	1100	9257606	47
14	3783794	2692	6216206	2·6428502	4087713	2·4463559	1·0803212	0743494	1101	9256506	46
15	3786486	2692	6213514	2·6409710	4091108	2·4443256	1·0804497	0744595	1102	9255405	45
16	3789178	2692	6210822	2·6390946	4094504	2·4422982	1·0805784	0745697	1102	9254303	44
17	3791870	2692	6208130	2·6372211	4097901	2·4402736	1·0807071	0746799	1104	9253201	43
18	3794562	2691	6205438	2·6353506	4101299	2·4382519	1·0808360	0747903	1104	9252097	42
19	3797253	2691	6202747	2·6334828	4104697	2·4362331	1·0809650	0749007	1105	9250993	41
20	3799944	2690	6200056	2·6316180	4108097	2·4342172	1·0810942	0750112	1106	9249888	40
21	3802634	2690	6197366	2·6297560	4111497	2·4322041	1·0812234	0751218	1106	9248782	39
22	3805324	2690	6194676	2·6278969	4114898	2·4301938	1·0813528	0752324	1108	9247676	38
23	3808014	2690	6191986	2·6260406	4118300	2·4281864	1·0814823	0753432	1108	9246568	37
24	3810704	2689	6189296	2·6241872	4121703	2·4261819	1·0816119	0754540	1109	9245460	36
25	3813393	2689	6186607	2·6223366	4125106	2·4241801	1·0817417	0755649	1109	9244351	35
26	3816082	2688	6183918	2·6204888	4128510	2·4221812	1·0818715	0756758	1111	9243242	34
27	3818770	2689	6181230	2·6186439	4131915	2·4201851	1·0820015	0757869	1111	9242131	33
28	3821459	2688	6178541	2·6168018	4135321	2·4181918	1·0821316	0758980	1112	9241020	32
29	3824147	2687	6175853	2·6149624	4138728	2·4162013	1·0822618	0760092	1113	9239908	31
30	3826834	2688	6173166	2·6131259	4142136	2·4142136	1·0823922	0761205	1113	9238795	30
31	3829522	2687	6170478	2·6112922	4145544	2·4122286	1·0825227	0762318	1115	9237682	29
32	3832209	2686	6167791	2·6094613	4148953	2·4102465	1·0826533	0763433	1115	9236567	28
33	3834895	2687	6165105	2·6076332	4152363	2·4082672	1·0827840	0764548	1116	9235452	27
34	3837582	2686	6162418	2·6058078	4155774	2·4062906	1·0829149	0765664	1116	9234336	26
35	3840268	2685	6159732	2·6039852	4159186	2·4043168	1·0830458	0766780	1118	9233220	25
36	3842953	2686	6157047	2·6021654	4162598	2·4023457	1·0831769	0767898	1118	9232102	24
37	3845639	2685	6154361	2·6003484	4166012	2·4003774	1·0833081	0769016	1119	9230984	23
38	3848324	2684	6151676	2·5985341	4169426	2·3984118	1·0834395	0770135	1120	9229865	22
39	3851008	2685	6148992	2·5967225	4172841	2·3964490	1·0835709	0771255	1121	9228745	21
40	3853693	2684	6146307	2·5949137	4176257	2·3944889	1·0837025	0772376	1121	9227624	20
41	3856377	2683	6143623	2·5931077	4179673	2·3925316	1·0838342	0773497	1122	9226503	19
42	3859060	2684	6140940	2·5913043	4183091	2·3905769	1·0839661	0774619	1123	9225381	18
43	3861744	2683	6138256	2·5895037	4186509	2·3886250	1·0840980	0775742	1124	9224258	17
44	3864427	2683	6135573	2·5877058	4189928	2·3866758	1·0842301	0776866	1124	9223134	16
45	3867110	2682	6132890	2·5859107	4193348	2·3847293	1·0843623	0777990	1126	9222010	15
46	3869792	2682	6130208	2·5841182	4196769	2·3827855	1·0844947	0779116	1126	9220884	14
47	3872474	2682	6127526	2·5823284	4200190	2·3808444	1·0846271	0780242	1126	9219758	13
48	3875156	2681	6124844	2·5805414	4203613	2·3789060	1·0847597	0781368	1128	9218632	12
49	3877837	2681	6122163	2·5787570	4207036	2·3769703	1·0848924	0782496	1129	9217504	11
50	3880518	2681	6119482	2·5769753	4210460	2·3750372	1·0850252	0783625	1129	9216375	10
51	3883199	2681	6116801	2·5751963	4213885	2·3731068	1·0851582	0784754	1130	9215246	9
52	3885880	2680	6114120	2·5734199	4217311	2·3711791	1·0852913	0785884	1130	9214116	8
53	3888560	2680	6111440	2·5716462	4220738	2·3692540	1·0854245	0787014	1132	9212986	7
54	3891240	2679	6108760	2·5698752	4224165	2·3673316	1·0855578	0788146	1132	9211854	6
55	3893919	2679	6106081	2·5681069	4227594	2·3654118	1·0856912	0789278	1133	9210722	5
56	3896598	2679	6103402	2·5663412	4231023	2·3634946	1·0858248	0790411	1134	9209589	4
57	3899277	2678	6100723	2·5645781	4234453	2·3615801	1·0859585	0791545	1135	9208455	3
58	3901955	2678	6098045	2·5628176	4237884	2·3596683	1·0860924	0792680	1135	9207320	2
59	3904633	2678	6095367	2·5610599	4241316	2·3577590	1·0862263	0793815	1136	9206185	1
60	3907311		6092689	2·5593047	4244748	2·3558524	1·0863604	0794951		9205049	0
′	Cosine	Dif.	Vers.	Secant.	Cotan.	Tang.	Cosec.	Covers	Dif.	Sine.	′

′	Sine.	Dif.	Cosec.	Verseds.	Tang.	Dif.	Cotang.	Covers.	Secant.	D.	Cosine.	′
0	9·5735754	3126	10·4264246	8·8622277	9·6064096	3636	10·3935904	9·7961533	10·0328341	511	9·9671659	60
1	9·5738880	3123	10·4261120	8·8628774	9·6067732	3634	10·3932268	9·7959660	10·0328852	511	9·9671148	59
2	9·5742003	3120	10·4257997	8·8635265	9·6071366	3631	10·3928634	9·7957786	10·0329363	512	9·9670637	58
3	9·5745123	3117	10·4254877	8·8641752	9·6074997	3630	10·3925003	9·7955912	10·0329875	511	9·9670125	57
4	9·5748240	3116	10·4251760	8·8648233	9·6078627	3627	10·3921373	9·7954037	10·0330386	513	9·9669614	56
5	9·5751356	3112	10·4248644	8·8654710	9·6082254	3626	10·3917746	9·7952161	10·0330899	513	9·9669101	55
6	9·5754468	3110	10·4245532	8·8661181	9·6085880	3623	10·3914120	9·7950285	10·0331412	513	9·9668588	54
7	9·5757578	3107	10·4242422	8·8667648	9·6089503	3621	10·3910497	9·7948408	10·0331925	513	9·9668075	53
8	9·5760685	3105	10·4239315	8·8674109	9·6093124	3618	10·3906876	9·7946531	10·0332438	514	9·9667562	52
9	9·5763790	3102	10·4236210	8·8680566	9·6096742	3617	10·3903258	9·7944653	10·0332952	515	9·9667048	51
10	9·5766892	3099	10·4233108	8·8687018	9·6100359	3614	10·3899641	9·7942774	10·0333467	515	9·9666533	50
11	9·5769991	3097	10·4230009	8·8693464	9·6103973	3613	10·3896027	9·7940895	10·0333982	515	9·9666018	49
12	9·5773088	3095	10·4226912	8·8699906	9·6107586	3610	10·3892414	9·7939015	10·0334497	516	9·9665503	48
13	9·5776183	3092	10·4223817	8·8706342	9·6111196	3608	10·3888804	9·7937135	10·0335013	516	9·9664987	47
14	9·5779275	3089	10·4220725	8·8712774	9·6114804	3605	10·3885196	9·7935254	10·0335529	517	9·9664471	46
15	9·5782364	3086	10·4217636	8·8719201	9·6118400	3604	10·3881591	9·7933373	10·0336046	517	9·9663954	45
16	9·5785450	3085	10·4214550	8·8725623	9·6122013	3602	10·3877987	9·7931491	10·0336563	517	9·9663437	44
17	9·5788535	3081	10·4211465	8·8732040	9·6125615	3599	10·3874385	9·7929608	10·0337080	518	9·9662920	43
18	9·5791616	3079	10·4208384	8·8738452	9·6129214	3598	10·3870786	9·7927725	10·0337598	518	9·9662402	42
19	9·5794695	3077	10·4205305	8·8744859	9·6132812	3595	10·3867188	9·7925841	10·0338116	519	9·9661884	41
20	9·5797772	3073	10·4202228	8·8751261	9·6136407	3593	10·3863593	9·7923956	10·0338635	519	9·9661365	40
21	9·5800845	3072	10·4199155	8·8757658	9·6140000	3591	10·3860000	9·7922071	10·0339154	520	9·9660846	39
22	9·5803917	3069	10·4196083	8·8764051	9·6143591	3589	10·3856409	9·7920186	10·0339674	520	9·9660326	38
23	9·5806986	3066	10·4193014	8·8770438	9·6147180	3586	10·3852820	9·7918300	10·0340194	520	9·9659806	37
24	9·5810052	3064	10·4189948	8·8776821	9·6150766	3585	10·3849234	9·7916413	10·0340715	521	9·9659285	36
25	9·5813116	3061	10·4186884	8·8783198	9·6154351	3583	10·3845649	9·7914525	10·0341236	521	9·9658764	35
26	9·5816177	3059	10·4183823	8·8789571	9·6157934	3580	10·3842066	9·7912637	10·0341757	522	9·9658243	34
27	9·5819236	3056	10·4180764	8·8795939	9·6161514	3579	10·3838486	9·7910749	10·0342279	522	9·9657721	33
28	9·5822292	3053	10·4177708	8·8802303	9·6165093	3576	10·3834907	9·7908859	10·0342801	524	9·9657199	32
29	9·5825345	3052	10·4174655	8·8808661	9·6168669	3574	10·3831331	9·7906970	10·0343323	524	9·9656677	31
30	9·5828397	3048	10·4171603	8·8815014	9·6172243	3572	10·3827757	9·7905079	10·0343847	523	9·9656153	30
31	9·5831445	3046	10·4168555	8·8821363	9·6175815	3570	10·3824185	9·7903188	10·0344370	524	9·9655630	29
32	9·5834491	3044	10·4165509	8·8827707	9·6179385	3568	10·3820615	9·7901297	10·0344894	524	9·9655106	28
33	9·5837535	3041	10·4162465	8·8834046	9·6182953	3566	10·3817047	9·7899405	10·0345418	525	9·9654582	27
34	9·5840576	3039	10·4159424	8·8840380	9·6186519	3564	10·3813481	9·7897512	10·0345943	525	9·9654057	26
35	9·5843615	3036	10·4156385	8·8846710	9·6190083	3562	10·3809917	9·7895618	10·0346468	525	9·9653532	25
36	9·5846651	3034	10·4153349	8·8853034	9·6193645	3560	10·3806355	9·7893725	10·0346994	526	9·9653006	24
37	9·5849685	3031	10·4150315	8·8859354	9·6197205	3557	10·3802795	9·7891830	10·0347520	527	9·9652480	23
38	9·5852716	3029	10·4147284	8·8865669	9·6200762	3556	10·3799238	9·7889935	10·0348047	527	9·9651953	22
39	9·5855745	3026	10·4144255	8·8871980	9·6204318	3554	10·3795682	9·7888039	10·0348574	527	9·9651426	21
40	9·5858771	3024	10·4141229	8·8878285	9·6207872	3551	10·3792128	9·7886143	10·0349101	528	9·9650899	20
41	9·5861795	3021	10·4138205	8·8884586	9·6211423	3550	10·3788577	9·7884246	10·0349629	528	9·9650371	19
42	9·5864816	3019	10·4135184	8·8890882	9·6214973	3547	10·3785027	9·7882348	10·0350157	529	9·9649843	18
43	9·5867835	3016	10·4132165	8·8897173	9·6218520	3546	10·3781480	9·7880450	10·0350686	529	9·9649314	17
44	9·5870851	3014	10·4129149	8·8903460	9·6222066	3543	10·3777934	9·7878551	10·0351215	529	9·9648786	16
45	9·5873865	3011	10·4126135	8·8909742	9·6225609	3541	10·3774391	9·7876652	10·0351744	530	9·9648256	15
46	9·5876876	3009	10·4123124	8·8916019	9·6229150	3540	10·3770850	9·7874752	10·0352274	531	9·9647726	14
47	9·5879885	3007	10·4120115	8·8922291	9·6232690	3537	10·3767310	9·7872852	10·0352805	530	9·9647195	13
48	9·5882892	3004	10·4117108	8·8928559	9·6236227	3536	10·3763773	9·7870950	10·0353335	532	9·9646665	12
49	9·5885885	3001	10·4114104	8·8934822	9·6239763	3533	10·3760237	9·7869049	10·0353867	531	9·9646133	11
50	9·5888897	2998	10·4111103	8·8941080	9·6243296	3531	10·3756704	9·7867146	10·0354398	533	9·9645602	10
51	9·5891897	2996	10·4108103	8·8947334	9·6246827	3529	10·3753173	9·7865243	10·0354931	532	9·9645069	9
52	9·5894893	2995	10·4105107	8·8953583	9·6250356	3528	10·3749644	9·7863340	10·0355464	533	9·9644537	8
53	9·5897888	2992	10·4102112	8·8959827	9·6253884	3525	10·3746116	9·7861436	10·0355996	534	9·9644004	7
54	9·5900880	2989	10·4099120	8·8966066	9·6257409	3523	10·3742591	9·7859531	10·0356530	533	9·9643470	6
55	9·5903869	2987	10·4096131	8·8972301	9·6260932	3522	10·3739068	9·7857626	10·0357063	535	9·9642937	5
56	9·5906856	2985	10·4093144	8·8978532	9·6264454	3519	10·3735546	9·7855720	10·0357598	534	9·9642402	4
57	9·5909841	2982	10·4090159	8·8984757	9·6267973	3518	10·3732027	9·7853813	10·0358132	536	9·9641868	3
58	9·5912823	2980	10·4087177	8·8990978	9·6271491	3515	10·3728509	9·7851906	10·0358668	535	9·9641332	2
59	9·5915803	2977	10·4084197	8·8997194	9·6275006	3513	10·3724994	9·7849998	10·0359203	536	9·9640797	1
60	9·5918780		10·4081220	8·9003406	9·6278519		10·3721481	9·7848090	10·0359739		9·9640261	0

| ′ | Cosine. | Dif. | Secant. | Covers. | Cotang. | Dif. | Tang. | Verseds. | Cosec. | D. | Sine. | ′ |

′	Sine.	Dif.	Covers	Cosec.	Tang.	Cotang.	Secant.	Vers.	Dif.	Cosine	′
0	3907311	2678	6092689	2·5593047	4244748	2·3558524	1·0863604	0794951	1137	9205049	60
1	3909989	2677	6090011	2·5575521	4248182	2·3539483	1·0864946	0796088	1138	9203912	59
2	3912666	2677	6087334	2·5558022	4251616	2·3520469	1·0866289	0797226	1139	9202774	58
3	3915343	2676	6084657	2·5540548	4255051	2·3501481	1·0867634	0798365	1139	9201635	57
4	3918019	2676	6081981	2·5523101	4258487	2·3482519	1·0868979	0799504	1140	9200496	56
5	3920695	2676	6079305	2·5505680	4261924	2·3463582	1·0870326	0800644	1141	9199356	55
6	3923371	2676	6076629	2·5488284	4265361	2·3444672	1·0871675	0801785	1142	9198215	54
7	3926047	2675	6073953	2·5470915	4268800	2·3425787	1·0873024	0802927	1142	9197073	53
8	3928722	2675	6071278	2·5453571	4272239	2·3406928	1·0874375	0804069	1143	9195931	52
9	3931397	2674	6068603	2·5436253	4275680	2·3388095	1·0875727	0805212	1144	9194788	51
10	3934071	2674	6065929	2·5418961	4279121	2·3369287	1·0877080	0806356	1145	9193644	50
11	3936745	2674	6063255	2·5401694	4282563	2·3350505	1·0878435	0807501	1146	9192499	49
12	3939419	2674	6060581	2·5384453	4286005	2·3331748	1·0879791	0808647	1146	9191353	48
13	3942093	2673	6057907	2·5367238	4289449	2·3313017	1·0881148	0809793	1147	9190207	47
14	3944766	2673	6055234	2·5350048	4292894	2·3294311	1·0882506	0810940	1148	9189060	46
15	3947439	2672	6052561	2·5332883	4296339	2·3275630	1·0883866	0812088	1149	9187912	45
16	3950111	2672	6049889	2·5315744	4299785	2·3256975	1·0885226	0813237	1149	9186763	44
17	3952783	2672	6047217	2·5298630	4303232	2·3238345	1·0886589	0814386	1150	9185614	43
18	3955455	2672	6044545	2·5281541	4306680	2·3219740	1·0887952	0815536	1151	9184464	42
19	3958127	2671	6041873	2·5264478	4310129	2·3201160	1·0889317	0816687	1152	9183313	41
20	3960798	2670	6039202	2·5247440	4313579	2·3182606	1·0890682	0817839	1152	9182161	40
21	3963468	2671	6036532	2·5230426	4317030	2·3164076	1·0892050	0818991	1154	9181009	39
22	3966139	2670	6033861	2·5213438	4320481	2·3145571	1·0893418	0820145	1154	9179855	38
23	3968809	2670	6031191	2·5196475	4323933	2·3127092	1·0894788	0821299	1155	9178701	37
24	3971479	2669	6028521	2·5179537	4327386	2·3108637	1·0896159	0822454	1155	9177546	36
25	3974148	2670	6025852	2·5162624	4330840	2·3090206	1·0897531	0823609	1157	9176391	35
26	3976818	2668	6023182	2·5145735	4334295	2·3071801	1·0898904	0824766	1157	9175234	34
27	3979486	2669	6020514	2·5128871	4337751	2·3053420	1·0900279	0825923	1158	9174077	33
28	3982155	2668	6017845	2·5112032	4341208	2·3035064	1·0901655	0827081	1159	9172919	32
29	3984823	2668	6015177	2·5095218	4344665	2·3016732	1·0903032	0828240	1159	9171760	31
30	3987491	2667	6012509	2·5078428	4348124	2·2998425	1·0904411	0829399	1161	9170601	30
31	3990158	2667	6009842	2·5061663	4351583	2·2980143	1·0905791	0830560	1161	9169440	29
32	3992825	2667	6007175	2·5044923	4355043	2·2961885	1·0907172	0831721	1161	9168279	28
33	3995492	2666	6004508	2·5028207	4358504	2·2943651	1·0908554	0832882	1163	9167118	27
34	3998158	2667	6001842	2·5011515	4361966	2·2925442	1·0909938	0834045	1164	9165955	26
35	4000825	2665	5999175	2·4994848	4365429	2·2907257	1·0911323	0835209	1164	9164791	25
36	4003490	2666	5996510	2·4978204	4368893	2·2889096	1·0912709	0836373	1165	9163627	24
37	4006156	2665	5993844	2·4961586	4372357	2·2870959	1·0914097	0837538	1165	9162462	23
38	4008821	2665	5991179	2·4944991	4375823	2·2852846	1·0915485	0838703	1167	9161297	22
39	4011486	2664	5988514	2·4928421	4379289	2·2834758	1·0916876	0839870	1167	9160130	21
40	4014150	2664	5985850	2·4911874	4382756	2·2816693	1·0918267	0841037	1168	9158963	20
41	4016814	2664	5983186	2·4895352	4386224	2·2798653	1·0919659	0842205	1169	9157795	19
42	4019478	2663	5980522	2·4878854	4389693	2·2780636	1·0921053	0843374	1170	9156626	18
43	4022141	2663	5977859	2·4862380	4393163	2·2762643	1·0922448	0844544	1170	9155456	17
44	4024804	2663	5975196	2·4845929	4396634	2·2744674	1·0923845	0845714	1171	9154286	16
45	4027467	2662	5972533	2·4829503	4400105	2·2726729	1·0925243	0846885	1172	9153115	15
46	4030129	2662	5969871	2·4813100	4403578	2·2708807	1·0926642	0848057	1173	9151943	14
47	4032791	2662	5967209	2·4796721	4407051	2·2690909	1·0928042	0849230	1173	9150770	13
48	4035453	2661	5964547	2·4780366	4410526	2·2673035	1·0929444	0850403	1175	9149597	12
49	4038114	2661	5961886	2·4764034	4414001	2·2655184	1·0930846	0851578	1175	9148422	11
50	4040775	2661	5959225	2·4747726	4417477	2·2637357	1·0932251	0852753	1175	9147247	10
51	4043436	2660	5956564	2·4731442	4420954	2·2619554	1·0933656	0853928	1177	9146072	9
52	4046096	2660	5953904	2·4715181	4424432	2·2601773	1·0935063	0855105	1177	9144895	8
53	4048756	2660	5951244	2·4698943	4427910	2·2584016	1·0936471	0856282	1178	9143718	7
54	4051416	2659	5948584	2·4682729	4431390	2·2566283	1·0937880	0857460	1179	9142540	6
55	4054075	2659	5945925	2·4666538	4434871	2·2548572	1·0939291	0858639	1180	9141361	5
56	4056734	2659	5943266	2·4650371	4438352	2·2530885	1·0940702	0859819	1180	9140181	4
57	4059393	2658	5940607	2·4634227	4441834	2·2513221	1·0942116	0860999	1182	9139001	3
58	4062051	2658	5937949	2·4618106	4445318	2·2495580	1·0943530	0862181	1182	9137819	2
59	4064709	2657	5935291	2·4602008	4448802	2·2477962	1·0944946	0863363	1182	9136637	1
60	4067366		5932634	2·4585933	4452287	2·2460368	1·0946363	0864545		9135455	0
′	Cosine	Dif.	Vers.	Secant.	Cotan.	Tang.	Cosec.	Covers	Dif.	Sine.	′

′	Sine.	Dif.	Cosec.	Verseds.	Tang.	Dif.	Cotang.	Covers.	Secant.	D.	Cosine.	′
0	9·5918780	2975	10·4081220	8·9003406	9·6278519	3512	10·3721481	9·7848090	10·0359739	537	9·9640261	60
1	9·5921755	2973	10·4078245	8·9009613	9·6282031	3509	10·3717969	9·7846181	10·0360276	537	9·9639724	59
2	9·5924728	2970	10·4075272	8·9015816	9·6285540	3508	10·3714460	9·7844271	10·0360813	537	9·9639187	58
3	9·5927698	2968	10·4072302	8·9022013	9·6289048	3505	10·3710952	9·7842361	10·0361350	538	9·9638650	57
4	9·5930666	2965	10·4069334	8·9028207	9·6292553	3504	10·3707447	9·7840450	10·0361888	538	9·9638112	56
5	9·5933631	2963	10·4066369	8·9034395	9·6296057	3501	10·3703943	9·7838539	10·0362426	538	9·9637574	55
6	9·5936594	2961	10·4063406	8·9040579	9·6299558	3500	10·3700442	9·7836627	10·0362964	540	9·9637036	54
7	9·5939555	2958	10·4060445	8·9046759	9·6303058	3498	10·3696942	9·7834715	10·0363504	539	9·9636496	53
8	9·5942513	2956	10·4057487	8·9052934	9·6306556	3496	10·3693444	9·7832801	10·0364043	540	9·9635957	52
9	9·5945469	2953	10·4054531	8·9059104	9·6310052	3493	10·3689948	9·7830888	10·0364583	540	9·9635417	51
10	9·5948422	2951	10·4051578	8·9065270	9·6313545	3492	10·3686455	9·7828973	10·0365123	541	9·9634877	50
11	9·5951373	2949	10·4048627	8·9071431	9·6317037	3490	10·3682963	9·7827058	10·0365664	541	9·9634336	49
12	9·5954322	2946	10·4045678	8·9077588	9·6320527	3488	10·3679473	9·7825143	10·0366205	542	9·9633795	48
13	9·5957268	2944	10·4042732	8·9083740	9·6324015	3486	10·3675985	9·7823226	10·0366747	542	9·9633253	47
14	9·5960212	2942	10·4039788	8·9089887	9·6327501	3484	10·3672499	9·7821309	10·0367289	543	9·9632711	46
15	9·5963154	2939	10·4036846	8·9096030	9·6330985	3483	10·3669015	9·7819392	10·0367832	543	9·9632168	45
16	9·5966093	2937	10·4033907	8·9102169	9·6334468	3480	10·3665532	9·7817474	10·0368375	543	9·9631625	44
17	9·5969030	2935	10·4030970	8·9108303	9·6337948	3478	10·3662052	9·7815555	10·0368918	544	9·9631082	43
18	9·5971965	2932	10·4028035	8·9114432	9·6341426	3477	10·3658574	9·7813636	10·0369462	544	9·9630538	42
19	9·5974897	2930	10·4025103	8·9120557	9·6344903	3475	10·3655097	9·7811716	10·0370006	545	9·9629994	41
20	9·5977827	2927	10·4022173	8·9126678	9·6348378	3472	10·3651622	9·7809796	10·0370551	545	9·9629449	40
21	9·5980754	2925	10·4019246	8·9132794	9·6351850	3471	10·3648150	9·7807875	10·0371096	545	9·9628904	39
22	9·5983679	2923	10·4016321	8·9138905	9·6355321	3469	10·3644679	9·7805953	10·0371642	546	9·9628358	38
23	9·5986602	2921	10·4013398	8·9145012	9·6358790	3467	10·3641210	9·7804031	10·0372188	546	9·9627812	37
24	9·5989523	2918	10·4010477	8·9151115	9·6362257	3465	10·3637743	9·7802108	10·0372734	546	9·9627266	36
25	9·5992441	2916	10·4007559	8·9157213	9·6365722	3463	10·3634278	9·7800184	10·0373281	547	9·9626719	35
26	9·5995357	2913	10·4004643	8·9163306	9·6369185	3461	10·3630815	9·7798260	10·0373828	547	9·9626172	34
27	9·5998270	2911	10·4001730	8·9169396	9·6372646	3460	10·3627354	9·7796335	10·0374376	548	9·9625624	33
28	9·6001181	2909	10·3998819	8·9175480	9·6376106	3457	10·3623894	9·7794410	10·0374924	548	9·9625076	32
29	9·6004090	2907	10·3995910	8·9181561	9·6379563	3456	10·3620437	9·7792484	10·0375473	549	9·9624527	31
30	9·6006997	2904	10·3993003	8·9187636	9·6383019	3454	10·3616981	9·7790558	10·0376022	549	9·9623978	30
31	9·6009901	2902	10·3990099	8·9193708	9·6386473	3452	10·3613527	9·7788630	10·0376572	550	9·9623428	29
32	9·6012803	2900	10·3987197	8·9199775	9·6389925	3450	10·3610075	9·7786703	10·0377122	550	9·9622878	28
33	9·6015703	2897	10·3984297	8·9205837	9·6393375	3448	10·3606625	9·7784774	10·0377672	550	9·9622328	27
34	9·6018600	2895	10·3981400	8·9211895	9·6396823	3446	10·3603177	9·7782845	10·0378223	551	9·9621777	26
35	9·6021495	2893	10·3978505	8·9217949	9·6400269	3446	10·3599731	9·7780916	10·0378774	551	9·9621226	25
36	9·6024388	2890	10·3975612	8·9223999	9·6403714	3442	10·3596286	9·7778985	10·0379326	552	9·9620674	24
37	9·6027278	2888	10·3972722	8·9230043	9·6407156	3441	10·3592844	9·7777055	10·0379878	552	9·9620122	23
38	9·6030166	2886	10·3969834	8·9236084	9·6410597	3439	10·3589403	9·7775123	10·0380431	553	9·9619569	22
39	9·6033052	2884	10·3966948	8·9242120	9·6414036	3437	10·3585964	9·7773191	10·0380984	553	9·9619016	21
40	9·6035936	2881	10·3964064	8·9248152	9·6417473	3435	10·3582527	9·7771258	10·0381537	553	9·9618463	20
41	9·6038817	2879	10·3961183	8·9254179	9·6420908	3434	10·3579092	9·7769325	10·0382091	554	9·9617909	19
42	9·6041696	2877	10·3958304	8·9260202	9·6424342	3431	10·3575658	9·7767391	10·0382645	554	9·9617355	18
43	9·6044573	2875	10·3955427	8·9266221	9·6427773	3430	10·3572227	9·7765457	10·0383200	555	9·9616800	17
44	9·6047448	2872	10·3952552	8·9272235	9·6431203	3428	10·3568797	9·7763521	10·0383755	555	9·9616245	16
45	9·6050320	2870	10·3949680	8·9278245	9·6434631	3426	10·3565369	9·7761586	10·0384311	556	9·9615689	15
46	9·6053190	2867	10·3946810	8·9284251	9·6438057	3424	10·3561943	9·7759649	10·0384867	556	9·9615133	14
47	9·6056057	2866	10·3943943	8·9290252	9·6441481	3422	10·3558519	9·7757712	10·0385424	557	9·9614577	13
48	9·6058923	2863	10·3941077	8·9296249	9·6444903	3421	10·3555097	9·7755775	10·0385980	556	9·9614020	12
49	9·6061786	2861	10·3938214	8·9302242	9·6448324	3419	10·3551676	9·7753836	10·0386538	558	9·9613462	11
50	9·6064647	2859	10·3935353	8·9308231	9·6451743	3417	10·3548257	9·7751898	10·0387096	558	9·9612904	10
51	9·6067506	2856	10·3932494	8·9314215	9·6455160	3415	10·3544840	9·7749958	10·0387654	559	9·9612346	9
52	9·6070362	2854	10·3929638	8·9320194	9·6458575	3413	10·3541425	9·7748018	10·0388213	559	9·9611787	8
53	9·6073216	2852	10·3926784	8·9326170	9·6461988	3412	10·3538012	9·7746077	10·0388772	560	9·9611228	7
54	9·6076068	2850	10·3923932	8·9332141	9·6465400	3410	10·3534600	9·7744136	10·0389332	560	9·9610668	6
55	9·6078918	2847	10·3921082	8·9338108	9·6468810	3407	10·3531190	9·7742194	10·0389892	560	9·9610108	5
56	9·6081765	2846	10·3918235	8·9344070	9·6472217	3407	10·3527783	9·7740252	10·0390452	561	9·9609548	4
57	9·6084611	2843	10·3915389	8·9350029	9·6475624	3404	10·3524376	9·7738308	10·0391013	561	9·9608987	3
58	9·6087454	2840	10·3912546	8·9355983	9·6479028	3403	10·3520972	9·7736365	10·0391574	562	9·9608426	2
59	9·6090294	2839	10·3909706	8·9361933	9·6482431	3400	10·3517569	9·7734420	10·0392136	562	9·9607864	1
60	9·6093133		10·3906867	8·9367878	9·6485831		10·3514169	9·7732475	10·0392698		9·9607302	0

| ′ | Cosine. | Dif. | Secant. | Covers. | Cotang. | Dif. | Tang. | Verseds. | Cosec. | D. | Sine. | ′ |

′	Sine.	Dif.	Covers	Cosec.	Tang.	Cotang.	Secant.	Vers.	Dif.	Cosine	′
0	4067366	2658	5932634	2·4585933	4452287	2·2460368	1·0946363	0864545	1184	9135455	60
1	4070024	2657	5929976	2·4569882	4455773	2·2442796	1·0947781	0865729	1184	9134271	59
2	4072681	2656	5927319	2·4553853	4459260	2·2425247	1·0949201	0866913	1185	9133087	58
3	4075337	2656	5924663	2·4537848	4462747	2·2407721	1·0950622	0868098	1186	9131902	57
4	4077993	2656	5922007	2·4521865	4466236	2·2390218	1·0952044	0869284	1187	9130716	56
5	4080649	2656	5919351	2·4505905	4469726	2·2372738	1·0953467	0870471	1187	9129529	55
6	4083305	2655	5916695	2·4489968	4473216	2·2355280	1·0954892	0871658	1188	9128342	54
7	4085960	2655	5914040	2·4474054	4476708	2·2337845	1·0956318	0872846	1189	9127154	53
8	4088615	2654	5911385	2·4458163	4480200	2·2320433	1·0957746	0874035	1190	9125965	52
9	4091269	2654	5908731	2·4442294	4483693	2·2303043	1·0959174	0875225	1191	9124775	51
10	4093923	2654	5906077	2·4426448	4487187	2·2285676	1·0960604	0876416	1191	9123584	50
11	4096577	2653	5903423	2·4410624	4490682	2·2268331	1·0962036	0877607	1192	9122393	49
12	4099230	2653	5900770	2·4394823	4494178	2·2251009	1·0963468	0878799	1193	9121201	48
13	4101883	2653	5898117	2·4379045	4497675	2·2233709	1·0964902	0879992	1193	9120008	47
14	4104536	2653	5895464	2·4363289	4501173	2·2216432	1·0966337	0881185	1195	9118815	46
15	4107189	2652	5892811	2·4347555	4504672	2·2199177	1·0967774	0882380	1195	9117620	45
16	4109841	2651	5890159	2·4331844	4508171	2·2181944	1·0969212	0883575	1196	9116425	44
17	4112492	2652	5887508	2·4316155	4511672	2·2164733	1·0970651	0884771	1196	9115229	43
18	4115144	2651	5884856	2·4300489	4515173	2·2147545	1·0972091	0885967	1198	9114033	42
19	4117795	2650	5882205	2·4284844	4518676	2·2130379	1·0973533	0887165	1198	9112835	41
20	4120445	2651	5879555	2·4269222	4522179	2·2113234	1·0974976	0888363	1199	9111637	40
21	4123096	2649	5876904	2·4253622	4525683	2·2096112	1·0976420	0889562	1200	9110438	39
22	4125745	2650	5874255	2·4238044	4529188	2·2079012	1·0977866	0890762	1200	9109238	38
23	4128395	2649	5871605	2·4222488	4532694	2·2061934	1·0979313	0891962	1201	9108038	37
24	4131044	2649	5868956	2·4206954	4536201	2·2044878	1·0980761	0893163	1202	9106837	36
25	4133693	2649	5866307	2·4191442	4539709	2·2027843	1·0982211	0894365	1203	9105635	35
26	4136342	2648	5863658	2·4175952	4543218	2·2010831	1·0983662	0895568	1204	9104432	34
27	4138990	2648	5861010	2·4160484	4546728	2·1993840	1·0985114	0896772	1204	9103228	33
28	4141638	2647	5858362	2·4145038	4550238	2·1976871	1·0986568	0897976	1205	9102024	32
29	4144285	2647	5855715	2·4129613	4553750	2·1959923	1·0988023	0899181	1206	9100819	31
30	4146932	2647	5853068	2·4114210	4557263	2·1942997	1·0989479	0900387	1207	9099613	30
31	4149579	2647	5850421	2·4098829	4560776	2·1926093	1·0990936	0901594	1207	9098406	29
32	4152226	2646	5847774	2·4083469	4564290	2·1909210	1·0992395	0902801	1209	9097199	28
33	4154872	2645	5845128	2·4068132	4567806	2·1892349	1·0993855	0904010	1209	9095990	27
34	4157517	2646	5842483	2·4052815	4571322	2·1875510	1·0995317	0905219	1209	9094781	26
35	4160163	2645	5839837	2·4037520	4574839	2·1858691	1·0996779	0906428	1211	9093572	25
36	4162808	2645	5837192	2·4022247	4578357	2·1841894	1·0998243	0907639	1211	9092361	24
37	4165453	2644	5834547	2·4006995	4581877	2·1825119	1·0999709	0908850	1212	9091150	23
38	4168097	2644	5831903	2·3991764	4585397	2·1808364	1·1001175	0910062	1213	9089938	22
39	4170741	2644	5829259	2·3976555	4588918	2·1791631	1·1002644	0911275	1214	9088725	21
40	4173385	2643	5826615	2·3961367	4592439	2·1774920	1·1004113	0912489	1214	9087511	20
41	4176028	2643	5823972	2·3946201	4595962	2·1758229	1·1005584	0913703	1215	9086297	19
42	4178671	2642	5821329	2·3931055	4599486	2·1741559	1·1007056	0914918	1216	9085082	18
43	4181313	2643	5818687	2·3915931	4603011	2·1724911	1·1008529	0916134	1217	9083866	17
44	4183956	2641	5816044	2·3900828	4606537	2·1708283	1·1010004	0917351	1217	9082649	16
45	4186597	2642	5813403	2·3885746	4610063	2·1691677	1·1011480	0918568	1218	9081432	15
46	4189239	2641	5810761	2·3870685	4613591	2·1675091	1·1012957	0919786	1219	9080214	14
47	4191880	2641	5808120	2·3855645	4617119	2·1658527	1·1014436	0921005	1220	9078995	13
48	4194521	2640	5805479	2·3840625	4620649	2·1641983	1·1015916	0922225	1221	9077775	12
49	4197161	2640	5802839	2·3825627	4624179	2·1625460	1·1017397	0923446	1221	9076554	11
50	4199801	2640	5800199	2·3810650	4627710	2·1608958	1·1018879	0924667	1222	9075333	10
51	4202441	2639	5797559	2·3795694	4631243	2·1592476	1·1020363	0925889	1223	9074111	9
52	4205080	2639	5794920	2·3780758	4634776	2·1576015	1·1021849	0927112	1223	9072888	8
53	4207719	2639	5792281	2·3765843	4638310	2·1559575	1·1023335	0928335	1225	9071665	7
54	4210358	2638	5789642	2·3750949	4641845	2·1543156	1·1024823	0929560	1225	9070440	6
55	4212996	2638	5787004	2·3736075	4645382	2·1526757	1·1026313	0930785	1226	9069215	5
56	4215634	2638	5784366	2·3721222	4648919	2·1510378	1·1027803	0932011	1227	9067989	4
57	4218272	2637	5781728	2·3706390	4652457	2·1494021	1·1029295	0933238	1227	9066762	3
58	4220909	2637	5779091	2·3691578	4655996	2·1477683	1·1030789	0934465	1228	9065535	2
59	4223546	2637	5776454	2·3676787	4659536	2·1461366	1·1032283	0935693	1229	9064307	1
60	4226183		5773817	2·3662016	4663077	2·1445069	1·1033779	0936922		9063078	0
′	Cosine	Dif.	Vers.	Secant.	Cotan.	Tang.	Cosec.	Covers	Dif.	Sine.	′

'	Sine.	Dif.	Cosec.	Verseds.	Tang.	Dif.	Cotang.	Covers.	Secant.	D.	Cosine.	'
0	9·6093133	2836	10·3906867	8·9367878	9·6485831	3309	10·3514169	9·7732475	10·0392698	563	9·9607302	60
1	9·6095969	2834	10·3904031	8·9373819	9·6489230	3398	10·3510770	9·7730530	10·0393261	563	9·9606739	59
2	9·6098803	2832	10·3901197	8·9379756	9·6492628	3395	10·3507372	9·7728583	10·0393824	564	9·9606176	58
3	9·6101635	2830	10·3898365	8·9385689	9·6496023	3394	10·3503977	9·7726636	10·0394388	564	9·9605612	57
4	9·6104465	2828	10·3895535	8·9391618	9·6499417	3392	10·3500583	9·7724689	10·0394952	564	9·9605048	56
5	9·6107293	2825	10·3892707	8·9397542	9·6502809	3390	10·3497191	9·7722741	10·0395516	565	9·9604484	55
6	9·6110118	2823	10·3889882	8·9403462	9·6506199	3388	10·3493801	9·7720792	10·0396081	565	9·9603919	54
7	9·6112941	2821	10·3887059	8·9409378	9·6509587	3387	10·3490413	9·7718843	10·0396646	566	9·9603354	53
8	9·6115762	2818	10·3884238	8·9415290	9·6512974	3385	10·3487026	9·7716893	10·0397212	566	9·9602788	52
9	9·6118580	2817	10·3881420	8·9421197	9·6516359	3383	10·3483641	9·7714942	10·0397778	567	9·9602222	51
10	9·6121397	2814	10·3878603	8·9427101	9·6519742	3381	10·3480258	9·7712991	10·0398345	567	9·9601655	50
11	9·6124211	2812	10·3875789	8·9433000	9·6523123	3380	10·3476877	9·7711039	10·0398912	568	9·9601088	49
12	9·6127023	2810	10·3872977	8·9438895	9·6526503	3378	10·3473497	9·7709087	10·0399480	568	9·9600520	48
13	9·6129833	2808	10·3870167	8·9444785	9·6529881	3376	10·3470119	9·7707134	10·0400048	568	9·9599952	47
14	9·6132641	2805	10·3867359	8·9450672	9·6533257	3374	10·3466743	9·7705180	10·0400616	569	9·9599384	46
15	9·6135446	2804	10·3864554	8·9456554	9·6536631	3373	10·3463369	9·7703225	10·0401185	569	9·9598815	45
16	9·6138250	2801	10·3861750	8·9462433	9·6540004	3371	10·3459996	9·7701271	10·0401754	570	9·9598246	44
17	9·6141051	2799	10·3858949	8·9468307	9·6543375	3369	10·3456625	9·7699315	10·0402324	570	9·9597676	43
18	9·6143850	2797	10·3856150	8·9474177	9·6546744	3368	10·3453256	9·7697359	10·0402894	571	9·9597106	42
19	9·6146647	2794	10·3853353	8·9480042	9·6550112	3365	10·3449888	9·7695402	10·0403465	571	9·9596535	41
20	9·6149441	2793	10·3850559	8·9485904	9·6553477	3364	10·3446523	9·7693444	10·0404036	571	9·9595964	40
21	9·6152234	2790	10·3847766	8·9491761	9·6556841	3363	10·3443159	9·7691486	10·0404607	572	9·9595393	39
22	9·6155024	2788	10·3844976	8·9497615	9·6560204	3360	10·3439796	9·7689528	10·0405179	573	9·9594821	38
23	9·6157812	2787	10·3842188	8·9503464	9·6563564	3359	10·3436436	9·7687568	10·0405752	573	9·9594248	37
24	9·6160599	2783	10·3839401	8·9509309	9·6566923	3357	10·3433077	9·7685608	10·0406325	573	9·9593675	36
25	9·6163382	2782	10·3836618	8·9515150	9·6570280	3356	10·3429720	9·7683648	10·0406898	574	9·9593102	35
26	9·6166164	2780	10·3833836	8·9520987	9·6573636	3353	10·3426364	9·7681687	10·0407472	574	9·9592528	34
27	9·6168944	2777	10·3831056	8·9526820	9·6576989	3352	10·3423011	9·7679725	10·0408046	574	9·9591954	33
28	9·6171721	2775	10·3828279	8·9532648	9·6580341	3351	10·3419659	9·7677762	10·0408620	575	9·9591380	32
29	9·6174496	2774	10·3825504	8·9538473	9·6583692	3349	10·3416308	9·7675799	10·0409195	575	9·9590805	31
30	9·6177270	2771	10·3822730	8·9544294	9·6587041	3346	10·3412959	9·7673835	10·0409771	576	9·9590229	30
31	9·6180041	2768	10·3819959	8·9550110	9·6590387	3346	10·3409613	9·7671871	10·0410347	576	9·9589653	29
32	9·6182809	2767	10·3817191	8·9555922	9·6593733	3343	10·3406267	9·7669906	10·0410923	577	9·9589077	28
33	9·6185576	2765	10·3814424	8·9561731	9·6597076	3342	10·3402924	9·7667940	10·0411500	577	9·9588500	27
34	9·6188341	2762	10·3811659	8·9567535	9·6600418	3340	10·3399582	9·7665974	10·0412077	578	9·9587923	26
35	9·6191103	2761	10·3808897	8·9573335	9·6603758	3339	10·3396242	9·7664007	10·0412655	578	9·9587345	25
36	9·6193864	2758	10·3806136	8·9579131	9·6607097	3337	10·3392903	9·7662040	10·0413233	579	9·9586767	24
37	9·6196622	2756	10·3803378	8·9584923	9·6610434	3335	10·3389566	9·7660072	10·0413812	579	9·9586188	23
38	9·6199378	2754	10·3800622	8·9590711	9·6613769	3334	10·3386231	9·7658103	10·0414391	579	9·9585609	22
39	9·6202132	2752	10·3797868	8·9596495	9·6617103	3331	10·3382897	9·7656134	10·0414970	580	9·9585030	21
40	9·6204884	2750	10·3795116	8·9602275	9·6620434	3331	10·3379566	9·7654164	10·0415550	581	9·9584450	20
41	9·6207634	2748	10·3792366	8·9608051	9·6623765	3328	10·3376235	9·7652193	10·0416131	581	9·9583869	19
42	9·6210382	2745	10·3789618	8·9613823	9·6627093	3327	10·3372907	9·7650222	10·0416712	581	9·9583288	18
43	9·6213127	2744	10·3786873	8·9619591	9·6630420	3325	10·3369580	9·7648250	10·0417293	582	9·9582707	17
44	9·6215871	2741	10·3784129	8·9625355	9·6633745	3324	10·3366255	9·7646277	10·0417875	582	9·9582125	16
45	9·6218612	2739	10·3781388	8·9631114	9·6637069	3322	10·3362931	9·7644304	10·0418457	582	9·9581543	15
46	9·6221351	2737	10·3778649	8·9636870	9·6640391	3320	10·3359609	9·7642330	10·0419039	583	9·9580961	14
47	9·6224088	2736	10·3775912	8·9642622	9·6643711	3319	10·3356289	9·7640356	10·0419622	584	9·9580378	13
48	9·6226824	2733	10·3773176	8·9648370	9·6647030	3316	10·3352970	9·7638381	10·0420206	584	9·9579794	12
49	9·6229557	2730	10·3770443	8·9654114	9·6650346	3316	10·3349654	9·7636405	10·0420790	584	9·9579210	11
50	9·6232287	2729	10·3767713	8·9659854	9·6653662	3313	10·3346338	9·7634429	10·0421374	585	9·9578626	10
51	9·6235016	2727	10·3764984	8·9665590	9·6656975	3313	10·3343025	9·7632452	10·0421959	585	9·9578041	9
52	9·6237743	2725	10·3762257	8·9671322	9·6660288	3310	10·3339712	9·7630474	10·0422544	586	9·9577456	8
53	9·6240468	2722	10·3759532	8·9677050	9·6663598	3309	10·3336402	9·7628496	10·0423130	586	9·9576870	7
54	9·6243190	2721	10·3756810	8·9682774	9·6666907	3307	10·3333093	9·7626517	10·0423716	587	9·9576284	6
55	9·6245911	2718	10·3754089	8·9688494	9·6670214	3305	10·3329786	9·7624537	10·0424303	587	9·9575697	5
56	9·6248629	2717	10·3751371	8·9694210	9·6673519	3304	10·3326481	9·7622557	10·0424890	588	9·9575110	4
57	9·6251346	2714	10·3748654	8·9699922	9·6676823	3303	10·3323177	9·7620577	10·0425478	588	9·9574522	3
58	9·6254060	2712	10·3745940	8·9705630	9·6680126	3300	10·3319874	9·7618595	10·0426066	588	9·9573934	2
59	9·6256772	2711	10·3743228	8·9711335	9·6683426	3299	10·3316574	9·7616613	10·0426654	589	9·9573346	1
60	9·6259483		10·3740517	8·9717035	9·6686725		10·3313275	9·7614630	10·0427243		9·9572757	0
'	Cosine	Dif.	Secant.	Covers.	Cotang.	Dif.	Tang.	Verseds.	Cosec.	D.	Sine.	'

′	Sine.	Dif.	Covers	Cosec.	Tang.	Cotang.	Secant.	Vers.	Dif.	Cosine	′
0	4226183	2636	5773817	2·3662016	4663077	2·1445069	1·1033779	0936922	1230	9063078	60
1	4228819	2636	5771181	2·3647265	4666618	2·1428793	1·1035277	0938152	1230	9061848	59
2	4231455	2635	5768545	2·3632535	4670161	2·1412537	1·1036775	0939382	1232	9060618	58
3	4234090	2635	5765910	2·3617826	4673705	2·1396301	1·1038275	0940614	1232	9059386	57
4	4236725	2635	5763275	2·3603136	4677250	2·1380085	1·1039777	0941846	1232	9058154	56
5	4239360	2634	5760640	2·3588467	4680796	2·1363890	1·1041279	0943078	1234	9056922	55
6	4241994	2634	5758006	2·3573818	4684342	2·1347714	1·1042783	0944312	1234	9055688	54
7	4244628	2634	5755372	2·3559189	4687890	2·1331559	1·1044289	0945546	1235	9054454	53
8	4247262	2633	5752738	2·3544581	4691439	2·1315423	1·1045795	0946781	1236	9053219	52
9	4249895	2633	5750105	2·3529992	4694988	2·1299308	1·1047303	0948017	1237	9051983	51
10	4252528	2633	5747472	2·3515424	4698539	2·1283213	1·1048813	0949254	1237	9050746	50
11	4255161	2632	5744839	2·3500875	4702090	2·1267137	1·1050324	0950491	1238	9049509	49
12	4257793	2632	5742207	2·3486347	4705643	2·1251082	1·1051836	0951729	1239	9048271	48
13	4260425	2631	5739575	2·3471838	4709196	2·1235046	1·1053349	0952968	1240	9047032	47
14	4263056	2631	5736944	2·3457349	4712751	2·1219030	1·1054864	0954208	1241	9045792	46
15	4265687	2631	5734313	2·3442881	4716306	2·1203034	1·1056380	0955449	1241	9044551	45
16	4268318	2631	5731682	2·3428432	4719863	2·1187057	1·1057898	0956690	1242	9043310	44
17	4270949	2630	5729051	2·3414002	4723420	2·1171101	1·1059417	0957932	1243	9042068	43
18	4273579	2629	5726421	2·3399593	4726978	2·1155164	1·1060937	0959175	1243	9040825	42
19	4276208	2630	5723792	2·3385203	4730538	2·1139246	1·1062458	0960418	1244	9039582	41
20	4278838	2629	5721162	2·3370833	4734098	2·1123348	1·1063981	0961662	1245	9038338	40
21	4281467	2628	5718533	2·3356482	4737659	2·1107470	1·1065506	0962907	1246	9037093	39
22	4284095	2628	5715905	2·3342152	4741222	2·1091611	1·1067031	0964153	1247	9035847	38
23	4286723	2628	5713277	2·3327840	4744785	2·1075771	1·1068558	0965400	1247	9034600	37
24	4289351	2628	5710649	2·3313548	4748349	2·1059951	1·1070087	0966647	1248	9033353	36
25	4291979	2627	5708021	2·3299276	4751914	2·1044150	1·1071616	0967895	1249	9032105	35
26	4294606	2627	5705394	2·3285023	4755481	2·1028369	1·1073147	0969144	1250	9030856	34
27	4297233	2626	5702767	2·3270790	4759048	2·1012607	1·1074680	0970394	1250	9029606	33
28	4299859	2626	5700141	2·3256575	4762616	2·0996864	1·1076214	0971644	1251	9028356	32
29	4302485	2626	5697515	2·3242381	4766185	2·0981140	1·1077749	0972895	1252	9027105	31
30	4305111	2625	5694889	2·3228205	4769755	2·0965436	1·1079285	0974147	1253	9025853	30
31	4307736	2625	5692264	2·3214049	4773326	2·0949751	1·1080823	0975400	1253	9024600	29
32	4310361	2625	5689639	2·3199912	4776899	2·0934085	1·1082363	0976653	1255	9023347	28
33	4312986	2624	5687014	2·3185794	4780472	2·0918437	1·1083903	0977908	1254	9022092	27
34	4315610	2624	5684390	2·3171695	4784046	2·0902809	1·1085445	0979162	1256	9020838	26
35	4318234	2623	5681766	2·3157615	4787621	2·0887200	1·1086989	0980418	1257	9019582	25
36	4320857	2624	5679143	2·3143554	4791197	2·0871610	1·1088533	0981675	1257	9018325	24
37	4323481	2622	5676519	2·3129513	4794774	2·0856039	1·1090079	0982932	1258	9017068	23
38	4326103	2623	5673897	2·3115490	4798352	2·0840487	1·1091627	0984190	1259	9015810	22
39	4328726	2622	5671274	2·3101486	4801932	2·0824953	1·1093176	0985449	1259	9014551	21
40	4331348	2622	5668652	2·3087501	4805512	2·0809438	1·1094726	0986708	1261	9013292	20
41	4333970	2621	5666030	2·3073536	4809093	2·0793942	1·1096277	0987969	1261	9012031	19
42	4336591	2621	5663409	2·3059588	4812675	2·0778465	1·1097830	0989230	1262	9010770	18
43	4339212	2620	5660788	2·3045660	4816258	2·0763007	1·1099385	0990492	1262	9009508	17
44	4341832	2621	5658168	2·3031751	4819842	2·0747567	1·1100940	0991754	1264	9008246	16
45	4344453	2619	5655547	2·3017860	4823427	2·0732146	1·1102498	0993018	1264	9006982	15
46	4347072	2620	5652928	2·3003988	4827014	2·0716743	1·1104056	0994282	1265	9005718	14
47	4349692	2619	5650308	2·2990134	4830601	2·0701359	1·1105616	0995547	1265	9004453	13
48	4352311	2619	5647689	2·2976299	4834189	2·0685994	1·1107177	0996812	1267	9003188	12
49	4354930	2618	5645070	2·2962483	4837778	2·0670646	1·1108740	0998079	1267	9001921	11
50	4357548	2618	5642452	2·2948685	4841368	2·0655318	1·1110304	0999346	1268	9000654	10
51	4360166	2618	5639834	2·2934906	4844959	2·0640008	1·1111869	1000614	1269	8999386	9
52	4362784	2617	5637216	2·2921145	4848552	2·0624716	1·1113436	1001883	1269	8998117	8
53	4365401	2617	5634599	2·2907403	4852145	2·0609442	1·1115004	1003152	1270	8996848	7
54	4368018	2616	5631982	2·2893679	4855739	2·0594187	1·1116573	1004422	1271	8995578	6
55	4370634	2617	5629366	2·2879974	4859334	2·0578950	1·1118144	1005693	1272	8994307	5
56	4373251	2615	5626749	2·2866286	4862931	2·0563732	1·1119716	1006965	1272	8993035	4
57	4375866	2616	5624134	2·2852618	4866528	2·0548531	1·1121290	1008237	1274	8991763	3
58	4378482	2615	5621518	2·2838967	4870126	2·0533349	1·1122865	1009511	1274	8990489	2
59	4381097	2614	5618903	2·2825335	4873726	2·0518185	1·1124442	1010785	1275	8989215	1
60	4383711		5616289	2·2811720	4877326	2·0503038	1·1126019	1012060		8987940	0
′	Cosine	Dif.	Vers.	Secant.	Cotan.	Tang.	Cosec.	Covers	Dif.	Sine.	′

′	Sine.	Dif.	Cosec.	Verseds.	Tang.	Dif.	Cotang.	Covers.	Secant.	D.	Cosine.	′
0	9·6259483	2708	10·3740517	8·9717035	9·6686725	3298	10·3313275	9·7614630	10·0427243	589	9·9572757	60
1	9·6262191	2706	10·3737809	8·9722731	9·6690023	3296	10·3309977	9·7612647	10·0427832	590	9·9572168	59
2	9·6264897	2704	10·3735103	8·9728424	9·6693319	3294	10·3306681	9·7610663	10·0428422	590	9·9571578	58
3	9·6267601	2702	10·3732399	8·9734113	9·6696613	3293	10·3303387	9·7608679	10·0429012	591	9·9570988	57
4	9·6270303	2700	10·3729697	8·9739797	9·6699906	3291	10·3300094	9·7606693	10·0429603	591	9·9570397	56
5	9·6273003	2698	10·3726997	8·9745478	9·6703197	3289	10·3296803	9·7604707	10·0430194	591	9·9569806	55
6	9·6275701	2696	10·3724299	8·9751155	9·6706486	3288	10·3293514	9·7602721	10·0430785	592	9·9569215	54
7	9·6278397	2693	10·3721603	8·9756828	9·6709774	3286	10·3290226	9·7600734	10·0431377	593	9·9568623	53
8	9·6281090	2692	10·3718910	8·9762497	9·6713060	3285	10·3286940	9·7598746	10·0431970	593	9·9568030	52
9	9·6283782	2690	10·3716218	8·9768163	9·6716345	3283	10·3283655	9·7596758	10·0432563	593	9·9567437	51
10	9·6286472	2688	10·3713528	8·9773824	9·6719628	3282	10·3280372	9·7594769	10·0433156	594	9·9566844	50
11	9·6289160	2685	10·3710840	8·9779482	9·6722910	3280	10·3277090	9·7592779	10·0433750	594	9·9566250	49
12	9·6291845	2684	10·3708155	8·9785135	9·6726190	3278	10·3273810	9·7590789	10·0434344	595	9·9565656	48
13	9·6294529	2682	10·3705471	8·9790785	9·6729468	3277	10·3270532	9·7588798	10·0434939	595	9·9565061	47
14	9·6297211	2679	10·3702789	8·9796431	9·6732745	3275	10·3267255	9·7586806	10·0435534	596	9·9564466	46
15	9·6299890	2678	10·3700110	8·9802073	9·6736020	3274	10·3263980	9·7584814	10·0436130	596	9·9563870	45
16	9·6302568	2675	10·3697432	8·9807711	9·6739294	3272	10·3260706	9·7582821	10·0436726	596	9·9563274	44
17	9·6305243	2674	10·3694757	8·9813346	9·6742566	3270	10·3257434	9·7580827	10·0437322	597	9·9562678	43
18	9·6307917	2672	10·3692083	8·9818976	9·6745836	3269	10·3254164	9·7578833	10·0437919	598	9·9562081	42
19	9·6310589	2669	10·3689411	8·9824603	9·6749105	3267	10·3250895	9·7576838	10·0438517	597	9·9561483	41
20	9·6313258	2668	10·3686742	8·9830226	9·6752372	3266	10·3247628	9·7574843	10·0439114	599	9·9560886	40
21	9·6315926	2668	10·3684074	8·9835845	9·6755638	3265	10·3244362	9·7572847	10·0439713	598	9·9560287	39
22	9·6318591	2665	10·3681409	8·9841460	9·6758903	3262	10·3241097	9·7570850	10·0440311	600	9·9559689	38
23	9·6321255	2664	10·3678745	8·9847072	9·6762165	3261	10·3237835	9·7568852	10·0440911	599	9·9559089	37
24	9·6323916	2661	10·3676084	8·9852679	9·6765426	3260	10·3234574	9·7566854	10·0441510	600	9·9558490	36
25	9·6326576	2660	10·3673424	8·9858283	9·6768686	3258	10·3231314	9·7564856	10·0442110	601	9·9557890	35
26	9·6329233	2657	10·3670767	8·9863883	9·6771944	3257	10·3228056	9·7562856	10·0442711	601	9·9557289	34
27	9·6331889	2656	10·3668111	8·9869480	9·6775201	3255	10·3224799	9·7560856	10·0443312	601	9·9556688	33
28	9·6334542	2653	10·3665458	8·9875072	9·6778456	3253	10·3221544	9·7558854	10·0443913	602	9·9556087	32
29	9·6337194	2652	10·3662806	8·9880661	9·6781709	3252	10·3218291	9·7556855	10·0444515	603	9·9555485	31
30	9·6339844	2650	10·3660156	8·9886246	9·6784961	3250	10·3215039	9·7554853	10·0445118	602	9·9554882	30
31	9·6342491	2647	10·3657509	8·9891827	9·6788211	3249	10·3211789	9·7552850	10·0445720	604	9·9554280	29
32	9·6345137	2646	10·3654863	8·9897404	9·6791460	3248	10·3208540	9·7550847	10·0446324	603	9·9553676	28
33	9·6347780	2643	10·3652220	8·9902992	9·6794708	3245	10·3205292	9·7548843	10·0446927	604	9·9553073	27
34	9·6350422	2642	10·3649578	8·9908548	9·6797953	3245	10·3202047	9·7546839	10·0447531	605	9·9552469	26
35	9·6353062	2640	10·3646938	8·9914114	9·6801198	3242	10·3198802	9·7544833	10·0448136	605	9·9551864	25
36	9·6355699	2637	10·3644301	8·9919676	9·6804440	3242	10·3195560	9·7542828	10·0448741	606	9·9551259	24
37	9·6358335	2636	10·3641665	8·9925235	9·6807682	3239	10·3192318	9·7540821	10·0449347	606	9·9550653	23
38	9·6360969	2634	10·3639031	8·9930790	9·6810921	3239	10·3189079	9·7538814	10·0449953	606	9·9550047	22
39	9·6363601	2632	10·3636399	8·9936341	9·6814160	3236	10·3185840	9·7536806	10·0450559	607	9·9549441	21
40	9·6366231	2630	10·3633769	8·9941888	9·6817396	3236	10·3182604	9·7534798	10·0451166	607	9·9548834	20
41	9·6368859	2628	10·3631141	8·9947432	9·6820632	3233	10·3179368	9·7532789	10·0451773	608	9·9548227	19
42	9·6371484	2625	10·3628516	8·9952972	9·6823865	3233	10·3176135	9·7530779	10·0452381	608	9·9547619	18
43	9·6374108	2624	10·3625892	8·9958508	9·6827098	3230	10·3172902	9·7528769	10·0452989	609	9·9547011	17
44	9·6376731	2623	10·3623269	8·9964041	9·6830328	3229	10·3169672	9·7526758	10·0453598	609	9·9546402	16
45	9·6379351	2620	10·3620649	8·9969569	9·6833557	3228	10·3166443	9·7524746	10·0454207	609	9·9545793	15
46	9·6381969	2618	10·3618031	8·9975056	9·6836785	3226	10·3163215	9·7522734	10·0454816	610	9·9545184	14
47	9·6384585	2616	10·3615415	8·9980616	9·6840011	3225	10·3159989	9·7520721	10·0455426	611	9·9544574	13
48	9·6387199	2614	10·3612801	8·9986134	9·6843236	3223	10·3156764	9·7518708	10·0456037	611	9·9543963	12
49	9·6389812	2613	10·3610188	8·9991648	9·6846459	3222	10·3153541	9·7516694	10·0456648	611	9·9543352	11
50	9·6392422	2610	10·3607578	8·9997158	9·6849681	3220	10·3150319	9·7514679	10·0457259	612	9·9542741	10
51	9·6395030	2608	10·3604970	9·0002665	9·6852901	3219	10·3147099	9·7512663	10·0457871	612	9·9542129	9
52	9·6397637	2607	10·3602363	9·0008168	9·6856120	3218	10·3143880	9·7510647	10·0458483	613	9·9541517	8
53	9·6400241	2604	10·3599759	9·0013667	9·6859338	3215	10·3140662	9·7508630	10·0459096	613	9·9540904	7
54	9·6402844	2603	10·3597156	9·0019163	9·6862553	3215	10·3137447	9·7506613	10·0459709	614	9·9540291	6
55	9·6405445	2601	10·3594555	9·0024655	9·6865768	3213	10·3134232	9·7504595	10·0460323	614	9·9539677	5
56	9·6408044	2599	10·3591956	9·0030144	9·6868981	3211	10·3131019	9·7502576	10·0460937	615	9·9539063	4
57	9·6410640	2596	10·3589360	9·0035628	9·6872192	3210	10·3127808	9·7500556	10·0461552	615	9·9538448	3
58	9·6413235	2595	10·3586765	9·0041109	9·6875402	3209	10·3124598	9·7498536	10·0462167	615	9·9537833	2
59	9·6415828	2593	10·3584172	9·0046587	9·6878611	3207	10·3121389	9·7496516	10·0462782	616	9·9537218	1
60	9·6418420	2592	10·3581580	9·0052061	9·6881818		10·3118182	9·7494494	10·0463398	616	9·9536602	0
′	Cosine.	Dif.	Secant.	Covers.	Cotang.	Dif.	Tang.	Verseds.	Cosec.	D.	Sine.	′

64 Deg.

′	Sine.	Dif.	Covers	Cosec.	Tang.	Cotang.	Secant.	Vers.	Dif.	Cosine	′
0	4383711	2615	5616289	2·2811720	4877326	2·0503038	1·1126019	1012060	1275	8987940	60
1	4386326	2614	5613674	2·2798124	4880927	2·0487910	1·1127599	1013335	1276	8986665	59
2	4388940	2613	5611060	2·2784546	4884530	2·0472800	1·1129179	1014611	1277	8985389	58
3	4391553	2613	5608447	2·2770987	4888133	2·0457708	1·1130761	1015888	1278	8984112	57
4	4394166	2613	5605834	2·2757445	4891737	2·0442634	1·1132345	1017166	1279	8982834	56
5	4396779	2613	5603221	2·2743921	4895343	2·0427578	1·1133929	1018445	1279	8981555	55
6	4399392	2612	5600608	2·2730415	4898949	2·0412540	1·1135516	1019724	1280	8980276	54
7	4402004	2611	5597996	2·2716927	4902557	2·0397519	1·1137103	1021004	1281	8978996	53
8	4404615	2612	5595385	2·2703457	4906166	2·0382517	1·1138692	1022285	1282	8977715	52
9	4407227	2611	5592773	2·2690005	4909775	2·0367532	1·1140282	1023567	1282	8976433	51
10	4409838	2610	5590162	2·2676571	4913386	2·0352565	1·1141874	1024849	1283	8975151	50
11	4412448	2611	5587552	2·2663155	4916997	2·0337615	1·1143467	1026132	1284	8973868	49
12	4415059	2609	5584941	2·2649756	4920610	2·0322683	1·1145062	1027416	1285	8972584	48
13	4417668	2610	5582332	2·2636376	4924224	2·0307769	1·1146658	1028701	1285	8971299	47
14	4420278	2609	5579722	2·2623012	4927838	2·0292873	1·1148255	1029986	1287	8970014	46
15	4422887	2609	5577113	2·2609667	4931454	2·0277994	1·1149854	1031273	1287	8968727	45
16	4425496	2608	5574504	2·2596339	4935071	2·0263133	1·1151454	1032560	1287	8967440	44
17	4428104	2608	5571896	2·2583020	4938689	2·0248289	1·1153056	1033847	1289	8966153	43
18	4430712	2607	5569288	2·2569736	4942308	2·0233462	1·1154659	1035136	1289	8964864	42
19	4433319	2608	5566681	2·2556461	4945928	2·0218654	1·1156263	1036425	1290	8963575	41
20	4435927	2607	5564073	2·2543204	4949549	2·0203862	1·1157869	1037715	1291	8962285	40
21	4438534	2606	5561466	2·2529964	4953171	2·0189088	1·1159476	1039006	1291	8960994	39
22	4441140	2606	5558860	2·2516741	4956794	2·0174331	1·1161084	1040297	1292	8959703	38
23	4443746	2606	5556254	2·2503536	4960418	2·0159592	1·1162694	1041589	1293	8958411	37
24	4446352	2605	5553648	2·2490348	4964043	2·0144869	1·1164306	1042882	1294	8957118	36
25	4448957	2605	5551043	2·2477178	4967669	2·0130164	1·1165919	1044176	1295	8955824	35
26	4451562	2605	5548438	2·2464025	4971297	2·0115477	1·1167533	1045471	1295	8954529	34
27	4454167	2604	5545833	2·2450889	4974925	2·0100806	1·1169148	1046766	1296	8953234	33
28	4456771	2604	5543229	2·2437770	4978554	2·0086153	1·1170766	1048062	1297	8951938	32
29	4459375	2603	5540625	2·2424669	4982185	2·0071516	1·1172384	1049359	1297	8950641	31
30	4461978	2603	5538022	2·2411585	4985816	2·0056897	1·1174004	1050656	1299	8949344	30
31	4464581	2603	5535419	2·2398517	4989449	2·0042295	1·1175625	1051955	1299	8948045	29
32	4467184	2602	5532816	2·2385468	4993082	2·0027710	1·1177248	1053254	1300	8946746	28
33	4469786	2602	5530214	2·2372435	4996717	2·0013142	1·1178872	1054554	1300	8945446	27
34	4472388	2602	5527612	2·2359419	5000352	1·9998590	1·1180498	1055854	1302	8944146	26
35	4474990	2601	5525010	2·2346420	5003989	1·9984056	1·1182124	1057156	1302	8942844	25
36	4477591	2601	5522409	2·2333438	5007627	1·9969539	1·1183753	1058458	1302	8941542	24
37	4480192	2600	5519808	2·2320474	5011266	1·9955038	1·1185383	1059760	1304	8940240	23
38	4482792	2600	5517208	2·2307526	5014906	1·9940554	1·1187014	1061064	1304	8938936	22
39	4485392	2600	5514608	2·2294595	5018547	1·9926087	1·1188647	1062368	1306	8937632	21
40	4487992	2599	5512008	2·2281681	5022189	1·9911637	1·1190281	1063674	1305	8936326	20
41	4490591	2599	5509409	2·2268783	5025832	1·9897204	1·1191916	1064979	1307	8935021	19
42	4493190	2599	5506810	2·2255903	5029476	1·9882787	1·1193553	1066286	1308	8933714	18
43	4495789	2598	5504211	2·2243039	5033121	1·9868387	1·1195191	1067594	1308	8932406	17
44	4498387	2597	5501613	2·2230192	5036768	1·9854003	1·1196831	1068902	1309	8931098	16
45	4500984	2598	5499016	2·2217362	5040415	1·9839636	1·1198472	1070211	1309	8929789	15
46	4503582	2597	5496418	2·2204548	5044063	1·9825286	1·1200115	1071520	1311	8928480	14
47	4506179	2596	5493821	2·2191752	5047713	1·9810952	1·1201759	1072831	1311	8927169	13
48	4508775	2597	5491225	2·2178971	5051363	1·9796635	1·1203405	1074142	1312	8925858	12
49	4511372	2595	5488628	2·2166208	5055015	1·9782334	1·1205051	1075454	1312	8924546	11
50	4513967	2596	5486033	2·2153460	5058668	1·9768050	1·1206700	1076766	1314	8923234	10
51	4516563	2595	5483437	2·2140730	5062322	1·9753782	1·1208350	1078080	1314	8921920	9
52	4519158	2595	5480842	2·2128016	5065977	1·9739531	1·1210001	1079394	1315	8920606	8
53	4521753	2594	5478247	2·2115318	5069633	1·9725296	1·1211653	1080709	1316	8919291	7
54	4524347	2594	5475653	2·2102637	5073290	1·9711077	1·1213308	1082025	1316	8917975	6
55	4526941	2594	5473059	2·2089972	5076948	1·9696874	1·1214963	1083341	1317	8916659	5
56	4529535	2593	5470465	2·2077323	5080607	1·9682688	1·1216620	1084658	1318	8915342	4
57	4532128	2593	5467872	2·2064691	5084267	1·9668518	1·1218278	1085976	1319	8914024	3
58	4534721	2592	5465279	2·2052075	5087929	1·9654364	1·1219938	1087295	1320	8912705	2
59	4537313	2592	5462687	2·2039476	5091591	1·9640227	1·1221600	1088615	1320	8911385	1
60	4539905	2592	5460095	2·2026893	5095254	1·9626105	1·1223262	1089935		8910065	0
′	Cosine	Dif.	Vers.	Secant.	Cotan.	Tang.	Cosec.	Covers	Dif.	Sine.	′

′	Sine.	Dif.	Cosec.	Verseds.	Tang.	Dif.	Cotang.	Covers.	Secant.	D.	Cosine.	′
0	9·6418420	2589	10·3581580	9·0052061	9·6881818	3205	10·3118182	9·7494404	10·0463398	617	9·9536602	60
1	9·6421009	2587	10·3578991	9·0057531	9·6885023	3204	10·3114977	9·7492472	10·0464015	616	9·9535985	59
2	9·6423596	2586	10·3576404	9·0062997	9·6888227	3203	10·3111773	9·7490449	10·0464631	618	9·9535369	58
3	9·6426182	2583	10·3573818	9·0068460	9·6891430	3201	10·3108570	9·7488426	10·0465249	617	9·9534751	57
4	9·6428765	2582	10·3571235	9·0073920	9·6894631	3200	10·3105369	9·7486402	10·0465866	619	9·9534134	56
5	9·6431347	2579	10·3568653	9·0079375	9·6897831	3199	10·3102169	9·7484377	10·0466485	618	9·9533515	55
6	9·6433926	2578	10·3566074	9·0084827	9·6901030	3196	10·3098970	9·7482352	10·0467103	619	9·9532897	54
7	9·6436504	2576	10·3563496	9·0090276	9·6904226	3196	10·3095774	9·7480326	10·0467722	620	9·9532278	53
8	9·6439080	2574	10·3560920	9·0095721	9·6907422	3194	10·3092578	9·7478299	10·0468342	620	9·9531658	52
9	9·6441654	2572	10·3558346	9·0101162	9·6910616	3193	10·3089384	9·7476272	10·0468962	620	9·9531038	51
10	9·6444226	2570	10·3555774	9·0106600	9·6913809	3191	10·3086191	9·7474244	10·0469582	621	9·9530418	50
11	9·6446796	2569	10·3553204	9·0112034	9·6917000	3189	10·3083000	9·7472216	10·0470203	622	9·9529797	49
12	9·6449365	2566	10·3550635	9·0117465	9·6920189	3189	10·3079811	9·7470186	10·0470825	622	9·9529175	48
13	9·6451931	2565	10·3548069	9·0122891	9·6923378	3187	10·3076622	9·7468156	10·0471447	622	9·9528553	47
14	9·6454496	2562	10·3545504	9·0128315	9·6926565	3185	10·3073435	9·7466126	10·0472069	623	9·9527931	46
15	9·6457058	2561	10·3542942	9·0133735	9·6929750	3184	10·3070250	9·7464095	10·0472692	623	9·9527308	45
16	9·6459619	2559	10·3540381	9·0139151	9·6932934	3183	10·3067066	9·7462063	10·0473315	624	9·9526685	44
17	9·6462178	2557	10·3537822	9·0144564	9·6936117	3181	10·3063883	9·7460030	10·0473939	624	9·9526061	43
18	9·6464735	2555	10·3535265	9·0149973	9·6939298	3180	10·3060702	9·7457997	10·0474563	624	9·9525437	42
19	9·6467290	2554	10·3532710	9·0155378	9·6942478	3178	10·3057522	9·7455963	10·0475187	625	9·9524813	41
20	9·6469844	2551	10·3530156	9·0160781	9·6945656	3177	10·3054344	9·7453926	10·0475812	626	9·9524188	40
21	9·6472395	2550	10·3527605	9·0166179	9·6948833	3176	10·3051167	9·7451893	10·0476438	626	9·9523562	39
22	9·6474945	2547	10·3525055	9·0171574	9·6952009	3174	10·3047991	9·7449857	10·0477064	626	9·9522936	38
23	9·6477492	2546	10·3522508	9·0176965	9·6955183	3172	10·3044817	9·7447821	10·0477690	627	9·9522310	37
24	9·6480038	2544	10·3519962	9·0182353	9·6958355	3172	10·3041645	9·7445784	10·0478317	628	9·9521683	36
25	9·6482582	2542	10·3517418	9·0187738	9·6961527	3170	10·3038473	9·7443746	10·0478945	627	9·9521055	35
26	9·6485124	2541	10·3514876	9·0193119	9·6964697	3168	10·3035303	9·7441707	10·0479572	629	9·9520428	34
27	9·6487665	2538	10·3512335	9·0198496	9·6967865	3167	10·3032135	9·7439668	10·0480201	628	9·9519799	33
28	9·6490203	2537	10·3509797	9·0203870	9·6971032	3166	10·3028968	9·7437628	10·0480829	630	9·9519171	32
29	9·6492740	2534	10·3507260	9·0209240	9·6974198	3165	10·3025802	9·7435588	10·0481459	630	9·9518541	31
30	9·6495274	2533	10·3504726	9·0214607	9·6977363	3163	10·3022637	9·7433547	10·0482088	630	9·9517912	30
31	9·6497807	2531	10·3502193	9·0219970	9·6980526	3161	10·3019474	9·7431505	10·0482718	631	9·9517282	29
32	9·6500338	2530	10·3499662	9·0225330	9·6983687	3160	10·3016313	9·7429462	10·0483349	631	9·9516651	28
33	9·6502868	2527	10·3497132	9·0230687	9·6986847	3159	10·3013153	9·7427419	10·0483980	631	9·9516020	27
34	9·6505395	2525	10·3494605	9·0236039	9·6990006	3158	10·3009994	9·7425375	10·0484611	632	9·9515389	26
35	9·6507920	2524	10·3492080	9·0241389	9·6993164	3156	10·3006836	9·7423331	10·0485243	633	9·9514757	25
36	9·6510444	2522	10·3489556	9·0246735	9·6996320	3154	10·3003680	9·7421286	10·0485876	632	9·9514124	24
37	9·6512966	2520	10·3487034	9·0252077	9·6999474	3154	10·3000526	9·7419240	10·0486508	634	9·9513492	23
38	9·6515486	2518	10·3484514	9·0257416	9·7002628	3152	10·2997372	9·7417193	10·0487142	634	9·9512858	22
39	9·6518004	2517	10·3481996	9·0262752	9·7005780	3150	10·2994220	9·7415146	10·0487776	634	9·9512224	21
40	9·6520521	2514	10·3479479	9·0268084	9·7008930	3150	10·2991070	9·7413099	10·0488410	634	9·9511590	20
41	9·6523035	2513	10·3476965	9·0273412	9·7012080	3147	10·2987920	9·7411050	10·0489044	636	9·9510956	19
42	9·6525548	2511	10·3474452	9·0278738	9·7015227	3147	10·2984773	9·7409001	10·0489680	635	9·9510320	18
43	9·6528059	2509	10·3471941	9·0284059	9·7018374	3145	10·2981626	9·7406951	10·0490315	636	9·9509685	17
44	9·6530568	2507	10·3469432	9·0289378	9·7021519	3144	10·2978481	9·7404901	10·0490951	637	9·9509049	16
45	9·6533075	2506	10·3466925	9·0294692	9·7024663	3142	10·2975337	9·7402850	10·0491588	637	9·9508412	15
46	9·6535581	2503	10·3464419	9·0300004	9·7027805	3141	10·2972195	9·7400798	10·0492225	637	9·9507775	14
47	9·6538084	2502	10·3461916	9·0305312	9·7030946	3140	10·2969054	9·7398745	10·0492862	638	9·9507138	13
48	9·6540586	2500	10·3459414	9·0310616	9·7034086	3139	10·2965914	9·7396692	10·0493500	639	9·9506500	12
49	9·6543086	2498	10·3456914	9·0315917	9·7037225	3137	10·2962775	9·7394638	10·0494139	638	9·9505861	11
50	9·6545584	2497	10·3454416	9·0321215	9·7040362	3135	10·2959638	9·7392584	10·0494777	640	9·9505223	10
51	9·6548081	2494	10·3451919	9·0326509	9·7043497	3135	10·2956503	9·7390529	10·0495417	639	9·9504583	9
52	9·6550575	2493	10·3449425	9·0331800	9·7046632	3133	10·2953368	9·7388473	10·0496056	641	9·9503944	8
53	9·6553068	2491	10·3446932	9·0337088	9·7049765	3132	10·2950235	9·7386416	10·0496697	640	9·9503303	7
54	9·6555559	2489	10·3444441	9·0342372	9·7052897	3130	10·2947103	9·7384359	10·0497337	641	9·9502663	6
55	9·6558048	2488	10·3441952	9·0347652	9·7056027	3129	10·2943973	9·7382301	10·0497980	642	9·9502022	5
56	9·6560536	2485	10·3439464	9·0352930	9·7059156	3128	10·2940844	9·7380243	10·0498620	642	9·9501380	4
57	9·6563021	2484	10·3436979	9·0358204	9·7062284	3126	10·2937716	9·7378184	10·0499262	643	9·9500738	3
58	9·6565505	2482	10·3434495	9·0363474	9·7065410	3125	10·2934590	9·7376124	10·0499405	643	9·9500005	2
59	9·6567987	2481	10·3432013	9·0368741	9·7068535	3124	10·2931465	9·7374063	10·0500548	643	9·9499452	1
60	9·6570468		10·3429532	9·0374005	9·7071659		10·2928341	9·7372002	10·0501191		9·9498809	0

′	Cosine.	Dif.	Secant.	Covers.	Cotang.	Dif.	Tang.	Verseds.	Cosec.	D.	Sine.	′

Y

′	Sine.	Dif.	Covers	Cosec.	Tang.	Cotang.	Secant.	Vers.	Dif.	Cosine	′
0	4339905	2592	5460095	2·2026893	5095254	1·9626105	1·1223262	1089935	1321	8910065	60
1	4542497	2591	5457503	2·2014326	5098919	1·9612000	1·1224927	1091256	1321	8908744	59
2	4545088	2591	5454912	2·2001775	5102585	1·9597910	1·1226592	1092577	1323	8907423	58
3	4547679	2590	5452321	2·1989240	5106252	1·9583837	1·1228259	1093900	1323	8906100	57
4	4550269	2590	5449731	2·1976721	5109919	1·9569780	1·1229928	1095223	1324	8904777	56
5	4552859	2590	5447141	2·1964219	5113588	1·9555739	1·1231598	1096547	1325	8903453	55
6	4555449	2589	5444551	2·1951733	5117259	1·9541713	1·1233269	1097872	1325	8902128	54
7	4558038	2589	5441962	2·1939262	5120930	1·9527704	1·1234942	1099197	1327	8900803	53
8	4560627	2589	5439373	2·1926808	5124602	1·9513711	1·1236616	1100524	1327	8899476	52
9	4563216	2588	5436784	2·1914370	5128275	1·9499733	1·1238292	1101851	1327	8898149	51
10	4565804	2588	5434196	2·1901947	5131950	1·9485772	1·1239969	1103178	1329	8896822	50
11	4568392	2587	5431608	2·1889541	5135625	1·9471826	1·1241648	1104507	1329	8895493	49
12	4570979	2587	5429021	2·1877150	5139302	1·9457896	1·1243328	1105836	1330	8894164	48
13	4573566	2587	5426434	2·1864775	5142980	1·9443981	1·1245010	1107166	1331	8892834	47
14	4576153	2586	5423847	2·1852417	5146658	1·9430083	1·1246693	1108497	1332	8891503	46
15	4578739	2586	5421261	2·1840074	5150338	1·9416200	1·1248377	1109829	1332	8890171	45
16	4581325	2585	5418675	2·1827746	5154019	1·9402333	1·1250063	1111161	1333	8888839	44
17	4583910	2586	5416090	2·1815435	5157702	1·9388481	1·1251750	1112494	1334	8887506	43
18	4586496	2584	5413504	2·1803139	5161385	1·9374645	1·1253439	1113828	1334	8886172	42
19	4589080	2585	5410920	2·1790859	5165069	1·9360825	1·1255130	1115162	1335	8884838	41
20	4591665	2583	5408335	2·1778595	5168755	1·9347020	1·1256821	1116497	1337	8883503	40
21	4594248	2584	5405752	2·1766346	5172441	1·9333231	1·1258514	1117834	1336	8882166	39
22	4596832	2583	5403168	2·1754113	5176129	1·9319457	1·1260209	1119170	1338	8880830	38
23	4599415	2583	5400585	2·1741895	5179818	1·9305699	1·1261905	1120508	1338	8879492	37
24	4601998	2582	5398002	2·1729693	5183508	1·9291956	1·1263603	1121846	1339	8878154	36
25	4604580	2582	5395420	2·1717506	5187199	1·9278228	1·1265302	1123185	1340	8876815	35
26	4607162	2582	5392838	2·1705335	5190891	1·9264516	1·1267003	1124525	1341	8875475	34
27	4609744	2581	5390256	2·1693180	5194584	1·9250819	1·1268705	1125866	1341	8874134	33
28	4612325	2581	5387675	2·1681040	5198278	1·9237138	1·1270408	1127207	1342	8872793	32
29	4614906	2580	5385094	2·1668915	5201974	1·9223472	1·1272113	1128549	1343	8871451	31
30	4617486	2580	5382514	2·1656806	5205671	1·9209821	1·1273819	1129892	1343	8870108	30
31	4620066	2580	5379934	2·1644712	5200368	1·9196186	1·1275527	1131235	1345	8868765	29
32	4622646	2579	5377354	2·1632633	5213067	1·9182565	1·1277237	1132580	1345	8867420	28
33	4625225	2579	5374775	2·1620570	5216767	1·9168960	1·1278948	1133925	1345	8866075	27
34	4627804	2578	5372196	2·1608522	5220468	1·9155370	1·1280660	1135270	1347	8864730	26
35	4630382	2578	5369618	2·1596489	5224170	1·9141795	1·1282374	1136617	1347	8863383	25
36	4632960	2578	5367040	2·1584471	5227874	1·9128236	1·1284089	1137964	1348	8862036	24
37	4635538	2577	5364462	2·1572469	5231578	1·9114691	1·1285806	1139312	1349	8860688	23
38	4638115	2577	5361885	2·1560482	5235284	1·9101162	1·1287524	1140661	1350	8859339	22
39	4640692	2577	5359308	2·1548510	5238990	1·9087647	1·1289244	1142011	1350	8857989	21
40	4643269	2576	5356731	2·1536553	5242698	1·9074147	1·1290965	1143361	1351	8856639	20
41	4645845	2575	5354155	2·1524611	5246407	1·9060663	1·1292687	1144712	1352	8855288	19
42	4648420	2576	5351580	2·1512684	5250117	1·9047193	1·1294412	1146064	1352	8853936	18
43	4650996	2575	5349004	2·1500772	5253829	1·9033738	1·1296137	1147416	1354	8852584	17
44	4653571	2574	5346429	2·1488875	5257541	1·9020299	1·1297864	1148770	1354	8851230	16
45	4656145	2574	5343855	2·1476993	5261255	1·9006874	1·1299593	1150124	1354	8849876	15
46	4658719	2574	5341281	2·1465127	5264969	1·8993464	1·1301323	1151478	1356	8848522	14
47	4661293	2573	5338707	2·1453275	5268685	1·8980068	1·1303055	1152834	1356	8847166	13
48	4663866	2573	5336134	2·1441438	5272402	1·8966688	1·1304788	1154190	1357	8845810	12
49	4666439	2573	5333561	2·1429615	5276120	1·8953322	1·1306522	1155547	1358	8844453	11
50	4669012	2572	5330988	2·1417808	5279839	1·8939971	1·1308258	1156905	1359	8843095	10
51	4671584	2572	5328416	2·1406015	5283560	1·8926635	1·1309996	1158264	1359	8841736	9
52	4674156	2571	5325844	2·1394238	5287281	1·8913313	1·1311735	1159623	1360	8840377	8
53	4676727	2571	5323273	2·1382475	5291004	1·8900006	1·1313475	1160983	1361	8839017	7
54	4679298	2571	5320702	2·1370726	5294727	1·8886713	1·1315217	1162344	1361	8837656	6
55	4681869	2570	5318131	2·1358993	5298452	1·8873436	1·1316961	1163705	1362	8836295	5
56	4684439	2570	5315561	2·1347274	5302178	1·8860172	1·1318706	1165067	1364	8834933	4
57	4687009	2569	5312991	2·1335570	5305906	1·8846924	1·1320452	1166431	1363	8833569	3
58	4689578	2569	5310422	2·1323880	5309634	1·8833690	1·1322200	1167794	1365	8832206	2
59	4692147	2569	5307853	2·1312205	5313364	1·8820470	1·1323950	1169159	1365	8830841	1
60	4694716		5305284	2·1300545	5317094	1·8807265	1·1325701	1170524		8829476	0

| ′ | Cosine | Dif. | Vers. | Secant. | Cotan. | Tang. | Cosec. | Covers | Dif | Sine. | ′ |

'	Sine.	Dif.	Cosec.	Verseds.	Tang.	Dif.	Cotang.	Covers.	Secant.	D.	Cosine.	'
0	9·6570468	2478	10·3429532	9·0374005	9·7071659	3122	10·2928341	9·7372002	10·0501191	644	9·9498809	60
1	9·6572946	2477	10·3427054	9·0379265	9·7074781	3121	10·2925219	9·7369940	10·0501835	644	9·9498165	59
2	9·6575423	2475	10·3424577	9·0384522	9·7077902	3120	10·2922098	9·7367878	10·0502479	645	9·9497521	58
3	9·6577898	2473	10·3422102	9·0389776	9·7081022	3119	10·2918978	9·7365814	10·0503124	646	9·9496876	57
4	9·6580371	2471	10·3419629	9·0395026	9·7084141	3117	10·2915859	9·7363750	10·0503770	645	9·9496230	56
5	9·6582842	2470	10·3417158	9·0400273	9·7087258	3116	10·2912742	9·7361686	10·0504415	647	9·9495585	55
6	9·6585312	2468	10·3414688	9·0405517	9·7090374	3114	10·2909626	9·7359621	10·0505062	646	9·9494938	54
7	9·6587780	2466	10·3412220	9·0410757	9·7093488	3113	10·2906512	9·7357555	10·0505708	647	9·9494292	53
8	9·6590246	2464	10·3409754	9·0415994	9·7096601	3112	10·2903399	9·7355488	10·0506355	647	9·9493645	52
9	9·6592710	2463	10·3407290	9·0421228	9·7099713	3111	10·2900287	9·7353421	10·0507003	648	9·9492997	51
10	9·6595173	2460	10·3404827	9·0426458	9·7102824	3109	10·2897176	9·7351353	10·0507651	649	9·9492349	50
11	9·6597633	2460	10·3402367	9·0431685	9·7105933	3108	10·2894067	9·7349284	10·0508300	649	9·9491700	49
12	9·6600093	2457	10·3399907	9·0436908	9·7109041	3107	10·2890959	9·7347215	10·0508949	649	9·9491051	48
13	9·6602550	2455	10·3397450	9·0442129	9·7112148	3106	10·2887852	9·7345145	10·0509598	650	9·9490402	47
14	9·6605005	2454	10·3394995	9·0447345	9·7115254	3104	10·2884746	9·7343074	10·0510248	651	9·9489752	46
15	9·6607459	2452	10·3392541	9·0452559	9·7118358	3103	10·2881642	9·7341003	10·0510899	651	9·9489101	45
16	9·6609911	2450	10·3390089	9·0457769	9·7121461	3101	10·2878539	9·7338931	10·0511550	651	9·9488450	44
17	9·6612361	2449	10·3387639	9·0462976	9·7124562	3100	10·2875438	9·7336858	10·0512201	652	9·9487799	43
18	9·6614810	2447	10·3385190	9·0468180	9·7127662	3099	10·2872338	9·7334785	10·0512853	652	9·9487147	42
19	9·6617257	2445	10·3382743	9·0473380	9·7130761	3098	10·2869239	9·7332711	10·0513505	653	9·9486495	41
20	9·6619702	2443	10·3380298	9·0478578	9·7133859	3097	10·2866141	9·7330636	10·0514158	653	9·9485842	40
21	9·6622145	2441	10·3377855	9·0483771	9·7136956	3095	10·2863044	9·7328561	10·0514811	654	9·9485189	39
22	9·6624586	2440	10·3375414	9·0488962	9·7140051	3094	10·2859949	9·7326485	10·0515465	654	9·9484535	38
23	9·6627026	2438	10·3372974	9·0494149	9·7143145	3092	10·2856855	9·7324408	10·0516119	654	9·9483881	37
24	9·6629464	2436	10·3370536	9·0499333	9·7146237	3092	10·2853763	9·7322331	10·0516773	655	9·9483227	36
25	9·6631900	2435	10·3368100	9·0504514	9·7149329	3090	10·2850671	9·7320252	10·0517428	656	9·9482572	35
26	9·6634335	2433	10·3365665	9·0509691	9·7152419	3089	10·2847581	9·7318174	10·0518084	656	9·9481916	34
27	9·6636768	2431	10·3363232	9·0514865	9·7155508	3087	10·2844492	9·7316094	10·0518740	656	9·9481260	33
28	9·6639199	2429	10·3360801	9·0520036	9·7158595	3087	10·2841405	9·7314014	10·0519396	657	9·9480604	32
29	9·6641628	2428	10·3358372	9·0525204	9·7161682	3085	10·2838318	9·7311933	10·0520053	658	9·9479947	31
30	9·6644056	2426	10·3355944	9·0530368	9·7164767	3084	10·2835233	9·7309852	10·0520711	658	9·9479289	30
31	9·6646482	2424	10·3353518	9·0535529	9·7167851	3082	10·2832149	9·7307769	10·0521369	658	9·9478631	29
32	9·6648906	2423	10·3351094	9·0540687	9·7170933	3081	10·2829067	9·7305686	10·0522027	659	9·9477973	28
33	9·6651329	2420	10·3348671	9·0545842	9·7174014	3080	10·2825986	9·7303603	10·0522686	659	9·9477314	27
34	9·6653054	2420	10·3346251	9·0550993	9·7177094	3079	10·2822906	9·7301519	10·0523345	660	9·9476655	26
35	9·6656168	2419	10·3343832	9·0556141	9·7180173	3078	10·2819827	9·7299434	10·0524005	660	9·9475995	25
36	9·6658586	2418	10·3341414	9·0561286	9·7183251	3076	10·2816749	9·7297348	10·0524665	661	9·9475335	24
37	9·6661001	2415	10·3338999	9·0566428	9·7186327	3075	10·2813673	9·7295262	10·0525326	661	9·9474674	23
38	9·6663415	2414	10·3336585	9·0571566	9·7189402	3074	10·2810598	9·7293175	10·0525987	661	9·9474013	22
39	9·6665828	2413	10·3334172	9·0576701	9·7192476	3073	10·2807524	9·7291087	10·0526648	663	9·9473352	21
40	9·6668238	2410	10·3331762	9·0581833	9·7195549	3072	10·2804451	9·7288999	10·0527311	662	9·9472689	20
41	9·6670647	2409	10·3329353	9·0586962	9·7198620	3070	10·2801380	9·7286910	10·0527973	663	9·9472027	19
42	9·6673054	2405	10·3326946	9·0592088	9·7201690	3069	10·2798310	9·7284820	10·0528636	664	9·9471364	18
43	9·6675459	2404	10·3324541	9·0597212	9·7204759	3068	10·2795241	9·7282729	10·0529300	664	9·9470700	17
44	9·6677863	2402	10·3322137	9·0602329	9·7207827	3066	10·2792173	9·7280638	10·0529964	664	9·9470036	16
45	9·6680265	2400	10·3319735	9·0607445	9·7210893	3065	10·2789107	9·7278546	10·0530628	665	9·9469372	15
46	9·6682665	2399	10·3317335	9·0612558	9·7213958	3064	10·2786042	9·7276454	10·0531293	665	9·9468707	14
47	9·6685064	2397	10·3314936	9·0617668	9·7217022	3063	10·2782978	9·7274361	10·0531958	665	9·9468042	13
48	9·6687461	2395	10·3312539	9·0622774	9·7220085	3062	10·2779915	9·7272267	10·0532624	666	9·9467376	12
49	9·6689856	2394	10·3310144	9·0627877	9·7223147	3060	10·2776853	9·7270172	10·0533290	667	9·9466710	11
50	9·6692250	2392	10·3307750	9·0632977	9·7226207	3059	10·2773793	9·7268077	10·0533957	667	9·9466043	10
51	9·6694642	2390	10·3305358	9·0638074	9·7229266	3058	10·2770734	9·7265981	10·0534624	668	9·9465376	9
52	9·6697032	2388	10·3302968	9·0643168	9·7232324	3057	10·2767676	9·7263885	10·0535292	668	9·9464708	8
53	9·6699420	2387	10·3300580	9·0648258	9·7235381	3055	10·2764619	9·7261787	10·0535960	669	9·9464040	7
54	9·6701807	2385	10·3298193	9·0653346	9·7238436	3054	10·2761564	9·7259689	10·0536629	669	9·9463371	6
55	9·6704192	2384	10·3295808	9·0658430	9·7241490	3053	10·2758510	9·7257591	10·0537298	670	9·9462702	5
56	9·6706576	2382	10·3293424	9·0663511	9·7244543	3052	10·2755457	9·7255491	10·0537968	670	9·9462032	4
57	9·6708958	2380	10·3291042	9·0668589	9·7247595	3051	10·2752405	9·7253391	10·0538638	670	9·9461362	3
58	9·6711338	2378	10·3288662	9·0673663	9·7250646	3049	10·2749354	9·7251290	10·0539308	671	9·9460692	2
59	9·6713716	2377	10·3286284	9·0678735	9·7253695	3049	10·2746305	9·7249189	10·0539979	672	9·9460021	1
60	9·6716093		10·3283907	9·0683803	9·7256744		10·2743256	9·7247087	10·0540651		9·9459349	0
'	Cosine.	Dif.	Secant.	Covers.	Cotang.	Dif.	Tang.	Verseds.	Cosec.	D.	Sine.	'

′	Sine.	Dif.	Covers	Cosec.	Tang.	Cotang.	Secant.	Vers.	Dif.	Cosine	
0	4694716	2568	5305284	2·1300545	5317094	1·8807265	1·1325701	1170524	1366	8829476	60
1	4697284	2568	5302716	2·1288899	5320826	1·8794074	1·1327453	1171890	1367	8828110	59
2	4699852	2567	5300148	2·1277267	5324559	1·8780898	1·1329207	1173257	1367	8826743	58
3	4702419	2567	5297581	2·1265651	5328293	1·8767736	1·1330962	1174624	1369	8825376	57
4	4704986	2567	5295014	2·1254048	5332020	1·8754588	1·1332719	1175993	1369	8824007	56
5	4707553	2566	5292447	2·1242460	5335765	1·8741455	1·1334478	1177362	1369	8822638	55
6	4710119	2566	5289881	2·1230887	5339503	1·8728336	1·1336238	1178731	1371	8821269	54
7	4712685	2565	5287315	2·1219328	5343242	1·8715231	1·1337999	1180102	1371	8819898	53
8	4715250	2565	5284750	2·1207783	5346981	1·8702141	1·1339762	1181473	1372	8818527	52
9	4717815	2565	5282185	2·1196253	5350723	1·8689065	1·1341527	1182845	1373	8817155	51
10	4720380	2564	5279620	2·1184737	5354465	1·8676003	1·1343293	1184218	1373	8815782	50
11	4722944	2564	5277056	2·1173235	5358208	1·8662955	1·1345060	1185591	1374	8814409	49
12	4725508	2563	5274492	2·1161746	5361953	1·8649921	1·1346829	1186965	1375	8813035	48
13	4728071	2563	5271929	2·1150274	5365699	1·8636902	1·1348600	1188340	1376	8811660	47
14	4730634	2563	5269366	2·1138815	5369446	1·8623896	1·1350372	1189716	1377	8810284	46
15	4733197	2562	5266803	2·1127371	5373194	1·8610905	1·1352146	1191093	1377	8808907	45
16	4735759	2562	5264241	2·1115940	5376943	1·8597928	1·1353921	1192470	1378	8807530	44
17	4738321	2561	5261679	2·1104523	5380694	1·8584965	1·1355697	1193848	1378	8806152	43
18	4740882	2561	5259118	2·1093121	5384445	1·8572015	1·1357476	1195226	1380	8804774	42
19	4743443	2561	5256557	2·1081733	5388198	1·8559080	1·1359255	1196606	1380	8803394	41
20	4746004	2560	5253996	2·1070359	5391952	1·8546159	1·1361030	1197986	1381	8802014	40
21	4748564	2560	5251436	2·1058998	5395707	1·8533252	1·1362819	1199367	1382	8800633	39
22	4751124	2559	5248876	2·1047652	5399464	1·8520358	1·1364603	1200749	1382	8799251	38
23	4753683	2559	5246317	2·1036320	5403221	1·8507479	1·1366389	1202131	1383	8797869	37
24	4756242	2559	5243758	2·1025002	5406980	1·8494613	1·1368176	1203514	1384	8796486	36
25	4758801	2558	5241199	2·1013698	5410740	1·8481761	1·1369965	1204898	1385	8795102	35
26	4761359	2558	5238641	2·1002408	5414501	1·8468923	1·1371755	1206283	1385	8793717	34
27	4763917	2557	5236083	2·0991131	5418263	1·8456099	1·1373547	1207668	1386	8792332	33
28	4766474	2557	5233526	2·0979869	5422027	1·8443289	1·1375341	1209054	1387	8790946	32
29	4769031	2557	5230969	2·0968620	5425791	1·8430492	1·1377135	1210441	1388	8789559	31
30	4771588	2556	5228412	2·0957385	5429557	1·8417709	1·1378932	1211829	1388	8788171	30
31	4774144	2556	5225856	2·0946164	5433324	1·8404940	1·1380730	1213217	1389	8786783	29
32	4776700	2555	5223300	2·0934957	5437092	1·8392184	1·1382529	1214606	1390	8785394	28
33	4779255	2555	5220745	2·0923764	5440862	1·8379442	1·1384330	1215996	1391	8784004	27
34	4781810	2554	5218190	2·0912584	5444632	1·8366713	1·1386133	1217387	1391	8782613	26
35	4784364	2555	5215636	2·0901418	5448404	1·8353999	1·1387937	1218778	1392	8781222	25
36	4786919	2553	5213081	2·0890265	5452177	1·8341297	1·1389742	1220170	1393	8779830	24
37	4789472	2554	5210528	2·0879127	5455951	1·8328610	1·1391550	1221563	1394	8778437	23
38	4792026	2553	5207974	2·0868002	5459727	1·8315936	1·1393358	1222957	1394	8777043	22
39	4794579	2552	5205421	2·0856890	5463503	1·8303275	1·1395169	1224351	1395	8775649	21
40	4797131	2552	5202869	2·0845792	5467281	1·8290628	1·1396980	1225746	1396	8774254	20
41	4799683	2552	5200317	2·0834708	5471060	1·8277994	1·1398794	1227142	1396	8772858	19
42	4802235	2551	5197765	2·0823637	5474840	1·8265374	1·1400608	1228538	1398	8771462	18
43	4804786	2551	5195214	2·0812580	5478621	1·8252767	1·1402425	1229936	1398	8770064	17
44	4807337	2551	5192663	2·0801536	5482404	1·8240173	1·1404243	1231334	1398	8768666	16
45	4809888	2550	5190112	2·0790506	5486188	1·8227593	1·1406062	1232732	1400	8767268	15
46	4812438	2549	5187562	2·0779489	5489973	1·8215026	1·1407883	1234132	1400	8765868	14
47	4814987	2550	5185013	2·0768486	5493759	1·8202473	1·1409706	1235532	1401	8764468	13
48	4817537	2549	5182463	2·0757496	5497547	1·8189932	1·1411530	1236933	1402	8763067	12
49	4820086	2548	5179914	2·0746519	5501335	1·8177405	1·1413356	1238335	1402	8761665	11
50	4822634	2548	5177366	2·0735556	5505125	1·8164892	1·1415183	1239737	1404	8760263	10
51	4825182	2548	5174818	2·0724606	5508916	1·8152391	1·1417012	1241141	1404	8758859	9
52	4827730	2547	5172270	2·0713670	5512708	1·8139904	1·1418842	1242545	1404	8757455	8
53	4830277	2547	5169723	2·0702746	5516502	1·8127430	1·1420674	1243949	1406	8756051	7
54	4832824	2546	5167176	2·0691836	5520297	1·8114969	1·1422507	1245355	1406	8754645	6
55	4835370	2546	5164630	2·0680940	5524093	1·8102521	1·1424342	1246761	1407	8753239	5
56	4837916	2546	5162084	2·0670056	5527890	1·8090086	1·1426179	1248168	1407	8751832	4
57	4840462	2545	5159538	2·0659186	5531688	1·8077664	1·1428017	1249575	1409	8750425	3
58	4843007	2545	5156993	2·0648328	5535488	1·8065256	1·1429857	1250984	1409	8749016	2
59	4845552	2544	5154448	2·0637484	5539288	1·8052860	1·1431698	1252393	1410	8747607	1
60	4848096		5151904	2·0626653	5543091	1·8040478	1·1433541	1253803		8746197	0
′	Cosine	Dif.	Vers.	Secant.	Cotan.	Tang.	Cosec.	Covers	Dif.	Sine.	′

′	Sine.	Dif.	Cosec.	Verseds.	Tang.	Dif.	Cotang.	Covers.	Secant.	D.	Cosine.	′
0	9·6716093	2375	10·3283907	9·0683803	9·7256744	3047	10·2743256	9·7247087	10·0540651	672	9·9459349	60
1	9·6718468	2373	10·3281532	9·0688869	9·7259791	3046	10·2740209	9·7244984	10·0541323	672	9·9458677	59
2	9·6720841	2372	10·3279159	9·0693931	9·7262837	3044	10·2737163	9·7242880	10·0541995	673	9·9458005	58
3	9·6723213	2370	10·3276787	9·0698990	9·7265881	3044	10·2734119	9·7240776	10·0542668	673	9·9457332	57
4	9·6725583	2369	10·3274417	9·0704046	9·7268925	3042	10·2731075	9·7238671	10·0543341	674	9·9456659	56
5	9·6727952	2367	10·3272048	9·0709099	9·7271967	3041	10·2728033	9·7236565	10·0544015	675	9·9455985	55
6	9·6730319	2365	10·3269681	9·0714148	9·7275008	3040	10·2724992	9·7234459	10·0544690	674	9·9455310	54
7	9·6732684	2363	10·3267316	9·0719195	9·7278048	3039	10·2721952	9·7232352	10·0545364	676	9·9454636	53
8	9·6735047	2362	10·3264953	9·0724238	9·7281087	3037	10·2718913	9·7230244	10·0546040	675	9·9453960	52
9	9·6737409	2360	10·3262591	9·0729279	9·7284124	3037	10·2715876	9·7228136	10·0546715	676	9·9453285	51
10	9·6739769	2359	10·3260231	9·0734316	9·7287161	3035	10·2712839	9·7226027	10·0547391	677	9·9452609	50
11	9·6742128	2357	10·3257872	9·0739350	9·7290196	3034	10·2709804	9·7223917	10·0548068	677	9·9451932	49
12	9·6744485	2355	10·3255515	9·0744381	9·7293230	3033	10·2706770	9·7221807	10·0548745	678	9·9451255	48
13	9·6746840	2354	10·3253160	9·0749409	9·7296263	3032	10·2703737	9·7219695	10·0549423	678	9·9450577	47
14	9·6749194	2352	10·3250806	9·0754434	9·7299295	3030	10·2700705	9·7217584	10·0550101	679	9·9449899	46
15	9·6751546	2350	10·3248454	9·0759455	9·7302325	3029	10·2697675	9·7215471	10·0550780	679	9·9449220	45
16	9·6753896	2349	10·3246104	9·0764474	9·7305354	3029	10·2694646	9·7213358	10·0551459	679	9·9448541	44
17	9·6756245	2347	10·3243755	9·0769490	9·7308383	3027	10·2691617	9·7211244	10·0552138	680	9·9447862	43
18	9·6758592	2345	10·3241408	9·0774502	9·7311410	3026	10·2688590	9·7209129	10·0552818	681	9·9447182	42
19	9·6760937	2344	10·3239063	9·0779511	9·7314436	3024	10·2685564	9·7207014	10·0553499	680	9·9446501	41
20	9·6763281	2344	10·3236719	9·0784518	9·7317460	3024	10·2682540	9·7204898	10·0554179	682	9·9445821	40
21	9·6765623	2342	10·3234377	9·0789521	9·7320484	3022	10·2679516	9·7202781	10·0554861	682	9·9445139	39
22	9·6767963	2340	10·3232037	9·0794521	9·7323506	3021	10·2676494	9·7200663	10·0555543	682	9·9444457	38
23	9·6770302	2339	10·3229698	9·0799518	9·7326527	3020	10·2673473	9·7198545	10·0556225	683	9·9443775	37
24	9·6772640	2338	10·3227360	9·0804512	9·7329547	3019	10·2670453	9·7196426	10·0556908	683	9·9443092	36
25	9·6774975	2335	10·3225025	9·0809503	9·7332566	3018	10·2667434	9·7194307	10·0557591	684	9·9442409	35
26	9·6777309	2334	10·3222691	9·0814491	9·7335584	3017	10·2664416	9·7192186	10·0558275	684	9·9441725	34
27	9·6779642	2333	10·3220358	9·0819476	9·7338601	3015	10·2661399	9·7190066	10·0558959	685	9·9441041	33
28	9·6781972	2330	10·3218028	9·0824458	9·7341616	3015	10·2658384	9·7187944	10·0559644	685	9·9440356	32
29	9·6784301	2329	10·3215699	9·0829437	9·7344631	3013	10·2655369	9·7185821	10·0560329	686	9·9439671	31
30	9·6786629	2328	10·3213371	9·0834413	9·7347644	3012	10·2652356	9·7183698	10·0561015	686	9·9438985	30
31	9·6788955	2326	10·3211045	9·0839386	9·7350656	3011	10·2649344	9·7181575	10·0561701	687	9·9438299	29
32	9·6791279	2324	10·3208721	9·0844356	9·7353667	3010	10·2646333	9·7179450	10·0562388	687	9·9437612	28
33	9·6793602	2323	10·3206398	9·0849322	9·7356677	3008	10·2643323	9·7177325	10·0563075	687	9·9436925	27
34	9·6795923	2321	10·3204077	9·0854286	9·7359685	3008	10·2640315	9·7175193	10·0563762	689	9·9436238	26
35	9·6798243	2320	10·3201757	9·0859247	9·7362693	3006	10·2637307	9·7173072	10·0564451	688	9·9435549	25
36	9·6800560	2317	10·3199440	9·0864204	9·7365699	3006	10·2634301	9·7170945	10·0565139	689	9·9434861	24
37	9·6802877	2317	10·3197123	9·0869159	9·7368705	3004	10·2631295	9·7168817	10·0565828	690	9·9434172	23
38	9·6805191	2314	10·3194809	9·0874111	9·7371709	3003	10·2628291	9·7166688	10·0566518	690	9·9433482	22
39	9·6807504	2313	10·3192496	9·0879059	9·7374712	3002	10·2625288	9·7164559	10·0567208	600	9·9432792	21
40	9·6809816	2312	10·3190184	9·0884005	9·7377714	3001	10·2622286	9·7162429	10·0567898	691	9·9432102	20
41	9·6812126	2310	10·3187874	9·0888948	9·7380715	2999	10·2619285	9·7160298	10·0568589	691	9·9431411	19
42	9·6814434	2308	10·3185566	9·0893887	9·7383714	2999	10·2616286	9·7158166	10·0569280	692	9·9430720	18
43	9·6816741	2307	10·3183259	9·0898824	9·7386713	2997	10·2613287	9·7156034	10·0569972	693	9·9430028	17
44	9·6819046	2305	10·3180954	9·0903758	9·7389710	2997	10·2610290	9·7153901	10·0570665	692	9·9429335	16
45	9·6821349	2303	10·3178651	9·0908688	9·7392707	2995	10·2607293	9·7151768	10·0571357	694	9·9428643	15
46	9·6823651	2302	10·3176349	9·0913616	9·7395702	2994	10·2604298	9·7149633	10·0572051	694	9·9427949	14
47	9·6825952	2301	10·3174048	9·0918541	9·7398696	2993	10·2601304	9·7147498	10·0572745	694	9·9427255	13
48	9·6828250	2298	10·3171750	9·0923462	9·7401689	2992	10·2598311	9·7145362	10·0573439	695	9·9426561	12
49	9·6830548	2298	10·3169452	9·0928381	9·7404681	2991	10·2595319	9·7143226	10·0574134	695	9·9425866	11
50	9·6832843	2295	10·3167157	9·0933297	9·7407672	2990	10·2592328	9·7141089	10·0574829	695	9·9425171	10
51	9·6835137	2294	10·3164863	9·0938210	9·7410662	2988	10·2589338	9·7138951	10·0575524	697	9·9424476	9
52	9·6837430	2293	10·3162570	9·0943120	9·7413650	2988	10·2586350	9·7136812	10·0576220	696	9·9423779	8
53	9·6839720	2290	10·3160280	9·0948027	9·7416637	2986	10·2583362	9·7134673	10·0576917	697	9·9423083	7
54	9·6842010	2290	10·3157990	9·0952931	9·7419624	2985	10·2580376	9·7132533	10·0577614	698	9·9422386	6
55	9·6844297	2287	10·3155703	9·0957832	9·7422609	2985	10·2577391	9·7130392	10·0578312	698	9·9421688	5
56	9·6846583	2286	10·3153417	9·0962730	9·7425594	2983	10·2574406	9·7128250	10·0579010	699	9·9420990	4
57	9·6848868	2285	10·3151132	9·0967625	9·7428577	2982	10·2571423	9·7126108	10·0579709	699	9·9420291	3
58	9·6851151	2283	10·3148849	9·0972517	9·7431559	2981	10·2568441	9·7123965	10·0580408	699	9·9419592	2
59	9·6853432	2281	10·3146568	9·0977406	9·7434540	2980	10·2565460	9·7121822	10·0581107	700	9·9418893	1
60	9·6855712	2280	10·3144288	9·0982293	9·7437520		10·2562480	9·7119677	10·0581807		9·9418193	0
′	Cosine.	Dif.	Secant.	Covers.	Cotang.	Dif.	Tang.	Verseds.	Cosec.	D.	Sine.	′

′	Sine.	Dif.	Covers	Cosec.	Tang.	Cotang.	Secant.	Vers.	Dif.	Cosine	′
0	4848096	2544	5151904	2·0626653	5543091	1·8040478	1·1433541	1253803	1411	8746197	60
1	4850640	2544	5149360	2·0615836	5546894	1·8028108	1·1435385	1255214	1411	8744786	59
2	4853184	2543	5146816	2·0605031	5550698	1·8015751	1·1437231	1256625	1412	8743375	58
3	4855727	2543	5144273	2·0594239	5554504	1·8003408	1·1439078	1258037	1413	8741963	57
4	4858270	2542	5141730	2·0583460	5558311	1·7991077	1·1440927	1259450	1413	8740550	56
5	4860812	2542	5139188	2·0572695	5562119	1·7978759	1·1442778	1260863	1415	8739137	55
6	4863354	2541	5136646	2·0561942	5565920	1·7966454	1·1444630	1262278	1415	8737722	54
7	4865895	2541	5134105	2·0551203	5569730	1·7954162	1·1446484	1263693	1416	8736307	53
8	4868436	2541	5131564	2·0540476	5573551	1·7941883	1·1448339	1265109	1416	8734891	52
9	4870977	2540	5129023	2·0529762	5577364	1·7929616	1·1450196	1266525	1417	8733475	51
10	4873517	2540	5126483	2·0519061	5581179	1·7917362	1·1452055	1267942	1418	8732058	50
11	4876057	2540	5123943	2·0508373	5584994	1·7905121	1·1453915	1269360	1419	8730640	49
12	4878597	2539	5121403	2·0497698	5588811	1·7892893	1·1455776	1270779	1420	8729221	48
13	4881136	2538	5118864	2·0487036	5592629	1·7880678	1·1457639	1272199	1420	8727801	47
14	4883674	2538	5116326	2·0476386	5596449	1·7868475	1·1459504	1273619	1421	8726381	46
15	4886212	2538	5113788	2·0465750	5600269	1·7856285	1·1461371	1275040	1422	8724960	45
16	4888750	2538	5111250	2·0455126	5604091	1·7844107	1·1463238	1276462	1422	8723538	44
17	4891288	2537	5108712	2·0444515	5607914	1·7831943	1·1465108	1277884	1423	8722116	43
18	4893825	2536	5106175	2·0433916	5611738	1·7819790	1·1466979	1279307	1424	8720693	42
19	4896361	2536	5103639	2·0423330	5615564	1·7807651	1·1468852	1280731	1425	8719269	41
20	4898897	2536	5101103	2·0412757	5619391	1·7795524	1·1470726	1282156	1425	8717844	40
21	4901433	2535	5098567	2·0402197	5623219	1·7783409	1·1472602	1283581	1426	8716419	39
22	4903968	2535	5096032	2·0391649	5627048	1·7771307	1·1474479	1285007	1427	8714993	38
23	4906503	2535	5093497	2·0381114	5630879	1·7759218	1·1476358	1286434	1428	8713566	37
24	4909038	2534	5090962	2·0370592	5634710	1·7747141	1·1478239	1287862	1428	8712138	36
25	4911572	2533	5088428	2·0360082	5638543	1·7735076	1·1480121	1289290	1429	8710710	35
26	4914105	2533	5085895	2·0349585	5642378	1·7723024	1·1482005	1290719	1430	8709281	34
27	4916638	2533	5083362	2·0339100	5646213	1·7710985	1·1483890	1292149	1431	8707851	33
28	4919171	2533	5080829	2·0328628	5650050	1·7698958	1·1485777	1293580	1431	8706420	32
29	4921704	2532	5078296	2·0318168	5653888	1·7686943	1·1487665	1295011	1432	8704989	31
30	4924236	2531	5075764	2·0307720	5657728	1·7674940	1·1489555	1296443	1433	8703557	30
31	4926767	2531	5073233	2·0297286	5661568	1·7662950	1·1491447	1297876	1433	8702124	29
32	4929298	2531	5070702	2·0286863	5665410	1·7650972	1·1493340	1299309	1435	8700691	28
33	4931829	2530	5068171	2·0276453	5669254	1·7639007	1·1495235	1300744	1435	8699256	27
34	4934359	2530	5065641	2·0266056	5673098	1·7627053	1·1497132	1302179	1435	8697821	26
35	4936889	2530	5063111	2·0255670	5676944	1·7615112	1·1499030	1303614	1437	8696386	25
36	4939419	2529	5060581	2·0245297	5680791	1·7603183	1·1500930	1305051	1437	8694949	24
37	4941948	2528	5058052	2·0234937	5684639	1·7591267	1·1502831	1306488	1438	8693512	23
38	4944476	2529	5055524	2·0224589	5688488	1·7579362	1·1504734	1307926	1438	8692074	22
39	4947005	2527	5052995	2·0214253	5692339	1·7567470	1·1506638	1309364	1440	8690636	21
40	4949532	2528	5050468	2·0203920	5696191	1·7555590	1·1508544	1310804	1440	8689196	20
41	4952060	2527	5047940	2·0193618	5700045	1·7543722	1·1510452	1312244	1441	8687756	19
42	4954587	2526	5045413	2·0183318	5703899	1·7531866	1·1512361	1313685	1441	8686315	18
43	4957113	2526	5042887	2·0173031	5707755	1·7520023	1·1514272	1315126	1443	8684874	17
44	4959639	2526	5040361	2·0162756	5711612	1·7508191	1·1516185	1316569	1443	8683431	16
45	4962165	2525	5037835	2·0152494	5715471	1·7496371	1·1518099	1318012	1444	8681988	15
46	4964690	2525	5035310	2·0142243	5719331	1·7484564	1·1520015	1319456	1444	8680544	14
47	4967215	2525	5032785	2·0132005	5723192	1·7472768	1·1521932	1320900	1445	8679100	13
48	4969740	2524	5030260	2·0121779	5727054	1·7460984	1·1523851	1322345	1446	8677655	12
49	4972264	2523	5027736	2·0111564	5730918	1·7449213	1·1525772	1323791	1447	8676209	11
50	4974787	2523	5025213	2·0101362	5734783	1·7437453	1·1527694	1325238	1448	8674762	10
51	4977310	2523	5022690	2·0091172	5738649	1·7425705	1·1529618	1326686	1448	8673314	9
52	4979833	2522	5020167	2·0080994	5742516	1·7413969	1·1531543	1328134	1449	8671866	8
53	4982355	2522	5017645	2·0070828	5746385	1·7402245	1·1533470	1329583	1450	8670417	7
54	4984877	2522	5015123	2·0060674	5750255	1·7390533	1·1535399	1331033	1450	8668967	6
55	4987399	2521	5012601	2·0050532	5754126	1·7378833	1·1537329	1332483	1451	8667517	5
56	4989920	2521	5010080	2·0040402	5757999	1·7367144	1·1539261	1333934	1452	8666066	4
57	4992441	2520	5007559	2·0030283	5761873	1·7355468	1·1541195	1335386	1453	8664614	3
58	4994961	2520	5005039	2·0020177	5765748	1·7343803	1·1543130	1336839	1453	8663161	2
59	4997481	2519	5002519	2·0010083	5769625	1·7332149	1·1545067	1338292	1454	8661708	1
60	5000000		5000000	2·0000000	5773503	1·7320508	1·1547005	1339746		8660254	0
′	Cosine	Dif.	Vers.	Secant.	Cotan.	Tang.	Cosec.	Covers	Dif.	Sine.	′

′	Sine.	Dif.	Cosec.	Verseds.	Tang.	Dif.	Cotang.	Covers.	Secant.	D.	Cosine.	′
0	9·6855712	2279	10·3144288	9·0982293	9·7437520	2079	10·2562480	9·7119677	10·0581807	701	9·9418193	60
1	9·6857991	2276	10·3142009	9·0987176	9·7440499	2079	10·2559501	9·7117532	10·0582508	701	9·9417492	59
2	9·6860267	2275	10·3139733	9·0992057	9·7443476	2077	10·2556524	9·7115387	10·0583209	701	9·9416791	58
3	9·6862542	2274	10·3137458	9·0996934	9·7446453	2077	10·2553547	9·7113240	10·0583910	702	9·9416090	57
4	9·6864816	2272	10·3135184	9·1001809	9·7449428	2975	10·2550572	9·7111093	10·0584612	703	9·9415388	56
5	9·6867088	2271	10·3132912	9·1006681	9·7452403	2973	10·2547597	9·7108945	10·0585315	703	9·9414685	55
6	9·6869359	2269	10·3130641	9·1011549	9·7455376	2973	10·2544624	9·7106797	10·0586018	703	9·9413982	54
7	9·6871628	2267	10·3128372	9·1016415	9·7458349	2971	10·2541651	9·7104647	10·0586721	704	9·9413279	53
8	9·6873895	2266	10·3126105	9·1021278	9·7461320	2970	10·2538680	9·7102497	10·0587425	704	9·9412575	52
9	9·6876161	2264	10·3123839	9·1026138	9·7464290	2969	10·2535710	9·7100346	10·0588129	705	9·9411871	51
10	9·6878425	2263	10·3121575	9·1030995	9·7467259	2968	10·2532741	9·7098195	10·0588834	705	9·9411166	50
11	9·6880688	2261	10·3119312	9·1035850	9·7470227	2967	10·2529773	9·7096043	10·0589539	706	9·9410461	49
12	9·6882949	2260	10·3117051	9·1040701	9·7473194	2966	10·2526806	9·7093890	10·0590245	707	9·9409755	48
13	9·6885209	2258	10·3114791	9·1045550	9·7476160	2965	10·2523840	9·7091736	10·0590952	706	9·9409048	47
14	9·6887467	2256	10·3112533	9·1050395	9·7479125	2964	10·2520875	9·7089582	10·0591658	708	9·9408342	46
15	9·6889723	2255	10·3110277	9·1055238	9·7482089	2963	10·2517911	9·7087427	10·0592366	707	9·9407634	45
16	9·6891978	2255	10·3108022	9·1060078	9·7485052	2961	10·2514948	9·7085271	10·0593073	708	9·9406927	44
17	9·6894232	2252	10·3105768	9·1064915	9·7488013	2961	10·2511987	9·7083115	10·0593781	709	9·9406219	43
18	9·6896484	2250	10·3103516	9·1069749	9·7490974	2960	10·2509026	9·7080957	10·0594490	709	9·9405510	42
19	9·6898734	2249	10·3101266	9·1074580	9·7493934	2958	10·2506066	9·7078799	10·0595199	710	9·9404801	41
20	9·6900983	2248	10·3099017	9·1079408	9·7496892	2958	10·2503108	9·7076641	10·0595909	710	9·9404091	40
21	9·6903231	2245	10·3096769	9·1084234	9·7499850	2956	10·2500150	9·7074481	10·0596619	711	9·9403381	39
22	9·6905476	2245	10·3094524	9·1089056	9·7502806	2956	10·2497194	9·7072321	10·0597330	711	9·9402670	38
23	9·6907721	2243	10·3092279	9·1093876	9·7505762	2954	10·2494238	9·7070160	10·0598041	711	9·9401959	37
24	9·6909964	2241	10·3090036	9·1098693	9·7508716	2953	10·2491284	9·7067999	10·0598752	713	9·9401248	36
25	9·6912205	2240	10·3087795	9·1103507	9·7511669	2953	10·2488331	9·7065837	10·0599465	712	9·9400535	35
26	9·6914445	2238	10·3085555	9·1108318	9·7514622	2951	10·2485378	9·7063674	10·0600177	713	9·9399823	34
27	9·6916683	2236	10·3083317	9·1113126	9·7517573	2950	10·2482427	9·7061510	10·0600890	714	9·9399110	33
28	9·6918919	2236	10·3081081	9·1117932	9·7520523	2949	10·2479477	9·7059346	10·0601604	714	9·9398396	32
29	9·6921155	2233	10·3078845	9·1122735	9·7523472	2948	10·2476528	9·7057180	10·0602318	714	9·9397682	31
30	9·6923388	2232	10·3076612	9·1127534	9·7526420	2948	10·2473580	9·7055015	10·0603032	715	9·9396968	30
31	9·6925620	2231	10·3074380	9·1132331	9·7529368	2946	10·2470632	9·7052848	10·0603747	716	9·9396253	29
32	9·6927851	2229	10·3072149	9·1137126	9·7532314	2945	10·2467686	9·7050681	10·0604463	716	9·9395537	28
33	9·6930080	2228	10·3069920	9·1141917	9·7535259	2944	10·2464741	9·7048513	10·0605179	716	9·9394821	27
34	9·6932308	2226	10·3067692	9·1146705	9·7538203	2943	10·2461797	9·7046344	10·0605895	717	9·9394105	26
35	9·6934534	2224	10·3065466	9·1151491	9·7541146	2942	10·2458854	9·7044174	10·0606612	717	9·9393388	25
36	9·6936758	2223	10·3063242	9·1156274	9·7544086	2941	10·2455912	9·7042004	10·0607329	718	9·9392671	24
37	9·6938981	2222	10·3061019	9·1161054	9·7547029	2940	10·2452971	9·7039833	10·0608047	719	9·9391953	23
38	9·6941203	2220	10·3058797	9·1165831	9·7549969	2939	10·2450031	9·7037661	10·0608766	719	9·9391234	22
39	9·6943423	2219	10·3056577	9·1170606	9·7552908	2938	10·2447092	9·7035489	10·0609485	719	9·9390515	21
40	9·6945642	2217	10·3054358	9·1175377	9·7555846	2937	10·2444154	9·7033316	10·0610204	720	9·9389796	20
41	9·6947859	2215	10·3052141	9·1180146	9·7558783	2935	10·2441217	9·7031142	10·0610924	720	9·9389076	19
42	9·6950074	2214	10·3049926	9·1184912	9·7561718	2935	10·2438282	9·7028967	10·0611644	721	9·9388356	18
43	9·6952288	2213	10·3047712	9·1189675	9·7564653	2934	10·2435347	9·7026792	10·0612365	721	9·9387635	17
44	9·6954501	2211	10·3045499	9·1194436	9·7567587	2933	10·2432413	9·7024616	10·0613086	722	9·9386914	16
45	9·6956712	2211	10·3043288	9·1199193	9·7570520	2932	10·2429480	9·7022439	10·0613808	722	9·9386192	15
46	9·6958922	2210	10·3041078	9·1203948	9·7573452	2931	10·2426548	9·7020262	10·0614530	723	9·9385470	14
47	9·6961130	2208	10·3038870	9·1208700	9·7576383	2930	10·2423617	9·7018084	10·0615253	723	9·9384747	13
48	9·6963336	2206	10·3036664	9·1213449	9·7579313	2929	10·2420687	9·7015905	10·0615976	724	9·9384024	12
49	9·6965541	2205	10·3034459	9·1218196	9·7582242	2928	10·2417758	9·7013725	10·0616700	724	9·9383300	11
50	9·6967745	2204	10·3032255	9·1222939	9·7585170	2926	10·2414830	9·7011545	10·0617424	725	9·9382576	10
51	9·6969947	2202	10·3030053	9·1227680	9·7588096	2926	10·2411904	9·7009363	10·0618149	725	9·9381851	9
52	9·6972148	2201	10·3027852	9·1232419	9·7591022	2925	10·2408978	9·7007182	10·0618874	726	9·9381126	8
53	9·6974347	2199	10·3025653	9·1237154	9·7593947	2924	10·2406003	9·7004999	10·0619600	726	9·9380400	7
54	9·6976545	2198	10·3023455	9·1241887	9·7596871	2923	10·2403129	9·7002816	10·0620326	727	9·9379674	6
55	9·6978741	2196	10·3021259	9·1246617	9·7599794	2922	10·2400206	9·7000631	10·0621053	727	9·9378947	5
56	9·6980936	2195	10·3019064	9·1251344	9·7602716	2922	10·2397284	9·6998447	10·0621780	728	9·9378220	4
57	9·6983129	2193	10·3016871	9·1256068	9·7605637	2921	10·2394363	9·6996261	10·0622508	728	9·9377492	3
58	9·6985321	2190	10·3014679	9·1260790	9·7608557	2920	10·2391443	9·6994075	10·0623236	729	9·9376764	2
59	9·6987511	2189	10·3012489	9·1265508	9·7611476	2919	10·2388524	9·6991888	10·0623965	729	9·9376035	1
60	9·6989700		10·3010300	9·1270225	9·7614394	2918	10·2385606	9·6989700	10·0624694	729	9·9375300	0

| ′ | Cosine. | Dif. | Secant. | Covers. | Cotang. | Dif. | Tang. | Verseds. | Cosec. | D. | Sine. | ′ |

′	Sine.	Dif.	Covers	Cosec.	Tang.	Cotang.	Secant.	Vers.	Dif.	Cosine	′
0	5000000	2519	5000000	2·0000000	5773503	1·7320508	1·1547005	1339746	1455	8660254	60
1	5002519	2518	4997481	1·9989929	5777382	1·7308878	1·1548945	1341201	1455	8658799	59
2	5005037	2519	4994963	1·9979870	5781262	1·7297260	1·1550887	1342656	1457	8657344	58
3	5007556	2517	4992444	1·9969823	5785144	1·7285654	1·1552830	1344113	1457	8655887	57
4	5010073	2518	4989927	1·9959788	5789027	1·7274060	1·1554775	1345570	1457	8654430	56
5	5012591	2516	4987409	1·9949764	5792912	1·7262477	1·1556722	1347027	1459	8652973	55
6	5015107	2517	4984893	1·9939753	5796797	1·7250905	1·1558670	1348486	1459	8651514	54
7	5017624	2516	4982376	1·9929752	5800684	1·7239346	1·1560620	1349945	1460	8650055	53
8	5020140	2515	4979860	1·9919764	5804573	1·7227797	1·1562572	1351405	1461	8648595	52
9	5022655	2515	4977345	1·9909787	5808462	1·7216261	1·1564525	1352866	1461	8647134	51
10	5025170	2515	4974830	1·9899822	5812353	1·7204736	1·1566480	1354327	1462	8645673	50
11	5027685	2514	4972315	1·9889869	5816245	1·7193222	1·1568436	1355789	1463	8644211	49
12	5030199	2514	4969801	1·9879927	5820139	1·7181720	1·1570394	1357252	1464	8642748	48
13	5032713	2514	4967287	1·9869997	5824034	1·7170230	1·1572354	1358716	1464	8641284	47
14	5035227	2513	4964773	1·9860080	5827930	1·7158751	1·1574315	1360180	1465	8639820	46
15	5037740	2512	4962260	1·9850172	5831828	1·7147283	1·1576278	1361645	1466	8638355	45
16	5040252	2513	4959748	1·9840276	5835726	1·7135827	1·1578243	1363111	1466	8636889	44
17	5042765	2511	4957235	1·9830393	5839627	1·7124382	1·1580209	1364577	1467	8635423	43
18	5045276	2512	4954724	1·9820520	5843528	1·7112949	1·1582177	1366044	1468	8633956	42
19	5047788	2510	4952212	1·9810659	5847431	1·7101527	1·1584146	1367512	1469	8632488	41
20	5050298	2511	4949702	1·9800810	5851335	1·7090116	1·1586118	1368981	1470	8631019	40
21	5052809	2510	4947191	1·9790972	5855241	1·7078717	1·1588091	1370451	1470	8629549	39
22	5055319	2509	4944681	1·9781146	5859148	1·7067329	1·1590065	1371921	1471	8628079	38
23	5057828	2510	4942172	1·9771331	5863056	1·7055953	1·1592041	1373392	1471	8626608	37
24	5060338	2508	4939662	1·9761527	5866965	1·7044587	1·1594019	1374863	1473	8625137	36
25	5062846	2509	4937154	1·9751735	5870876	1·7033233	1·1595999	1376336	1473	8623664	35
26	5065355	2508	4934645	1·9741954	5874788	1·7021890	1·1597980	1377809	1474	8622191	34
27	5067863	2507	4932137	1·9732185	5878702	1·7010559	1·1599963	1379283	1474	8620717	33
28	5070370	2507	4929630	1·9722427	5882616	1·6999238	1·1601947	1380757	1475	8619243	32
29	5072877	2507	4927123	1·9712680	5886533	1·6987929	1·1603933	1382232	1476	8617768	31
30	5075384	2506	4924616	1·9702944	5890450	1·6976631	1·1605921	1383708	1477	8616292	30
31	5077890	2506	4922110	1·9693220	5894369	1·6965344	1·1607911	1385185	1478	8614815	29
32	5080396	2505	4919604	1·9683507	5898289	1·6954069	1·1609902	1386663	1478	8613337	28
33	5082901	2505	4917099	1·9673805	5902211	1·6942804	1·1611894	1388141	1479	8611859	27
34	5085406	2504	4914594	1·9664114	5906134	1·6931550	1·1613889	1389620	1479	8610380	26
35	5087910	2504	4912090	1·9654435	5910058	1·6920308	1·1615885	1391099	1481	8608901	25
36	5090414	2504	4909586	1·9644767	5913984	1·6909077	1·1617883	1392580	1481	8607420	24
37	5092918	2503	4907082	1·9635110	5917910	1·6897856	1·1619882	1394061	1482	8605939	23
38	5095421	2503	4904579	1·9625464	5921839	1·6886647	1·1621883	1395543	1482	8604457	22
39	5097924	2502	4902076	1·9615829	5925768	1·6875449	1·1623886	1397025	1484	8602975	21
40	5100426	2502	4899574	1·9606206	5929699	1·6864261	1·1625891	1398509	1484	8601491	20
41	5102928	2501	4897072	1·9596593	5933632	1·6853085	1·1627897	1399993	1484	8600007	19
42	5105429	2501	4894571	1·9586992	5937565	1·6841919	1·1629905	1401477	1486	8598523	18
43	5107930	2501	4892070	1·9577402	5941501	1·6830765	1·1631914	1402963	1486	8597037	17
44	5110431	2500	4889569	1·9567822	5945437	1·6819621	1·1633925	1404449	1487	8595551	16
45	5112931	2500	4887069	1·9558254	5949375	1·6808489	1·1635938	1405936	1488	8594064	15
46	5115431	2499	4884569	1·9548697	5953314	1·6797367	1·1637953	1407424	1488	8592576	14
47	5117930	2499	4882070	1·9539150	5957255	1·6786256	1·1639969	1408912	1489	8591088	13
48	5120429	2498	4879571	1·9529615	5961196	1·6775156	1·1641987	1410401	1490	8589599	12
49	5122927	2498	4877073	1·9520091	5965140	1·6764067	1·1644007	1411891	1490	8588109	11
50	5125425	2498	4874575	1·9510577	5969084	1·6752988	1·1646028	1413381	1492	8586619	10
51	5127923	2497	4872077	1·9501075	5973030	1·6741921	1·1648051	1414873	1492	8585127	9
52	5130420	2496	4869580	1·9491583	5976978	1·6730864	1·1650076	1416365	1492	8583635	8
53	5132916	2497	4867084	1·9482102	5980926	1·6719818	1·1652102	1417857	1494	8582143	7
54	5135413	2495	4864587	1·9472632	5984877	1·6708782	1·1654130	1419351	1494	8580649	6
55	5137908	2496	4862092	1·9463173	5988828	1·6697758	1·1656160	1420845	1495	8579155	5
56	5140404	2495	4859596	1·9453725	5992781	1·6686744	1·1658191	1422340	1496	8577660	4
57	5142899	2494	4857101	1·9444288	5996735	1·6675741	1·1660224	1423836	1496	8576164	3
58	5145393	2494	4854607	1·9434861	6000591	1·6664748	1·1662259	1425332	1496	8574668	2
59	5147887	2494	4852113	1·9425445	6004648	1·6653766	1·1664296	1426829	1497	8573171	1
60	5150381		4849619	1·9416040	6008606	1·6642795	1·1666334	1428327	1498	8571673	0
′	Cosine	Dif.	Vers.	Secant.	Cotan.	Taug.	Cosec.	Covers	Dif.	Sine.	′

′	Sine.	Dif.	Cosec.	Verseds.	Tang.	Dif.	Cotang.	Covers.	Secant.	D.	Cosine.	′
0	9·6989700	2187	10·3010300	9·1270225	9·7614394	2917	10·2385606	9·6989760	10·0624694	729	9·9375306	60
1	9·6991887	2186	10·3008113	9·1274938	9·7617311	2916	10·2382689	9·6987512	10·0625423	730	9·9374577	59
2	9·6994073	2185	10·3005927	9·1279649	9·7620227	2915	10·2379773	9·6985322	10·0626153	731	9·9373847	58
3	9·6996258	2183	10·3003742	9·1284356	9·7623142	2914	10·2376858	9·6983132	10·0626884	731	9·9373116	57
4	9·6998441	2181	10·3001559	9·1289062	9·7626056	2913	10·2373944	9·6980942	10·0627615	732	9·9372385	56
5	9·7000622	2180	10·2999378	9·1293764	9·7628969	2912	10·2371031	9·6978750	10·0628347	732	9·9371653	55
6	9·7002802	2179	10·2997198	9·1298464	9·7631881	2911	10·2368119	9·6976558	10·0629079	732	9·9370921	54
7	9·7004981	2177	10·2995019	9·1303161	9·7634792	2910	10·2365208	9·6974365	10·0629811	733	9·9370189	53
8	9·7007158	2176	10·2992842	9·1307855	9·7637702	2910	10·2362298	9·6972172	10·0630544	734	9·9369456	52
9	9·7009334	2174	10·2990666	9·1312547	9·7640612	2908	10·2359388	9·6969977	10·0631278	734	9·9368722	51
10	9·7011508	2173	10·2988492	9·1317235	9·7643520	2907	10·2356480	9·6967782	10·0632012	734	9·9367988	50
11	9·7013681	2171	10·2986319	9·1321921	9·7646427	2907	10·2353573	9·6965586	10·0632746	735	9·9367254	49
12	9·7015852	2170	10·2984148	9·1326605	9·7649334	2905	10·2350666	9·6963390	10·0633481	736	9·9366519	48
13	9·7018022	2168	10·2981978	9·1331286	9·7652239	2904	10·2347761	9·6961192	10·0634217	736	9·9365783	47
14	9·7020190	2167	10·2979810	9·1335964	9·7655143	2904	10·2344857	9·6958994	10·0634953	736	9·9365047	46
15	9·7022357	2166	10·2977643	9·1340639	9·7658047	2902	10·2341953	9·6956795	10·0635689	737	9·9364311	45
16	9·7024523	2164	10·2975477	9·1345311	9·7660949	2902	10·2339051	9·6954596	10·0636426	738	9·9363574	44
17	9·7026687	2162	10·2973313	9·1349981	9·7663851	2900	10·2336149	9·6952396	10·0637164	738	9·9362836	43
18	9·7028849	2162	10·2971151	9·1354648	9·7666751	2900	10·2333249	9·6950194	10·0637902	738	9·9362098	42
19	9·7031011	2159	10·2968989	9·1359313	9·7669651	2899	10·2330349	9·6947993	10·0638640	739	9·9361360	41
20	9·7033170	2159	10·2966830	9·1363975	9·7672550	2898	10·2327450	9·6945790	10·0639379	740	9·9360621	40
21	9·7035329	2157	10·2964671	9·1368634	9·7675448	2896	10·2324552	9·6943587	10·0640119	740	9·9359881	39
22	9·7037486	2155	10·2962514	9·1373290	9·7678344	2896	10·2321656	9·6941383	10·0640859	740	9·9359141	38
23	9·7039641	2154	10·2960359	9·1377944	9·7681240	2895	10·2318760	9·6939178	10·0641599	741	9·9358401	37
24	9·7041795	2152	10·2958205	9·1382595	9·7684135	2894	10·2315865	9·6936973	10·0642340	742	9·9357660	36
25	9·7043947	2152	10·2956053	9·1387244	9·7687029	2893	10·2312971	9·6934766	10·0643082	741	9·9356918	35
26	9·7046099	2149	10·2953901	9·1391889	9·7689922	2892	10·2310078	9·6932559	10·0643823	743	9·9356177	34
27	9·7048248	2149	10·2951752	9·1396532	9·7692814	2891	10·2307186	9·6930352	10·0644566	743	9·9355434	33
28	9·7050397	2146	10·2949603	9·1401173	9·7695705	2891	10·2304295	9·6928143	10·0645309	743	9·9354691	32
29	9·7052543	2146	10·2947457	9·1405811	9·7698596	2889	10·2301404	9·6925934	10·0646052	744	9·9353948	31
30	9·7054689	2144	10·2945311	9·1410446	9·7701485	2888	10·2298515	9·6923724	10·0646796	745	9·9353204	30
31	9·7056833	2142	10·2943167	9·1415078	9·7704373	2888	10·2295627	9·6921513	10·0647541	744	9·9352459	29
32	9·7058975	2141	10·2941025	9·1419708	9·7707261	2886	10·2292739	9·6919302	10·0648285	746	9·9351715	28
33	9·7061116	2140	10·2938884	9·1424335	9·7710147	2886	10·2289853	9·6917090	10·0649031	746	9·9350969	27
34	9·7063256	2138	10·2936744	9·1428960	9·7713033	2884	10·2286967	9·6914877	10·0649777	746	9·9350223	26
35	9·7065394	2137	10·2934606	9·1433581	9·7715917	2884	10·2284083	9·6912663	10·0650523	747	9·9349477	25
36	9·7067531	2136	10·2932469	9·1438201	9·7718801	2883	10·2281199	9·6910449	10·0651270	747	9·9348730	24
37	9·7069667	2134	10·2930333	9·1442817	9·7721684	2882	10·2278316	9·6908233	10·0652017	748	9·9347983	23
38	9·7071801	2132	10·2928199	9·1447431	9·7724566	2881	10·2275434	9·6906017	10·0652765	749	9·9347235	22
39	9·7073933	2131	10·2926067	9·1452042	9·7727447	2880	10·2272553	9·6903801	10·0653514	748	9·9346486	21
40	9·7076064	2130	10·2923936	9·1456651	9·7730327	2879	10·2269673	9·6901583	10·0654262	750	9·9345738	20
41	9·7078194	2129	10·2921806	9·1461257	9·7733206	2878	10·2266794	9·6899365	10·0655012	750	9·9344988	19
42	9·7080323	2127	10·2919677	9·1465861	9·7736084	2877	10·2263916	9·6897146	10·0655762	750	9·9344238	18
43	9·7082450	2125	10·2917550	9·1470461	9·7738961	2877	10·2261039	9·6894926	10·0656512	751	9·9343488	17
44	9·7084575	2124	10·2915425	9·1475060	9·7741838	2875	10·2258162	9·6892706	10·0657263	751	9·9342737	16
45	9·7086699	2123	10·2913301	9·1479655	9·7744713	2875	10·2255287	9·6890485	10·0658014	752	9·9341986	15
46	9·7088822	2121	10·2911178	9·1484248	9·7747588	2874	10·2252412	9·6888263	10·0658766	752	9·9341234	14
47	9·7090943	2120	10·2909057	9·1488838	9·7750462	2872	10·2249538	9·6886040	10·0659518	753	9·9340482	13
48	9·7093063	2119	10·2906937	9·1493426	9·7753334	2872	10·2246666	9·6883817	10·0660271	753	9·9339729	12
49	9·7095182	2117	10·2904818	9·1498011	9·7756206	2871	10·2243794	9·6881593	10·0661024	754	9·9338976	11
50	9·7097299	2116	10·2902701	9·1502594	9·7759077	2870	10·2240923	9·6879368	10·0661778	755	9·9338222	10
51	9·7099415	2114	10·2900585	9·1507174	9·7761947	2869	10·2238053	9·6877142	10·0662533	754	9·9337467	9
52	9·7101529	2113	10·2898471	9·1511751	9·7764816	2869	10·2235184	9·6874915	10·0663287	756	9·9336713	8
53	9·7103642	2111	10·2896358	9·1516326	9·7767685	2867	10·2232315	9·6872688	10·0664043	756	9·9335957	7
54	9·7105753	2110	10·2894247	9·1520898	9·7770552	2866	10·2229448	9·6870460	10·0664799	756	9·9335201	6
55	9·7107863	2109	10·2892137	9·1525467	9·7773418	2866	10·2226582	9·6868231	10·0665555	757	9·9334445	5
56	9·7109972	2108	10·2890028	9·1530034	9·7776284	2865	10·2223716	9·6866002	10·0666312	757	9·9333688	4
57	9·7112080	2106	10·2887920	9·1534599	9·7779149	2863	10·2220851	9·6863772	10·0667069	758	9·9332931	3
58	9·7114186	2104	10·2885814	9·1539161	9·7782012	2863	10·2217988	9·6861541	10·0667827	758	9·9332173	2
59	9·7116290	2103	10·2883710	9·1543720	9·7784875	2862	10·2215125	9·6859309	10·0668585	759	9·9331415	1
60	9·7118393		10·2881607	9·1548276	9·7787737		10·2212263	9·6857076	10·0669344		9·9330656	0
′	Cosine.	Dif.	Secant.	Covers.	Cotang.	Dif.	Tang.	Verseds.	Cosec.	D.	Sine.	′

′	Sine.	Dif.	Covers	Cosec.	Tang.	Cotang.	Secant.	Vers.	Dif.	Cosine	′
0	5150381	2493	4849619	1·9416040	6008606	1·6642795	1·1666334	1428327	1499	8571673	60
1	5152874	2493	4847126	1·9406646	6012566	1·6631834	1·1668374	1429826	1499	8570174	59
2	5155367	2492	4844633	1·9397262	6016527	1·6620884	1·1670416	1431325	1500	8568675	58
3	5157859	2492	4842141	1·9387889	6020490	1·6609945	1·1672459	1432825	1501	8567175	57
4	5160351	2491	4839649	1·9378527	6024454	1·6599016	1·1674504	1434326	1501	8565674	56
5	5162842	2491	4837158	1·9369176	6028419	1·6588097	1·1676551	1435827	1502	8564173	55
6	5165333	2491	4834667	1·9359835	6032386	1·6577189	1·1678599	1437329	1503	8562671	54
7	5167824	2490	4832176	1·9350505	6036354	1·6566292	1·1680649	1438832	1504	8561168	53
8	5170314	2490	4829686	1·9341185	6040323	1·6555405	1·1682701	1440336	1504	8559664	52
9	5172804	2489	4827196	1·9331876	6044294	1·6544529	1·1684755	1441840	1505	8558160	51
10	5175293	2489	4824707	1·9322578	6048266	1·6533663	1·1686810	1443345	1506	8556655	50
11	5177782	2488	4822218	1·9313290	6052240	1·6522808	1·1688867	1444851	1506	8555149	49
12	5180270	2488	4819730	1·9304013	6056215	1·6511963	1·1690926	1446357	1508	8553643	48
13	5182758	2488	4817242	1·9294746	6060192	1·6501128	1·1692986	1447865	1508	8552135	47
14	5185246	2487	4814754	1·9285490	6064170	1·6490304	1·1695048	1449373	1508	8550627	46
15	5187733	2486	4812267	1·9276244	6068149	1·6479490	1·1697112	1450881	1510	8549119	45
16	5190219	2486	4809781	1·9267009	6072130	1·6468687	1·1699178	1452391	1510	8547609	44
17	5192705	2486	4807295	1·9257784	6076112	1·6457893	1·1701245	1453901	1511	8546099	43
18	5195191	2485	4804809	1·9248570	6080095	1·6447111	1·1703314	1455412	1511	8544588	42
19	5197676	2485	4802324	1·9239366	6084080	1·6436338	1·1705385	1456923	1513	8543077	41
20	5200161	2485	4799839	1·9230173	6088067	1·6425576	1·1707457	1458436	1513	8541564	40
21	5202646	2484	4797354	1·9220990	6092054	1·6414824	1·1709531	1459949	1513	8540051	39
22	5205130	2483	4794870	1·9211817	6096043	1·6404082	1·1711607	1461462	1515	8538538	38
23	5207613	2483	4792387	1·9202655	6100034	1·6393351	1·1713685	1462977	1515	8537023	37
24	5210096	2483	4789904	1·9193503	6104026	1·6382630	1·1715764	1464492	1516	8535508	36
25	5212579	2482	4787421	1·9184362	6108019	1·6371919	1·1717845	1466008	1517	8533992	35
26	5215061	2482	4784939	1·9175230	6112014	1·6361218	1·1719928	1467525	1517	8532475	34
27	5217543	2481	4782457	1·9166110	6116011	1·6350528	1·1722013	1469042	1518	8530958	33
28	5220024	2481	4779976	1·9156999	6120008	1·6339847	1·1724099	1470560	1519	8529440	32
29	5222505	2481	4777495	1·9147899	6124007	1·6329177	1·1726187	1472079	1519	8527921	31
30	5224986	2480	4775014	1·9138809	6128008	1·6318517	1·1728277	1473598	1521	8526402	30
31	5227466	2479	4772534	1·9129729	6132010	1·6307867	1·1730368	1475119	1521	8524881	29
32	5229945	2479	4770055	1·9120659	6136013	1·6297227	1·1732462	1476640	1521	8523360	28
33	5232424	2479	4767576	1·9111600	6140018	1·6286597	1·1734557	1478161	1523	8521839	27
34	5234903	2478	4765097	1·9102551	6144024	1·6275977	1·1736653	1479684	1523	8520316	26
35	5237381	2478	4762619	1·9093512	6148032	1·6265368	1·1738752	1481207	1524	8518793	25
36	5239859	2477	4760141	1·9084483	6152041	1·6254768	1·1740852	1482731	1524	8517269	24
37	5242336	2477	4757664	1·9075464	6156052	1·6244178	1·1742954	1484255	1526	8515745	23
38	5244813	2477	4755187	1·9066456	6160064	1·6233599	1·1745058	1485781	1526	8514219	22
39	5247290	2476	4752710	1·9057457	6164077	1·6223029	1·1747163	1487307	1526	8512693	21
40	5249766	2475	4750234	1·9048469	6168092	1·6212469	1·1749270	1488833	1528	8511167	20
41	5252241	2476	4747759	1·9039491	6172108	1·6201920	1·1751379	1490361	1528	8509639	19
42	5254717	2474	4745283	1·9030522	6176126	1·6191380	1·1753490	1491889	1529	8508111	18
43	5257191	2474	4742809	1·9021564	6180145	1·6180850	1·1755603	1493418	1529	8506582	17
44	5259665	2474	4740335	1·9012616	6184166	1·6170330	1·1757717	1494947	1531	8505053	16
45	5262139	2474	4737861	1·9003678	6188188	1·6159820	1·1759833	1496478	1531	8503522	15
46	5264613	2472	4735387	1·8994750	6192211	1·6149320	1·1761951	1498009	1531	8501991	14
47	5267085	2473	4732915	1·8985832	6196236	1·6138829	1·1764070	1499541	1532	8500459	13
48	5269558	2472	4730442	1·8976924	6200263	1·6128349	1·1766191	1501073	1533	8498927	12
49	5272030	2472	4727970	1·8968026	6204291	1·6117878	1·1768314	1502606	1534	8497394	11
50	5274502	2471	4725498	1·8959138	6208320	1·6107417	1·1770439	1504140	1535	8495860	10
51	5276973	2470	4723027	1·8950259	6212351	1·6096966	1·1772566	1505675	1535	8494325	9
52	5279443	2471	4720557	1·8941391	6216383	1·6086525	1·1774694	1507210	1536	8492790	8
53	5281914	2469	4718086	1·8932532	6220417	1·6076094	1·1776824	1508746	1537	8491254	7
54	5284383	2470	4715617	1·8923684	6224452	1·6065672	1·1778956	1510283	1538	8489717	6
55	5286853	2469	4713147	1·8914845	6228488	1·6055260	1·1781089	1511821	1538	8488179	5
56	5289322	2468	4710678	1·8906016	6232527	1·6044858	1·1783225	1513359	1539	8486641	4
57	5291790	2468	4708210	1·8897197	6236566	1·6034465	1·1785362	1514898	1540	8485102	3
58	5294258	2468	4705742	1·8888388	6240607	1·6024082	1·1787501	1516438	1540	8483562	2
59	5296726	2467	4703274	1·8879589	6244650	1·6013709	1·1789642	1517978	1541	8482022	1
60	5299193		4700807	1·8870799	6248694	1·6003345	1·1791784	1519519		8480481	0
′	Cosine	Dif.	Vers.	Secant.	Cotan.	Taug.	Cosec.	Covers.	Dif.	Sine.	′

58 Deg.

′	Sine.	Dif.	Cosec.	Verseds.	Tang.	Dif.	Cotang.	Covers.	Secant.	D.	Cosine.	′
0	9·7118393	2102	10·2881607	9·1548276	9·7787737	2862	10·2212263	9·6857070	10·0669344	759	9·9330656	60
1	9·7120495	2101	10·2879505	9·1552831	9·7790599	2860	10·2209401	9·6854843	10·0670103	760	9·9329897	59
2	9·7122596	2099	10·2877404	9·1557382	9·7793459	2859	10·2206541	9·6852609	10·0670863	761	9·9329137	58
3	9·7124695	2097	10·2875305	9·1561931	9·7796318	2859	10·2203682	9·6850374	10·0671624	760	9·9328376	57
4	9·7126792	2097	10·2873208	9·1566477	9·7799177	2857	10·2200823	9·6848139	10·0672384	762	9·9327616	56
5	9·7128889	2094	10·2871111	9·1571021	9·7802034	2857	10·2197966	9·6845902	10·0673146	762	9·9326854	55
6	9·7130983	2094	10·2869017	9·1575562	9·7804891	2856	10·2195109	9·6843665	10·0673908	762	9·9326092	54
7	9·7133077	2092	10·2866923	9·1580101	9·7807747	2855	10·2192253	9·6841428	10·0674670	763	9·9325330	53
8	9·7135169	2091	10·2864831	9·1584637	9·7810602	2854	10·2189398	9·6839189	10·0675433	763	9·9324567	52
9	9·7137260	2089	10·2862740	9·1589171	9·7813456	2853	10·2186544	9·6836950	10·0676196	764	9·9323804	51
10	9·7139349	2088	10·2860651	9·1593702	9·7816309	2853	10·2183691	9·6834710	10·0676960	764	9·9323040	50
11	9·7141437	2087	10·2858563	9·1598230	9·7819162	2851	10·2180838	9·6832469	10·0677724	765	9·9322276	49
12	9·7143524	2085	10·2856476	9·1602756	9·7822013	2851	10·2177987	9·6830227	10·0678489	765	9·9321511	48
13	9·7145609	2084	10·2854391	9·1607280	9·7824864	2851	10·2175136	9·6827985	10·0679254	766	9·9320746	47
14	9·7147693	2083	10·2852307	9·1611800	9·7827713	2849	10·2172287	9·6825741	10·0680020	767	9·9319980	46
15	9·7149776	2081	10·2850224	9·1616319	9·7830562	2849	10·2169438	9·6823498	10·0680787	766	9·9319213	45
16	9·7151857	2080	10·2848143	9·1620835	9·7833410	2848	10·2166590	9·6821253	10·0681553	768	9·9318447	44
17	9·7153937	2078	10·2846063	9·1625348	9·7836258	2848	10·2163742	9·6819007	10·0682321	768	9·9317679	43
18	9·7156015	2077	10·2843985	9·1629859	9·7839104	2846	10·2160896	9·6816761	10·0683089	768	9·9316911	42
19	9·7158092	2076	10·2841908	9·1634367	9·7841949	2845	10·2158051	9·6814514	10·0683857	769	9·9316143	41
20	9·7160168	2075	10·2839832	9·1638873	9·7844794	2845	10·2155206	9·6812266	10·0684626	769	9·9315374	40
21	9·7162243	2073	10·2837757	9·1643376	9·7847638	2844	10·2152362	9·6810018	10·0685395	770	9·9314605	39
22	9·7164316	2071	10·2835684	9·1647876	9·7850481	2843	10·2149519	9·6807769	10·0686165	770	9·9313835	38
23	9·7166387	2071	10·2833613	9·1652374	9·7853323	2842	10·2146677	9·6805519	10·0686935	771	9·9313065	37
24	9·7168458	2068	10·2831542	9·1656870	9·7856164	2841	10·2143836	9·6803268	10·0687706	772	9·9312294	36
25	9·7170526	2068	10·2829474	9·1661363	9·7859004	2840	10·2140996	9·6801016	10·0688478	772	9·9311522	35
26	9·7172594	2066	10·2827406	9·1665854	9·7861844	2840	10·2138156	9·6798764	10·0689250	772	9·9310750	34
27	9·7174660	2065	10·2825340	9·1670342	9·7864682	2836	10·2135318	9·6796511	10·0690022	773	9·9309978	33
28	9·7176725	2064	10·2823275	9·1674828	9·7867520	2838	10·2132480	9·6794257	10·0690795	773	9·9309205	32
29	9·7178789	2062	10·2821211	9·1679311	9·7870357	2837	10·2129643	9·6792002	10·0691568	774	9·9308432	31
30	9·7180851	2061	10·2819149	9·1683791	9·7873193	2836	10·2126807	9·6789747	10·0692342	775	9·9307658	30
31	9·7182912	2059	10·2817088	9·1688269	9·7876028	2835	10·2123972	9·6787491	10·0693117	774	9·9306883	29
32	9·7184971	2059	10·2815029	9·1692745	9·7878863	2835	10·2121137	9·6785234	10·0693891	776	9·9306109	28
33	9·7187030	2056	10·2812970	9·1697218	9·7881696	2833	10·2118304	9·6782976	10·0694667	776	9·9305333	27
34	9·7189086	2056	10·2810914	9·1701689	9·7884529	2833	10·2115471	9·6780717	10·0695443	776	9·9304557	26
35	9·7191142	2054	10·2808858	9·1706157	9·7887361	2832	10·2112639	9·6778458	10·0696219	777	9·9303781	25
36	9·7193196	2053	10·2806804	9·1710623	9·7890192	2831	10·2109808	9·6776198	10·0696996	778	9·9303004	24
37	9·7195249	2051	10·2804751	9·1715086	9·7893023	2831	10·2106977	9·6773937	10·0697774	778	9·9302226	23
38	9·7197300	2050	10·2802700	9·1719547	9·7895852	2829	10·2104148	9·6771676	10·0698552	778	9·9301448	22
39	9·7199350	2049	10·2800650	9·1724005	9·7898681	2827	10·2101319	9·6769413	10·0699330	779	9·9300670	21
40	9·7201399	2048	10·2798601	9·1728461	9·7901508	2827	10·2098492	9·6767150	10·0700109	779	9·9299891	20
41	9·7203447	2046	10·2796553	9·1732914	9·7904335	2826	10·2095665	9·6764886	10·0700888	780	9·9299112	19
42	9·7205493	2045	10·2794507	9·1737365	9·7907161	2826	10·2092839	9·6762622	10·0701668	780	9·9298332	18
43	9·7207538	2043	10·2792462	9·1741813	9·7909987	2824	10·2090013	9·6760356	10·0702449	781	9·9297551	17
44	9·7209581	2042	10·2790419	9·1746259	9·7912811	2824	10·2087189	9·6758090	10·0703230	781	9·9296770	16
45	9·7211623	2041	10·2788377	9·1750703	9·7915635	2823	10·2084365	9·6755823	10·0704011	781	9·9295989	15
46	9·7213664	2040	10·2786336	9·1755144	9·7918458	2822	10·2081542	9·6753555	10·0704793	782	9·9295207	14
47	9·7215704	2038	10·2784296	9·1759582	9·7921280	2821	10·2078720	9·6751287	10·0705576	783	9·9294424	13
48	9·7217742	2037	10·2782258	9·1764018	9·7924101	2820	10·2075899	9·6749017	10·0706359	783	9·9293641	12
49	9·7219779	2035	10·2780221	9·1768452	9·7926921	2820	10·2073079	9·6746747	10·0707143	784	9·9292857	11
50	9·7221814	2034	10·2778186	9·1772883	9·7929741	2819	10·2070259	9·6744476	10·0707927	784	9·9292073	10
51	9·7223848	2033	10·2776152	9·1777312	9·7932560	2818	10·2067440	9·6742203	10·0708711	785	9·9291289	9
52	9·7225881	2032	10·2774119	9·1781738	9·7935378	2817	10·2064622	9·6739932	10·0709496	786	9·9290504	8
53	9·7227913	2030	10·2772087	9·1786162	9·7938195	2816	10·2061805	9·6737659	10·0710282	786	9·9289718	7
54	9·7229943	2029	10·2770057	9·1790584	9·7941011	2816	10·2058989	9·6735385	10·0711068	787	9·9288932	6
55	9·7231972	2028	10·2768028	9·1795003	9·7943827	2814	10·2056173	9·6733110	10·0711855	787	9·9288145	5
56	9·7234000	2026	10·2766000	9·1799419	9·7946641	2814	10·2053359	9·6730835	10·0712642	787	9·9287358	4
57	9·7236026	2025	10·2763974	9·1803833	9·7949455	2813	10·2050545	9·6728558	10·0713429	788	9·9286571	3
58	9·7238051	2024	10·2761949	9·1808245	9·7952268	2813	10·2047732	9·6726281	10·0714217	789	9·9285783	2
59	9·7240075	2022	10·2759925	9·1812655	9·7955081	2811	10·2044919	9·6724003	10·0715006	789	9·9284994	1
60	9·7242097		10·2757903	9·1817061	9·7957892		10·2042108	9·6721725	10·0715795		9·9284205	0
′	Cosine.	Dif.	Secant.	Covers.	Cotang.	Dif.	Tang.	Verseds.	Cosec.	D.	Sine.	′

'	Sine.	Dif.	Covers	Cosec.	Tang.	Cotang.	Secant.	Vers.	Dif.	Cosine	'
0	5299193	2466	4700807	1·8870799	6245694	1·6003345	1·1791784	1519519	1542	8480481	60
1	5301659	2466	4698341	1·8862019	6252739	1·5992991	1·1793928	1521061	1542	8478939	59
2	5304125	2466	4695875	1·8853249	6256786	1·5982647	1·1796074	1522603	1544	8477397	58
3	5306591	2466	4693409	1·8844489	6260834	1·5972312	1·1798222	1524147	1544	8475853	57
4	5309057	2464	4690943	1·8835738	6264884	1·5961987	1·1800372	1525691	1544	8474309	56
5	5311521	2465	4688479	1·8826998	6268935	1·5951672	1·1802523	1527235	1546	8472765	55
6	5313986	2464	4686014	1·8818266	6272988	1·5941366	1·1804676	1528781	1546	8471219	54
7	5316450	2463	4683550	1·8809545	6277042	1·5931070	1·1806831	1530327	1547	8469673	53
8	5318913	2463	4681087	1·8800833	6281098	1·5920783	1·1808988	1531874	1547	8468126	52
9	5321376	2463	4678624	1·8792131	6285155	1·5910505	1·1811146	1533421	1549	8466579	51
10	5323839	2462	4676161	1·8783438	6289214	1·5900238	1·1813307	1534970	1549	8465030	50
11	5326301	2462	4673699	1·8774755	6293274	1·5889979	1·1815469	1536519	1549	8463481	49
12	5328763	2461	4671237	1·8766082	6297336	1·5879731	1·1817633	1538068	1551	8461932	48
13	5331224	2461	4668776	1·8757419	6301399	1·5869491	1·1819798	1539619	1551	8460381	47
14	5333685	2461	4666315	1·8748764	6305464	1·5859261	1·1821966	1541170	1552	8458830	46
15	5336145	2460	4663855	1·8740120	6309530	1·5849041	1·1824135	1542722	1552	8457278	45
16	5338605	2460	4661395	1·8731485	6313598	1·5838830	1·1826306	1544274	1554	8455726	44
17	5341065	2458	4658935	1·8722859	6317667	1·5828628	1·1828479	1545828	1554	8454172	43
18	5343523	2458	4656477	1·8714244	6321738	1·5818436	1·1830654	1547382	1554	8452618	42
19	5345982	2459	4654018	1·8705637	6325810	1·5808253	1·1832830	1548936	1556	8451064	41
20	5348440	2458	4651560	1·8699040	6329883	1·5798079	1·1835008	1550492	1556	8449508	40
21	5350898	2458	4649102	1·8688453	6333959	1·5787915	1·1837188	1552048	1557	8447952	39
22	5353355	2457	4646645	1·8679875	6338035	1·5777760	1·1839370	1553605	1557	8446395	38
23	5355812	2457	4644188	1·8671306	6342113	1·5767615	1·1841554	1555162	1559	8444838	37
24	5358268	2456	4641732	1·8662747	6346193	1·5757479	1·1843739	1556721	1559	8443279	36
25	5360724	2456	4639276	1·8654197	6350274	1·5747352	1·1845927	1558280	1559	8441720	35
26	5363179	2455	4636821	1·8645657	6354357	1·5737234	1·1848116	1559839	1561	8440161	34
27	5365634	2455	4634366	1·8637126	6358441	1·5727126	1·1850307	1561400	1561	8438600	33
28	5368089	2454	4631911	1·8628605	6362527	1·5717026	1·1852500	1562961	1562	8437039	32
29	5370543	2453	4629457	1·8620093	6366614	1·5706936	1·1854694	1564523	1563	8435477	31
30	5372996	2453	4627004	1·8611590	6370703	1·5696856	1·1856890	1566086	1563	8433914	30
31	5375449	2453	4624551	1·8603097	6374793	1·5686784	1·1859089	1567649	1564	8432351	29
32	5377902	2452	4622098	1·8594612	6378885	1·5676722	1·1861289	1569213	1565	8430787	28
33	5380354	2452	4619646	1·8586138	6382978	1·5666669	1·1863490	1570778	1565	8429222	27
34	5382806	2451	4617194	1·8577672	6387073	1·5656625	1·1865694	1572343	1566	8427657	26
35	5385257	2451	4614743	1·8569216	6391169	1·5646590	1·1867900	1573909	1567	8426091	25
36	5387708	2450	4612292	1·8560769	6395267	1·5636564	1·1870107	1575476	1568	8424524	24
37	5390158	2450	4609842	1·8552331	6399366	1·5626548	1·1872316	1577044	1568	8422956	23
38	5392608	2450	4607392	1·8543903	6403467	1·5616540	1·1874527	1578612	1569	8421388	22
39	5395058	2449	4604942	1·8535483	6407569	1·5606542	1·1876740	1580181	1570	8419819	21
40	5397507	2448	4602493	1·8527073	6411673	1·5596552	1·1878954	1581751	1570	8418249	20
41	5399955	2448	4600045	1·8518672	6415779	1·5586572	1·1881171	1583321	1571	8416679	19
42	5402403	2448	4597597	1·8510281	6419886	1·5576601	1·1883389	1584892	1572	8415108	18
43	5404851	2448	4595149	1·8501898	6423994	1·5566639	1·1885609	1586464	1573	8413536	17
44	5407298	2447	4592702	1·8493525	6428105	1·5556685	1·1887831	1588037	1573	8411963	16
45	5409745	2446	4590255	1·8485161	6432216	1·5546741	1·1890055	1589610	1574	8410390	15
46	5412191	2446	4587809	1·8476806	6436329	1·5536806	1·1892280	1591184	1575	8408816	14
47	5414637	2445	4585363	1·8468460	6440444	1·5526880	1·1894508	1592759	1575	8407241	13
48	5417082	2445	4582918	1·8460123	6444560	1·5516963	1·1896737	1594334	1576	8405666	12
49	5419527	2444	4580473	1·8451795	6448678	1·5507054	1·1898968	1595910	1577	8404090	11
50	5421971	2444	4578029	1·8443476	6452797	1·5497155	1·1901201	1597487	1577	8402513	10
51	5424415	2444	4575585	1·8435166	6456918	1·5487264	1·1903436	1599064	1579	8400936	9
52	5426859	2443	4573141	1·8426866	6461041	1·5477383	1·1905673	1600643	1579	8399357	8
53	5429302	2442	4570698	1·8418574	6465165	1·5467510	1·1907911	1602222	1579	8397778	7
54	5431744	2443	4568256	1·8410292	6469290	1·5457647	1·1910152	1603801	1581	8396199	6
55	5434187	2441	4565813	1·8402019	6473417	1·5447792	1·1912394	1605382	1581	8394618	5
56	5436628	2441	4563372	1·8393753	6477546	1·5437946	1·1914638	1606963	1582	8393037	4
57	5439069	2441	4560931	1·8385498	6481676	1·5428108	1·1916884	1608545	1582	8391455	3
58	5441510	2441	4558490	1·8377251	6485808	1·5418280	1·1919132	1610127	1583	8389873	2
59	5443951	2439	4556049	1·8369013	6489941	1·5408460	1·1921381	1611710	1584	8388290	1
60	5446390		4553610	1·8360785	6494076	1·5398650	1·1923633	1613294		8386706	0

'	Cosine	Dif.	Vers.	Secant.	Cotan.	Taug.	Cosec.	Covers.	Dif	Sine.	'

′	Sine.	Dif.	Cosec.	Verseds.	Tang.	Dif.	Cotang	Covers.	Secant.	D	Cosine.	
0	9·7242097	2021	10·2757903	9·1817061	9·7957892	2811	10·2042108	9·6721725	10·0715793	790	9·9284205	60
1	9·7244118	2020	10·2755882	9·1821466	9·7960703	2810	10·2039297	9·6719445	10·0716585	790	9·9283415	59
2	9·7246138	2018	10·2753862	9·1825868	9·7963513	2809	10·2036487	9·6717165	10·0717375	791	9·9282625	58
3	9·7248156	2018	10·2751844	9·1830268	9·7966322	2808	10·2033678	9·6714884	10·0718166	791	9·9281834	57
4	9·7250174	2015	10·2749826	9·1834665	9·7969130	2808	10·2030870	9·6712602	10·0718957	792	9·9281043	56
5	9·7252189	2015	10·2747811	9·1839060	9·7971938	2807	10·2028062	9·6710319	10·0719749	792	9·9280251	55
6	9·7254204	2013	10·2745796	9·1843452	9·7974745	2806	10·2025255	9·6708036	10·0720541	793	9·9279459	54
7	9·7256217	2012	10·2743783	9·1847812	9·7977551	2805	10·2022449	9·6705752	10·0721334	793	9·9278666	53
8	9·7258229	2011	10·2741771	9·1852230	9·7980356	2804	10·2019644	9·6703467	10·0722127	794	9·9277873	52
9	9·7260240	2009	10·2739760	9·1856615	9·7983160	2804	10·2016840	9·6701181	10·0722921	794	9·9277079	51
10	9·7262249	2008	10·2737751	9·1860998	9·7985964	2803	10·2014036	9·6698895	10·0723715	795	9·9276285	50
11	9·7264257	2007	10·2735743	9·1865378	9·7988767	2802	10·2011233	9·6696607	10·0724510	795	9·9275490	49
12	9·7266264	2005	10·2733736	9·1869756	9·7991569	2801	10·2008431	9·6694319	10·0725305	796	9·9274695	48
13	9·7268269	2004	10·2731731	9·1874132	9·7994370	2800	10·2005630	9·6692030	10·0726101	796	9·9273899	47
14	9·7270273	2003	10·2729727	9·1878505	9·7997170	2800	10·2002830	9·6689741	10·0726897	797	9·9273103	46
15	9·7272276	2002	10·2727724	9·1882876	9·7999970	2799	10·2000030	9·6687450	10·0727694	797	9·9272306	45
16	9·7274278	2000	10·2725722	9·1887245	9·8002769	2798	10·1997231	9·6685159	10·0728491	798	9·9271509	44
17	9·7276278	1999	10·2723722	9·1891611	9·8005567	2798	10·1994433	9·6682867	10·0729289	798	9·9270711	43
18	9·7278277	1998	10·2721723	9·1895974	9·8008365	2796	10·1991635	9·6680574	10·0730087	799	9·9269913	42
19	9·7280275	1996	10·2719725	9·1900336	9·8011161	2796	10·1988839	9·6678281	10·0730886	800	9·9269114	41
20	9·7282271	1996	10·2717729	9·1904695	9·8013957	2795	10·1986043	9·6675986	10·0731686	800	9·9268314	40
21	9·7284267	1993	10·2715733	9·1909051	9·8016752	2794	10·1983248	9·6673691	10·0732486	800	9·9267514	39
22	9·7286260	1993	10·2713740	9·1913406	9·8019546	2794	10·1980454	9·6671395	10·0733286	801	9·9266714	38
23	9·7288253	1991	10·2711747	9·1917758	9·8022340	2793	10·1977660	9·6669098	10·0734087	801	9·9265913	37
24	9·7290244	1990	10·2709756	9·1922107	9·8025133	2792	10·1974867	9·6666801	10·0734888	802	9·9265112	36
25	9·7292234	1989	10·2707766	9·1926454	9·8027925	2791	10·1972075	9·6664502	10·0735690	803	9·9264310	35
26	9·7294223	1988	10·2705777	9·1930799	9·8030716	2790	10·1969284	9·6662203	10·0736493	803	9·9263507	34
27	9·7296211	1986	10·2703789	9·1935142	9·8033506	2790	10·1966494	9·6659903	10·0737296	803	9·9262704	33
28	9·7298197	1985	10·2701803	9·1939482	9·8036296	2789	10·1963704	9·6657603	10·0738099	805	9·9261901	32
29	9·7300182	1983	10·2699818	9·1943819	9·8039085	2788	10·1960915	9·6655301	10·0738904	805	9·9261096	31
30	9·7302165	1983	10·2697835	9·1948155	9·8041873	2788	10·1958127	9·6652999	10·0739708	805	9·9260292	30
31	9·7304148	1981	10·2695852	9·1952488	9·8044661	2786	10·1955339	9·6650696	10·0740513	806	9·9259487	29
32	9·7306129	1980	10·2693871	9·1956819	9·8047447	2786	10·1952553	9·6648392	10·0741319	806	9·9258681	28
33	9·7308109	1978	10·2691891	9·1961147	9·8050233	2786	10·1949767	9·6646087	10·0742125	806	9·9257875	27
34	9·7310087	1977	10·2689913	9·1965473	9·8053019	2784	10·1946981	9·6643781	10·0742931	808	9·9257069	26
35	9·7312064	1976	10·2687936	9·1969797	9·8055803	2784	10·1944197	9·6641475	10·0743739	807	9·9256261	25
36	9·7314040	1975	10·2685960	9·1974118	9·8058587	2783	10·1941413	9·6639168	10·0744546	808	9·9255454	24
37	9·7316015	1974	10·2683985	9·1978437	9·8061370	2782	10·1938630	9·6636860	10·0745354	809	9·9254646	23
38	9·7317989	1972	10·2682011	9·1982754	9·8064152	2781	10·1935848	9·6634552	10·0746163	809	9·9253837	22
39	9·7319961	1971	10·2680039	9·1987068	9·8066933	2781	10·1933067	9·6632242	10·0746972	810	9·9253028	21
40	9·7321932	1970	10·2678068	9·1991380	9·8069714	2780	10·1930286	9·6629932	10·0747782	810	9·9252218	20
41	9·7323902	1968	10·2676098	9·1995690	9·8072494	2779	10·1927506	9·6627621	10·0748592	811	9·9251408	19
42	9·7325870	1967	10·2674130	9·1999997	9·8075273	2779	10·1924727	9·6625309	10·0749403	811	9·9250597	18
43	9·7327837	1966	10·2672163	9·2004302	9·8078052	2777	10·1921948	9·6622996	10·0750214	812	9·9249786	17
44	9·7329803	1965	10·2670197	9·2008605	9·8080829	2777	10·1919171	9·6620683	10·0751026	813	9·9248974	16
45	9·7331768	1963	10·2668232	9·2012906	9·8083606	2777	10·1916394	9·6618368	10·0751839	812	9·9248161	15
46	9·7333731	1962	10·2666269	9·2017204	9·8086383	2775	10·1913617	9·6616053	10·0752651	814	9·9247349	14
47	9·7335693	1961	10·2664307	9·2021499	9·8089158	2775	10·1910842	9·6613737	10·0753465	814	9·9246535	13
48	9·7337654	1960	10·2662346	9·2025793	9·8091933	2774	10·1908067	9·6611421	10·0754279	814	9·9245721	12
49	9·7339614	1958	10·2660386	9·2030084	9·8094707	2773	10·1905293	9·6609103	10·0755093	815	9·9244907	11
50	9·7341572	1957	10·2658428	9·2034373	9·8097480	2773	10·1902520	9·6606785	10·0755908	815	9·9244092	10
51	9·7343529	1956	10·2656471	9·2038660	9·8100253	2772	10·1899747	9·6604466	10·0756723	816	9·9243277	9
52	9·7345485	1955	10·2654515	9·2042944	9·8103025	2771	10·1896975	9·6602146	10·0757539	817	9·9242461	8
53	9·7347440	1953	10·2652560	9·2047226	9·8105796	2770	10·1894204	9·6599825	10·0758356	817	9·9241644	7
54	9·7349393	1952	10·2650607	9·2051506	9·8108566	2770	10·1891434	9·6597504	10·0759173	817	9·9240827	6
55	9·7351345	1951	10·2648655	9·2055783	9·8111336	2769	10·1888664	9·6595182	10·0759990	819	9·9240010	5
56	9·7353296	1950	10·2646704	9·2060058	9·8114105	2768	10·1885895	9·6592858	10·0760809	818	9·9239191	4
57	9·7355246	1949	10·2644754	9·2064331	9·8116873	2768	10·1883127	9·6590535	10·0761627	819	9·9238373	3
58	9·7357195	1947	10·2642805	9·2068602	9·8119641	2767	10·1880359	9·6588210	10·0762446	820	9·9237554	2
59	9·7359142	1946	10·2640858	9·2072870	9·8122408	2766	10·1877592	9·6585884	10·0763266	820	9·9236734	1
60	9·7361088		10·2638912	9·2077136	9·8125174		10·1874826	9·6583558	10·0764086		9·9235914	0
	Cosine.	Dif.	Secant.	Covers.	Cotang.	Dif.	Tang.	Verseds.	Cosec.	D.	Sine.	

′	Sine.	Dif.	Covers	Cosec.	Tang.	Cotang.	Secant.	Vers.	Dif.	Cosine	′
0	5446390	2440	4553610	1·8360785	6494076	1·5398650	1·1923633	1613294	1585	8386706	60
1	5448830	2439	4551170	1·8352565	6498212	1·5388848	1·1925886	1614879	1585	8385121	59
2	5451269	2438	4548731	1·8344354	6502350	1·5379054	1·1928142	1616464	1586	8383536	58
3	5453707	2438	4546293	1·8336152	6506490	1·5369270	1·1930399	1618050	1587	8381950	57
4	5456145	2438	4543855	1·8327959	6510631	1·5359494	1·1932658	1619637	1588	8380363	56
5	5458583	2437	4541417	1·8319774	6514774	1·5349727	1·1934918	1621225	1588	8378775	55
6	5461020	2436	4538980	1·8311599	6518918	1·5339969	1·1937181	1622813	1589	8377187	54
7	5463456	2436	4536544	1·8303432	6523064	1·5330219	1·1939446	1624402	1589	8375598	53
8	5465892	2436	4534108	1·8295274	6527211	1·5320479	1·1941712	1625991	1591	8374009	52
9	5468328	2435	4531672	1·8287125	6531360	1·5310746	1·1943980	1627582	1591	8372418	51
10	5470763	2435	4529237	1·8278985	6535511	1·5301023	1·1946251	1629173	1591	8370827	50
11	5473198	2434	4526802	1·8270854	6539663	1·5291308	1·1948523	1630764	1593	8369236	49
12	5475632	2434	4524368	1·8262731	6543817	1·5281602	1·1950796	1632357	1593	8367643	48
13	5478066	2433	4521934	1·8254617	6547972	1·5271904	1·1953072	1633950	1594	8366050	47
14	5480499	2433	4519501	1·8246512	6552129	1·5262215	1·1955350	1635544	1594	8364456	46
15	5482932	2433	4517068	1·8238416	6556287	1·5252535	1·1957629	1637138	1596	8362862	45
16	5485365	2432	4514635	1·8230328	6560447	1·5242863	1·1959911	1638734	1596	8361266	44
17	5487797	2431	4512203	1·8222249	6564609	1·5233200	1·1962194	1640330	1596	8359670	43
18	5490228	2431	4509772	1·8214179	6568772	1·5223545	1·1964479	1641926	1598	8358074	42
19	5492659	2431	4507341	1·8206118	6572937	1·5213899	1·1966767	1643524	1598	8356476	41
20	5495090	2430	4504910	1·8198065	6577103	1·5204261	1·1969056	1645122	1599	8354878	40
21	5497520	2430	4502480	1·8190021	6581271	1·5194632	1·1971346	1646721	1599	8353279	39
22	5499950	2429	4500050	1·8181985	6585441	1·5185012	1·1973639	1648320	1600	8351680	38
23	5502379	2428	4497621	1·8173958	6589612	1·5175400	1·1975934	1649920	1601	8350080	37
24	5504807	2429	4495193	1·8165940	6593785	1·5165796	1·1978230	1651521	1602	8348479	36
25	5507236	2427	4492764	1·8157930	6597960	1·5156201	1·1980529	1653123	1602	8346877	35
26	5509663	2428	4490337	1·8149929	6602136	1·5146614	1·1982829	1654725	1603	8345275	34
27	5512091	2427	4487909	1·8141937	6606313	1·5137036	1·1985131	1656328	1604	8343672	33
28	5514518	2426	4485482	1·8133953	6610492	1·5127466	1·1987435	1657932	1605	8342068	32
29	5516944	2426	4483056	1·8125977	6614673	1·5117905	1·1989741	1659537	1605	8340463	31
30	5519370	2425	4480630	1·8118010	6618856	1·5108352	1·1992049	1661142	1606	8338858	30
31	5521795	2425	4478205	1·8110052	6623040	1·5098807	1·1994359	1662748	1606	8337252	29
32	5524220	2425	4475780	1·8102102	6627225	1·5089271	1·1996671	1664354	1608	8335646	28
33	5526645	2424	4473355	1·8094161	6631413	1·5079743	1·1998985	1665962	1608	8334039	27
34	5529069	2423	4470931	1·8086228	6635601	1·5070224	1·2001300	1667570	1608	8332430	26
35	5531492	2423	4468508	1·8078304	6639792	1·5060713	1·2003618	1669178	1610	8330822	25
36	5533915	2423	4466085	1·8070388	6643984	1·5051210	1·2005937	1670788	1610	8329212	24
37	5536338	2422	4463662	1·8062481	6648178	1·5041716	1·2008258	1672398	1611	8327602	23
38	5538760	2422	4461240	1·8054582	6652373	1·5032229	1·2010582	1674009	1611	8325991	22
39	5541182	2421	4458818	1·8046691	6656570	1·5022751	1·2012907	1675620	1612	8324380	21
40	5543603	2421	4456397	1·8038809	6660769	1·5013282	1·2015234	1677232	1613	8322768	20
41	5546024	2420	4453976	1·8030935	6664969	1·5003821	1·2017563	1678845	1614	8321155	19
42	5548444	2420	4451556	1·8023070	6669171	1·4994367	1·2019894	1680459	1614	8319541	18
43	5550864	2419	4449136	1·8015213	6673374	1·4984923	1·2022226	1682073	1615	8317927	17
44	5553283	2419	4446717	1·8007365	6677580	1·4975486	1·2024561	1683688	1616	8316312	16
45	5555702	2419	4444298	1·7999524	6681786	1·4966058	1·2026898	1685304	1616	8314696	15
46	5558121	2418	4441879	1·7991693	6685995	1·4956637	1·2029236	1686920	1617	8313080	14
47	5560539	2417	4439461	1·7983869	6690205	1·4947225	1·2031577	1688537	1618	8311443	13
48	5562956	2417	4437044	1·7976054	6694417	1·4937822	1·2033919	1690155	1619	8309845	12
49	5565373	2417	4434627	1·7968247	6698630	1·4928426	1·2036264	1691774	1619	8308226	11
50	5567790	2416	4432210	1·7960449	6702845	1·4919039	1·2038610	1693393	1620	8306607	10
51	5570206	2415	4429794	1·7952658	6707061	1·4909659	1·2040958	1695013	1621	8304987	9
52	5572621	2415	4427379	1·7944876	6711280	1·4900288	1·2043308	1696634	1621	8303366	8
53	5575036	2415	4424964	1·7937102	6715500	1·4890925	1·2045660	1698255	1622	8301745	7
54	5577451	2414	4422549	1·7929337	6719721	1·4881570	1·2048014	1699877	1623	8300123	6
55	5579865	2414	4420135	1·7921580	6723944	1·4872223	1·2050370	1701500	1623	8298500	5
56	5582279	2413	4417721	1·7913831	6728169	1·4862884	1·2052728	1703123	1625	8296877	4
57	5584692	2413	4415308	1·7906090	6732396	1·4853554	1·2055088	1704748	1625	8295252	3
58	5587105	2412	4412895	1·7898357	6736624	1·4844231	1·2057450	1706372	1624	8293628	2
59	5589517	2412	4410483	1·7890633	6740854	1·4834916	1·2059814	1707998	1626	8292002	1
60	5591929		4408071	1·7882916	6745085	1·4825610	1·2062179	1709624	1626	8290376	0

′	Cosine	Dif.	Vers.	Secant.	Cotan.	Taug.	Cosec.	Covers.	Dif.	Sine.	′

′	Sine.	Dif.	Cosec.	Verseds.	Tang.	Dif.	Cotang.	Covers.	Secant.	D.	Cosine.	′
0	9·7361088	1944	10·2638912	9·2077136	9·8125174	2765	10·1874826	9·6583558	10·0764086	821	9·9235914	60
1	9·7363032	1944	10·2636968	9·2081400	9·8127939	2765	10·1872061	9·6581231	10·0764907	821	9·9235093	59
2	9·7364976	1942	10·2635024	9·2085661	9·8130704	2764	10·1869296	9·6578903	10·0765728	822	9·9234272	58
3	9·7366918	1941	10·2633082	9·2089920	9·8133468	2763	10·1866532	9·6576574	10·0766550	822	9·9233450	57
4	9·7368859	1940	10·2631141	9·2094177	9·8136231	2762	10·1863769	9·6574245	10·0767372	823	9·9232628	56
5	9·7370799	1938	10·2629201	9·2098432	9·8138993	2762	10·1861007	9·6571914	10·0768195	823	9·9231805	55
6	9·7372737	1938	10·2627263	9·2102684	9·8141755	2761	10·1858245	9·6569583	10·0769018	824	9·9230982	54
7	9·7374675	1936	10·2625325	9·2106934	9·8144516	2761	10·1855484	9·6567251	10·0769842	824	9·9230158	53
8	9·7376611	1935	10·2623389	9·2111182	9·8147277	2759	10·1852723	9·6564918	10·0770666	825	9·9229334	52
9	9·7378546	1933	10·2621454	9·2115428	9·8150036	2759	10·1849964	9·6562585	10·0771491	825	9·9228509	51
10	9·7380479	1933	10·2619521	9·2119671	9·8152795	2759	10·1847205	9·6560250	10·0772316	826	9·9227584	50
11	9·7382412	1931	10·2617588	9·2123912	9·8155554	2757	10·1844446	9·6557915	10·0773142	826	9·9226858	49
12	9·7384343	1930	10·2615657	9·2128151	9·8158311	2757	10·1841689	9·6555579	10·0773968	827	9·9226032	48
13	9·7386273	1928	10·2613727	9·2132388	9·8161068	2756	10·1838932	9·6553242	10·0774795	828	9·9225205	47
14	9·7388201	1928	10·2611799	9·2136622	9·8163824	2756	10·1836176	9·6550904	10·0775623	828	9·9224377	46
15	9·7390129	1926	10·2609871	9·2140854	9·8166580	2755	10·1833420	9·6548566	10·0776451	828	9·9223549	45
16	9·7392055	1925	10·2607945	9·2145084	9·8169335	2754	10·1830665	9·6546227	10·0777279	830	9·9222721	44
17	9·7393980	1924	10·2606020	9·2149311	9·8172089	2753	10·1827911	9·6543887	10·0778109	829	9·9221891	43
18	9·7395904	1923	10·2604096	9·2153537	9·8174842	2753	10·1825158	9·6541546	10·0778938	830	9·9221062	42
19	9·7397827	1921	10·2602173	9·2157760	9·8177595	2752	10·1822405	9·6539204	10·0779768	831	9·9220232	41
20	9·7399748	1920	10·2600252	9·2161981	9·8180347	2751	10·1819653	9·6536861	10·0780599	831	9·9219401	40
21	9·7401668	1919	10·2598332	9·2166199	9·8183098	2751	10·1816902	9·6534518	10·0781430	832	9·9218570	39
22	9·7403587	1918	10·2596413	9·2170416	9·8185849	2750	10·1814151	9·6532174	10·0782262	832	9·9217738	38
23	9·7405505	1916	10·2594495	9·2174630	9·8188599	2750	10·1811401	9·6529829	10·0783094	833	9·9216906	37
24	9·7407421	1916	10·2592579	9·2178842	9·8191348	2748	10·1808652	9·6527483	10·0783927	833	9·9216073	36
25	9·7409337	1914	10·2590663	9·2183052	9·8194096	2748	10·1805904	9·6525136	10·0784760	834	9·9215240	35
26	9·7411251	1913	10·2588749	9·2187259	9·8196844	2748	10·1803156	9·6522789	10·0785594	834	9·9214406	34
27	9·7413164	1911	10·2586836	9·2191464	9·8199592	2746	10·1800408	9·6520441	10·0786428	835	9·9213572	33
28	9·7415075	1911	10·2584925	9·2195668	9·8202338	2746	10·1797662	9·6518092	10·0787263	835	9·9212737	32
29	9·7416986	1909	10·2583014	9·2199868	9·8205084	2745	10·1794916	9·6515742	10·0788098	836	9·9211902	31
30	9·7418895	1908	10·2581105	9·2204067	9·8207829	2745	10·1792171	9·6513391	10·0788934	837	9·9211066	30
31	9·7420803	1907	10·2579197	9·2208263	9·8210574	2743	10·1789426	9·6511039	10·0789771	836	9·9210229	29
32	9·7422710	1906	10·2577290	9·2212458	9·8213317	2743	10·1786683	9·6508687	10·0790607	838	9·9209393	28
33	9·7424616	1904	10·2575384	9·2216650	9·8216060	2743	10·1783940	9·6506334	10·0791445	838	9·9208555	27
34	9·7426520	1903	10·2573480	9·2220839	9·8218803	2742	10·1781197	9·6503980	10·0792283	839	9·9207717	26
35	9·7428423	1902	10·2571577	9·2225027	9·8221545	2741	10·1778455	9·6501625	10·0793122	839	9·9206878	25
36	9·7430325	1901	10·2569675	9·2229212	9·8224286	2740	10·1775714	9·6499269	10·0793961	839	9·9206039	24
37	9·7432226	1900	10·2567774	9·2233396	9·8227026	2740	10·1772974	9·6496913	10·0794800	840	9·9205200	23
38	9·7434126	1898	10·2565874	9·2237577	9·8229766	2739	10·1770234	9·6494556	10·0795640	841	9·9204360	22
39	9·7436024	1897	10·2563976	9·2241755	9·8232505	2739	10·1767495	9·6492197	10·0796481	841	9·9203519	21
40	9·7437921	1896	10·2562079	9·2245932	9·8235244	2737	10·1764756	9·6489839	10·0797322	842	9·9202678	20
41	9·7439817	1895	10·2560183	9·2250106	9·8237981	2738	10·1762019	9·6487479	10·0798164	842	9·9201836	19
42	9·7441712	1894	10·2558288	9·2254279	9·8240719	2736	10·1759281	9·6485118	10·0799006	843	9·9200994	18
43	9·7443606	1892	10·2556394	9·2258449	9·8243455	2736	10·1756545	9·6482757	10·0799849	843	9·9200151	17
44	9·7445498	1892	10·2554502	9·2262617	9·8246191	2735	10·1753809	9·6480394	10·0800692	844	9·9199308	16
45	9·7447390	1890	10·2552610	9·2266782	9·8248926	2734	10·1751074	9·6478031	10·0801536	845	9·9198464	15
46	9·7449280	1889	10·2550720	9·2270946	9·8251660	2734	10·1748340	9·6475667	10·0802381	845	9·9197619	14
47	9·7451169	1887	10·2548831	9·2275107	9·8254394	2733	10·1745606	9·6473303	10·0803225	846	9·9196775	13
48	9·7453056	1887	10·2546944	9·2279266	9·8257127	2733	10·1742873	9·6470937	10·0804071	846	9·9195929	12
49	9·7454943	1885	10·2545057	9·2283423	9·8259860	2732	10·1740140	9·6468571	10·0804917	846	9·9195083	11
50	9·7456828	1884	10·2543172	9·2287578	9·8262592	2731	10·1737408	9·6466204	10·0805763	847	9·9194237	10
51	9·7458712	1883	10·2541288	9·2291731	9·8265323	2730	10·1734677	9·6463836	10·0806610	848	9·9193390	9
52	9·7460595	1882	10·2539405	9·2295881	9·8268053	2730	10·1731947	9·6461467	10·0807458	848	9·9192542	8
53	9·7462477	1881	10·2537523	9·2300029	9·8270783	2730	10·1729217	9·6459097	10·0808306	849	9·9191694	7
54	9·7464358	1879	10·2535642	9·2304175	9·8273513	2728	10·1726487	9·6456726	10·0809155	849	9·9190845	6
55	9·7466237	1878	10·2533763	9·2308319	9·8276241	2728	10·1723759	9·6454355	10·0810004	850	9·9189996	5
56	9·7468115	1877	10·2531885	9·2312461	9·8278969	2727	10·1721031	9·6451983	10·0810854	850	9·9189146	4
57	9·7469992	1876	10·2530008	9·2316601	9·8281696	2727	10·1718304	9·6449610	10·0811704	851	9·9188296	3
58	9·7471868	1875	10·2528132	9·2320738	9·8284423	2726	10·1715577	9·6447236	10·0812555	851	9·9187445	2
59	9·7473743	1874	10·2526257	9·2324874	9·8287149	2725	10·1712851	9·6444861	10·0813406	852	9·9186594	1
60	9·7475617		10·2524383	9·2329007	9·8289874		10·1710126	9·6442486	10·0814258		9·9185742	0

′	Cosine.	Dif.	Secant.	Covers.	Cotang.	Dif.	Tang.	Verseds.	Cosec.	D.	Sine.	′

'	Sine.	Dif.	Covers	Cosec.	Tang.	Cotang.	Secant.	Vers.	Dif.	Cosine	'
0	5591929	2411	4408071	1·7882916	6745085	1·4825610	1·2062179	1709624	1627	8290376	60
1	5594340	2411	4405660	1·7875208	6749318	1·481631]	1·2064547	1711251	1628	8288749	59
2	5596751	2411	4403249	1·7867508	6753553	1·4807021	1·2066917	1712879	1628	8287121	58
3	5599162	2410	4400838	1·7859817	6757790	1·4797738	1·2069288	1714507	1629	8285493	57
4	5601572	2410	4398428	1·7852133	6762028	1·4788463	1·2071662	1716136	1630	8283864	56
5	5603981	2409	4396019	1·7844457	6766268	1·4779197	1·2074037	1717766	1631	8282234	55
6	5606390	2409 2408	4393610	1·7836790	6770509	1·4769938	1·2076415	1719397	1631	8280603	54
7	5608798	2408	4391202	1·7829131	6774752	1·4760688	1·2078794	1721028	1632	8278972	53
8	5611206	2408	4388794	1·7821479	6778997	1·4751445	1·2081175	1722660	1632	8277340	52
9	5613614	2408	4386386	1·7813836	6783243	1·4742210	1·2083559	1724292	1634	8275708	51
10	5616021	2407	4383979	1·7806201	6787492	1·4732983	1·2085944	1725926	1634	8274074	50
11	5618428	2407	4381572	1·7798574	679174]	1·4723764	1.2088331	1727560	1634	8272440	49
12	5620834	2406 2405	4379166	1·7790955	6795993	1·4714553	1·2090720	1729194	1636	8270806	48
13	5623239	2406	4376761	1·7783344	6800246	1·4705350	1·2093112	]730830	1636	8269170	47
14	5625645	2404	4374355	1·7775741	6804501	1·4696155	1·2095505	1732466	1637	8267534	46
15	5628049	2404	4371951	1·7768146	6808758	1·4686967	1·2097900	1734103	1637	8265897	45
16	5630453	2404	4369547	1·7760559	6813016	1·4677788	1·2100297	1735740	1638	8264260	44
17	5632857	2403	4367143	1·7752980	6817276	1·4668616	1·2102696	1737378	1639	8262622	43
18	5635260	2403	4364740	1·7745409	6821537	1·4659452	1·2105097	1739017	1640	8260983	42
19	5637663	2403	4362337	1·7737845	6825801	1·4650296	1·2107500	1740657	1640	8259343	41
20	5640066	2401	4359934	1·7730290	6830066	1·4641147	1·2109905	1742297	1641	8257703	40
21	5642467	2402	4357533	1·7722743	6834333	1·4632007	1·2112312	1743938	1641	8256062	39
22	5644869	2401	4355131	1·7715204	6838601	1·4622874	1·2114721	1745580	1642	8254420	38
23	5647270	2400	4352730	1·7707672	6842871	1·4613749	1·2117132	1747222	1642	8252778	37
24	5649670	2400	4350330	1·7700149	6847143	1·4604632	1·2119545	1748865	1643 1644	8251135	36
25	5652070	2399	4347930	1·7692633	6851416	1·4595522	1·2121960	1750509	1644	8249491	35
26	5654469	2399	4345531	1·7685125	6855692	1·4586420	1·2124377	1752153	1645	8247847	34
27	5656868	2399	4343132	1·7677625	6859969	1·4577326	1·2126795	1753798	1646	8246202	33
28	5659267	2398	4340733	1·7670133	6864247	1·4568240	1·2129216	1755444	1647	8244556	32
29	5661665	2397	4338335	1·7662649	6868528	1·4559161	1·2131639	1757091	1647	8242909	31
30	5664062	2397	4335938	1·7655173	6872810	1·4550090	1·2134064	1758738	1648	8241262	30
31	5666459	2397	4333541	1·7647704	6877093	1·4541027	1·2136491	1760386	1649	8239614	29
32	5668856	2396	4331144	1·7640244	6881379	1·4531971	1·2138920	1762035	1649	8237965	28
33	5671252	2396	4328748	1·7632791	6885666	1·4522923	1·2141351	1763684	1650	8236316	27
34	5673648	2395	4326352	1·7625345	6889955	1·4513883	1·2143784	1765334	1651	8234666	26
35	5676043	2394	4323957	1·7617908	6894246	1·4504850	1·2146218	1766985	1651	8233015	25
36	5678437	2395	4321563	1·7610478	6898538	1·4495825	1·2148655	1768636	1652	8231364	24
37	5680832	2393	4319168	1·7603057	6902832	1·4486808	1·2151094	1770288	1653	8229712	23
38	5683225	2394	4316775	1·7595642	6907128	1·4477798	1·2153535	1771941	1654	8228059	22
39	5685619	2392	4314381	1·7588236	6911425	1·4468796	1·2155978	1773595	1654	8226405	21
40	5688011	2392	4311989	1·7580837	6915725	1·4459801	1·2158423	1775249	1655	8224751	20
41	5690403	2392	4309597	1·7573446	6920026	1·4450814	1·2160870	1776904	1656	8223096	19
42	5692795	2392	4307205	1·7566063	6924328	1·4441834	1·2163319	1778560	1656	8221440	18
43	5695187	2390	4304813	1·7558687	6928633	1·4432862	1·2165770	1780216	1657	8219784	17
44	5697577	2391	4302423	1·7551320	6932939	1·4423897	1·2168223	1781873	1658	5218127	16
45	5699968	2389	4300032	1·7543959	6937247	1·4414940	1·2170678	1783531	1658	8216469	15
46	5702357	2390	4297643	1·7536607	6941557	1·4405991	1·2173135	1785189	1659	8214811	14
47	5704747	2389	4295253	1·7529262	6945868	1·4397049	1·2175594	1786848	1660	8213152	13
48	5707136	2388	4292864	1·7521924	6950181	1·4388114	1·2178055	1788508	1660	8211492	12
49	5709524	2388	4290476	1·7514595	6954496	1·4379187	1·2180518	1790168	1662	8209832	11
50	5711912	2387	4288088	1·7507273	6958813	1·4370268	1·2182983	1791830	1661	8208170	10
51	5714299	2387	4285701	1·7499958	6963131	1·4361356	1·2185450	1793491	1663	8206509	9
52	5716686	2387	4283314	1·7492651	6967451	1·4352451	1·2187919	1795154	1663	8204846	8
53	5719073	2386	4280927	1·7485352	6971773	1·4343554	1·2190390	179C817	1664	8203183	7
54	5721459	2385	4278541	1·7478060	6976097	1·4334664	1·2192864	1798481	1665	8201519	6
55	5723844	2385	4276156	1·7470776	6980422	1·4325781	1·2195339	1800146	1665	8199854	5
56	5726229	2385	4273771	1·7463499	6984749	1·4316906	1·2197816	1801811	1666	8198189	4
57	5728614	2384	4271386	1·7456230	6989078	1·4308039	1·2200296	1803477	1667	8196523	3
58	5730998	2383	4269002	1·7448969	6993409	1·4299178	1·2202777	1805144	1667	8194856	2
59	5733381	2383	4266619	1·7441715	6997741	1·4290326	1·2205260	1806811	1669	8193189	1
60	5735764		4264236	1·7434468	7002075	1·4281480	1·2207746	1808480		8191520	0
'	Cosine	Dif.	Vers.	Secant.	Cotan.	Tang.	Cosec.	Covers	Dif	Sine.	'

′	Sine.	Dif.	Cosec.	Verseds.	Tang.	Dif.	Cotang.	Covers.	Secant.	D.	Cosine.	′
0	9·7475617	1872	10·2524383	9·2329007	9·8289874	2725	10·1710126	9·6442486	10·0814258	852	9·9185742	60
1	9·7477489	1871	10·2522511	9·2333138	9·8292599	2724	10·1707401	9·6440109	10·0815110	853	9·9184890	59
2	9·7479360	1870	10·2520640	9·2337267	9·8295323	2724	10·1704677	9·6437732	10·0815963	854	9·9184037	58
3	9·7481230	1870	10·2518770	9·2341393	9·8298047	2722	10·1701953	9·6435354	10·0816817	854	9·9183183	57
4	9·7483099	1869	10·2516901	9·2345518	9·8300769	2723	10·1699231	9·6432975	10·0817671	854	9·9182329	56
5	9·7484967	1868	10·2515033	9·2349640	9·8303492	2721	10·1696508	9·6430595	10·0818525	855	9·9181475	55
6	9·7486833	1866	10·2513167	9·2353761	9·8306213	2721	10·1693787	9·6428215	10·0819380	856	9·9180620	54
7	9·7488699	1864	10·2511302	9·2357879	9·8308934	2720	10·1691066	9·6425834	10·0820236	856	9·9179764	53
8	9·7490562	1863	10·2509438	9·2361995	9·8311654	2720	10·1688346	9·6423452	10·0821092	857	9·9178908	52
9	9·7492425	1862	10·2507575	9·2366109	9·8314374	2720	10·1685626	9·6421068	10·0821949	857	9·9178051	51
10	9·7494287	1861	10·2505713	9·2370221	9·8317093	2719	10·1682907	9·6418685	10·0822806	858	9·9177194	50
11	9·7496148	1859	10·2503852	9·2374330	9·8319811	2718	10·1680189	9·6416300	10·0823664	858	9·9176336	49
12	9·7498007	1859	10·2501993	9·2378438	9·8322529	2718	10·1677471	9·6413914	10·0824522	859	9·9175478	48
13	9·7499866	1857	10·2500134	9·2382543	9·8325246	2717	10·1674754	9·6411529	10·0825381	859	9·9174619	47
14	9·7501723	1856	10·2498277	9·2386647	9·8327963	2717	10·1672037	9·6409141	10·0826240	860	9·9173760	46
15	9·7503579	1855	10·2496421	9·2390748	9·8330679	2716	10·1669321	9·6406753	10·0827100	860	9·9172900	45
16	9·7505434	1853	10·2494566	9·2394847	9·8333394	2715	10·1666606	9·6404364	10·0827960	861	9·9172040	44
17	9·7507287	1853	10·2492713	9·2398944	9·8336109	2715	10·1663891	9·6401974	10·0828821	862	9·9171179	43
18	9·7509140	1851	10·2490860	9·2403038	9·8338823	2714	10·1661177	9·6399583	10·0829683	862	9·9170317	42
19	9·7510991	1851	10·2489009	9·2407131	9·8341536	2713	10·1658464	9·6397192	10·0830545	862	9·9169455	41
20	9·7512842	1849	10·2487158	9·2411222	9·8344249	2713	10·1655751	9·6394800	10·0831407	863	9·9168593	40
21	9·7514691	1847	10·2485309	9·2415310	9·8346961	2712	10·1653039	9·6392406	10·0832270	863	9·9167730	39
22	9·7516538	1847	10·2483462	9·2419396	9·8349673	2712	10·1650327	9·6390012	10·0833134	864	9·9166866	38
23	9·7518385	1846	10·2481615	9·2423481	9·8352384	2711	10·1647616	9·6387618	10·0833998	864	9·9166002	37
24	9·7520231	1844	10·2479769	9·2427563	9·8355094	2710	10·1644906	9·6385222	10·0834863	865	9·9165137	36
25	9·7522075	1844	10·2477925	9·2431643	9·8357804	2709	10·1642196	9·6382825	10·0835728	865	9·9164272	35
26	9·7523919	1842	10·2476081	9·2435721	9·8360513	2708	10·1639487	9·6380428	10·0836594	866	9·9163406	34
27	9·7525761	1841	10·2474239	9·2439797	9·8363221	2708	10·1636779	9·6378030	10·0837461	866	9·9162539	33
28	9·7527602	1840	10·2472398	9·2443871	9·8365929	2707	10·1634071	9·6375631	10·0838327	868	9·9161673	32
29	9·7529442	1838	10·2470558	9·2447942	9·8368636	2707	10·1631364	9·6373231	10·0839195	868	9·9160805	31
30	9·7531280	1838	10·2468720	9·2452012	9·8371343	2706	10·1628657	9·6370830	10·0840063	868	9·9159937	30
31	9·7533118	1836	10·2466882	9·2456079	9·8374049	2706	10·1625951	9·6368429	10·0840931	869	9·9159069	29
32	9·7534954	1836	10·2465046	9·2460145	9·8376755	2705	10·1623245	9·6366026	10·0841800	870	9·9158200	28
33	9·7536790	1834	10·2463210	9·2464208	9·8379460	2705	10·1620540	9·6363623	10·0842670	870	9·9157330	27
34	9·7538624	1833	10·2461376	9·2468269	9·8382164	2704	10·1617836	9·6361219	10·0843540	871	9·9156460	26
35	9·7540457	1831	10·2459543	9·2472328	9·8384867	2703	10·1615133	9·6358814	10·0844411	871	9·9155589	25
36	9·7542288	1831	10·2457712	9·2476385	9·8387571	2704	10·1612429	9·6356408	10·0845282	872	9·9154718	24
37	9·7544119	1830	10·2455881	9·2480440	9·8390273	2702	10·1609727	9·6354001	10·0846154	872	9·9153846	23
38	9·7545949	1828	10·2454051	9·2484493	9·8392975	2701	10·1607025	9·6351594	10·0847026	873	9·9152974	22
39	9·7547777	1827	10·2452223	9·2488544	9·8395676	2701	10·1604324	9·6349185	10·0847899	873	9·9152101	21
40	9·7549604	1827	10·2450396	9·2492593	9·8398377	2700	10·1601623	9·6346776	10·0848772	874	9·9151228	20
41	9·7551431	1825	10·2448569	9·2496640	9·8401077	2700	10·1598923	9·6344366	10·0849646	875	9·9150354	19
42	9·7553256	1824	10·2446744	9·2500684	9·8403776	2699	10·1596224	9·6341955	10·0850521	875	9·9149479	18
43	9·7555080	1822	10·2444920	9·2504727	9·8406475	2699	10·1593525	9·6339543	10·0851396	875	9·9148604	17
44	9·7556902	1822	10·2443098	9·2508767	9·8409174	2697	10·1590826	9·6337131	10·0852271	877	9·9147729	16
45	9·7558724	1820	10·2441276	9·2512806	9·8411871	2698	10·1588129	9·6334717	10·0853148	876	9·9146852	15
46	9·7560544	1820	10·2439456	9·2516842	9·8414569	2696	10·1585431	9·6332303	10·0854024	877	9·9145976	14
47	9·7562364	1818	10·2437636	9·2520876	9·8417265	2696	10·1582735	9·6329888	10·0854901	878	9·9145099	13
48	9·7564182	1817	10·2435818	9·2524909	9·8419961	2696	10·1580039	9·6327472	10·0855779	879	9·9144221	12
49	9·7565999	1816	10·2434001	9·2528939	9·8422657	2694	10·1577343	9·6325055	10·0856658	878	9·9143342	11
50	9·7567815	1815	10·2432185	9·2532967	9·8425351	2695	10·1574649	9·6322637	10·0857536	880	9·9142464	10
51	9·7569630	1814	10·2430370	9·2536993	9·8428046	2693	10·1571954	9·6320218	10·0858416	880	9·9141584	9
52	9·7571444	1812	10·2428556	9·2541017	9·8430739	2693	10·1569261	9·6317799	10·0859296	880	9·9140704	8
53	9·7573256	1812	10·2426744	9·2545039	9·8433432	2693	10·1566568	9·6315378	10·0860176	881	9·9139824	7
54	9·7575068	1810	10·2424932	9·2549059	9·8436125	2692	10·1563875	9·6312957	10·0861057	882	9·9138943	6
55	9·7576878	1809	10·2423122	9·2553077	9·8438817	2691	10·1561183	9·6310535	10·0861939	882	9·9138061	5
56	9·7578687	1808	10·2421313	9·2557093	9·8441508	2691	10·1558492	9·6308112	10·0862821	883	9·9137179	4
57	9·7580495	1807	10·2419505	9·2561107	9·8444199	2690	10·1555801	9·6305688	10·0863704	883	9·9136296	3
58	9·7582302	1806	10·2417698	9·2565119	9·8446889	2690	10·1553111	9·6303264	10·0864587	883	9·9135413	2
59	9·7584108	1805	10·2415892	9·2569128	9·8449579	2689	10·1550421	9·6300838	10·0865470	885	9·9134530	1
60	9·7585913		10·2414087	9·2573136	9·8452268		10·1547732	9·6298412	10·0866355		9·9133645	0
′	Cosine.	Dif.	Secant.	Covers.	Cotang.	Dif.	Tang.	Verseds.	Cosec.	D.	Sine.	′

′	Sine	Dif.	Covers	Cosec.	Tang.	Cotang.	Secant.	Vers.	Dif.	Cosine	′
0	5735764	2383	4264236	1·7434468	7002075	1·4281480	1·2207746	1808480	1668	8191520	60
1	5738147	2382	4261853	1·7427229	7006411	1·4272642	1·2210233	1810148	1670	8189852	59
2	5740529	2382	4259471	1·7419997	7010749	1·4263811	1·2212723	1811818	1670	8188182	58
3	5742911	2381	4257089	1·7412773	7015089	1·4254988	1·2215215	1813488	1671	8186512	57
4	5745292	2360	4254708	1·7405556	7019430	1·4246171	1·2217708	1815159	1672	8184841	56
5	5747672	2381	4252328	1·7398347	7023773	1·4237362	1·2220204	1816831	1672	8183169	55
6	5750053	2379	4249947	1·7391145	7028118	1·4228561	1·2222702	1818503	1673	8181497	54
7	5752432	2379	4247568	1·7383951	7032464	1·4219766	1·2225202	1820176	1673	8179824	53
8	5754811	2379	4245189	1·7376764	7036813	1·4210979	1·2227703	1821849	1675	8178151	52
9	5757190	2378	4242810	1·7369585	7041163	1·4202200	1·2230207	1823524	1675	8176476	51
10	5759568	2378	4240432	1·7362413	7045515	1·4193427	1·2232713	1825199	1676	8174801	50
11	5761946	2377	4238054	1·7355248	7049869	1·4184662	1·2235222	1826875	1676	8173125	49
12	5764323	2377	4235677	1·7348091	7054224	1·4175904	1·2237732	1828551	1677	8171449	48
13	5766700	2376	4233300	1·7340941	7058581	1·4167153	1·2240244	1830228	1678	8169772	47
14	5769076	2376	4230924	1·7333798	7062940	1·4158409	1·2242758	1831906	1678	8168094	46
15	5771452	2375	4228548	1·732 663	7067301	1·4149673	1·2245274	1833584	1680	8166416	45
16	5773827	2375	4226173	1·7319535	7071664	1·4140943	1·2247793	1835264	1680	8164736	44
17	5776202	2374	4223798	1·7312414	7076028	1·4132221	1·2250313	1836944	1680	8163056	43
18	5778576	2374	4221424	1·7305301	7080395	1·4123506	1·2252836	1838624	1681	8161376	42
19	5780950	2373	4219050	1·7298195	7084763	1·4114799	1·2255361	1840305	1682	8159695	41
20	5783323	2373	4216677	1·7291096	7089133	1·4106098	1·2257887	1841987	1683	8158013	40
21	5785696	2373	4214304	1·7284005	7093504	1·4097405	1·2260416	1843670	1683	8156330	39
22	5788069	2371	4211931	1·7276921	7097878	1·4088718	1·2262947	1845353	1684	8154647	38
23	5790440	2372	4209560	1·7269844	7102253	1·4080039	1·2265480	1847037	1685	8152963	37
24	5792812	2371	4207188	1·7262774	7106630	1·4071367	1·2268015	1848722	1685	8151278	36
25	5795183	2370	4204817	1·7255712	7111009	1·4062702	1·2270552	1850407	1687	8149593	35
26	5797553	2370	4202447	1·7248657	7115390	1·4054044	1·2273091	1852094	1686	8147906	34
27	5799923	2369	4200077	1·7241609	7119772	1·4045393	1·2275633	1853780	1688	8146220	33
28	5802292	2369	4197708	1·7234568	7124157	1·4036749	1·2278176	1855468	1688	8144532	32
29	5804661	2369	4195339	1·7227534	7128543	1·4028113	1·2280722	1857156	1689	8142844	31
30	5807030	2367	4192970	1·7220508	7132931	1·4019483	1·2283269	1858845	1689	8141155	30
31	5809397	2368	4190603	1·7213489	7137320	1·4010860	1·2285819	1860534	1691	8139466	29
32	5811765	2367	4188235	1·7206477	7141712	1·4002245	1·2288371	1862225	1691	8137775	28
33	5814132	2366	4185868	1·7199472	7146106	1·3993636	1·2290924	1863916	1691	8136084	27
34	5816498	2366	4183502	1·7192475	7150501	1·3985034	1·2293480	1865607	1692	8134393	26
35	5818864	2366	4181136	1·7185484	7154898	1·3976440	1·2296039	1867299	1693	8132701	25
36	5821230	2365	4178770	1·7178501	7159297	1·3967852	1·2298599	1868992	1694	8131008	24
37	5823595	2364	4176405	1·7171525	7163698	1·3959272	1·2301161	1870686	1694	8129314	23
38	5825959	2364	4174041	1·7164556	7168100	1·3950698	1·2303725	1872380	1695	8127620	22
39	5828323	2364	4171677	1·7157594	7172505	1·3942131	1·2306292	1874075	1696	8125925	21
40	5830687	2363	4169313	1·7150639	7176911	1·3933571	1·2308861	1875771	1697	8124229	20
41	5833050	2362	4166950	1·7143691	7181319	1·3925019	1·2311432	1877468	1697	8122532	19
42	5835412	2362	4164588	1·7136750	7185729	1·3916473	1·2314004	1879165	1698	8120835	18
43	5837774	2362	4162226	1·7129817	7190141	1·3907934	1·2316579	1880863	1698	8119137	17
44	5840136	2361	4159864	1·7122890	7194554	1·3899401	1·2319156	1882561	1699	8117439	16
45	5842497	2360	4157503	1·7115970	7198970	1·3890876	1·2321736	1884260	1700	8115740	15
46	5844857	2360	4155143	1·7109058	7203387	1·3882358	1·2324317	1885960	1701	8114040	14
47	5847217	2360	4152783	1·7102152	7207806	1·3873847	1·2326900	1887661	1701	8112339	13
48	5849577	2359	4150423	1·7095254	7212227	1·3865342	1·2329486	1889362	1702	8110638	12
49	5851936	2358	4148064	1·7088362	7216650	1·3856844	1·2332074	1891064	1702	8108936	11
50	5854294	2358	4145706	1·7081478	7221075	1·3848353	1·2334664	1892766	1704	8107234	10
51	5856652	2358	4143348	1·7074601	7225502	1·3839869	1·2337256	1894470	1704	8105530	9
52	5859010	2357	4140990	1·7067730	7229930	1·3831392	1·2339850	1896174	1704	8103826	8
53	5861367	2357	4138633	1·7060867	7234361	1·3822922	1·2342446	1897878	1706	8102122	7
54	5863724	2356	4136276	1·7054010	7238793	1·3814458	1·2345044	1899584	1706	8100416	6
55	5866080	2355	4133920	1·7047160	7243227	1·3806001	1·2347645	1901290	1706	8098710	5
56	5868435	2355	4131565	1·7040318	7247663	1·3797551	1·2350248	1902996	1708	8097004	4
57	5870790	2355	4129210	1·7033482	7252101	1·3789108	1·2352852	1904704	1708	8095296	3
58	5873145	2354	4126855	1·7026653	7256540	1·3780672	1·2355459	1906412	1709	8093588	2
59	5875499	2354	4124501	1·7019831	7260982	1·3772242	1·2358069	1908121	1709	8091879	1
60	5877853		4122147	1·7013016	7265425	1·3763819	1·2360680	1909830		8090170	0
′	Cosine	Dif.	Vers.	Secant.	Cotan.	Taug.	Cosec.	Covers	Dif.	Sine.	′

′	Sine.	Dif.	Cosec.	Verseds.	Tang.	Dif.	Cotang.	Covers.	Secant.	D.	Cosine.	′
0	9·7585913	1804	10·2414087	9·2573136	9·8452268	2688	10·1547732	9·6298412	10·0866355	885	9·9133645	60
1	9·7587717	1802	10·2412283	9·2577142	9·8454956	2688	10·1545044	9·6295985	10·0867240	885	9·9132760	59
2	9·7589519	1802	10·2410481	9·2581145	9·8457644	2688	10·1542356	9·6293557	10·0868125	886	9·9131875	58
3	9·7591321	1800	10·2408679	9·2585147	9·8460332	2686	10·1539668	9·6291128	10·0869011	887	9·9130989	57
4	9·7593121	1799	10·2406879	9·2589147	9·8463018	2687	10·1536982	9·6288698	10·0869898	887	9·9130102	56
5	9·7594920	1798	10·2405080	9·2593144	9·8465705	2685	10·1534295	9·6286267	10·0870785	887	9·9129215	55
6	9·7596718	1797	10·2403282	9·2597140	9·8468390	2685	10·1531610	9·6283836	10·0871672	888	9·9128328	54
7	9·7598515	1796	10·2401485	9·2601133	9·8471075	2685	10·1528925	9·6281403	10·0872560	889	9·9127440	53
8	9·7600311	1795	10·2399689	9·2605125	9·8473760	2684	10·1526240	9·6278970	10·0873449	889	9·9126551	52
9	9·7602106	1793	10·2397894	9·2609114	9·8476444	2683	10·1523556	9·6276536	10·0874338	890	9·9125662	51
10	9·7603899	1793	10·2396101	9·2613102	9·8479127	2683	10·1520873	9·6274101	10·0875228	890	9·9124772	50
11	9·7605692	1791	10·2394308	9·2617087	9·8481810	2682	10·1518190	9·6271665	10·0876118	891	9·9123882	49
12	9·7607483	1791	10·2392517	9·2621071	9·8484492	2682	10·1515508	9·6269228	10·0877009	892	9·9122991	48
13	9·7609274	1789	10·2390726	9·2625052	9·8487174	2681	10·1512826	9·6266791	10·0877901	892	9·9122099	47
14	9·7611063	1788	10·2388937	9·2629032	9·8489855	2681	10·1510145	9·6264352	10·0878793	892	9·9121207	46
15	9·7612851	1787	10·2387149	9·2633009	9·8492536	2680	10·1507464	9·6261913	10·0879685	893	9·9120315	45
16	9·7614638	1786	10·2385362	9·2636985	9·8495216	2680	10·1504784	9·6259473	10·0880578	894	9·9119422	44
17	9·7616424	1784	10·2383576	9·2640958	9·8497896	2679	10·1502104	9·6257031	10·0881472	894	9·9118528	43
18	9·7618208	1784	10·2381792	9·2644929	9·8500575	2678	10·1499425	9·6254589	10·0882366	895	9·9117634	42
19	9·7619992	1783	10·2380008	9·2648899	9·8503253	2678	10·1496747	9·6252147	10·0883261	895	9·9116739	41
20	9·7621775	1781	10·2378225	9·2652866	9·8505931	2677	10·1494069	9·6249703	10·0884156	896	9·9115844	40
21	9·7623556	1781	10·2376444	9·2656832	9·8508608	2677	10·1491392	9·6247258	10·0885052	897	9·9114948	39
22	9·7625337	1779	10·2374663	9·2660795	9·8511285	2676	10·1488715	9·6244813	10·0885949	896	9·9114051	38
23	9·7627116	1778	10·2372884	9·2664757	9·8513961	2676	10·1486039	9·6242367	10·0886845	898	9·9113155	37
24	9·7628894	1777	10·2371106	9·2668716	9·8516637	2675	10·1483363	9·6239919	10·0887743	898	9·9112257	36
25	9·7630671	1776	10·2369329	9·2672674	9·8519312	2675	10·1480688	9·6237471	10·0888461	899	9·9111359	35
26	9·7632447	1775	10·2367553	9·2676629	9·8521987	2674	10·1478013	9·6235022	10·0889540	899	9·9110460	34
27	9·7634222	1774	10·2365778	9·2680583	9·8524661	2674	10·1475339	9·6232573	10·0890439	900	9·9109561	33
28	9·7635996	1773	10·2364004	9·2684534	9·8527335	2673	10·1472665	9·6230122	10·0891339	900	9·9108661	32
29	9·7637769	1771	10·2362231	9·2688484	9·8530008	2672	10·1469992	9·6227670	10·0892239	901	9·9107761	31
30	9·7639540	1771	10·2360460	9·2692431	9·8532680	2672	10·1467320	9·6225218	10·0893140	901	9·9106860	30
31	9·7641311	1769	10·2358689	9·2696377	9·8535352	2671	10·1464648	9·6222765	10·0894041	902	9·9105959	29
32	9·7643080	1769	10·2356920	9·2700321	9·8538023	2671	10·1461977	9·6220311	10·0894943	902	9·9105057	28
33	9·7644849	1767	10·2355151	9·2704262	9·8540694	2671	10·1459306	9·6217855	10·0895845	904	9·9104155	27
34	9·7646616	1766	10·2353384	9·2708202	9·8543365	2669	10·1456635	9·6215400	10·0896749	903	9·9103251	26
35	9·7648382	1765	10·2351618	9·2712140	9·8546034	2670	10·1453966	9·6212943	10·0897652	904	9·9102348	25
36	9·7650147	1764	10·2349853	9·2716075	9·8548704	2668	10·1451296	9·6210485	10·0898556	905	9·9101444	24
37	9·7651911	1763	10·2348089	9·2720009	9·8551372	2669	10·1448628	9·6208026	10·0899461	905	9·9100539	23
38	9·7653674	1762	10·2346326	9·2723941	9·8554041	2667	10·1445959	9·6205567	10·0900366	906	9·9099634	22
39	9·7655436	1761	10·2344564	9·2727871	9·8556708	2668	10·1443292	9·6203107	10·0901272	907	9·9098728	21
40	9·7657197	1760	10·2342803	9·2731799	9·8559376	2666	10·1440624	9·6200645	10·0902179	906	9·9097821	20
41	9·7658957	1758	10·2341043	9·2735725	9·8562042	2666	10·1437958	9·6198183	10·0903085	908	9·9096915	19
42	9·7660715	1758	10·2339285	9·2739649	9·8564708	2666	10·1435292	9·6195720	10·0903993	908	9·9096007	18
43	9·7662473	1756	10·2337527	9·2743571	9·8567374	2665	10·1432626	9·6193256	10·0904901	909	9·9095099	17
44	9·7664229	1756	10·2335771	9·2747491	9·8570039	2665	10·1429961	9·6190792	10·0905810	909	9·9094190	16
45	9·7665985	1754	10·2334015	9·2751409	9·8572704	2664	10·1427296	9·6188326	10·0906719	910	9·9093281	15
46	9·7667739	1753	10·2332261	9·2755325	9·8575368	2663	10·1424632	9·6185860	10·0907629	910	9·9092371	14
47	9·7669492	1752	10·2330508	9·2759239	9·8578031	2663	10·1421969	9·6183392	10·0908539	911	9·9091461	13
48	9·7671244	1752	10·2328756	9·2763151	9·8580694	2663	10·1419306	9·6180924	10·0909450	911	9·9090550	12
49	9·7672996	1750	10·2327004	9·2767062	9·8583357	2662	10·1416643	9·6178455	10·0910361	912	9·9089639	11
50	9·7674746	1748	10·2325254	9·2770970	9·8586019	2661	10·1413981	9·6175985	10·0911273	913	9·9088727	10
51	9·7676494	1748	10·2323506	9·2774876	9·8588680	2661	10·1411320	9·6173514	10·0912186	913	9·9087814	9
52	9·7678242	1747	10·2321758	9·2778781	9·8591341	2661	10·1408659	9·6171042	10·0913099	913	9·9086901	8
53	9·7679989	1746	10·2320011	9·2782683	9·8594002	2659	10·1405998	9·6168569	10·0914012	915	9·9085988	7
54	9·7681735	1745	10·2318265	9·2786584	9·8596661	2660	10·1403339	9·6166096	10·0914927	914	9·9085073	6
55	9·7683480	1743	10·2316520	9·2790483	9·8599321	2659	10·1400679	9·6163621	10·0915841	916	9·9084159	5
56	9·7685223	1743	10·2314777	9·2794380	9·8601980	2658	10·1398020	9·6161146	10·0916757	916	9·9083243	4
57	9·7686966	1741	10·2313034	9·2798274	9·8604638	2658	10·1395362	9·6158669	10·0917673	916	9·9082327	3
58	9·7688707	1741	10·2311293	9·2802167	9·8607296	2658	10·1392704	9·6156192	10·0918589	917	9·9081411	2
59	9·7690448	1739	10·2309552	9·2806058	9·8609954	2656	10·1390046	9·6153714	10·0919506	918	9·9080494	1
60	9·7692187		10·2307813	9·2809947	9·8612610		10·1387390	9·6151235	10·0920424		9·9079576	0

| ′ | Cosine. | Dif. | Secant. | Covers. | Cotang. | Dif. | Tang. | Verseds. | Cosec. | D. | Sine. | ′ |

54 Deg.

′	Sine.	Dif.	Covers	Cosec.	Tang.	Cotang.	Secant.	Vers.	Dif.	Cosine	′
0	5877853	2353	4122147	1·7013016	7265425	1·3763819	1·2360680	1909830	1710	8090170	60
1	5880206	2352	4119794	1·7006208	7269871	1·3755403	1·2363293	1911540	1711	8088460	59
2	5882558	2352	4117442	1·6999407	7274318	1·3746994	1·2365909	1913251	1712	8086749	58
3	5884910	2352	4115090	1·6992612	7278767	1·3738591	1·2368526	1914963	1712	8085037	57
4	5887262	2351	4112738	1·6985825	7283218	1·3730195	1·2371146	1916675	1713	8083325	56
5	5889613	2351	4110387	1·6979044	7287671	1·3721806	1·2373768	1918388	1713	8081612	55
6	5891964	2350	4108036	1·6972271	7292125	1·3713423	1·2376393	1920101	1714	8079899	54
7	5894314	2349	4105686	1·6965504	7296582	1·3705047	1·2379019	1921815	1715	8078185	53
8	5896663	2349	4103337	1·6958744	7301041	1·3696678	1·2381647	1923530	1716	8076470	52
9	5899012	2349	4100988	1·6951990	7305501	1·3688315	1·2384278	1925246	1716	8074754	51
10	5901361	2348	4098639	1·6945244	7309963	1·3679959	1·2386911	1926962	1717	8073038	50
11	5903709	2348	4096291	1·6938504	7314428	1·3671610	1·2389546	1928679	1718	8071321	49
12	5906057	2347	4093943	1·6931771	7318894	1·3663267	1·2392183	1930397	1718	8069603	48
13	5908404	2346	4091596	1·6925045	7323362	1·3654931	1·2394823	1932115	1719	8067885	47
14	5910750	2346	4089250	1·6918326	7327832	1·3646602	1·2397464	1933834	1720	8066166	46
15	5913096	2346	4086904	1·6911613	7332303	1·3638279	1·2400108	1935554	1720	8064446	45
16	5915442	2345	4084558	1·6904907	7336777	1·3629963	1·2402754	1937274	1721	8062726	44
17	5917787	2345	4082213	1·6898208	7341253	1·3621653	1·2405402	1938995	1722	8061005	43
18	5920132	2344	4079868	1·6891516	7345730	1·3613350	1·2408052	1940717	1723	8059283	42
19	5922476	2343	4077524	1·6884830	7350210	1·3605054	1·2410704	1942440	1723	8057560	41
20	5924819	2344	4075181	1·6878151	7354691	1·3596764	1·2413359	1944163	1724	8055837	40
21	5927163	2342	4072837	1·6871479	7359174	1·3588481	1·2416016	1945887	1724	8054113	39
22	5929505	2342	4070495	1·6864814	7363660	1·3580204	1·2418675	1947611	1725	8052389	38
23	5931847	2342	4068153	1·6858155	7368147	1·3571934	1·2421336	1949336	1726	8050664	37
24	5934189	2341	4065811	1·6851503	7372636	1·3563670	1·2423999	1951062	1727	8048938	36
25	5936530	2341	4063470	1·6844857	7377127	1·3555413	1·2426665	1952789	1727	8047211	35
26	5938871	2340	4061129	1·6838219	7381620	1·3547162	1·2429333	1954516	1728	8045484	34
27	5941211	2339	4058789	1·6831586	7386115	1·3538918	1·2432003	1956244	1728	8043756	33
28	5943550	2339	4056450	1·6824961	7390611	1·3530680	1·2434675	1957972	1729	8042028	32
29	5945889	2339	4054111	1·6818342	7395110	1·3522449	1·2437349	1959701	1730	8040299	31
30	5948228	2338	4051772	1·6811730	7399611	1·3514224	1·2440026	1961431	1731	8038569	30
31	5950566	2338	4049434	1·6805124	7404113	1·3506006	1·2442704	1963162	1731	8036838	29
32	5952904	2337	4047096	1·6798525	7408618	1·3497794	1·2445385	1964893	1732	8035107	28
33	5955241	2336	4044759	1·6791933	7413124	1·3489589	1·2448069	1966625	1733	8033375	27
34	5957577	2336	4042423	1·6785347	7417633	1·3481390	1·2450754	1968358	1733	8031642	26
35	5959913	2336	4040087	1·6778768	7422143	1·3473198	1·2453442	1970091	1734	8029909	25
36	5962249	2335	4037751	1·6772195	7426655	1·3465011	1·2456131	1971825	1735	8028175	24
37	5964584	2334	4035416	1·6765629	7431170	1·3456832	1·2458823	1973560	1735	8026440	23
38	5966918	2334	4033082	1·6759070	7435686	1·3448658	1·2461518	1975295	1736	8024705	22
39	5969252	2334	4030748	1·6752517	7440204	1·3440492	1·2464214	1977031	1737	8022969	21
40	5971586	2333	4028414	1·6745970	7444724	1·3432331	1·2466913	1978768	1737	8021232	20
41	5973919	2332	4026081	1·6739430	7449246	1·3424177	1·2469614	1980505	1739	8019495	19
42	5976251	2332	4023749	1·6732897	7453770	1·3416029	1·2472317	1982244	1738	8017756	18
43	5978583	2332	4021417	1·6726370	7458296	1·3407888	1·2475022	1983982	1740	8016018	17
44	5980915	2331	4019085	1·6719850	7462824	1·3399753	1·2477730	1985722	1740	8014278	16
45	5983246	2331	4016754	1·6713336	7467354	1·3391624	1·2480440	1987462	1741	8012538	15
46	5985577	2329	4014423	1·6706828	7471886	1·3383502	1·2483152	1989203	1741	8010797	14
47	5987906	2330	4012094	1·6700328	7476420	1·3375386	1·2485866	1990944	1742	8009056	13
48	5990236	2329	4009764	1·6693833	7480956	1·3367276	1·2488583	1992686	1743	8007314	12
49	5992565	2328	4007435	1·6687345	7485494	1·3359172	1·2491302	1994429	1744	8005571	11
50	5994893	2328	4005107	1·6680864	7490033	1·3351075	1·2494023	1996173	1744	8003827	10
51	5997221	2328	4002779	1·6674389	7494575	1·3342984	1·2496746	1997917	1745	8002083	9
52	5999549	2327	4000451	1·6667920	7499119	1·3334900	1·2499471	1999662	1745	8000338	8
53	6001876	2326	3998124	1·6661458	7503665	1·3326822	1·2502199	2001407	1746	7998593	7
54	6004202	2326	3995798	1·6655002	7508212	1·3318750	1·2504929	2003153	1747	7996847	6
55	6006528	2326	3993472	1·6648553	7512762	1·3310684	1·2507661	2004900	1748	7995100	5
56	6008854	2325	3991146	1·6642110	7517314	1·3302624	1·2510396	2006648	1748	7993352	4
57	6011179	2324	3988821	1·6635673	7521867	1·3294571	1·2513133	2008396	1749	7991604	3
58	6013503	2324	3986497	1·6629243	7526423	1·3286524	1·2515872	2010145	1750	7989855	2
59	6015827	2323	3984173	1·6622819	7530981	1·3278483	1·2518613	2011895	1750	7988105	1
60	6018150		3981850	1·6616401	7535541	1·3270448	1·2521357	2013645		7986355	0
′	Cosine	Dif.	Vers.	Secant.	Cotan.	Tang.	Cosec.	Covers	Dif.	Sine.	′

'	Sine.	Dif.	Cosec.	Verseds.	Tang.	Dif.	Cotang.	Covers.	Secant.	D.	Cosine.	'
0	9·7692187	1738	10·2307813	9·2809947	9·8612610	2657	10·1387390	9·6151235	10·0920424	918	9·9079576	60
1	9·7693925	1737	10·2306075	9·2813834	9·8615267	2656	10·1384733	9·6148755	10·0921342	918	9·9078658	59
2	9·7695662	1736	10·2304338	9·2817720	9·8617923	2655	10·1382077	9·6146275	10·0922260	920	9·9077740	58
3	9·7697398	1736	10·2302602	9·2821603	9·8620578	2655	10·1379422	9·6143793	10·0923180	919	9·9076820	57
4	9·7699134	1734	10·2300866	9·2825484	9·8623233	2654	10·1376767	9·6141311	10·0924099	921	9·9075901	56
5	9·7700868	1733	10·2299132	9·2829364	9·8625887	2654	10·1374113	9·6138827	10·0925020	921	9·9074980	55
6	9·7702601	1731	10·2297399	9·2833241	9·8628541	2654	10·1371459	9·6136343	10·0925941	921	9·9074059	54
7	9·7704332	1731	10·2295668	9·2837117	9·8631195	2653	10·1368805	9·6133858	10·0926862	922	9·9073138	53
8	9·7706063	1730	10·2293937	9·2840990	9·8633848	2652	10·1366152	9·6131372	10·0927784	923	9·9072216	52
9	9·7707793	1730	10·2292207	9·2844862	9·8636500	2652	10·1363500	9·6128885	10·0928707	923	9·9071293	51
10	9·7709522	1729	10·2290478	9·2848732	9·8639152	2651	10·1360848	9·6126397	10·0929630	924	9·9070370	50
11	9·7711249	1727	10·2288751	9·2852600	9·8641803	2651	10·1358197	9·6123908	10·0930554	924	9·9069446	49
12	9·7712976	1727 / 1726	10·2287024	9·2856466	9·8644454	2651	10·1355546	9·6121418	10·0931478	925	9·9068522	48
13	9·7714702	1724	10·2285298	9·2860330	9·8647105	2650	10·1352895	9·6118928	10·0932403	926	9·9067597	47
14	9·7716426	1724	10·2283574	9·2864192	9·8649755	2649	10·1350245	9·6116436	10·0933329	926	9·9066671	46
15	9·7718150	1722	10·2281850	9·2868053	9·8652404	2649	10·1347596	9·6113944	10·0934255	926	9·9065745	45
16	9·7719872	1721	10·2280128	9·2871911	9·8655053	2649	10·1344947	9·6111451	10·0935181	927	9·9064819	44
17	9·7721593	1721	10·2278407	9·2875768	9·8657702	2648	10·1342298	9·6108956	10·0936108	928	9·9063892	43
18	9·7723314	1719	10·2276686	9·2879622	9·8660350	2647	10·1339650	9·6106461	10·0937036	928	9·9062964	42
19	9·7725033	1718	10·2274967	9·2883475	9·8662997	2647	10·1337003	9·6103965	10·0937964	929	9·9062036	41
20	9·7726751	1717	10·2273249	9·2887326	9·8665644	2647	10·1334356	9·6101469	10·0938893	930	9·9061107	40
21	9·7728468	1717	10·2271532	9·2891175	9·8668291	2646	10·1331709	9·6098971	10·0939823	930	9·9060177	39
22	9·7730185	1715	10·2269815	9·2895022	9·8670937	2646	10·1329063	9·6096472	10·0940753	930	9·9059247	38
23	9·7731900	1714	10·2268100	9·2898867	9·8673583	2645	10·1326417	9·6093972	10·0941683	931	9·9058317	37
24	9·7733614	1713	10·2266386	9·2902711	9·8676228	2645	10·1323772	9·6091472	10·0942614	932	9·9057386	36
25	9·7735322	1712	10·2264673	9·2906552	9·8678873	2644	10·1321127	9·6088971	10·0943546	932	9·9056454	35
26	9·7737039	1710	10·2262961	9·2910392	9·8681517	2643	10·1318483	9·6086468	10·0944478	933	9·9055522	34
27	9·7738749	1710	10·2261251	9·2914229	9·8684160	2644	10·1315840	9·6083965	10·0945411	933	9·9054589	33
28	9·7740459	1709	10·2259541	9·2918065	9·8686804	2642	10·1313196	9·6081461	10·0946344	934	9·9053656	32
29	9·7742168	1708	10·2257832	9·2921899	9·8689446	2643	10·1310554	9·6078956	10·0947278	935	9·9052722	31
30	9·7743876	1707	10·2256124	9·2925731	9·8692089	2642	10·1307911	9·6076450	10·0948213	935	9·9051787	30
31	9·7745583	1705	10·2254417	9·2929561	9·8694731	2641	10·1305269	9·6073943	10·0949148	936	9·9050852	29
32	9·7747288	1705	10·2252712	9·2933390	9·8697372	2641	10·1302628	9·6071436	10·0950084	936	9·9049916	28
33	9·7748993	1704	10·2251007	9·2937216	9·8700013	2640	10·1299987	9·6068927	10·0951020	937	9·9048980	27
34	9·7750697	1702	10·2249303	9·2941041	9·8702653	2640	10·1297347	9·6066417	10·0951957	937	9·9048043	26
35	9·7752399	1702	10·2247601	9·2944863	9·8705293	2640	10·1294707	9·6063907	10·0952894	933	9·9047106	25
36	9·7754101	1700	10·2245899	9·2948684	9·8707933	2639	10·1292067	9·6061396	10·0953832	938	9·9046168	24
37	9·7755801	1700	10·2244199	9·2952503	9·8710572	2638	10·1289428	9·6058883	10·0954770	939	9·9045230	23
38	9·7757501	1698	10·2242499	9·2956320	9·8713210	2638	10·1286790	9·6056370	10·0955709	940	9·9044291	22
39	9·7759199	1698	10·2240801	9·2960136	9·8715848	2638	10·1284152	9·6053856	10·0956649	940	9·9043351	21
40	9·7760897	1696	10·2239103	9·2963949	9·8718486	2637	10·1281514	9·6051341	10·0957589	941	9·9042411	20
41	9·7762593	1696	10·2237407	9·2967760	9·8721123	2637	10·1278877	9·6048825	10·0958530	941	9·9041470	19
42	9·7764289	1694	10·2235711	9·2971570	9·8723760	2636	10·1276240	9·6046308	10·0959471	942	9·9040529	18
43	9·7765983	1693	10·2234017	9·2975378	9·8726396	2636	10·1273604	9·6043791	10·0960413	943	9·9039587	17
44	9·7767676	1693	10·2232324	9·2979184	9·8729032	2636	10·1270968	9·6041272	10·0961356	943	9·9038644	16
45	9·7769369	1691	10·2230631	9·2982988	9·8731668	2634	10·1268332	9·6038752	10·0962299	944	9·9037701	15
46	9·7771060	1690	10·2228940	9·2986790	9·8734302	2635	10·1265698	9·6036232	10·0963243	944	9·9036757	14
47	9·7772750	1689	10·2227250	9·2990591	9·8736937	2634	10·1263063	9·6033710	10·0964187	945	9·9035813	13
48	9·7774439	1689	10·2225561	9·2994389	9·8739571	2633	10·1260429	9·6031188	10·0965132	945	9·9034868	12
49	9·7776128	1687	10·2223872	9·2998186	9·8742204	2634	10·1257796	9·6028665	10·0966077	946	9·9033923	11
50	9·7777815	1686	10·2222185	9·3001981	9·8744838	2632	10·1255162	9·6026141	10·0967023	946	9·9032977	10
51	9·7779501	1685	10·2220499	9·3005774	9·8747470	2632	10·1252530	9·6023616	10·0967969	947	9·9032031	9
52	9·7781186	1684	10·2218814	9·3009565	9·8750102	2632	10·1249898	9·6021090	10·0968916	948	9·9031084	8
53	9·7782870	1683	10·2217130	9·3013355	9·8752734	2631	10·1247266	9·6018563	10·0969864	948	9·9030136	7
54	9·7784553	1682	10·2215447	9·3017142	9·8755365	2631	10·1244635	9·6016035	10·0970812	949	9·9029188	6
55	9·7786235	1681	10·2213765	9·3020928	9·8757996	2631	10·1242004	9·6013506	10·0971761	950	9·9028239	5
56	9·7787916	1680	10·2212084	9·3024712	9·8760627	2630	10·1239373	9·6010977	10·0972710	950	9·9027289	4
57	9·7789596	1679	10·2210404	9·3028494	9·8763257	2629	10·1236743	9·6008446	10·0973661	950	9·9026339	3
58	9·7791275	1678	10·2208725	9·3032274	9·8765886	2629	10·1234114	9·6005914	10·0974611	951	9·9025389	2
59	9·7792953	1677	10·2207047	9·3036052	9·8768515	2629	10·1231485	9·6003382	10·0975562	952	9·9024438	1
60	9·7794630	1677	10·2205370	9·3039829	9·8771144		10·1228856	9·6000849	10·0976514		9·9023486	0
'	Cosine.	Dif.	Secant.	Covers.	Cotang.	Dif.	Tang.	Verseds.	Cosec.	D.	Sine.	'

′	Sine.	Dif.	Covers	Cosec.	Tang.	Cotang.	Secant.	Vers.	Dif.	Cosine	′
0	6018150	2323	3981850	1·6616401	7535541	1·3270448	1·2521357	2013645	1751	7986355	60
1	6020473	2322	3979527	1·6609990	7540102	1·3262420	1·2524102	2015396	1751	7984604	59
2	6022795	2322	3977205	1·6603586	7544666	1·3254397	1·2526850	2017147	1753	7982853	58
3	6025117	2322	3974883	1·6597187	7549232	1·3246381	1·2529601	2018900	1753	7981100	57
4	6027439	2321	3972561	1·6590795	7553799	1·3238371	1·2532353	2020653	1753	7979347	56
5	6029760	2320	3970240	1·6584409	7558369	1·3230368	1·2535108	2022406	1755	7977594	55
6	6032080	2320	3967920	1·6578030	7562941	1·3222370	1·2537865	2024161	1755	7975839	54
7	6034400	2319	3965600	1·6571657	7567514	1·3214379	1·2540625	2025916	1755	7974084	53
8	6036719	2319	3963281	1·6565290	7572090	1·3206393	1·2543387	2027671	1757	7972329	52
9	6039038	2318	3960962	1·6558929	7576668	1·3198414	1·2546151	2029428	1757	7970572	51
10	6041356	2318	3958644	1·6552575	7581248	1·3190441	1·2548917	2031185	1757	7968815	50
11	6043674	2317	3956326	1·6546227	7585829	1·3182474	1·2551685	2032942	1759	7967058	49
12	6045991	2317	3954009	1·6539885	7590413	1·3174513	1·2554456	2034701	1759	7965299	48
13	6048308	2316	3951692	1·6533550	7594999	1·3166559	1·2557229	2036460	1760	7963540	47
14	6050624	2316	3949376	1·6527221	7599587	1·3158610	1·2560005	2038220	1760	7961780	46
15	6052940	2315	3947060	1·6520898	7604177	1·3150668	1·2562782	2039980	1761	7960020	45
16	6055255	2315	3944745	1·6514581	7608769	1·3142731	1·2565562	2041741	1762	7958259	44
17	6057570	2314	3942430	1·6508270	7613363	1·3134801	1·2568345	2043503	1762	7956497	43
18	6059884	2314	3940116	1·6501966	7617959	1·3126876	1·2571129	2045265	1763	7954735	42
19	6062198	2313	3937802	1·6495668	7622557	1·3118958	1·2573916	2047028	1764	7952972	41
20	6064511	2313	3935489	1·6489376	7627157	1·3111046	1·2576705	2048792	1764	7951208	40
21	6066824	2312	3933176	1·6483090	7631759	1·3103140	1·2579497	2050556	1766	7949444	39
22	6069136	2311	3930864	1·6476811	7636363	1·3095239	1·2582291	2052322	1765	7947678	38
23	6071447	2311	3928553	1·6470537	7640969	1·3087345	1·2585087	2054087	1767	7945913	37
24	6073758	2311	3926242	1·6464270	7645577	1·3079457	1·2587885	2055854	1767	7944146	36
25	6076069	2310	3923931	1·6458009	7650188	1·3071575	1·2590686	2057621	1768	7942379	35
26	6078379	2310	3921621	1·6451754	7654800	1·3063699	1·2593489	2059389	1768	7940611	34
27	6080689	2309	3919311	1·6445506	7659414	1·3055828	1·2596294	2061157	1769	7938843	33
28	6082998	2308	3917002	1·6439263	7664031	1·3047964	1·2599102	2062926	1770	7937074	32
29	6085306	2308	3914694	1·6433027	7668649	1·3040106	1·2601912	2064696	1771	7935304	31
30	6087614	2308	3912386	1·6426796	7673270	1·3032254	1·2604724	2066467	1771	7933533	30
31	6089922	2307	3910078	1·6420572	7677893	1·3024407	1·2607539	2068238	1772	7931762	29
32	6092229	2306	3907771	1·6414354	7682517	1·3016567	1·2610356	2070010	1772	7929990	28
33	6094535	2306	3905465	1·6408142	7687144	1·3008733	1·2613175	2071782	1773	7928218	27
34	6096841	2306	3903159	1·6401936	7691773	1·3000904	1·2615997	2073555	1774	7926445	26
35	6099147	2305	3900853	1·6395736	7696404	1·2993081	1·2618820	2075329	1775	7924671	25
36	6101452	2304	3898548	1·6389542	7701037	1·2985265	1·2621647	2077104	1775	7922896	24
37	6103756	2304	3896244	1·6383355	7705672	1·2977454	1·2624475	2078879	1776	7921121	23
38	6106060	2303	3893940	1·6377173	7710309	1·2969649	1·2627306	2080655	1776	7919345	22
39	6108363	2303	3891637	1·6370997	7714948	1·2961850	1·2630140	2082431	1777	7917569	21
40	6110666	2303	3889334	1·6364828	7719589	1·2954057	1·2632975	2084208	1778	7915792	20
41	6112969	2301	3887031	1·6358664	7724233	1·2946270	1·2635813	2085986	1779	7914014	19
42	6115270	2302	3884730	1·6352507	7728878	1·2938488	1·2638653	2087765	1779	7912235	18
43	6117572	2301	3882428	1·6346355	7733526	1·2930713	1·2641496	2089544	1780	7910456	17
44	6119873	2300	3880127	1·6340210	7738176	1·2922943	1·2644341	2091324	1780	7908676	16
45	6122173	2300	3877827	1·6334070	7742827	1·2915179	1·2647188	2093104	1781	7906896	15
46	6124473	2299	3875527	1·6327937	7747481	1·2907421	1·2650038	2094885	1782	7905115	14
47	6126772	2299	3873228	1·6321809	7752137	1·2899669	1·2652890	2096667	1783	7903333	13
48	6129071	2298	3870929	1·6315688	7756795	1·2891922	1·2655745	2098450	1783	7901550	12
49	6131369	2297	3868631	1·6309572	7761455	1·2884182	1·2658601	2100233	1784	7899767	11
50	6133666	2298	3866334	1·6303462	7766118	1·2876447	1·2661460	2102017	1785	7897983	10
51	6135964	2296	3864036	1·6297359	7770782	1·2868718	1·2664322	2103802	1785	7896198	9
52	6138260	2296	3861740	1·6291261	7775448	1·2860995	1·2667186	2105587	1786	7894413	8
53	6140555	2296	3859444	1·6285169	7780117	1·2853277	1·2670052	2107373	1786	7892627	7
54	6142852	2295	3857148	1·6279083	7784788	1·2845566	1·2672921	2109159	1787	7890841	6
55	6145147	2295	3854853	1·6273003	7789460	1·2837860	1·2675792	2110946	1788	7889054	5
56	6147442	2294	3852558	1·6266929	7794135	1·2830160	1·2678665	2112734	1789	7887266	4
57	6149736	2293	3850264	1·6260861	7798812	1·2822465	1·2681541	2114523	1789	7885477	3
58	6152029	2293	3847971	1·6254799	7803492	1·2814776	1·2684419	2116312	1790	7883688	2
59	6154322	2293	3845678	1·6248743	7808173	1·2807094	1·2687299	2118102	1790	7881898	1
60	6156615		3843385	1·6242692	7812856	1·2799416	1·2690182	2119892		7880108	0

′	Cosine	Dif.	Vers.	Secant.	Cotan.	Taug.	Cosec.	Covers	Dif.	Sine.	′

52 Deg.

′	Sine.	Dif.	Cosec.	Verseds.	Tang.	Dif.	Cotang.	Covers.	Secant.	D.	Cosine.	′
0	9·7794630	1676	10·2205370	9·3039829	9·8771144	2628	10·1228856	9·6000849	10·0976514	952	9·9023486	60
1	9·7796306	1675	10·2203694	9·3043604	9·8773772	2628	10·1226228	9·5998314	10·0977466	953	9·9022534	59
2	9·7797981	1674	10·2202019	9·3047376	9·8776400	2627	10·1223600	9·5995779	10·0978419	953	9·9021581	58
3	9·7799655	1673	10·2200345	9·3051148	9·8779027	2627	10·1220973	9·5993243	10·0979372	954	9·9020628	57
4	9·7801328	1672	10·2198672	9·3054917	9·8781654	2627	10·1218346	9·5990706	10·0980326	955	9·9019674	56
5	9·7803000	1671	10·2197000	9·3058684	9·8784281	2626	10·1215719	9·5988168	10·0981281	955	9·9018719	55
6	9·7804671	1670	10·2195329	9·3062450	9·8786907	2626	10·1213093	9·5985629	10·0982236	956	9·9017764	54
7	9·7806341	1669	10·2193659	9·3066214	9·8789533	2625	10·1210467	9·5983089	10·0983192	956	9·9016808	53
8	9·7808010	1667	10·2191990	9·3069976	9·8792158	2624	10·1207842	9·5980549	10·0984148	957	9·9015852	52
9	9·7809677	1667	10·2190323	9·3073736	9·8794782	2625	10·1205218	9·5978007	10·0985105	957	9·9014895	51
10	9·7811344	1666	10·2188656	9·3077494	9·8797407	2624	10·1202593	9·5975464	10·0986062	958	9·9013938	50
11	9·7813010	1665	10·2186990	9·3081251	9·8800031	2623	10·1199969	9·5972921	10·0987020	959	9·9012980	49
12	9·7814675	1664	10·2185325	9·3085006	9·8802654	2623	10·1197346	9·5970376	10·0987979	959	9·9012021	48
13	9·7816339	1663	10·2183661	9·3088759	9·8805277	2623	10·1194723	9·5967831	10·0988938	960	9·9011062	47
14	9·7818002	1662	10·2181998	9·3092510	9·8807900	2622	10·1192100	9·5965285	10·0989898	960	9·9010102	46
15	9·7819664	1662	10·2180336	9·3096259	9·8810522	2622	10·1189478	9·5962737	10·0990858	961	9·9009142	45
16	9·7821324	1660	10·2178676	9·3100007	9·8813144	2621	10·1186856	9·5960189	10·0991819	962	9·9008181	44
17	9·7822984	1659	10·2177016	9·3103752	9·8815765	2621	10·1184235	9·5957640	10·0992781	962	9·9007219	43
18	9·7824643	1658	10·2175357	9·3107496	9·8818386	2621	10·1181614	9·5955090	10·0993743	963	9·9006257	42
19	9·7826301	1657	10·2173699	9·3111238	9·8821007	2620	10·1178993	9·5952539	10·0994706	963	9·9005294	41
20	9·7827958	1656	10·2172042	9·3114979	9·8823627	2619	10·1176373	9·5949987	10·0995669	964	9·9004331	40
21	9·7829614	1654	10·2170386	9·3118717	9·8826246	2620	10·1173754	9·5947434	10·0996633	964	9·9003367	39
22	9·7831268	1654	10·2168732	9·3122454	9·8828866	2618	10·1171134	9·5944881	10·0997597	965	9·9002403	38
23	9·7832922	1653	10·2167078	9·3126189	9·8831484	2619	10·1168516	9·5942326	10·0998562	966	9·9001438	37
24	9·7834575	1652	10·2165425	9·3129922	9·8834103	2618	10·1165697	9·5939770	10·0999528	966	9·9000472	36
25	9·7836227	1651	10·2163773	9·3133654	9·8836721	2617	10·1163279	9·5937214	10·1000494	967	9·8999506	35
26	9·7837878	1650	10·2162122	9·3137383	9·8839338	2618	10·1160662	9·5934656	10·1001461	967	9·8998539	34
27	9·7839528	1649	10·2160472	9·3141111	9·8841956	2616	10·1158044	9·5932098	10·1002428	968	9·8997572	33
28	9·7841177	1647	10·2158823	9·3144837	9·8844572	2617	10·1155428	9·5929538	10·1003396	968	9·8996604	32
29	9·7842824	1647	10·2157176	9·3148561	9·8847189	2616	10·1152811	9·5926978	10·1004364	969	9·8995636	31
30	9·7844471	1646	10·2155529	9·3152284	9·8849805	2615	10·1150195	9·5924417	10·1005333	970	9·8994667	30
31	9·7846117	1645	10·2153883	9·3156005	9·8852420	2615	10·1147580	9·5921854	10·1006303	970	9·8993697	29
32	9·7847762	1644	10·2152238	9·3159724	9·8855035	2615	10·1144965	9·5919291	10·1007273	971	9·8992727	28
33	9·7849406	1643	10·2150594	9·3163441	9·8857650	2614	10·1142350	9·5916727	10·1008244	972	9·8991756	27
34	9·7851049	1642	10·2148951	9·3167156	9·8860264	2614	10·1139736	9·5914162	10·1009216	972	9·8990784	26
35	9·7852691	1641	10·2147309	9·3170870	9·8862878	2614	10·1137122	9·5911596	10·1010188	972	9·8989812	25
36	9·7854332	1640	10·2145668	9·3174582	9·8865492	2613	10·1134508	9·5909029	10·1011160	973	9·8988840	24
37	9·7855972	1639	10·2144028	9·3178292	9·8868105	2613	10·1131895	9·5906461	10·1012133	974	9·8987867	23
38	9·7857611	1638	10·2142389	9·3182000	9·8870718	2612	10·1129282	9·5903893	10·1013107	974	9·8986893	22
39	9·7859249	1637	10·2140751	9·3185706	9·8873330	2612	10·1126670	9·5901323	10·1014081	975	9·8985919	21
40	9·7860886	1636	10·2139114	9·3189411	9·8875942	2612	10·1124058	9·5898752	10·1015056	976	9·8984944	20
41	9·7862522	1635	10·2137478	9·3193114	9·8878554	2611	10·1121446	9·5896181	10·1016032	976	9·8983968	19
42	9·7864157	1634	10·2135843	9·3196815	9·8881165	2610	10·1118835	9·5893608	10·1017008	977	9·8982992	18
43	9·7865791	1633	10·2134209	9·3200515	9·8883775	2611	10·1116225	9·5891034	10·1017985	977	9·8982015	17
44	9·7867424	1632	10·2132576	9·3204213	9·8886386	2610	10·1113614	9·5888460	10·1018962	978	9·8981038	16
45	9·7869056	1631	10·2130944	9·3207909	9·8888996	2609	10·1111004	9·5885885	10·1019940	978	9·8980060	15
46	9·7870687	1630	10·2129313	9·3211603	9·8891605	2609	10·1108395	9·5883308	10·1020918	979	9·8979082	14
47	9·7872317	1629	10·2127683	9·3215295	9·8894214	2609	10·1105786	9·5880731	10·1021897	980	9·8978103	13
48	9·7873946	1628	10·2126054	9·3218986	9·8896823	2609	10·1103177	9·5878153	10·1022877	980	9·8977123	12
49	9·7875574	1628	10·2124426	9·3222675	9·8899432	2608	10·1100568	9·5875573	10·1023857	981	9·8976143	11
50	9·7877202	1626	10·2122798	9·3226362	9·8902040	2607	10·1097960	9·5872993	10·1024838	981	9·8975162	10
51	9·7878828	1625	10·2121172	9·3230048	9·8904647	2607	10·1095353	9·5870412	10·1025819	982	9·8974181	9
52	9·7880453	1624	10·2119547	9·3233731	9·8907254	2607	10·1092746	9·5867830	10·1026801	983	9·8973199	8
53	9·7882077	1624	10·2117923	9·3237413	9·8909861	2607	10·1090139	9·5865247	10·1027784	983	9·8972216	7
54	9·7883701	1622	10·2116299	9·3241094	9·8912468	2606	10·1087532	9·5862663	10·1028767	984	9·8971233	6
55	9·7885323	1621	10·2114677	9·3244772	9·8915074	2605	10·1084926	9·5860078	10·1029751	984	9·8970249	5
56	9·7886944	1621	10·2113056	9·3248449	9·8917679	2606	10·1082321	9·5857492	10·1030735	985	9·8969265	4
57	9·7888565	1619	10·2111435	9·3252124	9·8920285	2605	10·1079715	9·5854905	10·1031720	986	9·8968280	3
58	9·7890184	1618	10·2109816	9·3255797	9·8922890	2604	10·1077110	9·5852318	10·1032706	986	9·8967294	2
59	9·7891802	1618	10·2108198	9·3259469	9·8925494	2604	10·1074506	9·5849729	10·1033692	987	9·8966308	1
60	9·7893420		10·2106580	9·3263138	9·8928098		10·1071902	9·5847139	10·1034679		9·8965321	0
′	Cosine.	Dif.	Secant.	Covers.	Cotang.	Dif.	Tang.	Verseds.	Cosec.	D.	Sine.	′

′	Sine.	Dif.	Covers	Cosec.	Tang.	Cotang.	Secant.	Vers.	Dif.	Cosine	′
0	6156615	2292	3843385	1·6242692	7812856	1·2799416	1·2690182	2119892	1792	7880108	60
1	6158907	2291	3841093	1·6236648	7817542	1·2791745	1·2693067	2121684	1792	7878316	59
2	6161198	2291	3838802	1·6230609	7822229	1·2784079	1·2695955	2123476	1792	7876524	58
3	6163489	2291	3836511	1·6224576	7826919	1·2776419	1·2698845	2125268	1793	7874732	57
4	6165780	2289	3834220	1·6218549	7831611	1·2768765	1·2701737	2127061	1794	7872939	56
5	6168069	2290	3831931	1·6212528	7836305	1·2761116	1·2704632	2128855	1795	7871145	55
6	6170359	2289	3829641	1·6206513	7841002	1·2753473	1·2707529	2130650	1795	7869350	54
7	6172648	2288	3827352	1·6200504	7845700	1·2745835	1·2710429	2132445	1796	7867555	53
8	6174936	2288	3825064	1·6194500	7850400	1·2738204	1·2713331	2134241	1796	7865759	52
9	6177224	2287	3822776	1·6188502	7855103	1·2730578	1·2716235	2136037	1798	7863963	51
10	6179511	2287	3820489	1·6182510	7859808	1·2722957	1·2719142	2137835	1798	7862165	50
11	6181798	2286	3818202	1·6176524	7864515	1·2715342	1·2722052	2139633	1798	7860367	49
12	6184084	2286	3815916	1·6170544	7869224	1·2707733	1·2724963	2141431	1799	7858569	48
13	6186370	2285	3813630	1·6164569	7873935	1·2700130	1·2727877	2143230	1800	7856770	47
14	6188655	2284	3811345	1·6158600	7878649	1·2692532	1·2730794	2145030	1801	7854970	46
15	6190939	2285	3809061	1·6152637	7883364	1·2684940	1·2733712	2146831	1801	7853169	45
16	6193224	2283	3806776	1·6146680	7888082	1·2677353	1·2736634	2148632	1802	7851368	44
17	6195507	2283	3804493	1·6140728	7892802	1·2669772	1·2739557	2150434	1802	7849566	43
18	6197790	2283	3802210	1·6134783	7897524	1·2662196	1·2742484	2152236	1803	7847764	42
19	6200073	2282	3799927	1·6128843	7902248	1·2654626	1·2745412	2154039	1804	7845961	41
20	6202355	2281	3797645	1·6122908	7906975	1·2647062	1·2748343	2155843	1805	7844157	40
21	6204636	2281	3795364	1·6116980	7911703	1·2639503	1·2751276	2157648	1805	7842352	39
22	6206917	2281	3793083	1·6111057	7916434	1·2631950	1·2754212	2159453	1806	7840547	38
23	6209198	2280	3790802	1·6105140	7921167	1·2624402	1·2757151	2161259	1806	7838741	37
24	6211478	2279	3788522	1·6099228	7925902	1·2616860	1·2760091	2163065	1808	7836935	36
25	6213757	2279	3786243	1·6093323	7930640	1·2609323	1·2763034	2164873	1807	7835127	35
26	6216036	2278	3783964	1·6087423	7935379	1·2601792	1·2765980	2166680	1809	7833320	34
27	6218314	2278	3781686	1·6081528	7940121	1·2594267	1·2768928	2168489	1809	7831511	33
28	6220592	2278	3779408	1·6075640	7944865	1·2586747	1·2771878	2170298	1810	7829702	32
29	6222870	2276	3777130	1·6069757	7949611	1·2579232	1·2774831	2172108	1810	7827892	31
30	6225146	2277	3774854	1·6063879	7954359	1·2571723	1·2777787	2173918	1812	7826082	30
31	6227423	2275	3772577	1·6058008	7959110	1·2564219	1·2780744	2175730	1811	7824270	29
32	6229698	2276	3770302	1·6052142	7963862	1 2556721	1·2783705	2177541	1813	7822459	28
33	6231974	2274	3768026	1·6046281	7968617	1·2549229	1·2786667	2179354	1813	7820646	27
34	6234248	2274	3765752	1·6040426	7973374	1·2541742	1·2789632	2181157	1814	7818833	26
35	6236522	2274	3763478	1·6034577	7978134	1·2534260	1·2792600	2182981	1814	7817019	25
36	6238796	2273	3761204	1·6028734	7982895	1·2526784	1·2795570	2184795	1815	7815205	24
37	6241069	2273	3758931	1·6022896	7987659	1·2519313	1·2798543	2186610	1816	7813390	23
38	6243342	2272	3756658	1 6017064	7992425	1·2511848	1·2801518	2188426	1817	7811574	22
39	6245614	2271	3754386	1·6011237	7997193	1·2504388	1·2804495	2190243	1817	7809757	21
40	6247835	2271	3752115	1·6005416	8001963	1·2496933	1·2807475	2192060	1817	7807940	20
41	6250156	2271	3749844	1·5999600	8006736	1·2489484	1·2810457	2193877	1819	7806123	19
42	6252427	2269	3747573	1·5993790	8011511	1·2482040	1·2813442	2195696	1819	7804304	18
43	6254696	2270	3745304	1·5987986	8016288	1·2474602	1·2816430	2197515	1820	7802485	17
44	6256966	2269	3743034	1·5982187	8021067	1·2467169	1·2819419	2199335	1820	7800665	16
45	6259235	2268	3740765	1·5976394	8025849	1·2459742	1·2822412	2201155	1821	7798845	15
46	6261503	2268	3738497	1·5970606	8030632	1·2452320	1·2825407	2202976	1822	7797024	14
47	6263771	2267	3736229	1·5964824	8035418	1·2444903	1·2828404	2204798	1822	7795202	13
48	6266038	2267	3733962	1·5959048	8040206	1·2437492	1·2831404	2206620	1823	7793380	12
49	6268305	2266	3731695	1·5953276	8044997	1·2430086	1·2834406	2208443	1824	7791557	11
50	6270571	2266	3729429	1·5947511	8049790	1·2422685	1·2837411	2210267	1824	7789733	10
51	6272837	2265	3727163	1·5941751	8054584	1·2415290	1·2840418	2212091	1825	7787909	9
52	6275102	2264	3724898	1·5935996	8059382	1·2407900	1·2843428	2213916	1826	7786084	8
53	6277366	2265	3722634	1·5930247	8064181	1·2400515	1·2846440	2215742	1827	7784258	7
54	6279631	2263	3720369	1·5924504	8068983	1·2393136	1·2849455	2217569	1827	7782431	6
55	6281894	2263	3718106	1·5918766	8073787	1·2385762	1·2852472	2219396	1827	7780604	5
56	6284157	2263	3715843	1·5913033	8078593	1·2378393	1·2855492	2221223	1828	7778777	4
57	6286420	2262	3713580	1·5907306	8083401	1·2371030	1·2858514	2223051	1829	7776949	3
58	6288682	2261	3711318	1·5901584	8088212	1·2363672	1·2861539	2224880	1830	7775120	2
59	6290943	2261	3709057	1·5895868	8093025	1·2356319	1·2864566	2226710	1830	7773290	1
60	6293204		3706796	1·5890157	8097840	1·2348972	1·2867596	2228540		7771460	0
′	Cosine	Dif.	Vers.	Secant.	Cotan.	Taug.	Cosec.	Covers	Dif.	Sine.	′

′	Sine.	Dif.	Cosec.	Versed.	Tang.	Dif.	Cotang.	Covers.	Secant.	D.	Cosine.	
0	9·7893420	1616	10·2106580	9·3263138	9·8928098	2604	10·1071902	9·5847139	10·1034679	987	9·8965321	60
1	9·7895036	1616	10·2104964	9·3266806	9·8930702	2604	10·1069298	9·5844549	10·1035666	988	9·8964334	59
2	9·7896652	1614	10·2103348	9·3270473	9·8933306	2603	10·1066694	9·5841957	10·1036654	988	9·8963346	58
3	9·7898266	1614	10·2101734	9·3274137	9·8935909	2602	10·1064091	9 5839364	10·1037642	989	9·8962358	57
4	9·7899880	1613	10·2100120	9·3277800	9·8938511	2603	10·1061489	9·5836771	10·1038631	990	9·8961369	56
5	9·7901493	1611	10·2098507	9·3281461	9·8941114	2601	10·1058886	9·5834176	10·1039621	990	9·8960379	55
6	9·7903104	1611	10·2096896	9·3285121	9·8943715	2602	10·1056285	9·5831581	10·1040611	991	9·8959389	54
7	9·7904715	1610	10·2095285	9·3288778	9·8946317	2601	10·1053683	9·5828985	10·1041602	992	9·8958398	53
8	9·7906325	1608	10·2093675	9·3292434	9·8948918	2601	10·1051082	9·5826387	10·1042594	992	9·8957406	52
9	9·7907933	1608	10·2092067	9·3296089	9·8951519	2600	10·1048481	9·5823789	10·1043586	992	9·8956414	51
10	9·7909541	1607	10·2090459	9·3299741	9·8954119	2600	10·1045881	9·5821190	10·1044578	993	9·8955422	50
11	9·7911148	1606	10·2088852	9·3303392	9·8956719	2600	10·1043281	9·5818589	10·1045571	994	9·8954429	49
12	9·7912754	1605	10·2087246	9·3307041	9·8959319	2599	10·1040681	9·5815988	10·1046565	995	9·8953435	48
13	9·7914359	1604	10·2085641	9·3310688	9·8961918	2599	10·1038082	9·5813386	10·1047560	995	9·8952440	47
14	9·7915963	1603	10·2084037	9·3314334	9·8964517	2599	10·1035483	9·5810783	10·1048557	995	9·8951445	46
15	9·7917566	1602	10·2082434	9·3317978	9·8967116	2598	10·1032884	9·5808179	10·1049550	997	9·8950450	45
16	9·7919168	1601	10·2080832	9·3321620	9·8969714	2598	10·1030286	9·5805574	10·1050547	996	9·8949453	44
17	9·7920769	1600	10·2079231	9·3325261	9·8972312	2598	10·1027688	9·5802968	10·1051543	996	9·8948457	43
18	9·7922369	1599	10·2077631	9·3328900	9·8974910	2597	10·1025090	9·5800361	10·1052541	998	9·8947459	42
19	9·7923968	1598	10·2076032	9·3332537	9 8977507	2597	10·1022493	9·5797753	10·1053539	998	9·8946461	41
20	9·7925566	1597	10·2074434	9·3336172	9·8980104	2596	10·1019836	9·5795144	10·1054537	1000	9·8945463	40
21	9·7927163	1597	10·2072837	9·3339806	9·8982700	2596	10·1017300	9·5792534	10·1055537	999	9·8944463	39
22	9·7928760	1595	10·2071240	9·3343438	9·8985296	2596	10·1014704	9·5789923	10·1056536	1001	9·8943464	38
23	9·7930355	1594	10·2069645	9·3347068	9·8987892	2595	10·1012108	9·5787311	10·1057537	1001	9·8942463	37
24	9·7931949	1594	10·2068051	9·3350697	9·8990487	2595	10·1009513	9·5784698	10·1058538	1001	9·8941462	36
25	9·7933543	1592	10·2066457	9·3354323	9·8993082	2595	10·1006918	9·5782085	10·1059539	1003	9·8940461	35
26	9·7935135	1592	10·2064865	9·3357949	9·8995677	2594	10·1004323	9·5779470	10·1060542	1002	9·8939458	34
27	9·7936727	1590	10·2063273	9·3361572	9·8998271	2594	10·1001729	9·5776854	10·1061544	1004	9·8938456	33
28	9·7938317	1590	10·2061683	9·3365194	9·9000865	2594	10·0999135	9·5774237	10·1062548	1004	9·8937452	32
29	9·7939907	1589	10·2060093	9·3368814	9·9003459	2593	10·0996541	9·5771620	10·1063552	1004	9·8936448	31
30	9·7941496	1587	10·2058504	9·3372432	9·9006052	2593	10·0993948	9·5769001	10·1064556	1005	9·8935444	30
31	9·7943083	1587	10·2056917	9·3376049	9·9008645	2592	10·0991355	9·5766382	10·1065561	1006	9·8934439	29
32	9·7944670	1586	10·2055330	9·3379664	9·9011237	2593	10·0988763	9·5763761	10·1066567	1007	9·8933433	28
33	9·7946256	1585	10·2053744	9·3383278	9·9013830	2592	10·0986170	9·5761139	10·1067574	1007	9·8932426	27
34	9·7947841	1584	10·2052159	9·3386889	9·9016422	2591	10·0983578	9·5758517	10·1068581	1007	9·8931419	26
35	9·7949425	1583	10·2050575	9·3390499	9·9019013	2591	10·0980987	9·5755893	10·1069588	1008	9·8930412	25
36	9·7951008	1582	10·2048992	9·3394107	9·9021604	2591	10·0978396	9·5753269	10·1070596	1009	9·8929404	24
37	9·7952590	1581	10·2047410	9·3397714	9·9024195	2591	10·0975805	9·5750643	10·1071605	1010	9·8928395	23
38	9·7954171	1580	10·2045829	9·3401319	9·9026786	2590	10·0973214	9·5748017	10·1072615	1010	9·8927385	22
39	9·7955751	1579	10·2044249	9·3404922	9·9029376	2590	10·0970624	9·5745390	10·1073625	1010	9·8926375	21
40	9·7957330	1579	10·2042670	9·3408524	9·9031966	2589	10·0968034	9·5742761	10·1074635	1011	9 8924354	20
41	9·7958909	1577	10·2041091	9·3412124	9·9034555	2589	10·0965445	9·5740132	10·1075646	1012	9·8923342	19
42	9·7960486	1576	10·2039514	9·3415722	9·9037144	2589	10·0962856	9·5737502	10·1076658	1013	9·8923342	18
43	9·7962062	1576	10·2037938	9·3419319	9·9039733	2588	10·0960267	9·5734870	10·1077671	1013	9·8922329	17
44	9·7963638	1574	10·2036362	9·3422913	9·9042321	2589	10·0957679	9·5732238	10·1078684	1013	9·8921316	16
45	9·7965212	1574	10·2034788	9·3426507	9·9044910	2587	10·0955090	9·5729605	10·1079697	1014	9·8920303	15
46	9·7966786	1573	10·2033214	9·3430098	9·9047497	2588	10·0952503	9·5726970	10·1080711	1015	9·8919289	14
47	9·7968359	1571	10·2031641	9·3433688	9·9050085	2587	10·0949915	9·5724335	10·1081726	1016	9·8918274	13
48	9·7969930	1571	10·2030070	9·3437276	9·9052672	2587	10·0947328	9·5721699	10·1082742	1016	9·8917258	12
49	9·7971501	1570	10·2028499	9·3440863	9·9055259	2586	10·0944741	9·5719062	10·1083758	1016	9·8916242	11
50	9·7973071	1569	10·2026929	9·3444448	9·9057845	2586	10·0942155	9·5716423	10·1084774	1018	9·8915226	10
51	9·7974640	1568	10·2025360	9·3448031	9·9060431	2586	10·0939569	9·5713784	10·1085792	1017	9·8914208	9
52	9·7976208	1567	10·2023792	9·3451612	9·9063017	2586	10·0936983	9·5711144	10·1086809	1019	9·8913191	8
53	9·7977775	1566	10·2022225	9·3455192	9·9065603	2585	10·0934397	9·5708503	10·1087828	1019	9·8912172	7
54	9·7979341	1565	10·2020659	9·3458770	9·9068188	2585	10·0931812	9·5705861	10·1088847	1020	9·8911153	6
55	9·7980906	1564	10·2019094	9·3462347	9·9070773	2584	10·0929227	9·5703218	10·1089867	1020	9·8910133	5
56	9·7982470	1564	10·2017530	9·3465922	9·9073357	2584	10·0926643	9·5700573	10·1090887	1021	9·8909113	4
57	9·7984034	1562	10·2015966	9·3469495	9·9075941	2584	10·0924059	9·5697928	10·1091908	1021	9·8908092	3
58	9·7985596	1562	10·2014404	9·3473067	9·9078525	2584	10·0921475	9·5695282	10·1092929	1022	9·8907071	2
59	9·7987158	1560	10·2012842	9·3476637	9·9081109	2583	10·0918891	9·5692635	10·1093951	1023	9·8906049	1
60	9·7988718		10·2011282	9·3480205	9·9083692		10·0916308	9·5689987	10·1094974		9·8905026	0
′	Cosine.	Dif.	Secant.	Covers.	Cotang.	Dif.	Tang.	Versed.	Cosec.	D.	Sine.	′

′	Sine.	Dif.	Covers	Cosec.	Tang.	Cotang.	Secant.	Vers.	Dif.	Cosine	′
0	6293204	2260	3706796	1·5890157	8097840	1·2348972	1·2867596	2228540	1831	7771460	60
1	6295464	2260	3704536	1·5884452	8102658	1·2341629	1·2870628	2230371	1832	7769629	59
2	6297724	2259	3702276	1·5878752	8107478	1·2334292	1·2873663	2232203	1832	7767797	58
3	6299983	2259	3700017	1·5873058	8112300	1·2326961	1·2876700	2234035	1833	7765965	57
4	6302242	2258	3697758	1·5867369	8117124	1·2319634	1·2879740	2235868	1834	7764132	56
5	6304500	2258	3695500	1·5861685	8121951	1·2312313	1·2882782	2237702	1834	7762298	55
6	6306758	2257	3693242	1·5856007	8126780	1·2304997	1·2885827	2239536	1835	7760464	54
7	6309015	2257	3690985	1·5850334	8131611	1·2297687	1·2888875	2241371	1835	7758629	53
8	6311272	2256	3688728	1·5844667	8136444	1·2290381	1·2891925	2243206	1837	7756794	52
9	6313528	2256	3686472	1·5839005	8141280	1·2283081	1·2894977	2245043	1836	7754957	51
10	6315784	2255	3684216	1·5833348	8146118	1·2275786	1·2898032	2246879	1838	7753121	50
11	6318039	2254	3681961	1·5827697	8150958	1·2268496	1·2901090	2248717	1838	7751283	49
12	6320293	2254	3679707	1·5822051	8155801	1·2261211	1·2904150	2250555	1839	7749445	48
13	6322547	2253	3677453	1·5816411	8160646	1·2253932	1·2907213	2252394	1839	7747606	47
14	6324800	2253	3675200	1·5810776	8165493	1·2246658	1·2910278	2254233	1841	7745767	46
15	6327053	2253	3672947	1·5805146	8170343	1·2239389	1·2913346	2256074	1840	7743926	45
16	6329306	2251	3670694	1·5799521	8175195	1·2232125	1·2916416	2257914	1842	7742086	44
17	6331557	2252	3668443	1·5793902	8180049	1·2224866	1·2919489	2259756	1842	7740244	43
18	6333809	2250	3666191	1·5788289	8184905	1·2217613	1·2922564	2261598	1843	7738402	42
19	6336059	2251	3663941	1·5782680	8189764	1·2210364	1·2925642	2263441	1843	7736559	41
20	6338310	2249	3661690	1·5777077	8194625	1·2203121	1·2928723	2265284	1844	7734716	40
21	6340559	2249	3659441	1·5771479	8199488	1·2195883	1·2931806	2267128	1845	7732872	39
22	6342808	2249	3657192	1·5765887	8204354	1·2188650	1·2934892	2268973	1845	7731027	38
23	6345057	2248	3654943	1·5760300	8209222	1·2181422	1·2937980	2270818	1846	7729182	37
24	6347305	2248	3652695	1·5754718	8214093	1·2174199	1·2941071	2272664	1847	7727336	36
25	6349553	2247	3650447	1·5749141	8218965	1·2166982	1·2944164	2274511	1847	7725489	35
26	6351800	2246	3648200	1·5743570	8223840	1·2159769	1·2947260	2276358	1848	7723642	34
27	6354046	2246	3645954	1·5738004	8228718	1·2152562	1·2950359	2278206	1849	7721794	33
28	6356292	2245	3643708	1·5732443	8233597	1·2145359	1·2953460	2280055	1849	7719945	32
29	6358537	2245	3641463	1·5726887	8238479	1·2138162	1·2956564	2281904	1850	7718096	31
30	6360782	2244	3639218	1·5721337	8243364	1·2130970	1·2959670	2283754	1851	7716246	30
31	6363026	2244	3636974	1·5715792	8248251	1·2123783	1·2962779	2285605	1851	7714395	29
32	6365270	2243	3634730	1·5710252	8253140	1·2116601	1·2965890	2287456	1852	7712544	28
33	6367513	2243	3632487	1·5704717	8258031	1·2109424	1·2969004	2289308	1852	7710692	27
34	6369756	2242	3630244	1·5699188	8262925	1·2102252	1·2972121	2291160	1854	7708840	26
35	6371998	2242	3628002	1·5693664	8267821	1·2095085	1·2975240	2293014	1854	7706986	25
36	6374240	2241	3625760	1·5688145	8272719	1·2087924	1·2978362	2294868	1854	7705132	24
37	6376481	2240	3623519	1·5682631	8277620	1·2080767	1·2981487	2296722	1855	7703278	23
38	6378721	2240	3621279	1·5677123	8282523	1·2073615	1·2984614	2298577	1856	7701423	22
39	6380961	2240	3619039	1·5671619	8287429	1·2066468	1·2987743	2300433	1857	7699567	21
40	6383201	2239	3616799	1·5666121	8292337	1·2059327	1·2990876	2302290	1857	7697710	20
41	6385440	2238	3614560	1·5660628	8297247	1·2052190	1·2994011	2304147	1857	7695853	19
42	6387678	2238	3612322	1·5655141	8302160	1·2045058	1·2997148	2306004	1859	7693996	18
43	6389916	2237	3610084	1·5649658	8307075	1·2037932	1·3000288	2307863	1859	7692137	17
44	6392153	2237	3607847	1·5644181	8311992	1·2030810	1·3003431	2309722	1860	7690278	16
45	6394390	2236	3605610	1·5638708	8316912	1·2023693	1·3006576	2311582	1860	7688418	15
46	6396626	2236	3603374	1·5633241	8321834	1·2016581	1·3009724	2313442	1861	7686558	14
47	6398862	2235	3601138	1·5627779	8326759	1·2009475	1·3012875	2315303	1862	7684697	13
48	6401097	2235	3598903	1·5622322	8331686	1·2002373	1·3016028	2317165	1862	7682835	12
49	6403332	2234	3596668	1·5616871	8336615	1·1995276	1·3019184	2319027	1863	7680973	11
50	6405566	2233	3594434	1·5611424	8341547	1·1988184	1·3022343	2320890	1864	7679110	10
51	6407799	2233	3592201	1·5605982	8346481	1·1981097	1·3025504	2322754	1864	7677246	9
52	6410032	2232	3589968	1·5600546	8351418	1·1974015	1·3028667	2324618	1865	7675382	8
53	6412264	2232	3587736	1·5595115	8356357	1·1966938	1·3031834	2326483	1865	7673517	7
54	6414496	2232	3585504	1·5589689	8361298	1·1959866	1·3035003	2328348	1867	7671652	6
55	6416728	2230	3583272	1·5584268	8366242	1·1952799	1·3038175	2330215	1867	7669785	5
56	6418958	2231	3581042	1·5578852	8371188	1·1945736	1·3041349	2332082	1867	7667918	4
57	6421189	2229	3578811	1·5573441	8376136	1·1938679	1·3044526	2333949	1869	7666051	3
58	6423418	2229	3576582	1·5568035	8381087	1·1931626	1·3047706	2335817	1869	7664183	2
59	6425647	2229	3574353	1·5562634	8386041	1·1924579	1·3050888	2337686	1870	7662314	1
60	6427876		3572124	1·5557238	8390996	1·1917536	1·3054073	2339556		7660444	0
′	Cosine	Dif.	Vers.	Secant.	Cotan.	Taug.	Cosec.	Covers	Dif.	Sine.	′

′	Sine.	Dif.	Cosec.	Verseds.	Tang.	Dif.	Cotang.	Covers.	Secant.	D.	Cosine.	′
0	9·7988718	1560	10·2011282	9·3480205	9·9083692	2583	10·0916308	9·5689987	10·1094974	1023	9·8905026	60
1	9·7990278	1558	10·2009722	9·3483772	9·9086275	2583	10·0913725	9·5687338	10·1095997	1024	9·8904003	59
2	9·7991836	1558	10·2008164	9·3487337	9·9088858	2582	10·0911142	9·5684688	10·1097021	1025	9·8902979	58
3	9·7993394	1557	10·2006606	9·3490900	9·9091440	2582	10·0908560	9·5682037	10·1098046	1025	9·8901954	57
4	9·7994951	1556	10·2005049	9·3494462	9·9094022	2581	10·0905978	9·5679385	10·1099071	1026	9·8900929	56
5	9·7996507	1555	10·2003493	9·3498022	9·9096603	2582	10·0903397	9·5676732	10·1100097	1026	9·8899903	55
6	9·7998062		10·2001938	9·3501580	9·9099185		10·0900815	9·5674078	10·1101123		9·8898877	54
7	9·7999616	1554	10·2000384	9·3505137	9·9101766	2581	10·0898234	9·5671423	10·1102150	1027	9·8897850	53
8	9·8001169	1553	10·1998831	9·3508692	9·9104347	2581	10·0895653	9·5668766	10·1103178	1028	9·8896822	52
9	9·8002721	1552	10·1997279	9·3512246	9·9106927	2580	10·0893073	9·5666109	10·1104206	1028	9·8895794	51
10	9·8004272	1551	10·1995728	9·3515798	9·9109507	2580	10·0890493	9·5663451	10·1105235	1029	9·8894765	50
11	9·8005823	1551	10·1994177	9·3519348	9·9112087	2580	10·0887913	9·5660792	10·1106264	1029	9·8893736	49
12	9·8007372	1549	10·1992628	9·3522897	9·9114666	2579	10·0885334	9·5658132	10·1107294	1030	9·8892706	48
13	9·8008921	1549	10·1991079	9·3526444	9·9117245	2579	10·0882755	9·5655471	10·1108325	1031	9·8891675	47
14	9·8010468	1547	10·1989532	9·3529989	9·9119824	2579	10·0880176	9·5652809	10·1109356	1031	9·8890644	46
15	9·8012015	1547	10·1987985	9·3533533	9·9122403	2578	10·0877597	9·5650146	10·1110388	1032	9·8889612	45
16	9·8013561	1546	10·1986439	9·3537075	9·9124981	2578	10·0875019	9·5647482	10·1111420	1032	9·8888580	44
17	9·8015106	1545	10·1984894	9·3540615	9·9127559	2578	10·0872441	9·5644817	10·1112453	1033	9·8887547	43
18	9·8016649	1543	10·1983351	9·3544154	9·9130137	2578	10·0869863	9·5642151	10·1113487	1034	9·8886513	42
19	9·8018192	1543	10·1981808	9·3547691	9·9132714	2577	10·0867286	9·5639484	10·1114521	1034	9·8885479	41
20	9·8019735	1543	10·1980265	9·3551227	9·9135291	2577	10·0864709	9·5636816	10·1115556	1035	9·8884444	40
21	9·8021276	1541	10·1978724	9·3554761	9·9137868	2577	10·0862132	9·5634147	10·1116592	1035	9·8883408	39
22	9·8022816	1539	10·1977184	9·3558293	9·9140444	2576	10·0859556	9·5631477	10·1117628	1036	9·8882372	38
23	9·8024355	1539	10·1975645	9·3561824	9·9143020	2576	10·0856980	9·5628806	10·1118665	1037	9·8881335	37
24	9·8025894	1537	10·1974106	9·3565353	9·9145596	2576	10·0854404	9·5626134	10·1119702	1037	9·8880298	36
25	9·8027431	1537	10·1972569	9·3568880	9·9148171	2575	10·0851829	9·5623461	10·1120740	1038	9·8879260	35
26	9·8028968	1536	10·1971032	9·3572406	9·9150747	2575	10·0849253	9·5620787	10·1121779	1039	9·8878221	34
27	9·8030504	1534	10·1969496	9·3575930	9·9153322	2575	10·0846678	9·5618112	10·1122818	1039	9·8877182	33
28	9·8032038	1534	10·1967962	9·3579453	9·9155896	2574	10·0844104	9·5615436	10·1123858	1040	9·8876142	32
29	9·8033572	1533	10·1966428	9·3582974	9·9158471	2575	10·0841529	9·5612759	10·1124898	1040	9·8875102	31
30	9·8035105	1532	10·1964895	9·3586494	9·9161045	2574	10·0838955	9·5610080	10·1125939	1041	9·8874061	30
31	9·8036637	1531	10·1963363	9·3590011	9·9163618	2574	10·0836382	9·5607401	10·1126981	1042	9·8873019	29
32	9·8038168	1531	10·1961832	9·3593528	9·9166192	2574	10·0833808	9·5604721	10·1128023	1042	9·8871977	28
33	9·8039699	1529	10·1960301	9·3597042	9·9168765	2573	10·0831235	9·5602040	10·1129066	1043	9·8870934	27
34	9·8041228	1529	10·1958772	9·3600555	9·9171338	2573	10·0828662	9·5599358	10·1130110	1044	9·8869890	26
35	9·8042757	1527	10·1957243	9·3604067	9·9173911	2573	10·0826089	9·5596675	10·1131154	1044	9·8868846	25
36	9·8044284	1527	10·1955716	9·3607576	9·9176483	2572	10·0823517	9·5593991	10·1132199	1045	9·8867801	24
37	9·8045811	1525	10·1954189	9·3611084	9·9179055	2572	10·0820945	9·5591305	10·1133244	1045	9·8866756	23
38	9·8047336	1525	10·1952664	9·3614591	9·9181627	2572	10·0818373	9·5588619	10·1134290	1046	9·8865710	22
39	9·8048861	1524	10·1951139	9·3618096	9·9184198	2571	10·0815802	9·5585932	10·1135337	1047	9·8864663	21
40	9·8050385	1523	10·1949615	9·3621599	9·9186769	2571	10·0813231	9·5583244	10·1136384	1047	9·8863616	20
41	9·8051908	1522	10·1948092	9·3625101	9·9189340	2571	10·0810660	9·5580555	10·1137432	1048	9·8862568	19
42	9·8053430	1521	10·1946570	9·3628601	9·9191911	2570	10·0808089	9·5577864	10·1138481	1049	9·8861519	18
43	9·8054951	1521	10·1945049	9·3632100	9·9194481	2570	10·0805519	9·5575173	10·1139530	1049	9·8860470	17
44	9·8056472	1519	10·1943528	9·3635597	9·9197051	2570	10·0802949	9·5572481	10·1140580	1050	9·8859420	16
45	9·8057991	1519	10·1942009	9·3639092	9·9199621	2570	10·0800379	9·5569787	10·1141630	1051	9·8858370	15
46	9·8059510	1517	10·1940490	9·3642586	9·9202191	2569	10·0797809	9·5567093	10·1142681	1052	9·8857319	14
47	9·8061027	1517	10·1938973	9·3646079	9·9204760	2569	10·0795240	9·5564398	10·1143733	1052	9·8856267	13
48	9·8062544	1516	10·1937456	9·3649569	9·9207329	2569	10·0792671	9·5561701	10·1144785	1053	9·8855215	12
49	9·8064060	1515	10·1935940	9·3653058	9·9209898	2568	10·0790102	9·5559004	10·1145838	1054	9·8854162	11
50	9·8065575	1514	10·1934425	9·3656546	9·9212466	2568	10·0787534	9·5556306	10·1146891	1054	9·8853109	10
51	9·8067089	1513	10·1932911	9·3660032	9·9215034	2568	10·0784966	9·5553606	10·1147945	1055	9·8852055	9
52	9·8068602	1512	10·1931398	9·3663516	9·9217602	2568	10·0782398	9·5550906	10·1149000	1055	9·8851000	8
53	9·8070114	1512	10·1929886	9·3666999	9·9220170	2567	10·0779830	9·5548204	10·1150055	1056	9·8849945	7
54	9·8071626	1510	10·1928374	9·3670480	9·9222737	2567	10·0777263	9·5545502	10·1151111	1057	9·8848889	6
55	9·8073136	1510	10·1926864	9·3673959	9·9225304	2567	10·0774696	9·5542798	10·1152168	1057	9·8847832	5
56	9·8074646	1508	10·1925354	9·3677437	9·9227871	2566	10·0772129	9·5540094	10·1153225	1058	9·8846775	4
57	9·8076154	1508	10·1923846	9·3680914	9·9230437	2567	10·0769563	9·5537388	10·1154283	1058	9·8845717	3
58	9·8077662	1507	10·1922338	9·3684389	9·9233004	2566	10·0766996	9·5534681	10·1155341	1060	9·8844659	2
59	9·8079169	1506	10·1920831	9·3687862	9·9235570	2565	10·0764430	9·5531974	10·1156401	1059	9·8843599	1
60	9·8080675		10·1919325	9·3691334	9·9238135		10·0761865	9·5529265	10·1157460		9·8842540	0

′	Cosine.	Dif.	Secant.	Covers.	Cotang.	Dif.	Tang.	Verseds.	Cosec.	D.	Sine.	′

′	Sine.	Dif.	Covers	Cosec.	Tang.	Cotang.	Secant.	Vers.	Dif.	Cosine	′
0	6427876	2228	3572124	1·5557238	8390996	1·1917536	1·3054073	2339556	1870	7660444	60
1	6430104	2228	3569896	1·5551848	8395955	1·1910498	1·3057261	2341426	1870	7658574	59
2	6432332	2227	3567668	1·5546462	8400915	1·1903465	1·3060451	2343296	1872	7656704	58
3	6434559	2226	3565441	1·5541081	8405878	1·1896437	1·3063644	2345168	1872	7654832	57
4	6436785	2226	3563215	1·5535706	8410844	1·1889414	1·3066839	2347049	1873	7652960	56
5	6439011	2225	3560989	1·5530335	8415812	1·1882395	1·3070038	2348913	1873	7651087	55
6	6441236	2225	3558764	1·5524970	8420782	1·1875382	1·3073239	2350786	1874	7649214	54
7	6443461	2224	3556539	1·5519610	8425755	1·1868373	1·3076442	2352660	1875	7647340	53
8	6445685	2224	3554315	1·5514254	8430730	1·1861369	1·3079649	2354535	1875	7645465	52
9	6447909	2223	3552091	1·5508904	8435708	1·1854370	1·3082858	2356410	1876	7643590	51
10	6450132	2223	3549868	1·5503558	8440688	1·1847376	1·3086069	2358286	1876	7641714	50
11	6452355	2222	3547645	1·5498218	8445670	1·1840387	1·3089284	2360162	1878	7639838	49
12	6454577	2221	3545423	1·5492882	8450655	1·1833402	1·3092501	2362040	1878	7637960	48
13	6456798	2221	3543202	1·5487552	8455643	1·1826422	1·3095720	2363918	1878	7636082	47
14	6459019	2221	3540981	1·5482226	8460633	1·1819447	1·3098943	2365796	1879	7634204	46
15	6461240	2220	3538760	1·5476906	8465625	1·1812477	1·3102168	2367675	1880	7632325	45
16	6463460	2219	3536540	1·5471590	8470620	1·1805512	1·3105396	2369555	1881	7630445	44
17	6465679	2219	3534321	1·5466280	8475617	1·1798551	1·3108626	2371436	1881	7628564	43
18	6467898	2218	3532102	1·5460974	8480617	1·1791595	1·3111860	2373317	1881	7626683	42
19	6470116	2218	3529884	1·5455673	8485619	1·1784644	1·3115095	2375198	1883	7624802	41
20	6472334	2217	3527666	1·5450378	8490624	1·1777698	1·3118334	2377081	1883	7622919	40
21	6474551	2216	3525449	1·5445087	8495631	1·1770756	1·3121575	2378964	1884	7621036	39
22	6476767	2217	3523233	1·5439801	8500640	1·1763820	1·3124820	2380848	1884	7619152	38
23	6478984	2215	3521016	1·5434520	8505653	1·1756888	1·3128066	2382732	1895	7617268	37
24	6481199	2215	3518801	1·5429244	8510667	1·1749960	1·3131316	2384617	1886	7615383	36
25	6483414	2214	3516586	1·5423973	8515684	1·1743038	1·3134568	2386503	1886	7613497	35
26	6485628	2214	3514372	1·5418706	8520704	1·1736120	1·3137823	2388389	1887	7611611	34
27	6487842	2214	3512158	1·5413445	8525726	1·1729207	1·3141081	2390276	1887	7609724	33
28	6490056	2212	3509944	1·5408189	8530750	1·1722298	1·3144341	2392163	1888	7607837	32
29	6492268	2212	3507732	1·5402937	8533777	1·1715395	1·3147604	2394051	1889	7605949	31
30	6494480	2212	3505520	1·5397690	8540807	1·1708496	1·3150870	2395940	1890	7604060	30
31	6496692	2211	3503308	1·5392449	8545839	1·1701601	1·3154139	2397830	1890	7602170	29
32	6498903	2211	3501097	1·5387212	8550873	1·1694712	1·3157410	2399720	1891	7600280	28
33	6501114	2210	3498886	1·5381980	8555910	1·1687827	1·3160684	2401611	1891	7598389	27
34	6503324	2209	3496676	1·5376752	8560950	1·1680947	1·3163961	2403502	1892	7596498	26
35	6505533	2209	3494467	1·5371530	8565992	1·1674071	1·3167240	2405394	1893	7594605	25
36	6507742	2209	3492258	1·5366313	8571037	1·1667200	1·3170523	2407287	1893	7592713	24
37	6509951	2207	3490049	1·5361100	8576084	1·1660334	1·3173808	2409180	1894	7590820	23
38	6512158	2208	3487842	1·5355892	8581133	1·1653472	1·3177096	2411074	1894	7588926	22
39	6514366	2206	3485634	1·5350689	8586185	1·1646615	1·3180386	2412969	1895	7587031	21
40	6516572	2206	3483428	1·5345491	8591240	1·1639763	1·3183680	2414864	1896	7585136	20
41	6518778	2206	3481222	1·5340297	8596297	1·1632916	1·3186976	2416760	1897	7583240	19
42	6520984	2205	3479016	1·5335109	8601357	1·1626073	1·3190274	2418657	1897	7581343	18
43	6523189	2205	3476811	1·5329925	8606419	1·1619234	1·3193576	2420554	1898	7579446	17
44	6525394	2204	3474606	1·5324746	8611484	1·1612400	1·3196881	2422452	1898	7577548	16
45	6527598	2203	3472402	1·5319572	8616551	1·1605571	1·3200188	2424350	1899	7575650	15
46	6529801	2203	3470199	1·5314403	8621621	1·1598747	1·3203498	2426249	1899	7573751	14
47	6532004	2202	3467996	1·5309238	8626694	1·1591927	1·3206810	2428149	1900	7571851	13
48	6534206	2202	3465794	1·5304078	8631768	1·1585112	1·3210126	2430049	1900	7569951	12
49	6536408	2201	3463592	1·5298923	8636846	1·1578301	1·3213444	2431950	1901	7568050	11
50	6538609	2201	3461391	1·5293773	8641926	1·1571495	1·3216765	2433852	1902	7566148	10
51	6540810	2200	3459190	1·5288627	8647009	1·1564693	1·3220089	2435754	1902	7564246	9
52	6543010	2199	3456990	1·5283487	8652094	1·1557896	1·3223416	2437657	1903	7562343	8
53	6545209	2199	3454791	1·5278351	8657181	1·1551104	1·3226745	2439561	1904	7560439	7
54	6547408	2199	3452592	1·5273219	8662272	1·1544316	1·3230078	2441465	1904	7558535	6
55	6549607	2197	3450393	1·5268093	8667365	1·1537532	1·3233413	2443370	1905	7556630	5
56	6551804	2198	3448196	1·5262971	8672460	1·1530754	1·3236750	2445276	1906	7554724	4
57	6554002	2196	3445998	1·5257854	8677558	1·1523979	1·3240091	2447182	1906	7552818	3
58	6556198	2197	3443802	1·5252741	8682659	1·1517210	1·3243435	2449089	1907	7550911	2
59	6558395	2195	3441605	1·5247634	8687762	1·1510445	1·3246781	2450996	1907	7549004	1
60	6560590		3439410	1·5242531	8692867	1·1503684	1·3250130	2452904	1908	7547096	0

′	Cosine	Dif.	Vers.	Secant.	Cotan.	Taug.	Cosec.	Covers	Dif.	Sine.	′

′	Sine.	Dif.	Cosec.	Verseds.	Tang.	Dif.	Cotang.	Covers.	Secant.	D.	Cosine.	′
0	9·8080675	1505	10·1919325	9·3691334	9·9238135	2566	10·0761865	9·5529265	10·1157460	1061	9·8842540	60
1	9·8082180	1504	10·1917820	9·3694804	9·9240701	2565	10·0759299	9·5526555	10·1158521	1061	9·8841479	59
2	9·8083684	1504	10·1916316	9·3698272	9·9243266	2565	10·0756734	9·5523845	10·1159582	1061	9·8840418	58
3	9·8085188	1502	10·1914812	9·3701739	9·9245831	2565	10·0754169	9·5521133	10·1160643	1063	9·8839357	57
4	9·8086690	1502	10·1913310	9·3705205	9·9248396	2564	10·0751604	9·5518420	10·1161706	1062	9·8838294	56
5	9·8088192	1500	10·1911808	9·3708669	9·9250960	2564	10·0749040	9·5515706	10·1162768	1064	9·8837232	55
6	9·8089692	1500	10·1910308	9·3712131	9·9253524	2564	10·0746476	9·5512992	10·1163832	1064	9·8836168	54
7	9·8091192	1499	10·1908808	9·3715592	9·9256088	2564	10·0743912	9·5510276	10·1164896	1065	9·8835104	53
8	9·8092691	1498	10·1907309	9·3719051	9·9258652	2563	10·0741348	9·5507559	10·1165961	1065	9·8834039	52
9	9·8094189	1497	10·1905811	9·3722508	9·9261215	2563	10·0738785	9·5504841	10·1167026	1066	9·8832974	51
10	9·8095686	1496	10·1904314	9·3725965	9·9263778	2563	10·0736222	9·5502122	10·1168092	1067	9·8831908	50
11	9·8097182	1496	10·1902818	9·3729419	9·9266341	2563	10·0733659	9·5499402	10·1169159	1067	9·8830841	49
12	9·8098678	1494	10·1901322	9·3732872	9·9268904	2562	10·0731096	9·5496681	10·1170226	1068	9·8829774	48
13	9·8100172	1494	10·1899828	9·3736323	9·9271466	2562	10·0728534	9·5493959	10·1171294	1068	9·8828706	47
14	9·8101666	1493	10·1898334	9·3739773	9·9274028	2562	10·0725972	9·5491236	10·1172362	1070	9·8827638	46
15	9·8103159	1491	10·1896841	9·3743221	9·9276590	2562	10·0723410	9·5488511	10·1173432	1069	9·8826568	45
16	9·8104650	1491	10·1895350	9·3746668	9·9279152	2561	10·0720848	9·5485786	10·1174501	1071	9·8825499	44
17	9·8106141	1490	10·1893859	9·3750113	9·9281713	2561	10·0718287	9·5483060	10·1175572	1071	9·8824428	43
18	9·8107631	1490	10·1892369	9·3753557	9·9284274	2561	10·0715726	9·5480333	10·1176643	1072	9·8823357	42
19	9·8109121	1488	10·1890879	9·3756999	9·9286835	2561	10·0713165	9·5477604	10·1177715	1072	9·8822285	41
20	9·8110609	1487	10·1889391	9·3760440	9·9289396	2560	10·0710604	9·5474875	10·1178787	1073	9·8821213	40
21	9·8112096	1487	10·1887904	9·3763879	9·9291956	2560	10·0708044	9·5472145	10·1179860	1073	9·8820140	39
22	9·8113583	1486	10·1886417	9·3767316	9·9294516	2560	10·0705484	9·5469413	10·1180933	1075	9·8819067	38
23	9·8115069	1485	10·1884931	9·3770752	9·9297076	2560	10·0702924	9·5466681	10·1182008	1074	9·8817992	37
24	9·8116554	1484	10·1883446	9·3774186	9·9299636	2559	10·0700364	9·5463947	10·1183082	1076	9·8816918	36
25	9·8118038	1483	10·1881962	9·3777619	9·9302195	2560	10·0697805	9·5461212	10·1184158	1076	9·8815842	35
26	9·8119521	1482	10·1880479	9·3781050	9·9304755	2559	10·0695245	9·5458477	10·1185234	1077	9·8814766	34
27	9·8121003	1481	10·1878997	9·3784480	9·9307314	2558	10·0692686	9·5455740	10·1186311	1077	9·8813689	33
28	9·8122484	1481	10·1877516	9·3787908	9·9309872	2559	10·0690128	9·5453002	10·1187388	1078	9·8812612	32
29	9·8123965	1479	10·1876035	9·3791335	9·9312431	2558	10·0687569	9·5450264	10·1188466	1079	9·8811534	31
30	9·8125444	1479	10·1874556	9·3794760	9·9314989	2558	10·0685011	9·5447524	10·1189545	1079	9·8810455	30
31	9·8126923	1478	10·1873077	9·3798184	9·9317547	2558	10·0682453	9·5444783	10·1190624	1080	9·8809376	29
32	9·8128401	1477	10·1871599	9·3801606	9·9320105	2557	10·0679895	9·5442041	10·1191704	1081	9·8808296	28
33	9·8129878	1476	10·1870122	9·3805026	9·9322662	2558	10·0677338	9·5439298	10·1192785	1081	9·8807215	27
34	9·8131354	1475	10·1868646	9·3808445	9·9325220	2557	10·0674780	9·5436554	10·1193866	1082	9·8806134	26
35	9·8132829	1474	10·1867171	9·3811863	9·9327777	2557	10·0672223	9·5433809	10·1194948	1082	9·8805052	25
36	9·8134303	1474	10·1865697	9·3815279	9·9330334	2556	10·0669666	9·5431063	10·1196030	1083	9·8803970	24
37	9·8135777	1473	10·1864223	9·3818693	9·9332890	2556	10·0667110	9·5428316	10·1197113	1084	9·8802887	23
38	9·8137250	1473	10·1862750	9·3822106	9·9335446	2557	10·0664554	9·5425568	10·1198197	1084	9·8801803	22
39	9·8138721	1471	10·1861279	9·3825517	9·9338003	2556	10·0661997	9·5422818	10·1199281	1085	9·8800719	21
40	9·8140192	1471	10·1859808	9·3828927	9·9340559	2555	10·0659441	9·5420068	10·1200366	1086	9·8799634	20
41	9·8141662	1470	10·1858338	9·3832335	9·9343114	2556	10·0656886	9·5417317	10·1201452	1086	9·8798548	19
42	9·8143131	1469	10·1856869	9·3835742	9·9345670	2555	10·0654330	9·5414564	10·1202538	1087	9·8797462	18
43	9·8144600	1469	10·1855400	9·3839147	9·9348225	2555	10·0651775	9·5411811	10·1203625	1088	9·8796375	17
44	9·8146067	1467	10·1853933	9·3842551	9·9350780	2555	10·0649220	9·5409056	10·1204713	1088	9·8795287	16
45	9·8147534	1467	10·1852466	9·3845953	9·9353335	2554	10·0646665	9·5406301	10·1205801	1089	9·8794199	15
46	9·8148999	1465	10·1851001	9·3849354	9·9355889	2555	10·0644111	9·5403544	10·1206890	1089	9·8793110	14
47	9·8150464	1465	10·1849536	9·3852753	9·9358444	2554	10·0641556	9·5400786	10·1207979	1091	9·8792021	13
48	9·8151928	1464	10·1848072	9·3856151	9·9360998	2554	10·0639002	9·5398027	10·1209070	1090	9·8790930	12
49	9·8153391	1463	10·1846609	9·3859547	9·9363552	2553	10·0636448	9·5395268	10·1210160	1092	9·8789840	11
50	9·8154854	1463	10·1845146	9·3862942	9·9366105	2554	10·0633895	9·5392507	10·1211252	1092	9·8788748	10
51	9·8156315	1461	10·1843685	9·3866335	9·9368659	2553	10·0631341	9·5389745	10·1212344	1093	9·8787656	9
52	9·8157776	1461	10·1842224	9·3869727	9·9371212	2553	10·0628788	9·5386982	10·1213437	1093	9·8786563	8
53	9·8159235	1459	10·1840765	9·3873117	9·9373765	2553	10·0626235	9·5384218	10·1214503	1094	9·8785470	7
54	9·8160694	1459	10·1839306	9·3876506	9·9376318	2553	10·0623682	9·5381452	10·1215624	1095	9·8784376	6
55	9·8162152	1458	10·1837848	9·3879893	9·9378871	2552	10·0621129	9·5378686	10·1216719	1095	9·8783281	5
56	9·8163609	1457	10·1836391	9·3883278	9·9381423	2552	10·0618577	9·5375919	10·1217814	1096	9·8782186	4
57	9·8165066	1457	10·1834934	9·3886662	9·9383975	2552	10·0616025	9·5373151	10·1218910	1096	9·8781090	3
58	9·8166521	1455	10·1833479	9·3890045	9·9386527	2552	10·0613473	9·5370381	10·1220006	1098	9·8779994	2
59	9·8167975	1454	10·1832025	9·3893426	9·9389079	2552	10·0610921	9·5367611	10·1221104	1097	9·8778896	1
60	9·8169429	1454	10·1830571	9·3896806	9·9391631		10·0608369	9·5364839	10·1222201		9·8777799	0

| ′ | Cosine. | Dif. | Secant. | Covers. | Cotang. | Dif. | Tang. | Verseds. | Cosec. | D. | Sine. | ′ |

′	Sine.	Dif.	Covers	Cosec.	Tang.	Cotang.	Secant.	Vers.	Dif.	Cosine	′
0	6560590	2195	3439410	1·5242531	8692867	1·1503684	1·3250130	2452904	1909	7547096	60
1	6562785	2195	3437215	1·5237433	8697976	1·1496928	1·3253482	2454813	1909	7545187	59
2	6564980	2194	3435020	1·5232339	8703087	1·1490176	1·3256837	2456722	1910	7543278	58
3	6567174	2193	3432826	1·5227250	8708200	1·1483429	1·3260194	2458632	1911	7541368	57
4	6569367	2193	3430633	1·5222166	8713316	1·1476687	1·3263554	2460543	1911	7539457	56
5	6571560	2192	3428440	1·5217087	8718435	1·1469949	1·3266918	2462454	1912	7537546	55
6	6573752	2192	3426248	1·5212012	8723556	1·1463215	1·3270284	2464366	1913	7535634	54
7	6575944	2191	3424056	1·5206942	8728680	1·1456486	1·3273653	2466279	1913	7533721	53
8	6578135	2191	3421865	1·5201876	8733806	1·1449762	1·3277024	2468192	1914	7531808	52
9	6580326	2190	3419674	1·5196815	8738935	1·1443041	1·3280399	2470106	1914	7529894	51
10	6582516	2190	3417484	1·5191759	8744067	1·1436326	1·3283776	2472020	1915	7527980	50
11	6584706	2189	3415294	1·5186708	8749201	1·1429615	1·3287156	2473935	1916	7526065	49
12	6586895	2188	3413105	1·5181661	8754338	1·1422908	1·3290539	2475851	1916	7524149	48
13	6589083	2188	3410917	1·5176619	8759478	1.1416206	1·3293925	2477767	1917	7522233	47
14	6591271	2187	3408729	1·5171581	8764620	1·1409508	1·3297314	2479684	1918	7520316	46
15	6593458	2187	3406542	1·5166548	8769765	1·1402815	1·3300706	2481602	1918	7518398	45
16	6595645	2186	3404355	1·5161520	8774912	1·1396126	1·3304100	2483520	1919	7516480	44
17	6597831	2186	3402169	1·5156496	8780062	1·1389441	1·3307497	2485439	1920	7514551	43
18	6600017	2185	3399983	1·5151477	8785215	1·1382761	1·3310897	2487359	1920	7512641	42
19	6602202	2184	3397798	1·5146462	8790370	1·1376086	1·3314301	2489279	1921	7510721	41
20	6604386	2184	3395614	1·5141452	8795528	1·1369414	1·3317707	2491200	1921	7508800	40
21	6606570	2184	3393430	1·5136447	8800688	1·1362747	1·3321115	2493121	1922	7506879	39
22	6608754	2182	3391246	1·5131446	8805852	1·1356085	1·3324527	2495043	1923	7504957	38
23	6610936	2183	3389064	1·5126450	8811017	1·1349427	1·3327942	2496966	1923	7503034	37
24	6613119	2181	3386881	1·5121459	8816186	1·1342773	1·3331359	2498889	1924	7501111	36
25	6615300	2182	3384700	1·5116472	8821357	1·1336124	1·3334779	2500813	1925	7499187	35
26	6617482	2180	3382518	1·5111489	8826531	1·1329479	1·3338203	2502738	1925	7497262	34
27	6619662	2180	3380338	1·5106511	8831707	1·1322839	1·3341629	2504663	1926	7495337	33
28	6621842	2180	3378158	1·5101538	8836886	1·1316203	1·3345058	2506589	1927	7493411	32
29	6624022	2178	3375978	1·5096569	8842068	1·1309571	1·3348489	2508516	1927	7491484	31
30	6626200	2179	3373800	1·5091605	8847253	1·1302944	1·3351924	2510443	1928	7489557	30
31	6628379	2178	3371621	1·5086645	8852440	1·1296321	1·3355362	2512371	1928	7487629	29
32	6630557	2177	3369443	1·5081690	8857630	1·1289702	1·3358802	2514299	1929	7485701	28
33	6632734	2176	3367266	1·5076739	8862822	1·1283088	1·3362246	2516228	1929	7483772	27
34	6634910	2177	3365090	1·5071793	8868017	1·1276478	1·3365692	2518158	1930	7481842	26
35	6637087	2175	3362913	1·5066852	8873215	1·1269872	1·3369141	2520088	1931	7479912	25
36	6639262	2175	3360738	1·5061915	8878415	1·1263271	1·3372594	2522019	1932	7477981	24
37	6641437	2175	3358563	1·5056982	8883619	1·1256674	1·3376049	2523951	1932	7476049	23
38	6643612	2173	3356388	1·5052054	8888825	1·1250081	1·3379507	2525883	1933	7474117	22
39	6645785	2174	3354215	1·5047131	8894033	1·1243493	1·3382968	2527816	1933	7472184	21
40	6647959	2172	3352041	1·5042211	8899244	1·1236909	1·3386432	2529749	1934	7470251	20
41	6650131	2173	3349869	1·5037297	8904458	1·1230329	1·3389898	2531683	1935	7468317	19
42	6652304	2171	3347696	1·5032387	8909675	1·1223754	1·3393368	2533618	1936	7466382	18
43	6654475	2171	3345525	1·5027481	8914894	1·1217183	1·3396841	2535554	1936	7464446	17
44	6656646	2171	3343354	1·5022580	8920116	1·1210616	1·3400316	2537490	1936	7462510	16
45	6658817	2170	3341183	1·5017683	8925341	1·1204053	1·3403795	2539426	1938	7460574	15
46	6660987	2169	3339013	1·5012791	8930569	1·1197495	1·3407276	2541364	1937	7458636	14
47	6663156	2169	3336844	1·5007903	8935799	1·1190941	1·3410761	2543301	1939	7456699	13
48	6665325	2168	3334675	1·5003020	8941032	1·1184391	1·3414248	2545240	1939	7454760	12
49	6667493	2168	3332507	1·4998141	8946268	1·1177846	1·3417738	2547179	1940	7452821	11
50	6669661	2167	3330339	1·4993267	8951506	1·1171305	1·3421232	2549119	1940	7450881	10
51	6671828	2166	3328172	1·4988397	8956747	1·1164768	1·3424728	2551059	1942	7448941	9
52	6673994	2166	3326006	1·4983531	8961991	1·1158235	1·3428227	2553001	1941	7446999	8
53	6676160	2166	3323840	1·4978670	8967238	1·1151706	1·3431729	2554942	1942	7445058	7
54	6678326	2164	3321674	1·4973813	8972487	1·1145182	1·3435234	2556885	1942	7443115	6
55	6680490	2165	3319510	1·4968961	8977739	1·1138662	1·3438742	2558827	1944	7441173	5
56	6682655	2163	3317345	1·4964113	8982994	1·1132146	1·3442253	2560771	1944	7439229	4
57	6684818	2163	3315182	1·4959270	8988251	1·1125635	1·3445767	2562715	1945	7437285	3
58	6686981	2163	3313019	1·4954431	8993512	1·1119127	1·3449284	2564660	1946	7435340	2
59	6689144	2162	3310856	1·4949596	8998775	1·1112624	1·3452804	2566606	1946	7433394	1
60	6691306		3308694	1·4944765	9004040	1·1106125	1·3456327	2568552		7431448	0
′	Cosine	Dif.	Vers.	Secant.	Cotan.	Tang.	Cosec.	Covers	Dif.	Sine.	′

48 Deg.

′	Sine.	Dif.	Cosec.	Versed.	Tang.	Dif.	Cotang.	Covers.	Secant.	D.	Cosine.	′
0	9·8169429	1453	10·1830571	9·3896806	9·9391631	2551	10·0608369	9·5364839	10·1222201	1099	9·8777799	60
1	9·8170882	1452	10·1829118	9·3900184	9·9394182	2551	10·0605818	9·5362067	10·1223300	1099	9·8776700	59
2	9·8172334	1451	10·1827666	9·3903561	9·9396733	2551	10·0603267	9·5359293	10·1224399	1100	9·8775601	58
3	9·8173785	1450	10·1826215	9·3906936	9·9399284	2551	10·0600716	9·5356518	10·1225499	1100	9·8774501	57
4	9·8175235	1450	10·1824765	9·3910309	9·9401835	2550	10·0598165	9·5353742	10·1226599	1101	9·8773401	56
5	9·8176685	1448	10·1823315	9·3913682	9·9404385	2551	10·0595615	9·5350965	10·1227700	1102	9·8772300	55
6	9·8178133	1448	10·1821867	9·3917052	9·9406936	2550	10·0593064	9·5348187	10·1228802	1102	9·8771198	54
7	9·8179581	1447	10·1820419	9·3920421	9·9409486	2550	10·0590514	9·5345408	10·1229904	1103	9·8770096	53
8	9·8181028	1446	10·1818972	9·3923789	9·9412036	2549	10·0587964	9·5342628	10·1231007	1104	9·8768993	52
9	9·8182474	1445	10·1817526	9·3927155	9·9414585	2550	10·0585475	9·5339847	10·1232111	1104	9·8767889	51
10	9·8183919	1445	10·1816081	9·3930520	9·9417135	2549	10·0582865	9·5337065	10·1233215	1105	9·8766785	50
11	9·8185364	1443	10·1814636	9·3933883	9·9419684	2549	10·0580316	9·5334281	10·1234320	1106	9·8765680	49
12	9·8186807	1443	10·1813193	9·3937245	9·9422233	2549	10·0577767	9·5331497	10·1235426	1106	9·8764574	48
13	9·8188250	1442	10·1811750	9·3940605	9·9424782	2549	10·0575218	9·5328712	10·1236532	1107	9·8763468	47
14	9·8189692	1441	10·1810308	9·3943964	9·9427331	2548	10·0572669	9·5325925	10·1237639	1107	9·8762361	46
15	9·8191133	1440	10·1808867	9·3947321	9·9429879	2549	10·0570121	9·5323137	10·1238747	1108	9·8761253	45
16	9·8192573	1439	10·1807427	9·3950677	9·9432428	2548	10·0567572	9·5320349	10·1239855	1109	9·8760145	44
17	9·8194012	1438	10·1805988	9·3954031	9·9434976	2548	10·0565024	9·5317559	10·1240964	1109	9·8759036	43
18	9·8195450	1438	10·1804550	9·3957384	9·9437524	2548	10·0562476	9·5314768	10·1242073	1111	9·8757927	42
19	9·8196888	1437	10·1803112	9·3960735	9·9440072	2547	10·0559928	9·5311976	10·1243184	1110	9·8756816	41
20	9·8198325	1436	10·1801675	9·3964085	9·9442619	2547	10·0557381	9·5309183	10·1244294	1112	9·8755706	40
21	9·8199761	1435	10·1800239	9·3967434	9·9445166	2548	10·0554834	9·5306389	10·1245406	1112	9·8754594	39
22	9·8201196	1434	10·1798804	9·3970781	9·9447714	2547	10·0552286	9·5303594	10·1246518	1113	9·8753482	38
23	9·8202630	1433	10·1797370	9·3974126	9·9450261	2546	10·0549739	9·5300797	10·1247631	1113	9·8752369	37
24	9·8204063	1433	10·1795937	9·3977470	9·9452807	2547	10·0547193	9·5298000	10·1248744	1114	9·8751256	36
25	9·8205496	1431	10·1794504	9·3980813	9·9455354	2546	10·0544646	9·5295201	10·1249858	1115	9·8750142	35
26	9·8206927	1431	10·1793073	9·3984154	9·9457900	2547	10·0542100	9·5292402	10·1250973	1115	9·8749027	34
27	9·8208358	1430	10·1791642	9·3987493	9·9460447	2546	10·0539553	9·5289601	10·1252088	1117	9·8747912	33
28	9·8209788	1429	10·1790212	9·3990831	9·9462993	2546	10·0537007	9·5286799	10·1253205	1116	9·8746795	32
29	9·8211217	1429	10·1788783	9·3994168	9·9465539	2545	10·0534461	9·5283997	10·1254321	1118	9·8745679	31
30	9·8212646	1427	10·1787354	9·3997503	9·9468084	2546	10·0531916	9·5281193	10·1255439	1118	9·8744561	30
31	9·8214073	1427	10·1785927	9·4000837	9·9470630	2545	10·0529370	9·5278388	10·1256557	1118	9·8743443	29
32	9·8215500	1426	10·1784500	9·4004169	9·9473175	2545	10·0526825	9·5275582	10·1257675	1120	9·8742325	28
33	9·8216926	1425	10·1783074	9·4007500	9·9475720	2545	10·0524280	9·5272774	10·1258795	1120	9·8741205	27
34	9·8218351	1424	10·1781649	9·4010829	9·9478265	2545	10·0521735	9·5269966	10·1259915	1120	9·8740085	26
35	9·8219775	1423	10·1780225	9·4014157	9·9480810	2545	10·0519190	9·5267157	10·1261035	1121	9·8738965	25
36	9·8221198	1423	10·1778802	9·4017484	9·9483355	2544	10·0516645	9·5264346	10·1262156	1122	9·8737844	24
37	9·8222621	1421	10·1777379	9·4020809	9·9485899	2544	10·0514101	9·5261535	10·1263278	1123	9·8736722	23
38	9·8224042	1421	10·1775958	9·4024132	9·9488443	2544	10·0511557	9·5258722	10·1264401	1123	9·8735599	22
39	9·8225463	1420	10·1774537	9·4027454	9·9490987	2544	10·0509013	9·5255908	10·1265524	1124	9·8734476	21
40	3·8226883	1419	10·1773117	9·4030775	9·9493531	2544	10·0506469	9·5253094	10·1266648	1125	9·8733352	20
41	9·8228302	1419	10·1771698	9·4034094	9·9496075	2544	10·0503925	9·5250278	10·1267773	1125	9·8732227	19
42	9·8229721	1417	10·1770279	9·4037412	9·9498619	2543	10·0501381	9·5247461	10·1268898	1126	9·8731102	18
43	9·8231138	1417	10·1768862	9·4040728	9·9501162	2543	10·0498838	9·5244643	10·1270024	1127	9·8729976	17
44	9·8232555	1416	10·1767445	9·4044043	9·9503705	2543	10·0496295	9·5241823	10·1271151	1127	9·8728849	16
45	9·8233971	1415	10·1766029	9·4047356	9·9506248	2543	10·0493752	9·5239003	10·1272278	1128	9·8727722	15
46	9·8235386	1414	10·1764614	9·4050668	9·9508791	2543	10·0491209	9·5236182	10·1273406	1128	9·8726594	14
47	9·8236800	1413	10·1763200	9·4053978	9·9511334	2542	10·0488666	9·5233359	10·1274534	1129	9·8725466	13
48	9·8238213	1413	10·1761787	9·4057287	9·9513876	2543	10·0486124	9·5230536	10·1275663	1130	9·8724337	12
49	9·8239626	1411	10·1760374	9·4060595	9·9516419	2542	10·0483581	9·5227711	10·1276793	1131	9·8723207	11
50	9·8241037	1411	10·1758963	9·4063901	9·9518961	2542	10·0481039	9·5224885	10·1277924	1131	9·8722076	10
51	9·8242448	1410	10·1757552	9·4067206	9·9521503	2542	10·0478497	9·5222058	10·1279055	1132	9·8720945	9
52	9·8243858	1409	10·1756142	9·4070509	9·9524045	2542	10·0475955	9·5219230	10·1280187	1132	9·8719813	8
53	9·8245267	1409	10·1754733	9·4073811	9·9526587	2541	10·0473413	9·5216401	10·1281319	1133	9·8718681	7
54	9·8246676	1407	10·1753324	9·4077111	9·9529128	2542	10·0470872	9·5213571	10·1282452	1134	9·8717548	6
55	9·8248083	1407	10·1751917	9·4080410	9·9531670	2541	10·0468330	9·5210739	10·1283586	1135	9·8716414	5
56	9·8249490	1406	10·1750510	9·4083708	9·9534211	2541	10·0465789	9·5207907	10·1284721	1135	9·8715279	4
57	9·8250896	1405	10·1749104	9·4087004	9·9536752	2541	10·0463248	9·5205073	10·1285856	1136	9·8714144	3
58	9·8252301	1404	10·1747699	9·4090298	9·9539293	2541	10·0460707	9·5202239	10·1286992	1136	9·8713008	2
59	9·8253705	1404	10·1746295	9·4093591	9·9541834	2540	10·0458166	9·5199403	10·1288128	1137	9·8711872	1
60	9·8255109		10·1744891	9·4096883	9·9544374		10·0455626	9·5196566	10·1289265		9·8710735	0

′	Cosine.	Dif.	Secant.	Covers.	Cotang.	Dif.	Tang.	Versed.	Cosec.	D.	Sine.	′

′	Sine.	Dif.	Covers	Cosec.	Tang.	Cotang.	Secant.	Vers.	Dif.	Cosine	′
0	6691306	2162	3308694	1·4944765	9004040	1·1106125	1·3456327	2568552	1946	7431448	60
1	6693468	2160	3306532	1·4939940	9009309	1·1099630	1·3459853	2570498	1948	7429502	59
2	6695628	2161	3304372	1·4935118	9014580	1·1093140	1·3463382	2572446	1948	7427554	58
3	6697789	2159	3302211	1·4930301	9019854	1·1086653	1·3466914	2574394	1948	7425606	57
4	6699948	2160	3300052	1·4925488	9025131	1·1080171	1·3470449	2576342	1950	7423658	56
5	6702108	2158	3297892	1·4920680	9030411	1·1073693	1·3473987	2578292	1950	7421708	55
6	6704266	2158	3295734	1·4915876	9035693	1·1067219	1·3477528	2580242	1950	7419758	54
7	6706424	2158	3293576	1·4911076	9040979	1·1060750	1·3481072	2582192	1951	7417808	53
8	6708582	2157	3291418	1·4906280	9046267	1·1054284	1·3484619	2584143	1952	7415857	52
9	6710739	2156	3289261	1·4901489	9051557	1·1047823	1·3488168	2586095	1952	7413905	51
10	6712895	2156	3287105	1·4896703	9056851	1·1041365	1·3491721	2588047	1953	7411953	50
11	6715051	2155	3284949	1·4891920	9062147	1·1034912	1·3495277	2590000	1954	7410000	49
12	6717206	2155	3282794	1·4887142	9067446	1·1028463	1·3498836	2591954	1954	7408046	48
13	6719361	2154	3280639	1·4882369	9072748	1·1022019	1·3502398	2593908	1955	7406092	47
14	6721515	2153	3278485	1·4877599	9078053	1·1015578	1·3505963	2595863	1956	7404137	46
15	6723668	2153	3276332	1·4872834	9083360	1·1009141	1·3509531	2597819	1956	7402181	45
16	6725821	2152	3274179	1·4868073	9088671	1·1002709	1·3513102	2599775	1957	7400225	44
17	6727973	2152	3272027	1·4863317	9093984	1·0996281	1·3516677	2601732	1957	7398268	43
18	6730125	2151	3269875	1·4858565	9099300	1·0989857	1·3520254	2603689	1958	7396311	42
19	6732276	2151	3267724	1·4853817	9104619	1·0983436	1·3523834	2605647	1959	7394353	41
20	6734427	2150	3265573	1·4849073	9109940	1·0977020	1·3527417	2607606	1959	7392394	40
21	6736577	2150	3263423	1·4844334	9115265	1·0970609	1·3531003	2609565	1960	7390435	39
22	6738727	2149	3261273	1·4839599	9120592	1·0964201	1·3534593	2611525	1960	7388475	38
23	6740876	2148	3259124	1·4834868	9125922	1·0957797	1·3538185	2613485	1962	7386515	37
24	6743024	2148	3256976	1·4830142	9131255	1·0951397	1·3541780	2615447	1961	7384553	36
25	6745172	2147	3254828	1·4825420	9136591	1·0945002	1·3545379	2617408	1963	7382592	35
26	6747319	2147	3252681	1·4820702	9141929	1·0938610	1·3548980	2619371	1963	7380629	34
27	6749466	2146	3250534	1·4815988	9147270	1·0932223	1·3552585	2621334	1963	7378666	33
28	6751612	2145	3248388	1·4811278	9152615	1·0925840	1·3556193	2623297	1965	7376703	32
29	6753757	2145	3246243	1·4806573	9157962	1·0919460	1·3559803	2625262	1965	7374738	31
30	6755902	2144	3244098	1·4801872	9163312	1·0913085	1·3563417	2627227	1965	7372773	30
31	6758046	2144	3241954	1·4797176	9168665	1·0906714	1·3567034	2629192	1966	7370808	29
32	6760190	2143	3239810	1·4792483	9174020	1·0900347	1·3570654	2631158	1967	7368842	28
33	6762333	2143	3237667	1·4787795	9179379	1·0893984	1·3574277	2633125	1967	7366875	27
34	6764476	2142	3235524	1·4788111	9184740	1·0887624	1·3577903	2635092	1968	7364908	26
35	6766618	2142	3233382	1·4778431	9190104	1·0881269	1·3581532	2637060	1969	7362940	25
36	6768760	2141	3231240	1·4773755	9195471	1·0874918	1·3585164	2639029	1969	7360971	24
37	6770901	2140	3229099	1·4769084	9200841	1·0868571	1·3588800	2640998	1970	7359002	23
38	6773041	2140	3226959	1·4764417	9206214	1·0862228	1·3592438	2642968	1971	7357032	22
39	6775181	2139	3224819	1·4759754	9211590	1·0855889	1·3596080	2644939	1971	7355061	21
40	6777320	2139	3222680	1·4755095	9216969	1·0849554	1·3599725	2646910	1972	7353090	20
41	6779459	2138	3220541	1·4750440	9222350	1·0843223	1·3603372	2648882	1972	7351118	19
42	6781597	2137	3218406	1·4745790	9227734	1·0836896	1·3607023	2650854	1973	7349146	18
43	6783734	2137	3216266	1·4741144	9233122	1·0830573	1·3610677	2652827	1974	7347173	17
44	6785871	2136	3214129	1·4736502	9238512	1·0824254	1·3614334	2654801	1974	7345199	16
45	6788007	2136	3211993	1·4731864	9243905	1·0817939	1·3617995	2656775	1975	7343225	15
46	6790143	2135	3209857	1·4727230	9249301	1·0811628	1·3621658	2658750	1975	7341250	14
47	6792278	2135	3207722	1·4722600	9254700	1·0805321	1·3625324	2660725	1976	7339275	13
48	6794413	2134	3205587	1·4717975	9260102	1·0799018	1·3628994	2662701	1977	7337299	12
49	6796547	2134	3203453	1·4713354	9265506	1·0792718	1·3632667	2664678	1977	7335322	11
50	6798681	2132	3201319	1·4708736	9270914	1·0786423	1·3636343	2666655	1978	7333345	10
51	6800813	2132	3199187	1·4704123	9276324	1·0780132	1·3640022	2668633	1979	7331367	9
52	6802946	2132	3197054	1·4699514	9281738	1·0773845	1·3643704	2670612	1979	7329388	8
53	6805078	2131	3194922	1·4694910	9287154	1·0767561	1·3647389	2672591	1980	7327409	7
54	6807209	2130	3192791	1·4690309	9292573	1·0761282	1·3651078	2674571	1980	7325429	6
55	6809339	2130	3190661	1·4685713	9297996	1·0755006	1·3654770	2676551	1982	7323449	5
56	6811469	2130	3188531	1·4681120	9303421	1·0748734	1·3658464	2678533	1981	7321467	4
57	6813599	2129	3186401	1·4676532	9308849	1·0742467	1·3662162	2680514	1983	7319486	3
58	6815728	2128	3184272	1·4671948	9314280	1·0736203	1·3665863	2682497	1982	7317503	2
59	6817856	2128	3182144	1·4667368	9319714	1·0729943	1·3669567	2684479	1984	7315521	1
60	6819984		3180016	1·4662792	9325151	1·0723687	1·3673275	2686463		7313537	0
′	Cosine	Dif.	Vers.	Secant.	Cotan.	Tang.	Cosec.	Covers	Dif.	Sine.	′

47 Deg.

'	Sine.	Dif.	Cosec.	Verseds.	Tang.	Dif.	Cotang.	Covers.	Secant.	D.	Cosine.	
0	9·8255109	1403	10·1744891	9·4096883	9·9544374	2541	10·0455626	9·5196566	10·1289265	1138	9·8710735	60
1	9·8256512	1401	10·1743488	9·4100174	9·9546915	2540	10·0453085	9·5193728	10·1290403	1139	9·8709597	59
2	9·8257913	1401	10·1742087	9·4103462	9·9549455	2540	10·0450545	9·5190889	10·1291542	1139	9·870845x	58
3	9·8259314	1401	10·1740686	9·4106750	9·9551995	2540	10·0448005	9·5188049	10·1292681	1140	9·8707319	57
4	9·8260715	1399	10·1739285	9·4110036	9·9554535	2540	10·0445465	9·5185207	10·1293821	1140	9·8706179	56
5	9·8262114	1398	10·1737886	9·4113321	9·9557075	2540	10·0442925	9·5182365	10·1294961	1141	9·8705039	55
6	9·8263512	1398	10·1736488	9·4116604	9·9559615	2539	10·0440385	9·5179521	10·1296102	1142	9·8703898	54
7	9·8264910	1397	10·1735090	9·4119885	9·9562154	2540	10·0437846	9·5176677	10·1297244	1143	9·8702756	53
8	9·8266307	1396	10·1733693	9·4123166	9·9564694	2539	10·0435306	9·5173831	10·1298387	1143	9·8701613	52
9	9·8267703	1395	10·1732297	9·4126445	9·9567233	2539	10·0432767	9·5170984	10·1299530	1144	9·8700470	51
10	9·8269098	1395	10·1730902	9·4129722	9·9569772	2539	10·0430228	9·5168136	10·1300674	1144	9·8699326	50
11	9·8270493	1394	10·1729507	9·4132998	9·9572311	2539	10·0427689	9·5165287	10·1301818	1145	9·8698182	49
12	9·8271887	1392	10·1728113	9·4136273	9·9574850	2539	10·0425150	9·5162436	10·1302963	1146	9·8697037	48
13	9·8273279	1392	10·1726721	9·4139546	9·9577389	2538	10·0422611	9·5159585	10·1304109	1147	9·8695891	47
14	9·8274671	1392	10·1725329	9·4142818	9·9579927	2538	10·0420073	9·5156733	10·1305256	1147	9·8694744	46
15	9·8276063	1390	10·1723937	9·4146088	9·9582465	2538	10·0417535	9·5153879	10·1306403	1148	9·8693597	45
16	9·8277453	1390	10·1722547	9·4149357	9·9585004	2538	10·0414996	9·5151024	10·1307551	1148	9·8692449	44
17	9·8278843	1388	10·1721157	9·4152625	9·9587542	2538	10·0412458	9·5148168	10·1308699	1149	9·8691301	43
18	9·8280231	1388	10·1719769	9·4155891	9·9590080	2538	10·0409920	9·5145311	10·1309848	1150	9·8690152	42
19	9·8281619	1387	10·1718381	9·4159156	9·9592618	2537	10·0407382	9·5142453	10·1310998	1151	9·8689002	41
20	9·8283006	1387	10·1716994	9·4162419	9·9595155	2537	10·0404845	9·5139594	10·1312149	1151	9·8687851	40
21	9·8284393	1385	10·1715607	9·4165681	9·9597693	2537	10·0402307	9·5136734	10·1313300	1152	9·8686700	39
22	9·8285778	1385	10·1714222	9·4168942	9·9600230	2537	10·0399770	9·5133872	10·1314452	1152	9·8685548	38
23	9·8287163	1384	10·1712837	9·4172201	9·9602767	2538	10·0397233	9·5131009	10·1315604	1154	9·8684396	37
24	9·8288547	1383	10·1711453	9·4175459	9·9605305	2537	10·0394695	9·5128146	10·1316758	1154	9·8683242	36
25	9·8289930	1382	10·1710070	9·4178715	9·9607842	2536	10·0392158	9·5125281	10·1317912	1154	9·8682088	35
26	9·8291312	1382	10·1708688	9·4181970	9·9610378	2537	10·0389622	9·5122415	10·1319066	1155	9·8680934	34
27	9·8292694	1381	10·1707306	9·4185223	9·9612915	2537	10·0387085	9·5119548	10·1320221	1156	9·8679779	33
28	9·8294075	1379	10·1705925	9·4188475	9·9615452	2536	10·0384548	9·5116679	10·1321377	1157	9·8678623	32
29	9·8295454	1379	10·1704546	9·4191726	9·9617988	2537	10·0382012	9·5113810	10·1322534	1157	9·8677466	31
30	9·8296833	1379	10·1703167	9·4194975	9·9620525	2536	10·0379475	9·5110940	10·1323691	1158	9·8676309	30
31	9·8298212	1377	10·1701788	9·4198223	9·9623061	2536	10·0376939	9·5108068	10·1324849	1159	9·8675151	29
32	9·8299589	1377	10·1700411	9·4201470	9·9625597	2536	10·0374403	9·5105195	10·1326008	1159	9·8673992	28
33	9·8300966	1376	10·1699034	9·4204715	9·9628133	2536	10·0371867	9·5102321	10·1327167	1160	9·8672833	27
34	9·8302342	1375	10·1697658	9·4207959	9·9630669	2535	10·0369331	9·5099446	10·1328327	1161	9·8671673	26
35	9·8303717	1374	10·1696283	9·4211201	9·9633204	2536	10·0366796	9·5096570	10·1329488	1161	9·8670512	25
36	9·8305091	1373	10·1694909	9·4214442	9·9635740	2535	10·0364260	9·5093693	10·1330649	1162	9·8669351	24
37	9·8306464	1373	10·1693536	9·4217681	9·9638275	2536	10·0361725	9·5090814	10·1331811	1163	9·8668189	23
38	9·8307837	1372	10·1692163	9·4220920	9·9640811	2535	10·0359189	9·5087934	10·1332974	1163	9·8667026	22
39	9·8309209	1371	10·1690791	9·4224156	9·9643346	2535	10·0356654	9·5085054	10·1334137	1164	9·8665863	21
40	9·8310580	1370	10·1689420	9·4227392	9·9645881	2535	10·0354119	9·5082172	10·1335301	1165	9·8664699	20
41	9·8311950	1370	10·1688050	9·4230626	9·9648416	2535	10·0351584	9·5079289	10·1336466	1165	9·8663534	19
42	9·8313320	1368	10·1686680	9·4233858	9·9650951	2535	10·0349049	9·5076405	10·1337631	1166	9·8662369	18
43	9·8314688	1368	10·1685312	9·4237089	9·9653486	2534	10·0346514	9·5073519	10·1338797	1167	9·8661203	17
44	9·8316056	1367	10·1683944	9·4240319	9·9656020	2535	10·0343980	9·5070633	10·1339964	1168	9·8660036	16
45	9·8317423	1366	10·1682577	9·4243548	9·9658555	2534	10·0341445	9·5067745	10·1341132	1168	9·8658868	15
46	9·8318789	1366	10·1681211	9·4246775	9·9661089	2534	10·0338911	9·5064857	10·1342300	1169	9·8657700	14
47	9·8320155	1364	10·1679845	9·4250000	9·9663623	2534	10·0336377	9·5061967	10·1343469	1169	9·8656531	13
48	9·8321519	1364	10·1678481	9·4253225	9·9666157	2535	10·0333843	9·5059076	10·1344638	1170	9·8655362	12
49	9·8322883	1363	10·1677117	9·4256447	9·9668692	2533	10·0331308	9·5056183	10·1345808	1171	9·8654192	11
50	9·8324246	1363	10·1675754	9·4259669	9·9671225	2534	10·0328775	9·5053290	10·1346979	1172	9·8653021	10
51	9·8325609	1361	10·1674391	9·4262889	9·9673759	2534	10·0326241	9·5050396	10·1348151	1172	9·8651849	9
52	9·8326970	1361	10·1673030	9·4266108	9·9676293	2534	10·0323707	9·5047500	10·1349323	1173	9·8650677	8
53	9·8328331	1360	10·1671669	9·4269325	9·9678827	2533	10·0321173	9·5044603	10·1350496	1173	9·8649504	7
54	9·8329691	1359	10·1670309	9·4272541	9·9681360	2533	10·0318640	9·5041705	10·1351669	1175	9·8648331	6
55	9·8331050	1358	10·1668950	9·4275756	9·9683893	2534	10·0316107	9·5038806	10·1352844	1175	9·8647156	5
56	9·8332408	1358	10·1667592	9·4278969	9·9686427	2533	10·0313573	9·5035906	10·1354019	1175	9·8645981	4
57	9·8333766	1356	10·1666234	9·4282181	9·9688960	2533	10·0311040	9·5033005	10·1355194	1177	9·8644806	3
58	9·8335122	1356	10·1664878	9·4285392	9·9691493	2533	10·0308507	9·5030102	10·1356371	1177	9·8643629	2
59	9·8336478	1355	10·1663522	9·4288601	9·9694026	2533	10·0305974	9·5027198	10·1357548	1177	9·8642452	1
60	9·8337833		10·1662167	9·4291809	9·9696559		10·0303441	9·5024294	10·1358725		9·8641275	0

| ' | Cosine. | Dif. | Secant. | Covers. | Cotang. | Dif. | Tang. | Verseds. | Cosec. | D. | Sine. | ' |

′	Sine.	Dif.	Covers	Cosec.	Tang.	Cotang.	Secant.	Vers.	Dif.	Cosine	′
0	6819984	2127	3180016	1·4662792	9325151	1·0723687	1·3673275	2686463	1984	7313537	60
1	6822111	2126	3177889	1·4658220	9330591	1·0717435	1·3676985	2688447	1985	7311553	59
2	6824237	2126	3175763	1·4653652	9336034	1·0711187	1·3680699	2690432	1985	7309568	58
3	6826363	2126	3173637	1·4649089	9341479	1·0704943	1·3684416	2692417	1986	7307583	57
4	6828489	2124	3171511	1·4644529	9346928	1·0698702	1·3688136	2694403	1987	7305597	56
5	6830613	2125	3169387	1·4639973	9352380	1·0692466	1·3691859	2696390	1987	7303610	55
6	6832738	2123	3167262	1·4635422	9357834	1·0686233	1·3695586	2698377	1988	7301623	54
7	6834861	2123	3165139	1·4630875	9363292	1·0680004	1·3699315	2700365	1989	7299635	53
8	6836984	2123	3163016	1·4626331	9368753	1·0673779	1·3703048	2702354	1989	7297646	52
9	6839107	2122	3160893	1·4621792	9374216	1·0667558	1·3706784	2704343	1989	7295657	51
10	6841229	2121	3158771	1·4617257	9379683	1·0661341	1·3710523	2706332	1991	7293668	50
11	6843350	2121	3156650	1·4612726	9385153	1·0655128	1·3714266	2708323	1991	7291677	49
12	6845471	2120	3154529	1·4608198	9390625	1·0648918	1·3718011	2710314	1991	7289686	48
13	6847591	2120	3152409	1·4603675	9396101	1·0642713	1·3721760	2712305	1992	7287695	47
14	6849711	2119	3150289	1·4599156	9401579	1·0636511	1·3725512	2714297	1993	7285703	46
15	6851830	2118	3148170	1·4594641	9407061	1·0630313	1·3729268	2716290	1994	7283710	45
16	6853948	2118	3146052	1·4590130	9412545	1·0624119	1·3733026	2718284	1994	7281716	44
17	6856066	2118	3143934	1·4585623	9418033	1·0617929	1·3736788	2720278	1994	7279722	43
18	6858184	2116	3141816	1·4581120	9423523	1·0611742	1·3740553	2722272	1996	7277728	42
19	6860300	2116	3139700	1·4576621	9429017	1·0605560	1·3744321	2724268	1996	7275732	41
20	6862416	2116	3137584	1·4572127	9434513	1·0599381	1·3748092	2726264	1996	7273736	40
21	6864532	2115	3135468	1·4567636	9440013	1·0593206	1·3751867	2728260	1997	7271740	39
22	6866647	2114	3133353	1·4563149	9445516	1·0587035	1·3755645	2730257	1998	7269743	38
23	6868761	2114	3131239	1·4558666	9451021	1·0580867	1·3759426	2732255	1998	7267745	37
24	6870875	2113	3129125	1·4554187	9456530	1·0574704	1·3763210	2734253	1999	7265747	36
25	6872988	2113	3127012	1·4549712	9462042	1·0568544	1·3766998	2736252	2000	7263748	35
26	6875101	2112	3124899	1·4545241	9467556	1·0562388	1·3770789	2738252	2000	7261748	34
27	6877213	2112	3122787	1·4540774	9473074	1·0556235	1·3774583	2740252	2001	7259748	33
28	6879325	2110	3120675	1·4536311	9478595	1·0550087	1·3778380	2742253	2001	7257747	32
29	6881435	2111	3118565	1·4531852	9484119	1·0543942	1·3782181	2744254	2002	7255746	31
30	6883546	2109	3116454	1·4527397	9489646	1·0537801	1·3785985	2746256	2003	7253744	30
31	6885655	2110	3114345	1·4522946	9495176	1·0531664	1·3789792	2748259	2003	7251741	29
32	6887765	2108	3112235	1·4518498	9500709	1·0525531	1·3793602	2750262	2004	7249738	28
33	6889873	2108	3110127	1·4514055	9506245	1·0519401	1·3797416	2752266	2005	7247734	27
34	6891981	2108	3108019	1·4509616	9511784	1·0513275	1·3801233	2754271	2005	7245729	26
35	6894089	2106	3105911	1·4505181	9517326	1·0507153	1·3805053	2756276	2005	7243723	25
36	6896195	2107	3103805	1·4500749	9522871	1·0501034	1·3808877	2758281	2007	7241719	24
37	6898302	2105	3101698	1·4496322	9528420	1·0494920	1·3812704	2760288	2007	7239712	23
38	6900407	2105	3099593	1·4491898	9533971	1·0488809	1·3816534	2762295	2007	7237705	22
39	6902512	2105	3097488	1·4487478	9539526	1·0482702	1·3820367	2764302	2008	7235698	21
40	6904617	2104	3095383	1·4483063	9545083	1·0476598	1·3824204	2766310	2009	7233690	20
41	6906721	2103	3093279	1·4478651	9550644	1·0470498	1·3828044	2768319	2010	7231681	19
42	6908824	2103	3091176	1·4474243	9556208	1·0464402	1·3831887	2770329	2010	7229671	18
43	6910927	2102	3089073	1·4469839	9561774	1·0458310	1·3835734	2772339	2010	7227661	17
44	6913029	2102	3086971	1·4465439	9567344	1·0452221	1·3839584	2774349	2011	7225651	16
45	6915131	2101	3084869	1·4461043	9572917	1·0446136	1·3843437	2776360	2012	7223640	15
46	6917232	2100	3082768	1·4456651	9578494	1·0440055	1·3847294	2778372	2013	7221628	14
47	6919332	2100	3080668	1·4452262	9584073	1·0433977	1·3851153	2780385	2013	7219615	13
48	6921432	2099	3078568	1·4447878	9589655	1·0427904	1·3855017	2782398	2013	7217602	12
49	6923531	2099	3076469	1·4443497	9595241	1·0421833	1·3858883	2784411	2015	7215589	11
50	6925630	2098	3074370	1·4439120	9600829	1·0415767	1·3862753	2786426	2015	7213574	10
51	6927728	2097	3072272	1·4434748	9606421	1·0409704	1·3866626	2788441	2015	7211559	9
52	6929825	2097	3070175	1·4430379	9612016	1·0403645	1·3870503	2790456	2016	7209544	8
53	6931922	2096	3068078	1·4426013	9617614	1·0397589	1·3874383	2792472	2017	7207528	7
54	6934018	2096	3065982	1·4421652	9623215	1·0391538	1·3878266	2794489	2017	7205511	6
55	6936114	2095	3063886	1·4417295	9628819	1·0385489	1·3882153	2796506	2018	7203494	5
56	6938209	2095	3061791	1·4412941	9634427	1·0379445	1·3886043	2798524	2019	7201476	4
57	6940304	2094	3059696	1·4408592	9640037	1·0373404	1·3889936	2800543	2019	7199457	3
58	6942398	2093	3057602	1·4404246	9645651	1·0367367	1·3893832	2802562	2020	7197438	2
59	6944491	2093	3055509	1·4399904	9651268	1·0361333	1·3897733	2804582	2020	7195418	1
60	6946584		3053416	1·4395565	9656888	1·0355303	1·3901636	2806602		7193398	0

′	Cosine	Dif.	Vers.	Secant.	Cotan.	Tang.	Cosec.	Covers	Dif.	Sine.	′

'	Sine.	Dif.	Cosec.	Versed.	Tang.	Dif.	Cotang.	Covers.	Secant.	D.	Cosine.	'
0	9·8337833	1355	10·1662167	9·4291809	9·9696559	2532	10·0303441	9·5024294	10·1358725	1179	9·8641275	60
1	9·8339188	1353	10·1660812	9·4295015	9·9699091	2533	10·0300909	9·5021388	10·1359904	1179	9·8640096	59
2	9·8340541	1353	10·1659459	9·4298220	9·9701624	2533	10·0298376	9·5018480	10·1361083	1180	9·8638917	58
3	9·8341894	1352	10·1658106	9·4301424	9·9704157	2532	10·0295843	9·5015572	10·1362263	1180	9·8637737	57
4	9·8343246	1351	10·1656754	9·4304626	9·9706689	2532	10·0293311	9·5012663	10·1363443	1181	9·8636557	56
5	9·8344597	1351	10·1655403	9·4307827	9·9709221	2533	10·0290779	9·5009752	10·1364624	1182	9·8635376	55
6	9·8345948	1349	10·1654052	9·4311027	9·9711754	2532	10·0288246	9·5006840	10·1365806	1183	9·8634194	54
7	9·8347297	1349	10·1652703	9·4314225	9·9714286	2532	10·0285714	9·5003927	10·1366989	1183	9·8633011	53
8	9·8348646	1348	10·1651354	9·4317422	9·9716818	2532	10·0283182	9·5001013	10·1368172	1184	9·8631828	52
9	9·8349994	1347	10·1650006	9·4320617	9·9719350	2532	10·0280650	9·4998098	10·1369356	1184	9·8630644	51
10	9·8351341	1347	10·1648659	9·4323811	9·9721882	2531	10·0278118	9·4995182	10·1370540	1186	9·8629460	50
11	9·8352688	1345	10·1647312	9·4327004	9·9724413	2532	10·0275587	9·4992264	10·1371726	1186	9·8628274	49
12	9·8354033	1345	10·1645967	9·4330196	9·9726945	2532	10·0273055	9·4989345	10·1372912	1186	9·8627088	48
13	9·8355378	1344	10·1644622	9·4333386	9·9729477	2531	10·0270523	9·4986425	10·1374098	1188	9·8625902	47
14	9·8356722	1344	10·1643278	9·4336574	9·9732008	2531	10·0267992	9·4983504	10·1375286	1188	9·8624714	46
15	9·8358066	1342	10·1641934	9·4339762	9·9734539	2532	10·0265461	9·4980582	10·1376474	1188	9·8623526	45
16	9·8359408	1342	10·1640592	9·4342948	9·9737071	2531	10·0262929	9·4977658	10·1377662	1190	9·8622338	44
17	9·8360750	1341	10·1639250	9·4346133	9·9739602	2531	10·0260398	9·4974734	10·1378852	1190	9·8621148	43
18	9·8362091	1340	10·1637909	9·4349316	9·9742133	2531	10·0257867	9·4971808	10·1380042	1191	9·8619958	42
19	9·8363431	1340	10·1636569	9·4352498	9·9744664	2531	10·0255336	9·4968881	10·1381233	1191	9·8618767	41
20	9·8364771	1338	10·1635229	9·4355678	9·9747195	2531	10·0252805	9·4965953	10·1382424	1193	9·8617576	40
21	9·8366109	1338	10·1633891	9·4358858	9·9749726	2531	10·0250274	9·4963024	10·1383617	1193	9·8616383	39
22	9·8367447	1337	10·1632553	9·4362036	9·9752257	2530	10·0247743	9·4960093	10·1384810	1193	9·8615190	38
23	9·8368784	1337	10·1631216	9·4365212	9·9754787	2531	10·0245213	9·4957162	10·1386003	1194	9·8613997	37
24	9·8370121	1335	10·1629879	9·4368387	9·9757318	2531	10·0242682	9·4954229	10·1387197	1195	9·8612803	36
25	9·8371456	1335	10·1628544	9·4371561	9·9759849	2530	10·0240151	9·4951295	10·1388392	1196	9·8611608	35
26	9·8372791	1334	10·1627209	9·4374734	9·9762379	2530	10·0237621	9·4948360	10·1389588	1197	9·8610412	34
27	9·8374125	1333	10·1625875	9·4377905	9·9764909	2531	10·0235091	9·4945424	10·1390785	1197	9·8609215	33
28	9·8375458	1332	10·1624542	9·4381075	9·9767440	2530	10·0232560	9·4942486	10·1391982	1197	9·8608018	32
29	9·8376790	1332	10·1623210	9·4384243	9·9769970	2530	10·0230030	9·4939547	10·1393179	1199	9·8606821	31
30	9·8378122	1331	10·1621878	9·4387411	9·9772500	2530	10·0227500	9·4936608	10·1394378	1199	9·8605622	30
31	9·8379453	1330	10·1620547	9·4390576	9·9775030	2530	10·0224970	9·4933667	10·1395577	1200	9·8604423	29
32	9·8380783	1329	10·1619217	9·4393741	9·9777560	2530	10·0222440	9·4930724	10·1396777	1201	9·8603223	28
33	9·8382112	1329	10·1617888	9·4396904	9·9780090	2530	10·0219910	9·4927781	10·1397978	1201	9·8602022	27
34	9·8383441	1328	10·1616559	9·4400066	9·9782620	2529	10·0217380	9·4924836	10·1399179	1202	9·8600821	26
35	9·8384769	1327	10·1615231	9·4403227	9·9785149	2530	10·0214851	9·4921891	10·1400381	1203	9·8599619	25
36	9·8386096	1326	10·1613904	9·4406386	9·9787679	2530	10·0212321	9·4918944	10·1401584	1203	9·8598416	24
37	9·8387422	1325	10·1612578	9·4409544	9·9790209	2529	10·0209791	9·4915996	10·1402787	1204	9·8597213	23
38	9·8388747	1325	10·1611253	9·4412700	9·9792738	2530	10·0207262	9·4913046	10·1403991	1205	9·8596009	22
39	9·8390072	1324	10·1609928	9·4415855	9·9795268	2529	10·0204732	9·4910096	10·1405196	1205	9·8594804	21
40	9·8391396	1323	10·1608604	9·4419009	9·9797797	2529	10·0202203	9·4907144	10·1406401	1206	9·8593599	20
41	9·8392719	1322	10·1607281	9·4422162	9·9800326	2530	10·0199674	9·4904191	10·1407607	1207	9·8592393	19
42	9·8394041	1322	10·1605959	9·4425313	9·9802856	2529	10·0197144	9·4901237	10·1408814	1208	9·8591186	18
43	9·8395363	1321	10·1604637	9·4428463	9·9805385	2529	10·0194615	9·4898282	10·1410022	1208	9·8589978	17
44	9·8396684	1320	10·1603316	9·4431611	9·9807914	2529	10·0192086	9·4895326	10·1411230	1209	9·8588770	16
45	9·8398004	1319	10·1601996	9·4434758	9·9810443	2529	10·0189557	9·4892368	10·1412439	1210	9·8587561	15
46	9·8399323	1319	10·1600677	9·4437904	9·9812972	2529	10·0187028	9·4889409	10·1413649	1210	9·8586351	14
47	9·8400642	1317	10·1599358	9·4441049	9·9815501	2529	10·0184499	9·4886449	10·1414859	1212	9·8585141	13
48	9·8401959	1317	10·1598041	9·4444192	9·9818030	2529	10·0181970	9·4883488	10·1416071	1211	9·8583929	12
49	9·8403276	1317	10·1596724	9·4447334	9·9820559	2528	10·0179441	9·4880525	10·1417282	1213	9·8582718	11
50	9·8404593	1315	10·1595407	9·4450475	9·9823087	2529	10·0176913	9·4877562	10·1418495	1213	9·8581505	10
51	9·8405908	1315	10·1594092	9·4453614	9·9825616	2529	10·0174384	9·4874597	10·1419708	1214	9·8580292	9
52	9·8407223	1314	10·1592777	9·4456752	9·9828145	2528	10·0171855	9·4871631	10·1420922	1215	9·8579078	8
53	9·8408537	1313	10·1591463	9·4459889	9·9830673	2529	10·0169327	9·4868664	10·1422137	1215	9·8577863	7
54	9·8409850	1312	10·1590150	9·4463024	9·9833202	2528	10·0166798	9·4865696	10·1423352	1216	9·8576648	6
55	9·8411162	1312	10·1588838	9·4466158	9·9835730	2529	10·0164270	9·4862726	10.1424568	1217	9·8575432	5
56	9·8412474	1311	10·1587526	9·4469291	9·9838259	2528	10·0161741	9·4859755	10·1425785	1217	9·8574215	4
57	9·8413785	1310	10·1586215	9·4472422	9·9840787	2528	10·0159213	9·4856783	10·1427002	1219	9·8572998	3
58	9·8415095	1309	10·1584905	9·4475552	9·9843315	2529	10·0156685	9·4853810	10·1428221	1218	9·8571779	2
59	9·8416404	1309	10·1583596	9·4478681	9·9845844	2528	10·0154156	9·4850836	10·1429439	1220	9·8570561	1
60	9·8417713		10·1582287	9·4481808	9·9848372		10·0151628	9·4847860	10·1430659		9·8569341	0

'	Cosine.	Dif.	Secant.	Covers.	Cotang.	Dif.	Tang.	Versed.	Cosec.	D.	Sine.	'

′	Sine.	Dif.	Covers	Cosec.	Tang.	Cotang.	Secant.	Vers.	Dif.	Cosine	′
0	6946584	2092	3053416	1·4395565	9656888	1·0355303	1·3901636	2806602	2021	7193398	60
1	6948676	2091	3051324	1·4391231	9662511	1·0349277	1·3905543	2808623	2022	7191377	59
2	6950767	2091	3049233	1·4386900	9668137	1·0343254	1·3909453	2810645	2022	7189355	58
3	6952858	2091	3047142	1·4382574	9673767	1·0337235	1·3913366	2812667	2023	7187333	57
4	6954949	2090	3045051	1·4378251	9679399	1·0331220	1·8917283	2814690	2023	7185310	56
5	6957039	2089	3042961	1·4373932	9685035	1·0325208	1·3921203	2816713	2024	7183287	55
6	6959128	2089	3040872	1·4369616	9690674	1·0319199	1·3925127	2818737	2025	7181263	54
7	6961217	2088	3038783	1·4365305	9696316	1·0313195	1·3929054	2820762	2025	7179238	53
8	6963305	2087	3036695	1·4360997	9701962	1·0307194	1·3932985	2822787	2026	7177213	52
9	6965392	2087	3034608	1·4356693	9707610	1·0301196	1·3936918	2824813	2026	7175187	51
10	6967479	2086	3032521	1·4352393	9713262	1·0295203	1·3940856	2826839	2027	7173161	50
11	6969565	2086	3030435	1·4348097	9718917	1·0289212	1·3944796	2828866	2028	7171134	49
12	6971651	2085	3028349	1·4343805	9724575	1·0283226	1·3948740	2830894	2028	7169106	48
13	6973736	2085	3026264	1·4339516	9730236	1·0277243	1·3952688	2832922	2029	7167078	47
14	6975821	2084	3024179	1·4335231	9735901	1·0271263	1·3956639	2834951	2030	7165049	46
15	6977905	2083	3022095	1·4330950	9741569	1·0265287	1·3960593	2836981	2030	7163019	45
16	6979988	2083	3020012	1·4326672	9747210	1·0259315	1·3964551	2839011	2030	7160989	44
17	6982071	2082	3017929	1·4322399	9752914	1·0253346	1·3968512	2841041	2032	7158959	43
18	6984153	2081	3015847	1·4318129	9758591	1·0247381	1·3972477	2843073	2032	7156927	42
19	6986234	2081	3013766	1·4313863	9764272	1·0241419	1·3976445	2845105	2032	7154895	41
20	6988315	2081	3011685	1·4309600	9769956	1·0235461	1·3980416	2847137	2033	7152863	40
21	6990396	2080	3009604	1·4305342	9775643	1·0229506	1·3984391	2849170	2033	7150830	39
22	6992476	2079	3007524	1·4301087	9781333	1·0223555	1·3988369	2851204	2034	7148796	38
23	6994555	2078	3005445	1·4296836	9787027	1·0217608	1·3992351	2853238	2035	7146762	37
24	6996633	2078	3003367	1·4292588	9792724	1·0211664	1·3996336	2855273	2036	7144727	36
25	6998711	2078	3001289	1·4288345	9798424	1·0205723	1·4000325	2857309	2036	7142691	35
26	7000789	2077	2999211	1·4284105	9804127	1·0199786	1·4004317	2859345	2037	7140655	34
27	7002866	2076	2997134	1·4279868	9809833	1·0193853	1·4008313	2861382	2037	7138618	33
28	7004942	2076	2995058	1·4275636	9815543	1·0187923	1·4012312	2863419	2038	7136581	32
29	7007018	2075	2992982	1·4271407	9821256	1·0181997	1·4016315	2865457	2039	7134543	31
30	7009093	2074	2990907	1·4267182	9826973	1·0176074	1·4020321	2867496	2039	7132504	30
31	7011167	2074	2988833	1·4262961	9832692	1·0170155	1·4024330	2869535	2039	7130465	29
32	7013241	2073	2986759	1·4258743	9838415	1·0164239	1·4028343	2871574	2041	7128426	28
33	7015314	2073	2984686	1·4254529	9844141	1·0158326	1·4032360	2873615	2041	7126385	27
34	7017387	2072	2982613	1·4250319	9849871	1·0152418	1·4036380	2875656	2041	7124344	26
35	7019459	2072	2980541	1·4246112	9855603	1·0146512	1·4040403	2877697	2043	7122303	25
36	7021531	2070	2978469	1·4241909	9861339	1·0140610	1·4044430	2879740	2042	7120260	24
37	7023601	2071	2976399	1·4237710	9867079	1·0134712	1·4048461	2881782	2044	7118218	23
38	7025672	2069	2974328	1·4233514	9872821	1·0128817	1·4052494	2883826	2044	7116174	22
39	7027741	2070	2972259	1·4229323	9878567	1·0122925	1·4056532	2885870	2044	7114130	21
40	7029811	2068	2970189	1·4225134	9884316	1·0117038	1·4060573	2887914	2045	7112086	20
41	7031879	2068	2968121	1·4220950	9890069	1·0111153	1·4064617	2889959	2046	7110040	19
42	7033947	2067	2966053	1·4216769	9895825	1·0105272	1·4068665	2892005	2047	7107995	18
43	7036014	2067	2963986	1·4212592	9901584	1·0099394	1·4072717	2894052	2047	7105948	17
44	7038081	2066	2961919	1·4208418	9907346	1·0093520	1·4076772	2896099	2047	7103901	16
45	7040147	2066	2959853	1·4204248	9913112	1·0087649	1·4080831	2898146	2048	7101854	15
46	7042213	2065	2957787	1·4200082	9918881	1·0081782	1·4084893	2900194	2049	7099806	14
47	7044278	2064	2955722	1·4195920	9924654	1·0075918	1·4088958	2902243	2050	7097757	13
48	7046342	2064	2953658	1·4191761	9930429	1·0070058	1·4093028	2904293	2050	7095707	12
49	7048406	2063	2951594	1·4187605	9936208	1·0064201	1·4097100	2906343	2050	7093657	11
50	7050469	2063	2949531	1·4183454	9941991	1·0058348	1·4101177	2908393	2051	7091607	10
51	7052532	2062	2947468	1·4179306	9947777	1·0052497	1·4105257	2910444	2052	7089556	9
52	7054594	2061	2945406	1·4175161	9953566	1·0046651	1·4109340	2912496	2053	7087504	8
53	7056655	2061	2943345	1·4171020	9959358	1·0040807	1·4113427	2914549	2053	7085451	7
54	7058716	2060	2941284	1·4166883	9965154	1·0034968	1·4117517	2916602	2053	7083398	6
55	7060776	2059	2939224	1·4162749	9970953	1·0029131	1·4121612	2918655	2054	7081345	5
56	7062835	2059	2937165	1·4158619	9976756	1·0023298	1·4125709	2920709	2055	7079291	4
57	7064894	2059	2935106	1·4154493	9982562	1·0017469	1·4129810	2922764	2056	7077236	3
58	7066953	2058	2933047	1·4150370	9988371	1·0011642	1·4133915	2924820	2056	7075180	2
59	7069011	2057	2930989	1·4146251	9994184	1·0005819	1·4138024	2926876	2056	7073124	1
60	7071068		2928932	1·4142136	1·0000000	1·0000000	1·4142136	2928932		7071068	0
	Cosine	Dif.	Vers.	Secant.	Cotan.	Tang.	Cosec.	Covers	Dif.	Sine.	′

45 Deg.

′	Sine.	Dif.	Cosec.	Verseds.	Tang.	Dif.	Cotang.	Covers.	Secant.	D.	Cosine.	′
0	9·8417713	1308	10·1582287	9·4481806	9·9848372	2528	10·0151628	9·4847860	10·1430659	1220	9·8569341	60
1	9·8419021	1307	10·1580979	9·4484934	9·9850900	2528	10·0149100	9·4844883	10·1431879	1221	9·8568121	59
2	9·8420328	1306	10·1579672	9·4488059	9·9853428	2528	10·0146572	9·4841905	10·1433100	1222	9·8566900	58
3	9·8421634	1306	10·1578366	9·4491183	9·9855956	2528	10·0144044	9·4838926	10·1434322	1223	9·8565678	57
4	9·8422939	1305	10·1577061	9·4494305	9·9858484	2528	10·0141516	9·4835946	10·1435545	1223	9·8564455	56
5	9·8424244	1305	10·1575756	9·4497426	9·9861012	2528	10·0138988	9·4832964	10·1436768	1224	9·8563232	55
6	9·8425548	1304 / 1303	10·1574452	9·4500546	9·9863540		10·0136460	9·4829981	10·1437992	1224	9·8562008	54
7	9·8426851	1303	10·1573149	9·4503664	9·9866068	2528	10·0133932	9·4826997	10·1439216	1226	9·8560784	53
8	9·8428154	1302	10·1571846	9·4506781	9·9868596	2528	10·0131404	9·4824012	10·1440442	1226	9·8559558	52
9	9·8429456	1301	10·1570544	9·4509897	9·9871123	2528	10·0128877	9·4821026	10·1441668	1226	9·8558332	51
10	9·8430757	1300	10·1569243	9·4513011	9·9873651	2528	10·0126349	9·4818038	10·1442894	1228	9·8557106	50
11	9·8432057	1300 / 1299	10·1567943	9·4516124	9·9876179	2527	10·0123821	9·4815049	10·1444122	1228	9·8555878	49
12	9·8433356	1299	10·1566644	9·4519236	9·9878706	2527	10·0121294	9·4812059	10·1445350	1228	9·8554650	48
13	9·8434655	1298	10·1565345	9·4522346	9·9881234	2528	10·0118766	9·4809068	10·1446579	1229	9·8553421	47
14	9·8435953	1297	10·1564047	9·4525456	9·9883761	2527	10·0116239	9·4806075	10·1447808	1231	9·8552192	46
15	9·8437250	1297	10·1562750	9·4528564	9·9886289	2528	10·0113711	9·4803082	10·1449039	1231	9·8550961	45
16	9·8438547	1295	10·1561453	9·4531670	9·9888816	2527	10·0111184	9·4800087	10·1450270	1231	9·8549730	44
17	9·8439842	1295	10·1560158	9·4534776	9·9891344	2528	10·0108656	9·4797091	10·1451501	1233	9·8548499	43
18	9·8441137	1295	10·1558863	9·4537880	9·9893871	2527	10·0106129	9·4794093	10·1452734	1233	9·8547266	42
19	9·8442432	1293	10·1557568	9·4540982	9·9896399	2528	10·0103601	9·4791095	10·1453967	1234	9·8546033	41
20	9·8443725	1293	10·1556275	9·4544084	9·9898926	2527	10·0101074	9·4788095	10·1455201	1234	9·8544799	40
21	9·8445018	1292	10·1554982	9·4547184	9·9901453	2528	10·0098547	9·4785094	10·1456436	1235	9·8543564	39
22	9·8446310	1291	10·1553690	9·4550283	9·9903981	2527	10·0096019	9·4782092	10·1457671	1236	9·8542329	38
23	9·8447601	1290	10·1552399	9·4553380	9·9906508	2527	10·0093492	9·4779088	10·1458907	1237	9·8541093	37
24	9·8448891	1290	10·1551109	9·4556477	9·9909035	2527	10·0090965	9·4776083	10·1460144	1237	9·8539856	36
25	9·8450181	1289	10·1549819	9·4559572	9·9911562	2527	10·0088438	9·4773078	10·1461381	1238	9·8538619	35
26	9·8451470	1288	10·1548530	9·4562665	9·9914089	2527	10·0085911	9·4770070	10·1462619	1239	9·8537381	34
27	9·8452758	1287	10·1547242	9·4565758	9·9916616	2527	10·0083384	9·4767062	10·1463858	1240	9·8536142	33
28	9·8454045	1287	10·1545955	9·4568849	9·9919143	2527	10·0080857	9·4764052	10·1465098	1240	9·8534902	32
29	9·8455332	1286	10·1544668	9·4571939	9·9921670	2527	10·0078330	9·4761042	10·1466338	1241	9·8533662	31
30	9·8456618	1285	10·1543382	9·4575027	9·9924197	2527	10·0075803	9·4758030	10·1467579	1242	9·8532421	30
31	9·8457903	1285	10·1542097	9·4578115	9·9926724	2527	10·0073276	9·4755016	10·1468821	1243	9·8531179	29
32	9·8459188	1283	10·1540812	9·4581201	9·9929251	2527	10·0070749	9·4752002	10·1470064	1243	9·8529936	28
33	9·8460471	1283	10·1539529	9·4584286	9·9931778	2527	10·0068222	9·4748986	10·1471307	1244	9·8528693	27
34	9·8461754	1282	10·1538246	9·4587369	9·9934305	2527	10·0065695	9·4745969	10·1472551	1245	9·8527449	26
35	9·8463036	1282	10·1536964	9·4590451	9·9936832	2527	10·0063168	9·4742951	10·1473796	1245	9·8526204	25
36	9·8464318	1281	10·1535682	9·4593532	9·9939359	2527	10·0060641	9·4739932	10·1475041	1246	9·8524959	24
37	9·8465599	1280	10·1534401	9·4596612	9·9941886	2527	10·0058114	9·4736911	10·1476287	1247	9·8523713	23
38	9·8466879	1279	10·1533121	9·4599690	9·9944413	2527	10·0055587	9·4733889	10·1477534	1248	9·8522466	22
39	9·8468158	1278	10·1531842	9·4602767	9·9946940	2526	10·0053060	9·4730866	10·1478782	1248	9·8521218	21
40	9·8469436	1278	10·1530564	9·4605843	9·9949466	2527	10·0050534	9·4727841	10·1480030	1249	9·8519970	20
41	9·8470714	1277	10·1529286	9·4608918	9·9951993	2527	10·0048007	9·4724816	10·1481279	1250	9·8518721	19
42	9·8471991	1276	10·1528009	9·4611991	9·9954520	2527	10·0045480	9·4721789	10·1482529	1251	9·8517471	18
43	9·8473267	1276	10·1526733	9·4615063	9·9957047	2526	10·0042953	9·4718761	10·1483780	1251	9·8516220	17
44	9·8474543	1274	10·1525457	9·4618134	9·9959573	2527	10·0040427	9·4715732	10·1485031	1252	9·8514969	16
45	9·8475817	1274	10·1524183	9·4621203	9·9962100	2527	10·0037900	9·4712701	10·1486283	1252	9·8513717	15
46	9·8477091	1274	10·1522909	9·4624271	9·9964627	2527	10·0035373	9·4709669	10·1487535	1254	9·8512465	14
47	9·8478365	1272	10·1521635	9·4627338	9·9967154	2526	10·0032846	9·4706636	10·1488789	1254	9·8511211	13
48	9·8479637	1272	10·1520363	9·4630404	9·9969680	2527	10·0030320	9·4703602	10·1490043	1255	9·8509957	12
49	9·8480909	1271	10·1519091	9·4633468	9·9972207	2527	10·0027793	9·4700566	10·1491298	1256	9·8508702	11
50	9·8482180	1270	10·1517820	9·4636531	9·9974734	2526	10·0025266	9·4697530	10·1492554	1256	9·8507446	10
51	9·8483450	1270	10·1516550	9·4639593	9·9977260	2527	10·0022740	9·4694492	10·1493810	1257	9·8506190	9
52	9·8484720	1269	10·1515280	9·4642654	9·9979787	2527	10·0020213	9·4691452	10·1495067	1258	9·8504933	8
53	9·8485989	1268	10·1514011	9·4645713	9·9982314	2526	10·0017686	9·4688412	10·1496325	1258	9·8503675	7
54	9·8487257	1267	10·1512743	9·4648771	9·9984840	2527	10·0015160	9·4685370	10·1497583	1260	9·8502417	6
55	9·8488524	1267	10·1511476	9·4651828	9·9987367	2526	10·0012633	9·4682327	10·1498843	1260	9·8501157	5
56	9·8489791	1266	10·1510209	9·4654884	9·9989893	2527	10·0010107	9·4679283	10·1500103	1260	9·8499897	4
57	9·8491057	1265	10·1508943	9·4657938	9·9992420	2527	10·0007580	9·4676237	10·1501363	1262	9·8498637	3
58	9·8492322	1264	10·1507678	9·4660991	9·9994947	2526	10·0005053	9·4673190	10·1502625	1262	9·8497375	2
59	9·8493586	1264	10·1506414	9·4664043	9·9997473	2527	10·0002527	9·4670142	10·1503887	1263	9·8496113	1
60	9·8494850		10·1505150	9·4667093	10·0000000		10·0000000	9·4667093	10·1505150		9·8494850	0
′	Cosine.	Dif.	Secant.	Covers.	Cotang.	Dif.	Tang.	Verseds.	Cosec.	D.	Sine.	′

Course Pts.	D.	Dist. 1 Lat.	Dep.	Dist. 2 Lat.	Dep.	Dist. 3 Lat.	Dep.	Dist. 4 Lat.	Dep.	Dist. 5 Lat.	Dep.	D.	Course Pts.
	1	0·9998	0·0175	1·9997	0·0349	2·9995	0·0524	3·9994	0·0698	4·9992	0·0873	89	
	2	0·9994	0·0349	1·9988	0·0698	2·9982	0·1047	3·9976	0·1396	4·9970	0·1745	88	
0¼		0·9988	0·0491	1·9976	0·0981	2·9964	0·1472	3·9952	0·1963	4·9940	0·2453		7¾
	3	0·9986	0·0523	1·9973	0·1047	2·9959	0·1570	3·9945	0·2093	4·9931	0·2617	87	
	4	0·9976	0·0698	1·9951	0·1395	2·9927	0·2093	3·9903	0·2790	4·9878	0·3488	86	
	5	0·9962	0·0872	1·9924	0·1743	2·9886	0·2615	3·9848	0·3486	4·9810	0·4358	85	
0½		0·9952	0·0980	1·9904	0·1960	2·9856	0·2941	3·9807	0·3921	4·9759	0·4901		7½
	6	0·9945	0·1045	1·9890	0·2091	2·9836	0·3136	3·9781	0·4181	4·9726	0·5226	84	
	7	0·9925	0·1219	1·9851	0·2437	2·9776	0·3656	3·9702	0·4875	4·9627	0·6093	83	
	8	0·9903	0·1392	1·9805	0·2783	2·9708	0·4175	3·9611	0·5567	4·9513	0·6959	82	
0¾		0·9892	0·1467	1·9784	0·2935	2·9675	0·4402	3·9567	0·5869	4·9459	0·7337		7¼
	9	0·9877	0·1564	1·9754	0·3129	2·9631	0·4693	3·9508	0·6257	4·9384	0·7822	81	
	10	0·9848	0·1736	1·9696	0·3473	2·9544	0·5209	3·9392	0·6946	4·9240	0·8682	80	
	11	0·9816	0·1908	1·9633	0·3816	2·9449	0·5724	3·9265	0·7632	4·9081	0·9540	79	
1		0·9808	0·1951	1·9616	0·3902	2·9424	0·5853	3·9231	0·7804	4·9039	0·9755		7
	12	0·9781	0·2079	1·9563	0·4158	2·9344	0·6237	3·9126	0·8316	4·8907	1·0396	78	
	13	0·9744	0·2250	1·9487	0·4499	2·9231	0·6749	3·8975	0·8998	4·8719	1·1248	77	
	14	0·9703	0·2419	1·9406	0·4838	2·9169	0·7258	3·8812	0·9677	4·8515	1·2096	76	
1¼		0·9700	0·2430	1·9401	0·4860	2·9101	0·7289	3·8801	0·9719	4·8502	1·2149		6¾
	15	0·9659	0·2588	1·9319	0·5176	2·8978	0·7765	3·8637	1·0353	4·8296	1·2941	75	
	16	0·9613	0·2756	1·9225	0·5513	2·8838	0·8269	3·8450	1·1025	4·8063	1·3782	74	
1½		0·9569	0·2903	1·9139	0·5806	2·8708	0·8709	3·8278	1·1611	4·7847	1·4514		6½
	17	0·9563	0·2924	1·9126	0·5847	2·8689	0·8771	3·8252	1·1695	4·7815	1·4619	73	
	18	0·9511	0·3090	1·9021	0·6180	2·8532	0·9271	3·8042	1·2361	4·7553	1·5451	72	
	19	0·9455	0·3256	1·8910	0·6511	2·8366	0·9767	3·7821	1·3023	4·7276	1·6278	71	
1¾		0·9415	0·3369	1·8831	0·6738	2·8246	1·0107	3·7662	1·3476	4·7077	1·6844		6¼
	20	0·9397	0·3420	1·8794	0·6840	2·8191	1·0261	3·7588	1·3681	4·6985	1·7101	70	
	21	0·9336	0·3584	1·8672	0·7167	2·8007	1·0751	3·7343	1·4335	4·6679	1·7918	69	
	22	0·9272	0·3746	1·8544	0·7492	2·7816	1·1238	3·7087	1·4984	4·6359	1·8730	68	
2		0·9239	0·3827	1·8478	0·7654	2·7716	1·1481	3·6955	1·5307	4·6194	1·9134		6
	23	0·9205	0·3907	1·8410	0·7815	2·7615	1·1722	3·6820	1·5629	4·6025	1·9537	67	
	24	0·9135	0·4067	1·8271	0·8135	2·7406	1·2202	3·6542	1·6269	4·5677	2·0337	66	
	25	0·9063	0·4226	1·8126	0·8452	2·7189	1·2679	3·6252	1·6905	4·5315	2·1131	65	
2¼		0·9040	0·4276	1·8080	0·8551	2·7120	1·2827	3·6160	1·7102	4·5199	2·1378		5¾
	26	0·8988	0·4384	1·7976	0·8767	2·6964	1·3151	3·5952	1·7535	4·4940	2·1919	64	
	27	0·8910	0·4540	1·7820	0·9080	2·6730	1·3620	3·5640	1·8160	4·4550	2·2700	63	
	28	0·8829	0·4695	1·7659	0·9389	2·6488	1·4084	3·5318	1·8779	4·4147	2·3474	62	
2½		0·8819	0·4714	1·7638	0·9428	2·6458	1·4142	3·5277	1·8856	4·4096	2·3570		5½
	29	0·8746	0·4848	1·7492	0·9696	2·6239	1·4544	3·4985	1·9392	4·3731	2·4240	61	
	30	0·8660	0·5000	1·7321	1·0000	2·5981	1·5000	3·4641	2·0000	4·3301	2·5000	60	
2¾		0·8577	0·5141	1·7155	1·0282	2·5732	1·5423	3·4309	2·0564	4·2886	2·5705		5¼
	31	0·8572	0·5150	1·7143	1·0301	2·5715	1·5451	3·4287	2·0602	4·2858	2·5752	59	
	32	0·8480	0·5299	1·6961	1·0598	2·5441	1·5896	3·3922	2·1197	4·2402	2·6496	58	
	33	0·8387	0·5446	1·6773	1·0893	2·5160	1·6339	3·3547	2·1786	4·1934	2·7232	57	
3		0·8315	0·5556	1·6629	1·1111	2·4944	1·6667	3·3259	2·2223	4·1573	2·7779		5
	34	0·8290	0·5592	1·6581	1·1184	2·4871	1·6776	3·3162	2·2368	4·1452	2·7960	56	
	35	0·8192	0·5736	1·6383	1·1472	2·4575	1·7207	3·2766	2·2943	4·0958	2·8679	55	
	36	0·8090	0·5878	1·6180	1·1756	2·4271	1·7634	3·2361	2·3511	4·0451	2·9389	54	
3¼		0·8032	0·5957	1·6064	1·1914	2·4096	1·7871	3·2128	2·3828	4·0160	2·9785		4¾
	37	0·7986	0·6018	1·5973	1·2036	2·3959	1·8054	3·1945	2·4073	3·9932	3·0091	53	
	38	0·7880	0·6157	1·5760	1·2313	2·3640	1·8470	3·1520	2·4626	3·9401	3·0783	52	
	39	0·7771	0·6293	1·5543	1·2586	2·3314	1·8880	3·1086	2·5173	3·8857	3·1466	51	
3½		0·7730	0·6344	1·5460	1·2688	2·3190	1·9032	3·0920	2·5376	3·8650	3·1720		4½
	40	0·7660	0·6428	1·5321	1·2856	2·2981	1·9284	3·0642	2·5712	3·8302	3·2139	50	
	41	0·7547	0·6561	1·5094	1·3121	2·2641	1·9682	3·0188	2·6242	3·7735	3·2803	49	
	42	0·7431	0·6691	1·4803	1·3383	2·2294	2·0074	2·9726	2·6765	3·7157	3·3457	48	
3¾		0·7410	0·6716	1·4819	1·3431	2·2229	2·0147	2·9638	2·6862	3·7048	3·3578		4¼
	43	0·7314	0·6820	1·4628	1·3640	2·1941	2·0460	2·9254	2·7280	3·6568	3·4100	47	
	44	0·7193	0·6947	1·4387	1·3894	2·1580	2·0840	2·8774	2·7786	3·5967	3·4733	46	
4	45	0·7071	0·7071	1·4142	1·4142	2·1213	2·1213	2·8284	2·8284	3·5355	3·5355	45	4
Pts.	Deg.	Dep.	Lat.	Dep.	Lat.	Dep.	Lat.	Dep.	Lat.	Dep.	Lat.	Deg.	Pts.
		Dist. 1.		Dist. 2.		Dist. 3.		Dist. 4.		Dist. 5.			

Course.		Dist. 6.		Dist. 7.		Dist. 8.		Dist. 9.		Dist. 10.		Course.	
Pts.	D.	Lat.	Dep.	Lat.	Dep.	Lat.	Dep.	Lat.	Dep.	Lat.	Dep;	D.	Pts.
	1	5·9991	0·1047	6·9989	0·1222	7·9988	0·1396	8·9986	0·1571	9·9985	0·1745	89	
	2	5·9963	0·2094	6·9957	0·2443	7·9951	0·2792	8·9945	0·3141	9·9939	0·3490	88	
0¼		5·9928	0·2944	6·9916	0·3435	7·9904	0·3925	8·9892	0·4416	9·9880	0·4907		7¾
	3	5·9918	0·3140	6·9904	0·3664	7·9890	0·4187	8·9877	0·4710	9·9863	0·5234	87	
	4	5·9854	0·4185	6·9829	0·4883	7·9805	0·5581	8·9781	0·6278	9·9756	0·6976	86	
	5	5·9772	0·5229	6·9734	0·6101	7·9696	0·6972	8·9658	0·7844	9·9619	0·8716	85	
0½		5·9711	0·5881	6·9663	0·6861	7·9615	0·7841	8·9567	0·8822	9·9518	0·9802		7½
	6	5·9671	0·6272	6·9617	0·7317	7·9562	0·8362	8·9507	0·9408	9·9452	1·0453	84	
	7	5·9553	0·7312	6·9478	0 8531	7·9404	0·9750	8·9329	1·0968	9·9255	1·2187	83	
	8	5·9416	0·8350	6·9319	0·9742	7·9221	1·1134	8·9124	1·2526	9·9027	1·3917	82	
0¾		5·9351	0·8804	6·9242	1·0271	7·9134	1·1738	8·9026	1·3206	9·8918	1·4673		7¼
	9	5·9261	0·9386	6·9138	1·0950	7·9015	1·2515	8·8892	1·4079	9·8769	1·5643	81	
	10	5·9088	1·0419	6·8937	1·2155	7·8785	1·3892	8·8633	1·5628	9·8481	1·7365	80	
	11	5 8898	1·1449	6·8714	1·3357	7·8530	1·5265	8·8346	1·7173	9·8163	1·9081	79	
1		5·8847	1·1705	6·8655	1·3656	7·8463	1·5607	8·8271	1·7558	9·8079	1·9509		7
	12	5·8689	1·2475	6·8470	1·4554	7·8252	1·6633	8·8033	1·8712	9·7815	2·0791	78	
	13	5·8462	1·3497	6·8206	1·5747	7·7950	1·7996	8·7693	2·0246	9·7437	2·2495	77	
	14	5·8218	1·4515	6·7921	1·6935	7·7624	1·9354	8·7327	2·1773	9·7030	2·4192	76	
1¼		5·8202	1·4579	6·7902	1·7009	7·7602	1·9438	8·7303	2·1868	9·7003	2 4298		6¾
	15	5·7956	1·5529	6·7615	1·8117	7·7274	2·0706	8·6933	2·3294	9 6593	2·5882	75	
	16	5·7676	1·6538	6·7288	1·9295	7·6901	2·2051	8·6513	2·4807	9·6120	2·7562	74	
1½		5·7416	1·7417	6·6986	2·0320	7·6555	2·3223	8 6125	2·6126	9·5694	2·9028		6½
	17	5·7378	1·7542	6·6941	2·0466	7·6504	2·3390	8·6067	2·6313	9·5630	2·9237	73	
	18	5·7063	1·8541	6·6574	2·1631	7·6085	2·4721	8·5595	2·7812	9·5106	3·0902	72	
	19	5·6731	1·9534	6·6186	2·2790	7·5642	2·6045	8·5097	2·9301	9·4552	3·2557	71	
1¾		5·6493	2·0213	6·5908	2·3582	7·5324	2·6951	8·4739	3·0320	9·4154	3·3689		6¼
	20	5·6382	2·0521	6·5778	2·3941	7·5175	2·7362	8·4572	3·0782	9·3969	3·4202	70	
	21	5·6015	2 1502	6·5351	2·5086	7·4686	2·8669	8·4022	3·2253	9·3358	3·5837	69	
	22	5·5631	2·2476	6·4903	2·6222	7·4175	2·9969	8·3447	3·3715	9·2718	3·7461	68	
2		5·5433	2·2961	6·4672	2·6788	7·3910	3·0615	8·3149	3·4442	9·2388	3·8268		6
	23	5·5230	2·3444	6·4435	2·7351	7·3640	3·1258	8·2845	3·5166	9·2050	3·9073	67	
	24	5·4813	2·4404	6·3948	2·8472	7·3084	3·2539	8·2219	3·6606	9·1355	4·0674	66	
	25	5·4378	2·5357	6·3442	2·9583	7·2505	3·3809	8·1568	3·8036	9·0631	4·2262	65	
2¼		5·4239	2·5653	6·3279	2·9929	7·2319	3·4204	8·1359	3·8480	9·0399	4·2756		5¾
	26	5·3928	2·6302	6·2916	3·0686	7·1904	3·5070	8·0891	3·9453	8·9879	4·3837	64	
	27	5·3460	2·7239	6·2370	3·1779	7·1280	3·6319	8·0191	4·0859	8·9101	4·5399	63	
	28	5·2977	2·8168	6·1806	3·2863	7·0636	3·7558	7·9465	4·2252	8·8295	4·6947	62	
2½		5·2915	2·8284	6·1734	3·2998	7·0554	3·7712	7·9373	4·2426	8·8192	4·7140		5½
	29	5·2477	2·9089	6·1223	3·3937	6·9970	3·8785	7·8716	4·3633	8·7462	4·8481	61	
	30	5·1962	3·0000	6·0622	3·5000	6·9282	4·0000	7·7942	4·5000	8·6603	5 0000	60	
2¾		5·1464	3·0846	6·0041	3·5987	6·8618	4·1128	7·7196	4·6269	8·5773	5·1410		5¼
	31	5·1430	3·0902	6·0002	3·6053	6·8573	4·1203	7·7145	4·6353	8 5717	5·1504	59	
	32	5·0883	3·1795	5·9363	3·7094	6·7844	4·2394	7·6324	4·7093	8·4805	5·2992	58	
	33	5·0320	3·2678	5·8707	3·8125	6·7094	4·3571	7·5480	4·9018	8·3867	5·4464	57	
3		4·9888	3·3334	5·8203	3·8890	6·6518	4·4446	7·4832	5·0001	8·3147	5·5557		5
	34	4·9742	3·3552	5·8033	3·9144	6·6323	4·4735	7·4613	5·0327	8·2904	5·5919	56	
	35	4·9149	3·4415	5·7341	4·0150	6·5532	4·5886	7·3724	5·1622	8·1915	5·7358	55	
	36	4·8541	3·5267	5·6631	4·1145	6·4721	4·7023	7·2812	5·2901	8·0902	5·8779	54	
3¼		4·8192	3·5742	5·6225	4·1699	6·4257	4·7656	7·2289	5·3613	8·0321	5 9570		4¾
	37	4·7918	3·6109	5·5904	4·2127	6·3891	4·8145	7·1877	5·4163	7·9864	6·0182	53	
	38	4·7281	3·6940	5·5161	4·3096	6·3041	4·9253	7·0921	5·5409	7·8801	6·1566	52	
	39	4·6629	3·7759	5·4400	4·4052	6·2172	5·0346	6·9943	5·6639	7·7715	6·2932	51	
3½		4·6381	3·8064	5·4111	4·4408	6·1841	5·0751	6·9571	5·7095	7·7301	6·3439		4½
	40	4·5963	3·8567	5·3623	4·4995	6·1284	5·1423	6·8944	5·7851	7·6604	6·4279	50	
	41	4·5283	3·9363	5·2830	4·5924	6·0377	5·2485	6·7924	5·9045	7·5471	6·5606	49	
	42	4·4589	4·0148	5·2020	4·6839	5·9452	5·3530	6·6883	6·0222	7·4314	6·6913	48	
3¾		4·4457	4·0294	5·1867	4·7009	5·9276	5·3725	6·6686	6·0440	7·4095	6·7156		4¼
	43	4·3881	4·0920	5·1195	4·7740	5·8508	5·4560	6·5822	6·1380	7·3135	6·8200	47	
	44	4·3160	4·1680	5·0354	4·8626	5·7547	5·5573	6·4741	6·2519	7·1934	6·9466	46	
4	45	4·2426	4·2426	4·9497	4·9497	5·6509	5·6569	6·3640	6·3640	7·0711	7·0711	45	4
Pts.	Deg.	Dep.	Lat.	Dep.	Lat.	Dep.	Lat.	Dep.	Lat.	Dep.	Lat.	Deg.	Pts.
		Dist. 6.		Dist. 7.		Dist. 8		Dist. 9.		Dist. 10.			

D	Arc	De	Arc	De	Arc	′	Arc	″	Arc	‴	A
1	·0174533	61	1·0646508	121	2·1118484	1	2909	1	48	1	1
2	·0349066	62	1·0821041	122	2·1293017	2	5818	2	97	2	2
3	·0523599	63	1·0995574	123	2·1467550	3	8727	3	145	3	2
4	·0698132	64	1·1170107	124	2·1642083	4	11636	4	194	4	3
5	·0872665	65	1·1344640	125	2·1816616	5	14544	5	242	5	4
6	·1047198	66	1·1519173	126	2·1991149	6	17453	6	291	6	5
7	·1221730	67	1·1693706	127	2·2165682	7	20362	7	339	7	6
8	·1396263	68	1·1868239	128	2·2340214	8	23271	8	388	8	6
9	·1570796	69	1·2042772	129	2·2514747	9	26180	9	436	9	7
10	·1745329	70	1·2217305	130	2·2689280	10	29089	10	485	10	8
11	·1919862	71	1·2391838	131	2·2863813	11	31998	11	533	11	9
12	·2094395	72	1·2566371	132	2·3038346	12	34907	12	582	12	10
13	·2268928	73	1·2740904	133	2·3212879	13	37815	13	630	13	11
14	·2443461	74	1·2915436	134	2·3387412	14	40724	14	679	14	11
15	·2617994	75	1·3089969	135	2·3561945	15	43633	15	727	15	12
16	·2792527	76	1·3264502	136	2·3736478	16	46542	16	776	16	13
17	·2967060	77	1·3439035	137	2·3911011	17	49451	17	824	17	14
18	·3141593	78	1·3613568	138	2·4085544	18	52360	18	873	18	15
19	·3316126	79	1·3788101	139	2·4260077	19	55269	19	921	19	15
20	·3490659	80	1·3962634	140	2·4434610	20	58178	20	970	20	16
21	·3665191	81	1·4137167	141	2·4609142	21	61087	21	1018	21	17
22	·3839724	82	1·4311700	142	2·4783675	22	63995	22	1067	22	18
23	·4014257	83	1·4486233	143	2·4958208	23	66904	23	1115	23	19
24	·4188790	84	1·4660766	144	2·5132741	24	69813	24	1164	24	19
25	·4363323	85	1·4835299	145	2·5307274	25	72722	25	1212	25	20
26	·4537856	86	1·5009832	146	2·5481807	26	75631	26	1261	26	21
27	·4712389	87	1·5184364	147	2·5656340	27	78540	27	1309	27	22
28	·4886922	88	1·5358897	148	2·5830873	28	81449	28	1357	28	23
29	·5061455	89	1·5533430	149	2·6005406	29	84358	29	1406	29	23
30	·5235988	90	1·5707963	150	2·6179939	30	87266	30	1454	30	24
31	·5410521	91	1·5882496	151	2·6354472	31	90175	31	1503	31	25
32	·5585054	92	1·6057029	152	2·6529005	32	93084	32	1551	32	26
33	·5759587	93	1·6231562	153	2·6703538	33	95993	33	1599	33	27
34	·5934119	94	1·6406095	154	2·6878070	34	98902	34	1648	34	27
35	·6108652	95	1·6580628	155	2·7052603	35	101811	35	1697	35	28
36	·6283185	96	1·6755161	156	2·7227136	36	104720	36	1745	36	29
37	·6457718	97	1·6929694	157	2·7401669	37	107629	37	1794	37	30
38	·6632251	98	1·7104227	158	2·7576202	38	110538	38	1842	38	31
39	·6806784	99	1·7278760	159	2·7750735	39	113446	39	1891	39	32
40	·6981317	100	1·7453293	160	2·7925268	40	116355	40	1939	40	32
41	·7155850	101	1·7627825	161	2·8099801	41	119264	41	1988	41	33
42	·7330383	102	1·7802358	162	2·8274334	42	122173	42	2036	42	34
43	·7504916	103	1·7976891	163	2·8448867	43	125082	43	2085	43	35
44	·7679449	104	1·8151424	164	2·8623400	44	127991	44	2133	44	36
45	·7853982	105	1·8325957	165	2·8797933	45	130900	45	2182	45	36
46	·8028515	106	1·8500490	166	2·8972466	46	133809	46	2230	46	37
47	·8203047	107	1·8675023	167	2·9146999	47	136717	47	2279	47	38
48	·8377580	108	1·8849556	168	2·9321531	48	139626	48	2327	48	39
49	·8552113	109	1·9024089	169	2·9496064	49	142535	49	2376	49	40
50	·8726646	110	1·9198622	170	2·9670597	50	145444	50	2424	50	40
51	·8901179	111	1·9373155	171	2·9845130	51	148353	51	2473	51	41
52	·9075712	112	1·9547688	172	3·0019663	52	151262	52	2521	52	42
53	·9250245	113	1·9722220	173	3·0194196	53	154171	53	2570	53	43
54	·9424778	114	1·9896753	174	3·0368729	54	157080	54	2618	54	44
55	·9599311	115	2·0071286	175	3·0543262	55	159989	55	2666	55	44
56	·9773844	116	2·0245819	176	3·0717795	56	162897	56	2715	56	45
57	·9948377	117	2·0420352	177	3·0892328	57	165806	57	2763	57	46
58	1·0122910	118	2·0594885	178	3·1066861	58	168715	58	2812	58	47
59	1·0297443	119	2·0769418	179	3·1241394	59	171624	59	2860	59	48
60	1·0471976	120	2·0943951	180	3·1415927	60	174533	60	2909	60	48
D	Arc	De	Arc	De	Arc	′	Arc	″	Arc	‴	A

Tab. 12. COMMON AND HYP. LOGARITHMS.   361

CL	HYP. LO.	CL	HYP. LO.	CL	HYP. LO.	CL	HYP. LO.
·01	·02302585	·26	·59867212	·51	1·17431840	·76	1·74996467
·02	·04605170	·27	·62169798	·52	1·19734425	·77	1·77299052
·03	·06907755	·28	·64472383	·53	1·22037010	·78	1·79601637
·04	·09210340	·29	·66774968	·54	1·24339595	·79	1·81904222
·05	·11512925	·30	·69077553	·55	1·26642180	·80	1·84206807
·06	·13815511	·31	·71380138	·56	1·28944765	·81	1·86509393
·07	·16118096	·32	·73682723	·57	1·31247350	·82	1·88811978
·08	·18420681	·33	·75985308	·58	1·33549935	·83	1·91114563
·09	·20723266	·34	·78287893	·59	1·35852520	·84	1·93417148
·10	·23025851	·35	·80590478	·60	1·38155106	·85	1·95719733
·11	·25328436	·36	·82893063	·61	1·40457691	·86	1·98022318
·12	·27631021	·37	·85195648	·62	1·42760276	·87	2·00324903
·13	·29933606	·38	·87498234	·63	1·45062861	·88	2·02627488
·14	·32236191	·39	·89800819	·64	1·47365446	·89	2·04930073
·15	·34538776	·40	·92103404	·65	1·49668031	·90	2·07232658
·16	·36841362	·41	·94405989	·66	1·51970616	·91	2·09535243
·17	·39143947	·42	·96708574	·67	1·54273201	·92	2·11837829
·18	·41446532	·43	·99011159	·68	1·56575786	·93	2·14140414
·19	·43749117	·44	1·01313744	·69	1·58878371	·94	2·16442999
·20	·46051702	·45	1·03616329	·70	1·61180957	·95	2·18745584
·21	·48354287	·46	1·05918914	·71	1·63483542	·96	2·21048169
·22	·50656872	·47	1·08221499	·72	1·65786127	·97	2·23350754
·23	·52959457	·48	1·10524084	·73	1·68088712	·98	2·25653339
·24	·55262042	·49	1·12826670	·74	1·70391297	·99	2·27955924
·25	·57564627	·50	1·15129255	·75	1·72693882	1·00	2·30258509

Tab. 13. — A TABLE of Rumbs, showing the Degrees, Minutes, and Seconds, that every Point and Quarter-point of the Compass makes with the Meridian.

North	North	Pts.qr	Pts.qr	°	′	″	Pts.qr	Pts.qr	South	South
		0	1	2	48	45	0	1		
		0	2	5	37	30	0	2		
		0	3	8	26	15	0	3		
N b E	N b W	1	0	11	15	0	1	0	S b E	S b W
		1	1	14	3	45	1	1		
		1	2	16	52	30	1	2		
		1	3	19	41	15	1	3		
NNE	NNW	2	0	22	30	0	2	0	SSE	SSW
		2	1	25	18	45	2	1		
		2	2	28	7	30	2	2		
		2	3	30	56	15	2	3		
NE b N	NW b N	3	0	33	45	0	3	0	SE b S	SW b S
		3	1	36	33	45	3	1		
		3	2	39	22	30	3	2		
		3	3	42	11	15	3	3		
NE	NW	4	0	45	0	0	4	0	S S	SW
		4	1	47	48	45	4	1		
		4	2	50	37	30	4	2		
		4	3	53	26	15	4	3		
NE b E	NW b W	5	0	56	15	0	5	0	SE b E	SW b W
		5	1	59	3	45	5	1		
		5	2	61	52	30	5	2		
		5	3	64	41	15	5	3		
ENE	WNW	6	0	67	30	0	6	0	ESE	WSW
		6	1	70	18	45	6	1		
		6	2	73	7	30	6	2		
		6	3	75	56	15	6	3		
E b N	W b N	7	0	78	45	0	7	0	E b S	W b S
		7	1	81	33	45	7	1		
		7	2	84	22	30	7	2		
		7	3	87	11	15	7	3		
East	West	8	0	90	0	0	8	0	East	West

EQUIVALENT EXPRESSIONS FOR SIN. A, COS. A, AND TAN. A

VALUES OF SIN. A.

1. FORMULA.... cos. A tan. A.

2. $\dfrac{\cos. A}{\cot. A}$.

3. $\sqrt{(1 - \cos.^2 A)}$.

4. $\dfrac{1}{\sqrt{(1 + \cot.^2 A)}}$, or $\dfrac{1}{\text{cosec. A}}$.

5. $\dfrac{\tan. A}{\sqrt{(1 + \tan.^2 A)}}$.

6. $2 \sin. \frac{1}{2} A \cos. \frac{1}{2} A$.

7. $\sqrt{\dfrac{1 - \cos. 2 A}{2}}$.

8. $\dfrac{2 \tan. \frac{1}{2} A}{1 + \tan.^2 \frac{1}{2} A}$.

9. $\dfrac{2}{\cot. \frac{1}{2} A + \tan. \frac{1}{2} A}$.

10. $\dfrac{1}{\cot. A + \tan. \frac{1}{2} A}$.

11. $2 \sin.^2 (45° + \frac{1}{2} A) - 1$.

12. $1 - 2 \sin.^2 (45° - \frac{1}{2} A)$.

13. $\dfrac{1 - \tan.^2 (45° - \frac{1}{2} A)}{1 + \tan.^2 (45° - \frac{1}{2} A)}$.

14. $\dfrac{\tan. (45° + \frac{1}{2} A) - \tan. (45° - \frac{1}{2} A)}{\tan. (45° + \frac{1}{2} A) + \tan. (45° - \frac{1}{2} A)}$.

15. $\sin. (60° + A) - \sin. (60° - A)$.

VALUES OF COS. A.

16. $\dfrac{\sin. A}{\tan. A}$.

17. $\sin. A \cot. A$.

18. $\sqrt{(1 - \sin.^2 A)}$.

19. $\dfrac{1}{\sqrt{(1 + \tan.^2 A)}}$, or $\dfrac{1}{\text{sec. A}}$.

20. $\dfrac{\cot. A}{\sqrt{(1 + \cot.^2 A)}}$.

21. $\cos.^2 \frac{1}{2} A - \sin.^2 \frac{1}{2} A$.

22. $1 - 2 \sin.^2 \frac{1}{2} A$.

23. $2 \cos.^2 \frac{1}{2} A - 1$.

24. $\sqrt{\dfrac{1 + \cos. 2 A}{2}}$.

25. $\dfrac{1 - \tan.^2 \frac{1}{2} A}{1 + \tan.^2 \frac{1}{2} A}$.

26. $\dfrac{\cot. \frac{1}{2} A - \tan. \frac{1}{2} A}{\cot. \frac{1}{2} A + \tan. \frac{1}{2} A}$.

27. $\dfrac{1}{1 + \tan. A \tan. \frac{1}{2} A}$.

28. $\dfrac{2}{\tan. (45° + \frac{1}{2} A) + \cot. (45° + \frac{1}{2} A)}$.

29. $2 \cos. (45° + \frac{1}{2} A) \cos. (45° \sim \frac{1}{2} A)$.

30. $\cos. (60° + A) + \cos. (60° - A)$

VALUES OF TAN. A.

31. $\dfrac{\sin. A}{\cos. A}$.

32. $\dfrac{1}{\cot. A}$.

33. $\sqrt{\left(\dfrac{1}{\cos.^2 A} - 1\right)}$.

34. $\dfrac{\sin. A}{\sqrt{(1 - \sin.^2 A)}}$.

35. $\dfrac{\sqrt{(1 - \cos.^2 A)}}{\cos. A}$.

36. $\dfrac{2 \tan. \frac{1}{2} A}{1 - \tan.^2 \frac{1}{2} A}$.

37. $\dfrac{2 \cot. \frac{1}{2} A}{\cot.^2 \frac{1}{2} A - 1}$.

38. $\dfrac{2}{\cot. \frac{1}{2} A - \tan. \frac{1}{2} A}$.

39. $\cot. A - 2 \cot. 2 A$.

40. $\dfrac{1 - \cos. 2 A}{\sin. 2 A}$.

41. $\dfrac{\sin. 2 A}{1 + \cos. 2 A}$.

42. $\sqrt{\dfrac{1 - \cos. 2 A}{1 + \cos. 2 A}}$.

43. $\dfrac{\tan. (45° + \frac{1}{2} A) - \tan. (45° - \frac{1}{2} A)}{2}$.

TABLE XV. 363

FORMULÆ RELATIVE TO TWO ARCS OR ANGLES.

1. $\sin. (A + B) = \sin. A \cos. B + \cos. A \sin. B.$

2. $\sin. (A - B) = \sin. A \cos. B - \cos. A \sin. B.$

3. $\cos. (A + B) = \cos. A \cos. B - \sin. A \sin. B.$

4. $\cos. (A - B) = \cos. A \cos. B + \sin. A \sin. B.$

5. $\tan. (A + B) = \dfrac{\tan. A + \tan. B}{1 - \tan. A \tan. B}$

6. $\tan. (A - B) = \dfrac{\tan. A - \tan B}{1 + \tan. A \tan. B}.$

7. $\begin{cases} \sin. (45^\circ \pm B) = \\ \cos. (45^\circ \mp B) = \end{cases} \dfrac{\cos. B \pm \sin. B}{\sqrt{2}}$

8. $\tan. (45^\circ \pm B) = \dfrac{1 \pm \tan. B}{1 \mp \tan. B}.$

9. $\tan.^2 (45^\circ \pm \frac{1}{2} B) = \dfrac{1 \pm \sin. B}{1 \mp \sin. B}.$

10. $\dfrac{\sin. (A+B)}{\sin. (A-B)} = \dfrac{\tan. A + \tan. B}{\tan. A - \tan. B} = \dfrac{\cot. B + \cot. A}{\cot. B - \cot. A}.$

11. $\dfrac{\cos. (A+B)}{\cos. (A-B)} = \dfrac{\cot. B - \tan. A}{\cot. B + \tan. A} = \dfrac{\cot. A - \tan. B}{\cot. A + \tan. B}.$

12. $\dfrac{\sin. A + \sin. B}{\sin. A - \sin. B} = \dfrac{\tan. \frac{1}{2} (A + B)}{\tan. \frac{1}{2} (A - B)}.$

13. $\dfrac{\cos. B + \cos. A}{\cos. B - \cos. A} = \dfrac{\cot. \frac{1}{2} (A + B)}{\tan. \frac{1}{2} (A - B)}.$

14. $\sin. A \cos. B = \frac{1}{2} \sin. (A + B) + \frac{1}{2} \sin. (A - B).$

15. $\cos. A \sin. B = \frac{1}{2} \sin. (A + B) - \frac{1}{2} \sin. (A - B).$

16. $\sin. A \sin. B = \frac{1}{2} \cos. (A - B) - \frac{1}{2} \cos. (A + B).$

17. $\cos. A \cos. B = \frac{1}{2} \cos. (A + B) + \frac{1}{2} \cos. (A - B).$

18. $\sin. A + \sin. B = 2 \sin. \frac{1}{2} (A+B) \cos. \frac{1}{2} (A - B).$

19. $\cos. A + \cos. B = 2 \cos. \frac{1}{2} (A+B) \cos. \frac{1}{2} (A - B).$

20. $\tan. A + \tan. B = \dfrac{\sin. (A + B)}{\cos. A \cos. B}.$

21. $\cot. A + \cot. B = \dfrac{\sin. (A + B)}{\sin. A \sin. B}.$

22. $\sin. A - \sin. B = 2 \sin. \frac{1}{2}(A-B) \cos. \frac{1}{2} (A + B).$

23. $\cos. B - \cos. A = 2 \sin. \frac{1}{2}(A-B) \sin. \frac{1}{2} (A + B).$

24. $\tan. A - \tan. B = \dfrac{\sin. (A - B)}{\cos. A \cos. B}.$

25. $\cot. B - \cot. A = \dfrac{\sin. (A - B)}{\sin. A \sin. B}.$

26. $\begin{cases} \sin.^2 A - \sin.^2 B = \\ \cos.^2 B - \cos.^2 A = \end{cases} \sin. (A - B) \sin. (A + B).$

27. $\cos.^2 A - \sin.^2 B = \cos. (A - B) \cos. (A + B).$

28. $\tan.^2 A - \tan.^2 B = \dfrac{\sin. (A - B) \sin. (A + B)}{\cos.^2 A \cos.^2 B}.$

29. $\cot.^2 B - \cot.^2 A = \dfrac{\sin. (A - B) \sin. (A + B)}{\sin.^2 A \sin.^2 B}.$

DIFFERENCES OF TRIGONOMETRICAL LINES.

30. $\Delta \sin. B = 2 \sin. \frac{1}{2} \Delta B \cos. (B + \frac{1}{2} \Delta B).$

31. $- \Delta \cos. B = 2 \sin. \frac{1}{2} \Delta B \sin. (B + \frac{1}{2} \Delta B).$

32. $\Delta \tan. B = \dfrac{\sin. \Delta B}{\cos. B \cos. (B + \Delta B)}.$

33. $- \Delta \cot. B = \dfrac{\sin. \Delta B}{\sin. B \sin. (B + \Delta B)}.$

34. $\begin{cases} \Delta (\sin.^2 B) = \\ - \Delta (\cos.^2 B) = \end{cases} \sin. \Delta B \sin. (2 B + \Delta B).$

35. $\Delta (\tan.^2 B) = \dfrac{\sin. \Delta B \sin. (2 B + \Delta B)}{\cos.^2 B \cos.^2 (B + \Delta B)}.$

36. $\Delta (\cot.^2 B) = \dfrac{\sin. \Delta B \sin. (2 B + \Delta B)}{\sin.^2 B \sin.^2 (B + \Delta B)}.$

DIFFERENTIALS OF TRIGONOMETRICAL LINES.

37. $d \sin. B = d B \cos. B.$

38. $- d \cos. B = d B \sin. B.$

39. $d \tan. B = \dfrac{d B}{\cos.^2 B}.$

40. $- d \cot. B = \dfrac{d B}{\sin.^2 B}$

41. $\begin{cases} d (\sin.^2 B) = \\ - d (\cos.^2 B) = \end{cases} 2 d B \sin. B \cos. B.$

42. $d (\tan.^2 B) = \dfrac{2 d B \tan. B}{\cos.^2 B}.$

43. $- d (\cot.^2 B) = \dfrac{2 d B \cot. B}{\sin.^2 B}$

ANALYTICAL FORMULÆ, FOR SOLVING ALL THE CASES OF A RECTILINEAR TRIANGLE ABC, OF WHICH THREE PARTS ARE KNOWN.

VALUES OF AB.

1. FORMULA ..... $\dfrac{BC \sin. C}{\sin. A}$.

2. $\dfrac{AC \sin. C}{\sin. B}$.

3. $\dfrac{BC}{\cos. B + \sin. B \cot. C}$.

4. $\dfrac{AC}{\cos. A + \sin. A. \cot. C}$.

5. $BC \cos. B + BC \sin. B \cot. A$.

6. $AC \cos. A + AC \sin. A \cot. B$.

7. $\sqrt{(BC^2 + AC^2 - 2 BC \times AC \cos. C)}$.

8. $BC \cos. B \pm \sqrt{(AC^2 - BC^2 \sin.^2 B)}$.

9. $AC \cos. A \pm \sqrt{(BC^2 - AC^2 \sin.^2 A)}$.

VALUES OF AC.

10. $\dfrac{AB \sin. B}{\sin. C}$.

11. $\dfrac{BC \sin. B}{\sin. A}$.

12. $\dfrac{AB}{\cos. A + \sin. A \cot. B}$.

13. $\dfrac{BC}{\cos. C + \sin. C \cot. B}$.

14. $AB \cos. A + AB \sin. A \cot. C$.

15. $BC \cos. C + BC \sin. C \cot. A$.

16. $\sqrt{(BC^2 + AB^2 - 2 BC \times AB \cos. B)}$.

17. $AB \cos. A \pm \sqrt{(BC^2 - AB^2 \sin.^2 A)}$.

18. $BC \cos. C \pm \sqrt{(AB^2 - BC^2 \sin.^2 C)}$.

VALUES OF BC.

19. $\dfrac{AC \sin. A}{\sin. B}$.

20. $\dfrac{AB \sin. A}{\sin. C}$.

21. $\dfrac{AC}{\cos. C + \sin. C \cot. A}$.

22. $\dfrac{AB}{\cos. B + \sin. B \cot. A}$.

23. $AC \cos. C + AC \sin. C. \cot. B$.

24. $AB \cos. B + AB \sin. B \cot. C$.

25. $\sqrt{(AB^2 + AC^2 - 2 AB \times AC \cos. A)}$.

26. $AC \cos. C \pm \sqrt{(AB^2 - AC^2 \sin.^2 C)}$.

27. $AB \cos. B \pm \sqrt{(AC^2 - AB^2 \sin.^2 B)}$.

VALUES OF SIN. A.

28. $\dfrac{BC \sin. C}{AB}$.

29. $\dfrac{BC \sin. B}{AC}$.

30. $\sin. (B + C)$.

31. $\sin. B \cos. C + \cos. B \sin. C$.

32. $\dfrac{BC \sin. C}{\sqrt{(BC^2 + AC^2 - 2 BC \times AC \cos. C)}}$.

33. $\dfrac{BC \sin. B}{\sqrt{(BC^2 + AB^2 - 2 BC \times AB \cos. B)}}$.

34. $\sqrt{1 - \left(\dfrac{AB^2 + AC^2 - BC^2}{2 AB \times AC}\right)^2}$.

35. $\dfrac{\sin. C (AC \cos. C \pm \sqrt{AB^2 - AC^2 \sin.^2 C})}{AB}$.

36. $\dfrac{\sin. B (AB \cos. B \pm \sqrt{AC^2 - AB^2 \sin.^2 B})}{AC}$.

VALUES OF COS. A.

37. $\dfrac{\pm \sqrt{(AB^2 - BC^2 \sin^2 C)}}{AB}$.

38. $\dfrac{\pm \sqrt{(AC^2 - BC^2 \sin.^2 B)}}{AC}$.

39. $- \cos. (B + C)$.

40. $\sin. B \sin. C - \cos. B \cos. C$.

41. $\dfrac{AC - BC \cos. C}{\sqrt{(BC^2 + AC^2 - 2 BC \times AC \cos. C)}}$.

42. $\dfrac{AB - BC \cos. B}{\sqrt{(BC^2 + AB^2 - 2 BC \times AB \cos. B)}}$.

43. $\dfrac{AB^2 + AC^2 - BC^2}{2 AB \times AC}$.

44. $\dfrac{AC \sin.^2 C \mp \cos. C \sqrt{(AB^2 - AC^2 \sin.^2 C)}}{AB}$.

45. $\dfrac{AB \sin.^2 B \mp \cos. B \sqrt{(AC^2 - AB^2 \sin.^2 B)}}{AC}$.

VALUES OF TAN. A.

46. $\dfrac{BC \sin. C}{\pm \sqrt{(AB^2 - BC^2 \sin.^2 C)}}$.

47. $\dfrac{BC \sin. B}{\pm \sqrt{(AC^2 - BC^2 \sin.^2 B)}}$.

48. $- \tan. (B + C)$.

49. $\dfrac{\tan. B + \tan. C}{\tan. B \tan. C - 1}$.

50. $\dfrac{BC \sin. C}{AC - BC \cos. C}$.

51. $\dfrac{BC \sin. B}{AB - BC \cos. B}$.

52. $\pm \sqrt{\left(\dfrac{2 AB \times AC}{AB^2 + AC^2 - BC^2}\right)^2 - 1}$.

53. $\dfrac{AC \cos. C \pm \sqrt{(AB^2 - AC^2 \sin.^2 C)}}{AC \sin. C \mp \cot. C \sqrt{(AB^2 - AC^2 \sin.^2 C)}}$

54. $\dfrac{AB \cos. B \pm \sqrt{(AC^2 - AB^2 \sin.^2 B)}}{AB \sin. B \mp \cot. B \sqrt{(AC^2 - AB^2 \sin.^2 B)}}$

TABLE XVI. CONTINUED. 365

VALUES OF SIN. B.

55. FORMULA .... $\dfrac{AC \sin. A}{BC}$.

56. $\dfrac{AC \sin. C}{AB}$.

57. $\sin. (A + C)$.

58. $\sin. A \cos. C + \cos. A. \sin. C$.

59. $\dfrac{AC \sin. A}{\sqrt{(AB^2 + AC^2 - 2 AB \times AC \cos. A)}}$.

60. $\dfrac{AC \sin. C}{\sqrt{(BC^2 + AC^2 - 2 BC \times AC \cos. C)}}$.

61. $\sqrt{1 - \left(\dfrac{BC^2 + AB^2 - AC^2}{2 BC \times AB}\right)^2}$.

62. $\dfrac{\sin. A (AB \cos. A \pm \sqrt{BC^2 - AB^2 \sin.^2 A})}{BC}$.

63. $\dfrac{\sin. C (BC \cos. C \pm \sqrt{AB^2 - BC^2 \sin.^2 C})}{AB}$.

VALUES OF COS. B.

64. $\dfrac{\pm \sqrt{(BC^2 - AC^2 \sin.^2 A)}}{BC}$.

65. $\dfrac{\pm \sqrt{(AB^2 - AC^2 \sin.^2 C)}}{AB}$.

66. $- \cos. (A + C)$.

67. $\sin. A \sin. C - \cos. A \cos. C$.

68. $\dfrac{AB - AC \cos. A}{\sqrt{(AB^2 + AC^2 - 2 AB \times AC \cos. A)}}$.

69. $\dfrac{BC - AC \cos. C}{\sqrt{(BC^2 + AC^2 - 2 BC \times AC \cos. C)}}$.

70. $\dfrac{BC^2 + AB^2 - AC^2}{2 BC \times AB}$.

71. $\dfrac{AB \sin.^2 A \mp \cos. A \sqrt{(BC^2 - AB^2 \sin.^2 A)}}{BC}$.

72. $\dfrac{BC \sin.^2 C \mp \cos. C \sqrt{(AB^2 - BC^2 \sin.^2 C)}}{AB}$.

VALUES OF TAN. B.

73. $\dfrac{AC \sin. A}{\pm \sqrt{(BC^2 - AC^2 \sin.^2 A)}}$.

74. $\dfrac{AC \sin. C}{\pm \sqrt{(AB^2 - AC^2 \sin.^2 C)}}$.

75. $- \tan. (A + C)$.

76. $\dfrac{\tan. A + \tan. C}{\tan. A \tan. C - 1}$.

77. $\dfrac{AC \sin. A}{AB - AC \cos. A}$.

78. $\dfrac{AC \sin. C}{BC - AC \cos. C}$.

79. $\pm \sqrt{\left(\dfrac{2 BC \times AB}{BC^2 + AB^2 - AC^2}\right)^2 - 1}$.

80. $\dfrac{AB \cos. A \pm \sqrt{(BC^2 - AB^2 \sin.^2 A)}}{AB \sin. A \mp \cot. A \sqrt{(BC^2 - AB^2 \sin.^2 A)}}$.

81. $\dfrac{BC \cos. C \pm \sqrt{(AB^2 - BC^2 \sin.^2 C)}}{BC \sin. C \mp \cot. C \sqrt{(AB^2 - BC^2 \sin.^2 C)}}$.

VALUES OF SIN. C.

82. $\dfrac{AB \sin. B}{AC}$.

83. $\dfrac{AB \sin. A}{BC}$.

84. $\sin. (A + B)$.

85. $\sin. A \cos. B + \cos. A \sin. B$.

86. $\dfrac{AB \sin. B}{\sqrt{(BC^2 + AB^2 - 2 BC \times AB \cos. B)}}$.

87. $\dfrac{AB \sin. A}{\sqrt{(AB^2 + AC^2 - 2 AB \times AC \cos. A)}}$.

88. $\sqrt{1 - \left(\dfrac{BC^2 + AC^2 - AB^2}{2 BC \times AC}\right)^2}$.

89. $\dfrac{\sin. B (BC \cos. B \pm \sqrt{AC^2 - BC^2 \sin.^2 B})}{AC}$.

90. $\dfrac{\sin. A (AC \cos. A \pm \sqrt{BC^2 - AC^2 \sin.^2 A})}{BC}$.

VALUES OF COS. C.

91. $\dfrac{\pm \sqrt{(AC^2 - AB^2 \sin.^2 B)}}{AC}$.

92. $\dfrac{\pm \sqrt{(BC^2 - AB^2 \sin.^2 A)}}{BC}$.

93. $- \cos. (A + B)$.

94. $\sin. A \sin. B - \cos. A \cos. B$.

95. $\dfrac{BC - AB \cos. B}{\sqrt{(BC^2 + AB^2 - 2 BC \times AB \cos. B)}}$.

96. $\dfrac{AC - AB \cos. A}{\sqrt{(AB^2 + AC^2 - 2 AB \times AC \cos. A)}}$.

97. $\dfrac{BC^2 + AC^2 - AB^2}{2 BC \times AC}$.

98. $\dfrac{BC \sin.^2 B \mp \cos. B \sqrt{(AC^2 - BC^2 \sin.^2 B)}}{AC}$.

99. $\dfrac{AC \sin.^2 A \mp \cos. A \sqrt{(BC^2 - AC^2 \sin.^2 A)}}{BC}$.

VALUES OF TAN. C.

100. $\dfrac{AB \sin. B}{\pm \sqrt{(AC^2 - AB^2 \sin.^2 B)}}$.

101. $\dfrac{AB \sin. A}{\pm \sqrt{(BC^2 - AB^2 \sin.^2 A)}}$.

102. $- \tan. (A + B)$.

103. $\dfrac{\tan. A + \tan. B}{\tan. A \tan. B - 1}$.

104. $\dfrac{AB \sin. B}{BC - AB \cos. B}$.

105. $\dfrac{AB \sin. A}{AC - AB \cos. A}$.

106. $\pm \sqrt{\left(\dfrac{2 BC \times AC}{BC^2 + AC^2 - AB^2}\right)^2 - 1}$.

107. $\dfrac{BC \cos. B \pm \sqrt{(AC^2 - BC^2 \sin.^2 B)}}{BC \sin. B \mp \cot. B \sqrt{(AC^2 - BC^2 \sin.^2 B)}}$.

108. $\dfrac{AC \cos. A \pm \sqrt{(BC^2 - AC^2 \sin.^2 A)}}{AC \sin. A \mp \cot. A \sqrt{(BC^2 - AC^2 \sin.^2 A)}}$.

ANALYTICAL EXPRESSIONS IN REFERENCE TO A SPHERICAL TRIANGLE ABC.

1. FORMULA .... $\sin A = \dfrac{\sin BC \; \sin C}{\sin AB}$.

2. $\sin A = \dfrac{\sin BC \; \sin B}{\sin AC}$.

3. $\sin B = \dfrac{\sin AC \; \sin A}{\sin BC}$.

4. $\sin B = \dfrac{\sin AC \; \sin C}{\sin AB}$.

5. $\sin C = \dfrac{\sin AB \; \sin B}{\sin AC}$.

6. $\sin C = \dfrac{\sin AB \; \sin A}{\sin BC}$.

7. $\cos A = \dfrac{\cos BC - \cos AB \; \cos AC}{\sin AB \; \sin AC}$.

8. $\cos A = \cos BC \; \sin B \; \sin C - \cos B \; \cos C$.

9. $\cos B = \dfrac{\cos AC - \cos BC \; \cos AB}{\sin BC \; \sin AB}$.

10. $\cos B = \cos AC \; \sin A \; \sin C - \cos A \; \cos C$.

11. $\cos C = \dfrac{\cos AB - \cos BC \; \cos AC}{\sin BC \; \sin AC}$.

12. $\cos C = \cos AB \; \sin A \; \sin B - \cos A \; \cos B$.

13. $\tan A = \dfrac{\sin B}{\sin AB \; \cot BC - \cos AB \; \cos B}$.

14. $\tan A = \dfrac{\sin C}{\sin AC \; \cot BC - \cos AC \; \cos C}$.

15. $\tan B = \dfrac{\sin C}{\sin BC \; \cot AC - \cos BC \; \cos C}$.

16. $\tan B = \dfrac{\sin A}{\sin AB \; \cot AC - \cos AB \; \cos A}$.

17. $\tan C = \dfrac{\sin A}{\sin AC \; \cot AB - \cos AC \; \cos A}$.

18. $\tan C = \dfrac{\sin B}{\sin BC \; \cot AB - \cos BC \; \cos B}$.

19. $\sin BC = \dfrac{\sin AB \; \sin A}{\sin C}$.

20. $\sin BC = \dfrac{\sin AC \; \sin A}{\sin B}$.

21. $\sin AC = \dfrac{\sin BC \; \sin B}{\sin A}$.

22. $\sin AC = \dfrac{\sin AB \; \sin B}{\sin C}$.

23. $\sin AB = \dfrac{\sin AC \; \sin C}{\sin B}$.

24. $\sin AB = \dfrac{\sin BC \; \sin C}{\sin A}$.

25. $\cos BC = \dfrac{\cos A + \cos B \; \cos C}{\sin B \; \sin C}$.

26. $\cos BC = \cos A \; \sin AB \; \sin AC + \cos AB \; \cos AC$.

27. $\cos AC = \dfrac{\cos B + \cos A \; \cos C}{\sin A \; \sin C}$.

28. $\cos AC = \cos B \; \sin BC \; \sin AB + \cos BC \; \cos AB$.

29. $\cos AB = \dfrac{\cos C + \cos A \; \cos B}{\sin A \; \sin B}$.

30. $\cos AB = \cos C \; \sin BC \; \sin AC + \cos BC \; \cos AC$.

31. $\tan BC = \dfrac{\sin AB}{\sin B \; \cot A + \cos B \; \cos AB}$.

32. $\tan BC = \dfrac{\sin AC}{\sin C \; \cot A + \cos C \; \cos AC}$.

33. $\tan AC = \dfrac{\sin BC}{\sin C \; \cot B + \cos C \; \cos BC}$.

34. $\tan AC = \dfrac{\sin AB}{\sin A \; \cot B + \cos A \; \cos AB}$.

35. $\tan AB = \dfrac{\sin AC}{\sin A \; \cot C + \cos A \; \cos AC}$.

36. $\tan AB = \dfrac{\sin BC}{\sin B \; \cot C + \cos B \; \cos BC}$.

In working by logarithms it is often necessary to transform these expressions by the introduction of a subsidiary arc or angle. Thus, in theorem 26, find a dependent angle, s, by the expression $\tan s = \cos A \; \tan AB$. Substitute for $\cos A$, in theor. 25. its value in this, it will become $\cos BC = \cos AB \; \cos AC + \tan s \; \cos AB \; \sin AC = \cos AB \left(\dfrac{\cos AC \; \cos s + \sin AC \; \sin s}{\cos s} \right)$ $= \dfrac{\cos AB \; \cos (AC - s)}{\cos s}$. The latter expression is evidently logarithmic; and the transformation is equivalent to dividing the triangle into two right-angled triangles, by letting fall a perpendicular.

TABLE XVIII.—TRIGONOMETRICAL FORMULÆ FOR THE SOLUTION OF QUADRATICS AND CUBICS.

Equations of the 2d degree

$x^2 + px = q$.

SOLUTION.

$$\tan. A = \frac{2}{p}\sqrt{q}.$$

Root $x = \tan. \tfrac{1}{2} A \sqrt{q}.$

The other root $v = -\cot. \tfrac{1}{2} A \sqrt{q}.$

$x^2 - px = q$.

SOLUTION.

$$\tan. A = \frac{2}{p}\sqrt{q}.$$

$x = -\tan. \tfrac{1}{2} A \sqrt{q}.$

$x = \cot. \tfrac{1}{2} A \sqrt{q}.$

$x^2 + px = -q$. If $p^2 < 4q$, x is imaginary.

SOLUTION.

$$\sin. A = \frac{2}{p}\sqrt{q}.$$

$x = -\tan. \tfrac{1}{2} A \sqrt{q}.$

$x = -\cot. \tfrac{1}{2} A \sqrt{q}.$

$x^2 - px = -q$. If $p^2 < 4q$, x is imaginary.

SOLUTION.

$$\sin. A = \frac{2}{p}\sqrt{q}.$$

$x = \tan. \tfrac{1}{2} A \sqrt{q}.$

$x = \cot. \tfrac{1}{2} A \sqrt{q}.$

Equations of the 3d degree

$x^3 + px - q = 0.$

SOLUTION.

$$\tan. B = \frac{p}{3q} \times 2\sqrt{\tfrac{1}{3}}\, p.$$

$\tan. A = \sqrt[3]{\tan. \tfrac{1}{3} B}.$

Only real root $x = -\cot. 2 A \times 2\sqrt{\tfrac{1}{3}}\, p.$

$x^3 + px + q = 0.$ We suppose $4p^3 < 27 q^2.$

SOLUTION.

$$\tan. B = \frac{p}{3q} \times 2\sqrt{\tfrac{1}{3}}\, p.$$

$\tan. A = \sqrt[3]{\tan. \tfrac{1}{2} B}.$

$x = -\dfrac{2\sqrt{\tfrac{1}{3}}\, p}{\sin. 2 A}.$

$x^3 - px + q = 0.$ We suppose $4p^3 < 27 q^2.$

SOLUTION.

$$\sin. B = \frac{p}{3q} \times 2\sqrt{\tfrac{1}{3}}\, p.$$

$\tan. A = \sqrt[3]{\tan. \tfrac{1}{2} B}.$

$x = -\dfrac{2\sqrt{\tfrac{1}{3}}\, p}{\sin. 2 A}.$

$x^3 - px - q = 0.$ We suppose $4p^3 < 27 q^2.$

SOLUTION.

$$\sin. B = \frac{p}{3q} \times 2\sqrt{\tfrac{1}{3}}\, p.$$

$\tan. A = \sqrt[3]{\tan. \tfrac{1}{2} B}.$

$x = \dfrac{2\sqrt{\tfrac{1}{3}}\, p}{\sin. 2 A}.$

Equations of the 3d degree in the irreducible case

These solutions apply to the cases where $4 p^3 >$ or $= 27 q$

$x^3 - px + q = 0.$

SOLUTION.

$$\sin. 3 A = \frac{3 q}{p} \times \frac{1}{2\sqrt{\tfrac{1}{3}}\, p}.$$

Root $x = \sin. A \times 2\sqrt{\tfrac{1}{3}}\, p.$

Another root $x = \sin. (60° - A) \times 2\sqrt{\tfrac{1}{3}}\, p.$

Third root $x = -\sin. (60° + A) \times 2\sqrt{\tfrac{1}{3}}\, p.$

$x^3 - px - q = 0.$

SOLUTION.

$$\sin. 3 A = \frac{3 q}{p} \times \frac{1}{2\sqrt{\tfrac{1}{3}}\, p}.$$

$x = -\sin. A \times 2\sqrt{\tfrac{1}{3}}\, p.$

$x = -\sin. (60° - A) \times 2\sqrt{\tfrac{1}{3}}\, p.$

$x = \sin. (60° + A) \times 2\sqrt{\tfrac{1}{3}}\, p.$

TABLE XIX.—FORMULÆ USEFUL IN ASTRONOMY.

If $a =$ right ascension, $d =$ declination, $l =$ latitude, $\lambda =$ longitude, $p =$ angle of position (or, the angle at a heavenly body formed by two great circles, one passing through the pole of the equator, and the other through the pole of the ecliptic), $i =$ inclination, or obliquity of the ecliptic; then the following equations, obtain generally, for all the stars and heavenly bodies.

1. $\tan. a = \tan. \lambda . \cos. i - \tan. l . \sec. \lambda . \sin. i.$
2. $\sin. d = \sin. \lambda . \cos. l . \sin. i + \sin. l . \cos. i.$
3. $\tan. \lambda = \sin. i . \tan. d . \sec. a + \tan. a . \cos. i.$
4. $\sin. l = \sin. d . \cos. i - \sin. a . \cos. d . \sin. i.$
5. $\cot an. p = \cos. d . \sec. a . \cot. i + \sin. d . \tan. a.$
6. $\cot an. p = \cos. l . \sec. \lambda . \cot. i - \sin. l . \tan. \lambda.$
7. $\cos. a . \cos. d = \cos. l . \cos. \lambda.$
8. $\sin. p . \cos. d = \sin. i . \cos. \lambda.$
9. $\sin. p . \cos. \lambda = \sin. i . \cos. a.$
10. $\tan. a = \tan. \lambda . \cos. i.$
11. $\cos. \lambda = \cos. a . \cos. d.$ $\Big\}$ when $l = 0$, as is always the case with the sun.

NUMBERS OFTEN USED IN CALCULATIONS, WITH THEIR LOGARITHMS.

[For the circumference of a circle to diameter unity, to 25 decimal places, and its logarithm to 20 places, see p. xxiii.]

		Logarithms.
Circumference of a circle, to diameter 1 .... ⎫		
Surface of a sphere to diameter 1 .... ⎬ = 3·1415926		0·4971499
Area of a circle to radius 1 .... ⎭		
Area of a circle to diameter 1.................... = ·7853982		9·8950899
Capacity of a sphere to diameter 1............. = ·5235988		9·7189986
Capacity of a sphere to radius 1 .... = 4·1887902		0·6220886
3·14159, &c., nearly = $\sqrt[2]{(31·00628)}$, nearly = $\sqrt[4]{(97·409089)}$, nearly = $\sqrt[3]{(306·02)}$, nearly = $\sqrt[4]{(961·39)}$, nearly = $\sqrt[8]{(9488·53)}$, arc equal to radius .... = 57°·2957795		1·7581226
Arc equal to radius, expressed in seconds .. .. = 206264″·8		5·3144251
Length of 1 degree .... = ·01745329		8·2418773
sin 1″ .... = ·00000485		4·6855749
sin 2″ .... = ·00000970		4·9866049
sin. 3″ .... = ·00001454		5·1626961
Number whose Hyper. log. is = 1............... = 2·7182818		0·4342945
Modulus of the common logarithms............. = ·43429448		9·6377843
Complement to the same .. .. = 2·3025851		0·3622157
12 hours, expressed in seconds .... = 43200		4·6354837
Complement to the same .... = ·00002315		5·3645163
24 hours, expressed in seconds................. = 86400		4·9365137
Complement to the same .... = ·00001157		5·0634863
360 degrees, expressed in seconds .... = 1296000		6·1126050
Compression of the earth = $\frac{1}{300}$................ = ·0033333		7·5228787
Sidereal revolution of the earth, in days .... = 365·25638		2·5625978
Tropical revolution of ditto .... = 365·24226		2·5625910
Sidereal year, in seconds .... = 31558150″		7·4991116
Tropical year, in seconds .... = 31556929″		7·4990947
Passage of a comet, perihelion distance 1, from perihelion to extremity of latus-rectum........ = 109ᵈ·61545		2·0398718
Log. k, (Gauss. Theor. M. c. c. 1.) k .... = 0·0172021		8·2355814
k in seconds, (Gauss. Theor. M. c. c. 6.) ..... = 3548″·1876		3·5500065

Constant logarithms (always additive) for converting

Sidereal time into mean solar time	9·9988126
French toises into metres ⎱ at 61°·3 ⎰	0·2898200
———— feet into metres ⎰	9·5116687
———— toises into English feet ⎱ at 56°·3 ⎰	0·8058372
———— feet into English feet ⎰	0·0276860
———— metre into English feet	0·5159929
———— millimetres into English inches	8·5951741
Centes. degrees of quadrant into sexages degrees...........	9·9542425
———— minutes of do. into ———— minutes..........	1·7323938
———— seconds of do. into ———— seconds..........	3·5105450

The arithmetical complements of these last 10 logarithms will serve to convert mean solar time into sidereal time, metres into toises, &c., English feet into toises, &c., and the sexagesimal divisions of the quadrant into the centesimal divisions.

Printed in the United States
By Bookmasters